FLOWERING
PLANT FAMILIES
OF THE WORLD

AUTHORS	V. H. Heywood, R. K. Brummitt, A. Culham, O. Seberg
ADDITIONAL CONTRIBUTIONS BY	H. E. Ballard, Jr., Diane M. Bridson, M. R. Cheek, R. J. Chinnock, A. P. Davis, A. S. George, D. J. Goyder, P. S. Green, D. J. N. Hind, Petra Hoffmann, P. C. Hoch, W. L. Wagner, S. L. Jury, R. V. Lansdown, Eve J. Lucas, R. M. Polhill, B. L. Stannard, G. W. Staples, Margaret Tebbs, T. Upson, T. M. A. Utteridge, Juliet A. Wege, P. H. Weston, Melanie C. Wilmot-Dear, Elizabeth M. Woodgyer
CONSULTANT EDITOR	V. H. Heywood
ARTISTS	Judith Dunkley, Victoria Goaman, Christabel King

Flowering Plant Families of the World is the successor to *Flowering Plants of the World* (1978).

FLOWERING
PLANT FAMILIES
OF THE WORLD

V. H. Heywood
R. K. Brummitt
A. Culham
O. Seberg

FIREFLY BOOKS

CONTRIBUTORS

A FIREFLY BOOK

Published by Firefly Books Ltd. 2007

Copyright © 2007
The Brown Reference Group plc

Publisher Cataloging-in-Publication Data (U.S.)

Heywood, V. H. (Vernon Hilton), 1927-
    Flowering plant families of the world / V. H. Heywood,
R. K. Brummitt, A. Culham, O. Seberg.
Includes index.
[424] p. : col. ill. , maps ; cm.
Summary: Survey of 506 flowering plant families, which
describes distribution, diagnostic features, classification
and commercial uses and a distribution map.
ISBN-13: 978-1-55407-206-4
ISBN-10: 1-55407-206-9
1. Angiosperms — Identification. 2. Flowers —
Identification. I. Flowering plants of the world. II. Title.
582.13 dc22   QK495.A1F46 2006

Library and Archives Canada Cataloguing in Publication

    Flowering plant families of the world / V. H. Heywood ..
[et al.].
Includes index.
ISBN-13: 978-1-55407-206-4
ISBN-10: 1-55407-206-9
    1. Angiosperms. 2. Flowers. I. Heywood, V. H. (Vernon
Hilton), 1927-
QK495.A1F46 2006   582.13   C2006-900869-8

Published in the United States by
Firefly Books (U.S.) Inc.
P.O. Box 1338 Ellicott Station
Buffalo, New York 14205

Published in Canada by
Firefly Books Ltd.
66 Leek Crescent
Richmond Hill, Ontario L4B 1H1

The Brown Reference Group plc
(Incorporating Andromeda Oxford Limited)
8 Chapel Place
Rivington Street
London EC2A 3DQ
www.brownreference.com

For The Brown Reference Group plc
COPYEDITOR: Leon Gray
DESIGNERS: Ron Callow, Imran Ghoorbin,
Dave Goodman, Sarah Williams
CARTOGRAPHER: Martin Darlinson
INDEXER: Ann Barrett
MANAGING EDITOR: Bridget Giles
PRODUCTION DIRECTOR: Alastair Gourlay
EDITORIAL DIRECTOR: Lindsey Lowe

Printed in U.A.E.

Citation: Heywood, V. H., Brummitt, R. K., Culham, A.,
and Seberg, O. *Flowering Plant Families of the World.*
Firefly Books: Ontario, Canada (2007).

HEB   Harvey E. Ballard, Jr., Ohio
       University, Athens, OH (USA)
DMB   Diane M. Bridson, Royal Botanic
       Gardens, Kew, Richmond (UK)
RKB   R. K. Brummitt, Royal Botanic
       Gardens, Kew, Richmond (UK)
MRC   Martin R. Cheek, Royal Botanic
       Gardens, Kew, Richmond (UK)
RJC   R. J. Chinnock, State Herbarium,
       Adelaide (South Australia)
AC   Alastair Culham, The University
       of Reading (UK)
APD   Aaron P. Davis, Royal Botanic
       Gardens, Kew, Richmond (UK)
ASG   Alex S. George, Murdoch University,
       Perth (Western Australia)
DJG   David J. Goyder, Royal Botanic
       Gardens, Kew, Richmond (UK)
PSG   Peter S. Green, Royal Botanic
       Gardens, Kew, Richmond (UK)
VHH   Vernon H. Heywood, The University
       of Reading (UK)
DJNH   D. J. Nicholas Hind, Royal Botanic
       Gardens, Kew, Richmond (UK)
PCH   Peter C. Hoch, Missouri Botanical
       Garden, St. Louis, Missouri (USA)
PH   Petra Hoffmann, Royal Botanic
       Gardens, Kew, Richmond (UK)
SLJ   Stephen L. Jury, The University
       of Reading (UK)
RVL   R. V. Lansdown, Stroud (UK)
EJL   Eve J. Lucas, Royal Botanic Gardens,
       Kew, Richmond (UK)
RMP   Roger M. Polhill, Royal Botanic
       Gardens, Kew, Richmond (UK)
OS   Ole Seberg, University
       of Copenhagen (Denmark)
BLS   B. L. Stannard, Royal Botanic
       Gardens, Kew, Richmond (UK)
GWS   G. W. Staples, Bishop Museum,
       Honolulu, Hawaii (USA)
MT   Margaret Tebbs, Royal Botanic
       Gardens, Kew, Richmond (UK)
TU   Tim Upson, University Botanic
       Garden, Cambridge (UK)
TMAU   T. M. A. Utteridge, Royal Botanic
       Gardens, Kew, Richmond (UK)
WLW   Warren L. Wagner, Smithsonian
       Institution, Washington, D.C. (USA)
JAW   Juliet A. Wege, Department of
       Environment and Conservation
       (Australia)
PHW   Peter H. Weston, Royal Botanic
       Gardens, Sydney, NSW (Australia)
CMW-D Melanie C. Wilmot-Dear (Melanie
       Thomas), Royal Botanic Gardens,
       Kew, Richmond (UK)
EMW   Elizabeth M. Woodgyer, Royal
       Botanic Gardens, Kew, Richmond (UK)

In addition, the following initials are
the abbreviations used for authors who
contributed to *Flowering Plants of the World*
(1978), and whose contributions have been
drawn upon to some extent in preparing the
family accounts for *Flowering Plant Families
of the World* (2007). The initials of these
contributors are cited within square brackets
before the abbreviation of the name of the
current author, thus: [DJM] VHH

DB   David Bramwell, Jardín Botánico
       Canario "Viera y Clavijo,"
       Tafira Alta (Spain)
CDC   C. D. K. Cook, University
       of Zurich (Switzerland)
JC   James Cullen, University Botanic
       Garden, Cambridge (UK)
WDC   W. D. Clayton, Royal Botanic
       Gardens, Kew, Richmond (UK)
JME   Jennifer M. Edmonds, University
       of Leeds (UK)
CJH   Christopher J. Humphries, Natural
       History Museum, London (UK)
FKK   Frances K. Kupicha (UK)
BFM   B. F. Mathew, Royal Botanic
       Gardens, Kew, Richmond (UK)
DJM   David Mabberley, University
       of Washington Botanic Garden,
       Seattle (USA)
DMM   David M. Moore, The University
       of Reading (UK)
BM   B. Morley, Adelaide Botanic Garden,
       Adelaide (South Australia)
BP   B. Pickersgill, The University
       of Reading (UK)
IBKR   Ian B. K. Richardson, Reading (UK)
GDR   Gordon D. Rowley, The University
       of Reading (UK)
NWS   N. W. Simmonds, University
       of Edinburgh (UK)
WTS   William T. Stearn (deceased)
HPW   H. P. Wilkinson, Royal Botanic
       Gardens, Kew, Richmond (UK)

FOREWORD

It is a great privilege and pleasure to be able to contribute
the Foreword for this magnificent book. The forerunner of this
work, *Flowering Plants of the World* (1978), was a huge success.
It found a place on the shelves of plant systematists, and also
other scientists of all kinds. It became an essential reference
work for anyone interested in the variety of plant life, including
professional and amateur gardeners. And this was also a book
that was well used: because it was comprehensive, visually
stunning, and packed with useful information. *Flowering Plant
Families of the World*, builds on the success of its predecessor
and has all the same virtues, including a wonderful consistency
of treatment and magnificent color illustrations. But it goes
further by also incorporating significant advances in knowledge
that have accumulated over the last two decades. A particular
focus has been to include new information on the relationships
among plant families, which has emerged from analyses
of structural and molecular evidence. This book therefore
helps encapsulate what we have learned during a remarkable
renaissance in the study of plant diversity.

One manifestation of the new data and new ideas summarized
in this book is that *Flowering Plant Families of the World* (2007)
treats 506 families, compared to the 306 of its predecessor.
The scope of the book has not changed: but this more refined
approach ensures that the individual family treatments are less
heterogeneous. As a result, the information presented about
the different plant families is more specific, and even more
useful. This book could never have been completed without
the coordinated efforts of many people. It is a great pleasure
to be able to acknowledge the hard work and commitment
of all those involved, especially the main authors and all the
other contributors of individual family treatments. Together
they have created a work that will be used around the world
by students and teachers, and professionals and amateurs.
Anyone interested in the grand sweep of plant diversity will
see the value in this book. And they will find it a marvellous
companion as they seek to understand the exuberant variety
in the world of plants.

Finally, this Foreword is not only an important opportunity
to congratulate Vernon Heywood for his efforts in coordinating
this volume, but it is also a good moment to acknowledge, more
generally, his outstanding lifelong contributions to plant diversity
science. Through his own talents and energy, and through his
influence on students, the public, and on policy makers, spanning
more than 50 years, he has made a profound difference to the
way we view the variety of plant life. This book is a further
remarkable testament to his energy, his perspective on what is
truly important, and his ability to engage others in that vision.

Professor Sir Peter Crane, FRS
Former Director, Royal Botanic Gardens,
Kew, Richmond, UK

CONTENTS

PREFACE

Since the publication of the first English-language edition of *Flowering Plants of the World* in 1978, our knowledge of the flowering plants has been dramatically changed as a result of a flood of publications on morphology, anatomy, distribution, and molecular phylogeny. Our original intention was to prepare an updated edition of the book to reflect this new information but so extensive were the changes needed that we abandoned the original text in most cases and prepared what is in effect a new book.

The aim of this book is to provide a concise account of all the families that are considered by the authors to be worthy of recognition, taking into account all the latest evidence. The number of families has increased from 306 in *Flowering Plants of the World* to 506. For each family we have provided a summary of the main features of its distribution and where appropriate details of important genera or other points of interest and often an indication of the type of habitat in which the family grows. We have tried to make the descriptions as consistent in presentation as possible. The section on classification for each family includes where appropriate information on the subdivision of the family into subfamilies and/or tribes and indicates where the circumscription and content of the family has been changed compared with previous treatments. It also summarizes both traditional and the most recent views on relationships, including those suggested by molecular evidence. Significant economic uses are summarized. As a break from the previous book, references are given to key literature, but because of space limitations these have had to be restricted. A general distribution map is given for most families but because of the scale, little detail is possible. Accurate distribution maps are difficult to compile and are surprisingly rare in the literature.

Professor V. H. Heywood
THE UNIVERSITY OF READING, UK

ACKNOWLEDGMENTS

The authors would like to acknowledge the following, who have assisted in the preparation of this book:

Conny B. Asmussen, Peter Brewer, Neil A. Brummitt, Julie E. Culham, Elvira Hörandl, Chris Humphries, Irmgard Jäger-Zurn, Pete Lowry, Gitte Petersen, Susanne Renner, Yuri Roskov, Mark P. Simmons, Tim Sutton, Nicola Toomey, Jun Wen, Chris Yesson

We are grateful to Peter F. Stevens for permission to base our distribution maps for some of the monocot families on maps provided on his Angiosperm Phylogeny Website.

Christine A. Heywood has contributed significantly to the production of the maps and has helped in many other ways.

THE FLOWERING PLANTS

The flowering plants contribute massively to the world's primary productivity and are arguably the most important component of global biodiversity: Not only do they provide the crops that feed us, as well as ornamentals, medicines, poisons, fibers, oils, tannins, beverages, and stimulants, and herbs and spices but constitute the main structure of our terrestrial ecosystems and afford habitats for countless animals. It is not surprising, therefore, that they have held a fascination for people over the centuries and that their classification has attracted a great deal of attention.

When the predecessor of this book, *Flowering Plants of the World*, was first published in 1978, the classification of the flowering plants was in a period of transition. The widely used systems of Cronquist (1968), Takhtajan (1969), and Thorne (1976), avowedly incorporated evolutionary principles rather than using any explicit methodology and with little documentation of the processes involved or the information base used in reaching the conclusions adopted. At that time, the use of phenetics with its emphasis on quantification of characters and character states was part of an attempt to make the procedures of classification more explicit and reproducible, and numerical phenetics (numerical or Adansonian taxonomy) was being increasingly used to handle large data sets within the limitations of the then existing computing technology. The phylogenetic approach of Hennig, although vigorously advocated by some, was still little known and only slowly made impact in botany, and the use of biochemical data, principally the chemistry of secondary compounds, in classification was just beginning and was yet to enter the macromolecular phase. Hence, neither approach made much impact on the classification of the flowering plant families, although the circumscription and relationships of some families was affected to some extent, and no new system of classification was produced. For the monocotyledons, however, Dahlgren and Clifford's major review, *Monocotyledons* (1982), not only included a revised classification of the class but also presented the extensive base data from which it was produced and subsequently served as the foundation of Dahlgren *et al.*'s (1985) magnum opus *The Families of the Monocotyledons*.

The sequence of families used in *Flowering Plants of the World* (1978) followed that of Stebbins in his *Flowering Plants – Evolution above the Species Level* (1974), itself largely based on Cronquist's 1968 system, but with the prophetic remark in the preface, ". . . it is likely that future systems will be radically different." In the intervening years since 1978, the systems of Cronquist, Takhtajan, and Thorne have been revised, incorporating new data (but only that of Thorne has been regularly updated).

A major shift has taken place over the past 20 years with the publication of many papers detailing morphological, anatomical, and other data for various flowering plant groups on the one hand, and the development of DNA sequencing technologies on the other. The latter have increasingly been applied to plants, leading to the production of large amounts of DNA sequence data which have in turn been analyzed using cladistic, phyletic, phenetic, and other analytical procedures made possible by the availability of high speed computing capacity.

The great advantage of DNA data is that they are plentiful, usually easily accessible, intrinsically qualitative, and the information is generally less subjective than morphological data and provides a more direct way of measuring genetic divergence among lineages for phylogenetic analysis. The accumulation of large bodies of DNA sequence data by a team of researchers, led to the seminal publication by 41 authors of a tentative phylogenetic tree for 500 species of flowering plants based on the analysis of nucleotide sequences of the plastid gene *rbc*L (Chase *et al*. 1993). This was followed by the formation of the Angiosperm Phylogeny Group and the publication of an ordinal system of classification of the flowering plants called APG I (1998) and APG II (2003) with a subsequent partial revision by Soltis *et al*. (2005). To date (2006) far more than 4,000 papers containing *rbc*L or other gene analyses have been published, about 10,000 flowering plant species have been sequenced for *rbc*L alone and the pace of sequencing and publication, if anything, is increasing.

Although often referred to as the molecular age of systematics, the characteristics of the current phase are much wider than just the use of DNA sequence data on their own. What is remarkable is the production of large data sets of morphological as well as molecular information and the construction of tree diagrams from these in various combinations. Indeed the combination of disparate data sets is one of the strengths of today's systematics and classification as well as providing both philosophical and technical challenges. These phylogenetic analyses have undoubtedly led to a much greater understanding of the evolution of flowering plants and although molecular systematics is still in its infancy, there is general agreement as to the basic framework of a phylogenetic system of classification. It has led to major realignments of families, the association of families or parts of them not previously regarded as related, the splitting of some families and the merging of others.

In this book we have refrained from offering a new system of classification or a further modification of APG II. The aim has been to focus on the family and its content and we have tried to evaluate all the available evidence, both morphological (in the broad sense) and molecular in deciding which families to recognize, how they may be circumscribed, their division into subfamilies and tribes where appropriate, and their relationships. A total of 506 families is recognized in this volume, compared with 306 in *Flowering Plants of the World*. They are arranged alphabetically within the two traditional major groups, dicotyledons and monocotyledons, while recognizing that the dicotyledons include the stem group leading to both the monocotyledons and eu-dicotyledons.

The question of which groups to recognize at family level cannot be answered by perusing the details of cladograms. Like the other taxonomic ranks, a family is a human construct with a long and complex history, and while cladograms (like phenograms) may help us decide which groups it makes sense to recognize, they cannot tell us at what rank: Cladograms like phenograms are rank free. Rank is a matter for taxonomic decision and preference, and ultimately consensus and as a consequence, in many cases groups that may be recognized as families in APG II or other systems are not so considered in this book and of course the reverse is true.

We have not formally recognized orders, historically one of the most arbitrary of taxonomic ranks, although in our discussion of relationships we have often referred to the orders recognized by Cronquist or Takhtajan and indicated the order in which the family is placed by APG II.

Although enormous progress has been made in our understanding of the classification and relationships of the flowering plant families during the past 25 years, we are still far from achieving a stable system. Many families are still poorly known and much basic work is needed, from plant exploration to detailed morphological analysis. At a molecular level, sampling is still very patchy and we may expect many more insights as work in this area continues. Our book may be regarded as a summary of the current state of play and we are certain that the picture will look very different in another 25 years from now.

The coming decades are likely to bring new approaches in developmental genetics and genomics which will change our understanding and perception of the evolution of plant architecture, structure, and physiology beyond what is currently known in model organisms. This should significantly illuminate our views, for instance, of the evolution of key genes and gene families expressed in flowers, and test hypotheses about the origin of the flower and the evolution of biochemical pathways, thus enhancing our understanding of evolutionary change, which will in turn be reflected to some extent in our formal classification of the flowering plants.

At another level taxonomy has moved further into cyberspace. Large specimen databases such as those of Australia's Virtual Herbarium (c. 6 million records) and the Botanical Society of the British Isles (c. 9 million records) are contributing to the Global Biodiversity Information Facility (GBIF) and offer new opportunities to discover, map and interpret plant distributions at various scales as well as promoting the free public access of data. E-Floras and taxonomic databases and information systems will become the norm. Further developments in cyber-taxonomy offer the promise of dealing with large-scale phylogenies via dispersed computing, of archiving massive phylogenetic knowledge and syntheses, browsing the tree of life and operating phylogenetically driven data-mining. Such developments will lead to a deeper understanding of the systematics, classification and relationships of the flowering plants and in turn are almost certain to lead to further modifications to our classifications both at family and other levels.

REFERENCE WORKS

Specific references are given after each family to relevant literature, but, in addition, some major works of reference are frequently mentioned in accounts by author and date but without an endnote reference, and these references are listed below.

APG II. An update of the Angiosperm Phylogeny Group classification for orders and families of flowering plants: APG II. *Bot. J. Linn. Soc.* 141: 399–43 (2003).

Bentham, G. & Hooker, W. J. *Genera Plantarum.* London, L. Reeve & Co. (1862–1883).

Brummitt, R. K. *Vascular Plant Families and Genera.* Richmond, Royal Botanic Gardens, Kew (1992).

Cronquist, A. *The Evolution and Classification of Flowering Plants.* Boston, Houghton Mifflin (1968).

Cronquist, A. *An Integrated System of Classification of Flowering Plants.* Columbia University Press, New York (1981).

Dahlgren, R. M. T., Clifford, H. T., & Yeo, P. F. *The Families of the Monocotyledons: Structure, Evolution, and Taxonomy.* Berlin & New York, Springer (1985).

Hultén, E. *The Circumpolar Plants.* Uppsala, Almqvist & Wiksells (1971).

Hutchinson, J. *The Families of Flowering Plants Arranged According to a New System Based on their Probable Phylogeny.* 3rd edn. Oxford, Clarendon Press (1973).

Kubitzki, K. *et al.* (eds), *The Families and Genera of Vascular Plants.* Berlin, Springer-Verlag (1990 on).

Mabberley, D. J. *The Plant-Book. A Portable Dictionary of the Vascular Plants.* 2nd edn. Cambridge, Cambridge University Press, (1997).

Smith, N. *et al.* (eds), *Flowering Plants of the Neotropics.* Princeton, Princeton, University Press (2004).

Soltis, D. E. *et al. Phylogeny and Evolution of Angiosperms.* Sunderland, Mass, Sinauer Associates (2005).

Stevens, P. F. (2001 onward). Angiosperm Phylogeny Website. Version 7, May 2006. http://www.mobot.org/MOBOT/research/APweb/ *or* http://seedplants.org/

Takhtajan, A. *Diversity and Classification of Flowering Plants.* New York, Columbia University Press (1997).

Thorne, R. F. The classification and geography of the flowering plants: dicotyledons of the class Angiospermae (subclass Magnoliidae, Ranunculidae, Caryophyllidae, Dilleniidae, Rosidae, Asteridae, and Lamiidae). *Bot. Rev.* 66: 441–647 (2000).

Thorne. R. F. *An Updated Classification of the Class Magnoliopsida ("Angiospermae").* http://rsabg.org/angiosperms

CLASSIFICATION

As noted in the introduction, it is not our intention to present a new system of classification but to provide a synthesis of the latest information at family level. To help the reader relate the families we have recognized to the widely cited APG II scheme, we have provided this table, in which families are arranged in the sequence given in APG II as modified by Soltis *et al., Phylogeny and Evolution of Angiosperms*, pp. 243–248 (2005).

Family names not indented under another family name are fully accepted in at least one of the two works. Those accepted by both are given in Roman type. Those accepted by Soltis *et al.* but not in this book are given in *italic type*. Those accepted in this book but not mentioned in Soltis *et al.* are given in **bold type**.

Family names indented under another family name are those listed by Soltis *et al.* as optionally recognizable from the nonindented family name above. Of these indented names, those given in *italic type* are here sunk under the nonindented family above, and those given in **bold type** are recognized in this book.

Amborellaceae
Chloranthaceae
Nymphaeaceae
Cabombaceae

Austrobaileyales
Austrobaileyaceae
Schisandraceae
Illiciaceae
Trimeniaceae

Ceratophyllales
Ceratophyllaceae

MAGNOLIIDS

Canellales
Canellaceae
Winteraceae

Laurales
Atherospermataceae
Calycanthaceae
Gomortegaceae
Hernandiaceae
Lauraceae
Monimiaceae
Siparunaceae

Magnoliales
Annonaceae
Degeneriaceae
Eupomatiaceae
Himantandraceae
Magnoliaceae
Myristicaceae

Piperales
Aristolochiaceae
Hydnoraceae
Lactoridaceae
Piperaceae
Saururaceae

MONOCOTYLEDONS

Petrosaviaceae

Acorales
Acoraceae

Alismatales
Alismataceae
Aponogetonaceae
Araceae
Butomaceae
Cymodoceaceae
Hydrocharitaceae
Juncaginaceae
Limnocharitaceae
Najadaceae
Posidoniaceae
Potamogetonaceae
Ruppiaceae
Scheuchzeriaceae
Tofieldiaceae
Zosteraceae

Asparagales
Alliaceae
    Agapanthaceae
    Amaryllidaceae
Anemarrhenaceae
Anthericaceae
Asparagaceae
    Agavaceae
    Aphyllanthaceae
    Hesperocallidaceae (see Agavaceae)
    Hyacinthaceae
    Laxmanniaceae
    Ruscaceae
    Themidaceae
Asteliaceae
Behniaceae
Blandfordiaceae
Boryaceae
Convallariaceae

Dracaenaceae
Doryanthaceae
Eriospermaceae
Herreriaceae
Hostaceae
Hypoxidaceae
Iridaceae
Ixioliriaceae
Johnsoniaceae
Lanariaceae
Nolinaceae
Orchidaceae
Tecophilaeaceae
Xanthorrhoeaceae
    Asphodelaceae
    Hemerocallidaceae
Xeronemataceae

Dioscoreales
Burmanniaceae
Dioscoreaceae
Taccaceae
Nartheciaceae

Liliales
Alstroemeriaceae
Campynemataceae
Colchicaceae
Corsiaceae
Liliaceae
Luzuriagaceae
Melanthiaceae
Petermanniaceae
Philesiaceae
Rhipogonaceae
Smilacaceae

Pandanales
Cyclanthaceae
Pandanaceae
Stemonaceae
Triuridaceae
Velloziaceae

COMMELINIDS

Dasypogonaceae

Arecales
Arecaceae

Commelinales
Commelinaceae
Haemodoraceae
Hanguanaceae
Philydraceae
Pontederiaceae

Poales
Anarthriaceae
Bromeliaceae
Centrolepidaceae
Cyperaceae

Ecdeiocoleaceae
Eriocaulaceae
Flagellariaceae
Hydatellaceae
Joinvilleaceae
Juncaceae
Mayacaceae
Poaceae
Prioniaceae
Rapateaceae
Restionaceae
Thurniaceae
Typhaceae
Xyridaceae

Zingiberales
Cannaceae
Costaceae
Heliconiaceae
Lowiaceae
Marantaceae
Musaceae
Strelitziaceae
Zingiberaceae

EUDICOTS

Buxaceae
Didymelaceae
Meliosmaceae
Sabiaceae
Trochodendraceae
    Tetracentraceae

Proteales
Nelumbonaceae
Proteaceae
    Platanaceae

Ranunculales
Berberidaceae
Circaeasteraceae
    Kingdoniaceae
Eupteleaceae
Glaucidiaceae
Hydrastidaceae
Lardizabalaceae
Menispermaceae
Papaveraceae
    Fumariaceae
    Pteridophyllaceae
Ranunculaceae
Sargentodoxaceae

CORE EUDICOTS

Aextoxicaceae
Berberidopsidaceae
Dilleniaceae

Gunnerales

Gunneraceae
Myrothamnaceae

Caryophyllales

Achatocarpaceae
Aizoaceae
Amaranthaceae
Ancistrocladaceae
Asteropeiaceae
Barbeuiaceae
Basellaceae
Cactaceae
Caryophyllaceae
Chenopodiaceae
Didiereaceae
Dioncophyllaceae
Droseraceae
Drosophyllaceae
Frankeniaceae
Gisekiaceae
Halophytaceae
Hectorellaceae
Limeaceae
Molluginaceae
Mesembryanthemaceae
Nepenthaceae
Nyctaginaceae
Petiveriaceae
Physenaceae
Phytolaccaceae
Plumbaginaceae
Polygonaceae
Portulacaceae
Rhabdodendraceae
Sarcobataceae
Simmondsiaceae
Stegnospermataceae
Tamaricaceae

Santalales

Eremolepidaceae
Loranthaceae
Misodendraceae
Olacaceae
Opiliaceae
Santalaceae
Viscaceae

Saxifragales

Altingiaceae (see
    Hamamelidaceae)
Aphanopetalaceae
Cercidiphyllaceae
Crassulaceae
Daphniphyllaceae
Grossulariaceae
Haloragaceae
Penthoraceae
Tetracarpaeaceae
Hamamelidaceae
Iteaceae

Pterostemonaceae
Paeoniaceae
Peridiscaceae
Saxifragaceae

ROSIDS

Aphloiaceae
Geissolomataceae
Ixerbaceae
Picramniaceae
Strasburgeriaceae
Vitaceae

Crossosomatales

Crossosomataceae
Stachyuraceae
Staphyleaceae

Geraniales

Geraniaceae
Hypseocharitaceae
Ledocarpaceae (see
    Vivianiaceae)
Melianthaceae
Francoaceae
Vivianiaceae

Myrtales

Alzateaceae
Combretaceae
Crypteroniaceae
Heteropyxidaceae
    (see Myrtaceae)
Lythraceae
Melastomataceae
Memecylaceae
Myrtaceae
Oliniaceae
Onagraceae
Penaeaceae
Psiloxylaceae (see Myrtaceae)
Rhynchocalycaceae
Vochysiaceae

EUROSIDS I (FABIDS)

Zygophyllaceae
Krameriaceae
Huaceae

Celastrales

Celastraceae
Lepidobotryaceae
Parnassiaceae
    Lepuropetalaceae
Stackhousiaceae

Cucurbitales

Anisophylleaceae
Begoniaceae
Coriariaceae
Corynocarpaceae
Cucurbitaceae
Datiscaceae

Tetramelaceae (see
    Datiscaceae)

Fabales

Leguminosae [= Fabaceae]
Polygalaceae
Quillajaceae
Surianaceae

Fagales

Betulaceae
Casuarinaceae
Fagaceae
Juglandaceae
    Rhoipteleaceae
Myricaceae
Nothofagaceae
Ticodendraceae

Malpighiales

Achariaceae
Balanopaceae
Bonnetiaceae
Caryocaraceae
Chrysobalanaceae
    Dichapetalaceae
    Euphroniaceae
    Trigoniaceae
Clusiaceae
Ctenolophonaceae
Elatinaceae
Euphorbiaceae
Flacourtiaceae
Goupiaceae
Hugoniaceae
Humiriaceae
Hypericaceae
    (see Clusiaceae)
Irvingiaceae
Ixonanthaceae
Lacistemataceae
Linaceae
Lophopyxidaceae
Malpighiaceae
Ochnaceae
    Medusagynaceae
    Quiinaceae
Pandaceae
Passifloraceae
    Malesherbiaceae
    Turneraceae
Phyllanthaceae
Picrodendraceae
Podostemaceae
Putranjivaceae
Rhizophoraceae
    Erythroxylaceae
Salicaceae
Scyphostegiaceae
Violaceae

Oxalidales

Brunelliaceae
Cephalotaceae
Connaraceae
Cunoniaceae
Elaeocarpaceae
Oxalidaceae

Rosales

Barbeyaceae
Cannabaceae
Cecropiaceae
Dirachmaceae
Elaeagnaceae
Moraceae
Rhamnaceae
Rosaceae
Ulmaceae
Urticaceae

EUROSIDS II (MALVIDS)

Tapisciaceae

Brassicales

Akaniaceae
    Bretschneideraceae
Bataceae
Brassicaceae
Capparaceae
Caricaceae
Cleomaceae
Emblingiaceae
Gyrostemonaceae
Koeberliniaceae
Limnanthaceae
Moringaceae
Pentadiplandraceae
Resedaceae
Salvadoraceae
Setchellanthaceae
Tovariaceae
Tropaeolaceae

Malvales

Bixaceae
    Diegodendraceae
    Cochlospermaceae
Bombacaceae
Brownlowiaceae
Byttneriaceae
Cistaceae
Dipterocarpaceae
Durionaceae
Helicteraceae
Malvaceae
Muntingiaceae
Neuradaceae
Pentapetaceae
Sarcolaenaceae
Sparrmanniaceae
Sphaerosepalaceae
Sterculiaceae
Tepuianthaceae
Thymelaeaceae
Tiliaceae

Sapindales

Anacardiaceae
Biebersteiniaceae
Burseraceae
Cneoraceae
Kirkiaceae
Leitneriaceae
Meliaceae

Nitrariaceae
Peganaceae
Tetradiclidaceae
Ptaeroxylaceae
Rutaceae
Sapindaceae
Simaroubaceae

ASTERIDS

Cornales
Cornaceae
   Nyssaceae
Curtisiaceae
Grubbiaceae
Hydrangeaceae
Hydrostachyaceae
Loasaceae

Ericales
Actinidiaceae
Balsaminaceae
Clethraceae
Coridaceae
Cyrillaceae
Diapensiaceae
Ebenaceae
Ericaceae
Fouquieriaceae
Lecythidaceae
Lissocarpaceae
Maesaceae
Marcgraviaceae
Myrsinaceae
Pentaphylaceae
   Sladeniaceae
Polemoniaceae
Primulaceae
Roridulaceae
Samolaceae
Sapotaceae
Sarraceniaceae
Styracaceae
Symplocaceae
Ternstroemiaceae (see
   Pentaphylacaceae)
Tetrameristaceae
   Pellicieraceae
Theaceae
Theophrastaceae

EUASTERIDS I (LAMIIDS)

Boraginaceae
Hydrophyllaceae
Icacinaceae
Lennoaceae
Oncothecaceae
Vahliaceae
Metteniusaceae

Garryales
Eucommiaceae
Garryaceae
   Aucubaceae

Gentianales
Apocynaceae
Gelsemiaceae
Gentianaceae
Loganiaceae
Rubiaceae

Lamiales
Acanthaceae
Avicenniaceae
Bignoniaceae
Buddlejaceae
Byblidaceae
Calceolariaceae (see
   Scrophulariaceae)
Callitrichaceae
Carlemanniaceae
Cyclocheilaceae
Gesneriaceae
Globulariaceae
Hippuridaceae
Lamiaceae
Lentibulariaceae
Martyniaceae
Myoporaceae
Nesogenaceae
Orobanchaceae (see
   Scrophulariaceae)
Oleaceae
Pedaliaceae
Phrymaceae
Plantaginaceae
Plocospermataceae
Schlegeliaceae
Scrophulariaceae
Stilbaceae
Symphoremataceae
Tetrachondraceae
Thomandersiaceae
Trapellaceae
Verbenaceae

Solanales
Convolvulaceae
Hydroleaceae
Montiniaceae
Solanaceae
Sphenocleaceae

EUASTERIDS II (CAMPANULIDS)

Bruniaceae
Columelliaceae
   Desfontainiaceae
Escalloniaceae
Paracryphiaceae
Polyosmaceae
Quintiniaceae
Sphenostemonaceae
Tribelaceae

Apiales
Apiaceae
Araliaceae
Aralidiaceae (see
   Torricelliaceae)

Griseliniaceae
Mackinlayaceae (see Apiaceae)
Melanophyllaceae
Myodocarpaceae
Pennantiaceae
Pittosporaceae
Torricelliaceae

Aquifoliales
Aquifoliaceae
Cardiopteridaceae
Helwingiaceae
Phyllonomaceae
Stemonuraceae

Asterales
Alseuosmiaceae
Argophyllaceae
Asteraceae
Calyceraceae
Campanulaceae
   Lobeliaceae
Goodeniaceae
Menyanthaceae
Pentaphragmataceae
Phellinaceae
Rousseaceae
Stylidiaceae
   Donatiaceae

Dipsacales
Adoxaceae
Caprifoliaceae
   Diervillaceae
   Dipsacaceae
   Linnaeaceae
   Morinaceae
   Valerianaceae
Sambucaceae
Triplostegiaceae
Viburnaceae

FAMILIES OF UNCERTAIN POSITION

Apodanthaceae
Balanophoraceae
Cynomoriaceae
Cytinaceae
Dipentodontaceae
Haptanthaceae
Hoplestigmataceae
Leptaulaceae
Medusandraceae
Mitrastemonaceae
Pottingeriaceae
Pteleocarpaceae
Rafflesiaceae

GLOSSARY

abaxial On side facing away from stem or axis.

acaulescent Stemless or nearly so.

achene A small, dry, single-seeded fruit that does not split open (Plate III).

acuminate Narrowing gradually to a point (Plate VII).

acute Having a sharp point (Plate VII).

adaxial On the side facing the stem or axis.

adnate Joined to or attached to; applied to unlike organs, e.g., stamens adnate to perianth; cf connate.

adventitious Arising from an unusual position, e.g., roots from a stem or leaf.

aerial root A root that originates above the ground level.

aestivation The arrangement of the parts of a flower within the bud, usually referring to sepals and petals (Plate I).

alternate (of leaves) One leaf at each node

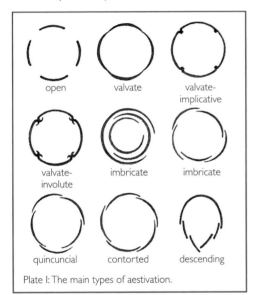

Plate I: The main types of aestivation.

of the stem (Plate VII); (of stamens) between the petals.

amphitropous (of ovules) Attached near its middle, half-inverted (Plate VIII).

anatropous (of ovules) Bent over through 180 degrees to lie alongside its stalk (funicle) (Plate VIII).

androecium All the male reproductive organs of a flower; the stamens; cf gynoecium (Plate IX).

androgynophore A column on which stamens and carpels are borne.

angiosperm A plant producing seeds enclosed in an ovary. A flowering plant.

annual A plant that completes its life cycle from germination to death within one year.

annular Ringlike.

anther The terminal part of the male organs (stamen), usually borne on a stalk (filament) and developing to contain pollen (Plate IX).

anthesis The period of flowering; from opening of the flower bud to the setting of the seed.

anthocyanin Pigment usually responsible for pink, red, purple, violet and blue in plants.

antipetalous Occurring opposite the petals, on the same radius, as distinct from alternating with the petals.

antisepalous Occurring opposite the sepals, on the same radius, as distinct from alternating with the sepals.

aperturate (of pollen) Having one or more apertures.

apetalous Without petals.

apex Tip of an organ; usually the growing point.

apical Pertaining to the apex.

apocarpous With carpels free from each other (Plate IV).

apomixis (adj. apomictic) Reproduction by seed formed without sexual fusion.

aquatic Living in water.

aril A fleshy or sometimes hairy outgrowth from the hilum or funicle of a seed.

asepalous Without sepals.

auricle (adj. auriculate) Small earlike projections at the base of a leaf or leaf blade or bract (Plate VII).

awn A stiff, bristlelike extension to an organ, usually at the tip.

axil The upper angle formed by the union of a leaf with the stem (Plate IV).

axile placentation A type of placentation in which the ovules are borne on placentas on the central axis of an ovary that has two or more locules (Plate XI).

axillary Pertaining to the organs in the axil, e.g., the buds, flowers, or inflorescence.

axis The main or central stem of a herbaceous plant or of an inflorescence.

baccate Berrylike.

basal Borne at or near the base.

basal placentation Having the placenta at the base of the ovary (Plate XI).

basifixed (of anthers) Attached at the base to the filament, and therefore lacking independent movement; cf dorsifixed (Plate IX).

berry A fleshy fruit without a stony layer, usually containing many seeds (Plate III).

betalains Red and yellow alkaloid pigments present in members of the Caryophyllales.

bi- A prefix meaning two or twice.

bicarpellate (of ovaries) Derived from two carpels.

biennial A plant that completes its life cycle in more than one, but less than two years and which usually flowers in the second year.

bifid Forked; having a deep fissure near center.

bilabiate Two-lipped (Plate IV).

bipinnate (of leaves) A pinnate leaf with the primary leaflets themselves divided in a pinnate manner; cf pinnate (Plate VII).

biseriate In two rows.

bisexual (of flowers) Containing both male and female reproductive organs in a single flower; cf unisexual (Plate IV).

blade The flattened part of a leaf; the lamina (Plate VII).

bostryx (of inflorescences) A cymose inflorescence with successive branches on one side only; normally coiled like a spring (Plate V).

bract A leaf, often modified or reduced, which subtends a flower or inflorescence in its axil (Plate V).

bracteole A small leaflike organ, occurring along the length of a flower stalk, between a true subtending bract and the calyx (Plate V).

bulb An underground organ comprising a short disklike stem, bearing fleshy scale leaves, buds and surrounded by protective scale leaves; it acts as a perennating organ and is a means of vegetative reproduction; cf corm, tuber.

bulbil A small bulb or bulblike organ often produced on above-ground organs.

caducous Falling off prematurely or easily.

calcicole A plant that favors soil containing lime.

calcifuge A plant that avoids soil containing lime.

calyculus A group of leaflike appendages below the calyx.

calyx Collective term for all a flower's sepals.

cambium A layer of cells that occurs within the stem and roots which divides to form secondary permanent tissues.

campanulate Bell shaped (Plate IV).

campylotropous (of ovules) Bent over through 90 degrees so that the stalk (funicle) appears to be attached to the side of the ovule (Plate VIII).

capitate Headlike.

capitulum An inflorescence consisting of a head of closely packed stalkless flowers (Plate V).

capsule A dry fruit that normally splits open to release its seeds (Plate III).

carnivorous plant A plant that is capable of catching and digesting small animals such as insects.

carpel One of the flower's female reproductive organs, comprising an ovary and a stigma, and containing one or more ovules (Plate IV).

caruncle (adj. carunculate) A fleshy, sometimes colored, outgrowth near hilum of some seeds.

caryopsis A dry fruit (achene) typical of grasses (Plate III).

catkin A pendulous inflorescence of simple, usually unisexual flowers (Plate V).

ciliate (of margins) Fringed with small hairs (Plate VII).

cincinnus A monochasial, cymose inflorescence with branches alternating from one side of the vertical axis to the other; normally curved to one side (Plate V).

circumscissile Opening all round by a transverse split (Plate III).

cladode A flattened stem which has assumed the form and function of a leaf.

claw The narrow basal part of some petals and sepals.

cleistogamic (cleistogamous) (of flowers) Self-pollinating, without the flower ever opening.

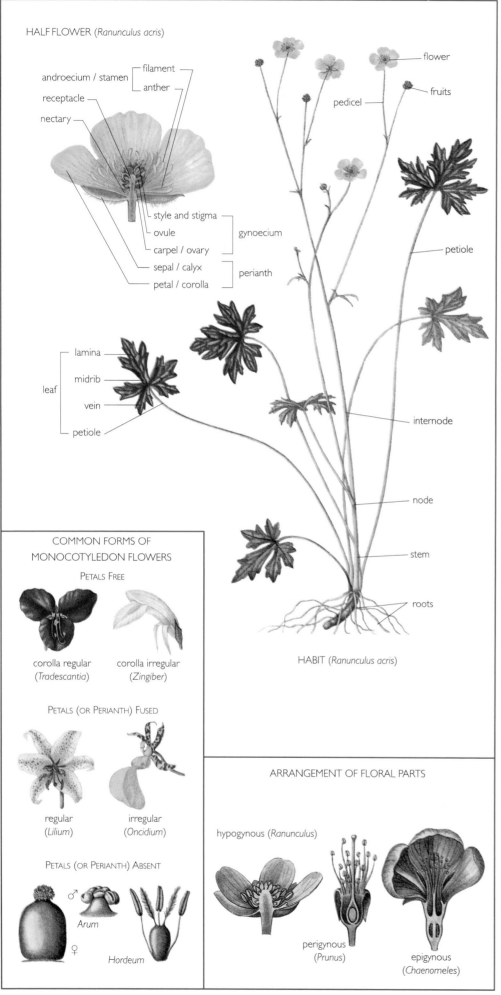

HALF FLOWER (*Ranunculus acris*)

androecium / stamen — filament — anther
receptacle
nectary
style and stigma
ovule — gynoecium
carpel / ovary
sepal / calyx — perianth
petal / corolla

flower
pedicel
fruits

petiole

lamina
midrib
leaf
vein
petiole

internode

node

stem

roots

HABIT (*Ranunculus acris*)

COMMON FORMS OF MONOCOTYLEDON FLOWERS

PETALS FREE

corolla regular (*Tradescantia*) corolla irregular (*Zingiber*)

PETALS (OR PERIANTH) FUSED

regular (*Lilium*) irregular (*Oncidium*)

PETALS (OR PERIANTH) ABSENT

Arum ♂ ♀ Hordeum

ARRANGEMENT OF FLORAL PARTS

hypogynous (*Ranunculus*)

perigynous (*Prunus*)

epigynous (*Chaenomeles*)

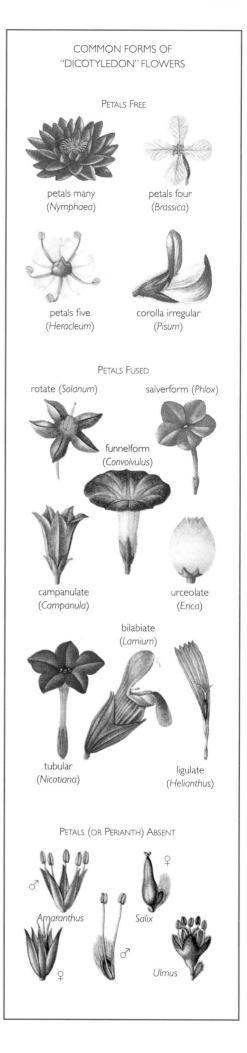

COMMON FORMS OF "DICOTYLEDON" FLOWERS

PETALS FREE

petals many (*Nymphaea*) petals four (*Brassica*)

petals five (*Heracleum*) corolla irregular (*Pisum*)

PETALS FUSED

rotate (*Solanum*) salverform (*Phlox*)

funnelform (*Convolvulus*)

campanulate (*Campanula*) urceolate (*Erica*)

bilabiate (*Lamium*)

tubular (*Nicotiana*) ligulate (*Helianthus*)

PETALS (OR PERIANTH) ABSENT

♂ ♀
Amaranthus Salix
♂
♀ Ulmus

Plate II: Structure of the flowering plant and flower, also showing the main flower forms.

colpate (of pollen) Having one or more colpi (oblong-elliptic apertures in the pollen-wall).

column (of a flower) Combined stamen and style (in orchids) or united staminal filaments (hibiscus).

compound Consisting of several parts, e.g., a leaf with several leaflets or an inflorescence with more than one group of flowers.

connate Joined or attached to; applied to similar organs fused during development, e.g., stamens fused into a tube; cf adnate.

connective (of stamens) The tissue connecting the pollen sacs of an anther (Plate IX).

contorted (of sepals and petals) Twisted in the bud so that they overlap on one side only; spirally twisted.

convolute Rolled together.

cordate (of leaves) Heart shaped (Plate VII).

coriaceous Leathery.

corm A bulbous, swollen, underground, stem base, bearing scale leaves and adventitious roots; cf bulb, tuber.

corolla All the petals of a flower; it is normally colored (Plate II).

corona A series of petal-like structures in a flower, either outgrowths from the petals, or modified from the stamens, e.g., a daffodil "trumpet."

corymb A rounded or flat-topped inflorescence of racemose type, in which the lower (outer) flower stalks (pedicels) are longer than the upper (inner) ones, so that all the flowers are at about the same level (Plate V).

corymbose Arranged in a corymb; corymb-like.

cotyledon(s) The first leaf, or pair of leaves, of an embryo within the seed.

crenate (of leaf margins) Round toothed (Plate VII).

crenulate Finely crenate (Plate VII).

cross-fertilization, cross-pollination *See* Fertilization, Pollination.

cupule (adj. cupulate) A cup-shaped sheath, surrounding some fruits.

cyme An inflorescence in which each terminal growing point produces a flower. Subsequent growth is therefore from a lateral growing point, the oldest flowers being at the apex, or center, if flat (Plate V).

cymose Arranged in a cyme; cymelike (Plate V).

cypsela A single-seeded fruit derived from a unilocular, inferior ovary (Plate III).

cystolith A crystal or deposit of lime, within a cell.

deciduous The shedding of leaves seasonally.

declinate (of stamens) Curving downward.

decussate (of leaves) Arranged in opposite pairs on the stem, with each pair at 90 degrees to the preceding pair (Plate VII).

dehiscence The method or act of opening.

dehiscent Opening to shed pollen or seeds.

dehiscing In the act of shedding pollen or seeds.

dentate Having a toothed margin (Plate VII).

denticulate Having a finely toothed margin (Plate VII).

di- A prefix meaning *two*.

dichasium (of inflorescences) A form of cymose inflorescence with each branch giving rise to two other branches; cf monochasium (Plate V).

didymous In pairs.

didynamous Having two long and two short stamens (Plate IX).

dimorphism (adj. dimorphic) Having two distinct forms.

dioecious Having male and female flowers borne on separate plants.

disk The fleshy outgrowth developed from the receptacle at the base of the ovary or from the stamens surrounding the ovary; it often secretes nectar.

distichous Arranged in two vertical rows.

dorsal Upper.

dorsifixed (of anthers) Attached at the back to the filament; cf basifixed (Plate IX).

drepanium (of inflorescences) A cymose inflorescence with successive branches on one side only; normally flattened in one plane and curved to one side (Plate V).

drupe A fleshy fruit containing one or more seeds, each of which is surrounded by a stony layer (Plate III).

elaiosome A fleshy outgrowth on a seed with oily substances attractive to ants.

elliptic(al) (of leaves) Oval shaped, with narrowed ends (Plate VII).

embryo The rudimentary plant within the seed.

embryo sac The central portion of the ovule; a thin-walled sac within the nucellus containing the egg nucleus (Plate VIII).

endocarp The innermost layer of the ovary wall (pericarp) of a fruit. In some fruits it becomes hard and "stony"; cf drupe (Plate III).

endosperm In some seeds, fleshy nutritive material derived from embryo sac (Plate III).

entire (of leaves) With an undivided margin (Plate VII).

epicalyx A whorl of sepal-like appendages resembling the calyx but outside the true calyx.

epidermis Usually a single layer of living cells forming the protective covering to many plant organs, particularly leaves, petals, sepals and herbaceous stems.

epigynous (of flowers) With sepals, petals, and stamens inserted near top of ovary (Plate II).

epipetalous Attached to petals or corolla.

epiphyte A plant that grows on the surface of another, without deriving food from its host.

erect (of an ovule) Upright, with its stalk at the base (Plate VIII).

exine Outer layer of a pollen grain's wall.

exocarp Outermost layer of fruit wall (Plate III).

exserted Protruding.

exstipulate Without stipules.

extrorse (of anthers) Opening away from the axis of growth towards the corolla (Plate IX).

fascicle A cluster or bundle.

female flower A flower containing functional carpels, but not stamens (Plate II).

fertilization The fusion of male and female reproductive cells (gametes) in the ovary after pollination. *Cross-fertilization* occurs between flowers from separate plants; *self-fertilization* occurs between flowers on the same plant or within the same flower.

filament The anther-bearing stalk of a stamen (Plate IX).

filiform Threadlike.

fimbriate (of margins) Fringed, usually with hairs.

flower The structure concerned with sexual reproduction in the Angiosperms. Essentially it consists of the male organs (androecium) comprising the stamens, and the female organs (gynoecium) comprising the ovary, style(s) and stigma(s), usually surrounded by a whorl of petals (the corolla) and a whorl of sepals (the calyx). The male and female parts may be in the same flower (bisexual) or in separate flowers (unisexual) (Plate II).

follicle A dry fruit which is derived from a single carpel and which splits open along one side only (Plate III).

free (of petals, sepals, etc.) Not joined to each other or to any other organ (Plate I, Plate IV).

free central placentation A type of placentation in which the ovules are borne on placentas on a free, central column within an ovary that has only one locule (Plate XI).

fruit Strictly the ripened ovary of a seed plant and its contents. Loosely, the whole structure containing ripe seeds, which may include more than the ovary; cf achene, berry, capsule, drupe, follicle, nut, samara (Plate III).

funicle The stalk of an ovule (Plate VIII).

gamopetalous / gamosepalous With petals / sepals fused, at least at the base.

glabrous Without hairs or projections.

gland (adj. glandular) Secreting organ producing oil, resin, nectar, water, etc.; cf hydathode, nectary.

glaucous With a waxy, grayish blue bloom.

globose Spherical, rounded.

gynobasic style A style that arises near the base of a deeply lobed ovary (Plate XI).

gynoecium All the female reproductive organs of a flower, comprising one or more free or fused carpels (Plate IV).

gynophore Stalk of a carpel or gynoecium.

habit The characteristic mode of growth or occurrence; the form and shape of a plant.

halophyte A plant that tolerates salty conditions.

hardy Able to withstand extreme conditions, usually of cold.

haustorium A peglike fleshy outgrowth from a parasitic plant, usually embedded in the host plant and drawing nourishment from it.

head A dense inflorescence of small, crowded, often stalkless, flowers; a capitulum (Plate V).

helicoid (of cymose inflorescences) Coiled like a spring.

herb (adj. herbaceous) A plant that does not develop persistent woody tissue above ground and either dies at the end of the growing season or overwinters by means of underground organs, e.g., bulbs, corms, rhizomes.

heterophylly Having leaves of more than one type on the same plant.

heterostyly Having styles (and usually stamens) of two or more lengths in different flowers within a species.

hilum The scar left on a seed marking the point of attachment to the stalk of the ovule.

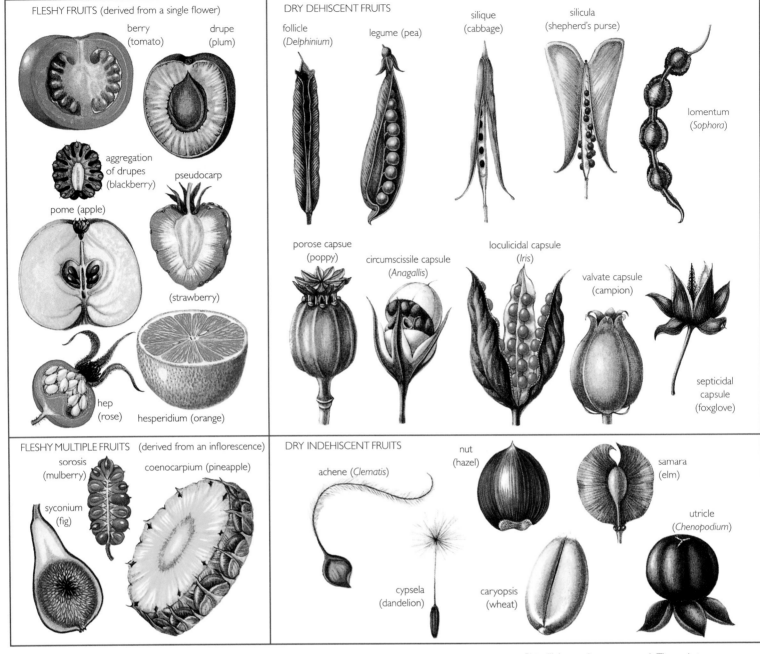

FLESHY FRUITS (derived from a single flower)

berry (tomato)

drupe (plum)

aggregation of drupes (blackberry)

pseudocarp

pome (apple)

(strawberry)

hep (rose)

hesperidium (orange)

DRY DEHISCENT FRUITS

follicle (*Delphinium*)

legume (pea)

silique (cabbage)

silicula (shepherd's purse)

lomentum (*Sophora*)

porose capsue (poppy)

circumscissile capsule (*Anagallis*)

loculicidal capsule (*Iris*)

valvate capsule (campion)

septicidal capsule (foxglove)

FLESHY MULTIPLE FRUITS (derived from an inflorescence)

sorosis (mulberry)

coenocarpium (pineapple)

syconium (fig)

DRY INDEHISCENT FRUITS

achene (*Clematis*)

nut (hazel)

samara (elm)

utricle (*Chenopodium*)

cypsela (dandelion)

caryopsis (wheat)

Plate III (ABOVE & OPPOSITE PAGE): The main types of fruit and their structure.

hirsute Covered in rough, coarse hairs.
honey-guide markings (e.g., lines or dots) on the perianth that direct insects to the nectar.
hydathode A specialized gland, usually found in leaves that exude water.
hydrophyte An aquatic plant.
hypanthium A cup-shaped enlargement of the floral receptacle or the bases of the floral parts, which often enlarges and surrounds the fruits, e.g., the fleshy tissue in rose-hips.
hypha The threadlike part of a fungal body.
hypogynous (of flowers) With the sepals, petals and stamens attached to the receptacle or axis, below and free from the ovary (Plate II).
imbricate (of sepals and petals) Overlapping, as in a tiled roof.
imparipinnate (of leaves) A pinnate leaf with an central unpaired terminal leaflet (Plate VII).
inaperturate (of pollen grains) Without an aperture; without any pores or furrows.
incised (of leaves) Sharply, deeply cut (Plate VII).
incompatible Describes plants between which hybrids cannot be formed.

indefinite (of flower parts) Of a number large enough to make a precise count difficult.
indehiscent Fruits not opening to release seeds; cf dehiscent.
indumentum A covering, usually of hairs.
inferior (of ovaries) An ovary with sepals, petals, and stamens attached to its apex (Plate II).
inflorescence Any arrangement of more than one flower, e.g., bostryx, capitulum, corymb, cyme, dichasium, fascicle, panicle, raceme, rhipidium, spadix, spike, and thyrse (Plate V).
infructescence A cluster of fruits, derived from an inflorescence.
integument (of ovules) The outer 1–2 protective covering of the ovule (Plate VIII).
internode The length of stem that lies between two leaf-joints (nodes) (Plate II).
introrse Directed and opening inward toward the center of the flower; cf extrorse (Plate IX).
involucel A whorl of bracteoles; cf bracteole.
involucre A whorl of bracts beneath an inflorescence; cf bract.

irregular (of flowers) Not regular; not divisible into halves by an indefinite number of longitudinal planes; zygomorphic.
lacerate (of leaves) Irregularly cut (Plate VII).
lamina Thin, flat blade of leaf or petal (Plate II).
lanceolate Narrow, as a lance, with tapering ends (Plate VII).
lateral Arising from the side of the parent axis or attached to the side of another organ.
latex A milky and usually whitish fluid that is produced by the cells of various plants and is the source of, e.g., rubber, gutta percha, chicle and balata.
laticiferous Producing a milky juice (latex).
leaf An aerial and lateral outgrowth from a stem that makes up the foliage of a plant. It typically consists of a stalk (petiole) and a flattened blade (lamina) (Plates II and VII).
leaflet Each separate lamina of a compound leaf (Plate VII).

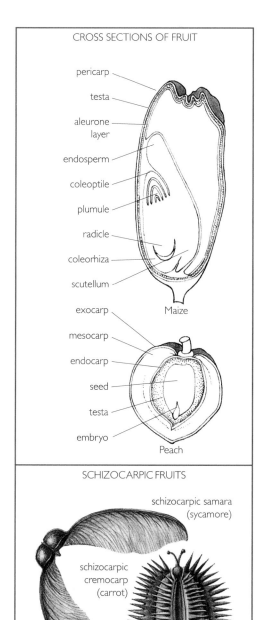

CROSS SECTIONS OF FRUIT

pericarp
testa
aleurone layer
endosperm
coleoptile
plumule
radicle
coleorhiza
scutellum

Maize

exocarp
mesocarp
endocarp
seed
testa
embryo

Peach

SCHIZOCARPIC FRUITS

schizocarpic samara (sycamore)

schizocarpic cremocarp (carrot)

liana A woody, climbing vine.

ligulate Strap or tongue shaped (Plate II).

ligule (of leaves) A scalelike membrane on the surface of a leaf; (of flowers) the strap-shaped corolla in some Compositae.

limb The upper, expanded portion of a calyx or corolla with fused parts; cf tube.

linear (of leaves) Elongated, and with parallel sides (Plate VII).

lithophyte A plant that grows on stones.

lobe (of leaves or perianths) A curved or rounded part.

lobed (of leaves) With curved or rounded edges (Plate VII).

locule The chamber or cavity of an ovary that contains the ovules, or, within an anther, that contains the pollen (Plate XI).

loculicidal Splitting open longitudinally along the dorsal suture (mid-rib) of each segment of the wall (Plate III).

male flower A flower containing functional stamens but no carpels.

marginal placentation A type of placentation in which the ovules are borne along the fused margins of a single carpel, eg pea seeds in a pod (Plate XI).

membranous Resembling a membrane; thin, dry and semitransparent.

mericarp A one-seeded portion of a fruit which splits up when the fruit is mature, e.g., the fruits of the Umbelliferae (Plate III).

meristem A group of cells capable of dividing indefinitely.

mesocarp The middle layer of the fruit wall (pericarp). It is usually fleshy, as in a berry (Plate III).

mesophyte A plant having moderate moisture requirements.

micropyle The opening through the integuments of an ovule, through which the pollen-tube grows after pollination (Plate VIII).

midrib The central or largest vein of a leaf or carpel (Plate II).

mono- A prefix meaning single, one, or once.

monocarpic Fruiting only once and then dying.

monochasium A cymose inflorescence in which there is a single terminal flower with below it a single branch bearing flower(s) (Plate V).

monocolpate (of pollen) Having a single colpus (an oblong-elliptic aperture).

monoecious Having separate male and female flowers on the same plant.

monogeneric (of a family) Of only one genus.

monopodial (of stems or rhizomes) With branches or appendages arising from a simple main axis; cf sympodial.

monotypic A genus or family of a single species.

mucilage A slimy secretion that swells on contact with water.

multiseriate (of flower parts) Borne in many series or whorls.

mycorrhiza The symbiotic association of the roots of some seed plants with fungi.

naked (of flowers) Lacking a perianth (Plate II).

nectar A sugary liquid secreted by some plants; it forms the principle raw material of honey.

nectary The gland in which nectar is produced.

node The point on a stem where one or more leaves are borne (Plate VII).

nucellus The central tissue of the ovule, containing the embryo sac and surrounded by, in angiosperms, 1–2 integuments (Plate VIII).

numerous (of floral parts) Usually meaning more than ten; cf indefinite.

nut A dry, single-seeded and nonopening (indehiscent) fruit with a woody pericarp (Plate III).

obligate (of parasite) Plant unable to grow on its own; entirely dependent on a host for nutrition.

obovate (of leaves) Having the outline of an egg, with the broadest part above the middle and attached at the narrow end (Plate VII).

ochrea A cup-shaped structure formed by the joining of stipules or leaf bases around a stem.

opposite (of leaves) Occurring in pairs on opposite sides of the stem; (of stamens) inserted in front of the petals (Plate VII).

orbicular More or less circular (Plate VII).

orthotropous (of ovules) Borne on a straight stalk (funicle); not bent over (Plate VIII).

ovary The hollow basal region of a carpel, containing one or more ovules and surmounted by the style(s) and stigma(s). It is made up of one or more carpels which may fuse together in different ways to form one or more chambers (locules). The ovary is generally above the perianth parts (superior) or below them (inferior) (Plate II).

ovate (of leaves) Having the outline of an egg with narrow end above middle (Plate VII).

ovoid (of leaves) Egg shaped (Plate VII).

ovule The structure in the chamber (locule) of an ovary containing the egg cell within the embryo sac which is surrounded by the nucellus. It is enclosed by 1–2 integuments. The ovule develops into the seed after fertilization (Plate VIII).

palmate (of leaves) With more than three segments or leaflets arising from a single point, as in the fingers of a hand (Plate VII).

panicle (of inflorescences) Strictly a branched raceme, with each branch bearing a further raceme of flowers. More loosely applies to any complex, branched inflorescence (Plate V).

paniculate Arranged in a panicle.

parasite Plant that usually obtains its food from another living plant to which it is attached.

parenchyma Tissue made up of thin-walled living photosynthetic or storage cells which is capable of division even when mature.

parietal placentation A type of placentation in which ovules are borne on placentas on the inner surface of ovary's outer wall (Plate XI).

paripinnate A pinnate leaf with all leaflets in pairs; cf imparipinnate.

pedate (of leaves) Palmately divided compound leaf, with the lateral lobes themselves divided. There may be a free central leaflet.

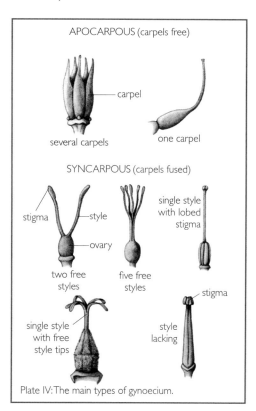

APOCARPOUS (carpels free)

carpel

several carpels

one carpel

SYNCARPOUS (carpels fused)

stigma
style
ovary

two free styles

five free styles

single style with lobed stigma

single style with free style tips

stigma

style lacking

Plate IV: The main types of gynoecium.

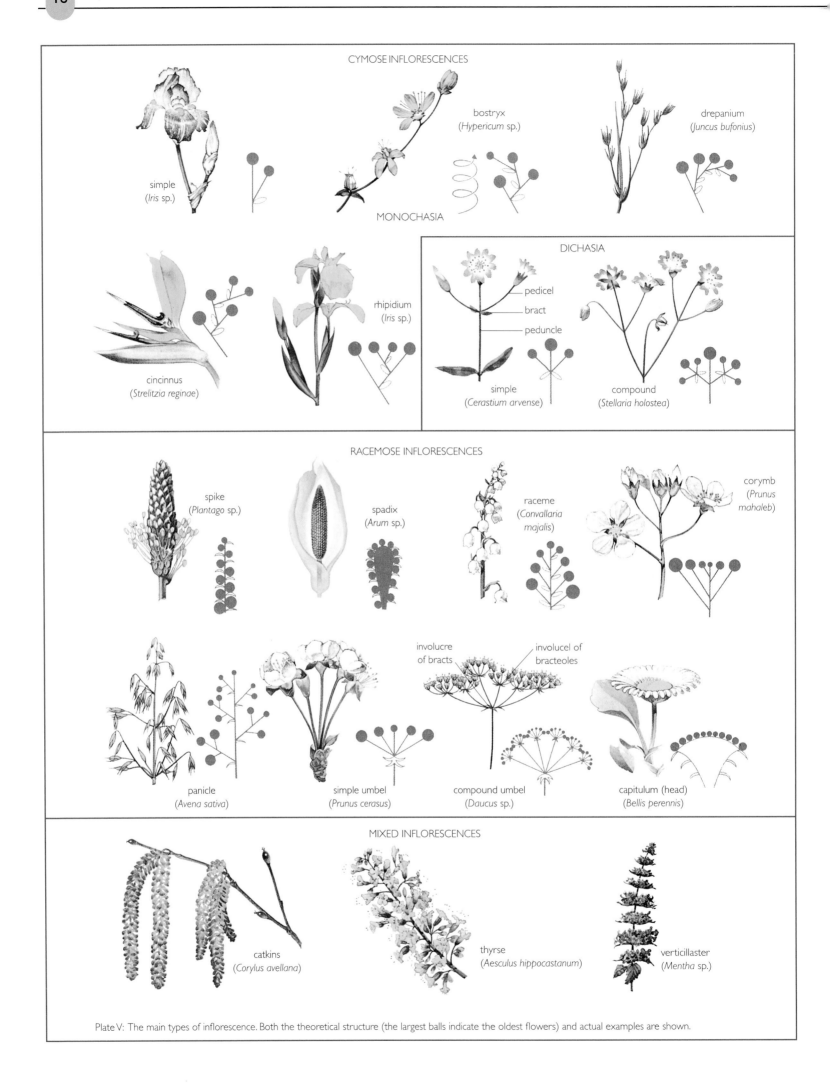

CYMOSE INFLORESCENCES

simple
(*Iris* sp.)

bostryx
(*Hypericum* sp.)

drepanium
(*Juncus bufonius*)

MONOCHASIA

cincinnus
(*Strelitzia reginae*)

rhipidium
(*Iris* sp.)

DICHASIA

pedicel
bract
peduncle

simple
(*Cerastium arvense*)

compound
(*Stellaria holostea*)

RACEMOSE INFLORESCENCES

spike
(*Plantago* sp.)

spadix
(*Arum* sp.)

raceme
(*Convallaria majalis*)

corymb
(*Prunus mahaleb*)

panicle
(*Avena sativa*)

simple umbel
(*Prunus cerasus*)

involucre
of bracts

involucel of
bracteoles

compound umbel
(*Daucus* sp.)

capitulum (head)
(*Bellis perennis*)

MIXED INFLORESCENCES

catkins
(*Corylus avellana*)

thyrse
(*Aesculus hippocastanum*)

verticillaster
(*Mentha* sp.)

Plate V: The main types of inflorescence. Both the theoretical structure (the largest balls indicate the oldest flowers) and actual examples are shown.

pedicel The stalk of a single flower (Plate II).

peduncle The stalk of an inflorescence (Plate V).

peltate (of leaves) More or less circular and flat with the stalk inserted in the middle (Plate VII).

pendulous Hanging down (Plate XI).

perennating Living over from season to season.

perennial A plant that persists for more than two years and normally flowers annually.

perfect flower A flower with functional male and female organs.

perianth The floral envelope whose segments are usually divisible into an outer whorl (calyx) of sepals, and an inner whorl (corolla) of petals. The segments of either or both whorls may fuse to form a tube (Plate II).

pericarp The wall of a fruit that encloses the seeds and develops from ovary wall (Plate II).

perigynous (of flowers) Having the stamens, corolla, and calyx inserted around the ovary, their bases often forming a cup-shaped disk (Plate II).

perisperm The nutritive storage tissue in some seeds, derived from the nucellus.

persistent Remaining attached, not falling off.

petal A nonreproductive (sterile) part of the flower, usually conspicuously colored; one of the units of the corolla (Plate II).

petaloid Petal-like.

petiole The stalk of a leaf (Plate VII).

phloem That part of the tissue of a plant that is concerned with conducting food material. In woody stems it is the innermost layer of the bark; cf xylem.

photosynthesis The process by which green plants manufacture sugars from water and carbon dioxide by converting the energy from light into chemical energy with the aid of the green pigment chlorophyll.

phyllode A flattened leaf stalk (petiole) that has assumed the form and function of a leaf blade.

pinnate (of leaves) Compound, with leaflets in pairs on opposite sides of the midrib; cf imparipinnate and paripinnate (Plate VII).

pinnatisect (of leaves) Pinnately divided, but not as far as the midrib (Plate VII).

pistil The female reproductive organ consisting of one or more carpels, comprising ovary, style and stigma; the gynoecium as a whole.

pistillate A flower that has only female organs.

pistillode A sterile, often reduced pistil.

placenta Part of the ovary wall to which the ovules are attached.

placentation The arrangement and distribution of the ovule-bearing placentas within the ovary; cf axile, basal, free central, marginal and parietal (Plate VI).

plumule The rudimentary shoot in an embryo.

pollen Collective name for the pollen grains, i.e., the minute spores (microspores) produced in the anthers.

pollen sac The chamber (locule) in an anther where the pollen is formed (Plate IX).

pollination The transfer of pollen grains from stamen to stigma. *Cross-pollination* occurs between flowers of different plants of the same species; *self-pollination* occurs between flowers of the same plant, or within one flower.

pollinium A mass of pollen grains produced

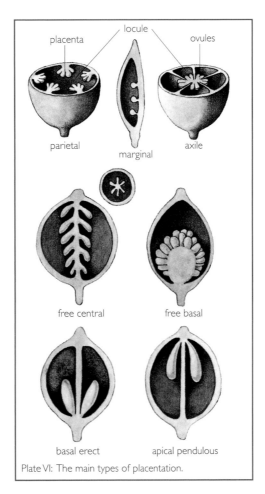
Plate VI: The main types of placentation.

by one anther-lobe, cohering together and transported as a single unit during pollination, as in the orchids.

polygamodioecious Having male and bisexual flowers on one individual plant and female and bisexual flowers on another.

polygamous Having separate male, female and bisexual flowers on the same plant.

polypetalous With petals free from each other.

pore A small hole.

protandrous (of flowers) The maturing of stamens and the consequent release of their pollen before the stigmas of the same flower become receptive.

protogynous (of flowers) The receptiveness of the stigmas before the stamens of the same flower mature and release their pollen.

pseudocopulation The attempted copulation by male insect visitors with a part of a flower that resembles the female of the insect species, as in some orchids.

pseudo-whorled (of leaves) Arising close together and so appearing to arise at the same level, although not in fact doing so.

pubescent Covered in soft, short hairs.

punctate Shallowly pitted or dotted, often with glands.

raceme An inflorescence consisting of a main axis, bearing single flowers alternately or spirally on stalks (pedicels) of approximately equal length. The apical growing point continues to be active so there is usually no terminal flower and the youngest branches or flowers are nearest the apex. This mode of growth is known as monopodial (Plate V).

racemose Arranged like a raceme; in general any inflorescence capable of indefinite prolongation, having lateral and axillary flowers (Plate V).

radical (of leaves) Arising from the base of a stem or from a rhizome; basal.

radicle The rudimentary root in an embryo.

raphe A ridge or tissue visible on the testa of seeds developed from ovules which are bent over through 180 degrees (anatropous). It results from the fusion of the stalk (funicle) with the rest of the bent-over ovules.

ray (of wood) Radial strands of living cells concerned with the transport of water and food.

receptacle Flat, concave or convex part of the stem from which all parts of a flower arise; the floral axis (Plate II).

recurved Curved backward.

reflexed Bent sharply backward at an angle.

regular (of flowers) Radially symmetrical, with more than one plane of symmetry; actinomorphic.

reticulate Marked with a network pattern, usually of veins.

rhachis The major axis of an inflorescence.

rhipidium (of inflorescences) A cymose inflorescence with branches alternating from one side of the vertical axis to the other; normally flattened in one plane and fan shaped (Plate V).

rhizome A horizontally creeping underground stem that lives over from season to season (perennates) and which bears roots and leafy shoots.

root The lower, usually underground, part of a plant (Plate II).

rosette A group of leaves arising closely together from a short stem, forming a radiating cluster on or near the ground.

rotate (of corollas) Wheel shaped; with the petals or lobes spreading out from the axis of a flower (Plate II).

ruminate (of endosperm in seeds) Irregularly grooved and ridged; having a "chewed" appearance.

sagittate (of leaves) Shaped like an arrow head; with two backward-directed barbs (Plate VII).

samara A dry fruit that does not split open and has part of the fruit wall extended to form a flattened membrane or wing (Plate III).

saponins A toxic, soaplike group of compounds that is present in many plants.

saprophyte A plant that cannot live on its own, but which needs decaying organic material as a source of nutrition.

scale A small, often membranous, reduced leaf frequently found covering buds and bulbs.

scaled Covered by scale leaves.

scape A leafless flower-stalk.

scarious Dry and membranous, with a dried-up appearance.

schizocarp A fruit derived from a simple or compound ovary in which the locules separate at maturity to form single-seeded units (Plate III).

sclerenchyma Tissue with thickened cell walls, often woody (lignified), and which give mechanical strength and support.

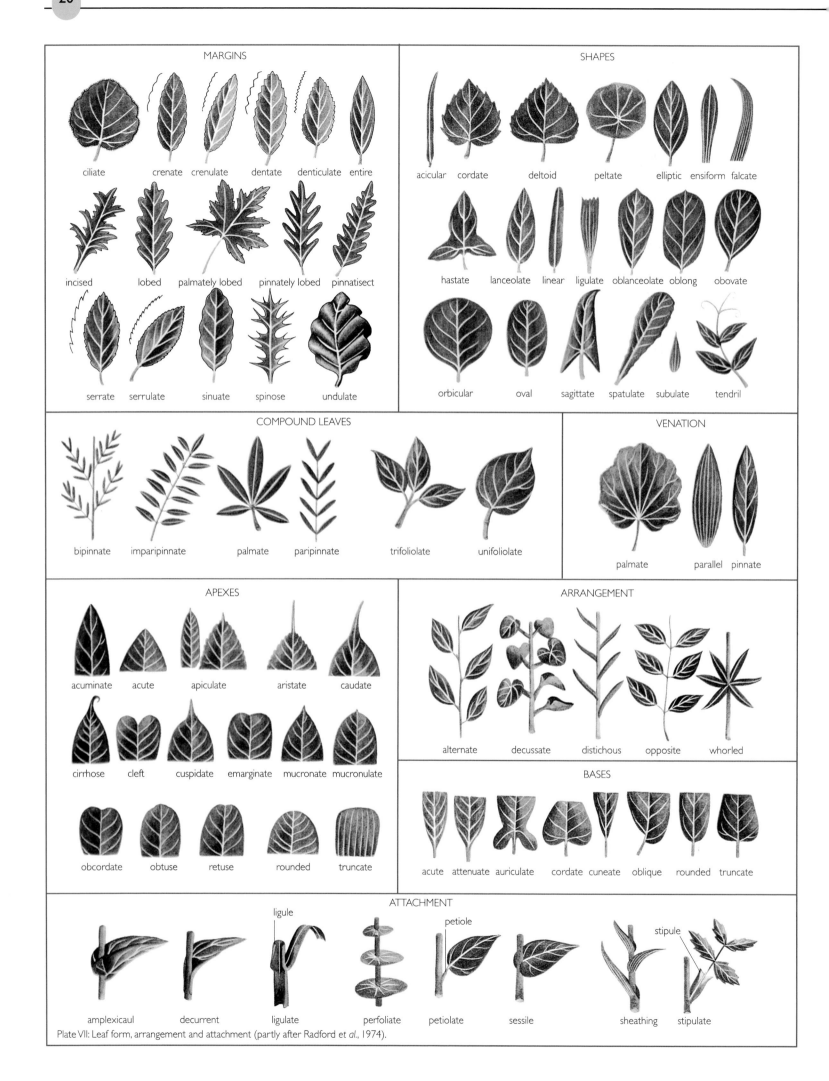

MARGINS

ciliate · crenate · crenulate · dentate · denticulate · entire
incised · lobed · palmately lobed · pinnately lobed · pinnatisect
serrate · serrulate · sinuate · spinose · undulate

SHAPES

acicular · cordate · deltoid · peltate · elliptic · ensiform · falcate
hastate · lanceolate · linear · ligulate · oblanceolate · oblong · obovate
orbicular · oval · sagittate · spatulate · subulate · tendril

COMPOUND LEAVES

bipinnate · imparipinnate · palmate · paripinnate · trifoliolate · unifoliolate

VENATION

palmate · parallel · pinnate

APEXES

acuminate · acute · apiculate · aristate · caudate
cirrhose · cleft · cuspidate · emarginate · mucronate · mucronulate
obcordate · obtuse · retuse · rounded · truncate

ARRANGEMENT

alternate · decussate · distichous · opposite · whorled

BASES

acute · attenuate · auriculate · cordate · cuneate · oblique · rounded · truncate

ATTACHMENT

ligule
petiole
stipule

amplexicaul · decurrent · ligulate · perfoliate · petiolate · sessile · sheathing · stipulate

Plate VII: Leaf form, arrangement and attachment (partly after Radford et al., 1974).

scorpioid (of cymose inflorescences) Curved to one side like a scorpion's tail.

scrambler A plant with a spreading, creeping habit usually anchored with the help of hooks, thorns or tendrils.

seed The unit of sexual reproduction developed from a fertilized ovule; an embryo enclosed in the testa which is derived from integument(s).

seedling The young plant that develops from a germinating seed.

self-fertilization See Fertilization.

self-incompatible Of plants incapable of self-fertilization, usually because the pollentube cannot germinate or grows very slowly.

self-pollination See Pollination.

semiparasite A plant that, although able to grow independently, is much more vigorous in a parasitic relationship with another plant.

sepal A floral leaf or individual segment of the calyx of a flower, usually green (Plate II).

septate (of ovaries) Divided into locules by walls.

septicidal (of fruits) Splitting longitudinally through septa so carpels are separated (Plate III).

septum (of ovaries) The wall between two chambers (locules) of an ovary made up of two or more fused carpels (syncarpous ovary).

seriate Arranged in a row.

serrate (of margins) Toothed, with the teeth pointing forward (Plate VII).

serrulate (of margins) Finely serrate (Plate VII).

sessile Without a stalk, e.g., leaves without petioles or stigmas without a style.

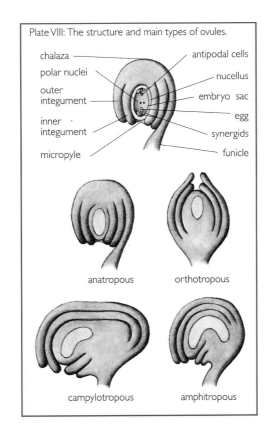

Plate VIII: The structure and main types of ovules.

chalaza — polar nuclei — outer integument — inner integument — micropyle — antipodal cells — nucellus — embryo sac — egg — synergids — funicle

anatropous orthotropous

campylotropous amphitropous

sheath (of leaves) The base of a leaf or leaf-stalk (petiole) that encases the stem.

sheathing (of leaves) With a sheath that encases the stem (Plate VII).

shoot The above-ground portions of a vascular plant, such as the stems and leaves; the plant part that develops from the embryo's plumule.

shrub A perennial woody plant with well developed side-branches that appear near the base, so that there is no trunk.

silicule or silicula A dry fruit that opens along two lines and has a central persistent partition; it is as broad as, or broader, than it is long, as in the Cruciferae (Plate III).

silique or siliqua A silicule-type of fruit that is longer than it is broad, as in the Cruciferae (Plate III).

simple (of leaves) Not divided or lobed in any way (Plate VII).

simple umbel (of inflorescences) An umbel in which the stalks (pedicels) arise directly from the top of the main stalk (Plate V).

sinuate (of margins) Divided into wide irregular teeth or lobes which are separated by shallow notches (Plate VII).

solitary (of flowers) Occurring singly in each axil.

spadix A spike of flowers on a swollen, fleshy axis (Plate V).

spathe A large bract subtending and often ensheathing an inflorescence. Applied only in the monocotyledons.

spatulate or spathulate (of leaves) Shaped like a spoon (Plate VII).

spicate Spikelike.

spike An inflorescence of simple racemose type in which the flowers are stalkless (sessile) (Plate V).

spikelet A small spike, as in grasses (Plate V).

spine The hard and sharply-pointed tip of a branch or leaf, usually round in cross section.

Plate IX: Stamen and androecium forms and structure.

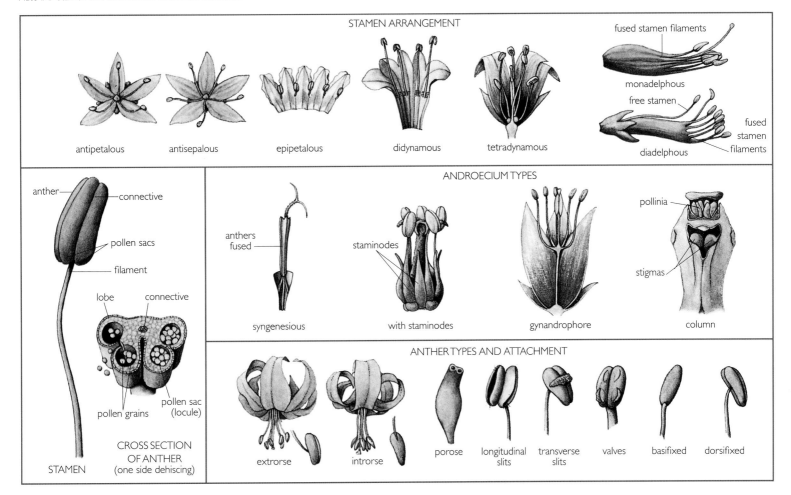

STAMEN ARRANGEMENT

fused stamen filaments

monadelphous

free stamen

fused stamen filaments

antipetalous antisepalous epipetalous didynamous tetradynamous diadelphous

anther — connective — pollen sacs — filament — lobe — connective — pollen grains — pollen sac (locule)

STAMEN

CROSS SECTION OF ANTHER (one side dehiscing)

ANDROECIUM TYPES

anthers fused — staminodes — pollinia — stigmas

syngenesious with staminodes gynandrophore column

ANTHER TYPES AND ATTACHMENT

extrorse introrse porose longitudinal slits transverse slits valves basifixed dorsifixed

spinose Spiny (Plate VII).

spur A hollow, usually rather conical, projection from the base of a sepal, petal, or fused corolla.

stamen The male reproductive organ of a flower. It consists of a usually bilobed anther born on a stalk (filament) (Plate IX).

staminate Having stamens (male organs), but no carpels (female organs); cf pistillate.

staminode A sterile, often reduced or modified stamen (Plate IX).

stellate Star shaped.

stem The main supporting axis of a plant. It bears leaves with buds in their axils. Usually aerial, it can however be subterranean (Plate II).

sterile Unable to reproduce sexually.

stigma The receptive part of the female reproductive organs on which pollen grains germinate; the apical part of carpel (Plate X).

stipitate Having a stalk or stipe.

stipulate Having stipules (Plate VII).

stipule A leafy appendage, often paired, and usually at the base of the leaf stalk.

stomata The pores that occur in large numbers in the epidermis of plants and through which gaseous exchange takes place.

stooling (of plants) Having several stems arising together at the base.

style The elongated apical part of a carpel or ovary bearing the stigma at its tip (Plate XI).

subapical Below the apex.

succulent With fleshy or juicy organs containing reserves of water.

suffrutescent (of herbaceous plants) Having a persistent woody stem base.

superior (of ovaries) Occurring above the level at which the sepals, petals and stamens are borne; cf inferior (Plate II).

suture A line of union; the line along which dehiscence often takes place in fruits.

symbiosis The nonparasitic relationship between living organisms to mutual benefit.

sympetalous With the petals united along their margins, at least at the base.

sympodial (of stems or rhizomes) With the apparent main stem consisting of a series of usually short axillary branches; cf cyme and monopodial.

syncarpous (of ovaries) Made up of two or more fused carpels.

tendril Part or all of a stem, leaf or petiole modified to form a delicate, threadlike appendage; a climbing organ with the ability to coil around objects (Plate VII).

tepal A perianth-segment that is not clearly distinguishable as either a sepal or a petal.

terminal Situated at the apex.

ternate (of leaves) Compound, divided into three parts more or less equally. Each part may itself be further subdivided.

tessellated (of leaves) Marked with a fine chequered pattern, like a mosaic.

testa Seed's outer protective covering (Plate III).

thallus A type of plant body that is not differentiated into root, stem or leaf.

theca One half of an anther containing two pollen sacs.

throat The site in a calyx or corolla of united parts where the tube and limbs meet.

thyrse (of inflorescences) Densely branched, broadest in the middle and in which the mode of branching is cymose.

tomentose Densely covered in short hairs.

tree A large perennial plant with a single branched and woody trunk and with a few or no branches arising from the base.

tri- A prefix meaning three.

trichome A hairlike outgrowth.

tricolpate (of pollen) Having three colpi.

trifoliolate (of leaves) Having three leaflets (Plate VII).

tube The united, usually cylindrical part of the calyx or corolla made up of united parts; cf limb.

tuber An underground stem or root that lives over from season to season (perennates) and which is swollen with food reserves; cf bulb, corm, rhizome.

turion A short, scaly branch produced from a rhizome.

umbel An umbrella-shaped inflorescence with all the stalks (pedicels) arising from the top of the main stem. Umbels are sometimes compound, with all the stalks (peduncles) arising from the same point and giving rise to several terminal flower stalks (pedicels) (Plate V).

undershrub A perennial plant with lower woody parts, but herbaceous upper parts that die back seasonally.

undulate (of leaves) With wavy margins.

unifoliolate With a single leaflet that has a stalk distinct from the stalk of the whole leaf (Plate VII).

unilocular (of ovaries) Containing one chamber (locule) in which the ovules or seeds occur.

uniseriate Arranged in a single row, series or layer, e.g., perianth-segments.

unisexual (of flowers) Of one sex.

utricle A small bladderlike, single-seeded dry fruit (Plate III).

valvate (of perianth-segments) With the margins adjacent without overlapping; cf imbricate.

vascular Possessing vessels; able to conduct water and nutrients.

vascular bundle A strand of tissue, consisting of phloem and xylem, involved in water and food transport.

vein Any of the visible strands of conducting and strengthening tissues running through a leaf (Plate VII).

venation The arrangement of the veins of a leaf (Plate VII).

ventral On the lower side.

vernation (of leaves) The manner and pattern of arrangement within the bud.

verticillate Arranged in whorls.

verticillaster (of inflorescences) Whorled dichasia at the nodes of an elongate rachis.

vessels Tubelike cells arranged end to end in the wood of flowering plants and which form the principal pathway in the transport of water and mineral salts.

whorl The arrangement of organs, such as leaves, petals, sepals and stamens so that they arise at the same level on the axis in an encircling ring.

xeromorphic Possessing characteristics such as reduced leaves, succulence, dense hairiness or a thick cuticle which are adaptations to conserve water and so withstand extremely dry conditions.

xerophyte A plant which is adapted to withstand extremely dry conditions.

xylem The woody fluid-conveying (vascular) tissue concerned with the transport of water about the plant; cf phloem; vessels.

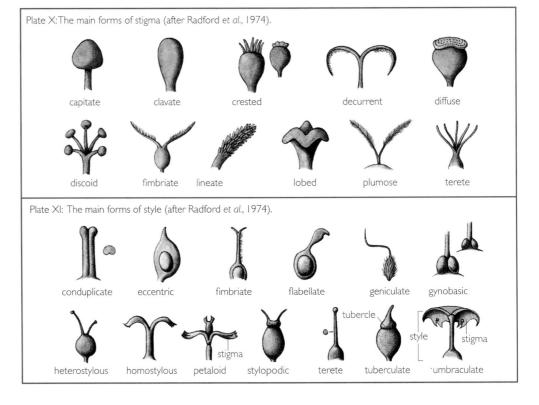

Plate X: The main forms of stigma (after Radford et al., 1974).

capitate   clavate   crested   decurrent   diffuse

discoid   fimbriate   lineate   lobed   plumose   terete

Plate XI: The main forms of style (after Radford et al., 1974).

conduplicate   eccentric   fimbriate   flabellate   geniculate   gynobasic

heterostylous   homostylous   petaloid   stigma   stylopodic   terete   tuberculate   tubercle   style   stigma   umbraculate

DICOTYLEDONS

ACANTHACEAE

ACANTHUS FAMILY

A mainly tropical or subtropical family of herbs, shrubs, or some trees, related to the Scrophulariaceae.

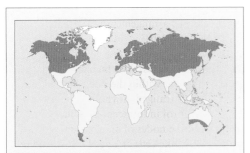

Genera c. 220 **Species** c. 3,000
Economic uses Many species cultivated as garden ornamentals

Distribution Tropical and subtropical, well represented in the New World, tropical Africa, Madagascar, and tropical Asia, but rather less so in Australia, and extending only sparingly into temperate regions. Only one genus, *Acanthus,* is native to Europe, and rather few species of the family occur in temperate Asia and North America. The family may be characteristic of both wet forests and arid regions.

Description This family comprises mostly annual to perennial herbs, shrubs, and climbers (subfamily Thunbergioideae), but occasionally they may be large trees such as *Bravaisia* in Mexico (more than 20 m high with a trunk of 40 cm diameter), while several species are mangroves (2 species of *Acanthus* in Asia and Australia, and species of *Bravaisia* and related genera in Central America). The leaves are opposite and usually entire (may be basal to subopposite and deeply divided in *Acanthus*), without stipules, and occasionally spiny. Except in subfamilies Nelsonioideae and Thunbergioideae and in the tribe Acantheae, the leaves have cystoliths that are visible with a magnifying lens as rod-shaped structures in the epidermis. Inflorescence bracts are often well developed and showy. The flowers are gamopetalous and zygomorphic, usually with a 2-lipped corolla or sometimes the upper lip lacking, and 2 or 4 (5) stamens. The anther structure is diverse, with the cells often being separated by a broad connective or reduced to 1 per anther. The ovary is bicarpellate and develops into a capsule with 2 to 8 (16) seeds, or in *Mendoncia* a single-seeded drupe. Except in the subfamilies Nelsonioideae and Thunbergioideae, each seed is borne on a retinaculum, which is a stiff hooklike structure arising from the axile placenta. This feature allows the great majority of the species to be readily referred to this family.

The family is well known for the remarkable diversity of its pollen in size, shape, pores, and exine structure and ornamentation. Many genera can be characterized by a distinctive pollen type. In some larger genera, however, such as *Justicia*, there is a wide range. *Crossandra stenostachya* from East Africa probably has the longest pollen of any terrestrial flowering plant, each javelin-shaped grain being over ½ mm long and clearly visible to the naked eye.

Classification The family has long been considered to have been derived from near the base of the Scrophulariaceae. Molecular evidence confirms this and also suggests that the mangrove family Avicenniaceae has arisen within the Acanthaceae, although it shares few potential family characters with that family apart from opposite leaves. Although the African genus *Thomandersia* has long been placed in Acanthaceae because of the hooklike structures subtending the seeds, these are not homologous with retinacula. Molecular evidence supports its recognition as a separate family, Thomandersiaceae, of uncertain affinity.

The Acanthaceae is a classical case of a polythetic family with no single character to define it, the obvious similarities of its major groups holding the family together. Recent reviews of

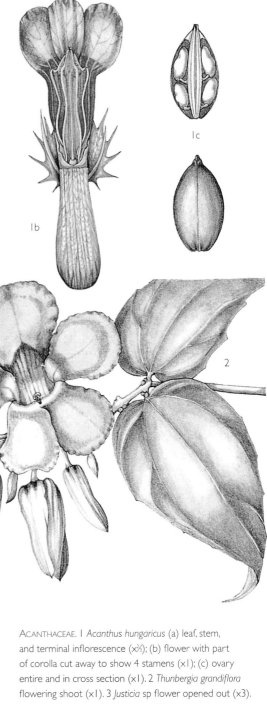

ACANTHACEAE. 1 *Acanthus hungaricus* (a) leaf, stem, and terminal inflorescence (×⅔); (b) flower with part of corolla cut away to show 4 stamens (×1); (c) ovary entire and in cross section (×1). 2 *Thunbergia grandiflora* flowering shoot (×1). 3 *Justicia* sp flower opened out (×3).

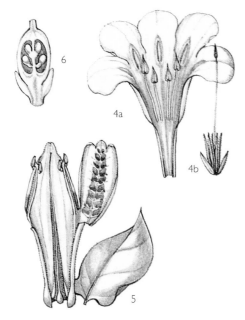

ACANTHACEAE. 4 *Ruellia dipteracantha* (a) corolla opened out to show epipetalous stamens (×1); (b) calyx and gynoecium (×1). 5 *Justicia brandegeeana* (*Beloperone guttata*) leaf and flower opened out showing 2 stamens with a broad connective (×2). 6 *Justicia patentiflora* vertical section of ovary with ovules on axile placentas (×9).

its classification[i,ii,iii] have resulted in major groupings very similar to those recognized by specialists in the family for the last century or more. Two small subfamilies, both lacking cystoliths and retinacula, are recognized outside the main group, the Nelsonioideae, coming fairly close to Scrophulariaceae, and the Thunbergioideae, which are predominantly climbers with a reduced calyx and poricidal anthers and so an apparently well-defined offshoot (including *Mendoncia* with fleshy fruits). In the subfamily Acanthoideae, the tribe Acantheae also lack cystoliths and have characteristic anthers, while the 2 larger tribes, Ruellieae and Justiceae, differ in their corolla aestivation. The larger genera are *Justicia* (600–700 spp., pantropical), *Strobilanthes* (c. 450 spp. in tropical Asia), *Ruellia* (c. 250 spp., pantropical), *Barleria* (more than 200 spp., mostly in Africa, some in Asia, but 1 in West Africa and the New World), *Blepharis* (c. 130 spp. in Africa and Asia) and *Thunbergia* (more than 100 spp., pantropical).

Economic uses The flowers, and often the bracts, are large and showy in many genera, making them popular ornamentals in tropical gardens or in conservatories in cooler regions. Among the most commonly grown are *Aphelandra, Asystasia, Barleria, Crossandra, Eranthemum, Hemigraphis, Hypoestes, Justicia* (including *Adhatoda, Beloperone,* and *Jacobinia*), *Mackaya, Odontonema, Pachystachys, Peristrophe, Pseuderanthemum, Ruellia, Ruspolia, Ruttya, Schaueria, Strobilanthes,* and *Thunbergia*. In temperate regions, 3 species of *Acanthus* (bears' breech) are commonly grown outdoors: *A. mollis* (the leaves of which are the motif of capitals of Corinthian columns in Greece), *A. hungaricus,* and *A. spinosus*. Several

species are grown as house plants, including *Aphelandra squarrosa* (Zebra Plant), *Fittonia albivenis* (Nerve Plant), *Justicia brandegeeana* (*Beloperone guttata*; Shrimp Plant), and *Thunbergia alata* (Black-eyed Susan). Other economic uses in the family are minimal.   RKB

i Scotland, R. W., Sweere, J. A., Reeves, P. A., & Olmstead, R. G. Higher-level systematics of Acanthaceae determined by chloroplast DNA sequences. *American J. Bot.* 82: 266–275 (1995).
ii McDade, L. A., Daniel, T. F., Masta S. E., & Riley, K. M. Phylogenetic relationships within the tribe Justiceae (Acanthaceae): evidence from molecular sequences, morphology and cytology. *Ann. Missouri Bot. Gard.* 87 (4): 435–458 (2000).
iii Scotland, R. W. & Vollesen, K. Classification of Acanthaceae. *Kew Bull.* 55: 513–589 (2000).

ACHARIACEAE

Three South African herbaceous, gamopetalous genera related to the Flacourtiaceae.

Distribution Confined to eastern parts of southern Africa. *Acharia* grows in scrub and woodland from the eastern Cape to Durban. *Ceratiosicyos* occurs in forests from the eastern Cape through Natal and Swaziland to northern Transvaal, up to 1,500 m altitude. *Guthriea* inhabits mountain grasslands from the eastern Cape to Lesotho and Natal up to 3,000 m altitude.

Description The 3 genera have diverse habits. *Acharia tragodes* is a herb to subshrub up to 45 cm, weakly woody at the base. *Ceratiosicyos*

laevis is a herbaceous climber up to 7 m or more. *Guthriea capensis* is a herb with a rosette of cordate leaves, resembling *Caltha palustris* in habit. The leaves are alternate, petiolate, palmately lobed or (*Guthriea*) unlobed but with crenate margins. Stipules are absent. The flowers are unisexual (plants monoecious), the male ones often maturing earlier, and borne solitary in leaf axils or in slender axillary racemes (*Ceratiosicyos*). The flowers are pedicellate, hypogynous, actinomorphic, and 3- to 5-merous. The 3 to 5 sepals are linear and may be adnate to the corolla. The corolla is green or

Genera 3 Species 3
Economic uses None

ACHARIACEAE. 1 *Guthriea capensis* (a) habit—a stemless herb with flowers of separate sexes on the same plant (×⅔); (b) female flower opened out, showing ovary surmounted by lobed stigma (×⅔); (c) cross section of ovary, showing single locule and parietal placentas (×1); (d) male flower opened out (×⅔); (e) fruit (a capsule) surmounted by the persistent corolla tube (×⅔). 2 *Acharia tragodes* (a) habit—a woody dwarf shrub (×⅔); (b) male flower opened out (×2); (c) female flower (×2); (d) capsule with one valve removed (×1). 3 (ABOVE RIGHT) *Ceratiosicyos ecklonii* (a) habit (×⅔); (b) young female flower (×2); (c) vertical section of gynoecium (×2); (d) male flower opened to show fleshy staminodes (×2); (e) dehiscing capsule (×⅔).

Genera 2 Species c.10
Economic uses None

white, campanulate with lobes shorter than the tube. There are 3 to 4 lobes in *Acharia*, 4 to 5 in *Ceratiosicyos*, and 5 in *Guthriea*. There are nectariferous glands at the base of the corolla tube. The stamens are usually equal in number to the corolla lobes, inserted at the base of the corolla (*Guthriea*) or in the throat of the corolla. The female flowers have a superior ovary with a single locule and 3 to 5 parietal placentas, each bearing numerous ovules. The style is 3- to 5-lobed, sometimes with 2-lobed stigmas. The fruit is a capsule, linear, and up to 8 cm long in *Ceratiosicyos*, or ellipsoid, up to 1.2 cm long, and enclosed in the persistent corolla in the other 2 genera. Seeds of *Acharia* and *Guthriea* are dispersed by ants.

Classification Although the 3 genera are so different in habit and show other important differences[i,ii], they are clearly closely related, sharing a herbaceous habit, unisexual flowers, strongly sympetalous corollas, and similar swollen trichomes on the anthers. Generally, they have been referred to the group of families united by their parietal placentation and often known as Parietales, and have most commonly been thought to be closely related to Passifloraceae and Cucurbitaceae and arising from a Flacourtiaceous stock[iii]. Molecular studies[iv] have found the 3 genera to be nested within the traditional Flacourtiaceae. As a result, the authors have combined one half of the latter

with the traditional Achariaceae as one family. Unfortunately, because of nomenclatural priority, the name *Achariaceae* has had to be taken up for this expanded family, and in recent years, a number of genera of traditional Flacourtiaceae have been referred to Achariaceae, causing confusion. The 3 genera are then referred to the tribe Acharieae. The close affinity of the 3 genera to the traditional Flacourtiaceae, particularly the tree genus *Kiggelaria*, is not questioned and has even been detected by butterflies feeding on the plants[ii]. However, these genera differ markedly in their general appearance (see published illustrations[v]), herbaceous habit, exstipulate leaves, and strongly gamopetalous flowers. They also differ from most Flacourtiaceae in having 3 to 5 instead of numerous stamens, though some Flacourtiaceae may have only 5. The 3 genera have apparently evolved sufficiently far from their Flacourtiaceous ancestors for family status to be well justified, and it seems unnecessary to sink them together taxonomically if the taxonomy is to reflect evolution of morphological characters and not conceal them. Those wishing to avoid a paraphyletic Flacourtiaceae should note that in the new classification, the tribe Acharieae makes the tribe Pangieae paraphyletic, so nothing is gained.

Economic uses No valuable properties are known. *Ceratiosicyos* has been recorded as a serious weed of sugarcane plantations. RKB

[i] Steyn, E. M. A., van Wyk, A. E., & Smith, G. F. A study of the ovule, embryo sac and young seed of *Guthriea capensis* (Achariaceae). *S. Afr. J. Bot.* 67: 206–213 (2001).
[ii] Steyn, E. M. A., van Wyk, A. E., & Smith, G. F. A study of ovule-to-seed development of *Ceratiosicyos* (Achariaceae) and the systematic position of the genus. *Bothalia* 32: 201–210 (2002).
[iii] van Wyk, A. E. Pp. 11–13. In: Dahlgren, R. & van Wyk, A. E. Structures and relationships of families endemic to or centred in southern Africa. In: Goldblatt, P & Lowry, P. P. (eds), *Modern Systematic Studies in African Botany*. Pp. 1–94. (1988).
[iv] Chase, M. W. *et al.* When in doubt, put it in Flacourtiaceae: a molecular phylogenetic analysis based on plastic *rbc*L DNA sequences. *Kew Bull.* 57: 141–181 (2002).
[v] Killick, D. J. B. Achariaceae. *Flora of Southern Africa* 22: 128–134 (1976).

ACHATOCARPACEAE

SNAKE EYES, LIMONACHO

A family of 2 woody genera with fleshy fruits from the tropics and subtropics of America.

Distribution Tropical and subtropical America. *Achatocarpus* comprises 5 to 10 species in Central America. *Phaulothamnus* is a monotypic genus from North America.

Description Small trees or shrubs, the side branches sometimes forming spines. Leaves alternate (sometimes fascicled), petiolate, and exstipulate; lamina entire and simple, with pinnate venation. Inflorescence small, a bracteate

panicle or raceme, the terminal unit cymose in *Phaulothamnus*; inflorescence without bracteoles in *Achatocarpus*. The flowers are unisexual (plants dioecious), the perianth sepaline; tepals 4 (*Phaulothamnus*) or 5 (*Achatocarpus*) in 1 whorl. Staminate flowers with 10 to 20 stamens, free from the calyx but basally connate; gynoecium is absent. Pistillate flowers with 2 carpels, forming a 1-locular superior ovary; 2 styles and stigmas; placentation basal. The fruit is a fleshy berry.

Classification Based on combined DNA sequence analysis[i], the family is a monophyletic sister to the Amaranthaceae and Chenopodiaceae but is often linked to the Phytolaccaceae or even included in that family[ii], although this assumption may be due to parallel evolution of the berries[iii].

Economic uses No major economic use, but the fruits of *Phaulothamnus spinescens* are part of the vegetable diet of the ring-tailed cat[iv]. AC

[i] Cuénoud, P. *et al.* Molecular phylogenetics of Caryophyllales based on nuclear 18S and plastid *rbc*L, *atp*B and *mat*K DNA sequences. *American J. Bot.* 89: 132–144 (2002).
[ii] Behnke, H.-D. Ultrastructure of sieve-element plastids in Caryophyllales (Centrospermae): evidence for the delimitation and classification of the order. *Pl. Syst. Evol.* 126: 31–54 (1976).
[iii] Bittrich, V. Achatocarpaceae. In: Kubitzki, K. *et al.* (eds), *The Families and Genera of Vascular Plants. II. Flowering Plants. Dicotyledons: Magnoliid, Hamamelid and Caryophyllid Families.* Pp. 35–36. Berlin, Springer-Verlag (1993).
[iv] Rodrigues-Estrella, R., Moreno, A. R., & Tam, K. G. Spring diet of the endemic Ring-tailed Cat (*Bassariscus astutus insulicola*) population on an island in the Gulf of California. *J. Arid Environments* 44: 241–246 (2000).

ACTINIDIACEAE

KIWI VINE, CHINESE GOOSEBERRY

A family of trees and some shrubs, previously thought to be related to Theaceae.

Distribution The family occurs mainly in tropical and subtropical America and Southeast Asia, with another center in temperate Asia.

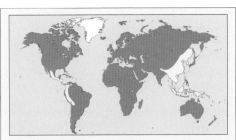

Genera 3 Species c. 350
Economic uses Some ornamentals; the fruits of *Actinidia chinensis* are edible

Saurauia is the largest genus with c. 250 spp. It is found in both tropical America and from the Himalayas to east Asia, Fiji, and Queensland. With 62 species, *Actinidia* occurs from Indonesia to Siberia with its center of distribution in China. *Clematoclethra* comprises 20 or a single species[i] endemic to montane forests of northwestern and central China.

Description Trees, shrubs, or woody lianas, frequently hairy on young growth. The leaves are alternate, simple, entire, or the margins serrate, exstipulate. The inflorescences are axillary, cymose, with few to numerous flowers (up to 500 in *Saurauia*), which are unisexual (plants monoecious in *Actinidia*, dioecious in *Clematoclethra*, and monoecious or functionally dioecious in *Saurauia*) or hermaphrodite, actinomorphic, and usually 5-merous. The sepals are (3–4)5(–9), free or adnate to the petals. The petals are (3–4)5(–8), usually white, sometimes red or brown-yellow, free or connate. The stamens are as few as 10–15 or numerous in fascicles. The ovary is superior, with 3 to 5 carpels, usually 5-locular, sometimes 3 or up to 20 or more; placentation axile. The fruit is a berry or a loculicidal capsule in some *Saurauia* spp. Seeds usually numerous.

Classification The genus *Sladenia* was previously included in the Actinidiaceae but is now treated as a separate family, Sladeniaceae (q.v.), or part of the Theaceae. Likewise, the genus *Saurauia* has sometimes been removed from the Actinidiaceae, but the balance of evidence indicates its correct placement along with *Clematoclethra* as a monophyletic group in this family. The tropical Asian–American disjunct distribution and cytological evidence suggest that *Saurauia* diverged from the common ancestor of *Actinidia* and *Clematoclethra*[ii].

In the past, the Actinidiaceae have been placed in or near the Theaceae or Dilleniaceae, but morphological[iii] and molecular data[iv,v] now suggest a basal placement in the Ericales. A Bayesian analysis of the Ericales suggests a sister-group relationship between the Actinidiaceae and Clethraceae, but further confirmation is needed[vi].

Economic uses The major economic species in the Actinidiaceae is *Actinidia chinensis*, which produces the Kiwi fruit or Chinese gooseberry. This fruit is widely cultivated in several parts of the world. The name Kiwi fruit refers to the fact that it was first comercialized by New Zealand fruit growers. It has also been studied extensively for its biochemical properties. Many species of *Saurauia* also have fruits with an edible pulp and are eaten locally. AC & VHH

i Tang, Y.-C. & Xiang, Q.-Y. A reclassification of the genus *Clematoclethra* (Actinidiaceae) and further note on the methodology of plant taxonomy. *Acta Phytotax. Sin.* 27: 81–95 (1989). [In Chinese with English abstract].

ii He, Z.-C., Li, J.-Q., Cai, Q., & Wang, Q. The cytology of *Actinidia*, *Saurauia* and *Clematoclethra* (Actinidiaceae). *Bot. J. Linn. Soc.* 147: 369–374 (2005).

iii Judd, W. S. & Kron, K. A. Circumscription of Ericaceae (Ericales) as determined by preliminary cladistic analyses based on morphological, anatomical and embryological features. *Brittonia* 45: 99–114 (1993).

iv Kron, K. A. & Chase, M. W. Systematics of the Ericaceae, Empetraceae, Epacridaceae and related taxa based on *rbc*L sequence data. *Ann. Missouri Bot. Gard.* 80: 735–741 (1993).

v Anderberg, A. A., Rydin, C., & Källerjö, M. Phylogenetic relationships in the order Ericales *s.l.*: Analyses of molecular data from five genes from the plastid and mitochondrial genomes. *American J. Bot.* 89: 677–687 (2002).

vi Geuten, K. *et al.* Conflicting phylogenies of balsamoid families and the polytomy in Ericales: Combining data in a Bayesian framework. *Molec. Phylog. Evol.* 31: 711–729 (2004).

ADOXACEAE
MOSCHATEL FAMILY

A small family of delicate herbs with 3-foliate or pinnate leaves and green flowers.

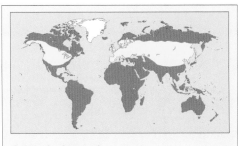

Genera 2   **Species** 3
Economic uses None known

Distribution *Adoxa moshatellina* is panboreal, extending south in Europe to central Spain, Corsica, and Bulgaria; in Asia to Kashmir, central China, and central Japan; and in North America to New Mexico and Illinois. *A. omeiensis* is known only from the mountains of Sichuan in southwestern China. *Sinadoxa corydalifolia* is known only from Quinghai province in northwestern China.

Description Delicate, glabrous herbs, measuring up to 25 cm from a slender rhizome or a short, vertical stock (*Sinadoxa*), with several long-petiolate basal leaves surrounding the erect flowering shoot, which has a single pair of opposite leaves below the inflorescence. Stipules are absent. The basal leaves are ternate to biternate or deeply 3-lobed (*A. omeiensis*), or pinnate with 3 to 4 pairs of pinnae (*Sinadoxa*), and each segment is shallowly to deeply incised but not serrate. The stem leaves are smaller and less divided, usually 3-lobed to 3-foliate. The inflorescence in *A. moschatellina* is a terminal glomerule of 5 sessile flowers, 1 terminal and 4 lateral in opposite pairs. In *A. omeiensis*, it is similar but lax, the 3 to 6 flowers having distinct pedicels. In *Sinadoxa*, there are several sessile glomerules of 3 to 5 sessile flowers at intervals on the stem, the lowermost being in the axils of the cauline leaves. There are no bracts or bracteoles. The flowers are up to 7 mm across, green, bisexual, and perigynous.

The calyx in *A. moschatellina* is 2-lobed in the terminal flower but 3-lobed in the lateral ones, while in *A. omeiensis* it is 4-lobed, and in *Sinadoxa*, it forms 3 saccate wings round the lower half of the ovary. The corolla is lobed almost to the base and rotate, in *A. moschatellina* with 4 lobes in the terminal flower and 5 (6) in the lateral flowers, but in *A. omeiensis*, it is 4-lobed, and in *Sinadoxa* 3- to 4-lobed, with small glands at the base of the lobes. The stamens are equal in number to the corolla lobes and inserted at the sinus between the lobes, but each stamen is divided longitudinally to the base so it appears that there are twice as many stamens. The ovary is semi-inferior, in *A. moschatellina* with 3 to 5 locules (usually 4 in the terminal flower and 5 in the lateral) and always 4 in *A. omeiensis*, while there is single locule in *Sinadoxa*. Each locule has a single, pendent ovule. There are as many styles as locules, each with a small, capitate stigma. In *Sinadoxa*, the upper half of the ovary is attenuate to a minute, sessile stigma. The fruit is a slightly fleshy drupe with 3 to 5 pyrenes in *Adoxa*, but the fruit is unknown in *Sinadoxa*.

Classification In the past, the family was sometimes thought to be related to Saxifragaceae or Araliaceae, but it is now found to be clearly related to Caprifoliaceae *sensu latissimo* on morphological and molecular grounds[i,ii]. Within that group, *Adoxa*, *Sambucus*, and *Viburnum* have been found to form a close group separate from the Caprifoliaceae and other families derived from it, differing in their partially inferior ovary, sessile or subsessile stigma, corolla without nectary at the base, and small, finely reticulate pollen grains. The closest relationship of *Adoxa* and *Sinadoxa* is with *Sambucus*, and both that and *Viburnum* have sometimes been placed in the Adoxaceae. A number of technical characters, including a distinctive and unusual embryology, suggest a close relationship with *Sambucus*. However, other morphological characters, including the habit and general appearance, are here considered sufficient to justify treating them as separate families. A particularly unusual and distinctive feature of the Adoxaceae is that the stamens are divided to the base, giving the appearance of being twice as many as they actually are.

Two discoveries in China in 1981 have given a new perspective to the family. First, the discovery of a second species of *Adoxa*, *A. omeiensis*, similar in general appearance to *A. moschatellina* but with a more lax terminal inflorescence, expands the concept of that genus. The placement of this in a separate genus *Tetradoxa*[iii] is difficult to uphold and was not followed by Hara (who originally described the species in *Adoxa*) in his later comprehensive account[iv]. Second, the discovery of a distinct new genus, *Sinadoxa*[v], with pinnate leaves but floral characters similar to those of *Adoxa*, confirms the distinctive characters of habit, inflorescence, and floral structure that define the family. A suggestion that *Sinadoxa*, for which molecular data are not available, may be not referable to Adoxaceae but nearer to Araliales[vi], seems unlikely given the floral similarities, including the green flowers with stamens divided to the base. RKB

[i] Bremer, K. *et al.* A phylogenetic analysis of 100+ genera and 50+ families of euasterids based on morphological and molecular data. *Pl. Syst. Evol.* 229: 137–169 (2001).
[ii] Bell, C. D. Preliminary phylogeny of Valerianaceae (Dipsacales) inferred from nuclear and chloroplast DNA sequence data. *Molec. Phylogen. Evol.* 31: 340–350 (2004).
[iii] Wu, C. Y. Another new genus of Adoxaceae with special references on the infrafamilial evolution and the systematic position of the family. *Acta Bot. Yunnan.* 3: 383–388 (1981).
[iv] Hara, H. A revision of Caprifoliaceae of Japan with reference to allied plants in other districts and the Adoxaceae. *Ginkgoana* 5: 1–336, t. 1–55 (1983).
[v] Wu, C. Y., Wu, Z. L., & Huang, R. F. *Sinadoxa* genus novum familiarum Adoxacearum. *Acta Phytotax. Sin.* 19: 203–210 (1981).
[vi] Donoghue, M. J., Olmstead, R. G., Smith, J. F., & Palmer, J. D. Phylogenetic relationships of Dipsacales based on *rbc*L sequences. *Ann. Missouri Bot. Gard.* 79: 333–345 (1992).

AEXTOXICACEAE
OLIVILLO

The family comprises 1 species, *Aextoxicon punctatum*, a large canopy or emergent tree up to 25 m high, from the temperate rain forests of Chile, where it forms a key element of the coastal evergreen forest from Limarí to Chiloé.

Genera 1 | Species 1
Economic uses
High-quality timber

The leaves are opposite; dark green on top and lighter below; covered in rusty, peltate scales; briefly petiolate; entire; and without stipules. The flowers are actinomorphic, unisexual (plants dioecious), in pendulous racemes, with prominent bracteoles. Sepals 5, deciduous; petals 5, broadly clawed. The staminate flowers have 5 stamens opposite the sepals, and vestigial carpels; the pistillate flowers have 2 carpels, fused to form a 2-locular ovary with 2 ovules in 1 locule, the other empty. The fruit is a single seeded, not fleshy, indehiscent drupe, resembling a small olive, hence the common name.

Aextoxicon has traditionally been placed in Euphorbiales but also in Sapindales and Celastrales. The species is now placed in the Berberidopsidales by APG II, forming a monophyletic group with Berberidopsidaceae[i]. Relationships are supported by wood anatomy[ii]. *Aextoxicon* is a source of high-quality timber. AC

[i] Soltis, P. S. & Chase, M. W. Angiosperm phylogeny inferred from multiple genes as a tool for comparative biology. *Nature* 402: 402–404 (1999).
[ii] Carlquist, S. C. Wood anatomy of Aextoxicaceae and Berberidopsidaceae is compatible with their inclusion in Berberidopsidales. *Syst. Bot.* 28: 317–325 (2003).

AGDESTIDACEAE
ROCKROOT

A monotypic family that comprises a perennial, vigorous climber, with large caudex weighing up to 68 kg, endemic to tropical South and Central America and the Greater Antilles. The leaves are alternate with long petioles and no stipules, and the lamina entire and cordate. The inflorescence is an axillary panicle with long peduncle and pedicels, giving a loose appearance. The flowers are hermaphrodite, with a single perianth whorl of 4 creamy white petaloid segments, each narrowly elliptic and borne on a perigynous disk. The stamens are 15 to 20, equal, and free. The carpels are 4, fused into a 4-celled, partly inferior ovary, bearing a single, short, terminal style, with 1 ovule per locule. The fruit is leathery, dry, and indehiscent, bearing wings formed from the perianth.

Genera 1 | Species 1
Economic uses
Anti-syphilitic; rarely, decorative climber in subtropical gardens

Agdestis belongs in the Caryophyllales[i], but opinion varies on the recognition of this family. Most authors place it in Phytolaccaceae[ii,iii], commenting on its pecularity. Wood anatomy has been used to propose recognition as a subfamily within Phytolaccaceae[iv]. The current balance of data supports treatment as a separate family. *Agdestis* root preparations are used against syphilis. AC

[i] Cuénoud, P. *et al.* Molecular phylogenetics of Caryophyllales based on nuclear 18S and plastid *rbc*L, *atp*B and *mat*K DNA sequences. *American J. Bot.* 89: 132–144 (2002).
[ii] Smith, N. *et al.* (eds), *Flowering Plants of the Neotropics.* Princeton, Princeton University Press (2004).
[iii] Brown, G. K. & Varadarajan, G. S. Studies in Caryophyllales I: Re-evaluation of classification of Phytolaccaceae *s.l.* *Syst. Bot.* 10: 49–63 (1985).
[iv] Carlquist, S. Wood anatomy of *Agdestis* (Caryophyllales): systematic position and nature of the successive cambia. *Aliso* 18: 35–43.

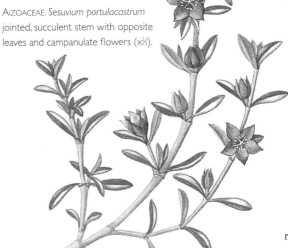

AIZOACEAE. *Sesuvium portulacastrum* jointed, succulent stem with opposite leaves and campanulate flowers (x⅔).

AIZOACEAE
CARPET WEED AND NEW ZEALAND SPINACH

A small family of subsucculent plants often found in coastal areas of the tropics.

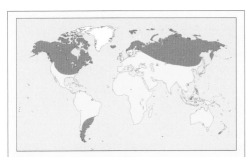

Genera 12 | Species c. 170
Economic uses *Tetragonia tetragonioides* has limited use as a leaf vegetable

Distribution Tropics and subtropics, extending into dry, warm, temperate Africa, Asia Minor, and North America. *Sesuvium portulacastrum* is a pantropical plant of beaches.

Description Annual herbs to perennial, woody shrubs, rarely tuberous, glabrous to hairy in all parts. Leaves entire, opposite, or alternate, generally small; sometimes tiny and imbricate, flat to cylindrical, the leaf surface smooth to papillate, sometimes hairy, with or without petiole. Inflorescence often leafy, with 1 to many flowers. Flowers actinomorphic, bisexual, 4- to 5-merous with a perianth of tepals that are green outside and green, white, yellow, or pink inside, usually triangular or linear when petaloid. The stamens are 4 to numerous. The carpels are 2 to 20, fused into a 1- to 10-locular ovary, each locule containing 1 ovule.

Classification The placement of the family in core Caryophyllales[i] is widely accepted. Exclusion of the Molluginaceae genera[i] is widely and consistently agreed, but there is ongoing controversy over the recognition of a broad, and monophyletic, Aizoaceae, including Mesembryanthemaceae, or a more narrowly defined family, excluding the main succulent plant radiations of southern Africa. Molecular evidence from a small sample of genera[ii] indicated Aizoaceae and Mesembryanthemaceae were sister groups, but a much more comprehensive study[iii] covering 51 genera and 4 DNA regions demonstrated that Aizoaceae was a paraphyletic mother group to Mesembryanthemaceae. Morphological evidence can be interpreted to support a broad Aizoaceae[iv] or a narrow one recognized by nectary morphology among other features[v]. Aizoaceae in this restricted sense comprises subfamilies Aizooideae (including Tetragonioideae) and Sesuvioideae with 7 and 5 genera respectively. Of the 170 species recognized, 85 are in *Tetragonia*.

Economic uses *Tetragonia tetragonioides* is grown as a leaf vegetable, known as New Zealand Spinach. It is invasive in several parts of the world and a noxious agricultural weed in several states of the USA and in Queensland, Australia. Some species have value in binding sand dunes in coastal regions. AC

i Bittrich, V. & Hartmann, H. E. K. The Aizoaceae – a new approach. *Bot. J. Linn. Soc.* 97: 239–254 (1988).
ii Cuénoud, P. *et al.* Molecular phylogenetics of Caryophyllales based on nuclear 18S and plastid *rbc*L, *atp*B and *mat*K DNA sequences. *American J. Bot.* 89: 132–144 (2002).
iii Klak, C., Khunou, A., Reeves, G., & Hedderson, T. A phylogenetic hypothesis for the Aizoaceae (Caryophyllales) based on four plastid DNA regions. *American J. Bot.* 90: 1433–1445 (2003).
iv Hartmann, H. E. K. Aizoaceae. In: Kubitzki, K., Rohwer, J. G., & Bittrich, V. (eds), *The Families and Genera of Vascular Plants. II. Flowering Plants. Dicotyledons: Magnoliid, Hamamelid and Caryophyllid Fam.* . Pp. 37–69. Berlin, Springer-Verlag (1993).
v Chesselet, P., Smith, G. F., & van Wyk, A. E. A new tribal classification of Mesembryanthemaceae: evidence from floral nectaries. *Taxon* 51: 295–308 (2002).

AKANIACEAE
TURNIPWOOD

The Akaniaceae comprises a single small tree species (*Akania bidwillii*), 8 to 12 m high, with dark brown or grey bark. It occurs in warm temperate eastern Australia in a broad coastal strip from Sydney north to Brisbane. Both the bark and freshly cut wood smell of turnip.

Genera 1 Species 1
Economic uses None

The leaves are usually paripinnate, up to 75 cm long, alternate; leaflets up to 30, lanceolate, with subulate stipules. The flowers are actinomorphic, bisexual, fragrant, in an axillary panicle. The sepals are 5, imbricate; the petals are 5, white or pink. The (5–)8(–10) stamens alternate with and are opposite the petals. The carpels are 3, fused to form a 3-celled ovary with a 3-lobed, stigma. The fruit is a dull red dehiscent capsule.

The genus is usually recognized in its own family and has been shown to have affinity with the Chinese Bretschneideraceae on the basis of wood anatomy and DNA sequence. The family has usually been placed in the Sapindales (for example, Cronquist 1988) but was later transferred to Brassicales (see APG II). It has no known economic uses. AC

i Royal Botanic Gardens Sydney. "New South Wales Flora Online" Demonstration Version. http://plantnet. rbgsyd.nsw.gov.au/PlantNet/NSWflora/index.html
ii Watson, L. & Dallwitz, M. J. The Families of Flowering Plants: Descriptions, Illustrations, Identification, Information Retrieval. http://delta-intkey.com (1992 onward).
iii Carlquist, S. Wood anatomy of Akaniaceae and Bretschneideraceae: a case of near-identity and its systematic implications. *Syst. Bot.* 21: 607–616 (1996).

iv Gadek, P. A. *et al.* Affinities of the Australian endemic Akaniaceae: new evidence from *rbc*L sequences. *Australian Syst. Bot.* 5: 717–724 (1992).

ALSEUOSMIACEAE

Australasian shrubs and subshrubs with usually serrate or fimbriate corolla lobes.

Distribution *Alseuosmia* has 5 species in New Zealand; *Wittsteinia* has 1 species in Victoria, Australia, and 1 in eastern and western New Guinea; *Periomphale* has 1 species in New Caledonia; and *Crispiloba* has 1 species in Queensland, Australia. All usually occur in the understories of montane

Genera 4 Species 9
Economic uses
Rarely cultivated as ornamentals

evergreen forests, up to an altitude of 3,000 m in New Guinea, or at sea level in New Zealand.

Description Subshrubs c. 40 cm tall or shrubs to small trees up to 5 m, sometimes scrambling. The leaves are alternate to subopposite or in a pseudowhorl of up to 6 leaves (*Crispiloba*), simple, and with entire or serrate margins. The inflorescence is an axillary or terminal cyme, or the flowers are solitary or fascicled in leaf axils. The flowers are actinomorphic, bisexual, (4)5(–7)-merous, perigynous, sweetly scented, sometimes cleistogamous. The calyx is shortly lobed. The corolla is funnel- or salver-shaped to campanulate or urceolate, 0.3–1 cm, with small teeth on the lobes[i], and greenish, yellowish, or red (*Alseuosmia*); or up to 5 cm, white, and with strongly fimbriate lobes in *Crispiloba*. The stamens are attached at the throat or at the base of the corolla. The ovary is inferior or three-quarters so, with 2 to 3 locules and 1 to several seeds per locule. The style is simple with a capitate to clavate 2- to 3-lobed stigma. The fruit is a berry with 1 to many seeds.

Classification *Alseuosmia* was often referred to Caprifoliaceae or the vicinity of Escalloniaceae until placed with *Periomphale* in the Alseuosmiaceae by Airy Shaw. *Wittsteinia* was thought to be an anomalous member of Ericaceae, until it was equated with *Periomphale* by van Steenis[ii], who also added *Crispiloba*. More recently, *Periomphale* has been reinstated as generically distinct from *Wittsteinia*[iii]. Molecular evidence[iv] confirms the close relationship of the 4 genera and places the family close to Argophyllaceae (eastern Australia, New Caledonia, and southern Pacific) and Phellinaceae (endemic in New Caledonia) in the Asterales of APG II.

Economic uses The colorful sweet-smelling flowers make *Alseuosmia* an attractive plant for horticulture. RKB

i Corrick, M. G. & Fuhrer, B. A. *Wildflowers of Victoria*: t. 6 (2001) [See color photograph of *Wittsteinia*].
ii van Steenis, C. G. G. J. A synopsis of Alseuosmiaceae in New Zealand, New Caledonia, Australia and New Guinea. *Blumea* 29: 387–394 (1984).
iii Tirel, C. Rétablissement de *Periomphale* Baill. (Alseuosmiaceae), genre endémique de Nouvelle Caledonie. *Bull. Mus. Nation. Hist. Nat., B, Adansonia* 18: 155–160 (1996).
iv Kårehed, J., Lundberg, J., Bremer, B., & Bremer, K. Evolution of the Australasian families Alseuosmiaceae, Argophyllaceae and Phellinaceae. *Syst. Bot.* 24: 660–682 (2000).

ALZATEACEAE

The family comprises 1 species, *Alzatea verticillata*, a shrub or medium tree growing in montane forests from Costa Rica and Panama to northern Bolivia, Colombia, Ecuador, and Peru. It had been included in a range of families before being placed in

Genera 1 Species 1
Economic uses None

its own family some 20 years ago[i] and apparently comfortably assigned to the Myrtales. The leaves are opposite, obovate to elliptical, or oblong. The flowers are actinomorphic, bisexual, and campanulate in paniculate cymes. Calyx lobes 5 or 6, persistent. Corolla absent. Stamens 5, opposite the sepals, with conspicuous heart-shaped connectives[ii]. The ovary is superior, the carpels 2, with 2 locules. The fruit is a loculicidal capsule. *Alzatea* is similar in appearance to Myrtaceae but differs in having a superior ovary. Its most closely related family appears to be Rhynchocalycaceae from South Africa, which is also a monotypic family. RKB

iGraham, S. A. Alzateaceae, a new family of Myrtales in the American tropics. *Ann. Missouri Bot. Gard.* 71: 757–779 (1985).
iiGraham, S. A. Alzateaceae. In: Smith, N. *et al.* (eds), *Flowering Plants of the Neotropics.* Pp. 11–13. Princeton/Oxford, Princeton University Press (2004).

AMARANTHACEAE
AMARANTHS, CELOSIAS, AND COCKSCOMBS

The Amaranthaceae is a large family of herbs and shrubs, containing the grain amaranths of Central and South America, as well as several species of horticultural importance, such as the cockscombs (*Celosia* spp.) and Love-lies-bleeding (*Amaranthus caudatus*).

Distribution Widespread in tropical, subtropical, and temperate regions. Most genera and species are tropical, occurring mainly in Africa and Central and South America.

Description Annual or perennial, largely non-succulent herbs or shrubs, rarely trees or climbers. Leaves entire, opposite or alternate, and lacking stipules. The flowers are actino-

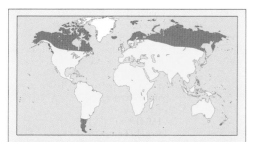

Genera 70 **Species** c. 1,000
Economic uses Several species used as cereals in Central and South America. Some species have horticultural importance as ornamentals

morphic, bisexual or rarely unisexual (then plants dioecious), solitary or in axillary dichasial cymes arranged in spikelike or headlike inflorescences, which are often conspicuously bracteate. The perianth is often dry, membranaceous, and colorless. Sometimes, the lateral flowers are sterile and develop spines, wings, or hairs that serve as a dispersal mechanism, as in *Froelichia*. The flowers are subtended by well-developed, dry, chaffy scales (bracteoles), and in the mass are often showy. Perianth segments 3 to 5, sometimes united. Stamens are 1 to 5, usually as many as perianth segments, free or often united at the base in a tube from which petaloid appendices arise between the stamens in some genera. The ovary is superior, comprising 2 to 3 fused carpels that are free from, or united with, the perianth, with 1 locule containing 1 to many ovules. The fruit may be a berry, pyxidium, or nut; the seeds have a shiny testa; and the embryo is curved.

Classification The boundaries of this family and Chenopodiaceae are controversial. Some authors consider the 2 distinct, based on molecular[i,ii] or morphological[iii] data, while others suggest one is nested within the other[iv] and that the two should be treated as a single family. There is consensus that the 2 groups are closely related, but the precise nature of the evolutionary relationship is still under debate.

Economic uses The edible seeds of some *Amaranthus* species (grain amaranths) were widely used, especially in Central and South America. Several are still grown today. Some species are used as pot herbs or vegetables, such as the fleshy leaves of *A. sessilis,* eaten in some tropical countries. A few species are grown as garden ornamentals, and some Celosias are grown as pot plants or tender bedding plants. The Cockscomb (*Celosia cristata*) is a tropical herbaceous annual. *Alternanthera* species from the New World tropics are grown for their ornamental leaves, and *Iresine herbstii* and *I. linderi*, both from South America, for their colorful (usually scarlet) leaves as house plants or tender bedding plants. *Gomphrena globosa*, a tropical annual with white, red, or purple heads, is grown as an "everlasting." Some species are reputed to have medicinal properties, and saponins are present widely in seeds of *Amaranthus* species. AC

AMARANTHACEAE.
1 *Amaranthus retroflexus* (a) leafy shoot with flowers in axillary tassels (x⅔); (b) fruit dehiscing by lid to disperse globular seed (x6). 2 *A. caudatus* (a) flower with reddish subtending bracteoles and reddish perianth (x14); (b) seed (x14); (c) leaf (x⅔); (d) vertical section of male flower (x12); (e) vertical section of female flower (x4). 3 *Deeringia amaranthoides* inflorescence with fruits at the base (x3). 4 *Froelichia gracilis* (a) shoot with large, lateral, hairy, sterile flowers (x⅔); (b) inflorescence (x⅔); (c) vertical section of sterile flower (x⅔).

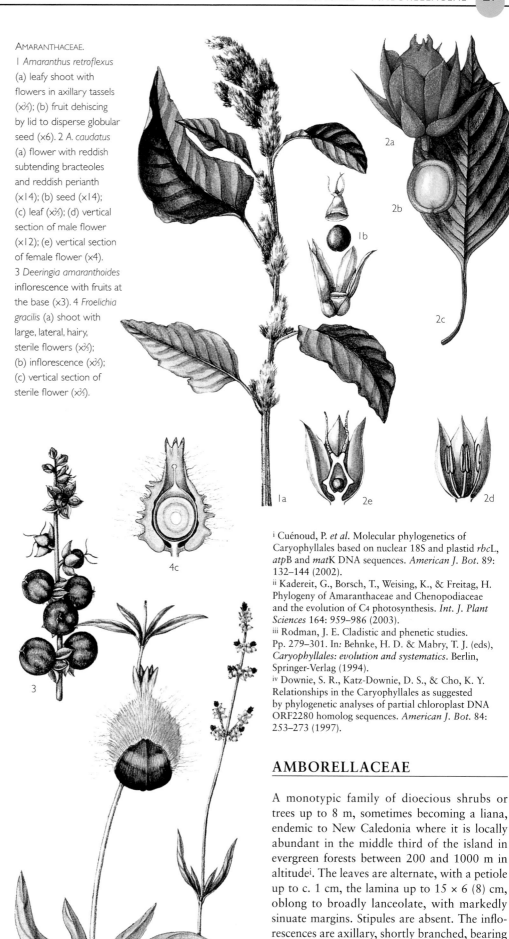

i Cuénoud, P. *et al.* Molecular phylogenetics of Caryophyllales based on nuclear 18S and plastid *rbc*L, *atp*B and *mat*K DNA sequences. *American J. Bot.* 89: 132–144 (2002).
ii Kadereit, G., Borsch, T., Weising, K., & Freitag, H. Phylogeny of Amaranthaceae and Chenopodiaceae and the evolution of C4 photosynthesis. *Int. J. Plant Sciences* 164: 959–986 (2003).
iii Rodman, J. E. Cladistic and phenetic studies. Pp. 279–301. In: Behnke, H. D. & Mabry, T. J. (eds), *Caryophyllales: evolution and systematics.* Berlin, Springer-Verlag (1994).
iv Downie, S. R., Katz-Downie, D. S., & Cho, K. Y. Relationships in the Caryophyllales as suggested by phylogenetic analyses of partial chloroplast DNA ORF2280 homolog sequences. *American J. Bot.* 84: 253–273 (1997).

AMBORELLACEAE

A monotypic family of dioecious shrubs or trees up to 8 m, sometimes becoming a liana, endemic to New Caledonia where it is locally abundant in the middle third of the island in evergreen forests between 200 and 1000 m in altitude[i]. The leaves are alternate, with a petiole up to c. 1 cm, the lamina up to 15 × 6 (8) cm, oblong to broadly lanceolate, with markedly sinuate margins. Stipules are absent. The inflorescences are axillary, shortly branched, bearing 2 to 30 flowers, with bracts in a spiral that sometimes intergrades into the tepals. The flowers are small, c. 3 mm, actinomorphic, the perianth not differentiated, the 5 to 8 tepals more or less triangular and rounded at the apex. The male flowers have 11 to 14 sessile stamens arranged in a spiral on the receptacle.

The female flowers have 5 to 8 free carpels and often have 1 to 2 staminodes. The carpels are not closed at their apex and each bears a single, pendent ovule. Each carpel develops into a shortly stipitate drupe, up to 5 mm, with a hard endocarp that turns red at maturity.

Although at first included in Monimiaceae, *Amborella trichopoda* has now been established in its own family on the basis of its ovules, its wood structure, and its alternate leaves. It has been thought to represent the basal lineage of all the flowering plants[ii,iii,iv]. This has been challenged recently[v], but that finding has been disputed[vi] with references to the many published papers on the family. No economic uses are known.   RKB

i Jérémie, J. Amborellaceae. *Flore de la Nouvelle-Calédonie et Dépendances* 11: 157–160 (1982).
ii Soltis, P. S., Soltis, D. E., & Chase, M. W. Angiosperm phylogeny inferred from multiple genes as a research tool for comparative biology. *Nature* 402: 402–404 (1999).
iii Qiu, Y.-L. *et al.* The earliest angiosperms: evidence from mitochondrial, plastid and nuclear genomes. *Nature* 402: 404–407 (1999).
iv Mathews, S. & Donoghue, M. J. The root of angiosperm phylogeny inferred from duplicate phytochrome genes. *Science* 286: 947–950 (1999).
v Goremykin, W., Bobrova, V., Pahnke, J., Troitsky, A., Antonov, A., & Martin, W. Analysis of the *Amborella trichopoda* chloroplast genome sequence suggests that *Amborella* is not a basal angiosperm. *Molec. Biol. Evol.* 20: 1499–1505 (2003).
vi Stefanović, S., Rice, D. W., & Palmer, J. D. Long branch attraction, taxon sampling, and the earliest angiosperms: *Amborella* or monocots? *BMC Evol. Biol.* 4: 19 (2004).

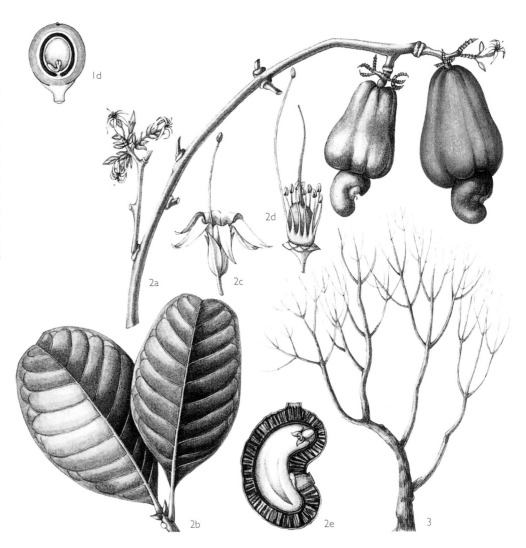

ANACARDIACEAE

CASHEW, MANGO, SUMACS, AND POISON IVY

An economically important, largely tropical and subtropical family of trees and shrubs producing exudates that often turn black, which may cause severe reactions in humans.

Distribution The family occurs in both the eastern and western hemispheres, with a more or less equal representation in South America, Africa, and Asia. Several genera are native to temperate North America and Eurasia (e.g., *Rhus*, *Pistacia*,

ANACARDIACEAE. 1 *Pistacia lentiscus* (a) shoot with imparipinnate leaves and male inflorescences (x⅔); (b) male flower with short, lobed calyx and stamens with short filaments (x10); (c) female flower with 3 spreading stigmas (x14); (d) vertical section of fruit (x4). 2 *Anacardium occidentale* (a) shoot with inflorescence and fruits, the latter swollen, with a pear-shaped stalk and receptacle with the kidney-shaped fruit below (x⅔); (b) simple leaves (x⅔); (c) male flower with a single stamen protruding (x3); (d) bisexual flower with petals removed showing all stamens except 1 to have short filaments (x4); (e) vertical section of fruit (x1). 3 *Rhus trichocarpa* habit.

and *Cotinus*). The family has diversified in Central and South America, southern and central Africa, Madagascar, Indochina, and Malaysia. The largest genus (see below), *Rhus*, with c. 200 species, grows in both temperate and tropical regions of both hemispheres. *Semecarpus* (c. 60 spp.) is Indo–Malaysian in distribution.

Description Trees and shrubs—occasionally scandent shrubs, herbaceous climbers, or lianas—with resin canals and clear to milky sap that may turn black on exposure to air and is often poisonous, causing severe contact dermatitis. The irritant chemicals may be distributed throughout the plant body or concentrated in particular organs, e.g., in the fruit wall of the cashew (*Anacardium occidentale*). The leaves are alternate, rarely opposite (*Abra-*

Genera 82 Species 700+
Economic uses *Cotinus, Pistacia, Schinopsis,* and *Rhus* are sources of tannins; fruits, such as cashew, mango, and pistachio; important ornamentals, such as smoke tree and sumac

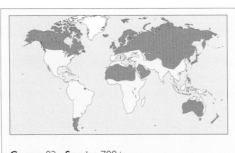

hamia, Blepharocarya, Bouea), exstipulate, often pinnate, sometimes simple (*Cotinus, Anacardium,* and *Mangifera*). The flowers are actinomorphic and bisexual, or sometimes unisexual (then plants dioecious), small and inconspicuous, in terminal or axillary panicles, thyrses, spikes, racemes, fascicles, or solitary. Sepals (3–)5(–7), usually connate at the base, but free above. Petals (3–)5(–7), rarely absent, and free. Stamens 5 to 10 or more, the filaments distinct or rarely connate at the base, forming a tube. An instrastaminal fleshy receptacle (disk) is usually present. Staminodes are present in female flowers. The ovary is usually superior (inferior in *Drimycarpus*) and comprises 1 or 3 to 5 united, very rarely free, carpels, with 1 to 5 locules, each containing a single ovule. There are 1 to 3 styles, sometimes none. The fruit is usually indehiscent and drupaceous but a diversity of unusual fruit types also occurs[i], such as samaras encircled by a marginal wing (*Campylopetalum, Dobinea, Laurophyllus*), samaroid with a single wing (*Faguetia, Loxopterygium*), dry syncarps (*Amphipterygium, Orthopterygium*), or achenial (*Apterokarpos*). In *Anacardium* and *Semecarpus,* the hypocarp subtending the drupe is enlarged and edible. The solitary seed may have thin endosperm, or none, and fleshy cotyledons.

Classification Traditionally, the Anacardiaceae has been subdivided into 5 tribes (Anacardieae, Dobineae, Rhoeae, Semecarpeae, and Spondiadeae). Recent molecular studies suggest that this arrangement is artificial and that 2 major clades should be recognized in the family, corresponding to the 2 subfamilies Anacardioideae and Spondioideae[i, ii]:

SUBFAM. ANACARDIOIDEAE Includes the tribes Anacardieae, Dobineae, Rhoeae, and Semecarpeae. Trees, shrubs, vines, or perennial herbs. Leaves simple or compound; stamens variable in number; carpels usually 1 or 3 and fused; producing dermatitis-causing catechols, resorcinols, and other compounds.

SUBFAM. SPONDIOIDEAE Includes the tribe Spondiadeae and a few members of the Rhoeae. Trees or shrubs. Leaves compound, rarely simple (*Haplospondias*) or simple and compound in a single individual (*Sclerocarya*); twice as many stamens as petals; no dermatitis-causing compounds produced.

As noted, in the broad sense the genus *Rhus* is widely considered to be the largest genus in the Anacardiaceae, but up to 7 segregate genera have been recognized in the *Rhus* complex: *Actinocheita, Cotinus, Malosma, Melanococca, Metopium, Searsia,* and *Toxicodendron*. Recent molecular data[iii] indicate that, in the narrow sense, *Rhus* is monophyletic and includes 2 subgenera, *Lobadium* (c. 25 spp.) and *Rhus* (c. 10 spp.), while *Actinocheita, Cotinus, Malosma, Searsia,* and *Toxicodendron* are distinct from it. The Anacardiaceae belongs in the Sapindales and is closely related to several other families in the order, most notably the Burseraceae of which it is a sister group. Molecular data

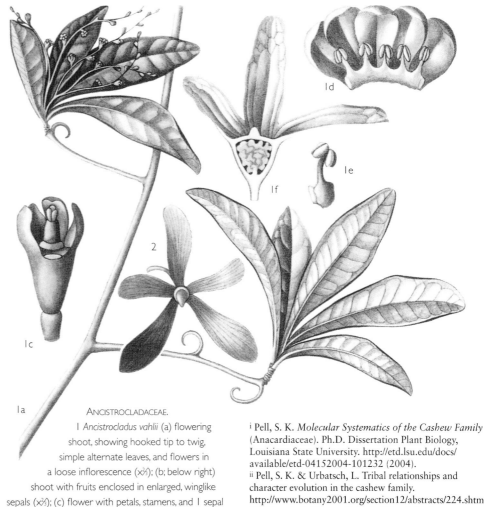

ANCISTROCLADACEAE.

1 *Ancistrocladus vahlii* (a) flowering shoot, showing hooked tip to twig, simple alternate leaves, and flowers in a loose inflorescence (×⅔); (b; below right) shoot with fruits enclosed in enlarged, winglike sepals (×⅔); (c) flower with petals, stamens, and 1 sepal removed to show thickened style surmounted by 3 stigmas (×8); (d) corolla opened out, showing stamens with short, fleshy filaments (×8); (e) stamen with basifixed anthers (×12); (f) half fruit containing single seed (×1). 2 *A. heyneanus* fruit (nut) surrounded by the persistent calyx (×⅔).

indicate the monophyly of the family and support its separation from the Burseraceae[i,ii].

Economic uses *Cotinus, Pistacia, Schinopsis,* and *Rhus* are major sources of tannins for the leather industry. The resin of *Rhus verniciflua,* native to China, is the basis of lacquer. Mastic and pistachio turpentine are produced from *Pistacia*. The family yields important fruits and nuts, e.g., cashew nuts (the swollen hypocarps) and cashew apples (*Anacardium occidentale*), pistachio nuts (*Pistacia vera*), Dhobi's nut (*Semecarpus anacardium*), the Mango (*Mangifera indica*), and the Otaheite apple, hog plum, and Jamaica plum (fruits of *Spondias* spp.). The small berrylike fruits of *Schinus terebinthifolius* (Brazilian Pepper Tree) are used as pink peppercorns (baies roses) in cooking, and the species is a serious invasive in some parts of the tropics and subtropics. *S. molle* (Pepper Tree) is widely cultivated as an ornamental and as a shade tree in warm climates. Other common ornamental trees include the sumacs (*Rhus* spp.) and the Smoke Tree or Wig Tree (*Cotinus coggygria*). Some genera, such as *Astronium, Myracrodruon,* and *Schinopsis,* include useful timber species. [FKK] VHH

[i] Pell, S. K. *Molecular Systematics of the Cashew Family* (Anacardiaceae). Ph.D. Dissertation Plant Biology, Louisiana State University. http://etd.lsu.edu/docs/available/etd-04152004-101232 (2004).
[ii] Pell, S. K. & Urbatsch, L. Tribal relationships and character evolution in the cashew family. http://www.botany2001.org/section12/abstracts/224.shtm
[iii] Miller, A. J., Young, D. A. & Wen, J. Phylogeny and biogeography of *Rhus* (Anacardiaceae) based on ITS sequence data. *Int. J. Plant Sci.* 162: 1401–1407 (2001).

ANCISTROCLADACEAE

This monogeneric family of tropical climbers gained a high profile when anti-HIV activity was discovered in *Ancistrocladus korupensis*.

Distribution Tropical Africa, then disjunct to Sri Lanka, Burma, eastern Himalayas, and southern China, extending to western Malaysia.

Description Mostly large woody lianas up to 20 m, climbing by twining stems; sometimes scrambling, sympodially branched, with each branch ending in a coiled hook. The leaves are simple, petiolate, alternate, and entire, with small

ANCISTROCLADACEAE.

3 *Ancistrocladus heyneanus*
(a) flowering shoot (x⅔); (b)
flower (x2); (c) flower opened
out (x2); (d) stamen (x6).

stipules that soon drop; sometimes the stipules are absent. The flowers are small, scentless, regular, and bisexual with articulated pedicels arranged in dichotomous cymes with branches curved back. The sepals are 5, overlapping, with a short calyx tube fused to the base of the ovary; in fruit, the sepals become unequally enlarged and winglike. The petals are 5, more or less fleshy, contorted, slightly joined or cohering (joined but apparently free). The stamens are 10, in a single series, rarely 5, with short, fleshy filaments joined beneath, and anthers with 2 locules dehiscing lengthwise. The ovary comprises 3 fused carpels, semi-inferior. It has 1 locule, containing 1 erect, basal ovule. The styles are 3, free or joined, and with 3 stigmas. The fruit is dry, woody, and surrounded by winglike, enlarged, and persistent (accrescent) calyx lobes. The seeds have a ruminate endosperm, and the cotyledons are deeply folded.

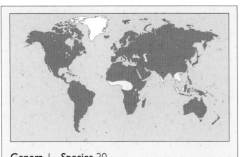

Genera 1   Species 20
Economic uses 1 species shows anti-HIV activity

Classification Traditionally, the affinities of the Ancistrocladaceae have been uncertain. It has been included in the Combretaceae, Malpighiaceae, and Dipterocarpaceae. Pollen evidence supports relationships with the Dioncophyllaceae[i]. DNA sequence evidence reveals a close relationship to the Dioncophyllaceae[ii].

Economic uses The major use has been the identification of naphthylisoquinoline alkaloids in *Ancistrocladus korupensis*, showing strong anti-HIV activity[iii,iv].   VHH & AC

[i] Erdtman, G. A note on the pollen morphology in the Ancistrocladaceae and Dioncophyllaceae. *Veroff. Geobot. Inst. Rubel Zurich* 33: 47–49 (1958).
[ii] Cuénoud, P. *et al.* Molecular phylogenetics of Caryophyllales based on nuclear 18S and plastid *rbc*L, *atp*B and *mat*K DNA sequences. *American J. Bot.* 89: 132–144 (2002).
[iii] Manfredi, K. P. *et al.* Novel alkaloids from the tropical plant *Ancistrocladus abbreviatus* inhibit cell killing by HIV-1 and HIV-2. *J. Med. Chem.* 34: 3402–3405 (1991).
[iv] Boyd, M. R. *et al.* Anti-HIV michellamines from *Ancistrocladus korupensis*. *J. Med. Chem.* 37: 1740–1745 (1994).

ANISOPHYLLEACEAE

A small, somewhat nondescript family of trees and shrubs with alternate leaves, often previously included in the Rhizophoraceae.

Distribution The Anisophylleaceae comprises 4 woody genera and 35 species growing in both Old and New World tropics. The largest genus *Anisophyllea* (c. 30 spp.) grows throughout most of the tropics but with the greatest number of species in the eastern part of its range. *Poga* (1 sp.) is endemic to western Africa, from Nigeria to the Congo; *Polygonanthus* (2 spp.) to Amazonia; and *Combretocarpus* (1 to 2 spp.) to Borneo. All occur in wet lowland primary forest, except *Combretocarpus*, which grows only in peat swamp forests.

Description Trees or shrubs. Leaves alternate, in 2 rows, or 4 with the upper reduced in size (*Anisophyllea*), simple, the lamina somewhat coriaceous, often asymmetrical at the base, the margins entire, exstipulate. Flowers actinomorphic, unisexual (plants dioecious) or bisexual (*Combretocarpus*), small, inconspicuous, in axillary racemes or panicles.

ANISOPHYLLEACEAE. 1 *Anisophyllea griffithii*
(a) shoot with alternate leaves and axillary inflorescences (x⅔); (b) flower with 4 sepals (x4); (c) flower with sepals removed (x6); (d) half flower, showing inferior ovary with free styles (x4); (e) petal and stamen (x14); (f) stamen (x14); (g) cross section of ovary (x5).

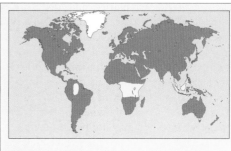

Genera 4   Species c. 35
Economic uses Some timber trees and fruits

The sepals are 3 to 5, free, persistent, valvate. The petals are 3 to 5, clawed, lobed, divided or laciniate (entire in *Polygonanthus*). Twice as many stamens as petals, (6–)8(–10), free. The ovary is inferior, syncarpous, of 3 to 4 carpels; 3 to 4 locules with axile placentation; styles free; 1 or 2 ovules in each locule. The fruit may or may not be fleshy: drupes (*Anisophyllea*), dry woody winged capsules (*Polygonanthus*), or samara (*Combretocarpus*).

Classification This somewhat enigmatic family has variously been placed in the Rosales, Santalales, Saxifragales, and Malpighiales. It has also been included in the Rhizophoraceae (Malpighiales), from which it is now considered to be quite distinct. Recent morphological and molecular data indicate that the Anisophylleaceae belong in the Cucurbitales[i,ii], and in a recent molecular analysis[ii] the Anisophylleaceae appear as sister to all other Cucurbitales. Recent studies of comparative floral

structure and systematics in the Cucurbitales[iii,iv] conclude that the Anisophylleaceae appear more similar to the Cunoniaceae in the Oxalidales and leaves open the question of whether the family belongs in the Cucurbitales or Oxalidales. *Polygonanthus* differs markedly from the other members of the family (entire petals, fruits dry, strongly winged, with persistent calyx lobes) and may not belong here.

Economic uses *Poga oleosa* is a valued timber tree, and its seeds yield edible oil. *Combretocarpus rotundatus* also produces timber used for heavy construction and paneling.   VHH

[i] Schwarzbach, A. E. & Ricklefs, R. E. Systematic affinities of Rhizophoraceae and Anisophylleaceae, and intergeneric relationships within Rhizophoraceae, based on chloroplast DNA, nuclear ribosomal DNA, and morphology. *American J. Bot.* 87: 547–564 (2000).
[ii] Zhang, L.-B. & Renner, S. S. Phylogeny of Cucurbitales inferred from seven chloroplast and mitochondrial loci. AIBS *Botany 2003* Abstract (2003).
[iii] Matthews, M. L. & Endress, P. K., Schönenberger, J., & Friis, E. M. A comparison of floral structures of Anisophylleaceae and Cunoniaceae and the problem of their systematic position. *Ann. Bot.* N.S. 88: 439–455 (2001).
[iv] Matthews, M. L. & Endress, P. K. Comparative floral structure and systematics in Cucurbitales (Corynocarpaceae, Coriariaceae, Tetramelaceae, Datiscaceae, Begoniaceae, Cucurbitaceae, Anisophylleaceae). *Bot. J. Linn. Soc.* 145: 129–185 (2004).

ANNONACEAE
SWEETSOP AND SOURSOP

A large family of mainly tropical trees and shrubs, some of which are cultivated for their large, edible fruits.

Distribution Annonaceae is pantropical in distribution, but the largest concentration of species (c. 1,150) is found in the Old World tropics, followed by the neotropics with 900 species, and 450 in Africa. Two genera, *Asimina* (8 spp.) and *Deeringothamnus* (2 spp., peninsular Florida), are restricted to subtropical and temperate USA and Canada. The most widespread genus is *Xylopia*, which occurs in tropical Amertica, Africa (including South Africa), Madagascar, Australia, and some Pacific islands[i] The Annonaceae are especially characteristic of lowland evergreen forest. The Neotropical genus *Guatteria,* with more than 260 species, is one of the largest genera of woody plants in the world.

Description Aromatic trees or shrubs, sometimes lianas, deciduous or evergreen. Leaves alternate, simple, usually in 2 ranks, without stipules, and with a glaucous or metallic sheen. The flowers are fragrant, often pendulous, actinomorphic, hermaphrodite or rarely unisexual, in terminal or axillary inflorescences. The sepals are (2)3(4); free, partially connate at the base, or wholly connate; unequal in size. The petals are usually 6 or in 2 whorls of 3, or 6–12(–15), equal or variable in size, often fleshy. Stamens usually numerous and spirally arranged, rarely fewer (10 to 20 in *Deeringothamnus*; 6 to 11 in *Bocagea*), the filaments short, with the connective expanded at the apex; staminodes usually absent. Ovary superior, apocarpous, with numerous carpels, rarely 1 to 3; placentation marginal or basal; stigma usually sessile. Fruit usually apocarpous, consisting of an aggregate of 2 to numerous berries, rarely a follicle (*Anaxagorea*), or syncarpous (e.g., *Annona*) and coalescing with an edible fleshy receptacle. Seeds 1 to numerous, with ruminate endosperm, sometimes arillate, or pachychalazal, i.e., the apical growth of the integuments is halted, and an intercalary

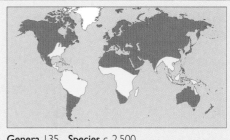

Genera 135   Species c. 2,500
Economic uses Cultivated for their fruits (cherimoya, sweetsop, soursop) and aromatic oils

growth beneath them gives rise to the bulk of the seed (*Cananga*). The flower in some species (e.g., *Duguetia* and some *Annona*) acts as a trap (pollination chamber) for scarab beetles.

Classification Although its limits are well defined, and despite the considerable attention it has attracted during the last century, the family is notoriously difficult to divide into subfamilies or tribes. Until recently, there has been a tendency to focus efforts on neotropical groups. The major classification proposed by Fries[ii] was based mainly on floral characters but has not been supported by either pollen ultrastructure or the limited molecular studies that have been undertaken. On the other hand, phylogenetic analyses based on morphological, palynological, and molecular data agree that *Anaxagorea* is sister to the rest of the family, and to the ambavioid group (*Ambavia, Cleistopholis, Tetrameranthus, Cananga*). Most of the other genera comprise a large group that includes 2 major clades. Members of the first, which include *Cymbopetalum, Guatteria, Trigynaea,* and *Xylopia,* have inaperturate pollen. The second clade contains genera with monosulcate and disulculate pollen such as *Crematosperma, Malmea,* and *Oxandra.* The details of the taxonomic groups to be recognized are still not clear, and multidisciplinary research teams are gradually clarifying these. The Annonaceae is one of the group of "primitive" families of flowering plants, once termed the Annoniflorae, often with indefinite numbers of free floral parts and spirally arranged stamens. It is a member of the Magnoliales and differs from the Magnoliaceae in having ruminate seeds and no stipules, and from the Myristicaceae by having free stamens. Annonaceae is sister to the Eupomatiaceae, which was previously a subfamily within it [iii,iv].

Economic uses Some species, such as *Annona cherimola* (Cherimoya), *A. muricata* (Guanabana, Soursop), *A. squamosa* (Sugar Apple, Sweet Apple), and *Rollinia mucosa* (Biribá) are cultivated throughout the tropics, and even subtropics, for their edible fruit. The fruits of the Old World *Artabotrys* are also edible. Although some *Xylopia* woods are used, annonaceous timber is generally of little importance, except for firewood and poles. Recently, pharmaeuctical research has been carried out on some

ANNONACEAE. 1 *Annona squamosa* (a) shoot with leaves in 2 ranks and axillary flower (x⅔); (b) young fruit (x1); (c) vertical section of fruit (sweetsop), comprising an aggregate of numerous berries with the fleshy receptacle (x⅔). 2 *Monodora myristica* (a) leafy shoot and flower (x⅔).

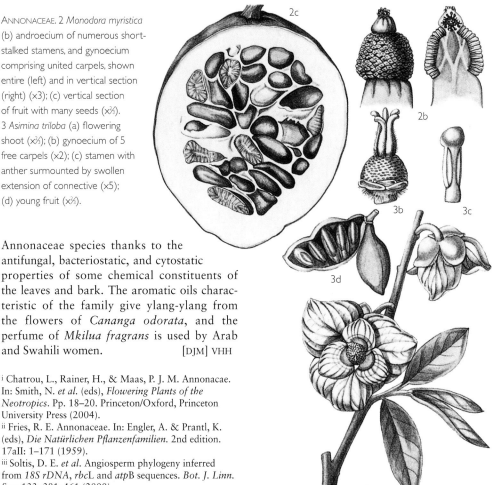

ANNONACEAE. 2 *Monodora myristica*
(b) androecium of numerous short-
stalked stamens, and gynoecium
comprising united carpels, shown
entire (left) and in vertical section
(right) (×3); (c) vertical section
of fruit with many seeds (×⅔).
3 *Asimina triloba* (a) flowering
shoot (×⅔); (b) gynoecium of 5
free carpels (×2); (c) stamen with
anther surmounted by swollen
extension of connective (×5);
(d) young fruit (×⅔).

Annonaceae species thanks to the antifungal, bacteriostatic, and cytostatic properties of some chemical constituents of the leaves and bark. The aromatic oils characteristic of the family give ylang-ylang from the flowers of *Cananga odorata*, and the perfume of *Mkilua fragrans* is used by Arab and Swahili women. [DJM] VHH

i Chatrou, L., Rainer, H., & Maas, P. J. M. Annonacae. In: Smith, N. *et al.* (eds), *Flowering Plants of the Neotropics*. Pp. 18–20. Princeton/Oxford, Princeton University Press (2004).
ii Fries, R. E. Annonaceae. In: Engler, A. & Prantl, K. (eds), *Die Natürlichen Pflanzenfamilien*. 2nd edition. 17aII: 1–171 (1959).
iii Soltis, D. E. *et al.* Angiosperm phylogeny inferred from *18S rDNA*, *rbc*L and *atp*B sequences. *Bot. J. Linn. Soc.* 133: 381–461 (2000).
iv Doyle, J. A. & Endress, P. K. Morphological phylogenetic analyses of basal angiosperms: Comparison and combination with molecular data. *Int. J. Plant Sci.* 161(6, suppl.): S121–S153 (2000).

APHANOPETALACEAE
AUSTRALIAN GUM VINE FAMILY

Distribution *Aphanopetalum clematideum* occurs in western Australia, from the Geraldton area to Shark Bay and inland. *A. resinosum* occurs from southeastern Queensland through subcoastal New South Wales to eastern Victoria.

Description Sometimes suberect shrubs c. 2 m high, but more often woody stragglers or climbers up to 15 m or more. The leaves are opposite, subsessile to shortly petiolate, linear with entire margins in *A. clematideum*, but in *A. resinosum* ellipti-

Genera 1 Species 2
Economic uses Limited horticultural use

cal, with serrate margins, and often gland-tipped teeth. Stipules are small and caducous or absent, sometimes interpetiolar. The flowers are in lax, few-flowered cymes or solitary in the leaf axils, with slender pedicels. The flowers are actinomorphic, bisexual, and 4-merous. The sepals are 4, more or less free, conspicuous, up to 12 mm, obovate to oblong, greenish to cream, accrescent in fruit. The

petals are absent or, if present, 4 and small and inconspicuous. The stamens are 8, in 2 whorls. The ovary is semi-inferior, conical above, and attenuate into a bifid stigma with each lobe again bifid. There are 4 locules, each with a single ovule pendent from an axile placenta. The fruit is a single-seeded nut.

Classification Until recently, the genus has been included in Cunoniaceae. A morphological cladistic study[i] placed it well nested within that family, close to *Bauera*. A later detailed anatomical study[ii] concluded that it should be excluded from Cunoniaceae and placed with woody families related to Saxifragaceae. The climbing habit is not found in Cunoniaceae, and pollen distinguishes it from both that family and Saxifragaceae *sensu lato*. Molecular data[iii] have placed it in Saxifragales, close to Tetracarpaeaceae and Haloragaceae, with Cunoniaceae referred to Oxalidales. Whatever its closest affinities, it seems that separate family status is preferable.

Economic uses None apart from limited horticultural value in Australia. RKB

i Hufford, L. & Dickison, W. C. A phylogenetic analysis of Cunoniaceae. *Syst. Bot.* 17: 181–200 (1992).
ii Dickison, W. C., Hils, M. H., Lucansky, T. W., & Stern, W. L. Comparative anatomy and systematics of woody Saxifragaceae. *Aphanopetalum* L. *Bot. J. Linn. Soc.* 114: 167–182 (1994).
iii Savolainen, V. *et al.* Phylogeny of the eudicots: a nearly complete familial analysis based on *rbc*L gene sequences. *Kew Bull.* 55: 257–309 (2000).

APHLOIACEAE

A monotypic family from East Africa to the Transvaal, Madagascar, Comoros, Mascarene Islands (to Rodriguez), and Seychelles, usually in montane evergreen forest and riverbanks, up to 2,900 m. It is a glabrous shrub or tree up to 14(–20) m with alternate, simple, elliptic to oblanceolate leaves up to 7(–10) cm, cuneate, finely crenate to serrate at least in the distal part, with small, inconspicuous stipules. The flowers are actinomorphic, bisexual and

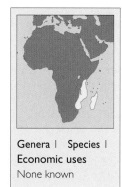

Genera 1 Species 1
Economic uses
None known

sweetly scented, solitary or fascicled in the leaf axils, on long pedicels that often bear 1–3 minute bracts in the lower half. The sepals are (4)5(6), unequal, free except at the base, suborbicular and imbricate, the 2 inner ones usually petaloid and whitish, deflexed in fruit. Petals are absent. The stamens are 65–110, inserted on the margin of a slightly concave receptacle. The ovary is slightly stipitate, apparently unicarpellate, with a single locule with 1 parietal placenta bearing around 6 ovules in 2 vertical rows. The stigma is subsessile, broad, peltate, with a median furrow, persistent on the fruit. The fruit is a white, fleshy berry through which the 6 orange seeds are visible at maturity.

Aphloia theiformis has usually been placed in Flacourtiaceae, but it differs in leaf and wood anatomy, pollen morphology, and embryology and was placed in its own family alongside Flacourtiaceae by Takhtajan[i]. Molecular evidence confirms the distinctness of the family and indicates that it is not closely related to Flacourtiaceae. A first analysis[ii] placed it close to Ixerbaceae and Staphyleaceae, while another[iii] has placed it far removed from those families and sister to Icacinaceae. In APG II, Aphloiaceae is unplaced under the Rosids. More recent analysis has confirmed a position close to Geissolomataceae, Strasburgeriaceae, and Ixerbaceae in the Crossosomatales[iv], although the claimed morphological similarities are far from easy to see. The species is variable, particularly in the Mascarene area, and a number of variants have been given names, although it seems unlikely that any of these can be maintained[v]. RKB

i Takhtajan, A. L. Three new families of flowering plants. *Bot. Zhurn.* 70: 1691–1693 (1985).
ii Soltis, D. E. *et al.* Angiosperm phylogeny inferred from *18S rDNA*, *rbc*L and *atp*B sequences. *Bot. J. Linn. Soc.* 133: 381–461 (2000).
iii Savolainen, V. *et al.* Phylogeny of the eudicots: a nearly complete familial analysis based on *rbc*L gene sequences. *Kew Bull.* 55: 257–309 (2000).
iv Matthews, M. I. & Endress, P. K. Comparative floral structure and systematics in Crossosomatales. *Bot. J. Linn. Soc.* 147: 1–46 (2005).
v Sleumer, H. *Aphloia*. *Flore des Mascareignes*. 42: 12–14 (1980).

APIACEAE (UMBELLIFERAE)

CARROT FAMILY

The Apiaceae is one of the best-known families of flowering plants, thanks to the characteristic inflorescences and fruits and the distinctive chemistry reflected in the odor, flavor, and even toxicity of many of its members. The Apiaceae seems to have been the first flowering plant family to be recognized as such by botanists, about the end of the 16th century, although only the temperate Old World species were then known. It was also the first group of plants to be the subject of a systematic study, published by Robert Morison in 1672.

Distribution The Apiaceae are found in most parts of the world, although most common in temperate upland areas and relatively rare in tropical latitudes. The 4 groups into which it may be divided (see below) have characteristic distributions. The largest, the Apioideae, is bipolar but mainly developed in the northern hemisphere in the Old World. The Saniculoideae is also bipolar but better represented in the southern hemisphere than the Apioideae. The Azorelloideae is found in the southern hemisphere in South America, Australia and New Zealand, and Antarctic and sub-Antarctic islands, with an extension to the Canary Islands and Somalia if the genus *Drusa* is included. The Mackinlayoideae are found mainly in the South Pacific Rim, with *Centella* mainly in South Africa and 3 more species occurring in southern tropical Africa, but *C. asiatica* is cosmopolitan and pantropical.

About two-thirds of the species of Apiaceae are native to the Old World, but the distribution of the subfamilies in the Old and New Worlds is different, 80% of the Apioideae being found in the Old World, and 60% of the Azorelloideae in the New World (almost 90% of these occurring in South America), where they form a significant component of the flora of temperate southern zones. The subfamily Saniculoideae is almost evenly split between the Old and New worlds. This pattern reflects the long evolutionary history and differentiation of this near cosmopolitan family. Many curious distributions are found in the Umbelliferae. Until very recently considered to be endemic to the Canary Islands, *Drusa glandulosa* has also now been identified in Morocco and Somalia and is most closely allied to Chilean species of *Bowlesia* and *Homalocarpus*, although no explanation for such a large geographical disjunction has been offered as yet. *Naufraga balearica*, from Majorca in Spain, was once thought to have its closest affinities with South American genera but has now been shown to be closely related to the predominantly Mediterranean genus *Apium*.

Description Most of the Apiaceae are herbaceous annuals, biennials, or perennials, with hollow internodes; sometimes they may be stoloniferous (*Schizeilema*), rosette plants (*Gingidia*), or cushion plants (*Azorella*). Several of the herbaceous species develop some degree of woodiness, but genuinely woody, treelike, or shrubby species also occur. Examples include *Eryngium bupleuroides*, *E. sarcophyllum*, and *E. inaccessum* of the Juan Fernández Islands, which develop a woody trunk; *Myrrhidendron* species from mountain summits above 3,000 m in Central and South America; and several shrubby species of *Bupleurum* (e.g., *B. fruticosum*). Several species are spiny, such as the thistlelike *Eryngium* species and the New Zealand species of *Aciphylla*, with rigid leaf- and bract-segments tipped by needle-sharp spines.

The leaves are alternate, without stipules, and the leaves usually dissected (ternate or variously pinnate). Entire leaves are found in *Bupleurum*, often with parallel venation and frequently resembling monocotyledons.

Genera 400–450 **Species** 3,500–3,700
Economic uses Important foods, herbs, spices, or flavoring plants (e.g., Angelica, Anise, Carrot, Celery, Dill, Fennel, Parsley, Parsnip); source of gum resins, medicines, and perfumes; a few ornamentals

The main inflorescence found in the Umbelliferae is a simple or compound umbel, although sometimes modified and reduced to a single flower, as in some *Azorella* spp. Less frequent are panicles or racemes (Mackinlayoideae). In *Eryngium* (Saniculoideae), the flowers are stalkless and crowded into a dense head surrounded by spiny bracts. Dichasia are found in the monotypic *Petagnia*, endemic to Sicily, Italy. The characteristic umbel is a flat-topped inflorescence in which each flower stalk (pedicel) arises from the same point on the rays (peduncles), being of different lengths to raise all the flowers to the same height. A compound umbel is one in which the ultimate umbels (*umbellets* or *umbellules*) are themselves arranged in umbels. Bracts are often present at the base of the rays of a compound umbel, forming an involucre, and bracteoles are present at the base of the umbellets, where they form an involucel. The bracts and bracteoles vary in number and size.

The flowers of an umbel and of the component umbellules open in sequence from the outer whorls to the center. Most umbels are protandrous, although a few genera, such as *Sanicula*, are protogynous. Sexual differentiation in the umbels is marked in some cases, varying from genus to genus according to the degree of protandry, ranging from a few male (staminate) flowers per umbel, to umbels that are composed of nothing but male (staminate) flowers. Moreover, the percentage of perfect (bisexual) flowers in the latter cases is higher in the primary umbels and progressively lower in the successive umbels, until the last umbels produced are almost entirely composed of male flowers. The degree of organization of the umbel is highly developed in some cases and comparable to that found in the capitula of the Compositae. In some umbels, whole flowers, which are functionally unisexual, take on the role of stamens or pistils (e.g., *Astrantia*, *Petagnia*, *Sanicula*), or the umbels themselves take on the role of stamens or pistils in more complex inflorescences.

The marginal flowers in the umbel are sometimes irregular, as in *Daucus carota* (Carrot), *Turgenia latifolia*, and *Artedia squamata*, thus serving as an attraction to insect pollinators. The visual impact is also enhanced by an

APIACEAE. 1 *Eryngium biscuspidatum* shoot with spiny leaves and bracts, and flowers in a dense head (x⅔). 2 *Eryngium maritimum* barbed fruit—a schizocarp (x6). 3 *Petroselinum crispum* (a) schizocarp—comprising 2 mericarps (x8); (b) cross section of a single mericarp with a central seed and canals (vittae) in the fruit wall (x12). 4 *Psammogeton canescens* schizocarp (x8).

increase in the number and size of the umbellules and a closer spacing of the individual flowers. The bracts forming the involucre may also become enlarged, colored, and showy as in various *Eryngium* and *Bupleurum* species. One of the most remarkable examples is the Mexican species *Mathiasella bupleuroides*, which has a showy involucre and involucels that are reminiscent of a malvaceous corolla, surrounding umbellules of staminate flowers with petals and naked pistillate flowers.

Most of the Apiaceae species are "promiscuous" plants in that they are pollinated by a wide variety of insects—mostly flies, mosquitoes, or gnats, but also some of the unspecialized bees, butterflies, and moths. Self fertilization is the normal situation, and self-sterile plants are rare. Pollination is often by geitonogamy, which means the pistils may be pollinated by the anthers of adjacent flowers in the same umbel. A curious feature of the family is the small amount of hybridization—few attested records of interspecific hybrids have been recorded.

The flower of the umbellifers is basically uniform, almost monotonously so, usually regular, consisting of 5 white, yellow, blue, or purplish petals; 5 free stamens; a greatly reduced calyx (except in *Eryngium*); an inferior ovary with 2 carpels and 2 locules; and a stylopodium supporting 2 styles. There is a single pendulous, anatropous ovule in each locule. Variations on this basic theme are limited: irregular corollas, the outer petals being sometimes larger and radiate, and unisexuality. A feature showing considerable variation that has been largely overlooked until recently is the stylopodium—the swollen, often colorful, nectar-secreting base of the styles, which is characteristic of the family. This organ varies widely in shape, size, color, and nectar secretion.

The fruit shows quite a remarkable range of variation. Basically, it is a dry schizocarp that splits down a septum (commisure) into 2 single-seeded mericarps that normally remain for some time suspended from a common forked stalk—the carpophore (absent in the Saniculoideae)—finally separating at maturity. The fruits are often compressed, either dorsally or laterally. The endocarp may be woody (Azorelloideae, Mackinlayoideae) or not (Apioideae, most Saniculoideae). The outer surface of the mericarp normally has 5 primary ridges—1 dorsal, 2 lateral, and 2 commisural—and between them 4 secondary vallecular ridges, all of which run longitudinally from the base to the stylar end of the fruit. In the furrows between the primary ridges, in the ridges themselves, or all over the fruit, oil ducts/cavities or resin canals (vittae) are often found. Crystals of calcium oxalate may be present in the pericarp. The fruit surface may bear spines, hooks, hairs, or tubercles of various kinds; in some fruits, the lateral ridges are extended into wings. All these

APIACEAE. 5 *Heracleum sphondylium* (a) leaf shoot bearing large inflorescences—note the outer flowers are irregular and have deeply cut petals (×⅔); (b) regular flower from center of inflorescence (×6). 6 *Daucus carota* schizocarp with spines on the ridges (×6). 7 *Sanicula europaea* schizocarp (×6). 8 *Peucedanum ostruthium* (a) winged schizocarp (×4); (b) cross section of schizocarp (×4).

features are related to their dispersal strategy: variations in shape, size, color, wings, and spines are numerous; some fruits are remarkable constructions and scarcely bear any resemblance to the basic umbelliferous type, such as those of *Petagnia, Scandix,* and *Thecocarpus.* The seeds (2; 1 per mericarp) have an oily endosperm and a small embryo.

Classification Traditionally, the Apiaceae has been divided into 3 subfamilies—Apioideae, Saniculoideae, and Hydrocotyloideae—and several tribes, following the system proposed by Drude (1897–1898), based largely on fruit characters, some of which are ambiguous or difficult to interpret. Recent extensive molecular and phylogenetic studies, which are congruent with later morphological and anatomical research, have led to major changes, both in the classification of the family Apiaceae, and its relationship to the Araliaceae and related families. Later work on the fruits indicates that taxonomically important characters are wing configuration; presence or absence of intrajugal oil ducts and vittae and their size; presence or absence of crystals and their positions; possible lignification of the endocarp; and the arrangement and position of ventral bundles that may or may not form free carpophores[i,ii]. Some authors adopt a wide

circumscription of the family to include the Araliaceae, but the balance of evidence suggests keeping them as separate families. Within the Apiaceae, perhaps the greatest change has been the recognition that the subfamily Hydrocotyoideae is polyphyletic, comprising 3 unrelated clades[iii,iv,v,vi], one of which approximates largely to Drude's original circumscription. However, since it excludes *Hydrocotyle* itself, which has been transferred to the Araliaceae, it has been renamed subfamily Azorelloideae by Plunkett *et al.*[iii], whose classification syntheses much of this recent work and is followed here. They also recognize three members of the former Hydrocotyloideae that remain in the Apiaceae, referred to as the "Azorella clade" by Downie *et al.*[vii], as the Azorelloideae, and establish a fourth subfamily, the Mackinlayoideae, for a clade made up of a number of former hydrocotyloid genera plus 2 genera, *Mackinlaya* and *Apiopetalum,* usually placed in the Araliaceae. The 4 subfamilies recognized are:

SUBFAM. APIOIDEAE Fruit with a soft endocarp, sometimes hardened by woody subepidermal layers; vittae vallecular and commissural; style on apex of disk; carpophore bifid; stipules absent. Between 350 and 400 genera; c. 3,000 species. The Apioideae were subdivided by Drude into 8 tribes, based largely on fruit characters, but recent molecular and systematic studies have revealed that most of these tribes are not monophyletic and have gone a long way to resolving relationships. Likewise, several of the genera as conventionally recognized are polyphyletic. A recent review by Downie et al.[viii] summarizes the results of recently published molecular cladistic analyses and presents a provisional classification of the subfamily based on taxonomic congruence among the data sets. Ten tribes (Aciphylleae, Bupleureae, Careae, Echinophoreae, Heteromorpheae, Oenantheae. Pleurospermeae, Pyramidoptereae, Scandiceae, and Smyrnieae) are proposed or confirmed as monophyletic. A further 7 clades are recognized but not yet treated formally. However, since only about half the known genera have so far been analyzed, and the placement of at least 23 of the genera examined is still uncertain, much further sampling and revisional work is needed before an acceptable comprehensive tribal classification can be proposed.

SUBFAM. SANICULOIDEAE Fruit with a soft parenchymatous endocarp, rarely somewhat woody (*Steganotaenia, Polemanniopsis*); vittae often absent or poorly developed; intrajugal oil ducts prominent; crystals dispersed in the mesocarp; base of style surrounded by a ringlike disk; carpophore absent. *Arctopus, Astrantia, Eryngium, Hacquetia, Lichtensteinia, Marlothiella, Petagnia, Polemanniopsis, Sanicula,* and *Steganotaenia.* The circumscription of this subfamily has been modified to take into account molecular research and morphological findings. Thus, the genus *Lagoecia* has been transferred to the Apioideae, while 3 genera, *Lichtensteinia, Polemanniopsis,* and *Steganotaenia,* previously placed in the Apioideae, have been transferred to the Saniculoideae, as well as *Arctopus,* formerly in the Hydrocotyloideae.

SUBFAM. AZORELLOIDEAE Fruits terete or dorsally compressed; vittae absent; endocarp woody; carpels splitting at maturity. At least 12 genera; c. 150 species in the genera *Azorella, Bolax, Bowlesia, Dichosciadium, Dickinsia, Diplapsis, Eremocharis, Gymnophyton, Huanaca, Mulinum, Schizeilema, Spananthe, Stilbocarpa,* and *?Klotzschia.*

Syn.: Hydrocotyloideae Link 1829, *pro parte,* excl. *Hydrocotyle.* This subfamily comprises a large part of Drude's Hydrocotyloideae but with *Hydrocotyle* itself removed to the Araliacaeae; *Centella, Micropleura, Actinotus, Platysace,* and *Xanthosia* removed to form part of the Mackinlayoideae; and with the addition of *Stilbocarp*a, and possibly *Klotzschia,* formerly in the Apioideae.

SUBFAM. MACKINLAYOIDEAE Fruit laterally compressed (terete in *Apiopetalum*); vittae absent; intrajugal canals minute or obsolete; endocarp becoming woody: *Mackinlaya, Apiopetalum, Centella, Micropleura, Actinotus, Platysace, Xanthosia.* Molecular studies by Plunkett and Lowry[ix] suggested that the *Mackinlaya, Apiopetalum,* and 4 hydrocotyloid genera (*Centella, Micropleura, Actinotus,* and

Xanthosia) form a separate tribe, Mackinlayeae. Later, the group was proposed as a subfamily of the Apicaceae by Plunkett et al[iii], supported by morphological features such as fruit structure.

The Apiaceae has frequently been included with the mainly tropical family Araliaceae in the order Umbellales, or sometimes united with it into a single family. No clear dividing line can be drawn between them, and nearly every vegetative or floral feature that characterizes the Umbelliferae can be found in the Araliaceae; there are also similarities in chemistry and pollen characters. Both families probably both arose from a common ancestral stock and have evolved separately and in parallel. However, a consensus is emerging to maintain the 2 families as separate within the same order, although with some transfer of genera between them as noted above. The Pittosporaceae (q.v.) and the Myodocarpaceae (q.v.), comprising 3 genera segregated from the Araliaceae, are also close to both the Apiaceae and Araliaceae and are included with them in a revised order Apiales by Plunkett *et al.*[iii].

Economic uses One of the remarkable features of umbellifers is the wide range of uses made of different species, ranging from food and fodder to spices, poisons, and perfumery. However, only the Carrot (*Daucus carota*) is a major vegetable crop and also used as animal feed, with a world production of 23.3 million megatons. The Carrot and Parsnip (*Pastinaca sativa*) are the only umbellifers of international

APIACEAE. 9 *Centella asiatica* creeping leafy stem bearing axillary flowers (x⅔). 10 *Sanicula europaea* leafy shoot and inflorescences (compound umbels) (x⅔). 11 *Artedia squamata* winged schizocarp (x3). 12 *Mackinlaya macrosciadea* (a) petal and stamen (x12); (b) flower from above (x12).

repute as root crops, but other members of the family have been so used, whether cultivated or not, such as the tubers of the Great Earthnut (*Bunium bulbocastanum*) and the Pignut (*Conopodium majus*). In South America, Arracacha or Peruvian Parsnip (*Arracacia xanthorrhiza*), one of the "Lost crops of the Incas," is domesticated in the Andes, where it sometimes replaces potato, and elsewhere in South and Central America[x]. Species of

Lomatium, the largest genus of umbellifers in the USA, have been true staple foods for several groups of Native Americans in the northwestern part of the country and in western Canada. Stems, petioles, and leaves may be used for food as in Angelica (*Angelica* and *Archangelica* spp.), Celery (*Apium graveolens*), and Lovage (*Levisticum officinale*). Herbs used for flavoring include Chervil (*Anthriscus cerefolium*), Fennel (*Foeniculum vulgare*, also used as a salad vegetable), and Parsley (*Petroselinum crispum*). Spices derived from fruits or seeds, which contain essential oils, are numerous in the Apiaceae. Examples include Anise (*Pimpinella anisum*), Caraway (*Carum carvi*), Coriander (*Coriandrum sativum*), Cumin (*Cuminum cyminum*), and Dill (*Anethum graveolens*). Several of these are used as flavorings for alcoholic beverages, especially anise. Many umbellifers have medicinal uses, for gastrointestinal complaints, cardiovascular ailments, and as stimulants, sedatives, antispasmodics, etc. They are also a source of gum resins and resins such as asafetida, derived from *Ferula asafoetida* and other species, the exudate being collected from cuts made at the base of the stem or at the top of the root; another is galbanum, an oleogum resin obtained from *Ferula galbaniflua*. There are many poisonous species, the most celebrated being Hemlock (*Conium maculatum*), which was responsible for the death of Socrates but is also used for its therapeutic value. A few umbellifers are grown in gardens for their ornamental value. Examples include *Eryngium giganteum* and various cultivars, *Astrantia* (Masterwort), *Bupleurum fruticosum, Ferula communis*, and *F. tingitana*, a variegated form of *Aegopodium podagraria*, and *Heracleum* species (hogweeds), especially the spectacular *H. mantegazzianum* (Giant Hogweed).   VHH

[i] Liu, M. R. A taxonomic evaluation of fruit structure in the family Apiaceae. Ph.D. Thesis, Rand Afrikaans University, Johannesburg, South Africa (2004) (unpublished but available on the Internet at http://etd.rau.ac.za/theses/available/etd-09012004-085620/).

[ii] Liu, M., Van Wyk, B.-E., & Tilney, P. M. The taxonomic value of fruit structure in the subfamily Saniculoideae and related African genera (Apiaceae). *Taxon* 52: 261–270 (2003).

[iii] Plunkett, G. M., Chandler, G. T., Lowry II, P. P., Pinney, S. M., & Sprenkle, T. S. Recent advances in understanding Apiales and a revised classification. *South African J. Bot.* 70: 371–381 (2004).

[iv] Plunkett, G. M., Soltis, D. E., & Soltis, P. S. Clarification of the relationship between Apiaceae and Araliaceae based on *mat*K and *rbc*L sequence data. *American J. Bot.* 84: 565–580 (1997).

[v] Plunkett, G. M., Wen, J., & Lowry II, P. P. Infrafamilial classifications and characters in Araliaceae: insights from the phylogenetic analysis of nuclear (ITS) and plastid (*trn*L-*trn*F) sequence data. *Plant Syst. Evol.* 245: 1–39 (2004).

[vi] Chandler, G. T. & Plunkett, G. M. Evolution in Apiales: nuclear and chloroplast markers together in (almost) perfect harmony. *Bot. J. Linn. Soc.* 144: 123–147 (2004).

[vii] Downie, S. R., Katz-Downie, D. S., & Watson, M. F. A phylogeny of the flowering plant family Apiaceae based on chloroplast DNA *rpl*16 and *rpo*C1 intron sequences: towards a suprageneric classification of subfamily Apioideae. *American J. Bot.* 87: 273–292 (2000).

[viii] Downie, S. R., Plunkett, G. M., Watson, M. F., Spalik, K., Katz-Downie, D. S., Valiejo-Roman, C. M., Terentieva, E. I., Troitsky, A. V., Lee, B.-Y., Lahham, J., & El-Oqlah, A. Tribes and clades within Apiaceae subfamily Apioideae: the contribution of molecular data. *Edinburgh J. Bot.* 58: 301–330 (2001).

[ix] Plunkett, G. M. & Lowry II, P. P. Relationships among "ancient araliads" and their significance for the systematics of Apiales. *Molecular Phylogenetics and Evolution* 19: 259–276 (2001).

[x] National Research Council. *Lost Crops of the Incas: Little-Known Plants of the Andes with Promise for Worldwide Cultivation*. National Academy Press, Washington, D.C. (1989).

APOCYNACEAE
OLEANDER AND MILKWEED FAMILY

Apocynaceae is a large, widespread family of woody and herbaceous plants, now including the families Asclepiadaceae and Periplocaceae, which were recognized as separate families until very recently but are now interpreted as highly specialized members of Apocynaceae as demonstrated by detailed morphological and molecular studies[i].

Distribution Pantropical and subtropical, with the 2 predominantly woody subfamilies most diverse in wetter biomes, such as the tropical rain forests and swamps of India and Malaya, and the 3 largely herbaceous subfamilies most diverse in seasonally dry environments; a few genera, such as *Vinca*, extend into temperate regions.

Description Trees, shrubs, woody lianas, vines, and herbs, mostly perennial but occasionally annual or even ephemeral, sometimes strongly succulent. Latex usually white, but occasionally red or yellow (e.g., *Aspidosperma*, *Strophanthus*), or clear, most notably in the succulent *Ceropegieae* (formerly *Stapelieae*). Leaves are simple and almost always entire; most commonly opposite, but frequently whorled (e.g., *Allamanda, Nerium*); alternate in several genera of the subfamily Rauvolfioideae and sometimes in the subfamily Apocynoideae (e.g., *Adenium, Pachypodium*). Flowers are actinomorphic, bisexual, mostly 5-merous, except for the carpels. The petals are fused at least at the base, the corolla varying from rotate or shallowly campanulate to tubular or funnel-shaped, the lobes often twisted in bud and overlapping to the left or right, but sometimes valvate. Coronal structures are often present either on the corolla or on the gynostegium, which is formed from the fusion of the stamens and the stylar head (the stylar head is free from the stamens in Rauvolfioideae). Stamens are inserted on the corolla tube or at its base, the filaments free or forming a tube, and the anthers often have specialized marginal wings or guide rails. Pollen is shed as single grains, tetrads, or pollinia. The ovary is superior to subinferior, syncarpous, with 2 carpels, or more commonly apocarpous with 2 carpels (rarely more) that fuse apically to form a stylar head with the stigmatic surfaces on the underside. Secretions from the stylar head are sticky and amorphous, or formed into 5 discrete spoonlike or cliplike structures, which assist in pollen transport. Generally, the fruits are a pair of ventrally dehiscent follicles, but follicles are commonly single through abortion and, in *Lepinia*, up to 5 slender follicles remain fused apically to form a cage. Drupes, berries, and capsules occur in the subfamily Rauvolfioideae.

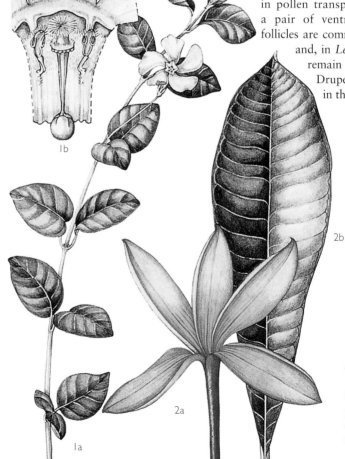
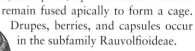

APOCYNACEAE. I *Vinca minor* (Periwinkle) (a) shoot with opposite leaves and solitary flower (x⅔); (b) part of dissected flower showing epipetalous stamens and thickened hairy stylar head (x3); (c) paired fruits (x1). 2 *Plumeria rubra* (Frangipani) (a) flower (x1); (b) leaf (x1).

Classification Related to Gelsemiaceae, Gentianaceae, Loganiaceae, and Rubiaceae—families characterized by their opposite entire leaves, interpetiolar ridges or stipules, and gamopetalous corollas. Within the Apocynaceae, the infrafamilial classification of Endress & Bruyns[ii] recognizes 5 subfamilies of increasing floral synorganization. The degree of fusion of the gynoecium and the androecium provide significant characters, as do structures associated with pollen transport. The often bizarre floral coronas present good characters for recognizing species but are less useful at higher taxonomic levels.

SUBFAM. RAUVOLFIOIDEAE (84 genera; c. 1,000 spp.). Old and New world trees, shrubs, woody lianas, and vines, with alternate, opposite, or whorled leaves; rarely herbs. Corolla aestivation almost always sinistrorse. Anthers are mostly fertile for their entire length (not in Tabernaemontaneae) and free from the stylar head; pollen shed as monads. Fruits dehiscent or indehiscent, with dry or fleshy pericarp, the endocarp sometimes forming a stone around the seed; seeds generally lacking a coma of hairs. Nine tribes are recognized: Alstonieae, Alyxieae, Carisseae, Hunterieae, Melodineae, Plumerieae, Tabernaemontaneae, Vinceae, and Willughbeeae. However, the relationships are still poorly understood.

SUBFAM. APOCYNOIDEAE (77 genera; c. 850 spp.). Old and New world woody lianas or vines, trees, or shrubs; occasionally perennial herbs. Leaves almost always opposite (alternate in *Adenium* and *Pachypodium*), and corolla aestivation generally dextrorse. Anthers usually have sterile basal portions, lignified marginal guide rails, and are weakly or strongly attached to the stylar head; pollen mostly shed as monads. Fruit usually a pair of dehiscent follicles; seeds normally have a tuft of hairs at one end, but occasionally absent or present at both ends. Five tribes are recognized: Apocyneae, Echiteae, Malouetieae, Mesechiteae, and Wrightieae.

SUBFAM. PERIPLOCOIDEAE (31 genera; c. 180 spp.). Entirely Old World, most diverse in Africa. Woody lianas, shrubs, rarely small trees (*Utleria*), or erect or twining herbs with annual stems arising from tuberous rootstocks. Leaves always opposite, sometimes with an irregularly lobed interpetiolar ridge. Corolla aestivation dextrorse or valvate, and the open flower is generally rotate, with a short or well-developed corolla tube. Corona lobes may be present in the corolla lobe sinus, or on the ridgelike foot on the corolla tube from which the staminal filaments arise. Anthers connivent over the stylar head and fused to it; anther margins not lignified. Pollen shed in tetrads or aggregated into pollinia (2 per locule), which lack a waxy outer wall. Five spoonlike translators are secreted from grooves in the stylar head—pollen or pollinia are deposited onto these translators, which have a sticky disk at the base, facilitating removal from the flower. Fruit generally a pair of dehiscent follicles, but one may be lost through abortion (e.g.,

Raphionacme); seeds are flattened, mostly with a tuft of hairs at one end, the hairs occasionally extending around the margin (*Finlaysonia*, *Raphionacme namibiana*).

SUBFAM. SECAMONOIDEAE (7 genera; c. 180 spp.). Old World, mostly occurring on Madagascar. Lianas, twining perennial herbs, or small shrubs, always with opposite leaves. Corolla aestivation dextrorse or valvate but sinistrorse in *Genianthus*. Coronas may occur as fleshy ridges on the corolla and as fleshy lobes on the stamens. Staminal filaments fused to form a tube; anther margins are lignified, forming guide rails. Stamens fused to the stylar head, forming a gynostegium. Pollen aggregated into tiny pollinia (2 per locule, 4 per pollinarium) lacking a waxy outer wall. Pollinia are attached to the

Genera 380–425 **Species** 5,000–6,000
Economic uses Arrow poisons, cultivated ornamentals, tropical timbers, traditional and Western medicines

soft, grooved corpusculum by 1 or 2 translator arms; the compound entity is removed from the flower as a unit. Fruit is a pair of dehiscent follicles; seeds are flattened, ovate, with a tuft of hairs at one end.

SUBFAM. ASCLEPIADOIDEAE (c. 180 genera; 2,500–3,500 spp.). Old and New world[iii] shrubs (*Calotropis*), twiners, succulents, and herbs. Mostly perennial but sometimes annual or even ephemeral (*Conomitra*). Leaves usually opposite, occasionally whorled, or reduced to a scale or spine. Corolla aestivation valvate or dextrorse. Corolline coronas are rare; the corona is mostly found on the gynostegium, in 1, 2, or rarely 3 whorls, although it is absent in some groups. Staminal filaments fused to form a tube; anther margins lignified, forming guide rails. Stamens fused to the stylar head, forming a gynostegium. Pollen is aggregated into pollinia (1 per locule, 2 per pollinarium) bound by a waxy outer wall. The pollinia are attached to the hard, brown or black, grooved corpusculum by 2 translator arms, forming a pollinarium extracted from the flower as a unit. Fruit is paired dehiscent follicles or frequently single by abortion. Seeds are generally flattened, ovate, mostly with a narrow marginal rim, and a tuft of hairs at one end. Three largely Old World tribes are recognized: Ceropegieae; Fockeeae; and Marsdenieae (*Marsdenia* itself is pantropical), with erect pollinia. Subtribes are recognized in Ceropegieae (Anisotominae, Heterostemminae,

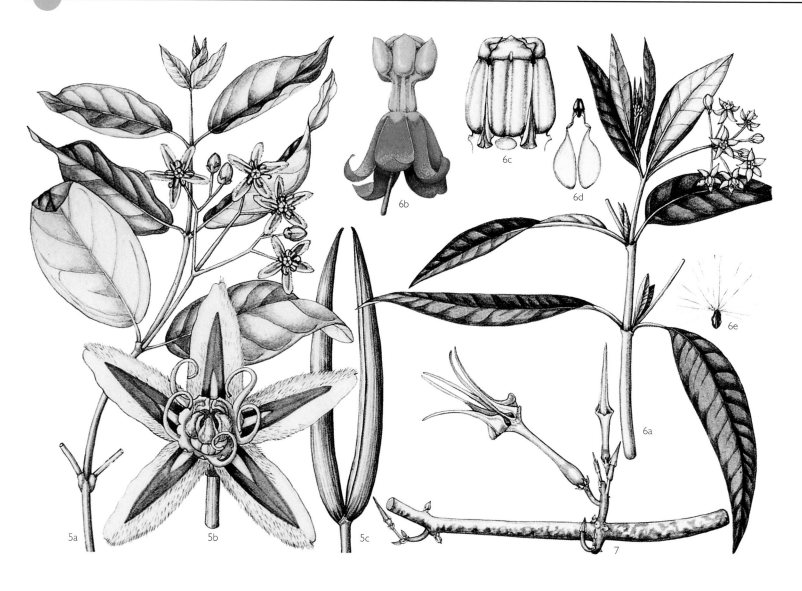

APOCYNACEAE. 5 *Periploca graeca* (a) shoot with opposite leaves and axillary inflorescence (×⅔); (b) flower with coiled corona lobes and hairy anthers (×3); (c) fruit, comprising a pair of follicles (×⅔). 6 *Asclepias curassavica* (Butterfly Plant) (a) leafy shoot with opposite leaves and extra-axillary inflorescence (×⅔); (b) flower with orange reflexed corolla lobes and yellow staminal corona obscuring the gynostegium (×3); (c) upper part of gynostegium with corona lobes removed to show anthers with thickened lateral margins forming guide rails, and triangular anther appendages curved over top of stylar head (×10); (d) pollinarium, formed of 2 pollinia linked by a pair of slender translator arms to the cliplike corpusculum (×14); (e) seed with tuft of hairs (×10). 7 *Ceropegia stapeliiformis* semisucculent shoot with a cluster of tubular trap-flowers (×⅔).

Leptadeniinae, and Stapeliinae[iv]). A fourth tribe, Asclepiadeae, which has mostly pendent or laterally disposed pollinia, contains 2 parallel radiations: one predominantly Old World (Asclepiadinae, Cynanchinae, and Tylophorinae), but with significant representation (*Asclepias*) in North America; the other centered on Central and South America (Gonolobinae, Metastelmatinae, Orthosiinae, and Oxypetalinae)[v,vi]. The small subtribe Astephaninae, from the Cape region of southern Africa, is considered to be sister to both radiations.

Economic uses Cardiac glycosides are obtained from many genera, e.g., *Strophanthus*, used locally as an arrow poison. *Rauvolfia* produces reserpine and rescinnamine, which are used to treat certain mental illnesses. *Catharanthus roseus* (Madagascar Periwinkle) yields 2 important cancer drugs—vinblastine and vincristine—used to treat childhood leukaemia and Hodgkins' disease. *Hoodia gordonii*, used by Kalahari bushmen as an appetite suppressant, has been commercialized for the treatment of obesity. Many members of the family are used in folk medicines. Latex or India rubber is valuable and is extracted from *Hancornia speciosa*, *Funtumia elastica*, and species of *Landolphia*. *Alstonia scholaris* is an important timber tree in Southeast Asia, and wood from *Aspidosperma* is particularly resistant to decay. Ornamentals widely planted in tropical regions include *Allamanda cathartica*, *Cryptostegia*, *Hoya* (Wax plant), *Plumeria* (Frangipani), and *Thevetia peruviana*. *Asclepias* (milkweeds), *Araujia* (Cruel Plant), *Mandevilla*, *Nerium oleander* (Oleander), *Stephanotis*, and *Vinca* (periwinkles) are popular in temperate regions, in addition to succulent genera such as *Caralluma*, *Ceropegia*, *Orbea*, and *Stapelia*.   DJG

i Sennblad, B. & Bremer, B. The familial and subfamilial relationships of Apocynaceae and Asclepiadaceae evaluated with *rbc*L data. *Pl. Syst. Evol.* 202: 153–175 (1996).

ii Endress, M. E. & Bruyns, P. V. A revised classification of the Apocynaceae *s.l. Bot. Rev.* 66: 1–56 (2000).
iii Goyder, D. J. An overview of Asclepiad biogeography. In: Ghazanfar, S. A. & Beentje, H. J. (eds), *Taxonomy and ecology of African plants and their conservation and sustainable use.* Pp. 205–214. (2006).
iv Meve, U. & Liede, S. Subtribal division of Ceropegieae (Apocynaceae-Asclepiadoideae). *Taxon* 53: 61–72 (2004).
v Rapini, A., Chase, M. W., Goyder, D. J., & Griffiths, J. Asclepiadeae classification: evaluating the phylogenetic relationships of New World Asclepiadoideae (Apocynaceae). *Taxon* 52: 33–50 (2003).
vi Liede-Schumann, S., Rapini, A., Goyder, D. J., & Chase, M. W. Phylogenetics of the New World subtribes of Asclepiadeae (Apocynaceae–Asclepiadoideae): Metastelmatinae, Oxypetalinae and Gonolobinae. *Syst. Bot.* 30: 183–194 (2005).

APODANTHACEAE

A small family of parasitic herbs of uncertain affinity.

Distribution *Apodanthes* (1 but possibly several species), is distributed from Mexico to northern parts of South America. *Pilostyles* (c. 20 spp.) grows in the neotropics from California south to Chile and Argentina, and in southwestern Asia and subtropical southwestern Australia. *Berlinianche* (2 spp.) is endemic to tropical Africa.

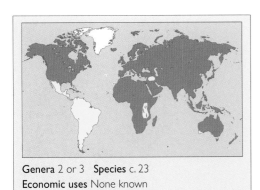

Genera 2 or 3 **Species** c. 23
Economic uses None known

Genera 1 **Species** More than 400
Economic uses Beverages (yerba maté),
ornamental trees and shrubs, and timber

Description Endoparasitic, achlorophyllous plants with the vegetative body resembling a fungal mycelium, consisting entirely of endophytic tissues residing inside the host. Only the flowering shoots are visible outside the host. *Pilostyles* is parasitic on legumes (*Astragalus, Dalea, Daviesia, Mimosa*), *Apodanthes* on *Casearia* and *Xylosma* (Flacourtiaceae) and representatives of Burseraceae and Meliaceae, while *Berlinianche* is restricted to Amherstieae of Leguminosae[i]. The stem is absent and the leaves are reduced to scalelike bracts subtending the flowers. The flowers are small, unisexual (plants monoecious or dioecious), each with 3 (or 4) whorls of 2–6 scales (bracts) that are 2- or 4-merous (*Apodanthes, Pilostyles*) or 3- and 6-merous (*Berlinianche*), of which the inner 1 or 2 correspond to a perianth[i]. In *Pilostyles,* the scales are free, red to brown, emerging from the sides of a rounded receptacle in the male flowers, or the outside parts of the ovary in the female flowers. In *Apodanthes*, the scales are cream and yellow to orange, brown, or red, free in the inner and outer whorl of scales and connate in the middle whorls. In *Berlinianche*, the scales are reddish and free. In the staminate flowers, the stamens are connate into a tube (synandrium) surrounding a sterile gynoecium; the individual stamens cannot be distinguished. In the female flowers, the gynoecium is syncarpous with 4 united carpels; the ovary is inferior or semi-inferior, 1-locular, with 4 or more parietal

placentae; there is no rudiment of an androecium. A nectary disk is present in both male and female flowers. In *Berlinianche*, there is a hair cushion on the inner perianth organs. The fruit is baccate, irregularly dehiscent, with numerous small seeds.

Nickrent's Parasitic Plant Connection[ii] is a valuable source of information and illustrations.

Classification The genus *Berlinianche* is sometimes included within *Pilostyles*[iii] but may be distinguished by its 3- and 6-merous scales and by its unique cushion of hairs on the inner perianth organs[i]. The position and relationships of Apodanthaceae are uncertain. It is often included in Rafflesiaceae or retained as a family positioned in the Rafflesiales within which it forms a small-flowered clade[iv]. In APG II, it is unplaced as a taxon of uncertain position. A recent study suggests that is either related to Malvales or Cucurbitales[v].   VHH

[i] Blarer, A., Nickrent, D. L., & Endress, P. K. Comparative floral structure and systematics in Apodanthaceae (Rafflesiales). *Plant Syst. Evol.* 245: 119–142 (2004).
[ii] Nickrent, D. L. The Parasitic Plant Connection. http://www.science.siu.edu/parasitic-plants/
[iii] Bouman, F. & Meijer, W. Comparative structure of ovules and seeds in Rafflesiaceae. *Plant Syst. Evol.* 193: 187–212 (1994).
[iv] Blarer, A., Nickrent, D. L., Bänziger, H., Endress, P. K., & Qiu, Y.-L. Phylogenetic relationships among genera of the parasitic family Rafflesiaceae *s. l.* based on nuclear ITS and SSU rDNA, mitochondrial LSU and SSU rDNA, atp1 and matR sequences. [Abstract]. *American J. Bot.* 87 (6 suppl.): 171 (2000).
[v] Nickrent, D. L., Blarer, A., Qiu, Y.-L., Vidal-Russell, R., & Anderson, F. E. Phylogenetic inference in Rafflesiales: the influence of rate heterogeneity and horizontal gene transfer. *BMC Evol. Biol.* 4: 40 (2004). See: http://www.biomedcentral.com/1471-2148/4/40

AQUIFOLIACEAE
HOLLIES AND YERBA MATÉ

A widespread, but mainly tropical, family of trees and shrubby plants with characteristic toothed leaves.

Distribution The single genus *Ilex*, with more than 400 species, has a more or less worldwide

distribution, but with the greatest diversity of species in tropical South America and Asia (mainly in Southeast Asia) and with a poor representation in Africa (1 sp.) and northern tropical Australia (1 sp.). The family extends into temperate regions of North America, Europe, and Asia, and a few species occur on the islands of the Azores, the Canaries, the Caribbean, Hawaii, Madeira, and Tahiti.

Description Shrubs or small trees, sometimes scandent or epiphytic, usually evergreen, rarely deciduous. Leaves alternate; rarely opposite or nearly so; simple; often leathery; with prickly dentate margins, although heterophylly is found in some species, whereby both dentate or entire leaves occur, either on the same or on separate plants. Flowers are actinomorphic; unisexual (plants dioecious); small and inconspicuous; white, cream, greenish-white, and yellow, in axillary cymose, racemose, or sub-umbellate inflorescences. The sepals are 4–5 or 6–8, valvate, more or less connate at the base. The petals are 4–5, rarely 6–9 or more, usually connate the base, less frequently distinct. The stamens are usually as many as the petals. Staminodes occur in the pistillate flowers and a pistillode in the staminate flowers. The ovary is superior, of (2–3)4–5 or more united carpels; as many locules as carpels, the locules with a single terminal, sometimes minute, style; the placentation is apical-axile, with 1 or rarely 2 pendulous, usually anatropous, ovules per locule. The fruit is a drupe containing 1–6 or more hard pyrenes; seeds 1–6, with copious endosperm.

Classification *Nemopanthus*, which comprises a single species, *N. mucronatus*, native to eastern North America, differs from *Ilex*, having flowers with a reduced *calyx* and narrow petals, but it is now included in *Ilex*[i,ii], showing a close relationship with *I. amelanchier*. *Phelline* and *Sphenostemon*, which are sometimes included in the Aquifoliaceae, are recognized as separate families. The molecular phylogeny of the family is still not fully clarified[ii,iii], and further sampling is needed. The Aquifoliaceae was previously placed in the Celastrales but is now considered on the basis of molecular evidence to belong in the Asterids, occupying a position basal to many of the Asterales.

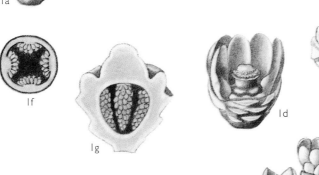

APODANTHACEAE. **1** *Berlinianche aethiopica* (a) male flower (×4); (b) section of host branch, showing flowers and flower buds (×2); (c) host branch bearing flowers (×⅔); (d) male flower with part of calyx removed (×4); (e) female flower with perianth removed (×3); (f) cross section of ovary (×3); (g) vertical section of ovary (×4½).

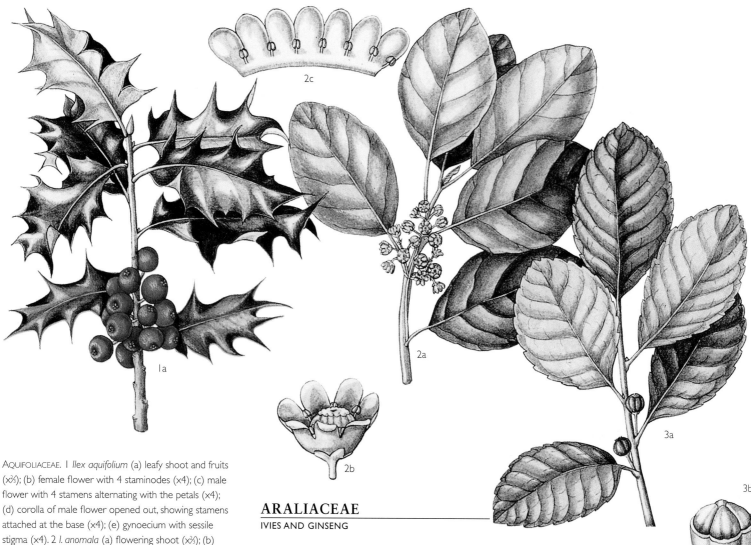

AQUIFOLIACEAE. 1 *Ilex aquifolium* (a) leafy shoot and fruits (x⅔); (b) female flower with 4 staminodes (x4); (c) male flower with 4 stamens alternating with the petals (x4); (d) corolla of male flower opened out, showing stamens attached at the base (x4); (e) gynoecium with sessile stigma (x4). 2 *I. anomala* (a) flowering shoot (x⅔); (b) bisexual flower (x3); (c) perianth opened out, showing stamens fused to the base of the perianth tube and alternating with the lobes (x3); (d) fruit—a berry (x4). 3 *I. paraguaensis* (a) leafy shoot with fruits (x⅔); (b) fruit with wall cut away, showing 4 hard pyrenes each containing a single seed (x4).

Economic uses Many *Ilex* species are grown as ornamentals (hollies) for their foliage and berries. Beverages are made from the leaves and other aerial parts of many species, most notably *I. paraguariensis* which is used in Argentina, southern Brazil, Paraguay, and Uruguay to prepare a popular beverage called yerba maté. The appeal of the beverage lies in its stimulative properties, which is due to the plant's caffeine and theobromine content. Other species are used in traditional medicines in Africa, Asia, Europe, and North America. The wood of a limited number of species of *Ilex* is valued for carving, furniture, and tools.   VHH

i Powell, M., Savolainen, V., Cuénoud, P., Manen, J.-F., & Andrews, S. The mountain holly (*Nemopanthus mucronatus*, Aquifoliaceae) revisited with molecular data. *Kew Bull.* 55: 341–347 (2000).
ii Manen, J.-F., Boulter, M. C., & Naciri-Graven, Y. The complex history of the genus *Ilex* L. (Aquifoliaceae): evidence from the comparison of plastid and nuclear DNA sequences and from fossil data. *Pl. Syst. Evol.* 235: 79–98 (2002).
iii Cuénoud, P., del Pero Martínez, M. A., Loizeau, P.-A., Spichiger, R., Andrews, S., & Manen, J.-F. Molecular phylogeny and biogeography of the genus *Ilex* L. (Aquifoliaceae). *Ann. Bot. London* 85: 111–122 (2000).

ARALIACEAE
IVIES AND GINSENG

The Araliaceae is a medium-sized, largely tropical, family of mainly trees or shrubs, characterized by their often compound leaves, with broad, more or less sheathing, leaf-bases and compound inflorescences that are usually capitate or umbellate, with small regular flowers and drupaceous fruits.

Distribution The Araliaceae is distributed throughout much of the world in both temperate and tropical regions, although the family is mainly tropical. In terms of generic diversity, the chief centers are located in eastern and southeastern Asia, the Pacific and Indian Ocean basins and, in terms of species, the Americas and Indo-Pacific regions.

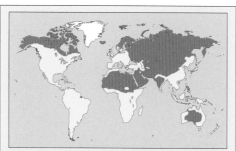

Genera 41–50 **Species** c. 1,450
Economic uses Ginseng and some medicinal products, rice paper, some species (e.g., *Hedera*) grown as ornamentals

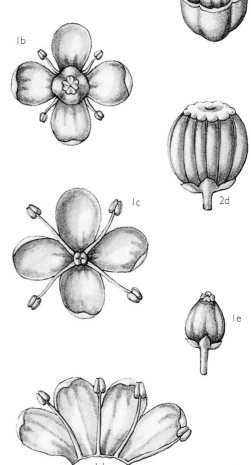

Description The Araliacae consists of small to medium, usually evergreen, trees or shrubs, occasionally climbers (*Hedera*) or rhizomatous herbs (*Aralia*). The leaves are usually alternate and often large and pinnately to palmately compound (simple in *Hydrocotyle* and *Oreopanax*), and sometimes crowded toward the end of the shoots in flushes; frequently covered with stellate or dendroid hairs; the stipules are small and poorly developed or conspicuous; the leaf-bases are sometimes sheathing, often leaving conspicuous scars. In those species with a climbing habit, aerial roots are modified for clinging to the supporting structures. The inflorescences are umbels, heads, or racemes, (*Munroidendron racemosum*) or spikes (*Oreopanax fontiana*). The flowers are actinomorphic, small, often greenish or whitish, bisexual or unisexual (then plants dioecious, rarely with male, female, and bisexual on the same plant). The calyx is small, with 4 or 5 teeth, fused to the ovary. Petals 5–10(–12), free or partially fused. The stamens are free, equal in number to the petals (c. 100 in *Plerandra*), and alternate with them, and attached to a more or less fleshy, nectariferous disk, which surmounts the ovary. The ovary is inferior (rarely superior, as in *Tetraplasandra*), with the disk surrounding the styles, which are free or fused into a column. The ovary has (2–)5–10 (–12) fused carpels (up to 15 in *Plerandra*), with (1)2–5(–10) locules, each with a single, pendulous ovule. The fruit is drupaceous, or less frequently baccate, containing 2–5 seeds, each with copious endosperm and a small embryo, rarely dry.

Classification Together with the Apiaceae, the Araliaceae has been placed in the order Apiales (or Araliales) by several authors, such as Cronquist and Takhtajan, while others, such as Judd *et al*[i,ii], have merged the two into a single family. As discussed under the Apiacaeae, here the two families are kept separate, but the Araliaceae is circumscribed so as to exclude the Myodocarpeae, which is recognized as a separate family, and the Mackinlayeae, which is referred to the Apiaceae, but with the inclusion of the genus *Hydrocotyle*, which is removed from the Apiaceae where it has traditionally been placed. Some authors, such as Frodin[iii], regard such a circumscription as premature and keep Myodocarpeae and Mackinlayeae as tribes with Hedereae (Schefflereae) and Aralieae and keep *Hydrocotyle* in the Apiaceae.

The subdivision of the family in its restricted sense, as adopted here, is still largely unresolved, although a recent phylogenetic analysis of data from molecular markers by Plunkett *et al*[iv] and Lowry[v] indicates that there are 3 major lineages in the family, corresponding generally with its centers of diversity: (1) the *Aralia–Panax* group centered in eastern and Southeast Asia; (2) the Asian palmate group (including *Hedera*, *Fatsia*, and part of the polyphyletic *Schefflera*) also centered in eastern and Southeast Asia; and (3), the

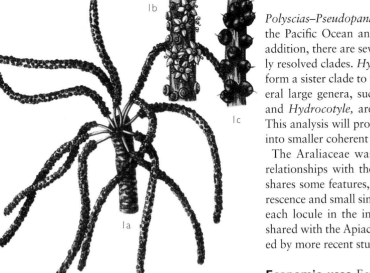

Polyscias–Pseudopanax group found throughout the Pacific Ocean and Indian Ocean basins. In addition, there are several other smaller or poorly resolved clades. *Hydrocotyle* and *Trachymene* form a sister clade to the large araliad clade. Several large genera, such as *Schefflera*, *Polyscias*, and *Hydrocotyle*, are in need of further study. This analysis will probably lead to their splitting into smaller coherent genera.

The Araliaceae was earlier thought to show relationships with the Cornaceae with which it shares some features, such as the compact inflorescence and small simple flowers with 1 ovule in each locule in the inferior ovary, features also shared with the Apiaceae, but this is not supported by more recent studies.

Economic uses Economically, the family is important for the plants of the ginseng group, which include American Ginseng (*Panax quinquefolia*), Oriental Ginseng (*P. pseudoginseng*), and Tienchi-ginseng (*P. notoginseng*), whose

ARALIACEAE. 1 *Cussonia kirkii* (a) portion of stem crowned by fruiting head (x⅔); (b) part of inflorescence (x⅔); (c) fruit (x⅔). 2 *Tetraplasandra hawaiensis* (a) flower with caplike corolla, which falls off (x2); (b) cross section of fruit (x2). 3 *Acanthopanax henryi* (a) shoot with stipulate trifoliolate leaves, and young and mature fruits (x⅔); (b) vertical section of ovary (x3). 4 *Aralia scopulorum* pinnate leaf (x⅔). 5 *Hedera helix* (a) climbing shoot with juvenile leaves and adventitious roots (x⅔); (b) shoot with adult leaves and flowers in umbels (x⅔); (c) cross section of fruit (x2); (d) flower (x3). 6 *Hydrocotyle vulgaris* (a) schizocarp (x10); (b) cross section of schizocarp, showing flattened appearance, and a narrow wall (commisure) between the 2 single-seeded mericarps, each with prominent ridges (x10).

thickened roots are the source of complex mixtures of triterpenoid saponins, which possess hormonal activity. Extracts from these roots are reputed to have stimulant, tonic, and aphrodisiac properties. The word *Panax* is derived from the Greek *panakeia*, which means "universal remedy," hence "panacea." Siberian ginseng, also known as eleuthero, is obtained from *Acanthopanax senticosus* and has similar properties to the extracts from other ginseng species. Chinese rice paper is obtained from the pith of *Tetrapanax papyrifera*. Medicinal extracts have also been extracted from a number of *Aralia* spp., e.g., *A. cordata* and *A. racemosa*. The family includes a number of ornamentals with attractive foliage, such as *Schefflera* (*Dizygotheca*) *elegantissima*, *S. arboricola*, and some ivies, in particular the ornamental cultivars of *Hedera helix* and the Canary Island ivy (*H. canariensis*). *Fatsia japonica* is also an attractive house plant and outdoor ornamental, with its glossy green leaves. First discovered in 1912 growing in a nursery in Nantes, France, Bush ivy (× *Fatshedera lizei*) is a bigeneric hybrid of the houseplant Japanese fatsia (*Fatsia japonica* "Moseri") and Irish ivy (*Hedera helix* var. *hibernica*). It is grown as a house plant or outdoor plant, often used for ground cover. Shrubby species from the genera *Polyscias* and *Acanthopanax* are grown as garden ornamentals. *Munroidendron racemosum*, an endangered tree endemic to the island of Kauai, may be grown as a landscape plant. Several ivies are grown outdoors for ground cover. [SRC] VHH

i Judd, W. S., Sanders, R. W., & Donoghue, M. J. Angiosperm family pairs: preliminary phylogenetic analyses. *Harvard Papers in Botany* 5 (1994).
ii Judd, W. S. *et al. Plant Systematics. A Phylogenetic Approach*. 2nd ed. Sunderland, Massachusetts, Sinauer Associates (2002).
iii Frodin, D. Araliaceae (Ginseng or Ivy Family). In: Smith, N. *et al.* (eds), *Flowering Plants of the Neotropics*. Pp. 28–31. Princeton/Oxford, Princeton University Press (2004).

ARGOPHYLLACEAE | *Corokia buddlejoides* (a) flowering shoot (×⅔); (b) flowerbud (×3); (c) perianth, opened out (×3); (d) vertical section of gynoecium (×3); (e) fruit, entire (left), and in vertical section (right) (×3); (f) fruits (×⅔).

iv Plunkett, G. M., Wen, J., & Lowry II, P. P. Infrafamilial classifications and characters in Araliaceae: insights from the phylogenetic analysis of nuclear (ITS) and plastid (*trn*L-*trn*F) sequence data. *Plant Syst. Evol.* 245: 1–39 (2004).
v Lowry II, P. P., Plunkett, G. M., & Wen, J. Generic relationships in Araliaceae: looking into the crystal ball. *South African J. Bot.* 70: 382–392 (2004).

ARGOPHYLLACEAE

This family of small trees or shrubs with 2-armed trichomes is found in the southwestern Pacific region.

Distribution *Argophyllum* comprises some 15 spp. from New Caledonia and eastern Australia. *Corokia* contains 6 spp.: 1 sp. in rain forests in northern New South Wales, Australia; 1 sp. endemic to Lord Howe Island; 2 spp. to New Zealand; 1 sp. to the Chatham Islands; and 1 sp. on Rapa Island.

Genera 2 Species c. 21
Economic uses Some ornamental shrubs

Description Mainly evergreen shrubs, to a height of 4 m or more, or small trees (some *Corokia* spp.), with intricate branching patterns (*Corokia*), often silky, pubescent, or tomentose with characteristic T-shaped trichomes. Leaves alternate, simple, the margins entire or dentate, exstipulate, often densely hairy, and white or silvery-gray beneath. The flowers are small, yellow, white, or green, hermaphrodite, actinomorphic in axillary or terminal panicles or corymbs, or solitary. The sepals 4–5(–8). The petals 4–5(–8), valvate, connate at the base, or adjacent lobes sometimes united for most of their length (*Corokia*), usually with a small ligulate appendage on the inner surface near the base. The stamens are 5. The ovary is inferior (*Corokia*) or semisuperior (*Argophyllum*), with 1–2(6) locules.

The fruit is a single-seeded drupe (*Corokia*) or a loculicidal capsule with numerous seeds (*Argophyllum*).

Classification The family comprise 2 genera that show appreciable differences in ovary placentation and fruit and seed structure and, as such, have not always been placed together. *Corokia* has been included in the Cornaceae by several authors, an affinity that has not subsequently been confirmed, and has been recognized as a separate family by Takhtajan. *Argophyllum* has been included in the Saxifragaceae or Escalloniaceae or again treated as a separate family. On the other hand, they share the characteristic T-shaped trichomes and ligulate petals. Evidence from floral morphology, pollen and molecular data support their association[i] and a probable position in the Asterales, although this is far from clearcut.

Economic uses Several *Corokia* spp., notably *C. cotoneaster* and *C. buddlejoides*, are grown as ornamental shrubs. VHH

i Kårehed, J., Lundberg, J., Bremer, B., & Bremer, K. Evolution of the Australasian families Alseuosmiaceae, Argophyllaceae and Phellinaceae. *Syst. Bot.* 24: 660–682 (2000).

ARISTOLOCHIACEAE
BIRTHWORT FAMILY

A primitive family with 3-merous, regular to strongly irregular and conspicuously tubular flowers, often with complex insect interactions.

Distribution Widespread in northern temperate (Canada and Scandinavia southward) and tropical regions but poorly represented in southern temperate regions, where it extends only to the north of Australia and is absent from southern Africa and most of southern South America. Species occur in rain forests, woodland, open grassy places, and dry Mediterranean-type vegetation.

Description Rhizomatous herbs (especially *Asarum*) to subshrubs, shrubs, or high-climbing lianes up to 50 m. The leaves are alternate, simple, often cordate at the base and palmately veined, occasionally trilobed. The flowers are solitary or in terminal or axillary cymes or racemes, in *Asarum* usually on the rhizomes hidden below the leaves. The flowers are actinomorphic to strongly zygomorphic, bisexual, in *Aristolochia* often spectacular and more than 10 cm long in *A. grandiflora*. The calyx is strongly developed, fused into a short tube with 3 equal lobes, or a longer complex tube forming an S-shaped pitcher with a swollen base constricted above and with 1 lobe usually much expanded, rarely with 4 or 5 lobes. The corolla is absent or rarely represented by 3 vestigial petals, or these are well developed in *Saruma*. There are usually 6 or 12 stamens but sometimes up to 40, in 1 to 4 whorls, usually fused with the gynoecium to

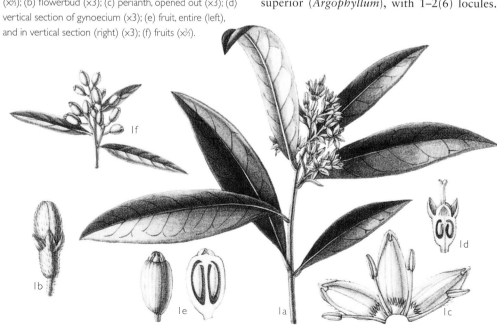

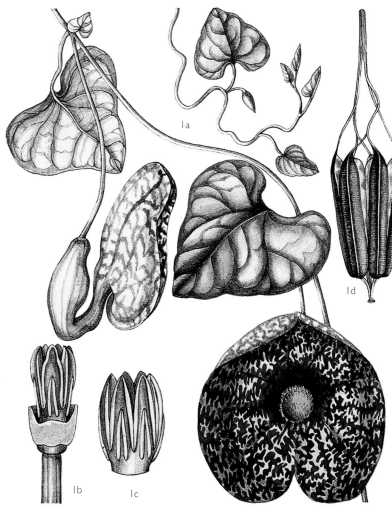

ARISTOLOCHIACEAE. 1 *Aristolochia elegans* (a) twining leafy shoot with alternate leaves and solitary flowers having a colorful calyx (x⅔); (b) receptacle and gynostemium formed from 6 sessile anthers below and 6 stigmatic lobes above (x1); (c) gynostemium (x1½); (d) fruit—a capsule opening from the base (x⅔). 2 *Asarum europaeum* (BELOW RIGHT) (a) habit, showing creeping aerial stems, scale leaves, and pairs of foliage leaves and single flower arising at joint of creeping stem (x⅔); (b) flower with perianth removed to show numerous stamens and thick style with lobed stigma (x4); (c) half flower with regular perianth lobes and an inferior ovary with ovules on axile placentas (x3).

Tribe Bragantieae comprise a group of subshrubs and shrubs, not climbing, with the calyx actinomorphic, stamens in 1 to 4 whorls and ovary 4-locular. *Thottea* (27 spp.) from India to Malaysia, and optionally *Asiphonia* with 1 species from Malaya, Sumatra, and Borneo.

Tribe Aristolochieae comprise a group of subshrubs to high climbers, with the calyx usually strongly zygomorphic, stamens in 1 whorl, and ovary 5- to 6-locular. *Isotrema* (c. 50 spp.) in tropical and temperate Asia to North and Central America; *Endodeca* (2 spp.) in North America; *Pararistolochia* (18 spp.) in tropical Africa and Malaysia; and *Aristolochia sensu lato* (c. 490 spp.) widespread, including optional genera *Holostylis*, *Euglypha*, and *Einomeia*.

Economic uses The family has long been associated with traditional medicines concerning childbirth (giving the name *Aristolochia* and the vernacular name Birthwort) and abortion. Numerous species of *Aristolochia* are cultivated as ornamentals or curiosities in the tropics or in temperate glasshouses, especially *A. sipho* (Dutchman's Pipe), *A. grandiflora*, and *A. elegans* (Hanging Baskets). *Asarum* species are sometimes grown for ground cover.   RKB

i Kelly, L. M. & González, F. A. Phylogenetic relationships in Aristolochiaceae. *Syst. Bot.* 28: 236–249 (2003).
ii Huber, H. Aristolochiaceae. In: Kubitzki, K. (ed.), *The Families and Genera of Vascular Plants. II.* Pp. 129–137. Berlin, Springer-Verlag (1993).
iii González, F. A. & Stevenson, D. W. A phylogenetic analysis of the subfamily Aristolochioideae (Aristolochiaceae). *Revista Acad. Colomb. Cienc.* 26: 25–60 (2002).

form a gynostemium. The ovary is semi-inferior to inferior with 4 to 6 usually fused carpels, but superior and the carpels almost free in *Saruma*; locules 4 to 6, each with usually many ovules. The fruit may be indehiscent (*Pararistolochia*), or a capsule breaking up irregularly (*Asarum*), or in *Aristolochia* often dehiscing from the base, with the ribs persisting to keep the dehisced fruit attached to the pedicel (the "hanging-basket" fruit of *A. elegans*).

The flowers are usually adapted to fly pollination, and they often smell strongly of carrion. In *Aristolochia*, the flies are attracted to the mouth of the tube, the swollen base of which often has a translucent window that attracts the flies down to the gynostemium. The inside wall of the tube bears stiff, downward-projecting hairs that prevent the flies from escaping until pollination is

effected. The hairs then relax and permit the flies to escape. The larvae of Swallowtail butterflies feed exclusively on leaves of the family.

Classification Despite the highly developed flowers adapted to fly pollination in *Aristolochia*, the 3-merous condition throughout the family has always suggested a position among the more primitive flowering plants. The precise relationships with other dicot families, and with the monocots (also 3-merous), have been much discussed recently—for summary and references see Kelly & Gonzalez[i]. A position in the Piperales seems to be generally agreed. The number of genera recognized in the family has varied considerably. In FGVP, Huber[ii] has recognized 12 genera, but some of these have been sunk on cladistic grounds[iii], especially into *Aristolochia*, despite some very distinctive characters. Two subfamiles are generally recognized.

SUBFAM. ASAROIDEAE Nonclimbing perennial herbs. The flowers are solitary and terminal, actinomorphic, with the calyx not constricted above the ovary. *Saruma* (1 sp. in southwestern China, with well developed petals and free carpels), and *Asarum* (c. 100 spp., northern temperate part of the range).

SUBFAM. ARISTOLOCHIOIDEAE. Subshrubs to high climbers. Flowers not solitary and terminal, actinomorphic to zygomorphic, with the calyx constricted above the ovary.

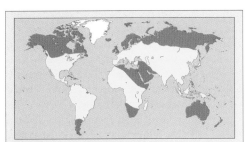

Genera 7–12   Species c. 600
Economic uses Traditional medicines, used especially in abortion and childbirth, and limited horticultural value

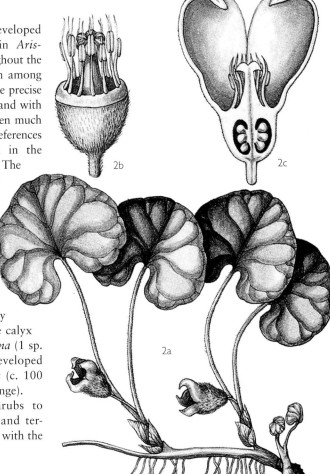

ASTERACEAE (COMPOSITAE)
SUNFLOWER FAMILY

The Asteraceae (Compositae) is one of the largest and best-known families of flowering plants and certainly the largest of the dicotyledonous families.

Distribution The family has a worldwide distribution, although absent from the Antarctic mainland. They are a significant element in the floras of the semiarid regions of the tropics and subtropics, such as the Mediterranean region; Mexico; southern Africa; and the woodland, wooded grassland, grassland, and bush ecosystems of Africa, South America, and Australia. They are also abundant in arctic, arctic-alpine, temperate, and montane floras throughout the world but poorly represented in tropical rain forests. About half the species of the Asteraceae are native to the Old World and half to the New World. Generally speaking, the Barnadesieae are restricted to South America, the Liabeae to the Neotropics, the Arctotideae are mostly South African, the Heliantheae and Eupatorieae are predominantly New World, and the highest concentrations of the Cardueae are in the Old World and predominantly Eurasian.

Description Most of the Asteraceae are annual or perennial herbs; subshrubs or shrubs; climbers (e.g., *Mikania* spp. and *Mutisia* spp.) or lianes; small or rarely large trees; epiphytes (e.g., *Gongrostylus*, *Neomirandea*); true aquatics (e.g., *Sclerolepis*, *Shinnersia*, *Trichocoronis*) are rare. Rootstocks fibrous and fleshy, sometimes with a distinct woody perennating rootstock (xylopodium). Stem usually unarmed, rarely spiny (e.g., in most members of the tribe Barnadesioideae and in few Mutisieae); spiny leaves are relatively common and typical in the tribe Cardueae. The Asteraceae are characterized by the presence of resin canals (in all except most members of the tribe Lactuceae) or laticifers (in all Lactuceae) and latex sacs (e.g., a few genera of Cardueae and Arctotideae). Characteristic biochemical features include the presence of the polysaccharide inulin, instead of starch, in the subterranean parts (e.g., *Helianthus tuberosus*) and fatty oils in the seeds. The leaves are alternate or opposite, rarely whorled, and without stipules; they are usually simple, rarely compound, pinnately or palmately veined, and either sessile or petiolate, sometimes with an expanded, sheathing or auriculate, base; they are often lobed or variously toothed, rarely succulent, rarely ending in a tendril (e.g., *Mutisia* spp.), and rarely reduced to scales and falling rapidly (e.g., *Aphylloclados*).

The headlike capitulum is one of the most characteristic features of the Asteraceae and is an inflorescence made up of many small individual flowers, called florets, and surrounded by an involucre of protective bracts; this type of inflorescence is constant throughout the family but sometimes highly modified. The whole structure resembles a single flower and is usually referred

Genera c. 1,600   Species c. 25,000
Economic uses Food plants (e.g., lettuce, globe and Jerusalem artichokes, sunflower), oil producers (e.g., sunflower, Niger seed, safflower), herbs (tarragon), ornamentals (e.g., asters, chrysanthemums, cinerarias, dahlias, cosmos, marigolds) insecticide (pyrethrum) and medicines and herbal remedies (e.g., arnica, calendula, chamomile, tansy), and drugs (artemisenin from *Artemisia annua*)

ASTERACEAE (COMPOSITAE). 1 *Gazania linearis,* a low-growing perennial herb, producing a basal rosette of leaves and terminal radiate (i.e., outer ray and inner disk florets) flower heads (capitula) that are subtended by a series of green phyllaries (x⅔). 2 *Mutisia oligodon* (a) a scrambling perennial herb, showing alternate leaves with tendrils and a terminal capitulum (x⅔); (b) hermaphrodite bilabiate floret with a 3-toothed outer lip and a 2-toothed inner lip (x3).

to as such by the layman. The capitula themselves, however, are cymosely arranged in the
overall inflorescence, which is variable in size,
shape, construction, number of capitula, and disposition on the plant. Commonly, the capitula
are arranged in terminal or terminal and upper
cymes (these sometimes scorpioid), corymbose
panicles, or rarely spicate (e.g., *Liatris*). In some
genera, the capitula are secondarily aggregated
into spikes, glomerules (e.g., *Eremanthus* spp.)
or compound capitula possessing a secondary
receptacle (synflorescences, e.g., *Chresta, Paraly-
chnophora, Pithecoseris*); sometimes each
individual capitulum in such a compound cluster
is reduced to the single-flowered condition (e.g.,
Echinops, Mexianthus, Neohintonia) and the
whole is sometimes surrounded by a secondary
involucre of its own (e.g., *Lagascea, Elephantopus*, and *Sphaeranthus* spp.).

Several types of capitula can be distinguished
by the types, arrangement, and sex of florets
within them. Homogamous capitula contain all
hermaphrodite, or more rarely, all male (staminate) or all female (pistillate) florets. Unisexual
capitula may be on the same plant (e.g.,
Ambrosia, Xanthium) or on different plants
(e.g., *Baccharis, Gochnatia* spp., some Inuleae).
Homogamous capitula may be discoid (with all
florets tubular and actinomorphic), ligulate (with
all florets ligulate), or bilabiate (with all florets
bilabiate), and the florets are all hermaphrodite,
all male, or all female. Heterogamous capitula
contain outer florets that are female (pistillate) or
sterile and inner florets, which are usually
hermaphrodite; such capitula are usually radiate
(with an outer whorl, or whorls, of rayed florets
that are female or neuter, and a central disk of
tubular, usually hermaphrodite, disk florets),
disciform (possessing at least 2 types of eradiate
florets, usually with filiform female outer florets
and central tubular hermaphrodite florets), or
radiant (possessing inner hermaphrodite florets
and outer enlarged sterile marginal florets, e.g.,
in many Cardueae). The involucre may be of 1 or
more rows of variously overlapping bracts (phyllaries); only rarely is it absent, and rarely is it
accompanied by an outer series of calycular
bracts, forming a calyculus (e.g., commonly in
members of the Senecioneae and Heliantheae).
The receptacle is solid (e.g., *Cynara*) or hollow
and may be concave, flat, or convex; sometimes
even conical or cylindrical (e.g., *Acmella, Ratibida, Rudbeckia, Spilanthes*); variously ornamented and areolate, alveolate, or fimbriate
around the achene attachment points; and chaffy
bracts (paleae), scales, bristles, or hairs may
occur between the florets; commonly, however,
the receptacle is naked.

Within the capitula, the florets are arranged
racemosely or indeterminately, the outer opening
first; this sequence may be unclear in compound
and secondary inflorescences. Morphologically,
florets can be divided into 2 types, those with
actinomorphic corollas that are typically disk
florets in radiate capitula, or florets found
throughout discoid capitula, and those with
zygomorphic corollas. Florets with zygomorphic

corollas are
of several types:
bilabiate (with a
3-toothed outer lip
and a 2-lobed inner lip,
e.g., many Mutisieae),
pseudobilabiate (with a 4-
toothed outer lip and a 1-lobed
inner lip, e.g., Barnadesieae),
rayed (lacking adaxial lobes but
with a limb with 1 to 4 apical
teeth), or ligulate (with a flat, strap-
shaped limb with 5 apical teeth,
e.g., all Lactuceae).

Predominantly, members of the
Asteraceae have 5 connate anthers,
forming a tube (surrounding the style),
and dehisce their pollen introrsely (i.e.,
into the tube); rarely are there genera in
which the anthers are free. The anthers may
be dorsifixed (fixed to the filaments by their
upper surface), e.g., subfamily Cichorioideae,
with pollen grains within the anthers below the
level of attachment; or basifixed (fixed at their
bases), e.g., subfamily Asteroideae, with all
the pollen grains above the insertion. In many
cases, the anthers are elongated below the
point of filament insertion and have sterile
tails at the base of the thecae. The filaments
are inserted basally inside the corolla tube,

ASTERACEAE (COMPOSITAE). 3 *Cichorium intybus*
(endive) (a) flowering shoot with capitula of
ligulate florets only (x⅔); (b) ligulate floret (x2);
(c) ligulate floret with corolla removed to show
stamens inserted in the corolla tube and anthers
united in a tube around the style (x4). 4 *Liatris
graminifolia* (a) flowering shoot bearing discoid
capitula (i.e., only with disk florets) (x⅔); (b) disk
floret with a regular 5-lobed corolla (x4). 5 *Centaurea
montana* leaf and terminal radiant capitulum of disk
florets, the outer ones being sterile and enlarged (x⅔).

higher up in the tube or sometimes just beneath the sinus of the corolla lobes, and are usually glabrous, rarely papillose or even hairy; rarely are the filaments fused together (e.g., some *Barnadesia* spp.). A filament, or anther, collar (found just below the point of insertion on the anther) is formed from thicker-walled cells and may be conspicuous and diagnostic in many groups (e.g., the Eupatorieae).

Typically, the style is divided into 2 style arms, with the stigmatic surface on their inner surfaces. Both the style and style arms may be hairy, papillose, or glabrous externally. The anthers ripen and dehisce the pollen into the anther cylinder before the styles. As they mature and elongate, thick styles usually push the pollen out of the anther tube; thinner and hairy or papillose styles and style arms brush, or sweep, the pollen through the tube. The style arms may have characteristic sterile appendages, which may be well developed (e.g., the Eupatorieae). The stigmatic papillae are often conspicuous and arranged along the margins or the inner surfaces of the style arms, or they are inconspicuous and either marginal or all over the inner surface. The style arms commonly separate in maturity to expose the stigmatic surfaces and may curve back to the pollen on top of the anther tube of the same floret to enable self-pollination if they are self-compatible.

The single-seeded fruit (the seed has no endosperm and a straight embryo) is indehiscent, nearly always dry, and is termed a cypsela as it develops from an inferior ovary and is thus surrounded by other floral tissues in addition to the ovary wall, although commonly referred to as an achene. It may be angular, rounded, variously compressed, or curved, ornamented or winged in various ways, glabrous or variously pubescent or glandular; rarely it is a drupe, with fleshy endocarp (e.g., *Chrysanthemoides*). Basally cypselas commonly have a carpopodium, the basal attachment area to the receptacle, whose form is a useful diagnostic character in some groups (e.g., the Eupatorieae); rarely (e.g., some Cardueae) is it replaced by an elaiosome. It often has an apical pappus (considered a modified calyx) made up of smooth, barbellate or plumose fine hairs, bristles, scales, or awns (sometimes more or less fused together), which acts as an aid to fruit dispersal, although it is sometimes caducous or deciduous; it may be completely lacking.

Classification The classification of the Asteraceae is still in a state of transition. Just as biochemistry, pollen analysis, and micromorphology, anatomy, and cytology have in their turn been held up as the ultimate method for proposing the ideal classification, so has molecular systematics. Molecular studies have been carried out on many taxa in the last few years[i,ii,iii,iv,v,vi,vii,viii]. The results have taken infrafamiliar division to the extreme, especially as increased generic sampling yields more data for analysis. Several recent proposals have seen monotypic subfamilies in the form of the Corymboideae, Gymnarrhenoideae, and

6

7

ASTERACEAE (COMPOSITAE). 6 *Ursinia speciosa* shoot bearing pinnatisect (deeply cut) leaves and radiate capitulum (x⅔). 7 *Bellis perennis* showing basal rosette of leaves and solitary radiate capitula (x⅔).

Hecastocleioideae (the 2 latter are also monospecific), and their corresponding tribes[iii]. There is every indication that this will continue with the formal position of several clades still to be recognized. What is admitted is the varying coverage of the "tribes" at generic level (generic diversity), which ranges from c. 10% to c. 75%; at the specific level the coverage is poor. To formalize the status of several clades, and indeed grades in the numerous trees provided, a number of new ranks have been proposed. In an effort not to merge some well recognized tribes 2 approaches have been taken. Jeffrey[ix] had considered this problem and had created the new infrafamiliar rank of supersubtribe, creating the Eupatoriodinae, to avoid the Eupatorieae being submerged into the Heliantheae. Robinson[x], apparently not recognizing or not aware of Jeffrey's proposal, subsequently proposed the recognition of 2 supertribes (likened to the situation in the Gramineae) to avoid merging the Eupatorieae within the Heliantheae. None of this is practical, nor desirable, and, until far higher percentages of species and genera are sampled, it is a guide only to the possible relationships of taxa but does not help identify plants. The positioning of some genera is still in a state of flux.

Three subfamilies recognized here from a practical point of view are:

SUBFAM. BARNADESIOIDEAE Capitula homogamous or heterogamous, discoid, disciform, or radiate; florets actinomorphic and usually deeply 5-lobed, or pseudobilabiate with more or less deeply 4-lobed and sometimes expanded limb with single adaxial lobe, or rarely pseudoligulate with deeper sinus between 2 of the lobes; corolla throat pubescent inside; corollas red, pink, violet, purple, white, or yellow; central florets sometimes variable, actinomorphic or pseudobilabiate, rarely bilabiate or ligulate, 5- or 4-lobed, lobes densely long-pubescent toward apices, sometimes male or female or neuter; anthers calcarate and distinctly caudate, filaments rarely united into a tube (some *Barnadesia* spp.); style more or less short-bilobed, glabrous or papillose in upper part, never pilose; pollen smooth, spiny, or ridged without spines (psilophate); cypselas brown, usually densely pubescent; pappus setae uniseriate, villous, or plumose; shrubs or trees, rarely herbs, often with axillary spines; leaves alternate or rarely rosulate or opposite, entire. The subfamily comprises a single tribe, Barnadesieae, with 9 genera, c. 95 spp. *Barnadesia*, *Chuquiraga*, *Dasyphyllum* (South America).

SUBFAM. CICHORIOIDEAE Capitula homogamous, ligulate, bilabiate, or discoid; less often heterogamous, radiate, radiant, or disciform; phyllaries usually multiseriate, free; disk florets usually with long, narrow lobes, purplish, pinkish, whitish, less often yellow; anthers dorsifixed; style arms usually with single stigmatic area on inner surface; pollen ridged, ridged and spiny, or spiny. There are 6 tribes:

Mutisieae, with capitula homogamous or heterogamous, bilabiate or rarely ligulate, rarely radiate, subradiate, or discoid; phyllaries usually imbricate, multiseriate; receptacle usually naked, sometimes paleaceous; latex ducts absent; corolla often bilabiate, sometimes actinomorphic or ligulate; corollas white or yellow; corolla lobes long; anther bases often conspicuously tailed; style arms short obtuse or truncate and fringed; cypselas usually brown; pappus usually of barbellate setae, rarely plumose; herbs, shrubs, or trees, stems rarely armed; leaves alternate, rarely opposite. With 10/11 subtribes, with 82 genera, c. 980 spp. Genera include *Chaptalia* (the Americas), *Gerbera* (Africa and Asia), *Gochnatia* (Central and South America), *Mutisia* (South America), *Nassauvia* (southern South America), *Perezia* (New World), *Stifftia* (tropical South America). One genus, *Amblysperma*, in Australia.

Cardueae, with capitula discoid, homogamous or heterogamous, and radiant with sterile outer florets; capitula rarely clustered into secondary heads (*Echinops*); phyllaries imbricate, multiseriate, often spiny or appendaged; receptacle usually densely bristly or fimbriate; latex ducts usually absent; corollas purple, white, or yellow; corollas rarely zygomorphic; corolla lobes usually long; anthers acute, often tailed; pollen spiny, not ridged; style arms usually short, often with papillose or pubescent zone below them; cypselas not black, rarely with a basal elaiosome; pappus usually of hairs, sometimes plumose, or of scales; herbs or rarely shrubs; leaves alternate, often thistlelike. With 4 subtribes (Bremer[xi]), 74 genera, c. 2,500 spp. Genera include *Arctium* (temperate Eurasia), *Carduus* (Eurasia), *Carlina* (Eurasia), *Carthamus* (Mediterranean), *Centaurea* (Eurasia, Africa), *Cirsium* (temperate northern hemisphere), *Cnicus* (Mediterranean), *Cousinia* (Asia), *Cynara* (Mediterranean, southwestern Asia), *Echinops* (Eurasia, Africa), *Jurinea* (Eurasia), *Onopordum* (Mediterranean, southwestern Asia), *Saussurea* (temperate Asia), *Silybum* and *Xeranthemum* (Mediterranean).

Lactuceae, with capitula homogamous, ligulate; phyllaries imbricate, often biseriate or multiseriate; receptacle sometimes paleaceous; latex ducts present; resin canals mostly absent; phyllaries imbricate, 1- to several-seriate; receptacle naked or scaly; corollas usually yellow or purplish, all ligulate, apically 5-toothed, corolla lobes usually long; anther bases acute; pollen usually ridged and spiny or spiny; style arms slender, tapered, or obtuse; cypselas not black; pappus usually of hairs, sometimes plumose; herbs or shrubs; leaves alternate. With 11 subtribes, 98 genera, c. 1,550 spp., excluding microspecies in genera such as *Taraxacum*, *Hieracium*, etc. Genera include *Catananche* (Mediterranean), *Chondrilla* (Eurasia), *Cichorium* (Mediterranean), *Crepis* (northern hemisphere, Africa), *Hieracium* (temperate, except Australasia), *Hypochaeris* (northern hemisphere, South America), *Lactuca* (northern hemisphere, Africa), *Lapsana* (Eurasia), *Picris* (Eurasia, Mediterranean), *Prenanthes* (temperate northern hemisphere), *Scolymus* (Mediterranean), *Scorzonera* (Eurasia), *Sonchus* (Old World), *Taraxacum* (mostly northern hemisphere), *Tragopogon* (Eurasia).

Vernonieae, with capitula homogamous, discoid, rarely with outer corollas liguliform (*Stokesia*); phyllaries imbricate, few- to multiseriate; receptacle rarely paleaceous; latex ducts absent; corollas regular, sometimes zygomorphic, rarely pseudoligulate (*Stokesia*); corolla lobes long; corollas purple, sometimes white, rarely yellow; anthers obtuse to acute, rarely tailed; pollen spiny, ridged and spiny, or ridged; style arms elongate, gradually attenuate, acute or obtuse, hairy; cypselas not black; pappus usually of hairs, often paleaceous, of 2 distinct series; herbs or shrubs, rarely trees; leaves alternate or sometimes opposite. With c. 6 subtribes, c. 90 genera, c. 1,300 spp. Genera include *Elephantopus* (pantropical), *Piptocarpha* (Central and South America), *Stokesia* (southeastern USA), *Vernonia* (pantropical).

Liabeae, with capitula homogamous, radiate or discoid; receptacle rarely paleaceous; latex frequently present but ducts absent; phyllaries imbricate, multiseriate; receptacle usually naked; corollas regular; corolla lobes long; corollas purple, sometimes white, rarely yellow; anthers acute or short-tailed; pollen spiny; style arms as in Vernonieae; cypselas not black; pappus setae usually biseriate, usually of hairs, sometimes paleaceous; herbs, shrubs, or small trees; leaves usually opposite, strongly 3-veined and white-tomentose beneath. With c. 4 subtribes, 16 genera, c. 180 spp. Genera include *Liabum* and *Munnozia* (Central and South America).

Arctotideae, with capitula radiate, rarely discoid; receptacle epaleaceous; latex ducts usually absent; phyllaries imbricate, multiseriate, often spiny or membranous; receptacle often alveolate, with fringed pits, rarely scaly; florets yellow; corollas actinomorphic; corolla lobes long; anthers obtuse to acute, not tailed; pollen spiny; style arms short usually obtuse, with swollen papillose zone below the style arms, usually with collar of hairs at its base; cypselas not black; pappus usually of scales; herbs or shrubs; leaves alternate, often thistlelike and white beneath. With 2 subtribes, 16 genera, c. 200 spp. Genera include *Arctotis* (South Africa), *Berkheya* (Africa), *Gazania* (South Africa).

SUBFAMILY ASTEROIDEAE Capitula heterogamous, radiate, or disciform, less often discoid; disk florets usually with short, broad lobes, usually yellow; anthers basifixed; style arms usually with 2 distinct stigmatic areas; pollen spiny; latex ducts absent. There are 7 tribes:

Inuleae, with capitula radiate, disciform and toroid, with inner bisexual and narrow outer female corollas, or discoid; phyllaries imbricate in several series, often papery and white or colored in upper part; receptacle naked, sometimes paleaceous; corolla lobes short; disk corollas yellow, rays usually yellow, rarely purplish; anthers usually tailed; style arms truncate and apically hairy, rounded or variously appendaged; cypselas not black; pappus usually of hairs; herbs or shrubs; leaves alternate, sometimes decurrent on the stems. With 3 subtribes, including the Plucheae and Gnaphalieae, c. 247 genera, c. 2,821 spp. Genera include *Anaphalis* (temperate northern hemisphere),

Antennaria (arctic and temperate northern hemisphere), *Blumea* (Old World tropics), *Gamochaeta* (worldwide), *Gnaphalium* (worldwide), *Helichrysum* (Africa, Madagascar, Eurasia), *Helipterum* (southern Africa, Australia), *Inula* (Eurasia, Africa), *Leontopodium* (Eurasia), *Ozothamnus* (Australasia), *Pluchea* (Asia, North and South America, Africa, Australia), *Pseudognaphalium* (worldwide), *Pulicaria* (Eurasia, North Africa), *Raoulia* (Australasia), *Sphaeranthus* (Old World tropics), *Xerochrysum* (Australasia).

Calenduleae, with capitula radiate; phyllaries 1- to 3-seriate; receptacle naked; corollas yellow, orange, or purple; corolla lobes short; anthers acute, more or less tailed; style arms truncate, with apical hairs; pappus absent; cypselas usually not black, large, often irregular or winged, or if black then fleshy (*Chrysanthemoides*), sometimes heteromorphic; pappus absent; herbs or rarely shrubs; leaves alternate or opposite. With 8 genera, c. 110 spp. Genera include *Calendula* (mostly Mediterranean), *Dimorphotheca* (South Africa), *Osteospermum* (Africa, southwest Asia).

Astereae, with capitula radiate, disciform and toroid, with inner bisexual and narrow outer female corollas, or discoid; phyllaries imbricate, 2- to multiseriate; the receptacle is naked, rarely paleaceous, rarely glandular; corolla lobes short; rays corollas commonly purplish, bluish, or white, sometimes yellow, and

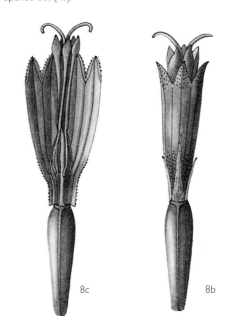

ASTERACEAE (COMPOSITAE).
8 *Helianthus giganteus* (a) female ray floret (×4); (b) hermaphrodite disk floret (×6); (c) disk floret opened out (×6).

8a

8c   8b

disk corollas yellow; anther bases usually obtuse, not tailed; style arms flattened, with hairy apical triangular to subulate appendages; cypselas not black, often flattened with 2 marginal veins; pappus often of hairs, sometimes reduced, coroniform, or absent; herbs, subshrubs, shrubs, or rarely small trees; leaves alternate or less often opposite. With c. 14 subtribes, c. 175 genera, c. 3,000 spp. Genera include *Aster* (northern hemisphere), *Baccharis* (New World), *Bellis* (Eurasia), *Brachycome* (Australasia), *Callistephus* (eastern Asia), *Conyza* (tropical), *Erigeron* (northern hemisphere), *Felicia* (Africa), *Grindelia* (temperate North and South America), *Haplopappus* (New World), *Olearia* (Australasia), *Solidago* (northern hemisphere, New World), *Symphyotrichum* (North America, Mexico, South America).

Anthemideae, with capitula radiate, disciform or discoid; phyllaries imbricate, 2- to multiseriate, with dry often brownish membranous or scarious margins; receptacle naked or paleaceous; corolla lobes short; ray corollas yellow or white, disk corollas yellow; anthers obtuse to acute, not tailed; style arms truncate, fringed with short hairs; cypselas not black; pappus not of bristles, usually cuplike, ear-shaped, or crownlike, or absent, rarely of scales; annual or perennial herbs, or less often shrubs; leaves alternate, very rarely opposite, often much divided, strongly aromatic. With 12 subtribes, c. 110 genera, c. 1,740 spp. Genera include *Achillea* (temperate northern hemisphere), *Anacyclus* (Mediterranean), *Anthemis* (Europe, Mediterranean, southwestern

Asia), *Argyranthemum* (Macaronesia), *Artemisia* (mostly northern hemisphere), *Chrysanthemum* (florists' chrysanthemum; mostly eastern Asia), *Leucanthemum* (Europe, Mediterranean, and southwestern Asia), *Lonas* (Mediterranean), *Matricaria* (Eurasia), *Santolina* (Mediterranean), *Tanacetum* (Europe, southwestern Asia).

Senecioneae, with capitula radiate or discoid, rarely disciform; phyllaries usually uniseriate, often with an outer series of reduced bracts; receptacle naked; corolla lobes usually short; corollas yellow, purplish blue, or white; anthers rounded to acute, sometimes tailed; style arms usually truncate, apically minutely hairy, less often variously appendaged; cypselas not black; pappus usually of hairs; herbs or shrubs, sometimes more or less succulent; leaves alternate. With 3 subtribes, c. 140 genera, c. 3,000 spp. Genera include *Adenostyles* (central and

ASTERACEAE (COMPOSITAE). 9 *Helianthus angustifolius* flowering shoot (x⅔). 10 *Leontopodium haplophylloides* flowering shoot (x2).

southern Europe), *Cineraria* (southern Africa, Madagascar), *Crassocephalum* (Africa), *Cremanthodium* (China, Himalayas), *Dendrophorbium* (South America) *Doronicum* (Eurasia), *Emilia* (Old World tropics), *Euryops* (Africa, Arabia), *Gynoxys* (South America), *Gynura* (Africa, Asia), *Kleinia* (Africa), *Ligularia* (East Asia), *Monticalia* (South America), *Othonna* (South Africa), *Pentacalia* (South America), *Pericallis* (Macronesia), *Petasites* (Eurasia), *Senecio* (worldwide), *Tussilago* (Eurasia).

Heliantheae, with capitula often radiate, sometimes disciform or discoid rays usually broad; phyllaries 1- to 2- to multiseriate, usually free, when uniseriate sometimes connate, often conspicuously gland-dotted; receptacle glabrous or paleaceous; corolla lobes short; corollas usually yellow, sometimes purple, disk corollas sometimes dark-colored; anthers obtuse to acute, not tailed; style arms truncate or appendiculate; cypselas usually black with phytomelanin in walls; pappus usually of awns or scales, sometimes absent, rarely of capillary hairs, sometimes plumose; herbs, subshrubs, or shrubs, rarely treelike; leaves opposite or alternate, frequently scabrid, sometimes gland-dotted and strongly aromatic. With c. 18 subtribes, although recently divided into several additional tribes, c. 300 genera, c. 3,330 spp. Genera include *Acanthospermum* (South America), *Acmella* (pantropical), *Ambrosia* (mostly New World), *Argyroxiphium* (Hawaii), *Arnica* (northern temperate, Arctic), *Bidens* (worldwide), *Coreopsis* (North and South America, Africa), *Cosmos* (Americas), *Dahlia* (Central America), *Echinacea* (North America), *Eriophyllum* (North America), *Espeletia* (South America), *Flaveria* (mostly Central America), *Gaillardia* (New World), *Galinsoga* (Central and South America), *Guizotia* (Africa), *Helenium* (North America), *Helianthus* (North America), *Heliopsis* (North America), *Lagascea* (New World tropics), *Madia* (western USA, Chile), *Parthenium* (New World), *Pectis* (New World), *Ratibida* (North and Central America), *Rudbeckia* (North America), *Sanvitalia* (Central America), *Silphium* (North America), *Spilanthes* (pantropical), *Tagetes* (New World), *Tithonia* (Central America), *Tridax* (Central and South America), *Xanthium* (worldwide), *Zinnia* (New World).

Eupatorieae, with capitula discoid, homogamous; latex ducts absent; phyllaries imbricate, or distant, 2- to multiseriate; receptacle usually naked; corollas purple, blue or white, never yellow, actinomorphic, corolla lobes short; anthers obtuse to acute, not tailed; pollen spiny; style arms obtuse, more or less club-shaped, often conspicuous and long-exserted; cypselas black, with phytomelanin in walls; pappus usually of hairs; herbs, shrubs, climbers, rarely trees; leaves usually opposite, sometimes whorled, sometimes alternate. With 17 subtribes, 181 genera, c. 2,000 spp. Genera include *Adenostemma* (pantropical), *Ageratina* (Americas), *Ageratum* (New World), *Ayapana* (New World), *Chromolaena* (New World), *Eupatorium* (temperate northern hemisphere), *Fleischmannia* (Central America, Andean South America), *Koanophyllon* (West Indies, Central and South America), *Liatris* (North America), *Mikania* (mostly New World tropics), *Stevia* (New World).

Morphological and molecular evidence would suggest that the Asteraceae is part of a group of families in the Asterales that include the Calyceraceae, Goodeniaceae, and Menyanthaceae[xii]. The closest relatives of the Asteraceae are now regarded as the Goodeniaceae and Calyceraceae[xiii,xiv,xv,xvi], and the latter are regarded as the sister group[i,xvii]. The basal lineages of the family (the Barnadesioideae and Mutisieae), as well as the sister group, are mostly southern South American. Funk *et al.*[xviii] have suggested that the separation of the lineages leading to the Goodeniaceae and Calyceraceae are believed to be about 50 million years ago, and that the basal radiation in the Asteraceae may be linked to the uplift of the southern Andes c. 50–90 million years ago.

Economic uses The importance of the Asteraceae is largely from an incalculable indirect economic value based on its contribution to the biodiversity of drier vegetation types throughout the world, especially in the temperate zones, subtropics, and tropics, where it at times approaches 10 to 15% of some floras. The direct economic importance of the family is relatively small and is largely based on a few food plants, limited sources of raw materials, medicinal and drug plants, several ornamentals and plants of horticultural interest, contrasted with the economics of often pernicious weeds and poisonous plants. Commercially, *Lactuca sativa* (lettuce) is the most important food plant, although several others are relatively widely eaten, extracts drunk, or used as flavorings, including *Cichorium endivia* (endive), *C. intybus* (chicory), *Cynara cardunculus* (cardoon, globe artichoke), *Helianthus tuberosus* (Jerusalem artichoke), *Scorzonera hispanica*, and *Tragopogon porifolius* (salsify); *Artemisia dracunculus* (tarragon) is used as a culinary herb. Oil seed crops include *Carthamus tinctorius* (safflower), *Guizotia abyssinica* (Niger seed), and *Helianthus annuus* (sunflower). The family is widely used in folk medicine, and herbal remedies are common, including extracts from *Anthemis nobilis* (chamomile), *Arnica montana* (arnica), *Calendula officinalis* (calendula), *Echinacea purpurea*, *Tanacetum vulgare* (tansy), among many others. The source of natural pyrethrum is *Tanacetum cinerariifolium*, and *Artemisia annua* is the source of artemisenin, a useful antimalarial. *Artemisia absinthium* is the source of the essential oil used to flavor the liqueur absinthe. Both *Parthenium argentatum*

(guayule) and *Taraxacum bicorne* have been used as minor alternative sources of latex to produce rubber.

The Asteraceae feature extensively in gardens throughout the world as ornamentals[xix]. A wide range of horticultural species is grown both under glass, or as herbaceous or shrubby garden plants throughout the world, many also important as cut flowers. These include *Achillea* (yarrow, sneezewort), *Ageratum conyzoides*, *A. houstonianum*, *Arctotis*, *Argyranthemum*, *Artemisia* (Wormwood), *Bellis* (daisy), *Bidens*, *Brachyglottis*, *Calendula officinalis* (Pot Marigold), *Callistephus* (China Aster), *Centaurea* (Cornflower, Knapweed), *Chrysanthemum*, *Coreopsis*, *Cosmos*, *Dahlia*, *Echinops* (Globe Thistle), *Eupatorium*, *Euryops*, *Felicia*, *Gaillardia*, *Gazania*, *Gerbera*, *Helenium*, *Helianthus* (Sunflower), *Helichrysum*, *Leucanthemum* (Ox-eye Daisy, Shasta Daisy), *Liatris*, *Ligularia*, *Olearia*, *Osteospermum*, *Pericallis* (Florists' Cineraria), *Ratibida*, *Rudbeckia* (cone flower), *Santolina* (Cotton Lavender), *Solidago* (Golden Rod), *Stokesia* (Stokes' Aster), *Symphiotrichum* (garden asters, Michaelmas Daisy), *Tagetes* (African and French Marigolds), *Tanacetum* (Feverfew, Tansy), *Xeranthemum*, *Xerochrysum* (Everlastings), and *Zinnia*. Thousands of cultivars can be found in *Callistephus*, *Chrysanthemum*, and *Dahlia*. Succulent plant enthusiasts grow several species of *Kleinia*, *Othonna*, and *Senecio*.

The family contains a number of weedy species, some of which are fast becoming pantropical in distribution. A number are considered pernicious or noxious, including *Ageratina riparia*, *Ageratum conyzoides*, *Ambrosia artemisiifolia*, *Bidens pilosa* (Black Jack), *Chondrilla juncea* (Skeleton Weed), *Chrysanthemoides monilifera*, *Cirsium arvense*, *C. vulgare* and *Carduus nutans* (thistles), *Chromolaena odorata* (Siam Weed), *Mikania micrantha* (Mile-a-minute Weed), *Crassocephalum crepidioides*, *Helichrysum kraussii*, various *Senecio* spp., *Silybum marianum*, *Sonchus oleraceus* (Sowthistle), *Taraxacum* spp. (dandelions), *Xanthium spinosum* and *X. strumarium* (Cocklebur). Several *Senecio* species are poisonous and serious weeds of pasture. They are responsible for more deaths of domestic stock than all other poisonous plants together. The windborne pollen of the ragweeds *Ambrosia artemisiifolia* and *A. trifida* is one of the main causes of hay fever in the regions of North America where these species occur. It is also apparent that some New World plants are potential weeds, having relatively recently gained a foothold in Australia. These include *Praxelis clematidea* and *Synedrellopsis grisebachii*. DJNH

ASTERACEAE (COMPOSITAE). 11 *Argyranthemum broussonetti* flowering shoot (×2).

i Baldwin, B. G., Wessa, B. L., & Panero, J. L. Nuclear rDNA evidence for major lineages of Helenioid Heliantheae (Compositae). *Syst. Bot.* 27: 161–198 (2002).
ii Keeley, S. C. & Jansen, R. K. Evidence from chloroplast DNA for the recognition of a new tribe, the Tarchonantheae, and the tribal placement of *Pluchea* (Asteraceae). *Syst. Bot.* 16(1): 173–181 (1991).
iii Panero, J. L. & Funk, V. A. Toward a phylogenetic subfamilial classification for the Compositae

(Asteraceae). *Proc. Biol. Soc. Washington* 115(4): 909–922 (2002).
iv Jansen, R. K. & Kim, K.-J. Implications of chloroplast DNA data for the classification and phylogeny of the Asteraceae. In Hind, D. J. N. & Beentje, H. (eds), *Compositae: Systematics. Proceedings of the International Compositae Conference, Kew, 1994.* (D. J. N. Hind, Editor-in-Chief). Volume 1. Pp. 317–339. Richmond, Royal Botanic Gardens, Kew (1996).
v Kim, H.-G. Keeley, S. C., Vroom, P. S., & Jansen, R. K. Molecular evidence for an African origin of the Hawaiian endemic *Hesperomannia* (Asteraceae). *Proc. Natl. Acad. Sci. USA* 95: 15,440–15,445 (2001c).
vi Kim, H.-G., Loockerman, D. J., & Jansen, R. K. Systematic implications of *ndh*F sequence variation in the Mutisieae (Asteraceae). *Syst. Bot.* 27(3): 598–609 (2002).
vii Schilling, E. E., Panero, J. L., & Cox, P. B. Chloroplast DNA restriction site data support a narrowed interpretation of *Eupatorium* (Asteraceae). *Pl. Syst. Evol.* 219: 209–223 (1999).
viii Ito, M. *et al.* Molecular phylogeny of Eupatorieae (Asteraceae) estimated from cpDNA RFLP and its implication for the polyploid origin hypothesis of the tribe. *J. Pl. Res.* 113: 91–96.
ix Jeffrey, C. (Systematics of *Compositae* at the beginning of the 21st century). *Botanicheskii Zhurnal* 87(11): 1–15 (2002). [Orig. in Russian].

x Robinson, H. New supertribes, Helianthodae and Senecionodae, for the subfamily Asteroideae (Asteraceae). *Phytologia* 86(3): 116–120 (2004).
xi Bremer, K. *Asteraceae: cladistics & classification.* Portland, Oregon, Timber Press (1994).
xii Gustafsson, M. H. G. Phylogenetic hypotheses for Asteraceae relationships. In Hind, D. J. N. & Beentje, H. (eds), *Compositae: Systematics. Proceedings of the International Compositae Conference, Kew, 1994.* (D. J. N. Hind, Editor-in-Chief). Volume 1. Pp. 9–19. Richmond, Royal Botanic Gardens, Kew (1996).
xiii Gustafsson, M. H. G., Backlund, A., & Bremer, B. Phylogeny of the Asterales *sensu lato* based on *rbc*L sequences with particular reference to the Goodeniaceae. *Pl. Syst. Evol.* 199: 217–242 (1996).
xiv Gustafsson, M. H. G., Grafström, E., & Nilsson, S. Pollen morphology of the Goodeniaceae and comparisons with related families. *Grana* 36: 185–207 (1997).
xv Lammers, T. G. Circumscription and phylogeny of the Campanulales. *Ann. Missouri Bot. Gard.* 79: 388–413 (1992).
xvi Takhtajan, A. L. *Systema Magnoliophytorum.* Leningrad, Nauka (1987).
xvii De Vore, M. L. & Stuessy, T. F. The place and time of origin of the Asteraceae, with additional comments on the Calyceraceae and Goodeniaceae. In: Hind, D. J. N., Jeffrey, C., & Pope, G. V. (eds), *Advances in Compositae Systematics.* Pp. 23–40. Richmond, Royal Botanic Gardens, Kew (1995).
xviii Funk, V. A. *et al.* Everywhere but Antarctica: Using a supertree to understand the diversity and distribution of the Compositae. *Biologiske Skrifter* 55: 343–374 (2005).
xix Sutton, J. *The Plantfinder's Guide to Daisies.* Portland, David & Charles, Newton Abbott & Timber Press (2001).

ASTEROPEIACEAE

MANOKA JAUNE

A monogeneric family of 7 or 8[i] species of trees or scrambling shrubs, endemic to Madagascar, distributed primarily in the east and north of the island[ii]. The leaves are alternate, shortly petiolate, exstipulate; lamina entire, simple, and the flowers are

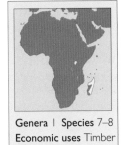

Genera 1 Species 7–8
Economic uses Timber

small, in axillary or terminal panicles, with the terminal unit cymose. Sepals and petals 5, free. Stamens 9 to 15, connate at the base. The ovary is superior, with 2 to 3 carpels, and thick-walled fruit is indehiscent.

Traditionally, the family is placed with the Theaceae and sometimes included within it[iii]. It is placed sister to the core Caryophyllales in molecular analyses[iv,v]. The wood of *Asteropeia rhopaloides* (Manoka Jaune) and other species is used locally.   AC

i http://www.mobot.org/phillipson/catalogue/pdf/asteropeiaceae
ii http://www.gbif.net/portal
iii Cronquist, A. *The Evolution and Classification of Flowering Plants*, 2nd ed. New York, New York Botanical Garden (1988).
iv Cuénoud, P. *et al.* Molecular phylogenetics of Caryophyllales based on nuclear 18S and plastid *rbc*L, *atp*B and *mat*K DNA sequences. *American J. Bot.* 89: 132–144 (2002).
v Morton, C. M., Karol, K. G., & Chase, M. W. Taxonomic affinities of *Physena* (Physenaceae) and *Asteropeia* (Theaceae). *Bot. Rev.* 63: 231–239 (1997).

ATHEROSPERMATACEAE

SOUTHERN OR TASMANIAN SASSAFRAS

A small family of trees and shrubs with opposite coarsely serrate leaves, found in the southern hemisphere.

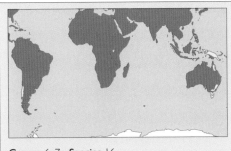

Genera 6–7 Species 16
Economic uses Timber and herbal tea

Distribution The family occurs in the southern hemisphere, with 2 species in Chile and 12 or more species in Australasia, from New Guinea to New Zealand and New Caledonia. *Atherosperma* contains 1 species, *A. moschatus*, from Tasmania, Victoria, and New South Wales in Australia. The 2 Chilean species are *Laurelia sempervirens* and *Laureliopsis philippiana*, and a second species of *Laurelia* (*L. novae-zelandiae*) grows in New Zealand. *Nemuaron* (2 spp.) is confined to New Caledonia, *Daphnandra* (5 spp.) grows in New Guinea, while *Dryadodaphne* (2 spp.) occurs in New Guinea and Australia, and *Doryophora* (2 spp.) grows in southeastern and northeastern Australia. Ecologically, there is a marked association between Atherospermataceae and Southern Beeches (*Nothofagus*) in the temperate forests of the southern hemisphere)[i].

Description Trees or shrubs. The leaves are opposite, petiolate, simple, coarsely serrate, often gland-dotted, without stipules. Flowers actinomorphic or slightly zygomorphic, hermaphrodite or unisexual (plants then monoecious or dioecious, or polygamodioecious), solitary or in cymose inflorescences. The calyx and corolla are distinct or the perianth sepaline, or absent. The sepals are 2 (enclosing the bud in *Atherosperma*). The petals are 7–20. The stamens are 4–6 (to numerous). The ovary is superior to inferior, the carpels 3 to numerous, free, sometimes embedded in the receptacle. The fruit is an aggregate of achaenial fruitlets.

Classification The Atherospermataceae are often included in the Monimiaceae (Laurales), but recent morphological and molecular data[ii] indicate that their relationship to that family is distant and that they are sister to *Gomortega keule* (Gomortegaceae), a tree endemic to Chile, also in the Laurales.

Economic uses The timber of *Atherosperma moschatus* (Southern or Tasmanian Sassafras)

is used commercially for furniture, toys, paneling, and musical instruments, and the bark is used to make herbal tea.   VHH

i Renner, S. S., Foreman. D. B., & Murray, D. Timing transantarctic disjunctions in the Atherospermataceae (Laurales): evidence from coding and noncoding chloroplast sequences. *Syst. Biol.* 49: 579–591 (2000).
ii Renner, S. S. Circumscription and phylogeny of the Laurales: evidence from molecular and morphological data. *American J. Bot.* 86: 1301–1325 (1999).

AUCUBACEAE

SPOTTED LAUREL FAMILY

A family comprising one genus of usually robust, evergreen, dioecious shrubs to small trees, native to forests from the eastern Himalayas to Taiwan and Japan. The leaves are up to 25 cm and commonly broadly elliptical, strongly serrate but may be almost linear or broadly obovate and conspicuously

Genera 1 Species 1–12
Economic uses Ornamentals grown for foliage and fruits

serrate-truncate, and may be both entire and toothed on the same plant. The flowers are up to 3 mm across, with 4 sepals and 4 inconspicuous green or red petals. In male flowers, there are 4 short stamens alternating with the petals. In female flowers, the ovary is inferior and 1-locular, with a single ovule suspended from the top of the locule, and a broad capitate stigma surrounded by the small calyx lobes. The fruit is a colorful, single-seeded ovoid berry, up to 2 cm.

Aucuba was for a long time treated as a member of the Cornaceae, with which it shares the regular 4-merous flowers and inferior ovary, but differs from it in many ways, especially ovary and fruit morphology, and has recently been usually excluded from that family. Molecular analyses[i,ii] now consistently place it well removed from the Cornaceae but near to Garryaceae and Eucommiaceae near the base of the Asterid group. The intectate pollen is

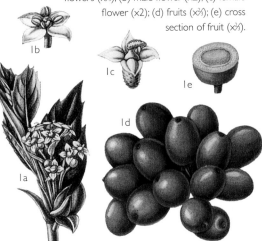

AUCUBACEAE. 1 *Aucuba japonica* (a) leafy shoot and female flowers (×⅔); (b) male flower (×2); (c) female flower (×2); (d) fruits (×⅔); (e) cross section of fruit (×⅔).

distinctive and unlike that of Garryaceae or related families[iii]. Variation in habit and leaf shape in *Aucuba* is considerable, and up to 12 spp. have been described. However, plants seem to be inherently variable and variation in leaf shape appears more or less continuous; some have suggested that there may be just 1 variable species. A broadly based taxonomic revision is needed. *Aucuba japonica* is widely grown in urban areas for its foliage (often variegated) and red or orange berries. Many cultivars are commercially available. RKB

[i] Savolainen, V. *et al.* Phylogeny of the eudicots: a nearly complete familial analysis based on *rbc*L gene sequences. *Kew Bull.* 55: 257–309 (2000).
[ii] Bremer, K. *et al.* A phylogenetic analysis of 100+ genera and 50+ families of euasterids based on morphological and molecular data with notes on possible higher level synapomorphies. *Pl. Syst. Evol.* 229: 137–169 (2001).
[iii] Eyde, R. H. Comprehending *Cornus*: puzzles and progress in the systematics of the dogwoods. *Bot. Rev.* 54: 311–312 (1988).

AUSTROBAILEYACEAE

A "primitive" family, comprising a single genus, *Austrobaileya*, of evergreen woody lianas up to 15 m, with 1 or 2 species, *A. scandens* and *A. maculata*, endemic to the rainforest of northeastern Queensland, Australia. The leaves are opposite, entire, with small stipules. The flowers

Genera 1 Species 1–2
Economic uses None

are quite large, pendulous, borne solitary in the leaf axils, with a putrescent odor. The perianth consists of c. 12 pale green, free, petals. The stamens are 6 to 11, laminar, the innermost sterile. The ovary is superior, of (4–)6–8(–14) free carpels, each with several ovules in 2 rows. The fruits are oblong, ellipsoid berrylike, apricot-colored, containing several seeds.

It was placed in the Magnoliales by Cronquist, but molecular evidence indicates that it belongs in a separate order, Austrobaileyales, one of the most basal lineages of the angiosperms, forming part of the so-called ANITA grade[i,ii] along with the Amborellaceae, Nymphaeaceae, Illiciaceae, Trimeniaceae, and Schisandraceae. VHH

[i] Qiu, Y.-L. *et al.* The earliest angiosperms: evidence from mitochondrial, plastid and nuclear genomes. *Nature* 402: 404–407 (1999).
[ii] Mathews, S. & Donoghue, M. J. The root of angiosperm phylogeny inferred from duplicate phytochrome genes. *Science* 286: 947–950 (1999).

AVICENNIACEAE

BLACK MANGROVE

Distribution Tropical coasts, extending north as far as southern China, southern Sinai, Egypt, and southern USA, and south to Australia, KwaZulu-Natal, and southern Brazil.

Genera 1 Species c. 8
Economic uses Sources of fuelwood, timber, and tanning materials

Description Tree mangroves with stilt roots and pneumatophores. The leaves are opposite, simple and entire, usually elliptical or ovate. The inflorescences are axillary or terminal, often densely crowded, usually pubescent to tomentose, with each flower subtended by a bract and a pair of bracteoles. The flowers in some species are regular, but in others are clearly zygomorphic with a weakly 2-lipped corolla. The sepals are 5, distinct, ovate, and usually imbricate. The corolla has a short tube with 4 (rarely 5) lobes exceeding the tube in length. The stamens are attached to the short corolla-tube and alternate with the lobes, and are sometimes slightly unequal. The ovary is superior, of 2 fused carpels, and has free-central placentation with 4 pendulous ovules. The fruit is a capsule, somewhat inflated and often beaked at the apex, including a single seed.

Classification *Avicennia* was formerly referred by many authors to Verbenaceae, but it differs markedly from that family in seedling morphology, stem and leaf anatomy, placentation, pollen, and other characters, leading to placement in its own family, Avicenniaceae. Recent molecular evidence[i] has placed it close to the basal groups of Acanthaceae, especially the Nelsonioideae and Thunbergioideae, and suggests that it might be placed within Acanthaceae, but it lacks the retinacula and cystoliths characteristic of most of that family, and the strictly actinomorphic 4-lobed flowers of the *A. marina* type[ii] are quite unlike anything in the Acanthaceae. With 4 pendulous ovules and the single-seeded fruit, the ovary is also unlike anything in Acanthaceae, and family status for Avicenniaceae seems justified. Nonetheless, the apparently independent occurrence of mangroves in *Acanthus* in the tribe Acantheae and in *Bravaisia* in the tribe Ruellieae of the Acanthaceae, the former in the Indian and Pacific Oceans and the other in Central America, may be seen by some as a further indication of affinity of *Avicennia* with that family.

Tomlinson[ii] has divided the genus informally into 3 groups according to corolla forms, which relate to different floral mechanisms. His *A. germinans* group has a distinctly zygomorphic white corolla with lobes at first imbricate, his *A. officinalis* group has a slightly zygomorphic yellow corolla with imbricate lobes, while

his *A. marina* group has strictly actinomorphic yellow or orange corolla with valvate lobes.

Economic uses *Avicennia* is one of the most common and characteristic components of mangrove vegetation and so is very important in coastal ecology and protection. The wood is sometimes used as fuel or for construction, and the bark is used in tanning leather. RKB

[i] Schwarzbach, A. E. & McDade, L. A. Phylogenetic relationships of the mangrove family Avicenniaceae based on chloroplast and nuclear ribosomal DNA sequences. *Syst. Bot.* 27: 84–98 (2002).
[ii] Tomlinson, P. B. *The Botany of Mangroves*. Pp. 186–207. Cambridge, Cambridge University Press (1986).

BALANOPACEAE

The Balanopaceae contains a single genus (*Balanops*) of tall, evergreen trees that are restricted to northern Queensland, Australia, New Caledonia, and Fiji. The leaves are alternate or pseudoverticillate, toothed, coriaceous, and without stipules.

Genera 1 Species 9
Economic uses None

The flowers are unisexual (plants dioecious), with a vestigial perianth, or absent. The male flowers are catkinate, each subtended by a scale and have (2–)5–6(–14) stamens; the female flowers are solitary and surrounded by a cupule of many bracts or, in addition, there are several lateral flowers. Each one is situated in the axil of a bract and surrounded by its own set of cupular bracts[i]. The flowers consist of 2 or 3 carpels, with the ovary superior, syncarpous, and with 2 superimposed ovules per carpel but only 1 of them commonly developing into a seed. The fruit is an acornlike drupe, with 1 or 2 seeds.

The relationships of the Balanopaceae remain problematic. In the past, affinities have been suggested with the Fagales, Salicales, and Euphorbiales, while molecular phylogenetic studies[ii,iii] have placed the family in the Malpighiales (*sensu* APG II) as a sister to Chrysobalanaceae/Euphroniaceae and Dichapetalaceae/Trigoniaceae. A study of the female flower and cupule structure[i] indicates that the Balanopaceae agrees more with Euphorbiaceae in the broad sense than with other families in the Malpighiales. There are no reported economic uses for the family. VHH

[i] Sutter, D. M. & Endress, P. K. Female flower and cupule structure in Balanopaceae, an enigmatic rosid family. *Ann. Bot.* 92: 459–469 (2003).
[ii] Litt, A. J. & Chase, M. W. The systematic position of *Euphronia*, with comments on the position of *Balanops*: An analysis based on *rbc*L sequence data. *Syst. Bot.* 23: 401–409 (1999).
[iii] Chase, M. W. *et al.* When in doubt, put it in Flacourtiaceae: a molecular phylogenetic analysis based on plastic *rbc*L DNA sequences. *Kew Bull.* 57: 141–181 (2002).

BALANOPHORACEAE

A largely tropical family of achlorophyllous root parasites.

Distribution The family is largely pantropical with concentrations of species in the neotropics and Asia. It is less well represented in Africa. The largest genus is *Balanophora* (15 spp.), which grows in the Old World tropics. *Chlamydophytum* (1 sp., *C. aphyllum*) and *Thonningia* (1 sp., *T. sanguinea*) are endemic to tropical Africa; *Sarcophyte* (2 spp.) occurs in tropical and southeastern Africa; *Ditepalanthus* (1 sp., *D. afzelii*) is endemic to Madagascar; *Mystropetalon* (2 spp.) is confined to the southwestern Cape area of South Africa; *Dactylanthus* (1 sp., *D. taylorii*) is restricted to New Zealand; *Exorhopala* (1 sp., *E. ruficeps*) is endemic to the Malay Peninsula; *Hachettea* (1 sp., *H. autro-caledonica*) is confined to New Caledonia; *Rhopalocnemis* (1 sp., *R. phalloides*) to Indomalaysia; *Helosis* (1 sp., *H. cayennensis*) occurs in Mexico and Central and South America; *Lathrophytum* (1 sp., *L. peckoltii*) is known from Rio de Janeiro, Brazil; *Corynaea* (1 or 2 spp.) grows in tropical American montane cloud forests; *Ombrophytum* (4 spp.) grows in warm South America and the Galápagos Islands; *Scybalium* (4 spp.) grows in tropical America; *Lophophytum* (3 spp.) grows in lowland rain forests of tropical South America. *Langsdorffia* has a highly disjunct distribution with 1 sp. in tropical America, one in Madagascar, and a third in New Guinea.

Description Achlorophyllous, obligate root parasites attaching to the host trees and shrubs, rarely herbs, by haustorial connections from an amorphous, underground "tuber." The tuber, which may reach the size of a baby's head, may

be entirely composed of parasite tissue as in *Dactylanthus, Helosis, Lophophytum,* and *Scybalium*. In others, it is part parasite, part host. The stem is absent, and the above-ground part of the plant consists of the inflorescence, which develops inside the tuber, rupturing the tuber tissue which remains as a "volva" at the base. In *Ombrophyton*, the volva becomes large and spathelike. The inflorescence may be terminal, racemose, spicate, globose or club-shaped, yellowish to purple, and sometimes conspicuously branched, the stalk bearing scales or bracts or naked. The bracts vary in shape from peltate to triangular or clavate; in *Helosis* they are strikingly geometrical and hexagonal, covering the inflorescence. The flowers are numerous or very numerous, small or minute (among the smallest in the angiosperms), actinomorphic, unisexual (plants monoecious with separate male and female inflorescences or mixed ones with the males toward the base, or plants dioecious), with or without a single whorl of perianth segments. In the staminate flowers, the perianth is either lacking (*Ombrophyton*) or may be valvate and with 3–4(–8) lobes that are free or connate at the base. The stamens are 1 or 2 in flowers without a perianth and equal in number and opposite to the segments in those with a perianth. The anthers have 2, 4, or many locules. In *Hachettea*, the filament is contracted, and the single stamen has 1 locule. In some genera (e.g., *Langsdorffia, Helosis,* and *Scybalium*), the lower parts of the stamens are variously united into a tube to form a synandrium with discrete anthers. In other genera, the tube is tipped with pollen sacs. In the pistillate flowers, the perianth is lacking or reduced and the carpels, placentation, and ovules are not easy to distinguish; the ovary is usually superior (inferior in *Mystropetalon*) with 2–3 carpels, syncarpous, without a discernible locule, and the placentas are reduced, bearing 1–3 ovules that are represented by embryo sacs without integuments. There are 1 or 2 styles, which are entire or 2-lobed, with a terminal stigma. The stigma is occasionally sessile. The fruits are tiny, indehiscent, 1-seeded achenes, sometimes surrounded by the swollen perianth tube (*Mystropetalon*) or swollen and aggregated into a fleshy multiple fruit. The stalks of some female flowers become elaiosomes, attractive to ants, which are reported to disperse the seeds of *Mystropetalon*. In general, the reproductive and dispersal biology is poorly understood.

BALANOPHORACEAE. 1 *Helosis cayensis* (a) habit showing irregular, underground tuber, spreading rhizomes, and aerial, club-shaped inflorescences (x2); (b) male flower with stamens fused into a tube (x20); (c) female flower with part of wall removed to show pendulous ovule (x20). 2 *Balanophora involucrata* (a) tuber and inflorescence (x⅔); (b) male flower (x20); (c) vertical section of ovary (x20). 3 *Lophophytum weddellii* (a) habit (x⅔); (b) vertical section of male inflorescence (x⅔). 4 *Thonningia sanguinea* (a) male inflorescence (x⅔); (b) female inflorescence (x⅔); (c) vertical section of female inflorescence (x⅔).

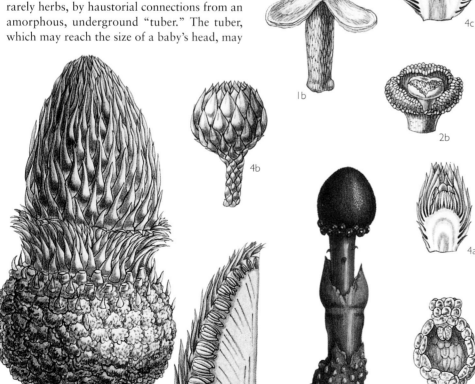

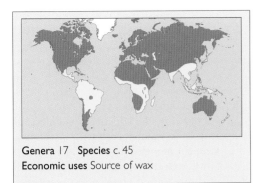

Genera 17 **Species** c. 45
Economic uses Source of wax

Economic uses *Langsdorffia* and *Balanophora* accumulate a waxy substance called balanophorin, not starch. Balanophorin is extracted from some species in Asia and used as a lighting fuel. [DJM] VHH

[i] Nickrent, D. L. Orígenes filogenéticos de las plantas parásitas. Capitulo 3, pp. 29–56. In: López-Sáez, J. A., Catalán, P., & Sáez, L. (eds), *Plantas Parásitas de la Península Ibérica e Islas Baleares*. Mundi-Prensa Libros, S. A., Madrid (2002).

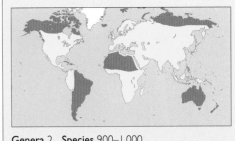

Genera 2 **Species** 900–1,000
Economic uses Widely grown ornamentals

Classification The Balanophoraceae has been split into 8 discrete families by Takhtajan (1997): Mystropetalaceae, Dactylanthaceae, Lophopytaceae, Sarcophytaceae, Scybaliaceae, Helosidaceae, Langsdorffiaceae, and Balanophoraceae itself. Some or all of these may be recognized as subfamilies. However, the differences between them do not appear to merit such recognition. The possible exception is *Mystropetalon*, with pollen unique among the flowering plants in being cuboid with pores at the 8 corners as well as its inferior ovary. The affinities of the Balanophoraceae remain unclear, caused partly by the difficulties of understanding morphology due to their parasitic habit. The family has been regarded as a member of the order Santalales (Cronquist 1988). Molecular studies also suggest this position[i]. It is treated by Takhtajan (1997) as the only family of the order Balanophorales in the superorder Balanophoranae in the Magnoliidae, along with the Cynomoriales, containing only the Cynomoriaceae, which he regarded as closely related and as having a common origin with the Balanophoraceae. Both the Balanophoraceae and Cynomoriaceae are considered unplaced eudicots in APG II.

BALSAMINACEAE
BALSAM, JEWEL WEED, TOUCH-ME-NOT

A family of annual and perennial herbs with watery, translucent stems.

Distribution *Impatiens* (900–1,000 spp.) is widespread in temperate and tropical regions, with centers of diversity in Southeast Asia (c. 250 spp.), China (220 spp.[i]), southern India and Sri Lanka (c. 150 spp.), Madagascar (c. 120 spp.), and tropical Africa (c. 110 spp.)[ii]. The Americas have fewer than 10 native species, but several Old World species have become invasive. *Hydrocera triflora* occurs from southern India and Sri Lanka, through Southern China, to Malaysia. An origin in Southeast Asia has been proposed for *Impatiens*[ii].

Description Annual or perennial herbs, sometimes with rhizomes or tubers (e.g., *I. tuberosa, I. etindensis*), shrubby with woody stems to 4 m high in some African and Indian species [iii], while *Hydrocera* is an erect marsh plant, about 1 m high. The leaves are alternate, whorled or opposite, toothed or crenate, and usually with petiole, often bearing capitate glands, but without stipules. The inflorescence is axillary, of 1 flower or several in a raceme or pseudoumbel, 1–3 together in *Hydrocera*. The flowers are bisexual and zygomorphic. The sepals are usually 3, (5 in *Hydrocera* and few *Impatiens*), free, the posterior 1 petaloid and usually spurred, and in addition sometimes 2 small or aborted anterior sepals. The petals are 5, unequal, the lower 4 of which are connate in lateral pairs. The stamens are 5, with short, flat filaments and introrse anthers that are more or less united to form a cap over the ovary. The ovary is superior, consisting of 5 fused carpels, with 5 locules containing numerous ovules on axile placentas. Stigmas are 1–5, more or less sessile. The fruit is usually an explosive capsule; rarely a berry (*Hydrocera*).

Classification *Impatiens* is distinguished from the genus *Hydrocera* by its lateral petals, which are connate in pairs, and the fruit a capsule. In *Hydrocera*, the lateral petals are free, and the fruit is a berry. The Indo-Malaysian

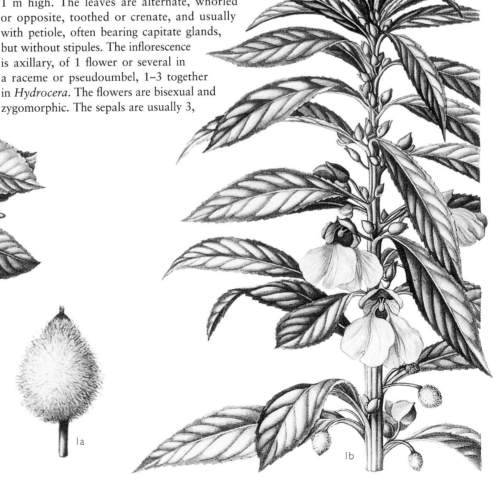

BALSAMINACEAE. I *Impatiens balsamina* (a) fruit (×1⅓); (b) leafy shoot and axillary flowers (×⅔). 2 *I. walleriana* leafy shoot and flowers (×⅔).

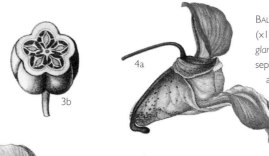

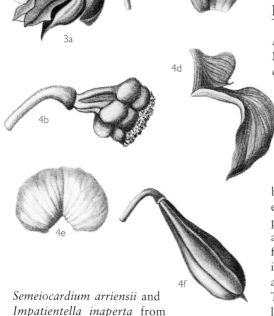

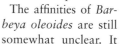

BALSAMINACEAE. 3 *Hydrocera triflora* (a) irregular flower (×1); (b) section of fruit—a berry (×1). 4 *Impatiens glandulifera* (a) irregular flower with posterior spurred sepal and small anterior sepals (×1); (b) stamens fused around the ovary (×3); (c) anterior sepal (×3); (d) lateral petal (×1); (e) anterior petal (×1); (f) fruit—a capsule (×1½).

Semeiocardium arriensii and *Impatientella inaperta* from Madagascar are now included in *Impatiens*. The presence of a spur in the flowers of the Balsaminaceae has led people to associate the family with the Geraniaceae and Tropaeolaceae (e.g., Cronquist 1981), but it is strictly an outgrowth of the calyx. In the other families, there is evidence that receptacle tissue is involved in spur formation. Molecular evidence from several genes and spacers indicates Balsaminaceae forms a clade with Marcgraviaceae, Pellicieraceae, and Tetrameristaceae sister to other Ericales families[iv,v].

Economic uses The economic value of the Balsaminaceae lies in the cultivation of *Impatiens* species as pot plants or garden ornamentals. The plants known commercially as "busy lizzies" are hybrids between *Impatients holstii* and *I. sultanii* and have white, pink, red, or orange flowers and green, red, or bronze foliage.   [SAH] AC

i Chen Yiling, Shinobu Akiyama, and Hideaki Ohba. Flora of China. Balsaminaceae [Draft]. http://flora.huh.harvard.edu/china/mss/volume12/Balsaminaceae-A-GH_edited.htm
ii Yuan, Y.-M. *et al*. Phylogeny and biogeography of Balsaminaceae inferred from ITS sequences. *Taxon* 53: 391–403 (2004).
iii Fischer, E. Balsaminaceae. Pp. 20–25. In: Kubitzki, K. (ed.), *The Families and Genera of Vascular Plants. VI. Flowering Plants. Dicotyledons: Celastrales, Oxalidales, Rosales, Cornales, Ericales*. Berlin, Springer-Verlag (2004).

iv Schönenberger, J., Anderberg, A. A., & Sytsma, K. J. Molecular phylogenetics and patterns of floral evolution in the Ericales. *Int. J. Pl. Sci.* 166: 265–288 (2005).
v Anderberg, A. A., Rydin, C., & Källersjö, M., Phylogenetic relationships in the order Ericales *s.l.*: analyses of molecular data from five genes from the plastid and mitochondrial genomes. *American J. Bot.* 89: 677–687 (2002).

BARBEUIACEAE

An obscure endemic Madagascan family comprising 1 sp. of liana, *Barbeuia madagascariensis*.

Genera 1 | Species 1
Economic uses None

The leaves are alternate, petiolate, simple, lacking stipules but articulated at the base; the lamina is entire, ovate, with pinnate veins. The inflorescence is racemose, axillary, with long pedicels supporting bracteate flowers. The flowers are actinomorphic, with 5 imbricate sepals with orbicular lobes; petals absent; stamens 30–100, free, and in 2–4 whorls. The ovary is superior, of 2 fused carpels, with 2 locules. Stigma sessile on ovary or on a short style. Ovules are 1 per locule. The fruit is a hard woody capsule containing 1–2 seeds.

Barbeuia is often included in the Phytolaccaceae[i], but phylogenetic work refutes this[ii]. The species is on the verge of extinction. No economic uses are reported.   AC

i Rohwer, J. G. Phytolaccaceae. Pp. 507–515. In: Kubitzki, K., Rohwer, J. G., & Bittrich, V. (eds), *The Families and Genera of Vascular Plants. II. Flowering Plants. Dicotyledons: Magnoliid, Hamamelid and Caryophyllid Families*. Berlin, Springer-Verlag (1993).
ii Cuénoud, P. *et al*. Molecular phylogenetics of Caryophyllales based on nuclear 18S and plastid *rbc*L, *atp*B and *mat*K DNA sequences. *American J. Bot.* 89: 132–144 (2002).

BARBEYACEAE

A monotypic family comprising *Barbeya oleioides*—a small evergreen tree from Ethiopia, Somalia, and the Arabian Peninsula (Yemen and Saudi Arabia), with simple, opposite, lanceolate leaves, densely silver with unicellular trichomes on the abaxial surface, with the margins entire, supervolute, and without stipules. The flowers are small, in axillary, fasciculate cymes, actinomorphic, unisexual (plants dioecious). The perianth is uniseriate, of 3–4 sepaline tepals, somewhat connate at the base. The staminate flowers have 6–12 stamens with short filaments. The ovary is superior, of 1–2(3) free or basally connate carpels, with 2 or 3 locules, each with 1 pendulous subapical ovule. The fruit is a dry nutlet with accrescent sepals.

Genera 1 | Species 1
Economic uses None

The affinities of *Barbeya oleoides* are still somewhat unclear. It was tentatively related to the Urticales by Takhtajan, although he later transferred it to the superorder Hamamelidanae. Molecular evidence places it in the redefined Rosales[i], where it belongs in a subclade along with Dirachmaceae, Elaeagnaceae, and Rhamnaceae[ii,iii].   VHH

i Qiu, Y.-L. *et al*. Phylogenetics of the Hamamelidae and their allies: Parsimony analyses of nucleotide sequences of the plastid gene *rbc*L. *Int. J. Plant Sci.* 159: 891–905 (1998).
ii Thulin, M. *et al*. Family relationships of the enigmatic rosid genera *Barbeya* and *Dirachma* from the Horn of Africa region. *Plant Syst. Evol.* 213: 103–119 (1998).
iii Sytsma, K. J. *et al*. Urticalean rosids: Circumscription, rosid ancestry, and phylogenetics based on *rbc*L, *trn*L-F, and *ndh*F sequences. *American J. Bot.* 89: 1531–1546 (2002).

BASELLACEAE
MADEIRA VINE

A small family of climbing vines, most of which are native to the tropical Americas.

Distribution The family is predominantly found in the Americas, particularly the Andes. Some species are found in tropical Africa, Madagascar, and southern Asia.

Description Short-lived, usually hairless and fleshy, perennial vines, or sometimes spreading herbs, often with a tuberous rootstock. The leaves are mostly simple, alternate, broadly ovate and fleshy, without stipules. The flowers are small, regular, bisexual or rarely unisexual (plants monoecious), perigynous, with a pair of bracts and arranged in spikes or in racemes in the leaf axils. The perianth has 5 segments that may be partially fused at the base and in some species may be colored. There are 5 stamens, each opposite a perianth segment that may be free or partially fused to the base of the perianth segment. The ovary is superior and comprises 3 fused carpels, with 1 locule containing 1 basal ovule. There is usually 1 style and 3 stigmas. The fruit is a drupe and is usual-

i Cuénoud, P. *et al.* Molecular phylogenetics of Caryophyllales based on nuclear 18S and plastid *rbc*L, *atp*B and *mat*K DNA sequences. *American J. Bot.* 89: 132–144 (2002).
ii National Research Council. Ad Hoc Panel of the Advisory Committee on Technology Innovation, Board on Science and Technology for International Development. *Lost Crops of the Incas: Little-Known Plants of the Andes with Promise for Worldwide Cultivation.* The National Academies Press (1989).
iii Csurhes, S. & Edwards, R. *Potential environmental weeds in Australia: Candidate species for preventative control.* Queensland Department of Natural Resources (1998).

Genera 4 **Species** 20–25
Economic uses Edible leaves and tubers; a pigment used to color food; some ornamental use

ly enclosed within a persistent fleshy perianth; otherwise it is dry. The seed usually has copious endosperm and a twisted or ringlike embryo.

Classification The 4 genera can be grouped into 2 tribes. The first, Baselleae, has the filaments erect and straight in bud—*Basella* (Malabar Nightshade), *Tournonia,* and *Ullucus* (Ulluco). The second, Anredereae, has the filaments curved outward in bud—*Anredera* (*Boussingaultia*), Madeira Vine. The family is a member of the core Caryophyllales[i], related to other fleshy familes such as Portulacaceae, Didiereaceae, and Cactaceae.

Economic Uses Tubers of *Ullucus tuberosus* are a staple starch source to people living in the Andes[ii]. The fleshy leaves of *Basella rubra* and *B. alba*, known as Malabar Spinach, are widely eaten in the tropics. *Anredera* (*Boussingaultia*) is grown in the tropics as an ornamental and has become a weed in Australasia and a serious invasive in some Pacific Islands[iii]. AC

BATACEAE
SALTWORT FAMILY

A monogeneric family of halophytic subshrubs with fleshy leaves.

Distribution Coasts of warm temperate and tropical Americas (*Batis maritima*) and tropical Australia to New Guinea (*Batis argillicola*), in tidal marshes and behind mangroves. *B. maritima* is recorded in Hawaii as an invasive alien[i]. Both species disperse by floating infructescences that disintegrate to release the floating fruit.

Genera 1 **Species** 2
Economic uses Occasional use as salad leaves

Description Dioecious (*B. maritima*) or monoecious (*B. argillicola*) succulent halophytic shrubs to 1.5 m, with opposite, decussate leaves. The leaves are obovoid or linear, simple, entire and fleshy, with no petiole and minute, caducous stipules or colleters[ii]. The inflorescence is a lax spike (*B. argillicola*) or a condensed catkinlike structure (*B. maritima*)— the male inflorescence with overlapping bracts and the female inflorescence with the flowers embedded. The male flowers are 4-merous, with the sepals joined and notably unequal in size. The large sepal forms 1 lobe, the 3 smaller ones form a second lobe. The petals are equal, clawed, and the stamens have a short, thick connective. The female flowers are highly reduced, with 2 carpels each with 2 locules fully embedded in the inflorescence tissue, and with a stigmatic cleft opening to the surface. The fruit is a drupe (*B. argillicola*) or drupelike syncarp[iii] (*B. maritima*), with a woody endocarp.

Classification The relationships of *Batis* have been disputed for many years. It has been placed in at least 12 different families or on its own in Bataceae, sometimes isolated in its own

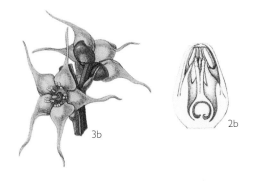

BASELLACEAE. **1** *Anredera cordifolia* (a) twining stem bearing alternate broadly ovate simple leaves and flowers in panicles (x⅔); (b) flower partly open, showing pair of bracteoles below the flower (x4); (c) fully open flower with 5 perianth segments and 5 stamens (x4); (d) perianth opened out to show stamens attached to its base (x6); (e) gynoecium with terminal style deeply divided into 3 stigmas (x3). **2** *Basella alba* (a) twining stem (x⅔); (b) half flower bud, showing ovary with single locule containing 1 ovule (x6). **3** *Ullucus tuberosus* (a) twining stem (x⅔); (b) flowers with brightly colored bracteoles and 5 greenish perianth segments (x2); (c) tuberous roots (x⅔); (d) gynoecium (x6).

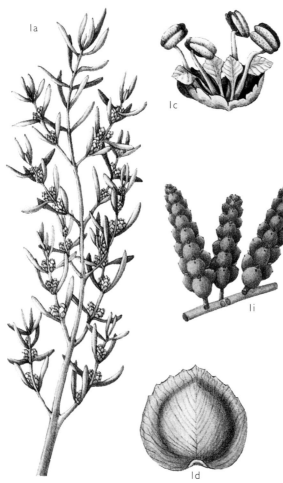

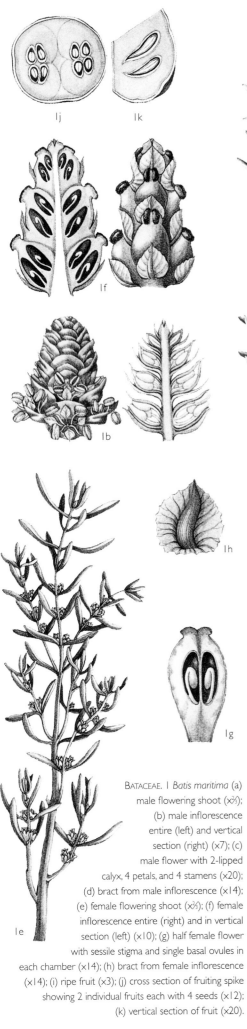

BATACEAE. I *Batis maritima* (a) male flowering shoot (×⅔); (b) male inflorescence entire (left) and vertical section (right) (×7); (c) male flower with 2-lipped calyx, 4 petals, and 4 stamens (×20); (d) bract from male inflorescence (×14); (e) female flowering shoot (×⅔); (f) female inflorescence entire (right) and in vertical section (left) (×10); (g) half female flower with sessile stigma and single basal ovules in each chamber (×14); (h) bract from female inflorescence (×14); (i) ripe fruit (×3); (j) cross section of fruiting spike showing 2 individual fruits each with 4 seeds (×12); (k) vertical section of fruit (×20).

order, Batidales. Most often it has been associated with families of the Centrospermae[iv], often with Gyrostemonaceae, that make up much of the modern Caryophyllales. Both Gyrostemonaceae and Bataceae were recognized as distinct from this group due to the presence of mustard oils. Evidence from DNA sequence studies indicates Bataceae to be nested within a broad Brassicales sister to Salvadoraceae[v] and Koeberliniaceae[vi].

Economic uses The salty leaves are sometimes eaten in salad. AC

[i] Rauzon, M. J. & Drigot, D. C., Red mangrove eradication and pickleweed control in a Hawaiian wetland, waterbird responses, and lessons learned. Pp. 240–248. In: Veitch, C. R. & Clout, M. N. (eds), *Turning the tide: the eradication of invasive species.* IUCN, Gland, Switzerland.
[ii] Ronse De Craene, L. P. Floral developmental evidence for the systematic position of *Batis* (Bataceae). *American J. Bot*, 92: 752–760 (2005).
[iii] Bayer, C. & Appel, O. Bataceae. Pp. 30–32. In: Kubitzki, K. & Bayer, C. (eds), *The Families and Genera of Vascular Plants. V. Flowering Plants. Dicotyledons: Malvales, Capparales and Non-betalain Caryophyllales.* Berlin, Springer-Verlag (2002).
[iv] Eckardt, T. Batales. Pp. 192–193. In Melchior, E. (ed.), *Engler's Syllabus der Pflanzenfamilien* 2. Berlin, Gebrüder Borntreger (1964).
[v] Chandler, G. T. & Bayer, R. J. Phylogenetic placement of the enigmatic Western Australian genus *Emblingia* based on *rbc*L sequences. *Plant Species Biol.* 15: 67–72 (2000).
[vi] Hall, J. C., Iltis, H. H., & Sytsma, K. J. Molecular phylogenetics of core Brassicales, placement of orphan genera *Emblingia, Forchhammeria, Tirania,* and character evolution. *Syst. Bot.* 29: 654–669 (2004).

BEGONIACEAE
BEGONIA FAMILY

A family of widely cultivated, mainly semi-succulent herbs, with asymmetrical leaves.

Distribution With some 1,400 spp., *Begonia* is largely confined to the tropics. It is absent from the Americas north of Mexico (although 1 sp. is naturalized in Florida), from Europe and North Africa, from northern temperate Asia (extending to the Ryukyu Islands but not to the main islands of Japan), and from Australasia. There are about 650 spp. in Asia, about 600 spp. in the Americas, and about 140 spp. in tropical and southern Africa. The single species of *Hillebrandia* is confined to Hawaii. They are commonly in wet habitats such as forest understory and are often epiphytic or lithophytic.

Description Usually semisucculent herbs up to 1 m high, sometimes acaulescent, sometimes softly woody and up to 4 m, occasionally climbing to more than 10 m, with woody stems and aerial roots. The leaves are alternate, with conspicuous stipules. The lamina is variable in shape but is characteristically asymmetrical and is sometimes shallowly to deeply lobed or rarely with separate leaflets, the margins serrate. The inflorescence is a compact to lax and broad (to 45 cm) cyme. The flowers are unisexual (plants monoecious), actinomorphic or zygomorphic, and the perianth is not clearly

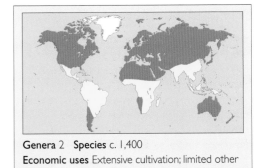

Genera 2 Species c. 1,400
Economic uses Extensive cultivation; limited other local uses

differentiated into sepals and petals. The male flowers have (2)4(5) tepals, usually in unequal opposite pairs, and many stamens or only 4 in sect. *Begoniella*, united into a column in sect. *Symbegonia*. The female flowers have 4–5(–9) tepals or 10 in *Hillebrandia*, in sect. *Symbegonia* fused into a tube for most of their length. The ovary is inferior or in *Hillebrandia* semi-inferior, with (2)3–8 locules, usually sharply angled and often winged. Female flowers have 2–6 styles that are free or connate and often twisted and often bifid or multifid. Placentation is axile and lobed or not, or rarely parietal (*Hillebrandia*), and there are many ovules. The fruit is a capsule or berry, with regular or irregular dehiscence to release the numerous minute seeds.

BEGONIACEAE. I *Begonia rex* (a) habit showing leaves with stipules and axillary inflorescences (x⅓); (b) young leaf with one side larger than the other—a characteristic feature (x1); (c) male flower buds (x1); (d) male flower showing 4 perianth segments and cluster of stamens each with an elongated connective (x1); (e) female flowers showing 5 perianth segments (x⅔); (f) young winged fruit with persistent styles that are fused at the base and bear twisted, papillose stigmatic surfaces (x2); (g) cross section of young winged fruit with 2 chambers and numerous seeds on branched, axile placentas (x2).

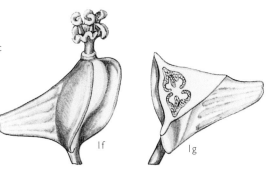

If   Ig

Classification The family has usually been considered close to Cucurbitaceae and Datiscaceae and is included with these in the Cucurbitales of APG II. The Anisophylleaceae have also been included in this group, and it is interesting to note the marked similarity of the ovary of the Malaysian forest tree *Combretocarpus rotundatus* to that of a *Begonia*. However, molecular data do not place the 2 families close together within the Cucurbitales, and a direct evolutionary connection may not be assumed. Many sections have been recognized within the genus *Begonia*[i]. Among many proposed segregates from the genus, 4 have achieved widespread acceptance. Three of these, *Begoniella* and *Semibegoniella* from South America, and *Symbegonia* from New Guinea, have been based largely on various degrees of fusion of the parts of the male and female flowers. Generally, the first 2 have sunk back into *Begonia* in recent years, but *Symbegonia*, which has the stamens united into a column and the tepals of the female flowers united into a tube, has persisted, as for example in the extensive recent global checklist and key to species[ii]. A recent molecular analysis[iii] has found this group to be nested in *Begonia*, and the 12 spp. have been formally transferred to the latter as sect. *Symbegonia*. Those who prefer to emphasise the morphological distinctness of this group may wish to maintain *Symbegonia*, but it seems likely that many will not accept it as a genus. On the other hand, the fourth segregate, *Hillebrandia*, has been found to fall outside *Begonia* and is regarded as the only palaeoendemic genus in Hawaii[iv]. It is thought to be a relict of a group that colonized these volcanic islands some millennia ago and hopped from island to island as the volcanoes rose and fell, while becoming extinct in the regions in which it originated.

Economic uses *Begonia* is extensively cultivated in temperate regions as a house or greenhouse plant or for summer bedding out of doors. It is estimated that some 10,000 cultivars have been recognized. Leaves have been cooked as a vegetable in Malaysia, and some local medicinal uses have been recorded.   RKB

i Doorenbos, J., Sosef, M. S. M., & de Wilde, J. J. F. E. The sections of *Begonia*, including descriptions, keys and species lists. *Wageningen Agric. Univ. Pap.* 98: 1–266 (1998).
ii Golding, J. & Wasshausen, D. C. Begoniaceae, edition 2. Part I: Annotated species list. Part II: Illustrated key, abridgement and supplement. *Contrib. U.S. Nat. Herb.* 43: 1–289 (2002).
iii Forrest, L. L. & Hollingsworth, P. M. A recircumscription of *Begonia* based on nuclear ribosomal sequences. *Pl. Syst. Evol.* 241: 193–211 (2003).
iv Clement, W. L. *et al*. Phylogenetic position and biogeography of *Hillebrandia sandwicensis* (Begoniaceae) a rare Hawaiian relict. *American J. Bot.* 91: 905–917 (2004).

BERBERIDACEAE
BARBERRY, SACRED BAMBOO, AND MAY APPLE

A small family of shrubs and herbs, many of which have ornamental value.

Distribution The herbaceous species are native to north temperate regions in both the Old and New Worlds, with several species pairs disjunct between the Far East and North America[i]; *Berberis* (including *Mahonia*) extends to Tierra del Fuego in South America.

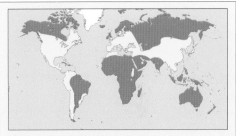

Genera 12–16 **Species** 650–700
Economic uses Garden ornamentals, dyes, and some medicinal use

Ib

Ic

Id

Ie

Ia

BERBERIDACEAE. 1 *Berberis darwinii* (a) flowering shoot (×⅔); (b) flower dissected to show small petaloid sepals, large petals, numerous stamens, and a single carpel (×4). 2 *Berberis* (*Mahonia*) *aquifolium* berries (×⅔). 3 *Jeffersonia dubia* (*Plagiorhegma dubium*) (a) habit (×⅔); (b) stamens (×4); (c) gynoecium entire (left) and in vertical section (right) (×4); (d) cross section of ovary (×4). 4 *Epimedium perralderianum* (a) habit (×⅔); (b) flower (×1); (c) gynoecium entire (left) and in vertical section (right) (×3); (d) stamen (×3); (e) nectariferous petal (×3); (f) fruit—a capsule (×3).

Description Woody shrubs (*Berberis* and *Nandina*) or herbs with rhizomes or tubers. The leaves are normally alternate (almost opposite in *Podophyllum*), basal in some herbaceous species, simple to pinnately or ternately compound (peltate in *Podophyllum* and *Diphylleia*), and usually without stipules. *Nandina* and some species of *Berberis*, including all of the *Mahonia* group, have evergreen, persistent leaves. In *Berberis*, the leaves of long shoots are often transformed into spines, while the short shoots bear simple leaves. The flowers are bisexual, actinomorphic, borne in panicles (e.g., *Nandina*), with reduction through racemes to solitary flowers on leafless basal peduncles (e.g., *Jeffersonia*). The perianth consists of several whorls of usually 6 or 4 segments variously differentiated, the innermost (petals) being usually distinct from the outer ones (outer and inner sepals) in being petaloid and bearing nectaries or else being reduced to small nectariferous sacs or scales. However, *Achlys* has no perianth and consequently no

nectaries. *Podophyllum, Diphylleia,* and *Nandina* have perianth segments without nectaries. There are 6 stamens opposite the innermost perianth segments, except in *Epimedium* (with 4) and *Podophyllum* (with up to 18). The anthers open by lengthwise slits in *Nandina* and *Podophyllum*; by valves hinged at the top in others. The ovary is superior with a single carpel and 1 locule with several or many ovules on the suture or 1 at the base. The style is short or rarely the stigma is sessile. The fruit is succulent in *Nandina, Berberis, Mahonia,* and *Podophyllum* but forms a 2-valved capsule in *Epimedium* and *Vancouveria*, and a papery bladder that is wind-dispersed in *Leontice*. *Caulophyllum* is remarkable in that the enlarging berrylike seeds burst through the wall of the ovary and develop in a completely exposed state; *Gymnospermium* also has exposed seeds.

Classification Berberidaceae is notable for the contrast between the large genus *Berberis* (including *Mahonia*), with more than 600 spp., *Epimedium* (c. 44 spp.), and the remaining genera all with fewer than 10 spp. The classification adopted by Hutchinson (1973) put the woody species into 2 distinct families: Nandinaceae (*Nandina*) and Berberidaceae (*Berberis* and *Mahonia*), while all the herbaceous perennials were put in the Podophyllaceae, including *Leontice* and its related species, sometimes regarded as a separate family, Leonticaceae. Strong evidence exists[ii] for the inclusion of *Mahonia* in *Berberis* with which it has been

consistently grouped in the past and shares most morphological features. The family has been the subject of several phylogenetic studies, including morphological[iii] and molecular[iv] analyses, that show clear relationships among the genera, identifying *Nandina* as distinct from the other genera in subfamily Nandinoideae, with the remaining genera in subfamily Berberidoideae. The relationship of Berberidaceae to Ranunculaceae and to other Ranunculales is well established and is supported by both morphological and molecular[v] evidence.

Economic uses Many of the Berberidaceae are prized ornamental plants, e.g., *Berberis buxifolia, B. darwinii, B. calliantha, B. stenophylla* and *B. thunbergii, Berberis* (*Mahonia*) *aquifolium, B. bealei, B. lomariifolia,* and *Nandina domestica* (Sacred Bamboo). *Berberis* fruit (Barberry) are barely edible, and the roots and wood contain a basic yellow dyestuff. The rhizomes of the May Apple, sometimes incorrectly called Mandrake (*Podophyllum peltatum* and *P. hexandrum*), yield a resin with drastic purgative and emetic properties. This resin is incorporated into certain types of commercial laxative pills.   [WTS] AC

i Liu, J. Q., Chen, Z. D., & Lu, A. M. Molecular evidence for the sister relationship of the eastern Asia–North American intercontinental species pair in the *Podophyllum* group (Berberidaceae). *Bot. Bull. Acad. Sinica* 43: 147–154 (2002).
ii Kim, Y. D., Kim, S. H., & Landrum, L. R. Taxonomic and phytogeographic implications from its phylogeny in *Berberis* (Berberidaceae). *J. Plant Res.* 117: 175–182 (2004).

iii Loconte, H. & Estes, J. R. Phylogenetic systematics of Berberidaceae and Ranunculales (Magnoliidae). *Syst. Bot.* 14: 565–579 (1989).
iv Kim, Y. D. & Jansen, R. K. Chloroplast DNA restriction site variation and phylogeny of the Berberidaceae. *American J. Bot.* 85: 1766–1778 (1998).
v Hoot, S. B., Magallon-Puebla, S., & Crane, P. R. Phylogeny of basal eudicots based on three molecular data sets: *atp*B, *rbc*L and 18S nuclear ribosomal DNA sequences. *Ann. Missouri Bot. Gard.* 86: 119–131 (1999).

BERBERIDOPSIDACEAE

Perennial climbers with glossy green leaves and beautiful red flowers found in temperate rain forests of Chile and eastern Australia.

Distribution *Berberidopsis corallina* (the Chilean Coral Vine) occurs in the temperate rain forests of Chile and is under threat of extinction in the wild[i]. *B. beckleri* and *Streptothamnus moorei* are endemic to rain forests of northern New South Wales and southern Queensland, the latter in warmer areas.

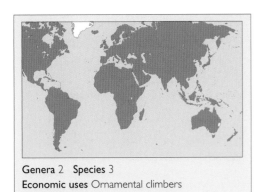

Genera 2 Species 3
Economic uses Ornamental climbers

Description Climbing evergreen glabrous to subglabrous perennials, becoming woody with age. The leaves are alternate, entire or spiny-toothed, ovate (dimorphic in *B. beckleri*), shiny above, glaucous below, petiolate, exstipulate. The flowers are bisexual, actinomorphic, usually solitary on a long peduncle arising from the leaf axil. The perianth consists of 9–12(–15) spirally arranged petaloid segments, usually red. The stamens are 8–15 (*Berberidopsis*) or c. 70 (*Streptothamnus*). The carpels are 3 or 5, fused to form a 1-locular ovary with 2 to many ovules. The fruit is a fleshy red or black berry.

Classification The 2 genera are often placed in Flacourtiaceae[ii] but molecular data support family recognition in the Berberidopsidales (APG II), forming a monophyletic group with Aextoxicaceae[iii]. These relationships are supported by wood anatomy[iv] and pollen[v].

Economic uses *Berberidopsis corallina* has limited use as an ornamental climber.   AC

i Etisham-Ul-Haq, M., Allnutt, T. R., Smith-Ramirez, C., Gardner, M. F., Armesto, J. J., & Newton, A.C. Patterns of genetic variation in *in* and *ex situ* populations of the threatened Chilean vine *Berberidopsis corallina*, using RAPD markers. *Ann. Bot.* 87: 813–821 (2001).

ii New South Wales Flora Online Demonstration Version. Royal Botanic Gardens Sydney. http://plantnet.rbgsyd.nsw.gov.au/ PlantNet/NSWflora/index.html
iii Soltis, P. S., Soltis, D. E., & Chase, M. W. Angiosperm phylogeny inferred from multiple genes as a tool for comparative biology. *Nature* 402: 402–404 (1999).
iv Carlquist, S. C. Wood anatomy of Aextoxicaceae and Berberidopsidaceae is compatible with their inclusion in Berberidopsidales. *Syst. Bot.* 28: 317–325 (2003).
v Van Heel, W. A. Flowers and fruit in Flacourtiaceae. V. The seed anatomy and pollen morphology of *Berberidopsis* and *Streptothamnus*. *Blumea* 30: 31–37 (1984).

BETULACEAE
ALDERS, BIRCHES, HAZELS, AND HORNBEAMS

A family of trees and shrubs that includes the alders (*Alnus*), birches (*Betula*), hazels (*Corylus*), and hornbeams (*Carpinus*).

Distribution The family occurs mainly in boreal and cool temperate regions. Some species grow on tropical mountain ranges through Central America and the Andes to Argentina.

Description Small, anemophilous trees or shrubs, often with smooth bark exfoliating in large, thin layers. The leaves are simple, alternate, 2- to 3-ranked, deciduous and with stipules, the lamina sometimes lobed, and the margins dentate to nearly entire. The flowers are unisexual (plants dioecious), in 1- to 3-flowered clusters, adhering to their bracts. The staminate flowers are in pendulous, conspicuously bracteate, cylindrical catkins comprising 1- to 3-flowered clusters. The pistillate flowers are borne on a stiff axis, in erect to pendulous, bracteate catkins or in 2- to 3-flowered clusters subtended by a leafy involucre. The perianth, when present, is of a variable number of scalelike

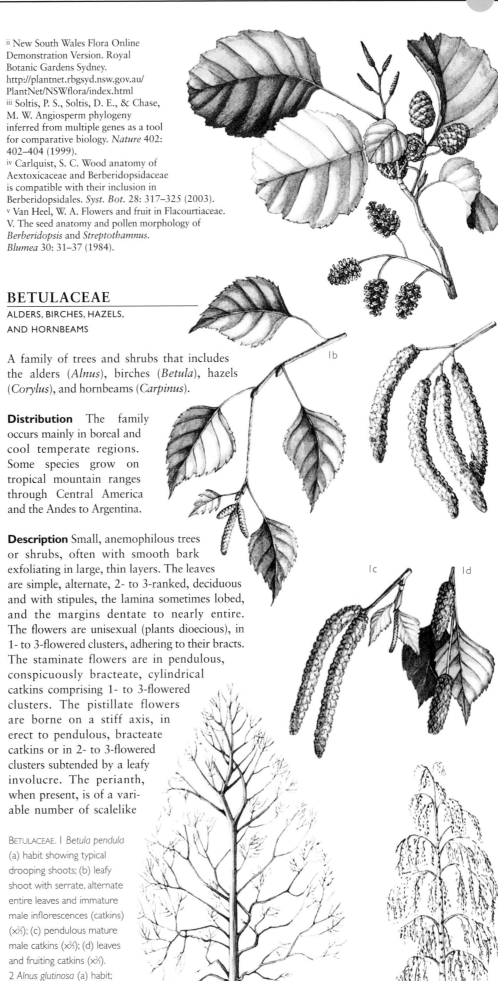

BETULACEAE. **I** *Betula pendula* (a) habit showing typical drooping shoots; (b) leafy shoot with serrate, alternate entire leaves and immature male inflorescences (catkins) (x⅔); (c) pendulous mature male catkins (x⅔); (d) leaves and fruiting catkins (x⅔). **2** *Alnus glutinosa* (a) habit; (b) male catkins (x⅔); (c) shoot with (from base) old fruiting cones, immature cones and young male catkins (x⅔).

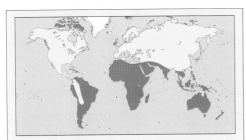

BETULACEAE. 3 Barks of various birch trees (a) *Betula pendula*, (b) *B. humilis*, (c) *Betula sp.* (×⅔).

Genera 6 Species 130[i]
Economic uses Timber; nuts from *Corylus* species (cobnuts, filberts, hazlenuts); ornamental trees

segments. The staminate flowers have (1–)4–6 (–12) stamens, the anthers 2-celled, and without traces of vestigial carpels. The pistillate flowers are highly reduced, consisting of a single inferior ovary of 2 fused carpels and completely lacking vestigial stamens. The fruit is a single-seeded nut, which is often winged for wind-dispersal, maturing in late summer or the fall. The seeds have no endosperm and a straight embryo.

Classification Previously the Betulaceae and Corylaceae have been treated as separate families, but recent work[ii,iii] upholds their recognition as 2 subfamilies that correspond to 2 major lineages detected through molecular studies[iv,v]. Cladistic and molecular work has also clarified to some degree the relationships between the genera and the family's relationships within the Fagales.

SUBFAM. BETULOIDEAE The male flowers are borne in 3-flowered groups and have a perianth; the female flowers lack a perianth. The single tribe, Betuleae, has 2 genera: *Betula* (c. 50 spp.) and *Alnus* (c. 30 spp.).

SUBFAM. CORYLOIDEAE The male flowers are solitary, the inflorescence therefore a simple spike; perianth absent; female flowers with a perianth. The tribe Coryleae (male flowers with an average of 2 stamens per flower) has 2 genera: *Corylus* (c. 15 spp.), and *Ostryopsis* (2 spp.). The tribe Carpineae (male flowers with an average of 6 stamens per flower) has 2 genera: *Carpinus* (c. 30 spp.) and *Ostrya* (7 spp.).

Economic uses The birches provide valuable hardwood timbers. *Betula lutea* and *B. lenta* are important in North America, providing wood for doors, floors, and furniture, etc. *B. papyrifera* is used for boxes, plywood, and in turnery. The bark of *B. papyrifera* is also used by Native Americans for making canoes and fancy goods. Branchlets of *Betula* spp. are used to make the besom brushes used by gardeners. *Alnus rubra* also provides a valuable timber, being a good imitation of mahogany. Both genera provide high-grade charcoal. *Ostrya* has extremely hard wood used for making mallets. Cobnuts, filberts, or hazelnuts, are produced by species of *Corylus*. Birches, hazels, and hornbeams are cultivated as ornamentals, and *Alnus* is used to aid soil nitrification. SLJ

[i] Chen, Z.-D. Phylogeny and phytogeography of the Betulaceae. *Acta Phytotax. Sinica* 29: 464–475 (1994).
[ii] Crane, P. R. Early fossil history and evolution of the Betulaceae. In: Crane, P. R. & Blackmore, S. (eds), *Evolution, Systematics, and Fossil History of the Hamamelidae*, vol. 1, 105–128. Oxford, Clarendon Press (1989).
[iii] Kubitzki, K. Betulaceae. Pp. 152–157. In: Kubitzki, K., Rohwer, J. G., & Bittrich, V. (eds), *The Families and Genera of Vascular Plants. II. Flowering Plants. Dicotyledons: Magnoliid, Hamamelid and Caryophyllid Families.* Berlin, Springer-Verlag (1993).
[iv] Bousquet, J., Strauss, S. H. & Li, P. Complete congruence between morphological and *rbc*L-based molecular phylogenies in birches and related species (Betulaceae). *Mol. Biol. Evol.* 9: 1076–1088 (1992).
[v] Chen, Z.-D., Manchester, S. R., & Sun. H.-Y. Phylogeny and evolution of the Betulaceae inferred from DNA sequences, morphology, and paleobotany. *American J. Bot.* 86: 1168–1181 (1999).

BIEBERSTEINIACEAE

This family comprises a single genus of 5 spp. of woody rhizomatous or tuberous perennial herbs, with alternate pinnately compound leaves. They occur in montane, semiarid areas of central and western Asia and southeastern Europe. *B. orphanidis* is the only European species (but also in central southern Turkey)—long thought to be extinct in its *locus classicus* on Mount Killini in northern Pelipponisos, Greece, but was recently discovered on a neighboring mountain, Saitas[i]. The flowers are hermaphrodite, actinomorphic in spikelike racemes or panicles, with 5 free sepals, and 5 free petals, which are sometimes clawed and alternating with 5 fleshy extrastaminal nectary glands; stamens 10 and connate at the base. The ovary is superior, deeply lobed, with a short gynophore; carpels 5, with a single ovule in each locule; the styles are free, connate apically to form a slender column with a capitate stigma. The fruit is a schizocarp of five 1-seeded mericarps.

The affinities of *Biebersteinia* are somewhat uncertain. Previously, it has been included in the Geraniaceae or recognized as a separate

Genera 1 Species 5
Economic uses An anti-inflammatory

family near the Geraniales, but molecular work suggests it should be placed in the Sapindales[ii]. Whatever its correct phylogenetic position, there does seem to be a case for recognizing it as a separate family. VHH

[i] Tan, K., Perdetzoglou. D. K., & Roussis, V. *Bieberstenia orphanidis* (Geraniaceae) from southern Greece. *Ann. Bot. Fennici* 34: 41–45 (1997).
[ii] Bakker, F. T., Vassiliades, D. D., & Morton, C. Phylogenetic relationships of *Bieberstenia* Stephan (Geraniaceae) inferred from *rbc*L and *atp*B sequence comparisons. *Bot. J. Linn Soc.* 127: 149–158 (1998).

BIGNONIACEAE
TRUMPET VINES, JACARANDAS, CATALPAS

A family of trees, shrubs, and climbers, often with tendrils, rarely herbaceous, mostly tropical and subtropical, and allied to Scrophulariaceae.

Distribution The family is pantropical and subtropical with a few genera in temperate parts of North America, Asia and the southern hemisphere. They often occupy a variety of wet forest habitats but also occur in wet or dry woodland or even subdesert conditions. The herbaceous genera *Incarvillea* and *Argylia* are found in montane grassy habitats in temperate Asia and the Andes, respectively. The tribes show interesting continental preferences: thus the Tecomeae are widespread in both Old and New World, whereas the mostly climbing Bignonieae are confined to the New World. The Coleeae are found in Africa, and especially Madagascar, and the Crescentieae are in Central and South America. The small tribes, Eccremocarpeeae and Tourretieae are monogeneric and both in the New World only, while the Oroxyleae have 4 genera in tropical Asia.

Description About half of all species are trees or shrubs while most of the others are woody climbers. *Incarvillea* and *Argylia* include perennial herbs, and *Tourrettia* is a herbaceous climber. Leaves are usually opposite but sometimes whorled or in a few genera alternate and compound or sometimes (e.g., *Catalpa*) simple. Compound leaves are either palmate or pinnate (often appearing 3-foliate), and in climbing South American genera the terminal leaflets of pinnate leaves are often modified to simple or branched tendrils. The flowers are solitary or in axillary or terminal racemes. The calyx is composed of 5 fused sepals, sometimes bilabiate or spathaceous to truncate or calyptrate, occasionally campanulate with a double margin. The corolla consists of 5 fused petals, usually tubular and bilabiate, or with rotate subequal lobes. Stamens are usually 5 and attached in the corolla tube, rarely 2 fertile and 3 staminodial. The ovary is bicarpellate and 2-locular or rarely (Tourrettieae) 4-locular with axile placentation, or rarely (Eccremocarpeae, Crescentieae) 1-locular with parietal placentation. The fruit is a capsule dehiscing by 2 valves either loculicidally or (Oroxyleae) septicidally, or sometimes (e.g., *Kigelia*) fleshy and indehiscent. The seeds

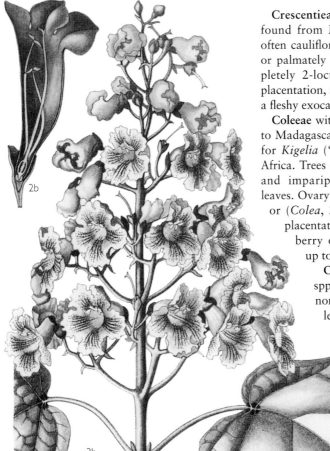

2b

3b

1a

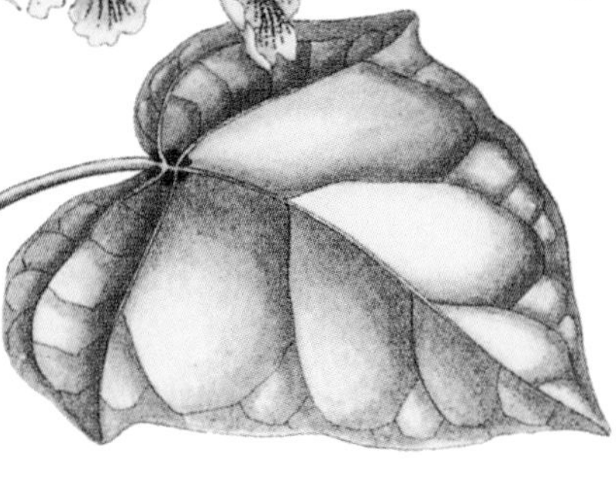

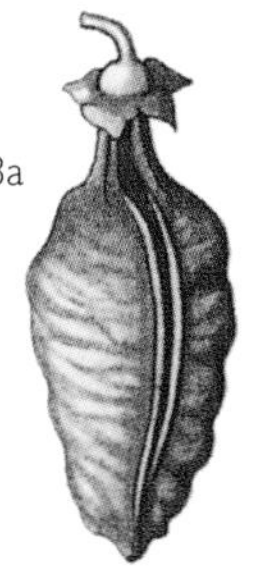

Crescentieae with 3 genera and 33 spp., found from Mexico to Brazil. Mostly trees, often cauliflorous, often with alternate, simple or palmately compound leaves. Ovary incompletely 2-locular or 1-locular with parietal placentation, and the fruit indehiscent and with a fleshy exocarp or endocarp.

Coleeae with 6 genera and 52 spp., confined to Madagascar and surrounding islands except for *Kigelia* ("sausage trees"), with 2 spp. in Africa. Trees and shrubs with often verticillate and imparipinnately compound or simple leaves. Ovary 2-locular, with axile placentation or (*Colea*, 21 spp.) 1-locular with parietal placentation, and the fruit an indehiscent berry or (*Kigelia*) fibrous and woody, up to 50 cm long.

Oroxyleae with 4 genera and 6 spp. in tropical Asia. Trees or woody nontendrilled climbers, with pinnate leaves. Ovary 2-locular with axile placentation and the fruit dehiscing septicidally.

3a

Genera c. 105 **Species** 850 or more
Economic uses Many cultivated ornamentals, especially in the tropics; condiments, dyes, medicinal derivatives, ropes, timber

Tourrettieae with a single species, *Tourrettia lappacea*, from Mexico to Argentina. A herbaceous climber with biternate leaves and a subterminal tendril. Ovary 4-locular, with axile placentation, the fruit dehiscing septicidally.

Eccremocarpeae with 1 genus and 3 spp. in South America. Herbaceous to woody climbers with bi- to tripinnatisect leaves, with the terminal leaflets modified to a branched tendril. Ovary 1-locular with parietal placentation, the fruit woody and flattened, dehiscent, with the valves remaining united at the apex.

Economic uses Bignoniaceae provides many showy trees and climbers often cultivated as ornamentals. A key to 34 spp. commonly grown in the tropics has been provided by Bidgood[iii]. In temperate regions, *Catalpa* (Indian

are numerous, usually flat and winged, but sometimes thicker and wingless.

Classification The family is close to the Scrophulariaceae and constant characters to separate them are difficult to find. Seven tribes are recognized[i,ii,iv]:

Tecomeae with 44 genera and c. 340 spp., including *Tabebuia* (c. 100 spp.) and *Jacaranda* (c. 50 spp.), the only tribe occurring both in the New and Old World (all the others tribes being confined to single continents). Trees, shrubs, and non-tendrilled climbers with pinnate, palmate, or simple leaves. Ovary 2-locular with axile placentation, the fruit dehiscing loculicidally. Apparently basal to the rest of the family. When *Tecomaria* is separated on androecium characters from *Tecoma* and other South American genera, none of the tropical American genera occur also in Africa. However, *Campsis* and *Catalpa* show striking disjunctions between China and temperate North America.

Bignonieae with 45 genera and c. 415 spp., including *Arrabidaea* (c. 100 spp.), *Anemopaegma* and *Adenocalymma* (c. 60 spp. each), and *Memora* (35 spp.) are confined to the New World tropics apart from the monotypic *Bignonia* from the southeastern USA. Woody climbers or rarely (*Sphingiphila*) a shrub to small tree, with pinnate leaves and the terminal pinna often modified to form a simple or trifid tendril, or occasionally with simple leaves. Ovary 2-locular with axile placentation, and the fruit dehisces septicidally.

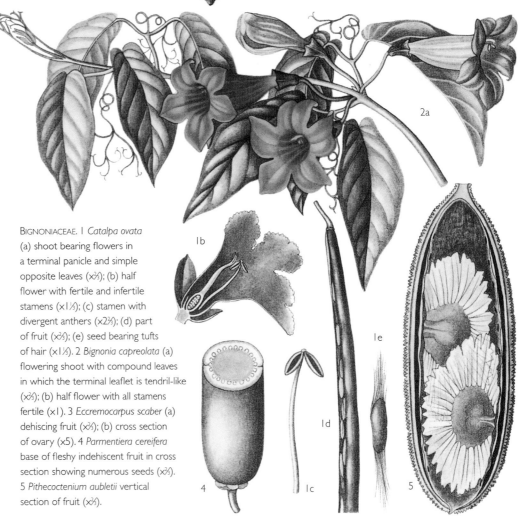

2a

1b

BIGNONIACEAE. 1 *Catalpa ovata* (a) shoot bearing flowers in a terminal panicle and simple opposite leaves (x⅔); (b) half flower with fertile and infertile stamens (x1⅓); (c) stamen with divergent anthers (x2⅔); (d) part of fruit (x⅔); (e) seed bearing tufts of hair (x1⅓). 2 *Bignonia capreolata* (a) flowering shoot with compound leaves in which the terminal leaflet is tendril-like (x⅔); (b) half flower with all stamens fertile (x1). 3 *Eccremocarpus scaber* (a) dehiscing fruit (x⅔); (b) cross section of ovary (x5). 4 *Parmentiera cereifera* base of fleshy indehiscent fruit in cross section showing numerous seeds (x⅔). 5 *Pithecoctenium aubletii* vertical section of fruit (x⅔).

1d

1e

4

1c

5

Bean Trees) and the climbers *Campsis radicans* (Trumpet Vine) and *Bignonia capreolata* (Trumpet Flower) from North America, *Pandorea* (Bower of Beauty) from Australia, and *Podranea* (Port St. John Creeper) and the shrub *Tecomaria* (Cape Honeysuckle) from South Africa, are often cultivated. Tropical Africa provides *Spathodea* (Flame of the Forest), *Fernandoa, Markhamia,* and *Stereospermum,* while *Kigelia* ("sausage trees") is well known to tourists. From Asia come other *Catalpa* spp., *Millingtonia* (Indian Cork Tree), *Radermachera,* and *Tecomanthe,* and from South America *Jacaranda mimosifolia, Tecoma* (Golden Bells), and *Tabebuia,* including some of the most spectacular flowering trees of the neotropics, as well as the climbing genera *Pyrostegia* (Flaming Vine), *Clytostoma* (Argentine Trumpet Vine), *Pithecoctenium* (Monkey's Hairbrush), *Crescentia* (Calabash), and *Eccremocarpus* (Glory Flower). Species of *Incarvillea* are sometimes grown as herbaceous perennials. Other useful products include timber from *Tabebuia* and ropes from some of the woody climbers. *Crescentia* fruits provide calabashes for carrying water or for use as maracas. Other genera produce condiments, red dyes for baskets, and various plant extracts used against major diseases, e.g., Pau d'arco, an extract of the bark of *Tabebuia impetiginosa* from tropical South America, commonly used as a food supplement and herbal remedy.   RKB

[i] Gentry, A. H. Bignoniaceae part 1: tribes Crescentiae and Tourrettiae. *Flora Neotropica* 25 (1): 1–131 (1980).
[ii] Fischer, E., Theisen, I., & Lohmann, L. G. Bignoniaceae. Pp. 9–38. In: Kadereit, J. W. (ed.), *Families and Genera of Vascular Plants. VII.* (2004).
[iii] Bidgood, S. Bignoniaceae: key to the cultivated species. *Flora Zambesiaca* 8 (3): 61–64 (1988).
[iv] Spangler, R. E. & Olmstead, R. G. Phylogenetic analysis of Bignoniaceae. *Ann. Missouri Bot. Gard.* 86: 33–46 (1999).

BIXACEAE

ANNATTO

A neotropical family, comprising a single genus of shrubs or trees with a yellow, orange, or reddish sap.

Distribution *Bixa* is confined to the neotropics. *B. orellana* is widely cultivated and often naturalized throughout the tropics and may no longer be found in its native tropical America.

Genera	1	Species	5

Economic uses
Annatto dye obtained from *Bixa orellana*

Description Evergreen trees or shrubs, the stems and leaves containing a yellowish or reddish exudate. The leaves are alternate, simple, entire, palmately nerved, covered with peltate scales on the undersurface, stipulate, with long petioles. The flowers are actinomorphic, hermaphrodite, in terminal showy thyrsoid inflorescences. The sepals are 5, free, caducous; the petals are 5, free, white, or pinkish. The stamens are numerous, with free filaments. The ovary is superior, bicarpellate, and 1-locular. The fruit is a loculicidal capsule, often covered with soft or prickly hairs. The seeds are covered with a reddish sarcotesta.

Classification The separation of the Bixaceae from the Cochlospermaceae is problematic. They were united by Cronquist, a position also adopted by Heald in FPN[i]. The morphological and chemical evidence is not conclusive and molecular data, indicating that *Bixa* and *Diegodendron* form a clade with *Cochlospermaceae* as a sister group[ii,iii], can be interpreted as support for merging them all in 1 family Bixaceae or recognizing 3 separate families: Bixaceae, Diegodendraceae, and Cochlospermaceae.

Economic uses An orange-yellow or reddish dye called annatto, which contains mainly bixin, is extracted from the seeds of *Bixa orellana*. It is used as a food colorant, and annatto

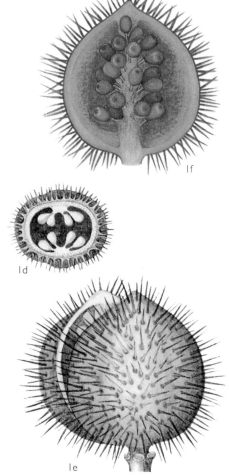

BIXACEAE. 1 *Bixa orellana* (a) shoot with alternate leaves and flowers in a terminal panicle (×1); (b) half base of flower showing numerous stamens and ovary with numerous ovules on parietal placentas (×2); (c) stamens dehiscing by short slits and ovary crowned by long style and bilobed stigma (×3); (d) cross section of ovary showing ovules on parietal placentas (×8); (e) capsule dehiscing by 2 valves (×1); (f) vertical section of fruit with numerous red seeds (×1).

paste is also used as a dye for cloth and wool and for paint and varnishes in industry. It is also used as a food supplement and as a folk medicine. It appears to have been cultivated in tropical America since before the time of Columbus and used as a body paint. Today, it is widely cultivated and grown on a plantation scale in some tropical countries. VHH

i Heald, S. W. Bixaceae. Pp. 54–55. In: Smith, N. *et al.* (eds), *Flowering Plants of the Neotropics*. Princeton: Princeton University Press (2004).
i Fay, M. F., Bayer, C., Alverson, W. S., de Bruijn, A. Y., & Chase, M. W. Plastid *rbc*L sequence data indicate a close affinity between *Diegodendron* and *Bixa*. *Taxon* 47: 43–50 (1998).
ii Alverson, W. S., Karol, K. G., Baum, D. A., Chase, M. W., Swensen, S. M., McCourt, R., & Systma, K. J. Circumscription of the Malvales and relationships to other Rosidae: Evidence from *rbc*L sequence data. *American J. Bot.* 85: 876–887 (1998).

BOMBACACEAE

BALSA, BAOBAB, KAPOK, AND SILK COTTONS

A family of trees mainly from South America, closely related to Malvaceae.

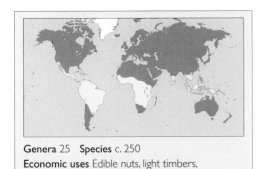

Genera 25 **Species** c. 250
Economic uses Edible nuts, light timbers, ornamental trees, specialist fibers, spices

Distribution Apart from the Baobab, *Adansonia* (Madagascar, Africa, and western Australia), and *Bombax* (Africa to Southeast Asia), all genera and most species are centered in Central and South America.

Description Trees, sometimes spiny, swollen-trunked, and deciduous; trunks with mucilaginous exudates and fibrous bark. The leaves are alternate, spiralled, palmately lobed or compound (simple in *Matisia* and *Quaraibea*), entire, petioles pulvinate, stipules caducous and inconspicuous. The hairs are usually stellate or tufted. Inflorescences are of 1 or 2 axillary flowers, rarely (e.g., *Bernoullia*) many-flowered. The flowers are actinomorphic or slightly zygomorphic, bisexual, showy, usually with an inconspicuous and caducous epicalyx of 3 bracts. The calyx is shortly cylindrical, with 5 sepal lobes at the apex, or more usually irregularly lobed or more or less truncate, sometimes splitting. The petals are 5, imbricate, joined at the base, and falling off with the androecium as 1 unit. The stamens are numerous, usually monothecal, the filaments united, usually in their lower half, into a cylinder around the style; the cylinder may be lobed, with the stamens in 5 antipetalous phalanges with the filaments completely united, numerous anthers, then sessile on each of the 5 entire phalanges. In some genera, the anthers on each phalange appear to have united into a "super-anther." These are septate dithecal (e.g., *Fremontodendron*, *Phragmotheca*, *Septotheca*), 5 non-septate dithecal (*Ceiba*), and tetrathecal septate (*Spirotheca* and *Huberodendron*). Pollen is usually spheroidal, more or less smooth, and reticulate. The ovary is superior, usually with 5 locules, each with 2 or more axile, anatropous ovules. The style is 5-lobed. The fruits are juicy, indehiscent, and 1- or 5-seeded (*Matisia* and *Quaraibea*), hard and indehiscent with numerous seeds (e.g., *Adansonia*) or dehiscent and 5-valved, the seeds often then embedded in endocarp hairs (kapok). Less usually the fruits or seeds are winged.

Bombacaceae are best known as emergent trees (*Ceiba pentandra* is reputed to be the tallest growing tree in Africa), usually of drier forests. However, *Matisia* and *Quaraibea* are understory trees of evergreen forest. *Spirotheca* is an epiphytic strangler. Pollination by bats, birds, and moths is known, even in a single genus (*Adansonia*)[i].

Classification Bombacaceae is part of the core Malvales (with Byttneriaceae, Brownlowiaceae, Durionaceae, Helicteraceae, Malvaceae, Pentapetaceae, Sparrmanniaceae, Sterculiaceae, and Tiliaceae, q.v.). Although closely related to Malvaceae, molecular data (other references under Malvaceae, Tiliaceae) supports their separation. Only pollen and habit seem to provide a morphological basis for the separation, and a few monotypic genera (*Lagunaria*, *Camptostemon*) are trees and so appear to connect the families. No satisfactory infrafamilial classification exists. The Durian group of Southeast Asia, formerly included in the Bombacaceae, are here treated as Durionaceae (q.v.)

Economic uses The lightweight, spongy wood of some species has specialist uses e.g., in model airplane making (balsa, *Ochroma*) or as a sandwich in plywood (fromagier, *Ceiba pentandra*). The internal fruit hairs of the latter, as kapok, have been used in stuffing life jackets and as insulation. Several species of *Pachira* are cultivated for their large, edible seeds ("nuts"), while the fruits of some *Quararibea* and *Matisia* are spices. *Ceiba* (*Chorisia*) *speciosa* is widely planted as an ornamental in the tropics and *Fremontodendron californicum* in warm temperate and Mediterranean gardens. MRC

i Baum, D. A. The comparative pollination and floral biology of Baobabs (*Adansonia*-Bombacaceae.) *Ann. Missouri Bot. Gard.* 82: 322–348 (1995).
ii Bayer, C. *et al.* Support for an expanded family concept of Malvaceae within a recircumscribed order Malvales: a combined analysis of plas tid *atp*B and *rbc*L DNA sequences. *Bot. J. Linn. Soc.* 129: 267–303 (1999).

BOMBACACEAE. 1 *Bombax ceiba* (a) habit; (b) digitate leaf (x⅔); (c) flower dissected to show petals and numerous bundles of stamens (x⅔); (d) gynoecium with superior ovary, simple style and divided stigma (x⅔); (e) cross section of ovary (x3); (f) fruit with seeds embedded in hairs (x⅔).

1a

1b

1c

1d

1e

1f

BONNETIACEAE

A small family of evergreen trees and shrubs producing yellow or white resinous sap.

Distribution *Ploiarium* (3 spp.) grows in Southeast Asia, western Malaysia, the Moluccas, and New Guinea. The other 2 genera are neotropical: both *Archytaea* (3 spp.) and *Bonnetia* (29 spp., of which 23 spp. are endemic to the tepuis of Venezuelan Guayana[i]) occur in Bolivia, Brazil, Colombia, and Venezuela (*Bonnetia* also in Peru and Cuba). *Neblinaria*, which may be part of *Bonnetia*, is endemic to the Cerro de la Neblina in southern Venezuela.

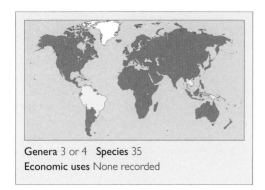

Genera 3 or 4 **Species** 35
Economic uses None recorded

Description Glabrous evergreen trees or shrubs, which secrete a resinous sap. The leaves are alternate, simple, usually crowded toward the tip of the shoots, without stipules, the margins entire or dentate. The flowers are actinomorphic, hermaphrodite, in terminal or axillary cymose inflorescences, or flowers solitary. The sepals are 5, unequal, persistent; the petals are 5, red or pink, contorted, and free. The stamens are numerous, free or basally connate, or in epipetalous fascicles of 5. The ovary is superior, with 3 (*Bonnetia*) or 5 (*Ploiarium, Archytaea*) carpels and 3–5 locules; the placentation is axile; the ovules are numerous. The fruit is a septicidal capsule with many seeds.

Classification The Bonnetiaceae was included in the Theaceae by Cronquist (1981) and was considered by Takhtajan (1997) to occupy an intermediate position between that family and the Clusiaceae. The family is included in the Malpghiales in APG II, in accord with molecular phylogenetic studies[ii]. VHH

[i] Stevenson, D. E. Bonnetiaceae (Bonnetia Family). Pp. 58–59. In: Smith, N. *et al.* (eds), *Flowering Plants of the Neotropics*. Princeton, Princeton University Press (2004).
[ii] Gustafsson, M. H. G., Bittrich, V., & Stevens, P. F. Phylogeny of Clusiaceae based on *rbc*L sequences. *Int. J. Plant Sci.* 163: 1045–1054 (2002).

BORAGINACEAE

BORAGE AND FORGET-ME-NOT FAMILY

As now circumscribed, the Boraginaceae ranges from large trees in tropical regions to annual herbs, usually with conspicuous hairs with bulbous bases.

Distribution The tree genera are pantropical and subtropical, while the herbaceous genera are worldwide although particularly abundant in the Mediterranean region and warm-temperate Asia. They occur in a wide range of habitats from sea level to more than 4,000 m.

Description Annual or perennial herbs to shrubs, occasionally woody climbers, or large forest trees. Vegetative parts characteristically bear conspicuous stiff hairs that are usually unicellular and mounted on a swollen multicellular base embedded in the epidermis, glabrous plants being rare (e.g., *Cerinthe*, some woody genera). The leaves are alternate or sometimes the lower ones opposite, or sometimes (some *Tournefortia*) all opposite, simple with entire or subentire margins, rarely serrate (some woody species), shallowly lobed (*Coldenia*) or serrate-spinulose (*Halgania*), rarely succulent (*Heliotropium curassavicum*). The inflorescence is characteristically a lax to condensed, 1-sided, terminal or axillary, cincinnate cyme, or occasionally the flowers are solitary in leaf axils. The flowers are hypogynous, gamopetalous, and actinomorphic or sometimes (*Echium*) slightly zygomorphic, usually 5-merous, sometimes unisexual (plants dioecious, *Cordia*), frequently heterostylous. The calyx lobes are free or fused and sometimes conspicuously accrescent (in *Sacellium, Auxemma, Cordia, Trichodesma*). The corolla is usually tubular to trumpet-shaped, rarely campanulate, sometimes bearing 5 appendages in the throat. The stamens are attached to the corolla tube, and the anthers are rarely (*Halgania*) connivent around the style. The ovary is bicarpellate, entire to deeply 4-lobed, with 1, 2, or 4 locules and with a terminal or gynobasic (subfamily Boraginoideae) style with 1, 2, or 4 stigmas. The ovules are usually 1 per locule and subbasal. The fruit may be a drupe with 1–4 pyrenes but is more often a schizocarp splitting into four 1-seeded nutlets, or is rarely a dehiscent capsule (see individual subfamilies). Two species from New Guinea, formerly referred to the genus *Zoelleria*, have 8–10 ovules and nutlets but are otherwise typical of *Trigonotis* and are included in that genus.

Classification Based on molecular evidence, the family is usually placed near to the Solanaceae and Convolvulaceae rather than to the Gentianaceae or Scrophulariaceae[i,ii] but is of uncertain ordinal placement in APG II. Six subfamilies are here recognized, 1 of which is divided into several tribes.

SUBFAM. CODONOIDEAE[iii] Includes only *Codon* with 2 spp. from western South Africa and Namibia. Annual herbs with spinelike trichomes up to 10 mm and also glandular trichomes (only occasionally found in other subfamilies). Flowers (calyx, corolla, and stamens) 10- to 12-merous, the corolla broadly cylindrical. Ovary pyriform with a long terminal style divided to half way. Fruit a dehiscent capsule with many seeds. *Codon* has until recently been regarded as a member of the Hydrophyllaceae.

SUBFAM. WELLSTEDIOIDEAE Includes only *Wellstedia*, with 7 spp. from northeastern Africa and South Africa. Dwarf subshrubs with a silky indumentum. Flowers 4-merous. Ovary 2-locular with a terminal bifid style and 1 ovule per cell. Fruit a capsule dehiscing to release the 1–2 seeds.

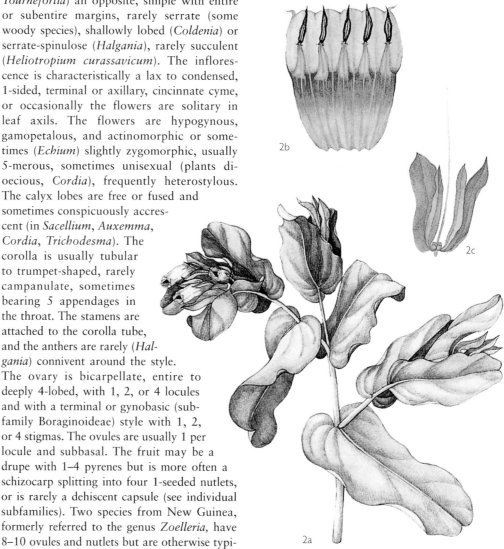

BORAGINACEAE. | *Anchusa officinalis* leafy shoot and inflorescence of regular flowers (x⅔). 2 *Cerinthe major* (a) leafy shoot and inflorescence (x⅔); (b) corolla opened out to show 5 epipetalous stamens (x2); (c) calyx and 4-lobed gynoecium with a gynobasic style, that is, arising from the base of the ovary between the lobes (x2). 3 *Echium vulgare* (a) inflorescence of irregular flowers (x⅔); (b) flower dissected to show epipetalous stamens and 4-lobed ovary with a thin, gynobasic style which is forked at its tip (x2). 4 *Heliotropium* sp. flower opened out with 5 arrow-shaped stamens and the ovary with a terminal style and an umbrella-shaped expansion below the stigma (x2).

SUBFAM. CORDIOIDEAE Includes 4 genera, the pantropical (but mainly South American) *Cordia* with c. 320 spp., *Auxemma* with 2 spp. in Brazil, *Patagonula* with 2 spp. in Brazil and Argentina, and *Saccellium* with 2 spp. from Peru and Brazil to Argentina. Shrubs to trees or rarely woody climbers. Flowers usually 5-merous. Ovary scarcely lobed, with a slender terminal style with a twice bifid style and 4 stigmas, with 4 ovules in 1, 2, or 4 locules. Fruit either dry and hard on the outside or thinly fleshy outside with a hard endocarp, developing 1, 2, or 4 pyrenes. Cotyledons plicate.

SUBFAM. EHRETIOIDEAE Recently defined[iv] as including *Ehretia* (c. 50 spp. of which 3 spp. are native to the New World and others in temperate and tropical Asia, Africa, and Australia), *Bourreria* (c. 30 spp. in the northern New World and 5 spp. perhaps congeneric in northeastern Africa), *Halgania* (17 spp. in Australia), *Coldenia* (1 sp. from the Old World now widely established as a weed), *Carmona* and *Rotula* (1 sp. each in the Old World), and *Cortesia*, *Rochefortia*, *Tiquilia*, and *Lepidocordia* (some 42 spp. from the New World). Herbs, shrubs and large trees. Flowers 5-merous except in *Coldenia*, where they are 4-merous. Ovary more or less entire with a slender terminal style divided once slightly to deeply with 2 stigmas, and producing 4 ovules in 2 or 4 locules. Fruit dry or thinly fleshy and breaking up into four 1-seeded or two 2-seeded nutlets or pyrenes.

SUBFAM. HELIOTROPIOIDEAE Traditionally with 2 main genera, *Heliotropium* with 250–300 spp. in tropical and warm-temperate regions, and *Tournefortia* with c. 100–150 spp., which are pantropical but mainly in the New World, with several smaller genera comprising c. 10 spp. Generic concepts are currently in a state of flux[v]. Mostly herbs to subshrubs, with some woody climbers which may be derived from herbaceous stock. Flowers 5-merous. Style terminal, short, and terminated by a broad stigmatic ring, with a conical extension that may be shallowly 2-lobed (the whole appearing like a mushroom or umbrella). Fruit dry or fleshy, dividing into two 2-seeded or four 1-seeded nutlets.

SUBFAM. BORAGINOIDEAE Includes more than half of the family and is developed mostly in warm-temperate areas. Herbs to occasionally subshrubs and distinguished from the other subfamilies by the style being gynobasic. Fruit breaks up into 4 dry, 1-seeded nutlets. Divided into a number of tribes, reflecting how the style is attached to the ovary lobes, there has been wide divergence of opinion on details. The following tribes have been variously recognized by recent authors:

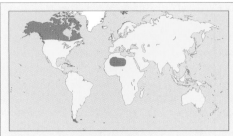

Genera c. 150   **Species** c. 2,700
Economic uses Dyestuffs, forage plants, garden ornamentals, tropical timber trees; some species serious weeds

Trigonotideae Gynobase elongate bearing tetrahedral nutlets attached basally; includes *Trigonotis*, *Mertensia*, etc.

Eritrichieae Gynobase elongate and ventrally keeled nutlets; includes *Eritrichium*, *Plagiobothrys*, *Lappula*, *Amsinckia* (but the latter similar in pollen to *Heliotropium*), etc.

Cynoglosseae Nutlets attached apically and basally divergent; includes *Cynoglossum*, *Lindelophia*, etc.

Trichodesmeae[vi] Nutlets attached over their whole adjacent side and anthers with contorted apical appendages; *Trichodesma* only.

Lithospermeae Gynobase flat, without cavities; includes *Moltkia*, *Onosma*, *Cerinthe*, *Arnebia*, *Lithospermum*, *Buglossoides* (the latter showing similarity to *Heliotropium* in stigma shape), etc.

Boragineae Gynobase flat, with 4 cavities; includes *Borago*, *Symphytum*, *Pulmonaria*, *Anchusa*, etc.

Echieae[vii] Flowers zygomorphic; includes *Echium*, *Lobostemon*.

Echiochileae[viii] Gynobase flat to elongate, submedial attachment of nutlets, stigmas 2; includes *Echiochilon*, *Ogastemma*, *Sericostoma* and *Antiphytum*.

Myosotideae Corolla lobes contorted instead of imbricate; *Myosotis* only.

The delimitation of the Boraginaceae is problematic. Early accounts tended to give the largely woody Cordioideae and Ehretioideae, which have a terminal style, family status, but the Heliotropioideae also have a terminal style and are mostly herbaceous. While some of the Old World *Cordia* species look rather unlike the herbaceous Boraginoideae, some of the South American species of the same genus have a Boraginaceous appearance. Inclusion of all of these 4 subfamilies in 1 family is now widely accepted. The Wellstedioideae have often been given family status, and with their 4-merous flowers and dehiscent capsules with 1–2 seeds, they do stand apart rather conspicuously, though *Coldenia* in Ehretioideae also has 4-merous flowers. With the transfer of *Codon* from Hydrophyllaceae to Boraginaceae, the position becomes even more disputatious, since this genus has 10- to 12-merous flowers, a broadly cylindrical corolla, and a dehiscent capsule with many seeds, unlike most Boraginaceae. Once this is included, a question mark hangs over the Hydrophyllaceae, which are

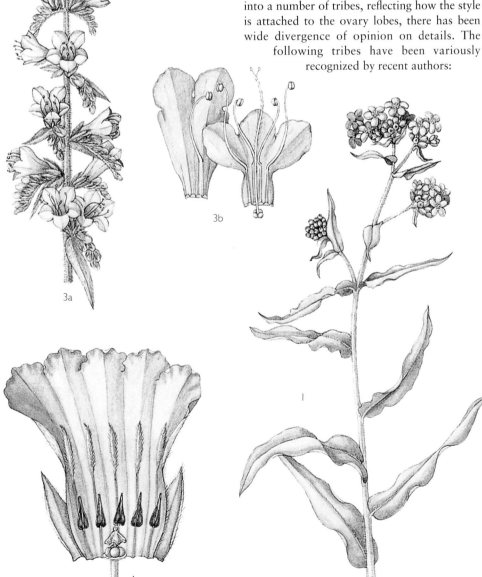

3a

3b

1

4

distinguished from Boraginaceae largely by their dehiscent many-seeded capsules. Both families share the rather distinctive and characteristic cincinnate inflorescence. But while *Codon* has a generally Boraginaceous superficial appearance (coarsely hairy herb with hairs typical of the Boraginaceae, as well as the striking spinulose hairs unmatched in the family), many Hydrophyllaceae have deeply divided leaves which is unlike Boraginaceae. Furthermore, the 10- to 12-merous flowers of *Codon* are reminiscent of the Lennoaceae, also clearly related to these families. The parasitic habit and highly modified morphology of the Lennoaceae argue for retaining family status for it but a good case can be made for treating Hydrophyllaceae as merely a subfamily of Boraginaceae, as is done by APG II. However, when too broad a family concept is adopted, the characterization of the family becomes more difficult and a more conservative approach is adopted here. One could take an even narrower view and give family status to both of the above capsule-bearing subfamilies, Codonoideae and Wellstedioideae, which also differ from much of the Boraginaceae in having a different number of floral parts, but since it seems that *Codon* and *Wellstedia* are phylogenetically basal to all the other subfamilies above, even a strongly cladistic approach could scarcely argue in principle with such a course of action. Then an order Boraginales would include 5 families.

Economic uses The tropical trees of the Cordioideae and Ehretioideae provide timber for construction and edible fruits, and some species are cultivated as decorative shrubs. Rhizomes of *Alkanna tinctoria* (Dyers' Alkanet), and *Lithospermum* and *Onosma* spp. have been used for red and purple dyes. Medicinal uses are uncommon, though *Lobostemon* is used in dressings for sores in South Africa and *Lithospermum ruderale* has been used locally as a contraceptive. *Borago* (Borage) is grown as a pot herb for its edible leaves, which are used in salads or flavoring drinks. *Symphytum* may also be eaten and is used as a forage plant. Several genera are valued by bee-keepers for honey production. Many herbaceous genera cultivated as garden ornamentals, especially *Myosotis* (Forget-Me-Not), *Symphytum* (Comfrey), *Pulmonaria*, *Lithospermum*, *Omphalodes*, *Onosma*, *Heliotropium,* and others. *Echium plantagineum* (Salvation Jane or Paterson's Curse) is an aggressive weed in Australia.   RKB

i Savolainen, V. *et al.* Phylogenetics of flowering plants based on combined analysis of plastid *atp*B and *rbc*L gene sequences. *Syst. Bot.* 49: 306–362 (2000).
ii Olmstead, R. G. *et al.* The phylogeny of the Asteridae *sensu lato* based on chloroplast *ndh*F gene sequences. *Molec. Phylogen. Evol.* 16: 96–112 (2000).
iii Retief, E. & van Wyk, A. E. Codonoideae, a new subfamily based on *Codon. Bothalia* 35: 78–80 (2005).
iv Miller, J. S. A revision of the New World species of *Ehretia* (Boraginaceae). *Ann. Missouri Bot. Gard.* 76: 1050–1076 (1989).
v Hilger, H. H. & Diane, N. A systematic analysis of Heliotropioideae (Boraginales) based on *trn*L and ITS1 sequence data. *Bot. Jahrb. Syst.* 125: 19–51 (2003).

vi Riedl, H. Boraginaceae. *Flora Iranica* 48: 1–281 (1967).
vii Retief, E. Boraginaceae. In: Leistner, O. A. (ed.), *Seed Plants of Southern Africa*: 178–183 (2000).
viii Långström, E. & Chase, M. W. Tribes of Boraginoideae (Boraginaceae) and placement of *Antiphytum, Echiochilon, Ogastemma* and *Sericostoma*: a phylogenetic analysis based on *atp*B plastid DNA sequence data. *Pl. Syst. Evol.* 224: 137–153 (2002).

BRASSICACEAE (CRUCIFERAE)
CABBAGE, MUSTARD, RAPESEED, TURNIP, AND WOAD

A cosmopolitan, mainly herbaceous family, often recognized by the 4-petalled cross-shaped (cruciform) flowers and the astringent mustard taste of the leaves.

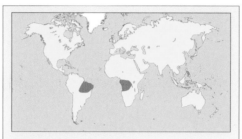

Genera c. 340 **Species** c. 3,350
Economic uses Edible and industrial oil, food, and fodder; stem, leaf, and inflorescence crops, mustard and wasabi, as well as many ornamentals. *Arabidopsis thaliana* was the first flowering plant for which the entire genome sequence was established

Distribution Members of the family are found throughout the world but are mainly concentrated in the north temperate region and more especially in the countries surrounding the Mediterranean basin and in southwestern and central Asia, where more genera occur than anywhere else in the world. The family is only sparingly represented in the southern hemisphere, and there are a few species in tropical regions, mostly at high altitude. Two tribes are confined to South Africa—the Chamireae containing a single species, *Chamira circaeoides*, and the Heliophileae comprising mainly the genus *Heliophila* with about 70 spp. mostly confined to the winter rainfall area around the Cape. Another tribe, the Pringleeae, contains as its sole member the species *Pringlea antiscorbutica*, the Kerguelen Island cabbage, found only on the remote islands of Kerguelen and Crozet in the southern hemisphere. A recently discovered genus, *Shangrilaia*, from high altitude, is endemic to Shangri-La[i].

Description Usually annual to perennial herbs, some geophytes (e.g., few *Cardamine*) rarely small shrubs (e.g., *Alyssum spinosum, Vella* spp.), tall shrubs to 2 m high or small trees (e.g., *Heliophila glauca* and *Farsetia somalensis*), rarely climbers (e.g., *Heliophila scandens, Lepidium scandens, Cremolobus peruvianus*[ii]), aquatics flowering above the water surface (e.g., *Subularia aquatica, Neobeckia aquatica,* and *Nasturtium officinale*), hummock plants

BRASSICACEAE. I *Iberis pinnata* leafy shoot and inflorescence with flowers having outer petals longer than the inner (x⅔).

(e.g., *Xerodraba pycnophylloides* from the Andes of Argentina), spiny plants (*Vella* spp., some *Alyssum, Lepidium, Moreira*), and whole-plant seed dispersal (*Anastatica hierochuntica*, Rose of Jericho). The leaves are usually alternate, simple and entire or variously lobed (but compound in *Nasturtium*, some *Cardamine* and *Yinshania*) sometimes strongly heterophyllous (e.g., *Lepidium perfoliatum*), and the mature leaves lack stipules. The hairs of the indumentum vary from simple to forked, many-branched, starlike or peltate, features that are useful for identification of genera and species. The inflorescence is usually a raceme or corymb, usually without bracts or bracteoles. The basic floral structure is highly characteristic and constant: 4 sepals, 4 cruciform petals, 6 stamens, (4 long and 2 short), and an ovary with 2 parietal placentas. There are, however, some exceptions to this plan. The flowers are usually bisexual, regular, and hypogynous. Sometimes the inner sepals are swollen and convex at the base and contain nectar secreted by the nectaries at the base of the stamens. The 4 petals are arranged in the form of a cross ("cruciform," hence Cruciferae), rarely absent as in *Pringlea* and some *Lepidium* and *Coronopus* spp., free, often clawed, imbricate or contorted. In a few genera, such as *Teesdalia* and *Iberis* (candy-tuft), the outer petals are radiate and larger than the inner petals. The stamens are typically 6 and tetradynamous, with 1 outer pair with short filaments, and 2 inner pairs, 1 posterior, 1 anterior, with long filaments. There may be only 4 stamens in some species of *Cardamine* and up to 16 in *Megacarpaea*. The filaments are sometimes winged or with toothlike appendages. The shape and disposition of the nectaries at the base of the stamens is variable and widely used in the classification of the family. The nectaries appear as swellings or little cushions. The ovary is superior, of 2 carpels, and is syncarpous, with 2 parietal placentas, usually with 2 locules through the formation of a membranous false septum or replum by the union of outgrowths from the placentas; sometimes the ovary is transversely plurilocular. The stigma is capitate to bilobed. As characteristic as the flower is the fruit, which is basically a 2-locular capsule with a false septum (replum), usually dehiscent, opening by 2 valves from below. When it is at least 3 times as long as wide it is called a siliqua, and when less than 3 times as long as wide, a silicula. The fruit may sometimes be indehiscent, breaking into single-seeded portions; rarely it is transversely articulate with dehiscent and indehiscent segments, sometimes breaking at maturity into single-seeded portions (a lomentum). The fruits range from linear-oblong to ovate to spherical; they may or may not be winged and stalked; the seeds may be in 1 or 2 rows. The range of variation shown in fruit types is vast and fruit characters have been relied upon extensively in the classification of the family at tribal, generic, and specific level. Examples of unusual or anomalous fruits are *Cakile,* which has siliquas that divide into two 1-seeded joints, the lower sterile and forming a thick stalk, the upper indehiscent, globose, and 1-seeded; *Lunaria* (Honesty) in which the silicula is flattened laterally to give a broad septum; or the silicula may be compressed anterior-posteriorly to give a narrow septum as in *Capsella* (Shepherd's Purse). *Geococcus pusillus* (Australia), *Morisia monanthos* (Sardinia and Corsica), *Cardamine chenopodiifolia* (South America), *Pegaeophyton* (the Himalayas), and *Lignarella* spp. (Japan) show geocarpy (the peduncles bend downward after flowering and bury the closed pod in the ground). The seeds of all species are non-endospermous, and the testa often contains mucilaginous cells of various types that swell up when wetted and produce a halo of mucilage. The ovules are campylotropous, the embryo being curved with the

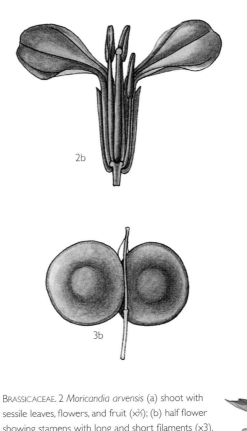

BRASSICACEAE. 2 *Moricandia arvensis* (a) shoot with sessile leaves, flowers, and fruit (×⅔); (b) half flower showing stamens with long and short filaments (×3). 3 *Biscutella didyma* var. *leiocarpa* (a) shoot with leaves, flowers, and fruit (×⅔); (b) fruit–a silicula (×4).

radicle in one half of the seed, and the cotyledons in the other. Great taxonomic importance has been attached to the shape of the embryo and the position of the radicle relative to the cotyledons.

Classification The delimitation and key diagnostic features of this family have remained stable for many decades[iii], and its placement in Capparales[iv], with Capparaceae and Cleomaceae, and the glucosinolate-containing families, is well established[v]. Morphological cladistic analyses have suggested Brassicaceae to be nested within a broad Capparaceae leading to a proposal to merge these 2 families[vi], but molecular evidence[vii] indicates 3 identifiable groups: Brassicaceae, Cleomaceae, and Capparaceae as an alternative scenario. There have been several proposed tribal classifications of the family of which one recognizing 19 tribes and 30 subtribes[viii] is the most widely used (although highly unnatural and splits similar genera into different tribes). Although numerous papers deal with aspects of the phylogeny of genera or generic groups in Brassicaceae, a comprehensive phylogeny of the family has yet to be constructed, despite the vast amount of data available from the *Arabidopsis* genome project[ix]. Generic limits are often based on a few characters and many closely allied morphological genera do not show correlation with phylogenetic analysis of sequence data[x].

Economic uses The Brassicaceae contains a considerable number and diversity of crop plants but not the diversity seen in the

Leguminosae or Poaceae. Although the crop species are mainly grown as food plants they do not form a substantial part of staple diets. Many crucifers are used as condiments or garnishes, such as mustard and cress, and many are collected from the wild rather than cultivated. Many cruciferous crops have been cultivated since ancient times. *Brassica oleracea*, the ancestral cabbage, was cultivated around 8,000 years ago in coastal areas of northern Europe. The first selection of sprouting broccoli was probably made in Greece and Italy in the pre-Christian era. All important cruciferous crops are propagated from seed; only minor crops such as watercress, horseradish, and sea kale are vegetatively propagated.

The seed crops can be divided into oils and mustard condiments; forage and fodder crops; and salads and vegetables for human consumption. Rapeseed oil now ranks second in world production behind soybeans and ahead of cottonseed oil[xi], providing 13% of the world's supply. The main oil crops are derived from *Brassica rapa* (Oilseed Rape) and *B. napus* (Oilseed Rape, Colza), while *B. campestris* and *B. juncea* are also important. Mustard is obtained from the ground seed of *Brassica juncea, B. nigra*, and *Sinapis alba*.

Animal feed is supplied by cruciferous crops in the form of silage—seed meal left over after oil extraction—forage crops grazed in the field, and stored root fodder, used as winter feeds. The characteristic glucosinolates produced by Cruciferae affect the economic use of many species. Glucosinolates are the precursors of the mustard oils, which are responsible for the pungency of most crucifers. Desirable

in the case of some crops, such as mustard, radish, and horseradish, these substances may also be responsible for toxic manifestations when used as animal feed or in human nutrition. As a result, the production of silage from crucifers is limited. Seed meal is obtained from species such as *Brassica napus*, especially races with a low glucosinolate content. Forage and fodder crops of crucifers are restricted mainly to countries such as Britain, the Netherlands, and New Zealand, all of which specialize in intensive small-scale farming of ruminants. The range of species used as forage crops includes *Brassica oleracea* (Kale, Cabbage), *B. campestris, B. napus* (Rape), and *Raphanus sativus* (Fodder Radish); fodder crops include those with swollen stems or root storage organs such as *B. oleracea* (Kohlrabi), *B. campestris* (Turnip), and *B. napus* (Swede).

Considerable proportions of the vegetable crop acreage in Eurasia are formed by cruciferous species. There are some curious geographical differences in what is cultivated, reflecting more national taste rather than the geographical origins of the crops. The Brussels sprout is very much a British crop, with British production equalling that of the rest of Europe and ten times that of the USA. Likewise, cauliflowers are a European crop. The most important species are *Brassica oleracea*, cultivars of which produce Kale, Brussels Sprouts, Kohlrabi, Cabbage, Broccoli, Calabrese, and Cauliflower, and *B. campestris*, producing Turnip, Chinese Cabbage, etc. A form of spiced and *Lactobacillus*-fermented cabbage, known as kimchi, is eaten at most meals in Korea and is considered an an integral part of the culture and cuisine[xii]. Kimchi may share a common origin with the European sauerkraut, which is widespread in northern Europe.

Ornamental genera include *Erysimum* (Wallflower), *Lunaria* (Honesty), *Iberis* (Candytuft), *Lobularia maritima* (Sweet Alysson), *Alyssum* spp. (Golden Alyssum), (*Matthiola* (stocks), *Hesperis* (Rocket), *Arabis* (Rock cress), *Draba, Aethionema*, and *Aubrieta*.

The short-lived annual *Arabidopsis thaliana* has now become one of the most studied flowering plants in molecular biology thanks to its small genome size and its rapid (6–10 week) life cycle. *Arabidopsis thaliana* was the first plant for which the entire genome was sequenced (completion announced in December 2000[xiii]).   VHH & AC

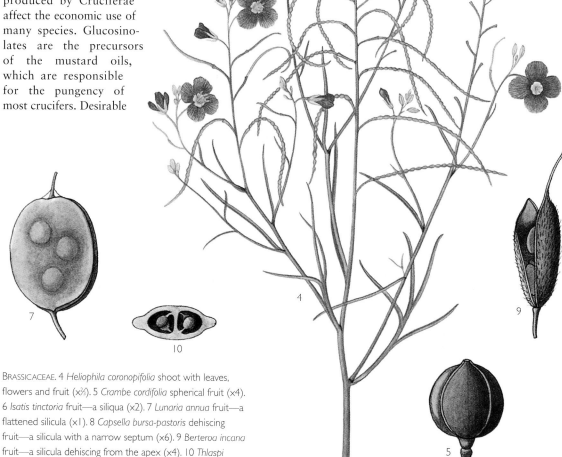

BRASSICACEAE. 4 *Heliophila coronopifolia* shoot with leaves, flowers and fruit (x⅔). 5 *Crambe cordifolia* spherical fruit (x4). 6 *Isatis tinctoria* fruit—a siliqua (x2). 7 *Lunaria annua* fruit—a flattened silicula (x1). 8 *Capsella bursa-pastoris* dehiscing fruit—a silicula with a narrow septum (x6). 9 *Berteroa incana* fruit—a silicula dehiscing from the apex (x4). 10 *Thlaspi arvense* cross section of 2-locular ovary showing false septum (x12). 11 *Erysimum cheiri* fruit—a siliqua (x1).

i Al-Shehbaz, I. A., Yue, J., & Sun, H. *Shangrilaia* (Brassicaceae), a new genus from China. *Novon* 14: 271–274 (2004).
ii Warwick S. I., Francis A., & La Fleche J. *Guide to Wild Germplasm of Brassica and Allied Crops (Tribe Brassiceae, Brassicaceae)*, 2nd edition (2000).
iii Warwick, S. I. & Sauder, C. A. Phylogeny of the tribe Brassiceae (Brassicaceae) based on chloroplast restriction site polymorphisms and nuclear ribosomal internal transcribed spacer and chloroplast *trn*L intron sequences. *Canadian J. Bot.* 83: 467–483 (2005).
iv Brummitt, R. K. *Vascular Plant Families and Genera.* Richmond, Royal Botanic Gardens, Kew (1992).
v Appel, O. & Al-Shehbaz, I. Cruciferae. Pp. 75–174. In: Kubitzki, K. & Bayer, C. (eds), *The Families and Genera of Vascular Plants. V. Flowering Plants. Dicotyledons: Malvales, Capparales and Non-betalain Caryophyllales.* Berlin, Springer-Verlag (2002).
vi Judd, W. S., Sanders, R. W., & Donoghue, M. J. Angiosperm family pairs: preliminary phylogenetic analyses. *Harvard Papers in Botany* 5: 1–51 (1994).
vii Hall, J. C., Sytsma, K. J., & Iltis, H. H. Phylogeny of Capparaceae and Brassicaceae based on chloroplast sequence data. *American J. Bot.* 89: 1826–1842 (2002).
viii Schulz, O. E. Cruciferae. Pp. 227–658. In: Engler, A. & Prantl, K. (eds), *Die natürlichen Pflanzenfamilien*, ed. 2, Vol. 17b. Leipzig, Germany, W. Engelmann (1936).
ix The Arabidopsis Information Resource (TAIR) http://www.arabidopsis.org/
x Mitchell-Olds, T., Al-Shehbaz, I. A., Koch, M. A., & Sharbel, T. F. Crucifer evolution in the post-genomic era. Pp. 119–137. In: Henry, R. J. (ed.), *Plant Diversity and Evolution: Genotypic and Phenotypic Variation in Higher Plants.* Wallingford, CABI International (2005).
xi Ash, M. & Dohlman, E. *Oil Crops Situation and Outlook Yearbook.* U.S. Department of Agriculture (2005).
xii http://www.kimchi.or.kr/eng/main.jsp
xiii Walbot, V. A green chapter in the book of life. *Nature* 408 (6814): 794–795 (2000).

BRETSCHNEIDERACEAE

A family comprising a single tree species, *Bretschneidera sinensis*, from southern China, Taiwan, northern Thailand, northern Vietnam, and probably Laos and Myanmar[i]. It grows 10–20 m tall and has gray-brown bark and alternate leaves. The leaves are impari-

Genera 1   Species 1
Economic uses None

pinnate, the leaflets are entire and narrowly elliptic to oblong, each with a petiolule; stipules are absent. The inflorescence is a loose, terminal raceme of several flowers on distinct pedicels. The flowers are hermaphrodite, zygomorphic, and 5-merous. The sepals are 5, fused to form a cup enclosing the petal bases. The petals are 5, free, white or pink, with dark pink veins, broadly spathulate, 1 petal forming a hood above the stamens. The stamens are 8 and free. The ovary is superior, of 3 fused carpels, each forming a locule containing 2 ovules, surmounted by an elongated style and 6-lobed stigma.

Bretschneidera contains glucosinolates and is most closely related to *Akania* (Akaniaceae) based on wood anatomy[ii], embryology[iii], fruit and seed anatomy[iv], and DNA sequence analysis[v]. It has consistently been placed in the Sapindales but was transferred to Brassicales,

along with other glucosinolate-containing genera, in APGII on the basis of molecular evidence supported by floral similarity[vi]. It is recognized in its own family or in Akaniaceae with *Akania*. No economic uses are recorded.   AC

i Lianli L. & Boufford, D. E. Bretschneideraceae. Pp. 197. In: Wu, Z.-Y. & Raven, P. (eds), *Flora of China*, Vol. 8, *Brassicaceae through Saxifragaceae.* Beijing, Science Press, and Missouri Botanical Garden Press (2001).
ii Carlquist, S. Wood anatomy of Akaniaceae and Bretschneideraceae: a case of near-identity and its systematic implications. *Syst. Bot.* 21: 607–616 (1996).
iii Tobe, H. & Peng, C. The embryology and taxonomic relationships of *Bretschneidera* (Bretschneideraceae). *Bot. J. Linn. Soc.* 103: 139–152 (1990).
iv Doweld, A. B. The systematic relevance of fruit and seed anatomy and morphology of *Akania* (Akaniaceae). *Bot. J. Linn. Soc.* 120: 379–389 (1996).
v Gadek, P. A. *et al.* Affinities of the Australian endemic Akaniaceae: new evidence from *rbc*L sequences. *Australian Syst. Bot.* 5: 717–724 (1992).
vi De Craene, L. P. R.,Yang, T. Y. A., Schols, P., & Smets, E. F. Floral anatomy and systematics of *Bretschneidera* (Bretschneideraceae). *Bot. J. Linn. Soc.* 139: 29–45 (2002).

BROWNLOWIACEAE

CHRISTIANA AND BROWNLOWIAS

A family of tropical trees.

Distribution Restricted to lowland, often coastal, evergreen forest in Southeast Asia and the Pacific Islands, from China and India to Australia and Fiji–Tahiti, with a secondary center in the Caribbean (most *Carpodiptera*) and South America (most *Christiana*), and with a single species of both *Carpodiptera* and *Christiana* in Africa. Many species are restricted to mangrove and freshwater swamps, or inundated forest.

Description Small to medium-sized trees 5–40 m tall, with fibrous bark and stellate-hairy indumentum, as in most core Malvales; peltate scales occur in several genera, e.g., most *Brownlowia*. The leaves are alternate, simple, and either entire or with the secondary nerves terminating in marginal, short, glandular teeth; venation palmate to pinnate; the leaves are peltate in several taxa of *Brownlowia*. The petioles are pulvinate at the ends, and the stipules usually caducous. The inflorescences are usually axillary panicles occurring in the distal 2–7 nodes. The bracts are usually accompanied by 2 bracteoles, and the pedicels are articulated near the base; no epicalyx is present. The flowers are actinomorphic and bisexual. The calyx is campanulate in the basal half, opening irregularly by 2–3 lobes and sometimes with a short stalk bearing the remaining flower parts. The petals are 5, free, often pink, slightly lobed, laciniate or emarginate at the apex, while the base is weakly stalked and lacks nectaries. The androecium consists of about 30 stamens grouped in 5 antipetalous bundles, the 2 anther thecae diverge at the base by about 90 degrees, while they touch at the apex and, after dehiscence, appear confluent. Staminodes 5, antisepalous,

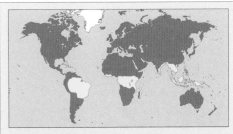

Genera 10   Species 81
Economic uses Hardwood timbers (small scale)

petaloid or ligulate (absent in *Christiana* and *Hainania*), inserted next to the ovary, each alternating with the bundles. The ovary is superior, globose or obovoid, and usually densely scaly or hairy, with 5 carpels that are more or less free (some *Christiana* and *Brownlowia*) to completely united (e.g., *Indagator*). The carpels usually have 2 axile to axile-pendulous, ± collateral, anatropous ovules. The styles are 5, linear, adhering to each other, giving the appearance of a single structure. The fruits are either apocarpous or syncarpous; either winged, papery, and indehiscent—often containing hairy seeds—or unwinged, leathery, or woody loculicidal capsules; often with tardy dehiscence; with smooth, marbled seeds. The embryos are usually foliose, endosperm sometimes being absent.

Classification Formerly included in a more broadly circumscribed Tiliaceae with what are here treated as the Sparrmanniaceae and Tiliaceae *sensu stricto*. It is 1 of 10 families in the core Malvales, differing by the campanulate, 2- to 3-lobed calyx (not divided to the base into 5 sepals); the 2 staminal thecae diverging by 90 degrees at the base while touching, and after dehiscence appearing confluent at the apex.

The traditional infrafamilial classification recognized three, probably artificial, tribes. Diplodisceae has syncarpous fruits, lacking wings (*Diplodiscus, Jarandersonia, Pityranthe, Hainania,* and *Indagator*). Berryeae has winged fruit (*Berrya, Carpodiptera,* and *Pentace*). Brownlowieae has apocarpous fruits lacking wings (*Brownlowia* and *Christiana*).

Economic uses *Berrya cordifolia* and *B. mollis* are occasionally cultivated in botanic gardens. Several species have valuable wood, such as the Indochinese *Brownlowia tabularis* (Béjaud) with durable timber that is used to make expensive furniture.   MRC

BRUNELLIACEAE

A monogeneric family of c. 60 spp. of evergreen trees from the neotropics, including the Antilles (*B. comocladiifolia*)[i], usually in humid montane regions from 600 to 3,800 m[ii], with many narrow endemics and a center of diversity in Colombia. The stems have prominent nodes. The leaves are opposite or whorled, simple or compound, then ternate, pinnate, or 1-foliate;

the lamina is coriaceous, with entire or dentate margins; the stipules are small, lateral, and caducous. The flowers are actinomorphic, hermaphrodite, or unisexual through abortion (plants dioecious or gynodioecious), in axillary inflorescences on new shoots. The flowers are 4- to 6-(–8)-merous, with a single perianth whorl of valvate sepals. The stamens are 8–10(–14), twice as many as sepals, usually in 2 whorls, the inner whorl sometimes branched, the filaments hairy. The ovary is apparently superior, of (2–)4–5(–8) free carpels, whose bases are submerged in the disk; carpels stylate with apical stigma, each bearing 2 ovules. The fruit is follicular, with red to yellow tomentum, often hispid.

The genus *Brunellia* is placed sister to *Cephalotus* (Cephalotaceae) in molecular and combined analyses[iii]. Most authors recommend recognition of the family but 1 paper[iv] states that the genus should be placed in Cunoniaceae. The family was placed in the Rosales (Cronquist 1988) but is now placed in the Oxalidales based on both molecular (see APG II) and morphological[iii] data. The leaves of a few species have traditional medicinal use[ii]. AC

Genera 1 **Species** c. 60
Economic uses Some medicinal value

[i] Kubitzki, K. Brunelliaceae. Pp. 26–28. In: Kubitzki, K. (ed.), *The Families and Genera of Vascular Plants. VI. Flowering Plants: Dicotyledons. Celastrales, Oxalidales, Rosales, Cornales, Ericales.* Berlin, Springer-Verlag (2004).
[ii] Zanoni, T. A. Brunelliaceae. P. 65. In: Smith, N. *et al.* (eds), *Flowering Plants of the Neotropics.* Princeton, Princeton University Press (2004).
[iii] Bradford, J. C. & Barnes, R. W. Phylogenetics and classification of Cunoniaceae (Oxalidales) using chloroplast DNA sequences and morphology. *Syst. Bot.* 26: 354–385 (2001).
[iv] Orozco, C. I. Pollen morphology of *Brunellia* (Brunelliaceae) and related taxa in Cunoniaceae. *Grana* 40: 245–255 (2001).

BRUNIACEAE

A small family of ericoid or microphyllous shrubs. Most species are native to the Western Cape of South Africa.

Distribution All of the genera occur in the Western Cape, some extending sparingly to the Eastern Cape, with 1 sp., *Raspalia trigyna*, recorded from extreme southern KwaZulu-Natal. All species are characteristic of the fynbos vegetation on the Table Mountain sandstones of the Eastern and Western Cape.

Genera 12 **Species** 77
Economic uses Dried flower arrangements

Description Subshrubs (often from an underground, woody rootstock) to shrubs or rarely small trees up to 3(4) m, often virgate or fastigiate. The leaves are small, hard-textured, ericoid to scalelike and imbricate, occasionally rhomboid, occasionally petiolate, with a black tip at least when young, usually with minute stipules. The inflorescence is a terminal dense spike to spherical capitulum with up to 400 flowers, the capitula themselves sometimes in panicles, or rarely solitary, with 1 or more bracts and usually 2 bracteoles per flower. The flowers are actinomorphic (rarely some lower flowers slightly irregular), bisexual, 5-merous, usually perigynous, often small (1 mm) and white but sometimes (*Thamnea*) up to 1.5 cm and brightly colored (*Audouinia*). The sepals are 5, sometimes vestigial. The petals are 5, free or in *Lonchostoma* their bases are connate with the stamens to form a basal tube. The stamens are 5. The ovary is semi-inferior to fully inferior or occasionally (some *Lonchostoma* and *Thamnea*) superior, of 1–3 fused carpels, with a simple bifid style, with 1–3(–5) locules or the septa incomplete, each locule with 1–16 ovules on a distal, parietal, or free central placenta. The fruit is indehiscent or a capsule dehiscing by 2–4 valves.

Classification The family has usually been placed in Rosales or near Saxifragaceae, but molecular evidence places it instead in the asterid line near the Dipsacales[i]. A close relationship to another South African family of ericoid shrubs, Grubbiaceae, suggested by Takhtajan (1997) and others, is not borne out by the molecular data. The shrubby ericoid habit shows notable parallel evolution with various other South African plants. The Stilbaceae are superficially similar to Bruniaceae but differ in many ways, including their whorled leaves and gamopetalous flowers. *Audouinia capitata* shows marked resemblance to *Erica margaritacea*, and *Staavia phylicoides* in Bruniaceae is so similar to a species of *Phylica* (Rhamnaceae) that dissection of the flowers may be needed to distinguish between them. Similarly, *Lonchostoma* spp. resemble the Thymelaeaceae, *Thamnea thesioides* resembles *Thesium* (Santalaceae), and *Nebelia* resembles *Lycopodium*. A recent generic revision[ii] has reduced the 12 genera to 9, but this has not been followed in a subsequent species checklist[iii]. The 12 genera seem well defined.

Economic uses Various species are sold locally in markets as dried flowers. RKB

[i] Savolainen, V. *et al.* Phylogeny of the eudicots: a nearly complete familial analysis based on *rbc*L sequences. *Kew Bull.* 55: 257–309 (2000).
[ii] Hall, A. V. Bruniaceae. In: Leistner, O. A. (ed.), *Seed plants of southern Africa: families and genera. Strelitzia* 10: 192–195 (2000).
[iii] Quint, M. Bruniaceae. In: Germishuizen, G. & Meyer, N. L. (eds), *Plants of southern Africa; an annotated checklist. Strelitzia* 14: 330–332 (2003).

BUDDLEJACEAE
BUDDLEJA FAMILY

A small family of shrubs, trees, and woody climbers in tropical and subtropical areas, distinguished from Scrophulariaceae mainly by the actinomorphic, 4-merous flowers.

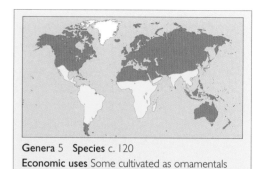

Genera 5 **Species** c. 120
Economic uses Some cultivated as ornamentals

Distribution *Buddleja* (including *Chilianthus* and *Nicodemia*) is widespread in tropical and South Africa, eastward from Arabia to India, China and Japan, and in southwestern USA to Brazil[i]. *Emorya* is in Texas and northern Mexico, *Gomphostigma* extends from Zaïre to South Africa, *Nuxia* is from Arabia to South Africa and the Mascarene Islands, and *Androya* is confined to southern Madagascar.

Description The family consists of mostly shrubs and small trees, occasionally big trees up to 30 m, sometimes woody climbers, sometimes (*Gomphostigma*, some *Buddleja*) perennial herbs from a woody base. The leaves are opposite, the pairs often connected by interpetiolar ridges, which may or may not represent stipules of other families, or sometimes (*Nuxia*) in whorls. The lamina is entire to dentate or (*Emorya*) deeply sinuate. Inflorescences are cymose or in *Gomphostigma* pseudoracemose by reduction of branches to single flowers. The flowers are often unisexual (plants often dioecious) in New World *Buddleja* species, but elsewhere the flowers are bisexual. The calyx consists of 4 fused and often slightly unequal sepals. The corolla is tubular, funnel-shaped to campanulate, with 4 short and more-or-less equal lobes. The stamens are 4, inserted on the corolla tube, the anthers exserted or not. The ovary is superior, bicarpellate, with 2 locules, and placentation is axile with many ovules. The fruit is a dry capsule or sometimes (*Buddleja* sect. *Nicodemia*) a fleshy berry.

Classification The family has often been included in Loganiaceae, but while the latter as defined here is close to the Gentianaceae, Buddlejaceae is more closely related to Scrophulariaceae[i]. It can be distinguished from the latter largely by its 4-merous, actinomorphic flowers. Circumscription of the family is still disputed[ii,iii]. *Polypremum* is here referred (with some doubts) to Tetrachondraceae, while *Sanango* and *Peltanthera* are referred to Gesneriaceae. However, *Androya*, which has been

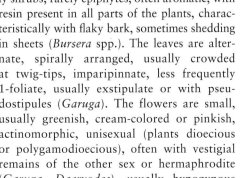

BUDDLEJACEAE. I *Buddleja crispa* (a) leafy shoot and terminal inflorescence (x⅔); (b) flower (x2⅔); (c) half flower (x3); (d) stamen (x8); (e) gynoecium (x5).

referred by some to Myoporaceae, and *Nuxia*, which has been referred to Stilbaceae, but which do not agree with those families in morphological characters, are here included in Buddlejaceae. *Nuxia* is strikingly different from Stilbaceae in lacking the ericoid appearance and not being confined to the Cape region. It seems, and looks, much better placed in Buddlejaceae with which it shares the unusual chromosome number of n=19. As noted in FGVP, it is similar

morphologically to *Buddleja* sect. *Chilianthus* from South Africa. It may be regarded as transitional toward Stilbaceae.

Economic uses A number of species of *Buddleja* are widely cultivated as ornamentals. The Chinese *B. davidii* is extremely well naturalized in Europe, sometimes to the extent of forming impenetrable thickets in native vegetation.   RKB

i Oxelman, B., Backlund, M., & Bremer, B. Relationship of the Buddlejaceae *s.l.* investigated using parsimony jack-knife and branch support analysis of chloroplast *ndh*F and *rbc*L sequence data. *Syst. Bot.* 24: 164–182 (1999).
ii Norman, E. M. Buddlejaceae. Pp. 66–67. In: Smith, N. *et al.* (eds), *Flowering Plants of the Neotropics.* Princeton, Princeton University Press (2004).
iii Oxelman, B., Kornhall, P., & Norman, E. M. Buddlejaceae. Pp. 39–44. In: Kadereit, J. W. (ed.), *Families and Genera of Vascular Plants. VII.* Berlin, Springer-Verlag (2004).

BURSERACEAE
FRANKINCENSE AND MYRRH

A tropical family of often aromatic, resinous trees and shrubs with usually imparipinnate leaves.

Distribution The family is found in the tropics of Africa, Asia, the Americas, West Indies, and Australia, extending also into subtropical and arid areas of southern North America. They grow in a wide range of habitats, ranging from montane and lowland rain forests (such as the dipterocarp forests of central and southern Malaysia) to dry forests and scrublands and desert regions. Trees and shrubs of the genus *Bursera* are particularly well represented and show great diversity in parts of Mexico (100 spp., most endemic) and, along with the genus *Protium* (c. 147 spp.), in South America such as the Amazon region. The largest genus, *Commiphora* (c.165 spp.)[i] is common in Africa and Madagascar. *Boswellia* comprises c. 17 spp., 6 spp. of which are endemic on Socotra (Yemen), 2 spp. to India, 1 sp. to Madagascar, and the remainder occur on mainland Africa and the Arabian peninsula, mostly in the Horn of Africa.

Description Large to small trees, less commonly shrubs, rarely epiphytes, often aromatic, with resin present in all parts of the plants, characteristically with flaky bark, sometimes shedding in sheets (*Bursera* spp.). The leaves are alternate, spirally arranged, usually crowded at twig-tips, imparipinnate, less frequently 1-foliate, usually exstipulate or with pseudostipules (*Garuga*). The flowers are small, usually greenish, cream-colored or pinkish, actinomorphic, unisexual (plants dioecious or polygamodioecious), often with vestigial remains of the other sex or hermaphrodite (*Garuga*, *Dacryodes*), usually hypogynous (perigynous in some *Bursera* and *Commiphora* spp.), in axillary or terminal to subterminal,

paniculate or racemose inflorescences. The sepals are 3–5(6), connate below and lobed, imbricate or valvate. The petals when present, are 3–5, free to partly connate, imbricate or valvate. The stamens are as many as or twice as many as petals, the filaments usually free or sometimes connate at the base (e.g., *Canarium* spp.); staminodes are present in female flowers. The ovary is superior with 2–5 carpels and locules; ovules 2 per locule; style simple. The fruits show considerable diversity[i]. They are basically compound, drupaceous or non-fleshy, capsular (pseudocapsules), with the pyrenes free or partly connate and dehiscent or remaining fused together. The dehiscent fruits in the tribes Protieae and Bursereae have as many valves as developed locules, and when fleshy, the exposed pyrenes are covered partially or almost completely by a pseudoaril. In the tribe Canarieae, the fruits are indehiscent and consist of a compound drupe with 2–3 locules, with a thin exocarp, an oily mesocarp, and a tough or bony endocarp.

Classification The family belongs in the Sapindales. Traditionally, it has been divided into 3 tribes—Bursereae, Protieae, and Canarieae—based partly on the degree of concrescence of the pyrenes. Recent molecular work[ii,iii] supports the monophyly of the latter 2, while the Bursereae appears not to be a natural grouping and is divisible into 2 clades that correspond to the subtribes Burserinae and Boswelliinae. The enigmatic monotypic Mexican genus *Beiselia* was only tentatively included in the Burseraceae[iv] but has recently been shown on molecular grounds to be a basal group and sister to the rest of the family.

Economic uses The aromatic resins produced by the Burseraceae have been used since ancient times for a variety of social, religious, cultural, and medicinal purposes. Frankincense is obtained from *Boswellia sacra* (Somaliland) and several other species. Myrrh, used in incense and perfumes, is obtained from *Commiphora myrryha* (True Myrrh) and other species. The wood of *Canarium littorale, Dacryodes costata, Santiria laevigata,* and *S. tomentosa* in Malaysia, and *Aucoumea* and *Canarium schweinfurthii* in Africa, is used for general building construction and carpentry. Varnish is obtained from several *Bursera* spp.,

Genera 18 **Species** c. 700
Economic uses Aromatic resins used to make incense, paints, perfumes, and soaps

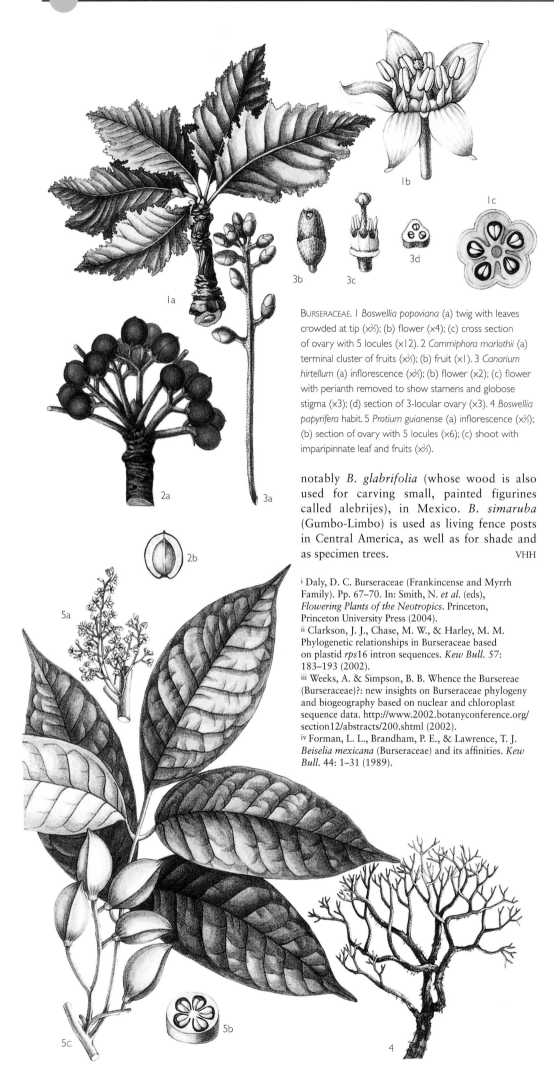

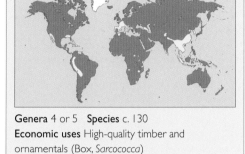

BURSERACEAE. 1 *Boswellia popoviana* (a) twig with leaves crowded at tip (x⅔); (b) flower (x4); (c) cross section of ovary with 5 locules (x12). 2 *Commiphora marlothii* (a) terminal cluster of fruits (x⅔); (b) fruit (x1). 3 *Canarium hirtellum* (a) inflorescence (x⅔); (b) flower (x2); (c) flower with perianth removed to show stamens and globose stigma (x3); (d) section of 3-locular ovary (x3). 4 *Boswellia papyrifera* habit. 5 *Protium guianense* (a) inflorescence (x⅔); (b) section of ovary with 5 locules (x6); (c) shoot with imparipinnate leaf and fruits (x⅔).

notably *B. glabrifolia* (whose wood is also used for carving small, painted figurines called alebrijes), in Mexico. *B. simaruba* (Gumbo-Limbo) is used as living fence posts in Central America, as well as for shade and as specimen trees. VHH

i Daly, D. C. Burseraceae (Frankincense and Myrrh Family). Pp. 67–70. In: Smith, N. *et al.* (eds), *Flowering Plants of the Neotropics*. Princeton, Princeton University Press (2004).
ii Clarkson, J. J., Chase, M. W., & Harley, M. M. Phylogenetic relationships in Burseraceae based on plastid *rps*16 intron sequences. *Kew Bull.* 57: 183–193 (2002).
iii Weeks, A. & Simpson, B. B. Whence the Bursereae (Burseraceae)?: new insights on Burseraceae phylogeny and biogeography based on nuclear and chloroplast sequence data. http://www.2002.botanyconference.org/section12/abstracts/200.shtml (2002).
iv Forman, L. L., Brandham, P. E., & Lawrence, T. J. *Beiselia mexicana* (Burseraceae) and its affinities. *Kew Bull.* 44: 1–31 (1989).

BUXACEAE
BOX FAMILY

A small, widespread family of evergreen shrubs or trees, including the common box.

Distribution The family is scattered throughout much of the world in temperate, subtropical, and tropical regions. Its main centers of species diversity are in eastern Asia (tribe Sarcococceae) and the Caribbean (especially Cuba), and Central America, where 50 spp. of *Buxus* are found (30 spp. endemic to Cuba). *Notobuxus* is confined to Africa.

Genera 4 or 5 **Species** c. 130
Economic uses High-quality timber and ornamentals (Box, *Sarcococca*)

Description Evergreen shrubs or medium-sized trees. The leaves are alternate (*Styloceras, Sarcococca, Pachysandra*) or opposite and decussate (*Buxus, Notobuxus*), simple, without stipules, often leathery, the margins entire or coarsely dentate (*Pachysandra*). The flowers are actinomorphic, unisexual (plants monoecious or, in most *Styloceras*, dioecious), subtended by bracts, in usually axillary, less commonly terminal, bracteate spikes, racemes, or clusters. In the staminate flowers, there are 4 weakly differentiated tepals (absent in *Styloceras*); the stamens are 4, usually with a central pistillode, 6–8 and without a pistillode (*Notobuxus*), or up to 45 (*Styloceras*), opposite the perianth segments. In the pistillate flowers, the tepals are 5–6 (*Buxus, Notobuxus*) or up to 20 (*Sarcococca, Pachysandra, Styloceras*), weakly differentiated; the ovary is superior, of 2 (*Sarcococca, Pachysandra, Styloceras*) or 3 (*Buxus, Notobuxus*) fused carpels, free above in the stylar region and spreading; the placentation is axile with 2 pendent, anatropous ovules per locule[i]. The fruit is a loculicidal capsule with the exocarp detaching and splitting septicidally, or drupaceous (*Sarcococca, Styloceras*). The seeds are black, shiny, usually with a caruncle or aril.

Classification The family comprises a number of groups of which only 2 are recognized here, in line with recent molecular studies that indicate that the family is divided into 2 clades[ii]: tribe Buxeae, comprising the genera *Buxus* (70 or more spp.) and *Notobuxus* (8 spp. sometimes included in *Buxus*), and the tribe Sarcococceae, comprising *Sarcococca* (11 spp.), *Styloceras* (4 spp.), and *Pachysandra* (4 spp.). *Pachsyandra* and *Styloceras* are sometimes recognized as distinct families (Pachysandraceae and Stylocerataceae). *Simmondsia* is sometimes included in the Buxaceae but is here treated as a separate family, Simmondsiaceae. The Buxaceae were placed in the Euphorbiales by Cronquist (1981) but are today considered, in light of recent work, including molecular analyses, to belong in their own order as sister to the Didymelaceae in a quite well-supported clade among families near the base of eudicots[iii,iv].

Economic uses *Buxus sempervirens* (Common Box) and numerous cultivars are widely grown as ornamentals, for hedging, and topiary. *Pachysandra procumbens* and *P. terminalis* are grown as ground cover. Species of *Sarcococca* have small, but strongly fragrant, flowers in winter. The wood of the Common Box and Cape Box (*B. macowanii*) is hard and dense and used in for carving, engraving, inlaying furniture, and for making rulers and instruments.　　VHH

i von Balthazar, M. & Endress, P. K. Reproductive structures and systematics of Buxaceae. *Bot. J. Linn. Soc.* 140: 193–228 (2002).
ii von Balthazar, M., Endress, P. K., & Qiu, Y.-L. Molecular phylogenetics of Buxaceae based on nuclear ITS and plastid *ndh*F sequences. *Int. J. Plant Sci.* 161: 785–792 (2000).
iii Qiu, Y.-L. *et al.* Phylogenetics of the Hamamelidae and their allies: Parsimony analyses of nucleotide sequences of the plastid gene *rbc*L. *Int. J. Plant Sci.* 159: 891–905 (1998).
iv Soltis, P. S., Soltis, D. E., & Chase, M. W. Angiosperm phylogeny inferred from 18S rDNA, *rbc*L, and *atp*B sequences. *Bot. J. Linn. Soc.* 133: 381–461 (2000).

BUXACEAE. **1** *Buxus sempervirens* (a) leafy shoot with axillary clusters of flowers (×⅔); (b) female flower surrounded by cluster of male flowers (×6); (c) male flower with 4 stamens having introrsely dehiscing anthers (×8); (d) female flower with 3 styles crowned by convolute stigmas (×8); (e) fruit—a capsule dehiscing by 3 valves (×2); (f) seed (×4). **2** *Pachysandra terminalis* (a) leafy shoot and inflorescence (×⅔); (b) half male flower (×3); (c) female flower (×6); (d) vertical section of ovary (×6); (e) fruit (×4); (f) seed (×4).

BYBLIDACEAE

A family comprising the single genus *Byblis* with 6 spp. of "carnivorous," shrubs or subshrubs to ephemeral herbs from northern and southwestern Australia and New Guinea. The leaves are alternate, simple, linear, entire, with circinnate vernation and covered with sticky glands that form a passive trapping mechanism. The flowers are hermaprodite, slightly zygomorphic, axillary, and solitary. The sepals are 5, connate at the base. The petals are 5, connate at the base, fimbriate apically, or entire. The stamens are 5, opposite the sepals, free, the anthers connivent and forming a cone. The ovary is superior, with 2 fused carpels, and a single elongate style; the placentation is axile with 10–50 ovules per locule. The fruit is a loculicidal capsule.

The affinities of this curious genus are still not entirely clear. A relationship with the Pittosporaceae has been suggested by several authors as discussed by Takhtajan (1997), who included Byblidaceae in the Byblidales next to the Pittosporales. The South African genus *Roridula* (here in the Roridulaceae, Ericales) has sometimes been included with *Byblis* in the Byblidaceae as by Cronquist (1981) and placed in the Rosales. An affinity with several families in the Scrophulariales, and in particular a sister relationship with Lentibulariaceae, has been

Genera 1 **Species** 6
Economic uses None

proposed[i,ii], but this is not supported in a recent study on the evolution of carnivory in the Lentibulariaceae and the Lamiales[iii]. *Byblis* is placed in the Lamiales in APG II. VHH

[i] Albert, V. A., Williams, S. E., & Chase, M. W. Carnivorous plants: phylogeny and structural evolution. *Science* 257 (5076): 1491–1495 (1992).
[ii] Conran, J. G. & Dowd, J. M. The phylogenetic relationships of *Byblis* and *Roridula* inferred from partial 18s ribosomal RNA sequences, *Pl. Syst. Evol.* 188: 73–86 (1993).
[iii] Müller, K. & Borsch, T., Legrende, L., Porembski, S., Theisen, I., & Barthlott, W. Evolution of carnivory in Lentibulariaceae and the Lamiales. *Plant Biol.* 6: 477–490 (2004).

BYTTNERIACEAE
CHOCOLATE FAMILY

A family of tropical shrubs, herbs, trees, and climbers that was formerly included in a more broadly circumscribed Sterculiaceae (q.v.).

Distribution The family is pantropical and subtropical, but mainly in Africa and South America, with a significant group of genera in southwestern Australia. They grow mainly in tropical lowland evergreen forest (most Byttnerieae) or in drier tropical scrub (most Hermannieae and Lasiopetaleae).

Description Mainly small shrubs, less usually small trees, lianas, and herbs, with highly fibrous bark and an indumentum mainly of stellate hairs, as in most core Malvales. The leaves are alternate, simple (palmately compound in *Herrania*), pinnately nerved, then often with the basal nerves most prominent ("triplinerved") or palmately nerved. The margins are usually serrate, the petiole ± pulvinate at base and apex, and the stipules moderately persistent, linear to cordate in shape. The inflorescence is axillary, rarely leaf-opposed, or terminal, usually a bracteate thyrse, but sometimes reduced to a single flower (e.g., some *Hermannia*). In *Theobroma* and *Herrania,* the flowers usually arise in fascicles from the trunk (cauliflorous). The flowers are actinomorphic or slightly zygomorphic and bisexual. An epicalyx is usually absent, while the calyx is valvate and more or less deeply divided into 5 sepals, less usually splitting into 2–3 lobes. The 5 petals are free, non-nectariferous, usually cupped at the base, often then with an apical, straplike appendage, or usually flat, with a weakly developed claw having involute margins (Hermannieae), or usually absent or highly reduced (Lasiopetaleae). The stamens are generally 5, in a single whorl, alternating with 5 often triangular staminodes, all united in a short tube. In several genera, the simple stamens are replaced by clusters of 2, or even 3, while in others (Hermannieae), staminodes are absent, and the staminal filaments are nearly free to the base. The anthers are usually basally attached, longitudinally dehiscent, extrorse, dithecal (trithecal in e.g., *Megatritheca* and *Ayenia*), and often sessile on the staminal tube where this is

well developed (*Byttnerieae*). The ovary is superior, syncarpous, of 5 carpels (1 in *Waltheria*), each with axile placentation and usually numerous anatropous ovules, reduced to as few as 2 in several genera. The fruits are loculicidally dehiscent, rarely indehiscent (e.g., *Theobroma* and *Herrania*); in several genera, notably *Byttneria*, mericarps are produced by septicidal dehiscence. The seeds are numerous in each locule or may be reduced to 1 (e.g., *Byttneria*) and frequently strophiolate.

Classification Formerly included in Sterculiaceae with the genera here included in Pentapetaceae, Helicteraceae and Sterculiaceae s.s., Byttneriaceae is 1 of 10 families in the core Malvales (see discussion under Malvaceae). Molecular investigations[i,ii,iii,iv] have placed Byttneriaceae as most closely related to the Sparrmanniaceae[ii]. Byttneriaceae is easily distinguished from the Sparrmanniaceae by the androecium being confined to a single whorl, with 5 (–15) fertile, usually basifixed, anthers, generally alternating with staminodes, the filaments often united into a tube. In Sparmanniaceae, fertile anthers are usually numerous, and the filaments are usually free to the base; where staminodes occur they are undifferentiated, merely lacking anthers, and on the outside of the staminal mass.

The Byttneriaceae have been divided into 3 tribes since the nineteenth century, and this is broadly supported by recent molecular work[v].

Byttnerieae (12 genera) includes the *Theobroma* alliance, concentrated in tropical Africa and South America, with outliers in Southeast Asia, mostly found in lowland evergreen or semideciduous forest, while the herbaceous *Glossostemon* is confined to arid habitat from Saudi Arabia to Iran. Usually with a well-defined staminal tube and petals with a constriction, and then a more or less elaborate apical appendage above the concave base.

Lasiopetaleae (9 genera) is confined to Australia, apart from genera with outlying species in Madagascar (*Thomasia*, *Keraudrenia*, and *Rulingia*) and *Commersonia*, which extends into Southeast Asia and the Pacific. Usually

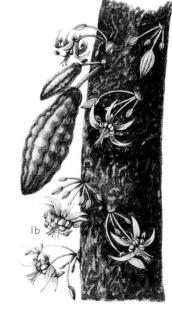

BYTTNERIACEAE. | *Theobroma cacao* (a) unripe fruit cut away to show seeds—the cocoa beans of commerce (x⅔); (b) flowers and young fruits, which form on old wood (x⅔).

characterized by the concave petals being reduced or absent. The last 2 genera are included by molecular evidence, having previously been placed in Byttnerieae.

Hermannieae (4 genera) is concentrated in seasonally arid scrub land in South America and Africa, except for the Australian monotypic *Dicarpidium*. The tribe is unusual in lacking both concave petals and staminodes, the 5 fertile stamens being free almost to the base.

Economic uses Locally used as stock fodder (some *Hermannia*), fiber (some *Leptonychia*), or medicinal (*Glossostemon* and some *Leptonychia*). However, the fermented seeds of *Theobroma cacao* have a vast international trade being in great demand for confectionery chocolate and cosmetic markets, mostly in the northern temperate zone. For this reason, trees are cultivated throughout the tropics, especially in cleared rain forest areas of West Africa. One or 2 spp. are pantropical weeds. MRC

[i] Baum, D. A. *et al.* Phylogenetic relationships of Malvatheca (Bombacoideae and Malvoideae; Malvaceae *sensu lato*) as inferred from plastid DNA sequences. *American J. Bot.* 91: 1863–1871 (2004).
[ii] Bayer, C. *et al.* Support for an expanded family concept of Malvaceae within a recircumscribed order Malvales: a combined analysis of plastid *atp*B and *rbc*L DNA sequences. *Bot. J. Linn. Soc.* 129: 267–303 (1999).
[iii] Alverson, W. S., Karol, K. G., Baum, D. A., Chase, M. W., Swensen, S. M., McCourt, R., & Systma, K. J. Circumscription of the Malvales and relationships to other Rosidae: Evidence from *rbc*L sequence data. *American J. Bot.* 85: 876–887 (1998).
[iv] Alverson, W. S. *et al.* Phylogeny of core Malvales: Evidence from *ndh*F sequence data. *American J. Bot.* 86: 1474–1486 (1999).
[v] Whitlock, B. A., Bayer, C., & Baum, D. A. Phylogenetic relationships and floral evolution of Byttnerioideae ("Sterculiaceae" or Malvaceae *s.l.*) based on sequences of the chloroplast gene, *ndh*F. *Syst. Bot.* 26: 420–437 (2001).

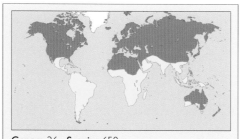

Genera 26 Species 650
Economic uses. Chocolate, food, fibers, medicines

CABOMBACEAE

A small cosmopolitan family of waterlilies.

Distribution The family is almost cosmopolitan but absent from Europe, where it is only naturalized. *Cabomba* has a wide distribution in both temperate and tropical regions of the western hemisphere. *Brasenia* is monotypic and *B. schreberi* has a worldwide distribution, occurring in East Asia, Australia, Africa, and North and Central America, and the Antilles, mainly in temperate regions except Europe and in tropical highlands.

Description Perennial, rhizomatous aquatic herbs. Leaves submerged or floating, alternate (floating) or opposite (submerged), heterophyllous (*Cabomba*) or not (*Brasenia*). The floating leaves are peltate, with the lamina elliptical to oval-elliptical, entire, and heavily covered with mucilage on the under surface. The underwater leaves (*Cabomba*) are finely subdivided, feathery, and fanlike, with capillary ultimate segments, coated with mucilage. The flowers are actinomorphic, hermaphrodite, axillary, solitary, usually 3-merous, white or yellow to violet or purplish. The sepals are 3 or rarely 2 (*Cabomba aquatica*), free, petaloid (*Cambomba*), or narrowly oblong to ovate and not

Genera 2 **Species** 6
Economic uses Aquarium plants

petaloid (*Brasenia)*. The petals are 3, free, clawed at the base, and with nectariferous auricles (*Cabomba*). The stamens are 3–6 (*Cabomba*), or 18–36 or more (*Brasenia*). The ovary is superior, with 1–4 (*Cabomba*) or 4–18 (*Brasenia*) free carpels. The stigmas are capitate or linear-decurrent. Ovules 1–2 (*Brasenia*) or 1–5 (*Cabomba*) in each carpel. The fruits are achene-like (*Brasenia*) or follicle-like (*Cabomba*) and indehiscent. Seeds 1–3, globose or ovoid, with scant endosperm and abundant perisperm.

Classification The 2 genera are quite distinct and are sometimes included in separate subfamilies—*Cabomba* in the Cabomboideae and *Brasenia* in the Hydropeltidoideae—with Takhtajan feeling compelled by the differences to recognize the latter as a distinct family the Hydropeltidaceae. However, molecular evidence suggests the family is monophyletic and that it is a sister group to the Nymphaeaceae[i]. It is a matter of judgment whether to keep the Cabombaceae and

Nymphaeaceae as separate families: Les *et al*[ii] support their distinction on the basis of both molecular and morphological data sets, while APGII leaves it as an open question.

Economic uses Both *Cabomba* and *Brasenia* are grown as aquarium plants. *Brasenia* shoots are eaten in soups or fried in China and Japan. Both genera may become invasive and can clog waterways and canals.   VHH

[i] Les, D. H. *et al*. Phylogeny, classification and floral evolution of water lilies (Nymphaeaceae; Nymphaeales): A synthesis of non-molecular, *rbc*L, *mat*K and 18S rDNA data. *Syst. Bot.* 24: 28–46 (1999).
[ii] Aoki, S., Uehara, K., Imafuku, M., Hasebe, M., & Ito, M. Phylogeny and divergence of basal angiosperms inferred from APETALA3- and PISTILLATA-like MADS-box genes. *J. Plant Res.* 117: 229–44 (2004).

CACTACEAE
CACTUS FAMILY

A large family of perennial, xerophytic trees, shrubs, climbers, and epiphytes, recognized by their more or less succulent and (except *Pereskia*) leafless appearance and many spines.

Distribution Cacti are mainly plants of semideserts of the warmer parts of the Americas, but *Rhipsalis baccifera* is native to Africa, Madagascar, and Sri Lanka; *Opuntia* is naturalized in Australia, Africa, the Mediterranean, and elsewhere. The characteristic habitats of cacti experience erratic rainfalls, with long drought periods in between, but night dews may be heavy when the temperature falls.

Description Perennial succulent trees, shrubs, climbers, and geophytes, mostly bearing spines. The spines, branches, and flowers arise from special sunken cushions or areoles, which may be regarded as condensed lateral branches— these are either set singly on tubercles or serially along raised ribs. Tufts of short barbed hairs (glochids) may also be present in the areoles. Photosynthesis is undertaken by the young green shoots, but with age these become corky and in the arborescent species develop into a hard, woody, unarmed trunk as in conventional trees. The vascular system forms a hollow, cylindrical, reticulated skeleton and lacks true vessels. The roots are typically superficial and in the larger species are widely spreading and adapted for rapid absorption near the soil surface. The flowers are solitary and sessile (*Pereskia* excepted), bisexual (with rare exceptions), and regular to oblique-limbed. Color range is from red and purple through various

shades of orange and yellow to white; blue is lacking. The stamens, petals, sepals, and bracts are numerous and spirally arranged, the last 3 in transitional series without sharp boundaries between them. The ovary is inferior, borne on an areole, and is commonly covered in hairs, bristles, or spines, consisting of 2 to numerous carpels and forming 1 locule with numerous ovules on parietal placentas. The style is simple. In *Opuntia*, the detached "fruit" (pseudocarp) grows roots and shoots, which form a new plant. The fruit is a berry, which is typically juicy, but may be dry and leathery, splitting open to release the seeds in various ways. The seeds have a straight to curved embryo and little or no endosperm.

Classification The Cactaceae is of especial interest to botanists for its survival under adverse conditions and drought and for its parallel

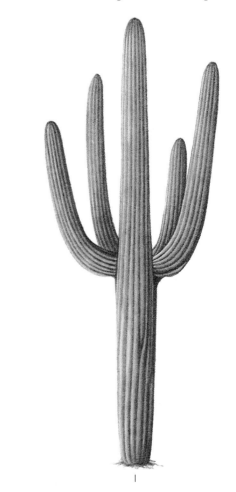

CACTACEAE. **1** *Carnegiea gigantea* showing the characteristic many-branched candelabra habit and ribbed stems (x⅟₆₀). **2** *Ariocarpus fissuratus*, a dwarf cactus with a many-ribbed, nonjointed stem bearing flowers on new areoles (x⅔).

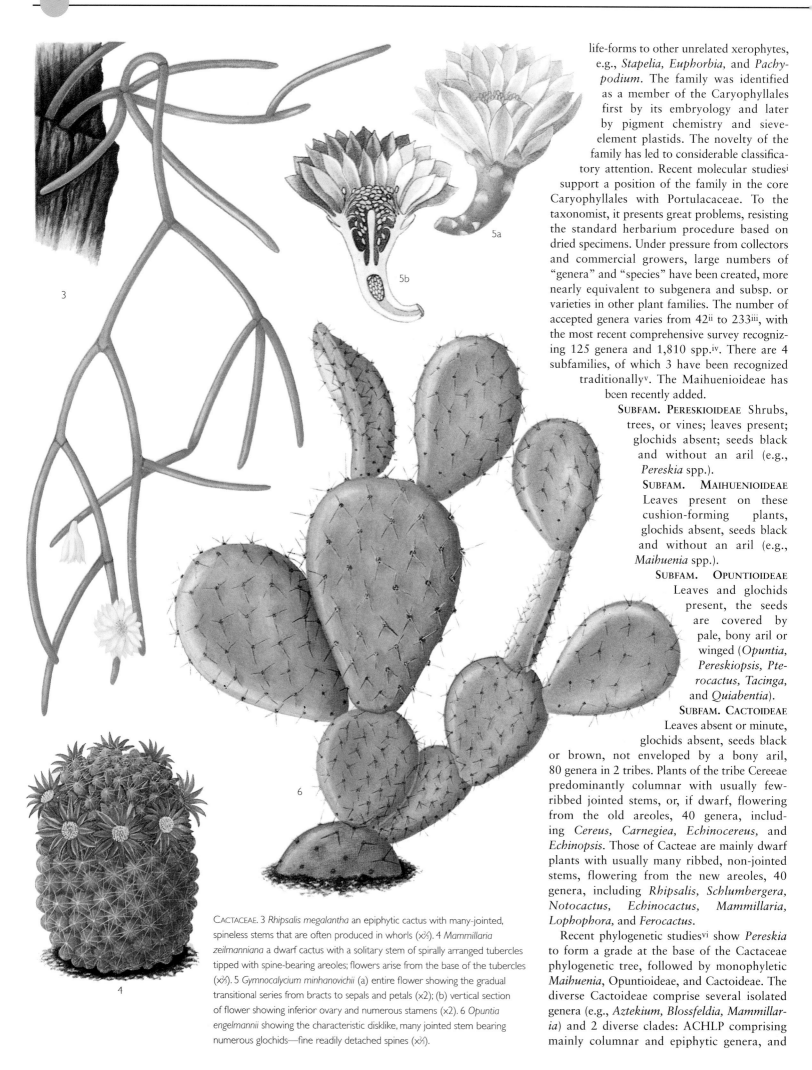

life-forms to other unrelated xerophytes, e.g., *Stapelia*, *Euphorbia*, and *Pachypodium*. The family was identified as a member of the Caryophyllales first by its embryology and later by pigment chemistry and sieve-element plastids. The novelty of the family has led to considerable classificatory attention. Recent molecular studies[i] support a position of the family in the core Caryophyllales with Portulacaceae. To the taxonomist, it presents great problems, resisting the standard herbarium procedure based on dried specimens. Under pressure from collectors and commercial growers, large numbers of "genera" and "species" have been created, more nearly equivalent to subgenera and subsp. or varieties in other plant families. The number of accepted genera varies from 42[ii] to 233[iii], with the most recent comprehensive survey recognizing 125 genera and 1,810 spp.[iv]. There are 4 subfamilies, of which 3 have been recognized traditionally[v]. The Maihuenioideae has been recently added.

SUBFAM. PERESKIOIDEAE Shrubs, trees, or vines; leaves present; glochids absent; seeds black and without an aril (e.g., *Pereskia* spp.).

SUBFAM. MAIHUENIOIDEAE Leaves present on these cushion-forming plants, glochids absent, seeds black and without an aril (e.g., *Maihuenia* spp.).

SUBFAM. OPUNTIOIDEAE Leaves and glochids present, the seeds are covered by pale, bony aril or winged (*Opuntia*, *Pereskiopsis*, *Pterocactus*, *Tacinga*, and *Quiabentia*).

SUBFAM. CACTOIDEAE Leaves absent or minute, glochids absent, seeds black or brown, not enveloped by a bony aril, 80 genera in 2 tribes. Plants of the tribe Cereeae predominantly columnar with usually few-ribbed jointed stems, or, if dwarf, flowering from the old areoles, 40 genera, including *Cereus*, *Carnegiea*, *Echinocereus*, and *Echinopsis*. Those of Cacteae are mainly dwarf plants with usually many ribbed, non-jointed stems, flowering from the new areoles, 40 genera, including *Rhipsalis*, *Schlumbergera*, *Notocactus*, *Echinocactus*, *Mammillaria*, *Lophophora*, and *Ferocactus*.

Recent phylogenetic studies[vi] show *Pereskia* to form a grade at the base of the Cactaceae phylogenetic tree, followed by monophyletic *Maihuenia*, Opuntioideae, and Cactoideae. The diverse Cactoideae comprise several isolated genera (e.g., *Aztekium*, *Blossfeldia*, *Mammillaria*) and 2 diverse clades: ACHLP comprising mainly columnar and epiphytic genera, and

CACTACEAE. 3 *Rhipsalis megalantha* an epiphytic cactus with many-jointed, spineless stems that are often produced in whorls (x⅔). 4 *Mammillaria zeilmanniana* a dwarf cactus with a solitary stem of spirally arranged tubercles tipped with spine-bearing areoles; flowers arise from the base of the tubercles (x⅔). 5 *Gymnocalycium minhanovichii* (a) entire flower showing the gradual transitional series from bracts to sepals and petals (x2); (b) vertical section of flower showing inferior ovary and numerous stamens (x2). 6 *Opuntia engelmannii* showing the characteristic disklike, many jointed stem bearing numerous glochids—fine readily detached spines (x⅔).

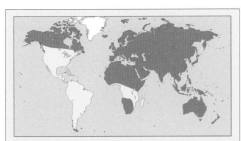

Genera 90–110 **Species** c. 2,000
Economic uses Garden and house ornamentals;
some edible fruits

Genera 1 **Species** c. 75
Economic uses Potential as food plant and
in water pollution estimation

i Cuénoud, P. *et al*. Molecular phylogenetics of
Caryophyllales based on nuclear 18S and plastid *rbc*L,
*atp*B and *mat*K DNA sequences. *American J. Bot.* 89:
132–144 (2002).
ii Mottram, R. *A contribution to a new classification
of the cactus family and index of suprageneric and
supraspecific taxa*. Thirsk, Britain, Whitestone Gardens
(1990).
iii Bakeberg, C. *Das Kakteenlexicon. Enumeratio
diagnostica Cactacearum*. VEB. Jena, Gustav Fischer
Verlag (1966).
iv Anderson, E. F. *The Cactus Family*. Portland, Timber
Press (2001).
v Barthlott, W. & Hunt, D. R. Cactaceae. Pp. 161–197.
In: Kubitzki, K., Rohwer, J. G., & Bittrich, V. (eds), *The
Families and Genera of Vascular Plants. II. Flowering
Plants. Dicotyledons: Magnoliid, Hamamelid and
Caryophyllid Families*. Berlin, Springer-Verlag (1993).
vi Nyffeler, R. Phylogenetic relationships in the cactus
family (Cactaceae) based on evidence from *trn*K/*mat*K
and *trn*L-F sequences. *American J. Bot.* 89: 312–326
(2002).

RNBCT, comprising Rhipsalideae, Notocacteae, and Browiningiae/Cereeae/Trichocereeae. The tribes Cereeae and Cacteae have no support in phylogenetic studies. Sampling of species is still sparse and much work is still needed to resolve a stable pattern of relationships.

Economic uses Apart from their wide appeal to specialist growers and collectors of the unusual, cacti have relatively few uses. The fleshy fruits of many are collected locally and eaten raw or made into jams or syrups. Some are used for hedging, while those with woody skeletons are used for rustic furniture and trinkets. Opuntias (prickly pears) are grown commercially in parts of Mexico and California for their large juicy fruits and are widely consumed in Mediterranean countries where they have been introduced. *Hylocereus undatus* (Dragon Fruit) is now widely seen in supermarkets in Europe. Some species, such as "peyote" (*Lophophora williamsii*), contain hallucinogenic substances and are used locally by Native Americans. Probably all genera are represented in cultivation, collectors being undeterred by the difficulties in growing them. The most popular genera among collectors are those that remain dwarf and combine attractive spine colors and rib formations with freedom of flowering e.g., *Rebutia, Lobivia, Echinopsis, Mammillaria, Notocactus, Parodia,* and *Neoporteria. Astrophytum* is valued for its prominent ribs and cottony white tufts and *Leuchtenbergia* for its extraordinarily long tubercles. More for cacti connoisseurs are the curiously squat, tuberculate, slow-growing species of *Ariocarpus, Pelecyphora,* and *Strombocactus. Melocactus* ("Turk's Cap Cactus"), one of the first cacti to reach Europe, is unique for the large, furry inflorescence terminating the short, stumpy axis. Even more widely grown are the epiphytes. The large-flowered epicacti are products of a long line of intergeneric crossings paralleled only in the orchids and are the only group grown primarily for flowers. Widespread collection has led all Cactaceae to be included on CITES appendix 1 or 2, but recently subfamilies Pereskioideae and Maihuenioideae have been proposed for exclusion from CITES on the basis that they have little trade value. The word *cactus* is commonly misapplied to a wide range of spiny or fleshy plants unrelated to the Cactaceae. [GDR] AC

CALLITRICHACEAE
WATER-STARWORT FAMILY

The family comprises 1 genus of submerged or amphibious aquatic, rarely terrestrial, herbs.

Distribution *Callitriche* is almost cosmopolitan, mostly in the temperate zones of both hemispheres, sporadic and largely montane in the tropics.

Description Delicate, annual or perennial herbs, often submerged aquatics, with elongate stems up to 1 m, sometimes amphibious when the aerial parts are more contracted, or terrestrial on damp ground with stems shorter, and erect or prostrate and rooting at the nodes. Leaves are decussate, often rosette-forming at the tips of floating stems, usually linear when submerged and often with a forked apex, but linear, elliptic, oblong, or spathulate when floating or aerial, with entire margins except for 1 sp. in Australia. Stipules are absent. The flowers are unisexual (plants monoecious), axillary, usually solitary or 1 male and 1 female (rarely up to 3 female) in the same leaf axil, in amphibious and some terrestrial species usually with 2 whitish bracts subtending the flower. There are no sepals or petals. The male flower consists of 1 stamen with a slender filament and an anther with 2 locules opening lengthwise, the slits confluent at the top. The female flower consists of a single naked ovary of 2 fused carpels, each longitudinally divided into 2 locules, each locule bearing a single pendulous, anatropous ovule, the whole structure 4-lobed, with each lobe winged or keeled, and with 2 elongate styles. At maturity, the fruit splits into four 1-seeded nutlets. The seeds are green, brownish, or black and have fleshy endosperms.

Callitriche has 2 quite different pollination syndromes, aerial plants being anemophilous while submerged plants are water-pollinated [i,ii]. Usually the pollen from flowers of submerged species is inaperturate while that from aerial flowers has 3–4 poorly developed colpi.

Classification The fruit divided into 4 nutlets has sometimes been taken to suggest that the family is an aquatic development from the Lamiaceae, but there the septum is median, while in the Callitrichaceae it is transverse. Molecular evidence shows the family in fact nesting in Scrophulariaceae, close to the Hippuridaceae, but the reduced flowers with no perianth, with the male flowers reduced to a single stamen and female flowers producing four 1-seeded nutlets, argue strongly for retention of Callitrichaceae as a separate family. Some classifications have divided the genus into

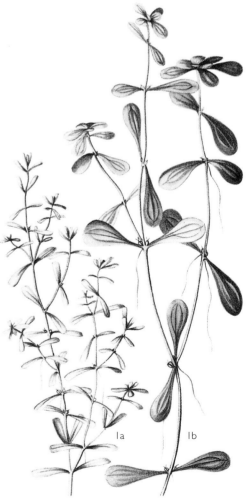

CALLITRICHACEAE. 1 *Callitriche heterophylla* (a) habit, showing narrow submerged leaves (×1⅓); (b) habit, with spatula-shaped aerial leaves (×1⅓).

3 sections[iii,iv] based largely on whether the plants are consistently submerged aquatics (sect. *Pseudocallitriche*) or amphibious (sect. *Callitriche*) or obligately terrestrial (sect. *Microcallitriche*). Although recent morphological and molecular analysis[v] suggests that this may be a good indication of the phylogenetic affinities of northern temperate species, it is not so when a wider range of species is examined.

Economic uses Early European explorers in the Antarctic ate *C. antarctica* as an antiscorbutic to avoid scurvy. Recent trials have been held in Australia on the use of this species as a commercial bulk food for use in times of famine. In Europe, some species have been used as a guide to water pollution.   RVL & RKB

[i] Philbrick, C. T. & Anderson, G. J. Pollination biology in the Callitrichaceae. *Syst. Bot.* 17: 282–292 (1992).
[ii] Philbrick, C. T. & Osborn, J. M. Exine reduction in underwater flowering *Callitriche* (Callitrichaceae): implications for the evolution of hypohydrophily. *Rhodora* 96: 370–381 (1994).
[iii] Fassett, N. C. *Callitriche* in the New World. *Rhodora* 53: 137–155 (1951).
[iv] Schotsman, H. D. *Les Callitriches: Espèces de France et Taxa Nouveaux d'Europe.* Paul Lechevalier (1967).
[v] Philbrick, C. T. & Les, D. H. Phylogenetic studies in *Callitriche*: implications for interpretation of ecological, karyological and pollination system evolution. *Aquatic Bot.* 68: 123–141 (2000).

CALLITRICHACEAE. 2 *Callitriche deflexa* habit showing solitary female flowers on long stalks (x2⅓). 3 *C. palustris* (a) habit showing creeping submerged and erect aerial shoots (x⅔); (b) male flower (left) comprising a single stamen and female flower (right) comprising an ovary with 2 styles (x20); (c) male and female flower in a single leaf axil (x20); (d) cross section of ovary (x34); (e) fruit with winged lobes (x23).

CALYCANTHACEAE
CAROLINA ALLSPICE, STRAWBERRY-SHRUB

A small family of aromatic, deciduous shrubs or small, evergreen trees.

Distribution *Sinocalycanthus* (1 sp.) and *Chimonanthus* (5 spp.) are endemic to eastern Asia, *Calycanthus* (4 spp.) is distributed in eastern and western North America, and *Idiospermum* (1 sp.) occurs in tropical rain forests of northern Queensland in Australia.

Description Deciduous or evergreen small trees or shrubs with aromatic bark. The leaves are opposite, simple, entire, without stipules. The often fragrant flowers are hermaphrodite, actinomorphic, and solitary. The perianth consists of 10–40 fleshy tepals, which are spirally inserted on outside of hypanthium, the outer bractlike and the remainder petaloid and often showy, secreting nectar. The stamens are 5–30, spirally inserted at top of the cuplike hypanthium; the staminodes are 10–25 on the inner surface of hypanthium. The ovary is superior, with 5–35 or 1–2 (3) (*Idiospermum*) carpels inserted on the inside of the receptacle, the carpels 1-locular with 1 or 2 ovules per style, elongate, filiform; stigma dry, decurrent. The fruits are achenes, 1–35, attached to the enlarged, persistent hypanthium (pseudocarp). The seeds are without endosperm but with a large embryo and 2 spirally twisted cotyledons or 3–4, massive and peltate (*Idiospermum*).

Genera 4   Species c. 11
Economic uses Ornamentals, local medicines, spice (Carolina Allspice)

Classification *Calycanthus* are deciduous shrubs with reddish or purple flowers; *Chimonanthus* are deciduous or evergreen shrubs with small, yellowish flowers; *Sinocalycanthus* is a deciduous shrub with large, white, magnolia-like flowers. The distinctive genus *Idiospermum*—a large evergreen tree with laminar stamens and an ovary of 1 or 2, sometimes 3, carpels—is included here because of the features it shares with the Calycanthaceae, such as phyllotaxy, undifferentiated tepals, the cuplike hypanthium, and achaenial fruits, although it could almost equally well be maintained a separate family (as done by Cronquist and Takhtajan). Molecular phylogenetic studies suggest that that *Chimonanthus* is basal and sister to the *Calycanthus*–*Sinocalycanthus* clade. The Calycanthaceae is a monophyletic group and belongs in the Laurales, where it forms a basal clade[i,ii].

Economic uses *Calycanthus floridus* (Carolina Allspice) and *C. occidentalis* are grown as ornamental shrubs for their sweetly fragrant summer flowers. *Calycanthus* contains calycanthine—an alkaloid similar to the poison strychnine and toxic to humans and livestock. *Chimonanthus fragrans* (*C. praecox*) is widely cultivated and is often known under the name Wintersweet.   VHH

[i] Renner, S. S. Circumscription and phylogeny of the Laurales: Evidence from molecular and morphological data. *American J. Bot.* 86: 1301–1315 (1999).
[ii] Li, J., Ledger, J., Ward, T., & del Tredici, P. Phylogenetics of Calycanthaceae based on molecular and morphological data with a special reference to divergent paralogues of the nrDNA ITS region. *Harvard Papers Bot.* 9: 69–82 (2004).

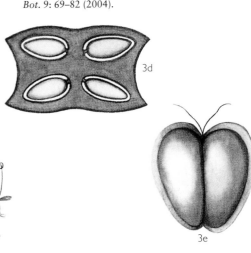

CALYCERACEAE

A small South American family of herbs with capitate inflorescences.

Distribution The family is restricted to South America, being most abundant in the Andes south from Bolivia, but extending eastward through Paraguay to Uruguay and southern Brazil, and throughout much of Argentina to southern Patagonia. *Calycera* (c. 15 spp.), *Nastanthus* (c. 10 spp.), *Acarpha* (1 sp.), *Gamocarpha* (6 spp.), and *Moschopsis* (8 spp.) grow mostly in dry open scrub or steppe vegetation. *Acicarpha* (5 spp.) grow in tropical South America, and *Boopis* (13 spp.) in the Andes, Argentina, and southern Brazil.

Genera 6	
Species c. 60	
Economic uses None	

CALYCANTHACEAE. 1 *Chimonanthus praecox* (a) shoot bearing solitary flowers, surrounded by numerous bracts in bud, which open before the leaves appear (x⅔); (b) half flower with bractlike outer and petaloid inner perianth segments (x3); (c) half false fruit comprising a fleshy receptacle surrounding the true fruits—achenes (x⅔); (d) leaf (x⅔). 2 *Calycanthus occidentalis* (a) shoot bearing leaves and axillary flowers with numerous petaloid perianth segments (x1); (b) half flower showing numerous perianth segments, stamens with short filaments, and the ovary comprising free carpels with free styles and stigmas, all inserted on a cuplike receptacle (x2).

Description Annual to perennial herbs, rarely woody at the base. The leaves are alternate, simple, often in a basal rosette, entire to pinnatisect, without stipules. The flowers are actinomorphic to slightly zygomorphic, bisexual (or rarely the central flowers of the head functionally unisexual), in capitate infloresecences, which are solitary, or in cymose panicles; the capitula are stalked or sessile and surrounded by 1 or 2 series of involucral bracts. The sepals are (4)5(6), connate, with small valvate lobes or teeth. The petals are (4)5(6), fused and cylindrical-tubular, with valvate lobes. The stamens are as many as, and alternating with, the corolla lobes, the filaments somewhat connate basally, and the anthers more or less connate into a tube. The ovary is inferior, with 2 fused carpels, and a single locule, with a solitary, anatropous, apical, pendulous ovule. The style is solitary and terminal; the stigma is capitate. The fruits are achenes with persistent calyx lobes at the apex, which are spiny in *Acicarpha* and *Calycera* and may be free or united with each other (*Acicarpha*). The seed is pendulous with fleshy endosperm and a straight embryo.

Classification Although the capitate inflorescences with involucral bracts and the form of pollen presentation in the Calyceraceae show similarities to the Asteraceae (Compositae), there are marked differences, such as the ovule position, the lack of a pappus, and the capitate stigma. However, the 2 families are closely related, as confirmed by phylogenetic analyses of morphological and molecular data[i,ii], and placed in the Asterales. The Calyceraceae together with the Asteraceae, Goodeniaceae, and Menyanthaceae form a strongly supported monophyletic group[ii,iii], but molecular data are not clear as to whether Asteraceae[iv,v] or Goodeniaceae[vi] are sister to Calyceraceae, although the morphological and molecular data combined support the former relationship[ii]. VHH

[i] Gustafsson, M. H. G. & Bremer, K. Morphology and phylogenetic interrelationships of the Asteraceae, Calyceraceae, Campanulaceae, Goodeniaceae and related families (Asterales). *American J. Bot.* 82: 250–265 (1995).
[ii] Lundberg, J. & K. Bremer, K. A phylogenetic study of the order Asterales using one morphological and three molecular data sets. *Int. J. Plant Sci.* 164: 553–578 (2003).
[iii] Bremer, B. *et al.* Phylogenetics of asterids based on three coding and three non-coding chloroplast DNA markers and the utility of non-coding DNA at higher taxonomic levels. *Molec. Phyl. Evol.* 24: 274–301 (2002).
[iv] Kårehed, J., Lundberg, J., Bremer, B., & Bremer K. Evolution of the Australasian families Alseuosmiaceae, Argophyllaceae, and Phellinaceae. *Syst. Bot.* 24: 660–682 (1999).
[v] Olmstead, R. G. *et al.* The phylogeny of the Asteridae *sensu lato* based on chloroplast *ndh*F gene sequences. *Molec. Phyl. Evol.* 16: 96–112 (2000).
[vi] Soltis, P. S., Soltis, D. E., & Chase, M. W. Angiosperm phylogeny inferred from 18S rDNA, *rbc*L, and *atp*B sequences. *Bot. J. Linn. Soc.* 133:381–461 (2000).

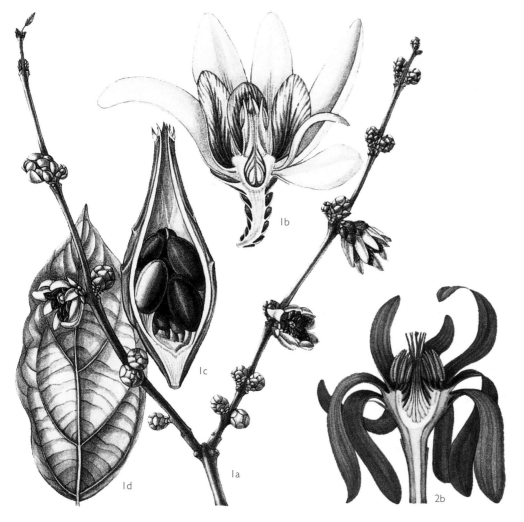

CALYCERACEAE. 1 *Calycera crassifolia* (a) leafy shoot and flowers in a head surrounded by involucre of bracts (×⅔); (b) floret immediately after opening (×2); (c) old floret (×2); (d) fruit—a ribbed achene with a persistent calyx (×⅔). 2 *Acicarpha spathulata* (a) habit (×⅔); (b) fertile marginal floret with spiked calyx-lobes and parts of the corolla tube and ovary wall removed to show stamens in tube around the style and single pendulous ovule (×4); (c) vertical section of achene (×4). 3 *Moschopsis rosulata* shoot (×⅔). 4 *Calycera herbacea* var. *sinuata* achene (×2). 5 *Nastanthus patagonicus* (a) habit (×⅔); (b) vertical section of capitulum (×2); (c) achene (×6).

CAMPANULACEAE
BELLFLOWER AND LOBELIA FAMILY

A family of herbs, sometimes climbing, or occasionally secondarily woody pachycauls.

Distribution Widespread from the arctic to the southern hemisphere in a variety of habitats (a few species aquatic), except the major deserts. The Campanuloideae are predominantly in the north temperate Old World, with a few genera confined to South Africa and *Canarina* in the Canaries and tropical Africa, *Campanula* also extending across North America, and *Wahlenbergia* extending commonly into the southern hemisphere; otherwise poorly represented in the tropics and New World. In contrast, the Lobelioideae are found mostly in the tropics and southern hemisphere and are poorly represented in the northern hemisphere apart from *Lobelia*, which is worldwide, and several genera endemic in Hawaii. Of the smaller subfamilies, the Cyphioideae are African, and the Nemacladoideae and Cyphocarpoideae are from the New World.

Description Annual to perennial herbs to herbaceous climbers (*Canarina* to 8 m high) or sometimes (giant *Lobelia* species and Hawaiian genera) pachycaul herbs to shrubs or trees several meters high. A white latex is often present. The leaves are alternate or sometimes opposite to occasionally (*Ostrowskia*) conspicuously whorled, simple or sometimes pinnatisect in Hawaii, sometimes 1 m or more long in pachycaul species, in South Africa sometimes ericoid (*Merciera*) or reduced to subulate scales (*Siphocodon*). The inflorescence is racemose or cymose, sometimes a capitulum (e.g., *Jasione*). The flowers are actinomorphic to strongly zygomorphic, bisexual or rarely unisexual, usually relatively large and showy, usually 5-merous but sometimes 6-merous (*Canarina*) or up to 9-merous (*Ostrowskia*), weakly to strongly epigynous or apparently rarely (*Cyananthus*) perigynous, resupinate in Lobelioideae. The sepals are usually elongate and acute. The corolla is regular to 2-lipped or deeply cleft down one side, cup-shaped to tubular, sometimes linear (up to 15 × 0.2 cm in *Hippobroma*). The stamens alternate with the corolla lobes and are attached to a disk or to the corolla; the filaments are free or connate. Pollen is shed into the space between the connivent or connate anthers, and the expanding stigma pushes through it with a fringe of hairs that draw the grains into the style. The ovary is usually obviously semi-inferior but more or less superior in *Cyananthus*, with 2 or 3 or 5–9 carpels (see separate subfamilies), usually with 1 locule per carpel but these sometimes divided by intrusive placentas, or the placentation rarely apical (*Siphocodon*) or basal (*Merciera*). The fruit is a capsule, dehiscing either by apical valves (above the calyx) or by lateral valves or pores (below the

Genera c. 90 **Species** c. 2,550
Economic uses Horticultural importance, alkaloid production, minor medical uses

calyx), or rarely (*Parishella*) circumscissile; or (*Canarina, Centropogon*) a fleshy berry or (e.g., *Gunillaea*) dry and indehiscent.

Classification Generally, the family has been placed in an order Campanulales together with related families such as Pentaphragmataceae and Goodeniaceae, or it has been included along with the latter families in a broader order Asterales including Asteraceae. In APG II, it is placed in Asterales, and on molecular evidence is perhaps nearest to Stylidiaceae. Some authors have preferred to recognize several families in what is here treated as Campanulaceae[i] or have adopted a broad family concept[ii]. The present account is based on that of Takhtajan (1997) but with lower taxonomic ranks. Numerous tribes have been listed by him in the Campanuloideae, but a comprehensive modern assessment is much needed. There is a risk, however, that strict emphasis on monophyletic taxa would vastly reduce the number of genera, particularly around *Campanula* and *Lobelia*, or lead to their being split into segregates of little value.

SUBFAM. CAMPANULOIDEAE With c. 56 genera and 1,260 spp., and 4 tribes:

Campanuleae has c. 50 genera, c. 1,200 spp., mostly north temperate Old World, except *Campanula* and *Wahlenbergia*. Herbs, sometimes twining; flowers usually 5-merous, actinomorphic; pollen zonoporate or pantoporate; ovary usually of 3 carpels; capsule dehiscing by apical valves or lateral pores, or indehiscent. Includes *Campanula* (415 spp.) and *Wahlenbergia* (c. 200 spp.).

Cyanantheae has 4 genera, 58 spp., from Afghanistan to East Asia. Herbs to small climbers; leaves alternate to subopposite or occasionally 3-whorled; flowers 5-merous, actinomorphic; pollen 5- to 6-colpate or 6- to 10-colpate; ovary more or less superior to inferior, 3- to 5-carpelled; capsule dehiscing by apical valves. Includes *Cyananthus* (22 spp.), *Codonopsis* (34 spp.), *Leptocodon*, and *Platycodon*.

Ostrowskieae comprises 1 genus, 1 sp., *Ostrowskia magnifica*, from the Tian Shan of Kazakhstan to Afghanistan. A robust fistulose herb to 1.8 m; most stem leaves in whorls of 4–6, subsessile, cuneate; flowers up to 10 × 10 cm, 5- to 9-merous in all parts; pollen grains with 5–7 poroid apertures; capsule dehiscing by a ring of lateral longitudinally elongate pores, which are 2–3 times as many as the sepals; seeds winged.

Canarineae comprises 1 genus (*Canarina*) and 3 spp.; Canary Islands and eastern Africa from Ethiopia to Malawi. Herbaceous high climbers to several meters or pendent epiphytes; leaves opposite or 3-whorled, cordate and petioled; flowers up to 10 cm, 6-merous in all parts; fruit a fleshy berry.

SUBFAM. CYPHOCARPOIDEAE Comprises 1 genus (*Cyphocarpus*), 3 spp., native to Chile. Small

annual herbs; leaves rigid, subspiny; flowers 5-merous, zygomorphic, with the corolla tubular, with 1 upper lobe and 4 lower lobes; stamens attached to the upper part of the corolla tube; pollen 3-colporate; ovary 2-locular with intrusive placentas developing to free-central.

SUBFAM. NEMACLADOIDEAE Comprises 3 genera and 15 spp. (*Nemacladus* 13 spp., *Pseudonemacladus*, and *Parishella*) from southwestern USA and northwestern Mexico. Delicate annuals or rarely perennials; flowers 5-merous, actinomorphic or the corolla sometimes with a 3-lobed upper lip and 2-lobed lower lip; stamens inserted on top of the ovary, the filaments free below but connate into a tube above; pollen 3-colpate or 3-colporate, spinulose; ovary 2-loculed; capsule dehiscing by 2 apical valves or rarely circumscissile.

SUBFAM. CYPHIOIDEAE Comprises 1 genus (*Cyphia*), 60 spp. southern tropical and South Africa. Erect or twining herbs; flowers

CAMPANULACEAE. I *Canarina eminii* shoot with opposite leaves and axillary, regular flowers (×⅔). 2 *Trachelium rumenlianum* shoots with alternate leaves and terminal inflorescences (×⅔). 3 *Lobelia cardinalis* "Red Flush" (×⅔).

CAMPANULACEAE. 4 *Campanula rapunculoides* (a) leafy shoot and racemose inflorescence (×⅔); (b) half flower showing free anthers to stamens, and inferior ovary surmounted by a single style with a lobed stigma (×1); (c) cross section of 3-locular ovary with ovules on axile placentas (×3). 5 *C. rapunculus* dehisced fruit—a capsule (×1⅓). 6 *Phyteuma orbiculare* inflorescence (×⅔).

zygomorphic, not resupinate; corolla with 5 subequal lobes or divided to the base into 3-lobed upper and 2-lobed lower lips; stamens free or the anthers sometimes loosely apically coherent; pollen 3-colporate, smooth; style topped by a fluid-filled stigmatic cavity; capsule dehiscing by 2 apical valves. Often regarded as an intermediate genus between Campanuloideae and Lobelioideae.

SUBFAM. LOBELIOIDEAE Includes c. 30 genera, c. 1,200 spp., mostly in the tropics and southern hemisphere except for *Lobelia*. Herbs to pachycaul giant herbs or small trees; flowers 5-merous, zygomorphic, resupinate (with 3- or 5-lobed lip appearing in lower position); stamens inserted on base of the corolla tube or on a disk, the filaments connate for at least part of their length, and the anthers connate into a tube around the style, often exserted through a slit in the corolla tube; pollen 3-colpate or 3-colporate; ovary of 2 carpels, the style with a brush of hairs below the stigma for collecting pollen; fruit a capsule, variously dehiscent. The pollination mechanism is highly specialized[ii]. Includes *Lobelia* (more than 400 spp.), *Siphocampylus* (c. 230 spp.), *Centropogon* (c. 220 spp.), and *Burmeistera* (c. 100 spp.).

Economic uses The showy flowers make the family attractive in horticulture as annuals, herbaceous perennials, and rock garden plants, especially the genera *Adenophora*, *Campanula*, *Codonopsis*, *Legousia*, *Lobelia*, *Michauxia*, *Phyteuma*, *Platycodon*, and *Wahlenbergia*. Leaves are rarely eaten as a vegetable (*Campanula rapunculus*, *Centropogon* species). Latex may contain poisonous alkaloids, but those in *Lobelia* species have been used in overcoming nicotine addiction and respiratory disorders.   RKB

[i] Lammers, T. G. Circumscription and phylogeny of Campanulales. *Ann. Missouri Bot. Gard.* 79: 388–413 (1992).
[ii] Lammers, T. G. Campanulaceae. Pp. 78–80. In: Smith, N. *et al.* (eds), *Flowering Plants of the Neotropics*. Princeton, Princeton University Press (2004).

CANELLACEAE
WHITE CINNAMON

A family of tropical aromatic trees or shrubs.

Distribution The family occurs in the Americas, West Indies, eastern tropical Africa, and Madagascar. In the western hemisphere, *Cinnamodendron* (6 spp.) grows in the West Indies, Venezuela, and Brazil; *Canella* (1 sp.) occurs in the West Indies and southern Florida in the USA; *Capsicodendron* (2 spp.) is confined to Brazil and *Pleodendron* (1 sp.) to the West Indies. In mainland Africa, *Warburgia* (3 spp.) occurs from Ethiopia to eastern Democratic Republic of Congo and northern South Africa, while *Cinnamosma* (3 spp.) is endemic to Madagascar.

Genera 6 **Species** c. 16
Economic uses Some species used in traditional medicine; *Canella winterana* is a source of white cinnamon

Description Aromatic evergreen trees or shrubs. The leaves are alternate, simple, often leathery, and dotted with pellucid glands, without stipules, the margins entire. The flowers are actinomorphic, hermaphrodite, in terminal or axillary cymes, or solitary. The sepals are 3, persistent, imbricate. The petals are (4) 5–12 in 1–2 or more whorls, usually free or slightly connate (*Cinnamosma*). The stamens are 6–12 (numerous in *Cinnamodendron*), with the filaments connate in a tube around the gynoecium. The ovary is superior, syncarpous, of 2–6 carpels, with a single locule and 2 to many ovules in 2 rows on parietal placentas. The fruit is a berry with 2 to many seeds.

Classification The family has been considered by most to be allied to Annonaceae or Myristicaceae and was placed by Cronquist (1981) in the Magnoliales. The structure of the seeds and other morphological features indicate a closer relationship with Winteraceae, a position supported by molecular studies (see references under Winteraceae). It differs notably from the latter in its wood, which contains vessel elements.

Economic uses Canella bark, or white cinnamon, is derived from *Canella winterana* and is used as a tonic and stimulant or as a condiment, although it is no longer exploited commercially. The bark of *Cinnamodendron corticosum* of the West Indies is also used as an aromatic tonic. *Warburgia ugandensis* has long been used for its medicinal properties, and the bark is sold in most major markets in Tanzania.   [DJM] VHH

CANNABACEAE
CANNABIS AND HOP FAMILY

A family of herbs and climbers with palmate leaves, related to Urticaceae and Moraceae.

Distribution *Cannabis sativa* probably originated in temperate Eurasia but has been widely cultivated and naturalized elsewhere, giving rise to recognition of local infraspecific variants, including subsp. *indica* from the tropics. *Humulus lupulus* is across all the northern temperate region and has also been divided into infraspecific variants. *H. scandens* is from East Asia but also widely cultivated.

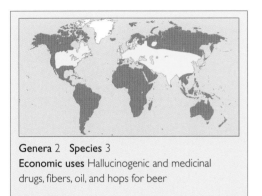

Genera 2   **Species** 3
Economic uses Hallucinogenic and medicinal drugs, fibers, oil, and hops for beer

Description *Cannabis* is an erect annual up to 3 m or more high, while *Humulus* species are herbaceous climbers up to 5 m or more. The leaves are opposite at least in the lower parts of the plant but often alternate above, heart-shaped, and palmately veined to deeply 3- to 7-lobed in *Humulus* and digitately compound in *Cannabis*, with long petioles and toothed margins. Stipules are lateral and persistent. The inflorescences are cymes in the axils of main leaves or the successively reduced upper leaves, clearly differentiated into male and female, the male ones diffuse and paniculate with small bracts, the female ones in *Cannabis* crowded among foliaceous bracts with each flower enveloped by a single, small, glandular bract, the female ones in *Humulus* conelike, with numerous, broad, papery, colored, persistent, laxly imbricate bracts. The flowers are small and greenish, unisexual (plants monoecious or usually dioecious), actinomorphic, with a uniseriate perianth. The tepals in male flowers are 5 and free, but in female flowers they are fused into a thin-textured tube adnate to the ovary or, in some cultivars, merely a ring at the base of the ovary. The stamens are 5, opposite the tepals. The ovary is superior, of 2 carpels united to form a 1-locular ovary with a bifid style, with a single apical ovule. The fruit is a small nut or achene closely invested by the persistent perianth.

Classification The family is clearly referable to the Urticales—see Urticaceae for references. The 2 genera are distinct in appearance but united by their palmate leaves, dimorphic inflorescences, and floral structure.

Economic uses *Cannabis* has been cultivated for thousands of years for its fiber used to make paper and rope (hemp) and for its use as a smoked drug (marijuana, pot, weed, dagga, ganja, bhang, etc.) or as an intoxicating liquid

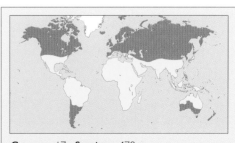

CAPPARACEAE. I *Capparis spinosa* (a) leafy shoot with spiny recurved stipules and large, solitary, axillary flowers (×⅔); fruit (b) entire; (c) in cross section (×⅔).

(hashish). It may have beneficial effects in treating various medical conditions but is dangerously addictive. Its seeds are used as a constituent of bird-seed, and the oil from them has been used in soap, cooking-oil, varnish, and even fuel for military tanks. *Humulus lupulus* provides hops (female inflorescences) used in the production of beer. *H. scandens* is cultivated as an ornamental, especially yellow-leaved variants.   CMW-D & RKB

CAPPARACEAE
CAPER FAMILY

A family of trees, shrubs, and herbs usually from seasonally dry areas of warm temperate and tropical regions.

Distribution Widespread in warm temperate and tropical regions, usually with seasonal drought. *Capparis* (c. 250 spp) is pantropical, extending from sea level to 2,500 m altitude with outlying species in subtropical regions and the Mediterranean. *Crataeva* (8 spp.) is pantropical but most diverse in Southeast Asia. *Cadaba* (c. 30 spp) is widespread throughout Africa and Asia into Australia. *Maerua* (c. 100 spp.) is predominantly African with outliers in Madagascar and Socotra through Arabia to tropical Asia. Seven genera are endemic to Africa and its offshore islands (but 1 sp. in

Arabia). Four genera are endemic from Mexico to South America. The monotypic *Dhofaria mackeishii* is endemic to Oman, and *Apophyllum anomalum* endemic to Australia.

Description Trees up to 30 m, shrubs, scramblers, or lianas (*Ritchea*), with alternate leaves, sometimes leafless with photosynthetic stems (*Apophyllum*). The leaves are simple in many genera or are (1–)3(–5) -foliate, only *Boscia*, *Cladostemon*, and *Crataeva* and without simple leaved or 1-foliate species; simple leaves usually without stipules, otherwise stipules minute except in *Apophyllum*, where the leaves are ephemeral and stipules to 7mm[i]. The inflorescence is usually a terminal or axillary raceme or single flowers borne in leaf axils, rarely a corymb or fascicle. The flowers are usually hermaphrodite but unisexual in *Apophyllum* (plants dioecious or polygamous) and *Capparis* (plants dioecious or monoecious), sometimes with male flowers by abortion of carpels (*Boscia*, *Maerua*). The flowers are 4-merous, actinomorphic to weakly or strongly zygomorphic (*Atamisquea*, *Cadaba*, *Capparis*), the receptacle sometimes forming a pronounced disk, cone or tube, sometimes with 4 peripheral nectaries or other appendages. The sepals are usually 4 (sometimes 3–5 or in *Buccholzia* 7), free, then unequal and sometimes petaloid (*Cadaba*, *Capparis*, *Euadenia*), opening early in floral

Genera c. 17   **Species** c. 470
Economic uses Capers eaten as a condiment; some ornamentals

development in some genera (*Bachmannia, Cladostemon, Crataeva, Euadenia*), often connate, then sometimes valvate or calyptrate (*Thilachium*), or rupturing into 2–4 lobes (e.g., *Morisonia*). The petals are usually 4 but sometimes 0 (*Bachmannia, Boscia, Thilachium, Buchholzia*), 0–4 (*Cadaba, Maerua*) or (0)4 or 8 (*Ritchiea*), when present pale and often white, cream, or yellow, subequal or unequal in size. The stamens and ovary are sometimes borne on a stalk (androgynophore), stamens (3–)4–8 (–many) but reduced to 1–3 sterile ones in female flowers of *Apophyllum*, staminodes usually absent but 2–7 when found (*Atamsiquea, Cladostemon, Euadenia, Morisonia*). The ovary is usually borne on a stalk (gynophore) and is superior, usually 1-locular, sometimes becoming 2-locular by the development of a septum; the stigma is usually sessile but sometimes a short style is present. The fruit is usually many-seeded, rarely 1- to 2-seeded (*Apophyllum, Buchholzia, Dhofaria*) or 4- seeded (some *Boscia, Cadaba*), but 8-seeded in *Bachmannia*.

Classification Capparaceae, placed in the Capparales by Cronquist (1981), forms a close knit group with Brassicaceae and Cleomaceae and the glucosinolate-containing families[ii]. Morphological cladistic analyses have suggested Capparaceae contains Brassicaceae, leading to a proposal to merge these 2 families[iii], but molecular evidence[iv] indicates 3 identifiable groups: Capparaceae, Cleomaceae, and Brassicaceae as an alternative scenario and suggests that several genera are not natural, especially *Capparis*, which is reflected in the high morphological variation seen in some of the larger genera. A thorough revision of the Capparaceae is needed.

Economic uses The pickled flower buds and fruits of *Capparis spinosa* are eaten (capers). Some species are grown as ornamentals.   AC

i Kers, L. E.. Capparaceae. Pp. 36–56. In: Kubitzki, K. & Bayer, C. (eds), *The Families and Genera of Vascular Plants. V. Flowering Plants. Dicotyledons: Malvales, Capparales and Non-betalain Caryophyllales*. Berlin, Springer-Verlag (2002).
ii Appel, O. & Al-Shehbaz, I. Cruciferae. Pp. 75–174. In: Kubitzki, K. & Bayer, C. (eds), *The Families and Genera of Vascular Plants. V. Flowering Plants. Dicotyledons: Malvales, Capparales and Non-betalain Caryophyllales*. Berlin, Springer-Verlag (2002).
iii Judd, W. S., Sanders, R. W., & Donoghue, M. J. Angiosperm family pairs: preliminary phylogenetic analyses. *Harvard Papers in Botany* 5: 1–51 (1994).
iv Hall, J. C., Sytsma, K. J., & Iltis, H. H. Phylogeny of Capparaceae and Brassicaceae based on chloroplast sequence data. *American J. Bot.* 89: 1826–1842 (2002).

CAPRIFOLIACEAE

HONEYSUCKLE FAMILY

A mainly northern temperate family of mostly shrubs or climbers, related to Dipsacaceae and Valerianaceae.

Distribution North temperate region, south to North Africa, Himalayas, central China, and Mexico.

Genera 12   Species c. 250
Economic uses Cultivated ornamental shrubs and climbers; some edible fruits

Description Mostly shrubs to 4 m, or sometimes woody climbers in *Lonicera*, rarely a small tree in *Lonicera* or *Heptacodium*, a slender wiry prostrate subshrub (*Linnaea*), or perennial herb (*Triosteum*). The leaves are opposite, sometimes with interpetiolar stipules in *Lonicera* and *Leycesteria*, sometimes connate in pairs (especially *Triosteum* and *Zabelia*), usually unlobed but occasionally deeply lobed (occasionally lobed and unlobed on the same plant), with entire to serrate margins. The inflorescence is a compact to diffuse terminal or axillary 2- to few-flowered cyme, or in *Heptacodium* a broad panicle of small capitula. Bracts and bracteoles are sometimes conspicuous, and in *Dipelta* are much accrescent. The sepals are usually 5 but sometimes 2–4 in *Abelia* and *Zabelia*, linear to obovate, and separate to the base, often persistent and sometimes accrescent. The corolla is tubular to funnel-shaped, sometimes more or less regular with 5 equal lobes (sometimes 4 in *Zabelia*), sometimes weakly to strongly zygomorphic and often 2-lipped, with a basal nectary and sometimes gibbous. The stamens are 4 or 5, attached to the corolla, sometimes 2 long and 2 short. The ovary is inferior, with a slender style and capitate stigma, and the upper part below the calyx is often narrowed into a sterile beak. In *Kolkwitzia* and some *Lonicera* species, the ovaries of adjacent flowers may be fused. The carpels are 2–5(–8), some of which may be abortive, with 1–5 fertile locules, and 1 to several ovules per locule. In *Weigela*, the ovary has a persistent central columella terminated by the persistent style. The fruit is a dry dehiscent or indehiscent capsule or a berry or fleshy drupe, with 1 to many seeds.

Classification The family as defined here is closely related to the Dipsacaceae, Morinaceae, Valerianaceae, and Triplostegiaceae, which probably make it paraphyletic, and less closely allied to *Sambucus, Viburnum*, and *Adoxa*, which have often been included within it in past classifications. Three subfamilies are recognized here, which have recently been treated as distinct families[i].

SUBFAM. DIERVILLOIDEAE Comprises *Diervilla*, with 2 spp. in North America (*D. lonicera* occurring from Saskatchewan and Newfoundland to North Carolina, and *D. sessilifolia* from the southern Appalachians), and *Weigela* with

c. 10 spp. from eastern Siberia, Korea, northeastern China, and Japan. Shrubs, stoloniferous in *Diervilla*. Calyx sometimes bilabiate in *Weigela*; corolla 5-lobed, trumpet-shaped, 2-lipped and yellowish in *Diervilla*, and sub-actinomorphic and pink to purple or yellow in *Weigela*, with a club-shaped nectary at the base; stamens 5; the ovary elongate, in *Diervilla* terminated by a narrow, sterile beak bearing the persistent calyx, with 2 carpels, and 2-locular, with numerous ovules in each locule. The fruit is dry, and in *Weigela* dehisces into 2 valves to reveal a central columella on which the style and stigma persist. In *Diervilla*, however, it is indehiscent or irregularly dehiscent below the sterile neck, and without a persistent columella. The seeds are many, often winged in *Weigela*. The latter 2 genera were for a long time treated as 1 genus under either name, and the nomenclature in horticultural literature has therefore been much confused.

SUBFAM. LINNAEOIDEAE Comprises 5 genera with 30 spp.. *Linnaea*, with 1 panboreal sp., is prostrate and wiry, while the other genera are shrubs up to 4 m. *Abelia* has c. 12 spp. centered in China and Japan and 3 spp. in Mexico. *Zabelia* has c. 10 spp. from Turkestan eastward to the Russian Far East and Japan. *Dipelta* has 3 spp. in western and central China. *Kolkwitzia* has 1 sp. in central and northern China. They have often conspicuous bracteoles below the flowers, especially in *Dipelta*. Sepals often large, obovate, and persistent; corolla subregular to zygomorphic; stamens 4, with 2 long and 2 short; ovary more or less globose, usually with a sterile neck above, and 3–4 locules, 2 of which are sterile, and the fertile locules each with 1 ovule. The fruit is dry and indehiscent with 1 seed, or in *Dipelta* fleshy with 2 seeds.

SUBFAM. CAPRIFOLIOIDEAE Comprises 5 genera and c. 210 spp.. *Lonicera* has c. 180 spp. more or less throughout the distribution of the family, including shrubs and woody climbers. *Symphoricarpos* has c. 15 spp. in North America and Mexico and 1 in China. *Leycesteria* has 5 shrubby species from Pakistan to western China. *Heptacodium* has 2 spp. of shrubs to small trees in central China. *Triosteum* has 3 rhizomatous spp. from the Himalayas to Japan and 2–3 spp. in eastern North America. Calyx small, with linear lobes; corolla regular to strongly zygomorphic; stamens 5; ovary more or less globose, with or without a sterile neck, with 2–4(5) locules or in *Heptacodium* only 1, and each locule has 3–8 ovules in 2 series. The fruit is a berry with 2 to several seeds, or in *Triosteum* a drupe with 3 stones. The herbaceous and drupaceous *Triosteum* has been maintained in a separate tribe Triosteeae by Hara[ii] (see below). *Heptacodium*, with heads of 6–7 flowers, each with 10–14 subtending bracts or bracteoles, all heads arranged in a broad terminal panicle, and with strikingly 3-nerved leaves, has also been assigned to a separate tribe Heptacodieae.

The molecular evidence[iii,iv] consistently places the 3 subfamilies adjacent to each other but with the Dipsacaceae and Valerianaceae (and no doubt Morinaceae and Triplostegiaceae) nested among them. While this is not here seen as a sufficient argument for treating the 3 subfamilies as families in order to avoid paraphyly, consideration must be given as to whether the morphological characters cited for the 3 groups would justify recognizing 3 separate families. There is wide variation in ovary and fruit characteristics, and the 4 unequal stamens of the Linnaeoideae rather strikingly define that group. However, 4 stamens are also found in *Symphoricarpos*, and the distinguishing characteristics quoted[i] for the Caprifoliaceae *sensu stricto*—entire leaves and 5 stamens—are neither constant in it, nor absent from the other groups. While the Caprifolioideae always have a fleshy fruit, so also does *Dipelta* in the Linnaeoideae, and while the 2 genera of Diervilloideae both have elongate dry fruits, those of *Diervilla* are indehiscent and break up below the conspicuous sterile neck and are more similar in this to the Linnaeoideae than to *Weigela*. The inflorescence bracts said to characterize the Linnaeoideae are apparently also well developed in *Heptacodium*, as already noted above. It seems that the similarities of these 3 groups heavily outweigh their differences, and subfamilial rank seems preferable. An alternative taxonomy seeking to avoid such paraphyly[v] has sunk the Dipsacaceae and Valerianaceae into Caprifoliaceae and would seem to indicate the impracticality of adopting a classification based purely on putative descent at the expense of characters.

Viburnum and *Sambucus* have usually been placed in Caprifoliaceae but differ in lacking a nectary in the corolla, in their short style, their fruits with 1–5 stones, small reticulate pollen grains, and large chromosomes. Molecular analysis places these genera nearer to Adoxaceae than to Caprifoliaceae, and they are here treated as separate families.

Economic uses Many species of *Diervilla* and *Weigela* (Bush Honeysuckle), *Abelia*, *Zabelia*, *Dipelta*, and *Kolkwitzia*, as well as the

well-known *Lonicera* (Honeysuckle) and *Leycesteria*, are cultivated for their showy flowers and fragrance, while *Symphoricarpos* (Snowberry) is grown more for its showy, fleshy fruits. The berries of *Lonicera edulis* closely resemble those of *Vaccinium myrtillus* (Bilberry) in appearance and taste and are gathered extensively in the wild in eastern Russia.   RKB

i Backlund, A. & Pyck, N. *Diervillaceae and Linnaeaceae*, two new families of caprifolioids. *Taxon* 47: 657–661 (1998).
ii Hara, H. A revision of Caprifoliaceae of Japan with reference to allied plants in other districts and the Adoxaceae. *Ginkgoana* 5: 1–336, plates 1–55 (1983).
iii Backlund, A. & Bremer, K. Phylogeny of the Asteridae *s. str.* based on *rbc*L sequences, with particular reference to the Dipsacales. *Pl. Syst. Evol.* 207: 225–254 (1997).
iv Bell, C. D. Preliminary phylogeny of Valerianaceae (Dipsacales) inferred from nuclear and chloroplast DNA sequence data. *Molec. Phylogen. Evol.* 31: 340–350 (2004).
v Judd, W. S. *et al. Plant Systematics. A Phylogenetic Approach*. 2nd ed. Sunderland, Sinauer Associates (2002).

CAPRIFOLIACEAE. 1 *Lonicera biflora* (a) twining stem with opposite leaves and flowers in pairs (x⅔); (b) flower opened out showing epipetalous stamens (x2); (c) paired fruit—berries (x4); (d) vertical section of fruit (x3). 2 *Weigela amabilis* (a) leafy shoot and inflorescence (x⅔); (b) corolla opened out (x1); (c) calyx, style, and stigma (x3); (d) section of ovary (x9).

CARDIOPTERIDACEAE

The content and circumscription of this family has been unclear. *Cardiopteris* is a twining herb from Southeast Asia to Australia and in China, with white latex and spirally arranged, alternate, entire, exstipulate leaves, and small, usually hermaphrodite, actinomorphic flowers, with 4–5 sepals and petals, connate at the base, and 4–5 epipetalous stamens, a superior 1-locular ovary, with 2 pendulous ovules, and dry, winged fruits. Recently, 4–5 other genera have been included in the Cardiopteridaceae, including *Citronella* (21 spp., Central and South America, Malaysia, Pacific), *Gonocaryum* (9–10 spp., Indo–Malaysia, Taiwan), *Dendrobangia* (3 spp., tropical South America), *Leptaulus* (5 spp., tropical Africa), and *Pseudobotrys* (2 spp., New Guinea)[i,ii], all of which are trees or shrubs with drupaceous fruits. These are here mostly referred to Leptaulaceae (q.v.). *Metteniusa*

(3 spp.) has also been suggested for inclusion here but has been recognized as a separate family Metteniusaceae (q.v.). Recent molecular work places it in the Euasterid I group near the Icacinaceae and Garryales[iii].   VHH

[i] Kårehed, J. *Evolutionary Studies in Asterids Emphasising Euasterids II.* University of Uppsala, Ph.D. thesis (2002). http://urn.kb.se/resolve?urn=urn:nbn:se:uu:diva-2696 (2005-07-07).
[ii] Kårehed, J. Multiple origin of the tropical forest tree family Icacinaceae. *American J. Bot.* 88: 2259–2259 (2001).
[iii] Chase, M. W. & Gonzales, F. Cited in Soltis, D. E., Soltis, P. S., Endress, P. K., & Chase, M. W. *Phylogeny and Evolution of Angiosperms.* Sunderland, Massachusetts, Sinauer Associates (2005).

CARICACEAE

PAPAYA OR PAWPAW

A tropical family of small trees celebrated for the edible fruits—pawpaw, or papaya—of *Carica papaya.*

Distribution Five genera are native to Mexico and extend to South America: *Carica* (1 sp.), *Horovitzia* (1 sp.), *Jacaratia* (7 spp.), *Jarilla* (3 spp.), and *Vasconcellea* (20 spp.), while *Cylicomorpha* (2 spp.) is native to Cameroon and eastern Africa.

Genera 6   Species 34
Economic uses *Carica papaya* grown for fruit (pawpaw, papaya) and latex containing the enzyme papain

Description Small, sparsely branched trees, with soft wood; all parts contain milky latex in articulated laticifers, which may be pungent. The stems are sometimes spiny and those of *Carica papaya* are unusual in that there is a little development of secondary xylem, the wood being formed largely from phloem, which gives the soft, large-pithed trunk much of its rigidity. The leaves are alternate, crowded at branch tips, digitately lobed or foliate, usually with long petioles and without stipules (or stipules spinelike in *Vasconcellea stipulata*). The inflorescences are axillary, 1-flowered, or a many-flowered thyrse. The flowers are unisexual (plants usually dioecious, very rarely monoecious or polygamous), rarely bisexual (sometimes in *Carica*). The flowers are regular, with the parts in fives; the sepals are more or less free; the corolla is contorted or valvate, with the corolla tube long in male, short in female, flowers. The stamens are attached to the petals, and the anthers open inward. The ovary is superior, of 5 fused carpels, with 1–5 locules, with many anatropous ovules on parietal placentas. The style is short and crowned by 5 stigmas. The fruit is a berry. The numerous seeds each have a gelatinous envelope, oily endosperm, and a straight embryo.

Classification Traditionally, the family has been included in the Violales[i], but molecular evidence identified a close relationship with *Moringa*[ii] and other mustard-oil–containing families. A species level analysis of relationships between Caricaceae and Moringaceae based on morphological and molecular data confirmed their monophyly and sister relationship[iii] and placed them adjacent to core Capparales. Analysis of *Carica* using morphological approaches supported the separation of the genus back into a monotypic *Carica,* with the remaining species in *Vasconcellea,* a view further supported by DNA fingerprint[v] and RFLP[vi] techniques. A sixth genus, *Horovitzia,* was described in 1993[vii], endemic to Oaxaca in Mexico.

Economic uses The family's economic importance resides largely in *Carica papaya,* which has a large, juicy fruit and is extensively cultivated throughout the tropics. This species has nocturnal, sweet-scented, moth-pollinated flowers. As with many other cultivated plants, its origin is unknown, and it may have arisen from hybridization. The fresh fruit is eaten with lemon or lime juice or in fruit salad; it may be tinned, crystallized, or made into jam, ice cream, jellies, pies, or pickles; when unripe, it is used like marrow or apple sauce. The green fruit produces latex that contains the proteolytic enzyme papain, which breaks down proteins. For this reason, the leaves are sometimes wrapped round meat to soften it. Commercial grade papain will digest 35 times its own weight of lean meat and is an important article of commerce for medicinal and industrial usages (e.g., in canned meat and leather tanning). A few species of *Vasconcellea* are cultivated in South America for their edible pericarp or the sweet, juicy seed envelopes, including *V. chrysophila* (higicho) and *V. pentagona* (babaco), both from Colombia and Ecuador, and *V. candicans* from Peru. The mountain pawpaw (*V. cundinamarcensis*) of the Andes has smaller fruits than the pawpaw and is successfully grown at altitudes in the tropics where *C. papaya* would fail. The fruits of *Jarilla caudata* and *Jacaratia mexicana* are also eaten locally.   [TWC] AC

[i] Brummitt, R. K. *Vascular Plant Families and Genera.* Richmond, Royal Botanic Gardens, Kew (1992).
[ii] Gadek, P. A. *et al.* Affinities of the Australian endemic Akaniaceae: new evidence from *rbc*L sequences. *Australian Syst. Bot.* 5: 717–724 (1992).
[iii] Olson, M. E. Intergeneric relationships within the Caricaceae-Moringaceae clade (Brassicales) and potential morphological synapomorphies of the clade and its families. *Int. J. Pl. Sci.* 163 (1): 51–65 (2002).
[iv] Badillo, V. M. *Carica* L. vs. *Vasconcella* St. Hil. (Caricaceae): con la rehabilitación de este último. *Ernestia* 10: 74–79 (2001).
[v] Van Droogenbroeck, B., Breyne, P., Goetghebeur, P., Romeijn-Peeters, E., Kyndt, T., & Gheysen, G. AFLP analysis of genetic relationships among papaya and its wild relatives (Caricaceae) from Ecuador. *Theor. Appl. Genet.* 105: 289–297 (2002).
[vi] Van Droogenbroeck, B., Kyndt, T., Maertens, I., Romeijn-Peeters, E., Scheldeman, X., Romero-Motochi, J. P., Van Damme, P., Goetghebeur, P., & Gheysen, G. Phylogenetic analysis of the highland papayas (*Vasconcellea*) and allied genera (Caricaceae) using PCR-RFLP. *Theor. Appl. Genet.* 108: 1473–1486 (2004).
[vii] Badillo, V. M. Caricaceae. Segundo esquema. *Rev. Fac. Agron. (Maracay)* 43: 1–111 (1993).

CARLEMANNIACEAE

A family of robust herbs to shrubs from the forests of Asia, with gamopetalous flowers with an inferior ovary, superficially reminiscent of a *Hydrangea.*

Distribution *Carlemannia* has 3 spp. from eastern Himalayas and Yunnan to Sumatra. *Silvianthus* has 2 species from Assam and Yunnan to Laos. All grow in evergreen forests, often on stream banks.

Genera 2 Species 5
Economic uses None

Description Both genera include robust herbs to soft-wooded shrubs, *Carlemaniia* being usually herbaceous and up to 1.5 m while *Silvianthus* is usually a shrub of 2–3.5 m. The leaves are opposite, without stipules, usually broadly elliptic and cuneate to a petiole up to 6 cm, in *Carlemannia* up to 18 × 9 cm and crenate to serrate, in *Silvianthus* up to 32 × 12 cm and entire to obscurely serrate. The inflorescence is a terminal or axillary cyme. The flowers are 4-merous in *Carlemannia* but 5-merous in *Silvianthus.* The calyx lobes are often markedly unequal. The corolla is funnel-shaped to narrowly campanulate, weakly to strongly zygomorphic. The stamens are 2, inserted in the corolla tube. The ovary is inferior, 2-carpellate and 2-locular, with numerous ovules—that of *Carlemannia* with basal placentas, that of *Silvianthus* with axile placentas in the middle of the septum. The fruit in *Carlemannia* is a dry capsule splitting into 2 valves, while that of *Silvianthus* is fleshy and white and splits into 5 valves. Seeds are many and small, probably dispersed by birds in *Silvianthus.*

Classification The 2 genera have been variously associated with Rubiaceae, Caprifoliaceae, Gesneriaceae, Verbenaceae (genera now in Lamiaceae), and Hydrangeaceae, but they are now placed by molecular evidence near the Oleaceae at the base of the Lamiaceae group[i]. In describing the family, Airy Shaw noted the rather strong differences between the 2 genera and wondered whether they were really closely related, but the anatomical evidence he quoted favors the close association and is supported by the molecular evidence. RKB

i Bremer, K. *et al*. A phylogenetic analysis of 100+ genera and 50+ families of euasterids based on morphological and molecular data with notes on possible higher level morphological synapomorphies. *Plant Syst. Evol.* 229: 137–169 (2001).

CARYOCARACEAE

A small family of evergreen tropical trees with drupaceous fruits.

Distribution Both genera *Caryocar* (16 spp.[i]) and *Anthodiscus* (9 spp.[i]) are confined to tropical America, mainly in the Amazon, but with a few species outside the region in coastal forests of Brazil. The Amazonian species grow in terra firme forests and in forests on white sand, while *Caryocar brasiliense* commonly occurs in Planalto cerrado, Brazil.

Genera 2
Species c. 25
Economic uses Edible fruits and seeds; durable timber; fish poisons

Description Small to large evergreen trees, sometimes with large buttresses, and occasionally shrublike. The leaves are opposite (*Caryocar*) or alternate (*Anthodiscus*), usually palmate and mostly 3-foliate, with usually caducous stipules (absent in *Anthodiscus*), the laminas serrate or serrulate. The flowers are actinomorphic, bisexual, large and showy, nocturnally pollinated by bats (*Caryocar*), in terminal racemes. The calyx has 5(6) lobes but is rudimentary in *Anthodiscus*. The petals are 4(5) free, slightly connate at the base (*Caryocar*), or fused at the apex (*Anthodiscus*). The stamens are 55–750, the innermost shorter and staminodial, the filaments often fused at the base. In *Caryocar,* the petals and stamens fall after anthesis, while in *Anthodiscus,* the petals fall as a unit. The ovary is superior, of 4–20 fused carpels, the locules 4(–6) (*Caryocar*) or 8–20 (*Anthodiscus*), with axile placentation and 1 ovule per locule. The fruit is drupaceous, separating into 1-seeded pyrenes when it matures. The seeds are 1–4 (*Caryocar*) or 8–20 (*Anthodiscus*).

Classification The Caryocaraceae has been monographed in *Flora Neotropica*[ii]. It was placed in the Theales (Cronquist 1981, Takhtajan 1997) and sometimes included in the Theaceae or with the latter in the Ternstroemiaceae. Molecular evidence[iii] now places the group in an expanded Malpighiales *sensu* APG II.

Economic uses The seeds of *Caryocar nuciferum* (Souari Nut) and other species are edible and may be used to produce cooking oil. It is sometimes cultivated in the tropics for its seeds.Fish poison is extracted from the fruits of *Caryocar* and *Anthodiscus* species. The wood of many Caryocaraceae species is hard and durable and may be used in construction and for shipbuilding. VHH

i Mori, S. A. Caryocaraceae. Pp. 87–88. In: Smith, N. *et al.* (eds), *Flowering Plants of the Neotropics.* Princeton, Princeton University Press (2004).
ii Prance, G. T. & Freitas da Silva, M. *Flora neotropica Monograph* No. 12 Caryocaraceae. New York, Hafner, (1973).
iii Savolainen, V. *et al.* Phylogeny of the eudicots: a nearly complete familial analysis based on *rbc*L sequences. *Kew Bull.* 55: 257–309 (2000).

CARYOPHYLLACEAE
CARNATION FAMILY

A large family of mainly temperate herbaceous plants with opposite and entire leaves. It includes the popular carnations and pinks and a number of widespread weeds.

Distribution The family is found from the Arctic to Antarctic in all temperate regions of the world and sparingly on mountains in the tropics; several species of *Stellaria* (chickweed) and *Cerastium* (mouse-ear) have become almost cosmopolitan weeds. The center of diversity is in the Mediterranean and adjoining parts of Europe and southwestern Asia. Representation in the temperate southern hemisphere is small in terms of genera and species. All the larger genera (*Silene, Dianthus, Arenaria*, etc.) are found in the northern hemisphere, with the strongest concentrations in the Mediterranean region.

Description Herbs, either annual or perennial, and dying back to the crown, rarely shrubby, with persistent woody stocks or small trees. The stems are often swollen at the nodes. The leaves are opposite (rarely alternate), always simple and entire, sometimes succulent, the bases of each pair often join around the stem to form a perfoliate base. Stipules are usually absent but usually scarious when present (subfamily Paronychioideae). The inflorescences are cymose, although varied in detail; the most complex is the dichasial panicle, in which each of the 2 bracteoles of the terminal flower sub-tends an inflorescence branch, itself bearing a terminal flower and 2 bracteoles, which repeats the structure. Suppression of individual flowers can lead to racemelike monochasia, and, ultimately, to a single-flowered inflorescence (as in the carnation). The flowers are usually regular and bisexual, less frequently unisexual (plants then dioecious or monoecious). The calyx consists of 4 or 5 free sepals or of united sepals with a 4- or 5-lobed apex; some subtending bracts are present at the base of the calyx in some genera, notably *Dianthus*. The petals are 4 or 5 (sometimes 0), free from each other and often sharply differentiated into limb and claw, often notched or deeply cut at the apex, and sometimes with 2 small outgrowths (ligules) at the junction of the 2 parts on the inner surface. The stamens are typically twice as many as the petals but may be reduced to as many or even fewer (e.g., *Stellaria media*), and usually free from each other and attached directly to the receptacle. In some apetalous species, the stamens are attached to the sepals, making the

CARYOPHYLLACEAE. I *Arenaria purpurascens* habit (x⅔).
2 *Silene dioica* inflorescence bearing flowers with deeply cleft limb and white claw to each petal (x⅔).

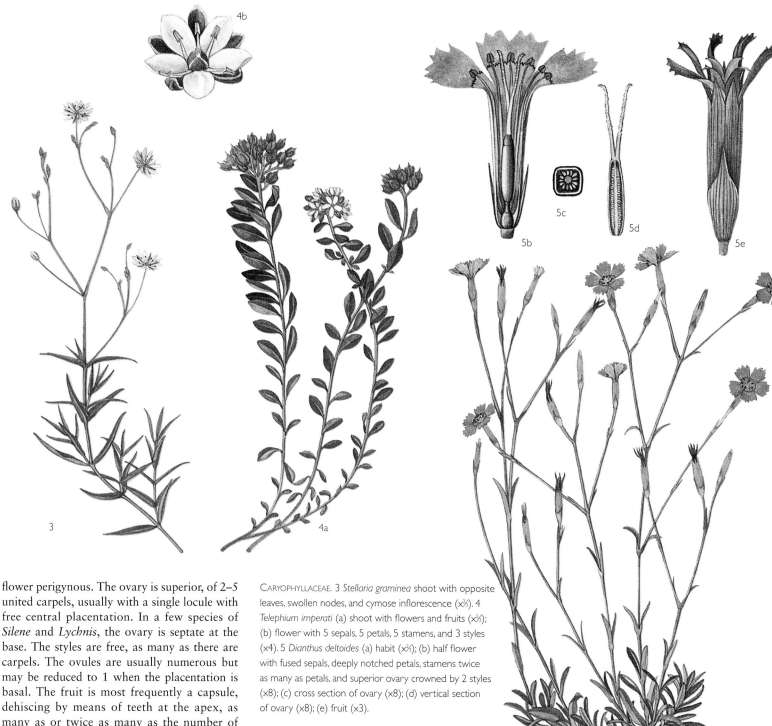

CARYOPHYLLACEAE. 3 *Stellaria graminea* shoot with opposite leaves, swollen nodes, and cymose inflorescence (×⅔). 4 *Telephium imperati* (a) shoot with flowers and fruits (×⅔); (b) flower with 5 sepals, 5 petals, 5 stamens, and 3 styles (×4). 5 *Dianthus deltoides* (a) habit (×⅔); (b) half flower with fused sepals, deeply notched petals, stamens twice as many as petals, and superior ovary crowned by 2 styles (×8); (c) cross section of ovary (×8); (d) vertical section of ovary (×8); (e) fruit (×3).

flower perigynous. The ovary is superior, of 2–5 united carpels, usually with a single locule with free central placentation. In a few species of *Silene* and *Lychnis*, the ovary is septate at the base. The styles are free, as many as there are carpels. The ovules are usually numerous but may be reduced to 1 when the placentation is basal. The fruit is most frequently a capsule, dehiscing by means of teeth at the apex, as many as or twice as many as the number of carpels. More rarely, in single-ovuled genera, the fruit is an achene or utricle. Seeds are usually numerous, with the embryo curved around the food-reserve material, which is usually perisperm rather than endosperm. In *Silene* and *Lychnis*, the petals, stamens, and ovary are separated from the calyx by a shortly extended internode (anthophore).

Classification The Caryophyllaceae groups with Amaranthaceae, Chenopodiaceae, and Achatocarpaceae in the core Caryophyllales[i], although the Caryophyllaceae does not contain betalains. Caryophyllaceae has traditionally been divided into 3 subfamilies[ii]:

SUBFAM. ALSINOIDEAE Stipules absent; sepals free from each other.

SUBFAM. SILENOIDEAE Stipules absent; sepals connate.

SUBFAM. PARONYCHIOIDEAE Stipules present, usually scarious; sepals free or connate. The Paronychioideae is more variable than the other two subfamilies and consists essentially of 2 groups of genera: those in which petals are present, the flower being hypogynous, and the fruit is usually a several- to many-seeded capsule, consisting of the tribes Sperguleae and Polycarpeae, similar to members of the Alsinoideae, differing only in the possession of stipules; and those in which the corolla is absent, the flower perigynous, and the fruit usually single-seeded and indehiscent. This latter group contains the tribe Paronychieae and 1 or 2 others and may be recognized as a separate family, the Illecebraceae, but this is not supported by DNA sequence data.

The infrafamilial classification is unstable and is not supported by the small amount of molecular evidence[iii,iv], which suggests that the

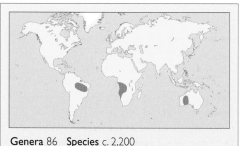

Genera 86 **Species** c. 2,200
Economic uses Cut flowers and ornamentals

subfamily Paronychioideae is a basal paraphyletic grade, with members of the tribe Corrigioleae (*Corrigiola* and *Telephium*) sister to the rest of the family. There is still the need for substantial research to establish the relationships within this family, and it is unlikely that the current infrafamilial classification will withstand scrutiny.

Economic uses The Caryophyllaceae provides a large number of widely cultivated garden ornamentals. The single most important species is the carnation (*Dianthus caryophyllus*), of which numerous garden forms (cultivars) exist; it is a specialized crop for the cut-flower market. Many other species of *Dianthus* (Pinks and Sweet William) are cultivated, including alpines, as well as species of *Silene* (Catchfly, Campion), *Gypsophila* (Baby's Breath), *Agrostemma* (Corn Cockle), *Lychnis*, and *Saponaria* (Soapwort). Several species, notably *Stellaria media*, are widespread annual weeds of fields, gardens, and other disturbed habitats. [JC] AC

i Cuénoud, P. *et al.* Molecular phylogenetics of Caryophyllales based on nuclear 18S and plastid *rbc*L, *atp*B and *mat*K DNA sequences. *American J. Bot.* 89: 132–144 (2002).
ii Rabeler, R. K. & Bittrich, V. Suprageneric nomenclature in the Caryophyllaceae. *Taxon* 42: 857–863 (1993).
iii Smissen, R. D., Clement, J. C., Garnock-Jones, P. J., & Chambers, G. K. Subfamilial relationships within Caryophyllaceae as inferred from 5'*ndh*F sequences. *American J Bot.* 89: 1336–1341 (2002).
iv Nepokroeff, M. *et al.* Relationships within Caryophyllaceae inferred from molecular sequence data. P. 105. In: *Botany 2002: Botany in the Curriculum*, Abstracts. Madison, Wisconsin (2002).

CASUARINACEAE

SHE OAK, RIVER OAK, BULL OAK, FOREST OAK, AND RED BEEFWOOD

A highly distinctive family of evergreen trees and shrubs, with slender wiry branches and reduced leaves vegetatively resembling *Equisetum*.

Distribution The family is distributed from Australia to the islands of the Pacific and Southeast Asia and has been widely planted and thought to be naturalized in the Mascarene Islands, Madagascar, and coastal regions of the Indian Ocean, Africa, and the Americas. It grows mainly in tropical climates, but *Casuarina* and *Allocasuarina* are also found in warm to cool temperate temperate zones of southern Australia.

Description Mostly tall trees, sometimes shrubs, with a characteristic weeping habit caused by their slender, jointed branches, which are more or less circular in outline and deeply grooved and with short internodes. The leaves are peculiar in structure, appearing as whorls of reduced, many-toothed sheaths surrounding the articulations of the jointed stems. The flowers are highly reduced, usually unisexual (plants monoecious or dioecious). The male flowers are borne in simple or branched terminal spikes, growing toward the tops of the plant, and the whole inflorescence is attached by short, green branchlets. The individual flowers tend to be aggregated into groups along the spike, giving the appearance of a cup with several stamens protruding out of it. Staminate flowers have 1 stamen and a perianth of 2 small lobes each subtended by 2 more small, leaflike scales or bracteoles. Pistillate flowers tend to grow on side branches lower down the tree and are borne in dense, spherical or oval-shaped heads; each flower is naked, growing out of a leaflike bract, and consists of a tiny ovary of 2 fused carpels, the posterior locule empty, the anterior locule containing 2 ovules. The style is short and divided into 2 long stigma branches. During development, the styles hang out well beyond the bracts. The fruit is a 1-seeded samaroid nut, enclosed in hard bracteoles that later open to release the fruit, so that mature inflorescences may resemble pine cones. The seeds have a straight embryo and no endosperm. Members of all 4 genera are nodulated by various clusters of *Frankia* strains[i].

Classification Four genera are recognized: *Allocasuarina* (58 spp.), *Casuarina* (17 spp.), *Ceuthostoma* (2 spp.), and *Gymnostoma* (18 spp.). Molecular studies have shown them to be monophyletic[ii]. The family appears to be distinct, with a combination of morphological characters unique in the flowering plants and independent from all other families, and it used to be thought that their reduced leaves and small, wind-pollinated flowers indicated an ancient origin and probable relationship with the gymnosperms.

Genera 4 **Species** 98
Economic uses Some *Casuarina* species are used as ornamentals, fuel and timber being extremely hard and suitable for furniture manufacture

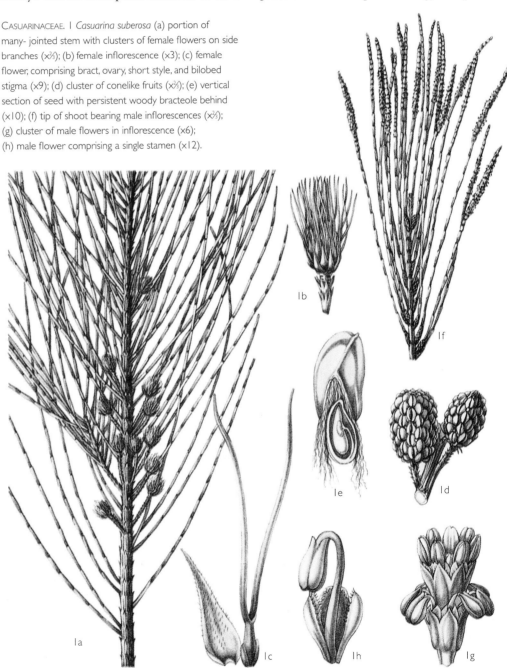

CASUARINACEAE. 1 *Casuarina suberosa* (a) portion of many-jointed stem with clusters of female flowers on side branches (x⅔); (b) female inflorescence (x3); (c) female flower, comprising bract, ovary, short style, and bilobed stigma (x9); (d) cluster of conelike fruits (x⅔); (e) vertical section of seed with persistent woody bracteole behind (x10); (f) tip of shoot bearing male inflorescences (x⅔); (g) cluster of male flowers in inflorescence (x6); (h) male flower comprising a single stamen (x12).

CASUARINACEAE. 2 *Casuarina* sp. portion of ribbed photosynthetic stem with whorl of reduced leaves (×12). 3 *Casuarina sumatrana* habit.

Subsequently, studies have shown that the family is quite derived and that its peculiar features are the result of extreme specializations in isolated conditions. It is now placed in the Fagales[iii] and is sister to Betulaceae and Myricaceae[iv]. In many of the important morphological features, the Casuarinaceae are similar to other petal-less flowering trees of the Hamamelidaceae[v], but do not indicate a close relationship. *Gymnostoma* is considered the most primitive genus and sister to the rest of the family, while *Allocasuarina* is the most specialized and diverse[vi,vii,viii].

Economic uses The timber obtained from several species is extremely hard and formerly greatly valued for furniture manufacture. *Casuarina equisetifolia* (Red Beefwood) is the most widespread cultivated species and important for dune stabilization and as an ornamental also useful as a shade tree in dry areas. It is also invasive in Hawaii, North and Central America, Réunion, and Japan. SLJ

i Maggia, L. & Bousquet, J. Molecular phylogeny of the actinorhizal Hamamelidae and relationships with host promiscuity towards *Frankia*. *Mol. Ecol.* 3: 459–467 (1994).
ii Steane, D. A., Wilson, K. L., & Hill, R. S. Using *mat*K sequence data to unravel the phylogeny of Casuarinaceae. *Mol. Phylogenet Evol.* 28: 47–59 (2003).
iii Chase, M. W. *et al.* Phylogenetics of seed plants: An analysis of nucleotide sequences from the plastid gene *rbc*L. *Ann. Missouri Bot. Gard.* 80: 528–580 (1993).
iv Soltis, D. E., Soltis, P. S., Endress, P. K., & Chase, M. W. *Phylogeny and Evolution of Angiosperms*. Sunderland, Sinauer Associates (2005).
v Nixon, K. C. Origins of Fagaceae. In: Crane, P. R. & Blackmore, S. (eds), *Evolution, Systematics and Fossil History of the Hamamelididae*. Systematics Association Special vol. no. 40A. 2: 117–145. Oxford, Clarendon Press (1989).

vi Johnson, L. A. S. & Wilson, K. L. Casuarinaceae. Pp. 237–242. In: Kubitzki, K., Rohwer, J. G., & Bittrich, V. (eds), *The Families and Genera of Vascular Plants. II. Flowering Plants. Dicotyledons: Magnoliid, Hamamelid and Caryophyllid Families*. Berlin, Springer-Verlag (1993).
vii Wilson, K. L. & Johnson, L. A. S. Casuarinaceae. In: George, A. S. (ed.), *Flora of Australia*. 3: 100–202 (1989).
viii Sogo, A., Setoguchi, H., Noguchi, J., Jaffre, T., & Tobe, H. Molecular phylogeny of Casuarinaceae based on *rbc*L and *mat*K gene sequences. *J. Plant Res.* 114: 459–464 (2001).

CECROPIACEAE
PARASOL TREE FAMILY

A family related to Urticaceae and Moraceae, with aerial adventitious roots, latex turning black, often palmate leaves, and amplexicaul stipules.

Distribution *Cecropia, Coussapoa,* and *Pourouma* have respectively 100, 45, and 25 spp. in the neotropics, while *Musanga* has 2 spp. in Guineo–Congolan Africa and *Myrianthus* has 7 spp. in tropical Africa. They occur mainly in rain forests. For *Poikilospermum* from Asia and Australia, see note below.

Description Trees and shrubs, or occasionally woody lianes, often with adventitious stilt roots and hemiepiphytic. The stems are often hollow and inhabited by ants, and the sap is mucilaginous and turns black on exposure to air. The leaves are alternate and spirally arranged, palmately lobed to digitately compound or simple (*Coussapoa*, some *Pourouma*), often large (60 cm or more), with entire margins, the stipules conspicuous and amplexicaul. The inflorescences are axillary or terminal, enclosed when young in conspicuous and sometimes (*Cecropia*) spathaceous bracts, dichotomously branched or, in female plants, globose. The flowers are unisexual (plants dioecious), actinomorphic, with a uniseriate perianth. The tepals in male flowers are 2–4, free; the perianth in female flowers is 2- to 4-lobed and tubular. The 1–4 stamens are straight in bud. The ovary is superior, with 1 stigma, which is tongue-shaped to peltate, with 1 locule and a single basal or sub-basal ovule. The fruit is a small nutlet or is large and either free or adnate to a fleshy perianth.

Classification The 5 genera are referable to the old Urticales and were placed in Moraceae until they were moved to Urticaceae in 1962 on account of their single basal ovule. They were separated as the family Cecropiaceae[i] in 1978, and this has been generally recognized since (see Urticaceae for references). Our description excludes *Poikilospermum*, which has generally been referred to Cecropiaceae but shows many anomalous characters so that this placement has been questioned. A genus of 20 spp. from tropical Asia to Australia, it differs from Cecropiaceae in having cystoliths, non-amplexicaul stipules and stamens inflexed, all characters found in Urticaceae (though the

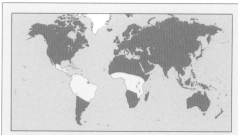

Genera 5 or 6 **Species** c. 180–200
Economic uses Edible fruits, medicinal products, wood for papermaking, some invasive weeds

stamens do not straighten abruptly) and *Poikilospermum* appears to be intermediate between the 2 families.

Economic uses *Pourouma cecropiifolia* is cultivated for its edible fruits. *Cecropia* also has edible fruits, as well as wood used for paper pulp, and the leaves are used medicinally, while some species of the genus have become invasive tree weeds. CMW-D & RKB

i Berg, C. C. Cecropiaceae a new family of the Urticales. *Taxon* 27: 39–44 (1978).

CELASTRACEAE
SPINDLE TREE FAMILY

A family of shrubs and trees or woody climbers with a pronounced floral disk and often characteristic capsule and arillate seeds.

Distribution Predominantly tropical and subtropical but extending sparingly into temperate regions. They grow in a wide range of habitats, including woodland, rain forest, grassland, coastal dunes and very dry areas.

Description Shrubs, trees, and non-tendrilled woody climbers, or rarely suffrutices, sometimes spiny. The leaves are opposite or alternate (sometimes in the same genus) or occasionally whorled, simple, usually petiolate, with entire to serrate margins. Stipules are small or absent. The inflorescence is axillary or terminal, cymose or racemose. The flowers are actinomorphic or, rarely (*Apodostigma*), somewhat zygomorphic, bisexual or unisexual (plants dioecious), (3)4–5-(6)-merous, with a marked cupular or occasionally lobed intrastaminal or extrastaminal disk. The sepals and petals are free, usually not particularly showy. The stamens are 3–5, inserted on or within the disk, sometimes with staminodes alternating with fertile stamens, or in the anomalous *Plagiopteron* the stamens are numerous in 2 series. The anthers dehisce either longitudinally or transversely or rarely apically (see subfamilies below). The ovary is superior or sometimes partly immersed in the disk, completely or incompletely 2- to 5-locular, with 1–12 or more ovules per locule, or in *Siphonodon* the locules are divided horizontally into 10 uniovulate chambers with axile placentation. The fruit is a loculicidal or sometimes (*Canotia*)

Genera c. 93 **Species** c. 1,200
Economic uses Wide range of medicinal,
construction, and food uses

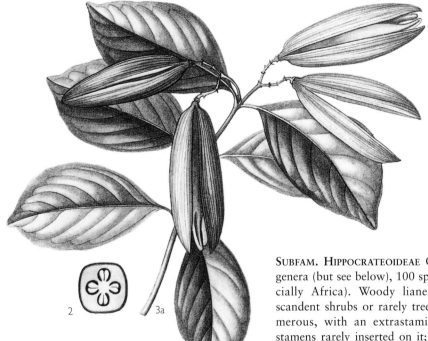

also septicidal capsule, or a fleshy drupe or
berry. The seeds are usually numerous and often
either winged or with an aril.

Classification Former concepts of the order
Celastrales[i] have not been supported by
molecular analysis, and in APG II the
Celastraceae are included in this order with
only the Lepidobotryaceae and Parnassiaceae,
placed in the major group Rosidae. The most
recent account[ii] has recognized 4 subfamilies, of
which Stackhousioideae are here treated as a
separate family. Of the other 3 subfamilies, the
Hippocrateoideae have often been treated as
a separate family but seem to be well placed
within Celastraceae. The subfamilies are:
SUBFAM. CELASTROIDEAE Comprises 68 genera,
832 spp., tropical and temperate. Trees and
shrubs, or rarely suffrutices, occasionally scandent; flowers bisexual or unisexual, 4- to

5-merous; disk intrastaminal or extrastaminal or
rarely 0, or rarely stamens inserted on it; stamens
3–5, with the anthers usually dehiscing longitudinally; the ovary 2- to 5-(10-) locular; fruit a
capsule, berry, or drupe; seeds arillate or not.
The subfamily includes *Maytenus* (c. 200 spp.),
Euonymus (c. 130 spp.), *Gymnosporia* (c. 80
spp.), *Microtropis* (66 spp.), *Elaeodendron* (40
spp.), and *Celastrus* (c. 30 spp.).

SUBFAM. HIPPOCRATEOIDEAE Comprises c. 19
genera (but see below), 100 spp., tropics (especially Africa). Woody lianes or sometimes
scandent shrubs or rarely trees; flowers (3-)5-
merous, with an extrastaminal disk or the
stamens rarely inserted on it; stamens usually
3(5) with anthers dehiscing transversely, or
(*Plagiopteron*) stamens many with apical dehiscence; ovary 3-locular; fruit a transversely
flattened deeply 3-lobed capsule; seeds 2 to
many with a basal wing and no aril. The
subfamily includes *Pristimera* (24 spp.). If
Hippocratea is treated in its traditional broad
sense, as is still favored by many authors, the
number of genera is much reduced.
SUBFAM. SALACIOIDEAE Comprises 6 genera,
260 spp.; tropics. Woody lianes or scandent
shrubs, rarely trees; flowers 5-merous with an

CELASTRACEAE. 1 *Euonymus myrianthus* (a) leafy shoot
bearing fruits—loculicidal capsules (x⅔); (b) flower with 4
distinct petals inserted on a fleshy disk (x2); (c) half flower
showing stamens inserted on the disk and gynoecium
with ovules on axile placentas (x4); (d) stamen (x12).
2 *E. vagans* cross section of ovary (x14). 3 *Hippocratea
welwitschii* (a) shoot with dehiscing fruit (x⅔); (b) seed with
aril (x⅔). 4 *Celastrus articulatus* (a) leafy shoot and cymose
inflorescences (x⅔); (b) flower (x2); (c) 2 stamens and
gynoecium (x2); (d) fruit—a capsule (x2). 5 *Elaeodendron
aethiopicum* (a) flower (x6); (b) half flower (x8).

extrastaminal disk and often a well-developed androgynophore; stamens (2) 3 (5), the anthers usually dehiscing transversely; ovary with (2) 3 (5) locules; fruit a berry; seeds 1 to many, arillate. Including *Salacia* (c. 200 spp.).

The genus *Nicobariodendron*, a tree from the Nicobar Islands, has been appended with doubt to the Celastraceae in FGVP. It has only 2 stamens and a fleshy fruit with a single basal seed and may be referable to a separate family. Molecular data are awaited.

Economic uses The family has widespread uses for medicinal extracts, insecticides, oil from seeds, edible fruits and seeds, wood for timber, stems for basket weaving, and bark for arrow poisons[ii]. *Catha edulis*, known as Kat (Khat, Gat), is commonly used in tropical countries as a stimulant, the leaves being chewed or used for an infusion drunk as a tea[iii]. Numerous species of *Celastrus* and *Euonymus* are cultivated in temperate regions for their foliage, particularly for autumnal color, and brightly colored fruits and seeds. *E. europaeus* (Spindle Tree) produces a fine-grained wood, and its seeds have been used in manufacture of soap and a dye for coloring butter.   RKB

[i] Savolainen, V., Manen, J. F., Douzery, E., & Spichiger, R. Molecular phylogeny of families related to Celastrales based on *rbc*L 5' flanking sequences. *Molec. Phylogen. Evol.* 3: 27–37 (1994).
[ii] Simmons, M. P. Celastraceae. Pp. 29–68. In: Kubitzki, K., Rohwer, J. G., & Bittrich, V. (eds), *The Families and Genera of Vascular Plants. VI. Flowering Plants. Dicotyledons: Celastrales, Oxalidales, Rosales, Cornales, Ericales.* Berlin, Springer-Verlag (2004).
[iii] Margetts, E. L. Miraa and myrrh in East Africa— clinical notes about *Catha edulis*. *Econ. Bot.* 21: 358–362 (1967).

CEPHALOTACEAE
ALBANY PITCHER PLANT

A single species of carnivorous evergreen herb bearing strongly dimorphic leaves, endemic to peaty swamps of the southern coast of the southwestern corner of Western Australia. The plant is a rosette-forming perennial herb, with ramifying underground rhizomes by which it forms clumps. The leaves are exstipulate, either entire and simple with ill-defined petiole and slightly fleshy, or forming a pitcher 2–6 cm high filled to half way with water and the opening surrounded by a pronounced rim of stiff ridges bearing inward pointing teeth overhung by a lid bearing

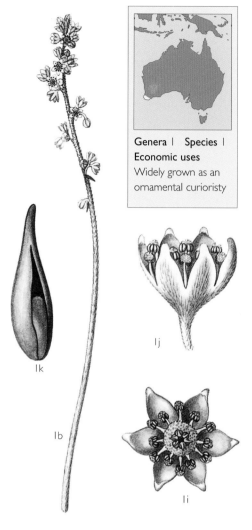

Genera 1
Species 1
Economic uses
Widely grown as an ornamental curioristy

CEPHALOTACEAE. 1 *Cephalotus follicularis* (a) habit showing foliage leaves, pitchers, and the leafless stalk to the inflorescence (×⅔); (b) inflorescence—a raceme of small cymes (×⅔); (c) pitcher and lid (×1⅓); (d) half section of pitcher showing downcurved spikes at rim (×1⅓); (e) section of flower showing hooded sepals, stamens of 2 lengths, a broad papillose disk, and the gynoecium consisting of 6 free carpels (×8); (f) long stamen with swollen glandular connective at apex (×24); (g) short stamen (×24); (h) vertical section of a carpel with single basal ovule (×24); (i) flower from above (×4); (j) flower from the side (×6); (k) fruit with part of wall removed to show single seed (×24); (l) seed (×32).

clear windowlike patches. Leaf type varies with season; both are commonly present. The inflorescence is terminal, arising from the basal rosette, forming an erect scape to c. 45cm high, bearing cymose clusters of small, white flowers that are actinomorphic, hermaphrodite, with 6 valvate sepals and no petals. The stamens are 12, 6 long and 6 short, in 2 whorls. The carpels are 6, free, hairy, each with a terminal stigma. The fruit is of 1- to 2-seeded follicles.

Cephalotus is placed sister to *Brunellia* (Brunelliaceae) in molecular and combined analyses[i]. The family was placed in the Rosales by Cronquist (1988) or Saxifragales by Takhtajan (1997), but it is now placed in the Oxalidales based on both molecular (APG II) and morphological[i] data. The plant is widely grown as an ornamental curiosity.   AC

[i] Bradford, J. C. & Barnes, R. W. Phylogenetics and classification of Cunoniaceae (Oxalidales) using chloroplast DNA sequences and morphology. *Syst. Bot.* 26: 354–385 (2001).

CERATOPHYLLACEAE
HORNWORT FAMILY

Submerged aquatics with whorled leaves divided dichotomously into linear segments.

Distribution *Ceratophyllum* has a worldwide distribution except in particularly dry areas. It grows in fresh or sometimes brackish still or slow-flowing water, where species float just below the surface.

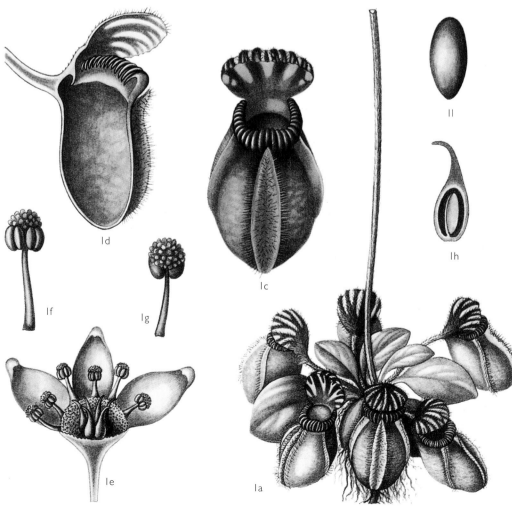

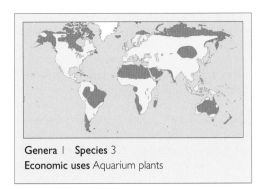

Description Submerged, free-floating aquatics, with slender, branched stems and finely divided leaves. The leaves are in whorls of 3–10, 1–4 times dichotomous, with the apical segments truncate, 2-spined, and with a central reddish glandular projection. The flowers are actinomorphic, unisexual (plants monoecious), axillary, with 1–4 male or 1 female sessile flowers at each node, the males and females often at different nodes. The flowers have (8)9–12(–15) perianth-lobes united at the base or higher, strap-shaped to obovate, each margin often with a single spine or lacerate, and with the apex as in the leaves. The male flowers have up to 45 stamens in several whorls around a pistillode. The female flowers have a superior ovary, tapering to a long style, and a single pendulous ovule. The fruit is a variously spiny achene, ovoid or ellipsoid, and slightly laterally flattened.

Classification The family has often been associated with the Nymphaeales or the Ranunculales, and in APG II it is placed in its own order among the early dicots. Many species have been described in the past. Recent authors have recognized 3 groups, comprising approximately 10 taxa but have differed in the ranks assigned. Wilmot-Dear[i] at first recognized only 2 spp., *C. submersum* and *C. demersum*. Les recognized 3 sections in the genus including 6 spp.[ii] Later Wilmot-Dear followed Les in recognizing *C. muricatum* at species rank[iii]. The 3 spp. of one system thus coincide more or less with the 3 sections of the other. Species recognized here are: *C. demersum* (cosmopolitan), with 1–2 times dichotomous coarse leaves and large fruits, with long terete apical and basal spines, or (in 3 less widespread variants) flattened spines, and often also additional spines; *C. submersum* (Old World and North America), with 3–4 times dichotomous, softer, longer bright green leaves and large fruits, with 1 short apical spine or (the variant *echinatum* as defined by Les[iv]) also with basal and marginal spines; and *C. muricatum* (pantropical), with similar 3–4 times dichotomous but yellow-green leaves and smaller fruits, with basal and marginal spines, with 3 variants in Central Europe and eastern Asia distinguished by fruit surface ornamentation and spine morphology.

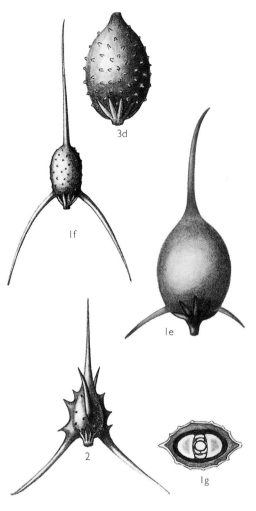

CERATOPHYLLACEAE. 1 *Ceratophyllum demersum* (a) a submerged aquatic herb with whorls of linear leaves and no roots (x⅔); (b) single twice-dichotomous leaf (x4); (c) anther with spurlike extension of connective (x8); (d) stem node with female (left) and male (right) flowers in the leaf axils (x3); (e, f) fruits—a 1-seeded nut tipped by the persistent, curved spinelike style and with 2 basal spines, (e) smooth form (x4), (f) papillose form (x2⅔); (g) cross section of papillose fruit (x8). 2 *C. demersum* var. *platyacanthum* fruit (x2⅔). 3 *C. submersum* (a) female flower (x10); (b) male flower (x10); (c) stamen (x16); (d) fruit (x4).

Economic uses Some species are grown as aquarium plants; sometimes a nuisance in choking waterways. CMW-D

i Wilmot-Dear, C. M. *Ceratophyllum* revised – a study in fruit and leaf variations. *Kew Bull.* 40: 243–271 (1985).
ii Les, D. H. The evolution of achene morphology in *Ceratophyllum* 4. *Syst. Bot.* 14: 254–262 (1989).
iii Wilmot-Dear C. M. *FSA* contributions 8: Ceratophyllaceae. *Bothalia* 27: 125–128 (1997).
iv Les, D. H. The taxonomic significance of plumule morphology in *Ceratophyllum*. *Syst. Bot.* 10: 338–346 (1985).

CERCIDIPHYLLACEAE

KATSURA FAMILY

A monogeneric family comprising 2 spp. of deciduous trees: *Cercidiphyllum japonicum*, which occurs on all 4 major islands of Japan (where it forms pure stands of large trees in the northern lowlands), southern Korea, and China as far west as Sichuan; and *C. magnificum*, which is confined to higher altitudes on mountains of Honshu in Japan (see map[i]). They are usually large trees, up to 35 m or more, with branches usually ascending, but sometimes pendent (in the wild and several cultivars)[ii], but sometimes remaining shrubby with many stems from the base (especially in cultivation). They produce long shoots (in the first year of growth) and short shoots (in subsequent years), the short shoots producing inflorescences and a single leaf. The leaves and

short shoots are opposite or subopposite or some alternate on the same plant. The leaves are ovate (especially on the long shoots) to suborbicular and strongly cordate at the base, petiolate, with 3–5 main veins from the base, the margins crenate. The stipules are deciduous. Inflorescences, each a condensed raceme on a short shoot, are usually borne before the leaves appear. The flowers are unisexual (plants dioecious) and lack a perianth. In male plants, each inflorescence includes up to 40 stamens with various bracts, but the individual flowers are difficult to delimit. In female plants, the inflorescence includes 2–7 carpels, each of which is apparently a single flower subtended by a bract. The carpel is slightly stipitate, elongate, with a long terminal style, bearing up to 20 ovules in 2 rows down one suture. The fruit is a follicle with many flattened winged seeds.

Originally placed close to the Magnoliales, the family was later included in the Hamamelididae. Molecular evidence[iii] places it close to Hamamelidaceae in the Saxifragales. The short shoots of *Cercidiphyllum* are reminiscent of Tetracentraceae, which have been placed basal to the Saxifragales. The 2 species are not always easy to distinguish, with *C. magnificum* being sometimes regarded as merely a high altitude variant of *C. japonicum*, although a majority of recent authors favor specific rank[ii]. *Cercidiphyllum* is cultivated as a botanical curiosity; the wood used for cabinet-making.   RKB

[i] Lindquist, B. Notes on *Cercidiphyllum magnificum* Nakai. *Bot. Tidsskr.* 51: 212–219 (1954).
[ii] Dosmann, M., Andrews, S., Del Tredici, P., & Li, J. H. Classification and nomenclature of weeping katsuras. *The Plantsman*, n.s. 2: 21–27 (2003).
[iii] Savolainen, V. *et al.* Phylogeny of the eudicots: a nearly complete familial analysis based on *rbc*L sequences. *Kew Bull.* 55: 257–309 (2000).

CHENOPODIACEAE

GOOSE FOOT, SUGAR BEET, BEETROOT, AND SPINACH

A largely herbaceous family with insignificant flowers, many species of which are important cash crops.

Distribution Cosmopolitan but especially in the Old World and in desert, semidesert, and saline areas. *Atriplex* (c. 300 spp.), *Chenopodium* (c. 100 spp.), and *Salicornia* (c. 30 spp.) are distributed globally. Twelve genera are native only to the Americas (e.g., *Cycloloma*, 1 sp., *Allenrolfea*, 3 spp., and *Nitrophila*, 7 spp.), and 16 genera are endemic to Australia (e.g., *Halosarcia*, 23 spp., *Sclerolaena*, 64 spp., and *Maireana*, 57 spp.).

Genera 1 Species 2
Economic uses
Sometimes cultivated as a botanical curiosity; in cabinet-making

The family is a major floristic constituent of xeric habitats of the world. Several genera are halophytes (e.g., *Salicornia*, 28 spp., *Kalidium*, 5 spp., *Halosarcia*, and *Allenrolfea*). Other genera are major components of ruderal floras (e.g., *Chenopodium*, *Atriplex*).

Description Annual or perennial herbs or shrubs, often with succulent stems (especially Salicornioideae), small trees (*Haloxylon*, 25 spp.), rarely lianas (*Holmbergia tweedii* and *Hablitzia tamnoides*), typically with deep penetrating or thickened taproots. The leaves are alternate or opposite, usually simple and entire, flat or often thickened, and succulent, sometimes reduced in size, rarely lobed, dentate or pinnatifid, usually glabrous, mealy textured or with glandular (often salt secreting) or eglandular hairs, sometimes spiny, lacking stipules. The leaves are absent in some genera, such as *Salicornia*, where fleshy green stems are the photosynthetic organ. Inflorescence cymose, a little to much-branched thyrse, usually axillary. The flowers are inconspicuous, usually bisexual, but sometimes unisexual (plants monoecious or dioecious), pentamerous, and actinomorphic. The tepals are usually imbricate, with 5 distinct lobes or sometimes connate almost to tip, rarely 0–1 (e.g., *Salicornia*) or 6–8, membranous or green. The stamens are usually isomerous with tepals, and opposite them, sometimes with interstaminal lobes, sometimes adnate to tepals. Nectaries are often present in bisexual flowers as a ring at the inner base of the stamens. The carpels are 2–3(–5), forming a 1-locular superior ovary (semi-inferior in *Beta*), with 1 basal ovule; the style is single and apical. The fruit is usually a small, round nut but sometimes an achene or utricle, often with a persistent perianth.

Classification The recognition of the family is controversial and relationships of some genera are still to be resolved. The most comprehensive study[i] indicates the Amaranthaceae and Chenopodiaceae to be sister groups with the tribe Beteae unresolved at the base of the 2 families. A review of Caryophyllales based on combined DNA sequence analysis[ii] also found the 2 families to be sister groups, but with *Beta* in the Chenopodiaceae. A survey of Caryophyllales based on chloroplast DNA sequence[iii] suggested Amaranthaceae to be nested within Chenopodiaceae. Whether or not the family is recognized, the component genera are placed in the core Caryophyllales[ii].

The Chenopodiaceae has been divided into 8 subfamilies[iv], but these were sometimes poorly defined. The modern treatments recognize 4 distinct subfamilies[v]:

SUBFAM. CHENOPODIOIDEAE Well-developed leaves, many flowered inflorescences, and flowers with 5 tepals that persist around the fruit. The subfamily is divided into 6 tribes:

Beteae (5 genera, 18 spp.) includes *Beta vulgaris*.

Chenopodieae (7 genera, c. 170 spp.) includes the ruderal weeds in *Chenopodium*.

Atripliceae (12 genera, c. 330 spp.) includes *Atriplex*, the largest genus of the family.

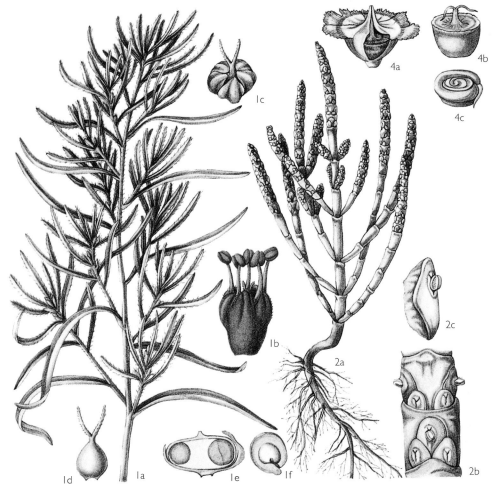

CHENOPODIACEAE. 1 *Kochia scoparia* (a) leafy shoot with inconspicuous flowers (x⅔); (b) bisexual flower with 5 perianth-segments and conspicuous stamens (x8); (c) female flower (x8); (d) gynoecium of female flower with 2 styles (x8); (e) vertical section of fruit (x8); (f) section of seed with circular embryo surrounding endosperm (x16). 2 *Salicornia europaea* (a) habit showing fleshy, leafless stem (x⅔); (b) part of flowering shoot (x4); (c) flower sunk in stem (x8). 3 *Atriplex triangularis* (a) flowering shoot (x⅔); (b) fruiting stem (x⅔); (c) enlarged bracteoles enclosing the fruit (x4); (d) fruit (x4). 4 *Salsola kali* (a) fruit with 1 segment of persistent bracteole removed (x4); (b) fruit (x6); (c) seed (x6).

Camphorosmeae (6 genera, c. 40 spp.) includes shrubby genera *Chenolea* and *Bassia*.

Sclerolaeneae (11 genera, c. 160 spp.) is endemic to Australia.

Corispermeae (3 genera, 68 spp.), which may be recognized by dendroid stellate hairs.

SUBFAM. SALICORNIOIDEAE Includes succulent plants, often with articulate stems and tiny leaves that appear cactuslike in growth form. The flowers are sunken into cavities in the fleshy inflorescence. There are 2 tribes: Halopeplideae (2 genera, 8 spp.); and Salicornieae (12 genera, c. 70 spp.).

SUBFAM. SALSOLOIDEAE Succulent or herbaceous plants, with 1–3 flowers in the axil of each bract, including 2 tribes: Suaedeae (*Suaeda* c. 100 spp. and 3 monotypic genera) *Suaeda* typically growing along coasts and in salt steppe; and Salsoleae (32 genera, 210–220 spp.) comprising many shrubby genera but the

Genera 97 **Species** c. 1,200
Economic uses Edible leaves and seeds; several grown as garden ornamentals

largest genus, *Salsola* (c. 120 spp.), including many herbaceous species. *Sarcobatus*, sometimes treated as tribe Sarcobateae, is here recognized at family level.

SUBFAM. POLYCNEMOIDEAE Comprises the single tribe, Polycnemeae (3 genera, 16–18 spp.), notable for the geographic separation of the genera: *Polycnemum* (Europe, North Africa, and Asia); *Nitrophila* (Americas), and *Hemichroa* (Australia).

Molecular data conflict with elements of this classification. The tribe Corispermeae is supported at subfamily level while the Chenopodieae is paraphyletic to Atripliciae[i].

Economic uses *Beta vulgaris* is a major agricultural crop (Sugar and Fodder Beet cultivars). It also includes the deep-red Beetroot and Mangel-wurzels, the large leaved Spinach Beet or Perpetual Spinach, and Sea Kale Beet or Swiss Chard. *Spinacia oleracea* is the source of Spinach. Quinoa, *Chenopodium quinoa*, is a staple in the Andes for its edible leaves and seeds, and *C. ambrosioides* var. *anthelminticum* is used as a vermifuge. *Chenopodium bonus-henricus* is a leaf vegetable but requires cooking due to the saponin and oxalic acid content. *Kochia scoparia* is grown for its brightly colored autumnal foliage. The woody genera (*Arthrophyton* and *Haloxylon*) are used for carving and as firewood. The desert genus *Cornulaca* is an important source of camel fodder[vi]. AC

i Kadereit, G., Borsch, T., Weising, K., & Freitag, H. Phylogeny of Amaranthaceae and Chenopodiaceae and the evolution of C4 photosynthesis. *Int. J. Plant Sci.* 164: 959–986. (2003).
ii Cuénoud, P. *et al.* Molecular phylogenetics of Caryophyllales based on nuclear 18S and plastid *rbc*L, *atp*B and *mat*K DNA sequences. *American J. Bot.* 89: 132–144 (2002).
iii Downie, S. R., Katz-Downie, D. S., & Cho, K. Y. Relationships in the Caryophyllales as suggested by phylogenetic analyses of partial chloroplast DNA ORF2280 homolog sequences. *American J. Bot.* 84: 253–273 (1997).
iv Ulbrich, E. Chenopodiaceae. Pp. 379–584. In: Engler, A. & Prantl, K. (eds), *Die natürlichen Pflanzenfamilien*. Vol. 16c. Leipzig, W. Engelmann (1934).
v Kuhn, U. *et al.* Chenopodiaceae. Pp. 253–281. In: Kubitzki, K., Rohwer, J. G., & Bittrich, V. (eds), *The Families and Genera of Vascular Plants. II. Flowering Plants. Dicotyledons: Magnoliid, Hamamelid and Caryophyllid Families*. Berlin, Springer-Verlag (1993).
vi Mabberley, D. J. *The Plant Book*. 2nd edn. Cambridge, Cambridge University Press (1997).

CHLORANTHACEAE

A small family of tropical trees, shrubs, and herbs, with usually aromatic, simple, and opposite leaves.

Distribution Pantropical (except Africa), with 3 Old World genera and 1 in the New World (with a single species in Asia)[i], mostly in high, wet, montane forests.

Genera 4 **Species** 65–70
Economic uses Limited uses as ornamentals, occasionally harvested locally for wood; consumed as teas and in traditional medicine

Description Trees (often with prop roots), shrubs, or herbs, including even annuals. The leaves are usually pungently aromatic when crushed, opposite, decussate, simple with dentate margins, petiole bases often meeting and fusing distinctively[ii]. Stipules are present, petiolar. Inflorescences are terminal or axillary, variable. The flowers are small, green, reduced, unisexual (plants usually dioecious, rarely monoecious), or bisexual. The male flowers are arranged in large numbers on a racemose inflorescence, without sepals and with 1 stamen or 3 fused together. Female and bisexual flowers are arranged in small numbers on a short raceme or spike, with a 3-toothed calyx fused to the 1-locular inferior ovary, which contains only 1 pendulous ovule inserted at the top of the locule and is surmounted by a stigma and short style. The fruit is a small, ovoid or globose, drupe with a thin exocarp and fleshy mesocarp, containing a single seed with a minute embryo surrounded by a well-developed, oily, starchy, endosperm.

Classification The Chloranthaceae consists of 4 genera: *Chloranthus* (10 spp.), *Sarcandra* (2 spp., both with bisexual flowers, the first with 3 stamens, the latter with one), *Ascarina* (10 spp., including *Ascarinopsis*), and *Hedyosmum* (40–50 spp.). Both *Ascarina* and *Hedyosmum* have unisexual flowers, with bracteate male flowers and the female flowers enclosed by a cuplike bract. Recent morphological and anatomical studies[iii,iv] relate the Chloranthaceae to the Trimeniaceae in the Laurales. Molecular studies[v,vii] have so far failed to resolve relationships (see APG II), although suggesting the family is sister to the magnoliids and eudicots[vi] and that it could belong to its own order, Chloranthales. It is not now thought to be closely related to the Piperaceae, differing by

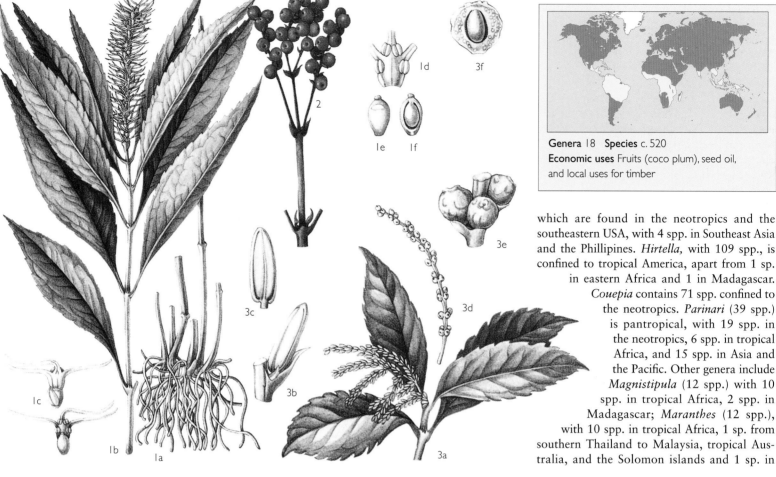

CHLORANTHACEAE. 1 *Chloranthus angustifolius* (a) roots, rhizome, and aerial stem-bases (x⅔); (b) leafy shoot and inflorescence (x⅔); (c) bisexual flower rear (upper) and front (lower) view (x2); (d) androecium, comprising central anther with 2 locules and 2 lateral anthers with 1 locule (x3); gynoecium (e) entire and (f) in vertical section (x3). 2 *Sarcandra glabra* fruiting shoot (x1). 3 *Ascarina lanceolata* (a) leafy shoot with male inflorescence (x⅔); (b) bracteate male flower (x6); (c) anther (x6); (d) female inflorescence (x⅔); (e) fruits—drupes (x4); (f) vertical section of fruit (x6). 4 *Hedyosmum brasiliense* (a) shoot with male inflorescence (x⅔); (b) female flowers (x4); gynoecium (x4); (d) male flower (x8).

its opposite leaves, united petiole bases, and inferior ovary with single ovule in a single locule. The Chloranthaceae is among the oldest lineages of flowering plants alive today[vii].

Economic uses *Hedyosmum* leaves are used to make medicinal teas and even sometimes as tea and coffee substitutes. They may also be applied externally to reduce swellings from the bites of insects and snakes. *H. scabrum* wood is of local importance. The fruits of *H. mexicanum* are edible. *Chloranthus glaber* is grown as an ornamental shrub. *C. officinalis* leaves are sometimes used to make a beverage in Malaysia and Indonesia.   SLJ

[i] Todzia, C., Chloranthaceae. Pp. 99–100. In: Smith, N. *et al.* (eds), *Flowering Plants of the Neotropics*. Princeton, Princeton University Press (2004).
[ii] Todzia, C.A., Chloranthaceae. Pp. 281–187. In: Kubitzki, K., Rohwer, J. G., & Bittrich, V. (eds), *The Families and Genera of Vascular Plants. II.*

Flowering Plants. Dicotyledons: Magnoliid, Hamamelid and Caryophyllid Families. Berlin, Springer-Verlag (1993).
[iii] Endress, P. K. Reproductive structures and phylogenetic significance of extant primitive angiosperms. *Pl. Syst. Evol.* 152: 1–28 (1986).
[iv] Endress, P. K. The Chloranthaceae: reproductive structures and phylogenetic position. *Bot. Jahrb. Syst.* 109: 153–226 (1987).
[v] Doyle, J. A., Eklund, H., & Herendeen, P. S. Floral evolution in Chloranthaceae: Implications of a morphological phylogenetic analysis. *Int. J. Pl. Sci.* 164 (Supplement): S365–S382 (2003).
[vi] Davies, T. J. *et al.* Darwin's abominable mystery: insights from a supertree of the angiosperms. *Proc. Nat. Acad. Sci. USA* 101: 1904–1909 (2004).
[vii] Zhang, L.-B. & Renner, S. The deepest splits in Chloranthaceae as resolved by chloroplast sequences. *Int. J. Pl. Sci.* 164 (Supplement): S383–S392 (2003).

CHRYSOBALANACEAE

COCOA PLUM FAMILY

The Chrysobalanaceae is a pantropical family of trees and shrubs, some of which are locally important as fruit trees.

Distribution Most of the species from the family are restricted to the lowlands of tropical and subtropical regions, growing mainly in rain forests. The family occurs in South and Central America, the Caribbean, and southeastern parts of the USA; it is found through much of tropical Africa, and in Asia it ranges from southern India and Myanmar to Indonesia and New Guinea, with an extension to the Pacific where it occurs on many of the islands. The largest genus is *Licania* with 218 spp., 214 spp. of

Genera 18 **Species** c. 520
Economic uses Fruits (coco plum), seed oil, and local uses for timber

which are found in the neotropics and the southeastern USA, with 4 spp. in Southeast Asia and the Phillipines. *Hirtella,* with 109 spp., is confined to tropical America, apart from 1 sp. in eastern Africa and 1 in Madagascar. *Couepia* contains 71 spp. confined to the neotropics. *Parinari* (39 spp.) is pantropical, with 19 spp. in the neotropics, 6 spp. in tropical Africa, and 15 spp. in Asia and the Pacific. Other genera include *Magnistipula* (12 spp.) with 10 spp. in tropical Africa, 2 spp. in Madagascar; *Maranthes* (12 spp.), with 10 spp. in tropical Africa, 1 sp. from southern Thailand to Malaysia, tropical Australia, and the Solomon islands and 1 sp. in

Central America; and *Dactyladenia* (30 spp.) confined to tropical Africa. *Parinari excelsa* and *Chrysobalanus icaco* occur in both tropical America and Africa.

Description Trees and shrubs. The leaves are simple, alternate, entire, with stipules that are often caducous. The flowers are bisexual, rarely unisexual (plants monoecious or polygamous), actinomorphic or zygomorphic, always with a disk and markedly perigynous, in racemose, paniculate, spicate, or cymose inflorescences. There are 5 sepals and usually 5 petals. The stamens are 2–100 (300 in *Couepia*), the filaments

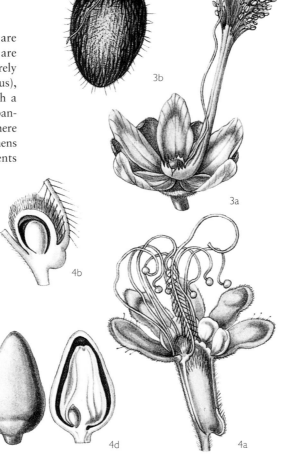

CHRYSOBALANACEAE. 1 *Couepia canomensis* (a) shoot with axillary inflorescences (x⅔); (b) flower with 5 free sepals and petals and numerous stamens (x3); (c) vertical section of ovary with 2 erect ovules (x3). 2 *Licania incana* half flower showing epipetalous stamens and ovary with single basal ovule (x8). 3 *Dactyladenia pallescens* (a) flower with bundle of stamens and filamentous style (x2); (b) fruit (x⅔). 4 *Hirtella zanzibarica* (a) flower opened out showing numerous epipetalous stamens with long filaments and globose anthers and lateral ovary (x3); (b) vertical section of ovary with single basal ovule and base of feathery style (x6); fruit (c) entire and (d) in vertical section (x1).

filiform, free or connate at the base, the anthers dehiscing introrsely. The ovary is superior with 3 carpels, but usually only 1 developing, gynobasic, 1-locular with 2 erect basal ovules or 2-locular with 1 ovule per locule. The style is simple, with a 3-lobed stigma. The fruit is a fleshy drupe, rarely dry, with a hard, fibrous or thin, bony endocarp (stone). The seeds are without endosperm.

Classification The Chrysobalanaceae has been revised for *Species Plantarum*[i]. Traditionally, it was considered to be closely allied to the Rosaceae and formerly included in it, sometimes as a subfamily, although differing in many features. Recent molecular studies indicate a placement in the Rosidae, together with the Dichapetalaceae and Trigoniaceae, in the order Malpighiales[ii]. Other molecular studies show the Chrysobalanaceae to be a sister group to the Euphroniaceae (q.v.)[iii].

Economic uses The fruits of several species are eaten locally, the most important being the Coco (Cocoa) Plum, *Chrysobalanus icaco* (Icaco), which is widely consumed in Venezuela and Colombia. *Parinari excelsa* is the Guinea Plum, which is rather dry and mealy; *P. curatellifolia* is the strawberry-flavored Mobola Plum, which is used in beer-making to extract a red dye; and *Neocarya macrophylla* is the Gingerbread Plum, all eaten in Africa. Oil may be extracted from the seeds of many species such as *Licania rigida* (Oiticica), which is cultivated for this purpose in Brazil, the oil being used as a substitute for tung oil, while the oil of *L. arborea* is inflammable and used in candle making and soapmaking. The wood of members of the Chrysobalanaceae contains silica, making it hard and difficult to work but extremely resistant. *Licania ternatensis* is used in underwater and underground construction, as well as for charcoal.   [DJM] VHH

i Prance, G. T. & Sothers, C. A. *Species Plantarum. Flora of the World.* Part 9. Chrysobalanaceae 1: *Chrysobalanus* to *Parinari*. Australian Biological Resources, Canberra (2003). Part 10. Chrysobalanaceae 2: *Acioa* to *Magnistipula*. Canberra, Australian Biological Resources (2003).
ii Savolainen, V. *et al.* Phylogeny of the eudicots: a nearly complete familial analysis based on *rbc*L sequences. *Kew Bull.* 55: 257–309 (2000).
iii Litt, A. J. & Chase, M. W. 1999. The systematic position of *Euphronia*, with comments on the position of *Balanops*: An analysis based on *rbc*L sequence data. *Syst. Bot.* 23: 401–409 (2003).

CIRCAEASTERACEAE

A family of small herbs with dichotomously veined leaves.

Distribution Himalayas to central China. *Circaeaster* grows in damp, rocky places from western Himalaya (Kumaon) to northern Yunnan and Shaanxi provinces of China at altitudes of between 2,200 and 3,800 m. *Kingdonia* grows in deep shade in damp gullies in coniferous forest from 2,800 to 3,800 m altitude in Yunnan, Sichuan, Ganzu, and Shaanxi provinces, reaching 3,950 m in *Rhododendron* forest in Yunnan.

Genera 2 Species 2
Economic uses None

Description *Circaeaster* is a slender, erect annual up to 20 cm tall, with a persistent hypocotyl surmounted by a rosette of leaves surrounding the inflorescence. *Kingdonia* is a perennial, with slender rhizomes bearing a single leaf with an erect petiole and a single-flowered peduncle up to 15 cm high. The leaves are flabellate, with dichotomous venation, simple and spathulate, or rarely bilobed (*Circaeaster*)[i], or more or less orbicular, but pedately divided almost to the base into around 5 lobes (*Kingdonia*). The inflorescence is either a compact, compound fascicle, or a solitary, terminal flower. The flowers are bisexual, actinomorphic, with a uniseriate perianth. The tepals are 2–3 and inconspicuous (*Circaeaster*) or 5–7 and petaloid (*Kingdonia*). The stamens are 1–3 in *Circaeaster*, while *Kingdonia* has 8–12 spirally arranged staminodes surrounding 3–6 stamens. The ovary is superior and of 1–3 or 5–9 free carpels, respectively, each carpel with a single ovule. The fruiting carpels in *Circaeaster* are linear-oblong and develop many conspicuous hooked hairs, while in *Kingdonia* they have the style sharply deflexed to form a recurved beak.

Classification Both genera have been referred to Ranunculaceae, and some recent accounts have retained *Kingdonia* within that family. Despite the many morphological dissimilarities between the 2 genera, they share the same curious dichotomous venation[ii], usually unilacunar nodes[i] and striate pollen, which is rare in related families. Molecular analysis has placed both genera together, either sister to Papaveraceae and Berberidaceae[iii], or close to Sargentodoxaceae[iv], and not close to Ranunculaceae. Affinity of both with Ranunculales is not in doubt. *Kingdonia* retains what is probably a plesiomorphic growth form for this group of families, while *Circaeaster* has been considered to be a highly derived neotenic plant[v]. If emphasis is given to the differences, there may be a good case for recognizing each genus in its own family, but it is preferred here to combine them in 1 family. RKB

[i] Ren, Y. & Hu, Z. H. The discovery of bilacunar nodes and bilobed leaves in *Circaeaster* and their systematic significance. *Acta Bot. Bor.-Occid. Sin.* 18: 566–569 (1998).
[ii] Yi, R., Xiao, Y. P., & Hu, Z. H. The morphological nature of the open dichotomous leaf venation of *Kingdonia* and *Circaeaster* and its systematic implication. *J. Plant Res.* 111: 225–230 (1998).
[iii] Oxelman, B. & Lidén, M. The position of *Circaeaster*—evidence from nuclear ribosomal DNA. In: Jensen, U. & Kadereit, J. W. Systematics and Evolution of the Ranunculiflorae. *Pl. Syst. Evol.*
(Supplement 9): 189–193 (1995).
[iv] Hoot, S. B. & Crane, P. R. Inter-familial relationships in the Ranunculidae based on molecular systematics. In: Jensen, U. & Kadereit, J. W. Systematics and Evolution of the Ranunculiflorae. *Pl. Syst. Evol.* (Supplement 9): 189–193 (1995).
[v] Ren, Y., Li, Z. J., & Hu, Z. H. Approaches to the systematic position of *Circaeaster* based on the morphological data. *Acta Bot. Bor.-Occid. Sin.* 23: 1091–1097 (2003).

CISTACEAE
ROCKROSES

A family of shrubs or herbs, often with showy flowers.

Distribution Most species grow in temperate or warm-temperate regions of the northern hemisphere, notably in southern Europe and the Mediterranean basin (*Fumana* 10 spp., *Cistus* 17 spp., *Halimium* 9 spp., *Tuberaria* 10 spp., and *Helianthemum* 80 spp.) and in the eastern USA. *Crocanthemum* (24 spp.) occurs from North America (with a concentration of species in the southeastern part of the continent) to Mexico, with a disjunct group of 3 species in southern South America. *Hudsonia* (3 spp.) is endemic to the Atlantic coast of North America. *Lechea* (18 spp.) grows in North and Central America and the Caribbean. Only a few species of *Crocanthemum*, *Helianthemum*, and *Lechea* occur in tropical America. *Cistus* and *Halimium* are characteristic of Mediterranean shrub communities, while *Helianthemum* species are often found in base-rich grasslands or open rocky habitats of the Mediterranean region.

Description Mainly shrubs or subshrubs, some annual to perennial herbs. The leaves are alternate or opposite, simple, entire, with or without stipules, sometimes ericoid. The flowers are solitary or in terminal or axillary cymes, actinomorphic, often showy. The sepals are 5, the outer 2 smaller and bracetolelike. The petals are 5, 3 (*Lechea*), or absent in cleistogamous flowers, often crumpled in bud; ephemeral. The stamens are 3–10 or more, usually numerous, free. The ovary is superior, usually with 3 carpels (6–12 in *Cistus*), the carpels usually 1-locular but some-

CISTACEAE (BELOW LEFT & RIGHT). 1 *Fumana procumbens* (a) habit showing simple opposite leaves and solitary flowers (x⅔); (b) flower with petals removed showing outer whorl of articulated sterile stamens and inner whorl of fertile stamens surrounding the elongated style with curved base and discoid stigma (x6); (c) dehiscing capsule showing 3 valves containing seeds (x2⅔). 2 *Tuberaria guttata* habit (x⅔). 3 *Cistus ladanifer* var. *maculatus* (a) habit (x⅔); (b) dehiscing capsule with 10 valves (x2); (c) cross section of capsule showing seed (x2⅔). 4 *C. symphytifolius* cross section of ovary showing projecting placentas bearing numerous ovules (x4). 5 (right) *Lechea mexicana* half flower with ovary containing 2 ascending ovules (x8).

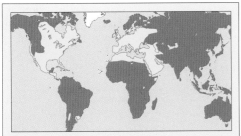

Genera 8 **Species** 170–180
Economic uses Several species of *Cistus*, *Halimium*, and *Helianthemum* are grown as ornamentals; some produce resin that has limited medicinal and cosmetic use

times incompletely 5- to 12-locular, with parietal placentation. The fruit is a loculicidal capsule, with 3 or 5–12 valves, and few to many seeds.

Classification The family is well circumscribed. It has previously been placed in the Violales or the Cistales, but molecular analysesi,ii,iii now indicate a placement with the Malvales, perhaps forming a sister group with the Dipterocarpaceae. The validity of the 8 presently recognized genera is supported by a recent molecular studyiv, which indicates that *Fumana* is the least derived genus in the family and forms a basal clade with *Lechea* and *Helianthemum*, while the other 5 genera form a terminal clade.

1b

5

3a

Economic uses Several species of *Cistus*, *Halimium*, and *Helianthemum* are widely cultivated as ornamentals for their showy flowers. Traditionally, *Cistus creticus* produced the aromatic gum ladanum, but the source of the modern commercial ladanum used in the perfumery industry is *C. ladanifer*. Extracts of several species are used in folk medicine in some Mediterranean countries. VHH

i Fay, M. F., Bayer, C., Alverson, W. S., de Bruijn, A. Y., & Chase, M. W. Plastid *rbc*L sequence data indicate a close affinity between *Diegodendron* and *Bixa*. *Taxon* 47: 43–50 (1998).
ii Alverson, W. S., Karol, K. G., Baum, D. A., Chase, M. W., Swensen, S. M., McCourt, R., & Systma, K. J. Circumscription of the Malvales and relationships to other Rosidae: Evidence from *rbc*L sequence data. *American J. Bot.* 85: 876–887 (1998).
iii Bayer, C. *et al.* Support for an expanded family concept of Malvaceae within a recircumscribed order Malvales: a combined analysis of plastid *atp*B and *rbc*L DNA sequences. *Bot. J. Linn. Soc.* 129: 267–303 (1999).
iv Arrington, J. M. & Kubitzki, K. Cistaceae. Pp. 62–70. In: Kubitzki, K. & Bayer, C. (eds), *The Families and Genera of Vascular Plants. V. Flowering Plants. Dicotyledons: Malvales, Capparales and Non-betalain Caryophyllales*. Berlin, Springer-Verlag (2002).

CLEOMACEAE
SPIDER FLOWER AND AFRICAN CABBAGE

A largely tropical family of plants that contain mustard oils.

Distribution Widespread throughout tropical and warm-temperate regions, with the greatest diversity in America. The largest genus, *Cleome* (c. 250 spp), is pantropical and extends into warm-temperate regions. All other genera are restricted to the Americas, except *Dipterygium* (1 sp.) in northeastern Africa through Arabia to Pakistan, and *Puccionia* (1 sp.), endemic to Somalia. Most genera and species are found in arid areas, but *Padandrogyne* and some *Cleome* grow in damp to wet conditions. Some genera (e.g., *Oxystylis* and *Wislizenia*) are predoninantly desert dwelling and show high tolerance to drought and saline conditions.

Description Annual or perennial herbs, subshrubs, or shrubs (*Isomeris*) with alternate leaves. The leaves are usually 3-foliate, but sometimes simple or to 7(–13) foliate (*Cleome*), usually petiolate, stipules, when present, from minute to large, and dissected or forming thorns. The inflorescence is usually a terminal bracteate raceme, sometimes ebracteate, rarely axillary clusters (*Oxystylis*) or single axillary flowers (some *Cleome*). The sepals are (2–)4(–6), free or fused at base. The petals are 4 (2 in *Haptocarpum*), free or fused at base, clawed and variously colored.

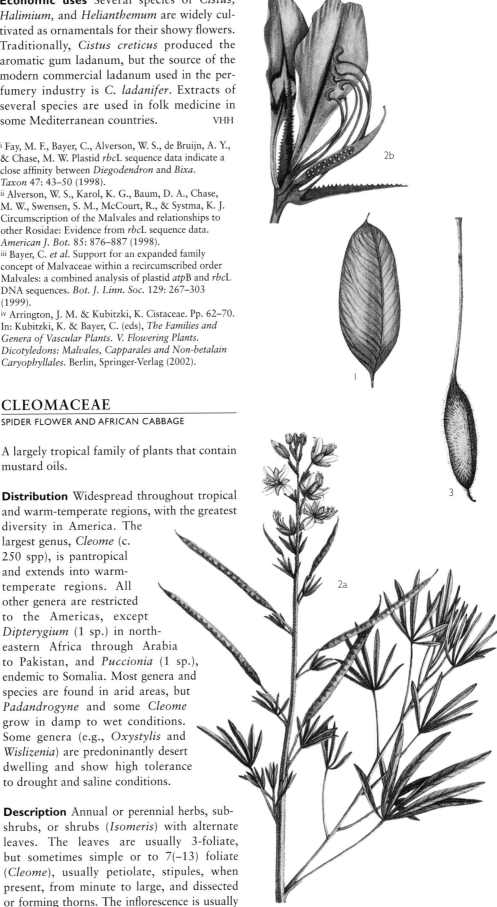

2b

1

3

2a

CLEOMACEAE. 1 *Dipterygium glaucum* winged seed (×6). 2 *Cleome hirta* (a) leafy shoot with flowers and capsular fruits (×⅔); (b) half flower with toothed sepals, 2 petals, 6 curved stamens, and ovary with numerous ovules (×2½). 3 *Podandrogyne brachycarpa* fruit—a capsule (×1).

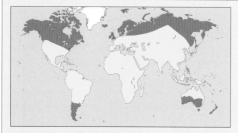

The stamens are usually 6 or (1, 2, 4–)6–many (*Cleome*), or up to 15 (*Polanisia*); or fertile stamens 4 and staminodes 2 (*Haptocarpum*) or fertile stamen 1 and staminodes 3–4 (*Dactylaena*). The ovary is superior, of 2 fused carpels borne on a gynophore, with 1 locule or 2 with a false septum (replum), and ten to many ovules. The fruit is a dry capsule, either a siliqua or silicula. Mustard oils (glucosinolates) are present in some species.

Classification The family is usually included in the Capparaceae, in Capparales[i], with Brassicaceae, and the other glucosinolate-containing families. Morphological cladistic analysis suggests an intimate relationship with Brassicaceae and Capparaceae, leading to a proposal to merge these families[ii], but molecular evidence[iii] indicates 3 identifiable groups: Brassicaceae, Cleomaceae, and Capparaceae as an alternative

scenario. Within the family, it is evident that much work is needed to resolve the relationships of the genera and to evaluate the generic limits of *Cleome*.

Economic uses *Cleome spinosa* and some other species are grown for their ornamental flowers. *Cleome gynandra* is a leaf vegetable grown in South Africa for its high mineral and vitamin content[iv]. AC

[i] Brummitt, R. K. *Vascular Plant Families and Genera.* Richmond, Royal Botanic Gardens, Kew (1992).
[ii] Judd, W. S., Sanders, R. W., & Donoghue, M. J. Angiosperm family pairs: preliminary phylogenetic analyses. *Harvard Papers in Botany* 5: 1–51 (1994).
[iii] Hall, J. C., Sytsma, K. J., & Iltis, H. H. Phylogeny of Capparaceae and Brassicaceae based on chloroplast sequence data. *American J. Bot.* 89: 1826–1842 (2002).
[iv] Van Wyk, B.-E. & Gericke, N. *People's Plants.* Pretoria, Briza Publications (2000).

CLETHRACEAE
LILY-OF-THE-VALLEY TREE

This family comprises 2 genera of tropical and subtropical evergreen or deciduous shrubs.

Distribution *Clethra* (c. 120 spp.) is found in tropical and subtropical Asia and America and in Macaronesia (Madeira). *Purdiaea* (10–15 spp.) occurs from Belize to Peru, with a center of diversity in Cuba.

Description Large shrubs or sometimes trees (*Clethra arborea*), with simple leaves and hairy shoots. The leaves are alternate, entire or toothed, with or without petioles, and without stipules. The inflorescence is a terminal or axillary raceme, condensed umbel-like raceme or panicle. The flowers are bisexual, actinomorphic or with irregular sepals (*Purdiaea*), white (*Clethra*) or mauve (*Purdiaea*), and borne in racemes or panicles, without bracteoles. The sepals are 5, free or fused, the petals 5, free or rarely fused, and the stamens in 2 whorls of 5, with the anthers bent outward in bud and opening by pores. The ovary is superior, occasionally with basal

nectaries, of 3 (*Clethra*) or 3–5 (*Purdiaea*) fused carpels, and contains 3 (*Clethra*) or 3–5 (*Purdiaea*) locules and numerous (*Clethra*) or a single (*Purdiaea*) ovule on axile placentas in each locule. The style is 3-lobed or undivided. The fruit is a loculicidal capsule enveloped by the enlarged calyx in *Purdiaea*. The seeds are often winged. *Clethra* is an ancient Greek word meaning "alder," and it is applied to this genus on account of the resemblance of some species of *Clethra* to those of alder (*Alnus*).

Classification Past classifications have placed Clethraceae with Ericales, with the exception of Thorne (1983), who placed the family in Theales. Phylogenetic studies have varied in the placement of *Clethra* with Styracaceae or Theaceae[i], or with Ericales[ii], the latter position becoming established as more sequence data have been added. The family is closely related to the Ericaceae and Cyrillaceae, forming a group with Cyrillaceae in 1 analysis[iii]. The inclusion of *Purdiaea* in this family rather than Cyrillaceae is still in need of further investigation to resolve the details of relationship among Clethraceae, Cyrillaceae, and Ericaceae[iv]. Based on seed morphology, *Clethra* is subdivided into two sections: sect. *Clethra* and sect. *Cuellaria*.

Economic uses *C. arborea*, the Lily-of-the-valley Tree, with fragrant white bell-like flowers, of Madeira, is cultivated as an ornamental. Several other species, including *C. alnifolia* (Sweet Pepper Bush), *C. acuminata* (White Alder), *C. monostachya*, and *C. tomentosa*, are cultivated as fragrant-flowered ornamental shrubs. There are several named cultivars[v] of the species. [SAH] AC

[i] Olmstead, R. G., Bremer, B., Scott, K. M., & Palmer, J. D. A parsimony analysis of the Asterideae *sensu lato* based on *rbc*L sequences. *Ann. Missouri Bot. Gard.* 80: 700–722 (1993).
[ii] Kron, K. A. Phylogenetic relationships of Empetraceae, Epacridaceae, Ericaceae, Monotropaceae, and Pyrolaceae: evidence from nuclear ribosomal 18s sequence data *Ann. Bot.* 77: 293–303 (1996).
[iii] Anderberg, A. A. & Zhang, X. Phylogenetic relationships of Cyrillaceae and Clethraceae (Ericales) with special emphasis on the genus *Purdiaea* Planch. *Organisms Diversity and Evolution* 2: 127–137 (2002).
[iv] Fior, S., Karis, P. O., & Anderberg, A. A. Phylogeny, taxonomy, and systematic position of *Clethra* (Clethraceae, Ericales) with notes on biogeography: evidence from plastid and nuclear DNA sequences. *Int. J. Plant Sci.* 164: 997–1006 (2003).
[v] Reed, S. M. Interspecific hybridization in *Clethra*. *HortScience* 37: 393–397 (2002).

CLETHRACEAE. 1 *Purdiaea nutans* (a) leafy shoot and terminal inflorescence (x⅔); (b) flower partly closed showing unequal sepals (x1½); (c) flower fully open showing 10 stamens and slender, undivided style (x1½); (d) young fruit with persistent calyx (x1½); (e) vertical section of gynoecium showing each locule with a single pendulous ovule (x14); (f) stamen (x2⅔).

CLUSIACEAE

MANGOSTEEN FAMILY

An economically important family whose circumscription has been much modified in recent years.

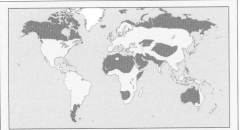

Genera 36 **Species** c. 1,630
Economic uses Drugs, dyes, fruits, gums, resins, and timber, and in horticulture

Distribution The Clusiaceae are wholly pantropical and lowland except for the Hypericoideae–Hypericeae, in which *Hypericum* is widespread (especially in the northern hemisphere) and mainly lowland to upland in temperate regions but montane in tropical and warm temperate regions, *Lianthus* is a monotypic genus from montane northeastern Yunnan, *Triadenum* has a mainly temperate disjunct East Asia/eastern North America distribution, and *Thornea* occurs in the mountains of tropical Mexico to Nicaragua. The other Hypericoidean tribes are Afro-American (Vismieae) and Madagascar–tropical eastern Asian (Cratoxyleae), respectively, and both are lowland to submontane. Genera of the Kielmeyeroideae are either American or tropical Asian, except for *Mammea* (pantropical), *Calophyllum* (pantropical but mainly in Asia and the Pacific region), and the 2 genera of the Endodesmieae, *Endodesmia*, and *Lebrunia*, all of which are African. In the Clusioideae, the Clusieae and the genus *Clusiella* are native to the USA, the Garcineae predominantly Old World but with some species of *Garcinia* in America, and the Symphonieae are pantropical.

Description Trees, shrubs, or perennial to annual herbs, evergreen or deciduous, with glands or canals in most parts of the plant secreting latex, essential oils, or resins. The leaves are simple and exstipulate, but occasionally with paired, stipuliform structures at or just below the leaf insertion point, opposite or more rarely alternate or whorled, with the margins entire or rarely fringed with glands. The flowers are actinomorphic, hermaphrodite or unisexual (plants dioecious), terminal and often axillary, solitary or in cymose or thyrsoid inflorescences. The sepals are 2–20 but usually 4–5, free or fused, imbricate in bud, persistent or deciduous after flowering. The petals are (3)4–5(–8) but may be absent, and free, equal, imbricate or contorted in bud, persistent or deciduous; epipetalous ligules are sometimes present in

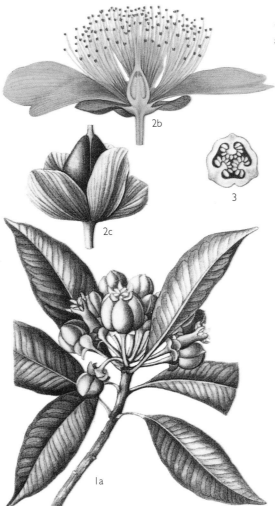

Cratoxylum and *Hypericum*. Essentially, the androecium consists of 2 whorls, each of 5 stamen bundles (fascicles), with filaments free almost to the base; the outer antisepalous whorl is sterile, with the members (fasciclodes) free or more or less united or absent, whereas the inner whorl is always fertile (except in female flowers) and may be modified by fusion of fascicles (most often 2+2+1) and/or stamens, or reduction to form masses or a ring of apparently free stamens. (In *Hypericum gentianoides*, each stamen fascicle is often reduced to a single stamen.) The ovary is superior and consists of usually 2–5 (–20) united carpels, each containing 1 to many ovules on axile to parietal, or more rarely apical or basal placentas. The stigmas are as many as the placentas, and the styles may be free, united, or totally lacking. Nectaries are absent; the fasciclodal

CLUSIACEAE. **1** *Symphonia globulifera* (a) leafy shoot and terminal inflorescence (x⅔); (b) androecium, comprising tube of fused stamens surrounding a 5-lobed stigma (x3); (c) fruit—a berry (x1½). **2** *Hypericum calycinum* (a) shoot with decussate leaves and terminal solitary flower (x⅔); (b) half flower with stamens in bundles and numerous ovules on axile placentas (x1); (c) fruit—a capsule (x1). **3** *H. frondosum* cross section of ovary showing single locule with ovules on 3 parietal placentas (x3). **4** *Calophyllum inophyllum* (a) shoot with inflorescences in axils of terminal leaves (x⅔); (b) gynoecium showing long, curved style (x4); (c) stamen (x4).

bodies at the base of the ovary sometimes swell (similar to lodicules in grasses) to expand the flower. The fruit is often a capsule but may be a dry or fleshy berry or a drupe. The seeds are sometimes winged or with an aril or rarely a caruncle, and often lack endosperm at maturity, and with a usually straight embryo, and the cotyledons well developed and free or united or sometimes much reduced and replaced in function by an enlarged hypocotyl (tigellus).

Classification The 5 subfamilies into which the family was once divided by Engler (1925) have now been reduced by Stevens[ii,iv] to 3, based on sex-distribution, androecium, ovary, fruit, and seeds:

SUBFAM. KIELMEYEROIDEAE Hermaphrodite to unisexual flowers, an androecium that lacks fasciclodes, and in which the stamen fascicles are rarely apparent, and styles that are wholly united (except in *Poeciloneuron* from southern India). The subfamily comprises 2 unequal tribes—Calophylleae (pantropical) and Endodesmieae—with 2 monotypic African genera, *Endodesmia* and *Lebrunia*.

SUBFAM. HYPERICOIDEAE Hermaphrodite flowers in which specialization for pollination has resulted in heterostyly, not unisexuality, fasciclodes are sometimes present, stamen fascicles are free or variously united, and styles are elongate and usually free. It comprises the tribes:

Vismieae, with regularly 5-merous flowers, the fruit a berry or drupe.

Cratoxyleae, with 5-merous perianth, 3-merous, 2-whorled androecium, and 3 carpels, the fruit a capsule with winged seeds.

Hypericeae, with 4- to 5-merous perianth and androecium, usually without fasciclodes, the fruit a capsule or rarely fleshy.

SUBFAM. CLUSIOIDEAE Hermaphrodite to unisexual flowers, with or without fasciclodes or apparent stamen fascicles, styles that are short or nearly always absent, and cotyledons that are nearly always much reduced in size. It comprises 4 tribes of which the first (confined to *Clusiella* but not yet validly named) has an apparently afasciculate androecium, fruit a berry, and elongate cotyledons.

Clusieae, with 4 genera in America, has unisexual or rarely polygamous flowers without fasciclodes, variable androecium and fruit a capsule or berry.

Garcinieae, with 2 genera—*Garcinia*, mostly Old-World, and *Allanblackia* in Africa—with usually unisexual flowers with or without fasciclodes, the androecium fasciculate or variously united, and fruit a berry.

Symphonieae, with 7 genera in America, Africa, and New Caledonia, the flowers hermaphrodite and regularly 5-merous, with fasciclodes free or (*Symphonia*) united in a ring, the androecium fasciculate and the fruit a berry.

The Clusiaceae are related to the Bonnetiaceae, differing from them in their entire or gland-fringed leaves and the absence of a secretory system. The nearest genus to the Clusiaceae in that family is *Ploiarium*,

which has regularly 5-merous flowers with fasciclodes. Both families are included in the Malpighiales in APG II. Gustafson *et al.*[ii] and some other molecular studies have placed the Podostemaceae, a family of tropical water plants, as a sister group to the Hypericoideae or to *Hypericum* itself, thus making the Clusiaceae paraphyletic. If this is indeed so, it seems more satisfactory to accept the paraphyly than to split the Clusiaceae.

Economic uses The Clusiaceae has been used as a source of hard and/or durable wood (species of *Mesua*, *Calophyllum*, *Cratoxylum*, *Platonia*, and *Montrouziera*), easily worked wood (species of *Calophyllum* and *Harungana*), drugs or dyes from bark (species of *Vismiea* and *Calophyllum*), gums, pigments, and resins from stems, including species of *Garcinia* (Gamboge) and *Clusia* (healing gums), drugs from leaves and flowers (species of *Hypericum* and *Harungana madagascariensis*), drugs and cosmetics from flowers (*Mesua ferrea*), edible fruits such as species of *Garcinia*, including *G. mangostana* (Mangosteen), *Mammea*, including *M. americana* (Mammey Apple), and *Platonia insignis*, and fats and oils from seeds (species of *Calophyllum*, *Mammea*, *Allanblackia*, *Garcinia*, and *Pentadesma*). NKBR

[i] Gustafson, M. H. G., Bittrich, V., & Stevens, P. F. Phylogeny of Clusiaceae based on *rbc*L sequences. *Int. J. Plant Sci.* 163: 1045–1054 (2002).

[ii] Robson, N. K. B. Studies in the genus *Hypericum* L. (Guttiferae). Parts 1–3, 7–8. *Bull. Br. Mus. (Nat. Hist.), Bot.* 5: 291–355 (1977), 8: 55–226 (1981), 12: 163–325 (1985), 16: 1–106 (1987), 20: 1–151 (1990); Parts 4 (1–2), 6. *Bull. Nat. Hist. Mus. Lond.* 26: 75–217 (1996), 31: 37–88 (2001), 33: 61–123 (2002); Part (3). *Syst. & Biodiv.* 4(1): 80 pp. (2006).

[iii] Stevens, P. F. Clusiaceae. In: Kubitzki, K. (ed.), *The Families and Genera of Vascular Plants. IX. Flowering Plants. Eudicots.* Berlin, Springer-Verlag (2007).

CNEORACEAE

A family of shrubs from the western Mediterranean region and Canaries, with a questionably native locality in Cuba.

Genera 2   Species 2
Economic uses Medicinal, especially as a purgative

Distribution *Cneorum tricoccon* occurs on calcareous rocky slopes in Spain, Baleares, southern France, Sardinia, and Italy. It is also known from a rather remote locality in Cuba, where its status as a native or introduced species is uncertain. *Neochamaelea pulverulenta* occurs in the Canaries (Gran Canaria, Tenerife, Gomera, Palma, and Hierro), where it is locally common and often dominant in *Euphorbia* communities.

Description Shrubs up to 1.5 m high. The leaves are alternate, entire, linear-oblanceolate, obtuse to retuse, cuneate at the base, and subsessile, without stipules. The inflorescence is a solitary flower or few-flowered, small, dense, few-flowered cyme on a short peduncle in the axil of a leaf, or often with the peduncle adnate to the base of the leaf. The flowers are actinomorphic, bisexual, 3-merous in *Cneorum*, and 4-merous in *Neochamaelea*. The sepals are small and ovate. The petals are narrowly oblong and yellow. There is a short, cushionlike androgynophore bearing the stamens, each of which is inserted in a lateral depression, and the terminal ovary. The ovary is superior, in *Cneorum* with 3 carpels attached to a central axis (resembling Euphorbiaceae), with a conspicuous terminal style, in *Neochamaelea* similar but with usually 4 carpels, with 2 pendulous ovules in each locule. The fruit is a schizocarp, with 3 mericarps in *Cneorum* and 1–4 mericarps developing in *Neochamaelea*, each mericarp with 1 or 2 seeds.

Classification The family has usually been placed near to Rutaceae, or sometimes Zygophyllaceae, and in APG II is sunk into Rutaceae along with Ptaeroxylaceae. Molecular evidence places Cneoraceae nested in Ptaeroxylaceae as defined here—for a discussion, see the latter family. However, Cneoraceae differs markedly in their slender shrubby habit, simple linear leaves, reduced axillary inflorescences, often 3-merous flowers, striate or verrucate pollen, and schizocarpic fruit and can be readily recognized as a separate family. The 2 genera are similar in appearance, but *Neochamaelea* differs from *Cneorum* in having minute bifid hairs, 4-merous flowers, and different pollen (3-colporate, with a striate exine in *Cneorum*, 4- to 6-colporate with a verrucate-gemmate exine in *Neochamaelea*).

The occurrence of the genus in Cuba has often been quoted in phytogeographical literature as a remarkable geographical discontinuity. The Cuban population was described as a third species, *Cneorum trimerum*, but this has recently convincingly been treated as conspecific with the western Mediterranean *C. tricoccon*[i]. The locality in Cuba is remote from habitation, and it might be argued that it is unlikely to have been introduced there by people. However, no satisfactory explanation for this discontinuity seems to be available at present.

Economic uses *C. tricoccon* is recorded as being used as a purgative.   RKB

[i] Lobreau-Callen, D. & Jérémie, J. L'espèce *Cneorum tricoccon* (Cneoraceae, Rutales) représentée à Cuba. *Grana* 25: 155–158 (1986).

COCHLOSPERMACEAE

A small pantropical family of trees, shrubs, or herbs, often with conspicuous flowers.

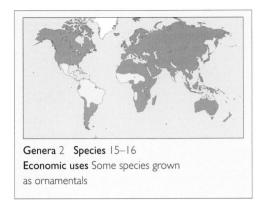

Genera 2   Species 15–16
Economic uses Some species grown as ornamentals

Distribution *Cochlospermum* subg. *Cochlospermum* is pantropical, occurring in tropical America, Africa, Asia, and northern Australia while subg. *Diporandra* is neotropical, confined to South America. The main center of distribution of *Amoreuxia* (3 or 4 spp.) is in Mexico and Central America, but it also occurs in parts of the southwestern USA, northwestern South America, and parts of the Caribbean.

Description Trees, shrubs, or low-growing herbs, with woody rootstocks (*Amoreuxia*). The leaves are alternate, palmatisect or

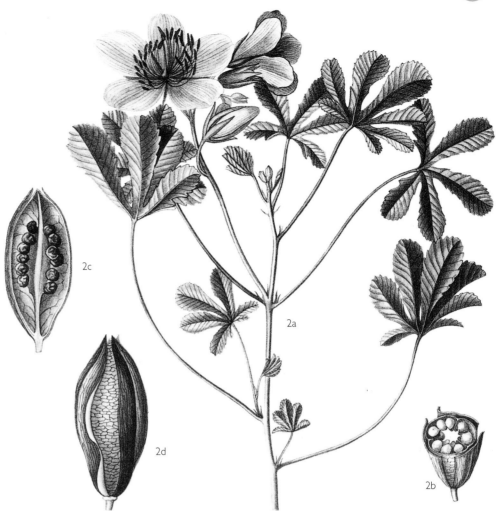

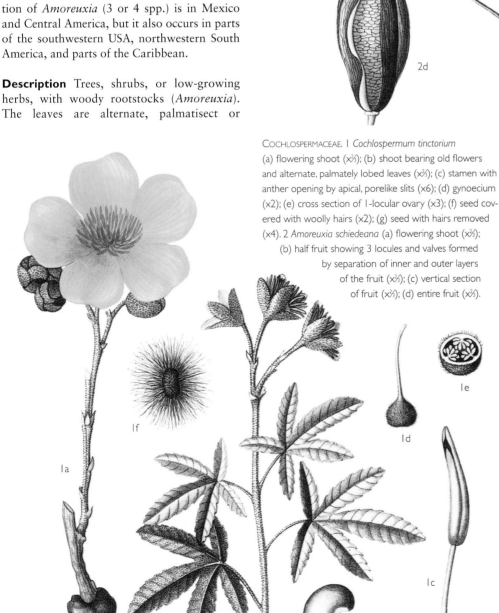

COCHLOSPERMACEAE. I *Cochlospermum tinctorium* (a) flowering shoot (x⅔); (b) shoot bearing old flowers and alternate, palmately lobed leaves (x⅔); (c) stamen with anther opening by apical, porelike slits (x6); (d) gynoecium (x2); (e) cross section of 1-locular ovary (x3); (f) seed covered with woolly hairs (x2); (g) seed with hairs removed (x4). 2 *Amoreuxia schiedeana* (a) flowering shoot (x⅔); (b) half fruit showing 3 locules and valves formed by separation of inner and outer layers of the fruit (x⅔); (c) vertical section of fruit (x⅔); (d) entire fruit (x⅔).

palmatilobed, sometimes nearly entire, with stipules. The flowers are large, hermaphrodite, actinomorphic (*Cochlospermum*) or slightly zygomorphic (*Amoreuxia*), in terminal thyrsoid inflorescences. The sepals are 5, free, deciduous. The petals are 5, free. The stamens are numerous, free (*Cochlospermum*) or in 2 bundles (*Amoreuxia*), the anthers deshiscing by pores. The ovary is superior, with 3–5 carpels, with 1 or 3 locules. The fruit is a 3- to 5-valved loculicidal capsule, with a single locule or with 3 locules and partly dehiscent. Seeds large, with whitish or reddish hairs in *Cochlospermum*, glabrous or hairy in *Amoreuxia*. Endosperm not starchy.

Classification In subg. *Cochlospermum*, the anthers have a single apical pore, while in subg. *Diporanda* there are 2. The Cochlospermaceae may be included in the Bixaceae but, as discussed under that family, the evidence is equivocal.

Economic uses Some species of *Cochlospermum* are grown as ornamentals, especially *C. vitifolium* (Rose Imperial), which also has a cultivar with double flowers. The seed hairs are occasionally used as a stuffing and some local medicinal uses have been reported.   VHH

COLUMELLIACEAE

A family consisting of the single genus *Columellia*, which comprises shrubs up to 2 m to trees up to 7 m, rarely up to 20 m, or with a trunk 2 m across[i], recorded from Colombia to Bolivia in open places (sometimes among cacti), rocky hillsides, or in montane forest patches, from 1,600 up to 3,900 m in altitude. The leaves are opposite, simple, subsessile, more or less oblanceolate, the margins entire or serrate in the distal part, the apex apiculate or with a short dark blunt spine or

| Genera | Species 2–4 |

Economic uses Some medicinal properties, wood used for small objects, stems for baskets

gland, often sericeous beneath, with interpetiolar stipules that fall to leave conspicuous raised nodes. The inflorescence is a usually terminal, dichasial cyme of (1) few to many flowers, with pedicels up to 1 cm. The flowers are slightly zygomorphic, bisexual, perigynous, 5-merous or rarely recorded as 4- to 8-merous. The sepals are normally 5, triangular, with a dark apiculum like that of the leaf apex, persistent on the sides of the dehiscing fruit. The corolla is 0.7 to 1 cm long, fused for half its length, funnel-shaped to campanulate, with usually 5 spreading suborbicular lobes, which are yellow. The stamens are 2 (rarely 3), inserted on the mid to upper part of the corolla tube, with a short, broad filament that is expanded above into a broad reniform connective, bearing 2 highly contorted plicate anther thecae on its margins, the 2(3) anthers occupying the mouth of the corolla. The ovary is semi-inferior, of 2 fused carpels, with a stout style and capitate stigma, with 2 incompletely separated locules and many ovules said to be inserted on intruded parietal placentas. The fruit is a many-seeded capsule.

Few if any other families have had so many diverse relationships postulated for them on morphological grounds, including Ebenaceae, Ericaceae, Escalloniaceae, Gentianaceae, Gesneriaceae, Hydrangeaceae, Lythraceae, Oleaceae, Onagraceae, and Saxifragaceae[ii]. Even the molecular evidence gives widely divergent suggestions, different analyses placing it in Dipsacales or close to Icacinaceae *sensu lato* or included in Desfontainiaceae. Columelliaceae is easily recognized by its raised nodes, its gamopetalous flowers with 2 stamens with unusual contorted anthers in the throat of the corolla, and its largely inferior ovary, and separate family status seems essential. In APG II, it is unplaced as to order in Euasterids II.

The use of the leaves against tertian fever and stomach complaints was first recorded by Ruiz and Pavón in 1798. The hard wood is used for making handles and various utensils. RKB

[i] Brizicky, G. A synopsis of the genus *Columellia* (Columelliaceae). *J. Arnold Arbor.* 42: 363–372 (1961).
[ii] Struwe, L. Columelliaceae. Pp. 108–109. In: Smith, N. *et al.* (eds), *Flowering Plants of the Neotropics*. Princeton, Princeton University Press (2004).

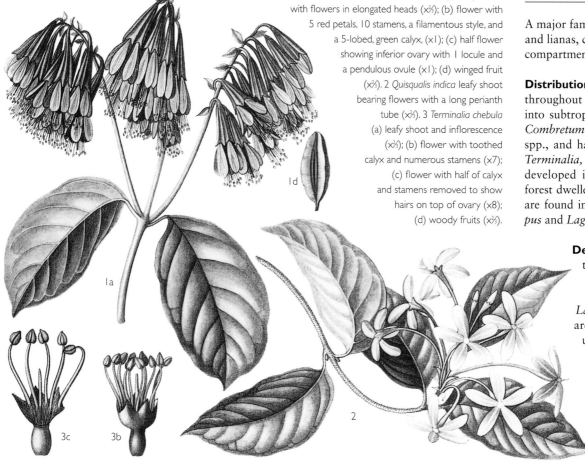

COMBRETACEAE. 1 *Combretum grandiflorum* (a) leafy shoot with flowers in elongated heads (×⅔); (b) flower with 5 red petals, 10 stamens, a filamentous style, and a 5-lobed, green calyx, (×1); (c) half flower showing inferior ovary with 1 locule and a pendulous ovule (×1); (d) winged fruit (×⅔). 2 *Quisqualis indica* leafy shoot bearing flowers with a long perianth tube (×⅔). 3 *Terminalia chebula* (a) leafy shoot and inflorescence (×⅔); (b) flower with toothed calyx and numerous stamens (×7); (c) flower with half of calyx and stamens removed to show hairs on top of ovary (×8); (d) woody fruits (×⅔).

COMBRETACEAE

A major family of largely tropical trees, shrubs, and lianas, characterized by unique, unicellular, compartmented hairs.

Distribution The Combretaceae is distributed throughout the tropics, with some extensions into subtropical and warm–temperate regions. *Combretum* is the largest genus, with some 250 spp., and has its center of diversity in Africa. *Terminalia*, with about 150–200 spp., is best developed in Asia. The trees and lianas are forest dwellers, while the shrubs and subshrubs are found in grasslands. *Lumnitzera*, *Conocarpus* and *Laguncularia* are mangroves.

Description Evergreen or deciduous trees up to 50 m, shrubs, low shrubs or lianas up to 30 m or more. Pneumatophores are found in *Laguncularia racemosa*. The leaves are opposite or alternate, simple, usually with unicellular hairs, with distinctive basal compartments, and with multicellular stalked glands; domatia sometimes occur mainly in the axils of the leaf veins; the leaf margins are entire or subentire; the stipules absent or vestigial. The flowers are

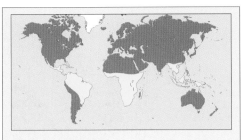

Genera 13–20   Species c. 500
Economic uses Timber trees; some ornamentals; local medicinal uses

small, actinomorphic or slightly zygomorphic, hermaphrodite or sometimes unisexual (plants dioecious or andromonoecious), in racemose, spicate, or paniculate inflorescences, often densely congested. A hypanthium is present at the top of the ovary, often nectariferous. The sepals are 4–5(–8), often small, and much reduced, free and attached toward the mouth of the hypanthium, or connate at the base, valvate. The petals are 4–5(–8), sometimes absent (*Terminalia, Buchenavia*), usually inserted near the mouth of the upper part of the hypanthium, valvate or imbricate. The stamens are 4–5, or 8–10, in 1 or 2 whorls. The ovary is inferior or rarely semi-inferior (*Strephonema*), of 2–5 carpels, 1-locular, the placentation apical, usually with 2–5 ovules, or sometimes up to 7 or more, pendulous and apical; style 1. The fruit is drupaceous, nutlike, or a false samara; 1-seeded. In the mangrove genera, the seeds germinate on the parent plant.

Classification The family may be divided into 2 subfamilies: Combretoideae, which contains most genera and comprises 2 tribes: Laguncularieae and Combreteae[i]; and Strephonematoideae, which contains a single genus *Strephonema* of 6 spp. from West Africa, characterized by having semi-inferior ovaries and seeds with massive hemispherical, hypogeal cotyledons. Strephonematoideae is sometimes treated as a separate family, Strephonemataceae.

The Combretaceae has long been included in the Myrtales despite the distinctive features, including the unique diagnostic "combretaceous hairs." This placement has been supported by molecular studies[ii,iii], which indicate that it is closest to Melastomataceae, occupying a basal position within the order.

Economic uses Many species of *Terminalia* are sources of timber, some of which is exported. Several climbers with attractive flowers are grown as ornamentals, such as *Combretum indicum* (*Quisqualis indica*) from Asia and *C. grandiflorum* from tropical West Africa. The kernels of *Terminalia catappa* from tropical Asia are edible (Indian Almond). Many Combretaceae are used locally as medicines and for tanning such as the myrabolans (myrobalans) obtained from the dried fruits rich in tannins of *Terminalia* species.   VHH

[i] Tan, F., Shi, S., Zhong, Y., Gong, X., & Wang, Y. Phylogenetic relationships of Combretoideae (Combretaceae) inferred from plastid, nuclear gene and spacer sequences. *J. Plant. Res.* 115: 475–481 (2002).
[ii] Conti, E., Litt, A., & Sytsma, K. J. Circumscription of Myrtales and their relationships to other Rosids: Evidence from *rbcL* sequence data. *American J. Bot.* 83: 221–233 (1996).
[iii] Conti, E. *et al*. Interfamilial relationships in Myrtales: Molecular phylogeny and patterns of morphological evolution. *Syst. Bot.* 22: 629–647 (1997) [1998].

CONNARACEAE
ZEBRA WOOD

The Connaraceae is a pantropical family of trees and climbing shrubs.

Distribution The family is pantropical. The most important genera are *Connarus* (c. 80 spp., distributed in Africa, Asia, the Pacific, Australasia, and tropical America), *Rourea* (pantropical, 40–70 spp.), *Cnestis* (c. 13 spp, 12 of which are native to Africa and Madagascar, and 1 in Malaysia), and *Agelaea* (8 spp. in tropical Africa, Madagascar, Southeast Asia, and Malaysia). It grows in lowland rain forest, rarely in mountains, and in forest patches in savannas.

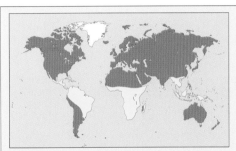

Genera c. 12   Species 160–190
Economic uses Medicines, timber (zebra wood); tannins and fibers used locally

Description Commonly evergreen, but sometime deciduous, trees (*Connarus, Jollydora*), shrubs (*Burttia, Connarus, Ellipanthus, Hemandradenia, Vismianthus*), or climbers (*Agelaea, Cnestidium, Cnestis, Connarus, Manotes, Pseudoconnarus, Rourea*) with alternate leaves. The leaf is usually imparipinnate or 3-foliate, a few species 1-foliate, the stipules absent, the petiole with a basal pulvinus; leaflets entire with a pulvinate petiolule. The flowers[i] are actinomorphic and hermaphrodite (but unisexual flowers have been reported in *Connarus*[ii]), in an axillary panicle, the panicles sometimes clustered near the shoot tips (e.g., *Agelaea* and *Connarus*), but sometimes cauliflorous (*Jollydora*, some *Cnestis*, and very few other species). The sepals are 5, free or fused, imbricate or sometimes valvate (*Manotes*). The petals are 5, free or fused, imbricate, sometimes with surface glands or hairs (*Connarus*). The stamens are in 2 equal whorls of 5

CONNARACEAE. 1 *Rourea foenumgraecum* (a) flowering shoot (x⅔); (b) half flower (x6). 2 *Rourea blanchetiana* (a) flowering shoot with imparipinnate leaves and flowers (x⅔); (b) vertical section of part of ovary (x14).

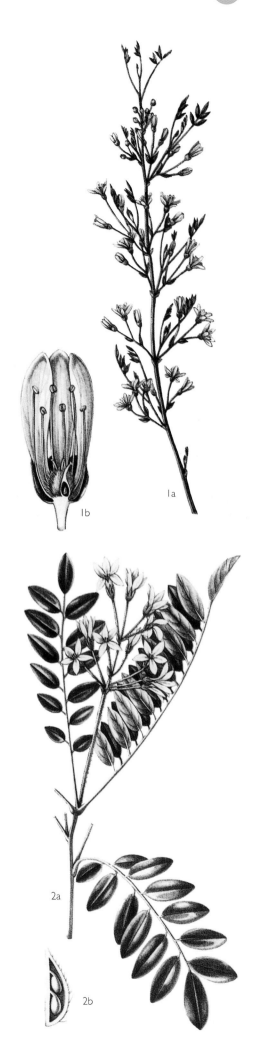

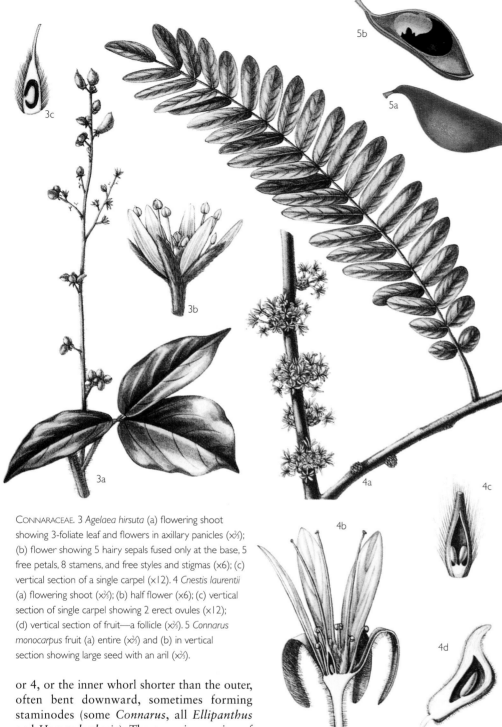

CONNARACEAE. 3 *Agelaea hirsuta* (a) flowering shoot showing 3-foliate leaf and flowers in axillary panicles (x⅔); (b) flower showing 5 hairy sepals fused only at the base, 5 free petals, 8 stamens, and free styles and stigmas (x6); (c) vertical section of a single carpel (x12). 4 *Cnestis laurentii* (a) flowering shoot (x⅔); (b) half flower (x6); (c) vertical section of single carpel showing 2 erect ovules (x12); (d) vertical section of fruit—a follicle (x⅔). 5 *Connarus monocarpus* fruit (a) entire (x⅔) and (b) in vertical section showing large seed with an aril (x⅔).

or 4, or the inner whorl shorter than the outer, often bent downward, sometimes forming staminodes (some *Connarus*, all *Ellipanthus* and *Hemandradenia*). The ovary is superior, of 1–5 free carpels each with 2 seeds. The fruits are usually 1-seeded follicles (2-seeded in *Jollydora*), the seeds often with a brightly colored, fleshy coat. The fruits are dispersed by birds such as pigeons and hornbills.

Classification The family has variously been placed in Rosales (Cronquist 1992), Rutales Thorne 1992), Sapindales[iii], Fabales[iv], and Connarales (Takhtajan 1997), in each case with comments about anomalous characters, but phylogenetic studies show a relationship to Oxalidaceae in the Oxalidales (APG II) based on DNA sequence analysis. This grouping was corroborated by morphological, biochemical, and anatomical features[v]. There is considerable variation in the estimates of species numbers (between 180 and 390) and in the generic circumscription (12–24).

Economic uses Economically, the family is important for Zebra Wood, which is obtained from *Connarus guianensis*, a native of Guyana. The seeds of the African *C. africanus* are used as anthelmintics. The leaves of the West African tree *Agelaea villosa* are used to treat dysentery, while those of *A. emetica* (native to Madagascar) yield an essential oil that promotes vomiting. The leaves of the West African species, *Cnestis corniculata* and *C. ferruginea*, are the source of an astringent and a laxative respectively. The oil from seeds of *Connarus* spp. have been used in the manufacture of soap in India[ii].   AC

[i] Matthews, M. L. & Endress, P. K. Comparative floral structure and systematics in Oxalidales (Oxalidaceae, Connaraceae, Brunnelliaceae, Cephalotaceae, Cunoniaceae, Elaeocarpaceae, Tremandraceae). *Bot. J. Linn. Soc.* 140: 321–381 (2002).

[ii] Lemmens, R. H. M. J., Breteler, F. J., & Jongkind, C. C. H. Connaraceae. Pp 74–81. In: Kubitzki, K. (ed.), *The Families and Genera of Vascular Plants. VI. Flowering Plants. Dicotyledons: Celastrales, Oxalidales, Rosales, Cornales, Ericales.* Berlin, Springer-Verlag (2004).
[iii] Dahlgren, R. M. T. General aspects of angiosperm evolution and macrosystematics. *Nordic J. Bot.* 3: 119–149 (1983).
[iv] Beddell, H. G. & Reveal, J. L. Amended outlines and indices for 6 recently published systems of angiosperm classification. *Phytologia* 51: 65–156 (1982).
[v] Nandi, O., Chase, M. W., & Endress, P. K. A combined cladistic analysis of angiosperms using *rbc*L and non-molecular data sets. *Ann. Missouri Bot. Gard.* 85: 137–212 (1998).

CONVOLVULACEAE
MORNING GLORIES AND BINDWEEDS

A family of climbers, herbs, shrubs, and occasionally trees, with usually regular flowers, with an unlobed corolla with 5 conspicuous midpetaline bands.

Genera 56 **Species** c. 1,840
Economic uses Edible crops, ornamentals, medicinal plants; some serious weeds

Distribution Widespread in tropical and temperate regions. In temperate and drier tropical places they are often low-growing herbs or shrubs in grassy or rocky places, sometimes on sand dunes (*Calystegia* and *Ipomoea* spp.) or salt marshes (*Cressa, Wilsonia*), or in semideserts (especially *Convolvulus*), or are climbers over other vegetation. *Ipomoea aquatica* grows in still water. In the wetter tropical areas they are often herbaceous climbers in thickets or at forest margins, or often high-climbing, woody lianes reaching the forest canopy at 35 m or more.

Description Annual to perennial herbs to shrubs or often herbaceous or woody climbers up to 35 m or more, or occasionally freestanding trees (*Humbertia*, some *Erycibe* and New World *Ipomoea* species), or leafless parasites (*Cuscuta*). The climbers twine sinistrally and have no tendrils or other climbing aids. The leaves are alternate, usually with a cordate base in the herbaceous climbers, but often coriaceous and cuneate in the woody lianes, usually simple but occasionally compound with palmate or pinnate segments (*Ipomoea* and *Merremia* species), the margins entire or weakly to deeply lobed (spinulose in *Hyaloscystis*), or leaves are absent (*Cuscuta*). The inflorescence is usually an axillary cyme and may be capitate (some *Jacquemontia, Convolvulus* etc.) or reduced to a single axillary flower (e.g., most *Calystegia*) but is sometimes racemose or a terminal thyrse (*Humbertia*). The bracts may sometimes be strongly accrescent in fruit (e.g., *Neuropeltis*, where they are adnate to the pedicel). The pedicels bear 2 bracteoles that may be large and enclosing the sepals (*Hewittia*, some *Calystegia*) or may be scarcely visible (*Neuropeltis*). The flowers are actinomorphic except in *Humbertia* (see first tribe below) and *Mina lobata*, bisexual or, in *Cladostigma* and *Hildebrandtia*, functionally unisexual (plants dioecious in *Hildebrandtia*), and 5-merous except in some 4-merous *Cladostigma* and *Hildebrandtia* species. The sepals are usually free and overlapping in a quincuncial arrangement, sometimes unequal or accrescent. The corolla is campanulate to funnel- or salver-shaped and 0.2 to 9.5 cm long, or (*Lepistemon*, *Mina lobata*) urceolate, or in some moth-pollinated *Ipomoea* species, the tube is narrowly

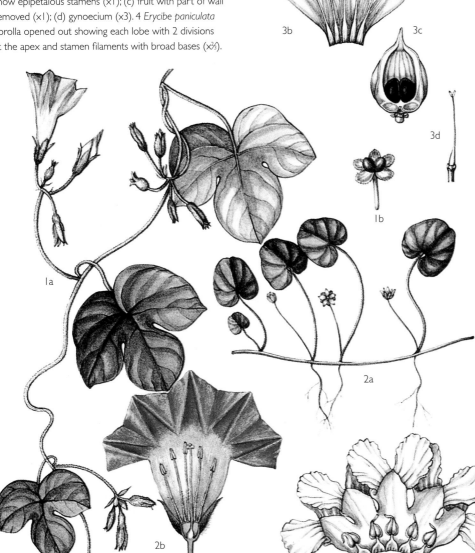

CONVOLVULACEAE. 1 *Ipomoea purpurea* (a) twining stem with axillary flowers (x⅔); (b) flower with corolla opened out showing stamens inserted at its base with a superior ovary surmounted by a thin style and lobed stigma (x1). 2 *Dichondra repens* (a) habit (x⅔); (b) fruit comprising 2 mericarps (x2). 3 *Calystegia sepium* (a) twining stem with solitary axillary flowers (x⅔); (b) corolla opened out to show epipetalous stamens (x1); (c) fruit with part of wall removed (x1); (d) gynoecium (x3). 4 *Erycibe paniculata* corolla opened out showing each lobe with 2 divisions at the apex and stamen filaments with broad bases (x⅔).

cylindrical and up to 16 cm long, always characterized by conspicuous mid-petaline bands (see below). The stamens are attached to, and usually included in, the corolla, but occasionally strongly exserted (*Humbertia*, some small-flowered genera). Pollen is variable, and provides important tribal characters. The ovary is superior, of 2 (or rarely 3–5) carpels, with a simple or forked terminal style or 2 distinct styles or, in *Falkia* and *Dichondra*, the ovary is deeply 2- to 4-lobed, with a gynobasic style; the stigma is variously divided or swollen. The ovary has 1–4 locules and usually 4 basally attached ovules, but *Humbertia* has many ovules (only 1–2 or 4 developing). The fruit is typically a dehiscent 4-seeded capsule but may be a single-seeded utricle or a fleshy or mealy berry (*Argyreia*) or become woody and nutlike, or may break up irregularly. The seeds typically have 2 flat faces and a rounded back and may bear long hairs.

When the flower is in bud, a band occupying the longitudinal middle third to fifth of each of the 5 segments of the corolla (tapering to a point at the apex) is exposed to the exterior, while the rest of the corolla is folded into the middle of the flower. These bands are clearly seen in the opened flower and are generally known as the mid-petaline bands. They usually differ markedly from the other areas of the corolla in color, texture, or indumentum and may include a clearly visible vein. They are characteristic of the family and may provide characters of generic significance (e.g., pubescent in *Convolvulus* and glabrous in *Calystegia*).

Classification The family is allied to other families that have alternate leaves and actinomorphic, gamopetalous flowers and is placed in the Solanales of APG II. Traditional subdivision of the family[i] has been based largely on characters of the ovary, number of styles, shape of stigmas, pollen characters, and dehiscence of the fruit, as well as the obvious differences in habit and leaf form. Recent molecular analyses[ii,iii] have largely reflected this, except in some details, and 12 tribes are here recognized:

Humbertieae, from Madagascar (*Humbertia*, 1 sp.).

Erycibeae, 1 genus and 74 spp., tropical Asia to Queensland. *Erycibe* (74 spp.).

Cardiochlamyeae, 6 genera and 21 spp., Madagascar, tropical Asia, and central Australia.

Poraneae, 4 (7) genera and 38 (54) spp., pantropical, *Calycobolus* (30 spp.), *Dipteropeltis* (2 spp.), *Porana* (2 spp.), *Rapona* (1 sp.). Of these genera, *Porana* is placed by the molecular data closer to the Dichondreae, to which it shows little morphological similarity. *Metaporana* (5 spp.) is placed as sister to *Porana* but has close morphological similarities to the Cresseae. In addition, *Neuropeltis* (13 spp.) and *Neuropeltopsis* (1 sp.) are placed by molecular data sister to *Calycobolus*, *Dipteropeltis*, and *Rapona* and resemble them in adaptation to wind for seed dispersal, but their accrescent parts are bracts rather than the calyx.

Maripeae, 3–4 genera, 35–36 spp., tropical America, *Dicranostyles* (15 spp.), *Lysiostyles* (1 sp.), *Maripa* (19 spp.). In addition, *Itzaea* (1 sp.) fits closely with Maripeae in morphological characters but is placed by molecular data near to *Calycobolus* of the Poraneae.

Dichondreae, 4 genera of which 1 genus is widespread, 1 genus is from Yemen to South Africa, 1 genus is confined to Ethiopia, and 1 genus in Mexico, with 20 spp., *Dichondra* (15 spp.), *Falkia* (3 spp.), *Nephrophyllum* (1 sp.), *Petrogenia* (1 sp.). **Cresseae**, 8 genera, with 204 spp., widespread, *Bonamia* (56 spp.), *Cladostigma* (3 spp.), *Cressa* (4 spp.), *Evolvulus* (97 spp.), *Hildebrandtia* (12 spp.), *Seddera* (23 spp.), *Stylisma* (6 spp.), and *Wilsonia* (3 spp.).

Cuscuteae, 1 genus (*Cuscuta*) with 183 spp., widespread.

Aniseieae, 4 genera with 7 spp., tropical America, *Aniseia* (3 spp.), *Iseia* (1 sp.), *Odonellia* (2 spp.), and *Tetralocularia* (1 sp.). The last of these is basal to the other genera in molecular position and differs markedly in its inflorescence and 4-lobed fruit and may be better referred to a different tribe.

Convolvuleae, 4 genera and 375 spp., temperate and tropical regions, *Calystegia* (26 spp.), *Convolvulus* (221 spp.), *Jacquemontia* (120 spp.), *Polymeria* (8 spp.). *Jacquemontia* is placed somewhat removed from the other genera in the molecular analysis but is so close to *Convolvulus* morphologically that it is difficult to define clearly.

Merremieae, 7 genera with 93 spp., pantropical, *Decalobanthus* (1 sp.), *Hewittia* (2 spp.), *Hyalocystis* (2 spp.), *Merremia* (70 spp.), *Operculina* (15 spp.), *Remirema* (1 sp.), and *Xenostegia* (2 spp.).

Ipomoeeae, 10 genera and c. 770 spp., pantropical, *Argyreia* (125 spp.), *Astripomoea* (12 spp.), *Ipomoea* (c. 600 spp.), *Lepistemon* (7 spp.), *Lepistemonopsis* (1 sp.), *Mina* (1 sp.), *Paralepistemon* (2 spp.), *Rivea* (3 spp.), *Stictocardia* (12 spp.), and *Turbina* (12 spp.).

Economic uses Root tubers of *Ipomoea batatas* are extensively eaten in the tropics (Sweet Potato) and are used as a source of industrial alcohol and sugar, or as livestock feed. Leaves of *I. aquatica* (Water Spinach) are eaten as a pot-herb (Kangkong, Ong choi) in the tropics. Minor medicinal products, mostly purgatives, are obtained from the roots of a few species of *Ipomoea* (especially *I. purga*, Jalap.), *Convolvulus* (*C. scammonia*, Scammony), *Operculina*, and others produce ergoline alkal-oids in their seeds and have been used in religious rituals. Species of *Ipomoea*, *Convolvulus*, *Evolvulus*, *Poranopsis*, and *Turbina* are grown as ornamentals, and *Argyreia* and *Merremia* also provide capsules and accrescent calyces used in dried flower arrangements. *Dichondra* species are sometimes used as a lawn substitute. *Cuscuta* (Dodder) may be a serious parasite of crop plants, while *Convolvulus arvensis* and *Calystegia* species (Bindweeds) may also be persistent aggressive weeds of gardens and agricultural land.   GWS & RKB

i Austin, D. F. The American Erycibeae (Convolvulaceae), *Maripa, Dicranostyles* and *Lysiostyles*: 1. Systematics. *Ann. Missouri Bot. Gard.* 60: 306–412 (1973).
ii Stefanoviæ, S., Krueger, L., & Olmstead, R. G. Monophyly of the Convolvulaceae and circumscription of their major lineages based on DNA sequences of multiple chloroplast loci. *American J. Bot.* 89: 1510–1522 (2002).
iii Stefanoviæ, S., Austin, D. F., & Olmstead, R. G. Classification of Convolvulaceae: a phylogenetic approach. *Syst. Bot.* 28: 791–806 (2003).

CORIARIACEAE

A family containing a single genus renowned for its interesting distribution[i].

Genera 1 **Species** 12–16
Economic uses Cultivated ornamentals, fly poisons, tannins, ink, and medicinal uses

Distribution *Coriaria myrtifolia* is confined to the western Mediterranean (Europe and North Africa) and is rare in Greece. *C. nepalensis* (probably including *C. kweichovensis*), *C. terminalis,* and *C. sinica* grow in the Himalayas and China, *C. japonica* is found in Japan, *C. intermedia* (optionally only a subsp. of *C. japonica*) is found in Taiwan and the Philippines, *C. papuana* is found in eastern New Guinea, and 8 spp. are currently recognized from New Zealand. The montane species *C. ruscifolia* has subsp. *ruscifolia* in the South Pacific islands, Chile, and Argentina, and subsp. *microphylla* from Peru to central and southern Mexico. However, if one takes a broad view of *C. ruscifolia*[ii] it may include also *C. papuana*[iii] and some of the New Zealand taxa. The species occur in coastal situations, in shrubby places, or (in New Guinea and New Zealand) often in subalpine areas, where they are on stream banks and are primary colonizers of alluvium on volcanic slopes, recorded up to 3,500 m in New Guinea[iii]. A molecular study[i] divides the genus into 2 main clades: the European and Asian species on the one hand, and the Pacific species on the other.

Description Rhizomatous perennial herbs or subshrubs to shrubs or small trees to 6 m. The stems, if woody, are markedly angled. The leaves are opposite, subsessile, entire, usually lanceolate to broadly ovate, and acute and up to 10 × 8 cm, but in some New Zealand and American species sometimes linear-lanceolate

CORIARIACEAE. 1 *Coriaria terminalis* (a) leafy shoot with fruits (pseudodrupes) surrounded by fleshy petals (x⅔); (b) inflorescence (x⅔); (c) fruit with 2 petals removed (x2); (d) vertical section of achene (x3); (e) flower showing large anthers (x2); (f) flower with 2 sepals removed to show small petals (x2); (g) gynoecium with ovary surrounded by 5 petals (x3); (h) stamen (x3).
2 *C. ruscifolia* (a) young protogynous flower with stigmas fully emerged (x6); (b) fertilized flower with mature stamens (x6); (c) part of perianth showing 3 sepals and 2 small petals (x6); (d) anthers (x 12); (e) vertical section of ovary (x8); (f) young flower with some of the petals and sepals removed to show free carpels (x6).

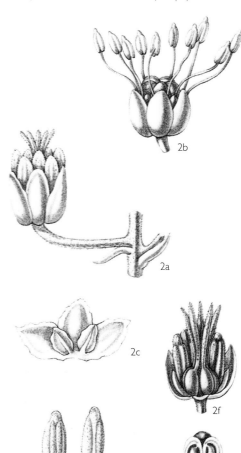

or (*C. angustissima*) linear and as small as 4 × 0.2 mm. The larger-leaved species characteristically have 3–7(–9) veins radiating from the base of the midrib, but in the small-leaved species these may be difficult to see. The flowers are borne in linear racemes 3–25 cm long, which may be terminal on main branches (*C. terminalis, C. pottsiana*) or simple and axillary, with small leaves on their lower part, or leafy and branched, with sometimes up to 12 racemes per node. The flowers are actinomorphic, hypogynous, 5-merous, male or female or hermaphrodite (but plants not dioecious), and markedly protogynous[iv]. The sepals are 5, free, usually exceeding the petals. The petals are 5, accrescent and fleshy, folded round a carpel in fruit, and turning black or red or yellow. The stamens are 10, long-exserted and dangling from the flower before dehiscing. The ovary is superior, with usually 5 free carpels, but in 2

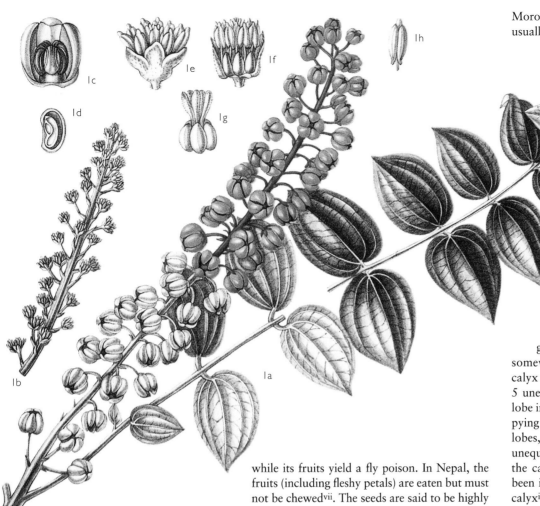

New Zealand species either 5 or 10 in 1 whorl or in 2 other New Zealand species always 10[iv]. Each carpel bears a long filiform style, which is exserted 2 to 4 mm from the flower and is stigmatic over its entire length and includes a single pendulous ovule. The fruit consists of 5 or 10 achenes enclosed within the persistent fleshy petals.

Classification *Coriaria* has long been regarded as an isolated genus best referred to its own family. It has often been placed in the Sapindales or Rutales, but Cronquist (1981) preferred to put it in Ranunculales (where the free carpels fit well). Surprisingly, the molecular evidence places it in the Cucurbitales of APG II, most families of which have an inferior ovary and show no obvious morphological resemblance to *Coriaria*. The nearest of these families in morphological terms[v] would be Corynocarpaceae, which at least has a superior ovary with a single, pendent ovule, but even there the similarities are difficult to find. Species level taxonomy is uncertain, the very broad concept of *C. ruscifolia*[ii] being queried by later authors[vi].

Economic uses Several species are occasionally cultivated as shrubs or perennial herbs with some curiosity value, attractive autumnal coloration, and attractive fruits. The leaves of the Mediterranean *C. myrtifolia* have been used to produce tannin for curing leather or for ink,

while its fruits yield a fly poison. In Nepal, the fruits (including fleshy petals) are eaten but must not be chewed[vii]. The seeds are said to be highly poisonous to people, producing convulsions, and even an elephant in a zoo in New Zealand is thought to have died from eating them. *C. japonica* is said to be one of the most poisonous plants of Japan.   RKB

i Yokoyama, J., Suzuki, M., Iwatsuki, K., & Hasebe, M. Molecular phylogeny of *Coriaria*, with special emphasis on the disjunct distribution. *Molec. Phylogen. Evol.* 14: 11–19 (2000).
ii Skog, L. E. The genus *Coriaria* (Coriariaceae) in the western hemisphere. *Rhodora* 74: 249 (1972).
iii Conn, B. J. Coriariaceae. In: Henty, E. E. (ed.), *Handbook of the Flora of Papua New Guinea* 2: 31–33 (1981).
iv Thomson, P. N. & Gornall, R. J. Breeding systems in *Coriaria* (Coriariaceae). *Bot. J. Linn. Soc.* 117: 293–304 (1995).
v Matthews, M. L. & Endress, P. K. Comparative floral structure and systematics in Cucurbitales (Corynocarpaceae, Coriariaceae, Tetramelaceae, Datiscaceae, Begoniaceae, Cucurbitaceae, Anisophylleaceae). *Bot. J. Linn. Soc.* 145: 129–185 (2004).
vi Oginuma, K., Nakata, M., Suzuki, M., & Tobe, H. Karyomorphology of *Coriaria* (Coriariaceae): taxonomic implications. *Bot. Mag. Tokyo* 104: 297–308 (1991).
vii Kanai, H. *Coriaria nepalensis*, an edible plant. *J. Jap. Bot.* 62: 15–16 (1987).

CORIDACEAE

Mediterranean subshrubs related to Primulaceae, with spiny, irregular flowers.

Distribution *Coris* occurs in southern Europe, from Spain to the Balkans (only recently recorded in Greece), North Africa, from

Morocco to Egypt, and in northern Somalia. It usually grows in dry places near the coast but is found in *Juniper* forest above 2,000 m in Somaliland.

Description Dwarf subshrubs up to 20(–30) cm high, slightly woody at the base. The leaves are alternate, ericoid in appearance, linear, rarely more than 10 × 1 mm, with the margins often inrolled below, and usually with black gland-dots alongside the midrib. The margins of the upper leaves often have long, spinose teeth that are longer than the width of the leaf itself. The inflorescence is a dense elongate spike up to 15 cm. The flowers are somewhat zygomorphic and bisexual. The calyx is obliquely cup-shaped to globose, with 5 unequal lobes that are valvate in bud, each lobe including a conspicuous black gland occupying most of its surface. Outside the 5 calyx lobes, there is a ring of up to 21 prominent, unequal spines that continue the main veins of the calyx cup and their branches, which has been interpreted as an epicalyx fused with the calyx[i]. The corolla, which is purple or pink to white, is strongly to sometimes weakly zygomorphic and 2-lipped, variable in shape and dissection[ii], with 5 bifid lobes longer or shorter than the tube. The stamens are inserted on the corolla tube. The ovary is superior, of 5 fused carpels, and is 1-locular, mostly filled by a tree-shaped placenta, with a trunk from the base of the locule surmounted by a spherical crown made up of starch-filled parenchymatous cells on top of which 5–6 ovules lie embedded[i]. The fruit is a globose capsule dehiscing by 5 valves.

Classification *Coris* has usually been placed in Primulaceae, where its many anomalous features have caused it to be placed in its own tribe. Suggestions that it might be allied to Lythraceae have been strongly opposed[iii], but at the same time numerous differences from Primulaceae have been emphasized. Separate family status as Coridaceae has been recognized by, e.g., Airy Shaw in *Willis's Dictionary* (1973), Ali in *Flora of Libya* 6 (1977), and posthumously by R. Dahlgren[iv]. A paper on floral anatomy[i] has also suggested that family status might be appropriate. Other classifications, such as Schwarz's revised system of the

Genera 1 Species 1–5
Economic uses
Rarely cultivated

Primulaceae[v], have excluded *Coris* and referred it to the affinity of the Lythraceae. Molecular analysis has placed it basal to one half of the Primulaceae[vi], where, for reasons of cladistic classification not accepted here, it was referred to the Myrsinaceae. The many significant characters that separate *Coris* from the Primulaceae—the subshrubby ericoid habit, the often spiny leaves, the slightly zygomorphic flowers, the spiny epicalyx fused with the calyx, the conspicuously black-glanded calyx lobes, the usually strongly 2-lipped corolla, large pollen grains with unique reticulate tectum, the ovules 5 rather than many, a chromosome number of 2n=18 rarely recorded in Primulaceae—indicate that it has diverged so far from the primulaceous stock that separate family status is indeed well justified. For further discussion and references, see Primulaceae. Species level taxonomy of the genus is problematic[vii], with many taxa having been described, but in *Med-Checklist* 4 (1989) only a single species, *C. monspeliensis*, was recognized, with numerous infraspecific variants.

Economic uses The genus is occasionally cultivated. RKB

i Ronse Decrane, L. P., Smets, E. F., & Clinckemaillie, D. The floral development and floral anatomy of *Coris monspeliensis*. *Canad. J. Bot.* 73: 1687–1698 (1995).
ii Gauba, E. Eine ägyptische *Coris* mit aktinomorphen Korollen. *Österr. Bot. Zeitschr.* 83: 266–272 (1934).
iii Anderberg, A. A., Trift, I., & Källersjö, M. On the systematic position of the genus *Coris* (Primulaceae). *Nordic J. Bot.* 18: 203–207 (1998).
iv Dahlgren, G. The last Dahlgrenogram: system of classification of the dicotyledons. Pp. 249–260. In: Tan, K. (ed.), *The Davis and Hedge Festschrift*. Edinburgh, Edinburgh University Press (1989).
v Schwarz, O. Die Gattung *Vitaliana* Sesl. und ihre Stellung innerhalb der Primulaceen. *Feddes Repert.* 67: 16–41 (1963).
vi Källersjö, M., Bergqvist, G., & Anderberg, A. A. Generic realignment in primuloid families of the Ericales *s.l.*: a phylogenetic analysis based on DNA sequences from three chloroplast genes and morphology. *American J. Bot.* 87: 1325–1341 (2000).
vii Airy Shaw, H. K. *Coris* (Primulaceae ?) in Somaliland. *Kew Bull.* 1951: 29–31 (1951).

CORNACEAE

DOGWOOD FAMILY

Perennial herbs to large trees, usually with broad, entire leaves, and terminal inflorescences, the latter sometimes capitate and subtended by showy, white bracts. The flowers are regular, with inferior ovaries that develop into fleshy drupes.

Distribution Mostly in northern temperate regions and tropical Asia, less commonly in South America, Africa, and Australia.

Description Most commonly shrubs to small trees, but sometimes large trees and occasionally (*Cornus* sect. *Chamaepericlymenum*) small herbs with woody rhizomes. The leaves are alternate or opposite, usually entire, but rarely

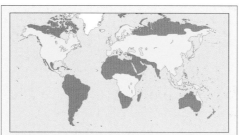

Genera 7 Species c. 105
Economic uses Ornamental shrubs and trees, limited timber, and edible fruits

serrate (*Davidia* and sometimes *Camptotheca*). The inflorescence is usually a terminal, but sometimes axillary, thyrse, which occasionally may be capitate and subtended by showy, white bracts. The flowers are epigynous and actinomorphic, sometimes (e.g., *Cornus* sect. *Afrocrania*) unisexual. The calyx has 4–5, or rarely 10, small teeth. The petals are 4–10, usually short and broad, but conspicuously long and narrow in *Alangium*. The stamens are usually as many as, or twice as many as, the petals, often in 2 whorls, or up to 40 in 1 whorl in *Alangium*. The ovary is inferior and usually 1- or 2-locular, but up to 6- to 10-locular in *Davidia*, with 1 pendulous ovule in each locule. The fruit is drupaceous, with a fleshy exocarp enclosing a stone dehiscing by germination valves except in *Alangium*.

Classification The concept of the family adopted is basically that developed by R. Eyde[i] but with the inclusion here of *Alangium* over which he hesitated. It is consistent with molecular

CORNACEAE. 1 *Cornus canadensis* (a) leafy shoot with inflorescence of small flowers surrounded by large bracts (×⅔); (b) fruits (×⅔).

analyses[ii,iii,iv] and coincides closely with that adopted in FGVP[v]. The precise delimitation of the family and the rank given to the groupings varies considerably, and recent cladistic considerations[vi] have suggested that 4 families—Cornaceae (including *Alangium*), Mastixiaceae (including *Diplopanax*), Nyssaceae (including *Camptotheca*), and monotypic Davidiaceae—should be recognized, echoing the views of Takhtajan (1997). Thorne (2001) has placed Mastixiaceae and Davidiaceae as subfamilies of Nyssaceae, with Cornaceae and Alangiaceae both monogeneric. APG II, however, has adopted the same family circumscription as is adopted here. Molecular evidence also places the small African families Curtisiaceae and Grubbiaceae close to, or nested within, the present concept of the family. Other closely allied families include Hydrangeaceae, Hydrostachyaceae, and Loasaceae. In this classification, 3 subfamilies are recognized under Cornaceae.

SUBFAM. CORNOIDEAE Includes only *Cornus*, here adopted in a broad sense to include *Swida*, *Chamaepericlymenum*, *Afrocrania*, and *Dendrobenthamia*, and totalling c. 60 spp. Leaves opposite except in 2 spp. (sect. *Alternifolia*); flowers 4-merous and petals short and without an inflexed apex;. ovary of 2(3) locules and the stone with germination valves.

SUBFAM. NYSSOIDEAE Including *Nyssa* (c. 8 spp. in tropical Asia and southeastern USA to Costa Rica), *Camptotheca* (1 sp., China), *Davidia* (1 sp., China), *Mastixia* (19 spp., from India to the Solomon Islands), and *Diplopanax* (2 spp., southern China and Vietnam). Leaves are alternate; petals 4–10 and short with an inflexed apex; stamens 8–26, usually in 2 whorls; ovary with 1 locule, and the stone with germination valves.

SUBFAM. ALANGIOIDEAE *Alangium* only, with 21 spp. widespread in the Old World tropics. Leaves alternate; petals 4–10, elongate, and strap-shaped, conspicuous; stamens 4–40 in a single whorl; ovary with 1 (2) locule, the fruit stone without germination valves.

Economic uses Many species of *Cornus* (Dogwood) are cultivated as ornamentals. *C. alba* and *C. stolonifera* provide autumnal color from their leaves, and winter color from their often bright red or yellow stems; *C. mas* is an early spring-flowering small tree; and *C. florida*, *C. kousa*, and *C. nuttallii* have striking, large white bracts subtending the inflorescences. *Davidia involucrata* has even larger white bracts, giving rise to the name of Handkerchief Tree. The wood of other *Cornus* species is used for a variety of purposes, and some species have edible fruits used for jam. *Nyssa* species (Tupelo) are grown in gardens, and *N. sylvatica* provides timber for constructing wharfs. RKB

i Eyde, R. H. Comprehending *Cornus*: puzzles and progress in the systematics of the dogwoods. *Bot. Rev.* 54: 233–351 (1988).

ii Hempel, A. L., Reeves, P. A., Olmstead, R. G., & Jansen, R. K. Implications of *rbc*L sequence data for higher order relationships of the *Loasaceae* and the anomalous aquatic plant *Hydrostachys* (*Hydrostachyaceae*). *Pl. Syst. Evol.* 194: 25–37 (1995).

iii Xiang, Q.-Y., Soltis, D. E., & Soltis P. S. Phylogenetic relationships of Cornaceae and close relatives inferred from *mat*K and *rbc*L sequences. *American. J. Bot.* 85: 285–297 (1998).

iv Xiang, Q.-Y. Systematic affinities of Grubbiaceae and Hydrostachyaceae within Cornales—insights from *rbc*L sequences. *Harvard Pap. Bot.* 4: 527–542 (1999).

v Kubitzki, K. Cornaceae. Pp. 82–90. In: Kubitzki, K. (ed.), *The Families and Genera of Vascular Plants. VI. Flowering Plants. Dicotyledons: Celastrales, Oxalidales, Rosales, Cornales, Ericales.* Berlin, Springer-Verlag (2004).

vi Xiang, Q.-Y., Moody, M. L, Soltis, D. E., Fan, C. Z., & Soltis, P. S. Relationships within Cornales and circumscription of Cornaceae—*mat*K and *rbc*L sequence data. *Molec. Phylog. Evol.* 24: 35–57 (2002).

CORYNOCARPACEAE

Trees and shrubs of the southwestern Pacific.

Distribution *Corynocarpus similis* extends from western New Guinea, through the Solomons, to Vanuatu; *C. cribbianus* grows in New Guinea and northern Queensland; *C. rupestris* is found in northeastern New South Wales and (subsp. *arborescens*) adjacent southeastern Queensland; while *C. dissimilis* is endemic to New Caledonia and *C. laevigatus* occurs in New Zealand, including the Kermadec and Chatham Islands. They are mostly found in forests or on river banks, often near the sea but up to 2,750 m in New Guinea.

Genera 1 Species 5
Economic uses Tree trunks for canoes; fruit and seeds edible

Description Trees up to 40 m, or sometimes shrubs. The leaves are alternate (sometimes clustered toward ends of branches), petiolate, simple, entire, thick, and leathery, up to 25 × 12 cm, with caducous cataphylls (? stipules), leaving a crescent-shaped scar above the leaf scar. The inflorescence is a terminal panicle of simple, linear, fairly dense cymes, and the flowers have a short pedicel subtended by a bract. The flowers are small (c. 2–4 mm), actinomorphic, bisexual, hypogynous, and 5-merous. The sepals and petals are 5, free, and imbricate. There are 5 fertile stamens, adnate to the petals, and 5 petaloid staminodes alternating with them, each staminode with a well-developed rectangular to ovoid nectary at its base. The ovary is superior and of 2 carpels, but one of these aborts to leave a single locule. There are 1 or 2 asymmetrically placed styles, with capitate stigmas. There is a single ovule pendent from the apex of the locule and filling the cavity. The fruit is a single-seeded drupe, with a fleshy mesocarp and woody endocarp.

Classification The affinities of the family have been obscure, but until recently it has usually been referred to the Celastrales. However, molecular studies have placed it—surprisingly—in the Cucurbitales. A recent detailed study of the family[i] has noted that there appear to be few morphological synapomorphies that unite all members of the Cucurbitales. In its superior ovary with apical placentation, the Corynocarpaceae differs markedly from Cucurbitaceae, Begoniaceae, and Datiscaceae, which are now its near neighbors. However, it does show more morphological resemblance to the Coriariaceae, which is apparently its nearest relative[ii].

Economic uses In New Zealand, the prunelike fruits of *C. laevigata* are eaten for their fleshy mesocarp and their seeds, but the seeds are highly poisonous unless steamed or steeped in saltwater—see T. F. Cheeseman, *Man. New Zealand Fl.* (1925), and associated references. The trunks of trees are well known for bearing glyphs left by the Moriori people on Chatham Island, and have also been used for making canoes.   RKB

i Wagstaff, S. J. & Dawson, M. I. Classification, origin, and patterns of diversification of *Corynocarpus* (Corynocarpaceae) inferred from DNA sequences. *Syst. Bot.* 25: 134–149 (2000).

ii Matthews, M. L. & Endress, P. K. Comparative floral structure and systematics in Cucurbitales (Corynocarpaceae, Coriariaceae, Tetrameleaceae, Datiscaceae, Begoniaceae, Cucurbitaceae, Anisophylleaceae). *Bot. J. Linn. Soc.* 145: 129–185 (2004).

CRASSULACEAE
STONECROPS AND HOUSELEEKS

A family of succulent herbs and small shrubs.

Distribution The Crassulaceae are distributed throughout the world, mainly in warm, dry regions, but have centers of diversity in South Africa, Madagascar, Macaronesia, East Asia, and Mexico. Like the other 2 great families of succulents, Mesembryanthemaceae and Cactaceae, they are characteristic of hot, exposed, rocky habitats subjected to long periods of drought. However, Crassulaceae has a wider range of adaptability: species of *Sempervivum* and some *Sedum* are frost-hardy, and some species of *Crassula* and *Sedum* live in a plentiful supply of water, 1 *Crassula* being an aquatic.

CRASSULACEAE. 1. *Echeveria nodulosa* (a) habit (×⅔); (b) half flower showing fleshy petals and carpels joined at base (×4½).

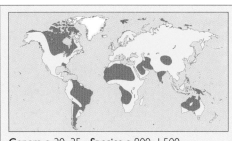

Genera c. 30–35 Species c. 900–1,500
Economic uses Ornamentals and some ethnobotanical uses

Description Perennial (rarely annual or biennial), soft-wooded herbs, or small shrubs, many with sessile rosettes, and 1 species (*Crassula helmsii*) is an emergent aquatic. The leaves are always more or less fleshy and mostly entire, without stipules, and commonly packed tightly in rosettes, which may reach the extreme of surface reduction in a sphere (*Crassula columnaris*). A full range of xerophytic specializations is found, including surfaces covered in papillae, hairs, bristles, or wax. Vegetative reproduction from offsets, adventitious buds, and fallen leaves is common. The inflorescence is a terminal or lateral cyme, sometimes a showy corymb or panicle. The flowers are small and structurally simple, with (3–)5(–30) mostly perigynous, free sepals, petals, and carpels, and twice the number of stamens in 1 (*Crassula* clade) or 2 whorls. The ovary is superior and

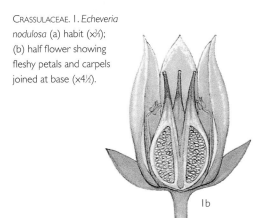

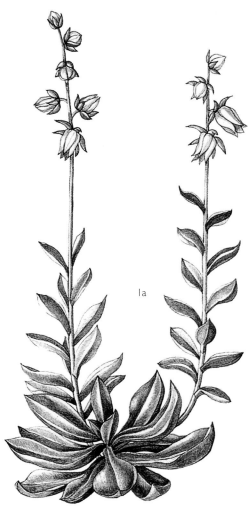

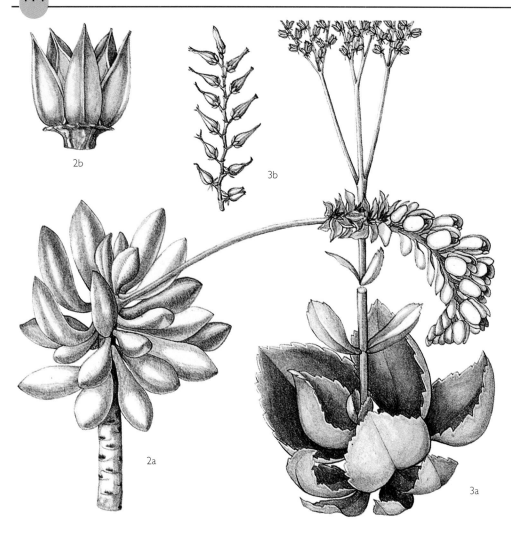

CRASSULACEAE. 2 *Pachyphytum longifolium* (a) shoot tip showing clusters of fleshy leaves and lateral inflorescence (x⅔); (b) fruit (x3). 3 *Kalanchöe crenata* (a) habit showing opposite succulent leaves arranged on a fleshy stem (x⅔); (b) tip of inflorescence bearing clusters of tubular flowers (x1).

the carpels (the same number as the petals) may be slightly joined at the base. The ovules are numerous (rarely few) and inserted on the adaxial suture. The style is short or elongated. The fruit is a group of follicles, each with minute seeds, with little or no endosperm.

Classification Circumscription of the ranks above the species is highly controversial, with the generic boundaries often blurred due to overlap in morphological characters[i]. The major genera recognized are: *Sedum* (c. 420 spp.); *Crassula* (c. 195 spp.); *Kalanchoe* (c. 145 spp.); and *Echeveria* (c. 140 spp.), of which at least *Sedum* is polyphyletic. The most widely used classification[ii] recognizes 6 subfamilies, but these are certainly not natural. In general terms, the family can be divided into 2 groups: those in which the stamens are as many as the petals and leaves are opposite, e.g., *Crassula* and *Rochea*; and those in which the stamens are twice as many as the petals and the leaves alternate or opposite, e.g., *Kalanchoe* (flower parts in 4s); *Umbilicus*, *Cotyledon* (corolla gamopetalous), *Sedum*, *Pachyphytum*, and *Echeveria* (flower parts in 5s); and *Monanthes*, *Sempervivum*, and *Greenovia* (flower parts in 6s or more). Analysis

of DNA sequences suggests *Crassula* (including *Tillaea*) to have split off before the major radiation in the family[iii]; 6 other major groups are then recognized. The family is placed in Saxifragales in many systems, and molecular evidence supports a close relationship to Haloragaceae[iv].

Economic uses The family is valued mainly for ornamental use. The hardy stonecrops (*Sedum* species) and houseleeks (*Sempervivum* species) are cultivated in the rock garden or alpine house, the tender species in the succulent collection, *Echeveria* for summer bedding, and *Kalanchoe*, *Rochea*, and a few others as house plants and florists' flowers. *Aeonium* is cultivated in warm regions. Some species are used in traditional medicine, including anti-inflammatories[v]. [GDR] AC

[i] van Ham, R. C. H. J. Phylogenetic relationships in the Crassulaceae inferred from chloroplast DNA variation. Pp. 16–29. In: 't Hart, H. & Eggli, U. *Evolution and Systematics of the Crassulaceae*. Leiden, Backhuys Publishers (1995).
[ii] Berger, A., Crassulaceae. Pp 352–485. In: Engler, A. & Prantl, K. (eds), *Die natürlichen Pflanzenfamilien*, Vol. 18a, edn. 2, Leipzig, W. Engelmann (1930).
[iii] Mort, M. E., Soltis, D. E., Soltis, P. S., Francisco-Ortega, J., & Santos-Guerra, A. Phylogenetic relationships and evolution of Crassulaceae inferred from *mat*K sequence data. *American J. Bot.* 88: 76–91 (2001).
[iv] Fishbein, M. & Soltis, D. E. Further resolution of the rapid radiation of Saxifragales (Angiosperms, Eudicots) supported by mixed-model Bayesian analysis. *Syst. Bot.* 29: 883–891 (2004).
[v] De Melo, G. O. *et al.* Phytochemical and pharmacological study of *Sedum dendroideum* leaf juice. *J. Ethnopharmacol.* 102: 217–220 (2005).

CROSSOSOMATACEAE
GREASE BUSH FAMILY

A small family of New World shrubs.

Distribution *Crossosoma* has 2 species, from Nevada, Arizona, and California, adjacent states of Mexico, and Guadalupe Island. *Glossopetalon* comprises 5 species, from Washington and Montana, south to California, Oklahoma, and Texas, and the states of Coahuila and San Luis Potosí in Mexico.

Genera 4 Species 9
Economic uses Rarely cultivated

Apacheria has a single species in southeastern Arizona and adjacent New Mexico. *Velascoa* has 1 species from Querétaro in central Mexico. They mostly occur in deserts, dry canyons, and rocky places.

Description Mostly shrubs, prostrate or up to 2(3) m high, often with branch spines and small leaves up to 1(2) cm, but in *C. californicum* a shrub to small tree up to 5 m with leaves up to 9 × 3 cm. The leaves are alternate or fascicled and entire, except in *Apacheria* where they are opposite and sometimes 3-toothed at the tip. Stipules are small and caducous, possibly sometimes absent. The inflorescence is a solitary terminal flower on a short shoot, sometimes appearing axillary, usually with conspicuous pedicels. The flowers are bisexual, actinomorphic, perigynous, with a hypanthium that is short and cup-shaped or (*Velascoa*) long and tubular, 4-merous in *Apacheria* or usually 5-merous in other genera, up to 4.5 cm across in *C. californicum* but usually smaller. The sepals are usually 4 or 5, free, triangular to orbicular, the outer ones sometimes smaller. The petals are usually 4 or 5, elliptical to orbicular, white or in *C. bigelovii* purplish. The stamens are equal in number or 2–10 times as many as the petals, in 1 to several whorls, 4–10 in *Glossopetalon*, 8 in *Apacheria*, 10 in 2 whorls and almost sessile in *Velascoa*, and 15–50 in *Crossosoma*. A disk is present under the ovary, except in *Velascoa*. The ovary is superior, of 1–5(–9 in *C. californicum*) free carpels, which are stipitate or sessile, each terminated by a slightly bilobed capitate stigma, and with 1–2 ovules per carpel or many (*Crossosoma*). The fruit consists of 1–9 follicles that dehisce ventrally, and the seeds are compressed, dark, and shiny, with a broad or narrow aril.

Classification Although the free carpels have in the past suggested affinity with Dilleniaceae or Rosaceae, and the family was later placed in Celastraceae, recent molecular[i] and floral anatomical[ii] analyses have established a close relationship with the Stachyuraceae

(Himalayas to Japan) and Staphyleaceae (Asia and South America), the order Crossosomatales also including Aphloiaceae, Geissolomatceae, Ixerbaceae, and Strasburgeriaceae, which are mostly in the southern hemisphere. Dilleniaceae and Crossosomataceae share numerous characters in addition to the free carpels that might indicate a closer relationship[ii].

Economic uses *Crossosoma californicum* is rarely cultivated as an ornamental.   RKB

[i] Sosa, V. & Chase, M. W. Phylogenetics of Crossosomataceae based on *rbc*L sequence data. *Syst. Bot.* 28: 96–105 (2003).
[ii] Matthews, M. L. & Endress, P. K. Comparative floral structure and systematics in Crossosomatales (Crossosomataceae, Stachyuraceae, Staphyleaceae, Aphloiaceae, Geissolomataceae, Ixerbaceae, Strasburgeriaceae). *Bot. J. Linn. Soc.* 147: 1–46 (2005).

CRYPTERONIACEAE

Trees from tropical Asia, with linear spikes of small flowers.

Distribution Confined to tropical Asia. *Crypteronia* has 7 spp., from Assam to New Guinea; *Dactylocladus* has 1 sp. in Borneo; and *Axinandra* has 4 spp.—1 in Sri Lanka and 3 in Borneo, 1 of which is also recorded from the southern Malay Peninsula.

Genera **3** Species **12**
Economic uses
Limited use as timber

Description Shrubs or usually trees up to 30 m. The leaves are opposite, entire, up to 30 × 18 cm, with parallel, secondary veins running to a prominent submarginal vein, except in *Dactylocladus*, with short dark pulvinuslike petioles. Stipules are present but caducous, large, and prominently veined in *Axinandra*. The inflorescence is a linear, sometimes pendent, spike of small, subsessile flowers, or a panicle of such spikes. The flowers are bisexual or unisexual[i] (plants polygamodioecious), actinomorphic, 1–2 mm, 4- to 5-merous, and perigynous to epigynous. The sepals are 4–5 and inserted on the hypanthium rim. The petals in *Dactylocladus* are hooded over the stamens, and in *Axinandra* are connate apically, falling off as an umbrella, but absent in *Crypteronia*. The stamens are equal in number to the sepals, or in *Axinandra*, are twice as many as the sepals. The ovary has 2–5 fused carpels and is superior, with many ovules per cell locule in *Crypteronia*, semi-inferior with 3 ovules per cell locule in *Dactylocladus*, and inferior with 1–2 ovules per cell locule in *Axinandra*, and placentation is parietal or basal. The capsule is chartaceous and up to 3 mm long, except in *Axinandra*, where it is woody and up to 2.5 × 1.5 cm. It dehisces internally to the base into 2–6 valves, these being evident externally only in the upper part.

Classification The family is referable to the Myrtales[ii] and has been at times included in Melastomataceae or Lythraceae. It is now considered to be most closely related to Alzateaceae from western tropical America, Oliniaceae from eastern tropical Africa, and Rhynchocalycaceae and Penaeaceae from South Africa[iii,iv]. The Alzateaceae and Rhynchocalycaceae have until recent years often been included in Crypteroniaceae but are quite distinct. *Axinandra* differs conspicuously from the other genera in many ways as noted above, and notably in embryology[v], and the recognition of a tribe Axinandreae, as in *Flora Malesiana*, seems well justified.

Economic uses Limited local use as timber has been recorded.   RKB

[i] Pereira, J. T. & Wong, K. M. Three new species of *Crypteronia* from Borneo. *Sandakania* 6: 41–53 (1995).
[ii] Dahlgren, R. & Thorne, R. F. The order Myrtales: circumscription, variation and relationships. *Ann. Missouri Bot. Gard.* 71: 633–699 (1984).
[iii] Schoenenberger, J. & Conti, E. Molecular phylogeny and floral evolution of Penaeaceae, Oliniaceae, Rhynchcalycaceae and Alzateaceae (Myrtales). *American J. Bot.* 90: 293–309 (2003).
[iv] Rutschmann, J., Eriksson, T., Schoenenberger, J., & Conti, E. Did Crypteroniaceae really disperse out of India? Molecular evidence dating from *rbc*L, *ndh*F and *rpl*16 entron sequences. *Int. J. Plant Sci.* 165 (Supplement): S69–S83 (2004).
[v] Tobe, H. & Raven, P. H. The embryology and relationships of *Crypteronia* (Crypteroniaceae). *Bot Gaz.* 148: 96–102 (1987).

CTENOLOPHONACEAE

The Ctenolophonaceae comprises 2 spp., *Ctenolophon englerianus,* which occurs in lowland forests in West and West–Central Africa from Nigeria to Cabinda, and *C. parvifolius* in the Malaysian region from the Malay Peninsula and Sumatra to the Philippines and New Guinea, where it occurs in primary and swamp forest, heaths, and on sand, mostly at low altitudes but up to 1,650 m in Borneo[i]. They are big trees up to 40 m, with buttressed trunks up to 1.2 m in diameter. The leaves are opposite, simple, ovate to elliptic, up to 10(–15) × 6(–9) cm, acuminate, with parallel pinnate veins, coriaceous, glabrous, with a short petiole (less than 1 cm), with interpetiolar stipules. The inflorescence is a terminal or occasionally axillary, cymose panicle, with paired caducous bracts at conspicuous nodes. The flowers are actinomorphic, bisexual, 5-merous, often with minute stellate hairs. The sepals are 5, broadly elliptical to suborbicular, imbricate. The petals are 5, narrow, up to 12 mm, somewhat leathery, twisted in bud, and caducous. The stamens are 10, unequal, inserted inside a cup-shaped disk. The ovary is superior, of 2 carpels, covered with long straight hairs, attenuate above to a simple style with 2 capitate stigmas. There are 2 locules, each with 2 ovules on an axile placenta. The fruit is a woody 1-celled capsule, splitting longitudinally, with a single seed pendulous from the apex of a central columella. The seed is ellipsoid and up to 2 cm long, with a conspicuous reddish-brown arillode.

The discontinuity in distribution between West Africa and Malaysia is notable, but other comparable examples are known, e.g, the papilionoid legume genus *Airyantha*. The pollen is distinctive, and fossils suggest that the genus also occurred in South America in the Palaeocene[ii]. The affinities of the genus have long been controversial. Recent authors have placed it in or close to Linaceae. Molecular data places it closer to Lacistemataceae, Goupiaceae, and Violaceae[iii] In APG II, it is placed in Malpighiales. A good account, including illustrations, of *C. parvifolius* notes that the timber is used for house construction in Borneo[iv].   RKB

[i] van Hooren, A. M. N. & Nooteboom, H. P. Ctenolophonaceae. *Flora Malesiana* 10: 629–634 (1988).
[ii] van der Hamm, R. W. J. M. New observations on the pollen of *Ctenolophon* Oliver, with remarks on the evolutionary history of the genus. *Rev. Palaeobot. Palynol* 59: 153–160 (1989).
[iii] Savolainen, V. *et al.* Phylogeny of the eudicots: a nearly complete familial analysis based on *rbc*L sequences. *Kew Bull.* 55: 257–309 (2000).
[iv] Pungga, R. S. Ctenolophonaceae. In: Soepadmo, E., Wong, K. M., & Saw, L.G. (eds), *Tree Flora of Sabah and Sarawak* 2: 151–153 (1996).

CUCURBITACEAE

GOURD OR PUMPKIN FAMILY

A highly specialized family of mainly climbing plants, often with coarsely hairy leaves; of major importance as a food source.

Distribution The Cucurbitaceae is well represented in the moist and moderately dry tropics of both Old and New Worlds, particularly in rain forest areas of South America and woodland, grassland, and bushland areas of Africa. Some species occur in semidesert or even desert vegetation. Cucurbits are poorly represented in Australasia and all temperate regions.

Description Usually climbing perennial herbs with a swollen tuberous subterranean or wholly or partly superficial rootstock, formed by swelling of the hypocotyl; sometimes annual, occasionally softly woody lianas, rarely a shrub or tree (*Acanthosicyos, Dendrosicyos*). The few nonclimbing species are probably derivative. The stems of most are characterized by bicollateral vascular bundles with internal, as well as external, phloem. The leaves are alternate, simple or sometimes ternate, or palmately compound with 3 or more leaflets, palmately veined

Genera **1** Species **2**

and usually lobed, sometimes pinnatifid (*Citrullus*), rarely absent (*Seyrigia*), usually coarsely hairy. In most species a solitary, branched or simple tendril arises at each side of the petiole base, the tip coiling round any suitable nearby object, such as a plant stem; the rest of the tendril then coils in a springlike manner, drawing the stem in close to its support; or the tip is adhesive. The inflorescences are usually axillary, of solitary flowers or in cymes, racemes, or panicles. The flowers are actinomorphic, nearly always unisexual (plants monoecious or dioecious). The sepals and petals are usually 5, borne at the top of a cup- or tubelike expanded hypanthium; the petals are often more or less united at the base. The androecium is complex, with 1–5 stamens, commonly 3, 2 of which are double with 4 pollen sacs each, and 1 single with 2 pollen sacs inserted on the lower part of the hypanthium, united in various ways, sometimes the filaments more or less completely fused into a single central column, the anthers bent to twisted or convoluted, with longitudinal dehiscence. In the pistillate flowers, the ovary is inferior, of 3 fused carpels (1 in tribe Sicyeae) 1-locular, usually with parietal placentation, or 2- to 5-locular by intrusive placentas; ovules from 1 to many in each fruit, anatropous; the seeds without endosperm, usually large and more or less flattened. The fruit is berry, sometimes firm-walled (such as the melon), and known as pepos, or fleshy or dry capsules that may be explosively dehiscent (e.g., *Ecballium*, *Schizopepon*), or leathery and indehiscent.

Classification The most comprehensive classification of the family is that of Jeffrey[i]. His classification has been been substantially supported at tribal level by subsequent molecular studies[ii,iii] . A revised version of Jeffrey's classification, *Systema Cucurbitacearum Nova*, is in press (2006). The family has been traditionally divided into 2 subfamilies: Cucurbitoideae, which contains 7 or more tribes and is probably monophyletic[iv]; and the Nhandiroboideae (Zanonioideae), which contains a single tribe.

SUBFAM. CUCURBITOIDEAE Includes c. 100 genera and c. 745 spp. Commonly annuals; tendrils unbranched or 2- to 7-branched from the lower part, spiralling only above the point of branching; filaments inserted on the hypanthium, free from the disk when the latter present; pollen grains various, not striate; style 1; seeds not winged. Here, divided into 7 tribes:

Joliffieae, including *Telfairia* (tropical Africa), *Momordica* (Old World tropics).

Benincaseae, including *Acanthosicyos* (southern tropical Africa, with *Acanthosicyos horridus* a spiny shrub of the Namib desert dunes), *Ecballium* (Mediterranean), *Benincasa* (tropical Asia), *Bryonia* (Eurasia), *Coccinia* (Old World tropics), and *Citrullus* (Old World tropics and subtropics).

Melothrieae, including *Dendrosicyos* (Socotra, with *D. socotranus* a small succulent-

Genera c. 120   Species 750–850
Economic uses Major food plants, including chayote, cucumbers, gherkins, gourds, marrows, melons, pumpkins, squashes, and zucchini, with many other uses

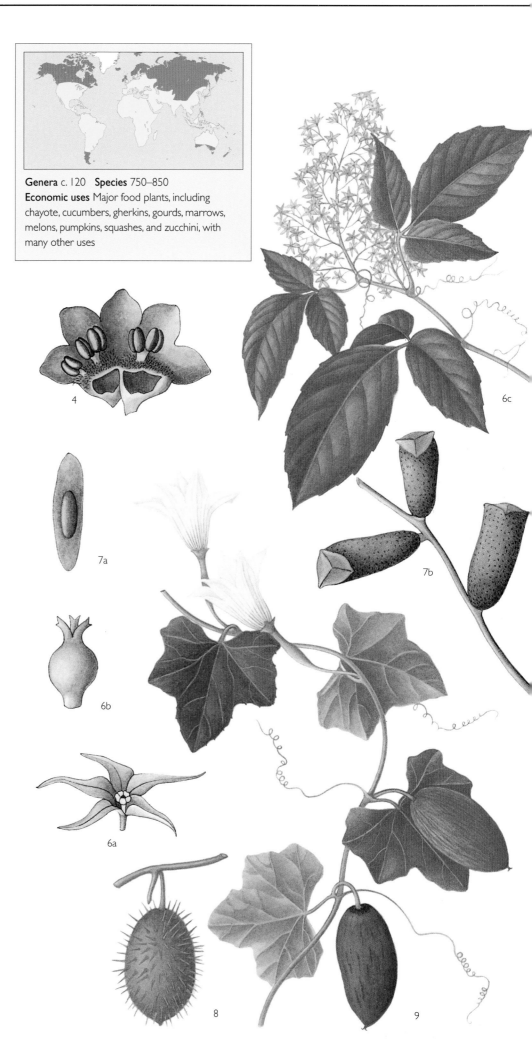

CUCURBITACEAE. 1 *Gurania speciosa* female flowers (×⅔).
2 *Curcurbita moschata* (a) male flower (×⅔); (b) cross
section of ovary (×⅔); (c) female flower with petals and
sepals removed (×⅔). 3 *Sechium edule* (a) female flower
with discoid stigma (×1⅓); (b) stamens partly joined in a
single column (×2); (c) vertical section of ovary with single
pendulous ovule (×2). 4 *Kedrostis courtallensis* male flower
opened out to show 2 double and 1 single epipetalous
stamen (×4). 5 *Trichosanthes tricuspidata* leaf, tendril, and
female flower (×⅔). 6 *Gynostemma pentaphyllum* (a) female
flower (×6); (b) young fruit with remains of styles (×8);
(c) leafy shoot with tendrils and inflorescence (×⅔).
7 *Zanonia indica* (a) winged seed (×¼); (b) fruits (×⅔).
8 *Echinocystis lobata* fruit (×⅔). 9 *Coccinea grandis* leaves,
tendrils, female flowers, and fruit (×⅔).

stemmed tree), *Trochomeria* (tropical Africa); *Corallocarpus* (Old World tropics), *Cucumeropsis* (tropical Africa), *Cucumis* (Africa and Asia), *Ibervillea* (southern North America), *Kedrostis* (Old World tropics), *Seyrigia* (Madagascar, leafless succulent lianas), *Zehneria* (Old World tropics), *Gurania* (New World tropics, lianas with a red or orange hypanthium and sepals, pollinated by hummingbirds).

Schizopeponeae, including *Schizopepon* (eastern Asia).

Cyclanthereae, including *Apatzingania* (Mexico, fruits geocarpic), *Cyclanthera* (New World), *Elateriopsis* (New World tropics), *Marah* (southwestern USA), and *Echinocystis* (North America).

Sicyoeae, including *Polakowskia* (Central America), *Sechium* (Central America), *Sicyos* (New World, Pacific, and Australia). Trichosantheae, including *Hodgsonia* (tropical Asia), *Peponium* (Africa and Madagascar), *Trichosanthes* (tropical Asia).

Cucurbiteae, including *Calycophysum* (tropical South America, pollinated by bats); *Cucurbita* (New World), composed of 12–14 spp. distributed from the United States to Argentina (pollinated by specialized solitary bees of the genera *Peponapis* and *Xenoglossa*); and *Sicana* (tropical New World).

SUBFAM. ZANONIOIDEAE This subfamily comprises 19 genera and c. 80 spp. Tendrils 2-branched from near the apex, spiraling above and below the point of branching; filaments inserted on or about the disk; styles 2–3; ovules pendulous; pollen small, striate, and uniform; seeds often winged. This subfamily includes *Fevillea* (tropical South America); *Alsomitra* (tropical Asia, with *Alsomitra macrocarpa*, a liana with large fruits and large, beautifully winged seeds); *Gerrardanthus* (tropical Africa); *Xerosicyos* (Madagascar, leaf succulents); *Cyclantheropsis* (tropical Africa and Madagascar, with the fruit being a single-seeded samara); and *Zanonia* (Indo–Malaysia).

The Cucurbitaceae has been considered closely related to Begoniaceae and Datiscaceae by Cronquist (1981) and Takhtajan (1997) and belong in the expanded Cucurbitales of APG and APG II, where they are sister to all other core group families of the order[v], although the relationships are still not fully resolved[vi].

Economic uses Major food crops are produced in tropical, subtropical, and temperate regions from *Cucurbita* species (squashes, pumpkins, and yellow-flowered gourds, vegetable marrows, vegetable spaghetti, zucchini), *Cucumis* species (*Cucumis melo*, Melon, Cantaloupe, honeydew, and *C. sativus*, Cucumber), and *Citrullus lanatus* (Watermelon). Other important crops include *Cucumis anguria* (West Indian Gherkin), *Lagenaria siceraria* (Calabash, Bottle Gourd), *Benincasa hispida* (Wax Gourd), *Sechium edule* (Chayote), *Luffa cylindrica* (Loofah) and *L. acutangula*, *Trichosanthes cucumerina* var. *anguina* (Snake Gourd), *Momordica charantia* (Bitter Melon; Balsam Apple), *Sicana odorifera* (Cassa-banana), *Cyclanthera pedata* (Achocha), *Hodgsonia heteroclita* (Lard Fruit), *Telfairia*

occidentalis (Oyster Nuts, the seeds yielding an edible oil), *Cucumeropsis mannii* (Egussi), and *Praecitrullus fistulosus* (Dilpasand, Tinda). *Luffa aegyptiaca* is the source of loofah sponges (dried vascular skeleton of the fruit), while dry fruits of *Lagenaria siceraria* (Bottle Gourd) have medicinal uses and have been used as containers since ancient times. The species is one of the earliest cultivated plants and the only one with an archaeologically documented prehistory in both Old and New Worlds. Fruits of wild *Citrullus lanatus* (Tsamma), *Acanthosicyos naudinianus*, and *A. horridus* (Narras) are food and water sources in the deserts of southern Africa. Bitter substances, called cucurbitacins, are widespread in the family. Many of the edible species occur in both bitter (inedible) and non-bitter (edible) variants.

As ornamentals, Cucurbitaceae are of minor importance: *Cucurbita pepo* produces the ornamental gourds, and species of *Momordica*, *Kedrostis*, *Corallocarpus*, *Ibervillea*, *Seyrigia*, *Gerrardanthus*, *Xerosicyos*, and *Cyclantheropsis* are sometimes cultivated by enthusiasts of succulent plants. [CJ] VHH

i Jeffrey, C. Appendix: an outline classification of the Cucurbitaceae. Pp. 449–463. In: Bates, D. M., Robinson, R. W., & Jeffrey, C. (eds), *Biology and Utilization of the Cucurbitaceae*. Ithaca, New York, Cornell University Press (2002).
ii Renner, S. S., Weerasooriya, A., & Olson, M. E. Phylogeny of Cucurbitaceae inferred from multiple chloroplast loci. P. 169. In: *Botany 2002: Botany in the Curriculum*, Abstracts. [Madison, Wisconsin.] (2002).
iii Kocyan, A., Zhang, L., & Renner, S. S. P. 783. A chloroplast phylogeny for 100 of the 122 genera of Cucurbitaceae reveals the family's spatiotemporal evolution. XVII International Botanical Congress, Vienna, Austria. Abstracts. (2005). http://www.ibc2005.ac.at/program/abstracts/ IBC2005_Abstracts.pdf
iv Volz, S., Kocyan, A., & Renner, S. S. Assessment of the value of ITS and *trn*L for a study of sexual system evolution in *Bryonia* and *Ecballium* (Cucurbitaceae). Abstracts Botany 2004. Systematics Section/ASPT. (2004). http://www.2004.botanyconference.org/engine/ search/index.php?func=detail&aid=628
v Soltis, D. E., Soltis, P. S., Endress, P. K., & Chase, M. W. *Phylogeny and Evolution of Angiosperms*. Sunderland, Sinauer Associates (2005).
vi Matthews, M. L. & Endress, P. K. Comparative floral structure and systematics in Cucurbitales (Corynocarpaceae, Coriariaceae, Tetrameleaceae, Datiscaceae, Begoniaceae, Cucurbitaceae, Anisoophyllaceae). *Bot. J. Linn. Soc.* 145: 129–185 (2004).

CUNONIACEAE

LIGHTWOOD, COACHWOOD, AND EUCRYPHIA

A family of evergreen trees and shrubs native to the southern hemisphere.

Distribution The main centers of distribution are Oceania and Australasia, but there are a few genera in South Africa and tropical America. The most important genus is *Weinmannia*, which has 160 spp. distributed throughout Madagascar, Malaysia, the Pacific, New Zealand, Chile, Mexico, and the West Indies. *Pancheria* has 25 spp. in New Caledonia. *Geissois* has 20 spp. in Australasia, New Caledonia, and Fiji. The 20 spp. of *Spiraeanthemum* are native to New Guinea and Polynesia. *Cunonia* and *Lamanonia* are smaller genera, consisting

Genera 27 **Species** c. 300
Economic uses Hard timber and ornamental trees and shrubs from *Eucryphia* spp.; lightweight timber from *Ceratopetalum apetalum*

of 15 and 10 spp., respectively. The former has a discontinuous distribution in South Africa and New Caledonia, while *Lamanonia* is native to Brazil and Paraguay. Of the 5 spp. of *Eucryphia*, 2 spp. are native to Chile, 1 sp. to New South Wales in Australia, and 2 spp. to Tasmania. Fossil evidence suggests the family once extended into the northern hemisphere[i].

Description Evergreen trees (*Eucryphia* to 25 m), shrubs, or stranglers (some *Weinmannia* spp.), with simple or complex hairs, some irritating (*Davidsonia*). The leaves are opposite or are rarely whorled (spiral in *Davidsonia*), leathery, often glandular, occasionally 1-foliate, but more often compound, 3-foliate or pinnate, the margins toothed or entire. The stipules may be large and united in pairs. The inflorescence is variously a panicle, thyrse, raceme, capitate, or rarely

CUNONIACEAE. 1 *Pancheria elegans* (a) shoot with whorls of simple leaves and flowers in compact heads (x⅔); (b) female flower showing 3 free sepals and petals and 2 free styles (×12); (c) male flower with 6 stamens (×12); (d) male flower opened out to show stamens with filaments of 2 lengths (×12); (e) bilobed fruit (×12). 2 *Cunonia capensis* shoot with pinnate leaf and flowers in a panicle (x⅔). 3 *Weinmannia hildebrandtii* (a) shoot with 3-foliate leaves and flowers in panicles (x⅔); (b) flower (×8); (c) half flower (×8). 4 *Geissois imthurnii* flower with 4 sepals, no petals, and numerous stamens inserted on a nectar-secreting disk (x2). 5 *Davidsonia prunens*, fruit (x⅔).

a solitary flower[ii]; inflorescences are positioned at the stem apex, in axils, or rarely cauliflorous. The flowers are regular, hermaphrodite, or unisexual in a few species (plants dioecious, *Hooglandia*[iii], *Pancheria*, *Vesselowskya*, and some species in other genera). The sepals are 3–6(–10), valvate or imbricate, either free or fused together at the base. The petals are alternate with the sepals and usually isomerous, free or united at the base, generally smaller than the sepals, and are absent in some species. The stamens are usually numerous but sometimes 4 or 5, alternating with the petals, or 8 or 10, usually inserted by their free filaments on a ringlike, nectar-secreting disk surrounding the ovary. The ovary is superior, comprising 1 (*Hooglandia*) 2–5 free or fused carpels (4–14 in *Eucryphia*), usually with 2 (sometimes 5) locules, each locule or free carpel containing sometimes 1–2, but usually numerous, ovules set in 2 rows on recurved, axile, or apical placentas. The styles are distinct, even in those species with fused carpels. The fruit is a woody or leathery capsule sometimes dehiscing along the ventral sutures or with winglike sepals or a nut; sometimes the carpels are bladderlike (*Platylophus*), or the seeds are winged (*Eucryphia*).

Classification The family has been placed in Rosales (Cronquist 1981), but phylogenetic studies show a relationship to *Cephalotus* and *Brunellia*[ii] in the Oxalidales (APG II) based on DNA sequence analysis. This is is corroborated by morphological and biochemical features[iv]. The genera have been comprehensively sampled in molecular phylogenetic studies[ii], and the most recent generic addition to the family, *Hooglandia* from New Caledonia, was placed there largely on molecular evidence[v]. Relationships of Cunoniaceae to Anisophylleaceae have been proposed based on floral structures[vi]. Both families are grouped in eurosids 1[i], but the latter is placed in Cucurbitales. The genera of Cunoniaceae were divided into 6 tribes based on molecular and morphological evidence[ii], but an additional tribe to accommodate *Hooglandia* was later added. Tribes include **Cunonieae** (*Cunonia*, 25 spp., 24 in New Caledonia, 1 in South Africa; *Pancheria*, c. 30 spp.; *Vesselowskya*, 2 spp.; *Weinmannia*, c. 150–160 spp.); **Codieae** (*Callicoma*, 1 sp.; *Codia*, c. 12 spp.; *Pullea*, c. 3 spp.); **Caldcluvieae** (*Ackama*, 4 spp.; *Caldcluvia*, 1 sp.; *Opocunonia*, 1 sp.; *Spiraeopsis*, 6 spp.); **Geissoieae** (*Geissois*, c. 18 spp.; *Lamanonia*, 5 spp.; *Pseudoweinmannia*, 2 spp.); **Schizomerieae** (*Anodopetalum*, 1 sp.; *Ceratopetalum*, c. 9 spp.; *Platylophus*, 1 sp.; *Schizomeria*, 10 spp.); **Spiraeanthemeae** (*Acsmithia*, 14 spp.; *Spiraeanthemum*, 6 spp.). In addition to these tribes, a "Basal Grade"[vii] comprising the genera *Aistopetalum* (2 spp.), *Hooglandia* (1 sp.), *Davidsonia* (3 spp.), and *Bauera* (4 spp.) is recognized together with a core Cunon[vii] group: *Eucryphia* (7 spp.), *Gillbea* (3 spp.), and *Acrophyllum* (1 sp.).

Economic uses The wood of various genera is used in construction work. The timber

(Lightwood) of *Ceratopetalum apetalum* from New South Wales is used in carpentry and cabinet-making, and there are reports of dermatitis caused by this timber. The timber of *Eucryphia cordifolia* from Chile has been used for a variety of purposes and the bark is a source of tannin. The timber of the Tasmanian species, *E. lucida*, is used for general building and cabinetmaking. *Eucryphia* spp. and hybrids are cultivated as small ornamentals. AC

[i] Schonenberger, J. *et al.* Cunoniaceae in the Cretaceous of Europe: Evidence from fossil flowers. *Ann. Bot.* 88: 423–437 (2001).
[ii] Bradford, J. C. & Barnes, R. W. Phylogenetics and classification of Cunoniaceae (Oxalidales) using chloroplast DNA sequences and morphology. *Syst. Bot.* 26: 354–385 (2001).
[iii] McPherson, G. & Lowry, P. P. *Hooglandia*, a newly discovered genus of Cunoniaceae from New Caledonia. *Ann. Missouri Bot. Gard.* 91: 260–265 (2004).
[iv] Nandi, O., Chase, M. W., & Endress, P. K. A combined cladistic analysis of angiosperms using *rbc*L and non-molecular data sets. *Ann. Missouri Bot. Gard.* 85: 137–212 (1998).
[v] Sweeney, P. W. *et al.* Phylogenetic position of the New Caledonian endemic genus *Hooglandia* (Cunoniaceae) as determined by maximum parsimony analysis of chloroplast DNA. *Ann. Missouri Bot. Gard.* 91: 266–274 (2004).
[vi] Matthews, M. L., Endress P. K., Schonenberger, J., & Friis, E. M. A comparison of floral structures of Anisophylleaceae and Cunoniaceae and the problem of their systematic position. *Ann. Bot.* 88: 439–455 (2001).
[vii] Bradford, J. C., Fortune Hopkins, H. C., & Barnes, R. W. Cunoniaceae. Pp. 91–111. In: Kubitzki, K. (ed.), *The Families and Genera of Vascular Plants. VI. Flowering Plants. Dicotyledons: Celastrales, Oxalidales, Rosales, Cornales, Ericales*. Berlin, Springer-Verlag (2004).

CURTISIACEAE
ASSEGAI FAMILY

| Genera | 1 | Species | 1 |
| Economic uses | Hard, close-grained timber |

The Curtisiaceae comprises a single species of evergreen tree, *Curtisia dentata*, up to 20 m high, found in the montane forests of the border of Zimbabwe and Mozambique and southward to Swaziland and South Africa to the Western Cape. The young parts are densely tomentose with brown hairs. The leaves are opposite, broadly ovate to elliptical, strongly dentate, prominently veined beneath. The inflorescence is a many-branched, terminal panicle. The flowers are subsessile, 4-merous, regular, and epigynous, with sepals and petals less than 2 mm long and inconspicuous. The ovary is inferior, with 4 locules and a single ovule in each one, developing into a drupe with a stone probably dehiscing by valves.

Curtisiaceae is close to Cornaceae and has been included in it by many authors. However, it has a rather different appearance, with its dense indumentum and serrate leaves, and differs in its 4-locular fruit with pyrenes with

CURTISIACEAE. I *Curtisia dentata* (a) leafy shoot with terminal inflorescence (x⅔); (b) flower (x6); (c) fruit (x3); (d) half section of fruit (x3); (e) seed (x3).

vascular bundles up the middle, which persuaded Eyde[i] that it did not fit in Cornaceae. Furthermore, its pollen is readily distinguishable from that of Cornaceae, being about half the size. Molecular analyses (see Cornaceae) place *Curtisia* basal to the Cornaceae and perhaps close to another South African family, Grubbiaceae, from which, however, it is very different morphologically. The tree is valuable for its timber, since it provides a heavy, hard-grained, durable wood that was much prized in earlier times in the manufacture of the axles and spokes of wheels of wagons, and as the shafts of spears (assegais)[ii]. RKB

[i] Eyde, R. H. Comprehending *Cornus*: puzzles and progress in the systematics of the dogwoods. *Bot. Rev.* 54: 233–351 (1988).
[ii] Palmer, E. & Pitman, N. *Trees of Southern Africa* 3: 1714–1716 (1972).

CYCLOCHEILACEAE

The family consists of stiff shrubs, with gamopetalous flowers lacking a calyx but enclosed by 2 opposed, cymbal-like, suborbicular, accrescent bracteoles, and is characteristic of dry habitats in the Horn of Africa and adjacent areas.

Distribution *Cyclocheilon* includes 3 species from southern Arabia, northern Somalia, and from eastern Ethiopia. *Asepalum* has 1 species in southern Arabia, Somalia and Ethiopia to northern Uganda, northern and eastern Kenya, and northeastern Tanzania.

Genera 2 Species 4
Economic uses Roots of both genera have been used to dye leather red[ii]

Description Shrubby, with stiff and sometimes almost spiny gray stems up to 2 m high. The leaves are opposite and without stipules, small, entire, and often rather hispid. The flowers are solitary in the leaf axils. The pedicel bears below the flower a pair of conspicuous bracteoles, broadly ovate to suborbicular, and accrescent and persistent in fruit, and also a pair of small bracts attached either below or alongside the bracteoles. The calyx is absent. The corolla is obliquely infundibuliform, 5-lobed, 1–3.5 cm long, and rather showy. The stamens are 4, inserted on the corolla tube. The ovary is a flattened disk consisting of 2 fused carpels and with 1 or 2 locules, with 2–10 ovules borne on long funicles. The fruit is a dry disk, which is either a dehiscent capsule or breaks up into 2 mericarps. The seeds have no endosperm.

Classification The family was described in 1981[i], having been formerly included in the broad concept of Verbenaceae or in the Chloanthaceae (which is now included in Lamiaceae). It was considered at the time to be close to Acanthaceae, being similar to the Thunbergioideae in the 2 conspicuous bracteoles and absence of a calyx, while the long funicles bearing the seeds are reminiscent of the retinacula of that family. Molecular data now place it close to the parasitic Scrophulariaceae[ii], and it has similarities in pollen and gynoecium structure to Nesogenaceae and Stilbaceae. *Cyclocheilon* has obovate to suborbicular bracts, inserted alongside the 2 bracteoles, the ovary is 1-locular or incompletely 2-locular, and the fruit is a dehiscent capsule, while *Asepalum* has smaller, linear bracts remote from the bracteoles, the ovary is 2-locular, and the fruit is a schizocarp. For quick identification, *Cyclocheilon* has pubescent bracteoles and a lilac to red corolla, while *Asepalum* has glabrous bracteoles and a white corolla. The accrescent bracteoles fall with the fruit and may help wind dispersal of the seeds.   RKB

[i] Marais, W. Two new gamopetalous families, Nesogenaceae and Cyclocheilaceae, for extra-Australian Dicrastylidaceae. *Kew Bull.* 35: 797–812 (1981).
[ii] Sebsebe, D. Cyclocheilaceae. Pp. 60–62. In: Kadereit, J. W. (ed.), *Families and Genera of Vascular Plants. VII. Flowering Plants. Dicotyledons: Lamiales (except Acanthaceae including Avicenniaceae)*. Berlin, Springer-Verlag (2004).

CYNOMORIACEAE

A family of fleshy, achlorophyllous root parasites, comprising the single genus *Cynomorium*, with 2 species.

Distribution *Cynomorium coccineum* occurs in dry areas in North Africa and the Mediterranean region. *C. songaricum* grows in the Middle East, extending into western and central Asia.

Genera 1 Species 2
Economic uses Eaten locally and widely used in traditional medicine

Description Fleshy, reddish brown to purplish obligate parasite, attaching to the roots of a range of species, including salt-tolerant plants, such as *Atriplex, Nitraria, Obione* and *Salsola, Tamarix, Limonium,* and various Cistaceae. The stem is represented by an underground rhizome bearing numerous haustoria. The aerial parts of the plant are the fleshy, clavate, spikelike inflorescence, which bears numerous, minute, tightly packed flowers interspersed with scalelike leaves, which bear stomata. The flowers are both unisexual and bisexual (plants polygamomonoecious). The perianth consists of (1–)4–5(–8) sepaloid, linear segments, which are free or partially connate. Staminate flowers have a single stamen adnate to the perianth; pistillate flowers have an inferior ovary, consisting of a single carpel with a single pendulous ovule. Fruit a single-seeded nut. Nickrent's Parasitic Plant Connection[i] is a valuable source of further information and illustrations.

Classification The Cynomoriaceae is often included in the Balanophoraceae or placed alongside the Balanophorales in the superorder Balanophoranae. The 2 families are now separated by a series of morphological characters, and molecular data do not support a close relationship. The family is treated as an unplaced eudicot in APGII. The molecular evidence so far available is not conclusive and suggests that the Cynomoriaceae may be sister to Saxifragales[ii].

Economic uses *Cynomorium* has been used for food and medicine since biblical times by various cultures, especially the Arabs, who call the plant "tarthuth." *C. coccineum* was known as Malta Mushroom (fungus melitensis) in the Middle Ages, and it was highly prized. It also has a reputation as an aphrodisiac. *C. songaricum* is an important constituent in traditional Chinese medicine.   VHH

[i] Nickrent, D. The Parasitic Plant Connection. http://www.science.siu.edu/parasitic-plants/
[ii] Nickrent, D. L. Orígenes filogenéticos de las plantas parásitas. Capitulo 3, pp. 29–56. In: López-Sáez, J. A., Catalán, P., & Sáez, L. (eds), *Plantas Parásitas de la Península Ibérica e Islas Baleares*. Mundi-Prensa Libros, S. A., Madrid (2002).

CYRILLACEAE
LEATHERWOOD AND BUCKWHEAT TREES

Two New World genera of deciduous or evergreen shrubs or small trees reproducing from root suckers.

Distribution *Cyrilla racemiflora* is spread from the southeast of North America to northern South America and the Caribbean. *Cliftonia monophylla* is native to southeastern North America.

Genera 2 Species 2
Economic uses Ornamental for decorative flowers

Description Evergreen or sometimes deciduous (*Cyrilla*), small, hairless trees or shrubs with simple leaves. The leaves are alternate, entire, with a short petiole or none, and without stipules. The inflorescence is of terminal or axillary racemes. The flowers are actinomorphic, hermaphrodite, with 5 free or partially fused persistent and imbricate sepals, often enlarged in the fruit. The petals are 5, equal, free or fused at the base, and imbricate. The stamens are 5 (*Cyrilla*) or 10, then in 2 whorls of 5 (*Cliftonia*), with free filaments inserted on the receptacle, surrounding a superior ovary of 2–4 (*Cyrilla*) or 3–5 (*Cliftonia*) fused carpels, each forming a locule containing 1 ovule. The style is short and 2–4 lobed (*Cyril-*

CYRILLACEAE. 1 *Cliftonia monophylla* (a) shoot with terminal inflorescences (x⅔); (b) winged fruit (x3).

la) or almost sessile with 3–5 lobes (*Cliftonia*). The fruit is a dry capsule and has 2–4 wings in *Cliftonia* and is often entirely seedless.

Classification *Cliftonia* is characterized by possessing 10 stamens, deciduous calyx (not enlarged after flowering), and a short style divided into 3 stigmas, while *Cyrilla* has only 5 stamens and a thick, short style ending in a 2- or 3-lobed stigma. *Cyrilla* has been treated as having up to 11 spp., but a revision of the genus[i] showed that 1 single and continuously variable species was the best interpretation. The family is often regarded as belonging to the Ericales (Dahlgren 1983). However, it is sometimes placed in the Celastrales (Melchior 1964) or Theales (Thorne 1983). Phylogenetic studies have consistently placed *Cyrilla* sister to Ericaceae[ii], the position becoming solidly established as more sequence data have been added[iii]. The family is closely related to the Ericaceae and Clethraceae[iv]. The proposed transfer of *Purdiaea* to Clethraceae[iv] needs further investigation to resolve the relationship among Clethraceae, Cyrillaceae, and Ericaceae[v].

Economic uses The Leatherwood (*Cyrilla racemosa*) and the Buckwheat Tree (*Cliftonia monophylla*) possess attractive white flowers and reddish-tinted autumnal foliage and are grown as ornamentals in gardens.   AC

[i] Thomas, J. L. The genera of the Cyrillaceae and Clethraceae of the Southeastern United States. *J. Arnold Arb.* 42: 96–106 (1961).
[ii] Kron, K. A. Phylogenetic relationships of Empetraceae, Epacridaceae, Ericaceae, Monotropaceae, and Pyrolaceae: evidence from nuclear ribosomal 18s sequence data *Ann. Bot.* 77: 293–303 (1996).
[iii] Anderberg, A. A., Rydin, C., & Källersjö, M. Phylogenetic relationships in the order Ericales *s.l.*: analyses of molecular data from five genes from the plastid and mitochondrial genomes. *American J. Bot.* 89: 677–687 (2002).
[iv] Anderberg, A. A. & Zhang, X. Phylogenetic relationships of Cyrillaceae and Clethraceae (Ericales) with special emphasis on the genus *Purdiaea* Planch. *Organisms Diversity and Evolution* 2: 127–137 (2002).
[v] Fior, S., Karis, P. O., & Anderberg, A. A. Phylogeny, taxonomy, and systematic position of *Clethra* (Clethraceae, Ericales) with notes on biogeography: evidence from plastid and nuclear DNA sequences. *Int. J. Plant Sci.* 164: 997–1006 (2003).

CYTINACEAE

A small family of achlorophyllous root parasites.

Distribution *Bdallophyton* (2–4 spp.) occurs in Mexico and Central America, while *Cytinus* grows in the Mediterranean region, South Africa, and Madagascar.

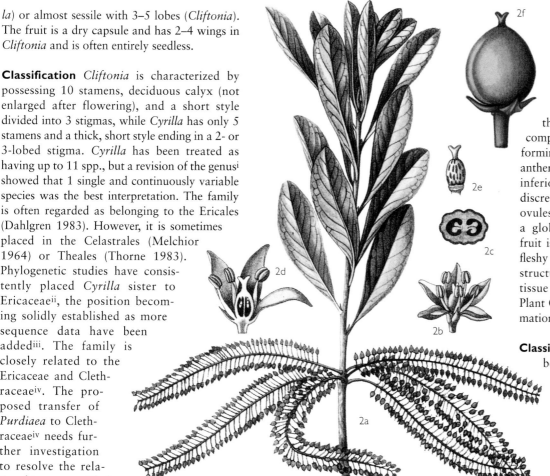

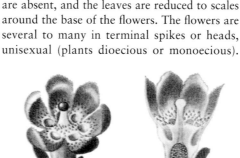

2f

CYRILLACEAE. 2 *Cyrilla racemosa* (a) shoot with axillary inflorescences (×⅔); (b) flower (×4); (c) cross section of ovary (×14); (d) half flower (×6); (e) gynoecium (×6); (f) fruit (×6).

The perianth is tubular, with 4–9 imbicate lobes, often brightly colored. In the staminate flowers, the androecium comprises 8–20 stamens, which are connate, forming a compact synandrium with extrorse anthers. The pistillate flowers have a 1-locular, inferior ovary, with 8–14 deeply intrusive, discrete parietal placentas bearing numerous ovules; the style is columnar and ends with a globose or capitate, viscous stigma. The fruit is an indehiscent or irregularly dehiscent fleshy berry (*Cytinus*) or compound berrylike structures made up of axial and ovarian tissue (*Bdallophyton*). Nickrent's Parasitic Plant Connection[i] is a valuable source of information and illustrations.

Classification The Cytinaceae has previously been included in the Rafflesiaceae but was recognized as a separate family by Takhtajan. It differs from the Rafflesiaceae and other families in the Rafflesiales by having an inflorescence of multiple flowers.

A study of the phylogenetic relationships among the holoparasitic members of the Rafflesiales using data derived from nuclear, mitochondrial, and chloroplast DNA, indicates that the Cytinaceae is strongly supported as a member of Malvales[ii].   VHH

[i] Nickrent, D. The Parasitic Plant Connection. http://www.science.siu.edu/parasitic-plants/
[ii] Nickrent, D. L., Blarer, A., Qiu, Y.-L., Vidal-Russell, R., & Anderson, F. E. Phylogenetic inference in Rafflesiales: the influence of rate heterogeneity and horizontal gene transfer. *BMC Evol. Biol.* 4: 40 (2004). http://www.biomedcentral.com/1471-2148/4/40

Description Achlorophyllous endoparasites with a filamentous or platelike endophyte resembling a fungal mycelium inside the host. *Cytinus hypocistis* is found on members of Cistaceae, while *Bdallophyton* is parasitic on the roots of plants of the Burseraceae. The stems are absent, and the leaves are reduced to scales around the base of the flowers. The flowers are several to many in terminal spikes or heads, unisexual (plants dioecious or monoecious).

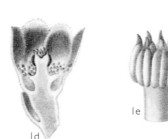

CYTINACEAE. 1 *Cytinus sanguineus* (a) inflorescences on surface of host stem (×⅔); (b) female flower (×⅔); (c) half female flower (×⅔); (d) half male flower (⅔); (e) staminal column (×2).

Genera 2   Species 10
Economic uses None

DAPHNIPHYLLACEAE

A monogeneric family of 10 spp. of evergreen trees or shrubs that grow in East and Southeast Asia. The leaves are alternate but appearing verticillate, simple, long-petiolate, entire. The flowers are small, actinomorphic, unisexual (plants dioecious) regular, in axillary or near-terminal racemes.

Genera 1 **Species** 10
Economic uses Some ornamental value

The sepals are (2)3–6, more or less connate; the petals are absent. Staminate flowers have 5–12(–24) distinct stamens with short filaments. Staminodes may be present. Pistillate flowers have a superior ovary of 2, rarely 4, united carpels, with 2(–4) locules and 2 apical to parietal ovules. The fruit is a 1-seeded drupe.

The taxonomic and phylogenetic position of *Daphniphyllum* remains somewhat unclear. It has sometimes been placed included in the Euphorbiaceae or in the Euphorbiales as a distinct family or placed in its own order, Daphniphyllales near the Hamamelidales (Cronquist 1981). Molecular evidence suggests inclusion in the Saxifragales *sensu* APG II although with some uncertainty about its position[i, ii]. Some species are occasionally cultivated as garden shrubs.   VHH

[i] Fishbein, M., Hibsch-Jetter, C., Soltis, D. E., & Hufford, L. Phylogeny of Saxifragales (angiosperms, eudicots): analysis of a rapid, ancient radiation. *Syst Biol.* 50: 817–847 (2001).
[ii] Soltis, D. E., Fishbein, M., & Kuzoff, R. K. Reevaluating the evolution of epigyny: Data from phylogenetics and floral ontogeny *Int. J. Pl. Sci.* 164 (Supplement): S251–S264 (2003).

DATISCACEAE

This family of massive trees and coarse herbs is an evolutionary puzzle

Distribution *Datisca cannabina* extends from the eastern Mediterranean (Crete, Samos, and Cyprus) to Kazakhstan and Nepal. *C. glomerata* is confined to California and northern Baja California, both species growing on riverbanks, in woodland, and in open places. *Tetrameles*

Genera 3 **Species** 4
Economic uses Canoes and poor quality timber

nudiflora occurs from western India, Nepal, and Yunnan, through Indo–China and Malaysia, to northern Queensland, in deciduous lowland forest. *Octomeles sumatrana* occurs through Malaysia from Sumatra to the Solomon Islands, but excluding Java and the Lesser Sunda Islands, a conspicuous emergent in lowland rain forest often on riverbanks.

Description Either robust, erect perennial herbs to 2(3) m (*Datisca*) or large, buttressed trees up to 60 m and with a bole up to 5 m in diameter. Of the trees, *Tetrameles* is deciduous but *Octomeles* is evergreen. The leaves are alternate and without stipules but different in form. The tree genera have simple, long-petioled, ovate to heart-shaped leaves, up to 45 × 25 cm, occasionally with 3–5 shallow acuminate lobes in *Tetrameles*, with an entire to weakly serrate margin. In *Datisca*, the leaves are imparipinnate to deeply pinnatifid or the upper ones successively reduced to ternate or unlobed leaves or bracts, with acute apices and serrate margins. The flowers are actinomorphic, unisexual (plants dioecious or, in *Datisca glomerata*, androdioecious), in long, pendent spikes or (in male *Tetrameles*) panicles, either fascicled at the branch tips or axillary, up to 35 cm long in the tree genera. The staminate male flowers in *Datisca* and *Tetrameles* have a short calyx tube 0.5–3 mm long, with 3–10 teeth (4 in *Tetrameles*), no petals, and 6–15(–25) stamens, while those of *Octomeles* have a campanulate calyx tube with 6 to numerous short, erect, greenish lobes, narrowly triangular greenish petals between the calyx teeth, and stamens opposite the calyx lobes. The pistillate or hermaphrodite flowers are epigynous, with 3–4(5) short calyx lobes on a small rim surmounting the ovary in *Datisca* and *Tetrameles*, or with 6–8 such calyx lobes in *Octomeles*, without petals. The ovary is inferior of 3–5 fused carpels and single locule with 3–8 parietal placentas. The styles are 3–4 and deeply bifid, with no distinct stigma in *Datisca*, 4 with oblique stigmas in *Tetrameles*, and 6–8 with capitate stigmas in *Octomeles*. The fruits of *Datisca* and *Tetrameles* are capsules, dehiscing by the apical valves curving inward to leave an apical pore, while that of *Octomeles* is complex, accrescent to a length of 13 mm or more and barrel-shaped, the exocarp splitting longitudinally from the base to be cast off with the calyx, revealing a hard endocarp that splits longitudinally from the apex into 6–8 light brown elliptic segments forming a persistent basket round the seeds. The seeds are numerous, small, sometimes spindle-shaped or winged.

Classification The family has long been considered close to Begoniaceae or Cucurbitaceae in the old Parietales and are placed with these in the Cucurbitales of APG II. Perhaps surprisingly, Coriariaceae and Corynocarpaceae are also closely related. The question remains as to whether the 2 tree genera should be taken out of the otherwise herbaceous Datiscaceae and referred to Tetramelaceae. An early molecular

analysis[i] placed *Datisca* slightly removed from the tree genera, only *Coriaria* falling between them, this group being sister to Begoniaceae and Cucurbitaceae. A later more detailed analysis[ii] produced 4 cladograms with differing results. In the first, *Datisca* was sister to the Begoniaceae, but only one step apart from the other 2 genera, which were sisters to the Coriariaceae and Cucurbitaceae. In the second, *Datisca* was sister to *Tetrameles* and *Octomeles* with Begoniaceae one step further removed. In the third, the 2 tree genera were sisters to *Datisca* and to Begoniaceae. In the fourth, Begoniaceae were sister to a group of *Tetrameles*, *Octomeles*, and then *Datisca*. A later analysis including all the Cucurbitales[iii] essentially placed all 3 genera included here in Datiscaceae as sister to the Begoniaceae. A broad phylogeny of the dicots[iv] with a much lesser sampling of the Cucurbitales placed *Datisca* next to *Combretocarpus* in Anisophylleaceae several steps distant from *Tetrameles* and *Octomeles*.

The molecular evidence is thus inconclusive as to the exact relationships of the genera here included in the family, but most analyses place them close to each other in 2 groups of 2 (1 being the 2 spp. of *Datisca* and the other the 2 tree genera). If we look closely at the morphological evidence, it will be clear from the above description that while *Tetrameles* and *Octomeles* are very similar in their massive tree habit, they are very different in floral and fruit characters. In these characters, *Tetrameles* closely resembles *Datisca* except in the stigmas, while *Octomeles* has many unique features. A detailed study of floral morphology of the Cucurbitales[v] has included only *Datisca cannabina* and *Octomeles sumatrana*, thus failing to consider the close similarity between *Datisca* and *Tetrameles*. Nonetheless, it has concluded that *Datisca* and *Octomeles* have a close relationship and share many special features.

The very close floral and fruit similarity between *Tetrameles* and *Datisca* has been disregarded by those who recently have sought to recognize the family Tetramelaceae for the 2 big tree genera. A monographic study by Davidson[vi] had resulted in the conclusion that evidence from anatomy and morphology of fruits and flowers does not support the separation of *Octomeles* and *Tetrameles* into their own family, and the differences do not even necessitate subfamilial rank. The same author later concluded[vii] that, despite the differences in habit, evidence from anatomy and morphology supports the inclusion of these three genera in a single family. A study of ovule and seed structure[viii] came to a similar conclusion. A cursory examination of the male flowers and the fruits of *Datisca* and *Tetrameles* shows they are so similar that it would seem difficult to put them in different families.

Economic uses The wood of the fast-growing *Tetrameles* and *Octomeles* is of poor quality and is used for short-term construction only or products such as boxes. Nonetheless, they have been planted commercially for quick supplies of wood, and *O. sumatrana* established in plantations in Nigeria some time in the twentieth century had reached a height of 35 m and girth of 6 m by 1966 (specimen in herbarium at Kew). In their native areas the trunks of both *Tetrameles* and *Octomeles* have been used in the past for making large canoes. RKB

i Swensen, S. M., Mullin, B. C., & Chase, M. W. Phylogenetic affinities of Datiscaceae based on an analysis of nucleotide sequences from the plastid *rbc*L gene. *Syst. Bot.* 19: 157–168 (1994).
ii Swensen, S. M., Luthi, J. N., & Rieseberg, L. H. Datiscaceae revisited: monophyly and the sequence of breeding system evolution. *Syst. Bot.* 23: 157–169 (1998).
iii Wagstaff, S. J. & Dawson, M. I. Classification, origin and patterns of diversification of *Corynocarpus* (Corynocarpaceae) inferred from DNA sequences. *Syst. Bot.* 25: 134–149 (2000).
iv Savolainen, V. *et al.* Phylogeny of the eudicots: a nearly complete familial analysis based on *rbc*L gene sequences. *Kew Bull.* 55: 257–309 (2000).
v Matthews, M. L. & Endress, P. K. Comparative floral structure and systematics in Cucurbitales (Corynocarpaceae, Coriariaceae, Tetrameleaceae, Datiscaceae, Begoniaceae, Cucurbitaceae, Anisophylleaceae). *Bot. J. Linn. Soc.* 145: 129–185 (2004).
vi Davidson, C. An anatomical and morphological study of Datiscaceae. *Aliso* 8: 49–110 (1973).
vii Davidson, C. Anatomy of xylem and phloem of Datiscaceae. *Contrib. Sci. Nat. Hist. Mus. Los Angeles Co.* 280: 1–28 (1976).
viii Boesewinkel, F. D. Ovule and seed structure in Datiscaceae. *Act Bot. Neerland.* 33: 419–429 (1984).

DEGENERIACEAE

The family Degeneriaceae comprises 1 or 2 spp. of large trees endemic to Fiji—*Degeneria vitiensis* from Suva and *D. roseiflora* from Levu Island. The latter is sometimes considered to be a subsp. of *D. vitiensis*. The leaves are alternate, simple, entire, petiolate, and without stipules. The flowers are solitary, hermaphrodite, and actinomorphic. The sepals are 3, free. The petals are 12–13(–18) in 3–5 whorls, free. The stamens are numerous, arranged in several series, and associated with staminodes. The gynoecium consists of a single, superior carpel with a ventral, elongate, spreading stigma; it is incompletely sealed when young and contains numerous ovules set in a double row on 2 almost marginal placentas. The fruit is large, indehiscent, and leathery, containing numerous flattened seeds with copious, oily, ruminate endosperm and a minute embryo.

Although the family is unique because of its incompletely sealed carpels during anthesis, it is closely allied to the Himantandraceae and Magnoliaceae within the Magnolialesi,ii. No economic uses are known. VHH

i Soltis, D. E. *et al.* Angiosperm phylogeny inferred from 18S rDNA, *rbc*L, and *atp*B sequences. *Bot. J. Linn. Soc.* 133: 381–461 (2000).
ii Sauquet, H. *et al.* Phylogenetic analysis of Magnoliaceae and Myristicaceae based on multiple data sets: Implications for character evolution. *Bot. J. Linn. Soc.* 142: 125–186 (2003).

DESFONTAINIACEAE

The family comprises 1 genus (*Desfontainia*) of holly-leaved erect shrubs and small trees up to 7 m, or occasionally scrambling or prostrate, that occur from Costa Rica through the Andes to southern Chile and Argentina at altitudes between 300 and 4,000 m, usually in cloud forest or wet mossy montane woodland or rocky terrain.

Genera 1
Species (1)3–5
Economic uses
Hallucinogenic properties, medicinal cures, yellow dye, often cultivated for ornament

The leaves are opposite, simple, usually 2–6 cm long, hollylike, with spiny margins of between 1 and 11 spines on each side, occasionally (especially in Colombia) obovate with 1 scarcely spinous crenation on each side near the apex, and with small caducous interpetiolar stipules that leave conspicuously raised nodes. The inflorescence is a 1- to few-flowered terminal or axillary cyme, with pedicels up to 1.5(–2) cm. The flowers are actinomorphic, bisexual, hypogynous, 5-merous. The sepals are triangular to oblong, usually with a distinct apiculum (cf. Columelliaceae). The corolla has a long tube, up to 3.5 cm, which is cylindrical or expanded upward and is deep red to orange outside and often yellow inside, with a waxy texture, much exceeding the calyx, with 5 rounded to acuminate lobes that are red to yellow and 5–15 mm long. The stamens are 5, inserted in the upper part of the corolla tube. The ovary is superior, syncarpous, of 5 carpels, incompletely divided into 5 locules, with many ovules on 5 axile placentas. The fruit is a globose to ovoid many-seeded berry, becoming white when mature.

The genus was formerly included in Loganiaceaei, but a number of micromorphological characters, as listed by Takhtajan (1997), set the genus apart from that family. Molecular evidence at first placed it in Dipsacalesii, but later basal to Dipsacales and closely allied to *Columellia*iii. A broader analysisiv again placed *Desfontainia* in the Dipsacales, but placed *Columellia* basal to the Gentianales close to Icacinaceae. In APG II, Desfontainiaceae is placed in Euasterids II, optionally included in Columelliaceae. The relationship with Columelliaceae may be supported by the Andean distribution of both, the wood anatomy, the opposite leaves with characteristic raised nodes, and perhaps even the apiculate sepals resembling the spines of the leaves. However, in its slightly zygomorphic flowers, 2 unique stamens, largely inferior ovary and capsular fruit, *Columellia* differs markedly from *Desfontainia*.

Although some 13 spp. have been described in *Desfontainia*, most recent authors since Leeuwenberg's revision have regarded it as

including only 1 sp., *D. spinosa*. However, a revision by Weigend, based on field and herbarium studiesv, recognizes 3 spp., even though the characters used to recognize taxa are not always obvious and intermediates may occur. *D. fulgens* from Chile and Argentina, *D. spinosa* confined to cloud forest in Peru, and *D. splendens* from Costa Rica to Bolivia are divided into 5 regional variants, some of which may deserve specific rank. The small-leaved variants in Colombia in particular look different from the larger-leaved spinier plants.

The hallucinogenic properties of *Desfontainia* have been well documented. The genus also provides cures for stomach complaints and a yellow dye for textile manufacture in Chile. The plant cultivated as a colorful flowering shrub in temperate countries as *D. spinosa* is more likely to be *D. fulgens*. RKB

i Leeuwenberg, A. J. M. Notes on American Loganiaceae, 4. Revision of *Desfontainia*, Ruiz & Pav. *Acta Bot. Neerl.* 18: 669–679 (1969).
ii Bremer, B., Olmstead, R. G., Struwe, L., & Sweere, J. A. *rbc*L sequences support exclusion of *Retzia, Desfontainia* and *Nicodemia* from the Gentianales. *Pl. Syst. Evol.* 190: 213–230 (1994).
iii Backlund, A. & Bremer, B. Phylogeny of the Asteridae s. str. based on *rbc*L sequences, with particular reference to the dipsacales. *Pl. Syst. Evol.* 207: 225–254 (1997).
iv Savolainen, V. *et al.* Phylogeny of the eudicots: a nearly complete familial analysis based on *rbc*L gene sequences. *Kew Bull.* 55: 257–309 (2000).
v Weigend, M. *Desfontainia* Ruiz & Pav. (Desfontainiaceae) revisited – a first step back towards α-diversity. *Bot. Jahrb. Syst.* 123: 281–301 (2001).

DIAPENSIACEAE

DIAPENSIA FAMILY

A small family of mat- or rosette-forming montane or subarctic herbs.

Distribution The family has an interesting distribution with centers of diversity in the eastern Himalayas and adjacent mountains of China, in Japan to Taiwan, and in the Appalachians of North America, plus 1 circumboreal species (see tribes below). They occur in montane to subarctic habitats or woods or pine barrens.

Genera 5 Species 13
Economic uses Some horticultural value

Description Perennial herbs, either prostrate and mat-forming, or rhizomatous and subacaulous. The leaves are alternate, simple, with entire to serrate margins, either small and obovate and crowded in mat-forming genera, or

Galacineae consists of 1 genus, 1 sp., native to the Appalachian Mountains (see map[iv]). Rhizomatous herbs with a rosette of long-petioled broad cordate leaves surrounding a long-peduncled spike of many smallish flowers up to 70 cm high; petals free; stamens and staminodes connate at their base and adnate to the base of the petals; anthers 1-locular; style short. *Galax urceolata.*

Shortieae consists of 2 genera, 7 spp., native to Asia and North America. Rhizomatous herbs with a loose rosette of long-petioled broad cuneate to cordate leaves; inflorescence a lax raceme of 1–15 relatively large, showy flowers on a long scape; corolla tubular; anthers 2-locular; staminodes inserted at the base of the corolla tube opposite the lobes; style elongate. *Berneuxia* (1 sp. from Tibet to Yunnan), *Shortia* (including *Schizocodon*), 1 sp. in Yunnan, 3 spp. in Japan (1 endemic in the Ryukyu Islands), 1 sp. in Taiwan, and 1 sp. in the Appalachians).

Diapensieae consists of 2 genera, 5 spp., native to western China and eastern Himalayas,

DIAPENSIACEAE. I *Diapensia himalaica* (a) creeping shoot bearing small, simple overlapping leaves and solitary flowers (x⅔); (b) fruiting shoot (x⅔); (c) 5-lobed perianth opened out to reveal 5 stamens fused to corolla (x2); (d) cross section of ovary (x6); (e) dehiscing fruit (x3); (f) detail of stamens (x8). 2 *Shortia soldanelloides* (a) habit (x⅔); (b) part of corolla opened out to show fertile stamens inserted on corolla tube and linear staminodes at the base (x2); (c) fruit enclosed in persistent bracts and calyx (x2); (d) gynoecium (x2); (e) cross section of ovary (x3); (f) stamens, dorsal view (left) and ventral view (right) (x6). 3 *Galax urceolata* habit (x⅔).

long-petioled and broad with a cuneate to cordate base. The inflorescence is an elongate, many-flowered spike, or a 1- to 15-flowered raceme on a long scape, or the flowers are solitary and terminal on short branches (see tribes). The flowers are actinomorphic, bisexual, 5-merous, sometimes showy. The sepals are free or shortly connate. The petals are free (*Galax*) or fused into a corolla tube with 5 lobes, which are conspicuously fringed in *Shortia*. The stamens are attached to the petals or corolla tube, sometimes alternating with 5 staminodes. In *Pyxidanthera* and *Galax,* the anthers are appendaged. The ovary is superior, of 3 fused carpels with a simple style and 3-lobed stigma. There are 3 locules, with few to many ovules on swollen axile placentas, or these become parietal during development. The fruit is a loculicidal capsule.

Classification The family has sometimes been considered close to Ericaceae, the rhizomatous species showing superficial similarity to *Pyrola*. Molecular evidence places them in the Ericales, but nearer to the Styracaceae than to the Ericaceae. Within the family, *Galax* consistently appears sister to the other genera[i]. The 5 genera fall at first sight into three very clear groups which are not contradicted by molecular evidence, and it is perhaps surprising that these have not all always been given formal rank[ii,iii]. Three tribes are recognized here.

North America and circumboreal. Mat-forming perennial herbs with small narrow crowded leaves; flowers solitary and terminal on branch stems; corolla tubular; anthers 2-locular; staminodes absent. *Pyxidanthera* (1 sp., New Jersey pine barrens and adjacent areas), *Diapensia* (1 sp. circumboreal, 3 spp. western China and eastern Himalaya). *Diplarche*, sometimes referred to Diapensiaceae, is now placed in Ericaceae.

Economic uses *Shortia* and *Diapensia* species are popular as rock-garden plants. *Galax* is sometimes grown on stream banks for its interesting architecture and has been the source of a major local industry in the Appalachians around the town of Galax in Virginia, where millions of leaves have been collected each year for use in flower arrangements[v].　　RKB

[i] Rönblom, K. & Anderberg, A. A. Phylogeny of the Diapensiaceae based on molecular data and morphology. *Syst. Bot.* 27: 383–385 (2002).
[ii] Scott, P. J. & Day, R. T. Diapensiaceae, a review of the taxonomy. *Taxon* 32: 417–423 (1983).
[iii] Scott, P. J. Diapensiaceae. Pp. 117–121. In: Kubitzki, K. (ed.), *The Families and Genera of Vascular Plants. VI. Flowering Plants. Dicotyledons: Celastrales, Oxalidales, Rosales, Cornales, Ericales.* Berlin, Springer-Verlag (2004).
[iv] Nesom, G. L. *Galax* (Diapensiaceae): geographical variation in chromosome numbers. *Syst. Bot.* 8: 1–14 (1983).
[v] Nelson, T. C. & Williamson, M. J. Decorative plants of Appalachia. *U.S. Dept. Agric. Inform. Bull.* 342: 1–31 (1970).

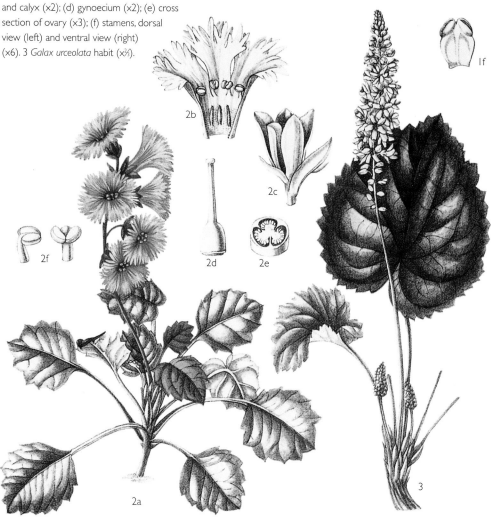

DICHAPETALACEAE

A mainly tropical family of trees, shrubs, and lianas, some of which are poisonous.

Distribution The family occurs in the neotropics, Africa, and tropical Asia. The largest genus, *Dichapetalum* (c. 130 spp.), is pantropical but strongly represented in Africa, including 1 sp. in South Africa (*D. cymosum*). *Stephanopodium* (12 spp.) is confined to Central America (1 sp., *S. costaricense*, endemic to Costa Rica) and South America, and the Caribbean, while *Tapura* (21 spp.) grows in the neotropics and in Africa.

Genera 3 Species c. 220
Economic uses A number of *Dichapetalum* spp. are poisonous

Description Trees, shrubs, or lianas. The leaves are alternate, simple, entire, with stipules. The inflorescence consists of axillary cymes with the stalk often united to the leaf petiole below (some species of *Dichapetalum* and *Tapura*), fascicles, or panicles. The flowers are actinomorphic or somewhat zygomorphic (*Tapura*), herma-phrodite or rarely unisexual (plants monoecious—some *Stephanopodium* spp.). The sepals are 5, imbricate, often unequal, free or connate. The petals are 5, imbricate, equal, free or connate into a tube and then with equal or unequal lobes, the apex deeply bifid, often drying black. The stamens are 3 or 5, free or epipetalous. The ovary is superior, syncarpous, of 2 or 3 carpels, with 2 or 3 terminal styles; locules 2 or 3 with 2 pendulous ovules. The fruit is a dry or fleshy drupe, often covered with soft hairs, with a single seed in each locule.

Classification The Dichapetalaceae has at various times been referred to the Rosales, Celastrales, or Thymelaeales. Recent molecular studies indicate a placement in the Rosidae and a close relationship with the Chrysobalanaceae and Trigoniaceae in the order Malpighiales[i,ii]. Dichapetalaceae, along with some other Malpighiales, is unusual in having tenuinucellate ovules.

Economic uses The leaves and seeds of *Dichapetalum* are poisonous and a source of the vertebrate pesticide 1080. The leaves of *Dichapetalum stuhlmannii* are used to poison

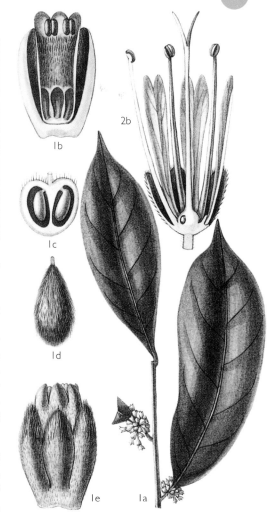

DICHAPETALACEAE. 1 *Stephanopodium peruvianum* (a) leafy shoot and axillary inflor_.scences united to the leaf stalk (×⅔); (b) half flower with sessile epipetalous anthers (×8); (c) vertical section of ovary showing pendulous ovules (×14); (d) fruit—a hairy drupe (×⅔); (e) flower (×8). 2 *Dichapetalum mombongense* (a) flowering and fruiting shoot (×⅔); (b) half flower (×8); (c) cross section of ovary (×21); (d) cross section of fruit (×1½). 3 *Dichapetalum toxicarium* (a) flower with gynoecium removed (×14); (b) gynoecium (×14). 4 *Tapura ciliata* (a) leafy shoot with small stipules and axillary inflorescences (×⅔); (b) flower (×4); (c) corolla opened out (×6); (d) hypogynous gland (×12); (e) ovary (×12).

wild pigs, monkeys, and rats in East Africa, where an extract is said to have been used in arrow poisons. The seeds of *D. toxicarium* are similarly used in West Africa, particularly as an effective rat poison. *D. cymosum* of the high veldt of southern Africa begins growth before the veldt grasses, and is therefore eaten by cattle, giving rise to "gifblaar" poisoning with disastrous effects. Similar effects are reported from other parts of Africa. The toxic principle is fluoracetic acid, which disrupts the tricarboxylic acid cycle of respiration.   VHH

i Savolainen, V. *et al.* Phylogeny of the eudicots: a nearly complete familial analysis based on *rbc*L gene sequences. *Kew Bull.* 55: 257–309 (2000).
ii Chase, M. W. *et al.* When in doubt, put it in *Flacourtiaceae*: a molecular phylogenetic analysis based on plastid *rbc*L DNA sequences. *Kew Bull.* 57: 141–181 (2002).

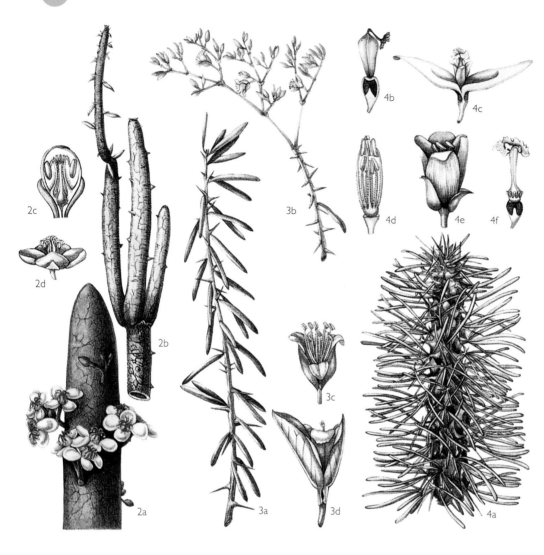

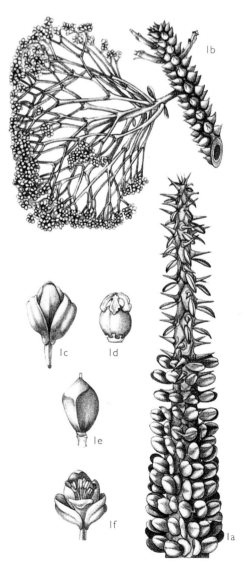

C. *taitensis*; and Portulacarioideae, comprising *Portulacaria* (2 spp.) and *Ceraria* (5 spp.). The strangest species is perhaps *Alluaudia procera*, which resembles a bent, thorny telegraph pole up to 15 m high, with flowers produced in incongruous apical clusters. The 2 spp. of *Didierea* are reminiscent of certain arborescent euphorbias. *Didierea madagascariensis* has erect branches 4–6 m tall, while *D. trollii* is smaller with horizontally spreading branches in the juvenile stage and erect branches only in the adult stage. The internal structure of the stem is divided into chambers by transverse diaphragms of pith that confer light weight but rigidity to the member. The single species of *Decaryia* is characterized by its spreading branches and thorny zigzag twigs.

DIDIEREACEAE. 1 *Alluaudia procera* (a) upper portion of flowering shoot (x⅓); (b) shoot bearing inflorescence of male flowers (x⅓); (c) female flower (x4); (d) gynoecium (x7); (e) fruit (x7); (f) male flower (x2). 2 *A. dumosa* (a) upper portion of flowering shoot (x⅔); (b) spiny branches (x⅓); (c) vertical section of male flower bud (x3½); (d) male flower (x1½). 3 *Alluaudiopsis fiherenensis* (a) spiny shoot (x⅓); (b) shoot with male flowers (x⅓); (c) male flower (x1⅔); (d) female flower (x1½). 4 *Didierea madagascariensis* (a) upper portion of branch (x⅓); (b) fruit (x3), (c) female flower opened out (x2); (d) androecium (x2); (e) male flower (x2); (f) gynoecium (x3½).

DIDIEREACEAE

The Didiereaceae is a curious family of chiefly branched columnar cactuslike plants from Madagascar, eastern Africa, Namibia, and South Africa.

Distribution The subfamily Didiereoideae (4 genera, 11 spp.) is confined to semidesert areas of Madagascar; subfamily Calyptrothecoideae (1 genera, 2 spp.) is endemic to tropical East Africa; subfamily Portulacarioideae (2 genera, 7 spp.) in South Africa and Kenya.

Description Succulent to woody xerophytic shrubs and trees, looking superficially like cacti. The stems are markedly dimorphic with long shoots starting succulent and becoming woody with age, and short shoots bearing spines[i]. The leaves are alternate, simple, entire and without stipules, and in some species wither and fall off to expose the spiny stems. The inflorescence is a thyrse. The flowers are unisexual (plants dioecious, or gynodioecious—*Decaryia*).

Genera 7 Species 20
Economic uses
Ornamental curiosity

The staminate flowers consist of 2 opposite petal-like sepals and 4 overlapping petals surrounding the 8–10 stamens, which are shortly united at the base; the pistillate flowers have a perianth like the male ones and a superior ovary of 3 fused carpels comprising 3 locules, only 1 of which is fertile, containing a single basal ovule. Pistillate flowers are like the unisexual flowers but contain both 8–10 stamens and a superior ovary. The style is single and usually expanded into 3 or 4 irregular stigmatic lobes. The triangular, dry fruits do not dehisce; the seed has a folded embryo, fleshy cotyledons, and little or no endosperm.

Classification The Didiereaceae is one of the core Caryophyllales containing betalain pigments rather than anthocyanins. Its affinities with this group have been apparent for more than a century despite its highly xeromorphic specialisations[ii]. Molecular evidence indicates a close relationship with Portulacaceae, Basellaceae, Halophytaceae, and Cactaceae[iii]. Some authors[iv,v] suggest an expanded family containing the African genera *Calyptrotheca*, *Ceraria*, and *Portulacaria*, which are usually placed in Portulacaceae. The 7 genera of the expanded family are recognized in 3 subfamilies. They are: Didieroideae, comprising *Alluaudia* (6 spp.), *Alluaudiopsis* (2 spp.), *Decaryia* (1 sp.), and *Didierea* (2 spp.); Calyptrothecoideae, comprising *Calyptrotheca somalensis* and

Economic uses The plants occasionally appear in succulent collections. Due to their rarity in the wild, all species are listed on appendix 2 of CITES; only *Allaudia procera* appears on the IUCN red list[vi]. AC

i Rauh, W. The morphology and systematic position of the Didieriaceae of Madagascar. *Bothalia* 14: 839–843. (1983).
ii Rowley, G. D. Didieraceae: *Cacti of the Old World*. Richmond, London, BCSS (1992).
iii Cuénoud, P. *et al*. Molecular phylogenetics of Caryophyllales based on nuclear 18S and plastid *rbcL*, *atpB* and *matK* DNA sequences. *American J. Bot.* 89: 132–144 (2002).
iv Applequist, W. L. & Wallace, R. S. Phylogeny of the Portulacaceous cohort based on *ndhF* sequence data. *Syst. Bot.* 26: 406–419 (2001).
v Applequist, W. L. & Wallace, R. S. Expanded circumscription of Didiereaceae and its division into three subfamilies. *Adansonia* sér. 3, 25: 13–16 (2003).
vi *IUCN Red List of Threatened Species*. IUCN. (2004).

DIDYMELACEAE

A monogeneric family of evergreen trees endemic to Madagascar and the Comores. The leaves are alternate, extipulate, glabrous, simple, the margins entire. The flowers are small, actinomorphic, unisexual (plants dioecious), subtended by 0–4 bracts. The staminate flowers are in

Genera 1 Species 2
Economic uses None

axillary racemes and have no perianth and 2 stamens with extrorse anthers. The pistillate flowers are in spikes and have possibly a single scalelike petal[i]; the ovary is superior with a single carpel; ovule usually one with the integument greatly elongated at the base. The fruit is drupaceous. The floral morphology and inflorescence structure are still not fully understood[i,ii].

The affinities of *Didymeles* have been a matter of some debate. It has been included as a separate order in the Hamamelididae by Cronquist (1981) and placed near the Euphorbiaceae, Daphniphyllaceae, and Buxaceae by other authors. Takhtajan (1997) placed it in its own order in the superorder Buxanae. A close relationship to the Buxaceae was proposed on chemical and anatomical grounds[iii], and this was later confirmed by molecular analyses that place the Didymelaceae as sister to the Buxaceae in a quite well-supported clade among families near the base of eudicots[iv,v]. VHH

i von Balthazar, M., Schatz, G. E., & Endress, P. K. Female flowers and inflorescences of Didymelaceae. *Pl. Syst. Evol.* 237: 199–208 (2003).
ii von Balthazar, M. & Endress, P. K. Development of inflorescences and flowers in Buxaceae and the problem of perianth interpretation. *Int. J. Plant Sci.* 163: 847–876 (2002).
iii Sutton, D. A. The Didymelales: a systematic review. In: Crane, P. R. & Blackmore, S. (eds), *Evolution, Systematics and Fossil History of the Hamamelidaceae* 1: 279–284. Systematics Association. Oxford, Clarendon Press (1989).

iv Qiu, Y.-L. *et al*. Phylogenetics of the Hamamelidae and their allies: Parsimony analyses of nucleotide sequences of the plastid gene *rbcL*. *Int. J. Plant Science* 159: 891–905 (1998).
v Soltis, D. E. *et al*. Angiosperm phylogeny inferred from 18S rDNA, *rbcL*, and *atpB* sequences. *Bot. J. Linn. Soc.* 133: 381–461 (2000).

DIEGODENDRACEAE

This family consists of a single species, *Diegodendron humbertii*, which is endemic to Madagascar. It is an aromatic small tree or shrub with alternate, petiolate, entire leaves and caducous intra-petiolar stipules. The large, fragrant, pink flowers are actinomor-

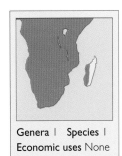

Genera 1 Species 1
Economic uses None

phic, hermaphrodite, in terminal thyrsoid inflorescences, with 5 or 6 free sepals and petals, numerous stamens, and an ovary with 2, rarely 4, carpels and a gynobasic style. The fruit consists of an indehiscent, single-seeded mericarp (sometimes up to 4) with a leathery pericarp.

Diegodendron was originally considered to be closest to the Ochnaceae, Sarcolaenaceae, or Sphaerosepalaceae[i], sharing a gynobasic style and several other features with the latter. It was later placed in the Malvales, within which molecular evidence places it closest to the Bixaceae[ii] in which it is sometimes included along with the Cochlospermaceae. In this classification, the 3 groups are tentatively retained as separate families. VHH

i Capuron, R. *Diegogodendon*. R. Capuron gen. nov., type de la nouvelle famille Diegodendraceae (Ochnales *sensu* Hutchinson). *Adansonia* 2: 385–392 (1963).
ii Fay, M. F. *et al*. Plastid *rbcL* sequence data indicate a close affinity between *Diegodendron* and *Bixa*. *Taxon* 47: 43-50 (1998).

DILLENIACEAE
DILLENIA FAMILY

Dilleniaceae is a large family of mainly tropical trees and shrubs.

Distribution Widespread in the tropics and warm-temperate zones, growing in lowland forests or savannas. *Hibbertia* (c. 115 spp.) ranges from Madagascar to Australia (to which most species are endemic) and Fiji. *Tetracera* is pantropical in distribution. Four genera are endemic to the neotropics—*Curatella, Davilla, Doliocarpus,* and *Pinzona*.

Description Medium-sized trees, shrubs, lianas, and perennial herbs. The leaves are alternate, simple, persistent or caducous, without stipules, the lamina serrate or dentate and often with prominent lateral veins. The inflorescences are axillary or terminal, cymose,

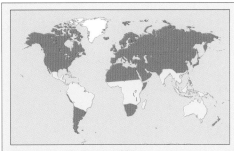

Genera 11–12 Species 335
Economic uses Some ornamental value

paniculate or fasciculate. The flowers are actinomorphic, bisexual or unisexual (plants monoecious, or functionally dioecious in neotropical species of *Tetracera*), sometimes large and conspicuous. The sepals are (3–)5(–20), imbricate, persistent. The petals are 3–5, imbricate, deciduous, often crumpled in bud. The stamens are numerous (up to 500 in *Tetratheca*), developing centrifugally, the anthers often with a prolonged and well-developed connective. The ovary is superior of 1–5 or more carpels, free or variously connate, sometimes compound (*Curatella* and *Pinzona*), with marginal placentation and 1–5 locules with 1 to many seeds. The styles are separate

DILLENIACEAE. 1 *Tetracera masuiana* (a) flowering shoot (x⅔); (b) vertical section of gynoecium with free carpels (x6).

1a   1b

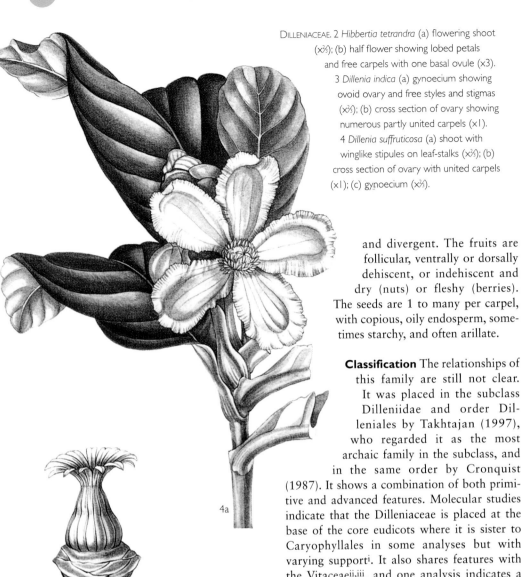

DILLENIACEAE. 2 *Hibbertia tetrandra* (a) flowering shoot (x⅔); (b) half flower showing lobed petals and free carpels with one basal ovule (x3). 3 *Dillenia indica* (a) gynoecium showing ovoid ovary and free styles and stigmas (x⅔); (b) cross section of ovary showing numerous partly united carpels (x1). 4 *Dillenia suffruticosa* (a) shoot with winglike stipules on leaf-stalks (x⅔); (b) cross section of ovary with united carpels (x1); (c) gynoecium (x⅔).

and divergent. The fruits are follicular, ventrally or dorsally dehiscent, or indehiscent and dry (nuts) or fleshy (berries). The seeds are 1 to many per carpel, with copious, oily endosperm, sometimes starchy, and often arillate.

Classification The relationships of this family are still not clear. It was placed in the subclass Dilleniidae and order Dilleniales by Takhtajan (1997), who regarded it as the most archaic family in the subclass, and in the same order by Cronquist (1987). It shows a combination of both primitive and advanced features. Molecular studies indicate that the Dilleniaceae is placed at the base of the core eudicots where it is sister to Caryophyllales in some analyses but with varying support[i]. It also shares features with the Vitaceae[ii,iii], and one analysis indicates a sister relationship with that family although it is not strongly supported[iv], but further work is needed to establish its correct relationships.

Economic uses *Hibbertia scandens* and *Dillenia indica* are occasionally grown as ornamentals. The timber of *Dillenia* spp. is used for general construction and boatbuilding. The cut stems of *Tetracera arborescens* yield potable water. The scabrid stems of several liana species, such as *Doliocarpus dentatus* and *Davilla kunthii*, known as *Cipó-de-fogo*, have irritant hairs that cause a burning sensation if touched.   VHH

[i] Savolainen, V. *et al.* Phylogenetics of flowering plants based upon a combined analysis of plastid *atp*B and *rbc*L gene sequences. *Syst. Biol.* 49: 306–362 (2000).

[ii] Nandi, O. I., Chase, M. W., & Endress, P. K. A combined cladistic analysis of angiosperms using *rbc*L and nonmolecular data sets. *Ann. Missouri Bot. Gard.* 85: 137–212 (1998).

[iii] Metcalfe, C. R. & Chalk, L. Anatomy of the dicotyledons. Leaves, stem, and wood in relation to taxonomy with notes on economic uses, vols. 1–2, Oxford, Clarendon Press (1950).

[iv] Hilu, K. *et al.* Inference of angiosperm phylogeny based on *matK* sequence information. *American J. Bot.* 90: 1758–1776 (2003).

DIONCOPHYLLACEAE

A family of 3 monospecific genera of climbers and scramblers from tropical Africa, 1 sp. of which is seasonally carnivorous, with antimalarial activity.

Distribution The family is restricted to the Guineo-Congolian rain forests of Africa. *Dioncophyllum* is found in primary and secondary forest in Congo and Gabon. *Habropetalum* occurs in drier rain forests and coastal thickets on sandy ground on the borders of Liberia and Sierra Leone. The carnivorous *Triphyophyllum* occurs in both primary forest and secondary areas of regeneration in Ivory Coast, Liberia, and Sierra Leone.

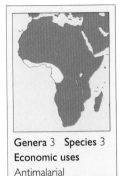

Genera 3 Species 3
Economic uses
Antimalarial compounds are found in 1 sp.

Description Large woody lianas or scandent shrubs, stems with conventional secondary thickening or with wood developing from concentric cambial rings. The leaves are alternate, leathery, with pinnate, or rarely parallel, veins, sometimes heterophyllous, simple, linear, and carnivorous, or hooked and climbing (*Triphyophyllum*). The leaves are without stipules, the margins entire or crenate, vernation rarely circinnate (*Triphyophyllum*). The inflorescence is cymose, more or less supra-axillary, lax. The flowers have large or small bracts, regular, hermaphrodite. The sepals are 5, fused at the base in to a tube or free, valvate. The petals are 5, free, alternating with the sepals, contorted, white, sometime fleshy, deciduous. The stamens are 10 (*Habropetalum*, *Triphyophyllum*) or 25–30 (*Dioncophyllum*), free, equal or strongly unequal. The carpels are 2 or 5, ovary of 1 locule, superior, styles 2 or 5, isomerous with carpels, capitate, punctate or feathery, ovules 30 to numerous, placentation parietal. The fruit is a dry dehiscent capsule, splitting longitudinally, often well before the seeds are mature. The seeds are borne on long funicles, large and winged, with starchy endosperm.

Classification The family has been linked to Flacourtiaceae and Nepenthaceae and, following the discovery of carnivory in *Triphyophyllum*, to Droseraceae. Palynological evidence supports a close affinity of Dioncophyllaceae with Ancistrocladaceae that has since been confirmed with molecular evidence[i].

Economic uses The naphthylisoquinoline alkaloids in *Triphyophyllum peltatum* have antimalarial activity[ii], and the species is used in folk medicine to treat malaria, elephantiasis and some other diseases.   AC

i Cameron, K. M., Chase, M. W., & Swensen, S. M. Molecular evidence for the relationships of *Triphyophyllum* and *Ancistrocladus*. *American J. Bot.* 82: 117–118 (1995).
ii Francois, G., Bringmann, G., Dochez, C., Schneider, C., Timperman, G., & Assi, L. A. Activities of extracts and naphthylisoquinoline alkaloids from *Triphyophyllum peltatum*, *Ancistrocladus abbreviatus* and *Ancistrocladus barteri* against plasmodium-berghei (anka strain) in-vitro. *J. Ethnopharmacol.* 46: 115–120 (1995).

DIPENTODONTACEAE

A monogeneric family of uncertain affinity, comprising the single species, *Dipentodon sinicus*. Small deciduous tree or shrub native to southern China and adjacent Myanmar and northeastern India. The leaves are alternate, petiolate, simple,

Genera 1 Species 1
Economic uses None

with ovate dentate laminas. The flowers are small, bisexual, actinomorphic, in umbellate inflorescences. The sepals and petals are 5–7, very similar, connate at the base. The stamens are 5–7 with a similar number of staminodial nectaries. The ovary is superior, of 3 fused carpels, 1-locular with free central placentation, or incompletely 3-locular at the base; ovules 2 in each locule but only 1 developing. The fruit is a tardily dehiscent single-seeded capsule.

The taxonomic position of Dipentodontaceae has been uncertain since it was first described and placed in the Celastraceae. It was subsequently proposed as a separate family, in the Rosales between the Hamamelidaceae and Rosaceae, and then regarded by Cronquist (1981), on the basis of ovary features, as belonging in the Santalales. Takhtajan (1997) placed Dipentodontaceae in the Violales. It was given its own order (Dipentodontales)[i], and a recent molecular study[ii] suggests that *Dipentodon* is sister to *Tapiscia* (Tapisciaceae), both unplaced Eurosid II in APG II, forming a clade that is closest to the Malvales and Sapindales. Stephens (2001) places both families in the Huerteales (=Dipentodontales).   VHH

i Wu, Z.-Y., Lu, A.-M., Tang, Y.-C., Chen, Z.-D., & Li, D.-Z. Synopsis of a new "polyphyletic-polychronic-polytopic" system of the angiosperms. *Acta Phytotax. Sinica* 40: 298–322 (2002). [In Chinese.]
ii Peng, Y., Chen, Z., Gong, X., Zhong, Y., & Shi, S. Phylogenetic position of *Dipentodon sinicus*: evidence from DNA sequences of chloroplast *rbc*L, nuclear ribosomal 18S, and mitochondria *mat*R genes. *Bot. Bull. Acad. Sinica* 44: 217–222 (2003).

DIPSACACEAE
SCABIOUS AND TEASEL FAMILY

A family resembling Asteraceae in having flowers in a head but related to Caprifoliaceae and characterized by the involucel subtending each flower.

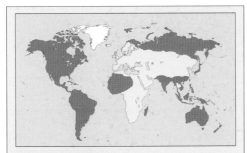

Genera 10 Species c. 250
Economic uses Cultivated herbaceous perennials and fulling cloth

Distribution Confined to the Old World (occasionally introduced in North America), centered in the Mediterranean region but extending westward to the Canary Islands (*Pterocephalus*), southward to South Africa (*Scabiosa* and *Cephalaria* both well developed there), and eastward sparingly to Japan (*Dipsacus*, *Scabiosa*). It is unusual to find an essentially Mediterranean family with strong extensions into tropical Africa (*Dipsacus*, *Cephalaria*, *Scabiosa*, *Succisa*, *Pterocephalus*).

Description Annual or perennial herbs to low subshrubs. The leaves are opposite or in a basal rosette, without stipules, and entire to deeply pinnately or bipinnately divided. The inflorescence is a terminal compact capitulum surrounded by involucral bracts, the receptacle sometimes rather elongate (*Dipsacus*). The capitulum receptacle may bear prominent scales or (*Knautia*) hairs between the flowers. Each flower is subtended by an 8-nerved involucel derived from the fusion of 2 bracteoles, except in a few species of *Cephalaria* and *Succisa*, in which it is reported to be absent. The involucel surrounds the inferior ovary and may be variously developed into a corona or setae (see below). The flowers are regular to slightly zygomorphic, those at the margin of the capitulum often markedly bigger than the inner ones and with enlarged outer lobes. The calyx either forms a scarious cup or is divided into 4 or 5 setae or, in *Knautia*, up to 16(–24) setae. The corolla is tubular or trumpet-shaped, with 4 or 5 lobes, and is either regular or 2-lipped. The stamens are 4, sometimes 2 long and 2 short, usually exserted. The ovary is bicarpellary, but 1 carpel aborts to leave a single locule that bears 1 ovule pendulous at its apex. The style is elongate and often exserted. The fruit is a hard, dry cypsela, often surmounted by the persistent scarious calyx, or enveloped by the involucel.

The striking variation in inflorescence and calyx characters is probably related to dispersal mechanisms of the single-seeded indehiscent fruits. In *Dipsacus*, the prominent spinescent long outer bracts are members of the involucre of the capitulum, while the acute bracts separating and exceeding the flowers are scales on the capitulum receptacle, the involucel and calyx being inconspicuous. In other genera, the cup-shaped involucel surrounding the inferior ovary may be conspicuous and developed in various ways. In *Succisella*, it is urceolate, closely enfolding the fruit. In *Scabiosa*, the involucel has a prominent, translucent, orbicular corona persisting with the fruit. In *Pterocephalus*, the involucel may bear a corona of setae. In *Pterocephalidium*, the involucel develops on its rim one very long curved tooth far exceeding the length of the rest of the involucel. On the other hand, in *Knautia*, the involucel is inconspicuous, but the calyx develops many terminal awns. The capitulate inflorescences are superficially very similar to those of the Asteraceae, in which, however, the anthers are always connate round the style. The Dipsacaceae can be readily distinguished by their free anthers that are usually prominently exserted.

Classification The family is apparently derived from ancestors referable to Caprifoliaceae, some present-day members of which have a similar inferior ovary that is indehiscent and single-seeded. However, Dipsacaceae is distinctive with its capitulate inflorescences and the development of the involucel by the fusion of the bracteoles (but see also Morinaceae, which has sometimes been included in Dipsacaceae). An analysis of the likely evolution of morphological characters[i] has placed *Dipsacus* and *Cephalaria* (which may sometimes be difficult to distinguish) basal to other genera, with Morinaceae, Triplostegiaceae, and Valerianaceae successively related outside the family. A further discussion[ii] has stressed the importance (often overlooked by European botanists) of the Asian species in the evolution of the family. For further discussion of variant family concepts, see Caprifoliaceae.

The larger genera are widespread in Africa and Eurasia, *Scabiosa* (c. 80 spp.), *Knautia* (c. 60 spp.), *Cephalaria* (c. 60 spp.), *Pterocephalus* (c. 25 spp.), and *Dipsacus* (c. 15 spp.). The smaller genera tend to be locally confined. *Pseudoscabiosa* comprises 4 spp. confined to central and western Mediterranean; *Pterocephalidium* has 1 sp. confined to the Iberian Peninsula; *Succisella* has 5 spp. of which 1 spp. is widespread in Europe but 3 spp. are narrow endemics in Spain; and *Succisa* also has 1 sp. widespread in Europe, 1 sp. in the northwest of the Iberian Peninsula, and another in Nigeria and Cameroon. The single Asian endemic genus, recently described as *Bassecoia* (an anagram of Scabiosae), has 2 spp. One is native to western China and the other is found in Thailand.

Economic uses The genera *Scabiosa* (Scabious), *Pterocephalus*, and *Cephalaria* are grown as ornamental herbaceous perennials or rock-garden plants. The inflorescences of *Dipsacus* species were formerly used for raising the nap on cloth (Fullers' Teasel) and although largely replaced by other means, are still used when a high-quality finish is needed.   RKB

i Caputo, P. & Cozzolino, S. A cladistic analysis of Dipsacaceae (Dipsacales). *Pl. Syst. Evol.* 189: 41–61 (1994).
ii Burtt, B. L. The importance of some Far Eastern species of Dipsacaceae in the history of the family. In: Tandon, R. K. & Prithipalsingh (eds), *Biodiversity, Taxonomy and Ecology: Professor K. M. M. Dakshini Festschrift.* Pp. 131–139 (1999).

DIPTEROCARPACEAE

A pantropical family comprising some of the best-known tropical forest trees, characterizing major formations, especially in Southeast Asia, providing a main source of hardwood timber.

Distribution Subfamily Dipterocarpoideae is confined to the tropics of Asia (Sri Lanka, India, Myanmar, Thailand, Indo–China, southern China, Yunnan, Malaysia, Sumatra, Java, Bali, Borneo, New Guinea, the Philippines, and the Seychelles). Subfamily Monotoideae grows in Africa and Madagascar and has recently been discovered in the neotropics, in Colombia[i]. Subfamily Pakaraimaeoideae is endemic to the Guyana highlands of South America.

DIPSACACEAE. 1 *Scabiosa anthemifolia* var *rosea* (a) leafy shoot and inflorescences (×⅔); (b) inner flower with bristlelike calyx-segments (×3); (c) larger outer flower opened out (×2); (d) fruit with epicalyx (involucel) expanded into an umbrella-shaped extension and crowned by spines (×3); (e) vertical section of fruit (×2). 2 *Dipsacus fullonum* (a) dense flower head surrounded by spiny bracts (×⅔); (b) flower (×3); (c) fruit (×3); (d) cross section of fruit (×5). 3 *Pterocephalus perennis* flowering shoot (×⅔).

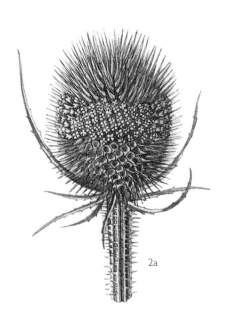

Description Trees, mainly evergreen, some of them reaching a height of up to 70 m, often with buttressed bases, rarely shrubs (*Pakaraimaea*). The leaves are alternate, usually coriaceous or chartaceous, simple, with margins entire. The flowers are actinomorphic, bisexual, in terminal and axillary racemes or panicles, rarely cymes. The sepals are 5, accrescent, persistent, mostly enlarged and winglike in fruit, distinct or connate at the base. The petals are 5, free or connate at the base. The stamens are 5 to numerous, the filaments free or connate below, the anthers dorsifixed or basifixed, often appendaged with the connective extended into a sterile tip. The ovary is superior to semi-inferior, with (2)3–4(5) carpels and locules, each with 1 (*Pseudomonotes*) or 2–4 ovules that are pendulous or laterally attached. The fruit is a single-seeded nut or a capsule, usually with persistent winged and membranous sepals. The seeds usually lack endosperm.

Classification A general review of the systematics and biogeography of the Dipterocarpaceae is given by Maury-Lechon & Curtet[ii]. The classification of Ashton[iii], which is followed here, divides the family into 3 monophyletic subfamilies:

SUBFAM. DIPTEROCARPOIDEAE Comprises 13 genera and 470 spp. that dominate the canopy of lowland equatorial forests in Southeast Asia. The main genera are *Shorea* (194 spp.), *Hopea* (102 spp.), *Dipterocarpus* (69 spp.), and *Vatica* (65 spp.). Contains 2 tribes, Dipterocarpeae with 8 genera, and Shoreae with 5 genera, including the 2 largest in the family, *Shorea* and *Hopea*.

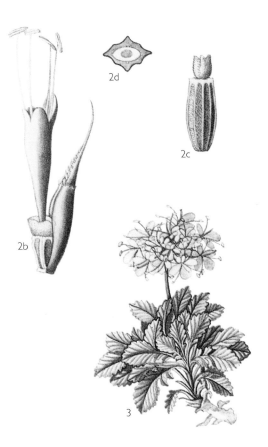

SUBFAM. MONOTOIDEAE Comprises 3 genera and 34 spp. *Monotes* (30 spp.) grows in tropical Africa and Madagascar, *Marquesia* (3 spp.) in tropical Africa, and *Pseudomonotes tropinbosii* in South America (Arraracuara, Colombia).

SUBFAM. PAKARAIMAEOIDEAE Comprises the single species *Pakaraimaea dipterocarpacea* from the Guyana highlands.

The Dipterocarpaceae was included in the Theales by Cronquist and in the Malvales by Takhtajan, who separated off the Monotoideae and Pakaraimaeoideae as a separate family, the Monotaceae, comprising 3 subfamilies: Monotoideae, Pakaraimaeoideae, and Pseudomonotoideae. Its inclusion in the Malvales is confirmed by recent molecular phylogenetic analyses using *rbcL* chloroplast data; these also indicate that Sarcolaenaceae is the family's closest relative[iii,iv].

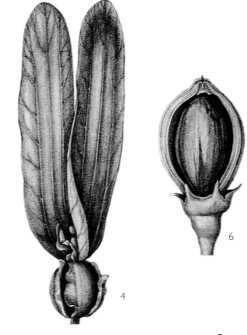

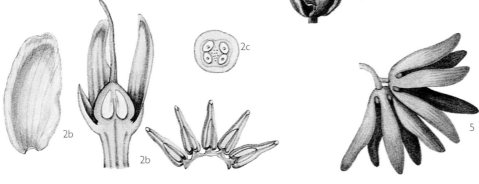

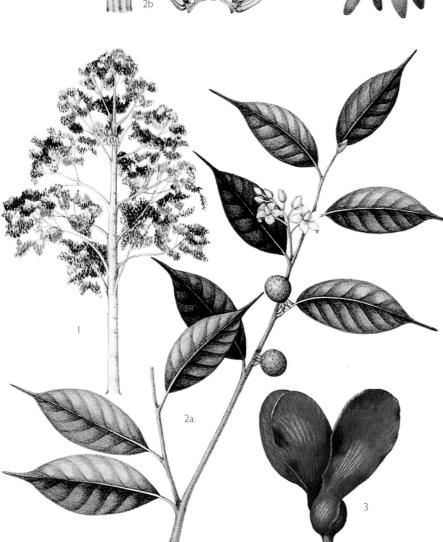

Genera 17 Species c. 500
Economic uses Major sources of hardwood; oleoresins and cocoa butter substitutes

Economic uses Dipterocarps, especially species of *Dipterocarpus, Hopea, Shorea,* and *Vatica* are of major importance as a source of hardwood and currently dominate the international tropical timber market[v]. Many species are now at risk due to overexploitation and habitat loss. Dipterocarps provide many nonwoody forest products[vi], which are mainly used as by village people but the products from a few species have gained commercial importance in industry and trade due to their properties and chemical constituents. Some species are a source of oleoresins (especially *Dipterocarpus*), resins (damar), "butter fat," obtained from *Shorea* spp., and camphor (*Dryobalanops aromatica* formerly a main source). VHH

[i] Londono, A. C., Alvarez, E., Forero, E., & Morton, C. M. A new genus and species of Dipterocarpaceae from the Neotropics. I. Introduction, taxonomy, ecology and distribution. *Brittonia* 47: 225–236 (1995).
[ii] Maury-Lechon, G. & Curtet, L. Biogeography and evolutionary systematics of Dipterocarpaceae. In: Appanah, S. & Turnbull, J. M. (eds), *A Review of Dipterocarps: Taxonomy, ecology and silviculture.* Pp. 5–44. Bogor, Indonesia, Center for International Forestry Research (1998).
[iii] Dayanandan, S., Ashton, P. S., Williams, S. M., & Primack, R. B. Phylogeny of the tropical tree family Dipterocarpaceae based on nucleotide sequences of the chloroplast *rbcL* gene. *American J. Bot.* 86: 1182–1190 (1999).
[iv] Ducousso, M. *et al.* The last common ancestor of Sarcolaenaceae and Asian dipterocarp trees was ectomycorrhizal before the India-Madagascar separation, about 88 million years ago. *Mol. Ecol.* 13: 231–236 (2004).
[v] Appanah, S. & Turnbull, J. M. (eds), *A Review of Dipterocarps: Taxonomy, ecology and silviculture.* Pp. 187–197. Bogor, Indonesia, Center for International Forestry Research (1998).
[vi] Shiva, M. P. & Jantan, I. Non-timber forest products from Dipterocarps. In: Appanah, S. & Turnbull, J. M. (eds), *A Review of Dipterocarps: Taxonomy, ecology and silviculture.* Pp. 187–197. Bogor, Indonesia, Center for International Forestry Research (1998).

DIPTEROCARPACEAE. 1 *Shorea ovalis*, tree in leaf. 2 *Monoporandra elegans* (a) leafy twig with fruit and inflorescence (x⅔); (b) flower dissected to show sepal (left), petals, and gynoecium in vertical section (center), and stamens fused at the base and anthers with long connectives (right) (x6); (c) cross section of ovary (x6). The fruits are often enclosed in winged extensions of the calyx. Shown here are: 3 *Dipterocarpus incanus* (x⅔). 4 *Shorea ovalifolia* (x1½) and 5 *Monotes tomentellus* (x⅔). 6 *Dipterocarpus oblongifolia* vertical section of single-seeded fruit (x1).

DIRACHMACEAE

The family Dirachmaceae comprises the single genus *Dirachma*, with 2 spp. of spindly trees or shrubs (*D. socotrana*, endemic to Socotra, and *D. somalensis* from Somalia). The leaves are small, alternate, clustered on the short shoots, serrate-

Genera 1 | Species 2
Economic uses Wood is aromatic if burned

dentate, with subulate persistent stipules. The flowers are terminal, borne singly, actinomorphic, bisexual, with an epicalyx of 4–8 lobes. The sepals are 5–8, connate at the base. The petals are 5–8, free, inserted on the calyx-tube, with fleshy nectariferous appendages near the base or nectaries at the base. The stamens are 5–8, antepetalous, inserted on the calyx tube, the anthers with extrorse dehiscence. The ovary is superior, of 5–8 fused carpels, deeply lobed with 5–8 locules, and 1 ovule per locule. The fruit is a septicidal capsule, densely woolly inside. The seeds are flattened.

The Dirachmaceae used to be considered allied to the Geraniaceae or even included in it but recent anatomical[i], morphological and molecular evidence[ii] supports its recognition as a separate family and its placement in the Rosales *sensu* APG II. The wood of *D. socotrana* is aromatic when burned.   VHH

[i] Baas, P., Jansen, S., & Smets, E. Vegetative anatomy of *Dirachma socotrana* (Dirachmaceae). *Syst. Bot.* 26: 231–241 (2001).
[ii] Richardson, J. E. *et al.* A phylogenetic analysis of Rhamnaceae using *rbcL* and *trnL-trnF* plastid DNA sequences. *American J. Bot.* 87: 1309–1324 (2000).

DONATIACEAE

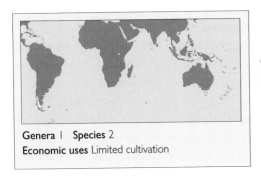

Genera 1 | Species 2
Economic uses Limited cultivation

The family comprises a single genus of perennial herbs, with 2 spp.: *Donatia novae-zelandiae*, which is restricted to New Zealand and Tasmania, and *D. fascicularis* endemic to southern parts of South America. Both species are common in montane and alpine cushion bogs and herbfields, occasionally present in bogs near sea level. They form compact, often broad, cushions and are glabrous except for dense tufts of white, eglandular hairs in the leaf axils. The stems are short, densely branched, and closely packed, with adventitious roots, and covered in overlap-

ping, long-persistent leaves. Secondary growth is absent. The leaves are simple, linear-subulate, subacute with entire margins, sessile and exstipulate. The flowers are solitary and sessile at the end of the branches, bisexual, and actinomorphic. The calyx-tube is turbinate and adnate to the ovary with 3–7 free, acute lobes. The petals are 5–10 free, white. The stamens are 2–3 and inserted on a nectiferous epigynous disk immediately adjacent to, and alternating with, an equivalent number of styles. The anthers are extrorse, opening by longitudinal slits. The ovary is inferior, with 2 or 3 locules with numerous ovules in the upper portion. Placentation is axile. The fruit is an indehiscent capsule containing few, small seeds.

Donatia has been included within Saxifragaceae, Stylidiaceae, and the monogeneric family Donatiaceae. Most molecular studies place Donatiaceae within the order Asterales and sister to Stylidiaceae[i]. A close relationship between Donatiaceae and Stylidiaceae is supported by imbricate corolla aestivation, a reduced number of stamens (2–3), extrorse anthers, and extrastaminal nectaries. Unlike Stylidiaceae, however, in which the corolla is sympetalous and the stamens united with the style to form a unique floral column, Donatiaceae is characterized by free petals, stamens, and styles. There are also notable differences in vegetative anatomy[ii]. Reuniting these 2 families as recently suggested[i] would unnecessarily deprive Stylidiaceae of its defining synapomorphies. The Donatiaceae have some use in cultivation as glasshouse or rock-garden ornamental plants.   JAW

[i] Lundberg, J. & Bremer, K. A phylogenetic study of the order Asterales using one morphological and three molecular data sets. *Int. J. Plant Sci.* 164: 553–578 (2003).
[ii] Rapson, L. J. Vegetative anatomy in *Donatia*, *Phyllachne*, *Forstera* and *Oreostylidium* and its taxonomic significance. *Trans. Roy. Soc. N.Z.* 80: 399–402 (1953).

DROSERACEAE

SUNDEWS, WATERWHEEL PLANT, AND VENUS' FLYTRAP

This well-known family of carnivorous plants is dominated by the diverse sundew genus, *Drosera*, characterized by its sticky tentacles.

Distribution The family is widespread and has colonized a range of habitats, from fully aquatic (*Aldrovanda*) to seasonally dry (*Drosera* sect. *Ergaleium*), and from sea level to high altitude. The Venus Flytrap (*Dionaea*) is restricted to the Green Swamp (borders of North and South Carolina), *Aldrovanda* is widespread but infrequent from northern temperate Old World to tropical and subtropical Australia. The type genus, *Drosera* (150–180 spp.), is cosmopolitan with centers of diversity in southwestern Australia, South Africa, and tropical South America, absent only from Antarctica.

Description Perennial or rarely annual, acaulescent or caulescent, carnivorous herbs to 3 m, evergreen or dormant by stem or root tubers in dry climates, or by apical bud in cold climates. The leaves are spirally arranged or

DROSERACEAE. 1. *Drosera capensis* (a) habit showing basal rosette of leaves covered in stalked glands (×⅔); (b) perianth opened out to reveal stamens (×2); (c) gynoecium (×2); (d) half section of ovary (×10). 2 *Dionaea muscipula* (a) habit showing leaves modified to form a trap (×⅔); (b) inflorescence (×⅔); (c) vertical section of base of flower showing ovary with basal ovules (×4).

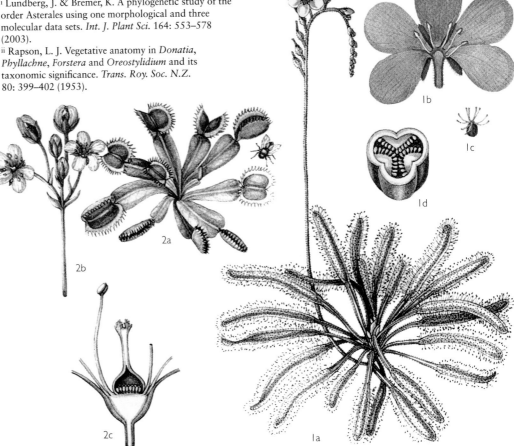

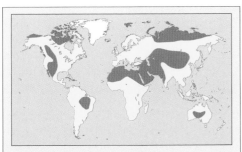

Genera 3 Species 150–180
Economic uses Medical naphthoquinones widespread; horticultural curiosities

whorled (*Aldrovanda*, some *Drosera* sect. *Stolonifera*), with or without stipules, with or without petiole, lamina attached basally or peltate, hinged in *Dionaea* and *Aldrovanda* to form a bilobed trap, upper surface of lamina bearing glandular tentacles in *Drosera*, often pubescent below. Inflorescence is a terminal or apparently lateral cyme, usually scorpioid, 1- to many-flowered. The flowers are actinomorphic, hermaphrodite, usually 5-merous, (rarely 4-merous: *D. pygmaea*; or up to 12-merous: *D. heterophylla*), the sepals free or fused at base, the petals free, often brightly colored, alternating with the sepals. The stamens are usually 5, free. The ovary is superior, usually of 3 carpels, sometimes 5, fused, 1-locular. The styles are as many as the carpels, often much branched and causing confusion in number; the stigmas are often highly branched, crested, or papillate. The fruit is a dehiscent capsule containing few to many seeds.

Classification The Droseraceae have often been allied to the Sarraceniaceae and Nepenthaceae in the Sarraceniales or Nepenthales, although others have separated it into the Rosales with Saxifragaceae (Thorne 1992) or in Theales with Violaceae[ii]. Molecular investigations have identified clear affinities in the Caryophyllales[i] and particularly with Nepenthaceae[ii]. The family currently comprises 2 monotypic genera, *Aldrovanda* and *Dionaea*, and the large *Drosera* that has been the subject of phylogenetic investigation[iii]. The monotypic genera have each sometimes been recognized in their own families Aldrovandaceae and Dionaeaceae. *Drosophyllum* has long been considered part of the Droseraceae, but its molecular affinities[iv], as well as its glandular structure, ally it more closely with Dioncophyllaceae, and it is now treated in its own family Drosophyllaceae (q.v.).

Economic uses The major economic use of the species is in herbal medicine, where the widespread naphthoquinones[v] show a range of activity including antispasmodic action, as a treatment for coughs, and anti-inflammatory properties[vi], as well as exhibiting antibacterial action[vii]. The family is also of economic value, particularly *Dionaea muscipula*, as ornamental curiosities sold in many countries and represented by societies for growing them[viii].   AC

i Williams, S. E., Albert, V. A., & Chase, M. W. Relationships of Droseraceae; a cladistic analysis of *rbc*L sequence and morphological data. *American J. Bot.* 81: 1027–1037 (1994).
ii Meimberg, H., Dittrich, P., Bringmann, G., Schlauer, J., & Heubl, G. Molecular phylogeny of Caryophyllidae *s.l.* based on *mat*K sequences with special emphasis on carnivorous taxa. *Plant Biology* 2: 218–228 (2000).
iii Rivadavia, F., Kondo, K., Kato, M., & Hasebe, M. Phylogeny of the sundews, *Drosera* (Droseraceae), based on chloroplast *rbc*L and nuclear 18S ribosomal DNA sequences. *American J. Bot.* 90: 123–130 (2003).
iv Cameron, K. M., Wurdack, K. J., & Jobson, R. W. Molecular evidence for the common origin of snap-traps among carnivorous plants. *American J. Bot.* 89: 1503–1509 (2002).
v Culham, A. & Gornall, R. J. The taxonomic significance of Naphthoquinones in the Droseraceae. *Biochem. Systematics Ecol.* 22: 507–515 (1994).
vi Krenn, L., Beyer, G., Pertz, H. H., Karall, E., Kremser, M., Galambosi, B., & Melzig, M. E. In vitro antispasmodic and antiinflammatory effects of *Drosera rotundifolia*. *Arzneimittel-Forschung-Drug Research* 54: 402–405 (2004).
vii Didry, N., Dubreuil, L., Trotin, F., & Pinkas, M. Antimicrobial activity of aerial parts of *Drosera peltata*. Smith on oral bacteria. *J. Ethnopharmacol.* 60: 91–96 (1998).
viii http://www.carnivorousplants.org/
http://thecps.org.uk/
http://www.acps.org.au/

DROSOPHYLLACEAE
PORTUGUESE DEWY PINE

This family comprises 1 sp. of carnivorous perennial subshrub, *Drosophyllum lusitanicum*, smelling strongly of honey, which is endemic to southern Portugal, Spain, and northern Morocco, where it grows on acid or basic soils in seasonally dry open pine and oak forest habitats. It is readily recognized by its leaves with sticky tentacles on the underside and its sulfur-yellow flowers. The stem is short and woody, usually to 60 cm, sometimes branched. It has a well-developed and persistent taproot. The leaves are spirally arranged in a closely packed rosette, unrolling with reverse circination, the glandular hairs on the underside tightly packed and inrolled until the leaf unfurls, the older leaves drying out and reflexed to form a sheathing skirt around the stem. The leaves are linear, slightly grooved on the upper surface, convex, and with numerous secretory tentacles below. The inflorescence is terminal, a branched cymose panicle covered in numerous glandular secretory hairs. The flowers are actinomorphic, hermaphrodite, 5-merous, the sepals elliptical, fused at the base, glandular on the outer surface, alternating with the sulfur-yellow petals. The stamens are 10, in 2 whorls of 5, free. The ovary is superior, of 5 fused carpels, and 1-locular with basal placentation. The fruit is a dry, terminally dehiscent capsule. The seeds are several, with thick testa.

Although traditionally placed in the Droseraceae, several recent molecular studies have shown this plant to be more closely related to Nepenthaceae, Dioncophyllaceae, and Ancistrocladaceae[i,ii]. Only 1 sp. is recognized in the genus, and it is morphologically consistent throughout its range. The rarity of this species has recently caused concern, and ecological surveys have addressed the status of the wild populations[iii,iv]. The genus is of ornamental value to specialist growers.   AC

Genera 1 Species 1
Economic uses Some ornamental use by specialist growers

i Cuénoud, P. *et al*. Molecular phylogenetics of Caryophyllales based on nuclear 18S and plastid *rbc*L, *atp*B and *mat*K DNA sequences. *American J. Bot.* 89: 132–144 (2002).
ii Meimberg, H., Dittrich, P., Bringmann, G., Schlauer, J., & Heubl, G. Molecular phylogeny of Caryophyllidae *s.l.* based on *mat*K sequences with special emphasis on carnivorous taxa. *Plant Biology* 2: 218–228 (2000).
iii Correia, E. and Freitas, H. *Drosophyllum lusitanicum*, an endangered West Mediterranean endemic carnivorous plant: threats and its ability to control available resources. *Bot. J . Linn. Soc.* 140: 383–390 (2002).
iv Garrido, B., Hampe, A., Marañón, T., and Arroyo, J. Regional differences in land use affect population performance of the threatened insectivorous plant *Drosophyllum lusitanicum* (Droseraceae). *Diversity and Distributions* 9: 335–350 (2003).

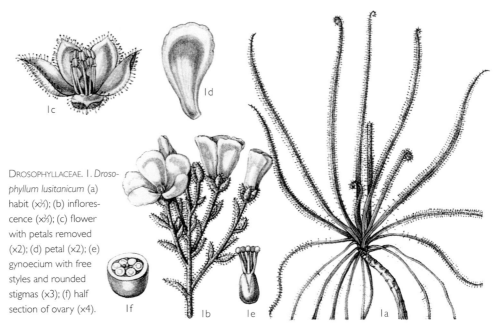

DROSOPHYLLACEAE. 1. *Drosophyllum lusitanicum* (a) habit (x⅔); (b) inflorescence (x⅔); (c) flower with petals removed (x2); (d) petal (x2); (e) gynoecium with free styles and rounded stigmas (x3); (f) half section of ovary (x4).

DURIONACEAE
DURIANS

A family of tropical trees, formerly included in Bombacaceae (q.v.).

Distribution Southeast Asia, usually trees of lowland evergreen forest, with a few species in swamp forest.

Description Ever-green trees, often tall, with an indu-mentum mainly of lepidote scales, but stellate hairs also occur. The leaves are alternate, simple, usu-ally oblong, pinnately nerved and entire, the lower surface is dense-ly covered in scales (sparsely in *Neesia* and also mixed with stellate hairs), the petioles pulvinate, and the stipules generally caducous and linear. The inflorescence is usually axillary, of 1–2 flowers, but often fasciculate or subfasciculate, ramiflorous or cauliflorous. The flowers are actinomorphic, bisexual. An epicalyx is present, usually entirely enclosing the flower in bud and splitting into 2–5 lobes. The calyx itself is shallowly divided into 5 valvate lobes. The corolla is calyptrate in *Neesia* and *Coelostegia* but otherwise is divided into 5 free linear to spathulate petals. An andro-gynophore is absent. The stamens are numerous; either free, united at the base, or unit-ed into 5 phalanges that are either free or connate into a staminal tube, the filaments becoming free at various heights; the anthers are some-times dithecal and longitudinally dehiscent, but are more usually monothecal and borne in clusters, dehiscing by terminal pores or longi-tudinally. The pollen is suboblate, 3-colporate, lacking spines but with low, large, smooth verrucae. The ovary is superior or partly embedded in the receptacle (*Coeloste-gia*), subovoid or slightly 5-angled, clothed in peltate scales (except *Neesia*), 5-locular with 2 to numerous axile anatropous ovules in 2 ranks

Genera 6
Species c. 45
Economic uses Some *Durio* spp. are cultivated throughout Southeast Asia for the production of the fruits, comestibly highly esteemed, although foetid

per locule. The style is usually well developed (± absent in *Kostermansia*), and the stigma peltate or capitate. The fruits are globose to ellipsoid, spiny to subspiny, loculicidally dehiscent cap-sules, each of the 5 locules with 1 to several large raphe-funicular arillate seeds. Rarely are the fruits indehiscent or the seeds exarillate.

The pollinators of durians are not recorded but it is speculated that some species are bird-pollinated. Their fruits are highly sought after for the sweet flesh of the seed arils that they contain. Species as varied as hornbills and ele-phants consume the fruits.

Classification The durians were included in the Bombacaceae on the basis of their monothecal anthers (or, at least, dithecal anthers with nearly separate thecae in *Coelostegia* and *Kostermansia*) together with their arborescent habit. Molecular evidence[i,ii,iii,iv] from several genes, however, has placed them in the same clade as the Helicter-aceae, itself formerly included in Sterculiaceae. Characters that distinguish the durians within the core Malvales are their usually oblong–elliptic leaves with pinnate nerves and lepidote lower surface; their spiny, often large, dehiscent fruit with arillate seeds; and South-east Asian distribution. Within Durionaceae, *Neesia*, with dithecal anthers and sparsely lepidote leaves, has been shown to be sister to the remaining, usually monothecal, genera each with densely lepidote leaves. No satisfactory infra-familial classification is available.

Economic uses There are numerous and varied local uses attached to durians, particularly to *Durio zibethinus*, the most widespread and well known species of the family. Its roots are used to treat fever, its wood for cheap furniture, its seeds for food, the ashes for bleaching silk and to promote abortions and menstruation, while the valves are used as remedies for constipation and skin diseases[v]. However, it is the foul-smelling fruits that are highly esteemed and traded internationally within Southeast Asia, China being a major importer of these spiny comestibles. The seed arils are delicious. The fruits are banned on some international airlines due to their stench when ripening.   MRC

i Baum, D. A., Alverson W. S., & Nyffeler R. A durian by any other name: taxonomy and nomenclature of the core Malvales. *Harvard Pap. Bot.* 3: 315–330 (1998).
ii Bayer, C. *et al.* Support for an expanded family concept of Malvaceae within a recircumscribed order Malvales: a combined analysis of plastid *atp*B and *rbc*L DNA sequences. *Bot. J. Linn. Soc.* 129: 267–303 (1999).
iii Alverson, W. A. *et al.* Phylogeny of the core Malvales: evidence from *ndh*F sequence data. *American J. Bot.* 86 (10): 1474–1486 (1999).
iv Nyffler, R. & Baum, D. A. Phylogenetic relationships of the durians (Bombacaceae-Durioneae or Malvaceae/Helicteroideae/Durioneae) based on chloroplast and nuclear ribosomal DNA sequences. *Plant Syst. Evol.* 224: 55–82 (2000).
v Kostermans, A. J. G. H. The genus *Durio. Reinwardtia* 4: 75–76 (1958).

EBENACEAE
EBONY AND PERSIMMON

A tropical and warm-temperate family of trees or shrubs known for the high-value dark heart-wood, ebony.

Distribution Lowland tropics and subtropics, with a few warm-temperate species. *Diospyros* (c. 500 spp.) is most diverse in Asia and the Pacific, with c. 100 spp. each in Madagascar, Africa, and the USA, and another 13 in Aus-tralia. *Euclea* (c. 15 spp.) is restricted to Africa, Arabia, and the Comoro and Socotra archipela-gos, with a center of endemism in the Cape. *Tetraclis* (2 spp.) is endemic to Madagascar.

Description Small evergreen trees or shrubs with a single leading shoot and flattened foliage sprays. The outer bark is usually black, gritty, and charcoal-like. The leaves are alternate (rarely opposite), simple, entire, usually petio-late, and without stipules, sometimes bearing extrafloral nectaries on the underside. The inflorescences are short and determinate in the leaf axils, sometimes reduced to a single flower, especially in the females. The flowers are regular, usually unisexual (plants generally dioecious; occasionally monoecious), rarely structurally bisexual, and jointed at the base, 3- or 5-merous (rarely 6- or 7-merous). The sepals are fused, with lobes valvate or imbricate. The petals are fused into a tube with as many lobes as there are sepals, the lobes contorted, white, cream, or suffused pink. In male flowers, the

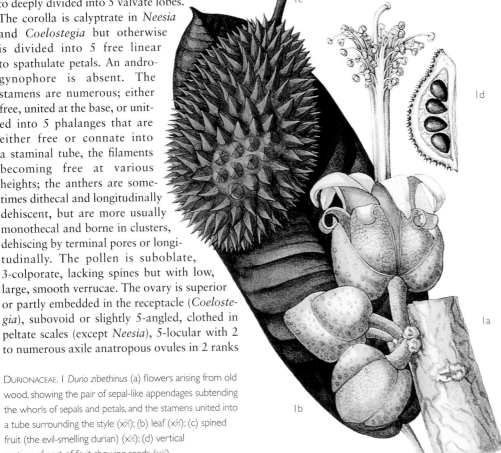

DURIONACEAE. **1** *Durio zibethinus* (a) flowers arising from old wood, showing the pair of sepal-like appendages subtending the whorls of sepals and petals, and the stamens united into a tube surrounding the style (x⅔); (b) leaf (x⅔); (c) spined fruit (the evil-smelling durian) (x⅓); (d) vertical section of part of fruit showing seeds (x¾).

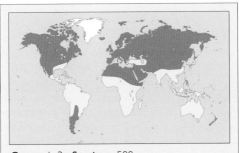

Genera 1–3 **Species** c. 500
Economic uses High-value timber and edible fruit

stamens are usually epipetalous, short and flattened, in 2 whorls, 2 to 4 times the number of the corolla lobes and fused in radial pairs. The female flowers usually bear staminodes in a single whorl. The ovary is superior and sessile, with as many locules as there are petals and sepals. Each locule has 2 pendent ovules attached at the apex but is divided by a false septum, with 1 ovule in each half, the halves being connate at the apex. The styles are fused at least at the base; there are as many stigmas as locules. The male flowers usually have a pistillode. The fruit is a berry, the pericarp pulpy to fibrous, with a stony inner part, only rarely dehiscent, seated on the persistent, often enlarged, calyx. The seeds are several, with endosperm, which is sometimes ruminate.

Classification Traditionally, the family has been placed in the Ebenales, but since the development of APGII (2003) Ebenaceae has been included in Ericales. It is sister to Primulaceae/Myrsinaceae/Theophrastaceae[i] and basal to the Primuloid group of Ericales[ii] in molecular phylogenetic studies. The inclusion of *Lissocarpa* in *Diospyros* on morphological grounds has been refuted[iii], although *Lissocarpa* was found to be the closest relative of Ebenaceae in a molecular study[iv] that recommended its inclusion in the family although it is sometimes recognized at subfamilial level[v]. This research identified a monophylum including *Euclea* nested within *Diospyros* and these sister to *Lissocarpa*. The recognition of the genus *Euclea* is supported by calyx, fruit, and seed characters. The recognition of *Lissocarpa* (q.v.) at family level has been promoted on the basis of floral symmetry, pollen, and seeds[vi]. These data allow us to recognize 2 families: Ebenaceae, 3- or 5-merous flowers, ovary superior and locules divide by a false septum, and Lissocarpaceae (q.v.), 4-merous flowers, ovary inferior, lacking a false septum, the system followed here. *Tetraclis* is usually included in *Diospyros*[vii], and it is possible that *Euclea* should also be included but more study is needed.

Economic uses The family is best known for the black, hard, commercially valuable heartwood ebony, which is produced by most, but not all, species of *Diospyros*. *D. reticulata* (in Mauritius) and *D. ebenum* (in Sri Lanka) are among the finest producers of ebony.

The fruits of several species are eaten, and a few have been brought into cultivation, the best known being the persimmons, a group of outlying warm north-temperate species: *Diospyros kaki* of eastern Asia (Kaki, Chinese or Japanese Date Plum, or Persimmon), extensively cultivated in China and Japan, and known through to the Mediterranean; *D. lotus* of Eurasia (Date Plum); and *D. virginiana* of North America. Crushed seeds of certain Malaysian species are used as fish poison. In all species, the fruit is extremely astringent until it is ripe. AC

i Morton, C., Chase, M. W., Kron, K., & Swensen, S. M. A molecular evaluation of the monophyly of the order Ebenales based upon *rbc*L sequence data. *Syst. Bot.* 21: 567–586 (1996).
ii Schönenberger, J., Anderberg, A. A., & Sytsma, K. J. Molecular phylogenetics and patterns of floral evolution in the Ericales. *Int. J. Pl. Sci.* 166: 265–288 (2005).
iii Berry, P. E. A synopsis of the family Lissocarpaceae. *Brittonia* 51: 214–216 (1999).
iv Berry, P. E., Savolainen, V., Sytsma, K. J., Hall, J. C., & Chase, M. W. *Lissocarpa* is sister to *Diospyros* (Ebenaceae). *Kew Bulletin* 56: 725–729 (2001).
v Wallnöfer, B. A revision of *Lissocarpa* Benth. (Ebenaceae subfam. Lissocarpoideeae (Gilg in Engler) B. Walln). *Ann. Naturhist. Mus. Wien, B*, 105: 515–564 (2004).
vi Wallnöfer, B. Lissocarpaceae. Pp. 236–238. In: Kubitzki, K. (ed.), *The Families and Genera of Vascular Plants. VI. Flowering Plants. Dicotyledons: Celastrales, Oxalidales, Rosales, Cornales, Ericales.* Berlin, Springer-Verlag (2004).
vii Wallnöfer, B. Ebenaceae. Pp. 125–130. In: Kadereit, J. W. (ed.), *Families and Genera of Vascular Plants. VII. Dicotyledons: Lamiales.* Berlin, Springer-Verlag (2004).

ELAEAGNACEAE
OLEASTER AND SEA BUCKTHORN

A smallish family of many-branched shrubs or small trees, which are covered with silvery or golden scales.

Distribution *Elaeagnus* (45 spp.) is wide-spread in temperate eastern Asia with a single species, *E. triflora*, extending from Southeast Asia to Queensland in northeastern Australia, 1 sp. in southern Europe, and another in North America. *Hippophaë* (3 spp.) grows in temperate Eurasia, and *Shepherdia* (3 spp.) occurs in North America.

Description Shrubs, woody climbers, or small trees, often thorny and densely covered with silver, gold, or brown peltate scales and/or stellate hairs. The leaves are deciduous or evergreen, leathery, alternate, opposite or whorled, simple, entire, without stipules. The flowers are axillary; solitary in clusters, racemes, or spikes; bisexual (*Elaeagnus*) or unisexual (plants monoecious in

ELAEAGNACEAE. 1 *Shepherdia argentea* (a) flowering branch showing thorns and male flowers (x⅔); (b) leafy shoot with female flowers (x⅔); (c) female flower (x6); (d) male flower (x6); (e) fruit—a drupelike structure cut away here to reveal the single seed (x3). 2 *Hippophaë rhamnoides* (a) thorny, leafy, shoot bearing fruits (x⅔); (b) female flower with bilobed calyx-tube (x3); (c) gynoecium (x3); (d) fruit (x2); (e) fruit with part of fleshy calyx cut away (x2); (f) male flower (x3); (g) male flower opened out (x2).

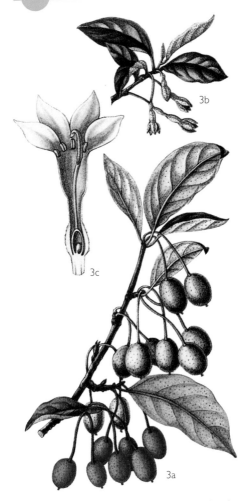

ELAEAGNACEAE. 3 *Elaeagnus multiflora* (a) shoot bearing fleshy fruits (x⅔); (b) shoot bearing bisexual flowers (x⅔); (c) flower opened out with vertical section of 1-locular ovary (x2).

Elaeagnus, dioecious in *Shepherdia* and *Hippophaë*); strongly perigynous. The perianth is in 1 whorl, 2- to 4-lobed, rarely more, often petaloid, with a tubular hypanthium, often constricted above the level of the ovary (erroneously giving the appearance of an inferior ovary), sometimes flat in male flowers. The stamens are inserted on the hypanthium throat, 4 in a single whorl, alternating with the perianth segments, or 8 in 2 whorls (1 alternate and 1 opposite), 2-locular, with longitudinal dehiscence. A nectary disk is often present on inner hypanthium surface. The ovary is superior, 1-locular, and the ovule 1, anatropous, placentation basal. The style is long with a basal to capitate stigma. The fruit is a single-seeded achene, drupelike

Genera 3 **Species** c. 50
Economic uses Ornamental shrubs; limited uses of fruits and wood

through enclosure by the thickened lower part of persistent hypanthium. The seeds have little or no endosperm.

Classification Recently, the Elaeagnaceae has usually been included in the Proteales in accordance with Cronquist's 1989 classification, because of its perigynous flowers, lack of petals, and single carpel. It has also been referred to the Rhamnales because of its golden or silvery hairy indumentum, the nature of the fruits, and by the presence of a basally inserted ovule (e.g., Thorne 1992), the Thymelaeales (Melchior 1964), or even placed in its own monotypic order (Takhtajan 1997). Recent morphological and molecular studies have shown it to be a sister group to *Barbeya* (a small deciduous tree similar in many respects to *Hippophaë*) in the Barbeyaceae[i,ii,iii] and both families to be members of a revised and expanded Rosales (APG II).

Economic uses A number of species are grown as ornamental shrubs, such as *Elaeagnus angustifolia*, *E. pungens*, *E. umbellata*, and *E. macrophylla*, their hybrids and cultivars, and *Hippophaë rhamnoides* (Sea Buckthorn), whose female plants produce bright-orange berries. The fruits of some species are edible, e.g., *Shepherdia argentea* (Silver Buffalo Berry) used to make jelly and also eaten dried with sugar in various parts of North America. The berries of *S. canadensis* (Russet Buffalo Berry) when dried or smoked are used as food by the Inuit. The wood of this species is fine-grained and is used for turnery. The fruits of the Japanese shrub *Elaeagnus multiflora* (Cherry Elaeagnus) are used as preserves and are used in an alcoholic beverage. SLJ

i Thulin, M. *et al.* Family relationships of the enigmatic rosid genera *Barbeya* and *Dirachma* from the Horn of Africa region. *Pl. Syst. Evol.* 213: 103–119 (1998).
ii Soltis, D. E. *et al.* Angiosperm phylogeny inferred from 18S rDNA, *rbc*L, and *atp*B sequences. *Bot. J. Linn. Soc.* 133: 381–461 (2000).
iii Richardson, J. E. *et al.* A phylogenetic analysis of Rhamnaceae using *rbc*L and *trn*L–*trn*F plastid DNA sequences. *American J. Bot.* 87: 1309–1324 (2000).

ELAEOCARPACEAE

A family of often showy tropical and subtropical trees and shrubs and Australian herbs.

Distribution Members of the family are found in eastern Asia, Indo–Malaysia, Australasia, the Pacific area, Madagascar, South America, and the West Indies. Of the largest 2 genera, *Elaeocarpus* (c. 200 spp.) is from eastern Asia, Indo–Malaysia, Australasia, and the Pacific area, and *Sloanea* (c. 120 spp) is from tropical Asia and America. *Tetratheca* (c. 40 spp.), *Tremandra* (2 spp.), and *Platytheca* (2 spp.) are confined to Australia. Many species show only a very narrow distribution and consequently have small populations that are prone to extinction. In *Tetratheca* alone, 20 of the 42 spp. listed for Western Australia show some level of threat.

Description Trees, shrubs, ericoid shrubs, or herbs with glandular or rarely star-shaped hairs (*Tetratheca*, *Tremandra*, and *Platytheca*), the trees sometimes with buttresses[i]. The leaves are alternate, spirally arranged, opposite or rarely whorled (some *Tetratheca*), simple (but rarely pinnate or pinnatisect in juvenile plants), with entire or serrate margins, with stipules or mucilaginous hairs. Inflorescence compound or simple, racemose or cymose, axillary or terminal, sometimes of solitary axillary flowers (*Tetratheca*, *Tremandra*, and *Platytheca*). The flowers are regular, mostly hermaphrodite, and have 3–5 usually valvate sepals, free or partly united. The petals are 4 or 5, or absent (some *Sloanea* spp.), usually free, valvate, but sometimes overlapping at tip, often fringed or lacerated at their tips. The stamens are 4 to numerous, free, the anthers with 2 locules that release pollen through 2 apical pores. The receptacle is sometimes swollen and glandular, lobed between the stamens and petals. The ovary is superior and contains 2 to many locules (rarely 1); each locule contains 1 to many pendulous ovules. The style is simple and sometimes lobed at the tip or may be slender. The fruit is a capsule or drupe. Seeds 1 to many per locule.

Classification The family has been placed consistently in Malvales in most major classifications, but phylogenetic studies show a relationship to Cunoniaceae and Brunelliaceae in the Oxalidales (APG II) based on DNA sequence analysis. This grouping was corroborated by morphological and biochemical features[ii]. Relationship of Elaeocarpaceae to Tremandraceae has been confirmed based on floral structures[iii]. *Sloanea* differs from *Elaeocarpus* in not having a succulent fruit but a hard capsular one covered with rigid bristles. The Australian genera *Tremandra* (2 spp.), *Tetratheca* (39 spp.), and *Platytheca* (2 spp.) are often recognized in Tremandraceae[iv] (Cronquist 1981), but phylogenetic evidence shows *Platytheca* to be placed within Elaeocarpaceae[v]. The 3 genera appear to be a specialized group, defined by ericoid vegetative features. The broader view of Elaeocarpaceae, including Tremandraceae, is followed here.

Economic uses Several species of *Elaeocarpus*, *Crinodendron*, and *Aristotelia* are cultivated. *Elaeocarpus reticulatus* (*E. cyaneus*), a native of

Genera 15 **Species** c. 400–600
Economic uses Ornamental shrubs and limited local use of fruits

ELAEOCARPACEAE. 1 *Aristotelia racemosa* (below left) (a) shoot with axillary racemes of flowers (×⅔); (b) male flower with three-lobed petals and numerous stamens (×3); (c) gynoecium with free curled styles (×6); (d) fruits (×⅔). 2 *Elaeocarpus dentatus* (a) shoot with inflorescence (×⅔); (b) bisexual flower (×2); (c) 2-locular ovary in vertical (left) and cross (right) section (×6). 3 *Sloanea jamaicensis* (a) half flower with numerous stamens having short filaments and large anthers (×1½); (b) cross section of ovary with 4 locules and numerous ovules on axile placentas (×3); (c) dehiscing fruit (×⅓).

Australia, and *E. dentatus* from New Zealand are both cultivated in Europe. Two ornamental evergreen species of *Crinodendron* are well known in cultivation: *C. hookerianum*, with pendulous, urnlike, crimson flowers, about 2.5cm long on crimson stalks to 7cm long; and *C. patagua*, with white and bell-shaped flowers. *Aristotelia chilensis* produces Maqui berries, which are said to have medicinal properties and are made into a local wine in Chile. The fruits and seeds of several *Elaeocarpus* species are eaten, e.g., *E. calomala* (Philippines), *E. dentatus* (New Zealand), and *E. serratus* (Sri Lanka). The West Indian *Sloanea berteriana* (Motillo) and *S. woollsii* (Gray or Yellow Carabeen) yield a heavy and a light timber, respectively. The New Zealand *Aristotelia racemosa* and Australian *Elaeocarpus grandis* produce wood used for cabinet-making.   AC

i Coode, M. J. E. Elaeocarpaceae. Pp. 135–142. In: Kubitzki, K. (ed.), *The Families and Genera of Vascular Plants. VI. Flowering Plants. Dicotyledons: Celastrales, Oxalidales, Rosales, Cornales, Ericales*. Berlin, Springer-Verlag (2004).

ii Nandi, O. I., Chase, M. W., & Endress, P. K. A combined cladistic analysis of angiosperms using *rbc*L and nonmolecular data sets. *Ann. Missouri Bot. Gard.* 85: 137–212 (1998).
iii Matthews, M. L. & Endress, P. K. Comparative floral structure and systematics in Oxalidales (Oxalidaceae, Connaraceae, Brunelliaceae, Cephalotaceae, Cunoniaceae, Elaeocarpaceae, Tremandraceae). *Bot. J. Linn. Soc.* 140: 321–381 (2002).
iv Dahlgren, R. M. T. General aspects of angiosperm evolution and macrosystematics. *Nordic J. Bot.* 3: 119–149 (1983).
v Soltis, D. E. *et al.* Angiosperm phylogeny inferred from 18s rDNA, *rbc*L, and *atp*B sequences. *Bot. J. Linn. Soc.* 133: 381–461 (2000).

ELATINACEAE

A small family of herbs and shrubby plants found in moist or wet habitats or in dry habitats worldwide.

Distribution The genus *Elatine* (12–25 spp.) is almost cosmopolitan, with the majority of species in temperate regions of both hemispheres. *Bergia* (c. 25 spp.) is found mainly in the paleotropics with 10–20 spp. in eastern and southern Africa, 10 in Australia, 8 in South and Southeast Asia, and 3 in the neotropics; only a few species grow in temperate zones. Most species are adapted to fluctuating water levels and are found in shallow water that seasonally dries out. They are particularly common in rice paddies and periodically drained fish ponds.

Genera 2 Species c. 33
Economic uses Weeds in rice paddies and irrigation ditches

Description Annual or short-lived perennial herbs or subshrubs (*Bergia*). The leaves are opposite or whorled (*Elatine alsinastrum*), simple, and with small, scarious, interpetiolar stipules; the lamina is linear to elliptical or ovate, with entire or serrate margins. The flowers are usually small and inconspicuous, actinomorphic, bisexual, either solitary in leaf axils or in cymes. The sepals are 2–5(–6), free or partly connate at the base. The petals are 2–5, free, membranous. The ovary is superior, of 2–3(–5) carpels, with 2–5 locules, and numerous ovules on axile or basal placentas. The fruit is a septicidal capsule. The seeds are numerous with the surface usually finely reticulate, sometimes smooth.

Classification The Elatinaceae has been linked with families such as the Caryophyllaceae, Frankeniaceae, Tamaricaceae, and with Clusiaceae. This latter affinity has been supported by molecular evidence[i,ii]. Elatinaceae and Clusiaceae are placed in the Malpighiales in APG II but with some ambiguity. A recent study indicates strong support for the Elatinaceae as sister to Malpighiaceae[iii].

Economic uses *Elatine* is considered beneficial, as it effectively consolidates mud and the leaves and seeds are eaten by waterfowl. *Elatine* and *Bergia* are frequently found as weeds in rice paddies and irrigation ditches. VHH

[i] Savolainen, V. *et al.* Phylogenetics of flowering plants based upon a combined analysis of plastid *atp*B and *rbc*L gene sequences. *Syst. Biology* 49: 306–362 (2000).
[ii] Chase, M. W. *et al.* When in doubt, put it in *Flacourtiaceae*: a molecular phylogenetic analysis based on plastid *rbc*L DNA sequences. *Kew Bull.* 57: 141–181 (2002).
[iii] Davis, C. C. & Chase, M. W. Elatinaceae are sister to Malpighiaceae; Peridiscaceae belong to Saxifragales. *American J. Bot.* 91: 262–273 (2004).

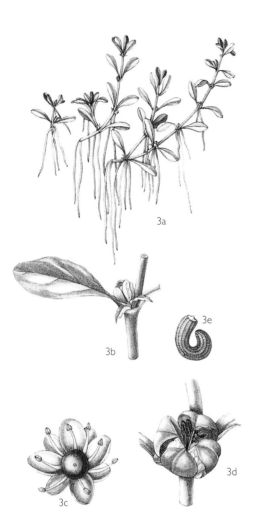

ELATINACEAE. 1 *Bergia ammannioides* (a) habit (x⅔); (b) inflorescences in axils of leaves (x1½); (c) flower viewed from above with 3 each of sepals, petals, and stamens and a 4-lobed ovary (x8); (d) dehiscing capsule with seeds exposed (x20); (e) cross section of fruit (x20). 2 *B. capensis* (a) part of creeping stem with adventitious roots (x⅔); (b) fruit (x20). 3 *Elatine hydropiper* (a) habit showing long adventitious roots (x⅔); (b) solitary flower in leaf axil (x4); (c) flower with 4 sepals and petals, 8 stamens, and globose, superior ovary (x8); (d) dehiscing fruit (x8); (e) curved seed (x20).

EMBLINGIACEAE

Emblingia calceoliflora is a strange and rare plant confined to western Australia in the areas between Eneabba and Shark Bay and in the Cape Range north of Carnarvon, where it occurs among shrubs on sandy flats usually over limestone. It is

Genera 1 Species 1
Economic uses None

probably a short-lived perennial prostrate subshrub, up to 1 m across. Its younger parts are densely covered with stiff, spreading hairs. The leaves are more or less opposite, said to be minutely stipulate, subsessile with an elliptical rather brittle lamina up to 5 × 2 cm, with entire or somewhat undulate margins. The flowers are solitary in the axils of leaves, and up to 1.2 cm long, strongly zygomorphic, and hypogynous. The calyx consists of 5 fused sepals, with the median lobe somewhat larger than the others, and the whole calyx slit to the base opposite the median lobe. The petals are 2, hooded at the apex, and connate along their adjacent sides to make a slipper-shaped hairy corolla. The stamens and ovary are borne on a broad, flat androgynophore that arises in the slit of the calyx and arches over into the hood of the corolla. This structure bears 4 or 5 staminodes at its apex and a line of 4 short fertile stamens below them. The ovary is attached to the underside of the arched androgynophore and is held pendent in the center of the flower. Anthers dehisce onto the stigmatic surface, ensuring automatic self-pollination[i]. The ovary has 2 wings near its apex and a sessile stigma. It was originally described as being unicarpellate with a single basal ovule, but was later said to have 3 fused carpels with axile placentation with 1 ovule in each locule. The fruit is dry and indehiscent, the thin pericarp adhering to the single seed, which is flattened and has a laciniate funicle.

The plant was described by Müller in 1861 in great detail and with excellent illustration with analyses but has baffled successive generations since as to its affinities. Müller placed it in Capparaceae because of its conduplicate embryo and its stamens and ovary being borne on an androgynophore. In the *Pflanzenfamilien*, Pax made it a separate subfamily of Capparaceae, but it differs from that family in so many characters that others have looked for affinities elsewhere. When 4 authors presented simultaneous reports on the plant[ii], Erdtman proposed a link to Polygalaceae based on pollen (note also bilateral flowers, 8 stamens, some vegetative anatomical similarities, funiculate seeds), Leins preferred a connection with Sapindaceae based on flower morphology, Melville suspected affinity with Goodeniaceae through floral vasculature (supposing the androgynophore to be a modified third petal) and geography, while

Metcalfe also preferred affinity with Goodeniaceae based on stem and leaf anatomy but could also see connections with Polygalaceae. Airy Shaw made it a separate family of uncertain affinity. Since then, Cronquist and Takhtajan have placed it in, or close to, Polygalaceae, while Thorne has made it subfamily of Sapindaceae. However, molecular evidence[iii] has found *Emblingia* to be referable to the Capparales, confirming Müller's first impression. The question posed by the four joint authors "Will it turn out that all of us have failed?" seems to be answered in the affirmative.   RKB

[i] Keighery, G. J. The breeding system of *Emblingia* (Emblingiaceae). *Pl. Syst. Evol.* 137: 63–65 (1981).
[ii] Erdtman, G., Leins, P., Melville, R., & Metcalfe, C. R. On the relationships of *Emblingia. Bot. J. Linn. Soc.* 62: 169–186 (1969).
[iii] Chandler, G. T. & Bayer, R. J. Phylogenetic placement of the enigmatic Western Australian genus *Emblingia* based on *rbc*L sequences. *Plant Sp. Biol.* 15: 67–72 (2000).

EREMOLEPIDACEAE
CATKIN-MISTLETOE FAMILY

Small, shrubby, parasitic plants with small flowers; mainly neotropical.

Distribution Western South America, from Chile to Venezuela and Colombia, then northward to Mexico and the Greater Antilles, mainly in temperate to tropical montane areas. *Antidaphne* occurs on a wide range of hosts, the other genera (*Eubrachion* and *Lepidoceras*) mostly grow on Myrtaceae.

Genera 3 Species 12
Economic uses None recorded

Description Small, shrubby parasitic plants generally on the branches of dicotyledonous angiosperms, rarely on gymnosperms, attached by a sucker (haustorium), sometimes with runners over the bark of the host. The leaves are generally alternate, simple, entire, occasionally reduced to scales, without stipules. The flowers are small, unisexual (plants monoecious or dioecious), in small catkinlike spikes or racemes, rarely single in the leaf axils. The perianth segments are 2–4, occasionally lacking in male flowers. The stamens are as many as the petals and opposite them. The ovary is partly or entirely inferior, without evident chambers. The fruit has a single seed with a sticky covering.

Classification The 3 genera were formerly included in Loranthaceae but associated into a single family in recent decades[i]. Current molecular studies reaffirm the general affinity of the genera and indicate that they can also be regarded as a basal grouping within Santalaceae[ii] in which they are included in APG II.   RMP

[i] Kuijt, J. Monograph of Eremolepidaceae. *Syst. Bot. Monogr.* 18: 1–60 (1988).
[ii] Nickrent, D. L. & Malécot, V. A molecular phylogeny of Santalales. In: Fer, A., Thalouarn, P., Joel, D. M., Musselman, L. J., Parker, C., & Verkleij, J. A. C. (eds), *Proceedings of the 7th International Parasitic Weed Symposium.* Pp. 69–74. Nantes, Faculté des Sciences, Université de Nantes (2001).

ERICACEAE
BILBERRIES, BLUEBERRIES, CRANBERRIES, HEATHERS, HEATHS, RHODODENDRONS, WINTERGREENS

A large family containing many well-known ornamental shrubs.

Distribution The family is cosmopolitan with several genera (e.g., *Empetrum*) whose species are widely disjunct. The disjunction of genera and species across major oceans is a feature of the family. *Erica, Rhododendron,* and *Agapetes* account for more than 50% of the species, but 50% of the genera are of 5 spp. or fewer. Notable areas of diversity are the southwestern cape of South Africa, where c. 450 *Erica* spp. occur; Southeast Asia and Malaysia with more than 400 *Agapetes* spp. and almost 300 spp. of *Rhododendron* sect. *Vireya;* and the area of western China, Tibet, Myanmar, and Assam, where *Rhododendron* is most diverse (c. 700 spp.)[i]. Epacrideae has a center of diversity in eastern Australia.

Description Evergreen or deciduous shrubs, sometimes scramblers or climbers (e.g., some *Agapetes*), herbs or trees (*Arbutus*), sometimes epiparasites without chlorophyll (*Monotropa*), with simple leaves. The leaves are alternate and sprially arranged or opposite or whorled, entire or serrate, with or without a petiole, sometimes strongly revolute and forming narrow ericoid needles, and without stipules. The inflorescence is of terminal or axillary racemes usually with paired bracteoles (prophylls), rarely single-flowered. The flowers are regular, usually hermaphrodite, with (2–)4–5(–9) sepals fused at the base. The petals are (3–)4–5(–9), equal, and usually fused, except for pointed tips. The stamens are (2–)5(–8) or 10(–16) free or rarely adnate to the petals. The ovary is usually superior but inferior in the Vaccineae, of (1–)4–5(–14) fused carpels, containing 1 to many ovules. The style is usually as long as the corolla tube, hollow, and without lobes. The fruit is a capsule, berry, or drupe bearing small seeds.

Classification Ericaceae is nested within its order sister to Cyrillaceae then Clethraceae[ii]. The family has long been accepted but has undergone a major expansion due to molecular and morphological phylogenetic studies[iii] to include Empetraceae, Epacridaceae, Monotropaceae, and Pyrolaceae, all of which were previously recognized as separate families by many authors (e.g., Cronquist 1981, Dahlgren 1983). The enlarged family is split into 8 subfamilies and 20 tribes[ii], totaling some 4,050 spp.[iv] in 124 genera.

SUBFAM. ENKIANTHOIDEAE Evergreen or deciduous shrubs, with simple alternate or whorled leaves, with entire or serrate margins, and a distinctive indumentum of elongate, nonglandular hairs, and scaly winter buds. Inflorescence lacks bracteoles; calyx-lobes deciduous; anthers with a pair of awns. One genus, *Enkianthus* (16 spp.).

SUBFAM. MONOTROPOIDEAE Evergreen herbs or subshrubs or lacking chlorophyll and growing in association with fungal mycorrhizae, with the leaves alternate or spirally arranged. Divided into three tribes:

Monotropeae comprises mycotrophic herbs without chlorophyll and usually without indumentum (9 genera, probably over split, 11 spp.).

Pyroleae are evergreen herbs or subshrubs with spirally arranged leaves, the flowers with free petals, a 5-locular ovary, and a peltate or lobed stigma (4 genera, 38 spp., of which *Pyrola* comprises 30 spp.).

Pterosporeae comprises mycotrophic herbs without chlorophyll, with an indumentum of glandular multiseriate-stalked hairs (2 genera, 2 spp.).

SUBFAM. ARBUTOIDEAE Mostly evergreen trees or shrubs with alternate leaves and a terminal inflorescence. Calyx-lobes small, with distinctive tomentum of long unicelluar hairs on the inner surface; stamens with filaments strongly dilated at the base; ovary 4- to 10-locular, with 1 or few ovules per locule; fruit fleshy, usually with 1 to several seeds (4 genera, 81 spp., of which *Arctostaphylos* comprises 60 spp.).

SUBFAM. CASSIOPOIDEAE Low-growing evergreen shrubs with decussate, ericoid leaves and single-flowered axillary inflorescences bearing 4–6 basal bracteoles. One genus, *Cassiope*, with 12 spp.

SUBFAM. ERICOIDEAE Evergreen or deciduous shrubs or trees, with alternate, whorled or decussate leaves that are ericoid or not. Inflorescences varied, flowers sometimes held erect. Fruit is a loculicidal or septicidal capsule, drupe, or indehiscent pod. It is subdivided into 5 tribes:

Ericeae is a tribe of evergreen shrubs or trees usually with 3- to 4-whorled ericoid leaves; inflorescences always with a bract; flowers usually 4-merous and actinomorphic with usually persistent calyx-lobes; ovary of (1–)4(–8) carpels capped by a gradually tapered or sunken style (3 genera, c. 860 spp., of which *Erica* c. 860 spp.).

Empetreae consists of evergreen shrubs, sometimes forming mats; leaves ericoid, alternate to almost verticillate; inflorescence axillary with small, solitary, unisexual or bisexual actinomorphic flowers that are wind pollinated, the petals

Genera 124 Species 4,050
Economic uses Ornamental for decorative flowers

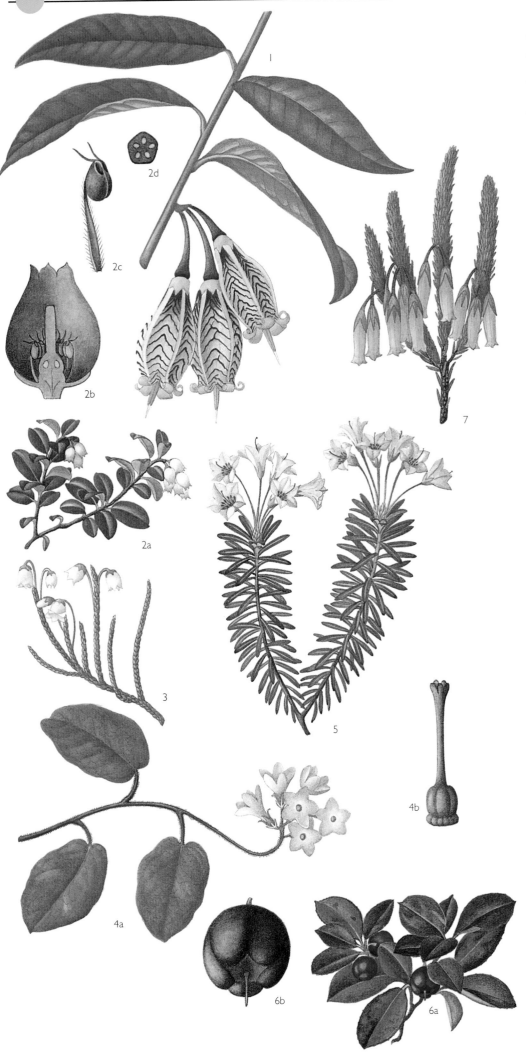

ERICACEAE. 1 *Agapetes macrantha* part of leafy shoot with axillary inflorescence (x⅔). 2 *Arctostaphylos uva-ursi* (a) leafy shoot with terminal inflorescences (x⅔); (b) half flower (x4); (c) stamen with broad, hairy filament and anthers crowned by recurved arms and opening by terminal pores (x10); (d) cross section of ovary (x4). 3 *Cassiope selaginoides* stem covered with small clasping leaves (x⅔). 4 *Epigaea repens* (a) leafy stem and inflorescence (x1); (b) gynoecium with lobed ovary and stigma (x4). 5 *Phyllothamnus erectus* flowering shoot (x⅔). 6 *Gaultheria* sp. (a) leafy shoot and berries (x⅔); (b) berry (x2⅔). 7 *Erica vallis-aranearum* flowering shoot (x⅔).

free and the sepals inconspicuous; style short, with deeply divided stigmatic lobes; ovary ripening to form a dry or fleshy fruit (3 genera, 5 spp.).

Bejarieae comprises low evergreen shrubs, with decussate, whorled or alternate, ericoid or flat, leaves; infloresence terminal or axillary racemes or panicles; flowers with free petals; ovary of 4–7 carpels, bearing a jointed style, capped with a truncate or capitate stigma; fruit a septicidal capsule sometimes with a warty surface (3 genera, 9 spp.).

Phyllodoceae consists of evergreen shrubs with spirally arranged, decussate or whorled, leaves that may be ericoid; inflorescence usually terminal, forming a raceme, spike, or corymb; flowers 4- to 6-merous with free or fused petals; ovary of 2–6 carpels, sometimes with a sunken style; the stigma truncate; fruit a septicidal or partially loculicidal capsule (6 genera, 28 spp.).

Rhodoreae comprises evergreen or deciduous small trees or shrubs, with alternate leaves; inflorescence usually terminal; flowers 4- to 5- merous and weakly zygomorphic, often with blotched petals; ovary of 3–14 carpels bearing a sunken or erect style capped by a truncate or expanded stigma; fruit a distinctive cylindrical to ovoid septicidal capsule (4 genera, c. 860 spp., of which *Rhododendron* c. 850 spp).

SUBFAM. **HARRIMANELLOIDEAE** Mat-forming evergreen shrubs with ericoid leaves and an indumentum of unicellular hairs; inflorescence terminal, of one 5-merous flower; no bracts or bracteoles; the petals fused; ovary of 5 carpels capped by an indented style that is short and stout and bears a truncate stigma; fruit a capsule. One genus: *Harrimanella* (2 spp.).

SUBFAM. **STYPHELIOIDEAE** Evergreen trees, shrubs or climbers with alternate leaves; inflorescence terminal or axillary; flowers 4- to 5-merous with persistent petals fused in a tube with large or small lobes; stamens 2 or 4–5, usually attached to the petals; ovary of 1–11 carpels; fruit fleshy or dry. The subfamily consists of 7 tribes:

Prionoteae comprises shrubs or climbers with alternate leaves and axillary single-flowered inflorescences borne at the end of long branches; flowers 5-merous with fused petals; ovary of 5 fused carpels, bearing a long style with a tapered stigma (2 genera, 2 spp.).

Archerieae comprises shrubs with alternate leaves; inflorescences axillary; single-flowered or racemose; flowers 5-merous, with fused petals; ovary of 5 fused carpels, capped by a sunken style (*Archeria*, 7 spp.).

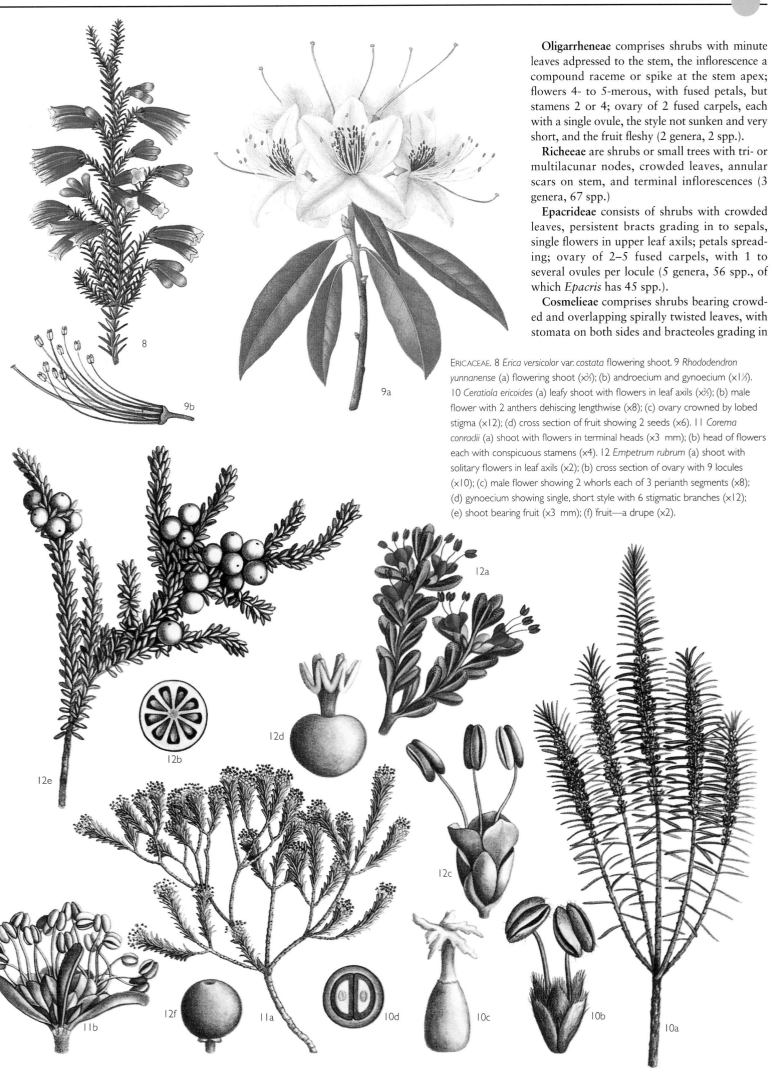

Oligarrheneae comprises shrubs with minute leaves adpressed to the stem, the inflorescence a compound raceme or spike at the stem apex; flowers 4- to 5-merous, with fused petals, but stamens 2 or 4; ovary of 2 fused carpels, each with a single ovule, the style not sunken and very short, and the fruit fleshy (2 genera, 2 spp.).

Richeeae are shrubs or small trees with tri- or multilacunar nodes, crowded leaves, annular scars on stem, and terminal inflorescences (3 genera, 67 spp.)

Epacrideae consists of shrubs with crowded leaves, persistent bracts grading in to sepals, single flowers in upper leaf axils; petals spreading; ovary of 2–5 fused carpels, with 1 to several ovules per locule (5 genera, 56 spp., of which *Epacris* has 45 spp.).

Cosmelieae comprises shrubs bearing crowded and overlapping spirally twisted leaves, with stomata on both sides and bracteoles grading in

ERICACEAE. 8 *Erica versicolor* var. *costata* flowering shoot. 9 *Rhododendron yunnanense* (a) flowering shoot (×⅔); (b) androecium and gynoecium (×1⅓). 10 *Ceratiola ericoides* (a) leafy shoot with flowers in leaf axils (×⅔); (b) male flower with 2 anthers dehiscing lengthwise (×8); (c) ovary crowned by lobed stigma (×12); (d) cross section of fruit showing 2 seeds (×6). 11 *Corema conradii* (a) shoot with flowers in terminal heads (×3 mm); (b) head of flowers each with conspicuous stamens (×4). 12 *Empetrum rubrum* (a) shoot with solitary flowers in leaf axils (×2); (b) cross section of ovary with 9 locules (×10); (c) male flower showing 2 whorls each of 3 perianth segments (×8); (d) gynoecium showing single, short style with 6 stigmatic branches (×12); (e) shoot bearing fruit (×3 mm); (f) fruit—a drupe (×2).

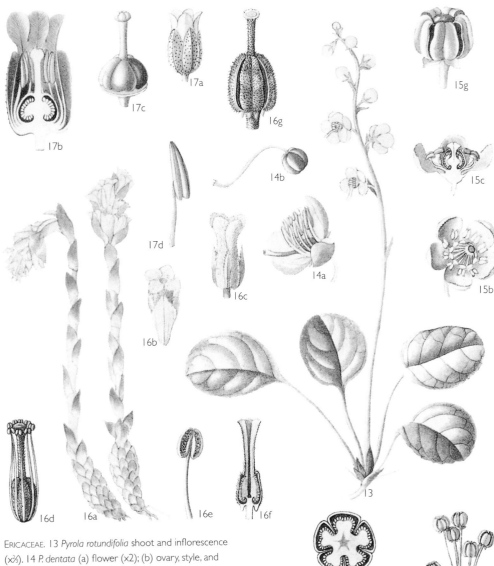

ERICACEAE. 13 *Pyrola rotundifolia* shoot and inflorescence (x⅔). 14 *P. dentata* (a) flower (x2); (b) ovary, style, and stigma (x3). 15 *Chimaphila umbellata* (a) shoot and inflorescence (x⅔); (b) flower (x1⅓); (c) half flower (x2); (d) stamen side (left) and front (right) views (x4); (e) cross section of ovary (x4); (f) fruits (x⅔); (g) dehisced fruit (x3). 16 *Monotropa hypopithys* (a) habit (x⅔); (b) flower (x2); (c) flower (x2); (d) gynoecium and stamens (x3); (e) stamen (x12); (f) vertical section of gynoecium (x3); (g) fruit (x4). 17 *Sarcodes sanguinea* (a) flower (x1); (b) half flower (x2); (c) gynoecium (x2); (d) stamen (x3).

to sepals in the single, terminal flowers (3 genera, 41 spp., of which *Andersonia* has 35 spp.)

Styphelieae comprises shrubs to small trees, inflorescences terminal, axillary, or on stem; flowers with stamens attached to the petals and 1 apical ovule per locule; fruit a drupe (19 genera, 370 spp., of which *Leucopogon* has 230 spp.).

SUBFAM. VACCINIOIDEAE Evergreen or deciduous trees, shrubs, or climbers; the apical bud of vegetative shoots aborting. There are five tribes:

Oxydendreae contains deciduous trees with terminal inflorescences; anthers without an appendage and splitting for half their length. There is 1 genus, *Oxydendrum*, with 1 sp.

Lyonieae comprises trees, shrubs, occasionally vines, with woody stems, flowers various; stamens with the filaments bent (4 genera, 48 spp., of which *Lyonia* 35 spp.).

Andromedeae consists of small shrubs with hairs only on the leaf margins, inflorescence

terminal; flowers with awned anthers (2 genera, 3 spp.).

Gaultherieae consists of shrubs with usually racemose inflorescences, usually urn-shaped; stamens with filaments swollen and anthers with short slits (6 genera, 246 spp., of which *Gaultheria* 130 spp. and *Diplycosia* 100 spp.)

Vaccinieae comprises evergreen or deciduous shrubs or small trees, sometimes lianas, often epiphytic; leaves leathery; inflorescences axillary, the flowers with a jointed pedicel; fruits fleshy (35 genera, 1,275 spp., including *Agapetes* 400 spp., *Vaccinium* 140 spp., *Cavendishia* 130 spp., and *Dimorphanthera* 80 spp).

The classification adopted here is based on molecular phylogenetic study of 124 spp. of Ericaceae and of morphological phylogenetic study of 80 spp[ii]. Genera seem to be defined by breeding barriers, and there are no authenticated intergeneric hybrids in the family known in cultivation[v]. The parasitic Monotropoideae show high specificity between species and fungal symbiont and, unusually for Ericaceae, associate with Basidiomycete fungi[vi].

Economic uses The family is economically important for the many species in horticulture, *Rhododendron* particularly commanding high prices. *Erica*, *Arbutus*, *Kalmia*, and *Pieris* are all also widely grown for their bright flowers and usually glossy, evergreen foliage. The major fruit crops are Cranberry (*Vaccinium* sect. *Oxycoccus*, especially the American Cranberry, *Vaccinium macrocarpon*) and the Blueberry (*V.* sect. *Cyanococcus*, especially *V. corymbosum* and *V. angustifolium*). Some species are highly toxic, and the honey from *Rhododendron* can cause illness or even fatality thanks to the effects of a grayanotoxin (andromedotoxin)[vii]. According to Pliny the Elder, this honey was used to poison Xenophon's troops in 401 BCE, although it is now commercially available and used in recipes[viii]. AC

[i] Heads, M. Ericaceae in Malaysia: vicariance biogeography, terrane tectonics and ecology. *Telopea* 10: 311–449 (2003).
[ii] Anderberg, A. A., Rydin, C., & Källersjö, M. Phylogenetic relationships in the order Ericales *s.l.*: analyses of molecular data from five genes from the plastid and mitochondrial genomes. *American J. Bot.* 89: 677–687 (2002).
[iii] Kron, K. A. *et al.* Phylogenetic classification of Ericaceae: molecular and morphological evidence. *Bot. Rev.* 68: 335–423 (2002).
[iv] Stevens, P. F. *et al.* Ericaceae. Pp. 145–194. In: Kubitzki, K. (ed.), *The Families and Genera of Vascular Plants. VI. Flowering Plants. Dicotyledons: Celastrales, Oxalidales, Rosales, Cornales, Ericales.* Berlin, Springer-Verlag (2004).
[v] Grant, M. L., Toomey, N. H., & Culham, A. Is there such a thing as *Kalmia* × *Rhododendron*? *J. American Soc. Hort. Sci.* 129: 517–522 (2004).
[vi] Bidartondo M. I. & Bruns, T. D. Extreme specificity in epiparasitic Monotropoideae (Ericaceae): widespread phylogenetic and geographical structure. *Molecular Ecology* 10: 2285–2295 (2001).
[vii] http://www.chm.bris.ac.uk/webprojects2001/gerrard/home.html
[viii] Nischan, M. & Goodbody, M. *Taste Pure and Simple: Irresistible Recipes for Good Food and Good Health.* San Francisco, Chronicle Books (2003).

ERYTHROXYLACEAE
COCA AND COCAINE

A tropical and subtropical family of trees and shrubs, including the important cocaine-producing coca plant.

Distribution *Erythroxylum* is pantropical, but most species are neotropical, centered mainly in the Andes and Amazonian basin. The other 3 genera are confined to Africa: *Aneulophus* (2 spp.) in west tropical Africa, *Nectaropetalum* (8 spp.) in east and west tropical Africa down to the Cape, with 1 sp. in Madagascar, and *Pinacopodium* (2 spp.) in tropical Africa (Congo and Gabon).

Description Evergreen or deciduous, small trees and shrubs, often with persistent cataphylls. The leaves are alternate, rarely opposite (*Aneulophus*), usually 2-ranked, simple, the margins entire, the stipules intrapetiolar, often caducous. The flowers are small, actinomorphic, bisexual or infrequently unisexual (plants dioecious), hypogynous, in axillary fascicles, or flowers solitary. The sepals are 5, small, basally connate, valvate or imbricate, persistent. The petals are 5, free, caducous, imbricate, deciduous, with a ligular appendage near the base. The stamens are 10, in 2 whorls of 5, connate at the base, usually forming a short tube or free except at the very base (*Nectaropetalum, Pinacopodium*). The ovary is superior, of (2)3(4) fused carpels, with 2–4 locules, only 1 or 2 of them fertile, each containing 1 or 2 apical pendulous, anatropous ovules. The styles are 3, free or connate at the base. The fruit is an ovoid single-seeded drupe, with the persistent calyx at the base. The seeds have copious endosperm and a straight embryo.

Classification Within the Erythroxylaceae, *Aneulophus* is distinctive in having opposite leaves, a capsular fruit, and arillate seeds. It is treated as a taxon of uncertain position in APG II. *Nectaropetalum* is sometimes distinguished in a separate family. The Erythroxylaceae is included in the Linales by Cronquist (1981) and Takhtajan (1997), but recent morphological and molecular studies indicate that it is sister to the Rhizophoraceae within the Malpighiales[i,ii].

Economic uses The leaves of several species of *Erythroxylum* produce alkaloids, the most economically important and potent of which are *E. coca* and *E. novagranatense* (coca), which yield the important alkaloid cocaine, a narcotic that has long been used as a local anaesthetic and subsequently as a template for the production of several synthetic anaesthetics used in modern medicine and dentistry. Coca has been used in the Andes by Native South Americans for 3,000 years as a tonic and stimulant to alleviate hunger and altitude sickness as well as for medicinal purposes. *E. coca* and *E. novogranatense* have been extensively cultivated in South America, Sri Lanka, and Java, and concentrated and purified or semipurified forms of cocaine are used globally as recreational drugs. Other species of *Erythroxylum* produce valuable alkaloids, such as catuabine, which is used locally as an aphrodisiac and for the production of catuabine A, B, and C, stimulants of the nervous system. A major ingredient of the soft drink Coca-Cola® is an extract of coca leaves, although today without cocaine. Other species are of local importance for their wood, bark dye, wood tar, and essential oils.   VHH

[i]Schwarzbach, G. & Ricklefs, R. E. Systematic affinities of Rhizophoraceae and Anisophylleaceae, and intergeneric relationships within Rhizophoraceae, based on chloroplast DNA, nuclear ribosomal DNA, and morphology. *American J. Bot.* 87: 547–564 (2000).
[ii] Setoguchi, H., Kosuge, K., & Tobe, H. Molecular phylogeny of Rhizophoraceae based on *rbc*L gene sequences. *J. Plant Res.* 112: 443–455 (1999).

ESCALLONIACEAE
ESCALLONIA, FEATHERWOOD

A family most known for the garden ornamentals in *Escallonia*.

Distribution The family is scattered throughout the southern hemisphere with some remarkable disjunctions: sister genera *Valdivia gayana* (1 sp., Chile) with *Forgesia racemosa* (1 sp., Reunion) and *Eremosyne pectinata* (1 sp., southwestern coast of Western Australia) with *Escallonia* (c. 45 spp., centered in South America but extending into North America) and within *Anopterus* (2 spp.), *A. glandulosus* (Tasmania) and *A. macleayanus* (within 100 km of the coast at the border of Queensland and New South Wales). The remaining two genera are *Polyosoma* (c. 85 spp.) that extends from the eastern Himalayas through Malaysia to tropical Australia and *Tribeles australis* that is endemic to temperate South America.

Description Trees, shrubs, subshrubs, sometimes creeping (*Tribeles*), rarely an annual herb (*Eremosyne*), with opposite (*Polyosma*) or alternate leaves. The leaves are usually simple, usually evergreen, margins entire or variously toothed, usually petiolate. The inflorescence is usually terminal, sometimes on lateral shoots or axillary, cymose or racemose (*Polyosma*), of many to 1 flower (*Tribeles*). The flowers are 4-merous (*Polyosma*), or 5- to 9-merous. The sepals are attached to a pronounced hypanthium, joined at the base but forming distinct valvate lobes. The petals are as many as the sepals, free, often colorful. The stamens are mostly 5, alternating with the petals. The ovary is superior or wholly or partly inferior, of 2–4 fused carpels, with 1 to many seeds per carpel. The style is often long, and the stigmas can be wet or dry. The fruit is a capsule or drupe. The morphology of this family is highly variable.

Classification The family has undergone major changes in its circumscription and placement. Many systems put all the component genera in Saxifragaceae (e.g., Thorne 1992) but some authors separate selected genera out into small satellite families (e.g., Dahlgren 1983), while Cronquist (1981) recognized the genera as part of a broad Grossulariaceae. A DNA sequence study of 2 genes[i] demonstrated the extreme diversity of the broadly circumscribed Saxifragaceae and gave strong support for its division into smaller and better defined families including Escalloniaceae. The Escalloniaceae sits next to a clade comprising Apiales and Dipsacales in an analysis of *rbc*L only[ii], but support was weak. Molecular evidence appears to support a family comprising the 7 genera listed above[iii] but with *Valdivia* and *Forgesia* placed within *Escallonia*, and they could easily be included within the genus given their morphology. *Quintinia*, traditionally placed in the family and sharing many of the features, is now recognized to group with *Paracryphia* (Paracryphiaceae) and *Sphenostemon* (Sphenostemonaceae) in a well-supported lineage. It is sometimes (as here) recognized in its own family, Quintiniaceae (q.v.).

Economic uses *Escallonia* is well known for garden ornamental shrubs with glossy evergreen leaves and bright flowers. It also has limited local use as a fuelwood, for charcoal and construction, and as a source of red dye[iv]. *Anopterus* has brightly colored flowers, and these trees have ornamental use.   AC

[i] Soltis, D. E. & Soltis, P. S. Phylogenetic relationships in Saxifragaceae *sensu lato*: a comparison of topologies based on 18S rDNA and *rbc*L sequences. *American J. Bot.* 84: 504–522 (1997).

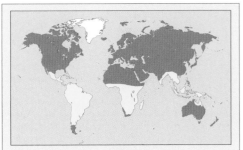

Genera 4 Species c. 250
Economic uses Coca used as a tonic and stimulant and source of cocaine

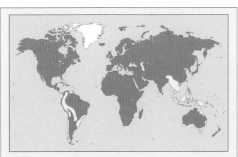

Genera 5–7 Species c. 140
Economic uses Ornamental, used locally for fuel and timber; source of red dye

ii Savolainen, V. *et al.* Phylogeny of the eudicots: a nearly complete familial analysis based on *rbc*L gene sequences. *Kew Bull.* 55: 257–309 (2000).
iii Lundberg, J. Phylogenetic studies in the Euasterids II with particular reference to Asterales and Escalloniaceae. Chapter V: A well resolved and supported phylogeny of Euasterids II based on a Bayesian inference, with special emphasis on Escalloniaceae and other incertae sedis. *Acta Univ. Ups. Comprehensive Summaries of Uppsala Dissertations from the Faculty of Science and Technology* 676. 38 pp. Uppsala (2001).
iv Mori, S. A. Escalloniaceae. Pp. 145–146. In: Smith, N. *et al.* (eds), *Flowering Plants of the Neotropics.* Princeton, Princeton University Press (2004).

i Bremer, K. *et al.* A phylogenetic analysis of 100+ genera and 50+ families of euasterids based on morphological and molecular data with notes on possible higher level morphological synapomorphies. *Pl. Syst. Evol.* 229: 137–169 (2001).

EUCOMMIACEAE

The family comprises a single species, *Eucommia ulmoides*. This deciduous tree can be found in the montane forests of China, and it is noted in horticultural literature that it may reach up to 20 m tall. *Eucommia ulmoides* has been recorded in Gansu, Guizhou,

Genera 1 | Species 1
Economic uses Latex

Hubei, Hunan, Shaanxi, and Sichuan provinces but is rare and possibly no longer found in the wild. The leaves are spirally arranged, petiolate, ovate or elliptic, crenate to serrate, and usually with a long, acute apex. When a leaf is broken and slowly pulled apart, the veins exude a latex that hardens and holds the parts together. The flowers are subsessile to shortly pedicelled and borne in the axils of small bracts toward the base of a shoot terminated above the flowers by several leaves. The flowers are unisexual (plants dioecious). Sepals and petals are absent. In male plants, the flower consists of a cluster of 5–12 stamens, each up to 1.5 cm long, the filaments short, but the long anther cells terminated by a short prolongation of the connective. The flowers of female plants consist of a flattened, bicarpellate, 1-locular ovary terminated by 2 styles. There are at first 2 ovules suspended from the apex of the locule, but one soon aborts. The fruit is a flat samara, 2.5–3 × 0.8 cm, with a wing c. 2 mm broad surrounding the single seed.

The family was long considered to be referable either to Urticales or Hamamelidales and was regarded by Cronquist as intermediate between them, but Takhtajan has preferred an affinity with Cornales. Molecular evidence[i] now places Eucommiaceae close to the monogeneric Garryaceae and Aucubaceae, both formerly included in Cornaceae but now considered remote from that in the Asterid line. All are dioecious woody plants, and some similarities may be seen in their ovary structure.

The latex derived from the tree, called gutta percha, has been used to line oil pipelines, insulate electric cables, and as a filling for teeth. The species is extensively cultivated in China for its medicinal properties and sometimes grown as an ornamental.   RKB

EUPHORBIACEAE

SPURGE FAMILY

A large, diverse family with unisexual flowers, superior syncarpous ovaries, and 1 ovule per locule. The family has undergone major taxonomic revision following molecular phylogenetic studies.

Distribution Cosmopolitan, except Antarctica, with greatest diversity in the tropics; often abundant in low to medium altitudes, dominant in some ecosystems.

Description Trees, shrubs, herbs, climbers, or succulents, with hairs ranging from simple to dendritic and T-shaped to stellate and lepidote, urticating in *Cnidoscolus* and in the unrelated tribe Plukenetieae. Clear or milky latex is present in many taxa. It can be variously colored but is usually white or red; the white milky latex of subfamily Euphorbioideae is usually caustic or toxic. Also notable is the secretion of resin from floral bracts in *Dalechampia*[i]. The leaves are simple or palmately compound, entire to toothed or deeply lobed. Stipules are generally, but not always, present and can be minute and caducous to large and foliaceous. Leafless succulent species are often spiny, superficially resembling cacti. Inflorescences are diverse. The tribe Euphorbieae is characterized by highly specialized pseudanthial synflorescences (cyathia), consisting of bracts subtending 4–5 staminate inflorescences (reduced to a single stamen), and a terminal pistillate flower (reduced to a gynoecium). The flowers are unisexual (plants monoecious or dioecious). Sepals, petals, disks, staminodes, and pistillodes may be absent or present. The stamens range from 1–1000 (variously branched in *Ricinus*) and are free or variously fused. The ovary is superior, of (1–)2–5(–20) fused carpels, forming as many locules, each with a single apical-axile, anatropous and epitropous ovule provided with an obturator. The fruit is characteristically an explosive schizocarp, with typical dehiscence and persistent columella, a unique character confined to many taxa of Euphorbiaceae *sensu stricto*, Phyllanthaceae, and Picrodendraceae (Pandaceae and Putranjivaceae are indehiscent), which has high recognition value in a group that generally lacks striking morphological synapomorphies. Whether the fruit type is plesiomorphic for the 3 families, however, cannot be shown by the molecular phylogenetic analyses so far[ii,iii,iv] due to poorly supported deep nodes. In many taxa, the fruit is indehiscent, and in critical taxa, such as Peroideae, it is atypically nonexplosive. A parallel origin of this structure in different euphorbiaceous lineages cannot be excluded.

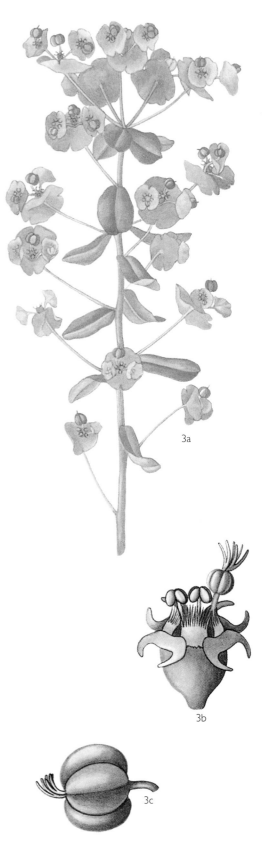

EUPHORBIACEAE. 1 *Euphorbia stapfii* a cactuslike species (x⅔). 2 *Acalypha* sp. (a) leafy shoot and lateral inflorescence (x⅔); (b) female flower with large, branched styles (x6). 3 *Euphorbia amygdaloides* (a) flowering shoot showing inflorescences (cyathia) condensed to resemble a single flower (x⅔); (b) the cyathium consisting of an outer cup-shaped structure bearing horseshoe-shaped glands on the rim, within which is a ring of male flowers each consisting of a single stamen and in the center the female flower that consists of a stalked ovary and branched stigmas (x6); (c) 3-lobed fruit (x4). 4 *Croton fothergillifolius* (a) flowering shoot (x⅔); (b) fruit (x4).

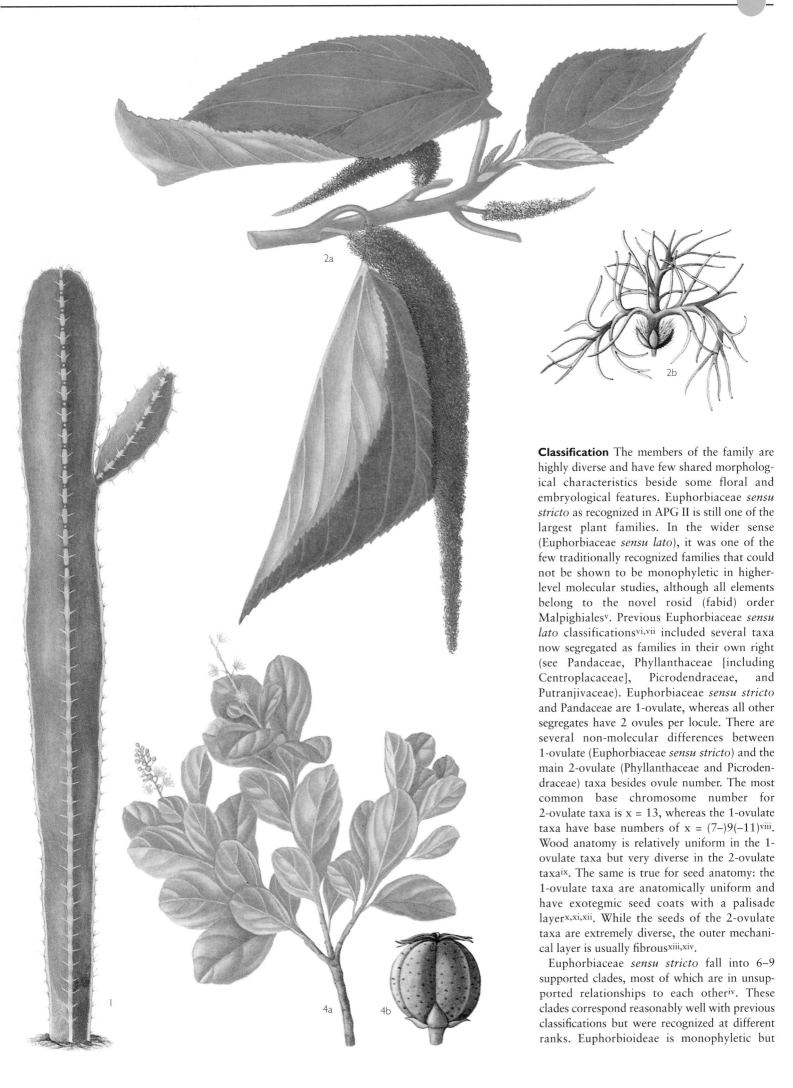

Classification The members of the family are highly diverse and have few shared morphological characteristics beside some floral and embryological features. Euphorbiaceae *sensu stricto* as recognized in APG II is still one of the largest plant families. In the wider sense (Euphorbiaceae *sensu lato*), it was one of the few traditionally recognized families that could not be shown to be monophyletic in higher-level molecular studies, although all elements belong to the novel rosid (fabid) order Malpighiales[v]. Previous Euphorbiaceae *sensu lato* classifications[vi,vii] included several taxa now segregated as families in their own right (see Pandaceae, Phyllanthaceae [including Centroplacaceae], Picrodendraceae, and Putranjivaceae). Euphorbiaceae *sensu stricto* and Pandaceae are 1-ovulate, whereas all other segregates have 2 ovules per locule. There are several non-molecular differences between 1-ovulate (Euphorbiaceae *sensu stricto*) and the main 2-ovulate (Phyllanthaceae and Picrodendraceae) taxa besides ovule number. The most common base chromosome number for 2-ovulate taxa is x = 13, whereas the 1-ovulate taxa have base numbers of x = (7–)9(–11)[viii]. Wood anatomy is relatively uniform in the 1-ovulate taxa but very diverse in the 2-ovulate taxa[ix]. The same is true for seed anatomy: the 1-ovulate taxa are anatomically uniform and have exotegmic seed coats with a palisade layer[x,xi,xii]. While the seeds of the 2-ovulate taxa are extremely diverse, the outer mechanical layer is usually fibrous[xiii,xiv].

Euphorbiaceae *sensu stricto* fall into 6–9 supported clades, most of which are in unsupported relationships to each other[iv]. These clades correspond reasonably well with previous classifications but were recognized at different ranks. Euphorbioideae is monophyletic but

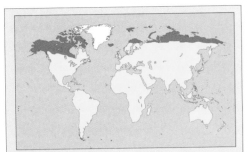

Genera c. 250 **Species** c. 6,300
Economic uses Major starch crop, seed oil, latex, purgatives, ornamentals, some weeds

relationships between the different crotonoid clades remain unclear; Acalyphoideae is clearly paraphyletic as predicted by Webster[xv].

SUBFAM. PEROIDEAE Fruit septa membranous, fragile and without visible vascularization, the valves often remaining attached to the base of the columella after deshiscence. Seeds shiny, black, smooth, with a substantial caruncle sometimes covering more than half of the seed surface. Seed coat exotestal with a tracheoidal exotegmen (except *Pogonophora*).

SUBFAM. CHEILOSOIDEAE Pollen-tectum echinate. Seeds with a sarcotesta. Marginal foliar glands situated at abaxial side of teeth.

Erismantheae Stipules interpetiolar due to unequal internode growth and reduction of 1 stipule.

SUBFAM. ACALYPHOIDEAE *sensu stricto* Latex is absent, laticifers rare (if present, then inarticulate). Leaves always simple. Petals are present or absent. Pollen grains are mostly 3-colporate or 3-porate.

SUBFAM. CROTONOIDEAE *sensu lato* Latex is reddish or yellowish to milky (sometimes scanty or absent); laticifers are articulate or inarticulate. Leaves simple or compound. Indumentum simple, stellate, dendritic, or lepidote. Sepals imbricate to valvate and covering the anthers in bud. Petals present (at least in staminate flowers) or absent. Pollen grains 3-colporate or more often porate or inaperturate, with "crotonoid" pattern of exinous processes. There are 4 groups:

Crotonoideae *sensu stricto* The pollen grains are inaperturate. Petals usually present. Laticifers articulated or non-articulated. Contains *Croton*, the second largest genus in the family with more than 1,200 spp.

"Articulated crotonoids" The pollen grains are aperturate (apertures cryptic in *Elateriospermum*). Latex present and laticifers are articulated. Petals absent and plants mostly monoecious. Basic chromosome number x = 9. Inner integuments are thick, multiplicative, with vascular bundles.

Gelonieae The pollen grains are pantoporate. Leaves pellucid-punctate. Latex not evident and laticifers nonarticulate. Petals absent. Plants mostly dioecious. Basic chromosome number x = 11. Inner integuments thin, without vascularization (except in *Klaineanthus*).

Adenoclineae (including Omphaleae) The pollen grains are 3-colpate. Leaves not pellucid-punctate. Latex present (clear or colored) with nonarticulated laticifers. Petals absent. Plants mostly dioecious. Basic chromosome number x = 11. Inner integuments thin, without vascularization.

SUBFAM. EUPHORBIOIDEAE Latex always present, whitish (often caustic or toxic); laticifers inarticulate. Leaves always simple. Indumentum simple or absent, rarely dendritic, never stellate or lepidote. Bracts often 2-glandular at base. Sepals imbricate or reduced; petals absent. Anthers usually not covered in bud. Pollen grains 3-colporate with the sexine mostly perforate-reticulate. *Euphorbia*, with more than 2,000 spp., is the largest genus of the family and one of the largest of all plant genera.

Economic uses *Manihot esculenta* (cassava, manioc, tapioca, yuca) is the fourth most important food plant in the tropics. This starch crop of neotropical origin is known to have been in cultivation from 2000 BCE. It grows in poor, dry soils and is nearly immune to locust attack. The starchy, tuberous roots contain HCN, which is removed by correct preparation. Another economically important species is *Hevea brasiliensis* (Rubber Tree). It originated in the Amazon Basin. In 1873, the English forester Henry Wickham brought 70,000 seeds from Brazil to England from where they were shipped to Southeast Asia, which is now the main center of cultivation. Natural rubber has played a major role in the industrial revolution, and still accounts for around one-third of the world's consumption of tires and tire accessories. *Ricinus communis* (Castor Oil Plant) is grown as a source of oil. The seeds have been found in Egyptian tombs dating back to 4000 BCE. The seed oil was probably used as an illuminant and unguent, later medicinally as a purgative, but is also an excellent industrial lubricant. The oil is free of the water-soluble compound ricin, a compound highly toxic to humans and animals, which has made news headlines in connection with terrorist threats. *Ricinus* is also grown as an ornamental as are *Euphorbia pulcherrima* (Poinsettia), *Codiaeum variegatum* ("Croton"), *Acalypha hispida* (Cat's Tail, Chenille Plant), *A. wilkesiana* (Copper Leaf), and *Euphorbia milii* (Crown-of-thorns). Many succulent species of *Euphorbia*, mainly from Africa and Madagascar, are horticulturally desirable and threatened by illegal collecting, whereas many herbaceous *Euphorbia* species of temperate regions are well-known garden weeds.   PH

i Armbruster, W. S. The role of resin in angiosperm pollination: ecological and chemical considerations. *American. J. Bot.* 71: 1149–1160 (1984).
ii Wurdack, K. J. *et al.* Molecular phylogenetic analysis of Phyllanthaceae (Phyllanthoideae pro parte, Euphorbiaceae *sensu lato*) using plastid *rbc*L DNA sequences. *American J. Bot.* 91: 1882–1900 (2004).
iii Kathriarachchi, H., Hoffmann, P., Samuel, R., Wurdack, K. J., & Chase, M. W. Molecular phylogenetics of Phyllanthaceae inferred from 5 genes (plastid *atp*B, *mat*K, 3' *ndh*F, *rbc*L) and nuclear *PHYC*. *Mol. Phyl. Evol.* 36: 112–134 (2005).
iv Wurdack, K. J., Hoffmann, P., & Chase, M. W. Molecular phylogenetic analysis of uniovulate Euphorbiaceae (Euphorbiaceae *sensu stricto*) using plastid *rbc*L and *trn*L-*trn*F DNA sequences. *American J. Bot.* 92: 1397–1420 (2005).
v Wurdack, K. J. & Davis, C. C., unpublished manuscript.
vi Webster, G. L. Synopsis of the genera and suprageneric taxa of Euphorbiaceae. *Ann. Missouri Bot. Gard.* 81: 33–144 (1994).
vii Radcliffe-Smith, A. *Genera Euphorbiacearum*. Richmond, Royal Botanic Gardens, Kew (2001).
viii Hans, A. S. Chromosomal conspectus of the Euphorbiaceae. *Taxon* 22: 591–636 (1973).
ix Mennega, A. M. W. Wood anatomy of the Euphorbiaceae, in particular of the subfamily Phyllanthoideae. *Bot. J. Linn. Soc.* 94: 111–126 (1987).
x Tokuoka, T. & Tobe, H. Ovules and seeds in Crotonoideae (Euphorbiaceae): structure and systematic implications. *Bot. Jahrb. Syst.* 120: 165–196 (1998).
xi Tokuoka, T. & Tobe, H. Ovules and seeds in Euphorbioideae (Euphorbiaceae): structure and systematic implications. *J. Pl. Res.* (Tokyo) 115: 361–374 (2002).
xii Tokuoka, T. & Tobe, H. Ovules and seeds in Acalyphoideae (Euphorbiaceae): structure and systematic implications. *J. Pl. Res.* (Tokyo) 116: 355–380 (2003).
xiii Corner, E. J. H. *The Seeds of the Dicotyledons*. Cambridge, Cambridge University Press (1976).
xiv Stuppy, W. *Systematische Morphologie und Anatomie der Samen der biovulaten Euphorbiaceen*. Ph.D. dissertation. Kaiserslautern, University of Kaiserslautern (1996).
xv Webster, G. L. The saga of the spurges: a review of classification and relationships in the Euphorbiales. *Bot. J. Linn. Soc.* 94: 3–46 (1987).

EUPHRONIACEAE

A small tropical family consisting of a single genus, *Euphronia*, and 3 spp. from the tropical Americas. *E. guianensis* is confined to southeastern Venezuela and adjacent Guyana (the Guyana Shield) and *E. acuminatissima* is restricted to southwestern Venezuela in the border

Genera 1 **Species** 3
Economic uses None

area with Colombia. *E. hirtelloides* is known from scattered collections in Venezuela, Colombia, and northwestern Brazil[i]. They are mostly shrubs but some trees, with alternate, simple, leaves covered with dense white or grayish hairs on the under surface, with small stipules. The flowers are zygomorphic, hermaphrodite, in axillary or terminal racemes; the calyx has 5 unequal lobes, and there are 3 free, spathulate petals; the stamens are 4 and usually 1 staminode, all connate at the base and forming a tube; the ovary is superior, syncarpous, with 3 carpels and 3 locules; the style is terminal; the fruit is a cylindrical, 3-locular, septicidal capsule, with 1 seed in each locule.

Euphronia is sometimes included in the Trigoniaceae (q.v.) or placed in the Vochysiaceae from both of which it differs in a several morphological and anatomical characters. On the basis of molecular evidence, it is a sister to the Chrysobalanaceae (q.v.) but differs from the latter in its terminal style, lack of a disk, capsular fruit, and opposite leaves. Until it is better known, it seems preferable to keep it as a separate family.   VHH

i Litt, A. Euphroniaceae. Pp. 150–151. In: Smith, N. et al. (eds), *Flowering Plants of the Neotropics*. Princeton, Princeton University Press (2004).

EUPOMATIACEAE

Genera 1 Species 3
Economic uses Fruits, timber, ornamentals

The Eupomatiaceae comprises a single genus (*Eupomatia*) of glabrous aromatic shrubs, small trees, or rhizomatous herbs, native only to eastern Australia (Queensland, New South Wales, and Victoria) and New Guinea, (formerly recorded from Sulawesi but record not confirmed and unlikely), mainly in rain forests. The leaves are alternate, simple, with entire margins, shortly petiolate, and lack stipules. The flowers are showy, solitary or rarely paired in the leaf axils, with a well developed pedicel, strongly perigynous, actinomorphic and hermaphrodite, strongly scented, and without a perianth. They possess a conical calyptra formed from a modified amplexicaul bract, which falls off to reveal an expanded thickened concave receptacle. The stamens are numerous, arising from the edge of the receptacle; the innermost 3–5 rows are sterile and petaloid staminodes that arch over. The ovary consists of many carpels that are fused at the base and spirally arranged and sunk into the conical receptacle but free above; each carpel contains 2 or more ovules attached at 1 side. The carpels are not sealed above and are terminated by the feathery stigmas, which are sessile and decurrent on the surface. The berrylike fruit is flat-topped, with oil glands; each of the several locules contains 1–3 angular seeds. The seeds contain copious endosperm and a small embryo.

Eupomatia contains 3 spp. *E. bennettii* from southeastern Queensland and northeastern New South Wales is a shrub with fleshy roots, flowers to 2.5 cm in diameter, and having yellow central staminodes, orange marginal staminodes, and fertile stamens reflexed onto the flower stalk. *E. laurina* is a tree from Victoria, New South Wales, Queensland, and New Guinea, growing to about 10 m, with white flowers and spreading instead of reflexed stamens. A third species, *E. barbata*, was recently described from tropical Queensland.

The characteristic calyptra of the flowers appears to be best interpreted as a modified bract[i]. The Eupomatiaceae was formerly included in the Annonaceae but is now regarded as a separate family and indeed as forming a sister pair with it in the Magnoliales[ii].

The wood of *E. laurina* is used locally for its decorative qualities. Both *E. laurina* and *E. bennettii* are sometimes grown as ornamentals. The fruits are edible and were apparently eaten by aborigines.   RKB & VHH

i Endress, P. K. Early floral development and nature of the calyptra in Eupomatiaceae (Magnoliales). *Int. J. Plant Sci.* 164: 489–503 (2003).
ii Soltis, D. E. et al. Angiosperm phylogeny inferred from 18S rDNA, *rbc*L, and *atp*B sequences. *Bot. J. Linn. Soc.* 133: 381–461 (2000).
iii Doyle, J. A. & Endress, P. K. Morphological phylogenetic analyses of basal angiosperms: Comparison and combination with molecular data. *Int. J. Pl. Sci.* 161 (Supplement): S121–S153 (2000).

EUPTELEACEAE

Genera 1 Species 2
Economic uses None

The family is monogeneric, comprising 2 spp. of deciduous often multitrunked trees with spirally arranged leaves: *Euptelea pleiosperma* from China, northeastern India, and Bhutan, and *E. polyandra* from Japan. The leaves are simple, elliptical, with a dentate margin and an extended tip, petiolate but without stipules. The flowers are axillary, in clusters on short lateral shoots, hermaphrodite and disymmetric, surrounded by bracts but lacking both sepals and petals, probably wind pollinated. The stamens are 6–19, free, the prominent red anthers accounting for more than one-half the total length. The carpels are 8–31, superior and free, each with 1–3(–4) ovules and an elongated lateral stigmatic region[i]. The fruit is dry and winged (a samara), retaining the stigmatic region.

The family has most often been included in the Hamamelid group of dicotyledons due to its lack of perianth and wind-pollination syndrome. DNA sequence data for three genes showed *Euptelea* to diverge after Papaveraceae at the base of the Ranunculales clade[ii], but a subsequent 4-gene analysis[iii] suggested that Eupteleaceae was the first family to diverge from the Ranunculales lineage. *Euptelea polyandra* has been used in traditional medicine and has been shown to possess potent gastroprotective saponins[iv].   AC

i Endress, P. K. Eupteleaceae. Pp. 299–301. In: Kubitzki, K., Rohwer, J. G., & Bittrich, V. (eds), *The Families and Genera of Vascular Plants. II. Flowering Plants. Dicotyledons: Magnoliid, Hamamelid and Caryophyllid Families*. Berlin, Springer-Verlag (1993).
ii Soltis, D. E. et al. Angiosperm phylogeny inferred from 18S rDNA, *rbc*L, and *atp*B sequences. *Bot. J. Linn. Soc.* 133: 381–461 (2000).

iii Kim, S., Soltis, D. E., Soltis, P. S., Zanis, M. J., & Suh, Y. Phylogenetic relationships among early-diverging eudicots based on four genes: were the eudicots ancestrally woody? *Mol. Phylogen. Evol.* 31: 16–30 (2004).
iv Murakami, T. et al. Bioactive saponins and glycosides. XVIII. Nortriterpene and triterpene oligoglycosides from the fresh leaves of *Euptelea polyandra* Sieb. et Zucc. (2): Structures of euptelea saponins VI, VI acetate, VII, VIII, IX, X, XI, and XII. *Chem. Pharm. Bull.* 49: 741–746 (2001).

FAGACEAE

BEECHES, OAKS, AND SWEET CHESTNUTS

A commercially important family of hardwood timber trees, more rarely shrubs with edible fruits, and ornamentals.

Distribution The Fagaceae are mostly found in the northern hemisphere, with a few species crossing the equator in Southeast Asia. Beeches (*Fagus*), oaks (*Quercus*), and sweet chestnuts (*Castanea*) figure prominently and are frequently the dominant members of the broad-leaved forests that cover, or used to cover, vast areas of North America and Eurasia at midlatitudes. Evergreen oaks are important members of the forests around the Gulf of Mexico and in southern China and southern Japan. In Southeast Asia, the structure of the mixed mountain forest is largely determined by evergreen members of the family, particularly oaks. In total, therefore, the Fagaceae produces a colossal biomass, possibly exceeded only by the conifers. The Fagaceae have a long fossil record, suggesting an origin by at least the middle Cretaceous, about 90 million years ago[i].

Description Deciduous or evergreen trees, rarely shrubs, with alternate (rarely whorled) simple, entire to pinnately lobed leaves, and scarious, usually deciduous stipules. The flowers are unisexual (plants monoecious) and usually arranged in catkins or small spikes that may comprise only flowers of one sex, as in oaks, or may have female flowers at the base of otherwise male inflorescences, a more ancestral condition found e.g., in chestnuts. The perianth is bractlike, with 4–7 lobes. The male flowers have as many or twice as many stamens as perianth segments, occasionally up to 40, with the filaments free, with or without a pistillode. The female flowers are in groups of 1–3, each group being surrounded by a basal involucre. The

Genera 8 Species 620–750
Economic uses Timber, edible fruits, ornamentals

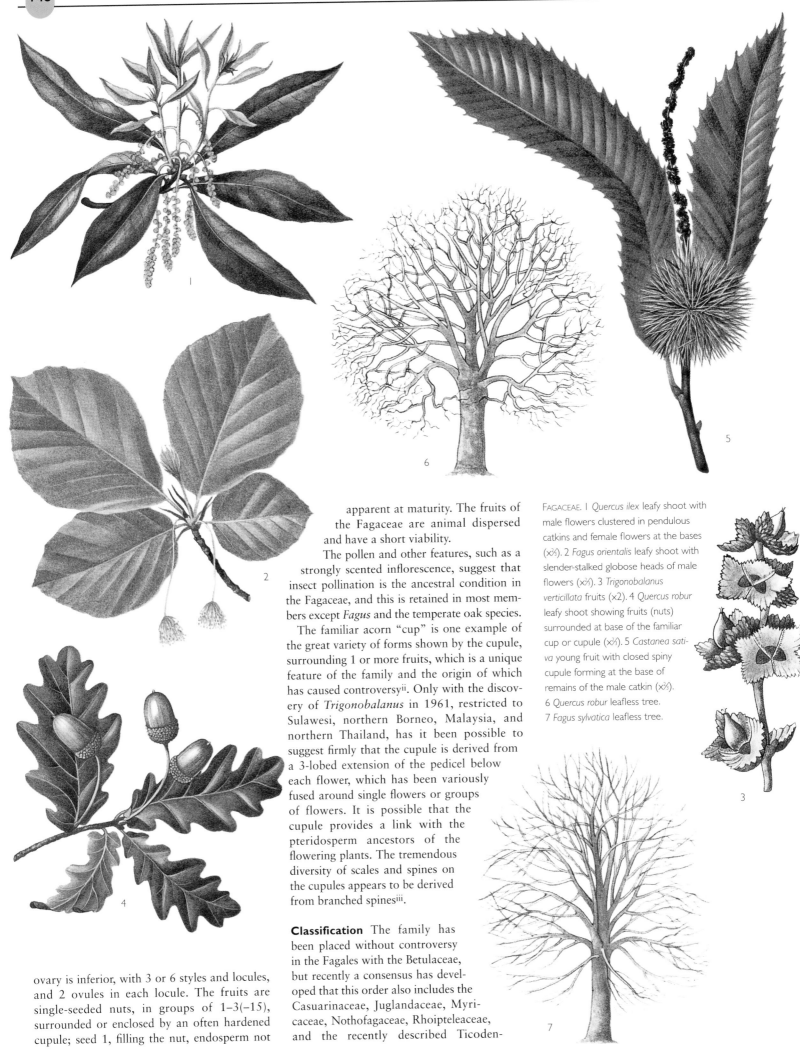

apparent at maturity. The fruits of the Fagaceae are animal dispersed and have a short viability.

The pollen and other features, such as a strongly scented inflorescence, suggest that insect pollination is the ancestral condition in the Fagaceae, and this is retained in most members except *Fagus* and the temperate oak species.

The familiar acorn "cup" is one example of the great variety of forms shown by the cupule, surrounding 1 or more fruits, which is a unique feature of the family and the origin of which has caused controversy[ii]. Only with the discovery of *Trigonobalanus* in 1961, restricted to Sulawesi, northern Borneo, Malaysia, and northern Thailand, has it been possible to suggest firmly that the cupule is derived from a 3-lobed extension of the pedicel below each flower, which has been variously fused around single flowers or groups of flowers. It is possible that the cupule provides a link with the pteridosperm ancestors of the flowering plants. The tremendous diversity of scales and spines on the cupules appears to be derived from branched spines[iii].

Classification The family has been placed without controversy in the Fagales with the Betulaceae, but recently a consensus has developed that this order also includes the Casuarinaceae, Juglandaceae, Myricaceae, Nothofagaceae, Rhoipteleaceae, and the recently described Ticoden-

FAGACEAE. 1 *Quercus ilex* leafy shoot with male flowers clustered in pendulous catkins and female flowers at the bases (x⅔). 2 *Fagus orientalis* leafy shoot with slender-stalked globose heads of male flowers (x⅔). 3 *Trigonobalanus verticillata* fruits (x2). 4 *Quercus robur* leafy shoot showing fruits (nuts) surrounded at base of the familiar cup or cupule (x⅔). 5 *Castanea sativa* young fruit with closed spiny cupule forming at the base of remains of the male catkin (x⅔). 6 *Quercus robur* leafless tree. 7 *Fagus sylvatica* leafless tree.

ovary is inferior, with 3 or 6 styles and locules, and 2 ovules in each locule. The fruits are single-seeded nuts, in groups of 1–3(–15), surrounded or enclosed by an often hardened cupule; seed 1, filling the nut, endosperm not

draceae[iv] (APG II). There is also agreement that *Nothofagus,* the southern beeches, are a separate family in their own right, based on a cladistic analysis of morphology, but now backed up by molecular studies. The order constitutes the monophyletic Higher Hamamelididae clade of the subclass Rosidae 5. With *Nothofagus* removed, *Fagus* is sister to the remainder of the family and makes up the Fagoideae. Subfamily Castaneoideae holds together reasonably well with *Castanea, Lithocarpus, Castanopsis,* and *Chrysolepis.* The problem occurs with the genus *Trigonobalanus,* where *Colombobalanus* and *Formanodendron* may be separated and which at present are best placed with *Quercus* (including *Cyclobalanopsis*) in the Quercoideae:

SUBFAM. FAGOIDEAE Inflorescence a 1- to many-flowered axillary cluster; contains only *Fagus,* (male inflorescences, long-stalked, many-flowered; styles long).

SUBFAM. CASTANEOIDEAE Catkinlike inflorescences, flowers usually with 12 stamens and dorsifixed anthers about 0.25 mm long; contains *Chrysolepis* (cupule divided into free valves), *Castanea* (cupule valves joined when young; styles 6 or more; leaves deciduous), *Castanopsis* (cupule valves joined when young; styles 3; leaves evergreen), and *Lithocarpus* (cupule without valves).

SUBFAM. QUERCOIDEAE Catkinlike inflorescences, flowers usually with 6 stamens and more or less basifixed anthers 0.5–1 mm long; contains *Quercus* (female flowers borne singly in the inflorescence; fruit round in transverse section, cupule not lobed) and *Trigonobalanus* (female flowers in clusters of 3, sometimes up to 7; fruit 3-angled; cupule lobed).

A world checklist of the Fagales[vi] lists 7 genera and 1,013 spp. for the Fagaceae.

Economic uses The Fagaceae is the source of some of the world's most important hardwood timbers, the most notable being oak (particularly the North American white oaks), beech, and chestnut. Together with clearance for agriculture, this has resulted in the destruction of large areas dominated by the family. Although the timber is of good quality, the tropical members of *Castanopsis* and *Lithocarpus* have as yet been little exploited. Taken as a whole the timber of the family exhibits a wide range of properties and thus many uses, from floorboards and furniture to charcoal and whisky barrels. Commercial cork is derived from the bark of the Mediterranean cork oak (*Quercus suber*), and in southeastern Europe and Asia Minor the oak galls were an important source of tannin. Many species of chestnut, but principally the Sweet Chestnut (*Castanea sativa*) of southern Europe, are grown for their large edible nuts, from which are made purees, stuffings, stews, and the French delicacy *marrons glacés.* The nuts of beech (beech-mast) are rich in oil (46%) and in many regions constitute an important food for pigs, as also the acorns of oaks.

The form and rich autumnal coloring in many deciduous species results in many Fagaceae being grown as ornamentals in parks and larger gardens, particularly oaks, chestnuts, and beech The only American species of *Castanopsis* (*C. crysophylla*) and *Lithocarpus* (*L. densiflorus*) may be cultivated in warmer regions.   SLJ

i Van Steenis, C. G. G. J. *Nothofagus,* key genus of plant geography, in time and space, living and fossil, ecology and phylogeny. *Blumea* 9: 65–98 (1971).
ii Fey, B. S. & Endress, P. K. Development and morphological interpretation of the cupule in Fagaceae. *Flora* 173: 451–468 (1983).
iii Nixon, K. C. & Crepet, W. L. *Triganobalanus* (Fagaceae): taxonomic status and phylogenetic relationships. *American J. Bot.* 76: 828–841 (1989).
iv Nixon, K. C. Fagaceae. Pp. 156–158. In: Kubitzki, K., Rohwer, J. G., & Bittrich, V. (eds), *The Families and Genera of Vascular Plants. II. Flowering Plants. Dicotyledons: Magnoliid, Hamamelid and Caryophyllid Families.* Berlin, Springer-Verlag (1993).
v Nixon, K. C. Origins of Fagaceae. In: Crane, P. R. & Blackmore, S. (eds), *Evolution, Systematics and Fossil History of the Hamamelididae.* Vol. 1. *Introduction and lower Hamamelididae.* Systematics Association. Special vol. no. 40B, 2: 23–43. Oxford, Clarendon Press (1989).
vi Govaerts, R. & Frodin, D. G. *World Checklist and Bibliography of Fagales* (Betulaceae, Corylaceae, Fagaceae and Ticodendraceae). Richmond, Royal Botanic Gardens, Kew (1998).

FLACOURTIACEAE

A fairly large family of trees and shrubs, often recognizable by their regular flowers with many conspicuous stamens.

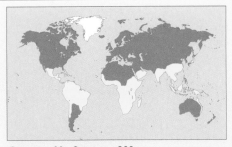

Genera c. 80 **Species** c. 900
Economic uses Some timber, seeds used for skin diseases, some ornamentals cultivated

Distribution Widespread in tropics and subtropics, absent from Europe and the Mediterranean, 1 sp. just reaching the USA in southern Texas, 2 spp. reaching Japan, several genera in Australia but confined to forests of Queensland.

Description Shrubs and trees, sometimes with axillary spines, sometimes dioecious. The leaves are simple, usually alternate, entire to crenate or serrate, the teeth sometimes glandular. The inflorescence is a subterminal or usually axillary raceme or cyme, or occasionally capitulum, or reduced to fascicles, or the flowers solitary in leaf axils. The flowers are bisexual or unisexual, actinomorphic, hypogynous or perigynous, the perianth sometimes spirally arranged and undifferentiated. The sepals are usually 4–7 sometimes appear to be up to 15, free or fused at the base. The petals are 3–8(–15), free and often inserted on a receptacular disk, or sometimes absent. The stamens are usually many but sometimes as few as 4–5, sometimes in bundles, sometimes the outer series sterile. The ovary is superior or occasionally semi-inferior, of 2–10 united carpels, usually with 1 locule with parietal placentation, but sometimes incompletely or completely divided into more than 1 locule by intrusion of the placentas. There are 2 to many ovules on each placenta. The styles are short and as many as the carpels. The fruit is usually a berry, sometimes a capsule or drupe. The seeds are 1 to many, often arillate.

Classification The family has generally been recognized as being referable to the Parietales, alias Violales, and falls within the Malpighiales of APG II. However, recent molecular analysis[i] has found the Salicaceae to nest within the family, leading the authors to sink the 2 families together, and unfortunately the name *Salicaceae* is the earlier name that has to be used. At the same time the traditional Flacourtiaceae has divided in the cladogram into two parts, with the other part uniting with the Achariaceae. In some recent

FLACOURTIACEAE. 1 *Oncoba spinosa* (a) flowering shoot armed with axillary spines (x⅔); (b) cross section of fruit (x⅔); (c) entire fruit with remains of flower below (x⅔).

FLACOURTIACEAE. I *Idesia polycarpa* (a) pendulous inflorescence (x⅔); (b) cross section of I-locular ovary with ovules on 5 large parietal placentas (x2); (c) vertical section of female flower showing branches style (x2); (d) male flower with numerous short stamens (x2); (e) ripe fruit (x⅔). 2 *Azara microphylla* (a) flowering shoot (x⅔); (b) vertical section of flower with single style and large stamens (x9); (c) part of inflorescence (x3); (d) cross section of I-locular ovary (x18).

literature, therefore, the name Flacourtiaceae has been dropped from usage and the genera previously referred there have been referred instead to the Salicaceae and Achariaceae respectively. This classification is not followed here. The Achariaceae forms a well-distinguished family with herbaceous habit, exstipulate leaves, and strongly gamopetalous flowers. The Scyphostegiaceae, which has also been sunk into former Flacourtiaceae, appears even more distinct, and is also here treated as a separate family. Once these 2 groups are maintained as separate families, it is easy to argue for the Salicaceae also to be maintained distinct on account of their catkinate inflorescences, lack of a perianth, and plumose seeds. The question then arises as to whether the traditional family should be split into two based solely on the molecular evidence, when a sample of only 30 genera out of a total of c. 80 was taken. The basal genus of the new Salicaceae in the cladogram, *Casearia*, is a large genus of 180 spp., only one of which was included in the molecular study. This is the only genus of the 13 referred to the same tribe that was included in the study. As noted here under Achariaceae and Salicaceae, sinking of these families does not get rid of paraphyly and merely transfers it to a different rank. Examination of the characters given in the paper for the several tribes shows that there is no character that separates the traditional family into two groups. It seems premature to make major taxonomic and nomenclatural changes that are not supported by morphological evidence.

Economic uses Some trees provide timber, especially *Gossypiospermum praecox* (West Indian Boxwood). Seeds of *Hydnocarpus* and other genera provide chaulmoogra oil, which is used to treat skin diseases such as leprosy. Some shrubs and trees are cultivated as ornamentals, especially *Idesia* and *Azara*. RKB

i Chase, M. W. *et al*. When in doubt, put it in *Flacourtiaceae*: a molecular phylogenetic analysis based on plastid *rbc*L DNA sequences. *Kew Bull.* 57: 141–181 (2002).

FOUQUIERIACEAE

OCOTILLO AND BOOJUM

The Fouquieriaceae comprises just one genus of spiny succulents endemic to the southwestern USA, where *Fouquieria splendens* (Ocotillo), is a characteristic shrub of the Mojave and Colorado deserts and Mexico, showing superficial resemblance to Didiereaceae of Madagascar. The 11 spp. are all now included in *Fouquieria*, although the Boojum has often been placed in the segregate genus *Idria*[i]. They are tall, columnar or branching shrubs or small trees with succulent or woody

Genera I Species II
Economic uses Of little importance

trunks bearing lateral branches, with numerous spines and small simple leaves. The leaves are succulent and borne in groups or singly and are without stipules. The inflorescence is a terminal or axillary spike, raceme, or panicle. The showy flowers are yellow (subgen. *Idria*) or red (subgen. *Fouquieria*) and actinomorphic, bisexual, with 5 imbricate sepals, 5 hypogynous petals fused into a tube at the base, and 10 or more hypogynous stamens in 1 or more series, with free or slightly fused filaments. The ovary is superior comprising 3 fused carpels and has a single locule with axile placentation below, but with septiform parietal placentation above, dividing it incompletely into 3 locules. The style is branched (subgen. *Fouquieria* and subgen. *Bronnia*) or entire (subgen. *Idria*). The fruit is a capsule with oblong, winged, and compressed seeds containing thin endosperm and a straight embryo with thick, flat cotyledons.

Three subgenera[ii] are recognized in *Fouquieria*: *Fouquieria* (8 spp.), *Bronnia* (2 spp.), and *Idria* (1 sp.). They are supported by molecular evidence, but the relationships among them are equivocal[iii]. The family is placed sister to Polemoniaceae in the Ericales based on molecular evidence[iv]. Traditionally its placement has been very variable including in Solanales (Thorne 1983) Fouquieriales (Dahlgren 1983) and Violales (Cronquist 1988).

The family has limited importance as ornamental curiosities, although trade in many species is restricted by CITES due to the small natural populations. AC

i Kubitzki, K. Fouquieriaceae. In: Kubitzki, K., (ed.), *The Families and Genera of Vascular Plants. VI. Flowering Plants. Dicotyledons: Celastrales, Oxalidales, Rosales, Cornales, Ericales*. Berlin, Springer-Verlag (2004).
ii Henrickson, J. A taxonomic revision of Fouquieriaceae. *Aliso* 7: 439–537 (1972).
iii Schultheis, L. M. & Baldwin, B. G. Molecular phylogenetics of Fouquieriaceae: evidence from nuclear rDNA ITS studies. *American J. Bot.* 86: 578–589 (1999).
iv Schönenberger, J., Anderberg, A. A., & Sytsma, K. J. Molecular phylogenetics and patterns of floral evolution in the Ericales. *Int. J. Pl. Sci.* 166: 265–288 (2005).

FRANKENIACEAE

Shrubs and herbs of saline, gypseous, or calcareous substrates in Mediterranean-type climate regions around the world, recognized by small, inrolled, heathlike leaves.

Distribution *Frankenia* is most diverse in Mediterranean-type climate regions but is distributed throughout the warm-temperate and subtropical regions. The family is well represented in arid and maritime environments and shows pronounced salt and gypsum tolerance.

Description Sprawling or clump-forming shrubs or cushionlike subshrubs, rarely with a caudex, or rarely annual. Stems with salt glands. The leaves are opposite and decussate, simple and entire, without stipules, united by a

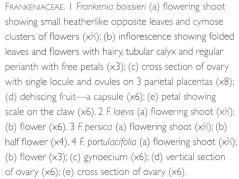

FRANKENIACEAE. 1 *Frankenia boissieri* (a) flowering shoot showing small heatherlike opposite leaves and cymose clusters of flowers (x⅔); (b) inflorescence showing folded leaves and flowers with hairy, tubular calyx and regular perianth with free petals (x3); (c) cross section of ovary with single locule and ovules on 3 parietal placentas (x8); (d) dehiscing fruit—a capsule (x6); (e) petal showing scale on the claw (x6). 2 *F. laevis* (a) flowering shoot (x⅔); (b) flower (x6). 3 *F. persica* (a) flowering shoot (x⅔); (b) half flower (x4). 4 *F. portulacifolia* (a) flowering shoot (x⅔); (b) flower (x3); (c) gynoecium (x6); (d) vertical section of ovary (x6); (e) cross section of ovary (x6).

sheathing membrane around stem, often inrolled at margin, the petiole short or absent. The inflorescence is terminal or axillary with solitary to many flowers in a dichasial cyme, with bracts 2 free, or 4 fused at base. The flow-

ers are regular and hermaphrodite, or unisexual (*F. triandra* is gynodieocious). The calyx is of 4–5(–6) fused sepals, tubular, campanulate or urceolate. The petals are equal in number to sepals, free, with a long claw in the calyx tube but expanded and spreading above, and each with a scale at the base of this expanded limb, which is continued down the sides of the claw. Stamens 3–6(–24 in *F. persica*) in 2 whorls, reduced to staminodes in pistillate flowers (*F. triandra*), sometimes united at the base of the ovary, sometimes free. The ovary is superior, the carpels (1–)2–4 fused, with a single locule containing several parietal placentas. The ovules are few to many, anatropous, ascending or with a long recurved stalk. Style simple with a usually 2- or 3-lobed stigma. The fruit is a longitudinally dehiscent capsule surrounded by a persistent calyx. The seeds are numerous, except *F. triandra*, which has a single placenta and 1 seed.

Classification The family is allied to Tamaricaceae, differing by the presence of opposite (decussate) leaves in Frankeniaceae, not alternate as in the Tamaricaceae. The family is remarkable for the fairly constant number of recognized species, but the highly variable number of genera ranging from 4[i] through 2[ii] to only one[iii] in recent years. DNA sequence evidence places the family in the non-core Caryophyllales[iv], sister to Tamaricaceae[v]. When split into 4 genera: *Frankenia* (73 spp.) is widespread but *Anthobryum* (4 spp.) and *Niederleinia* (3 spp.) are restricted to South America. The monotypic *Hypericopsis* occurs in southern Iran.

Economic uses Some species are occasionally grown as ornamentals. The largest member of the family is a relatively spectacular shrub *Frankenia portulacifolia* (St. Helena Tea) some 60 cm high, native to St. Helena. *Frankenia ericifolia* is used locally in the Macronesian islands as a fish poison, and *F. berteroana*, a small shrub in Chile, is burnt and the ash used as a source of salt. AC

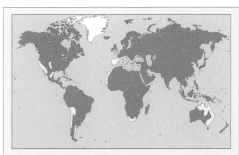

Genera 1–4   **Species** 80
Economic uses Limited use as ornamental oddities and local use of *Frankenia* spp. for fish poison

[i] Watson, L. & Dallwitz, M. J. *The Families of Flowering Plants: Descriptions, Illustrations, Identification, Information retrieval.* (1992 onwards) Version: 13th January 2005. http://delta-intkey.com
[ii] Kubitzki, K. Frankeniaceae. In: Kubitzki, K. & Bayer, C. (eds), *The Families and Genera of Vascular Plants. V. Flowering Plants. Dicotyledons: Malvales, Capparales and Non-betalain Caryophyllales.* Berlin, Springer-Verlag (2002).
[iii] Olson, M .E., Gaskin, J. F., & Ghahremani-nejad, F. Stem anatomy is congruent with molecular phylogenies placing *Hypericopsis persica* in *Frankenia* (Frankeniaceae): Comments on vasicentric tracheids. *Taxon* 52: 525–532. (2003).
[iv] Cuénoud, P. *et al.* Molecular phylogenetics of Caryophyllales based on nuclear 18S and plastid *rbc*L, *atp*B and *mat*K DNA sequences. *American J. Bot.* 89: 132–144 (2002).
[v] Gaskin, J. F., Ghahremani-nejad, F., Zhang, D.-Y., & Londo, J. P. A systematic overview of Frankeniaceae and Tamaricaceae from nuclear rDNA and plastid sequence data. *Ann. Missouri Bot. Gard.* 91: 401–409 (2004).

GARRYACEAE

SILK TASSEL FAMILY

The family comprises 1 genus of evergreen dioecious shrubs and trees often with catkinate inflorescences.

Distribution Western parts of North and Central America, from Washington State to Panama and eastward to Texas and the Caribbean, from coastal chaparral to mountain slopes at 4,000 m. Subgen. *Garrya* extends from Washington State to Baja California, with 1 aberrant species isolated in Guatemala, while subgen. *Fadyenia* is in Central America (especially Mexico) and the Caribbean.

Description Dioecious evergreen shrubs and trees up to 13 m. The leaves are opposite, simple, usually coriaceous and entire or nearly so, subsessile to petiolate, with the petioles connate in pairs. The inflorescences are terminal or axillary, linear, sometimes branching, sometimes (e.g., *G. elliptica*) catkinlike and pendent, those of male plants up to 30 cm long, those of female plants shorter and broader. The inflorescences have opposing bracts connate at each node, those of the male plants forming a conspicuous cup-shaped involucre surrounding the small flowers in their axils, while those of the female plants are less conspicuously connate, and each bract has 1–3 subsessile or pedicellate flowers in its axil. The male flowers are regular with a single perianth whorl of 4 tepals, with 4 stamens alternating with them. The female flowers are essentially inferior but the perianth is reduced and often merely a pair of teeth beneath the 2(3) free terminal styles. The ovary is 2-carpellate or rarely 3-carpellate, with a single locule bearing 2(3) pendulous ovules from a subapical parietal placenta. The fruit is a thin-walled globose 2-seeded berry, which becomes dry at maturity with its thin wall breaking up irregularly to release the seeds.

Classification *Garrya* was placed in the Cornaceae and had been thought to have particular affinities with *Aucuba*, likewise placed in Cornaceae. With the successive trimming of Cornaceae as a result of anatomical, palynological, and chemical investigation, *Garrya* has become widely referred to its own family. Recent molecular evidence supports this and confirms the suspected affinity with Aucubaceae (see the latter for references) and remoteness from Cornaceae. These 2 families and Eucommiaceae form a group toward the base of the Asterid line. The genus is divided into 2 subgenera on the basis largely of inflorescence characters[i]. Subgen. *Garrya* has catkinlike inflorescences, with the connate bracts modified into cups and never leaflike. Subgen. *Fadyenia* has inflorescences branching at the base and those of the female ones usually leaflike, so looking much less catkinlike.

Economic uses A number of species are cultivated as architectural plants or botanical curiosities. The pendent inflorescences, up to 30 cm long, of male plants of *G. elliptica* are often conspicuous during winter periods in the northern hemisphere. The bark of some species is used in traditional medicine. RKB

[i] Dahling, G. V. Systematics and evolution of *Garrya*. *Contrib. Gray Herb.* 209: 1–104 (1978).

GEISSOLOMATACEAE

The family comprises a single species, *Geissoloma marginatum*, a bushy shrub up to 1 m high growing on the Langeberg Mountains between Swellendam and Riversdale, 200 km east of Cape Town. The leaves are opposite and decussate, simple, small (usually 1–3 × 0.7–2.5 cm), entire and with the margins slightly recurved beneath, ovate to suborbicular, stiffly coriaceous, subsessile, with small stipules, and with leaf bases decurrent down the stem. The inflorescence is a solitary terminal flower on lateral short shoots that bear 3 pairs of leaflike bracts, often becoming petaloid and pink. The flowers are bisexual, actinomorphic, 4-merous, with a single perianth whorl. The 4 tepals are basally connate, pink and showy, ovate and acute. The stamens are 8 in 2 whorls of 4, attached to the base of the perianth tube. The ovary is superior and consists of 4 carpels that are fused centrally and form 4 narrow vertical wings running above into the 4 styles that are twisted in a spiral. The locules are 4, each with 2 collateral pendulous ovules. The fruit is a loculicidal capsule enclosed in the persistent perianth, with 4 seeds.

Genera 1 Species 1
Economic uses None

The family was first considered part of another South African family, Penaeaceae, in the Myrtales. This affinity has seemed increasingly untenable[i] and affinity with Celastraceae has been suggested. After analysis of floral morphology and molecular data, Geissolomataceae has been placed in the Crossosomatales with particular association with the Afro-Madagascan Aphloiaceae, Ixerbaceae from New Zealand, and Strasburgeriaceae from New Caledonia[ii]. RKB

[i] Dahlgren, R. Structures and relationships of families endemic to or centered on southern Africa: Geissolomataceae. In: Goldblatt, P. & Lowry, P. P., *Modern Systematic Studies in African Botany* (*Proc. 11th Plenary AETFAT meeting*). Pp. 16–21. St. Louis (1988).
[ii] Matthews, M. L. & Endress, P. K. Comparative floral structure and systematics in Crossosomatales (Crossosomataceae, Stachyuraceae, Staphyleaceae, Aphloiaceae, Geissolomataceae, Ixerbaceae, Strasburgeriaceae). *Bot J. Linn. Soc.* 147: 1–46 (2005).

GELSEMIACEAE

JESSAMINE FAMILY

Shrubs and twining climbers with stipulate, opposite, simple leaves and regular, trumpet-shaped, white, pink, or yellow corollas.

Distribution *Gelsemium* has 1 sp. from Assam to southeastern China, Borneo, and Sumatra, and 2 spp. from southern and southeastern USA to Guatemala. *Mostuea* has 7 spp. from tropical Africa and Madagascar and 2 in the Guianas and Brazil. Most species occur in rain forests, the more widespread species also extending into woodland or occasionally savannah, from coasts to altitudes of 2,000 m.

Description *Gelsemium* is usually a twining climber to several meters, often woody in lower parts, while *Mostuea* is a low to vigorous shrub up to 7 m, which may become scandent. The leaves are opposite and with interpetiolar stipules or at least raised lines representing them, and shortly petiolate; the lamina is simple and entire, or almost so, and usually narrowly lanceolate to broadly ovate, with a long acute apex, but in some *Mostuea* spp. it may be small and suborbicular. The inflorescence consists of 1- to few-flowered axillary cymes, but these may be aggregated into large terminal panicles (especially the Asiatic *G. elegans*). The flowers are actinomorphic or very nearly so, 5-merous, and usually heterostylous[i]. The sepals are small, sometimes fused at the base. The corolla is trumpet-shaped and rotate with 5 suborbicular lobes, bright yellow in *Gelsemium* and white to pinkish or orange in *Mostuea*. The stamens are attached near the base of the corolla tube, and the anthers held near the mouth of the corolla. The ovary is superior, 2-carpellate, 2-locular, with many ovules on the axile placentas, and the stigma is twice dichotomously branched. The fruit is a capsule, ovoid in *Gelsemium*, and laterally compressed in the American species, 2-lobed and much flattened in *Mostuea*.

Classification The gamopetalous regular flowers have usually placed the 2 genera in Loganiaceae[ii], but in their own tribe. Molecular evidence has shown that they are somewhat removed from the main Loganiaceae and perhaps most closely related to Apocynaceae[iii]. Family recognition now seems widely accepted[iv].

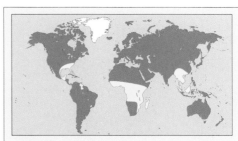

Genera 2 Species 12
Economic uses Ornamental climbers; some local medicinal uses and stimulants, poisonous in excess

Economic uses *Gelsemium* has been used in Asia and America for medicinal cures of various ailments but is poisonous in excess. *Mostuea* has been used as a stimulant, keeping party-goers active in Africa, and also for curing stomachaches and vomiting. The American *Gelsemium* spp. make good climbing ornamentals, with masses of yellow flowers, often used to cover trellises. The Old Government Building in Canberra, Australia, has a display over 200 m long in front of it.   RKB

[i] Ornduff, R. The systematics and breeding system of *Gelsemium* (Loganiaceae). *J. Arnold Arbor.* 51: 1–17 (1970).
[ii] Leeuwenberg, A. J. M. The Loganiaceae of Africa. 2. A revision of *Mostuea* Didr. *Meded. Landb. Wageningen* 61 (4): 1–31 (1961).
[iii] Backlund, M., Oxelman, B., & Bremer, B. Phylogenetic relationships within the Gentianales based on *ndh*F and *rbc*L sequences, with particular reference to the Loganiaceae. *American J. Bot.* 87: 1029–1043 (2000).
[iv] Struwe, L. & Albert, V. A. Gelsemiaceae. Pp. 164–166. In: Smith, N. *et al.* (eds), *Flowering Plants of the Neotropics*. Princeton, Princeton University Press (2004).

GENTIANACEAE
GENTIAN FAMILY

Gentianaceae has commonly been considered to comprise only annual to perennial herbs with a few shrubs, but recent work has resulted in the inclusion of some massive tropical trees and woody lianes, formerly included in Loganiaceae. It is mostly characterized by opposite entire leaves and regular gamopetalous flowers with a superior ovary. A peculiarity of the family is that plants are usually completely glabrous.

Distribution Cosmopolitan, from arctic tundra to tropical forests, but perhaps most often associated with open grassy places, such as chalk grassland and alpine meadows.

Description Mostly annual to perennial herbs (twining in *Crawfurdia* and *Tripterospermum*), but in tropical genera occasionally woody and sometimes (especially *Anthocleista* and *Fagraea*) large trees up to 35 m or woody lianes. Saprophytes without chlorophyll are rare but have evolved separately several times in different tribes (*Voyria*, *Voyriella*, *Sebaea*, *Bartonia*,

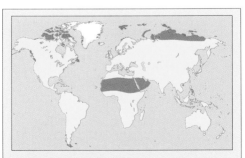

Genera 87   Species c. 1,650
Economic uses Used in traditional medicines worldwide; many cultivated ornamentals

GENTIANACEAE. 1 *Chironia purpurascens* (a) flowering shoot (x⅔); (b) corolla opened out showing epipetalous stamens with coiled anthers (x1); (c) cross section of ovary with 2 parietal placentas (x3). 2 *Voyria primuloides* (a) habit (x⅔); (b) flower from above (x⅔); (c) half flower with epipetalous stamens and globose stigma (x3). 3 *Gentiana depressa* (a) habit (x⅔); (b) ovary with part of wall cut away (x3); (c) corolla opened out (x1). 4 *Sabatia campestris* (a) flowering shoot (x⅔); (b) section of ovary (x2); (c) corolla opened out (x1).

Obolaria, *Cotylanthera*). The leaves are opposite or rarely (*Saccifolium*, some *Swertia*, *Voyriella*) alternate or whorled, sometimes perfoliate (*Blackstonia*), usually with raised interpetiolar lines at the nodes. The lamina nearly always has an entire margin, and is rarely (*Saccifolium* from Guyana Highlands) saccate. The inflorescence is a terminal or axillary cyme or rarely raceme, spike, or head, and flowers are rarely solitary. The flowers are bisexual except in *Veratrilla*, hypogynous, 4- or 5-merous, or rarely 3- or 6- to 16-merous, nearly always actinomorphic, occasionally heterostylous. The sepals are fused at the base, and the calyx may be zygomorphic in some *Exacum* species. The corolla is tubular to campanulate or funnel-shaped, may be slightly zygomorphic in *Chelonanthus*, *Symbolanthus*, and *Macrocarpaea*, and may have 4 nectariferous spurs on the outside (*Halenia*). The stamens are the same in number as the corolla lobes and inserted on the corolla tube or rarely between the corolla lobes, and may be rarely asymmetrical. The anthers open by terminal pores in *Exacum*. The ovary is superior, 2-carpellate, 1-locular to 2-locular, with a variety of placentation types. The fruit is a dry capsule, or a berry in the herbaceous *Chironia* and *Tripterospermum* and the woody *Anthocleista*, *Fagraea*, and *Potalia*.

Classification The family is related to other families that are characterized by opposite entire leaves and actinomorphic gamopetalous flowers, notably the Loganiaceae, Apocynaceae, Rubiaceae, and Gelsemiaceae. The comprehensive monograph of Struwe & Albert and others[i] has presented a much needed new classification of the family, and recognizes 6 tribes and some further subtribes. Most of these groups show wide character variation, and easy morphological diagnosis of them is apparently not possible.

Saccifolieae is a somewhat heterogeneous tribe of 5 genera with 16–20 spp. from tropical America, mostly from the Guiana Highlands. It is basal to other tribes. *Saccifolium*, with alternate extraordinary saccate leaves, has been treated as a separate family (Saccifoliaceae).

Exaceae comprises 6 genera and up to 185 spp. in the Old World, 3 genera being confined to Madagascar but the 2 larger genera *Exacum* (65 spp.) and *Sebaea* (more than 100 spp.) are widespread from tropical Africa to Australasia.

Chironieae tribe is further divided into 3 subtribes: Chironiinae, with 12 genera and 106 spp., widespread, largest genus *Centaurium* (c. 50 spp.); Canscorinae, with 7 genera and 24 spp. in the Old World; and Coutoubeinae, with 5 genera and 29 spp. from South America, the largest genus being *Schultesia* (15 spp.).

Helieae consists of 22 genera and 184 spp. in tropical America, the only large genus being *Macrocarpaea* (c. 90 spp.) and all other genera having 10 or fewer species.

Potalieae is further divided into three subtribes: Faroinae, with 9 genera and 31 spp., mostly from tropical Africa, but rarely in South America and Malaysia, predominantly herbaceous and often tiny annuals, *Faroa* having 19 spp.; Lisianthinae, with 1 genus *Lisianthus* from Central America and West Indies varying from annual herb to small tree; and Potaliinae, with 3 genera with berry fruits, *Anthocleista* with 14 spp., making huge trees and lianes in Africa, *Fagraea* with up to 70 spp. similarly in tropical Asia, and *Potalia* with 9 spp. of shrubs and small trees in South America.

Gentianeae is further divided into 2 subtribes: Gentianinae, with 3 genera and c. 400 spp., widespread, mostly in *Gentiana* with 360; and Swertiinae, with 14 genera and c. 550 spp., including *Gentianella* with 250 spp. and *Halenia* with 80 spp. (of which 3 are in temp-erate Asia, 1 in North America, and 76 in alpine Central and South America).

Economic uses Many medicinal treatments are based on the unique bitter seco-iridoids and xanthones found in the family. *G. lutea* (Yellow Gentian) is the source of "gentianroot" of the Pharmacopeia and the main commercial source of the bitter tonic, gentian bitter, and used in aperitifs such as Gentiane and Suze. A number of genera are grown as ornamentals, including *Exacum* as a house plant. By far the most important commercially is *Gentiana*, many species of which are cultivated as rock garden or herbaceous border perennials.　　　RKB

[i] Struwe, L. & Albert, V. A. (eds), *Gentianaceae: Systematics and Natural History*. 652 pp. Cambridge, Cambridge University Press (2002).

GERANIACEAE. 1 *Geranium malviflorum* (a) shoot with compound leaves and inflorescences (x⅔); (b) vertical section of flower showing bilobed petals (x1⅓).
2 *G. sanguineum* fruit with persistent calyx and with 1 awn separating from the central axis to disperse a seed (x1⅓). 3 *Erodium romanum* (a) tip of leafy shoot with inflorescence and fruits (x⅔); (b) fruit before dehiscence (x1⅓). 4 *Sarcocaulon patersonii* (a) fleshy stem with thorns (remains of leaf stalks) and solitary flowers (x1); (b) flower with petals removed showing 5 pointed sepals and 15 stamens of 2 lengths and with fused filament bases (x3).

GERANIACEAE
GERANIUMS AND PELARGONIUMS

The Geraniaceae is a family of annual to perennial herbs or shrubs with usually colorful flowers, with most genera containing horticulturally valuable plants.

Distribution The Geraniaceae are widely distributed reaching from Arctic to sub-Antarctic regions but are commonest in temperate regions and largely absent from the wet tropics and much of Australasia. Of the major genera, *Geranium* is cosmopolitan, while *Pelargonium* has a center of diversity in the South African Cape, *Erodium* is most diverse in the Mediterranean region, and *Monsonia* is African in distribution. *Hypseocharis*, by contrast, is endemic to the Andes.

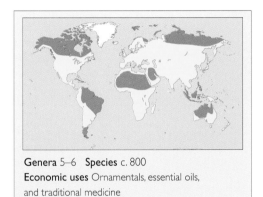

Genera 5–6 **Species** c. 800
Economic uses Ornamentals, essential oils, and traditional medicine

Description Shrubs, geophytes, and herbs, mostly perennial but some annual or biennial, usually with a pronounced taproot. Stems, when present, are sometimes woody or succulent. The leaves are alternate, with stipules and usually petiolate, the lamina simple or lobed to deeply divided and pinnate. The inflorescences are terminal or axillary, umbellate, cymose, or of a single flower. The flowers are bisexual, radially (*Monsonia, Hypseocharis*) or bilaterally symmetrical, strongly so in *Pelargonium*, 5-merous, bracteolate, with a usually brightly colored perianth of sepals and petals, the petals clawed, highly variable in form in *Pelargonium*. Nectaries are present (in a hypanthial tube in *Pelargonium*). The stamens are 5, 10, or 15, sometimes modified to staminodes (commonly in *Pelargonium*). The ovary is superior, of 5 fused carpels, developing into a distinctive schizocarpic fruit that separates into single-seeded mericarps, each ususally with a hygroscopic awn.

Classification The delimitation of Geraniaceae has varied, and its relationships remain uncertain. The core genera are *Geranium*, *Pelargonium*, *Erodium*, and *Monsonia* (optionally including *Sarcocaulon*). *Hypseocharis* seems best placed in this family but has several morphological anomalies. The genera are well defined on both morphological and molecular evidence[i]. Several genera have been segregated into other families.

Economic uses *Pelargonium* is of major economic importance in horticulture as a summer bedding plant and as an ornamental houseplant. *Pelargonium triste* was introduced to Europe from the Cape before 1631 and later introductions led to extensive hybridization in the eighteenth century[ii]. Oils from *Pelargonium* are used in perfume and as a flavoring. Many Geraniaceae have traditional medicinal use. AC

[i] Bakker, F. T., Culham, A., Hettiarachi, P., Toloumenidou, T., & Gibby, M. Phylogeny of *Pelargonium* (Geraniaceae) based on DNA sequences from three genomes. *Taxon* 53: 17–28 (2004).
[ii] Miller, D. *Pelargonium – A Gardener's Guide to the Species and Cultivars and Hybrids*. London, BT Batsford (1996).

GESNERIACEAE
AFRICAN VIOLETS AND GLOXINIAS

Mostly tropical herbs and shrubs, sometimes lianes or trees, often epiphytic or epilithic, related to Scrophulariaceae.

Distribution The family is pantropical, with extensions to southern temperate areas and with 6 spp. isolated in the mountains of southern Europe (Pyrenees, Balkans). It is much less represented in the Afro–Madagascan region than in the neotropics and Asia. Species grow from sea level to high alpine places in the Himalayas and Andes, most preferring a moist situation in forests, where some 25% of species are epiphytic or on wet rocks. Distributions of the major genera correspond closely with different continents, and as a broad generalization, genera that occur in 1 region tend to be closely related to each other. Thus, subfamily GESNERIOIDEAE occurs in the New World, DIDYMOCARPOIDEAE in the Old World, and CORONANTHEROIDEAE in Australasia and temperate South America. The distributions of these groups may be related to the break-up of Gondwanaland[i].

GESNERIACEAE. 1 *Chrysothemis pulchella* shoot with opposite leaves and inflorescences (x⅔). 2 *Aeschynanthus microtrichus*, fruit—an elongate capsule (x⅔). 3 *Columnea crassifolia* shoot with alternate leaves and solitary, 2-lipped flower (x⅔).

Description The family varies in habit from herbs (rarely annual) to shrubs, lianes, or trees to 15 m, usually polycarpic, sometimes monocarpic. In the Old World, many genera comprise plants consisting only of 1 much enlarged cotyledon (sometimes more than 1 m long) and an inflorescence, with no normal foliage leaves. When leaves are present, they are opposite or sometimes whorled or alternate, sometimes forming a basal rosette, entire or variously toothed but rarely deeply divided, often anisophyllous, occasionally peltate, usually clothed with glandular or egandular hairs. The inflorescence is a specialized "pair-flowered" cyme found elsewhere only in a few genera of Scrophulariaceae[ii]. The 5 sepals are

Genera c. 150 **Species** c. 3,500
Economic uses Ornamentals for cultivation under glass in temperate regions, and medicinal uses especially as an antidote to snake bites

free or fused into a tube, sometimes 2-lipped, sometimes fleshy and colored. The corolla has a short or more often long tube (often varying greatly in 1 genus), more or less actinomorphic to strongly zygomorphic. The fertile stamens are 2 (some Old World genera) or most commonly 4 or sometimes (in regular flowers) 5. A nectary below the ovary is often well developed. The ovary consists of 2 fused carpels and may be superior or (in the New World) semi-inferior or inferior, usually 1-locular with parietal placentae, but may have an incomplete or sometimes complete septum dividing it into 2 locules with axile placenta; the ovules are many. Occasionally only 1 carpel is fertile. In the Old World, the fruit is usually a dry capsule and may be short, globose, and dehiscing in its upper part, sometimes circumscissile, or much elongate and podlike, with many small seeds. In the New World, the fruit is often fleshy and usually indehiscent.

The family shows many adaptations to specialized habitats in both vegetative and floral characters. Many epiphytic or epilithic species have the ability to dry out for short periods and revive themselves in wetter conditions. In the Old World, the cotyledons develop unequally, and in some genera the only adult vegetative structure is one much-enlarged cotyledon that lies flat against a tree-trunk or rock. Floral adaptations are associated with pollinating agents, involving often brightly colored corolla and sometimes calyx. In some species, the pollinator is attracted by extra-floral structures, such as colored hairs or leaves. Many New World genera have developed brightly colored fleshy fruits, which are adapted to seed dispersal by birds, bats, monkeys, or ants.

Classification Gesneriaceae falls in the complex surrounding the Scrophulariaceae, and constant characters to separate it from that family are not easy to find. For the most part, the 1-locular ovary with parietal placentation separates Gesneriaceae (and perhaps suggests a relationship with Martyniaceae), but some members of the family may have a divided ovary with more or less axile placentation. Nonetheless, the family is well defined with few genera of doubtful inclusion. It is usually divided into 3 subfamilies[iii], but the comprehensive

account in FGVP by Weber suggests that the *Epithema* group (Old World) might also be given subfamily rank. The definition of tribes within the subfamilies seems also to be in question at present.
SUBFAM. CORONANTHEROIDEAE
Comprises 9 genera and 20 spp., 13 spp. of which are in the 2 genera in New Caledonia, the remainder in monotypic

GESNERIACEAE. 4 *Ramonda myconi* basal rosette of leaves and inflorescences (x⅔). 5 *Gesneria cuneifolia* (a) basal rosette of leaves and solitary flowers (x⅔); (b) flower with part of calyx and corolla cut away to show 2 stamens with anthers cohering together (x2). 6 *Aeschynanthus pulcher* flowering shoot (x⅔). 7 *Streptocarpus caulescens* leafy shoot and inflorescences (x⅔). 8 *Aeschynanthus pulcher* half flower showing corolla tube constricted at the base, 4-lobed ovary crowned by a long style and stamens with curved filaments (x1).

genera from northeastern and southeastern Australia, Lord Howe Island, New Zealand, and Chile–Argentina. Mostly trees or epiphytic shrubs or climbers, differing from other subfamilies in having a nectary adnate to the ovary. SUBFAM. DIDYMOCARPOIDEAE Comprises 85 genera with more than 1,900 spp., from Africa (especially *Streptocarpus* with 140 spp. from East Africa to South Africa and Madagascar, and *Saintpaulia* up to 22 spp. from Kenya and Tanzania), Europe (*Ramonda, Haberlea, Jancaea*), Asia (especially *Cyrtandra* with c. 640 spp. extending through the Pacific Islands as far as Hawaii, *Aeschynanthus* 198 spp., *Henckelia* 180 spp., *Chirita* 140 spp., *Agalmyla* 90 spp., and *Paraboea* 90 spp.), and Australia (1 sp. each of *Boea* and *Cyrtandra*). Surprisingly, 1 sp. of the essentially Old World subfamily, *Rhynchoglossum azureum*, occurs from Mexico to Peru, but molecular analysis places it close to a southern Indian species of the same genus, and it may have been transported to the New World in recent times by migrations of early Polynesians. They have cotyledons of markedly different sizes, the larger one sometimes persisting to maturity as the only foliar organ. The ovary is always superior and, except in *Rhynchotechum* and *Cyrtandra*, the fruit is a dry capsule, often elongate and splitting into much twisted valves. The distinctive *Epithema* group (tribe Klugieae) are usually succulent and have a globose rather than elongate ovary with triangular rather than lamelliform placentas bearing spirally reticulate seeds.

SUBFAM. GESNERIOIDEAE Comprises 7 genera with more than 1,500 spp., confined to the tropical and subtropical New World, the larger genera being *Besleria* c. 170 spp., *Drymonia* 140 spp., *Dalbergaria* 90 spp., and *Columnea* 75 spp.. They have equal cotyledons, and the ovary is often semi-inferior or inferior and develops into a dry or fleshy capsule that is more or less globose and neither elongate nor splitting into twisted valves.

Economic uses The colorful flowers and comparative ease of cultivation have made many genera important in horticulture, but few are hardy in the temperate regions. The 3 European genera are occasionally grown as alpines on rock gardens. Species and hybrids of *Streptocarpus* and *Saintpaulia* (inappropriately called Cape Primrose and African Violet respectively) are extensively marketed as house plants. Many thousands of cultivars are known. In glasshouses in temperate regions, or outside in the tropics, *Sinningia* (the florists' "Gloxinia"), *Columnea, Kohleria, Episcia, Nematanthus, Aeschynanthus*, and other genera are extensively grown; special-interest groups of gesneriad growers thrive. The family also provides many medicinal plants[iv].   RKB

i Burtt, B. L. Climatic accommodation and phytogeography of the Gesneriaceae of the Old World. In: Mathew, P. & Sivadasan, M. (eds), *Diversity and Taxonomy of Tropical Flowering Plants*: 1–27 (1998).
ii Weber, A. Gesneriaceae. Pp. 63–158. In: Kubitzki, K. (ed.), *Families and Genera of Vascular Plants 7* (2004).
iii Burtt, B. L. & Wiehler, H. Classification of the family Gesneriaceae. *Gesneriana* 1: 1–4 (1995).
iv Wiehler, H. Medicinal gesneriads. *Gesneriana* 1: 98–120 (1995).

GISEKIACEAE

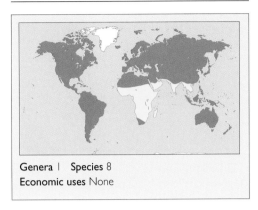

Genera | Species 8
Economic uses None

A monogeneric family of erect or prostrate, much-branched, fleshy herbs from Africa and southern Asia. *Gisekia pharnaceoides* is widespread, found in drier areas of Africa, Madagascar, and the Mascarines, tropical Asia from Arabia to Indo–China[i]. *G. pierrei* is found in China[ii], but other species are endemic to Africa, particularly Somalia. The leaves are opposite or pseudoverticillate, entire, exstipulate, petiole short. The flowers are actinomorphic, bisexual, 5-merous, in a dense, nearly sessile, axillary umbel-like cyme or a diffuse terminal cyme. The tepals are 5, ovate to lanceolate, green to white, or sometimes pink to red. The stamens are 5–20, free; the carpels are (3–)4 or 5(–15), joined only at the base, each with 1 ovule. The fruit is a cluster of mericarps, heterocarpic, thin-walled smooth or thicker-walled and ornamented with papillae, warts or spines or with prominent white wings[i].

Gisekia is in the core Caryophyllales[iii], but its relationships within that group are still equivocal. Opinion varies on the recognition of this family, with many authors placing it in Phytolaccaceae[iv,v], commenting on the peculiarity of the genus. Wood anatomy has been used to propose its recognition as a subfamily within Phytolaccaceae[vi]. In many African floras, it is placed in Aizoaceae[i] but others place it in Molluginaceae[ii]. The current balance of data supports treatment as a separate family.

Gisekia pharnaceoides is a weed of cultivated land, becoming established in the Americas.   AC

i Gilbert, M. G. A review of *Gisekia* (Gisekiaceae) *Kew Bull.* 48: 343–356 (1993).
ii Lu, D. & Hartmann, H. E. K. Molluginaceae. Pp. 437–439. In: *Flora of China*. Vol. 5. *Ulmaceae through Basellaceae*. Missouri Botanical Garden (2003).
iii Cuénoud, P. *et al.* Molecular phylogenetics of Caryophyllales Based on Nuclear 18S and Plastid *rbc*L, *atp*B and *mat*K DNA sequences. *American J. Bot.* 89: 132–144 (2002).
iv Smith, N. *et al.* (eds), *Flowering Plants of the Neotropics*. Princeton, Princeton University Press (2004).
v Brown, G. K. & Varadarajan, G. S. Studies in Caryophyllales I: Re-evaluation of classification of Phytolaccaceae *s.l. Syst. Bot.* 10: 49–63 (1985).
vi Carlquist, S. Wood anatomy of *Agdestis* (Caryophyllales): systematic position and nature of the successive cambia. *Aliso* 18: 35–43 (1999).

GLAUCIDIACEAE

Genera | Species |
Economic uses Rare in cultivation

Glaucidium palmatum is endemic to Japan, where it grows in Hokkaido and northern and central Honshu in the understory of deciduous woods. It is a perennial herb, with rhizomes producing at their apex persistent bud scales and a flowering shoot up to 50 cm high, usually bearing 3 leaves in the upper part and a single terminal flower, sometimes also with 1 or 2 long, petiolate basal leaves. The leaves are palmate and petiolate, up to 30 cm diameter, divided for one-third to two-thirds of their radii into (5–)7–11(–13) lobes that may be secondarily lobed, with serrate margins, or the sessile upper cauline leaf may be reniform and unlobed but unevenly serrate. The flowers are bisexual and actinomorphic, with a 1-seriate perianth. The tepals are 4, large and showy, up to 5.5 × 4.5 cm, 2 outer overlapping 2 inner, pale blue to white. The stamens are 350–500 spirally arranged and developing centrifugally. The carpels are normally 2, inserted opposite the outer tepals, with a short style and a stigma folded down each side. Each carpel has 15–20 ovules along its ventral margin. The carpels are at first erect and slightly connate at the base, but the receptacle beneath them elongates during maturation of the fruit to push each carpel into a more horizontal position with the dorsal suture uppermost, making a T- or Y-shaped fruit. The fruiting follicles are rectangular to diamond-shaped and dehisce along both margins to reveal a tightly packed row of flat and broadly winged overlapping seeds hanging from the dorsal suture.

Glaucidium is clearly referable to the Ranunculales and has been variously referred to Ranunculaceae or Berberidaceae (rarely also to Paeoniaceae) or has been associated with *Hydrastis*, which is here referred to Hydrastidaceae. In its palmate leaves and rhizomatous habit, with terminal solitary flowers, it resembles genera in a wide range of families in this order, e.g., *Calathodes* in Ranunculaceae, *Podophyllum* in Berberidaceae, *Hylomecon* in Papaveraceae, *Hydrastis* in Hydrastidaceae, and *Kingdonia* in Circaeasteraceae. It seems likely that this is merely a plesiomorphic lifeform for the group rather than an apomorphy. *Glaucidium* differs from Ranunculaceae in many anatomical, cytological, embryological, and chemical characters, and from Berberidaceae and Hydrastidaceae also in its pollen and many centrifugal stamens. In its 4 tepals and many stamens, and in some molecular analyses, it shows affinity with Papaveraceae. However, its unique ovary and T-shaped fruit marks it off from all other families. A range of morphological and molecular papers in the 1994 Bayreuth symposium on Ranunculiflorae[i] consistently placed *Glaucidium* and *Hydrastis* apart from Ranunculaceae and Berberidaceae, usually as separate families but united as Hydrastidaceae with the subfamily Glaucidiaceae in one case (*l.c.*, p. 105). It seems likely, in fact, that both genera are independently derived from a basal stock of Ranunculales, sharing a basic growth habit, and their floral morphology is so different that there seems to be no case for uniting them. Neither Glaucidiaceae nor Hydrastidaceae is recognized in APG II.　　RKB

[i] Jensen, U. & Kadereit, J. W. (eds.), Systematics and Evolution of the Ranunculiflorae. *Pl. Syst. Evol.* (Supplement 9) Berlin, Springer-Verlag (1995).

GLOBULARIACEAE

A small family of 2 genera of subshrubs with woody bases to low, slender shrubs, centered in the Mediterranean area and related to Scrophulariaceae.

GLOBULARIACEAE. 1 *Globularia trichosantha* (a) leafy shoot with an erect capitulate inflorescence (×⅔); (b) lower deeply three-lobed portion of corolla with 4 epipetalous stamens (×6); (c) upper petal (×6); (d) calyx opened out and gynoecium (×6). 2 *G. salicina* flowering shoot (×⅔). 3 *Poskea socotrana* (a) flowering shoot (×⅔); (b) corolla opened out showing 5 petal lobes, 4 epipetalous stamens, and ovary with a single style crowned by a forked stigma (×16).

Distribution *Globularia* occurs from Madeira, Canary Islands, and Cape Verde to the Mediterranean, the Alps, extending north to the Baltic Sea and east to the Transcaucasus and northern Iraq, with 1 sp. in northern and western Arabia and northern Somalia. *Poskea* has 2 spp. confined to Socotra and Somalia.

Genera 2
Species Up to 30
Economic uses
Few rock-garden ornamentals

Description Perennials, woody at least at the base, often with a rosette of leaves and erect flowering shoots, sometimes creeping and rooting, sometimes forming a caespitose subshrub or an erect slender shrub up to 1 m. Leaves are alternate, often differentiated into large spathulate lower leaves forming rosettes, and much smaller elliptic-lanceolate cauline leaves, but rosette leaves are often lacking. They are entire or minutely toothed, rarely (*G. spinosa*) strongly spinose on the margins, emarginate to spinulose at the tip. Flowers are in capitate inflorescences with an involucre, with the heads relatively large (to 1.5 cm) and solitary to much smaller (3 mm) and arranged in a slender leafless "raceme" (*G. sintenisii*), or in dense terminal spikes with no involucre (*Poskea*). The calyx is 5-lobed and deeply divided. The corolla is tubular, 5-lobed, and 2-lipped, usually blue with the 2 lips unequal and the lobes more or less linear in *Globularia*, but with the lips subequal and the lobes rounded in *Poskea*. Stamens are 4, inserted on the corolla tube. The ovary is derived from 2 carpels but the adaxial one does not develop and the mature ovary is 1-locular with 1 pendulous ovule. The 1-seeded fruit is indehiscent and remains within the persistent calyx.

Classification The family is evidently derived from the Scrophulariaceae stock[i] and has been considered close to *Selago* in Scrophulariaceae because of the few-seeded ovary in that genus. This may, however, be a case of parallel evolution, and it may be closer to the *Antirrhinum* group[ii]. The quite different ovary and fruit structure of Globulariaceae argue for keeping the family distinct, and the exact relationships to Scrophulariaceae are still unclear.

Economic uses Several species of *Globularia* are cultivated as rock-garden plants.　　RKB

[i] Wagenitz, G. Globulariaceae. Pp. 159–162. In: Kadereit, J. W. (ed.), *Families and Genera of Vascular Plants. VII. Flowering Plants. Dicotyledons: Lamiales.* Berlin, Springer-Verlag (2004).
[ii] Oxelman, B., Backlund, M., & Bremer, B. Relationships of the Buddlejaceae s.l. investigated using parsimony jackknife and branch support analysis of chloroplast *ndh*F and *rbc*L sequence data. *Syst. Bot.* 24: 164–182 (1999).

GOMORTEGACEAE

The Gomortegaceae comprises just one species, *Gomortega keule*, which is an aromatic tree native to Chile with simple, opposite, entire, exstipulate leaves. The flowers are smallish, actinomorphic, in terminal and axillary racemes. The perianth is sepaline, the tepals (5–)7(–9), free, more or less spirally arranged. The androecium consists of 5–10(–13) free stamens, with basal glandular appendages; staminodes sometimes present. The ovary is syncarpous and inferior, a condition that is unique in Laurales, with 2–6 spirally arranged carpels, with 2–3 locules. The fruits are fleshy inferior coenocarpous drupelets, giving the appearance of a syncarpous fruit[i]. The species is known from only 22 populations, growing on the coastal cordillera and is highly endangered.[ii] Within the Laurales the Gomortegaceae is most closely related to the Atherospermataceae (q.v.). VHH

Genera 1 Species 1 Economic uses Fruits reputedly used locally as a hallucinogen

[i] Doweld, A. B. Carpology and phermatology of Gomortega (Gomortegaceae): Systematic and evolutionary implications. *Acta Bot. Malacitana* 26: 19–37 (2001).
[ii] Villegas, P., Le Quesne, C., & Lusk, C. H. Estructura y dinámica de una población de *Gomortega keule* (Mol.) Baillon en un rodal antiguo de bosque valdiviano, cordillera de Nahuelbuta, Chile. *Gayana Bot.* 60: 107–113 (2003).

GOODENIACEAE

GOODENIAS, LECHENAULTIAS, AND FAN-FLOWERS

A mainly Australian family characterized by a cup-shaped indusium below the apex of the style, which acts as a pollen presenter, often superficially resembling the Lobelioid Campanulaceae in their corolla.

Distribution All genera and some 385 spp. occur in Australia, 200 of them confined to western Australia. *Scaevola* has the greatest number of species outside Australia, estimates

Genera 12 Species c. 420
Economic uses Some ornamentals

ranging from 25 in a genus of 96 spp.[i] to 40 in a genus of 130 spp.[ii], most of these in Southeast Asia and the Pacific. *S. socotraensis* is endemic to Socotra, and *S. wrightii* is endemic to Cuba. The only other genera known outside Australia are *Goodenia*, with 3 out of 179 Australian species extending to New Guinea and 1 sp. to Malaysia and China, while 1 is endemic to Java; *Lechenaultia* and *Velleia* each with 1 of numerous Australian species extending to New Guinea; and the monotypic *Selliera* from the coasts of southeastern Australia, New Zealand, and Chile.

Description Annual to perennial, glabrous or pubescent to tomentose herbs or shrubs, rarely scrambling or making small trees. The inflorescence is a cyme, thyrse, raceme, spike or subumbel, or the flowers may be solitary. The flowers are weakly to strongly zygomorphic, or rarely actinomorphic (*Brunonia*), bisexual, usually 5-merous, often showy. The sepals are (3)5 or sometimes obsolete. The corolla tube is often split longitudinally in 1 or 2 places on the upper side, the 5 lobes sometimes almost regularly radially arranged, or with 2 making an upper lip and 3 a lower, or all 5 making a lower lip, each lobe usually differentiated into a median part and pronounced lateral wings or auricles. The stamens are 5, free or attached to the base of the corolla tube, sometimes with the anthers connate into a tube round the style. The style has a cup-shaped indusium below the apex, which collects the pollen from the anthers (which dehisce in bud) and present it to visiting pollinating agents. The ovary is inferior to semi-inferior or (*Velleia*) superior, apparently of 2 carpels but perhaps derived from 4, with a short basal 2-locular zone, a middle 1-locular zone, and small upper 2-locular zone, with 1 to many ovules on axile or basal placentas. The fruit is a drupe, nut, or capsule, or in *Lechenaultia* may break transversely into 1-seeded woody segments.

Classification Like the Campanulaceae, the Goodeniaceae have long been considered close to the Asteraceae and are included in Asterales in recent classifications[iii],[iv]. The similarities to the Campanulaceae, and particularly the subfamily Lobelioideae, may be superficial, however, since the structure and vasculature of the 2 families is "so different as to preclude a very close relationship[i]." The genus *Brunonia* has often been placed in its own family Brunoniaceae characterized by its actinomorphic flowers borne in a capitate head, its seeds without endosperm, and its "rubiaceous" stomata, but it has the stylar indusium of the Goodeniaceae and is now included in that family (the separation of Brunoniaceae in the *Flora of Australia* was dictated by Cronquist having recognized the family).

Recent accounts have not recognized formal infrafamilial taxa in the family. Genera standing morphologically somewhat isolated are *Brunonia* and *Lechenaultia*[v]. Carolin has recognized a "*Dampiera* group" and a "*Goodenia* group."

Economic uses Many species are showy with brightly colored flowers and may be grown under glass or outside in warm regions, popular genera being *Goodenia*, *Dampiera*, and particularly *Lechenaultia*[vi]. RKB

[i] Carolin, R. C., Rajput, M. T. M., & Morrison, D. A. Brunoniaceae, Goodeniaceae. *Flora of Australia* 35. 351 pp. (1992).
[ii] Howarth, D. G., Gustafsson, M. H. G., Baum, D. A., & Motley, T. J. Phylogenetics of the genus *Scaevola* (Goodeniaceae): implication for dispersal patterns across the Pacific Basin and colonization of the Hawaiian islands. *American J. Bot.* 90: 915–923 (2003).
[iii] Gustafsson, M. H. G., Backlund, A., & Bremer, B. Phylogeny of the Asterales *sensu lato* based on rbcL sequences with particular reference to the Goodeniaceae. *Pl. Syst. Evol.* 199: 217–242 (1996).
[iv] Olmstead, R. G., Kim, K.-J., Jansen, R. K., & Wagstaff, S. J. The phylogeny of the Asteridae *sensu lato* based on chloroplast ndhF gene sequences. *Mol. Phylogen. Evol.* 16: 96–112 (2000).
[v] Gustafsson, M. H. G., Grafström, E., & Nilsson, S. Pollen morphology of the Goodeniaceae and comparison with related families. 36: 185–207 (1997).
[vi] Morrison, D. & George, A. S. The genus *Lechenaultia*. *Curtis Bot. Mag.* 21: 106–131 (2004).

GOUPIACEAE

A poorly known monogeneric family of evergreen trees, with its 2 spp. endemic to the Amazon-Guyanan region of tropical northeastern South America. The leaves are alternate, coriaceous, simple, entire or toothed, with long, caducous stipules. The flowers are actinomorphic, bisexual, small, in axillary umbellate inflorescences. The sepals are 5, connate; the petals are 5, long and subulate, induplicate-valvate. The stamens are 5, on the edge of the nectary disk, with short filaments. The ovary is superior, more or less enclosed by the disk, with 5 locules each containing several ovules; styles 5, free. The fruit is a globose berrylike drupe with 2–3 seeds.

Genera 1 Species 2 Economic uses Timber

The taxonomic and phylogenetic position of *Goupia* is not fully resolved. The genus has been included in the Celastraceae or as a separate family in the Celastrales, but molecular evidence indicated that it is unrelated to that family and is more closely related to the Euphorbiaceae[i],[ii],[iii], which is included, as is the Goupiaceae, in the Malpighiales in APG II. The wood of *Goupia glabra* is used as timber. VHH

[i] Simmons, M. P. & Hedin, J. P. Relationships and morphological character changes among genera of Celastraceae *sensu lato* (including Hippocrateaceae). *Ann. Missouri Bot. Gard.* 86: 723–757 (1999).
[ii] Savolainen, V., Spichiger, R., & Manen, J.-F. Polyphyletism of Celastrales deduced from a non-coding chloroplast DNA region. *Mol. Phyl. Evol.* 7: 145–157 (1997).
[iii] Simmons, M. P. *et al.* Phylogeny of the Celastraceae inferred from phytochrome B gene sequence and morphology. *American J. Bot.* 88: 313–325 (2000).

GRISELINIACEAE

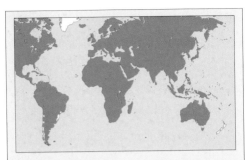

Genera 1   Species 7
Economic uses Some species are cultivated as ornamentals

The Griseliniaceae is a monotypic family of dioecious, small evergreen trees or shrubs, sometimes epiphytic, with simple, entire or dentate leaves, and the flowers in racemes or panicles. The single genus *Griselinia* contains 7 spp. that show a trans-Pacific disjunct distribution between southern South America (Argentina, Chile, and southeastern Brazil) and New Zealand. The flowers are actinomorphic, the staminate ones with 5 petals and 5 stamens, and the pistillate ones with 5 petals, sometimes absent, and 3, 2-locular carpels. The fruit is a fleshy single-seeded berry.

The placement of this family has varied. It has previously been included by several authors in the Cornaceae, also showing a possible relationship with *Garrya*, or treated as a separate monotypic family, showing relationships with the Hydrangeales. Evidence from morphology, anatomy, and floral vasculature suggested links also with the Araliales and Pittosporales[i,ii]. Recent molecular studies suggest its treatment as a separate family within the Apiales, near the genera *Melanophylla* (recognized here as the Melanophyllaceae), *Torricellia*, and *Araldidium* (recognized as the Torricelliaceae) and *Pennantia* (also placed in a separate family Pennantiaceae)[iii,iv] although the morphological support for this is not at all clear[v].   VHH

i Philipson, W. R. *Griselinia* Forst. fil. - Anomaly or link. *New Zealand J. Bot.* 5: 134–165 (1967).
ii Philipson, W. R. Ovular morphology and the classification of the dicotyledons. *Plant. Syst. Evol.* (Supplement 1): 123–140 (1977).
iii Chandler, G. T. & Plunkett, G. M. Evolution in Apiales: nuclear and chloroplast markers together in (almost) perfect harmony. *Bot. J. Linn. Soc.* 144: 123–147 (2004).
iv Plunkett, G. M., Chandler, G. T., Lowry II, P. P., Pinney, S. M., & Sprenkle, T. S. Recent advances in understanding Apiales and a revised classification. *South African J. Bot.* 70: 371–381(2004).
vPlunkett, G. M. Relationships of the order Apiales to subclass Asteridae: a reevaluation of morphological characters based on insights from molecular data. *Edinburgh J. Bot.* 58: 183–200 (2001).

GROSSULARIACEAE
CURRANTS, GOOSEBERRIES

Plants of northern temperate and Andean high altitudes, comprising the single genus *Ribes*.

Distribution The family occurs mainly in the northern hemisphere, reaching high latitudes and altitudes. In the southern hemisphere, the family occurs along the Andes (subgen. *Parilla* and *Berisia*) and south into Tierra Del Fuego (*Ribes magellanicum*).

Description Shrubs, sometimes prostrate or scrambling, with alternate leaves and usually with multicellular glandular hairs. The leaves are usually evergreen, sometimes deciduous, simple, often toothed, usually petiolate, often palmately lobed, stipules attached to petiole and usually winged or fimbriate. The inflorescence is a raceme with the flowers typically borne on a short lateral shoot. The flowers are actinomorphic, bisexual or unisexual (plants dioecious—subgen. *Parilla* and some *Berisia*), with a pronounced tubular to rotate hypanth-ium. The sepals are usually (4–)5, free, sometimes strongly reflexed at anthesis. The petals are as many as the sepals and usually shorter, free, sometimes minute, obovate. The stamens are 4–5, usually as many as the sepals, the filaments short, attached to top of hypanthial tube, free. The ovary is inferior, of 2 joined carpels, with 1 locule. The fruit is a berry, often covered with dense, glandular hairs.

Classification Grossulariaceae has often been treated as part of Saxifragaceae or, by later authors, as a closely related family in the Rosales[i]. APG II states its position is still uncertain, and one analysis places it sister to the eurosid group[ii] while another places it sister to all eudicots except Gunnerales[iii], but it is consistently sister to Saxifragaceae. The family has been treated as having 1 genus, *Ribes,* or 2 when *Grossularia,* with highly reflexed sepals, is recognized. Both morphological and molecular data show *Grossularia* to be a distinct group within a broad *Ribes*[iv].

Economic uses The family is commercially important for the many edible fruits[v]: Blackcurrants (*R. nigrum*), Redcurrants (*R. rubrum*), white currants (*R. sanguineum*), Alpine Currant (*R. alpinum*), American Blackcurrant (*R. americanum*), Golden Currant (*R. aureum*), and Gooseberries (*R. uva-crispa*), as well as the hybrid Jostaberry (*R. × culverwellii*). Flowering currants (*R. sanguineum, R. speciosum*) are grown for their ornamental flowers.   AC

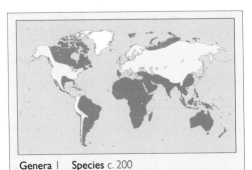

Genera 1   Species c. 200
Economic uses Soft fruit and some ornamentals

i Brummitt, R. K. *Vascular Plant Families and Genera.* Richmond, Royal Botanic Gardens, Kew (1992).
ii Soltis, D. E. *et al.* Angiosperm phylogeny inferred from 18S rDNA, *rbc*L, and *atp*B sequences. *Bot. J. Linn. Soc.* 133: 381–461 (2000).
iii Soltis, D. E. *et al.* Gunnerales are sister to other core eudicots: Implications for the evolution of pentamery. *American J. Bot.* 90: 461–470 (2003).
iv Senters, A. E. & Soltis, D. E. Phylogenetic relationships in *Ribes* (Grossulariaceae) inferred from ITS sequence data. *Taxon* 52: 51–66 (2003).
v Lord, T. (Consultant Editor), Armitage, J., Cubey, J., Grant, M., Lancaster, N., & Whitehouse, C. (RHS eds). *RHS Plant Finder 2005–6.* London, Dorling Kindersley (2005).

GRUBBIACEAE

Evergreen, ericoid shrubs from the Cape region of South Africa, closely related to the Curtisiaceae and Cornaceae.

Distribution *Grubbia* is confined to the Cape region, extending from the southeastern part of the Western Cape to Eastern Cape. It occurs in damp habitats on sandstone.

Description Ericoid shrubs up to 1.5 m high and often more across, either from a taproot or a massive lignotuber, with stiffly ascending branches. The leaves are opposite, mostly linear-lanceolate to linear, with the margins inrolled beneath, but occasionally those on

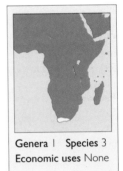

Genera 1   Species 3
Economic uses None

the main stems may be more or less flat and 2 cm across. The flowers are small, usually less than 1 mm across, in compact sessile axillary dichasia of 2–20 flowers, the ovaries of adjacent flowers coalescing to form a syncarpium. In sect. *Grubbia*, there are only 2–3 flowers per axil, and these are subtended by brown scarious bracts through which the whitish floral hairs protrude, but in sect. *Strobilocarpus* there are up to 20 flowers per axil and subtending bracts are not conspicuous. The flowers are actinomorphic, 4-merous and epigynous, with a single perianth whorl of 4 members that are densely hairy inside and out. The stamens are 8, 4 opposite and more or less attached to the perianth members, and 4 alternating with them. The ovary is inferior and bicarpellate, at first 2-locular but later 1-locular by breakdown of the septum, with 2 ovules pendent from the upper part of the septum. The fruits are single-seeded with a fleshy exocarp and hard endocarp, and those of adjacent flowers in an inflorescence coalesce to form a fleshy syncarpium, which is globose and up to 8 mm in diameter.

Classification Various affinities have been suggested in the past for the family, Santalaceae, Pittosporaceae, Cunoniaceae, and Ericaceae having been seriously proposed. Recent molecular evidence, however, has established a close relationship with Curtisiaceae and through that with Cornaceae (see the latter for references). A recent combination of the Grubbiaceae and-Curtisiaceae has been described in FGVP[i] as a shot-gun marriage, with good reason, for although there are many characters to support such a close relationship, there are also many characters quite apart from the ericoid habit to justify keeping the families apart. The 3 spp. have been divided among 2 sections[ii], see description above. RKB

[i] Kubitzki, K. Grubbiaceae. Pp. 199–201. In: Kubitzki, K. (ed.), *The Families and Genera of Vascular Plants. VI. Flowering Plants. Dicotyledons: Celastrales, Oxalidales, Rosales, Cornales, Ericales.* Berlin, Springer-Verlag (2004).
[ii] Carlquist, S. A revision of Grubbiaceae. *J. S. Afr. Bot.* 43: 115–128 (1977).

GUNNERACEAE

GUNNERA FAMILY

A monogeneric family of mostly rhizomatous or stoloniferous herbs, some of them gigantic with rosettes of large leaves.

Distribution *Gunnera* is widely distributed mainly in the southern hemisphere, in New Zealand, Tasmania (*G. cordifolia*), Malaysia, New Guinea, the Philippine Islands, the Juan Fernandez Islands (*G. masafuerae*), East–Central Africa, South Africa, Madagascar, Central and South America, and extending northward into Mexico and Hawaii.

Description Perennial rhizomatous or stoloniferous herbs, rarely annual (*G. herteri*), the rhizomes usually creeping, suberect in subg. *Panke*. Symbiotic cyanobacteria (*Nostoc*) occur in the parenchyma of the rhizomes. The stems are covered with large triangular scales between the leaves in subg. *Panke* and these are homologous with the bracts found in the stoloniferous species of subgenera *Misandra, Pseudogunnera,* and *Milligania*[i]. The leaves are alternate, simple, in rosettes, small to very large (up to 3 m wide in *G. manicata*), with petioles that are often robust and elongate (up to 2.5 m in *G. manicata*); the lamina is often peltate, sometimes cordate or cuneate at the base, and with dentate, crenate, or lobed margins and palmate venation, the veins prominent on the under surface; stipules intrapetiolar. The flowers are somewhat zygomorphic, small to minute, bisexual (subg. *Panke*) or

Genera 1 **Species** 50–70
Economic uses Ornamental herbs; minor uses

unisexual (plants monoecious, dioecious, or polygamomonoecious), in axillary or terminal spicate panicles. The sepals are 2(3), free, valvate, sometimes fleshy or scalelike, or virtually absent; the petals are 2(3), sometimes absent, free. The stamens are 1 or 2, rarely 3, free, with short filaments. The ovary is inferior, of 2 united carpels, 1-locular, with a single pendulous ovule. The fruit is usually a fleshy drupe or sometimes a membranous or coriaceous nut.

Classification *Gunnera* is divided into 6 monophyletic subgenera: *Gunnera, Milligania, Misandra, Ostenigunnera, Panke,* and *Pseudogunnera,* based on distribution, morphology, and molecular data[ii]. Traditionally, *Gunnera* has been considered an anomalous member of the Haloragaceae, but the differences in mor-

GUNNERACEAE. 1 *Gunnera magellanica* (a) habit (x⅔); (b) female inflorescence (x⅔); (c) tip of male inflorescence each flower comprising only 2 stamens (x4); (d) female flower (x4); (e) fruit (x4).

phology, anatomy, embryology, and palynology— reinforced by the unique *Nostoc* symbiotic relationship—have been used to justify recognition as a separate family. Molecular data support a sister relationship with *Myrothamnus*, which belongs with *Gunnera* in the Gunnerales or should optionally even be included within it despite the significant differences between the 2 genera[iii]. There is also strong support for placement of Gunnerales (*Gunnera+Myrothamnus*) as the sister group to all remaining core eudicots[iv,v].

Economic uses Some species of *Gunnera*, especially *G. manicata*, are grown for ornament or ground cover in gardens in Europe and North America. The stem of the Chilean *Gunnera tinctorea* (=*G. chilensis*) has been used on a small scale for tanning and dyeing and the petioles in confectionery. Some species have medicinal properties. VHH

[i] Wanntorp, L., Wanntorp, H.-E., & Rutishauser, R. On the homology of the scales in *Gunnera* (*Gunneraceae*). *Bot. J. Linn. Soc.* 142: 303–308 (2003).
[ii] Wanntorp, L., Wanntorp, H.-E., & Källersjö, M. Phylogenetic relationships of *Gunnera* based on nuclear ribosomal DNA ITS region, *rbc*L and *rps*16 intron sequences. *Syst. Bot.* 27: 512–521 (2002).
[iii] Wilkinson, H. P. A revision of the anatomy of Gunneraceae. *Bot. J. Linn. Soc.* 134: 233–266 (2000).
[iv] Savolainen, V. *et al.* Phylogenetics of flowering plants based upon a combined analysis of plastid *atp*B and *rbc*L gene sequences. *Syst. Biol.* 49: 306–362 (2000).
[v] Soltis, D. E. *et al.* Gunnerales are sister to other core eudicots: Implications for the evolution of pentamery. *American J. Bot.* 90: 461–470 (2003).

GYROSTEMONACEAE

An Australian family of short-lived (mostly fire-opportunist) herbs, shrubs, and small trees with corolla-less flowers.

Distribution Endemic in drier and southern regions of Australia, except a single, unconfirmed record from New Zealand.

Description Short-lived trees or shrubs, or annuals, dioecious or monoecious, glabrous or papillose, soft-wooded, pungently scented when crushed. The branchlets are commonly orange, red, or brown. The leaves are alternate, simple, entire, sessile, or petiolate, commonly succulent; the stipules are small. The flowers are small, actinomorphic, or nearly so, solitary, or in axillary or terminal racemes or panicles. The calyx is broadly cupular, 4- to 8-lobed or entire, persistent, the lobes imbricate in bud. There is no corolla. The male flower has 7–100 stamens, in 1 or several concentric series, when uniseriate with a central disk; the anthers are almost sessile, quadrangular, 2-locular, opening widely by longitudinal slits. In the female flowers, the ovary is superior, with 1 (*Cypselocarpus*) or 2 to many carpels, which are either connate around a central column, or fused (*Tersonia, Walteranthus*). There is 1 campylotropous ovule per carpel, and 1 commonly petaloid stigma per carpel; placentation

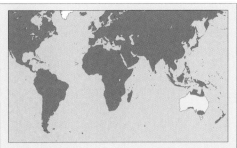

Genera 5 Species c. 20
Economic uses Three species may be toxic to stock

is axile. The fruit is a dry or succulent schizo-carp, a hard achene, or a syncarp; carpels commonly fall before shedding the seeds. The seeds are U-shaped, usually rugose, red-brown, with a prominent, translucent aril.

Classification Although Endlicher described the family in 1841, most authors treated the genera in the Phytolaccaceae until Heimerl (later supported by Hutchinson) recognized Gyrostemonaceae again as a distinct family. Debate has continued on its affinities, with Prijanto supported by Cronquist allying it with Bataceae on the basis of its pollen. A study of the pollen and cytology by Goldblatt *et al* indicated a placement within the Capparales, whereas Carlquist allied it with Sapindales on the basis of its anatomy. Recent molecular research by Rodman *et al*[i] and an embryological study by Tobe and Raven[ii] support its placement within the Capparales. The genera are closely knit morphologically (George[iii,iv]).

Economic uses Three species from the family (*Codonocarpus cotinifolius*, *C. attenuatus*, and *Gyrostemon australasicus*) may be responsible for the losses of stock (including camels in arid Australia).   ASG

[i] Rodman, J. E., Karol, K. G., Price, R. A., Conti, E., & Sytsma, J. D. Nucleotide sequences of *rbc*L confirm the Capparalean affinity of the Australian endemic Gyrostemonaceae. *Austral. Syst. Bot.* 7: 57–69 (1994).
[ii] Tobe, H. & Raven, P. H. The embryology and relationships of Gyrostemonaceae. *Austral. Syst. Bot.* 4: 407–420 (1995).
[iii] George, A. S. Gyrostemonaceae. *Fl. Australia* 8: 362–379 (1982).
[iv] George, A. S. Gyrostemonaceae. Pp. 213–217. In: Kubitzki, K. & Bayer, C. (eds), *The Families and Genera of Vascular Plants. V. Flowering Plants. Dicotyledons: Malvales, Capparales and Non-betalain Caryophyllales*. Berlin, Springer-Verlag (2002).

HALOPHYTACEAE

A family comprising 1 sp., *Halophytum ameghinoi*, a fleshy monoecious annual herb with spreading glabrous branches, endemic to scattered localities in dry temperate Argentina[i]. The leaves are alternate and without stipules. The small unisexual flowers are borne in axillary racemose inflorescences. The stem first produces female inflorescences and later a spike of male flowers above. The female flowers lack a perianth and consist of 1 small bract and 2 bracteoles surrounding a superior ovary of 3 fused carpels that are partially sunken into the cortex. The male flowers are also bracteate, arranged into spikes 8–10mm long, flowers bear 4 membranous tepals, with 4 alternating stamens. The fruits are single-seeded and enclosed in groups of 2 or 3 in a woody cortex.

The species was first described in *Tetragonia* (Aizoaceae) and later in Chenopodiaceae but was recognized as distinct as early as 1946[ii]. On the basis of molecular evidence it is placed in the core Caryophyllales[iii].   AC

Genera 1 Species 1
Economic uses None

[i] Zuloaga, F. O. & Morrone, O. (eds) *Catálogo de las plantas vasculares de la república Argentina* II. St. Louis, Missouri Botanical Garden (1999).
[ii] Soriana, A. Halophytaceae. Nueva familia del orden Centrospermae. *Notas Mus. La Plata* 11: 161–175 (1946).
[iii] Cuénoud, P. *et al.* Molecular phylogenetics of Caryophyllales based on nuclear 18S and plastid *rbc*L, *atp*B and *mat*K DNA sequences. *American J. Bot.* 89: 132–144 (2002).

HALORAGACEAE
HALORAGIS AND WATER MILFOIL FAMILY

Usually narrow-leaved annual herbs to shrubs or submerged or emergent aquatics, usually with 4-merous flowers and an inferior ovary

Distribution Six of the 8 genera in this family, and more than 100 spp., are native to Australia, but the family extends to every continent through its wetland members, especially the widespread *Myriophyllum*.

Description The dryland Australian genera are mostly erect annual to perennial herbs to subshrubs, or *Haloragodendron* may be a robust erect shrub up to 3 m, while the wetland genera are mostly prostrate to submerged herbs rooting at the nodes. The leaves are alternate or opposite to whorled (often variable in the same genus and sometimes even on the same plant); in the dryland genera they are usually narrow and entire to strongly serrate, but in wetland species they are often strongly dimorphic with finely pinnately dissected submerged leaves and broad entire to serrate emergent leaves, often with a gradation from one to the other on the same stem, and sometimes only finely dissected submerged leaves are present. The inflorescence is a spike or globular to flat-topped corymb. The flowers are bisexual or unisexual (plants monoecious or dioecious), actinomorphic, usually 2- to 4-merous, but in *Proserpinaca* 3-merous in all parts, occasionally showy (*Glischrocaryon*). The sepals are usually 4. The petals are usually 4 and hooded or boat-shaped but are absent in *Proserpinaca* and female flowers of *Myriophyllum* and *Laurembergia*. The stamens are usually 8 in 2 whorls, rarely 4. The ovary is inferior, with usually 4 free styles and usually capitate stigmas, completely or incompletely 4-locular with 1–2 pendent ovules per locule. The fruit is an indehiscent nut or a schizocarp splitting into 4 mericarps, variously winged or ornamented.

Classification The genus *Gunnera* has sometimes been included in Haloragaceae, but it is now not considered to be closely related and is referred to Gunneraceae. Otherwise the family has usually been considered somewhat isolated in the Rosid line. In APG II, it is placed in the Saxifragales, with Penthoraceae and Tetracarpaeaceae optionally included in it. These 2

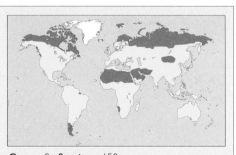

Genera 8 Species c. 150
Economic uses Aquarium plants; sometimes invasive weeds of waterways

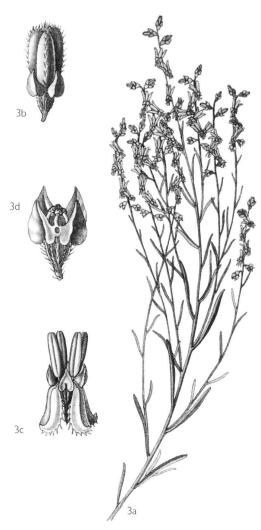

3b

3d

3c

3a

genus *Laurembergia* is pantropical (excluding Australia), with probably 4 spp. Of the Australian genera, *Haloragis* (28 spp.) extends to tropical Asia and the Pacific, as does *Gonocarpus* (41 spp.), which also extends to temperate Asia. *Glischrocaryon* (4 spp.), *Haloragodendron* (5 spp.), and *Meziella* (1 sp.) are confined to Australia[i]. Following the recent rediscovery of *Meziella*, which was for some 150 years known only from a single collection from southwestern Australia, 2 tribes have been re-circumscribed[ii]. The first tribe, Myriophylleae, comprises the genera *Myriophyllum* and *Meziella*; fruits (2–)4 single-seeded pyrenes. *Meziella* is somewhat intermediate between *Haloragis* of the following tribe and *Myriophyllum*. The other 6 genera belong to the tribe Halorageae; fruits indehiscent 1- to 4-seeded nut, not subdivided.

Economic uses Some *Myriophyllum* spp. (Water Milfoil) are cultivated as aquarium or pond plants, but they can become invasive weeds of waterways and may be poisonous to animals.   RKB

[i] Orchard, A. E. Haloragaceae. In: *Flora of Australia*. 18: 5–85 (1990).
[ii] Orchard, A. E. & Keighery, G. J. The status, ecology and relationships of *Meziella* (Haloragaceae). *Nuytsia* 9: 111–117 (1993).

HAMAMELIDACEAE
WITCH HAZEL AND LIQUIDAMBAR FAMILY

Temperate and tropical trees and shrubs, with petals usually linear to strap-shaped.

Distribution The family has a rather disjunct distribution, from eastern North America to Venezuela, eastern Mediterranean, Ethiopia to South Africa, temperate East Asia to Indo–Malaysia and northern Australia, and is absent or scarce in much of South America, Africa, and Australia. In temperate regions, they are usually found in open places or woodland, but in the tropics they are more often located in rain forest.

Description Trees and shrubs, sometimes with stellate hairs. The leaves are alternate or rarely opposite, simple, ovate or elliptical to heart-shaped or palmately lobed, with entire to serrate margins, usually with stipules. The inflorescence is usually aggregated into a dense spike or head, sometimes spherical with many flowers, the individual flowers sessile, rarely a lax spike with shortly pedicellate flowers. The flowers are actinomorphic or (*Rhodoleia*) zygomorphic, bisexual or unisexual (plants monoecious or, in Hamamelideae, andromonoecious), hypogynous to usually epigynous. The sepals are 4–5(–10) and small, or absent. The petals are 4 or 5, or absent, usually linear or strap-shaped when present. The stamens are 4–5(–10), alternating with the petals when present, sometimes with many staminodes. The

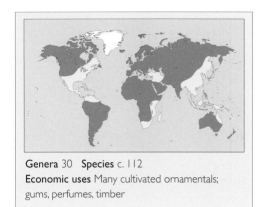

Genera 30   Species c. 112
Economic uses Many cultivated ornamentals; gums, perfumes, timber

ovary is superior to inferior, of 2 fused carpels each with 1 to many ovules with axile or rarely parietal placentation. The fruit is a capsule, usually woody, the seeds being ejected or not.

Classification Relationships formerly assumed in the subclass Hamamelididae of, e.g., Cronquist, have not been borne out by molecular data. Hamamelidaceae is placed in Saxifragales in APG II, most closely related to Daphniphyllaceae and Cercidiphyllaceae. The delimitation and classification of the family adopted here is that of Endress[i]. Evolutionary development has been considered by the same author[ii].

SUBFAM. HAMAMELIDOIDEAE Leaves usually ovate with pinnate venation, with mostly small caducous stipules; flowers mostly sessile in spikes or heads, usually bisexual; sepals usually present but petals present or absent; each carpel with a single ovule, and the seed ejected at maturity (unlike other subfamilies). There are 4 tribes:

Hamamelideae has 10 genera and 35 spp. confined to the Old World except for the North American species of *Hamamelis*. Flowers are bisexual and 4- to 5-merous with showy linear petals. There are 3 subtribes:

Hamamelidinae (*Hamamelis*); Loropetalinae (*Loropetalum, Tetrathyrium, Maingaya,* and *Embolanthera*); and Dicoryphinae (*Dicoryphe, Trichocladus, Ostrearia, Neostrearia,* and *Noahdendron*—the last three all monotypic Australian genera).

Corylopsideae has just a single genus, *Corylopsis*, with 7 spp. from Bhutan to Japan. Flowers bisexual with 5 elliptic clawed petals; fruits dehiscing with short curved beaks.

Eustigmateae has 3 genera: *Eustigma, Fortunearia,* and *Sinowilsonia*, and 4 spp. from China and Indochina. Flowers bisexual or unisexual; petals small and inconspicuous or absent; fruits large and with straight beaks.

Fothergilleae has 8 genera: *Molinadendron, Fothergilla, Parrotiopsis, Parrotia, Sycopsis, Distyliopsis, Distylium,* and *Matudaea*, and 33 spp. in North and Central America and Himalaya to tropical Asia. Flowers bisexual or male, apetalous and sometimes asepalous.

SUBFAM. EXBUCKLANDIOIDEAE There are 4 genera: *Disanthus, Mytilaria, Chunia,* and *Exbucklandia*, with 5 spp. native to temperate and tropical Asia. Leaves ovate to cordate and

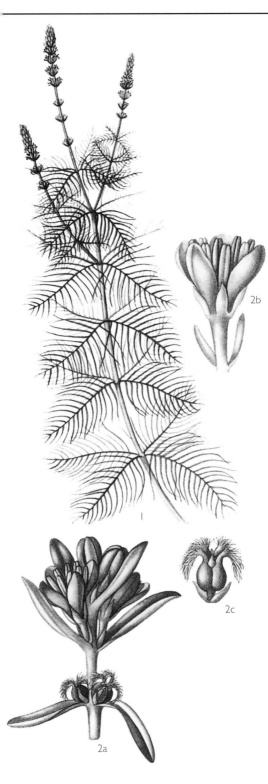

HALORAGACEAE. 1 *Myriophyllum spicatum* shoot bearing submerged much divided leaves and aerial inflorescences (x⅔). 2 *M. pedunculatum* (a) tip of inflorescence with separate male and female flowers (x6); (b) male flower (x8); (c) female flower with plumed stigmas (x8). 3 *Gonocarpus cordiger* (a) flowering shoot (x⅔); (b) flower (x6); (c) flower showing heart-shaped sepals, downward-curved petals and large anthers (x4); (d) flower with petals and stamens removed to show styles and stigmas (x8).

families have superior ovaries and seem quite distinct morphologically. The largest genus is the cosmopolitan aquatic or emergent genus *Myriophyllum*, with c. 60 spp. Closely similar to it in habitat and habit is *Proserpinaca*, with 3 spp. extending from Canada to Central America and the West Indies. The prostrate wetland

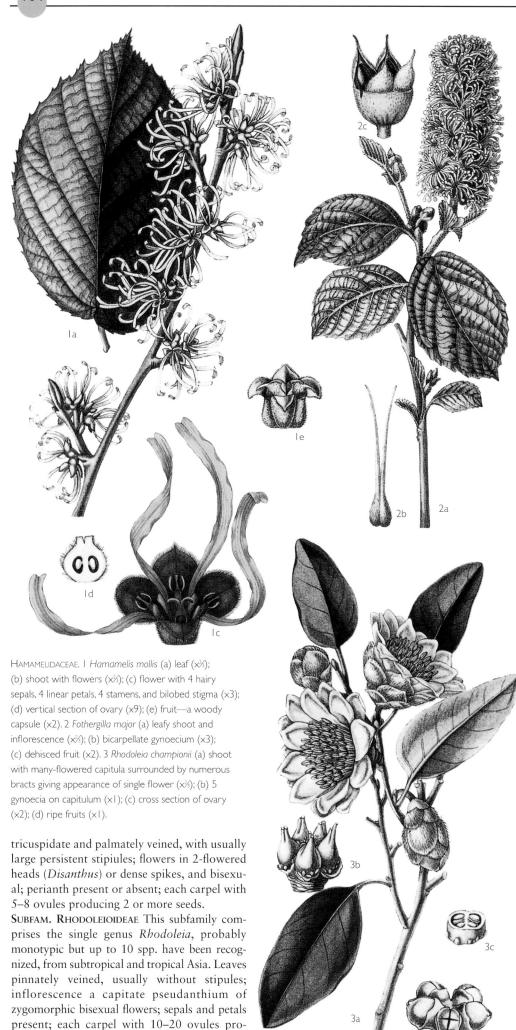

SUBFAM. ALTINGIOIDEAE The subfamily comprises *Liquidambar*, *Semiliquidambar*, and *Altingia*, with up to 16 spp. from Central and North America (*L. styraciflua*), eastern Mediterranean (*L. orientalis*, if different from the American species; see Meikle's *Flora of Cyprus*), and China to tropical Asia. Leaves tricuspidate to palmately lobed, with small caducous stipules; petiole with 3–5 separate vascular bundles (cylindrical or semicylindrical in other subfamilies); inflorescence a dense spherical head of unisexual flowers lacking sepals and petals (often resembling that of Casuarinaceae in fruit); each carpel with more than 20 ovules producing 2 or more seeds.

Economic uses *Altingia*, *Liquidambar*, and *Exbucklandia* provide high-quality timber. *Liquidambar* and *Hamamelis* yield gums used in perfumery or to provide medicinal products used as expectorants, inhalants, or for the treatment of skin complaints, e.g., *H. virginica* (Witch Hazel). Many species and cultivars of *Corylopsis*, *Fothergilla*, *Hamamelis*, *Liquidambar*, *Loropetalum*, *Parrotia*, and *Sycopsis* are grown as ornamentals (early sweet-scented flowers with an autumnal color) in temperate regions. Twigs of witch hazel have been used in water-divining ("witching"). RKB

[i] Endress, P. K. A suprageneric taxonomic classification of the Hamamelidaceae. *Taxon* 38: 371–376 (1989).
[ii] Endress, P. K. Aspects of evolutionary differentiation of the Hamamelidaceae of the lower Hamamelidae. *Pl. Syst. Evol.* 162: 193–211 (1989).

HAPTANTHACEAE

This enigmatic monoecious evergreen broad-leaved tree was collected from central Honduras in 1980 and described in 1989 as *Haptanthus hazlettii* without being assigned to a family[i]. The family Haptanthaceae was subsequently created for it. Several attempts have been made to recollect but without success. It has opposite leaves with entire margins and cymose axillary inflorescences. The structure of the staminate flowers has recently been reinterpreted. The 2 elliptical flattened structures that bear the anthers are borne on top of a solid, pedicel-like stalk and may be leaflike introrse stamens or tepals with adnate introrse stamens[ii]. The carpellate flowers are central, with a superior ovary of 3 fused carpels with parietal placentation; the styles are 3, spreading. Fruit is not known.

The affinities of *Haptanthus* are obscure. It has been placed in the Violales[iii], but until further material is available no firm decision can be made on its position. VHH

[i] Goldberg A. & Nelson C. S. *Haptanthus*, a new dicotyledonous genus from Honduras. *Syst. Bot.* 14: 16–19 (1989).
[ii] Doust, A. N. & Stevens, P. Systematic placement of the enigmatic genus *Haptanthus*, based on floral morphology. Systematics Section/ASPT. http://www.2005.botanyconference.org/engine/search/index.php?func=detail&aid=388
[iii] Shipunov, A. The system of flowering plants from a synthetic point of view. *Zhurn. Obshchei Biol.* 64: 501–510 (2003).

HAMAMELIDACEAE. 1 *Hamamelis mollis* (a) leaf (x⅔); (b) shoot with flowers (x⅔); (c) flower with 4 hairy sepals, 4 linear petals, 4 stamens, and bilobed stigma (x3); (d) vertical section of ovary (x9); (e) fruit—a woody capsule (x2). 2 *Fothergilla major* (a) leafy shoot and inflorescence (x⅔); (b) bicarpellate gynoecium (x3); (c) dehisced fruit (x2). 3 *Rhodoleia championii* (a) shoot with many-flowered capitula surrounded by numerous bracts giving appearance of single flower (x⅔); (b) 5 gynoecia on capitulum (x1); (c) cross section of ovary (x2); (d) ripe fruits (x1).

tricuspidate and palmately veined, with usually large persistent stipiules; flowers in 2-flowered heads (*Disanthus*) or dense spikes, and bisexual; perianth present or absent; each carpel with 5–8 ovules producing 2 or more seeds.

SUBFAM. RHODOLEIOIDEAE This subfamily comprises the single genus *Rhodoleia*, probably monotypic but up to 10 spp. have been recognized, from subtropical and tropical Asia. Leaves pinnately veined, usually without stipules; inflorescence a capitate pseudanthium of zygomorphic bisexual flowers; sepals and petals present; each carpel with 10–20 ovules producing 2 or more seeds.

HECTORELLACEAE

A family comprising 2 monotypic genera of alpine cushion plants of uncertain affinity. Both genera are endemic to South Island of New Zealand and Kerguelen Island (not shown on map). The species are densely hairy to glabrous subshrubs

Genera 2 Species 2
Economic uses None

with spirally arranged leaves. The leaves are simple, without stipules, small, overlapping, and leathery. The inflorescence has 1 flower, axillary. The flowers are actinomorphic, hermaphrodite (*Lyallia*) or unisexual (plants dioecious, but sometimes also with herma-phrodite flowers; *Hectorella*), with 2 opposite sepals, and 4 or 5 petals. The stamens are equal in number to the petals (*Hectorella*) or 1 fewer (*Lyallia*). The ovary is superior, of 2 fused carpels, 1-locular, with 4–7 ovules. Style short, with a 2-lobed stigma. The fruit is a capsule containing few seeds. The position of these genera in the Caryophyllales is long established[i], but there is still little evidence to confirm either their recognition as a separate family or their detailed relationships within this order. One account of the family implies a definite relationship with Portulacaceae but states the genera differ in their spirally arranged exstipulate leaves and axillary flowers[ii] and deserve family status on this basis. The genera were not sampled in recent molecular analyses of the order. AC

[i] Ng, S. Y., Phillipson, W. R., & Walker, J. R. L. Hectorellaceae—a member of the Centrospermae. *New Zealand J. Bot.* 13: 567–570 (1975).
[ii] Philipson, W. R. Hectorellaceae. Pp. 331–334. In: Kubitzki, K. & Bayer, C. (eds), *The Families and Genera of Vascular Plants. V. Flowering Plants. Dicotyledons: Malvales, Capparales and Non-betalain Caryophyllales.* Berlin, Springer-Verlag (2002).

HELICTERACEAE

HELICTERES FAMILY

A family of tropical trees and shrubs, formerly included in a more broadly circumscribed Sterculiaceae (q.v.).

Distribution Pantropical, but sparse in Africa. The dominant genus is *Helicteres*, with c. 60 spp. in deciduous and semideciduous thicket and scrub in Central and South America and tropical Asia to Australia, followed by *Reevesia*, with 25 spp. and a similar distribution.

Description Shrubs to large trees. The bark is usually fibrous, and the indumentum is mainly of stellate hairs, as in most core Malvales. The leaves are alternate, simple, rarely lobed, usually palmately nerved and cordate. The margins are serrate or entire, the petiole ± pulvinate at base and apex, and the stipules linear and caducous.

The inflorescences are cymose and axillary or terminal. Bracts are present, but epicalycular bracts are only recorded in *Triplochiton* and *Helicteres*, and these are foliaceous. The flowers are more or less zygomorphic in all genera except *Triplochiton*, this being most marked in *Helicteres*. The calyx in bud is large, campanulate to clavate, and valvately 5-lobed at the apex. In most genera, it opens only at the apex, so remaining largely cylindrical. In *Mansonia*, however, it splits along its length on one side, and in *Triplochiton* is 5 lobed to the base. The 5 contorted petals are free to the base and markedly clawed, in most genera with hair patches, pockets, or lateral flaps at the junction of blade and claw. A conspicuous androgynophore lifts the stamens and ovary usually well clear of the perianth. The androecium consists of 10–20(–30) stamens often arranged in clusters of 2–4 in one whorl. The 2 thecae are placed end to end in the stamen, often on a swollen connective. In *Triplochiton*, *Mansonia*, and *Neoregnellia*, monothecal anthers occur, probably by fusion of the 2 thecae. The pollen is peroblate, with 3(–5) pores at the angles, the surface being finely verrucate and/or reticulate. Ovate staminodes form an inner whorl to the androecium, alternating with and about as long as the outer anthers (absent or reduced in *Ungeria* and *Reevesia*). The gynoecium is apocarpous (except in *Ungeria* and *Reevesia*), of 5 carpels each tapering to an erect filiform style. The ovules are 1 to numerous with axile placentation. The fruits are dry and loculicidally dehiscent, or dehisce only at the apex in some *Helicteres*, often with numerous winged seeds. In *Triplochiton* and *Mansonia* the mericarps are samaroid.

Classification Helicteraceae were formerly included as a tribe in Sterculiaceae *s.l.*, with other groups here treated as Byttneriaceae, Pentapetaceae, and Sterculiaceae s.s. Molecular analysis of core Malvales (q.v.), shows that Helicteraceae forms a monophyletic clade with Durionaceae[i,ii,iii] (itself once included in Bombacaceae), from which it differs morphologically in its zygomorphic flowers; calyces clavate-campanulate in bud, cylindrical to subcylindrical at anthesis; clawed petals, often with pouches, hair patches, or lateral lobes at the junction with the blade; conspicuous androgynophore; the anther thecae arranged end-to-end; inner androecial whorl petaloid;

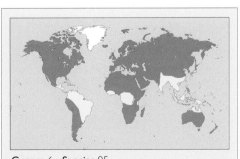

Genera 6 Species 95
Economic uses Timber trees and flowering shrubs

pollen peroblate, angulato-aperturate, microverrucate to reticulate. Genera traditionally included in the former tribe Helictereae, but which have on molecular grounds been placed elsewhere, are *Pterospermum* (now Pentapetaceae) and *Kleinhovia* (now Byttneriaceae). There is no formal infra-familial classification, but *Helicteres* and *Neoregnellia* are shrubs and subshrubs; anthers with short stalks; staminodes ovate, apocarpous capsules. *Ungeria* and *Reevesia* are trees and shrubs; anthers sessile, the filaments fused, forming a 5-dentate goblet around the ovary; staminodes reduced or absent; syncarpous capsules. *Mansonia* and *Triplochiton* are trees; anthers stalked, usually monothecal; apocarpous, samaroid mericarps.

Economic uses *Triplochiton scleroxylon* and *Mansonia altissima* are significant timber trees in western and central Africa. Members of the genus *Mansonia* are also timber trees in India and Indo–China. *Mansonia gagei* gives the scented and culturally significant kalamet wood of India. Some *Helicteres* and *Reevesia* are occasionally planted as ornamentals in tropical and temperate areas. MRC

[i] Baum, D. A., Alverson W. S., & Nyffeler R. A durian by any other name: taxonomy and nomenclature of the core Malvales. *Harvard Pap. Bot.* 3: 315–330 (1998).
[ii] Bayer, C. *et al.* Support for an expanded family concept of Malvaceae within a recircumscribed order Malvales: a combined analysis of plastid *atp*B and *rbc*L DNA sequences. *Bot. J. Linn. Soc.* 129: 267–303 (1999).
[iii] Whitlock, B. A., Bayer, C., & Baum, D. A. Phylogenetic relationships and floral evolution of Byttnerioideae ("Sterculiaceae" or Malvaceae *s.l.*) based on sequences of the chloroplast gene, *ndh*F. *Syst. Bot.* 26: 420–437 (2001).

HELWINGIACEAE

A monogeneric family comprising 3 spp. of evergreen shrubs or trees with epiphyllous inflorescences, distributed from the Himalayas to Japan. The leaves are alternate to subopposite, petiolate, simple, the margins serrate, ovate to linear–lanceolate, with more or less fringed, caducous stipules. The flowers are

Genera 1 Species 3
Economic uses
Edible leaves and medicinal properties

small, actinomorphic, unisexual (plants dioecious), in fasciculate cymose inflorescences on the adaxial leaf surfaces. The perianth is 1-whorled, consisting of 3–5 petals, free, valvate, greenish or reddish purple. In the staminate flowers, the stamens are 3–5. In the pistillate flowers, the ovary is inferior, of 2–4 fused carpels, with apical placentation and 1 ovule per locule. The fruit is a fleshy drupe with 3–4 separate pyrenes.

Helwingia was included by Cronquist (1981) in Cornaceae, from which it differs in a series of features. Takhtajan (1997) maintained the family and created an order for it, Helwingiales, placed next to the Araliales. Kårehed[i] proposes including Helwingiaceae along with Phyllonomaceae in the Aquifoliaceae. The results of molecular analyses are still not resolved. Some studies indicate a sister relationship between Helwingiaceae and Aquifoliaceae[ii] or the Aquifoliaceae as sister to Phyllonomaceae and Helwingiaceae together[iii] in an expanded Aquifoliales. The sister relationship of Helwingiaceae and Phyllonomaceae makes morphological sense in that both have epiphyllous inflorescences.

VHH

[i] Kårehed, J. Evolutionary Studies in Asterids Emphasising Euasterids II (2002). http://urn.kb.se/resolve?urn=urn:nbn:se:uu:diva-2696
[ii] Olmstead, R. G., Kim, K.-J., Jansen, R. K., & Wagstaff, S. J. The phylogeny of the Asteridae *sensu lato* based on chloroplast *ndh*F gene sequences. *Mol. Phylogenet. Evol.* 16: 96–112 (2000).
[iii] Bremer, B. *et al.* Phylogenetics of asterids based on 3 coding and 3 non-coding chloroplast DNA markers and the utility of non-coding DNA at higher taxonomic levels. *Molecular Phylog. Evol.* 24: 274–301 (2002).

HERNANDIACEAE

A family of tropical trees and shrubs or lianas.

Distribution The family is pantropical, especially in coastal areas. The largest genus, *Hernandia* (c. 20 spp.) is widespread in the tropics, with a concentration in the Indo–Malaysian region and extending eastward into Australia and the Pacific and extending westward to tropical Africa, Madagascar, Central America, northern South America, and the West Indies. *Illigera* (19 spp.) ranges from China and Indochina to Malaysia, Africa, and Madagascar. *Gyrocarpus* (3–5 spp.) is found in Sri Lanka, tropical Africa, and Central America with 1 sp., *G. americanus*, widespread in tropical coastal vegetation. *Sparattanthelium* (13 spp.) grows in the Guayanan and Amazonian regions.

Description Trees, shrubs, and some lianas, often evergreen, with aromatic oils in the stems and leaves. The leaves are alternate, usually simple, entire palmately lobed or compound, without stipules. The flowers are actinomorphic

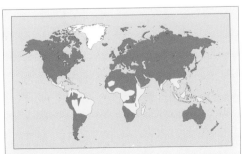

Genera 4   Species c. 60
Economic uses Timber used locally

or zygomorphic, bisexual (*Gyrocarpus, Sparattanthelium*) or unisexual (plants monoecious in *Hernandia*) in axillary, rarely terminal (*Hernandia*), corymbose or paniculate cymes. The tepals are (3–)4–8(–10), in 1 or 2 whorls, sepaloid. The stamens are 3–7, in 1 or 2 whorls, the filaments sometimes with nectariferous appendages; anthers dehiscing by longitudinal valves. The ovary is inferior, with a single carpel; ovule solitary. The fruits are dry indehiscent nuts, or drupes, longitudinally ribbed or with 2 lateral wings (*Illigera*), or with 2 apical spathulate wings (*Gyrocarpus*), or wingless and enclosed in bracteoles (*Hernandia*). A novel type of heterodichogamy not previously reported in the Angiosperms is reported in *Hernandia nymphaeifolia*: it has unisexual flowers and is monoecious, with 2 kinds of individual—1 with exclusively female flowers that open in the morning, and male flowers that open in the afternoon, and the other with the reverse behavior[i].

Classification There are 2 subfamilies.
SUBFAM. HERNANDIOIDEAE (*Hernandia, Illgera*) with thyrsoid infloresences; only a few of the fruits in the infructescence attain maturity.
SUBFAM. GYROCARPOIDEAE (*Gyrocarpus, Sparattanthelium*), with dichasial infloresences and many of the fruits maturing per infructesence[ii]. The Hernandiaceae was placed in the Laurales by Cronquist and within that order, morphological and molecular work indicates that Hernandiaceae, Lauraceae, and Monimiaceae are closely related. However, there appears to be a discrepancy between molecular and morphological evolution as regards the details of the relationships between these families[iii].

Economic uses Timber of *Hernandia* and *Gyrocarpu*s is used locally.

VHH

[i] Endress, P. K. & Lorence, D. H. Heterodichogamy of a novel type in *Hernandia* (Hernandiaceae) and its structural basis. *Int. J. Plant Sci.* 165: 753– 763 (2004).
[ii] Kubitzki, K. Hernandiaceae. Pp. 180–181. In: Smith, N. *et al.* (eds), *Flowering Plants of the Neotropics*. Princeton, Princeton University Press (2004).
[iii] Renner, S. S. & Chanderbali, A. S. What is the relationship among Hernandiaceae, Lauraceae and Monimiaceae, and why is this question so difficult to answer? *Int. J. Plant Sci.* 161 (Supplement 6): S109–119 (2000).

HIMANTANDRACEAE

A family of aromatic trees, covered with peltate scales, consisting of a single genus, *Galbulimima*, with only 2 spp.

Distribution The family is found in tropical rain forests of northeastern Australia and parts of Papua New Guinea, Indonesia, the Solomon Islands, and the Moluccas.

Description Aromatic trees with a covering of numerous, minute, peltate scales. The leaves are alternate, simple, petiolate, and without stipules. The flowers are regular, polypetalous, hermaphrodite, and usually solitary. The flower

bud is calyptrate with leathery sepals or bracts forming a cap round the bud. The sepals are 4 or 6. The petals are 7–9 (merging acropetally into the androecium), regular, white, cream, or reddish brown. The stamens are 15–40, petaloid in shape and texture, or laminoid, the anthers being at the edges of the flat, lanceolate blades.

Genera 1   Species 2
Economic uses
Used locally in traditional medicine

There are 8–10 staminodes. The ovary consists of 7–10, free, superior carpels, each with 1 locule and 1 or 2 pendulous ovules. The fruit is a syncarpous globose drupe. The seeds have oily endosperm and a small embryo.

Classification The family is considered one of the so-called primitive angiosperms because of petaloid stamens, carpel structure, vessels with porous perforations, and ladderlike pits. It is placed in Magnoliales complex and is closest morphologically to the Degeneriaceae with which it forms a sister pair according to molecular studies (see Degeneriaceae for references). However, a recent study notes that the morphology and anatomy of seed coats is quite distinct in structure and origin from both Magnoliaceae and Degeneriaceae and counters the traditional affiliation of Himantandraceae with Magnoliales, proposing instead that it should be placed in its own order. A great deal of work on the flower and fruit structure of this enigmatic family has been published but its interpretation is not always clear.

Economic uses The bark of *G. belgraviana* is used in Papua New Guinea as an local medicine and for its alleged psychoactive properties and Galbulimima alkaloids have aroused interest as a source of potential new drugs.

VHH

[i] Doweld, A. B. & Shevyryova, N. A. Carpology, anatomy and taxonomic relationships of *Galbulimima* (Himantandraceae). *Ann. Bot.* N.S. 81: 337–347 (1997).

HIPPURIDACEAE

Distribution *Hippuris* comprises 4 spp. of aquatic herbs found in shallow water in arctic to temperate wetlands of the northern hemisphere south to the Mediterranean, the Himalayas, China, and southern USA, and as far south as southern Chile and Argentina.

Description Perennial herbs with creeping rhizomes rooting beneath shallow water and producing erect shoots that emerge above water to flower. The leaves are borne in whorls of 4–12, the submerged ones linear, pale green, and flaccid; the emergent ones often shorter, linear, to broadly obovate or spathulate, dark green, and

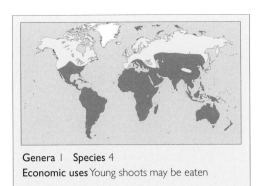

rigid. The flowers are borne in the axils of emergent leaves and are small and inconspicuous and variable within a single plant, being bisexual or unisexual, or rudimentary and apparently sterile. The perianth is reduced to a small rim around the top of the inferior ovary. There is 1 massive stamen also situated on top of the ovary. The ovary consists of 1 carpel that contains a single pendulous ovule. The style is long and slender, with stigmatic papillae along its length. The fruit is a small, ovoid, smooth, single-seeded nutlet.

Pollination is assumed to be by wind. The characteristic habit with unbranched stems bearing simple narrow leaves in symmetrical whorls is mimicked in numerous unrelated aquatic plants in a wide range of families, and has been called "the *Hippuris* syndrome"[i].

Classification The phylogenetic position of the Hippuridaceae has been much disputed in the past, though it has most often been considered to be related to the Haloragaceae, where *Myriophyllum* has a somewhat similar habit. Molecular evidence now shows this to be incorrect and places *Hippuris* instead next to the Callitrichaceae nested within the Scrophulariaceae. The highly reduced flowers with an inferior ovary argue against combining it with either of these families.

Some authors have preferred to treat the genus as comprising only 1 sp., *H. vulgaris*, while others have given specific rank to some easily recognizable variants such as *H. tetraphylla*, which has whorls of only 4 broadly obovate to spathulate leaves, unlike *H. vulgaris*, which has whorls of 6–12 linear leaves. Although in *Flora Europaea* it has been suggested it is only a form controlled by environmental factors, its immediately recognizable appearance and markedly coastal distribution, now well mapped around the northern hemisphere, scarcely coinciding with that of *H. vulgaris*, suggest that specific recognition is desirable. The diminutive *H. montana* from Alaska and adjacent areas also appears distinct. *H. spiralis* has recently been described from China.

Economic uses The submerged shoots of *Hippuris* remain green in winter and form a winter food for many animals. Native Americans also gather and eat the young shoots.   RKB

[i]Cook, C. D. K. The *Hippuris* syndrome. In: Street, H.E. (ed.), *Essays in Plant Taxonomy* 163–176. Academic Press, London (1978).

HOPLESTIGMATACEAE

The genus *Hoplestigma* includes 2 spp. from the Guineo–Congolian forests of Africa. *H. klaineanum* has a bimodal distribution in Ivory Coast[i] and Gabon, and *H. pierreanum* occurs in Cameroon. Both species occur in lowland rain forest and

are apparently rare, occurring as single isolated trees in their few known localities. Some sites have been lost to forest clearance, so the family may be regarded as endangered and conservation measures are needed[ii]. Both species are forest trees up to 25 m high. Young parts are usually tomentose but become glabrous at maturity. The leaves are alternate, large, up to 55 × 25 cm, oblanceolate, entire, coriaceous, shortly petiolate. The inflorescence is a short terminal cyme with stout branches and subsessile flowers. The flowers are actinomorphic, bisexual, subspherical, and c. 1–2 cm diameter[iii]. The calyx is subspherical, enclosing the bud, splitting irregularly into 2–5 lobes at anthesis. The corolla is tubular in the lower half, with c. 9–14 lobes that are strongly reflexed and much overlapping each other, white. The stamens are 20–35, inserted on the corolla tube and exserted in a tight, erect bunch. The ovary is superior, of 2 carpels, dome-shaped with 2 strongly geniculate styles free almost to their base. It is 1-locular, with 2 intruded forked parietal placentas, each bearing 2 ovules. The fruit is a spherical or ovoid, laterally grooved, drupe up to 3 cm long with a leathery exocarp and hard bony endocarp, with 2–4 seeds.

The affinities of the family have been disputed. Cronquist placed it in the Violales largely because of the parietal placentation, also mentioning possible affinity with Ebenales and observing that the many corolla lobes and stamens conflicted with Boraginaceae. It may be noted, however, that the South African herbaceous genus *Codon*, here referred to its own subfamily in Boraginaceae, has 10–12 corolla lobes and stamens. It seems likely that *Hoplestigma* is close to the base of the Boraginaceae, especially the subfamily Ehretioideae as suggested by the pollen[iv], but confirmation of this by molecular data is currently awaited. No economic uses are known, though *H. pierreanum* has potential as a flowering ornamental.   RKB & MRC

[i] Aké Assi, L. Flore de la Côte d'Ivoire. Hoplestigmataceae. *Boissiera* 57: 269 (2001).
[ii] Cheek, M. R. Hoplestigmataceae. In: Takahashi, K. (ed.), *The World of Plants*. Asahi shimbun 70: 6–312 (1995).
[iii] Aubréville, A. *Hoplestigma klaineanum. Fl. Forest. Côte d'Ivoire*, edn. 2, 3: t. 307 (1959).
[iv] Nowicke, J. W. & Miller, J. S. Pollen morphology and the relationships of the Hoplestigmataceae. *Taxon* 38: 12–16 (1989).

HUACEAE

Tropical African evergreen trees of uncertain affinity.

Distribution Tropical Africa.

Decription Shrubs, lianas, or herbs, aromatic (smelling of garlic). Leaves alternate, 2-ranked, simple, with caducous stipules, the lamina with entire margins, with small glands at the base, pinnately and finely reticulate-veined. The flowers are axillary, solitary or in clusters, actino-morphic, bisexual.

The sepals are (4)5 free, valvate (*Hua*) or connate to form a closed calyx (*Afrostyrax*). The petals are (4)5, free, with a long claw and peltate blade (*Hua*) or sessile, the blade obovate (*Afrostyrax*). The stamens are twice as many as the petals, in a single whorl. The ovary is superior, of 5 carpels, 1-locular, with a single basal ovule (*Hua*) or 4–6 (*Afrostyrax*). The style is single and the stigma small. The fruit is a capsule with valvular dehiscence (*Hua*) or a drupe (*Afrostyrax*). The seeds are large, solitary.

Classification The affinities of Huaceae remain obscure. The 2 genera have been placed in Sterculiaceae and Styracaceae and Huaceae was included in the Malvales by various authors including Takhtajan (1997), who noted, however, that its position was rather isolated, and in the Violales by Cronquist (1988). Molecular evidence is equivocal (see the discussion in Soltis *et al.*)[i].

Economic uses The leaves and seeds of *Afrostyrax lepidophyllus* are used as a condiment. *Hua gaboni* is used in local medicine. VHH

[i] Soltis, D. E., Soltis, P. S., Endress, P. K., & Chase, M. W. *Phylogeny and Evolution of Angiosperms*. Sunderland, Sinauer Associates (2005).

HUGONIACEAE

Trees and climbers related to Linaceae.

Distribution Pantropical. Four genera occur in the Old World: *Hugonia* with c. 30 spp. in tropical Africa and Madagascar and 4 spp. from Sri Lanka to Borneo; *Durandea* (sometimes treated as a section of *Hugonia*) with 4 spp. from New Guinea to Queensland, New Caledonia, and Fiji; *Philbornea* with 1 sp. from Sumatra, Borneo, and the Philippines; and *Indorouchera* with 2 spp. from the Nicobar Islands and Vietnam to Sumatra and Borneo. They occur mostly in or at the margins of evergreen forests.

Two genera are found in the New World: *Roucheria*, with 7 spp. from Nicaragua to Bolivia and central Brazil and *Hebepetalum* with 3 spp. from Colombia to Bolivia and central Brazil[i]. They are mostly in lowland seasonally inundated forest or on nutrient-poor sand, less commonly in *terra firme* forest.

Description The Old World genera are woody lianes, 3–35 m, climbing by woody recurved hooks or tendrils (modified stems) in opposite or subopposite pairs, or sometimes remaining as small trees without hooks until support for the climbing habit is available. The New World genera are trees up to 30 m, without hooks or tendrils. The leaves are alternate, simple, ovate to elliptical, usually crenate or serrate and somewhat undulate, with caducous stipules that may be dentate to palmately laciniate. The inflorescence is an axillary or terminal panicle or cyme, or rarely the flowers are solitary. The flowers are actinomorphic or the sepals unequal, bisexual, 5-merous, sometimes heterodistylous. The sepals are imbricate. The petals are free, rarely short-clawed, not showy. There are 10(–15) stamens, alternately unequal, the filament bases united into a tube surrounding the ovary. The ovary is superior, of 2–5 fused carpels, with 3–5 locules and as many styles, with 2 axile ovules per locule. The fruit is a fleshy drupe with 1 to many seeds, or splitting into 5 pyrenes (*Durandea*).

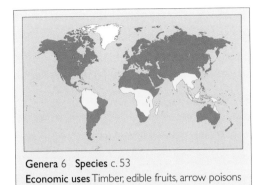

Genera 6   Species c. 53
Economic uses Timber, edible fruits, arrow poisons

Classification The family is closely related to Linaceae and is often included in it. It differs in its strongly woody habit, with small flowers (looking quite unlike even the shrubby Linaceae, which have large showy flowers), in their stamens twice as many as petals, and their fleshy drupaceous fruit. The Old World species are easily recognized by their woody recurved hooks, but these are not found in the New World genera. Leaf anatomy also supports separation of the 2 families[ii]. For further discussion of the family, see Jardim cited below.

Economic uses Species of *Roucheria* and *Hebepetalum* are used for timber in South America. The fleshy fruits of *Hugonia* are eaten in Africa. *Indorouchera contestiana* is recorded as containing saponin-like compounds used for arrow poisons.   RKB

[i] Jardim, A. A revision of *Roucheria* Planch. and *Hebepetalum* Benth. (Hugoniaceae). M.Sc. thesis, St. Louis, Missouri, University of Missouri (1999).
[ii] Van Welzen, P. C. & Baas, P. A leaf anatomical contribution to the Linaceae complex. *Blumea* 29: 453–479 (1984).

HUMIRIACEAE

A small family of evergreen trees and shrubs with a disjunct amphiatlantic distribution.

Distribution The family is predominantly tropical American, with all 8 genera and all but 1 sp. occurring there in a range of habitats, from forests to scrubland. *Sacoglottis gabonensis* is the only non-American species. It grows in the Guineo–Congolian forests of Africa.

Description[i] Evergreen buttressed trees or shrubs, the stems often angled or winged, with smooth, irregularly peeling bark. The leaves are alternate, simple, drying black, rolled inward

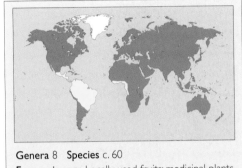

Genera 8   Species c. 60
Economic uses Locally used fruits; medicinal plants

when young, with short, often swollen, petioles; the lamina is entire to serrate, often with longitudinal striations and dotlike glands. The inflorescences are usually axillary, cymose. The flowers are actinomorphic, bisexual, usually smallish and inconspicuous, rarely brightly colored (*Vantanea guianensis*). The sepals are 5, connate at the base. The petals are 5, distinct. The stamens are 10–30, in 1 or 2 whorls, or numerous (*Vantanea*), the filaments more or less connate at the base, the anther connectives usually thick and well developed. An instrastaminal disk is present. The ovary is superior of 4–5(–8) fused carpels, 1-locular, with axile placentation and 1–2 ovules per carpel. The style is simple and the stigma capitate. The fruit is drupaceous, ovoid to globose, with a woody endocarp often with resin-filled cavities, often colored when ripe. The seeds are few with large embryos and and foliaceous cotyledons.

Classification The Humiriaceae were placed in the Linales by Cronquist (1981), Takhtajan (1997), close to the Ixonanthaceae, a relationship confirmed by a recent phylogenetic analysis[ii]. Molecular evidence, however, places Humiriaceae as sister to Salicaceae and Pandaceae[iii] in a large Malpighiales clade.

Economic uses The fruits of some species are consumed locally, the timber of *Humiria floribunda* is used for construction, and extracts from some species have medicinal value.   VHH

[i] Sabatier, D. Humiriaceae. In: Smith, N. *et al.* (eds), *Flowering Plants of the Neotropics*. Princeton, Princeton University Press (2004).
[ii] Bove, C. P. Phylogenetic analysis of Humiriaceae with notes on the monophyly of Ixonanthaceae. *J. Comp. Biol.* 2: 19–24 (1997).
[iii] Savolainen, V. *et al.* Phylogeny of the eudicots: a nearly complete familial anaylsis based on *rbc*L sequences. *Kew Bull.* 55: 257–309 (2000).

HUGONIACEAE. 1 *Hugonia castaneifolia* (a) leafy shoot showing hooklike modified branches (×1); (b) flower (×3); (c) stamens in 2 whorls of 5 surrounding the gynoecium which has three styles (×4); (d) vertical section of fruit (×1½).

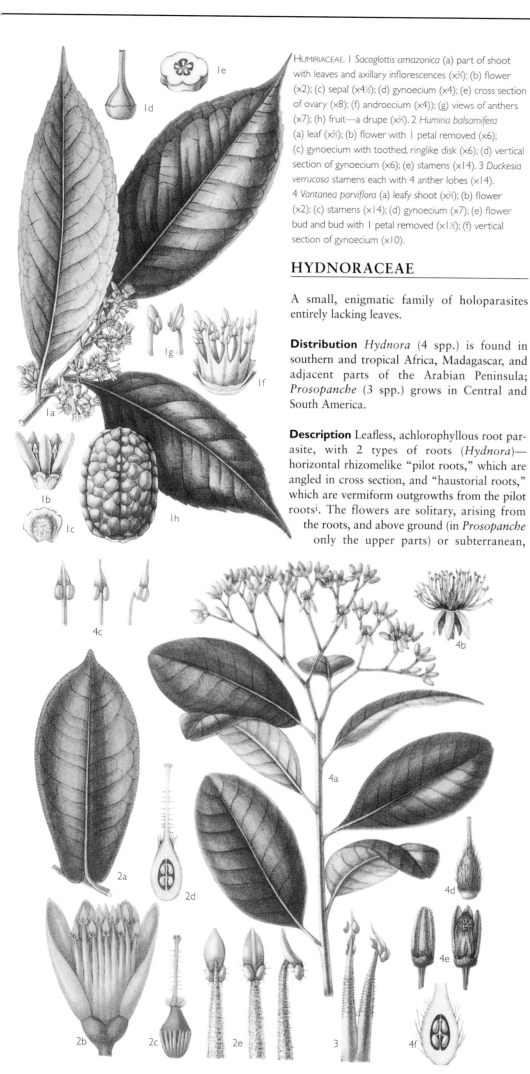

HUMIRIACEAE. I *Sacoglottis amazonica* (a) part of shoot with leaves and axillary inflorescences (x⅔); (b) flower (x2); (c) sepal (x4½); (d) gynoecium (x4); (e) cross section of ovary (x8); (f) androecium (x4); (g) views of anthers (x7); (h) fruit—a drupe (x⅔). 2 *Humiria balsamifera* (a) leaf (x⅔); (b) flower with I petal removed (x6); (c) gynoecium with toothed, ringlike disk (x6); (d) vertical section of gynoecium (x6); (e) stamens (x14). 3 *Duckesia verrucosa* stamens each with 4 anther lobes (x14). 4 *Vantanea parviflora* (a) leafy shoot (x⅔); (b) flower (x2); (c) stamens (x14); (d) gynoecium (x7); (e) flower bud and bud with I petal removed (x1½); (f) vertical section of gynoecium (x10).

HYDNORACEAE

A small, enigmatic family of holoparasites entirely lacking leaves.

Distribution *Hydnora* (4 spp.) is found in southern and tropical Africa, Madagascar, and adjacent parts of the Arabian Peninsula; *Prosopanche* (3 spp.) grows in Central and South America.

Description Leafless, achlorophyllous root parasite, with 2 types of roots (*Hydnora*)—horizontal rhizomelike "pilot roots," which are angled in cross section, and "haustorial roots," which are vermiform outgrowths from the pilot roots[i]. The flowers are solitary, arising from the roots, and above ground (in *Prosopanche* only the upper parts) or subterranean, actinomorphic, bisexual. The tepals are 3-4, fleshy, fused. The stamens in *Hydnora* are 3-4, with the filaments fused with the tepals to form a tube (tepalostemon) or, in *Prosopanche*, the andreoecium is fused into a domelike structure with a central opening, and sometimes with an innermost whorl of staminodes. The ovary is inferior, with a single locule and parietal placentation with numerous ovules. The style is single, and the stigma capitate. The fruit, which starts developing underground, is large, baccate, splitting irregularily and circumscissilely, containing numerous small seeds.

Classification Hydnoraceae was placed in the Rafflesiales by Cronquist (1981), who also noted its links with Aristolochiaceae recognized by previous authors. Takhtajan (1997) placed Hydnoraceae in its own order, Hydnorales, next to the Rafflesiales in the superorder Rafflesianae, apparently regarding both as related to the Aristolochiaceae. Molecular analyses do not support a close relationship with the Rafflesiaceae but indicate that *Hydnora* and *Prosopanche* form a clade associated with the Aristolochiaceae and Lactoridaceae in the Piperales[i], although its precise relationship is still to be determined. *Hydnora* has been called "the strangest plant in the world."[ii]   VHH

[i] Nickrent, D.L. *et al.* Molecular data place Hydnoraceae with Aristolochiaceae. *American J. Bot.* 89: 2809–1817 (2002).
[ii] Musselman, L. J. & Visser, L. J. The strangest plant in the world! *Veld and Flora* 71: 109–111 (1986).

HYDRANGEACEAE
HYDRANGEA AND MOCK ORANGE FAMILY

A family of mostly shrubs, often with dimorphic flowers, some of them sterile with an enlarged and showy calyx, usually with many stamens and an inferior ovary.

Distribution Mostly temperate and warm northern hemisphere, also in tropical Asia (especially *Dichroa*), but absent from Africa and South America. Habitats range from arid rocky places in southwestern USA through mesic woodland to wet tropical forests in Malaysia. The 3 largest genera—*Philadelphus* (80 spp.), *Deutzia* (60 spp.), and *Hydrangea* (29 spp.)—are all common to both Old and

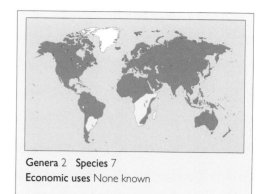

Genera 2 **Species** 7
Economic uses None known

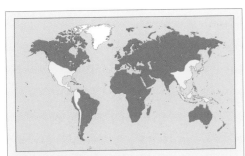

Genera 17 Species 220
Economic uses Dried roots provide drugs; many species are cultivated as ornamentals

New Worlds. Smaller genera are more restricted in distribution. *Deutzia* shows an interesting disjunction between Asia and Mexico.

Description Mostly shrubs, but *Kirengeshoma*, *Deinanthe*, and *Cardiandra* are robust rhizomatous herbs, and *Decumaria*, *Pileostegia*, *Schizophragma*, and some *Hydrangea* species are woody vines climbing by adventitious roots. The leaves are characteristically opposite, but rarely (*Cardiandra*) alternate, simple, sometimes small (1 cm) in xerophytic species but large (to 25 cm) in forest species, often crenate or serrate but sometimes entire, occasionally (*Kirengeshoma*) palmately lobed. The inflorescence is usually a terminal cyme, sometimes capitate in xerophytic species, but usually lax and paniculate. The flowers are sometimes conspicuously dimorphic, with perfect fertile flowers and sterile flowers with an enlarged calyx acting as an attraction to visitors. *Broussaisia*, from Hawaii, is dioecious. The sepals and petals are usually 4–5, but sometimes up to 8 in *Deinanthe*, and up to 12 in *Decumaria*, and are free or rarely fused at the base. The stamens are usually 2–5 times as many as the petals, very rarely (some *Dichroa*) equal in number, and in *Carpenteria* and *Deinanthe* there are more than 200 per flower. The ovary is syncarpous and partly to completely inferior, with (2)3–5(7) carpels, or in *Decumaria* up to 12, and with usually several locules but sometimes a single locule with intruding parietal placentas. There are usually several ovules on axile or intruding parietal placentas. The fruit is a capsule with septicidal to loculicidal or apical dehiscence, or a berry (*Dichroa* and *Broussaisia*).

Classification The family was at one time included in a broad concept of Saxifragaceae but is now widely accepted as having its relatives elsewhere. Molecular analysis[i] places it clearly with Loasaceae, close to Cornaceae. Interestingly, molecular analysis also makes the Hydrangeaceae paraphyletic with respect to the aquatic Afro–Madagascan family Hydrostachyaceae (see that family for references) with which it appears to have very few morphological characters in common. There seem to be no appropriate characters to maintain the 2 subfamilies recently recognized[ii], but 2 tribes may be distinguished:

Philadelpheae has 8 genera, including *Jamesia* and *Fendlera*, and 154 spp., confined to the New World except for *Philadelphus* and *Deutzia*. Sterile flowers are absent. The fruit is a loculicidal or septicidal capsule.

Hydrangeeae Nine genera, 67 spp., confined to Asia and the Pacific, except for *Hydrangea*. Sterile flowers are usually present and conspicuous. The fruit is a capsule with apical dehiscence or breaking up irregularly, or it may be a berry.

Economic uses Roots of *Dichroa* and *Hydrangea* are used locally as a source of drugs, and *Deutzia* is used for medicinal purposes in China. Many species are cultivated as ornamentals, especially in the genera *Hydrangea*, *Philadelphus* and *Deutzia*, and to a lesser extent *Schizophragma* and *Kirengeshoma*.   RKB

[i] Hufford, L., Moody, M. L., & Soltis, D. E. A phylogenetic analysis of Hydrangeaceae based on sequences of the plastid gene *mat*K and their combination with *rbc*L and morphological data. *Int. J. Pl. Sci.* 162: 835–846 (2001).
[ii] Hufford, L. Hydrangeaceae. Pp. 202–215. In: Kubitzki, K. (ed.), *The Families and Genera of Vascular Plants. VI. Flowering Plants. Dicotyledons: Celastrales, Oxalidales, Rosales, Cornales, Ericales.* Berlin, Springer-Verlag (2004).

HYDRASTIDACEAE

The family comprises a single monotypic genus, *Hydrastis canadensis*, widespread from Minnesota, southern Ontario, and Vermont, south to Oklahoma and Georgia, where it grows in the understory of deciduous woods. It is a rhizomatous perennial herb, with persistent bud scales and erect flowering stems to 50 cm high, bearing 2(3) cauline leaves in the upper part below a single terminal flower, and sometimes a long-petiolate basal leaf. The leaves are palmate and petiolate, divided into 5(–7) lobes that may be shallowly secondarily lobed and unevenly serrate. The flowers are bisexual, actinomorphic, and with a uniseriate perianth. The tepals are (2)3(4), petaloid, greenish white, caducous. The stamens are 30–50, spirally arranged, developing centripetally. The carpels are (8–)12–20, free, with a short style and discoid stigma, each 1-locular with 2 ovules inserted on the ventral margin. In fruit, the carpels are fleshy, red, forming a somewhat *Rubus*-like aggregate, each carpel 1-2-seeded and apparently tardily dehiscing along the ventral suture to release the black, shiny, obovoid seeds.

In most North American Floras, *Hydrastis* has been referred to Ranunculaceae, though more monographic treatments have referred it to Berberidaceae. In either family, it has often been associated with *Glaucidium* from Japan. In 1985,

Genera 1   Species 1
Economic uses
Medicinal properties

a strong case was made for placing it in its own monotypic family[i]. It differs from Ranunculaceae in its pollen, chromosome number, floral anatomy, vegetative anatomy, and embryological and chemical characters. Most of these features also separate it from Berberidaceae, from which it differs also in its undifferentiated perianth, many stamens, and numerous free carpels (the individual carpels somewhat similar in appearance to the whole fruits of Berberidaceae). Recent analyses have consistently placed it apart from both these families—for discussion and references see Glaucidiaceae. Like *Glaucidium*, *Hydrastis* seems to be an isolated genus derived from a common stock with the Ranunculaceae and Berberidaceae, perhaps closer to the latter but certainly worthy of separate family status. The similarity to *Glaucidium* is probably due to both genera retaining a plesiomorphic growth habit common to many families in this group. The family is not recognized in APG II.   RKB

[i] Tobe, H. & Keating, R. C. The morphology and anatomy of *Hydrastis* (Ranunculales): systematic reevaluation of the genus. *Bot. Mag.* (Tokyo) 98: 291–316 (1985).

HYDROLEACEAE

A family comprising 1 genus, *Hydrolea*, found in semiaquatic habitats across North America and the tropics, formerly included in Hydrophyllaceae.

Distribution USA (Texas to Chesapeake Bay), Central and South America to Argentina, tropical Africa (especially western Africa), tropical Asia to Taiwan, and northern Australia. All species are characteristic of wet habitats and are often rooted under water.

Description Herbs to subshrubs up to 2 m high, with stems erect to prostrate, succulent to woody, and sometimes with spongy aerenchyma. The stems often (in New World only) bear spines up to 3 cm long, which are borne 1–2 at nodes and are apparently modified shoots (rarely branching or producing small leaves). The leaves are alternate, ovate to linear, entire to serrulate, glabrous to pubescent, sometimes glandular. The flowers are borne in terminal corymbs, or leafy panicles, or in short axillary fascicles, or sometimes solitary in the leaf axils. The flowers are actinomorphic, bisexual, 5-merous, and hypogynous. The sepals are lanceolate to cordate. The corolla is campanulate, bright blue to purple or occasionally white, and the stamens are inserted on the tube. The ovary is 2-carpellate or rarely with 3–4 carpels, with 2–4 long styles and funnel-shaped stigmas, with axile placentation (see sections) and many ovules. The fruit is a globose capsule.

Classification *Hydrolea* was usually placed in its own family until 1875, when Asa Gray placed it as the last genus of Hydrophyllaceae in

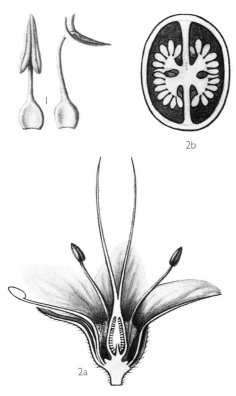

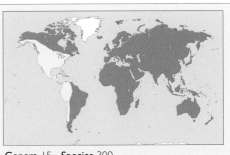

Genera 15 **Species** 300
Economic uses Garden ornamentals; limited medicinal value

HYDROLEACEAE. 1 *Hydrolea floribunda* stamen before (left) and after (right) dehiscence (×6). 2 *H. spinosa* (a) half flower (×4); (b) cross section of ovary showing ovules on placentas adnate to the septa (×10).

the tribe Hydroleae, which he attributed to Bentham, whose account of the family was published a year later. Subsequently, it has almost invariably been referred to a monogeneric tribe at the end of the Hydrophyllaceae. In the *Pflanzenreich*, Brand[i] treated it as a tribe separated from the other 2 tribes by its truly 2-locular ovary, with 2 styles rather than 1, and different placentation. In 1963, Constance said it had no close relationship with Hydrophyllaceae, but in a life-time's study of the latter he seems never to have formally excluded it. The molecular evidence[ii,iii], however, has consistently placed it outside Hydrophyllaceae and nearer to Sphenocleaceae or Solanaceae. Apart from the ovary characters, it lacks the typical cincinnate inflorescenes of Hydrophyllaceae and stands apart from that family in many ways, including its often spiny habit, aquatic habitat, and distribution extending outside the New World. Family status seems well justified.

The most recent revision[iv] divides the genus into 2 sections. Sect. *Hydrolea* (7 spp.) is confined to the New World, has the placenta

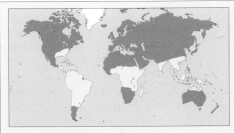

Genera 1 **Species** 11
Economic uses None

divided into 2 wings, and is commonly spiny. Sect. *Attaleria* (4 spp.) occurs in the Old World, has the placenta entire, and is never spiny.

Economic uses None known, but some species may be weeds in rice paddies.   RKB

i Brand, A. Tribus Hydroleae. In: Engler, A. (ed.), *Das Pflanzenreich* 4 (251): 174–185 (1913).
ii Chase, M. W. *et al*. Phylogenetics of seed plants: an analysis of nucleotide sequences from the plastid gene *rbc*L. *Ann. Missouri Bot. Gard.* 80: 528–580 (1993).
iii Ferguson, D. M. Phylogenetic analysis and relationships in Hydrophyllaceae based on *ndh*F sequence data. *Syst. Bot.* 23: 253–268 (1999).
iv Davenport, L. J. A monograph of *Hydrolea* (Hydrophyllaceae). *Rhodora* 90: 169–208 (1988).

HYDROPHYLLACEAE

PHACELIA FAMILY

A New World family related to Boraginaceae and usually resembling them in having a cincinnate inflorescence.

Distribution With the exclusion of *Hydrolea* and *Codon* (see below), the family is now considered to be restricted to the New World, from Alaska to Argentina. The center of generic diversity and greatest number of species is in the southwestern USA, especially California. Members of the family are mostly found in open, dry, sandy, or rocky habitats. A few species of *Nemophila*, *Phacelia*, and *Wigandia* are occasionally naturalized in the Old World.

Description Annual to perennial herbs to shrubs, or rarely (*Wigandia*) small trees to 5 m with massive leaves. The leaves are alternate or rarely (*Draperia*) opposite, simple and entire to palmate (*Romanzoffia*, some *Hydrophyllum*), or deeply pinnately divided, sometimes 3-pinnate, in *Wigandia* sometimes broadly ovate and up to 40 cm across. Irritant hairs are found in some genera, e.g., *Phacelia*, *Turricula*, and *Wigandia*. The inflorescences are characteristically cincinnate, the flowers opening as the coil unwinds, or sometimes flowers are solitary with long pedicels in the leaf axils. The flowers are more or less actinomorphic, bisexual, hypogynous, and 5-merous. The calyx has sepals fused at the base and is usually persistent and conspicuous and may be accrescent (especially *Tricardia*). The corolla is campanulate to funnel-shaped, with 5 usually short, rounded lobes, usually white to blue or purple, rarely yellow. The stamens are inserted on the corolla tube and may be exserted or included. The ovary is bicarpellate and 1-locular, with a single, often exserted, style and stigma (sometimes divided in *Phacelia*), and with parietal placentation and usually many ovules. The fruit is a capsule, dehiscing by 2–4 valves.

Classification As defined here, the family is a natural one, confined to the New World. The genus *Hydrolea*, with a pantropical semiaquatic distribution, until recently has usually been included as an aberrant tribe of its own but is

here excluded as the family Hydroleaceae. The genus *Codon* (2 spp., South Africa) has also usually been included but is surely to be excluded on molecular and morphological characters and referred either to Boraginaceae or a family of its own characterized by 10- to 12-merous flowers. The long-supposed relationship of Hydrophyllaceae with Polemoniaceae, of similar distribution, seems to be fictitious in the light of the molecular evidence placing the latter in the Ericales. Hydrophyllaceae makes Boraginaceae paraphyletic but differs from it by its parietal placentation and many-seeded dehiscent capsules. Easily distinguishable morphologically, it is much better treated as a separate family than sunk into Boraginaceae as in recent cladistic classifications (see APG II). Three tribes are supported by molecular data[i]. The first of these has been recognized by ovary characters since the *Pflanzenreich* account by Brand in 1913; the other tribes are less easy to diagnose morphologically.

Hydrophylleae consists of 5 genera and 25 spp. spread across North America and only *Pholistoma* extending slightly into Mexico: *Hydrophyllum* 8 spp., *Nemophila* 11 spp., *Ellisia* 1 sp., *Pholistoma* 3 spp., and *Eucrypta* 2 spp.

Phacelieae comprises 6 genera and 210 spp. (*Phacelia* 200 spp.) and extends from North America south to Argentina. The other genera are virtually confined to western North America. *Romanzoffia* has 5 spp. on wet cliffs and rocks from northern California to Alaska, *Hesperochiron* 2 spp. and *Draperia* and *Tricardia* 1 sp. each, all in the dry southwestern USA, and *Emmenanthe* with 1 sp. that jumps across the border from California to Guadalupe Island off Mexico.

Nameae contains 4 genera and 65 spp. from southwestern USA and West Indies to South America: the mostly herbaceous *Nama* (c. 50 spp.), predominately in Texas and Mexico but extending into the West Indies and south to Argentina; the herb to 3 m subshrub *Turricula* (1 sp.) in California and northern Mexico; the evergreen shrubby *Eriodictyon* (9 spp.) in California and northern Mexico; and the arborescent *Wigandia* (c. 5 spp.) from Mexico to Peru.

Economic uses Species of *Nemophila* and *Phacelia* are grown as ornamental annuals. *Wigandia* is widely cultivated as a large-leaved,

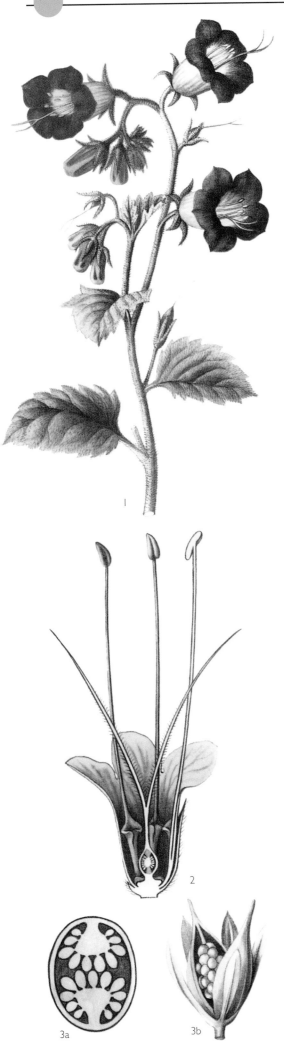

Genera 1 Species 22
Economic uses None

HYDROPHYLLACEAE. 1 *Phacelia minor* shoot with leaves and flowers (x⅔). 2 *P. tanacetifolia* half flower showing appendages on bases of filaments (x4). 3 *P. franklinii* (a) cross section of ovary with ovules on parietal placentas (x18); (b) dehiscing capsule with many seeds (x1½). 4 *Hydrophyllum virginianum* (a) shoot with pinnate leaf and flowers in a headlike cyme (x⅔); (b) dehiscing capsule with 2 seeds (x4).

fast-growing novelty tree in the tropics. Leaves of *Nama jamaicense* are said to be smoked as a tobacco in Argentina and are used to cure gastric ulcers in Mexico.   RKB

i Ferguson, D. M. Phylogenetic analysis and relationships in Hydrophyllaceae based on *ndh*F sequence data. *Syst. Bot.* 23: 253–268 (1999).

HYDROSTACHYACEAE

A monogeneric family of aquatic plants with reduced flowers.

Distribution *Hydrostachys* comprises 22 spp., of which 14 are endemic to Madagascar and the other 8 from continental Africa from the Democratic Republic of Congo and Tanzania to South Africa.

Description Almost stemless plants, with a basal holdfast usually 1–2 cm across attaching them to rocks in fast-flowing water by means of adventitious roots. The holdfast bears a rosette of narrow "leaves" up to 50(–100) cm long, which lie in the water flow, with several inflorescence spikes arising in the center of the rosette. The inflorescences are erect with a peduncle up to 30 cm and spike up to 15 cm, usually becoming exposed above the water as the level drops during the dry season but remaining submerged in some species. Each spike resembles

that of a *Plantago*. The "leaves" are composed of narrow cylindrical axes, up to 2 mm thick, which are sometimes unbranched but more often 1- to 2-pinnate in a regular pattern with up to 60 pairs of "fernlike" pinnae, the whole "leaf" then densely clothed with small scales ("emergences") of various shapes (cylindrical, linear, concave, flattened, strap-shaped, fimbriate), and usually up to 2 mm long but up to 5 mm in species with undivided axes (giving an appearance of a leafy liverwort). These emergences, unique in the flowering plants, contain bundles[i], and a clear distinction from the pinnae cannot be made. Indeed, since in *H. monoica* the male inflorescences and some roots are borne at the axils of the pinnae (see fig. 15A of Jäger-Zürn[i]), it may be that the "leaves" of *Hydrostachys* are in fact modified stems. The homologies of the various structures are unclear. Plants are dioecious or in *H. monoica* and sometimes *H. stolonifera* (both Madagascan) monoecious. The flowers are reduced, lacking sepals and petals, each one sessile and subtended closely by a small bract. The male flower consists of 2 more or less sessile thecae, variously regarded as derived from 1 stamen or 2. The female flower consists of a superior bicarpellate ovary with 2 terminal stylodia and a single locule with many ovules on parietal placentas. The fruit is a septicidal capsule and produces numerous small seeds. These are mucilaginous, which apparently enables them to attach themselves to rocks.

Classification The affinities of this peculiar family have been debated for many decades. A supposed relationship with the Podostemaceae, also characteristically growing on rocks in fast-flowing water, was rejected a century ago. Relationships with the Lamiaceae and Pittosporaceae have been suggested. However, in the last decade the molecular evidence, reviewed in FGVP[ii], has repeatedly placed *Hydrostachys* within the Hydrangeaceae, with its closest relatives likely to be the woody genera *Whipplea* and *Fendlerella* which grow in the dry country of southwestern USA[iii]. It appears that adaptation to the aquatic habitat has been accompanied by major changes in morphological characters.   RKB & BLS

i Jäger-Zürn, I. Anatomy of the Hydrostachyaceae. Pp. 129–196. In: Landolt, E., Jäger-Zürn, I., & Schnell, R. A. A., *Extreme adaptations in angiospermous hydrophytes*. Bornträger, Berlin & Stuttgart (1998).
ii Erbar, C. & Leins, P. Hydrostachyaceae. Pp. 216–220. In: Kubitzki, K. (ed.), *The Families and Genera of Vascular Plants. VI. Flowering Plants. Dicotyledons: Celastrales, Oxalidales, Rosales, Cornales, Ericales.* Berlin, Springer-Verlag (2004).
iii Albach, D. C., Soltis, D. E., Chase, M. W., & Soltis, P. S. Phylogenetic placement of the enigmatic angiosperm *Hydrostachys*. *Taxon* 50: 781–805 (2001).

ICACINACEAE

Small-flowered shrubs, trees, and lianes of tropical evergreen forests.

Distribution Pantropical and southern hemisphere, usually in evergreen forests.

Description Shrubs or trees up to 45 m (*Platea*) or high-climbing lianes up to 40 m (*Phytocrene, Sleumeria*), sometimes with tendrils that are modified branches or inflorescences, the lower parts sometimes forming a massive rootstock above ground (*Pyrenacantha*), and sometimes (e.g., *Phytocrene*) with irritating hairs. The leaves are alternate or, in the climbers, sometimes opposite, simple, usually entire

Genera 35 **Species** c. 300
Economic uses Minor local use for timber or edible fruits or tubers

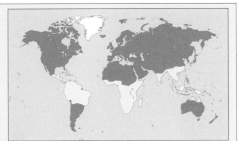

or rarely serrulate, sometimes palmately 3- to 7-lobed, often with prominent venation beneath. The inflorescence is racemose, sometimes compound, rarely cymose, sometimes a pendent linear spike. The flowers are actinomorphic, bisexual or unisexual (plants dioecious), (4-)5-merous, and small (mostly 3–5[7] mm). The calyx is shallowly cup shaped, rarely absent (*Pyrenacantha*). The petals are free or fused at the base or (e.g., *Iodes*) for most of their length. The stamens are as many as the petals or corolla lobes, free or inserted on the corolla, or with many congested anther-like chambers (*Polyporandra*). The ovary is superior, with a simple style or sessile stigma, normally 1-locular with 2 pendent apical ovules. The fruits are drupes, sometimes aggregated into large globose clusters (e.g., *Phytocrene*), rarely with a lateral fleshy colored appendage (*Apodytes*).

Classification The family was formerly included in Olacaceae, which is now placed in the Santalales. More recently, it has been placed in Celastrales, Theales, or Cornales. In APG II, it is referred to Euasterids I.

ICACINACEAE (LEFT AND ABOVE).
1 *Pyrenacantha volubilis* (a) leafy shoot with axillary inflorescences (x⅔); (b) female flower (x6); (c) male flower (x6); (d) vertical section of ovary (x6). 2 *Phytocrene bracteata* fruits (x⅔). 3 *Iodes usambarensis* (a) leafy shoot with tendril and inflorescence (x⅔); (b) female flower (x6); (c) fruits (x1); (d) male flower (x6).

The traditional concept of the family was difficult to define and recent molecular and morphological analysis[i] has segregated several families from it (see Stemonuraceae, Cardiopteridaceae, Leptaulaceae and Metteniusaceae, which are referred to Aquifoliales, and Pennantiaceae, which is referred to Apiales). The revised concept of Icacinaceae may be recognized by its racemose inflorescence of small flowers, glabrous stamens, 1-locular ovary with 2 pendent ovules, and drupaceous fruits.

Economic uses The wood is seldom of very good quality and is used only locally. Stems of lianes, such as *Phytocrene,* are said to hold fresh drinking water. In the New World, *Poraqueiba* has edible fruits and seeds, while starchy tubers of *Casimirella* are sold commercially. *Sarcostigma* seeds provide an oil used in India to treat rheumatism, and *Cassinopsis* leaves and bark are used to treat dysentery in Madagascar.   TMAU & RKB

[i] Kårehed, J. Multiple origin of the tropical forest tree family Icacinaceae. *American J. Bot.* 88: 2259–2274 (2001).

ILLICIACEAE
STAR ANISE

A small, disjunct family of aromatic shrubs and trees, comprising the single genus *Illicium.*

Distribution *Illicium* has a disjunct distribution from southeastern North America (*I. floridanum, I. parviflorum*), Mexico (*I. mexicanum*), Cuba (*I. cubense*), and Hispaniola (*I. ekmanii, I. hottense*) to East and Southeast Asia where most species occur.

Genera 1 **Species** c. 40
Economic uses Source of the spice star anise; limited medicinal value

Description Shrubs or small trees. The leaves are alternate, simple, entire, exstipulate, aromatic. The flowers are actinomorphic, bisexual, usually solitary or sometimes 2 or 3, axillary or subterminal, highly variable in number, size, and shape of perianth segments, stamens, and carpels. The perianth segments are 7–12 to numerous, free, spirally arranged, the outer sepaloid, intergrading with the inner more petaloid ones, which themselves are transitional to stamens. The stamens are 4–8 or numerous, with thick, short filaments. The ovary is superior, of (5–)8–15(–21) free carpels, sometimes appearing whorled; ovules are 1 per carpel. The fruit is a follicle, single or in aggregates of up to 13, with a well-developed system of ballistic dehiscence along the adaxial surface ejecting the seed. The seeds are 1 per follicle.

ILLICIACEAE. 1 *Illicium floridanum* leafy shoot and axillary flowers (x⅔). 2 *I. anisatum* dehiscing fruits—follicles (x1⅓).

Classification

The Illiciaceae was placed by Cronquist in the Illiciales, along with the Schisandraceae with which it is often united as a single family. The family is a member of the so-called ANITA grade, along with Amborellaceae, Nymphaeaceae, and Schisandraceae, and is one of the basal groups in the evolution of the angiosperms. It has been suggested that floral characters may have evolved in parallel in different groups of *Illicium* and, on their own, do not form a satisfactory basis for subdividing the genus. On the other hand, seed features may support a major division between New World and Old World species as indicated by molecular data[i].

Economic uses

Illicium verum (Star Anise), native to China and Japan, is widely cultivated in east Asia for its fruits, which are used as a culinary spice and as a flavoring for liqueurs along with or as a substitute for European aniseed. Anethole, which is extracted from the fruits, has many medicinal uses. Other species of *Illicium* are also used medicinally.　　VHH

[i]Hao, G., Saunders, R. M. K., & Chye, M.-L. A phylogenetic analysis of the Illiciaceae based on sequences of internal transcribed spacers (ITS) of nuclear ribosomal DNA. *Pl. Syst. Evol.* 223: 81–90 (2000).

IRVINGIACEAE

A small family of tropical trees, resembling *Ficus* in the extended stipules protecting the terminal bud.

Distribution

Desbordesia has 1 sp. from Nigeria to western Democratic Republic of Congo, *Klainedoxa* has 2 spp. from Guinea-Bissau to southern Sudan and Zambia, and *Irvingia* has 6 spp. from Sierra Leone to southern Sudan, Democratic Republic of Congo, and Angola, and 1 sp. from Thailand and Vietnam to Sumatra and Java. They occur in forests, on dry ground or occasionally on riverbanks or seasonally flooded forest.

Genera 3　Species 10
Economic uses Edible seeds and timber

Description

Large trees 10–55m, with buttressed trunks. The leaves are alternate, simple, entire, ovate to elliptical, shortly petiolate, with copious mucilage cells and canals. The stipules are large and conspicuous, fused in pairs, unequal and asymmetrically inserted, those at the stem apices long and curved and protecting the terminal bud from falling and leaving a scar. The inflorescence is a panicle that is terminal or axillary or on leafless branches. The flowers are actinomorphic, bisexual, up to 4 mm, 5-merous. The sepals are free or slightly fused. The petals are free, elliptical, widely spreading or reflexed. The stamens are 10, in 1 whorl, free, exceeding the petals, inserted below a prominent nectariferous disk that bears the ovary. The ovary is superior, either 2-locular (*Desbordesia*) or 5-locular, with a simple style and stigma, with 1 pendulous ovule per locule. The fruit is a 1(2)-seeded samara in *Desbordesia*, a drupe with 5 single-seeded pyrenes in *Klainedoxa*, and a drupe with a single pyrene in *Irvingia*[i].

Classification

The family has in the past been variously placed with the Linaceae *sensu lato* or Simaroubaceae but has been situated close to the Putranjivaceae and Euphorbiaceae by molecular data[ii] in Malpighiales of APG II.

Economic uses

Harris (*loc. cit.*) has recorded that the seeds of several species are eaten in Africa, while the mesocarp may produce a mucilaginous sauce used in cooking. Saplings are often used for building-poles.　　RKB

[i] Harris, D. J. Irvingiaceae. *Species Plantarum*, 1. Canberra, ABRS (1999).
[ii] Savolainen, V. *et al.* Phylogeny of the eudicots: a nearly complete familial analysis based on *rbc*L gene sequences. *Kew Bull.* 55: 257–309 (2000).

ITEACEAE

Two genera of evergreen trees and shrubs sometimes included in the Escalloniaceae.

Distribution

Itea (16 spp.) is found mainly in Southeast Asia and the Himalayas to China and Japan, with 1 sp. occurring in eastern North America. *Choristylis* (1 sp.) grows in eastern tropical and southern Africa.

Description

Trees and shrubs, the branches with chambered pith. The leaves are small, alternate, simple, petiolate, elliptical to lanceolate; the stipules are deciduous or persistent; the leaf blade margin is glandular-serrate, dentate, or spinose-dentate, rarely crenate or entire. The flowers are small, actinomorphic, bisexual or unisexual (plants polygamonoecious), in axillary racemose or paniculate inflorescences. The sepals are 5, connate below into a turbinate or obconical

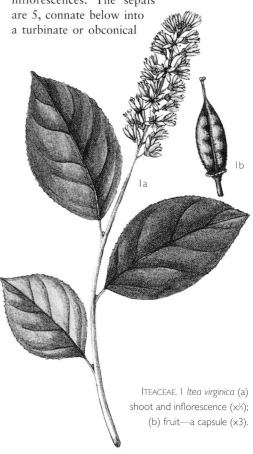

ITEACEAE. 1 *Itea virginica* (a) shoot and inflorescence (x⅔); (b) fruit—a capsule (x3).

Genera 2 Species c. 17
Economic uses Some species are used in herbal medicine; a few ornamentals

tube (hypanthium), with 5 lobes, persistent. The petals are 5, valvate, persistent. The stamens are 5, inserted on the margin of the nectary disk. The ovary is superior to semi-inferior, with 2 carpels that are fused or initially free and becoming connate (*Choristylis*); the locules are 2, placentation axile, with numerous ovules; the styles free or remaining partly joined at the stigma. The fruit is a septicidal capsule with persistent calyx lobes and petals, the valves usually remaining attached by the stigma. The seeds are numerous.

Classification *Itea* has been included in the Grossulariaceae or Escalloniaceae but is now widely recognized as a separate family in the Saxifragales *sensu* APG II. A sister relationship between Iteaceae and Pterostemonaceae (q.v.) is strongly supported by molecular data[i].

Economic uses Some species of *Itea* such as *I. virginica* (Sweetspire) are cultivated as ornamental shrubs, and several are used locally in traditional medicine in China and India.   VHH

[i] Fishbein, M., Hibsch-Jetter, C., Soltis, D. E., & Hufford, L. Phylogeny of Saxifragales (angiosperms, eudicots): analysis of a rapid, ancient radiation. *Syst Biol.* 50: 817–847 (2001).

IXERBACEAE

A monotypic family comprising the species *Ixerba brexioides*, an evergreen tree up to 15 m, from the low montane forests of the north of North Island, New Zealand, with alternate, dentate, exstipulate leaves. The flowers are large, waxy white, bisexual, strongly scented, in terminal subumbellate inflorescences. The sepals are 5, thick, the outer 2 shorter and broader at the base than the inner 3, which completely enclose the stamens and carpels[i]. The petals are 5, clawed. The stamens are 10, with long filaments. There is a 5-lobed nectary disk. The ovary is superior to slightly inferior, of 5 fused carpels, tapering into the persistent style; the placentation is axile, with 2 ovules per locule. The fruit is a coriaceous, loculicidal capsule, with woody valves. Seeds few, black, arillate.

Ixerba was originally associated with *Brexia* (of which it is an anagram) and placed in the Celastraceae. Later, it was included in the Saxifragaceae or Escalloniaceae, or as a monotypic family (Ixerbaceae), which was placed by Takhtajan (1997) in the Brexiales alongside the Brexiaceae. Molecular evidence has indicated the inclusion of Ixerbaceae in the Crossomatales[i,ii], and that inclusion has been confirmed by a detailed study of the floral structure and systematics of the order. A sister relationship with Strasburgeriaceae was suggested by *rbcL* analyses[ii,iii] and a 3-gene analysis[iv] and is supported by floral structure and wood anatomy[i]. Tawari honey obtained from bees that pollinate the flowers of *Ixerba brexioides* is highly prized.   VHH

[i] Matthews, M. L. & Endress, P. K. Comparative floral structure and systematics in Crossomotales (Crossomataceae, Stachyuraceae, Staphyleaceae, Aphloiaceae, Geissolomataceae, Ixerbaceae, Strasburrgeriaceae). *Bot. J. Linn Soc.* 147: 1–46 (2005).
[ii] Savolainen, V. *et al.* Phylogeny of the eudicots: a nearly complete familial analysis based on *rbcL* gene sequences. *Kew Bull.* 55: 257–309 (2000).
[iii] Sosa, V. & Chase, M. W. Phylogenetics of Crossosomataceae based on *rbcL* sequence data. *Syst. Bot.* 96–105 (2003).
[iv] Cameron, K. M. On the phylogenetic position of the New Caledonian endemic families Paracryphiaceae, Oncothecaceae, and Strasburgeriaceae: a comparison of molecules and morphology. *Bot. Rev.* 68: 428–443 (2003).

IXONANTHACEAE

A family of pantropical trees and shrubs, formerly referred to Linaceae.

Distribution *Ixonanthes* has 3 spp. from India to New Guinea (but not Java, Lesser Sunda, or Moluccas), *Allantospermum* has 1 sp. in Malaysia and Borneo and 1 sp. in Madagascar, *Phyllocosmus* has 9 spp. in tropical Africa, *Ochthocosmus* has 7 spp. in northern South America, and *Cyrillopsis* has 2 spp. in Venezuela and northeastern Brazil. All occur in forests, woodland (often on sand), or savanna.

Description Shrubs or trees up to 30 m. The leaves are alternate, simple, more or less elliptical, shortly petiolate with entire or serrate margins. Stipules are small, lateral (intrapetiolar in *Allantospermum*), inconspicuous, caducous. The inflorescence is a terminal or axillary cyme or raceme. The flowers are actinomorphic or nearly so, bisexual, 5-merous, small. The sepals are more or less imbricate. The petals are free, imbricate and often contorted, persistent in fruit. The stamens are 5, 10, 15, or 20, in 1 whorl, the base of the filaments expanded but not connate, inserted outside a shortly cylindrical disk and free from it in *Allantospermum*. The ovary is superior or slightly inferior, syncarpous, and of 5 carpels or in *Cyrillopsis* only 2, with a simple style and stigma, the locules often divided incompletely by outgrowths of the carpel wall (as in Linaceae), with 2 pendulous ovules in each locule or in *Allantospermum* only 1. The fruit is a capsule, in *Allantospermum* with the valves splitting away from a central column.

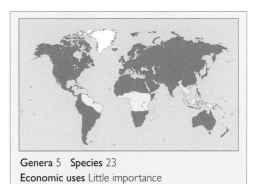

Genera 5 Species 23
Economic uses Little importance

Classification The relationships of the 5 genera have been disputed in recent decades. They were combined with Irvingiaceae in 1 family with 2 subfamilies by Forman[i], while Nooteboom[ii] thought *Allantospermum* to be nearer to Simarubaceae (in which Irvingiaceae has been included). For further review of these controversial relationships, see Kool[iii]. Molecular evidence places these putative relatives rather widely apart, with Ixonanthaceae and Irvingiaceae both well separated in the broad Mapighiales[iv], but Simarubaceae in Sapindales. At present the 5 genera seem to be generally treated together as 1 family, but the position of *Allantospermum* seems somewhat anomalous.

Economic uses None reported.   RKB

[i] Forman, L. L. A new genus of Ixonanthaceae with notes on the family. *Kew Bull.* 19: 517–526 (1965).
[ii] Nooteboom, H. P. The taxonomic position of Irvingioideae, *Allantospermum* Forman and *Cyrillopsis* Kuhlm. *Adansonia* II 7: 161–168 (1967).
[iii] Kool, R. Ixonanthaceae. *Flora Malaysiana* 10: 621–627 (1988).
[iv] Savolainen, V. *et al.* Phylogeny of the eudicots: a nearly complete familial analysis based on *rbcL* gene sequences. *Kew Bull.* 55: 257–309 (2000).

JUGLANDACEAE
WALNUTS, HICKORIES, PECAN NUTS

A family of chiefly deciduous, often aromatic, trees, well known for the edible fruits of its members.

Distribution The family is well-represented in temperate and subtropical regions of the northern hemisphere, with extensions reaching the tropics of Asia and America. Most *Juglans* (21 spp.) grow in southeastern Europe extending to East Asia and Japan, and from North America and the West Indies into northern South America. *Carya* (c. 16 spp., including *Annamocarya*) extends from eastern North America to southern Mexico, and from eastern China to Vietnam and eastern Pakistan. Genera that are restricted to the New World are *Alfaroa* (c. 7 spp.) from Mexico to Colombia, and *Oreomunnea* (3 spp.), from Mexico to Panama. *Cyclocarya* (1 sp.), *Engelhardia* (16 spp., including *Alfaropsis*[i]), *Platycarya* (1 sp.) and *Pterocarya* (6 spp.) are all found in the Far East, with only *Pterocarya* reaching as far west as the Caucasus.

Description Resinous trees, usually deciduous, sometimes evergreen (*Alfaroa, Oreomunnia*). The leaves are alternate, rarely opposite (*Alfaroa, Oreomunnea*), pinnate, usually aromatic, and lacking stipules. The flowers are bracteate and unisexual (plants monoecious); male flowers usually in catkinlike, pendulous inflorescences forming on the previous year's growth, the females in smaller erect or long pendulous (*Pterocarya*) spikes on the new growth. The perianth is typically 4-lobed but is often reduced or absent. The male flowers have 3–40 free stamens in 2 or more series, with short filaments and 2-locular anthers opening longitudinally. Pollination is by wind. The female flowers have an inferior ovary of 2 fused carpels forming a single locule containing 1 erect orthotropous ovule, lacking endosperm. The style is short with 2 stigmas. The fruit is a drupaceous nut or 3-winged and samaroid, tightly enclosed by the coriaceous or fibrous husk developed from the perianth, bracts, and bracteoles, dehiscent in most species of *Carya* and *Juglans*. The characteristic boat-shaped walnut halves, obtained on cracking open, do not correspond to the 2 carpels as the suture is along their midribs.

Classification The relationships of the family have recently been established (APG II) as belonging to the Fagales with a close sister-group of the family Rhoipteleaceae. Previously, the family had been placed close to the Anacardiaceae in the Rutales[ii] and even as an order, Juglandales, in the Hamamelididae[iii]. The subfamilial classification has been controversial, based on the morphology of the inflorescence and fruit, particularly the wings and husk, characters associated with the mode of dispersal. One classification[iv] isolates the Platycaryoideae with *Platycarya* from the Juglandoideae, the latter with 3 tribes, while another classification[v,vi,vii], based on the good fossil evidence available, realigns the genera in to 2 clades: a united Engelhardieae and Platycaryeae (*Platycarya, Engelhardia, Oreomunnea,* and *Alfaroa*), with the remaining genera in the Juglandoideae. This shows wind and animal dispersal has arisen several times. A more recent *cp*DNA cladistic analysis[viii] shows 2 major clades: the Engelhardieae (*Engelhardia, Oreomunnea,* and *Alfaroa*) and the rest. This last is incongruent with the fossil evidence.

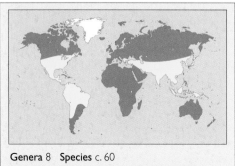

Genera 8 Species c. 60
Economic uses Edible nuts and oil (walnut, hickory, and pecan), timber, and ornamentals

Economic uses *Juglans*, and to a lesser extent *Carya*, produce valuable timber and are much prized for their fine grain and toughness, used extensively for veneers. Though timber may be the main use, the family is perhaps best known for its edible nuts, the walnuts (*Juglans regia* and other spp.), pecan nuts (*Carya pecan, C. illinoinensis*), and hickory nuts (*C. ovata*). The oil from the nuts has a high percentage of oleic acid and is also regarded as a healthy oil. As a result, it is used in foods, in the manufacture of cosmetics and soap, and as a drying agent in paints. It is too expensive to use compared to commercially produced olive oil except for specialist uses. Species of walnut, hickory, and wingnut (*Pterocarya*) are also often grown for their ornamental value.   SLJ

[i] Stone, D. E. Juglandaceae. Pp. 348–359. In: Kubitzki, K., Rohwer, J. G., & Bittrich, V. (eds), *The Families and Genera of Vascular Plants. II. Flowering Plants. Dicotyledons: Magnoliid, Hamamelid and Caryophyllid Families.* Berlin, Springer-Verlag (1993).
[ii] Thorne, R. F. "Hamamelididae": a commentary. In: Crane, P. R. & Blackmore, S. (eds), *Evolution, Systematics and Fossil History of the Hamamelidaceae.* Systematics Association. Oxford, Clarendon Press (1989).

JUGLANDACEAE. I *Juglans regia* (a) imparipinnate leaf (x⅔); (b) male, catkinlike inflorescence borne on old wood (x⅔); (c) tip of shoot with female flower with plumose stigmas (x⅔); (d) fruit (x⅔); (e) fruit with fleshy husk removed to show hard, sculptured endocarp (inner fruit wall) (x⅔); (f) fruit with endocarp removed to show seed with contorted cotyledons (x⅔); (g) habit of an old tree.

iii Wolfe, J. A. Leaf-architectural analysis of the Hamamelididae. In: Crane, P. R. & Blackmore, S. (eds), *Evolution, Systematics and Fossil History of the Hamamelidaceae*. Systematics Association. Oxford, Clarendon Press (1989).
iv Manning, W. E. The classification within the Juglandaceae. *Ann. Missouri Bot. Gard.* 65: 1058–1087 (1978).
v Manchester, S. R. The fossil history of the Juglandaceae. *Monogr. Ann. Missouri Bot. Gard.* 21: 1–137 (1987).
vi Gunther, L. E., Kochert, G., & Giannasi, D. E. Phylogenetic relationships of the Juglandaceae. *Pl. Syst. Evol.* 192: 11–29 (1994).
vii Smith, J. F. & Doyle, J. J. A cladistic analysis of chloroplast DNA restriction site variation and morphology for the genera of the Juglandaceae. *American J. Bot.* 82: 1163–1172 (1995).
viii Manos, P. & Stone, D. E. Evolution, phylogeny, and systematics of the Juglandaceae. *Ann. Missouri Bot. Gard.* 88: 231–269 (2001).

KIRKIACEAE

Distribution *Kirkia* occurs from Ethiopia and Somalia south to the Transvaal, and in Namibia, and a single species in west–central Madagascar (Bemara-ha) formerly referred to *Pleiokirkia* is here also included in *Kirkia*. All species are usually in dry habitats, often on limestone hills or open woodland.

Genera 1 | Species 6
Economic uses Timber, live electric fences, bark for fibers

Description Small to fairly large deciduous trees, up to 20 m or more high. The leaves are alternate but characteristically clustered at the ends of long or short shoots, leaving a conspicuous leaf scar on falling. Leaves are compound, imparipinnate, with 2–30 or more pairs of opposite leaflets that are narrowly lanceolate to suborbicular and serrate to entire. The inflorescence is an axillary thyrse of dichotomous cymes borne below the subterminal cluster of leaves. The flowers are actinomorphic, up to 4 mm long, 4-merous, usually dimorphic, and unisexual or at least functionally unisexual (plants monoecious or polygamomonoecious; *P. leandrii* from Madagascar is dioecious[i]), with some parts reduced or vestigial. The sepals are 4, slightly united at the base. The petals are 4, spreading or erect. The stamens are 4, alternating with the petals, inserted beneath an annular fleshy disk. The ovary is superior, angular or lobed, of 4–8 connate carpels, each attached to a central columella and each with a single locule that bears a single ovule or one fertile ovule and one reduced one aborting[i]. The style is terminal, with a capitate stigma, and at anthesis the stigma is cast off and the style divides into its 4–8 parts (1 per carpel) with the carpels separating from each other but remaining attached to the columella. The fruit is dry and breaks up into 4–8 single-seeded mericarps, remaining suspended from the apex of the columella and carrying on their dorsal side the portion of the style to which they were attached.

Classification *Kirkia* has been regarded as a subfamily of Simaroubaceae since Engler's account in the Pflanzenfamilien (1931). Recent studies of the fruit anatomy and molecular data have supported raising Engler's subfamilies to family rank (see Simaroubaceae for discussion and references). Kirkiaceae is retained with Simaroubaceae in the Sapindales of APG II. The single Madagascan species has been recognized as a separate genus *Pleiokirkia* said to be distin-guished from *Kirkia* by its united styles and 8 carpels, but since *K. dewinteri* from Namibia also has 8 carpels, and *K. tenuifolia* from East Africa may have 4–6, it seems preferable to combine the 2 genera into one.

Economic uses *K. acuminata* is planted as a live fence in southern tropical Africa, and pro-vides poles, planks and veneered wood. Swollen roots of this species are chewed to quench thirst, and also cure toothache, while the fibers from the bark are woven into cloth. The roots of *K. tenuifolia* in East Africa and of *K. wilmsii* in the Transvaal (tubers up to 30 cm across) are also eaten to quench thirst, and the bark of the latter species is used to make cord[ii]. RKB & BLS

i Schatz, G. Kirkiaceae. *Generic Tree Flora of Madagascar*: 220 (2001).
iiStannard, B. L. A revision of *Kirkia* (Simaroubaceae). *Kew Bull.* 35: 829–839 (1981).

KOEBERLINIACEAE
ALLTHORN, JUNCO

Xerophytic shrubs with spiny photo-synthetic stems and deeply penetrating roots[i]. There is 1 sp., *Koeberlinia spinosa*, found in arid regions of southwestern USA and northern Mexico and Bolivia in semi-arid inter-Andean chaco woodland. It is a spiny shrub or tree with photosynthetic stems to 5 m tall, bearing leaves in the

Genera 1 | Species 1
Economic uses None recorded

wet season only. The leaves are simple, elliptic, and entire. Inflorescences in short axillary racemes. The flowers are 4-merous, minutely bracteate. The sepals are 4, free, in 2 whorls, small. The petals are 4, in 1 whorl, free but overlapping and clawed, white. The stamens are 8, free, diplostemonous. There is a gyno-phore supporting an ovary of 2 fused carpels bearing a subulate style. The fruit is a berry.

The family Koeberliniaceae is probably most closely related to Bataceae[iii] and is in the peripheral Brassicales. It has been included in the Capparaceae in most previous systems. There are no economic uses recorded. AC

i Gile, L. H., Gibbens, R. P., & Lenz, J. M., Soils and sediments associated with remarkable, deeply penetrating roots of crucifixion thorn (*Koeberlinia spinosa* Zucc.). *J. Arid Envir.* 31: 137–151 (1995).
iihttp://www.nybg.org/botany/nee/ambo/Veg.html
iii Hall, J. C., Iltis, H. H., & Sytsma, K. J Molecular phylogenetics of core Brassicales, placement of orphan genera *Emblingia, Forchhammeria, Tirania*, and evolution. *Syst. Bot.* 29: 654–669 (2004).

KRAMERIACEAE
RHATANY FAMILY

Hemiparasitic small-leaved herbs and shrubs with bilateral flowers and spiny fruits.

Distribution From Nevada and Kansas through Central Am-erica and the Andes to Chile; from Flori-da, through the West Indies and Guyana, to eastern Brazil[i]; absent from most of the Amazon basin. Centers of diversity are in Mexico and eastern Brazil. They occur in arid or sea-sonally dry habitats, rising to an altitude of 3,600 m in the

Genera 1 | Species 18
Economic uses Some value in traditional medicine; also used in tooth-paste manufacture

Andes. They are very particular in their choice of hosts for their hemiparasitic habit.

Description *Krameria* includes prostrate perennial herbs to shrubs, up to 2 m high, with spinose branches in *K. grayi* of USA and Mexi-co, and with haustoria penetrating the roots of hosts. The leaves are alternate, simple or in *K. cytisoides* 3-foliate, rather small (3.5 cm) to minute (1 mm), usually narrowly elliptical to linear, sessile or petiolate. The inflorescence is a solitary axillary flower or a terminal raceme or panicle, each flower subtended by a pair of bracteoles. The flowers are bisexual, zygomor-phic, 4- to 5-merous. The sepals are 4 or 5, free, spirally arranged, unequal, with the lowermost, the largest, all showy and colored (pink to pur-ple or rarely the uppermost ones yellow), lanceolate and acute, concave to saccate, usual-ly spreading or reflexed. The petals are 5 or rarely 4, dimorphic: the 2 lowermost petals are modified into fleshy orbicular to rectangular oil-secreting glands that lie either side of the ovary, their outer surfaces striate to rugose or with saccate blisters in their upper part, while the upper 3 (rarely 2) petals are petaloid (pink to purple) but less showy than the sepals, insert-ed at the base of the uppermost sepal and forming a "flag" above the stamens and ovary, free or connate for up to three-quarters of their length. The stamens are 4, borne in 2 pairs, 1 pair above the other, or occasionally (*K. lap-pacea*) 3 with 2 below and 1 above, all inserted between the flag petals and the ovary. The ovary is superior, pyriform to ovoid, and curved

KRAMERIACEAE. 1 *Krameria lappacea* shoot with alternate leaves, flowers, and fruits (x⅔). 2 *K. tomentosa* (a) half flower (x2); (b) cross section of ovary (x4); (c) bristly, barbed, indehiscent fruit (x2⅔). 3 *K. cistoidea* (a) flowering shoot (x⅔); (b) leaf (x3); (c) flower with 2 opposite subtending bracts, 5 large sepals, 5 unequal petals, 4 stamens and a simple style (x2); (d) flower side view showing petals of 2 sizes (x2); (e) small anterior petal (x4); (f) large posterior petal (x4); (g) stamen (x4); (h) anther showing porose dehiscence (x12); (i) cross section of anther (x12); (j) gynoecium (x6). 4 *K. argentea* fruit (x1½).

upward into an arcuate style, essentially bicarpellate but with one carpel aborting early, the fertile carpel 1-locular with 2 pendulous ovules. The fruit is a single-seeded capsule (or often all seeds aborting and the capsule sterile), usually conspicuously adorned with radiating spines (or spineless in *K. grandiflora* only), the spines variously bearing retrorse barbs at their tip or along their distal part or sometimes without barbs.

Classification The very characteristic bilateral flowers have usually suggested a close relationship with Polygalaceae, or sometimes with Leguminosae. Cronquist and Takhtajan placed the family in Polygalales. Molecular data, however, place the family sister to Zygophyllaceae, a family to which they show little obvious similarity. In APG II, it is unplaced to order in Eurosids I.

Economic uses The Rhatany, *K. lappacea*, was claimed by Ruiz in 1797 to be second only to Cinchona in importance as a medicinal herb. The roots were extensively imported into Europe for a wide range of uses. Today its uses are confined to producing dyes, an ingredient for toothpaste because of its styptic qualities, and in cosmetics[ii]. RKB

[i] Simpson, B. B. *Flora Neotropica*. Monograph 49: Krameriaceae. 108 pp. New York, New York Botanical Garden (1989).
[ii] Simpson, B. B. Krameriaceae (Rhatany family). In: Smith, N. *et al.* (eds), *Flowering Plants of the Neotropics*. Princeton, Princeton University Press (2004).

LACISTEMATACEAE

A family of neotropical shrubs and small trees with catkinate inflorescences of apetalous unistaminate flowers.

Distribution *Lacistema* has c. 11 spp., from Mexico and the West Indies to northern Argentina, in low-altitude rain forest or dry forest. *Lozania* has 4 spp., from Costa Rica to Peru, in lowland or montane wet forests.

Genera 2
Species c. 15
Economic uses None recorded

Description Shrubs or small trees up to 6 m. The leaves are alter-nate, simple, shortly petiolate, coriaceous, more or less elliptical, usually up to 12 × 7 cm, with entire (*Lacistema*) or crenate (*Lozania*) margins, with small caducous stipules. Inflorescences in *Lacistema* are clustered in leaf axils, catkinlike, short (1–2 cm) and dense, in *Lozania* one to several in the axils, more lax, and up to 6 cm. The flowers are small (c. 1 mm), more or less zygomorphic, hermaphrodite or sometimes unisexual (plants monoecious or andromonoecious). The sepals are (1)2–6, somewhat unequal, or sometimes absent. The petals are absent. The stamen is single, inserted on a semilunar to annular disk, with the anther thecae separated by a widely divergent connective. The ovary is superior, 2- to 3-carpellate, 1-locular, with a single style and 2–3 stigmas. Placentation is parietal with 1–2 ovules on each placenta. The fruit is a capsule with 1–2 seeds with a brightly colored sarcotesta.

Classification The family has always been considered to be in the old Parietales, often thought to be close to Flacourtiaceae and Salicaceae. A recent molecular study of the Flacourtiaceae group has placed it nearer to Ctenolophonaceae, Goupiaceae and Violaceae, but the authors have surprisingly commented that they expect further data to place it in Salicaceae *sensu lato*[i]. Separate family status seems desirable. RKB

[i] Chase, M. W. *et al.* When in doubt, put it in *Flacourtiaceae*: a molecular phylogenetic analysis based on plastid *rbc*L DNA sequences. *Kew Bull.* 57: 141–181 (2002).

LACTORIDACEAE

Lactoridaceae comprises a single species, *Lactoris fernandeziana,* which is confined to the island of Masatierra in the Juan Fernandez archipelago. It is a much-branched subshrub, up to 1 m high, found in the understory of moist forests between 250 and 690 m altitude, or rarely in more open habitats, and about 1,000 plants are thought to exist[i]. Pollen resembling that of *Lactoris* has been found in deep sea deposits and around Australia, and it is thought to have been widespread in the southern hemisphere in the late Cretaceous (around 70–90 million years ago[ii]); it now just survives in a refuge on Juan Fernandez on a volcanic island only 3–4 million years old. The stems are brittle, up to 1 cm thick, often with zigzag branching. The leaves are alternate and distichous, up to 18 × 10 mm, obovate, truncate to retuse at the apex, with the 2 stipules fused into a conspicuous short ochrea. The flowers are actinomorphic, either hermaphrodite or female only (plants dioecious or polygamodioecious), borne terminally, though often appearing axillary, in small cymes of 3 flowers or solitary[iii]. They are small, green, up to 5 mm, with 3 perianth segments. In the hermaphrodite flowers, the stamens are 6, apparently in 2 whorls. The ovary is superior, with 3 carpels that are free or slightly connate at the base, each with 4–8 ovules on an intruded marginal placenta. Each carpel develops into a small beaked dehiscent follicle.

The family is referable to the Piperales, its 3-merous flowers suggesting affinity with Aristolochiaceae and the monocots. Its inclusion within Aristolochiaceae on molecular evidence has not been widely accepted. Affinities with Piperaceae are suggested by wood anatomy, and with that and Saururaceae by the stipules.   RKB

Genera 1   Species 1   Economic uses None

[i] Stuessy, T. F., Crawford, D .J., Anderson, G. J., & Jensen, R. J. Systematics, biogeography and conservation of Lactoridaceae. *Perspect. Pl. Ecol. Evol. Syst.* 1: 267–290 (1998).
[ii] Macphail, M. K., Partridge, A. D., & Truswell, E.M. Fossil pollen records of the problematical primitive angiosperm family Lactoridaceae in Australia. *Pl. Syst. Evol.* 214: 199–210 (1999).
[iii] Gonzlez, F. & Rudall, P. The questionable affinities of *Lactoris*: evidence from branching pattern, inflorescence morphology and stipule development. *American J. Bot.* 88: 2143–2150 (2001).

LAMIACEAE (LABIATAE)

MINT FAMILY

A large family of herbs, shrubs, or trees, usually with opposite leaves and markedly zygomorphic flowers. Its circumscription has changed considerably, as noted below, and it will come as a surprise to many to find genera such as *Tectona* (teak), *Gmelina,* *Vitex, Clerodendrum,* and *Caryopteris* included.

Distribution The family is cosmopolitan, ranging from tropical forests to arctic tundra, and from sea level to high altitude, with several main centers of diversification[i]. It is particularly characteristic of the Mediterranean region, where genera such as *Phlomis, Rosmarinus, Salvia, Sideritis, Micromeria,* and *Thymus* are common aromatic components of maquis, matorral, and garrigue communities.

Description Annual or perennial herbs, through shrubs to small or large trees (*Tectona grandis* to 30 m high or more) or rarely woody climbers. The stems are characteristically square in section in the more herbaceous genera. The leaves are opposite or very rarely alternate (some *Aeollanthus*) or whorled, usually simple, and often variously toothed on the margins, but sometimes deeply divided (e.g., some *Lavandula*) or 3-foliate (*Cedronella*) or digitately (*Vitex*), pinnately (*Petraeovitex*) or even bipinnately (*Perrierastrum,* now in *Plectranthus*) compound, sometimes simple and very large (*Tectona grandis* to 35 × 25 cm). The inflorescence is cymose, usually dichasial, but sometimes monochasial, lax or often contracted into axillary verticillasters, usually with many bracts and bracteoles. The flowers are gamopetalous and strongly zygomorphic, or occasionally secondarily actinomorphic. The calyx has a short tube, often with 10–15 conspicuous nerves, and (4)5 lobes or teeth, or 2-lipped (*Scutellaria*), rarely fleshy in fruit (*Hoslundia*). The corolla is usually strongly 2-lipped but sometimes subactinomorphic. Stamens are usually 4 (2 long and 2 short) and characteristically exserted from the corolla tube. There is often a nectariferous disk below the ovary. The ovary is superior, of 2 carpels, divided into 2 or 4 locules by intrusion of the carpel wall, often deeply 4-lobed and with a gynobasic style, with 4 basal or sub-basal ovules. When the ovary is entire, the fruit may be a drupe and have 4 pyrenes, but when the ovary is 4-lobed the fruit breaks up into 4 dry nutlets.

Classification The family was until recently restricted to herbs and shrubs (rarely trees), characterized by a 4-lobed ovary with a gynobasic style. It is now much enlarged[ii,iii] by the transfer to it of many genera formerly referred to Verbenaceae (including many large trees) and the inclusion of the Chloanthaceae (10 genera of shrubs endemic to Australia). The family now parallels the Boraginaceae in having many tropical tree genera, in which the ovary is entire with a terminal style, and many often temperate genera, which are herbaceous to shrubby and characterized by a deeply 4-lobed ovary and gynobasic style. The tropical trees are thought to represent a more primitive condition, while the often temperate herbaceous genera probably represent a derived condition. Athough the family is now much less easily characterizedthan formerly, it now forms a more coherent evolutionary unit. The family falls within the gamopetalous families related to Scrophulariaceae, but its closest relative is uncertain.

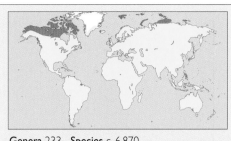

Genera 233   Species c. 6,870
Economic uses Timber (especially teak), many garden ornamentals, aromatic and culinary herbs, and plants yielding essential oils

LAMIACEAE. 1 *Stachys sylvatica* shoot with opposite leaves and terminal inflorescence (x⅔). 2 *Scutellaria indica* flowering shoot (x⅔).

Molecular evidence suggests Symphoremataceae, Verbenaceae, Acanthaceae, Pedaliaceae
and also Scro- phulariaceae[iv] as possibilities.
The family is divided into 6 subfamilies:

SUBFAM. VITICOIDEAE Includes 10 genera and
about 525 spp., including *Vitex* with c. 250 spp.
and *Premna*, with c. 200 spp. Pantropical or
rarely extending into temperate areas. Ovary
entire, developing into a drupe with 1–2 pyrenes.

SUBFAM. AJUGOIDEAE Includes 34 genera with c.
1,075 spp., including *Clerodendrum* with c. 500
spp., mostly woody and tropical but extending
into more temperate regions, particularly in
Teucrium (250 spp., especially in the Mediterranean region). They differ from the Viticoideae
in having usually actinomorphic or single-lipped
flowers and distinctive pollen.

SUBFAM. PROSTANTHEROIDEAE Includes 16 genera and 316 spp. confined to Australia, all of
which are non-aromatic shrubs to small trees
with albuminous seeds and often stellate or
branched hairs. There are 2 tribes: Chloantheae
(previously family Chloanthaceae), with an
entire ovary, and Westringeae, with a 4-lobed
ovary and gynobasic style.

SUBFAMILY SCUTELLARIOIDEAE Widespread
comprising 4 genera and 362 spp., including
Scutellaria with c. 360 species. Herbs or
shrubs, with a 2-lipped calyx with the lips
entire or only shallowly lobed, and the ovary

shallowly to deeply 4-lobed with the style terminal or rarely gynobasic.

SUBFAMILY LAMIOIDEAE Cosmopolitan with
63 genera and c. 1,215 spp., including *Stachys*
with c. 300 spp., *Sideritis* with c. 140 spp., and
Phlomis and *Leucas* each with c. 100 spp.
Mostly non-aromatic herbs or shrubs, with a 4-
lobed ovary and gynobasic style, and their
pollen usually 3-colpate.

LAMIACEAE. 3 *Salvia roemeriana* flowering shoot (×⅔). 4 *Salvia* sp. (a) section of flower showing stamen with much elongated connective (×2); (b) detail of stamens (×3). 5 *Plectranthus welwitschii* (a) flowering shoot showing the square stem characteristic of herbaceous members of the family (×⅔); (b) detail of flower (×2). 6 *Teucrium fruticans* (a) flowering shoot (×⅔); (b) detail of flower (×2). 7 *Rosmarinus officinalis* flower with stigma and stamens projecting from the corolla (×⅔). 8 *Lamium maculatum* 4-lobed ovary (a) entire (×2) and (b) in vertical section (×9) showing style attached to base of ovary (i.e. gynobasic). 9 *Clerodendrum thomsoniae* (a) shoot with cymose inflorescence of flowers which have an inflated, winged calyx (×⅔); (b) fruit—a 4-lobed drupe (×1). 10 *Vitex agnus-castus* (a) shoot bearing digitate leaves and cymose inflorescences (×⅔); (b) flower (×2); (c) corolla opened out to show stamens of 2 lengths (didynamous) (×3); (d) fruit (×4); (e) cross section of fruit showing 4 locules (×4).

SUBFAMILY NEPETOIDEAE Comprises 105 genera
and c. 3,190 spp., including *Salvia* widespread
with c. 900 spp., *Plectranthus* confined to the
Old World tropics, especially Africa, with c.
300 spp., *Hyptis* with about 280 spp., nearly all
in the New World tropics, *Thymus* with 220

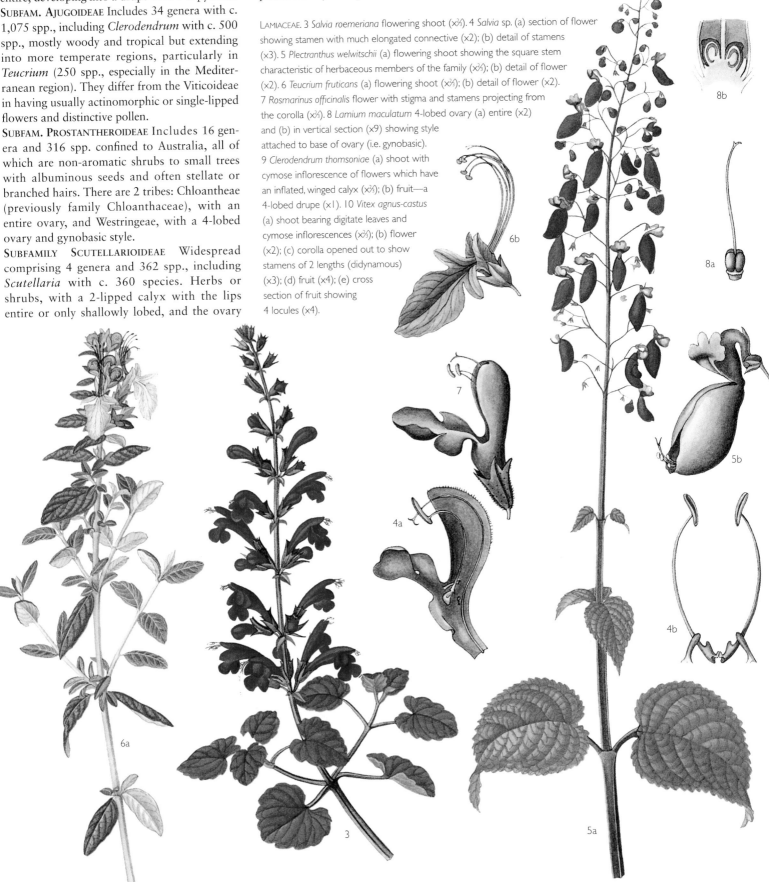

9a

9b

10b

10c

10d 10e

10a

i Hedge, I. C. A global survey of the biogeography of the Labiatae. Pp. 7–17. In: Harley, R. M. & Reynolds, T. (eds), *Advances in Labiate Science*. Richmond, Royal Botanic Gardens, Kew (1992).
ii Cantino, P. D., Harley, R. M., & Wagstaff, S. J. Genera of Labiatae: status and classification. Pp. 511–522. In: Harley, R. M. & Reynolds, T. (eds), *Advances in Labiate Science*. Richmond, Royal Botanic Gardens, Kew (1992).
iii Harley, R. M. *et al*. Lamiaceae. Pp. 167–275. In: Kadereit, J. W. (ed.), *Families and Genera of Vascular Plants. VII. Flowering Plants. Dicotyledons: Lamiales*. Berlin, Springer-Verlag (2004).
iv Olmstead, R. G., Kim, K.-J., Jansen, R. K., & Wagstaff, S. J. The phylogeny of the Asteridae *sensu lato* based on chloroplast *ndh*F gene sequences. *Mol. Phylogen. Evol.* 16: 96–112 (2000).
v Rivera Nuñez , D. & Obón de Castro, C. The ethnobotany of Old World Labiatae. Pp. 455–473. In: Harley, R. M. & Reynolds, T. (eds), *Advances in Labiate Science*. Richmond, Royal Botanic Gardens, Kew (1992).
vi Heinrich, M. Economic botany of American Labiatae. Pp. 475–488. In: Harley, R. M. & Reynolds, T. (eds), *Advances in Labiate Science*. Richmond, Royal Botanic Gardens, Kew (1992).

LARDIZABALACEAE

AKEBIA AND HOLBOELLIA FAMILY

Distribution Seven of the 9 genera are in temperate to subtropical Asia, from western Himalaya (Jammu–Kashmir) to northern Indo–China, Taiwan, Korea, and Japan (*Akebia* reaching Hokkaido). They occur in humid thickets and forests, often on riverbanks, up to an altitude of 2,000 m or *Decaisnea* up to 3,600m in the eastern Himalayas. The 2 other genera—*Lardizabala* and *Boquila*—both monotypic, occur in similar habitats in central Chile and adjacent western Argentina, with the former also on Juan Fernandez Islands.

Description Deciduous or evergreen, usually glabrous, twining woody climbers, usually recorded up to 5 m high but sometimes much higher and with a stem up to 39 cm diameter, or (*Decaisnea*) an erect shrub up to 4 m. The leaves are alternate or apparently fascicled on short shoots, compound, in *Decaisnea* pinnate, in other genera 3-foliate (rarely, in *Lardizabala*, bi- or tri-ternate) or palmate with up to 9 leaflets, usually with long petiolules that are swollen at the base. Stipules are absent except in *Lardizabala*. The inflorescence is a usually fairly lax axillary raceme, corymb, or umbel, or a terminal raceme in *Decaisnea*, or a spike of small, sessile flowers in *Sinofranchetia*, often bearing male flowers apically and female flowers basally. The flowers are usually unisexual (plants monoecious or dioecious), actinomorphic, 3-merous, usually 1–2 cm, except *Sinofranchetia*. The sepals are 3(4) or 6, free, petaloid, conspicuous, sometimes with a long acute apex. The petals are 6, free, in 2 whorls, petaloid or modified to nectaries, or (*Decaisnea, Archakebia, Akebia*, sometimes *Stauntonia*) absent. The stamens are (3–)6(–8), free or connate into a tube, with a well developed connective often conspicuously prolonged at the apex, female flowers usually having 6 small staminodes. The ovary is superior, of 3–6(–12) free carpels, each with a

spp. centered in the Mediterranean region, and *Isodon* with c. 100 spp. nearly all in tropical Asia. They are similar to the Lamioideae but are mostly strongly aromatic and usually have 6-colpate pollen.

Some 10 genera formerly included in Verbenaceae, including the pantropical *Callicarpa* with c. 140 spp. and *Tectona* with only 4 spp., but providing teak timber, are so far unplaced in these 6 subfamilies. The genera that were traditionally placed in the Lamiaceae are mostly now split between the Lamioideae and the Nepetoideae, but Ajuga and *Teucrium*, both lacking an upper lip to the corolla, are placed in the Ajugoideae together with many woody genera formerly placed in Verbenaceae, and *Scutellaria* with a 2-lipped calyx is referred to Scutellarioideae with 3 other genera.

Economic uses The family is important economically[v,vi]. As expanded, it now includes important tropical timber trees, especially teak

(*Tectona*), and species of *Vitex* and *Gmelina*. *G. arborea* is widely planted in the tropics as a source of quick-growing fuelwood. The subfamily Nepetoideae is well known for producing aromatic oils and includes numerous culinary or aromatic herbs such as sage (*Salvia*), Mint (*Mentha*), Marjoram or Oregano (*Origanum*), Thyme (*Thymus*), Lavender (*Lavandula*), Rosemary (*Rosmarinus*), Basil (*Ocimum*), Lemon Balm (*Melissa*), Savory (*Satureja*), Mountain Tea (*Sideritis*), and Catmint (*Nepeta*). Some of these are grown as commercial crops. Decorative garden plants include species of *Salvia, Phlomis, Stachys, Monarda, Ajuga, Lamium, Physostegia, Rosmarinus, Thymus, Agastache, Caryopteris, Vitex*, while *Plectranthus* (including *Coleus*) are popular house plants. In the tropics, *Plectranthus* provides hedge plants, garden ornamentals, edible tubers, pot herbs, and medicinal cures, while *Holmskioldia* is grown for its colorful accrescent calyx and *Callicarpa* for its colorful fleshy fruits. *Pogostemon* provides patchouli used in the perfume industry, and the oily nutlets of *Perilla* are crushed for oil for the paint industry. *Ocimum sanctum*, a holy plant for Hindus, is frequently grown near temples in India.   RKB

Genera 9 Species 35
Economic uses Drugs, medicinal uses, stems for cord, edible fruits, ornamental cultivated plants

short sessile or subsessile stigma and usually not completely closed. Each carpel has few to numerous ovules on laminar or (*Decaisnea, Sinofranchetia*) submarginal placentae. The fruit is either a dehiscent leathery follicle (*Decaisnea, Archakebia, Akebia*) or an indehiscent berry. The seeds are often hard and shiny.

Classification With its 3-merous flowers and free unsealed carpels, the family has always been placed among the more primitive families, close to Ranunculaceae, sometimes particularly close to the Menispermaceae, which are also climbers, and all molecular analyses confirm this position[i,ii]. *Sargentodoxa* is sometimes included in Lardizabalaceae[iii], but here it is treated as a separate but closely related family. The recent comprehensive revision of the family[iv] excludes *Sargentodoxa* and recognizes 4 tribes:

Decaisneeae contains 1 monotypic genus, *Decaisnea*, from Nepal to eastern China. They are erect shrubs with pinnate leaves with 6 or more pairs of opposite leaflets; racemes usually terminal; fruit a cylindrical leathery follicle.

Sinofranchetieae contains 1 monotypic genus, *Sinofranchetia*, from central and southwestern China. Deciduous vine with 3-foliate leaves; flowers small, subsessile, in linear spikes; sepals 6, broad and fleshy; stamens 6, free; carpels with numerous ovules arranged in 2 series; and the fruit a small berry.

Akebieae comprises 5 genera, *Archakebia* (1 sp.), *Akebia* (4 spp.), *Parvatia* (2 spp.), *Holboellia* (11 spp.), and *Stauntonia* (13 spp.), ranging from the Himalayas to East Asia. Evergreen or deciduous vines with 3-foliate or palmately (3–)5–7(–9)-foliate leaves; inflorescence a lax raceme with long pedicels; placentation laminar; fruit a leathery capsule or berry.

Lardizabaleae contains 2 monotypic genera, *Boquila* and *Lardizabala*, from South America. Evergreen vines with 3-foliate or ternate to 3-ternate leaves; inflorescence umbellate; ovules arranged in rows; fruit a small or large berry.

LARDIZABALACEAE. 1 *Akebia quinata* leafy shoot with separate female (large) and male (small) flowers (x⅔). 2 *Decaisnea fargesii* (a) male inflorescence (x⅔); (b) part of leaf showing swollen bases to leaflet stalks (x⅔); (c) female flower and young fruit (x⅔); (d) androecium from male flower with 6 partly fused stamens (x2); (e) gynoecium with vestigial stamens (x2); (f) vertical section of carpel with numerous ovules (x3); (g) cross section of carpel (x3); (h) fruits—follicles (x⅔).

Economic uses Many of the species in China are used as a source of drugs, and the stems of some are used for making strong cord. In South America, *Boquila* is said to be used to cure eye infections, and fruits of *Lardizabala* are edible. *Decaisnea insignis* is cultivated as a shrub, while other genera are cultivated as climbers on walls, especially *Akebia quinata, Holboellia latifolia,* and *Stauntonia hexaphylla.*   RKB

i Hoot, S. B., Culham, A., & Crane, P. R. Phylogenetic relationships of the Lardizabalaceae and Sargentodoxaceae: chloroplast DNA sequence evidence. *Pl. Syst. Evol.* (Suppl. 9): 195–199 (1995).
ii Wang, F., Li, D. Z., & Yang, J. B. Molecular phylogeny of the Lardizabalaceae based on *trn*L-*trn*F sequences and combined chloroplast data. *Acta Bot. Sin.* 44: 971–977 (2002).
iii Wu, C. Y. & Kubitzki, K. Lardizabalaceae. Pp. 361–365. In: Kubitzki, K., Rohwer, J. G., & Bittrich, V. (eds), *The Families and Genera of Vascular Plants. II. Flowering Plants. Dicotyledons: Magnoliid, Hamamelid and Caryophyllid Families.* Berlin, Springer-Verlag (1993).
iv Qin, H. N. A taxonomic revision of the Lardizabalaceae. *Cathaya* 8–9: 1–214 (1997).

LAURACEAE
AVOCADO, CINNAMON, BAY LAUREL

A large, mainly tropical family of aromatic trees and shrubs, with some parasitic climbers.

Distribution The family is mainly tropical in distribution, centered in tropical America, Southeast Asia, and Australasia in lowland to montane rain forests but extending into the subtropics and temperates zones. Genera reaching temperate areas include *Lindera, Persea,* and *Sassafras* and, in the relict laurel forests of the Canary Islands and Madeira, *Apollonias, Laurus, Ocotea,* and *Persea.*

Description Mostly large evergreen trees, some shrubs, or herbaceous climbing parasites (*Cassytha*). Leaves alternate to opposite (absent in *Cassytha*), simple, entire and coriaceous, often containing essential oils; stipules absent. The flowers are bisexual or unisexual (plants dioecious or monoecious), small, inconspicuous, greenish to cream colored, and radially symmetrical, in umbellate, racemose or cymose infloresences. The perianth consists of whorls of only slightly differentiated tepals; tepals (2)3(4) or in multiples. The stamens are (3–)9(–12), in 3 or 4 alternating whorls, 1 or more whorls reduced to staminodes or absent, those of the third whorl with 2 basal glands; the anthers with 2 or 4 locules, mostly splitting by valves on the inner face, but often those of the inner third whorl opening toward the outside. Receptacle, cup-shaped, well developed. The ovary is usually superior, often surrounded by the receptacle, of a single carpel with a single, pendulous, anatropous ovule. The style is simple and the stigma small. The fruit is a berry or drupelike, often enclosed by the receptacle, which may become a more or less fleshy cupule. The seed, without endosperm, contains a straight embryo.

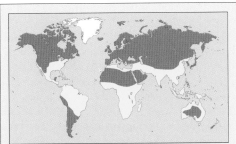

Genera c. 50 **Species** 2,500–2,750
Economic uses Avocado pear (*Persea americana*);
several important spices, such as bay laurel (*Laurus nobilis*), cinnamon, camphor (*Cinnamomum zeylanicum*), and sassafras oil (*Sassafras albidum*);
timber; some ornamentals

Classification Most species in this family are found in large genera such as *Cinnamomum* (350 spp.), *Cryptocarya* (350 spp.), *Litsea* (400 spp.), *Ocotea* (350 spp.), *Beilschmiedia* (250 spp.), and *Persea* (200 spp.). The Lauraceae is usually divided into 2 tribes: the mainly neotropical Perseeae, and the mainly palaeotropical or temperate Laureae. They are sometimes regarded as subfamilies. *Cassytha* (16–20 spp.), which is anomalous in its parasitic habit and leaves reduced to scales, is sometimes treated as a separate tribe or family Cassythaceae. However, molecular evidence confirms that it belongs in the Lauraceae, although its exact position there is not clear[i].

Lauraceae, together with Gomortegaceae, Hernandiaceae, Monimiaceae, Atherospermataceae, Siparunaceae, and Calycanthaceae, constitute the monophyletic core Laurales[ii], although the relationships between the families are still not fully elucidated[iii].

Economic uses The Lauraceae contains a number of economically important species. Avocado Pear (*Persea americana*) is extensively cultivated in the tropics and subtropics, where it is an important cash crop. The fruits are rich in monounsaturated oils, and several cultivars exist. *Cinnamomum* provides both Cinnamon and Camphor, and the aromatic leaves of the Bay Laurel, *Laurus nobilis*, are widely used as a flavoring. *Sassafras* yields oil of sassafras. Many genera contain valuable timber trees, such as species of *Beilschmiedia*, *Endiandra*, *Ocotea* (green heart), and *Litsea*, the last also being a source of many local medicines. *Laurus* and *Lindera* species are grown as ornamentals. VHH

i Chanderbali, A. S., van der Werff, H., & Renner, S. S. Phylogeny and historical biogeography of Lauraceae: Evidence from the chloroplast and nuclear genomes. *Ann. Missouri Bot. Gard.* 88: 104–134 (2001).
ii Renner, S. S. Circumscription and phylogeny of the Laurales: evidence from molecular and morphological data. *American J. Bot.* 86: 1301–1315 (1999).
iii Renner, S. S., Chanderbali, A. S. What is the relationship among Hernandiaceae, Lauraceae, and Monimiaceae, and why is this question so difficult to answer? *Int. J. Plant Sci.* 161 (Supplement 6): S109–S119 (2000).

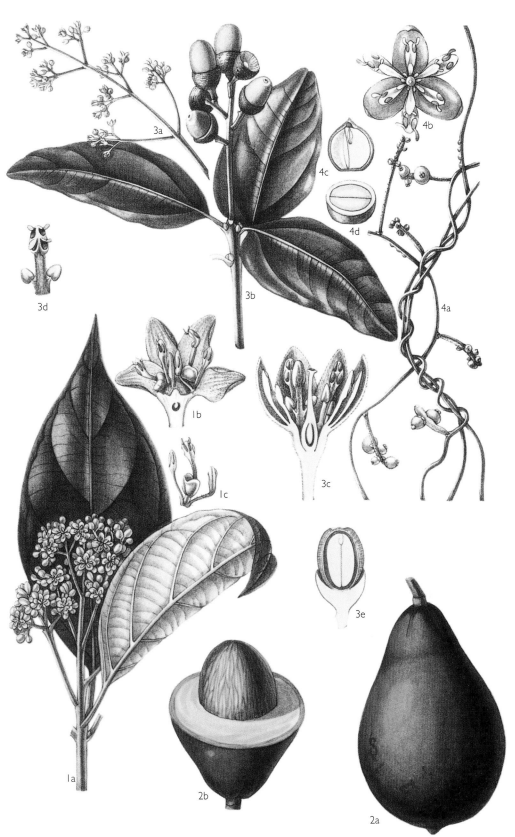

LAURACEAE. 1 *Hypodaphnis zenkeri* (a) shoot and inflorescence (x⅔); (b) half flower (x4); (c) stamen from second whorl, 2 from third whorl, and gland (x4). 2 *Persea gratissima* (avocado) fruit (a) entire and (b) cut away to reveal hard seed (x⅔). 3 *Cinnamomum litseifolium* (a) inflorescence (x⅔); (b) shoot with fruits enclosed in fleshy cupules (x⅔); (c) half flower showing introrse and extrorse dehiscence of anthers (x8); (d) stamen with glandular bases and anthers dehiscing by flaps (x2); (e) half fruit (x1). 4 *Cassytha filiformis* (a) habit of this parasitic plant (x⅔); (b) flower from above (x8); fruit in vertical (c) and cross (d) section (x2).

LECYTHIDACEAE

BRAZIL NUTS AND CANNONBALL TREE

A small family of tropical trees, the best known of which is *Bertholletia excelsa*, which yields Brazil nuts.

Distribution The Lecythidaceae is centered in the wet regions of tropical South America, with some genera in Africa and Asia. Subfamily Scytopetaloideae is native to tropical West Africa.

Description Small to very large trees or occasionally shrubs, with spirally arranged leaves in clusters at the tips of the twigs or borne in 2 ranks. Leaves large, simple, usually without gland-dots, sometimes toothed and asymmetrical at the base, with or without stipules[i]. The inflorescence is a spike up to 1 m long, as in *Couroupita guianensis*, or panicle borne terminally, or racemes in the axils of leaves on young stems on side shoots or on the older parts of the stems. The flowers are actinomorphic or zygomorphic, bisexual, often large and showy, in shades of red, pink, yellow, or white, and have an attractive, fluffy appearance due to numerous stamens (e.g., subfam. Lecythidoideae) or with long pedicels (e.g., subfam. Scytopetaloideae). They seldom last long, in *Barringtonia* remaining open for only 1 night before stamens and petals fall at dawn. The flowers are regular or irregular, with (2–)3–6 sepals fused into a cup-shaped structure, and 3–6(–18) petals (absent in *Foetidia* and subfam. Scytopetaloideae), free (subfam. Lecythidoideae) or united at the base or into a ribbed tube (subfam. Napoleonaeoideae). In subfam. Scytopetaloideae, the petals are replaced by staminodes[ii] that are sometimes thick and either become reflexed as the anthers mature or do not separate and fall off the flower as a cap. The stamens are joined to each other and to the petals or staminodes at their base in one or more rings. The ovary is inferior of 2–6 fused carpels, totally fused with the top of the receptacle or partially sunk into it, or superior[iii] (subfam.

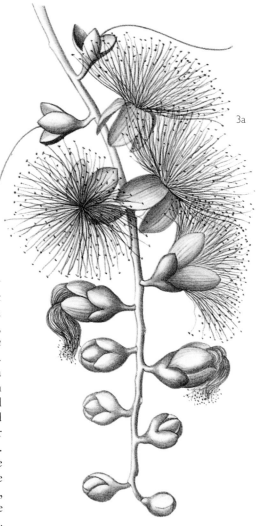

Scytopetaloideae); locules 2–6 or more, each with 1 to many ovules, a long simple style and a lobed or disk-shaped stigma, or locules 3–8 and 2 or more pendulous ovules per locule, borne on axile placentas, a simple style and simple stigma (subfam. Scytopetaloideae). Pollination is frequently by bats, attracted by the sweet-smelling flowers. The usually large fruits have fleshy outer layers, hard and woody inner layers, and are indehiscent, with a lid through which the seeds leave the fruit, or a woody capsule, but in a few cases is drupelike. The seed is large, woody and lacks endosperm except in subfam. Scytopetaloideae where the seed contains a narrow embryo surrounded by copious endosperm, which may have a rough or mottled appearance.

Classification The component genera have often been recognized as 2 or 3 distinct families: Lecythidaceae[iv], Napoleonaeaceae[ii], and Scytopetalaceae[v]. Inclusion of the latter in the family makes it heterogenous in floral and fruit morphology. Some classification schemes assign the African and Asian genera to a separate family, the Barringtoniaceae. The Lecythidaceae has been associated with the Myrtaceae in Myrtiflorae, or even included in it, but more recently has been placed in Theales by Thorne (1983) or Lecythidales (Takhtajan 1987 and Cronquist 1988). The Scytopetalaceae has been associated with the Malvales but was placed in Theales by Thorne (1983) and Cronquist (1988) or Ochnales (Takhtajan 1987). These different views seem to reflect the numerous floral differences seen within the expanded Lecythidaceae. Phylogenetic analysis of the group places it in Ericales[vi] sister to Sapotaceae. The recircumscription is based on analysis of one DNA region and morphological characters[vii] that led to the component genera being lumped into one heterogeneous family[viii].

Economic uses *Bertholletia excelsa* is the source of Brazil nuts or para nuts. Sapucaia or paradise nuts, from several species of *Lecythis*, such as *Lecythis zabucajo*, are reputedly superior to Brazil nuts. The fruits are woody fibrous capsules resembling pots with a terminal lid. After the nuts are shed, the empty pots are used as traps for wild monkeys (hence the common name "monkey-pots"). *Couroupita guianensis* (Cannonball Tree), a native of trop-

LECYTHIDACEAE. 1 *Gustavia pterocarpa* (a) flowering shoot (x⅔); (b) flower with stamens and petals removed to show simple style surmounted by a lobed stigma (x⅔). 2 *Napoleona imperialis* (a) petal-less flower showing several whorls of stamens and outer ring of staminodes, forming a corolla-like saucer-shaped body (x⅔); (b) half flower with calyx and staminodes removed, showing fertile outer whorls of stamens within which are outgrowths (disk) of the receptacle and a short flat-topped style (x1½); (c) berry-like fruit with apical lid (x⅔). 3 *Barringtonia racemosa* (a) spike-bearing flowers with many stamens (x⅔); (b) vertical section of flower base showing ovary fused into the receptacle (x2); (c) fruit (x⅔).

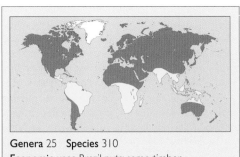

Genera 25 **Species** 310
Economic uses Brazil nuts; some timber

ical South America, is grown as an ornamental for its large 10 cm, waxy, sweet-smelling, red-and-yellow flowers, which are followed by spectacular spherical fruits measuring around 15–20 cm in diameter, the "cannonballs" borne on the trunks. Useful timber comes from *Lecythis grandiflora* (wadadura wood), *Careya* (tummy wood) from Malaysia and India, and from *Bertholletia excelsa*.   AC

i Breteler, F. J. Scytopetalaceae are stipulate. *Kew Bull.* 57: 759–761 (2002).
ii Prance, G. T. Napoleonaeaceae. Pp. 282–284. In: Kubitzki, K. (ed.), *The Families and Genera of Vascular Plants. VI. Flowering Plants. Dicotyledons: Celastrales, Oxalidales, Rosales, Cornales, Ericales.* Berlin, Springer-Verlag (2004).
iii Appel, O. Morphology and systematics of the Scytopetalaceae. *Bot. J. Linn. Soc.* 121: 207–227 (1996).
iv Prance, G. T. Lecythidaceae. Pp. 221–232. In: Kubitzki, K. (ed.), *The Families and Genera of Vascular Plants. VI. Flowering Plants. Dicotyledons: Celastrales, Oxalidales, Rosales, Cornales, Ericales.* Berlin, Springer-Verlag (2004).
v Appel, O. Scytopetalaceae. Pp. 426–430. In: Kubitzki, K. (ed.), *The Families and Genera of Vascular Plants. VI. Flowering Plants. Dicotyledons: Celastrales, Oxalidales, Rosales, Cornales, Ericales.* Berlin, Springer-Verlag (2004).
vi Anderberg, A. A., Rydin, C., & Källersjö, M. Phylogenetic relationships in the order Ericales *s.l.*: analyses of molecular data from five genes from the plastid and mitochondrial genomes. *American J. Bot.* 89: 677–687 (2002).
vii Morton, C. M., Mori, S. M., Prance, G. T., Karol, K. G., & Chase, M. W. Phylogenetic relationships of Lecythidaceae: A cladistic analysis using *rbc*L sequence and morphological data. *American J. Bot.* 84: 530–540 (1997).
viii Morton, C. M., Prance, G. T., Mori, S. A., & Thorburn, L. G. 1998. Recircumscription of the Lecythidaceae. *Taxon* 47: 817–827.

LEGUMINOSAE (FABACEAE)

PEAS, BEANS, MIMOSA

The legumes are the third largest family of flowering plants and second only to the cereals in their economic importance. They are important for enriching nutrient-poor soils through nitrogen-fixing symbioses with *Rhizobium* bacteria.

Distribution The family is cosmopolitan, absent only from Antarctica. Caesalpinioideae and especially Mimosoideae are predominantly tropical and warm-temperate, while Papilionoideae has major areas of diversity in temperate regions. Many genera are extremely widespread, while others are endemic to single

countries. The family has diversified in most major land biomes from arid to wet tropical, grassland, and coastal but is notably absent from marine environments (although some, e.g., some *Entada, Sophora,* and *Lathyrus japonicus*, have long-distance ocean dispersal of seeds) and are poorly represented in fresh water (e.g., *Neptunia oleracea* and others).

Description Herbs, shrubs, trees, woody (e.g., *Wisteria,* some *Entada*) or herbaceous climbers, geophytes, rarely freshwater aquatics (e.g., *Neptunia*), with usually alternate leaves. The leaves are sometimes simple (e.g., *Ulex*), sometimes absent, or present only in juveniles, then replaced by leaflike petioles (e.g., some *Acacia*) but usually 3-foliate (e.g., *Trifolium, Medicago*), palmate (*Lupinus, Lupinaster*), pinnate (e.g., *Vicia, Lathyrus*) or bipinnate (e.g., some *Leucaena* and *Acacia*), sometimes terminating in a tendril (e.g., *Vicia*), the petioles and leaflets with pulvini, sometimes reacting to touch (e.g., *Mimosa pudica*), many allowing leaflets to fold up at night. Stipules are present, sometimes as spines (e.g., some *Acacia*), large and leafy (e.g., *Pisum*) or otherwise modified, but usually small and green. Inflorescence terminal or axillary racemes, heads, spikes, or single flowers. The flowers are bilaterally or radially (Mimosoideae) symmetrical, hermaphrodite or unisexual (plants monoecious—some Mimosoideae), the whorls of 5 sepals, 5 petals, and 2 of 5 stamens alternating[i]. The sepals are 5, joined, or partially so; in bilaterally symmetrical flowers, there are sometimes 2 or 4 lobes by fusion. The petals are 5–4(–1) or 0, variable in size and symmetry, all free, or the lower 2 fused to form a keel (e.g., Papilionoideae). Stamens usually in 2 whorls of 5, 10 fused in a tube (mona-

delphous), or 9 fused and 1 free (diadelphous), or all free, and sometimes 50–100 or more (e.g., Acacieae, Ingeae), *Acacia myrtifolia* has up to 500 stamens. A nectary frequently surrounds the superior ovary. Carpel usually 1, but several (free) in some Ingeae. The fruit is usually a 1-locular dry or fleshy, dehiscent or indehiscent, pod that is inflated, compressed, winged, or not, variously dull or brightly colored, from 4 mm to 3 m in length, with one to many seeds. The seeds usually have a tough coat and vary greatly in size and color; some have a caruncle; all have little endosperm and 2 large cotyledons in proportion to the seed size. Many species form nodulating symbioses with nitrogen-fixing bacteria (usually *Rhizobium* spp.).

Classification The family has long been grouped with the Rosales or, as part of the Fabales, with the Rosideae[ii], a conclusion matched by its placement in the Eurosid I group of flowering plants[iii] using DNA sequence data. Leguminosae appears to be

LEGUMINOSAE. 1 *Spartium junceum* inflorescence—a raceme (x⅔). 2 *Piptanthus nepalensis* shooting bearing 3-foliate leaves with stipules, flowers and fruit (x⅔).

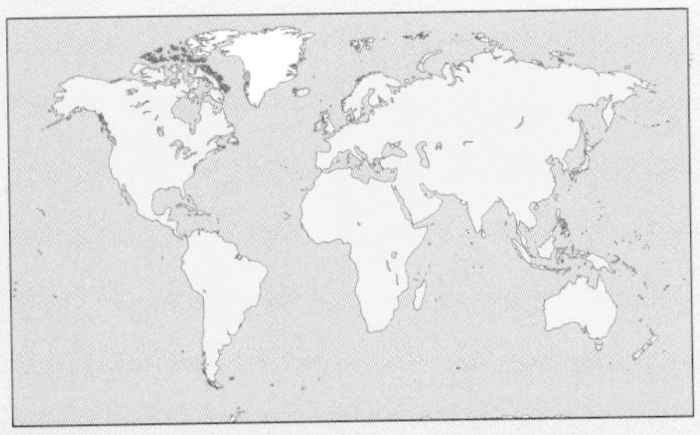

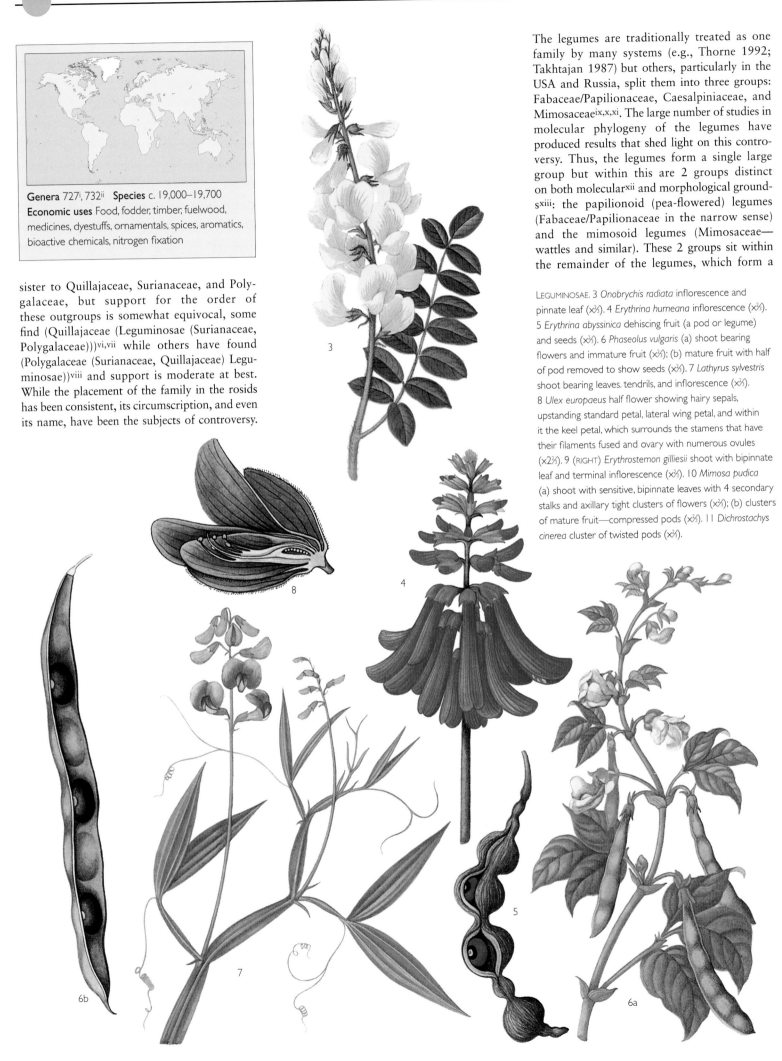

Genera 727[i], 732[ii] **Species** c. 19,000–19,700
Economic uses Food, fodder, timber, fuelwood, medicines, dyestuffs, ornamentals, spices, aromatics, bioactive chemicals, nitrogen fixation

sister to Quillajaceae, Surianaceae, and Polygalaceae, but support for the order of these outgroups is somewhat equivocal, some find (Quillajaceae (Leguminosae (Surianaceae, Polygalaceae)))[vi,vii] while others have found (Polygalaceae (Surianaceae, Quillajaceae) Leguminosae))[viii] and support is moderate at best. While the placement of the family in the rosids has been consistent, its circumscription, and even its name, have been the subjects of controversy.

The legumes are traditionally treated as one family by many systems (e.g., Thorne 1992; Takhtajan 1987) but others, particularly in the USA and Russia, split them into three groups: Fabaceae/Papilionaceae, Caesalpiniaceae, and Mimosaceae[ix,x,xi]. The large number of studies in molecular phylogeny of the legumes have produced results that shed light on this controversy. Thus, the legumes form a single large group but within this are 2 groups distinct on both molecular[xii] and morphological grounds[xiii]: the papilionoid (pea-flowered) legumes (Fabaceae/Papilionaceae in the narrow sense) and the mimosoid legumes (Mimosaceae—wattles and similar). These 2 groups sit within the remainder of the legumes, which form a

3

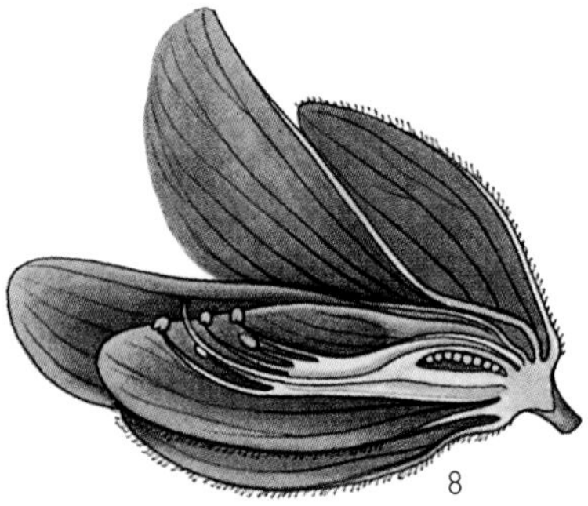
8

4

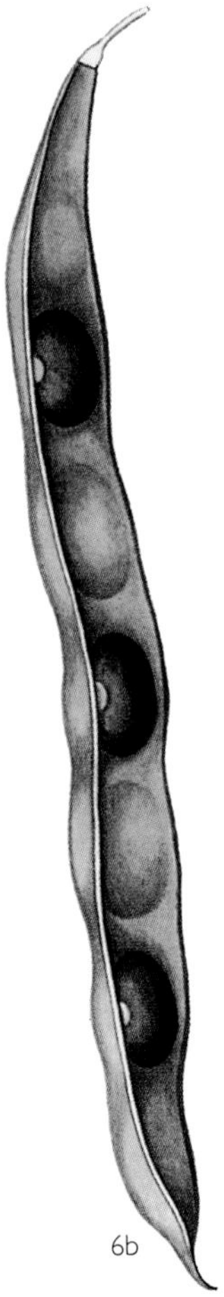
6b

7

5

6a

LEGUMINOSAE. 3 *Onobrychis radiata* inflorescence and pinnate leaf (x⅔). 4 *Erythrina humeana* inflorescence (x⅔). 5 *Erythrina abyssinica* dehiscing fruit (a pod or legume) and seeds (x⅔). 6 *Phaseolus vulgaris* (a) shoot bearing flowers and immature fruit (x⅔); (b) mature fruit with half of pod removed to show seeds (x⅔). 7 *Lathyrus sylvestris* shoot bearing leaves. tendrils, and inflorescence (x⅔). 8 *Ulex europaeus* half flower showing hairy sepals, upstanding standard petal, lateral wing petal, and within it the keel petal, which surrounds the stamens that have their filaments fused and ovary with numerous ovules (x2⅔). 9 (RIGHT) *Erythrostemon gilliesii* shoot with bipinnate leaf and terminal inflorescence (x⅔). 10 *Mimosa pudica* (a) shoot with sensitive, bipinnate leaves with 4 secondary stalks and axillary tight clusters of flowers (x⅔); (b) clusters of mature fruit—compressed pods (x⅔). 11 *Dichrostachys cinerea* cluster of twisted pods (x⅔).

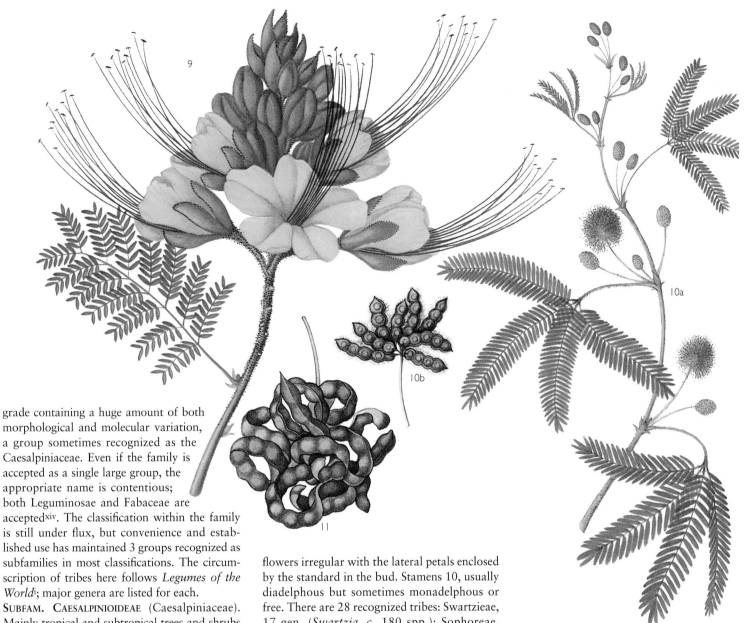

grade containing a huge amount of both morphological and molecular variation, a group sometimes recognized as the Caesalpiniaceae. Even if the family is accepted as a single large group, the appropriate name is contentious; both Leguminosae and Fabaceae are accepted[xiv]. The classification within the family is still under flux, but convenience and established use has maintained 3 groups recognized as subfamilies in most classifications. The circumscription of tribes here follows *Legumes of the World*[i]; major genera are listed for each.

SUBFAM. CAESALPINIOIDEAE (Caesalpiniaceae). Mainly tropical and subtropical trees and shrubs (c. 171 genera, 2,200–2,300 spp.). Leaves usually pinnate but sometimes bipinnate, and the flowers usually more or less irregular with the lateral petals (wings) covering the standard in the bud. Stamens 10 or fewer, free or monadelphous. There are 4 tribes: Cercideae, 12 genera (*Bauhinia*, c. 155 spp., *Phanera*, c. 125 spp.); Detarieae, 82 genera (*Cynometra*, c. 85 spp.); Cassieae, 21 genera (*Chamaecrista*, c. 330 spp., *Senna*, c. 300 spp.); and Caesalpinieae, 56 genera (*Tachigali*, c. 65 spp.).

SUBFAM. MIMOSOIDEAE (Mimosaceae). Mainly tropical and subtropical trees and shrubs (c. 78 genera and 3,200-3,300 spp.). The leaves are often bipinnate and the flowers are regular with the petals valvate in bud, and with 10 or more stamens. There are 4 tribes: Mimoseae, 40 genera (*Mimosa*, c. 500 spp.); Mimozygantheae, 1 genus (*Mimozyganthus carinatus*); Acacieae, 1 genus (*Acacia*, c. 1,450 spp., sometimes split into several genera); and Ingeae, 36 genera (*Inga*, c. 300 spp.).

SUBFAM. PAPILIONOIDEAE (Papilionaceae). Temperate, tropical, and subtropical in distribution, mostly herbs but some trees and shrubs (c. 478 genera and 13,600–14,060 spp.). Leaves usually pinnate but sometimes simple, and the flowers irregular with the lateral petals enclosed by the standard in the bud. Stamens 10, usually diadelphous but sometimes monadelphous or free. There are 28 recognized tribes: Swartzieae, 17 gen. (*Swartzia*, c. 180 spp.); Sophoreae, 45 gen. (*Ormosia*, c. 130 spp.); Dipterygeae, 3 gen. (*Dipteryx*, c. 12 spp.); Brongniartieae, 10 gen. (*Brongniartia*, c. 63 spp.); Euchresteae, 1 gen. (*Euchresta*, 4 spp.); Thermopsideae, 6 gen. (*Thermopsis*, c. 23 spp.); Podalyrieae, 8 gen. (*Amphithalea*, 42 spp.); Crotalarieae, 11 gen. (*Crotalaria*, c. 690 spp., *Aspalathus*, 278 spp.); Genisteae, 25 gen. (*Lupinus*, c. 225 spp.); Amorpheae, 8 gen. (*Dalea*, c. 165 spp.); Dalbergieae, 49 gen. (*Dalbergia*, c. 250 spp., *Adesmia*, c. 240 spp., *Aeschynomene*, c. 180 spp.); Hypocalypteae, 1 gen. (*Hypocalyptus*, 3 spp.); Mirbelieae, 25 gen. (*Daviesia*, 135 spp.); Bossiaeeae, 6 gen. (*Bossiaea*, c. 60 spp.); Indigofereae, 7 gen. (*Indigofera*, c. 700 spp.); Millettieae, 45 gen. (*Tephrosia*, c. 350 spp.); Abreae, 1 gen. (*Abrus*, c. 17 spp.); Phaseoleae, 89 gen. (*Rhynchosia*, c. 230 spp.); Desmodieae, 30 gen. (*Desmodium*, c. 275 spp.); Psoraleeae, 9 gen. (*Otholobium*, 61 spp.); Sesbanieae, 1 gen. (*Sesbania*, c. 60 spp.); Loteae, 22 gen. (*Lotus*, c. 125 spp.); Robinieae, 11 gen. (*Coursetia*, c. 35 spp.); Galegeae, 24 gen. (*Astragalus*, c. 2,400 spp., *Oxytropis*, c. 350 spp.); Hedysareae, 12 gen. (*Hedysarum*, c. 160 spp.); Cicereae, 1 gen. (*Cicer*, 43 spp.); Trifolieae, 6 gen. (*Trifolium*, c. 250 spp.); Fabeae, 5 gen. (*Vicia s.l.*, c. 160 spp.). There is a disporportionately high number of small genera in Caesalpinioideae, suggesting either a different genus concept, a lower rate of speciation, or relict lineages from which species have been lost. More than 20% of all species are found in the 2 genera *Astragalus* and *Acacia s.l.* alone.

Economic uses The family is of major economic importance. The Papilionoideae is especially important because the seeds and pods of many of the herbaceous species are sources of human and animal food and are of particular value in the protein-deficient areas of the world because they are rich in protein as well as mineral content. Certain species such as Clover (*Trifolium repens*), Lucerne (*Medicago sativa*), and Lupin (*Lupinus polyphyllus*) either can be used for feeding livestock or can be plowed into the soil, functioning as excellent fertilizer and greatly increasing the nitrogen levels of the soil. Among the better-known species used as human food are the Garden Pea (*Pisum sativum*); Chickpea (*Cicer arietinum*); French, Haricot,

Snap, String, Green, or Kidney Bean (*Phaseolus vulgaris*); Broad Bean (*Vicia faba*); Pigeon Pea (*Cajanus cajan*); Grass Pea (*Lathyrus sativus*); Lablab (*Dolichos lablab*); Jack Bean (*Canavalia ensiformis*); Lima Bean (*Phaseolus. lunatus*); Mung Bean (*P. aureus*); Scarlet Runner (*P. coccineus*); Lentil (*Lens culinaris*); Soybean (*Glycine max*); and Peanut, or Groundnut, (*Arachis hypogea*). In addition, Fenugreek (*Trigonella foenum-graecum*) is a widely used spice.

The Cowpea (*Vigna sinensis*) Clover (*Trifolium* spp.), Lucerne (*Medicago sativa*), and Vetch (*Vicia sativa*) are widely used as forage plants. Many genera contain species highly prized as ornamentals in both temperate and tropical countries. Some of the better-known in this category are Lupin (*Lupinus*), Broom (*Cytisus*), Golden Rain (*Laburnum*), Sweet Pea (*Lathyrus odorata*), Wild Indigo (*Baptisia*), Locust (*Robinia*), Silk Tree (*Albizia*), *Sophora*, *Wisteria*, and *Genista*. The twigs, leaves, and flowers of *Genista tinctoria* were the source of a yellow dye used for coloring fabrics. Species of *Indigofera* yield the dye indigo. In the Mimosoideae, *Acacia* yields a number of valuable products. The Australian Black Wattle (*Acacia decurrens*) and Golden Wattle (*A. pycnantha*) are the source of wattle bark, which is used in tanning. *A. dealbata* is the "Mimosa" of florists, widely cultivated in the Côte d'Azur. A number of species, including the Australian Blackwood (*A. melanoxylon*) and *A. visco*, are the source of useful timbers. Species including *A. stenocarpa* and *A. senegal* yield gum arabic, while the pods and beans of the Mexican Mesquite Tree (*Prosopis juliflora*) are ground up and used as an animal stock feed. A number of *Albizia* species are valuable timber trees. Caesalpinioideae also contains a number of useful species, including *Cassia acutifolia* and *C. angustifolia* native to the Middle East, whose dried leaves are the source of the purgative senna. Several *Caesalpinia* species are sources of dyes and timber. The pods of the Tamarind (*Tamarindus indica*) are used as a fresh fruit and for medicinal purposes in India, while a number of species, such as the Flamboyant Tree, *Delonix regia* and species of *Caesalpinia* (e.g., *Caesalpinia pulcherrima*, Pride of Barbados), are grown as ornamentals in the tropics and in greenhouses in temperate zones.   AC

i Lewis, G., Schrire, B., Mackinder, B., & Lock, M. *Legumes of the World*. Richmond, Kew Publishing (2005).
ii Roskov Y. R., Bisby F. A., Zarucchi J. L., Schrire, B. D., & White R. J. (eds), ILDIS World Database of Legumes: draft checklist, version 10. Reading, UK: CD-ROM. ILDIS (2005).
iii Tucker, S. C. Floral Development in Legumes. *Plant Physiology* 131: 911–926 (2003).

iv Brummitt, R. K. *Vascular Plant Families and Genera*. Richmond, Royal Botanic Gardens, Kew (1992).
v Soltis, D. E. *et al.* Angiosperm phylogeny inferred from 18S rDNA, *rbc*L, and *atp*B sequences. *Bot. J. Linn. Soc.* 133: 381–461 (2000).
vi Crayn, D. M., Fernando, E. S., Gadek, P. A., & Quinn, C. J. A reassessment of the familial affinity of the Mexican genus *Recchia* Mociño & Sessé ex DC. *Brittonia* 47: 397–402 (1995).
vii Forest, F., *Systematics of Fabales and Polygalaceae, with emphasis on* Muraltia *and the origin of the Cape Flora*. Ph.D. dissertation, University of Reading (2004).
viii Wojiekowski, M. F., Lavin, M., & Sanderson, M. J. A phylogeny of legumes (Leguminosae) based on analysis of the plastid *mat*K gene resolves many well-supported subclades within the family. *American J. Bot.* 91: 1845–1861 (2004).
ix Dahlgren, R. M. T. General aspects of angiosperm evolution and macrosystematics. *Nordic J. Bot.* 3: 119–149 (1983).
x Young, D. A. In: Bedell, H. G. & Reveal, J. L. Amended outlines and indices for six recently published systems of angiosperm classification. *Phytologia* 51: 65–156 (1982).
xi Cronquist, A. *An Integrated System of Classification of Flowering Plants*. New York, Columbia University Press (1981).
xii Doyle, J. J., Chappill, J. A., Bailey, C. D., & Kajita, T. Towards a comprehensive phylogeny of legumes: evidence from *rbc*L sequences and non-molecular data. In: Herendeen, P. S. & Bruneau, A. (eds.), *Advances in Legume Systematics* 9: 1–20. Richmond, Royal Botanic Gardens, Kew (2000).
xiii Doyle, J. J. & Luckow, M. A. The rest of the iceberg. Legume diversity and evolution in a phylogenetic context. *Plant Physiology* 131: 900–910 (2003).
xiv Greuter, W., McNeill, J., Barrie, F. R., Burdet, H.-M., Demoulin, V., Filgueiras, T. S., Nicolson, D. H., Silva, P. C., Skog, J. E., Trehane, P., Turland, N. J., & Hawksworth, D. L. *International Code of Botanical Nomenclature (St. Louis Code)*. *Regnum Vegetabile* 138. Koeltz Scientific Books, Königstein (2000).

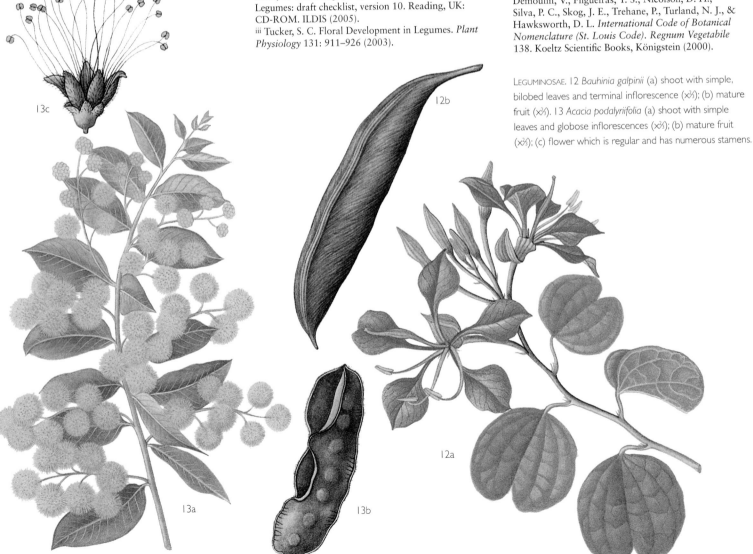

LEGUMINOSAE. 12 *Bauhinia galpinii* (a) shoot with simple, bilobed leaves and terminal inflorescence (x⅔); (b) mature fruit (x⅔). 13 *Acacia podalyriifolia* (a) shoot with simple leaves and globose inflorescences (x⅔); (b) mature fruit (x⅔); (c) flower which is regular and has numerous stamens.

13c

12b

13a

13b

12a

LEITNERIACEAE

CORKWOOD FAMILY

The family comprises a single species, *Leitneria floridana*, a shrub or tree, 1–4(–8) m high, occurring in swamps and wet thickets and on estuarine shores on the coastal plain from Georgia and Florida to eastern Texas (curiously absent from Alabama, Mississippi, and Louisiana) and up the Mississippi

Genera 1 | Species 1
Economic uses The wood is used for fishing-net floats

valley to southeastern Missouri. It spreads vegetatively by underground suckers. With its alternate, narrowly elliptical to linear-lanceolate leaves up to 20 × 5 cm, and catkinate inflorescences, it superficially resembles a *Salix* species. The flowers are actinomorphic, unisexual (plants usually dioecious, rarely polygamodioecious[i] or monoecious), the catkins borne on old wood before the leaves and stiffly erect or the male ones drooping. Interpretation of the structure of inflorescences and flowers is controversial. Male catkins are 2–6 × 1–1.5 cm with 40–50 bracts and clusters of up to 15 stamens, which are either said to constitute several flowers of 1–5 stamens, arranged in 3-flowered cymules, or are treated as a single flower; neither bracteoles nor perianth are present. Female catkins are 1–3 × 1 cm, with 10–15 bracts subtending a pair of bracteoles beneath each flower, each with an irregular ring of (3–)4(–8) minute structures, which may be either bracts or sepals surrounding the ovary. The ovary is superior, normally of 1 carpel, with 1 pendulous ovule. The fruits are leathery drupes up to 2.5 cm with a bony endocarp.

Widely different affinities have been suggested for the family, especially other catkinate families such as Juglandaceae. Both Cronquist and Takhtajan emphasized the uniqueness of *Leitneria* and placed it in its own order, Leitneriales, the former placing it in the Hamamelididae and the latter in the Rutanae. Chemical evidence first suggested an affinity with Simaroubaceae[ii], later confirmed by molecular evidence, and *Leitneria* has actually been placed in Simaroubaceae. However, it has been found to be near the base of that family, with only *Ailanthus* basal to it in a rather small sample of genera, and appears to have diverged very early from the Simaroubaceae[iii] and has become highly modified. Further sampling may give a different picture, but on present evidence there seems no reason to sink Leitneriaceae when it looks so unlike Simaroubaceae and is so different in its simple leaves and catkinate inflorescences with reduced apetalous flowers and unicarpellate ovary. The wood is extremely light and is used as floats for fishing nets. RKB

[i] Godfrey, R. K. Leitneriaceae (Corkwood family). *Trees Shrubs and Woody Vines of Northern Florida and adjacent Georgia and Alabama*. Pp. 448–450 University of Georgia Press (1988)
[ii] Petersen, F. P. & Fairbrothers, D. E. A serotaxonomic appraisal of *Amphipterygium* and *Leitneria*: two amentiferous taxa of Rutiflorae (Rosidae). *Syst. Bot.* 8: 134–148 (1983).
[iii] Fernando, E. S., Gadek, P. A., & Quinn, C. J. Simaroubaceae, an artificial construct: evidence from *rbcL* sequence variation. *American J. Bot.* 82: 92–1003 (1995).

LENNOACEAE

Annual to perennial parasitic herbs related to Boraginaceae, lacking chlorophyll.

Genera 2 | Species 4
Economic uses Roots formerly eaten by indigenous people in desert areas

Distribution *Lennoa madreporoides* occurs from northwestern Mexico to Nicaragua with 2 disjunct records from the north coast of Colombia and adjacent Venezuela, in a wide range of habitats[i]. *Pholisma* extends from southern California to the Sonoran Desert of northwestern Mexico. *P. arenarium* occurs from San Luis Obispo County, California, to western Arizona and northern Baja California in coastal sand dunes and in the Sierra Nevada. *P. sonorae* is restricted to the area around the intersection of the California, Arizona and Mexico boundaries, growing on sand dunes and in citrus groves. *P. culiacanum* is found on the west slopes of the Sierra Madre Occidental (Mexico).

Description Annual (*Lennoa*) or perennial (*Pholisma*) herbaceous succulent parasitic herbs, lacking chlorophyll, growing on the roots of other herbs or shrubs. The stems arise from haustoria attached to the roots of the host plant and may extend underground up to 1.5 m from the host, producing inflorescences at or a little above ground level. The leaves are reduced to linear to deltoid scales up to 25 × 3 mm arranged spirally and often densely around the stems. Inflorescences are dense cymose panicles or spikes, sometimes branched, sometimes flattened into convex or (*P. sonorae*) concave capitula to several centimeters across, with many crowded flowers, usually very colorful and showy[ii]. The flowers are actinomorphic, 8-merous in *Lennoa* or (4–)5–9(–10)-merous on the same plant in *Pholisma*. The calyx is divided almost to the base or may be tubular (*P. culiacanum*). The corolla is up to 10 mm long, cylindrical or salver-shaped, with short and sometimes reflexed lobes at its rim, bright mauve to bluish or purple. The stamens are attached to the corolla tube in 1 row (*Pholisma*) or 2 alternating rows (*Lennoa*), with short filaments. The ovary is superior and composed of 5–9(–16?) fused carpels, each

carpel 2-ovulate with a false partition between the ovules. Placentation is axile but appears free-central in fruit. The fruit, although somewhat fleshy at first, is a capsule, irregularly circumscissile to reveal a ring of up to 28 wedge-shaped seeds.

All species show some degree of host preference, and their distributions are often linked to those of the hosts. The most widespread species of *Pholisma*, *P. arenarium*, has been recorded as parasitic on Asteraceae, Euphorbiaceae, and Hydrophyllaceae, while *P. sonorae* is recorded from Asteraceae, Boraginaceae, and Polygonaceae, and *P. culiacanum* is known only parasitic on Euphorbiaceae. *Lennoa* has been recorded on Asteraceae, Nyctaginaceae, and Zygophyllaceae. The parasite has been recorded as weighing up to 40 times as much as its host, which it does not appear to harm, and there may even be some symbiotic benefit. For further discussion of the parasitism of the family, see Yatskievych 1985[ii], and for photographs of the attachment of the plants to their hosts see Thackery & Gilman[vi].

Classification The family was at one time thought to be related to Ericaceae, but more recent morphological and molecular studies [iii,iv] show it to be most closely related to the basal Boraginaceae. However, the sinking of the family into Boraginaceae in APG II on cladistic grounds ignores the many morphological differences between them. The latest taxonomic revision[v] includes *Ammobroma* in *Pholisma* and recognizes 2 variants within *Lennoa madreporoides* as formae. These differ markedly in flower size, the f. *caerulea* having smaller flowers and occurring mainly from southeastern Mexico to Venezuela, but also recorded rarely in northwestern Mexico, while the large-flowered f. *madreporoides* predominates in Central Mexico. These variants had been given subspecific rank earlier.

Economic uses Roots of *Pholisma* have been used by native Americans and others as a thirst-quenching, nourishing, and tasty food available even in the driest of desert conditions[vi]. They can be eaten raw or cooked or may be ground into flour. However, excessive use of the plants as food causes dental decay. RKB

[i] Yatskievych, G. Lennoaceae. Pp. 209–211. In: Smith, N. *et al.* (eds), *Flowering Plants of the Neotropics*. Princeton, Princeton University Press (2004).
[ii] Yatskievych, G. Notes on the biology of the Lennoaceae. *Cact. Succ. J.* (USA) 57: 73–79 (1985).
[iii] Smith, R. A. *et al.* Molecular phylogenetic evidence for the origin of Lennoaceae: a case of adelphoparasitism in the angiosperms. *American J. Bot.* 87: (Supplement 6): 158 (2000).
[iv] Gottschling, M., Hilger, H. H., Wolf, M., & Diane, N. Secondary structure of the ITS1 transcript and its application in a reconstruction of the phylogeny of Boraginales. *Pl. Biol.* 3: 629–636 (2001).
[v] Yatskievych, G. & Mason, C. T. A revision of the Lennoaceae. *Syst. Bot.* 11: 531–548 (1986).
[vi] Thackery, F. A. & Gilman, M. F. A rare parasitic food plant of the southwest. *Smithsonian Report* 1930: 409–416, plates 1–9 (1931).

LENTIBULARIACEAE
BUTTERWORTS AND BLADDERWORTS

A widespread family of carnivorous herbs found predominantly in wet habitats, sometimes largely submerged under water.

Distribution *Pinguicula* has up to 84[i,ii] terrestrial species spread across the north–temperate region, in the Himalayas, and through the West Indies and Central America down the Andes to Tierra del Fuego. *Genlisea* has about 21 terrestrial or occasionally aquatic species in tropical America, Africa, and Madagascar[iii,iv]. The largest genus, *Utricularia*, has probably about 220 spp. widely distributed in subarctic, temperate, and tropical regions, a large majority being terrestrial on damp ground and a minority aquatic[v].

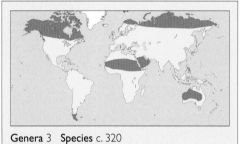

Genera 3   Species c. 320
Economic uses Cultivation by enthusiasts

Description The homologies of the vegetative parts with those of other families are not straight-forward[ii]. In *Genlisea* and *Utricularia,* the whitish organs that anchor the plants having no vascular bundles or root cap and are usually referred to as rhizoids. The leaves usually form a rosette at ground level, except in some aquatic and many terrestrial *Utricularia* species, and are variously modified for the carnivorous habit (see below). The flowers are borne on an erect scape arising in the middle of the rosette or rising above the surface of the water in aquatic species, bearing 1 flower in *Pinguicula* and usually a raceme of flowers in the other genera. Bracts and bracteoles of various shapes are present on the scape. The flowers are strongly zygomorphic and bisexual. The sepals are 2 in most *Utricularia* species, but 4 in the 3 spp. of subgen. *Polypompholyx*, while *Genlisea* and *Pinguicula* have 5. The corolla is strongly 2-lipped and usually has a raised boss at the throat, the lower lip sometimes much divided. There are 2 stamens attached in the corolla tube, and sometimes 2 staminodes. The ovary consists of 2 fused carpels with a single locule, bearing few to many ovules arranged in free central placentation. The fruits are capsules with irregular dehiscence or vertical valves (*Genlisea* subgen. *Tayloria*) or (*Genlisea* subgen. *Genlisea*) circumscissile with one to three rings of dehiscence (a mechanism apparently unique in the flowering plants)[iv]. The seeds are usually many and small, without endosperm.

The carnivorous organs differ markedly in the 3 genera. The leaves in *Pinguicula* are usually ovate to lanceolate and arranged in a rosette with secretory glands over the upper surface, some stalked and emitting shiny mucilage, the others more numerous but sunk into the leaf surface. The stalked glands may attract insects that are trapped by the mucilage when they alight, and they also secrete some preliminary digestive juices. Larger prey may be further enveloped by the margins of the leaves rolling inward. *Genlisea* has dimorphic leaves—the normal leaves are usually in a rosette, and the modified carnivorous ones (traps) are in a ring below the foliage leaves. Occasionally, both sorts of leaves are loosely scattered along a rhizome. The traps[vi] are downwardly directed in the damp soil or water surrounding them, and consist of a solid basal stalk, a short inflated bulb, and a long hollow cylinder terminated by 2 spirally twisted arms, usually as long as the cylinder itself. The mouth of the cylinder is an oblique slit that continues up each arm as a long groove. Inside, the cylinder is divided into backwardly directed, cone-shaped cells each bearing many retrorse hairs[vii] acting on the lobster pot principle to prevent prey from escaping. *Utricularia* (bladderworts), including *Polypompholyx*, has highly modified vegetative parts bearing complicated globose traps with oral appendages and a moving door to ensnare prey. The traps are 0.2–12 mm long and very variable in detailed structure[v], sometimes dimorphic or polymorphic on the same plant. They are borne on the stem or peduncle base, or from the nodes or internodes of stolons, or at leaf apices (an alternative view would be that the stalk of the trap is expanded and foliose), or laterally on the leaves, or occasionally on the rhizoids. The traps are saccate with a small to relatively large mouth, which may be close to or distal from the trap stalk. The mouth is closed by a lid that suddenly opens inward when the long trigger cells projecting outside it are stimulated by a passing organism, which is then washed into the trap and digested[vii]. It has been demonstrated recently[viii] that these traps chemically attract, ingest, and digest protozoa living in the film of water round the soil particles in which they occur.

Classification Both morphological and molecular evidence suggest that the family is closely related to Scrophulariaceae, especially to the *Antirrhinum* group of genera whose floral morphology is similar. The carnivorous habit, with

LENTIBULARIACEAE (BELOW & RIGHT). 1 *Pinguicula moranenis* (a) habit (x⅔); (b) spurred flower (x1); (c) calyx, ovary, and stamens (x3); (d) vertical section of ovary showing the large free-central placenta (x8). 2 *Genlisea africana* (a) habit (x1); (b) flower (x4½), (c) gynoecium with sessile stigma (x12); (d) fruit—a capsule (x8); (e) section of pitcher (x14). 3 *Utricularia subulata* (a) habit (x1); (b) flower, front view (x6); (c) flower, back view (x6); (d) gynoecium and calyx (x6); (e) fruit (x12); (f) trap with projecting bristles around the entrance (x40).

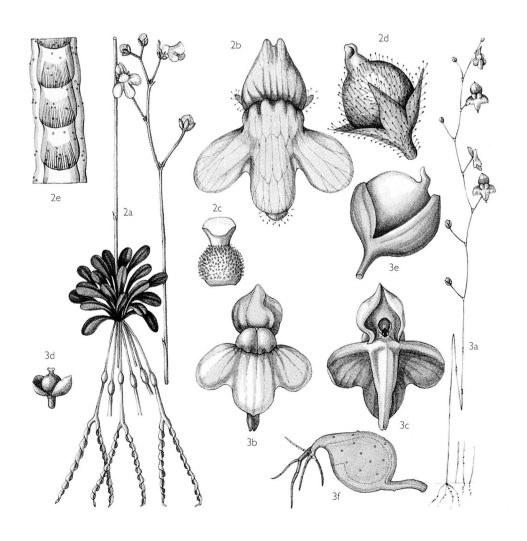

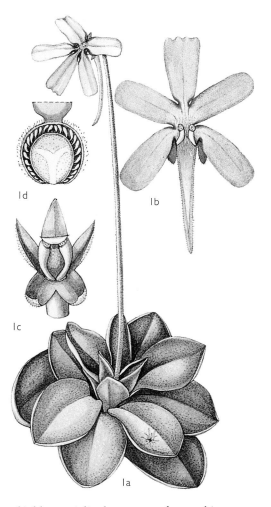

Id   Ib

Ic

Ia

highly specialized structures for catching prey, the often peculiar vegetative morphology, with rhizoids rather than true roots, the free central placentation, and non-endospermous seeds maintain it as a separate family. Recent molecular work[ix] largely confirms previous morphological classifications of the family.

Economic uses Widely cultivated in botanic gardens and by specialists because of their carnivorous habit. Some *Utricularia* species are found as weeds in rice paddies.   RKB

i Legendre, L. The genus *Pinguicula* (Lentibulariaceae): an overview. *Acta Bot. Gall.* 147: 77–95 (2000).

ii Gluch, O. Neuerungen in der Systematik der Gattung Fettkraut (*Pinguicula*). *Taublatt* 49: 14–18 (2004).

iii Taylor, P. The genus *Genlisea. Carniv. Pl. Newsl.* 20: 20–33 (1991).

iv Fischer, E., Porembski, S., & Barthlott, W. Revision of the genus *Genlisea* (Lentibulariaceae) in Africa and Madagascar with notes on ecology and phytogeography. *Nordic J. Bot.* 20: 291–318 (2000).

v Taylor, P. The genus *Utricularia* – a taxonomic monograph. *Kew Bull.* Addit. Ser. 14. 724 pp. London, HMSO (1989).

vi Reut, M. S. Trap structure of the carnivorous plant *Genlisea* (Lentibulariaceae). *Bot. Helv.* 103: 101–111 (1993).

vii Slack, A. *Carnivorous Plants*. Yeovil, Somerset, Marston House (2000).

viii Barthlott, W., Porembski, S., Fischer, E., & Gemmel, B. First protozoa-trapping plant found. *Nature* 392: 447 (1998).

ix Jobson, R. W., Playford, J., Cameron, K. M., & Albert V. A. Molecular phylogenetics of Lentibulariaceae inferred from plastid *rps*16 intron and *trn*L-*trn*F DNA sequences; implications for character evolution and biogeography. *Syst. Bot.* 28: 157–171 (2003).

LEPIDOBOTRYACEAE

Two monotypic genera of evergreen trees, one from tropical America, and the other from tropical Africa.

Distribution *Lepidobotrys staudtii* is confined to Guineo-Congolan Africa and *Ruptiliocarpon caracolito* to tropical Central and South America.

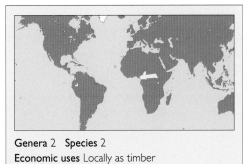

Genera 2   Species 2
Economic uses Locally as timber

Description Evergreen trees or large shrubs. The leaves are alternate, simple (?1-foliate), articulated at junction of petiole and petiolule which are pulvinate at the base, with caducous stipules, and a single elongate caducous stipel; lamina elliptical, the margins entire, the apex acute to acuminate. The flowers are small, greenish, actinomorphic, unisexual (plants dioecious) in terminal racemes or spicate panicles, appearing leaf-opposed. The sepals are 5, free (or slightly connate below), imbricate; the petals are 5, free, imbricate. A fleshy nectary disk is present (*Lepidobotrys*) or absent (*Ruptiliocarpon*). The stamens are 10 in 2 whorls, the inner shorter than the outer, the filaments connate at the base. The ovary is superior, with 2(–3) carpels, with axile placentation; ovules 2 per locule. The fruit is a septicidal or irregularly splitting capsule with 1 (sometimes 2) seeds, partly covered by a fleshy aril.

Classification *Lepidobotrys* was the only genus in the family until the monotypic *Ruptiliocarpon* was described in 1993[i] and included in the Lepidobotryaceae, although Takhtajan (1997) suggested that it better placed in the Meliaceae with which it shares similarities in wood anatomy. An affinity with the Oxalidaceae was suggested for *Lepidobotrys* by Hutchinson, who also included *Sarcotheca* and *Dapania* with it (here placed in the Oxalidaceae), and it was included in that family by Cronquist (1981). It is placed in the Celastrales by APG II although unplaced in APG I. More detailed work is needed to establish the affinities of these 2 genera more accurately.

Economic uses The timber of *Ruptiliocarpon* is used locally.   VHH

i Hammel, B. E. & Zamora, N. *Ruptiliocarpon* (Lepidobotryaceae): A new arborescent genus and tropical American link to Africa, with a reconsideration of the family. *Novon* 3: 408-417 (1993).

LEPTAULACEAE

Pantropical forest trees and shrubs recently segregated from the Icacinaceae.

Distribution Widespread in the tropics. The largest genus is *Citronella,* with 21 spp. in Malaysia, Australia, the Pacific, and tropical America; *Gonocaryum* has 12 spp. from China to New Guinea; *Leptaulus* has 6 spp. in tropical Africa and Madagascar; and *Pseudobotrys* has 2 spp. in New Guinea. All are typically found in wet forests.

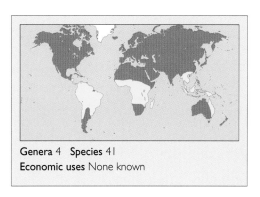

Genera 4   Species 41
Economic uses None known

Description Shrubs, occasionally scandent, to trees up to 36 m. The leaves are alternate, simple, usually entire but sometimes serrate in *Citronella*, shortly petiolate. The inflorescence is axillary and cymose, often in congested fascicles or (*Pseudobotrys*) cauliflorous. The flowers are actinomorphic, bisexual, usually small (c. 4 mm) but up to 1.3 cm in *Pseudobotrys*, 5-merous. The petals are 5, free or fused into a short tube in *Citronella*, fused into a narrow tube with 5 usually very short lobes in other genera. The stamens are 5, with filaments adnate to the corolla tube or sometimes free, not exserted. The ovary is superior, with a long and slender or short and conical style, with a single locule (perhaps pseudomonomerous), with 2 apical pendent ovules. The fruit is a single-seeded drupe up to 6.5 cm, those of *Gonocaryum* often found as "drift seeds."

Classification The family has not previously been recognized with this circumscription and may require reconsideration when further data are available. It was included in Icacinaceae until the recent analysis by Kårehed[i], who referred them to a heterogeneous Cardiopteridaceae, while saying that the several elements there might be given separate subfamily or family status. This group is placed on molecular evidence alongside Stemonuraceae, well removed from Icacinaceae. We have here separated 4 genera from *Cardiopteris*, which is probably the most distinctive genus in the whole complex and should comprise a monotypic family distinguished by its herbaceous climbing habit, white latex, laxly dichotomous inflorescence, and dry compressed suborbicular conspicuously winged fruit. The name *Leptaulaceae* is available for the 4 genera. *Metteniusa* is here also placed in its own family, but does show some similarity to

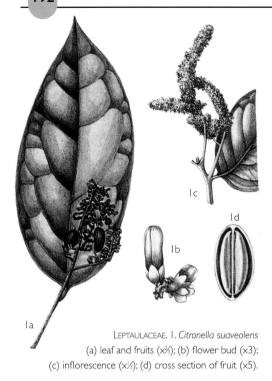

LEPTAULACEAE. 1. *Citronella suaveolens*
(a) leaf and fruits (x⅔); (b) flower bud (x3);
(c) inflorescence (x⅓); (d) cross section of fruit (x5).

Leptaulaceae, especially to the poorly known *Pseudobotrys*. On the other hand, *Dendrobangia*, placed with this group by Kårehed, is probably closely allied to *Platea* which was placed by him in Icacinaceae, though this position was later queried by him[ii] and further information is required.   TMAU & RKB

[i] Kårehed, J. Multiple origin of the tropical forest tree family Icacinaceae. *American J. Bot.* 88: 2259–2274 (2001).
[ii] Kårehed, J. Not just hollies – the expansion of Aquifoliales. In: *Evolutionary Studies in Asterids emphasising Euasterids II. Acta Univ. Uppsala, Comprehensive Summaries of Uppsala Dissertations from the Faculty of Science and Technology.* 676. 38 pp. Uppsala (2001).

LIMEACEAE

A small family of herbs and subshrubs, previously included in Molluginaceae.

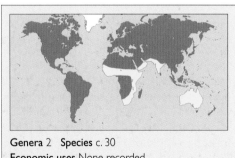

Genera 2 **Species** c. 30
Economic uses None recorded

Distribution The 2 genera in this family are geographically separated. *Limeum* (21 spp.) is distributed from southern Africa, through Ethiopia to southern Asia. *Macarthuria* (7–9 spp.) is endemic to Australia, distributed predominantly in both temperate and tropical coastal regions.

Description Herbs or subshrubs. The leaves are spirally arranged, without stipules. The flowers are hermaphrodite, small and regular in axillary cymose inflorescences of 1 to many flowers. The sepals are 5, green. The petals are 5, white or cream, alternating with the sepals, or sepaline perianth only. The stamens are 8, in 2 whorls. The ovary is superior, with 3 united carpels and 1 or 3 locules. Styles 3, simple. The fruit is a dehiscent capsule with few seeds.

Classification The 2 component genera were grouped until recently[i] with the Molluginaceae with which they share several morphological features, but are separated from it by molecular systematic analysis[ii]. Molluginaceae appears to be a polyphyletic grouping in its broad sense, but sampling of genera is still too poor to fully resolve the issue of relationship in this group.   AC

[i] Shipunov, A. B. The system of flowering plants from a synthetic point of view. *J. Gen. Biol.* 64: 501–510 (2003). [In Russian.]
[ii] Cuénoud, P. *et al.* Molecular phylogenetics of Caryophyllales based on nuclear 18S and plastid *rbc*L, *atp*B and *mat*K DNA sequences. *American J. Bot.* 89: 132–144 (2002).

LIMNANTHACEAE

POACHED EGG PLANT, MEADOWFOAM

Soft annual herbs with bright flowers and divided leaves.

Genera 2 **Species** 10
Economic uses
Ornamental flowers

Distribution *Limnanthes* (9 spp.) is found in vernal pools and other seasonally wet areas of Pacific Western North America, especially California. *Floerkia* (1 sp., *F. proserpinacoides*) grows in northern USA and Canada.

Description Soft annual herbs, usually without hairs. The leaves are alternate, pinnate, with lobed leaflets, or 1–3 pinnatisect; stipules are absent. The inflorescence is a raceme, sometimes with leafy bracts, of few to many flowers. The flowers are actinomorphic and hermaphrodite, usually 5-merous (*Limnanthes*) or 3-merous (*Floerkia*). The sepals are valvate and fused at the base. The petals are free, twisted together clockwise in bud, usually large and bright, but small in *Floerkia*. The stamens are twice as many as petals, free. The ovary is of 3 or 5 fused carpels, each forming a lobe. The fruit usually separates into single seeded mericarps.

Classification The family has been included in Geraniales (Cronquist 1981), Tropaeolales (Dahlgren 1983), and Limnanthales (Takhtajan 1987), but molecular evidence supports its

inclusion with Capparales. The 2 genera have been defined as differing in numbers of floral parts, but this distinction is blurred by variation within and among species. The 2 genera do have a clear ecological separation, and the petal length, much shorter than sepals in *Floerkia*, and equal or usually much longer in *Limnanthes*, provide distinguishing features.

Economic uses *Limnanthes douglasii* is widely cultivated as an ornamental.   AC

[i] Bayer, C. & Appel, O. Limnanthaceae. Pp. 220–224. In: Kubitzki, K. & Bayer, C. (eds), *The Families and Genera of Vascular Plants. V. Flowering Plants. Dicotyledons: Malvales, Capparales and Non-betalain Caryophyllales.* Berlin, Springer-Verlag (2002).
[ii] Rodman, J. E. Soltis, P. S., Soltis, D. E., Systma, K. J., & Karol, K. G. Parallel evolution of glucosinolate biosynthesis inferred from congruent nuclear and plastid gene phylogenies. *American J. Bot.* 85: 997–1007 (1998).

LINACEAE

FLAX FAMILY

Mostly temperate herbs but with 2 small woody genera

Distribution *Linum* has c. 160 spp. almost worldwide but mostly northern temperate. *Hesperolinon* has 13 spp. from Oregon to Mexico. The 2 woody genera, *Tirpitzia* and *Reinwardtia*, have 2 spp. each in China and tropical Asia. *Cliococca*, *Sclerolinon*, and *Radiola* are all monotypic, the first in South America from southeastern Brazil to Chile, the second in western USA, and the third in Europe and the African mountains south to Malawi. The anomalous *Anisadenia* has 2 spp. from the Himalaya to China and Thailand.

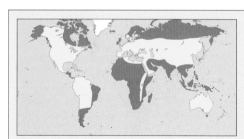

Genera 8 **Species** c. 180
Economic uses Fibers used for textiles; linseed oil for paints and preservatives; some garden ornamentals

Description Annual to perennial herbs—sometimes with extensive underground parts (*Cliococca*) or woody rootstock (*Linum* spp.)—or (*Reinwardtia* and *Tirpitzia*) subshrubs or shrubs to 1.5 m or the latter genus rarely recorded as up to 6 m and with a trunk of 15 cm. Plants are often completely glabrous. The leaves are alternate or opposite, usually sessile, with entire margins, with stipules small or absent. The inflorescence is a cymose panicle, or rarely (*Anisadenia*) a narrow spike of sessile flowers. The flowers are actinomorphic, bisex-

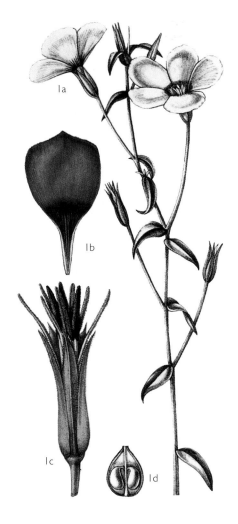

LINACEAE. 1 *Linum grandiflorum* (a) leafy shoot with cymose inflorescence (x⅔); (b) petal (x2); (c) flower with petals removed to show 5 blue stamens and 5 pink staminodes (x3); (d) vertical section of ovary showing axile, pendulous ovules (x4). 2 *Reinwardtia sinensis* (a) leafy shoot and inflorescence (x⅔); (b) fruit—a capsule (x⅔); (c) cross section of ovary (x3); (d) whorl of stamens and small staminodes (x3); (e) calyx and 4 styles (x3).

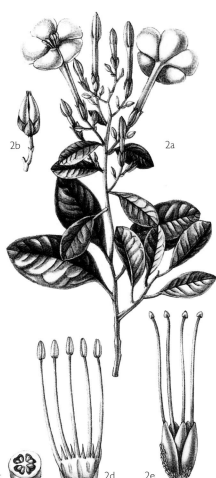

ual, 5-merous or (*Radiola*) 4-merous, often heterostylous. The sepals are 5, imbricate. The petals are 5, free, often clawed, usually showy, but caducous. The stamens are 5 (4 in *Radiola*) alternating with the petals or (*Anisadenia*) opposite them, sometimes with extra staminodes, and the bases of the filaments are fused into a short tube round the ovary. The ovary is superior of 2–5 carpels, with 3–5 locules, with each locule divided almost completely by a false septum from the carpel wall (except in *Anisadenia*), each cell with 2 ovules. The fruit is a septicidal capsule, or a schizocarp (*Anisadenia*) dividing into 2 single-seeded mericarps.

Classification The family was formerly much more widely interpreted, including Hugoniaceae, Ixonanthaceae, Ctenolophonaceae, and Erythroxylaceae, all of which are—like Linaceae—placed in the Malpighiales of APG II. The closest relationships, according to molecular data, seem to be with Hugoniaceae and Picrodendraceae. As noted in the above description, *Anisadenia* seems to be very anomalous in this restricted concept of Linaceae, and further evidence may lead to its exclusion from the family.

Economic uses *Linum usitatissimum* (Flax) has long been grown for its stem fibers used in making fabrics (linen, canvas, etc.) and paper. Its seeds are used to extract linseed oil, which has wide applications in making paint, varnish, and printing ink, and in preserving wood (e.g., cricket bats) and some foods, while the residue is fed to cattle. Characteristic blue fields of flax are still a common sight in many European countries. *L. catharticum* produces a purgative and diuretic. Several other species of *Linum* are grown as ornamentals. *Reinwardtia* is an attractive shrub grown in sheltered conditions.   RKB

LISSOCARPACEAE

Distribution *Lissocarpa* extends from Colombia and the Guianas to Bolivia, in lowland dry or seasonally flooded forest on sand or in montane cloud forest.

Description Shrubs or usually trees, up to 28 m, glabrous in all parts. The leaves are alternate, simple, entire, up to 35 × 9 cm, elliptical, with short pulvinus-like petioles. The inflorescence is a compact

| Genera 1 | Species 8 |
| Economic uses Fish |
| poison; ulcer treatment |

axillary cluster. The flowers are actinomorphic, unisexual (plants probably dioecious) or sometimes apparently sterile, opening at night, 4-merous. The calyx is 4-lobed and does not enlarge in fruit. The corolla is shortly tubular and 4-lobed, and most species have a corona at the throat. The stamens are 8 in a single series. The ovary is inferior, of 4 carpels with a single style and capitate stigma, with 4 locules and 2 pendent axile ovules in each. The fruit is a smooth drupe, with 1–2 seeds that have 6–12 prominent longitudinal vascular ridges.

Classification The family has long been considered close to Styracaceae and Ebenaceae in the Ericales. Molecular evidence places it as sister to Ebenaceae[i], and a recent revision[ii] has made it a subfamily of that family. Others, however, have given Lissocarpaceae family status[iii], which seems well justified[iv], even if the corona is now found to be not present in all species[ii]. It differs from Ebenaceae in its inferior ovary, with a single style and capitate stigma, non-accrescent calyx, corolla usually with a corona probably derived from staminodes, stamens in 1 series with basally connate filaments, 3-porate pollen with prominent reticulate sculpturing (not 3-colporate and smooth to finely warty), and vessels with simple and scalariform pores as in Styracaceae.

Economic uses The leaves may be boiled to prepare a cure for skin ulcers. They have also been recorded as having been used as a fish poison.   RKB

[i] Berry, P. E. *et al.* *Lissocarpa* is sister to Diospyros Ebenaceae). *Kew Bull.* 56: 725–729 (2001).
[ii] Wallnöfer, B. A revision of *Lissocarpa* Benth. (Ebenaceae subfam. Lissocarpoideae). *Ann. Naturhist. Mus. Wien* 105B: 515–564 (2004).
[iii] Gracie, C. Lissocarpaceae. P. 216. In: Smith, N. *et al.* (eds), *Flowering Plants of the Neotropics.* Princeton, Princeton University Press (2004).
[iv] White, F. The botany of the Guyiana Highlands: 9. Lissocarpaceae. *Mem. New York Bot. Gard.* 32: 329–330 (1981).

LOASACEAE

ROCK NETTLE FAMILY

A family of mainly herbs and shrubs, sometimes climbing, with rough and often stinging hairs, closely related to Hydrangeaceae.

Distribution The family occurs predominantly in western parts of the New World, from southern Canada (southern British Columbia, Manitoba) to Argentina and Chile, being especially common in the drier southwestern United States and Mexico and in the Andes, extending more sparingly east to southeastern USA and the Caribbean to western Brazil, also with outlying genera *Kissenia* with 1 sp. in southern Arabian Peninsula and northeastern Africa and 1 sp. in southwestern Africa and *Plakothira* with 3 spp. in the Pacific (Marquesas). Some genera are characteristic of subdesert areas, but others are predominantly in wet forests.

Description Mostly annual to perennial herbs or shrubs, sometimes annual to woody twining climbers, occasionally small trees to 4 m (*Mentzelia arborescens* from Mexico) or rarely to 10 m. A majority of species are covered with unicellular stinging hairs, which cause a skin rash in humans, but the genus *Mentzelia* (80 spp.) and some of the small other genera lack such hairs. Glochidiate and gland-tipped hairs also occur. The leaves are opposite or alternate (sometimes on the same plant), and entire to more often lobed or pinnatisect or palmate. The inflorescence is complex but usually a bracteate terminal or axillary raceme. The flowers are actinomorphic, usually with (4–)5(–8) sepals and petals which are free or rarely fused at the base. The stamens are usually many (up to 150 or more) except in the 2 smaller subfamilies (see below), sometimes some reduced to staminodes, which may be in clusters and modified to form colorful nectaries or petaloid scales. The ovary is inferior or sometimes partially superior, composed of 3–5 carpels, usually 1-locular with parietal placentae, which are simple or variously divided, usually (but see subfamilies) with many ovules. The fruit is a capsule dehiscing apically into 3–5 valves to release many small seeds, or in the smaller subfamilies is indehiscent and single-seeded.

Classification The family was often regarded as an isolated member of the Parietales nearest to Passifloraceae or Turneraceae, but it is now established as referable to Cornales and very close to Hydrangeaceae and particularly to the genera *Jamesia* and *Fendlera* from the southwestern USA[i]. Generic concepts have recently been overhauled by Weigend[ii], who recognized 4 subfamilies in FGVP.

SUBFAM. LOASOIDEAE Comprises 10 genera and 220 spp. mostly in South America. The petals are boat-shaped or fleshy and have 3–5 veins from the base. Stamens are many, and staminodes are present in groups opposite the sepals. The ovary has parietal placentae with usually many ovules. In the tribe Loaseae, the flowers are 5- to 8-merous and usually pendent, the capsule is 10-veined, and the larger genera, *Nasa*, *Loasa*, *Caiophora*, and *Blumenbachia*, all have stinging hairs but some of the small other genera do not. In the tribe Klaprotheae, the flowers are 4-merous and usually erect, the capsule is not veined, and stinging hairs are absent.

SUBFAM. MENTZELIOIDEAE Comprises 3 genera and c. 94 spp. mostly in dry areas of southwestern USA and Central America. Flowers erect and 5-merous, with flat and membranous petals with 3–5 veins from the base; stamens are many and rarely (some *Mentzelia*) in 2 whorls, and the staminodes, if present, are not in groups opposite the sepals; ovary with parietal placentae with usually many ovules. Stinging hairs present in most *Eucnide* species. The recent exclusion of *Eucnide* and *Schismocarpus* from Mentzelioideae to avoid paraphyly is not followed here[iii].

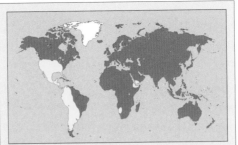

Genera 20 **Species** 330
Economic uses Medicinal purposes, edible seeds, occasional horticultural value

SUBFAM. GRONOVIOIDEAE Comprises 3 genera and 4 spp. from southwestern USA to Peru. Flowers erect and 5-merous, with membranous petals that have a single vein from the base; stamens 5; ovary with a single pendulous ovule; fruit indehiscent with a persistent accrescent calyx. Stinging hairs present in *Gronovia*, but *Fuertesia* and *Cevallia* have glochidiate hairs instead.

SUBFAM. PETALONYCHOIDEAE Comprises 1 genus (*Petalonyx*) and 5 spp. in southwestern USA and northern Mexico. Flowers erect and 5-merous with petals with 1 vein from the base; stamens 5; the ovary with a single pendulous ovule; fruit indehiscent and remaining attached to the bract. Stinging hairs absent but glochidiate hairs present. Further relationships have been clarified by molecular analysis[iv].

Economic uses Species of *Caiophora*, *Nasa* and *Mentzelia* have been used in local medicine. Seeds of *Mentzelia* have formerly been important as a local food in North America, and those of some species such as *M. laevicaulis* (Blazing Star) are sold commercially as garden annuals.   RKB

i Weigend, M. Loasaceae. Pp. 239–254. In: Kubitzki, K. (ed.), *The Families and Genera of Vascular Plants. VI. Flowering Plants. Dicotyledons: Celastrales, Oxalidales, Rosales, Cornales, Ericales*. Berlin, Springer-Verlag (2004).
ii Weigend, M. *NASA and the Conquest of South America*. München: M. Weigend (1997).
iii Moody, M. L., Hufford, L., Soltis, D. E., & Soltis, P. S. Phylogenetic relationships of Loasaceae subfamily Gronovioideae inferred from *mat*K and ITS sequence data. *American J. Bot.* 88: 326–336 (2001).
iv Hufford, L., *et al*. The major clades of Loasaceae: phylogenetic analysis using the plastid *mat*K and *trn*L-*trn*F regions. *American J. Bot.* 90: 1215–1228 (2003).

LOGANIACEAE

A diverse family, ranging from delicate annual herbs to large forest trees and woody climbers, closely related to Gentianaceae.

Distribution Tropics and subtropics and sparingly in temperate regions, in a wide range of habitats from arid areas (herbaceous genera) to rain forests (trees and lianes).

Description Annual to perennial herbs to shrubs and small to large trees and high-climbing lianes. The leaves are opposite with the petiole bases usually joined by a raised line; the lamina is simple with an entire margin. The inflorescence is usually cymose and terminal or axillary, often many-flowered, sometimes (*Spigelia*, *Mitreola*) conspicuously cincinnate. The flowers are actinomorphic, bisexual, sometimes unisexual (plants sometimes dioecious in *Logania* and *Labordea*; gynodioecious in *Geniostoma*), and 4- to 5-merous. The sepals are fused at least at the base and sometimes (*Usteria*) very unequal, or in *Phyllangium* the calyx is absent and replaced by a 2-lobed foliaceous involucre. The corolla forms a usually narrow tube with short lobes, and the stamens are inserted on the tube. The ovary is superior, except in *Mitreola* and *Phyllangium*, where it is semi-inferior, of 2 fused carpels, with (1)2(3) locules. Placentation is axile often with a peltate placenta bearing usually many ovules. The fruit is a capsule dehiscing either apically or basally, or (*Neuburgia*) a drupe, or berrylike or fleshy and coriaceous and sometimes (*Strychnos*) globose and up to 15 cm diameter or pear-shaped and up to 20 cm long. The seeds are 2-numerous and sometimes winged.

Classification The Loganiaceae falls clearly within the group of families with opposite, simple, entire leaves allied to the Gentianaceae. Earlier classifications adopted a broader concept than the one resulting from largely molecular analysis of the last decade[i]. Groups included in the relatively recent *Pflanzenfamilien* account of Loganiaceae by Leeuwenberg and Leenhouts[ii] but which are now excluded are Desfontainiaceae (not closely related); Buddlejaceae and Plocospermataceae (moved to more Scrophulariaceous affinity); *Polypremum* (transferred here with some hesitation to Tetrachondraceae); *Anthocleista*, *Fagraea* and *Potalia* (all transferred to Gentianaceae); *Retzia* (transferred to Stilbaceae); *Peltanthera* and *Sanango* (transferred to Gesneriaceae); and Gelsemiaceae (which some might prefer to keep in Loganiaceae). The remaining family may be divided into 4 tribes[iii]:

Spigelieae, comprises 1 genus, *Spigelia* (c. 70 spp.) from South and Central America and the USA. Herbaceous with a cincinnate inflores-

Genera 15 **Species** c. 420
Economic uses: Poisonous alkaloids, especially strychnine and nux-vomica, and various local medicinal products

cence, a valvate corolla, and a capsule with various dehiscence.

Loganieae comprises 7 genera and 144 spp. centered in Australia but extending through much of the warm Old World and with 2 spp. in South America: *Mitreola* (6 spp. pantropical and subtropical), *Mitrasacme* (54 spp. from India to China and Australia, 48 spp. in the latter), *Phyllangium* (5 spp. from Australia), *Schizacme* (3 spp. from Australia and New Zealand), *Logania* (35 spp. all Australian except 1 sp. in New Caledonia and 1 sp. in New Zealand), *Geniostoma* (24 spp. from the Mascarenes to Japan and the Pacific, only 1 sp. in Australia), and *Labordea* (15 spp. in Hawaii, arguably included in *Geniostoma*). Annual herbs to small trees with an imbricate corolla aestivation and dehiscent capsules. *Mitreola*, with cincinnate inflorescences and 2 spp. in South America, links closely to *Spigelia*.

Antonieae comprises 4 genera and 8 spp. across the tropics: *Bunyunia* (4 spp.) from the Guianas to Peru and Brazil, *Antonia* (1 sp.) from the Guianas to Brazil and Bolivia, *Norrisia* (2 spp.) from Malaysia to Borneo and the Philippines, and *Usteria* (1 sp.) from Senegal to eastern Democratic Republic of Congo and Cabinda. Shrubs and trees or (*Usteria*) lianes with valvate aestivation and capsules with flattened seeds.

Strychneae comprises 3 genera and c. 200 spp., pantropical: *Strychnos* (c. 170 spp.), widespread; *Neuburgia* (12 spp.), from the Philippines to the Pacific; and *Gardneria* (5 spp.), from India to Japan and sparingly to Malaysia. Shrubs to trees and especially lianes with curled tendrils, with valvate aestivation and fleshy fruits.

Economic uses *Strychnos* produces highly poisonous alkaloids such as strychnine and brucine as well as the arrow poison curare and fish poisons. More positively, the same genus has also provided many medicinal products used locally to cure fevers, stomach-ache, malaria, rheumatism, and other complaints. The fruits of some species are also edible. Other genera have similar properties, and a survey has been provided by Bisset[iv]. RKB

i Backlund, M., Oxelman, B., & Bremer, B. Phylogenetic relationships within the Gentianales based on *ndh*F and *rbc*L sequences, with particular reference to the Loganiaceae. *American J. Bot.* 87: 1029–1043 (2000).
ii Leeuwenberg, A. J. M. & Leenhouts, P. W. Taxonomy. In: Engler, A. & Prantl, K. (eds), *Pflanzenfamilien* 28 b 1: 8–96 (1980).
iii Struwe, L. Loganiaceae. Pp. 219–211. In: Smith, N. *et al.* (eds), *Flowering Plants of the Neotropics*. Princeton, Princeton University Press (2004).
iv Bisset, N. G. Useful plants. In: Engler, A. & Prantl, K. (eds), *Pflanzenfamilien* 28 b 1: 8–96 (1980).

LOPHOPYXIDACEAE

The family comprises the single species, *Lophopyxis maingayi*, a liana with leaf tendrils, which extends from Malaysia to the Solomon and Caroline Islands. The leaves are alternate, simple, stipulate, the margins serrate. The flowers are borne in clusters in axillary panicles and are actinomorphic, unisexual (plants monoecious). The sepals are 5, free or slightly connate, persistent. The petals are 5, free, small. In the staminate flowers, the stamens 5, opposite the sepals, alternating with petaloid staminodes. In the pistillate flowers, the ovary is superior, of (4–)5 carpels, with as many locules as carpels, the ovules 2 per locule, pendulous. The fruit is dry, 5-winged, samaroid, single-seeded.

The Lophopxyidcaeae is an isolated lineage. It was included in the Celastraceae by Cronquist (1981) and in the Celastrales by Takhtajan (1997). Recent analyses place it in the Malpighiales and close to the Pandaceae[i].

Genera 1	Species 1
Economic uses	
None recorded	

Molecular data also support a relationship with Putranjivaceae[ii]. No economic uses are reported. VHH

i Savolainen, V. *et al.* Phylogenetics of flowering plants based upon a combined analysis of plastid *atp*B and *rbc*L gene sequences. *Syst. Biology* 49: 306–362 (2000).
ii Wurdack K. J. Molecular systematics and evolution of Euphorbiaceae *sensu lato*. Ph.D. dissertation. Chapel Hill, North Carolina: University of North Carolina (2002).

LORANTHACEAE

MISTLETOE FAMILY

A family of shrubby and leafy parasitic plants, generally with showy flowers.

Distribution Mainly in the tropics to temperate regions of the southern hemisphere, extending north to southern North America, southern Europe, and southern Asia to Japan. Widely distributed in wooded places, almost invariably growing on dicotyledonous angiosperms. The species usually occur on a variety of hosts, but sometimes more or less restricted to one genus or species. They are mostly to be found on the branches of the host, but a few are root parasites, among which *Gaiodendron* in the New World and the Western Australian Christmas tree, *Nuytsia floribunda*, grow into trees.

Description Shrubs or rarely small trees on the roots, or more generally the branches of other dicotyledons, attached by suckers (haustoria) and sometimes with runners in or over the bark of the host, generally evergreen. In certain cases, the parasite may mimic the host in aspect and leaf shape, notably in Australia. The leaves are opposite, alternate or whorled, simple, entire, often leathery or rather fleshy, without stipules. The flowers are usually bisexual, variously borne in mostly 3-flowered cymes, in racemes, umbels, heads, or singly. Calyx rimlike to tubular, entire to shortly toothed. The petals are often brightly colored, (3–)4–5(–9), free or united, valvate, radially symmetrical or with a unilateral split. The stamens are as many as the petals and usually attached to them. The ovary is inferior with a single chamber or in more primitive genera several obscure chambers; the ovules are more or less immersed in the placenta. The fruit is usually a berry with 1 seed but occasionally dry and winged. The seed lacks a testa and is generally surrounded by a sticky layer developed from the fruit wall, by which the seed may be affixed to a new branch when wiped off the beak or excreted by birds.

Classification Related to Olacaceae[i]. The basal groups of about 15 genera with relatively primitive features, all small and often with only a single species, occur in small areas scattered almost exclusively in temperate or tropical montane forest zones in South America, New Zealand, and Australia. From there, more advanced groups can be traced, extending

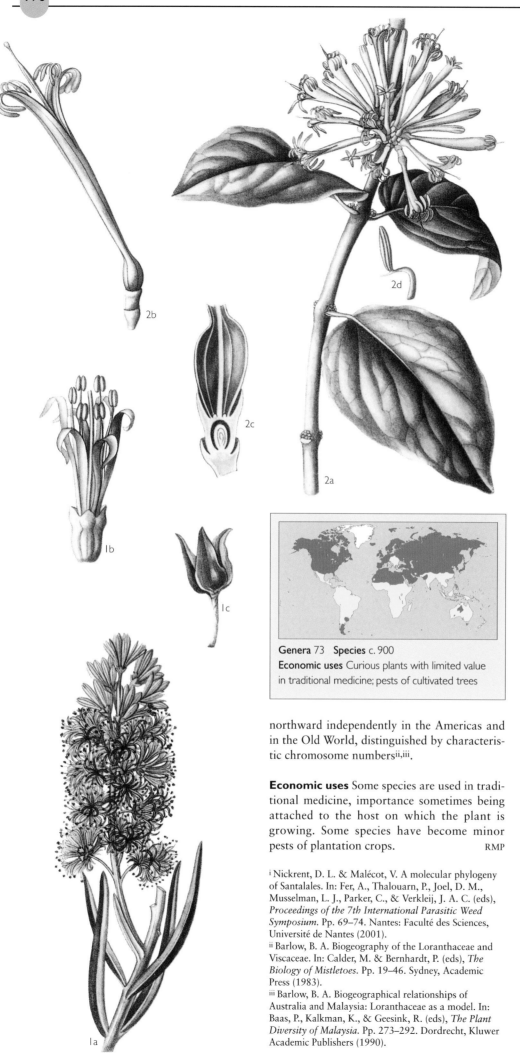

LORANTHACEAE. 1 *Nuytsia floribunda* (a) shoot with flowers in threes (x⅔); (b) flower (x4); (c) dry 3-winged fruit (x1). 2 *Tapinanthus constrictiflorus* (a) flowering branch (x⅔); (b) flower (x2); (c) half of flower base (x5); (d) epipetalous stamen (x4).

LYTHRACEAE

LOOSESTRIFE, POMEGRANATE, WATER CHESTNUT, HENNA, CREPE MYRTLE, MANGROVE APPLE

A family of trees, shrubs, and aquatics.

Distribution The family is primarily found throughout the tropics, extending into temperate regions of the world, but generally absent from African and Arabian deserts and from high latitudes. *Cuphea* (c. 260 spp.), by far the largest genus, is restricted to the New World tropics; *Diplusodon* (c. 74 spp.) is restricted to Brazil; *Lagerstroemia* (c. 53 spp.) occurs from tropical Asia to Australia; *Nesaea* (c. 56 spp.) through tropical and southern Africa; *Rotala* (c. 44 spp.) is temperate and pantropical; and *Lythrum* (c. 36 spp.) is probably the most widespread genus reaching from Europe to Australia and an introduced invasive in North America.

Description Trees, including mangroves (*Pemphis*), some with cone-shaped pneumatophores (*Sonneratia*), shrubs, or herbs, rarely aquatic (*Trapa*), with 4-angled stems on young growth. The leaves are usually opposite, rarely whorled or alternate, simple, dimorphic in amphibious species[i], usually entire, rarely with swollen pet-ioles for flotation (*Trapa*); stipules, when present, small and arising in the leaf axils. Inflorescences terminal or axillary, in racemes, panicles, cymes, or solitary. The flowers are radially symmetrical to strongly zygomorphic, usually hermaphrodite, with a distinctive campanulate to tubular hypanthium. An epicalyx is sometimes present, alternating with the sepals. The sepals are 4–6(–16), joined at the base, valvate, often with pronounced external ridges. The petals are 4–6(–16), crinkled in bud, often brightly colored. The stamens are usually twice as many as sepals, attached to the calyx tube in 2 whorls, free, usually alternating short and long. The ovary is usually superior, sometimes inferior, of 2–4(–many) fused carpels, each forming a locule, with 1 to many ovules per locule, the whole tipped with a simple and dry style. The fruit is variously a dry dehiscent capsule to fleshy-seeded capsule or berry. Distyly and tristyly are common in the family and appear to have arisen repeatedly[ii] as an aid to outcrossing, and bat pollination occurs in *Lafoensia* and *Sonneratia*. There has been much study of tristyly in *Lythrum salicaria* in its invasive populations because it is comomon that not all style morphs occur in these narrow gene pools, but this has not materially hindered invasive behavior.

Classification Lythraceae has been placed in the Myrtales (or Myrtineae) by all major

---

northward independently in the Americas and in the Old World, distinguished by characteristic chromosome numbers[ii,iii].

Economic uses Some species are used in traditional medicine, importance sometimes being attached to the host on which the plant is growing. Some species have become minor pests of plantation crops. RMP

[i] Nickrent, D. L. & Malécot, V. A molecular phylogeny of Santalales. In: Fer, A., Thalouarn, P., Joel, D. M., Musselman, L. J., Parker, C., & Verkleij, J. A. C. (eds), *Proceedings of the 7th International Parasitic Weed Symposium*. Pp. 69–74. Nantes: Faculté des Sciences, Université de Nantes (2001).

[ii] Barlow, B. A. Biogeography of the Loranthaceae and Viscaceae. In: Calder, M. & Bernhardt, P. (eds), *The Biology of Mistletoes*. Pp. 19–46. Sydney, Academic Press (1983).

[iii] Barlow, B. A. Biogeographical relationships of Australia and Malaysia: Loranthaceae as a model. In: Baas, P., Kalkman, K., & Geesink, R. (eds), *The Plant Diversity of Malaysia*. Pp. 273–292. Dordrecht, Kluwer Academic Publishers (1990).

Genera 73 **Species** c. 900
Economic uses Curious plants with limited value in traditional medicine; pests of cultivated trees

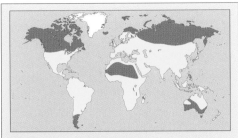

Genera c. 32 **Species** c. 600
Economic uses Fruit, dyes, timber, and ornamentals

traditional systems[iii] of classification and by APG II. It is the sister group of Onagraceae in multigene DNA sequence analyses[iv], but Onagraceae is nested within Lythraceae in an analysis of *rbcL* only[v]. There is debate on the extent of the family; some authors[vi] include *Duabanga*, *Punica*, and *Trapa*, as here, which are otherwise included in monogeneric families sister to Lythraceae. Each of these genera is nested within Lythraceae in a recent molecular analysis[vii] that finds the monotypic *Decodon verticillatus* sister to the rest of the family. In this analsysis, *Duabanga* and *Trapa* form sister pairs with more typical Lythraceae: *Duabanga* with *Lagerstroemia* and *Trapa* with *Sonneratia*, the 4 forming a monophylum in combined data analyses. *Punica* is also nested well within the family on molecular evidence in a group with *Galpinia*, *Capuronia*, and *Lafoensia*. The genera within the family are mostly well defined, although there is still some discussion about the limits of 3 generic pairs: *Ammannia-Nesaea*, *Lythrum-Peplis*, and *Ginoria-Haitia*.

Economic uses The pomegranate (*Punica granatum*) is an ancient fruit crop of the Middle East and Mediterranean. The water chestnut (*Trapa natans*) is grown in the far east as a food crop and is a problem invasive in the USA. The dye plant Henna (*Lawsonia inermis*) is the source of the most widely used products derived from this family, and work is being conducted to improve yields[viii]; the active constituent, Lawsone, gives an orange dye used especially on skin and hair. Crepe Myrtle (*Lagerstroemia indica*, *L. speciosa*) is grown widely in warm climates as a decorative tree due to the plentiful and colorful flowers borne in the summer. Seeds of *Cuphea* have been identified as a good source of medium chain triglycerides (fatty acids), and work is underway to develop cultivars that can

be harvested on a commercial scale. Some species provide good quality timber, particularly *Physocalymma scaberrima*, which has pink wood used in decorative products. *Cuphea* species, particularly *C. ignea* and hybrids, are grown for their ornamental flowers in warm climates, and *Lythrum salicaria* is grown as a pond-side plant in temperate regions for its decorative purple flowers, although it is replaced by *Decodon verticillatus* in North America, where *L. salicaria* has become a problem invasive.   AC

[i] Graham, S. A. Lythraceae. Pp 223–225. In: Smith, N. *et al.* (eds), *Flowering Plants of the Neotropics*. Princeton, Princeton University Press (2004).
[ii] Graham, S. A., Crisci, J. V., & Hoch, P. C. Cladistic analysis of the Lythraceae *sensu lato* based on morphological characters. *Bot. J. Linn. Soc.* 113: 1–33 (1993).
[iii] Brummitt, R. K. *Vascular Plant Families and Genera*. Richmond, Royal Botanic Gardens, Kew (1992).
[iv] Soltis, D. E. *et al.* Angiosperm phylogeny inferred from 18S rDNA, *rbcL*, and *atpB* sequences. *Bot. J. Linn. Soc.* 133: 381–461 (2000).
[v] Savolainen, V. *et al.* Phylogeny of the eudicots: a nearly complete familial analysis based on *rbcL* gene sequences. *Kew Bull.* 55: 257–309 (2000).
[vi] Dahlgren, R. & Thorne, R. F. The order Myrtales: circumscription, variation, and relationships. *Ann. Missouri Bot. Gard.* 71: 633–699 (1984).
[vii] Graham, S. A., Hall, J., Sytsma, K., & Shi, S.-H. Phylogenetic analysis of the Lythraceae based on four gene regions and morphology. *Int. J. Plant Sci.* 166: 995–1017. (2005).
[viii] Khandelwal, S. K., Gupta, N. K., & Sahu, M. P. Effect of plant growth regulators on growth, yield and essential oil production of henna (*Lawsonia inermis* L.). *J. Hort. Sci. & Biotech.* 77: 67–72 (2002).

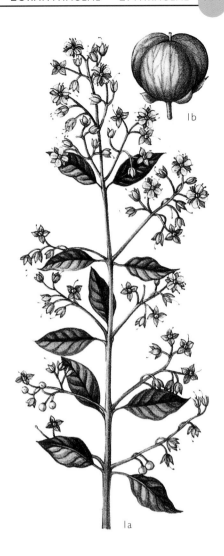

LYTHRACEAE. **1** *Lawsonia inermis* (a) leafy shoot with axillary and terminal inflorescences (x⅔); (b) fruit (x3); (c) cross section of fruit (x3). **2** *Peplis portula* (a) habit showing adventitious roots (x⅔); (b) vertical section of fruit (x4). **3** *Cuphea ignea* (a) leafy shoot with solitary axillary flower (x⅔); (b) vertical section of flower (x1½). **4** *Lythrum salicaria* produces 3 types of flowers (only 1 type on each individual), with the style and the 2 whorls of stamens at 3 levels in the floral tube (tristyly); seed-set is far higher when the stigma receives pollen from stamens of the same length as itself (shown as arrows) than when it is pollinated from longer or shorter stamens.

MAESACEAE

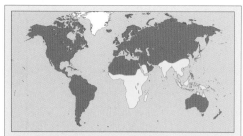

Genera 1 Species c. 150
Economic uses Seeds used for cooking oil; medicinal uses; fish poison and insecticides; some timber

A family differing from Myrsinaceae in their inferior ovary and many seeds.

Distribution *Maesa* is widespread in tropics and subtropics of the Old World. There are only 6 spp. in Africa, but *M. lanceloata* is common in many parts and extends to southern Arabia. The great majority of species are in tropical Asia and China, extending to northeastern Australia and the Pacific islands. The genus is usually characteristic of forest margins, understory of lowland forests and shrubby disturbed habitats.

Description Shrubs and small trees or sometimes scramblers not able to support themselves but not twining or possessing hooks or tendrils[i]. The leaves are alternate, simple, entire to serrate, often with glandular lines. The inflorescence is a simple or many-branched raceme, usually with distinct spreading pedicels that bear 2 bracteoles. The flowers are rather small, actinomorphic, bisexual but plants sometimes functionally dioecious[ii], perigynous. The calyx is fused with the ovary for most of its length, the (4)5 sepals being inserted near the top of the ovary. The corolla has a short tube and (4)5 lobes that have glandular canals. The stamens are inserted opposite the corolla lobes. The ovary is three-quarters inferior, with 1 locule and with free central placentation. The ovules are many, in 1–5 whorls, not or only slightly immersed in the placenta. The fruit is more or less spherical, a dry or more usually fleshy drupe with a crustaceous endocarp enclosing numerous angular seeds.

Classification *Maesa* has long been included in Myrsinaceae, but as its own subfamily Maesoideae differing in having a nearly completely inferior ovary, many angular seeds, glandular lines (rather than gland dots) in the leaves, a pair of bracteoles on the pedicel, and in wood anatomy. Molecular analysis places it outside the Myrsinaceae and basal to the complex including Coridaceae, Myrsinaceae, Primulaceae, Samolaceae, and Theophrastaceae. For further discussion and references, see Primulaceae.

MAESACEAE. 1 (LEFT) *Maesa alnifolia* vertical section of ovary with calyx (x6).

Economic uses The seeds of *M. lanceolata* are recorded as being used for cooking oil in Ethiopia, and the crushed fruits are used for treating intestinal parasites. Other species have been used as fish poisons and insecticides. The leaves of *M. indica* are used as an ingredient of curries in India. Trees may occasionally be used as timber.   RKB

i Utteridge, T. M. A & Saunders, R. M. K. The genus *Maesa* (Maesaceae) in the Philippines. *Bot. J. Linn. Soc.* 145: 17–43 (2004).
ii Utteridge, T. M. A & Saunders, R. M. K. Sexual dimorphism and functional dioecy in *Maesa perlarius* and *M. japonica* (Maesaceae/Myrsinaceae). *Biotropica* 33: 368–374 (2001).

MAGNOLIACEAE
MAGNOLIAS AND TULIP TREE

A family of trees and shrubs native to the Americas and Asia, usually with showy flowers, long thought to be among the earliest and least evolved of all the angiosperms.

Distribution Approximately 75% of species in the family are distributed in temperate eastern and tropical Southeast Asia, from the Himalayas eastward to Japan and southeastward through the Malay Archipelago to Papua New Guinea. The remainder are found in America, from temperate southeast North America to tropical and subtropical South America to Brazil, with most neotropical species occurring in Colombia. All the Ameri-

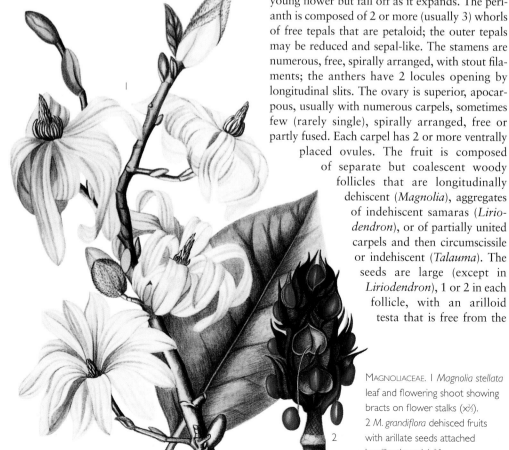

Genera 2 or 3 (or 10) Species c. 220
Economic uses Several important ornamental trees and shrubs

can species belong to the genera *Magnolia*, *Talauma*, and *Liriodendron*. *Talauma* is confined to the New World, but *Magnolia* and *Liriodendron* also occur in Asia and thus have independent discontinuous distributions. *Liriodendron* comprises 2 spp., *L. tulipifera* from eastern North America and *L. chinense* from China. Fossil records indicate that the family has a long evolutionary history of more than 100 million years and was formerly much more widely distributed in the northern hemisphere.

Description Trees or shrubs, deciduous or evergreen. The leaves are alternate, simple, petiolate, entire or lobed (*Liriodendron*), often with large stipules, which at first surround the stem, but fall off as the leaf expands, leaving a conspicuous scar around the node. The flowers are actinomorphic, hermaphrodite (rarely unisexual), terminal, usually solitary, often large and showy, with an elongate receptacle, the peduncle bearing 1 or more spathaceous bracts that enclose the young flower but fall off as it expands. The perianth is composed of 2 or more (usually 3) whorls of free tepals that are petaloid; the outer tepals may be reduced and sepal-like. The stamens are numerous, free, spirally arranged, with stout filaments; the anthers have 2 locules opening by longitudinal slits. The ovary is superior, apocarpous, usually with numerous carpels, sometimes few (rarely single), spirally arranged, free or partly fused. Each carpel has 2 or more ventrally placed ovules. The fruit is composed of separate but coalescent woody follicles that are longitudinally dehiscent (*Magnolia*), aggregates of indehiscent samaras (*Liriodendron*), or of partially united carpels and then circumscissile or indehiscent (*Talauma*). The seeds are large (except in *Liriodendron*), 1 or 2 in each follicle, with an arilloid testa that is free from the

MAGNOLIACEAE. 1 *Magnolia stellata* leaf and flowering shoot showing bracts on flower stalks (x⅔). 2 *M. grandiflora* dehisced fruits with arillate seeds attached by silky thread (x⅔).

MAGNOLIACEAE. 3 *M. heptapeta* half flower showing 2 whorls of perianth segments, numerous spirally arranged stamens and numerous free carpels on an elongate receptacle (×⅓). 4 *Liriodendron tulipifera* (a) flower and leaf (×⅓); (b) vertical section of carpel (×1⅓); (c) fruiting head (×⅔). 5 *Talauma ovata* fruit with upper portions of carpels falling away to reveal 1 or 2 seeds in each carpel locule (×⅓).

endocarp but attached by a silky threadlike funicle; in *Liriodendron* they adhere to the endocarp and are without an arilloid testa. The seeds have copious oily endosperm and a tiny embryo.

Classification The family is monophyletic and clearly delimited, comprising 2 groups that are sometimes recognized as subfamilies: Liriodendroideae (*Liriodendron*) and Magnolioideae (*Magnolia sensu lato*). However, there is no agreement on the number of genera that should be recognized. Frodin and Govaerts[i] recognize 7 genera: *Elmerrillia* (4 spp.), *Kmeria* (2 spp.), *Liriodendron* (2 spp.), *Magnolia* (128 spp.), *Manglietia* (29 spp.), *Michelia* (47 spp.), and *Pachylarnax* (2 spp.). Others recognize as many as 14 or as few as 2 genera, *Magnolia* and *Liriodendron*. Molecular data suggest that the genus *Magnolia sensu lato* is not monophyletic but rather substantially paraphyletic[ii,iii], and that the family is sister to the Myristicaceae[iv,v], along with the rest of the Magnoliales.

Economic uses The genus *Magnolia* is widely cultivated as an ornamental, with numerous species and hybrids available. *Liriodendron* is also often grown in temperate countries. The wood of the North American Tulip Tree (*Liriodendron tulipifera*) is a valuable timber product of the eastern USA. The bark and flower buds of *Magnolia officinalis* and other species yield a valuable drug or tonic exported from China for medicinal use.   VHH

i Frodin D. G. & Govaerts, R. *World Checklist and Bibliography of Magnoliaceae*. Richmond, Royal Botanic Gardens, Kew (1996).
ii Azuma H., Thien, L. B., & Kawano, S. Molecular phylogeny of *Magnolia* (Magnoliaceae) inferred from cpDNA sequences and evolutionary divergence of the floral scents. *J. Plant Research* 112: 291–291 (1999).
iii Kim, S., Park, C.-W., Kim, Y.-D., & Suh, Y. Phylogenetic relationships in family Magnoliaceae inferred from *ndhF* sequences. *American J. Bot.* 88: 717–728 (2001).
iv Soltis, D. E. *et al.* Angiosperm phylogeny inferred from 18S rDNA, *rbcL*, and *atpB* sequences. *Bot. J. Linn. Soc.* 133: 381–461 (2000).
v Sauquet, H. *et al.* Phylogenetic analysis of Magnoliaceae and Myristicaceae based on multiple data sets: Implications for character evolution. *Bot. J. Linn. Soc.* 142: 125–186 (2003).

MALESHERBIACEAE

The family comprises the single genus *Malesherbia* with 24 spp. confined to dry habitats in the Pacific coasts and adjacent arid Andean regions of South America, in Chile, Peru, and Argentina. They are perennial herbs (rarely annual), often woody at the base, to subshrubs or shrubs, sometimes foetid, with alternate simple linear or lanceolate to ovate or obovate leaves, the lamina entire or variously toothed or lobed or deeply pinnatiparite, sometimes with long narrow leaf-bases. The flowers are variously colored, actinomorphic, bisexual, in terminal racemes or panicles. The sepals and petals form a cylindrical, campanulate or funnel-shaped floral tube. The sepals and petals are 5; a corona is developed inside the floral tube, sometimes large and showy, sometimes reduced. There is a prominent central androgynophore. The stamens are 5. The ovary is superior, syncarpous with 3 carpels, with parietal placentation and numerous ovules; styles 3, distinct. The fruit is a loculicidal capsule with the persistent floral tube adhering; seeds 1 to many. Some species have local herbal and medicinal uses.

The family was placed in the Violales by Cronquist and Takhtajan but is now placed in the Malpighiales and considered close to the Turneraceae and Passifloraceae[i,ii].   VHH

i Chase, M. W. *et al.* When in doubt, put it in *Flacourtiaceae*: a molecular phylogenetic analysis based on plastid *rbcL* DNA sequences. *Kew Bull.* 57: 141–181 (2002).
ii Gengler-Nowak, K. Reconstruction of the biogeographical history of Malesherbiaceae. *Bot. Rev.* 68: 171–188 (2000).

MALPIGHIACEAE

A family of tropical lianas, trees, shrubs, and some herbs, with characteristic 2-branched unicellular hairs.

Distribution The family is found in the tropics and subtropics of the both the Old and New worlds but is mainly concentrated in tropical America, where species occur in most countries, with a center of diversity in Brazil. Some genera extend into the southern USA. There are 15 Old World genera, growing in tropical and southern Africa, and from the Indian subcontinent to China to Southeast Asia and northeast Australia and the Pacific. They grow mainly in forests or savannas.

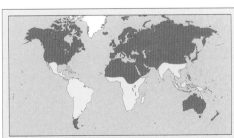

Description Trees, shrubs, subshrubs or herbs, or climbers, often with unicellular and 2-branched "Malpighian hairs," rarely the hairs stellate. The leaves are simple, usually opposite, sometimes whorled, rarely alternate. Stipules are usually present. The flowers are in axillary or terminal racemose or paniculate inflorescences or are solitary. The flowers are usually bisexual, sometimes unisexual (plants dioecious or functionally dioecious), actinomorphic to obliquely zygomorphic. The sepals are 5, united, usually imbricate, rarely valvate, with 2

MALPIGHIACEAE. 1 *Malpighia coccigera* (a) flowering shoot (x⅔); (b) flower with petals removed (x2⅓); (c) sepal dorsal view (x5⅓); (d) gynoecium (x5); (e) cross section of ovary (x6); (f) fruit (x2). 2 *Acridocarpus natalitius* (a) inflorescence (x⅔); (b) flower with petals removed (x2); (c) gynoecium (x2⅔); (d) vertical section of ovary (x3); (e) cross section of ovary (x2⅔); (f) winged fruit (x⅔). 3 *Sphedamnocarpus pruriens* (a) leafy shoot and terminal inflorescence (x⅔); (b) flower with filament bases united in a ring (x2); (c) gynoecium (x3).

(rarely 1) abaxial glands on 4 or 5 of the sepals at the base in neotropical species (much reduced or absent in Old World species). The petals are 5, usually imbricate, clawed at the base. The stamens are 10, sometimes fewer, in a single whorl, the filaments distinct or connate at the base, the anthers 2-locular, dehiscing longitudinally. The ovary is superior, of (2)3(4) separate to united carpels, with axile placentation, and 1 pendulous anatropous ovule per locule. The fruits are diverse: dry, usually schizocarpic, splitting into 2–3 mericarps, which are often winged, or fleshy (drupe or berry). The seeds are without endosperm.

Classification Both morphological and molecular evidence indicate that the Malpighiaceae is monophyletic and may be divided into 2 major clades[ii,iii] that correspond to 2 subfamilies: Byrsonimoideae and Malpighioideae.

SUBFAM. BYRSONIMOIDEAE Arborescent, pollen 3-colporate. The fruits are schizocarpic, unwinged, and dispersed by the water or animals. 10 genera.

SUBFAM. MALPIGHIOIDEAE Climbers, shrubs, or small trees. The fruits are samaroid, winged, and wind dispersed. 61 genera. Within these subfamilies there is still a lack of clarity as to the groups to be recognized, and their state of classification is in a state of flux according to leading expert, W. R. Anderson[i]. The family was placed in the Polygalales by Cronquist (1981) but has subsequently been shown to belong in the large and diverse Malpighiales[ii,iii]. Within the Malpighiales, data from plastid (*ndh*F and *rbc*L) and nuclear (PHYC) genes indicate strong support for Elatinaceae as sister to Malpighiaceae[iv], despite their obvious morphological dissimilarities.

Economic uses *Malpighia emarginata*, Barbados Cherry or Acerola, is locally cultivated for its edible fruits, which are a rich source of vitamin C[v]. Hallucinogenic compounds are obtained from species of *Banisteriopsis* and *Diplopterys*, notably *Banisteriopsis caapi*, which is widely cultivated by indigenous groups in the western Amazon, where it is used as a principal compound of a hallucinogenic beverage. Some species of *Banisteriopsis*, *Galphimia*, *Malpighia*, *Peixotoa*, and *Stigmaphyllon* are cultivated as ornamentals.   VHH

i Anderson, W. R. Malpighiaceae. Pp. 229–232. In Smith, N. *et al.* (eds), *Flowering Plants of the Neotropics*. Princeton, Princeton University Press (2004).
ii Davis, C. C. & Chase, M. W. Elatinaceae are sister to Malpighiaceae; Peridiscaceae belong to Saxifragales. *American J. Bot.* 91: 262–273 (2004).
iii Davis, C. C., Anderson, W. R., & Donoghue, M. J. Phylogeny of Malpighiaceae: Evidence from chloroplast *ndh*F and *trn*L-*trn*F nucleotide sequences. *American J. Bot.* 88: 1830–1846 (2001).
iv Cameron, K. M., Chase, M. W., Anderson, W. R., & Hills, H. G. Molecular systematics of Malpighiaceae: Evidence from plastid *rbc*L and *mat*K sequences. *American J. Bot.* 88: 1847–1862 (2001).
v Mezquita, P. C. & Vigoa, Y. G. The acerola: marginal fruit from America with a high level of ascorbic acid. *Alimentaria* 37: 113–125 (2000).

MALVACEAE

COTTON, MALLOWS, AND HOLLYHOCKS

An economically important family of mainly herbs and shrubs.

Distribution The Malvaceae is worldwide in distribution but primarily tropical and subtropical, extending into temperate areas. Around 75% of the species and 78 of the 115 genera occur in the New World. Estimates of species numbers vary from 1,800 to 2,300. Malvaceae are species of savanna, scrub, and forest edge and seem to be light demanders since they are absent from rain forest. A few species specialize in coastal habitats e.g., *Althaea officinalis* (saltmarshes of western Europe) and *Lavatera arborea* (cliffs of western Europe).

Description Subshrubs or herbs, less usually shrubs, and rarely trees, indumentum of stellate hairs, often mixed with simple hairs, rarely with peltate scales. Most parts contain mucilage. Stem bark usually very fibrous. The leaves are alternate, simple, usually orbicular in outline and then often more or less deeply digitately lobed. Species of drier climates often have oblong, more or less entire leaves. The leaf venation is usually digitate or at least 3-nerved at the base. The petioles are often dilated at base and apex, especially in the shrubbier species; stipules are present but are usually caducous. The inflorescences are either terminal and then often racemose, or axillary, then simple or cymose; they bear pedicellate, bisexual, usually protandrous flowers, with well developed persistent epicalyces (absent in *Abutilon* and the *Sida* group). The sepals are 5, valvate, and often nearly free. The petals are 5, convolute or imbricate, and showy, united at the base to the androecium. The androecium consists of numerous monothecal, longitudinally dehiscing, versatile stamens, their filaments united for most of their length into a cylinder around the style. The pollen is globose, long spiny and usually porate. The ovary is superior and syncarpous, usually with 5 or more locules, each with 1 to numerous axile ovules. At anthesis, the 5 (sometimes 10 or more) cohering styles and stigmas are initially concealed within the apex of the androecium. In the female phase, the stigmas emerge from the end of the androecium by extension of the styles. The

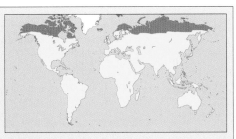

MALVACEAE. 1 *Malva sylvestris* (a) flowering shoot (x⅔); (b) gynoecium (x4); (c) androecium and base of corolla (x4); (d) fruit and persistent calyx viewed from above (x1½). 2 *Malope trifida* (a) flowering shoot (x⅔); (b) young fruit with remains of styles and stigmas removed (x2); (c) vertical section of young fruit (x2); (d) ripe fruit enclosed in calyx and epicalyx (x⅔); (e) flower (x1). 3 *Hibiscus schizopetalus* (a) leafy shoot bearing flower and fruit—a capsule (x⅔); (b) vertical section of lower part of flower showing ovary containing ovules on axile placentas (x1); (c) cross section of ovary showing 5 locules (x2).

Genera 115 **Species** c. 2,000

Economic uses Cotton, flowering herbs and shrubs, stem fibers

fruits are many-seeded dry capsules (e.g., *Abutilon*) or break into 5 or more 1-seeded mericarps, often equipped with awns (e.g., *Malva, Pavonia,* and *Lavatera*). Pollination is mainly by bees.

Classification The Malvaceae is 1 of 10 families now recognized in the Malvales *sensu stricto*, or core Malvales. Formerly only 4 families were recognized: Tiliaceae, Malvaceae, Bombacaceae, and Sterculiaceae. However, molecular evidence has shown indisputably that some parts of traditional

Tiliaceae and Sterculiaceae are more closely related to each other than to the other parts of their respective families. A change in classification has been unavoidable. Within Malvales there are ten groups that are morphologically and palynologically distinct, and, on molecular evidence, monophyletic in all cases but one. One approach is to lump them all into a "super" Malvaceae, recognizing them as subfamilies (FGVP)[i]. The other, taken here, is to recognize each of these ten groups as families, these being Sterculiaceae, Pentapetaceae, Byttneriaceae, Durionaceae, Tiliaceae, Brownlowiaceae, Sparrmanniaceae, Helicteraceae, Malvaceae, and Bombacaceae (q.v.). Although Malvaceae is closely related to Bombacaceae, and previous authors have questioned their distinctness, molecular evidence[i] (and see under Tiliaceae account for refs.) supports their traditional separation. Bombacaceae is distinguished from Malvaceae mainly in being trees, not shrubs or herbs, and in having smooth, not spiny, pollen.

The infrafamilial classification of Malvaceae has remained broadly unchanged since Bentham and Hooker in 1862. The following classification is derived from Hutchinson's *Genera of Flowering Plants* (1967), itself based on Kearney's treatment of the American genera of Malvaceae (1951) and accepted by Fryxell in his revision of Kearney's work (1997). Alternative classifications usually maintain the Ureneae, Hibisceae, and Malveae, the major part of the family, but demote one or both of the other tribes, replacing them with small, coherent groups such as Gossypieae and Decaschistieae (e.g., Mabberley 1997).

Malopeae Carpels numerous in 2 or more superimposed whorls; fruit a schizocarp; carpels uniovulate, the ovule ascending; stigmas apical; involucel absent. Genera include *Palava, Malope,* and *Kitaibelia.*

Ureneae Carpels in 1 whorl; style branches twice the number of the carpels, stigmas apical; fruit a schizocarp; carpels uniovulate, ovule ascending; epicalyx present except in most *Malachra.* Main genera include *Pavonia, Urena,* and *Malvaviscus.*

Hibisceae Carpels in 1 whorl; style branches equal in number to the carpels or styles undivided; fruit capsular, rarely baccate; carpels multiovulate (except *Kosteletzkya*) epicalyx usually present. Main genera include *Hibiscus, Thespesia, Gossypium,* and *Abelmoschus.*

Malveae Carpels in 1 whorl; style branches equal in number to the carpels; stigmas decurrent on style branches, the latter filiform to clavate, but not clearly expanded at the tip. There are two subtribes:

Corynabutilinae Carpels multiovulate, epicalyx absent; style branches thick, blunt. Genera include *Corynabutilon* and *Neobaclea.*

Malvinae Carpels uniovulate, epicalyx present or absent; style branches slender, pointed. Genera include *Althaea, Callirhoe, Lavatera, Malva, Napaea,* and *Sidalcea.*

Abutileae Carpels in 1 whorl; style branches equal in number to the carpels; stigmas apical, capitate, discoid or truncate. There are 2 subtribes:

Abutilinae Ovules 2+ per carpel, if 1 then erect or ascending; epicalyx present or absent. Genera include *Abutilon, Bakeridesia, Malvastrum, Modiola, Nototriche,* and *Sphaeralcea.*

Sidinae Ovules usually solitary, pendulous or horizontal; epicalyx absent. Genera include *Sida, Anoda, Cristaria,* and *Tetrasida.*

Economic uses Cotton is the major internationally traded commodity produced by the family, derived from the seed hairs mainly of 2 spp. of *Gossypium.* The stem fibers of *Hibiscus cannabinus* (Kenaf) also produce a fiber but of lesser importance. *Abelmoschus esculentus,* (Okra or Ladies' Fingers) is cultivated throughout the tropics and subtropics for the mucilaginous, edible, immature fruit and is also exported to temperate areas. *Hibiscus sabdariffa* (Roselle) is extensively cultivated for the fleshy calyces, rich in vitamin C, consumed as an infusion. In the floricultural industry, Malvaceae are significant for providing tropical shrubs, particularly *Hibiscus rosa-sinensis,* which is common throughout the tropics in the form of numerous cultivars and hybrids with related species. In northern temperate gardens, species of *Alcea* and, more recently, *Lavatera* are also cultivated for their flowers, while species of *Abutilon* are increasing in popularity as pot-plants.

MRC

[i] Bayer, C. & Kubitzki, K. Malvaceae. Pp. 225–311. In: Kubitzki, K. & Bayer, C (eds), *The Families and Genera of Vascular Plants. V. Flowering Plants. Dicotyledons: Malvales, Capparales and Non-betalain Caryophyllales.* Berlin, Springer-Verlag (2002).

MARCGRAVIACEAE

Tropical climbers, shrubs, or trees with a number of interesting features, including highly modified nectaries.

Distribution The family is restricted to tropical America, from southern Mexico to northern Bolivia and the West Indies, occurring in primary wet tropical forests. *Ruyschia* is restricted to montane cloud forest.

Genera 7
Species c. 130
Economic uses None

Description Terrestrial or often epiphytic climbers or shrubs, sometimes small trees (some *Sarcopera* and *Schwartzia*). The leaves are alternate, either spirally or distichously (*Marcgravia*) arranged, simple, often glabrous and leathery, with entire or minutely crenate margins and without stipules. In *Marcgravia* the climbing shoots bear different leaves from the mature shoots (dimorphic foliage). The inflorescences are terminal racemes, sometimes umbel-like (*Marcgravia, Marcgraviastrum*) or spicate (*Sarcopera*). The flowers are actinomorphic, bisexual, and borne in pendulous racemes or racemose umbels, with the bracts modified into variously shaped pitchers that secrete nectar. In *Marcgravia,* the nectaries subtend sterile flowers at the apex of the inflorescence. The differing positions of the bracts in the inflorescence help identify the genera in the family[i]. The sepals are 4 or 5, imbricate, and the petals 5, free or fused, forming a cap that falls off when the flower opens (*Marcgravia*). The stamens are 3 to many, free or variously fused. The ovary is superior and has at first 1, but later 2 to many locules with multiple ovules by ingrowth of the parietal placentas. The stigma has a short

MARCGRAVIACEAE. 1 *Marcgravia umbellata* (a) climbing stem with juvenile foliage and adventitious roots (x⅔); (b) nectar cup (x⅔); (c) flower with petals united into deciduous cap (x2). 2 *M. exauriculata* (a) tip of shoot with adult leaves and inflorescence (x⅔); (b) young flower (x1½).

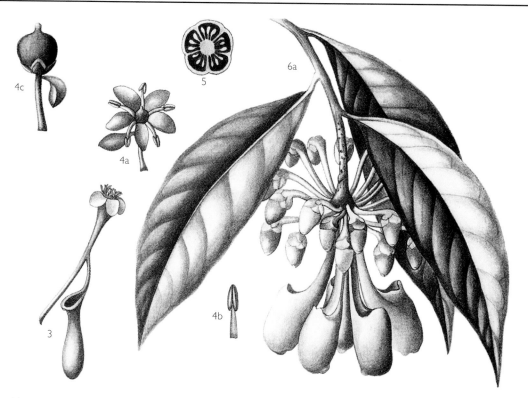

MARCGRAVIACEAE. 3 *Norantea peduncularis* flower and nectar cup (×⅔). 4 *Ruyschia clusiifolia* (a) flower (×2); (b) stamen (×4); (c) fruit (×1½). 5 *Souroubea* sp. cross section of ovary (×3). 6 *Marcgravia nepenthoides* (a) shoot with adult foliage and inflorescence with infertile flowers bearing conspicuous nectar cups (×⅔); (b) half flower (×1½); (c) gynoecium (×2); (d) cross section of ovary (×2). 7 *M. affinis* half flower (×1½).

style and radiates out in 5 lobes. The fruits are globose, fleshy and thick, often indehiscent, and contain many small seeds that lack endosperm. Each seed has a somewhat curved embryo, a large radicle, and small cotyledons.

Classification Study of 3 chloroplast DNA regions indicates support for 2 clades in the family: *Marcgravia*, and the rest[ii] with no support for monophyly of the other genera. Pollinators vary among species, some are hummingbird, bananaquit, or tody pollinated[iii], others utilize bats, moths, and even opossums[iv], others are visited by lizards and bees, still others are self-pollinated before the flower opens. The Marcgraviaceae was thought to be related to the Theaceae (Cronquist 1981 and others) but evidence from DNA sequence studies places it with the Pelicieraceae and Tetrameristaceae[v] in the broader Ericales.

Economic uses There are no major economic uses for this family but various folk medicinal uses have been reported[i]. AC

[i] Dressler, S. Marcgraviaceae. Pp. 258–265. In: Kubitzki, K. (ed.), *The Families and Genera of Vascular Plants. VI. Flowering Plants. Dicotyledons: Celastrales, Oxalidales, Rosales, Cornales, Ericales*. Berlin, Springer-Verlag (2004).
[ii] Ward, N. M. & Price, R. A. Phylogenetic relationships of Marcgraviaceae: Insights from three chloroplast genes. *Syst. Bot.* 27: 149–160 (2002).
[iii] Sazima, I., Buzatao, S., & Sazima, M. The bizarre inflorescence of *Norantea brasiliensis* (Marcgraviaceae): visits of hovering and perching birds. *Bot. Acta.* 106: 507–513 (1993).

[iv] Tschapka, M. & von Helversen, O. Pollinators of synoptic *Marcgravia* species in Costa Rican lowland rain forest: bats and opossums. *Plant Biol.* 1: 382–388 (1999).
[v] Anderberg, A. A., Rydin, C., & Källersjö, M. Phylogenetic relationships in the order Ericales *s.l.*: analyses of molecular data from five genes from the plastid and mitochondrial genomes. *American J. Bot.* 89: 677–687 (2002).

MARTYNIACEAE
UNICORN PLANTS

Mostly annual herbs, rarely a shrub, related to Scrophulariaceae, usually with showy flowers and large fruits with a marked prolongation at their apex.

Distribution Native only in the New World, from central USA to Argentina, usually in dry and disturbed habitats. *Martynia annua, Proboscidea louisianica,* and *Ibicella lutea* are Old World weeds.

Genera 5
Species c. 16
Economic uses Some grown as decorative annual herbs; some limited use for fruits

Description Herbs, usually annual but occasionally perennating from fleshy roots or tubers, or rarely (*Holoregmia* from northeastern Brazil) a shrub to 3 m, all covered with viscid glands. The leaves are opposite or occasionally alternate, usually broad and cordate, palmately lobed to suborbicular. The flowers are large and borne in lax terminal racemes, usually with long pedicels that bear 2 bracteoles. The calyx has 5 sepals that are free or fused into a short tube or (*Craniolaria*) spathe. The corolla has a tubular basal part (linear and up to 12 cm long in *Craniolaria*) and a campanulate upper part with 5 lobes, almost regular to distinctly 2-lipped. The stamens are 4, 2 long and 2 short, or in *Martynia* 2 with 2 staminodes inserted in the corolla tube. The pollen is inaperturate. The ovary is superior and surrounded at the base by a nectariferous disk. The ovary is of 2 fused carpels with a single locule with stout T-shaped parietal placentas on each side, bearing 4 to many ovules. The style is slender and the forked stigma is sensitive so that when an insect pollinator touches

MARTYNIACEAE. I *Proboscidea fragrans* leafy shoot covered with hairs and bearing irregular flowers in a terminal raceme (×⅔).

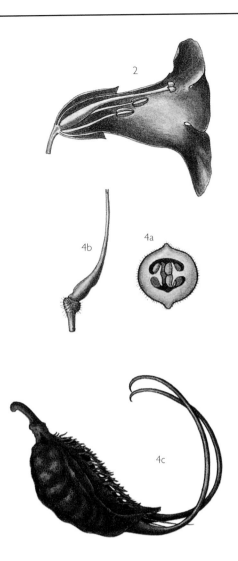

MARTYNIACEAE. 2 *Proboscidea louisianica* half flower showing stamens of 2 lengths (×1). 3 *Martynia annua* (a) flowering shoot (×⅔); (b) part of corolla with 2 fertile stamens with united anthers and 2 staminodes (×2); (c) young fruit (×⅔). 4 *Ibicella lutea* (a) cross section of ovary showing T-shaped parietal placentas (×4); (b) gynoecium (×1); (c) fruit (a capsule) showing horns that aid dispersal by animals (×⅔).

it, the 2 flat lobes close up. The fruit is a loculicidal capsule (sometimes only partially dehiscent) terminated by a short beak or often a long hooked prolongation that splits longitudinally when the capsule ruptures. The wall of the fruit consists of a woody or fibrous endocarp and a soft exocarp that usually flakes off to reveal the variously sculpted wall of the endocarp. *Proboscidea* and *Ibicella* produce many seeds per fruit; other genera have only 4–6 seeds.

Classification The family falls into the complex around Scrophulariaceae. It has often been included in the Pedaliaceae, probably only because of the superficial resemblance of the hooked fruits of some species. However, numerous significant characters separate them[i], especially in the inflorescence, pollen, parietal placentation and details of the fruit structure. The genus *Holoregmia* was rediscovered[ii] some time ago, but it is not mentioned in the most recent account[iii].

Economic importance Some species are grown as decorative annual herbs. The agricultural weeds introduced in the Old World are said to inhibit growth of crop plants, and the spiny fruits are a nuisance to livestock.　RKB

i Ihlenfeldt, H.-D. Martyniaceae. Pp. 283–288. In: Kadereit, J. W. (ed.), *Families and Genera of Vascular Plants. VII. Flowering Plants. Dicotyledons: Lamiales.* Berlin, Springer-Verlag (2004).
ii Harley, R. M., Giulietti, A. M., & Dos Santos, F. R. *Holoregmia* Nees, a recently rediscovered genus of Martyniaceae from Bahia, Brazil. *Kew Bull.* 58: 205–212 (2003).
iii Zanoni, T. A. Martyniaceae. Pp. 239–240. In: Smith, N. et al. (eds), *Flowering Plants of the Neotropics.* Princeton, Princeton University Press (2004).

MEDUSAGYNACEAE

Medusagyne oppositifolia is confined to about 4 localities in the intermediate forest zone between 150 and 500 m altitude on the island of Mahé in the Seychelles. For some years, it was thought to be extinct but was rediscovered in 1970. By 1989, it was believed that only about 10 trees were surviving[i]. Conservation measures have been implemented[ii]. Recently further localities have been discovered[iii], and about 40 plants are now known[iv]. It is a tree up to 10 m high with a rounded crown and a trunk up to 30 cm diameter. The leaves are opposite, simple (? 1-foliate), up to 8 × 4 cm, elliptical, coriaceous, with margins entire to sinuate or subcrenate, with tertiary veins forming a fine raised reticulum of more or less isodiametric cells, the petiole short (to 5 mm), and usually dark and pulvinus-like. Stipules are absent. The inflorescence is a lax terminal cyme, with bisexual and male flowers in the same inflorescence, the male flowers borne in the upper part. The flowers are actinomorphic, 5-merous, up to 1.3 cm across. The sepals are 5, broadly ovate, becoming reflexed. The petals are 5, free, broadly ovate to suborbicular, white or in one population pink[iii], reflexed at maturity. The stamens in male flowers are up to 200–300[iv] on a clavate receptacle, those in the bisexual flowers fewer and smaller and arranged in 3–4 whorls inserted below the ovary. The ovary is superior, barrel-shaped, composed of (17–)20–23(–25) carpels shaped like those of an orange, with their inner margin fused to a central columella and their dorsal surface verrucose with a row of rough papillae. There are 2 ovules per locule, on axile placentae. Each carpel is prolonged into a stout style with a capitate stigma, all of these forming a ring round the upper rim of the barrel, like the head of Medusa. The fruits are slightly woody and pendulous, c. 1.5 cm long, and dehisce from the base, with each carpel remaining attached to the summit of the columella and spreading in an umbrella-like structure. The seeds are winged and up to 3 mm.

Early opinions placed *Medusagyne* in the affinity of Theaceae, but molecular analysis[v] has given it a place close to Ochnaceae and Quiinaceae. It has subsequently sometimes been included in Ochnaceae on cladistic grounds (see Quiinaceae for discussion and references), but it is so distinctive in its floral morphology that it seems unnecessary and unwise to sink Medusagynaceae. In APG II, it is optionally included in Ochnaceae or treated as a separate family, and is placed in the Malpighiales. It is apparently a palaeoendemic surviving in the Seychelles, which are a fragment of the former Gondwanaland. The short pulvinus-like petioles, similar to those of most Ochnaceae and Quiinaceae, suggest that their common ancestor may have had compound leaves.　RKB

i Robertson, A., Wise, R., & White, F. *Medusagyne oppositifolia.* Medusagynaceae. *Kew Magazine* 6: 166–171 (1989).
ii Fay, M. F., Cowan, R. S., Beltran, G., & Allen, B. Genetic fingerprinting of two endemics from the Seychelles, *Medusagyne oppositifolia* (Medusagynaceae) and *Rothmannia annae* (Rubiaceae). *Phelsuma* 8: 11–22 (2000).
iii Wise, R. *A Fragile Eden: portraits of the endemic flowering plants of the granitic Seychelles.* Plate 27. Princeton, Princeton University Press (1998).
iv Friedmann, M. Medusagynaceae. In: *Flore des Seychelles: Dicotylédones.* Pp. 151–154. Paris, ORSTOM (1994).
v Fay, M. F., Swenson, S. M., & Chase, M. W. Taxonomic affinities of *Medusagyne oppositifolia* (Medusagynaceae). *Kew Bull.* 52: 111–120 (1997).

is a coriaceous, dehiscent, 3-valved, 1-seeded capsule with reflexed accrescent sepals.

The affinities of *Medusandra* are not clear. It has been placed in its own order Medusandrales by Takhtajan (1997) next to the Santalales and in the Santalales by Cronquist (1981). Affinities have also been suggested with the Olacaceae and Flacourtiaceae. Medusandraceae is treated as an unplaced eudicot in APG II.   VHH

MELANOPHYLLACEAE

The family comprises the single genus *Melanophylla*, with 7 spp. of glabrous trees up to 15(–20) m high confined to Madagascar, where they grow in humid evergreen forests up to 1,800 m. The leaves are alternate, often clustered at the branch tips, often somewhat succulent, up to 50 cm long or more, usually oblanceolate, the margins subentire to markedly crenate, at least in the distal part, petiolate, the petiole-base sometimes semi-amplexicaul. The inflorescence is a many-flowered terminal or subterminal raceme or panicle, with many bracts, and pedicels up to 1 cm. The flowers are bisexual, actinomorphic, 5-merous, up to 1 cm across, moderately showy. The calyx is tubular with 5 short teeth. The petals are 5, free, reflexed in flower, white to pink, rarely yellow. The stamens are 5, alternating with the petals. The ovary is inferior, with 3 styles, the locules 2–3, with 1 ovule in 1 of the locules and the others sterile. The fruit is a fleshy indehiscent single-seeded drupe, dark purple when ripe[i].

The genus has usually been considered part of, or close to, the broad concept of Cornaceae but is referred by molecular data to the Apiales, close to the Asian families Aralidiaceae and Torricelliaceae[ii]. The leaves often turn black as they dry (hence the generic name), which has suggested affinity with another Madagascan genus, *Kaliphora*, showing this feature, and that has sometimes been placed in Melanophyllaceae although the floral characters are different and it is now placed in Montiniaceae. *Melanophylla* has been revised, with color illustrations[iii]. No economic uses are known.   RKB

Genera 1 Species 7
Economic uses None

[i] Schatz, G. E. Melanophyllaceae. *Generic Tree Flora of Madagascar* 255. Kew & Missouri (2001).
[ii] Savolainen, V. *et al.* Phylogeny of the eudicots: a nearly complete familial analysis based on *rbc*L gene sequences. *Kew Bull.* 55: 257–309 (2000).
[iii] Schatz, G. E., Lowry, P. P., & Wolf, A.-E. Endemic families of Madagascar: 1. A synoptic revision of *Melanophylla* Baker (Melanophyllaceae). *Adansonia* III, 20: 233–242 (1998).

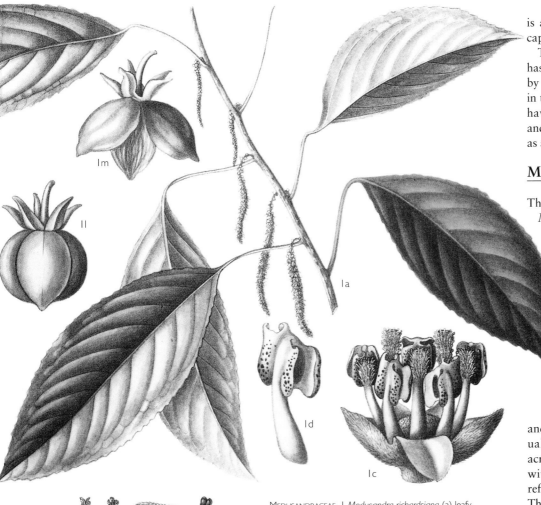

MEDUSANDRACEAE. 1 *Medusandra richardsiana* (a) leafy shoot bearing axillary, pendulous racemes of flowers (x⅔); (b) flower showing the 5 long, hairy staminodes (x6); (c) flower with petals removed to show the 5 short, fertile stamens and bases of the staminodes (x18); (d) stamen with dehisced anthers (x26); (e) tip of staminode (x26); (f) tip of staminode with vestigial anthers split open (x26); (g) cross section of anther (x26); (h) cross section of dehisced anther (x26); (i) gynoecium showing 3 short styles (x26); (j) cross section of ovary (x10); (k) vertical section of ovary showing pendulous ovules (x14); (l) fruit—a capsule (x1); (m) dehisced fruit showing the 3 valves (x1).

MEDUSANDRACEAE

A single genus of 2 spp. confined to Cameroon, of trees with alternate, apparently simple, pinnately veined leaves, with long petioles that are swollen distally forming a pulvinus, the lamina slightly serrate to crenulate. The flowers are small, actinomorphic, bisexual, borne in pendulous, catkinlike racemose inflorescences. The sepals and petals are 5, free, and the stamens are 5, opposite the petals, free, and there are 5 elongate densely hairy staminodes opposite the sepals. The ovary is superior, with 3–4 fused carpels, 1-locular, with central placentation, ovules pendulous. The fruit

Genera 1 Species 2
Economic uses None

MELASTOMATACEAE
MELASTOME FAMILY

Genera 150–166 **Species** c. 4,570
Economic uses Cultivation of some ornamental genera; edible fruit; dyes; local timber source

A large pantropical family of trees, shrubs, woody climbers, and herbs, characterized by the lateral leaf veins running parallel to the leaf margin from the base, converging toward the apex, and often purple and showy flowers.

Distribution The family occurs throughout the wet tropics in montane to lowland forests, savannas and disturbed vegetation[i]. Species are represented in Africa, Asia, and Australia, and two-thirds occur in the neotropics. In America, the genus *Rhexia* reaches as far north as eastern USA and Canada. In South America, 7 genera extend as far south as northern Argentina, and in Africa the genera *Antherotoma*, *Dissotis*, and *Melastomastrum* extend as far as southern Africa. *Medinilla*, *Melastoma*, and *Osbeckia* occur in Australia; and *Astronidium*, *Medinilla*, and *Melastoma* occur in the Pacific Islands.

Description Trees, shrubs, herbs, lianas, and epiphytes. The leaves are opposite, exstipulate, decussate, simple, entire or dentate, and usually petiolate, with 2–8 primary lateral vascular bundles running sub-parallel to the leaf margin, diverging from the base (or just above it) and converging toward the apex. Anisophylly is quite common (particularly in the Sonerileae and Miconieae) and ant domatia occur in c. 10 genera in the Miconieae and Blakeeae and are typically situated in the leaf bases. The diversity of vestiture is the greatest in the angiosperms: the epidermis may be glabrous, or possess an elaborate indumentum varying from simple, unicellular hairs to complex, multicellular, non-glandular or glandular trichomes. The inflorescences are usually terminal, axillary or cauline cymes, but flowers are occasionally solitary. The flowers are actinomorphic or often with zygomorphic androecia, often bracteolate, bisexual, and with (3)4–5(–8) calyx lobes and petals, and a campanulate or urn-shaped hypanthium that bears the calyx lobes, petals,

MELASTOMATACEAE. 1 *Melastoma malabathricum* (a) flowering shoot showing leaves with characteristic parallel veins (×⅔); (b) half flower showing stamens with elbow-shaped filaments and anthers dehiscing by a single pore (×1½); (c) cross section of ovary (×2⅔); (d) stamen showing lobed appendages at base of the connective (×3⅓); (e) fruit—a capsule (×2⅔); (f) seed (×8). 2 *Sonerila* sp. leafy shoot and cymose inflorescence (×⅔).

and stamens on a torus (vascular plexus) at its apex. The calyx lobes are usually regularly lobed and imbricate to valvate, sometimes indistinct or absent, with external teeth or fused to form a calyptra. The petals are free, right-contorted, usually white to purple (sometimes pink, seldom orange or yellow), and usually spreading. The stamens are usually twice as many as the petals, isomorphic or dimorphic with the inner set smaller than the outer one. The anthers are basifixed, obovate to subulate and open by 1–2(–4) apical pores (rarely slits). The connective is often prolonged below the anther and often variously appendaged ventrally and/or dorsally, and the filaments are often twisted, bringing all the anthers to one side of the flower. The ovary is superior or inferior (the hypanthium free or adhering to the ovary completely or partially), with (3)4–5(–14) fused carpels and the same number of locules, usually with axile placentation. The style is single and the stigma is punctiform to capitate. The fruit is either a fleshy berry or a loculicidal capsule enclosed by the persistent hypanthium. The seeds are usually numerous, small, lacking endosperm,

and may be straight with a smooth or tubulate surface, straight with a foveolate surface, or cochleate with a tubulate surface.

Classification The Melastomataceae are divided into 2 subfamilies[ii]:

SUBFAM. KIBESSIOIDEAE Characterized by fiber tracheids, radially and axially included phloem, and median-parietal placentation. Comprises a single monogeneric tribe, Kibessieae, consisting of the tree genus *Pternandra* (15 spp.), with most of its diversity in Borneo but also extending into peninsular Malaysia, Indochina, and New Guinea. Capsules fleshy with dorsal-median placentas and wood with interxylary phloem islands.

SUBFAM. MELASTOMATOIDEAE Characterized by libriform fibers, lacking included phloem and with axile-basal or axillary placentation.

Astronieae tribe is found in Southeast Asia (4 genera, 150 spp.). *Astronia* occurs from peninsular Malaysia to New Guinea; *Astronidium* from New Guinea and the Pacific Islands; *Beccarianthus* from Borneo, the Philippines, and New Guinea; and monotypic *Astrocalyx* from the Philippines. Woody habit, dry capsules with cuneate or linear seeds, basal to basal-axile placentation, anthers opening by 2 longitudinal slits, and leaf midribs with a complex vascular pattern[iii].

Sonerileae occurs mainly Southeast Asia and Madagascar, with some species in the neotropics (40 genera, c. 560–600 spp.). Mostly herbaceous plants of shady habitats, often with a basal whorl of large, somewhat turgescent leaves, and with the ovary apex often crowned by lobes. Old World Sonerileae usually have isomerous flowers, but New World species are typically anisomerous (5-merous androecia with 3-locular ovaries). Recent molecular work[i] points to the possible segregation of the Sonerileae into 2 more tribes: Bertolonieae, a neotropical group (13 genera, 90 spp.), mostly understory shrubs and herbs; and Dissochaeteeae, a paleotropical group, with several poorly circumscribed genera.

Merianieae is neotropical (16 genera, 220 spp.), almost entirely treelets, shrubs, or woody climbers with dry capsules, leathery leaves, and large flowers, with dorsally thickened and variously spurred stamens. The major genera are *Meriania* (c. 74 spp.), *Graffenrieda* (c. 44 spp.), and *Axinaea* (c. 30 spp.).

Microlicieae comprises 275–300 spp., of which more than 90% are endemic to the cerrado of Brazil. Shrubby, often microphyllous, capsules terete, glabrous ovaries, basally prolonged anther connectives, rostrate anther thecae, and oblong or reniform seeds with a foveolate testa. Recent studies[iv] have suggested that it comprises only 6, rather than the traditional 11–17, genera: *Chaetostoma, Lavoisiera, Microlicia* (c. 170 spp.), *Rhynchanthera, Stenodon,* and *Trembleya.*

Melastomeae is a pantropical tribe of some 48 genera and c. 550 spp. Mainly annual or perennial herbs, subshrubs or shrubs, distinguished by cochleate seeds, stamens with

pedoconnectives (basal-ventrally prolonged connectives), and ovaries typically crowned by persistent trichomes; flowers usually diplostemonous with unequal stamens, or 1 whorl staminodial or lacking; often with elaborate and conspicuous trichomes on their leaves and hypanthia. Major genera are the neotropical *Tibouchina* (c. 240 spp.) and *Brachyotum* (c. 58 spp.) and the palaeotropical *Dissotis* (c. 140 spp.), *Osbeckia* (c. 60 spp.), and *Melastoma* (c. 50 spp.). The genus *Rhexia* used to be formally recognized as a separate tribe (Rhexieae), but molecular evidence has shown that it should be included within Melastomeae[i,v].

Miconieae is pantropical, with c. 25 genera and 2,000 spp. Mostly shrubs, trees, and treelets with 4- or 5-merous flowers, short to elongate and persistent external calyx teeth, usually glabrous petals with blunt to acute apices, stamens usually twice the number of petals, with ovoid to elongated, and usually at least slightly curved anthers that open by a minute apical pore or longitudinal slits, and epigynous ovaries that develop into baccate fruits, typically dispersed by birds. *Miconia* (1,000 spp.) occurs throughout tropical America, especially in the Andes.

Blakeeae is neotropical, comprising the genera *Blakea* (100 spp.) and *Topobea* (c. 50+ spp.) only. Flowers 6-merous, axillary, subtended by 2 pairs of persistent floral bracts, conspicuously cross-venulate leaves, with often numerous, strictly parallel, lateral veins and baccate fruit.

Economic uses Some genera are cultivated for their showy flowers or striking leaves including *Tibouchina* (Glory Bushes), *Medinilla, Melastoma, Dissotis, Rhexia, Bertolonia,* and *Centradenia.* Some timber (*Astronia*), edible fruit (*Bellucia, Conostegia,* and *Heterotrichum*), and dyes (*Dionycha*) are produced locally. Introduced Melastomes can become aggressive weeds. *Miconia calvescens* is on the list of the world's 100 most noxious weeds, forming large monospecific stands in Hawaii and Tahiti[v]. *Clidemia hirta*, native to the Caribbean and Central and South America, has also spread throughout the Pacific islands.   EMW

[i] Clausing, G. & Renner, S. S. Molecular phylogenetics of Melastomataceae and Memecylaceae: implications for character evolution. *American J. Bot.* 88: 486–498 (2001).

[ii] Renner, S. S. Phylogeny and classification of the Melastomataceae and Memecylaceae. *Nordic J. Bot.* 13: 519–540 (1993).

[iii] Mentik, H. & Baas, P. Leaf anatomy of the Melastomataceae, Memecylaceae, and Crypteroniaceae. *Blumea* 37: 189–225 (1992).

[iv] Fritsch, P. W. *et al.* Phylogeny and circumscription of the near-endemic Brazilian tribe Microlicieae (Melastomataceae). *American J. Bot.* 91: 1105–1114 (2004).

[v] Renner, S. S. Melastomataceae. In: Smith, N. *et al.* (eds), *Flowering Plants of the Neotropics.* Princeton, Princeton University Press (2004).

MELIACEAE

MAHOGANY, NEEM

An economically important family of tropical and subtropical trees and shrubs, including the commercial mahoganies.

Distribution The family is essentially pantropical, with an extension into northern and western China (the genus *Toona*). They are particularly common as understory trees in rain forests but also grow in mangroves and even arid regions.

Description Small to large trees and occasionally shrubs, sometimes "tuft trees," rarely herbaceous with a woody stock (*Munronia*). The bark is often aromatic, smelling sweet when slashed. The leaves are alternate, spiral, usually pinnate (rarely 2-pinnate), 3-foliate or 1-foliate, exstipulate. In the Malaysian genus *Chisocheton*, the leaf-tip buds continue to produce new pinnae periodically for many years[i]; similar indeterminate growth is found in the American and African genus *Guarea.* The flowers are actinomorphic, functionally unisexual, but appearing bisexual (plants monoecious, dioecious, or polygamous), in cymose panicles, borne on the trunk or branches, or in the axils of undeveloped leaves, or terminal or rarely epiphyllous (*Chisocheton*) or cauliflorous. The inflorescences show considerable variation. In some *Aglaia* species the inflorescence shows indeterminate growth over a period of months; in the genus overall, it can range from one-third to two-thirds of a meter long, with profuse branching and abundant flowers all the way to being much reduced and few-flowered inflorescence only 1–2 cm long. The flowers have distinct sepals and petals. The sepals are (3)4–5(–8), free or partly united. The petals 3–5(–7), distinct or partly united. The stamens are 5–10, the filaments free (*Cedrela, Toona*) or partially to completely united into a tube. The ovary is superior, syncarpous, with 2–13 carpels and 2–13 locules with axile placentation and 1–2 or more ovules per locule. The stigma is usually capitate or discoid. The fruit is a capsule, berry, or drupe—rarely a nut. The seeds are 1 to many, winged or fleshy, or unwinged and then with a fleshy arillode or testa (sarcotesta).

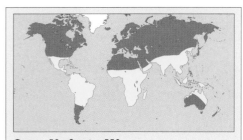

Genera 50   Species 550
Economic uses Timber trees, including mahogany and neem; edible fruits; some species have ornamental value

including *Aglaia* (c.110 spp., Southeast Asia, the Pacific, and Australia), the largest genus in the family[iii], *Trichilia* (85 spp.), *Turraea* (c.65 spp., Old World), *Melia* (5 spp., Old World), *Azadirachta* (2 spp., Indomalaysia and the Pacific), *Guarea* (50 spp., Africa and America), *Chisocheton* (50 spp., Indomalaysia), *Dysoxylum* (80 spp., Indomalaysia and Pacific), and *Xylocarpus* (2 or 3 spp. in mangrove and coastal forests of the Old World).

SUBFAM. SWIETENIOIDEAE Buds usually protected by scales; fruit a septifragal capsule; seeds usually winged with a corky or woody testa; plants monoecious. About 15 genera and 50 spp., including *Swietenia* (3 spp., New World tropics), *Entandrophragma* (15 spp., Africa), *Khaya* (8 spp., Africa), *Cedrela* (9 spp., America), and *Toona* (15 spp., Asia and Australasia).

The Meliaceae belongs in the Sapindales *sensu* Cronquist, as confirmed by molecular data and is most closely related to the Simaroubaceae and Rutaceae[iv,v], from which it differs in androecial characters.

Economic uses The Meliaceae contains many economically important species. It includes some of the most highly prized timber trees in the world, including the commercial mahoganies obtained from *Swietenia* and *Khaya*. Other important timbers include Sapele (*Entandrophragma cylindricum*), utile (*E. utile*), Omu (*E. candollei*), Spanish Cedar (*Cedrela odorata*), and species of *Melia, Carapa, Azadirachta,*

Guarea, Cedrela, Toona, Soymida, Chukrasia, Dysoxylum, Lovoa, Aglaia, and *Owenia*. Other uses of Meliaceae comprise shade and street trees; fruit trees, e.g., Langsat (*Lansium domesticum*) and Santol (*Sandoricum koetjape*), especially in Southeast Asia; and as sources of biologically active compounds. Many species have medicinal properties, e.g., Vinkohomba (*Munronia pumila*), which is used to bring down body temperature and in the treatment of malaria. Oils are extracted from the seeds of several species. The Neem Tree (*Azadirachta indica*) has been used for more than 2,000 years in Ayurvedic and Unani medicine and as a shade tree and is now commercially exploited as a source of pharmaceuticals and cosmetic products, pesticides, and as a fertilizer and in organic farming. VHH

i Fisher, J. B. Indeterminate leaves of *Chisocheton* (Meliaceae): survey of structure and development. *Bot. J. Linn. Soc.* 139: 207–221 (2002).
ii Muellner, A. N. *et al*. Molecular phylogenetics of Meliaceae (Sapindales) based on nuclear and plastid DNA sequences. *American J. Bot.* 90: 471–480 (2003).
iii Muellner, A.N. *et al*. *Aglaia* (Meliacae): An evaluation of taxonomic concepts based on DNA data and secondary metabolites. *American J. Bot.* 92: 534–543 (2005).
iv Gadek, P. A. *et al*. Sapindales: Molecular delimitation and infraordinal groups. *American J. Bot.* 83: 802–811 (1996).
v Chase, M. W., Morton, C. M., & Kallunki, J. A. Phylogenetic relationships of Rutaceae: A cladistic analysis of the subfamilies using evidence from *rbc*L and *atp*B sequence variation. *American J. Bot.* 86: 1191–1199 (1999).

MELIACEAE. **1** *Melia azedarach* (a) bipinnate leaf and axillary inflorescence (x⅔); (b) fruits (x⅔); (c) staminal tube opened out to show anthers attached and gynoecium with basal disk (x2); (d) vertical section of ovary with ovules on axile placentas (x6); (e) cross section of fruit (x⅔); (f) seed (x4). **2** *Swietenia mahagoni* (a) shoot with pinnate leaf and axillary inflorescence (x⅔); (b) half flower showing staminal tube (x8); (c) winged seed (x⅔). **3** *Cedrela australis* (a) flower opened out showing free stamens arising from the disk and superior ovary crowned by discoid stigma (x3½); (b) part of inflorescence (x⅔); (c) winged seeds (x1); (d) fruits surrounded by persistent sepals (x⅔).

Classification The family has been divided into 4 (often 5) subfamilies, of which 2 are small and restricted to Madagascar; the other 2, Melioideae and Swietenioideae, are pantropical. Molecular and phylogenetic analyses now support just 2 subfamilies, Melioideae and Swietenioideae[ii], which are sister groups.

SUBFAM. MELIOIDEAE Buds naked; fruit a loculicidal capsule, sometimes a berry, drupe, or nut; seeds unwinged and usually with a fleshy sarcotesta or aril; plants usually dioecious. About 35 genera and 550 spp., mainly Old World,

MELIANTHACEAE

A formerly entirely African family recently expanded to include Greyiaceae from South Africa and Francoaceae from Chile, now including 5 genera rather diverse in habit and leaf characters but showing similarities in inflorescence and flower structure.

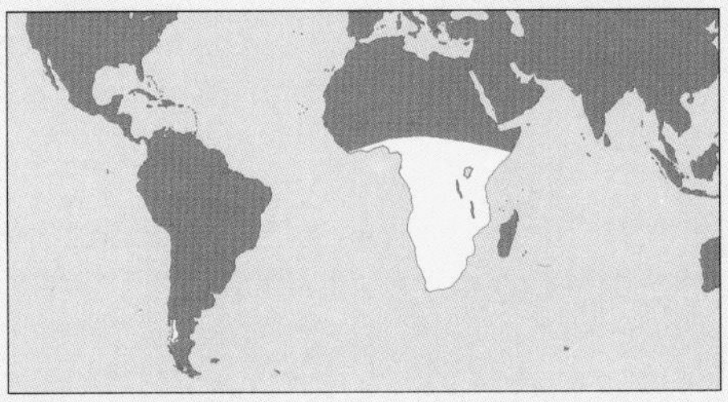

Genera 5 **Species** 18
Economic uses Medicinal purposes; timber; cultivation as ornamentals

Distribution *Bersama,* with 7 spp. in rain forest and woodland in tropical and South Africa, and *Melianthus,* with 6 spp. in both wet and dry habitats in South Africa, have long comprised Melianthaceae. *Greyia* has 3 spp. in eastern South Africa, growing in rocky places, on riverbanks, and at forest margins up to 1,500 m altitude. *Francoa,* with probably 1 highly polymorphic species, *F. appendiculata* (Bridal Wreath), and *Tetilla hydrocotylifolia,* are both confined to Chile in woods and rocky places from sea level to 1,500 m.

Description *Bersama* includes shrubs and trees up to 24 m; *Greyia* comprises shrubs and small spreading trees up to 5(–7) m high, with soft-wooded and sometimes almost succulent stems; *Melianthus* is a shrub or robust subshrub up to 2.3 m with leaves crowded in the lower parts; *Francoa* and *Tetilla* are rhizomatous erect herbs up to 1.5 or 0.3 m high, respectively, with leaves mostly crowded near the base. The leaves are alternate, pinnate, and sometimes with a winged rhachis in *Bersama* and *Melianthus,* lyrate to pinnatisect in *Francoa* with a large terminal lobe resembling the lamina of *Greyia,* simple and suborbicular with lobed and crenate margins and distinct petioles in *Greyia* and *Tetilla.* The stipules are small to very large and sheathing and in *Greyia* the sheath is fused to the stem as an easily shed pseudocortex. The inflorescence is a lax to dense terminal raceme or spike with pedicels spreading widely. The flowers are actinomorphic to weakly zygomorphic, resupinate at least in *Bersama* and *Melianthus,* bisexual or unisexual (plants polygamodioecious[i]), 5-merous in *Bersama* and 4- to 5-merous in the other genera. The sepals are 4 or 5, often unequal. The petals are 4 or 5, free, sometimes showy.

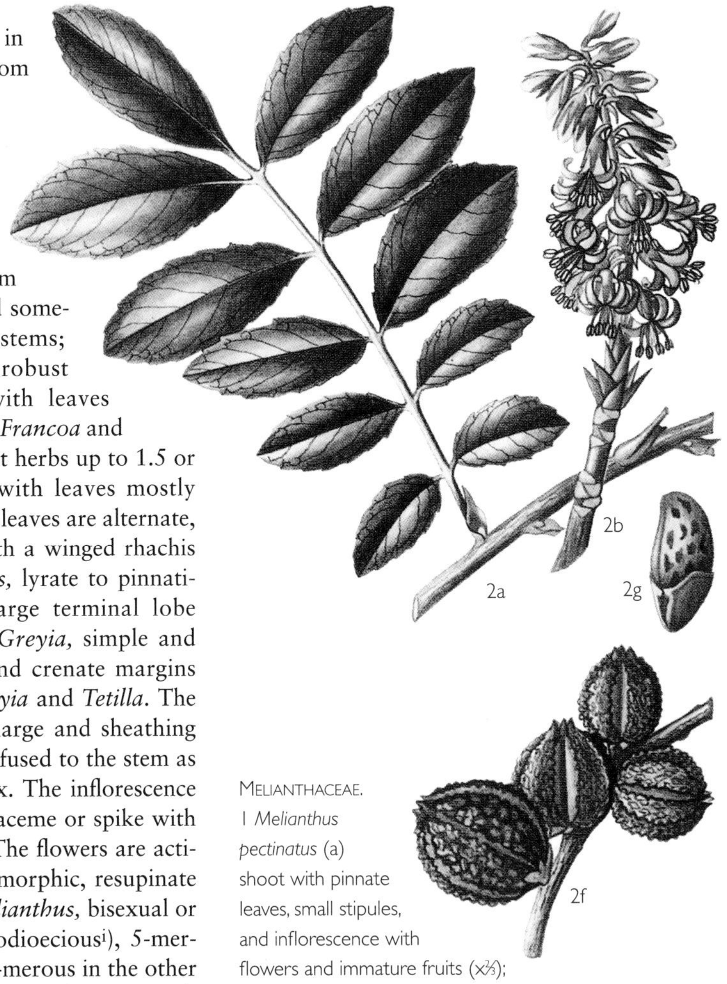

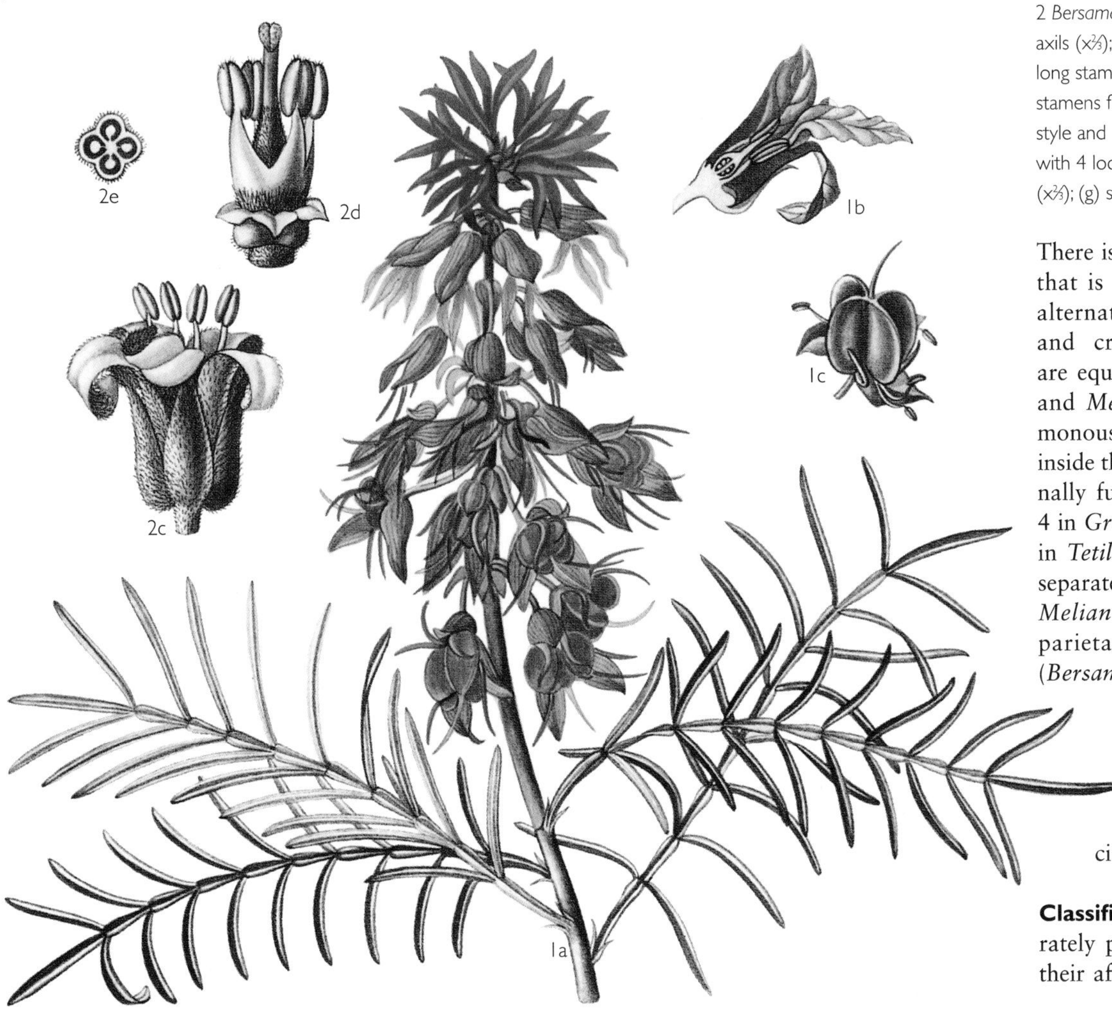

MELIANTHACEAE. 1 *Melianthus pectinatus* (a) shoot with pinnate leaves, small stipules, and inflorescence with flowers and immature fruits (x⅔); (b) half flower with irregular sepals and petals and swollen nectar-secreting disk (x1); (c) capsule (x⅔). 2 *Bersama tysoniana* (a) leafy shoot with stipules in the axils (x⅔); (b) inflorescence (x⅔); (c) mature flowers with long stamens (x3); (d) young androecium with 4 short stamens fused at the base and ovary crowned by simple style and lobed stigma (x4½); (e) cross section of ovary with 4 locules and ovules on axile placentas (x3); (f) fruits (x⅔); (g) seed with aril (x1).

There is a disk between the petals and stamens that is either annular with 8–10 acute lobes alternating with the stamens or unilateral and crescent- or V-shaped. The stamens are equal in number to the petals in *Bersama* and *Melianthus,* twice as many (obdiplostemonous[ii]) in the other genera, inserted on or inside the disk. The ovary is superior, longitudinally furrowed, with 4–5 carpels in *Bersama;* 4 in *Greyia, Melianthus,* and *Francoa;* and 2–4 in *Tetilla.* It is either completely divided into separate locules with axile placentas (*Bersama, Melianthus*) or is 1-locular with intrusive parietal placentas[ii], each carpel with few (*Bersama, Melianthus*) to many (other genera) ovules in 2 rows. The fruit is a capsule, usually elongate and chartaceous, but in *Bersama* spherical and woody, dehiscing loculicidally in *Bersama* and *Melianthus* or septicidally in other genera.

Classification *Greyia* and *Francoa* have separately puzzled botanists for a long time as to their affinities, and it has only been in the last decade that their close similarities to each other, and to the traditional Melianthaceae, have been appreciated, despite rather striking differences in habit and leaf form. *Greyia* was placed in the Melianthaceae by J. D. Hooker in 1873[iii] but has more recently been placed in Saxifragaceae *sensu lato* or close to Cunoniaceae or Escalloni-aceae and has been given separate family status in recent South African literature[iv]. *Francoa* and the related but comparatively little-known *Tetilla* have usually been placed in Saxifra-gaceae. In his broadly based study, van Wyk[iii] considered relationships of Greyiaceae with both Melianthaceae and *Francoa* but concluded that it was not close to either. However, over-looking the marked differences in habit and geographical distribution, floral characters show few obvious differences between *Greyia* and *Francoa*[ii] except in *Greyia* having a long slender style whereas *Francoa* and *Tetilla* hav-ing a more or less sessile stigma.

Molecular evidence has placed these genera together[v] and as a sister group to *Bersama* and *Melianthus*[vi]. The latter genera differ in having pinnate leaves (but those of *Francoa* are pinnati-sect), slightly zygomorphic flowers (but *Greyia* may also have slightly one-sided flowers), sta-mens equalling petals in number, 1–5 ovules per carpel, and few large seeds (see van Wyk[iii] for more detailed comparison). There are many obvi-ous similarities of *Greyia* with *Melianthus,* and inclusion of Francoaceae in a broad Meli-anthaceae now seems acceptable. Each of the 5 genera has distinctive significant features, and the extremes of *Bersama* and *Tetilla* are widely differ-ent, but there seems to be no convenient way of dividing up the whole group. In APG II, Melianthaceae, with Francoaceae optionally rec-ognizable, is included in Geraniales.

Economic uses Some species are used in tradi-tional medicine (e.g., *Bersama tysoniana* in South Africa)[i]. The root, bark, and leaves of *Melianthus comosus* are used in South Africa for treating snake bites, while a decoction of the leaves of *M. major* is used for healing wounds. *Bersama abyssinica* produces a hard, heavy wood that is used for house construction in West Africa. The soft wood of *Greyia* is used in South Africa for carving ornaments and household utensils. Species of *Melianthus* emit a strong scent, and the wood may be burned as incense. *Greyia sutherlandii* and *G. radlkoferi* are attrac-tive showy small trees that are increasingly cultivated in tropical countries or in temperate conservatories. *Francoa appendiculata* is widely cultivated under several different species names as a herbaceous perennial. AC & EJL & RKB

i http://www.hort.purdue.edu/newcrop/ proceedings1999/v4-160.html
ii Ronse-Decraene, L. P. & Smets, E. F. Floral ontogeny of *Francoa* and *Greyia*. *Int. J. Plant Sci.* 160: 377–393 (1999).
iii van Wyk, A. E. Greyiaceae. Pp 29–35. In: van Wyk, A. E. & Dahlgren, R. Structures and relationships of families endemic to or centred in southern Africa. In: Goldblatt, P & Lowry, P. P. (eds), *Modern Systematic Studies in African Botany.* [Proc. XI Meeting St. Louis, AETFAT] (1988).
iv Archer, R. H. Greyiaceae, Melianthaceae. Pp. 315–316, 356. In: Leistner, O. A. (ed.), *Seed Plants of Southern Africa: Families and Genera,* (Strelitzia 10) (2000).
v Morgan, D. R. & Soltis, D. E. Phylogentic relationships among members of Saxifragaceae *sensu lato* based on *rbc*L sequence data. *Ann. Missouri Bot. Gard.* 80: 631–660 (1993).
vi Savolainen, V. *et al.* Phylogeny of the eudicots: a nearly complete familial analysis based on *rbc*L gene sequences. *Kew Bull.* 55: 257–309 (2000).

MELIOSMACEAE

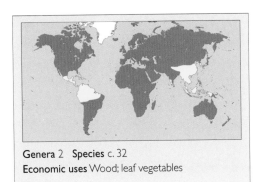

Genera 2 Species c. 32
Economic uses Wood; leaf vegetables

Distribution *Meliosma* has a bimodal distribu-tion, with 15 spp. from the western Himalayas and southwestern India to Japan and New Guinea, and about 10 from Mexico and the West Indies to the northern half of South Amer-ica. Even 1 sp., *M. alba,* occurs from northern Myanmar to central China and also in Mexico (with fossil records from eastern Europe and the Caucasus)[i]. *Ophiocaryon* has 7 spp. from the Guyanas to Brazil and Peru. The family is found in both wet and dry forests.

Description Evergreen or deciduous shrubs and trees up to 25 m or more high. The leaves are alternate, simple or (*Ophiocaryon, Meliosma* subsect. *Pinnatae*) pinnate, up to 45 cm long, with entire or serrate margins, the simple lamina or leaflets with a distinct pulvinus, without stip-ules. The inflorescence is a terminal or axillary pyramidal panicle, often large, the pedicels short or absent, with small caducous bracts and no bracteoles (sometimes sepals remote from the flower resemble bracteoles). The flowers are small, zygomorphic, bisexual. The sepals are (3)5, sometimes unequal, sometimes forming an involucre with the small bracts. The petals are 5, unequal with the 3 outer ones suborbicular and concave and the 2 inner ones variously reduced and scalelike (*Meliosma*), or subequal and elon-gate (*Ophiocaryon*). The fertile stamens are 2, opposite the inner petals, and in *Meliosma* attached to them at the base, and with 3 stamin-odes that are often fleshy and enveloping the fertile stamens and the ovary. The ovary is supe-rior, 2(3)-locular, with 2 ovules in each locule. The fruit is a globose to pyriform, usually single-seeded, drupe.

Classification The family is closely related to the Sabiaceae and often included in it. Further comments on relationships are made under that family (q.v.).

Economic uses Some Asian *Meliosma* species provide hard wood used to make wheels, carry-ing poles, or other implements. The leaves of other species may be eaten as vegetables. RKB

i van Beusekom, C. F. Revision of *Meliosma* (Sabiaceae), section *Lorenzanea* excepted, living and fossil, geography and phylogeny. *Blumea* 19: 357–523 (1971).

MEMECYLACEAE

A family resembling Melastomataceae but lack-ing their characteristic venation and with few large seeds.

Distribution Pantropical with 4 genera in the Old World and 2 genera in the New World. *Memecylon* has c. 300 spp. in Africa and tropical Asia to southwestern Pacific; *Lijndenia* has 10 spp. in tropical Africa and Asia; *Warneckea* 33 spp. in Africa and Madagascar; *Spathandra* has 2 spp. in Guineo-Congolan Africa; *Mouriri* has 81 spp. from Mexico to Bolivia and Brazil; and *Votomita* has 10 spp. from Panama to northern South America and in Cuba. Most species occur in lowland moist forests.

Description Shrubs or forest trees up to 35 m or more, usually glabrous. The leaves are opposite, simple, entire, pinnately veined or with 3 promi-nent veins from the base not or scarcely joining each other near the apex (see note below). The inflorescence is cymose to umbelloid or fascicu-late, sometimes cauliflorous, rarely flowers solitary in the axils. The flowers are actinomor-phic, bisexual, 4- to 5-merous. The sepals are free and minute. The petals are free, yellow to pink or bluish. The stamens are twice as many as

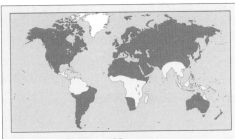

Genera 6 Species c. 435
Economic uses Wood for furniture, dyes

MEMECYLACEAE. 1 *Memecylon laurinum* half flower showing inferior ovary with sub-basal placentas (x3⅓). 2 *M. intermedium* fruits—berries (x⅔).

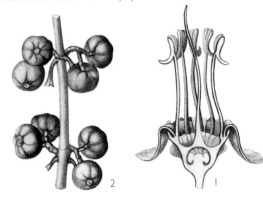

the petals, isomorphic or dimorphic, with basi-fixed anthers dehiscing by 2 short apical slits and the connective with a dorsal gland producing a terpenoid. The ovary is inferior, varying from 1 locule with free central placentation to up to 5(6) locules with basal to axile or parietal placenta-tion, with up to 16 ovules per locule but usually fewer and only 1–2 developing into seeds. The fruit is a berry with 1–12 large seeds.

Classification The family is referable to the Myrtales and has often been included within Melastomataceae. It differs from the latter in its leaf venation (see below), presence of leaf scle-reids, paracytic stomata, anthers usually dehiscing by longitudinal slits and with a gland on the connective, and the fruits with few large seeds which have cryptocotylar germination[i]. Furthermore the Memecylaceae are nearly always glabrous and are often large trees while the Melastomataceae nearly always have an indumentum and are seldom big trees. The Melastomataceae have characteristic acro-dromous leaf venation, with up to 9 veins radiating from the base and joining up again near the leaf apex. Memecylaceae can often be distinguished from them at a glance by their brochidodromous (loop-veined) leaf venation, which is often pinnate but sometimes (e.g., *Warneckea*) has 3 prominent veins radiating from the base of each leaf, in which case they usually stop short of joining up near the leaf apex. However, a clear distinction from Melas-tomataceae in this character is sometimes difficult to maintain. The family is probably also close to Crypteroniaceae and Lythraceae.

Economic uses *Memecylon* species produce wood suitable for making furniture or some-times a yellow dye. RKB & EMW

[i] Renner, S. S. Phylogeny and classification of the Melastomataceae and Memecylaceae. *Nordic J. Bot.* 13: 519–540 (1993).

MENISPERMACEAE

CURARE

The Menispermaceae is a family predominantly of lianas of tropical lowlands, including sources of curare (*Chondrodendron tomentosum* and *Sciadotenia toxifera*).

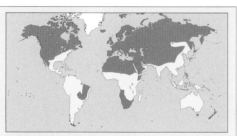

Genera c. 70 Species c. 420
Economic uses Medicines and drugs (curare); sarsaparilla substitute

Distribution Most species grow in tropical rain forest, but there are some subtropical and warm-temperate species. The genus *Menispermum* (Moonseed) is widespread from Atlantic North America and Mexico to eastern Asia. *Chondrodendron, Discipha-nia*, and *Hyperbaena* occur in South America. The large genus *Cocculus*, also present in Atlantic North America and Mexico, extends across Africa and India into Indoma-laysia, growing in rain forests and also in semidesert scrub and deciduous bushland in Africa, Madagascar, and Socotra. *Cissampelos* has a similar distribu-tion. Other African genera are *Jateorhiza* and the large genus *Triclisia* (also in Mada-gascar and Socotra), and *Tiliacora* which also extends into India, Myanmar, and Sri Lanka.

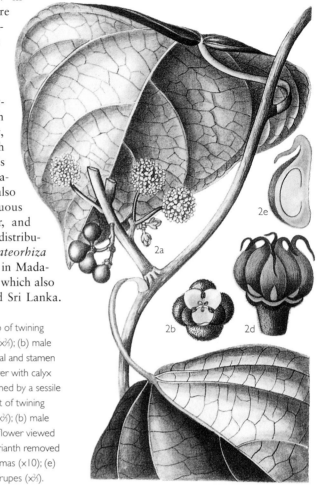

MENISPERMACEAE. 1 *Tinospora cordifolia* (a) tip of twining stem with flowers in lateral inflorescences (x⅔); (b) male flower with 6 sepals and petals (x4); (c) petal and stamen (x6); (d) female flower (x2); (e) female flower with calyx removed to show 3 free carpels each crowned by a sessile stigma (x6). 2 *Coscinium fenestratum* (a) part of twining stem with lateral inflorescences and fruits (x⅔); (b) male flower viewed from above (x4); (c) female flower viewed from above (x6); (d) female flower with perianth removed to show 6 free carpels crowned by thin stigmas (x10); (e) vertical section of carpel (x18); (f) fruits—drupes (x⅔).

Tinospora and *Pycnarrhena* (a large genus) also grow in Indian forests but extend into Thailand, Cambodia, the Philippines, Indomalaysia, and northern Australia.

Description Climbers (sometimes gigantic), less often shrubs or small trees. The leaves are without stipules, alternate and sometimes peltate. The inflorescence is a raceme, panicle, or cyme, usually with many flowers borne on a short, leafless, axillary shoot. The flowers are unisexual (plants dioecious), very small, often greenish-white. The sepals are usually in 2 rows of 3 (sometimes more), and the petals in 2 or 3 rows. In the staminate flowers, the 3–6(–many) stamens may be free, united, or in bundles. In the pistillate flowers, the 1–32 free carpels each contain 2 ovules that reduce to 1. The stigma is terminal, entire, or lobed. The fruit is drupaceous, usually curved, often to horseshoe-shape, and the endocarp has attractive sculpturing.

Classification The family was divided into 8 tribes by Diels[i] based largely on seed structure but subsequently reduced to 5[ii]:

Pachygoneae Sepals and petals distinguishable, carpels 3–12, endocarp curved, not sculptured, or straight and with a swelling, endosperm absent (e.g., *Hyperbaena*, *Penianthus*).

Anomospermeae Sepals and petals indistinguishable (none in *Abuta*), carpels 3, endocarp usually curved, hardly sculptured, endosperm ruminate (e.g., *Anomospermum*).

Tinosporeae Sepals and petals distinguishable (e.g., *Jateorhiza* and *Tinospora*) or not (*Arcangelisia*), carpels 3–6, endocarp straight, usually sculptured, endosperm ruminate.

Fibraureae Sepals and petals indistinguishable, usually 3, carpels sometimes 6, endocarp straight, usually sculptured, endosperm little (*Anamirta*, *Coscinium*, *Fibraurea*, *Tinomiscium*).

Menispermeae Sepals and petals distinguishable, petals sometimes absent, carpels 6, endocarp curved, ribbed and sculptured, slight endosperm present or absent (e.g., *Menispermum*, *Cocculus*, *Sinomenium*, *Stephania*, and *Cissampelos*).

Several research teams are currently working on Menispermaceae phylogeny at the family level[iii,iv,v,vi], and it is hoped that publication of their work will help stabilize the tribal classification. The Menispermaceae is consistently placed in Ranunculales[vii] sometimes sister to Berberidaceae and Ranunculaceae[viii].

Economic uses Curare (tubocurarine chloride) is obtained mainly from the bark of *Chondrodendron tomentosum* and is used as a muscle relaxant in surgical operations and in neurological conditions. The drupes of *Anamirta cocculus* ("fish berries") contain a poison (picrotoxin), which is used locally in fishing and in treating skin diseases. A tonic and febrifuge is prepared from the roots of *Jateorhiza palmata* (Calumba).   [HPW] AC

[i] Diels, L. Menispermaceae. P. 94. In: Engler, A. (ed.), *Pflanzenreich* IV. Leipzig, Engelmann (1910).
[ii] Kessler, P. J. A. Menispermaceae. Pp. 402–418. In: Kubitzki, K., Rohwer, J. G., & Bittrich, V. (eds), *The Families and Genera of Vascular Plants. II. Flowering Plants. Dicotyledons: Magnoliid, Hamamelid and Caryophyllid Families*. Berlin, Springer-Verlag (1993).
[iii] http://www.umsl.edu/services/kellogg/people/ortiz.html
[iv] White, P. J. & Stevenson, D. W. Phylogenetic relationships and character evolution in the Menispermaceae [abstract]. *Botany 2002, Botany in the Curriculum: Integrating Research and Teaching*. Madison, Wisconsin (2002).
[v] http://www.uwm.edu/People/hoot/Menisp.html
[vi] Jacques, F. Towards a new phylogeny of Menispermaceae (Ranunculales) [abstract]. *Cladistics* 20 (6): 597–597 (2004).
[vii] Brummitt, R. K. *Vascular Plant Families and Genera*. Richmond, Royal Botanic Gardens, Kew (1992).
[viii] Soltis, D. E. *et al.* Angiosperm phylogeny inferred from 18S rDNA, *rbc*L, and *atp*B sequences. *Bot. J. Linn. Soc.* 133: 381–461 (2000).

MENYANTHACEAE

BOG BEAN FAMILY

A family of wetland plants, including numerous floating aquatics.

Distribution Almost worldwide (except in dry areas), but especially in Australia. *Nymphoides* has about 40 spp., of which *N. indica* is pantropical, 13 spp. occur in Africa and Madagascar, up to 10 in tropical and subtropical Asia, and 20 in Australia. *Villarsia* has around 20 spp., including 3 in South Africa, 3 in Malaysia and Indochina, and 13 in Australia. *Nephrophyllidium* has 1 sp. in Japan and Alaska to Washinton State. *Menyanthes* has 1 sp. widespread in northern temperate regions. *Liparophyllum* has 1 sp. in Tasmania and New Zealand. All are found in wet habitats with still lying water, often rooting under shallow open water.

Description Perennial or occasionally annual herbs, sometimes rhizomatous, with well developed air canals in vegetative parts. All parts are glabrous (except the fimbriate corolla lobes). The leaves are alternate, often in a basal rosette, with petioles sheathing at the base, the lamina linear (*Liparophyllum*) to elliptical, reniform (*Nephrophyllidium*, some *Villarsia*), heart-shaped (*Nymphoides*), suborbicular to sagittate, or 3-foliate with elliptical leaflets (*Menyanthes*). In *Nymphoides,* and sometimes *Villarsia*, the leaves float on the surface of the water, like water lily leaves. The inflorescence is a simple or branched cyme or raceme or dense head, or the flowers are solitary. The flowers are actinomorphic, usually bisexual but sometimes unisexual (plants monoecious), often heterostylous[i]. The sepals are 5, sometimes somewhat connate with the base of the ovary. The corolla has 5 lobes about as long as the tube, often with conspicuous fimbria on the margins or upper surface. The stamens are 5, inserted on the corolla tube. The ovary is superior to semi-inferior, of 2 carpels, with a single locule, with 2 parietal placentas bearing few to many ovules. The fruit is a capsule dehiscing by 2–4 valves or irregularly, or is a fleshy berry in *Liparophyllum*.

Genera 5 **Species** c. 62
Economic uses Medicinal uses; leaves and rhizomes occasionally eaten; some cultivated aquatic species; aquatic weeds

MENYANTHACEAE. 1 *Menyanthes trifoliata* (a) habit showing 3-foliate leaves and hairy flowers in an erect raceme (x⅔); (b) part of hairy corolla showing epipetalous stamens (x2); (c) gynoecium and calyx with one sepal removed (x2); (d) cross section of 1-locular ovary with ovules on 2 parietal placentas (x4). 2 *Liparophyllum gunnii*, entire plant (a) with flower and (b) with fruit (x⅔); (c) vertical section of gynoecium (x4); (d) corolla opened out to show epipetalous stamens (x3). 3 *Nymphoides peltata* (a) shoot with peltate leaves and solitary flowers (x⅔); (b) gynoecium (x2); (c) part of corolla with basal crests of hairs and epipetalous stamens (x1); (d) fruit—a capsule; (e) seed (x3).

1a

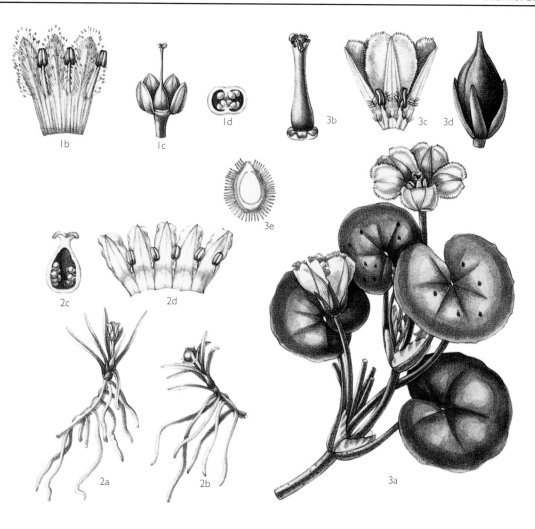

1b 1c 1d

3b 3c 3d

3e

2c 2d

2a 2b 3a

Classification The family used to be included in Gentianaceae and was later considered close-ly related to that family. However, molecular evidence places it in the Asterales rather than Gentianales. Seed characters have been found valuable in defining species[ii].

Economic uses Leaves of *Menyanthes trifolia-ta* yield the glucoside menyanthin, which is used medicinally to reduce fever or in larger doses as a purgative or for getting rid of worm parasites. The leaves have also been eaten as a vegetable, used in place of hops in making beer, and used as a substitute for tea. The rhizomes have been used in arctic regions for making bread. Species of *Nymphoides* are grown as aquatics, and some have become serious weeds of waterways in tropical countries. *Nephro-phyllidium crista-galli* is also grown beside streams in gardens.   RKB

[i] Ornduff, R. Distyly and monomorphism in *Villarsia* (Menyanthaceae): some evolutionary considerations. *Ann. Missouri Bot. Gard.* 75: 761–767 (1988).
[ii] Aston, H. I. Seed morphology of Australian species of *Nymphoides* (Menyanthaceae). *Muelleria* 18: 33–65 (2003).

MESEMBRYANTHEMACEAE
ICE PLANTS, LIVING STONES, VYGIES, AND MESEMBS

A large family of succulent plants with usually showy, daisylike flowers, often forming brilliant sheets of color when growing en-masse.

Distribution The majority of species are endemic to southern Africa, especially the Suc-culent Karoo. By contrast, most *Disphyma* are native to Australasia, and some *Carpobrotus* and *Sarcozona* species are restricted to this region. A few other outlying species occur in coastal regions of the Mediterranean, south-western South America, the Near East, and North Africa. Some species have naturalized extensively in mediterranean climate ecosys-tems of Europe, Australia, and America.

Description Annual or perennial herbs or small shrubs, usually more or less succulent. The leaves are alternate or opposite, mostly simple, entire, with or without stipules. The flowers are solitary or in cymes, and regular, with their parts in whorls, and usually bisexual. The sepals are 4–8 (usually 5) and are imbricate or rarely valvate, more or less united below. Petals are absent, their role fulfilled by whorls of petaloid staminodes, often brightly colored. The stamens are perigy-nous, usually numerous, sometimes with connate filaments. The ovary is superior or inferior, with between 1 and 20 (usually 5) stigmas and locules and usually numerous ovules. The fruit is a dry capsule, rarely a berry or nut. The seed has a large curved embryo surrounding a mealy endosperm.

Many of the features of the Mesembryan-themaceae are the result of adaptations to extremely dry climates (xeromorphy), and typi-cal members are able to survive long periods of extreme insolation and drought, e.g., in the desert regions of South Africa. The leaves are more or less succulent, and in addition some plants have succulent roots or caudices. Often the plant is reduced to a single annual pair of opposite leaves, which may be so condensed as to approach a sphere, with minimal surface in relation to volume, enabling the plant to resist desiccation. The internal tissues also show modi-fications. Large watery cells rich in sugars called pentosans are characteristic of succulents. In *Muiria*, these cells may be 1 mm in diameter and can retain their moisture for weeks when sepa-rated and exposed to dry air. The possession of 2 different leaf forms (heterophylly) is common (*Mitrophyllum, Monilaria*), the leaf pair formed at the start of the dormant season being more united and compact than that formed when in full growth and acting as a protective sheath to the stem apex. Other genera are partly subter-ranean, with only the clear "window" in each leaf tip exposed above soil. A type of optical sys-tem exists whereby a layer of apical tissue rich in calcium oxalate crystals acts as a filter to intense sunlight before it reaches the thin chlorophyllous layer below (*Fenestraria, Frithia, Conophytum* subgen. *Ophthalmophyllum*).

Other so-called mimicry plants show a strik-ing similarity to their background rocks and are difficult to detect when not in flower. These are the pebble plants or living stones (*Lithops*); each species is associated with one particular type of rock formation and occurs nowhere else. *Titanopsis*, with a white encrustation to the leaves, is confined to limestone outcrops. This is probably a rare case of protective col-oration in plants akin to examples of mimicry found in the animal kingdom.

The phenomenon of crassulacean acid metabolism occurs in members of the Mesem-bryanthemaceae, having evolved independently in many different families of succulent plants.

The mostly showy, diurnal flowers have a superficial resemblance to the flower heads of the Compositae. They are insect-pollinated, and most require full sunlight before they will expand. Several have set hours for opening and closing. *Carpobrotus* produces an edible berry, the Hottentot Fig, but the remainder form dry, dehiscent capsules operated by a hygroscopic mechanism that expands the valves in response to moistening, closing them again upon drying out. In desert conditions, this ensures germina-tion during the brief rainy periods. *Conicosia* and certain related genera have 3 different

Genera 123   **Species** 1,680
Economic uses Garden ornamentals and curiosities

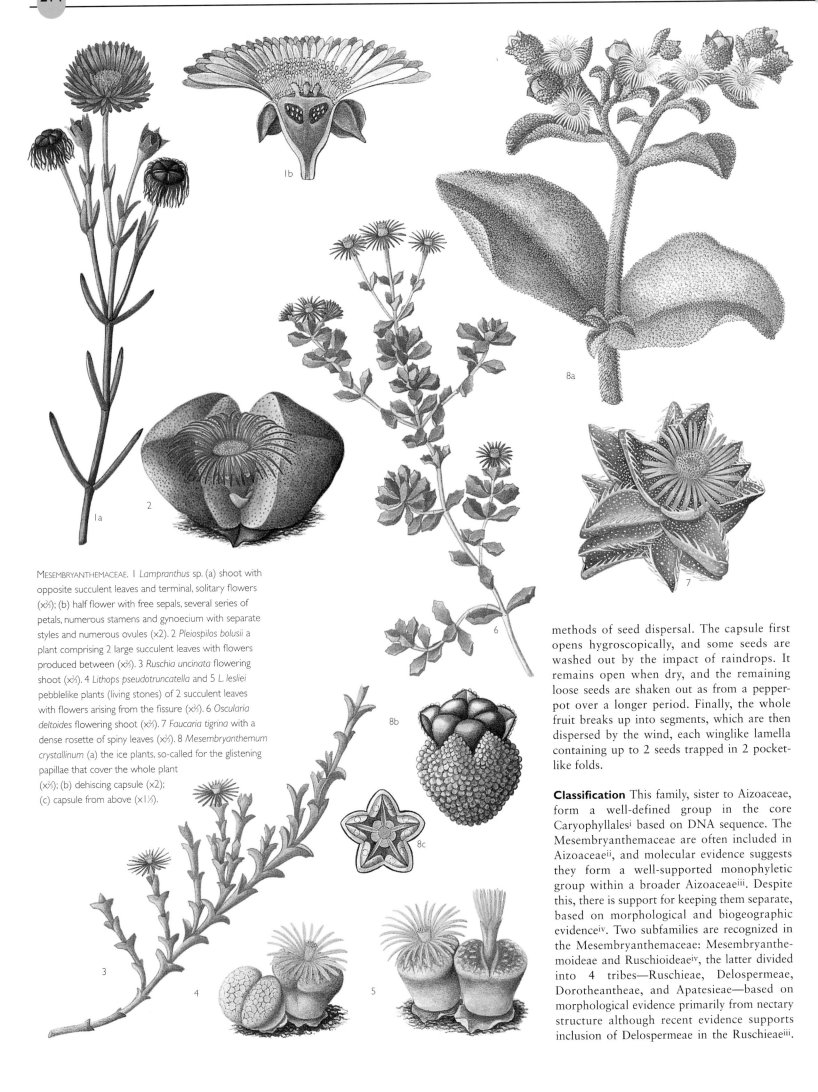

MESEMBRYANTHEMACEAE. I *Lampranthus* sp. (a) shoot with opposite succulent leaves and terminal, solitary flowers (x⅔); (b) half flower with free sepals, several series of petals, numerous stamens and gynoecium with separate styles and numerous ovules (x2). 2 *Pleiospilos bolusii* a plant comprising 2 large succulent leaves with flowers produced between (x⅔). 3 *Ruschia uncinata* flowering shoot (x⅔). 4 *Lithops pseudotruncatella* and 5 *L. lesliei* pebblelike plants (living stones) of 2 succulent leaves with flowers arising from the fissure (x⅔). 6 *Oscularia deltoides* flowering shoot (x⅔). 7 *Faucaria tigrina* with a dense rosette of spiny leaves (x⅔). 8 *Mesembryanthemum crystallinum* (a) the ice plants, so-called for the glistening papillae that cover the whole plant (x⅔); (b) dehiscing capsule (x2); (c) capsule from above (x1⅓).

methods of seed dispersal. The capsule first opens hygroscopically, and some seeds are washed out by the impact of raindrops. It remains open when dry, and the remaining loose seeds are shaken out as from a pepper-pot over a longer period. Finally, the whole fruit breaks up into segments, which are then dispersed by the wind, each winglike lamella containing up to 2 seeds trapped in 2 pocket-like folds.

Classification This family, sister to Aizoaceae, form a well-defined group in the core Caryophyllales[i] based on DNA sequence. The Mesembryanthemaceae are often included in Aizoaceae[ii], and molecular evidence suggests they form a well-supported monophyletic group within a broader Aizoaceae[iii]. Despite this, there is support for keeping them separate, based on morphological and biogeographic evidence[iv]. Two subfamilies are recognized in the Mesembryanthemaceae: Mesembryanthemoideae and Ruschioideae[iv], the latter divided into 4 tribes—Ruschieae, Delospermeae, Dorotheantheae, and Apatesieae—based on morphological evidence primarily from nectary structure although recent evidence supports inclusion of Delospermeae in the Ruschieae[iii].

Of the 123 genera, about 115 are separated out from the original Linnean genus *Mesembryanthemum*[v], indicating a substantial change in understanding of generic limits in the family.

Economic uses The Hottentot Fig, *Carpobrotus edulis,* produces edible fruit. Shrubby members of the Ruschioideae (*Lampranthus, Oscularia, Drosanthemum, Erepsia,* etc.) are half-hardy and grown for summer bedding, especially in southern Europe and California where they flower profusely. Only 1 sp., *Ruschia uncinata,* verges on complete hardiness, although *Carpobrotus* survives most winters in coastal areas and is much planted as a sandbinder. Several species of *Carpobrotus,* notably *C. edulis, C. acinaciformis,* and *C. chilensis,* have become naturalized and serious invasives in parts of the Mediterranean basin and California. Hybrids of the annual *Dorotheanthus* enjoy great popularity and have supplanted the original ice plant *Mesembryanthemum crystallinum,* which has glossy papillae-like water droplets covering the foliage, in popularity in gardens.   [GDR] AC

[i] Cuénoud, P. *et al.* Molecular phylogenetics of Caryophyllales Based on Nuclear 18S and Plastid *rbc*L, *atp*B and *mat*K DNA sequences. *American J. Bot.* 89: 132–144 (2002).
[ii] Bittrich, V. & Hartmann, H. E. K. The Aizoaceae – a new approach. *Bot. J. Linn. Soc.* 97: 239–254 (1988).
[iii] Klak, C., Khunou, A., Reeves, G., & Hedderson, T. A phylogenetic hypothesis for the Aizoaceae (Caryophyllales) based on four plastid DNA regions. *American J. Bot.* 90: 1433–1445 (2003).
[iv] Chesselet, P., Smith, G. F., & van Wyk, A. E. A new tribal classification of Mesembryanthemaceae: evidence from floral nectaries. *Taxon* 51: 295–308 (2002).
[v] Hartmann, H. E. K. Aizoaceae. Pp. 37–69. In: Kubitzki, K., Rohwer, J. G., & Bittrich, V. (eds), *The Families and Genera of Vascular Plants. II. Flowering Plants. Dicotyledons: Magnoliid, Hamamelid and Caryophyllid Families.* Berlin, Springer-Verlag (1993).

METTENIUSACEAE

Distribution In montane forests up to 2,000 m in the Andes of Colombia to Venezuela and Peru.

Genera 1 | Species 6
Economic uses None

Description Trees up to 20 m. The leaves are alternate, simple, entire, up to 25 × 15 cm, petiolate. The inflorescence is an axillary cyme of up to 17 long flowers, with short pedicels bearing up to 4 bracts. The flowers are actinomorphic, bisexual, up to 5.5 cm long, and fragrant. The sepals are 5, imbricate, slightly fused at the base. The corolla is fused for one-third or slightly more of its length into a narrow tube, with 5 linear lobes that are reflexed and red-tomentose on their inner surfaces. The stamens are 5, attached to the corolla throat, and strongly exserted, the anthers moniliform, and

each cell curled upward at the base (published illustrations of the anthers are conflicting). The ovary is superior, with a long filiform style, with 1 locule with a single ovule pendent from the apex. The fruit is a drupe up to 4.5 cm long, with a single seed up to 3 cm.

Classification *Metteniusa* has usually been included in the Icacinaceae, but its affinities have been long debated. On evidence of wood anatomy, the family specialist, Howard, considered it best placed outside Icacinaceae. Airy Shaw placed it in Alangiaceae, despite the inferior ovary of the latter, and Takhtajan placed it in its own order. In a revision of the genus in 1988[i], it was clearly placed in its own family. In a recent analysis of Icacinaceae *sensu lato* by Kårehed[ii], it was placed with some hesitation on morphological characters (molecular evidence not available) in a heterogeneous Cardiopteridaceae, a group here mostly referred to Leptaulaceae, and *Metteniusa* shows some close similarities to the larger-flowered members of the Leptaulaceae, such as *Pseudobotrys* from New Guinea, but differs in its larger flowers, strongly exserted stamens with long filaments, moniliform anthers with up-curved bases, and its uniovulate ovary. The leaf venation of *Metteniusa* is also distinctive, with closely arranged scalariform tertiary veins.   TMAU & RKB

[i] Lozano-C., G. & de Lozano, N. B. Metteniusaceae. Bogota, Colombia, *Flora of Colombia* 11: 1–56 (2001).
[ii] Kårehed, J. Multiple origin of the tropical forest tree family Icacinaceae. *American J. Bot.* 88: 2259–2274 (2001).

MISODENDRACEAE
FEATHERY MISTLETOE FAMILY

A family comprising 1 genus (*Misodendron*) of mistletoe-like parasitic plants, growing mostly on southern beeches (*Nothofagus*).

Distribution
Nothofagus forests of the Andes south of 33°S in Chile and Argentina.

Genera 1 | Species 8
Economic uses None

Description Small, shrubby plants growing mostly on the branches of *Nothofagus,* attached by a sucker (haustorium), which occasionally spreads under the bark to emit secondary shoots. The leaves are alternate, entire, deciduous, without stipules, sometimes reduced to scales. The flowers are small, generally unisexual (monoecious or dioecious), in racemes, spikes, or clusters. The male flowers have 2 or 3 stamens arising from a central cushion, without perianth segments. The female flowers have 3 perianth segments fused to the sides of the ovary and extending out as slight wings, leaving 3 deep grooves; a staminode is included in the groove at each corner of the ovary and grows

out to form a long barbed bristle. The ovary is superior and has 3 ovules borne on a free central placenta in a single chamber, which is divided into 3 compartments below the ovules. The fruit is an achene with 3 barbed awns and a single seed. The seedling develops a sticky pad on the host branch before penetration.

Classification Morphologically, Misodendraceae is considered to be a derivative of Olacaceae[i], a position reaffirmed by more recent molecular studies[ii].   RMP

[i] Kuijt, J. *The Biology of Parasitic Flowering Plants.* Berkeley and Los Angeles, University of California Press, (1969).
[ii] Nickrent, D. L. & Malécot, V. A molecular phylogeny of Santalales. Pp. 69–74. In: Fer, A. *et al.* (eds), *Proceedings of the 7th International Parasitic Weed Symposium.* Nantes, Faculté des Sciences, Université de Nantes (2001).

MITRASTEMONACEAE

A small family of root parasites.

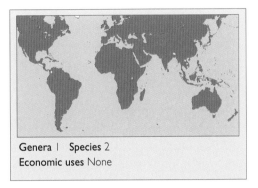

Genera 1 | Species 2
Economic uses None

Distribution The 2 spp. of *Mitrastemon* show a markedly disjunct distribution: *M. matudae* is confined to the neotropics, from Mexico to Guatemala and northwestern Colombia, while *M. yamamotoi* has a scattered distribution in Asia (Japan, Indochina, and Malaysia).

Description Achloropyllous endoparasitic herbs, with a filamentous endophyte inside the roots of Fagaceae. Stems are absent and the leaves scalelike, opposite, and decussate. The flowers are medium-sized (2–2.5 cm across), white, terminal, and occur singly. They are bisexual, actinomorphic. The calyx (perianth) is cupular, truncate, persistent, with 4 lobes. The stamens are connate into a tube (androphore), crowned by a fertile zone of pollen-bearing cavities (locules) that has an opening at the top. The ovary is superior, with 9–15(–20) carpels, 1-locular with 8–15(–20) parietal lobes; the style is short and thick. The fruit is a somewhat woody baccate capsule, opening by a transverse slit, containing numerous small seeds.

Nickrent's Parasitic Plant Connection[i] is a valuable source of information and illustrations.

Classification *Mitrastemon* has traditionally been included in the Rafflesiaceae but differs by its bisexual flowers, androecial structure, and superior ovary. The relationships suggested by recent molecular analyses are not fully resolved, with some indicating a link with the Malvales and others a placement in the Ericales[ii],[iii]. Further work is needed to clarify the position of this and the other groups once considered members of the Rafflesiales or Rafflesiaceae. VHH

[i] Nickrent, D. The Parasitic Plant Connection. http://www.science.siu.edu/parasitic-plants/
[ii] Barkman, T. J., Lim, S.-H., Salleh, K. M., & Nais, K. Mitochondrial DNA sequences reveal the photosynthetic relatives of *Rafflesia*, the world's largest flower. *Proc. National Acad. Sci. USA* 101: 787–792 (2004).
[iii] Nickrent, D. L. *et al*. Phylogenetic inference in Rafflesiales: the influence of rate heterogeneity and horizontal gene transfer. *BMC Evol. Biol.* 4: 40 (2004). http://www.biomedcentral.com/1471-2148/4/40

MOLLUGINACEAE
CARPET WEEDS

A family of largely weedy species living in tropical and warm-temperate regions.

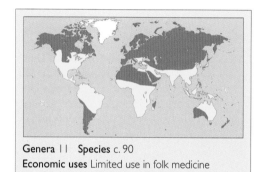

Genera 11 Species c. 90
Economic uses Limited use in folk medicine

Distribution Widespread in tropical, subtropical, and temperate regions. Most genera and species are tropical, occurring mainly in America and Africa, with a center of diversity in southern Africa.

Description Annual or perennial, slightly succulent, herbs or subshrubs. The leaves are simple, entire, alternate or occasionally opposite, sometimes apearing whorled, exstipulate or with membranous stipules. The inflorescences are usually terminal cymes that appear axillary by growth of lateral shoots, rarely axillary or solitary flowers. The flowers are hermaphrodite, rarely unisexual (plants dioecious), actinomorphic, individually inconspicuous or sometimes showy. The tepals are (4)5, green-white, sometimes pink or red, petaloid staminodes 0 or 5(–8), usually white. The stamens are (3)4–5(–8) fused at the base into a short cup-shaped cylinder. The carpels are (1)2–5, each with a short style or sessile stigma, usually free but rarely fused at the base, ovules 1–8 per locule, placentation axile. The fruit is a loculicidal capsule, seeds sometimes with aril.

Classification The Molluginaceae genera are recognized as core Caryophyllales based on molecular[i] and morphological[ii] data, but recognition and circumscription of this family

varies among authors[iii]. Some consider the family within Phytolaccaceae while others (e.g., Cronquist 1975) place it with Aizoaceae. Molluginaceae is now generally accepted due to the presence of anthocyanins compared with betalains in the other 2 families. *Limeum* and *Macarthuria* are now recognized in Limeaceae on molecular grounds[iv] and on morphological evidence. *Corrigiola* is placed in Caryophyllaceae[v]. *Corbichonia* (2 spp. from Africa and Asia), traditionally placed in Molluginaceae, is most closely related to *Lophiocarpus* on molecular evidence[i] and may be placed with it in a new family in due course. At present it is treated here for ease of reference.

Economic uses A few species of *Mollugo* and *Glinus* are used in folk medicine, particularly in India and Africa. AC

[i] Cuénoud, P. *et al*. Molecular phylogenetics of Caryophyllales based on nuclear 18S and plastid *rbc*L, *atp*B and *mat*K DNA sequences. *American J. Bot.* 89: 132–144 (2002).
[ii] Behnke, H.-D. *et al*. Ultrastructural, micromorphological and phytochemical evidence for a "central position" of *Macarthuria* (Molluginaceae) within the Caryophyllales. *Plant Syst. Evol.* 143: 151–161 (1983).
[iii] Endress, M. E. & Bittrich, V. Molluginaceae. Pp. 419–426. In: Kubitzki, K. *et al*. (eds), *The Families and Genera of Vascular Plants II. Flowering Plants. Dicotyledons: Magnoliid, Hamamelid and Caryophyllid Families*. Berlin, Springer-Verlag (1993).
[iv] Shipunov, A. B. The system of flowering plants from a synthetic point of view. *J. Gen. Biol.*, 64: 501–510 (2003). [In Russian.]
[v] Downie, S. R., Katz-Downie, D. S., & Cho, K.-J. Relationships in the Caryophyllales as suggested by phylogenetic analysis of partial chloroplast DNA ORF2280 homolog sequences. *American J. Bot.* 84: 253–273 (1997).

MONIMIACEAE

A pantropical family of small trees or shrubs.

Distribution The family is distributed in the tropics and subtropics. In the neotropics, it ranges from Mexico and Central America to Chile, and it occurs in southwestern and southeastern tropical Africa, Madagascar, Sri Lanka, Malaysia, tropical and Eastern Australia, Polynesia, and New Zealand. The largest genus, *Tambourissa* (50 spp.), occurs in Madagascar, the Mascarenes, and the Comores. *Mollinedia* (20 spp.) grows in lowland moist forest from Mexico down to the Amazon, and in gallery forests of Argentina, Brazil, and Paraguay[i]. *Monimia* (3 spp.) is endemic to the Mascarenes (Mauritius and Réunion), while *Hortonia* (1–3 spp.) is endemic to Sri Lanka.

Description Monoecious, dioecious, or polygamous trees, shrubs, some lianas. The leaves are opposite, rarely whorled, simple, the margins entire or serrate-dentate, without stipules. The flowers are actinomorphic, usually unisexual, or with rudiments of the other sex (*Peumus boldus*), rarely hermaphrodite (*Hortonia*), usually small, in axillary cymes or cauliflorous on

Genera c. 25 Species c. 200
Economic uses Herbal tea (boldo) from *Peumus boldus*

older wood. The perianth is sepaloid or petaloid, sometimes calyptrate, the tepals 3 to many, commonly 4–6(–8) in 2 whorls (10–20), spirally arranged in *Hortonia* and *Peumus*), free or connate. The stamens are few to numerous (up to 1,800), scattered in the hollow receptacle, filaments free. The ovary is superior, apocarpous, the carpels one to few or up to 1,000 (*Decarydendron*) or 2,000 (*Tambourissa*), often deeply embedded in the receptacular tissue. The fruit is an aggregate of stalked or sessile, juicy, dark blue or red drupelets, often embedded or enclosed in the conspicuous well-developed fleshy or woody receptacle; seeds 1 in each carpel.

Classification The family was placed in the Laurales by Cronquist, and this is confirmed by molecular data. It was divided into 6 subfamilies

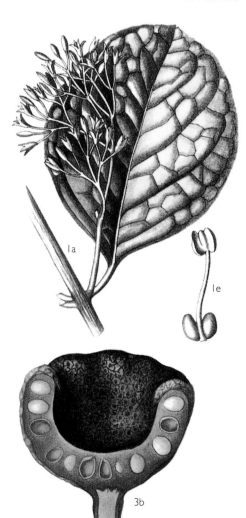

MONIMIACEAE. 1 *Monimia rotundifolia* (a) leaf and axillary inflorescence (×⅔) (b) female flower with hairy receptacle and many separate styles (×4); (c) vertical section of female flower showing free carpels (×4); (d) male flower with many stamens (×4); (e) stamen with basal glands (×18). 2 *Hedycarya arborea* (a) leafy shoot bearing axillary inflorescences of male flowers (×⅔); (b) leafy shoot with female flowers and fruits (×⅔); (c) male flower with uniform perianth segments (×2); (d) female flower with numerous free carpels (×2); (e) carpel (×6). 3 *Tambourissa elliptica* (a) flower (×2); (b) vertical section of fruit showing achenes deeply embedded in the receptacle (×2). 4 *Tambourissa* sp. fruit (×⅔).

by Philipson[ii]: Atherospermatoideae, Glossocalycoideae, Hortonoideae, Mollinedioideae, Monimioideae, and Siparunoideae. This is supported substantially by molecular data[iii], although Atherospermataceae and Siparunaceae q.v. are now accepted as separate families, and Glossocalycoideae is part of the Siparunaceae. The Monimiaceae contains 2 main lineages, 1 comprising the monospecific *Peumus* from Chile plus *Monimia* from the Mascarenes and *Palmeria* from eastern Australia and New Guinea. The other consists of *Hortonia* and the remaining genera[iv]. The division into genera is unclear. Within the Laurales, the Monimiaceae is most closely related to the Lauraceae and Hernandiaceae[v].

Economic uses The wood of several species is used locally and aromatic oil extracts from the leaves and bark may be used medicinally and as

perfumes. A tonic or herbal tea is obtained from *Peumus boldus* (Boldo) and is a traditional remedy used by the Araucanian Native Americans of Chile. VHH

i Renner, S. S. Monimiaceae. Pp. 252–253. In: Smith, N. *et al.* (eds), *Flowering Plants of the Neotropics.* Princeton, Princeton University Press (2004).
ii Philipson, W. R. Monimiaceae. Pp. 426–437. In: Kubitzki, K., Rohwer, J. G., & Bittrich, V. (eds), *The Families and Genera of Vascular Plants. II. Flowering Plants. Dicotyledons: Magnoliid, Hamamelid and Caryophyllid Families.* Berlin, Springer-Verlag (1993).
iii Renner, S. S. Circumscription and phylogeny of the Laurales: evidence from molecular and morphological data. *American. J. Bot.* 86: 1301–1315 (1999).
iv Renner, S. S. Phylogenetic affinities of Monimiaceae based on cpDNA gene and spacer sequences. *Perspect Plant Ecol Evol Syst.*1:61–77(1998).
v Renner, S. S. & Chanderbali, A. S. What is the relationship among Hernandiaceae, Lauraceae and Monimiaceae, and why is this question so difficult to answer? *Int. J. Plant Sci.* 161 (Supplement): S109–S119 (2000).

MONTINIACEAE

Distribution *Montinia caryophyllacea* occurs in dry country and woodland from South Africa northward just into Botswana and southwestern Angola. *Grevea* has 3 spp.: *G. madagascariensis* in dry deciduous forests in Madagascar, *G. eggelingii*[i] usually in riverine thickets in southeastern Kenya, southeastern Tanzania, and northeastern Mozambique, and *G. bosseri* in forest in western Congo, and the genus has also recently been discovered in Ghana. *Kaliphora madagascariensis* grows in subhumid to montane evergreen forest in Madagascar.

Description Shrubs and trees. *Montinia* is a stiff shrub up to 2(–4) m high, *Grevea bosseri* is a woody liane, while the other species of the genus are shrubs to trees up to 7.5 m, and *Kaliphora* is a many-branched shrub to small tree up to 5 m. The stems are usually ridged from the decurrent leaf bases, lenticellate, and with raised nodes that often bear a bifid bract. The leaves are simple, entire, small (up to 5(–7) × 1(–2.5) cm) and alternate in *Montinia,* but large (up to 25 × 20 cm) and opposite or subopposite in *Grevea*, while they are intermediate in size and appear alternate in *Kaliphora* but usually have a reduced leaf or bract opposite their base. They are subsessile or with short petioles in *Montinia* and *Kaliphora* but with longer petioles in *Grevea*. Leaves of *Grevea* and *Kaliphora* are recorded as turning black on drying and having a pungent or burning odor[ii]. In *Montinia* and *Grevea*, the inflorescence in the male plants is a diffuse terminal or axillary panicle, and in female plants is a solitary or paired erect terminal flower, while in *Kaliphora* both male and female inflorescences are short racemes in the axil of a bract or reduced leaf. The flowers are unisexual (plants dioecious), actinomorphic, 3–4(5)-merous, the male small, but the female larger. Male flowers have a short tubular calyx with 3 or 4(5) short lobes, and the same number of free petals that are small and white to yellowish and inconspicuous, and stamens alternating with the petals inserted below a fleshy disk. In *Montinia* and *Kaliphora,* the petals and stamens are usually 4, and in *Grevea* usually 3. Female flowers have an inferior ovary that is ovoid to elongate and fusiform, with usually 4 short calyx lobes and 4 petals inserted below a fleshy disk at the apex of the ovary. In *Montinia* and *Grevea,* there is a stout conical style in the middle of the conspicuous and persistent disk, and a bifid stigma, but in *Kaliphora* the bifid stigma is sessile. The ovary is inferior, of 2 fused carpels and 1–2

Genera 3 **Species** 5
Economic uses
Cure for colds
and headaches

locules, with (2)3–9(–12) ovules in 2 rows on an axile placenta in each locule (*Montinia* and *Grevea*), or with 1 ovule in each locule (*Kaliphora*). The fruit is a 2-valved capsule in *Montinia*, indehiscent and sometimes warty and spiny in *Grevea*, and a fleshy drupe with 2 single-seeded pyrenes in *Kaliphora*.

Classification The family as now defined comprises 3 rather disparate genera. The female flowers and fruits of *Montinia* and *Grevea* are similar and suggest close relationship despite marked differences in habit and leaf arrangement, but they are rather different in *Kaliphora*, which has sometimes been referred to another Madagascan family, Melanophyllaceae, although it agrees better with *Grevea* and was placed with it on morphological grounds in 1969 by Capuron. Pollen and wood anatomy have been found to support this. Molecular evidence has confirmed this affinity[iii], and surprisingly has suggested a placement of the family in the Solanales. However, studies of morphology and floral anatomy[iv] offer no support for an affinity with Solanaceae but suggest close relationship with Escalloniaceae.

Economic uses The pungent leaves of *Grevea* are used to soothe colds and headaches[ii]. RKB

[i] Verdcourt, B. A reappraisal of the *Grevea* growing on the Kenya coast. *Kew Bull.* 51: 771–773 (1996).
[ii] Schatz, G. E. *Generic Tree Flora of Madagascar.* Kew & Missouri (2001).
[iii] Fay, M. F. *et al.* Molecular data support the inclusion of *Duckeodendron cestroides* in Solanaceae. *Kew Bull.* 53: 203–212 (1998).
[iv] Ronse-Decraene, L. P., Linder, H. P., & Smets, E. F. The questionable relationship of *Montinia* (Montiniaceae): evidence from a floral ontogenetic and anatomical study. *American J. Bot.* 87: 1408–1424 (2000).

MORACEAE

MULBERRY AND FIG FAMILY

A family of woody plants with milky latex, conspicuous stipules, and fruits enclosed by a fleshy perianth or receptacle.

Distribution Pantropical, extending into warm-temperate regions, in lowland rain forest, woodland, and a variety of other habitats.

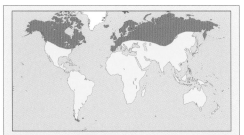

Genera 37 Species c. 1,050
Economic uses Edible fruits, timber, leaves for silkworms, fibers for paper, some medicinal and poisonous properties, some horticultural uses

Description Occasionally herbs, usually shrubs and trees, at times massive, or sometimes lianes, often hemiepiphytic ("stranglers"), sometimes with branch thorns (*Maclura*) or prickles (*Poulsenia*). Milky latex is present, sometimes turning brownish when exposed to air. The leaves are alternate, spiral or distichous, or rarely opposite (*Bagassa*), simple or occasionally palmately lobed, or rarely pinnately compound (*Artocarpus*), the margins entire to toothed, with stipules often amplexicaul and (especially *Ficus*) enclosing and exceeding the terminal bud. The inflorescences are unisexual or bisexual, forming racemes, spikes, or globose heads, or flattened (*Dorstenia*) to completely invaginated to form a syconium or fig enclosing many flowers with only a narrow terminal ostiole (*Ficus*), or rarely the flowers solitary. The flowers are unisexual (plants dioecious or monoecious), actinomorphic or, in reduced female flowers, one-sided, with the perianth

MORACEAE. 1 *Ficus religiosa* (a) shoot with drip tip leaf and figs (x⅔); (b) fruit from below (x⅔); (c) stalked and sessile female flowers (x5); (d) sterile female or gall flower (x5); (e) male flower (x5); 2 *Morus nigra* (a) shoot with infructescence (the mulberry)—an aggregation of achenes and the fleshy perianth (x⅓); (b) female inflorescence (x1½); (c) female flower (x3). 3 *Ficus benghalensis*—the banyan tree. 4 *Ficus carica*—the fig, a multiple fruit of numerous achenes inside the swollen receptacle (x⅓). 5 *Artocarpus communis* (a) infructescence comprising numerous achenes, swollen perianths and swollen receptacle (x1); (b) female inflorescence (x⅓); (c) leaf (x¼).

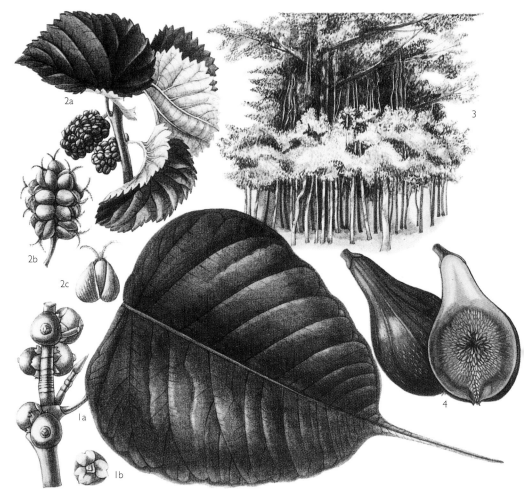

uniseriate or absent. The tepals are 2–6 (more in male *Naucleopsis* flowers) or often absent. The stamens are 1–4 opposite the tepals, straight or inflexed. The ovary is superior, of 2 carpels, with 1–2 stigmas, with a single locule and 1 apical ovule. The fruits are achenes or drupes often adnate to or enclosed in a fleshy perianth, often grouped into complex infructescences, in *Ficus* with many individual fruits enclosed within the infructescence (fig).

Classification Moraceae have always been associated with Urticaceae[i] and other families commonly placed in the Urticales, which are all placed in Rosales in APG II. The family is divided into 5 tribes:

Moreae 8 genera, c. 70 spp., tropical to northern temperate. Woody; inflorescences mostly racemose; stamens inflexed in bud and reflexing abruptly at anthesis (as in Urticaceae). Genera include *Morus* (mulberries, 13 spp.) and *Streblus* (23 spp.).

Dorstenieae 8 genera, c. 130 spp., tropical Africa and America, rare in Asia. Mostly herbaceous, with bisexual inflorescences and strongly reduced flowers. Genera include *Dorstenia* (105 spp.).

Castilleae 7 genera, c. 50 spp., mostly neotropical and few from Africa to Asia and Australia. Woody; unisexual capitate and involucrate inflorescences. Genera include *Naucleopsis* (20 spp.).

Ficeae comprises the genus *Ficus* only, with 700–750 spp., pantropical but mostly in Asia to Australia. Woody; inflorescences invaginated to enclose flowers.

Artocarpeae 10 genera, c. 80 spp., pantropical but mostly in Asia and America. Woody; inflorescences unisexual, the male ones spicate to racemose. Genera include *Artocarpus* (47 spp.).

Economic uses *Ficus* is important for its edible fruit (figs), timber, fibers for amate paper, cures for worms, and planting as large tropical shade-giving trees or smaller house plants. *Artocarpus* provides the well-known breadfruit and jackfruit. *Castilla* and *Ficus* yield rubber that was formerly important commercially. *Broussonetia papyrifera* has been planted for paper production in Africa. *Morus* produces mulberries and its leaves are fed to silkworms. *Antiaris* and *Naucleopsis* have been used as arrow poisons.   CMW-D & RKB

[i] Berg, C. C. Systematics and phylogeny of Urticales. In: Crane, P. R. & Blackmore, S. *Evolution, Systematics and Fossil History of the Hamamelidae.* 2: 193–217 (1989).

MORINACEAE

Usually spiny perennial herbs related to Dipsacaceae but differing in their verticillate inflorescences, different involucels, and the many technical characters shown by the flowers as well as their characteristic pollen.

Distribution From the western Balkan Peninsula to central Asia, the Himalayas, and southwestern China, in meadows, woods, and open places on mountains up to 4,900 m. *Acanthocalyx* (3 spp.) and *Cryptothladia* (8 spp.) are from Sino-Himalaya, the former occurring from Nepal eastward through the eastern Himalayas to the mountains of southwestern China, and the latter from Kirgizstan and eastern Kazakhstan through the Himalayas and Tibet to southwestern China. *Morina* (4 spp.) extends from Albania and Palestine eastward to Kirgizstan, Tibet, and the eastern Himalayas.

Genera 3
Species 13–15
Economic uses Few species are grown as garden perennials

Description Perennial herbs with short stout rhizomes, a basal rosette of leaves, and erect

MORINACEAE. I *Morina betonicoides* (a) leafy shoot and inflorescence (x⅔); (b) flower (x2); (c) corolla tube opened to show stamens of 2 lengths (x2); (d) gynoecium and calyx (x2).

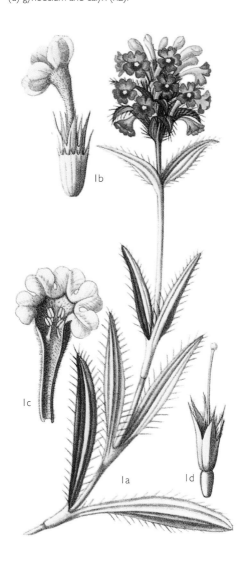

stems 20–130 cm high, sometimes forming several-stemmed clumps. The leaves are simple, usually oblong to linear, up to 50 cm, usually sinuate and conspicuously spiny, rarely with entire margins. The stem leaves are opposite in *Acanthocalyx*, but in *Cryptothladia* and *Morina* they are in whorls of 3–6. The flowers are borne in crowded verticels at the apex and at several nodes below it, or in *Acanthocalyx* the nodes may be few so that the inflorescence appears subcapitate. Each verticel is subtended by opposite or whorled more or less spiny bracts that usually equal or exceed the flowers. Each flower is subtended by a 12-veined tubular to campanulate involucel, which is formed from the fusion of 4 bracteoles. The involucel surrounds the inferior ovary and is persistent in fruit. It has few to many teeth on its margin, and in *Morina* 2 of these are much enlarged. The calyx has 4 or 5 lobes, and in *Acanthocalyx* is obliquely truncate, but 2-lipped in the other genera. The corolla in *Acanthocalyx* is slightly 2-lipped with 5 subequal lobes, while in *Cryptothladia* it is strongly 2-lipped with 1(2) lobes in the upper lip and 3 in the lower lip and the whole corolla scarcely equals the calyx in length, and in *Morina* it has a long narrow tube much exceeding the calyx and a 2-lobed upper lip and 3-lobed lower lip. In *Acanthocalyx*, the stamens are 4, attached near the corolla throat with 2 above the other 2, but in *Morina* the fertile stamens are 2 at the throat and 2 staminodes shortly below them, while in *Cryptothladia* the fertile stamens are 2, included in the corolla tube, and 2 stamens at or near the base of the corolla. It has been suggested that *Cryptothladia* may be cleistogamous. The ovary is inferior, 6-veined, and formed from 3 carpels, but with a single locule with 1 pendent ovule. The style is slender, with a simple stigma. The fruit is a single-seeded cypsela on which the calyx is persistent and is shed enclosed by the involucel.

Classification The family has often been included in Dipsacaceae and is undoubtedly closely related but differs most obviously by its verticillate inflorescence and spinose and usually whorled stem leaves, as well as by its zygomorphic calyx, 3-carpellate ovary, 12-nerved involucel derived from 4 bracteoles, distinctive and rather extraordinary pollen, chromosome number 2n=34, and various embryological characters.

Until the recent monograph[i], all species were included in *Morina*, but 3 genera can now be clearly recognized. Molecular studies have indicated the monophyly of the Morinaceae and its relationships in the Dipsacales[ii].

Economic uses The white- to pink-flowered *Morina longifolia* (Afghanistan to Bhutan) and *M. persica* (Albania to Pakistan) and the yellow-flowered *M. coulteriana* (Afghanistan to Tibet) are occasionally grown as spiny ornamentals in herbaceous borders. Seeds of *M. persica* are said to be eaten as rice in Iran.   RKB

i Cannon, M. J. & Cannon, J. F. M. A revision of the Morinaceae (Magnoliophyta – Dipsacales). *Bull. Brit. Mus. (Nat. Hist.), Bot. ser.* 12(1): 1–35 (1984).
ii Bell, C. D. & Donoghue M. J. Phylogeny and biogeography of Morinaceae (Dipsacales) based on nuclear and chloroplast DNA sequences. *Org. Divers. Evol.* 3: 227–237 (2003).

MORINGACEAE

HORSERADISH TREE, DRUMSTICK TREE, BEN-OIL TREE

Trees, bottle trees, and geophytes, with highly variable flowers and often large pinnate leaves.

Distribution *Moringa* is distributed primarily in the Horn of Africa but extends along the southern coast of the Arabian peninsula and through India with additional species in Namibia, Angola, and Madagascar.

Genera 1 Species 13
Economic uses
Edible leaves

Description Trees (often with swollen trunks), shrubs, and subshrubs, sometimes with a tuberous root-stock, all with soft and usually brittle wood. The leaves are compound, 1–3-imparipinnate with a petiole, the entire leaflets in opposite pairs, stipules represented by stipitate glands. The inflorescence is an axillary thyrse. The flowers are 5-merous; hermaphrodite; red, yellow, or white; and range from actinomorphic (e.g., *M. ovalifolia*) to strongly zygomorphic (e.g., *M. borziana*), with a cuplike to tubular receptacle. The sepals are 5, free and overlapping in bud, equal or unequal. The petals are 5, free and overlapping, equal or unequal. The stamens are 5, alternating with the 3–5 staminodes and opposite the petals. The ovary is superior on a stalk, formed of 3 fused carpels forming 1 locule with many ovules. The large fruit is an elongated capsule, with 3 valves, sometimes twisted.

Classification The family is included in Capparales in most systems[i], and a relationship with Caricaceae is suggested[ii]. The placement of the family in Brassicales is secure (APG II). The genus, and hence family, has been the subject of focused research on diversity and phylogeny[iii] that shows remarkable levels of floral and growth form diversification in such a small genus.

Economic uses Although all species have local economic uses, the species of major interest is *M. oleifera*, which has extensive use in folk medicine but is also under development as a major leaf vegetable crop in warm areas due to the high vitamin content of the leaves and the resistance of the plants to pests and poor conditions.   AC

i Brummitt, R. K. *Vascular Plant Families and Genera.* Richmond, Royal Botanic Gardens, Kew (1992).

ii Hall, J. C., Iltis, H. H., & Sytsma, K. J. Molecular phylogenetics of core Brassicales, placement of orphan genera *Emblingia, Forchhammeria, Tirania,* and character evolution. *Syst. Bot.* 29: 654–669 (2004).
iii Olsen, M. E. Combining data from DNA sequences and morphology for a phylogeny of Moringaceae (Brassicales). *Syst. Bot.* 27: 55–73 (2002).
iv Morton, J. F. The horseradish tree, *Moringa pterygosperma* (Moringaceae) – A boon to arid lands? *Econ. Bot.* 45: 318–333 (1991).

MUNTINGIACEAE

A small, recently recognized neotropical family with genera formerly included in Elaeo-carpaceae, Tiliaceae, or Flacourtiaceae.

Distribution The family is confined to the neotropics, and all the species are rare except *Muntingia calabura,* which is widespread in tropical America. *Neotessmannia uniflora* is known from only a single locality in Peru.

Genera 4 Species 4
Economic uses Edible fruits, fibers, some ornamentals

Description Shrubs or small trees, lacking mucilaginous cavities, with stellate, simple, and glandular hairs. The leaves are simple, alternate, asymmetric and cordate at the base, serrate, palmately nerved; stipule-like structures, usually dimorphic, occur at each leaf and are filiform or foliaceous and peltate or vestigal. The inflorescences are supra-axillary. The flowers are solitary or in clusters, actinomorphic, bisexual. The sepals are 5, valvate, fused at the base. The petals are 5, imbricate, crumpled in the bud, caducous. The stamens are numerous, with free, filiform filaments and dithecal, longitudinally dehiscent, usually ± basally attached anthers. The ovary is inferior (superior but embedded in receptacle in *Muntingia*), with 5 carpels and locules (rarely more), with pendulous placentae and numerous ovules. The style is short and thick, the stigma capitate to slightly lobed and grooved. The fruit is a berry with numerous seeds.

Classification Muntingiaceae was recognized only recently[i], having previously been grouped together, but regarded as anomalous, in various families, e.g., Tiliaceae. The inferior ovaries, dimorphic stipule-like structures, lack of mucilage cavities, and pendulous multiovulate placentae, are the main characters that delimit this family from, e.g., the core Malvales (see under Malvaceae).

Economic uses *Muntingia calabura* is sometimes cultivated in the tropics as an ornamental, and locally in South America for its edible fruits and useful bark fibers. Ecologically, it is important as a pioneer in forests.   MRC

i Bayer, C., Chase, M. W., & Fay, M. F. Muntingiaceae, a new family of dicotyledons with malvalean affinities. *Taxon* 47: 37–42 (1998).

MYODOCARPACEAE

A small family of trees and shrubs, with simple or compound leaves and compound umbels, segregated from the Araliaceae on the basis of morphological features, notably the anatomy of the fruits, which contain specialized oil vesicles in the

Genera 2 Species 19
Economic uses None

endocarp, and molecular studies. The family comprises 2 genera—*Delarbrea* (including *Pseudosciadium*) and *Myodocarpus*—centered in New Caledonia but with 1 sp., *Delarbrea michieana,* endemic to Queensland and another, *D. paradoxa,* occurring widely in the southwestern Pacific. The fruits are terete, fleshy or spongy drupes (*Delarbrea*) or dry, schizocarps (*Myodocarpus*).   VHH

i Lowry, P. P., II., A Systematic Study of Three Genera of Araliaceae Endemic to or Centered on New Caledonia: *Delarbrea, Myodocarpus,* and *Pseudosciadium.* Ph.D. dissertation. St. Louis, Missouri, Washington University, (1986).
ii Plunkett, G. M. & Lowry, P. P., II. Relationships among "ancient araliads" and their significance for the systematics of Apiales. *Mol. Phylogen. Evol.* 19: 259–276 (2001).
iii Plunkett, G. M. *et al.* Recent advances in understanding Apiales and a revised classification. *South African J. Bot.* 70: 371–381 (2004).

MYOPORACEAE

A family of herbs or woody plants found mostly in the southern hemisphere, especially in Australia, often resembling Scrophulariaceae but distinguished by their characteristic resin cavities and different ovary and fruit structure.

Distribution *Eremophila,* with about 215 spp., is confined to Australia, but 1 sp. is naturalized in New Zealand. *Myoporum* (c. 30 spp.) ranges from Timor to Australia and New Zealand and the islands of the Pacific with a single species in Mauritius and Rodrigues. *Pentacoelium* has

Genera 7 Species c. 250
Economic uses Ornamentals, hedge plants, a few useful timber trees

MYOPORACEAE. **1** *Myoporum petiolatum* (a) shoot with flowers in leaf axils (x⅔); (b) flower comprising a 5-lobed fused calyx and corolla, 4 stamens and a simple style (x1⅓); (c) corolla opened out showing epipetalous stamens alternating with the corolla lobes (x2); (d) calyx and gynoecium (x2⅔); (e) fruit—a drupe (x2); (f) cross section of fruit (x2). **2** *Eremophila bignoniiflora* (a) shoot bearing linear leaves, flower with irregular 2-lipped corolla, and fruits (x⅔); (b) corolla opened out showing stamens of 2 lengths (x⅔); (c) stamen with divergent anthers and longitudinal dehiscence (x2); (d) fruit (x⅔); (e) vertical section of fruit (x⅔). **3** *Eremophila glabra* leafy shoot with flowers and fruits (x⅔).

1 sp. from the coasts of northern Vietnam, China, Taiwan, and Japan, often in mangrove communities. The monotypic genus *Bontia* occurs in the Caribbean, markedly disjunct from the main center of distribution. Three further endemic Australian genera, widely recognized in herbaria through the work of R. J. C., have not yet been formally described.

Description Tall shrubs or small trees or rarely up to 10 m. Vegetative parts are characterized by unique secretory resin cavities, which often protrude through the epidermis as raised tubercles. The leaves are alternate or rarely opposite or whorled, entire or toothed, without stipules, often glandular, scaly or woolly. The flowers are solitary or in cymose clusters in leaf axils, and are usually zygomorphic, but in *Myoporum* often almost regular. The calyx comprises (4)5 sepals that are free or fused into a shallow to deep tube, and is often accrescent. The corolla

has 5 lobes and may be strongly 2-lipped or, in *Myoporum*, shortly tubular with 5 subequal rotate lobes. The stamens are 4(5), fused to the corolla tube, and alternating with the lobes. The anther locules run into one another. The ovary is superior, of 2 fused carpels with 2 locules or these divided into 4 or rarely up to 12

locules by intruding septal placentas, each cell with a single pendulous ovule (rarely up to 4 in *Eremophila*). The fruit is dry or fleshy, indehiscent or dividing toward the apex and semidehiscent, rarely (*Pentacoelium*) with 5 subapical pores, or rarely (*Eremophila tetraptera*) splitting into 4 single-seeded segments, with a thick woody endocarp and sometimes a fleshy mesocarp, with 1–4 locules each with 1–3 seeds.

Classification The family is derived from the basal stock of the Scrophulariaceae. It differs from Scrophulariaceae in its characteristic secretory resin cavities, its ovary structure, and its fruits with a woody endocarp and sometimes fleshy mesocarp. The flowers in *Eremophila* resemble those of Scrophulariaceae while those of *Myoporum* do not. The genera *Oftia* and *Ranapisoa* have sometimes been referred to Myoporaceae but lack the secretory cavities and are better referred to Scrophulariaceae. The latest generic account[i] includes a monotypic genus *Androya* from Madagascar based on limited molecular evidence[ii], but it does not have the morphological characters of Myoporaceae and is excluded here.

Economic uses A number of *Myoporum* spp., often known as Boobialla in Australia, are cultivated as ornamentals, hedges, or windbreaks. *M. sandwicense* from Hawaii provides useful timber. *Eremophila* spp. are extensively grown as ornamentals in Australia and increasingly elsewhere. The purplish fleshy fruits of many species are important food for birds. RKB & RJC

[i] Theisen, I. & Fischer, E. Myoporaceae. Pp. 289–292. In: Kadereit, J. W. (ed.), *Families and Genera of Vascular Plants. VII. Flowering Plants. Dicotyledons: Lamiales.* Berlin, Springer-Verlag (2004).
[ii] Oxelman, B., Backlund, M., & Bremer, B. Relationships of the Buddlejaceae s.l. investigated using parsimony jackknife and branch support analysis of chloroplast *ndh*F and *rbc*L sequence data. *Syst. Bot.* 24: 164–182 (1999).

MYRICACEAE

A small family of aromatic trees or shrubs.

Distribution The family is found on all major landmasses except Australia and New Zealand. Most species are temperate or subtropical with the main center of diversity in southern Africa.

Description Small trees or shrubs, usually aromatic, often evergreen. The roots have nitrogen-fixing nodules with the bacterium *Frankia*. Trichomes are often present. The leaves are alternate, simple or pinnatifid, the margins entire, serrate or irregularly dentate. Stipules are present in *Comptonia*. The flowers are unisexual (plants monoecious or dioecious), inconspicuous, borne on axillary catkinlike spikes. The male flowers usually have 2 bracteoles, the stamens (2)4(–20) with the filaments sometimes joined at the base;

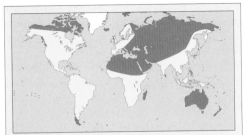

Genera 3–4 **Species** 57–62
Economic uses Fruits boiled to produce wax

sometimes with a disk at the base. The female flowers have 2–4 bracteoles and no disk; the ovary is superior, comprising 2 fused carpels that form a single locule with 1 erect ovule with basal placentation; the styles are 2, distinct or fused at base, with short style branches. The fruit is a small, rough, often waxy drupe with a hard endocarp and seed with little or no endosperm; embryo straight.

Classification The family was thought to be related to the Betulaceae but sufficiently distinctive to stand in an order of its own, the Myricales (Cronquist 1981). It has also been related to the Juglandaceae[i]. Thorne's view of a link to the Rutales (1973) has now been shown erroneous, and both Myricaceae and Juglandaceae are now thought to belong to the Fagales (APG II). The genus *Myrica* has been split up in various

ways, with even *Myrica gale* put into a monotypic genus, *Gale*. *Comptonia*, although close to *Myrica* and sometimes included in it, differs in its pinnatifid leaves, semicordate stipules, fruits nutlike, subtended by 6–8 linear-subulate, and persistent bracteoles, longer than the fruit. The monotypic *Canacomyrica* from New Caledonia differs considerably in floral structure, with bisexual flowers replacing the female flowers of *Myrica*, and in having 6 stamens that are fused below into a ring surrounding the ovary. Also surrounding the ovary is a 6-lobed disk that enlarges in fruit to enclose the ovary completely. *Canacomyrica* also has a certain amount of endosperm in the seeds, but despite this and other morphological differences, it is still thought to be a member of this otherwise homogeneous family, although a subfamily was created to include it[ii]. A molecular study[iii] indicates that *Myrica*, including *M. gale* and *M. hartwegii*, and *Comptonia*, including *C. peregrina*, belong to a phylogenetic cluster distinct from the other *Myrica* species that are transferred to the genus *Morella*. This grouping parallels the specificity

MYRICACEAE 1 *Myrica gale* (a) shoot with leaves and catkinlike male and female inflorescences (x⅔); (b) female flower (x10); (c) fruit (x5); (d) cross section of fruit (x5); (e) male flower (x5). 2 *Canacomyrica monticola* (a) leafy shoot (x⅔); (b) half fruit surrounded by enlarged disk (x10). 3 *Comptonia aspleniifolia* (a) flowering shoot (x⅔); (b) male catkin (x⅔); (c) half female flower (x2). 4 *Morella nagi* (a) shoot with fruits (x⅔); (b) female flower (x4); (c) ovary and styles (x4); (d) male flower (x4); (e) fruit cut away to reveal hard endocarp (x⅔).

of each cluster with *Comptonia-Myrica* and *Morella* being nodulated by 2 phylogenetically divergent clusters of *Frankia* strains.

Economic uses The fruits produce a white exudate of triglycerides (especially *Myrica cerifera*, Wax Myrtle) and were boiled to produce wax for candles. *Myrica gale* has been long used in northern Europe for flavorings and a source of a dye, and recent research is investigating it use as an insect repellent. The bark from *Myrica pennsylvanica* has been used in medicines as a tonic. SLJ

i Hjelmqvist, H. Studies on the floral morphology and phylogeny of the Amentiferae. *Bot. Not. Suppl.* 2. 1: 1–171 (1948).
ii Leroy, J.-F. De la morphologie florale et la classification des Myricaceae. *C. R. Acad. Sci. Paris* 229: 1162–1163 (1949).
iii Huguet, V. *et al.* Molecular phylogeny of Myricaceae: a reexamination of host-symbiont specificity. *Mol. Phylogen. Evol.* 34: 557–568 (2005).

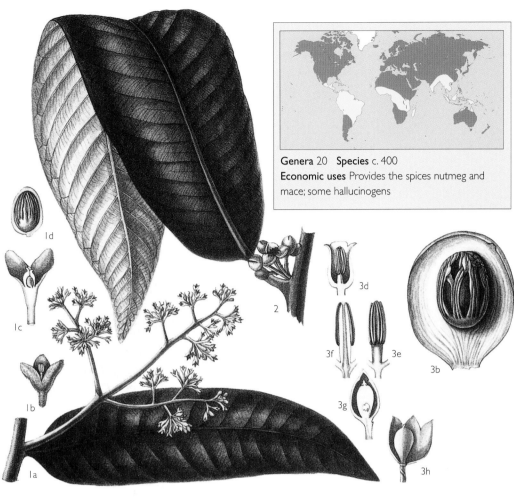

Genera 20 **Species** c. 400
Economic uses Provides the spices nutmeg and
mace; some hallucinogens

about 175 spp. centered in New Guinea. Other
large Asian genera are *Horsfieldia* (105 spp.),
which grows in south Asia from India to Papua
New Guinea, and *Knema* (85–90 spp.) in south
Asia from India to the Philippines and Papua
New Guinea. *Virola* (45 spp.) is confined
to tropical Central and South America, where
it is an important constituent of the Amazon
forests. Recent phylogenetic analyses of the
family, based on morphology and several plastid regions, reinforce the view that the ancestral
area was Africa-Madagascar and that Asian
taxa are derived[i].

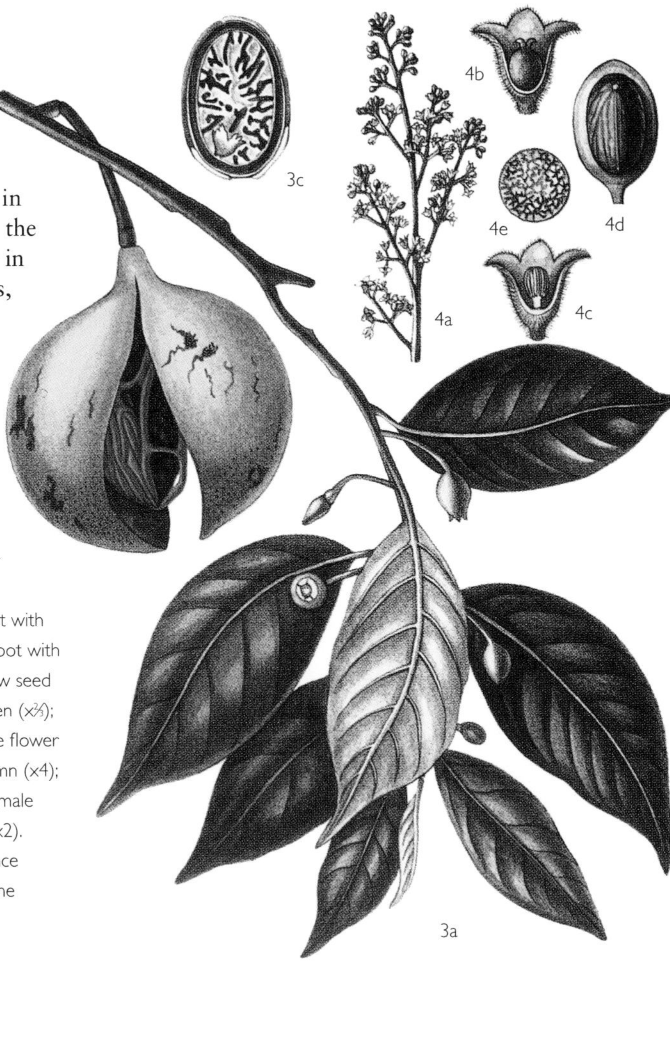

Description Trees, some shrubs, usually evergreen, the wood exuding a reddish sap when
wounded. The leaves are alternate, entire,
without stipules, often with glandular dots
containing aromatic oil. The flowers are unisexual (plants dioecious, sometimes monoecious—*Doyleanthus*), small and inconspicuous, actinomorphic, borne in terminal
racemose or corymbose inflorescences, sometimes cauliflorous. The perianth is in one series
consisting of (2)3(–5) partly connate, sepal-like
tepals. In the male flowers, the stamens are
fused into a synandrium, which consists of a
sterile column and a collection of 2–60 anthers
fused to various degrees to this column, exceptionally with short filaments (*Mauloutchia*)[ii];
the anthers shed their pollen through 2 longitudinal slits. The female flowers have a 1-locular
superior ovary containing 1 basal ovule. The
fruit becomes fleshy, and on maturity it splits
into 2 or 4 valves (a dehiscent berry), disclosing
the large seed, or is rarely leathery and indehiscent. The seed contains a small embryo and
much endosperm, rich in oil, ruminate in cross
section, and enveloped by an aril consisting of a
usually brightly colored network of tissue.

Classification For a long time, all members
were classified in the single genus *Myristica,*
probably because of their uniformity in flower
and fruit structure. However, later studies
revealed considerable variation in the structure
of the male flower, especially the androecium
and pollen, and in the inflorescence and aril.
Today, up to 20 genera are recognized. Relationships within the family remain uncertain,
although pollen aperture shape, tectal sculp-
ture, and infratectal structure have been used to
distinguish an Afro-Malagasy clade of 5 genera
(including *Mauloutchia*), which may or may
not be basal, from the rest of the family[iii].

The family is distinct and was placed in the
Magnoliales by Cronquist, a position supported
by recent molecular and morphological data
sets, indicating that the Myristicaceae not only
belongs in the Magnoliales but is the sister
group of all the other families in the order.

Economic uses *Myristica fragrans* is the
Nutmeg Tree from which the spices nutmeg
(the seed) and mace (the aril) are obtained.
They also have powerful narcotic properties.
Although nutmeg originated in the Moluccas, it

MYRISTICACEAE

NUTMEG AND MACE

A pantropical family of
mainly trees, best known
for the Nutmeg Tree.

Distribution The family is
exclusively tropical, mainly in
lowland rain forests, occurring in the
Indo-Malaysian region, especially in
New Guinea and the Phillipines,
in tropical America, especially
the Amazon basin, and in
Africa and Madagascar. Most
members inhabit lowland
rain forests. The largest
genus is *Myristica*, with

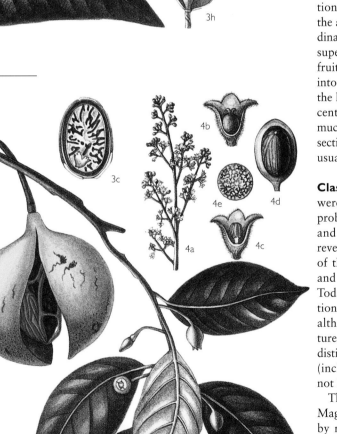

MYRISTICACEAE. 1 *Virola glaziovii* (a) leaf
and male inflorescence (×⅔); (b) male flower
(×4); (c) half female flower (×4); (d) vertical
section of fruit (×⅔). 2 *Knema pectinata* shoot with
male flowers (×⅔). 3 *Myristica fragrans* (a) shoot with
flowers and fruit (nutmeg) split open to show seed
covered by the red aril (×⅔); (b) fruit cut open (×⅔);
(c) vertical section of seed (×⅔); (d) half male flower
(×2); (e) androecium with stamens in a column (×4);
(f) vertical section of column (×4); (g) half female
flower (×2); (h) female flower opened out (×2).
4 *Horsfieldia macrocoma* (a) male inflorescence
(×⅔); (b) female and (c) male flowers with one
sepal removed (×4); (d) fruit cut open (×⅔);
(e) cross section of seed (×⅔).

is now widely cultivated, especially in the West Indies. The pericarp of the fruit is also used to make a jelly preserve, and inferior seeds are pressed to make "nutmeg butter" used in perfumery and making candles, as are the Asian *Gymnacranthera farquhariana* and the Brazilian *Virola surinamensis* (ucuúba), whose waxy seeds are also used as a source of "butter" and for making candles. The wood of this family makes poor-quality lumber with a high moisture content, and is little used with the exception of *Virola surinamensis,* which is widely used in carpentry and as plywood and is also the source of an indigenous hallucinogenic snuff throughout Amazonia.   VHH

i Doyle, J. A., Sauquet, H., Scharaschkin, T., & Le Thomas, A. Phylogeny, molecular and fossil dating, and biogeographic history of Annonaceae and Myristicaceae (Magnoliales). *Int. J. Plant Sci.* 165 (Supplement): S55–S67 (2004).
ii Sauquet, H. Androecium diversity and evolution in Myristicaceae (Magnoliales), with a description of a new Malagasy genus, *Doyleanthus* gen. nov. *American J. Bot.* 90: 1293–1305 (2003).
iii Sauquet, H. et al. Phylogenetic analysis of Magnoliales and Myristicaceae based on multiple data sets: implications for character evolution. *Bot. J. Linn. Soc.* 142: 125–125 (2003).

MYROTHAMNACEAE

An enigmatic mono-generic family that contains 2 spp. of glabrous, aromatic, resinous shrubs with narrowly winged stems from tropical and southern Africa, and Madagascar. The leaves are small, opposite, simple, flat or plicate, folding up in dry conditions and expanding again after rain, with a

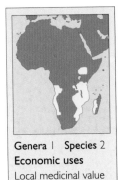

Genera 1 **Species** 2
Economic uses
Local medicinal value

long petiolar sheath at the base, dentate margins, and small interpetiolar stipules. The flowers are actinomorphic to slightly zygomorphic, sessile, unisexual (plants dioecious), in terminal spicate inflorescences that may have a terminal flower. The floral structure is not fully understood: there are no petals and in the staminate flowers with 3–4 free stamens (*M. moschata*) there are 4 scales (?or tepals) alternating with them, or when the stamens are 3–8 (*M. flabellifolia*), they are connate by their filaments. In the carpellate flowers the ovary is superior, with 3–4 carpels connate at the base, the locules 3–4, placentation axile, ovules numerous. The fruit is follicular, with 3–4 mericarps separating septicidally. The seeds are numerous.

Myrothamnus has been included in the Hamamelidaceae as a subfamily or recognized as a separate family within the Hamamelidales. It is made a superorder, Myrothamnanae, in the subclass Hamamelididae by Takhtajan (1997). A closer link with the Trochodendrales has also been suggested[i]. Molecular data indicate that *Myrothamnus* belongs with *Gunnera* in the Gunnerales or should optionally even be included within it despite the significant differences between the 2 genera[ii]. They also strongly support the placement of Gunnerales (*Gunnera* + *Myrothamnus*) as the sister group to all remaining core eudicots[iii,iv]. Both species are used medicinally and their essential oils have antimicrobial and antifungal properties.   VHH

i Endress, P. K. The systematic position of the Myrothamnaceae. Pp. 193–200. In: Crane, P. R. & Blackmore, S. *Evolution, Systematics and Fossil History of the Hamamelidae.* Systematics Association Special Volume 40A (1989).
ii Wilkinson, H. P. A revision of the anatomy of Gunneraceae. *Bot. J. Linn. Soc.* 134: 233–266 (2000).
iii Savolainen, V. et al. Phylogenetics of flowering plants based upon a combined analysis of plastid *atp*B and *rbc*L gene sequences. *Syst. Biol.* 49: 306–362 (2000).
iv Soltis, D. E. et al. Gunnerales are sister to other core eudicots: Implications for the evolution of pentamery. *American J. Bot.* 90: 461–470 (2003).

MYRSINACEAE

A family of evergreen trees and shrubs with resin canals and gland dots.

Distribution Widespread in tropics and subtropics, extending as far north as Florida, Macaronesia, the Himalayas, and Japan, and south sparingly to South Africa, New Zealand, and Argentina. The widespread species *Myrsine africana* extends to the Azores, and the mono-

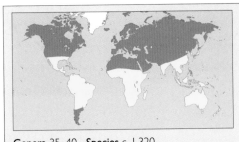

Genera 35–40 **Species** c. 1,320
Economic uses Few cultivated ornamentals; occasional medicinal use

typic genera *Pleiomeris* and *Heberdenia* are both endemic to the Canary Islands. Other monotypic genera are *Emblemantha* on Sumatera, *Elingamita* endemic to Three Kings Islands at the north end of New Zealand, *Synardisia* in Central America, *Vegaea* on Hispaniola, and *Solonia* on Cuba. By contrast, *Oncostemum* is endemic to Madagascar but has c. 90 spp. Most species of the family occur in lowland or upland forests. *Aegiceras* is a mangrove occurring from the Malaysian region to northern Australia and the Pacific.

Description Mostly evergreen trees and shrubs or woody climbers, rarely subshrubs, woody only at the base in *Ardisia* (*A. primulifolia* from China is acaulous with a rosette of leaves surrounding the inflorescence). Resin ducts and gland dots are usually conspicuous in vegetative

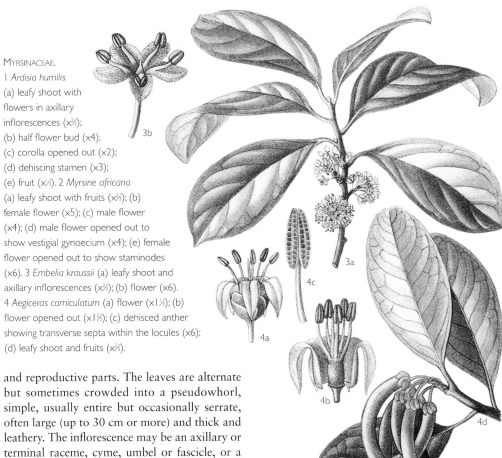

MYRSINACEAE.

1 *Ardisia humilis*
(a) leafy shoot with flowers in axillary inflorescences (x⅔);
(b) half flower bud (x4);
(c) corolla opened out (x2);
(d) dehiscing stamen (x3);
(e) fruit (x⅓). 2 *Myrsine africana*
(a) leafy shoot with fruits (x⅔); (b) female flower (x5); (c) male flower (x4); (d) male flower opened out to show vestigial gynoecium (x4); (e) female flower opened out to show staminodes (x6). 3 *Embelia kraussii* (a) leafy shoot and axillary inflorescences (x⅔); (b) flower (x6). 4 *Aegiceras corniculatum* (a) flower (x1½); (b) flower opened out (x1½); (c) dehisced anther showing transverse septa within the locules (x6); (d) leafy shoot and fruits (x⅔).

and reproductive parts. The leaves are alternate but sometimes crowded into a pseudowhorl, simple, usually entire but occasionally serrate, often large (up to 30 cm or more) and thick and leathery. The inflorescence may be an axillary or terminal raceme, cyme, umbel or fascicle, or a solitary axillary flower. Bracteoles are absent (cf. Maesaceae). The flowers are usually rather small, actinomorphic, bisexual or unisexual (the plants dioecious)[i], usually 5-merous but 3–4-merous sometimes in *Cybianthus*. The corolla is shortly tubular to campanulate, imbricate or valvate, sometimes divided almost to the base (*Embelia*). The stamens are inserted opposite the corolla lobes. The ovary is superior, derived from 5 fused carpels, 1-locular, and with free central placentation. The ovules are few to numerous, in 1 to several rows, embedded in the placenta. The fruit is a single-seeded fleshy drupe or berry, usually subspherical or drawn out to a terminal point, or (*Aegiceras*) an elongate viviparous capsule. The seeds are subglobose and not angular.

Classification The family is closely allied to the Primulaceae. The recent inclusion in Myrsinaceae of several herbaceous genera normally referred to Primulaceae is not followed here. Theophrastaceae, sometimes formerly included in Myrsinaceae, is here accorded separate family status. The genus *Maesa*, which has generally been treated as subfamily Maesoideae of Myrsinaceae, is here referred to its own family Maesaceae. Perhaps contrary to expectations, the molecular cladograms suggest that the essentially woody tropical Myrsinaceae has been derived from the essentially herbaceous temperate Primulaceae—see Gentianaceae for similar suggestion on evolutionary direction. See the Primulaceae for further discussion and references. The major genera are pantropical *Rapanea* with c. 300 spp. (but sometimes included in *Myrsine*), *Ardisia* (c. 250 spp.) in the tropics and sub-

tropics of the New World and Asia, *Cybianthus* (150 spp.) in the New World, and *Embelia* (130 spp.) in the Old World. The mangrove genus *Aegiceras* has sometimes been placed in its own family.

Economic uses *Myrsine africana*, unusual in the family in having very small leaves and with notable metallic-purple fruits, is sometimes cultivated as a hedge plant and has become naturalized in southern Britain. Several species of *Ardisia*, especially *A. crispa*, are grown in greenhouses or as house plants for their bright red fruits. Other *Ardisia* species have been found to have medicinal properties or have been used as a food that cures scurvy. RKB

[i] Heenan, P. B. Dioecism in *Elingamita johnsonii* (Myrsinaceae). *New Zealand J. Bot.* 38: 569–574 (2000).

MYRTACEAE
MYRTLE, EUCALYPTUS, CLOVE, AND GUAVA FAMILY

A family of mostly tropical, occasionally temperate, tall trees and shrubs, usually with numerous showy stamens, often with flaking or peeling bark.

Distribution The family has a pantropical distribution from southeastern North America throughout central and southern America, sub-Saharan Africa to the Cape, India, Indochina, to Japan, Australasia and the Pacific, with 1

genus (*Myrtus*) in the Mediterranean. In temperate regions, they are usually found in open places or woodlands, but in the tropics more often in rain forest or savanna.

Description Trees or shrubs. The leaves are mostly opposite, sometimes sub-opposite or spirally arranged, mostly simple, without conspicuous stipules, with pellucid gland dots containing ethereal oils, aromatic when crushed. The inflorescence is basically a leafy panicle, but many variations occur: solitary flowers, 3-flowered dichasia, racemose, cymose or thyrsoid. The flowers are actinomorphic or occasionally zygomorphic, bisexual or unisexual, epigynous or occasionally somewhat perigynous, often subtended by persistent or falling bracts and are mostly pedicellate. The sepals and petals are 4–5, usually free and simple, but occasionally the sepals are fused into a calyptra (e.g., *Calyptranthes*) or the sepals and petals are fused into an operculum (e.g., *Eucalyptus*) that detaches and falls as the flower opens. The stamens are usually many and showy. The ovary is inferior to semi-inferior with usually 2–5 locules, bearing 2 to many ovules with axile or rarely parietal placentation. In Australia, the fruit is commonly a dry, dehiscent capsule with many small, unspecialised seeds. In the Americas, a fleshy berry is more common with 2 to several seeds with a varying degree of embryo specialization. Throughout the rest of the distribution, fruits are dry or fleshy.

Classification The delimitation and classification of the family adopted here is that of Wilson *et al*[i]. It is divided into 2 subfamilies:
SUBFAM. PSILOXYLOIDEAE Flowers dioecious, male flowers with 10 or fewer stamens, female flowers with a superior ovary. Two tribes, Psiloxyleae and Heteropyxideae with one genus in each; *Psiloxylon* (Mauritius and Reunion) and *Heteropyxis* (southern Africa).
SUBFAM. MYRTOIDEAE Flowers bisexual, ovary almost always inferior or semi-inferior, 15 tribes: Backhousieae, 2 genera from Australia and New Guinea; Chamelaucieae, 23 genera from East and Southeast Asia, Australasia, and New Caledonia (23 genera); Eucalypteae, 7 genera (including *Eucalyptus*) from Southeast Asia, Australasia, and New Caledonia; Kanieae,

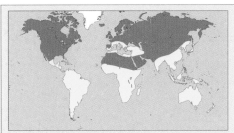

Genera c. 140 **Species** c. 5,800
Economic uses Timber, gums, essential oils, economically important fruits and spices, cultivated ornamentals

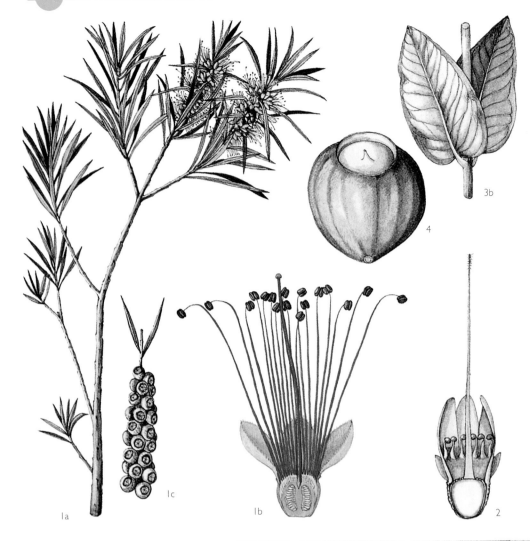

MYRTACEAE. 1 *Callistemon subulatus* (a) leafy shoot and inflorescences (x⅔); (b) half flower showing numerous stamens and inferior ovary crowned by a single style and containing ovules on parietal placentas (x3); (c) fruits (x⅔). 2 *Darwinia citriodora* half flower with epipetalous stamens and a long style crowned by a hairy stigma (x3). 3 *Eucalyptus melanophloia* (a) habit; (b) adult leaves (x⅔). 4 *Eugenia gustavioides* fruit—a berry with remains of the style (x1).

8 genera from Southeast Asia, Australasia, and New Caledonia; Leptospermeae, 8 genera, including *Leptospermum*, from Southeast Asia, Australasia, and New Zealand; Lindsayomyrteae, 1 genus from Australasia; Lophostemoneae: 4 genera from Sulawesi, Australasia; Melaleuceae, 9 genera, including *Callistemon* and *Melaleuca*, from India, Southeast Asia, Australasia, and New Caledonia; Metrosidereae, 5 genera, including *Metrosideros*, from South Africa, Southeast Asia, Australasia, New Zealand, the Pacific and Chile; Myrteae, 47 genera (including *Eugenia*, *Pimenta*, and *Psidium*), mainly in the Americas but also Asia, Australasia, the Pacific, sub-Saharan Africa, and the Mediterranean; Osbornieae, 1 genus from Southeast Asia and Australasia; Syncarpieae, 1 genus from Australia; Syzygieae, 5 genera, including *Syzygium*, from Arabia, Africa, India, eastern and Southeast Asia, Australasia and the Pacific; Tristanieae, 3 genera from Southeast Asia, Australasia, and New Caledonia; Xanthostemoneae, 3 genera from Southeast Asia, Australasia, New Caledonia, and the Pacific.

Economic uses *Eucalyptus* is widely cultivated for the timber and pulp industry. *Syzygium aromaticum* (Clove) and *Pimenta dioica* (Allspice) are well-known spices. *Pimenta racemosa* (Bay Rum), *Melaleuca* (Cajeput), and *Eucalyptus* provide oils important for the perfume industry, while antiseptic oils are extracted from *Eucalyptus*, *Leptospermum scoparium* (Tea Tree), *Callistemon,* and *Melaleuca.* A range of *Eucalyptus* spp. are cultivated as ornamentals as is *Acmena smithii* (Lily Pily), *Callistemon, Melaleuca* (the "bottle brush" plants), *Myrtus communis* (Common Myrtle), and *Acca sellowiana* (Pineapple Guava). Almost all fleshy fruited Myrtaceae are edible, and economically important species include *Psidium guajava* (guava) and *Syzygium aqueum* (Rose Apple). Many lesser-known species are locally important for jams, juice, and sweets, e.g., Jaboticaba (*Myrciaria cauliflora*) and Pitanga (*Eugenia uniflora*). Due to their high propensity for water uptake, some species of *Eucalyptus,* and in particular *Metrosideros quinquenervia,* are planted to drain boggy land although in some areas these are invasive weeds.   EJL

i Wilson, P. G., O'Brien, M. M., Heslewood, M. M., & Quinn, C. J. Relationships within Myrtaceae *sensu lato* based on a *mat*K phylogeny. *Plant Syst. Evol.* 251: 3–19 (2004).

NELUMBONACEAE
SACRED LOTUS

A family of aquatic herbs with peltate leaves, comprising the single genus *Nelumbo*.

Distribution *Nelumbo lutea* grows in eastern and central North America, while *N. nucifera* is native to tropical and subtropical East Asia and Australia. They are widely cultivated and naturalized.

Description Perennial aquatic herbs. The rhizomes are creeping, fleshy, becoming tuberous, and with adventitious roots. Well-developed air chambers (aerenchyma) are found in the vegetative parts of the plant. The leaves are alternate, arising from the rhizomes in groups of three, 2 of them scale-leaves and 1 emergent or floating peltate leaf with dichotomous main veins extending to the margin. The flowers are bisexual, actinomorphic, large and showy, solitary, borne above the water surface on long peduncles. The perianth consists of 2–5 outermost members (sepals), merging into the 20–30 spirally arranged petals that are yellow, pink, or red. The stamens are 200–400, spirally arranged, with long filaments, and the connective with a long appendage. The ovary is apocarpous with numerous free carpels immersed in a massive turbinate, spongy receptacle; locules 1 with parietal placentation and a single ovule. The fruit consists of indehiscent nutlets lying loosely in the

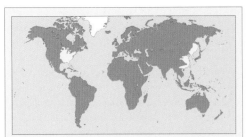

Genus 1 **Species** 2
Economic uses Edible tubers, rhizomes, young leaves, and seeds; religious and social significance of the sacred lotus

accresecent receptacle that distintegrates to release them. The seeds are large.

Classification Recently, *Nelumbo lutea* has been considered a subspecies of *N. nucifera*[i]. *Nelumbo* has traditionally been included in the Nymphaeales along with other water lilies, either included in the Nymphaeaceae or as a separate family, but new systematic and molecular data have situated *Nelumbo* with lower eudicots in the Proteales and sister to the Platanaceae[ii]. A molecular analysis of the Nymphaeaceae data indicated that *Nelumbo* was basal to the other genera with a bootstrap value of 100% and occupied an isolated position, justifying its separation in its own family, order Nelumbonales[iii]. Support for separating the Nelumbonaceae also comes from a study of root anatomy[iv].

Economic uses The edible tubers, rhizomes, young leaves, and seeds of *N. nucifera* are consumed in Asia and widely cultivated. It is also cultivated for its ornamental leaves and flowers. Known as the Sacred Lotus, it plays an important role in several religions in India, China, and other Asian countries and in Egypt.   VHH

[i] Borsch, T. & Barthlott. Classification and distribution of the genus *Nelumbo* Adans. (Nelumbonaceae). *Beitr. Biol. Pflanzen* 68: 421–450 (1994).
[ii] Chase, M. W. *et al*. Phylogenetics of seed plants: An analysis of nucleotide sequences from the plastid gene *rbc*L. *Ann. Missouri Bot. Gard*. 80: 528–580 (1993).
[iii] Liu, Y.-L., Xu, L.-M., Ni, X.-M., & Zhao, J.-R. Phylogeny of Nymphaeaceae inferred from ITS sequences. *Acta Phytotax. Sinica* 43: 22–30 (2005).
[iv] Seago, J. L. The root cortex of the Nymphaeaceae, Cabombaceae, and Nelumbonaceae. *J. Torrey Bot. Soc*. 129: 1–9 (2002).

NELUMBONACEAE. 1 *Nelumbo nucifera* receptacle containing the fruits is shown (a) entire (x⅖) and (b) in vertical section (x⅖).

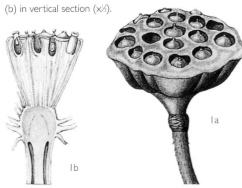

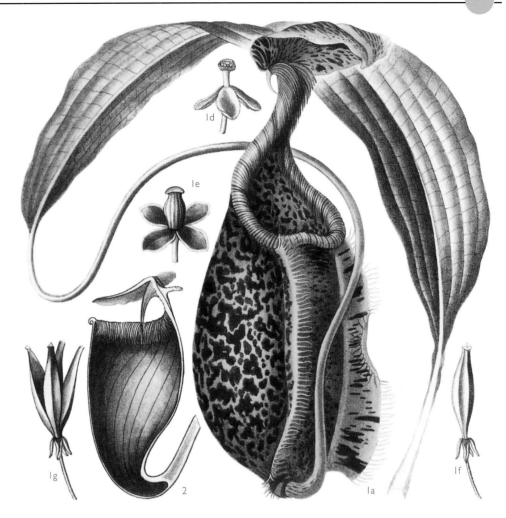

NEPENTHACEAE. 1 *Nepenthes rafflesiana* (a) leaf, the tip of which is modified to form a pitcher (x⅔); (b) male inflorescence (x⅔); (c) leaves with apical tendrils (x⅔); (d) male flower with filaments united into a column (x1½); (e) female flower with sessile stigma (x1½); (f) fruit—a capsule (x⅔); (g) fruit dehiscing by 4 valves (x⅔). 2 *N. bicalcarata* vertical section of pitcher (x⅔). 3 *N. fimbriata* cross section of ovary with 4 locules and numerous ovules on axile placentas (x2⅔).

NEPENTHACEAE
TROPICAL PITCHER PLANTS

The family consists of the genus *Nepenthes* only, well known because of their brightly colored pitchers.

Distribution The family is largely endemic to the Malaysian region, with a center of diversity in Borneo and outlying species around the Indian Ocean in Madagascar, Seychelles, Sri Lanka, northern India, and northern Australia. Species occur from sea level (*N. mirabilis*) to >2,500 m (*N. villosa*) in nutrient-poor soils, often in open or disturbed sites at the edge of woodland, although some species grow in open grassland (*N. pervillei* and *N. distillatoria*).

Description Perennial, dioecious, insectivorous climbers (to 40 m), scramblers or epiphytes, with alternate leaves. Young plants form a basal rosette from which 1 to many climbing stems develop. The leaves lack stipules and are often without distinct petioles. The lamina forms wings along an often prominent midrib, the

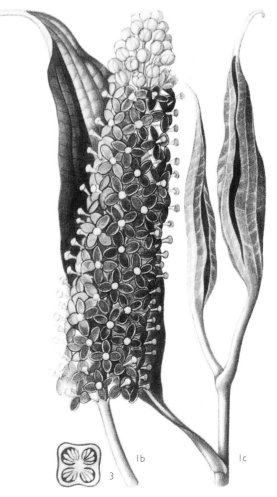

basal portion often clasping the stem. The plant climbs by means of tendrils that are prolongations of the leaf midrib. The end of the tendril generally develops into a pitcher, with a lid projecting over the mouth, which opens as the pitcher matures. The distinctive pitcher usually consists of an elongated cuplike structure,

Genera 1 | Species 85
Economic uses
Ornamental plants for specialist growers

with a pronounced rim of hard tissue, the upper opening capped by a lid, with basal spur that prevents entry of rainwater. Pitchers vary from 5 cm to as much as 40 cm in length, and some in Borneo are large enough to hold 2 liters of water. In some species, the young pitchers are clothed in tight, rusty, branched hairs. At the entrance of the pitcher there are secretory glands, below which the interior is slippery with fine wax scales. Insects are attracted to the pitcher by smell or bright color, which may be red or green and is often blotched. Animals entering the pitcher are unable to climb out because of the slippery surface and eventually drown in the liquid secretions in the base of the pitcher. The plant absorbs the products of decay.

Many members of the Nepenthaceae are epiphytic and have climbing stems up to 3 cm in diameter. The small red, yellow, or green flowers often smell strongly of stale sweat and are 4-merous, regular and unisexual (plants dioecious) and are borne in a spikelike inflorescence. The perianth consists of 3 or 4 tepals. In the male flower, the filaments of the 4–24 stamens are united into a column, and the anthers crowded into a mass. The female flower has a discoid stigma. The style is short or totally absent. The ovary is superior, of 4 fused carpels, and has 4 locules bearing numerous ovules on central placentas. The fruit is a leathery capsule, and the light seeds have hairlike projections on the end. The seeds have a minute embryo and fleshy endosperm.

Classification The affinities of the Nepenthaceae with Droseraceae, Ancistrocladaceae, and Dioncophyllaceae were first proposed by Airy-Shaw[i] and confirmed by phytochemistry. Molecular phylogenetic evidence indicates that the family is closely related to Droseraceae, Dioncophyllaceae, and Ancistrocladaceae[ii,iii]. One species, *Nepenthes pervillei*, is sometimes recognized as a separate genus, *Anurosperma*, based on its different seed anatomy and inflorescence branching. In molecular analysis[iv], this and the other non-Malaysian species form a basal grade from which the monophyletic Malaysian *Nepenthes* arise.

Economic uses The family are grown as ornamentals by specialist horticulturists. All species are listed by CITES, some are on Appendix 1.

The stems of some species (*N. distillatoria* in Sri Lanka, *N. reinwardtiana* in Malaysia) are used locally for basket-making and as a type of cordage. AC

i Airy-Shaw, H. K. On the Dioncophyllaceae, a remarkable new family of Flowering Plants. *Kew Bull.* 1951: 327–347 (1952).
ii Cuénoud, P. *et al.* Molecular phylogenetics of Caryophyllales based on nuclear 18S and plastid *rbc*L, *atp*B and *mat*K DNA sequences. *American J. Bot.* 89: 132–144 (2002).
iii Meimberg, H. *et al.* Molecular phylogeny of Caryophyllidae s.l. based on *mat*K sequences with special emphasis on carnivorous taxa. *Plant Biology* 2: 218–228 (2000).
iv Meimberg, H., Wistuba, A., Dittrich, P., & Heubl, G. Molecular phylogeny of Nepenthaceae based on cladistic analysis of plastid *trn*K intron sequence data. *Plant Biology* 3: 164–175 (2001).

NESOGENACEAE

A little-known family of small-flowered, gamopetalous, usually low-growing herbs and subshrubs in the affinity of Scrophulariaceae.

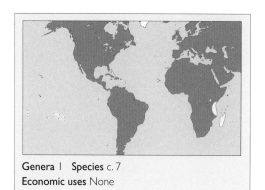

Genera 1 | Species c. 7
Economic uses None

Distribution *Nesogenes* has c. 6 spp. in the Mascarene region (Madagascar, Réunion, Rodrigues, and Aldabra Islands), with 1 sp. extending to inland northern Mozambique and southwestern Tanzania and the Kenya coast, and with 1 sp. in the South Pacific (Tuamotu to Ducie Island). Most species are characteristic of coral islands, while the inland African species occurs on limestone or granite outcrops.

Description Annual or perennial herbs to subshrubs, which are prostrate to erect. The leaves are opposite and without stipules and range from linear-lanceolate and entire to ovate and serrate or the lower leaves sometimes deeply pinnatifid with 5–6 narrow lobes. The flowers are solitary or in few-flowered to densely aggregated cymes in the leaf axils. The calyx is 5-lobed with a short 7- to 10-nerved tube. The corolla is 5-lobed, regular to weakly 2-lipped, up to 8 mm long, white to lilac. The stamens are 4, inserted on the corolla tube. There is no floral disk. The ovary consists of 2, 1-locular fused carpels, each with a single basal ovule, and has a simple terminal elongate stigma. The fruit includes a hard, indehiscent, woody, 2-seeded pyrene.

Classification The family was described in 1981[i], having been formerly included in the broad concept of Verbenaceae or in the Chloanthaceae (which is now included in Lamiaceae), and was considered close to Lamiaceae with which it shares a similar calyx. Molecular evidence places it close to the parasitic Scrophulariaceae[ii], and it may be close to Cyclocheilaceae with which it shares a similar pollen type. Its fruit, with a woody 2-seeded pyrene, argues against its inclusion in any of these families. The absence of a floral disk is also said to distinguish it from Lamiaceae. RKB

i Marais, W. Two new gamopetalous families, Nesogenaceae and Cyclocheilaceae, for extra-Australian 'Dicrastylidaceae'. *Kew Bull.* 35: 797–812 (1981).
ii Harley, R. M. Nesogenaceae. Pp. 293–295. In: Kadereit, J. W. (ed.), *Families and Genera of Vascular Plants. VII. Flowering Plants. Dicotyledons: Lamiales.* Berlin, Springer-Verlag (2004).

NEURADACEAE

A small family of subtropical to tropical desert herbs or subshrubs.

Distribution Dry and desert areas from Africa and the eastern Mediterranean to India. *Grielum* (5 spp.) grows in southern Africa in the Cape region, the Karoo, and on southern Namib inselbergs. *Neuradopsis* comprises 2 spp.: *N. austrafricana* from southwestern Africa and *N. bechuanaensis* known only from Botswana. *Neurada* consists of a single species, *N. procumbens*, although a new species, *N. al-eisawii*, has been described from Jordan[i].

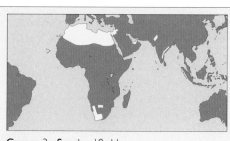

Genera 3 | Species 10–11
Economic uses Some ornamental value

Description Annual or perennial herbs to subshrubs. The leaves are alternate, toothed to pinnatifid, exstipulate. The flowers are apparently axillary, but in fact terminal and solitary, often showy. The sepals are 5, free with or without an "epicalyx" of spiny appendages. The petals are 5, free, imbricate or contorted. The stamens are 10, the inner whorl shorter than the outer. The carpels are usually 10, some smaller and less developed, making the gynoecium appear somewhat irregular; each carpel has a solitary ovule. The fruit is dry and indehiscent

Classification This still poorly understood family was placed in the Rosales, but affinities with the Malvales have long been suggested and this is supported by molecular studies[ii,iii].

Economic uses Some species of *Grielum*, such as *G. humifusum* (White-Eyed Duiker Root), may be cultivated as ornamentals.   VHH

i Barsotti, G., Borzatti von Löwenstern, A., & Garbari, F. *Neurada al-eisawii* (Neuradaceae), a new species from southern desert of Jordan. *Botanica Chronica* 13: 111–115 (2000).
ii Alverson, W. S. *et al.* Circumscription of the Malvales and relationships to other Rosidae: Evidence from *rbc*L sequence data. *American J. Bot.* 85: 876–887 (1998).
iii Bayer, C. *et al.* Support for an expanded family concept of Malvaceae with a recircumscribed order Malvales: a combined analysis of plastid *atp*B and *rbc*L DNA sequences. *Bot. J. Linn. Soc.* 129: 267–303 (1999).

NITRARIACEAE

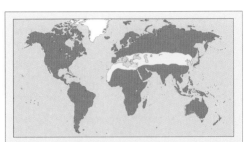

Genera 1   Species 10–13
Economic uses Revegetation of arid areas; fruits eaten locally

Distribution Dry regions of the Old World, from Senegal to North Africa, the Arabian Peninsula and Romania, and across temperate Asia to Korea, with a disjunct area in southern and western Australia[i]. Most species occur in sandy or saline places. The main center of diversity is in central Asia and northwestern China[ii].

Description Stiff shrubs to 1 m high, sometimes with lateral branches modified into spines. The leaves are alternate but often in pairs or fascicles at the nodes, small (to 3 cm), obtriangular to spathulate, entire or somewhat notched toward the apex, often fleshy or glaucous, usually with small stipules. The inflorescence is a few- to many-flowered axillary or terminal dichasial cyme. The flowers are actinomorphic, bisexual, 5-merous, small (usually 1–3 mm). The sepals are fused at the base. The petals are free, white to yellowish or greenish. The stamens are 10–15, 5 opposite the sepals and the others opposite the petals. The ovary is superior, usually 3-carpellate, with (2)3(–6) locules, each with a single pendulous ovule. There is a single style with as many stigma lobes as locules. The fruit is a single-seeded drupe, with a fleshy exocarp and stony endocarp.

NOTHOFAGACEAE. 1. *Nothofagus procera* dehisced 4-valved cupule (×2).

Classification *Nitraria* has usually been included in Zygophyllaceae but differs in leaf arrangement and morphology, 3-carpellate ovary, and fleshy single-seeded fruits, and is placed by molecular data together with Peganaceae and Tetradiclidaceae in Sapindales, well removed from the Zygophyllaceae—see under Peganaceae for discussion and references.

Economic uses May be used to revegetate degraded arid areas. Fruits may be eaten locally, and are vital food for emus in Australia.   RKB

i Pan, X. L., Shen, G. M., & Chen, P. A preliminary research on taxonomy and systematics of genus *Nitraria*. *Acta Bot. Yunnan.* 21: 287–295 (1999).
ii Pan, X. Y. *et al.* Polyploidy: classification, evolution and applied perspective of the genus *Nitraria* L. *Chinese Bull. Bot.* 20: 632–638 (2003).

NOTHOFAGACEAE
SOUTHERN BEECHES

A family of trees and shrubs from the southern hemisphere, many of them cultivated.

Distribution *Nothofagus* has a Gondwanaland distribution in the southern hemisphere, with a discontinuity between the Old and New Worlds. It has 10 temperate species in South America and 25 in Australia (including Tasmania), New Zealand, New Guinea, and New Caledonia. Many species are dominants or co-dominants in both cool temperate or tropical montane forests.

Description Trees or shrubs, evergreen or deciduous. Leaves are alternate, simple, entire or serrate, with peltate stipules. The flowers are unisexual (plants monoecious), the male flowers in sessile to shortly pedunculate 1- to 3-flowered dichasia, with numerous non-persistent bracts, the perianth campanulate, splitting irregularly. The stamens are 5–90. The female flowers are 1–3(–7) in a sessile or shortly stalked involucre; the perianth is minute, dentate; the styles are short; the fruits are surrounded by a (1–)4-valved cupule with lamellar appendages, the cupule lobes sometimes largely fused or missing. The fruits are 2- or 3-angled, usually winged nuts, and single-seeded with a glabrous endocarp.

Classification Traditionally, *Nothofagus* has been included in the Fagaceae[i], but in recent years it is believed that it should form a family on its own, based on morphological and molecular studies[ii,iii] (including APG II). The closest relationship is still with the genus *Fagus* and the Fagaceae in the Fagales[iv]. However, the subgeneric classification has been hotly debated as the genus has always been of such great interest to the biogeographers and phylogeneticists. *Nothofagus* has had a much greater distribution and diversity in the past, and this has led to a general lack of consensus in biogeographical and evolutionary reconstructions[iv,v,vi]. Currently four subgenera are

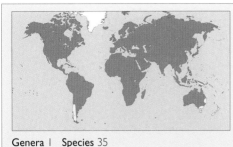

Genera 1   Species 35
Economic uses Widely grown for timber

preferred: subg. *Brassospora* (tropical with *N. grandis*, *N. parryi*, *N. brassii*, and *N. resinosa* in New Guinea, and *N. balansae* and *N. aequilateralis* in New Caledonia) sister to the temperate subg. *Nothofagus* (South America: *N. antarctica*, *N. pumilio*, *N. betuloides*, *N. nitida*, and *N. dombeyi*), subg. *Fuscospora* (New Zealand: *N. trucata*, *N. fusca*, and *N. solandri*; Tasmania: *N. gunnii*; South America *N. alessandri*) and subg. *Lophozonia* (South America: *N. alpina*, *N. glauca* and *N. obliqua*; Australia-Tasmania: *N. cunninghamii*; Australia: *N. moorei*; New Zealand: *N. menziesii*).

Economic uses Several of the temperate species (especially in South America) are used for hardwood timber in furniture, building construction and fencing, and are second only to *Eucalyptus* in the southern hemisphere. *Nothofagus procera* (Rauli), *N. obliqua* (Roble), *N. glauca*, and *N. alessandri* have the best timbers, with the last 2 used for boatbuilding in Chile. *Nothofagus antactica*, *N. obliqua*, and *N. procera* have good coloring during the autumn and are grown as ornamentals. Following the British forestation programmes in the 1930s, an increasing use has been made of *Nothofagus* for timber production, especially *N. obliqua* and *N. procera* and more recently with the demise of *Ulmus* species through Dutch elm disease.   SLJ

i Kubitzki, K. Fagaceae. Pp. 301–309. In: Kubitzki, K., Rohwer, J. G., & Bittrich, V. (eds), *The Families and Genera of Vascular Plants. II. Flowering Plants. Dicotyledons: Magnoliid, Hamamelid and Caryophyllid Families.* Berlin, Springer-Verlag (1993).
ii Hill, R. S. & Jordan, G. J. The evolutionary history of *Nothofagus* (Nothofagaceae). *Australian Syst. Bot.* 6: 111–126 (1993).
iii Manos, P. S. Systematics of *Nothofagus* (Nothofagaceae) based on RDNA spacer sequences (ITS): taxonomic congruence with morphology and plastid sequences. *American J. Bot.* 84: 1137–1155 (1997).
iv Philipson, W. R. & Philipson, M. N. A classification of the genus *Nothofagus* (Fagaceae). *Bot. J. Linn. Soc.* 98: 27–36 (1988).
v Humphries, C. J. Biogeographical methods and the southern beeches (Fagaceae: *Nothofagus*). In: Funk, V. A. & Brooks, D. R. (eds), *Advances in Cladistics: Proceedings of the First Meeting of the Hennigian Society.* Pp. 177–207. New York Botanical Garden (1981).
vi Humphries, C. J., Cox, J. M., & Nielsen, E. S. *Nothofagus* and its parasites: a cladistic approach to coevolution. Pp. 55–76. In: Stone, A. R. & Hawksworth, D. L. (eds), *Coevolution and Systematics.* Oxford, Clarendon Press (1986).

NYCTAGINACEAE
BOUGAINVILLEAS AND FOUR-O'CLOCKS

The family is best known for *Bougainvillea*, which is widely cultivated in the tropics and warm-temperate regions as an ornamental.

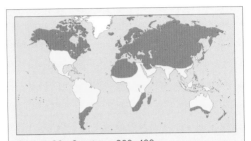

Genera 30 **Species** c. 300–400
Economic uses *Bougainvillea* and *Mirabilis* cultivated as ornamentals; *Pisonia* used as vegetables; several pantropical weeds

Distribution The family is found throughout the tropics and warm-temperate zones but more strongly represented in the Americas. Species most commonly occur at low altitudes[i], but habitats include beaches, forest understory, and hillsides.

Description Trees, shrubs, climbers, and herbs, with swollen nodes and sometimes fleshy roots. The shoot tip and terminal bud are often covered with rich brown indumentum. The leaves are alternate, opposite or whorled, petiolate, simple, often dark grey, and without stipules. The flowers are bisexual or sometimes unisexual (plants dioecious), usually actinomorphic, and sometimes surrounded by colored bracts that resemble a calyx, in usually cymose or paniculate inflorescences. The perianth is usually petal-like, tubular, and often the lower part persists into fruiting time. There are no petals and the 1 to many stamens, usually 5, alternate with the 5 perianth lobes, the filaments being free or fused together at the base, or sometimes branched above. The ovary is superior, comprising a single carpel with a single basal erect ovule, usually surmounted by a long style. The fruit is an indehiscent achene, sometimes enclosed by the persistent base of the calyx, which may serve to assist fruit dispersal. The seeds have endosperm, perisperm, and a straight or curved embryo.

Classification The family are related to Phytolaccaceae and Sarcobataceae in the core Caryophyllales[ii]. *Mirabilis*, including *Oxybaphus*, comprises about 60 American spp. In *Mirabilis*, a neotropical genus of about 45–60 spp., the involucre has taken on the role of a parachute-like fruit-dispersal structure. In *Bougainvillea*, a genus of 18 South American spp., the 3 decorative and colored "sepals" are in fact bracts that subtend groups of 3 inconspicuous tubular flowers. In *Pisonia*, which comprises 50–70 tropical and subtropical spp., the fruits

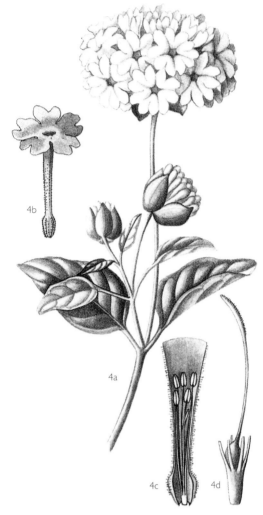

are glandular so that they adhere to animals and are dispersed. Other important genera include the monotypic *Nyctaginia* from southern North America and Mexico; the 40 tropical and subtropical species of herbs in the genus *Boerhavia*; the genus *Abronia*, with 35 spp. from North America; the neotropical *Guapira* with 55 spp.; and *Neea*, comprising some 50–80 tropical American spp.

Economic uses Bougainvilleas are often grown as defensive and decorative hedges in warmer climates and as greenhouse plants farther north. The 2 most commonly grown species are *Bougainvillea glabra* and *B. spectabilis*. From these, and from *B. peruviana* and *B. × buttiana*, many cultivars have been produced. *Mirabilis jalapa* and *M. coccinea* are among many species of *Mirabilis* cultivated for their ornamental value. The flow-

NYCTAGINACEAE. 1 *Bougainvillea spectabilis* (a) leafy shoot with flowers subtended by conspicuous bracts (x⅔); (b) bract and half flower showing petaloid tubular calyx and no petals (x2). 2 *Mirabilis jalapa* (a) leafy shoot with each flower enclosed by a calyxlike involucre of bracts (x⅔); (b) indehiscent fruit (x3⅓); (c) vertical section of fruit showing single seed (x3⅓). 3 *Pisonia aculeata* glandular fruit that is dispersed by birds (x1). 4 (ABOVE) *Abronia fragrans* (a) shoot bearing dense clusters of flowers (x⅔); (b) flower (x1); (c) section of base of perianth-tube showing stamens and style (x2); (d) gynoecium with elongate, hairy stigma (x4).

ers of *Mirabilis jalapa* open in the evening, which gives rise to one of the common names—"Four O'clock." Another common name, Marvel of Peru, relates to the polychromic flowers, which are white, yellow, or red. The tuberous roots of *M. jalapa* are the source of a purgative drug used as a substitute for jalap. The leaves of the Brown Cabbage Tree (*Pisonia grandis*) and the Lettuce Tree (*P. alba*) can be used as a vegetable. Decoctions of the leaves of *P. aculeata* and of the fruits of *P. capitata* are used medicinally to treat a range of complaints. Several species have become pantropical weeds, e.g., *Pisonia*.   [BM] AC

i Stemmerik, J. F. Nyctaginaceae. Pp. 450–469. In: van Steenis, C. G. G. J. (ed.), *Flora Malaysiana Ser. 1* Djakarta, Noordhoff-Kolff N. V., (1964).
ii Cuénoud, P. *et al.* Molecular phylogenetics of Caryophyllales based on nuclear 18S and plastid *rbc*L, *atp*B and *mat*K DNA sequences. *American J. Bot.* 89: 132–144 (2002).

NYMPHAEACEAE

WATER LILIES

A cosmopolitan family of large-flowered water plants, the water lilies.

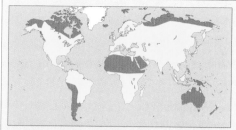

Genera 6   Species 55–60
Economic uses Ornamental aquatics such as the water lilies and edible rhizomes and seeds

Distribution The water lily family occurs worldwide in freshwater habitats (lakes and ponds, rivers and streams, springs, marshes, ditches, canals, and tidal waters) in both temperate and tropical regions. The largest genus *Nymphaea* (c. 40 spp.) is also the most widespread, occurring in both temperate and tropical regions. *Nuphar* (c. 14 spp.) grows in north-temperate regions of North America, Europe, and Asia, with 1 sp. (*N. advena*) extending into the neotropics (Mexico and Cuba). *Victoria* (2 spp.) is endemic to tropical and subtropical South America, *Barclaya* (3–4 spp.) is tropical Indo-Malaysian, *Euryale* (1 sp.) occurs in eastern Asia, and *Ondinea* (1 sp.) is endemic to western Australia.

Description Freshwater aquatic herbs, normally perennial with long horizontal rhizomes, sometimes tuberous, or short and erect. The leaves are submerged, floating or emergent, simple, peltate, cordate or sagittate, with or without stipules; the lamina is ovate to orbicular, entire

NYMPHAEACEAE. 1 *Nymphaea micrantha* (a) habit, showing floating leaves and aerial flowers (×⅔); (b) half flower with greenish sepals, many petals grading into numerous stamens and ovary sunk in receptacle (×1). 2 *Barclaya motleyi* (sometimes placed in the Barclayaceae) (a) habit (×⅔); (b) cross section of fruit containing many seeds (×⅔). 3 *Victoria amazonica* vertical section of flower (×⅔).

or dentate. In *Victoria amazonica*, the leaves may attain 2 m in diameter, with spines on the surface and petioles, and upturned rims. They are able to support a weight of 40–75 kg due to the network of tubular veins on their undersides. The flowers are often large and showy, actinomorphic, hermaphrodite, solitary, borne above the water surface on stout peduncles. The sepals are 4–9, green or colored, and petaloid (*Nuphar*). The petals are 3 to many (absent in *Ondinea*), white, yellow, pink, red, blue, or purplish, grading in some genera into the numerous stamens. The ovary is inferior or semi-inferior (*Barclaya, Euryale, Nymphaea, Victoria*) or superior (*Nuphar, Ondinea*). The carpels are 3–40, united or partly free. The fruit is a spongy berrylike capsule, dehiscing by the swelling of mucilage within or indehiscent. The seeds often have an aril (*Nymphaea, Ondinea*). The flowers are pollinated by insects, especially beetles.

Classification The circumscription of the Nymphaeaceae is contentious. The genus *Nuphar* has been separated as a subfamily (Nupharoideae) or even as a separate family (Nupharaceae), as by Takhtajan (1997), because of its superior ovary, small petals with an abaxial nectary, and exarillate seeds. The palaeotropical *Barclaya*, which has completely inferior ovaries and the petals connate into a lobed tube, is sometimes treated as a separate family (Barclayaceae) or as a third subfamily of the Nymphaeaceae. Here, however, it is included in the subfamily Nymphaeoideae. *Victoria* and *Euryale* are sometimes placed in a separate family, the Euryalaceae, but are retained here in the subfamily Nymphaeoideae as is *Ondinea*. A proposed classification of the Nymphaeaceae (and of the order Nymphaeales), based on both molecular and non-molecular data, by Les *et al.*[i], largely reflects that of Takhtajan, although with different ranks accepted for some groups and recognizes 3 subfamilies, the Nupharoideae, the Nymphaeoideae, and the Barclayoideae.

The Nymphaeaceae is a critical group for the understanding of the orgin and evolution of the early angiosperms. It is usually included in the Nymphaeales and is sister to the Cabombaceae (q.v.). The infrageneric classification of *Nuphar* remains controversial, and a phylogenetic study based on 3 independent data sets

(morphology, chloroplast DNA, and nuclear DNA) indicates that it comprises 2 lineages, one of New World taxa and the other primarily Old World[ii]. A phylogenetic study[iii] of the family inferred from ITS sequences supports the inclusion of *Nuphar* in the Nymphaeaceae and as a sister group to the other genera and suggests a number of subclades.

Economic uses Many species from the family are cultivated as ornamentals, especially *Nymphaea* (water lilies), *Nuphar* (yellow water lilies), *Euryale*, and *Victoria*. The seeds and rhizomes of *Nymphaea* spp. are sometimes eaten, as are the roasted seeds of *Victoria*. *Euryale* seeds yield arrowroot. VHH

[i] Les, D. H. *et al.* Phylogeny, classification and floral evolution of water lilies (Nymphaeaceae; Nymphaeales): A synthesis of non-molecular, *rbc*L, *mat*K, and rDNA data. *Syst. Bot.* 24: 28–46 (1999).
[ii] Padgett, D. J., Les, D. H., & Crow, G. E. Phylogenetic relationships in *Nuphar* (Nymphaeaceae): evidence from morphology, chloroplast DNA, and nuclear ribosomal DNA. *American J. Bot.* 86: 1316–1324 (1999).
[iii] Liu, Y.-L. *et al.* Phylogeny of Nymphaeaceae inferred from ITS sequences. *Acta Phytotax. Sinica* 43: 22–30 (2005).

OCHNACEAE

A family of tropical trees, shrubs, or herbs, usually with shiny leaves and with dense pinnate venation, toothed margins, and conspicuous stipules

Distribution Tropics of New and Old World, rarely extending into subtropics, and almost completely absent from temperate regions. They occur mostly in rain forest or woodland.

Description Low growing herbs (*Sauvagesia*) to shrubs or trees (up to 30 m in *Cespedesia*), rarely lianas in *Krukoviella*. The leaves are alternate, simple (? 1-foliate) and unlobed or in *Rhytidanthera* pinnate with large leaflets, usually shiny (coriaceous in the woody genera), with tertiary veins closely parallel and the cells between the finer veins more or less rectangular. Leaf margins are usually finely crenate to serrate or sometimes fimbriate, the lamina usually broad and up to 70 cm or more long in *Cespedesia* but sometimes ericoid in *Sauvagesia*. Petioles are usually characteristically very short or completely absent and when present resemble a pulvinus. Stipules are present and often conspicuous, persistent or caducous, sometimes fimbriate or glandular-ciliate. The inflorescence is a terminal or axillary raceme, panicle or corymb, or the flowers are sometimes solitary. Flowers are bisexual, actinomorphic or slightly zygomorphic, usually 5-merous (3–6-merous in *Elvasia*), often showy. The sepals

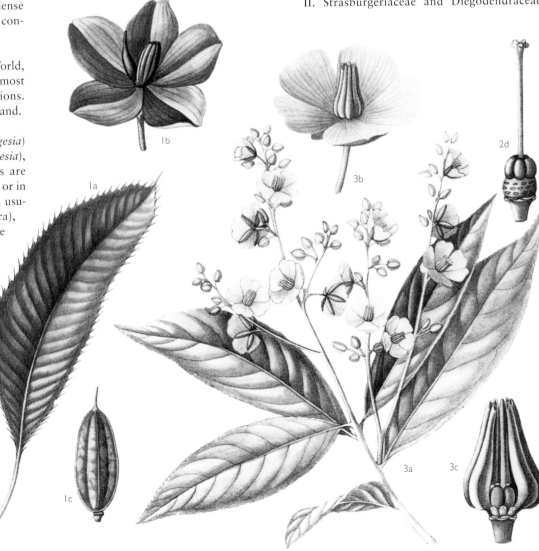

OCHNACEAE. 1 *Luxemburgia ciliosa* (a) leaf (×⅔); (b) flower (×1); (c) fruit (×1); (d) cross section of fruit (×2). 2 *Ochna atropurpurea* (a) flowering shoot (×⅔); (b) stamen with apical pores on anthers (×8); (c) fruit—a cluster of drupes on a fleshy receptacle, with a persistent colorful calyx (×1); (d) gynoecium (×6). 3 *Ouratea intermedia* (a) leafy shoot with terminal inflorescence (×⅔); (b) flower with cluster of 10 stamens (×2); (c) flower with perianth removed to show stamens (with filaments much shorter than anthers) and gynoecium (×4).

are (3–4)5(6–10), often enlarging in fruit. The petals are (3–4)5(6–12), free, often caducous. The stamens are 5 to many, in 1 to several whorls, sometimes the inner ones reduced to staminodes and fused into a short tube round the ovary. The ovary is superior, with 2–15 carpels, which are fused and with a terminal style or more or less free and with a gynobasic style and the receptacle becoming swollen and colored beneath them. The ovules are 1 to many per carpel, on axile or parietal placentas. The fruit is either a capsule or a berry or a collection of single-seeded drupelets.

Classification Although traditionally associated with the Theales, the Ochnaceae are now placed on molecular evidence close to Linaceae and allied families in the Malpighiales of APG II. Strasburgeriaceae and Diegodendraceae

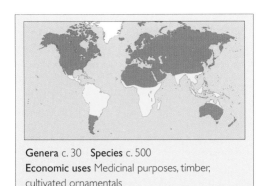

Genera c. 30 **Species** c. 500
Economic uses Medicinal purposes, timber, cultivated ornamentals

have been excluded and placed remote from Ochnaceae, but the South American Quiinaceae and the Seychelles endemic Medusagynaceae are found to be closely related to the family and have been included in it by some—see Quiinaceae for discussion. The family Sauvagesiaceae has sometimes been recognized as distinct from Ochnaceae but is better regarded as a subfamily[i]. The greater number of genera are referred to Sauvagesioideae, but the great majority of species are in the Ochnoideae.

SUBFAM. OCHNOIDEAE Trees and shrubs; carpels 1–10, single-seeded, and usually appearing free and black on a swollen red to purple receptacle, with a gynobasic style; fruit usually a collection of single-seeded drupelets; seeds without endosperm; staminodes absent. Several genera, including *Ouratea* (pantropical, c. 200 spp.), *Ochna* (Old World, c. 90 spp.), and *Brackenridgea* (Old World, 12 spp.). *Elvasia* (New World, 11 spp.) may be referred to the tribe Elvasieae with the carpels united into a dry capsular fruit. *Lophira* (western Africa, 2 spp.) is referred to the tribe Lophireae, differing in its 2 carpels with 5–10 ovules per carpel and the capsular fruit attached to 2 persistent winglike sepals like a dipterocarp.

SUBFAM. SAUVAGESIOIDEAE Low herbs to trees and shrubs; carpels 2–5, fused, each single- to many-seeded; fruit a many-seeded capsule; seeds with endosperm; 10 or more staminodes usually present in 1 or more whorls. Numerous genera. In the tribe Sauvagesieae, the carpels have more than 2 ovules and actinomorphic flowers, including *Sauvagesia* (c. 39 herbaceous spp., of which 2 are in Africa, 2 in tropical Asia, and the others in the New World). The Luxemburgieae differ in having slightly zygomorphic flowers (*Luxemburgia*, 17 spp. in Brazil, etc.). The Euthemideae differ in having 1–2 ovules per carpel and actinomorphic flowers with no staminodes (or rarely 5) (*Euthemis*, 2 spp. from tropical Asia).

Economic uses The flowers of many species are often showy, and some, such as the South African *Ochna atropurpurea,* are cultivated as ornamentals. The large leaves of *Cespedesia* are used in roofing and baskets in the neotropics. *Sauvagesia* has been used for curing eye disorders and to make a herbal tea. In West Africa, *Lophira* is important for antibacterial use, for cooking oil from its seeds, and for timber (so-called "African Oak") for wharves and railroad ties. RKB

[i] Amaral, M. C. E. Phylogenetische Systematik der Ochnaceae. *Bot. Jahrb. Syst.* 113: 105–196 (1991).

OLACACEAE
AMERICAN HOG PLUM AND AFRICAN WALNUT FAMILY

A family of tropical woody autotrophic or hemiparasitic plants, whose circumscription is still unclear.

Distribution The family occurs widely in the tropics of Africa, Asia, and America, often in lowland rain forests but also growing in deciduous forests and savanna[i]. In the Americas, it ranges from the southern USA, through Mexico and Central America, down to Argentina, and it extends into the Antilles.

OLACACEAE. 1 *Heisteria parvifolia* (a) leafy shoot with small flowers (x⅔); (b) half flower (x7); (c) cross section of ovary (x7); (d) fruit (x⅔). 2 *Ximenia caffra* (a) fruits (x⅔); (b) vertical section of fruit (x1). 3 *Olax obtusifolia* (a) flowering shoot (x⅔); (b) flower with calyx represented by a small rim below the recurved petals (x3); (c) part of flower showing gynoecium and stamens (x3). 4 *Schoepfia vacciniiflora* (a) flowering shoot (x⅔); (b) flower (x2); (c) flower opened out showing disk around the ovary with below it the reduced calyx (x2); (d) vertical section of ovary which is partly sunk in the disk (x4); (e) cross section of ovary (x4); (f) fruits (x⅔).

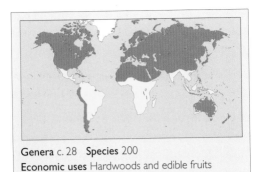

Genera c. 28 **Species** 200
Economic uses Hardwoods and edible fruits

Description Usually small to medium ever-green trees, sometimes shrubs (*Ximenia*), rarely lianas, and some genera are known to be root hemiparasites (e.g., *Ximenia*)[ii]. The leaves are simple, alternate, spiral, or distichous, without stipules, the lamina margin entire or sinuate-dentate. The inflorescence is axillary, racemose, or paniculate. The flowers are actinomorphic, bisexual, or rarely unisexual in *Ximenia* (plants dioecious), usually small and greenish. The calyx is small, cupular, 3- to 6-lobed. The petals are 3–6, free or connate below or forming a tube, valvate. The stamens are as many or twice as many (or more) as the petals, in a single whorl, the filaments free or connate. An intra- or extra-staminal nectary disk is sometimes present. The ovary is superior, semi-inferior, or inferior, of 2–5 carpels, with 1–5 locules, the placentation free central or axile, and with 1 ovule per locule. The fruit is a drupe or nut, often surrounded or enclosed by the fleshy accrescent calyx, usually single-seeded.

Classification The classification and circum-scription of the Olacaceae is unclear. Both morphological[iii] and molecular analyses[iv] indi-cate that Olacaceae in the broad sense is paraphyletic and contains 4 main clades[i], some of which are congruent with previously described tribes. Until further clarification is achieved, the subdivision of the family must remain tentative. Olacaceae belongs in the Santalales, where it is paraphyletic to the other families.

The position of *Schoepfia*, a genus of some 25 spp. from tropical America and Southeast Asia is also uncertain. It is distinctive in its flowers subtended by conspicuous bracteoles, its tubu-lar corolla, and in its wood anatomy. It is sometimes included in Olacaceae as a separate tribe or subfamily, but some recent analyses suggest that it is sister to *Misodendron* (Miso-dendraceae) and to all remaining families in the Santalales[iii]. Its recognition as a separate family (Schoepfiaceae) seems premature.

Economic uses Some Olacaceae provide useful timber, such as *Ximenia americana* (Tallow Wood), which is used as a substitute for sandal-wood, and species of *Heisteria, Minquartia*, and *Scorodocarpus*. The fruits of *Ximenia americana* (Hog Plum) are edible, but the flesh contains hydrocyanic acid, and the seeds are purgative. Some species are used in local medicines. VHH

[i] Pennington, T. Olacaceae. Pp. 276–277. In: Smith, N. *et al.* (eds), *Flowering Plants of the Neotropics*. Princeton, Princeton University Press (2004).
[ii] Nickrent, D. L. Phylogenetic origins of parasitic plants. Chapter 3. In: López-Sáez, J. A., Catalán, & Sáez, L. (eds), *Parasitic Plants of the Iberian Peninsula and Balearic Islands*. Madrid, Mundi-Prensa (2002).
[iii] Malécot, V. *et al.* A morphological cladistic analysis of Olacaceae. *Syst. Bot.* 29: 569–586 (2004).
[iv] Nickrent, D. L. & Malécot, V. A molecular phylogeny of Santalales. Pp. 69–74. In: Fer, A., Thalouarn, P., Joel, D. M., Musselman, L. J., Parker, C., & Verkleij, J. A. C. (eds), *Proceedings of the 7th International Parasitic Weed Symposium*. Nantes, Faculté des Sciences, Université de Nantes (2001).

OLEACEAE

OLIVE FAMILY

A widespread family of shrubs, trees, and woody climbers, placed near the base of the Scrophulariaceae complex.

Distribution Widespread in temperate and tropical regions. Several genera show an unusu-al disjunction between southern Europe and the Far East[i]. Most tropical genera have relatively narrow distributions in either the New World or Old World, but *Schrebera* and *Chionanthus* span both hemispheres. *Menodora* is in the New World, apart from 3 spp. in South Africa (one of which has its closest relatives in the USA and Mexico), and *Nestegis* is confined to New Zealand and Norfolk Island, except for 1 sp. in Hawaii. *Hesperelaea* was confined to Guadalupe Island off the Pacific coast of Mexi-co but is now extinct. *Fraxinus* is an important woodland genus in the temperate northern hemisphere but also extends to the tropics in Costa Rica and mountains of Malaysia.

OLEACEAE. 1 (OPPOSITE PAGE, ABOVE) *Forsythia viridissima* (a) shoot with flowers borne on previous year's side shoots (x⅔); (b) vertical section of ovary (x4); (c) part of corolla opened out to show epipetalous stamens (x1). 2 *Fraxinus platypoda* (a) winged fruit—samaras (x⅔); (b) vertical section of samara base (x2); (c) vertical section of seed (x2). 3 *Phillyrea vilmoriniana* fleshy fruits (x⅔). 4 (RIGHT) *Syringa vulgaris* (a) leaves and inflore-scence (x⅔); (b) half flower (x1½); (c) cross section of ovary (x6); (d) corolla opened out to show epipetalous stamens (x⅔); (e) dehisced fruits—2-locular capsules (x⅔).

Description The family comprises mostly trees and shrubs, but *Jasminum* and *Myxopyrum* include woody climbers to 12 m or more, and *Menodora* and *Dimetra* may be herbs with a woody rootstock. The leaves are opposite except in *Jasminum* sect. *Alternifolia*, and sim-ple to imparipinnate. The inflorescence is a compound cyme and is usually terminal. The flowers are actinomorphic, occasionally unisex-ual or the plants dioecious, usually sweetly scented. The calyx is 4-lobed, or in Jasmineae 5- to 15-lobed, or sometimes absent. The corolla is usually gamopetalous and actinomorphic but is absent in *Forestiera* and most *Fraxinus* and *Nestegis*, and usually 4-lobed but 5- to 12-lobed in Jasmineae. The stamens are characteristically 2 but rarely 4 in *Fraxinus, Chionanthus, Nestegis*, and *Priogymnanthus*. The ovary is 2-carpellate and 2-locular, each locule bearing 1, 2, 4, or (*Forsythia*) more ovules. The fruit is a capsule, samara, berry or drupe.

Classification The family has often been thought to be rather isolated in the bicarpellate gamopetalous families or sometimes close to the Gentianales, but molecular evidence places it near the base of the Scrophulariaceae complex. The main natural groupings within the family seem to be uncontroversial, but the rank given to them varies. *Nyctanthes* and *Dimetra* have often been excluded from the family and placed either in a broad Verbenaceae or their own fam-ily Nyctanthaceae, but molecular analysis places them basal to Oleaceae and near to *Myxopyrum*

4e   4a   4b   4c   4d

(though they are morphologically not similar to the latter). The remaining genera fall clearly into 2 groups, as demonstrated on the basis of chromosome number as well as other characters by Taylor[ii] and confirmed by Wallender and Albert[iii] by molecular and other characters. The recent treatment for FGVP[iv] reflects the same groupings. There seems to be some advantage now in recognizing a paraphyletic subfamily, the Oleoideae as recognized below nesting within Jasminoideae. When these are given subfamilial status, it seems also appropriate to group the 3 anomalous basal genera at the same rank. It is here preferred to recognize 3 subfamilies, each divided into tribes:

SUBFAM. MYXOPYROIDEAE Comprises 3 genera and 7 spp. Stems usually rectangular with cortical bundles at the angles. Calyx 4-lobed, and the corolla 4- to 8-lobed. Ovules 1–3 per locule, and basal and ascending. Chromosome numbers n=11, n=12, n=22, or n=23. Two morphologically very distinct tribes:

Myxopyreae *Myxoporum* has 4 spp. from India to New Guinea. Woody climbers to 12 m or more, with leaves with 2 prominent veins parallel with the midrib. Corolla campanulate to urceolate. Ovary with 1–3 ovules per locule and developing into a single- to 4-seeded drupe. Chromosomes n=22 or n=24.

Nyctantheae *Nyctanthes* comprises 2 spp. of shrubs or small trees from India to Indochina. Leaves not 3-nerved. Corolla funnel-shaped and rotate, unlike most Oleaceae. Ovary with a single basal ovule per locule and developing into a

schizocarp with flattened seeds. Chromosome number n=11, n=18, n=22, or n=23. *Dimetra*, 1 sp. from Thailand, is a herb from a woody rootstock, unlike *Nyctanthes* in general appearance. Its fruit is a flattened capsule. It seems somewhat little-known, and further investigation may suggest that it deserves recognition as its own tribe alongside Nyctantheae.

SUBFAM. JASMINOIDEAE Comprises 5 genera and c. 234 spp. Stems not rectangular. Calyx 4- to 15-lobed and corolla 4- to 5-lobed or more. Ovules 1–4 per locule and axile or pendulous. Chromosome number n=11, n=13, or n=14. There are 3 tribes.

Jasmineae *Jasminum* comprises c. 200 spp. in the Old World and the genus *Menodora* comprises 23 spp. in the New World. Calyx 5- to 15-lobed and corolla 5–12 lobed and funnel- or salver-shaped. Ovary with 2 or 4 ovules per locule and developing into a capsule or berry. Chromosome number n=11, n=12, or n=13.

Forsythieae *Abeliophyllum* comprises 1 sp. from Korea, and the genus *Forsythia* comprises 9 spp. from southern Europe and East Asia. Calyx and corolla 4-lobed, and the corolla campanulate with well-developed narrow lobes. Ovary with 1 pendulous or (*Forsythia*) many ovules per locule and developing into a samara or capsule. Chromosome number n=14.

Fontanesieae *Fontanesia* comprises 1 sp. with 2 subsp. from the Mediterranean to China. Calyx and corolla are 4-lobed, the corolla deeply so. Ovary with 1 pendulous ovule per locule and developing into a single- to 2-seeded samara. Chromosome number n=13.

SUBFAM. OLEOIDEAE Comprises 7 genera and c. 330 spp. Stems not rectangular. Calyx and corolla 4-lobed or the corolla 5- to 7-lobed in *Schrebera*. Ovules 2 per locule or 4 in *Schrebereae*, pendulous. Chromosomes number n=23 or in *Haenianthus* n=20. There are 4 tribes:

Fraxineae *Fraxinus* only, with 40–50 spp, widespread. Corolla absent or consists of 2 or 4 more or less free petals. Fruit a single-seeded samara.

Schrebereae *Schrebera* comprises 6 spp., pantropical, and *Comoranthus* comprises 3 spp. from Madagascar and the Comoros.

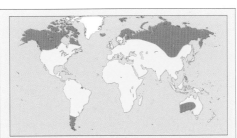

Genera 25 Species c. 570
Economic uses Olives and oil; timber; numerous ornamental shrubs, climbers, and hedge plants

Corolla hairy and funnel- to salver-shaped. Fruit a woody pear-shaped capsule with winged seeds.

Ligustreae *Syringa* comprises c. 20 spp. from Eurasia, and *Ligustrum* comprises c. 40 spp., mainly in Asia but 1 each in Europe and Australia. Corolla glabrous and cylindrical. Fruit a capsule (*Syringa*) or a fleshy berry or drupe.

Oleeae 12 genera and c. 210 spp., widespread, including *Chionanthus* (formerly *Linociera*) with c. 60 spp, *Noronhia* with c. 45 spp. in the Mascarene area, *Olea* with c. 32 spp. in Old World tropics and warm areas, and *Osmanthus* with c. 32 spp. Corolla glabrous or absent. Fruit a drupe, usually with a hard endocarp.

Economic uses *Olea europaea* subsp. *europaea* (Olive) has been cultivated by humans for thousands of years for its edible fruits and the oil extracted from them for multiple purposes. The genus *Fraxinus* (Ash) yields valuable timber. *Ligustrum ovalifolium*, native of Japan, is extensively planted as an urban hedge plant (Privet). Many others are cultivated as ornamental (often strongly scented) garden plants, especially *Jasminum* (jasmines), *Syringa* (lilacs), *Forsythia*, *Osmanthus*, and *Abeliophyllum*.

RKB & PSG

i Green, P. S. *Osmanthus decorus* and disjunct Asiatic-European distributions in the Oleaceae. *Kew Bull.* 26: 487–490 (1972).
ii Taylor, H. Cyto-taxonomy and phylogeny of the Oleaceae. *Brittonia* 5: 337–367 (1945).
iii Wallander, E. & Albert, V. A. Phylogeny and classification of the Oleaceae based on *rps*16 and *trn*L-*trn*F sequence data. *American. J. Bot.* 87: 1827–1841 (2000).
iv Green, P. S. Oleaceae. Pp. 296–306. In: Kadereit, J. W. (ed.), *Families and Genera of Vascular Plants. VII.* Berlin, Springer-Verlag (2004).

OLINIACEAE

A monotypic family of 5 spp. of trees or shrubs from montane and coastal forests of eastern, western, and southern Africa and on St. Helena where it was probably introduced. The leaves are simple, opposite or rarely ternate, coriaceous, with rudimentary stipules or none. The flowers are actinomorphic, hermaphrodite, in terminal or axillary cymes. The perianth is 5-merous and consists of 3 whorls whose interpretation is a matter of some debate[i]: an outer of small, toothlike organs inserted on top of the somewhat thickened hypanthial rim; a middle whorl of conspicuous pink to red, lingulate, petaloid organs, with a broad base and alternating with them 4–5 pubescent, colored scales. The stamens are 4–5, opposite, and just below the lobes of the inner perianth whorl on the

Genera 1 Species 5
Economic uses Some medicinal value; timber

adaxial side of the hypanthial rim. The ovary is inferior, of 3–5 carpels, with axile placentation; locules 3–5 with 2–3 ovules per locule. The fruit is drupaceous, 3- to 5-locular, with 1 seed (stone) per locule.

The Oliniaceae has been related by several authors to the Penaeaceae in the order Myrtales[ii]. Some molecular studies[iii,iv], confirm this relationship and indicate a sister group relationship of Oliniaceae and Penaeaceae but this is only weakly supported in another study[i], although it does confirm the Rhynchocalycaceae as sister to the clade formed by Oliniaceae and Penaeaceae. The timber of some species is used locally and some medicinal uses are reported. VHH

[i] Schönenberger, J. & Conti, E. Molecular phylogeny and floral evolution of Penaeaceae, Oliniaceae, Rhynchocalycaceae, and Alzateaceae (Myrtales). *American J. Bot.* 90: 293–309 (2003).
[ii] Johnson, L. A. S. & Briggs, B. G. Myrtales and Myrtaceae - a phylogenetic analysis. *Ann. Missouri Bot. Gard.* 71: 700–756 (1984).
[iii] Conti, E. *et al.* Interfamilial relationships in Myrtales: Molecular phylogeny and patterns of morphological evolution. *Syst. Bot.* 22: 629–647 (1997 [1998]).
[iv] Clausing, G. & Renner, S. S. Molecular phylogenetics of Melastomataceae and Memecylaceae: Implications for character evolution. *American J. Bot.* 88: 486–498 (2001).

ONAGRACEAE
EVENING PRIMROSE FAMILY

A well-defined family distributed nearly worldwide, but especially rich in genera and species in the New World.

Distribution Cosmopolitan, species occurring mostly in open habitats, ranging from dry to wet, even aquatic mostly in *Ludwigia*, from tropics to deserts of western North America, and to arctic tundra; a few species of *Epilobium*, *Ludwigia* and *Oenothera* can be weeds in wetlands or cultivated fields.

Description Mostly perennial to annual herbs, with some shrubs, rarely trees to 25 m. The leaves are usually cauline, often with a basal rosette, rarely basal only, simple, alternate or opposite, occasionally whorled or spirally arranged, petiolate to sessile; stipules present and intrapetiolar, usually small and caducous, or absent in tribes Epilobieae and Onagreae.

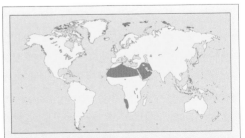

Genera 18 **Species** 656
Economic uses Used in traditional and commercial medicines; provides many cultivated ornamentals

The lamina is entire or toothed to pinnatifid. The flowers are axillary, in leafy spikes or racemes or solitary, or occasionally in panicles, hermaphroditic (protandrous in *Chamerion*, *Epilobium*, and *Clarkia*; protogynous in *Circaea* and *Fuchsia*) or occasionally unisexual (plants gynodioecious, dioecious, or subdioecious), actinomorphic, or zygomorphic, 2-7-merous (usually 4-merous), morning or evening opening, insect or bird pollinated or autogamous. A floral tube is usually present (absent in *Chamerion*, *Ludwigia*, and sometimes *Lopezia*), nectariferous within. The sepals are green or colored, valvate, in *Ludwigia* persistent after anthesis. The petals are as many as sepals or rarely absent, yellow, white, red, pink, or various shades of purple, imbricate or convolute in bud, occasionally clawed. Occasionally, the flowers are slightly to markedly zygomorphic, and the petals can be subequal and presented to the upper side of the flower (*Megacorax* and *Oenothera*) or unequal (*Lopezia*), with the upper and lower pairs often markedly dissimilar, the upper 2 petals united in some species, with the upper 3 sepals for part of their length. The stamens are as many as or twice as many as sepals and in 2 series, in *Lopezia* reduced to 2 or 1 plus 1 sterile staminode (± enclosing the fertile stamen and releasing it explosively). The anthers are versatile or basifixed, dithecal, septa present dividing sporogenous tissue, sometimes cross-partitioned at maturity, opening by longitudinal slits, pollen grains united by viscin threads, shed singly or in tetrads or polyads, pollen ektexine paracrystalline, beaded. The ovary is inferior, consisting of as many carpels and locules as sepals, the septa sometimes thin or absent at maturity, placentation axile or parietal, ovules 1 to numerous per locule, in 1 or several rows or clustered, anatropous, bitegmic. The style is single; the stigma globose, clavate, or lobed (with as many lobes as sepals), papillate and dry without secretions, or non-papillate and wet with sticky secretions. The fruit is a loculicidal capsule, sometimes woody, or indehiscent berry or nut, sessile to long-pedicellate. The seeds are usually small, smooth or variously sculptured, sometimes with a coma or wings, anatropous, with straight, oily embryo, 4-nucleate embryo sac, endosperm lacking.

Classification Recent molecular studies have shown the family to be most closely related to the Lythraceae, and more broadly related to other Myrtalean families Melastomataceae, Myrtaceae, and allied families, with Combretaceae basal in the order. The comprehensive generic monograph of Wagner *et al*[i] presents a revised classification of the family, and recognizes 2 subfamilies:

SUBFAM. LUDWIGIOIDEAE With 1 genus, *Ludwigia* (82 spp.), characterized by the sepals persistent after anthesis; floral tube absent; flowers usually 4- 5-merous, but may be 3- to 7-merous. *Ludwigia* is pantropical, currently

divided into 23 sections, especially well represented in North and South America, with endemic groups also in Africa and Asia. Diurnal and pollinated mainly by bees (also some flower flies and butterflies).

SUBFAM. ONAGROIDEAE With 6 tribes and 17 genera:

Hauyeae *Hauya* (2 spp.) only, small trees or shrubs from Hidalgo and Guerrero, Mexico, south to Costa Rica. Flowers large, white nocturnal, pollinated by hawk-moths and possibly bats. The thick, woody capsules contain large winged seeds.

Circaeeae Two genera: *Circaea* (8 spp.), perennials occurring throughout the northern hemisphere in moist, temperate, evergreen and/or deciduous and cool boreal forests, and *Fuchsia* (106 spp., 12 sections), shrubs, lianas, epiphytes, or rarely small trees, with most species in South America, where the largest section *Fuchsia* is centered, but extending to New Zealand (3 spp.) and Society Islands (1 sp.), and north to Central America, Mexico, and Hispaniola. Members of the tribe are characterized by indehiscent fruits, with hooked hairs in *Circaea* and a fleshy berry in *Fuchsia*. Flowers diurnal with pollination in *Circaea* by syrphid flies and small bees, and in *Fuchsia* by hummingbirds, passerine birds (in species of Oceania), bees, flies (tachinid and flower), and butterflies.

Lopezieae Two genera: *Megacorax* (1 sp.), endemic to Durango, Mexico, pollinated by bee flies, and *Lopezia* (22 spp., 6 sections) occurring mainly in Mexico and Guatemala, extending south to Panama. Characterized by zygomorphic flowers, in *Lopezia* with stamens reduced to 2 or 1 plus a staminode; flowers diurnal and pollinated by hummingbirds, syrphid flies, and rarely bees.

Gongylocarpeae *Gongylocarpus* (2 spp.) only, with 1 shrub species endemic to a small area of western Baja California, Mexico, and a self-pollinating annual species scattered across Mexico to Guatemala. The mature fruits become embedded in the stem at maturity, possibly an adaptation for seawater dispersal.

Epilobieae Two genera: *Chamerion* (fireweeds, 8 spp., 2 sections, 1 in western Asia and Europe, and the other widespread in the northern hemisphere) and *Epilobium* (164 spp., 7 sections, with 6 sections mainly in western North America, and the seventh sect. *Epilobium* [150 spp.] in cool temperate and montane

ONAGRACEAE. 1 *Fuchsia alpestris* (a) flowering shoot (x⅔); (b) half flower showing free sepals that are longer than the purple petals (x1½); (c) cross section of fruit (x3). 2 *Circaea cordata* (a) dehisced capsule (x6); (b) cross section of fruit (x6). 3 *Epilobium hirsutum* (a) flowering shoot (x⅔); (b) fruit (a capsule) with part of wall cut away (x2); (c) ripe fruit dehiscing to release plumed seeds (x4). 4 *Lopezia coronata* flower with upper petals marked with blotches that resemble nectar, 1 erect fertile stamen, and petaloid spoon-shaped staminodes—all adaptations to a specialized form of insect pollination (x3). 5 *Oenothera biennis* (a) flowering shoot (x⅔); (b) partly opened flower with petals removed (x2).

habitats on all continents except Antarctica). The key innovation of the tribe seems to be the comose, easily dispersed seeds from thin capsules. Mostly perennial, only a few annuals and subshrubs. Flowers diurnal and pollinators including bees, flies, and hummingbirds (1 sp.), although many species have relatively small flowers and commonly self-pollinate.

Onagreae Nine genera primarily of western North America: *Xylonagra*, with 1 shrub species with tubular red flowers endemic to central Baja California, Mexico; *Clarkia* (42 spp., 9 sections), mostly restricted to California, but extending to other western states of the USA and Baja California, with 1 sp. in southern South America; *Gayophytum* (9 spp.) of western North America, with 1 sp. in temperate South America and 1 sp. common to both continents; *Chylismiella* (1 sp.) of the northern desert areas of the western USA; *Taraxia* (4 spp.) in moist habitats in northwestern North America; *Camissonia* (37 spp., 5 sections), with most species restricted to California, but a few extending throughout western North America and northern Mexico, and 1 sp. in southern South America; *Eulobus*, with 4 spp. distributed from southern California to southern Arizona, Sonora, and Baja California, Mexico; *Chylismia* (16 spp., 2 sections), restricted to desert regions of western North America; and *Oenothera* (evening primroses, 149 spp., 16 sections now including *Calylophus, Stenosiphon,* and *Gaura*), a large genus widely distributed in temperate to subtropical areas of

North and South America, with several species naturalized worldwide, usually of open, often disturbed, habitats, with the center of diversity in southwestern North America. Tribe Onagreae comprises over 40% of the species in the family. Aneuploidy and hybridization (*Clarkia*), polyploidy (*Camissonia, Gayophytum*), and chromosomal translocations with the formation of permanent translocation heterozygotes that form a ring of chromosomes in meiosis (*Gayophytum, Oenothera*) are all important features of the evolution of the tribe.

Economic uses Species of *Clarkia, Oenothera,* and *Fuchsia* are widely and commonly grown horticulturally. Evening primrose oil (EPO) from *Oenothera* seeds contains an omega-6 essential fatty acid, gamma-linolenic acid (GLA), which is considered to be the active therapeutic ingredient. EPO has been proposed for the treatment of a variety of disorders, particularly those affected by metabolic products of essential fatty acids, although convincing evidence for its efficacy in treating most disorders is still lacking. *Clarkia* has been popular in ecological and evolutionary studies for decades with a large body of literature. One group of *Oenothera* species has been intensively studied due to its unusual chromosomal behaviour (translocation heterozygosity, resulting in rings of chromosomes in meiosis) and is currently widely used in research. WLW & PCH

[i] Wagner, W. L., Hoch, P. C., & Raven, P. H. A revised classification of the family Onagraceae based on molecular sequence data and morphology. *Syst. Bot. Monogr.* (unpublished as of 2006).

ONCOTHECACEAE

Oncotheca is confined to the main island of New Caledonia, where both species have been collected mostly in the southeast in forests at 0–800 m. They are shrubs to large trees up to 30 m. All parts are glabrous. The leaves are borne in lax clusters toward the ends of branches and are simple, entire, coriaceous, up to 18 × 8 cm, oblanceolate with a rounded apex, and the base gradually narrowed to a short petiole. The inflorescences are axillary and borne among the lax leaf clusters, each one with a stout main axis up to 8 cm bearing short side branches terminated by compact glomerules of several small flowers, the whole inflorescence triangular in shape. The bracts subtending the branches, and the 2 bracteoles per flower, are broadly triangular. The flowers are subsessile, c. 2 mm across, actinomorphic and bisexual. There are 5 suborbicular sepals. The corolla is up to 2 mm long and has a short tube with 5 broad white lobes. The 5 stamens are inserted in the sinuses between the corolla lobes, with short filaments, and *O. balansae* has the connectives prolonged and inflexed to form a cover over the ovary. The ovary is superior, made up of 5 carpels, and is more or less entire in *O. humboldtiana* but deeply furrowed

238

between the carpels in *O. balansae*. The locules are 5, with 2 pendent ovules in each. The stigmas are 5, more or less free, but sessile. Few fruits develop in each inflorescence, each being an ovoid to pyriform drupe up to 3.6 × 2.4 cm, with a thin slightly fleshy exocarp surrounding a thick-walled stone in which 5, single- to 2-seeded locules are embedded.

Oncotheca has been variously thought to be related to Aquifoliaceae or Ebenaceae, but a recent Flora account[i] has concluded that they are certainly related to Theaceae. However, molecular evidence[ii] has referred them (without strong support) to the Garryales, to which they have little obvious morphological resemblance. The Oncothecaceae are similar in habit and general appearance to another family endemic to New Caledonia, the Phellinaceae, but the molecular evidence places that far removed in the Asterales.   RKB

[i] Morat, P. & Veillon, J.-M. *Flore de la Nouvelle-Calédonie et Dépendances*. 15: 90–98 (1988).
[ii] Savolainen, V. *et al*. Phylogeny of the eudicots: a nearly complete familial analysis based on *rbc*L gene sequences. *Kew Bull*. 55: 257–309 (2000).

OPILIACEAE

A pantropical family of root-parasite trees and lianes.

Distribution Pantropical in evergreen or occasionally deciduous forests. The largest genus, *Agonandra* (10 spp.), occurs from Mexico to Argentina, but the other genera are all found in the Old World.

Description Evergreen shrubs and small trees up to 5(–13) m, or in South America up to 40 m or lianes up to 15(–30) m. Of the 10 genera, 8 have been confirmed as root parasites[i]. The leaves are alternate, simple, entire, and petiolate. The inflorescence is usually an axillary or cauliflorous panicle or umbel or spike. The flowers are actinomorphic, usually bisexual but unisexual (plants dioecious) in *Agonandra* and the monotypic genera *Melientha* and *Gjellerupia*, or gynodioecious in *Champereia*, sometimes 3-merous, but usually 4- to 5-merous, small and with the perianth inconspicuous or sometimes absent. The stamens are as many as the tepals. The ovary is superior, with a subsessile stigma, and is 1-celled

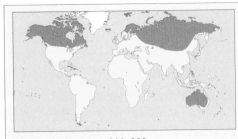

Genera 10 **Species** 33
Economic uses Fruits eaten locally; roots and bark used in medicine; seeds used as rubber substitute

with a pendulous ovule. The fruit is a drupe with a thin pericarp, fleshy mesocarp, and woody endocarp, with 1 large seed.

Classification The family has always been associated with, and often included in, the Olacaceae, and is placed in the Santalales in APG II. The New World genus *Agonandra* has sometimes been placed in a separate tribe.

Economic uses The fruits of several genera are eaten locally. In Malaysia, young shoots are boiled and eaten (but beware poisonous *Urobotrya*). The roots of several genera are used medicinally against fever or rheumatism or as a purgative. The bark of *Agonandra* provides a cure for colds and malaria[ii] and is used to kill mollusks, while the seeds produce an oil used to make a rubber substitute.   RKB

[i] Hiepko, P. Opiliaceae. *Species Plantarum* (2006).
[ii] Hiepko, P. Opiliaceae. *Flora Neotropica Monographs*. 82: 1–53 (2000).

OXALIDACEAE
WOOD SORREL, BERMUDA BUTTERCUP, OCA, STARFRUIT

This family is dominated by the weedy genus *Oxalis* but is also known for the Starfruit (*Averrhoa carambola*).

OXALIDACEAE. I *Oxalis adenophylla* (a) habit showing palmate leaves (×⅔); (b) gynoecium comprising 5 united carpels each with a capitate stigma (×4½); (c) flower with petals removed (×2); (d) androecium and gynoecium showing trimorphic heterostyly (i.e., stamens in 2 rows, each at different levels to the stigmas) (×3); (e) section of base of fruit (×2); (f) leaflet and short petiolule (×2).

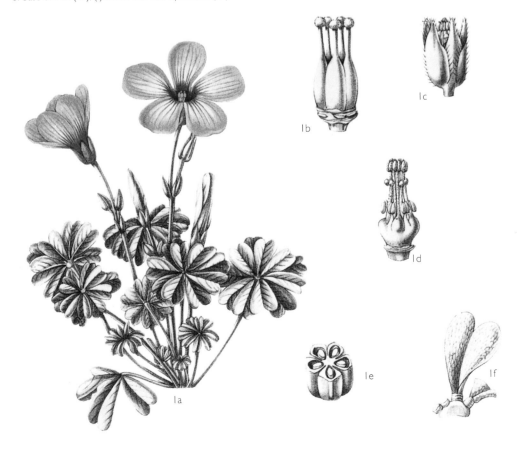

Genera 5 **Species** c. 800–900
Economic uses Tubers and leaves of some *Oxalis* species are eaten; some are ornamentals

Distribution Most of the family is native to the tropics, with many species at high altitudes, the remainder subtropical or sometimes temperate. The full extent of the natural distribution is blurred by the presence of several widespread and weedy species. The origin of *O. corniculata* is unknown, because it has become so widespread and established around the tropics.

Description Small trees, shrubs, climbers, or usually perennial herbs (rarely annual), with alternate leaves, often with underground storage bulbs, tubers, or fleshy roots. Plants are often clump-forming or spreading by stolons. The leaves are sometimes simple but often pinnately or palmately compound from a well-defined petiole, the leaflets articulate, and in many *Oxalis* species, folding downward at night and in cold weather. In some *Biophytum* species, the leaflets bend when touched. In a few *Oxalis* species (e.g., *O. bupleurifolia*), the ordinary leaves are replaced by phyllodes (leaflike petioles). The petioles are sometimes woody and persistent or rarely succulent

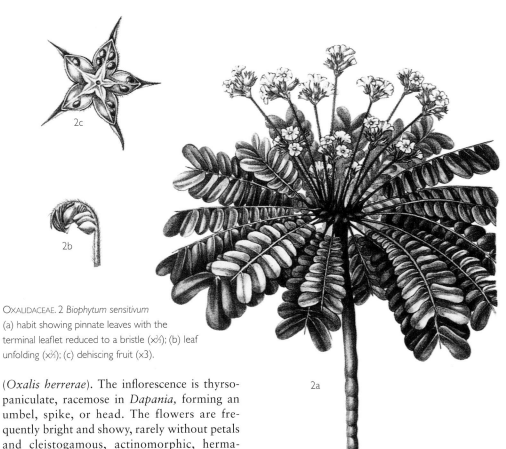

OXALIDACEAE. 2 *Biophytum sensitivum* (a) habit showing pinnate leaves with the terminal leaflet reduced to a bristle (x⅔); (b) leaf unfolding (x⅔); (c) dehiscing fruit (x3).

(*Oxalis herrerae*). The inflorescence is thyrso-paniculate, racemose in *Dapania*, forming an umbel, spike, or head. The flowers are frequently bright and showy, rarely without petals and cleistogamous, actinomorphic, hermaphrodite, but unisexual (plant androdioecious) in *Dapania*. The sepals are 5, persistent and overlapping. The petals are 5, usually twisted together in bud, and free or fused at the base, clawed at the base, often brightly colored. The stamens are 10, arranged in 2 whorls and connate at the base, the outer whorl of 5 lying opposite the petals. Sometimes the outer stamens are sterile (some *Averrhoa* species). The ovary is superior, consisting of 5 free (e.g., *Biophytum*) or fused (e.g., *Oxalis*) carpels with 5 free styles and capitate stigmas, and with 5 locules, each with 1 or 2 rows of ovules on axile placentas. Many *Oxalis* species display trimorphic heterostyly, i.e., flowers with long styles and medium and short stamens, medium styles and long and short stamens, and short styles and long and medium stamens. Fertile crosses are only possible between different flower-types. The Eurasian species *Oxalis acetosella* (Wood Sorrel) has flowers that exhibit cleistogamy in cold conditions. The fruit is a capsule. The seeds of some *Biophytum* and *Oxalis* species may have a fleshy aril at the base. The turgid inner cell layers of the aril turn inside out rapidly, separate from the testa, and the seed is explosively flung out.

Classification This is the largest family of the Oxalidales and is related to Connaraceae based on floral structure[i]. *Oxalis* (c. 800 spp.), with leaves with 1–20 or more leaflets and *Biophytum* (70 spp.), whose leaves possess a bristle representing the end leaflet, have capsular fruit. *Averrhoa* (2 spp.), with 5 or more leaflets and 3–7 ovules per locule, and *Dapania* (3 spp.) are lianas or small shrubs with 1-foliate or 3-foliate leaves and dehiscent fruit with 0–2

ovules per locule. *Sarcotheca* (11 spp.) are shrubs or trees with 1-foliate or 3-foliate leaves and indehiscent fruit with 0–2 ovules per locule. The tree genus *Averrhoa* is sometimes separated as the family Averrhoaceae.

Economic uses Some Andean *Oxalis* species have tubers that are boiled and eaten as a vegetable (Oca), and the leaves may be used in salads. The leaves of *O. acetosella* are sometimes used in salads instead of sorrel, and the bulbous stems of *O. pes-caprae* (Bermuda Buttercup) are sometimes used as a vegetable in southern France and North Africa. The tubers of the Mexican species *O. deppei* are also used as food and are cultivated in France and Belgium. A number of *Oxalis* species are cultivated as ornamentals, and some are troublesome weeds. The Starfruit (*Averrhoa carambola*) is eaten but also used as a bleaching agent[ii], and *A. bilimbi* (Cucumber Tree) produces somwhat sour fruit used in pickles and jams.   AC

[i] Matthews, M. L. & Endress, P. K. Comparative floral structure and systematics in Oxalidales (Oxalidaceae, Connaraceae, Brunnelliaceae, Cephalotaceae, Cunoniaceae, Elaeocarpaceae, Tremandraceae). *Bot. J. Linn. Soc.* 140: 321–381 (2002).
[ii] Coccucci, A. A. Oxalidaceae. Pp. 285–288. In: Kubitzki, K. (ed.), *The Families and Genera of Vascular Plants. VI. Flowering Plants. Dicotyledons: Celastrales, Oxalidales, Rosales, Cornales, Ericales.* Berlin, Springer-Verlag (2004).

PAEONIACEAE
PEONIES

A family comprising a single genus (*Paeonia*) of perennial, rhizomatous herbs or shrubs, with large showy flowers.

Distribution Native to northern-temperate regions, chiefly southern Europe, northern and east Asia, and western North America. Fifteen spp., 10 spp. of which are endemic, grow in China. One species, *P. coriacea*, occurs in northwestern Africa. The biogeography of the genus is discussed by Sang *et al*[ii].

Description Perennial herbs with fleshy roots to soft-wooded shrubs. The leaves are alternate, compound, without stipules, variously divided once or twice into 3 to many (*P. tenuifolia*) leaflets, segments, and lobes, the leaflets linear to broadly elliptical, entire, or lobed. The flowers are bisexual, actinomorphic, large, conspicuous, white, pink, red, purple, or yellow, solitary, and mostly terminal. Leafy floral bracts are present below the calyx. The sepals are (3–)5(–7), free, persistent, unequal. The petals are 5–9 or more. The stamens are numerous, arranged centrifugally. The ovary consists of (1)2–5(–8) free carpels borne on a conspicuous fleshy disk; ovules numerous and in 2 rows. The fruit consist of 1–5 large leathery divergent follicles. The seeds are large, fleshy, arillate, red initially, and turning black when mature.

Classification The genus *Paeonia* has attracted a great deal of interest, largely because of its great horticultural interest, and there is considerable disagreement between different authors[i,iii] as to the number of species and subsp. to be recognized, with new ones still being described[iii]. The taxonomic position of *Paeonia* is still not fully resolved. In the past, it has been placed in the Ranunculaceae and a relationship with *Glaucidium* has been suggested but now discounted on morphological and molecular grounds. It has been placed in its own order Paeoniales, but it is currently regarded as a member of the Saxifragales *sensu* APG II, within which some molecular studies suggest a link with the Crassulaceae[v].

Economic uses Species of *Paeonia* have long been the source of medicinally active compounds. The European *P. officinalis* has been

Genera 1 Species 25–33
Economic uses Garden ornamentals; some local medicinal use

iii e.g., Hong, D.-Y. & Pan, K.-Y. A revision of the *Paeonia suffruticosa* complex (Paeoniaceae). *Nordic J. Bot.* 19: 289–299 (1999).
iv Soltis, D. E., & Soltis, P. S. Phylogenetic relationships in Saxifragaceae *sensu lato*: A comparison of topologies based on 18S rDNA and *rbc*L sequences. *American J. Bot.* 84: 504–522 (1997).
v Fishbein M., Hibsch-Jetter C., Soltis, D. E., & Hufford, L. Phylogeny of Saxifragales (angiosperms, eudicots): analysis of a rapid, ancient radiation. *Syst. Biol.* 50: 817–847 (2001).

PANDACEAE

A small woody Old World family similar to Euphorbiaceae but always with drupaceous fruits.

Distribution Confined to the Old World. Asiatic *Galearia* ranges from Burma to the Solomon Islands. *Microdesmis* is distributed both in Southeast Asia (2 spp.) and tropical Africa (9 spp.), wheras *Panda* occurs only in West Africa (Liberia to Central African Republic).

Description Trees or shrubs with simple, alternate leaves on long shoots that can superficially resemble pinnate leaves, with simple hairs. The leaves are penninerved and entire or toothed at the margin, often unequal at the base, with small stipules and often 1 stipule inserted higher than the other (unequal nodes). The inflorescences are terminal or cauline, in long pendulous racemes (flagelliflory, *Galearia*), in axillary or supraaxillary (*Microdesmis*) or cauline to ramiflorous (*Panda*) fasciculate racemes. The bracts are minute. The flowers are actinomorphic, unisexual (plants dioecious). The calyx is 5-lobed or 5-toothed with imbricate segments (*Galearia* and *Microdesmis*) or truncate and open in bud (*Panda*). The petals are 5, longer than the calyx, imbricate or contorted in bud (*Microdesmis*), or valvate and concave with a median ridge, covering in staminate flowers 1 anther in each half (*Galearia*), or halfway between these states (*Panda*). The disk is absent or negligible. The stamens are 5–15, free, sometimes in 2 whorls, with the filaments of the whorls sometimes differing in length; anthers with longitudinal dehiscence. Pollen in Pandaceae is 3-colporate. A pistillode is present in the staminate flowers. The ovary is superior, 2- to 5-locular, with a single ovule per locule. Ovules lack an obturator and are anatropous in *Microdesmis* and *Galearia* (but may be orthotropous in *G. maingayi*) and orthotropous in *Panda*. The styles are short and the stigmas are simple, bifid, or variously lacerated. The fruit is a drupe with a woody, fleshy, or brittle pericarp. The seeds are ecarunculate and contain copious endosperm as well as broad, flat cotyledons. The large seeds of *Panda oleosa* are exclusively dispersed by elephants.

Classification Pandaceae contains the genera *Galearia* (6 spp.), *Microdesmis* (11 spp.), and *Panda* (1 sp.). Some authors[ii] also included the biovulate monospecific genus *Centroplacus*, but this is now excluded from the family (see Phyllanthaceae for *Centroplacus*). The pseudo-

PAEONIACEAE. 1 *Paeonia peregrina* (a) shoot with upper leaves and solitary terminal flower (x⅔); (b) young fruit comprising 3 follicles (x⅔); (c) dehisced fruit (x⅔); (d) vertical section of seed with copious endosperm and a small embryo (x2); (e) young leafy shoots (x⅓). 2 *P. wittmanniana* (a) cross section of carpel (x⅔); (b) vertical section of carpel (x⅔); (c) young fruit (x⅔). 3 *P. mascula* fruit of 5 follicles (x⅔). 4 *P. emodi* fruit (x⅔). 5 *P. tenuifolia* flower (x⅔).

cultivated and used in traditional herbal healing since the Middle Ages as an antispasmodic or a sedative. Radix Paeoniae is the dried root of *Paeonia lactiflora* from Siberia, Mongolia, and China. It has been cultivated in China for more than 2,500 years. It and 3 other Chinese species are used in Chinese herbal medicine. Many species, hybrids, and cultivars are grown as garden flowers, making attractive border plants. More than 20 spp. are grown in European gardens alone. The "tree paeonies," soft-woody shrubs such as *P. suffruticosa* (*P. moutan*), *P. delavayi*, *P. lutea*, and *P. potaninii*, which grow in some cases to 2.5 m, are also popular.   VHH

i Halda, J. J. & Waldick, J. W. The genus *Paeonia*. Portland, Oregon, Timber Press, (2004).
ii Sang, T., Crawford, D. J., & Stuessy, T. Chloroplast DNA phylogeny, reticulate evolution, and biogeography of *Paeonia* (Paeoniaceae). *American J. Bot.* 84: 1120–1136 (1997).

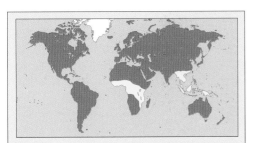

Genera 3 **Species** 18
Economic uses Cooking oil and wood for tools

pinnate appearance of the shoots had led to placement of members of this family in Burseraceae, Sapindaceae, and Anacardiaceae. Mostly, however, it has been variously assigned to Euphorbiaceae-Acalyphoideae as tribe Galearieae[iii,iv], or regarded as a separate family[v,vi]. Recent molecular studies confirm the family status of Pandaceae within the order Malpighiales in the fabid clade of the rosids[vii,viii]. *Galearia*+*Panda* are sister to *Microdesmis* despite the fact that pollen characters group *Galearia* and *Microdesmis* together. The main morphological differences of Pandaceae from Euphorbiaceae *sensu stricto* are the orthotropous ovules of the monotypic genus *Panda*, the divergent anatomy of the nodes and the absence of an obturator. Because of the similar petal shape in *Galearia* and biovulate *Dicoelia*, Webster moved *Dicoelia* from Euphorbiaceae-Phyllanthoideae[x] to Pandaceae[xi] and later to Euphorbiaceae-Acalyphoideae next to tribe Galearieae[iii]. *Dicoelia* has been shown to be a member of Phyllanthaceae.

Economic uses Seed oil of *Panda oleosa* is locally used in cooking. The hard wood of *Microdesmis* is used for tool handles and similar purposes. PH

i Forman, L. L. A synopsis of *Galearia* Zoll. & Mor. (Pandaceae). *Kew Bull.* 26: 153–165 (1971).
ii Govaerts, R., Frodin, D. G., & Radcliffe-Smith, A. *World checklist and bibliography of Euphorbiaceae (with Pandaceae)*. Vols. 1–4. Richmond, Royal Botanic Gardens, Kew (2000).
iii Webster, G. L. Synopsis of the genera and suprageneric taxa of Euphorbiaceae. *Ann. Missouri Bot. Gard.* 81: 33–144 (1994).
iv Radcliffe-Smith, A. *Genera Euphorbiacearum*. Richmond, Royal Botanic Gardens, Kew (2001).
v Pierre, J. B. L. Plantes du Gabon. *Bull. Soc. Linn. Paris* 2: 1249–1256 (1896).
vi Forman, L. L. The reinstatement of *Galearia* Zoll. & Mor. and *Microdesmis* Hook.f. in the Pandaceae. *Kew Bull.* 20: 309–321 (1966).
vii Wurdack, K. J. *et al.* Molecular phylogenetic analysis of Phyllanthaceae (Phyllanthoideae pro parte, Euphorbiaceae *sensu lato*) using plastid *rbc*L DNA sequences. *American J. Bot.* 91: 1882–1990 (2004).
viii Davis, C. C. *et al.* Explosive radiation of Malpighiales supports a mid-Cretaceous origin of tropical rain forests. *American Nat.* 165: E36–E65 (2005).
ix Nowicke, J. W. A palynological study of the Pandaceae. *Pollen et Spores* 26: 31–42 (1984).
x Webster, G. L. Conspectus of a new classification of the Euphorbiaceae. *Taxon* 24: 593–601 (1975).
xi Webster, G. L. The saga of the spurges: a review of classification and relationships in the Euphorbiales. *Bot. J. Linn. Soc.* 94: 3–46 (1987).

PAPAVERACEAE
POPPY, FUMITORY, BLEEDING HEART

A family of mainly herbaceous annuals or perennials, but with some shrubs, most of which produce latex. Current circumscriptions of the family now usually include Fumariaceae and Pteridophyllaceae.

Distribution The family occurs throughout northern temperate regions. Only a few are found south of the Equator: *Bocconia* occurs in Central and South America; a few species of *Corydalis* on mountains in East Africa; and *Papaver aculeatum* and the small genera *Phacocapnos*, *Cysticapnos*, *Trigonocapnos*, and *Discocapnos* in southern Africa. They often grow in open areas, mountain screes and disturbed ground.

PAPAVERACEAE. 1 *Eschscholzia californica* (a) leafy shoot and flower with 4 petals (×1); (b) flower bud with calyx forming cap (×1½); (c) fruit—a capsule dehiscing by 2 valves (×1); (d) cross section of fruit (×7). 2 *Glaucium flavum* (a) capsule (×⅔); (b) tip of opened capsule showing seeds and apical valve (×1½). 3 *Platystemon californicus* fruit—a group of follicles (×1½). 4 *Macleaya cordata* inflorescence (×⅔). 5 *Argemone mexicana* dehisced spiny capsule with seeds exposed (×⅔). 6 *Papaver dubium* (a) shoot with dissected leaves and solitary flowers (×⅔); (b) capsule dehiscing by apical pores (×1½); (c) vertical section of a capsule (×1½); (d) cross section of capsule (×1½).

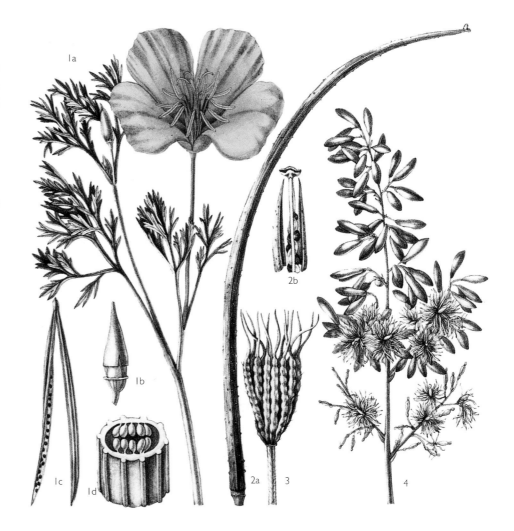

Genera 41 **Species** c. 760
Economic uses Opium, oils for soaps, ornamentals

Description Annual or perennial herbs, rarely shrubs[i] (e.g., *Dendromecon, Hunnemannia, Romneya*), geophytes (e.g., some *Corydalis*), or scramblers and climbers (e.g., *Dactylicapnos*) with alternate leaves. The stems, leaves, and other parts of many species contain a well-developed system of secretory canals that produce yellow, milky, or watery latex. All contain alkaloids. The leaves are entire but often lobed or deeply dissected (Papaveroideae), or usually pinnately or palmately compound (Fumarioideae, *Pteridophyllum*), without stipules. The inflorescence is usually racemose or cymose, sometimes a thyrse or solitary. The flowers are generally large and conspicuous, regular or disymmetric, bisexual and hypogynous (perigynous in *Eschscholzia*); in Fumarioideae, the flowers are of complex and unusual structure, whose derivation from the simpler papaveraceous type is, however, demonstrated by some of the smaller genera. In

Papaveroideae and *Pteridophyllum*, the sepals are 2, free, falling off before the flower opens, and the petals are usually in 2 whorls of 2, free, showy (absent in *Macleaya*) and often crumpled in the bud. The stamens are usually numerous, in several whorls, few in *Pteridophyllum*. The filaments are sometimes petaloid, while the anthers have 2 locules and dehisce longitudinally. The gynoecium consists of 2 to numerous fused carpels (separate except at the base in *Platystemon*). The ovary is superior and contains usually a single locule with intruding parietal placentas, as many as the number of carpels, and each bearing numerous ovules. The stigmas are as many as the carpels and are opposite to, or alternate with, the placentas. The fruit is a capsule, opening by valves or pores (follicular in *Platystemon*), and containing seeds with a small embryo and copious mealy or oily endosperm.

PAPAVERACEAE. 7 *Corydalis lutea* (a) shoot with much divided leaves and irregular flowers in a racemose infloresence (x⅔); (b) half flower showing spurred petal and elongate ovary (x4); (c) vertical section of fruit (x2). 8 *Pteridophyllum racemosum* (a) habit showing fernlike leaves (x⅔); (b) flower—the simplest form in this family (x2); (c) vertical section of ovary (x4). 9 *Fumaria muralis* (a) flowering shoot (x⅔); (b) half flower with spurred petal and stamens in 2 bundles (x3); (c) vertical section of bladderlike fruit (x6). 10 *Dicentra spectabilis* (a) leaf and inflorescence (x1); (b) flower dissected to show varied form of petals and stamens arranged in 2 bundles (x2).

In Fumarioideae, a more complex situation is found in *Hypecoum*, in which the 2 inner petals are prominently 3-lobed, with the middle lobe stalked and with an expanded, cuplike apex; the apices of the middle lobes wrap around the anthers and form a chamber into which the

pollen falls. Here again the stamens are 4, with nectar secreted at the filament bases. In *Dicentra*, there is further elaboration: all 4 petals are variably fused, particularly toward the apex; the outer petals are spurred at the base, and the apices of the inner petals are fused around the anthers. The stamens are arranged in 2 bundles opposite the inner petals; each bundle has a single filament that divides into 3 parts at the apex; the central division of each bears a complete anther, while the lateral divisions each bear half an anther. This complex structure appears to have evolved from the 4-staminate condition by the splitting of the stamens opposite the outer petals. The base of each compound filament is prolonged into the petal spur and secretes nectar there. In the other genera, only 1 of the outer petals is spurred, producing an unusual irregular flower; all have similar staminal arrangements to *Dicentra*, and all have a 2-carpellate ovary, which is usually-many-ovuled, though with 1 ovule in *Fumaria* and related genera. The fruit is usually a capsule, sometimes swollen and bladderlike; more rarely it is indehiscent, either a single-seeded nutlet, or many-seeded and breaking up into single-seeded indehiscent segments. *Ceratocapnos heterocarpa* has dimorphic fruits. The seed has a small embryo and fleshy endosperm.

Classification The largest genus[ii] is *Corydalis* (c. 320 spp.); *Papaver* is the second largest genus (c. 80 spp. or c. 130 if *Meconopsis* is included), followed by *Platystemon* (c. 60 spp.), *Fumaria* (c. 50 spp.), *Meconopsis* (48 spp.), *Argemone* (32 spp.), and *Glaucium* (25 spp.); other genera have fewer than 20 spp. The genera *Bocconia* (10 spp. from Asia and tropical America) and *Macleaya* (2 spp. from eastern Asia) are of interest as they possess apetalous flowers that are aggregated into compound racemes. *Pteridophyllum racemosum* has flowers with only 4 stamens and is sometimes regarded as constituting a separate family Pteridophyllaceae[iii]. The Papaveraceae is usually placed in its own order, Papaverales[iv], but extensive analysis of DNA sequences and morphology support inclusion in a broad Ranunculales[v]. Phylogeny of the Papaveraceae indicates a basal position for *Pteridophyllum*, and the Fumarioideae to be sister to the Papaveroideae[vi]. Phylogenetic analysis also supports the recognition of 3 families: Pteridophyllaceae; Fumariaceae, and Papaveraceae.

Economic uses Economically, the most important species in this family is *Papaver somniferum* (Opium Poppy), which yields opium. The seeds do not contain opium and are used in baking. They also yield an important drying oil. Likewise, the seeds of *Glaucium flavum* and *Argemone mexicana* yield oils that are important in the manufacture of soaps. Many species are cultivated as garden ornamental plants, e.g., *Dendromecon rigida* (Californian Bushy Poppy). *Eschscholzia californica* (Californian Poppy), *Papaver alpinum* (Alpine Poppy), *P. nudicaule* (Iceland Poppy), *P. orientalis* (Oriental Poppy), *Macleaya cordata* (Plume Poppy), and a few species of *Corydalis* and *Dicentra* (Dutchman's Breeches, Bleeding Heart). Some species of *Fumaria* are agricultural weeds, such as *Fumaria officinalis* (fumitory). AC

[i] Kadereit, J. W. Papaveraceae. Pp. 494–506. In: Kubitzki, K., Rohwer, J. G., & Bittrich, V. (eds), *The Families and Genera of Vascular Plants. II. Flowering Plants. Dicotyledons: Magnoliid, Hamamelid and Caryophyllid Families.* Berlin, Springer-Verlag (1993).
[ii] Lidén, M. Fumariaceae. Pp. 310–318. In: Kubitzki, K., Rohwer, J. G., & Bittrich, V. (eds), *The Families and Genera of Vascular Plants. II. Flowering Plants. Dicotyledons: Magnoliid, Hamamelid and Caryophyllid Families.* Berlin, Springer-Verlag (1993).
[iii] Lidén, M. Pteridophyllaceae. Pp. 556–557. In: Kubitzki, K., Rohwer, J. G., & Bittrich, V. (eds), *The Families and Genera of Vascular Plants. II. Flowering Plants. Dicotyledons: Magnoliid, Hamamelid and Caryophyllid Families.* Berlin, Springer-Verlag (1993).
[iv] Brummitt, R. K. *Vascular Plant Families and Genera.* Richmond, Royal Botanic Gardens, Kew (1992).
[v] Soltis, D. E. *et al.* Gunnerales are sister to other core eudicots: Implications for the evolution of pentamery. *American J. Bot.* 90: 461–470 (2003).

[vi] Hoot, S. B. *et al.* Data congruence and phylogeny of the Papaveraceae *s.l.* based on four data sets: *atp*B and *rbc*L sequences, *trn*K restriction sites, and morphological characters. *Syst. Bot.* 22: 575–590 (1997).

PARACRYPHIACEAE

The Paracryphiaceae comprises a single species, *Paracryphia alticola,* a small tree of uncertain affinity from New Caledonia. The leaves are subverticillate, simple, exstipulate. The flowers are small, inconspicuous, bisexual or staminate (plants andromonoecious) in branched terminal spikes. The perianth is poorly differentiated, with 4 tepals, the outermost larger and enclosing the others. The stamens are usually 8, the filaments somewhat flattened in the staminate flowers. The ovary is superior, of 8–15 laterally connate carpels, with axile placentation and 4 ovules per locule. The fruit is capsular, with the carpels separating at the base but remaining attached to the central column. The seeds are small and winged.

Paracryphia is an enigmatic plant. It combines both unspecialized wood anatomy and vegetative morphology with quite advanced reproductive features[i,ii]. It was located in the Theales by Cronquist (1981) and in the Paracryphiales in the superorder Theanae by Takhtajan (1997). It is treated as a Euasterid II but unplaced as to order by APG II, and molecular data suggest that it is sister to Sphenostemonales[iii]. Lundberg[iv] considers that *Sphenostemon, Paracryphia,* and *Quintinia* (formerly Escalloniaceae) belong in the same clade and should be merged into a single family Paracryphiaceae. VHH

[i] Dickison, W. C. & Baas, P. The morphology and relationships of *Paracryphia* (Paracryphiaceae). *Blumea* 23: 417–438 (1977).
[ii] Cameron, K. M. On the phylogenetic position of the New Caledonian endemic families Paracryphiaceae, Oncothecaceae, and Strasburgeriaceae: a comparison of molecules and morphology. *Bot. Rev.* 68: 428–443 (2003).
[iii] Savolainen, V. *et al.* Phylogeny of the eudicots: a nearly complete familial analysis based on *rbc*L gene sequences. *Kew Bull.* 55: 257–309 (2000).
[iv] Lundberg, J. Phylogenetic studies in the Euasterids II with particular reference to Asterales and Escalloniaceae. Chapter V: A well-resolved and supported phylogeny of Euasterids II based on a Bayesian inference, with special emphasis on Escalloniaceae and other incertae sedis. Uppsala, Sweden, *Acta Univ. Ups. Comprehensive Summaries of Uppsala Dissertations from the Faculty of Science and Technology* 676 (2001).

PARNASSIACEAE
GRASS OF PARNASSUS

A family of small perennials and 1 annual, of which almost 90% grow in China.

Distribution *Parnassia* occurs predominantly in China (63 spp., 49 endemic[i]), but *P. palustris* extends around the northern hemisphere and the *P. fimbriata* complex occurs in North America. *Lepuropetalon spathulatum* is disjunct, growing in the southeastern USA and Uruguay to central Chile, occurring on sandy soils and seasonally moist rock outcrops[ii].

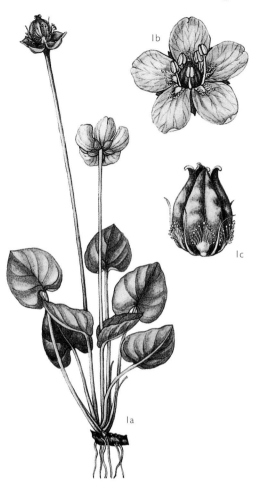

PARNASSIACEAE. *Parnassia palustris* (a) habit (x⅔); (b) flower (viewed from above) showing fanlike staminodes alternating with the stamens (x1½); (c) fruit—a capsule with persistent staminodes at base (x2).

Description Rosette-forming perennial herbs (*Parnassia*) or small, winter-growing annuals (*Lepuropetalon*) with alternate to almost opposite leaves. The basal leaves are petiolate (cauline leaves sometimes appearing sessile) and entire, the blade variously orbicular-ovate, reniform or spathulate, stipules absent. The inflorescence is a 1-flowered (rarely 2-flowered) scape. The flowers are 5-merous, appearing radially symmetrical. The sepals are usually 5, overlapping. The petals are usually 5 (*Lepuropetalon* sometimes without petals), minute to large, white or cream, margin entire, toothed or fringed. The stamens are 5, alternating with petals, free, and staminodes 5, aligned with petals, free. The ovary is superior or partially inferior, of 3–4(5) joined carpels, with 1 locule. The fruit is a dehiscent many-seeded capsule.

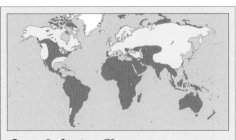

Genera 2 **Species** c. 70
Economic uses Garden ornamentals

Classification The inclusion of Parnassiaceae in Saxifragaceae (e.g., Cronquist 1981) or allied to Droseraceae (Diels 1906) was disputed on morphological grounds by several botanists. Parnassiaceae has been recognized as a family based on embryology[iii], floral anatomy[iv], and pollen structure[v]. The placement of Parnassiaceae in Celastrales was established by the use of combined molecular and morphological data[vi] and by the use of DNA sequence data[vii] approaches that also confirmed the relationship between *Parnassia* and *Lepuropetalon*. Re-examination of floral structure further supported the inclusion of the family in Celastrales[viii] based on joined carpels with stigmatic lobes along the junctions, and clusters of oxalate crystals in their floral parts. Species limits in *Parnassia* are based largely on floral structure, especially staminodes and secondarily on leaf shape.

Economic uses Some species are grown as ornamental curiosities. AC

i Gu, C. & Hultgard, U.-M. Parnassiaceae. Pp. 358–379. In: Wu, Z.-Y. & Raven, P. (eds), *Flora of China*. Beijing and Missouri Botanical Garden, St. Louis, Science Press (2001).

ii Simmons, M. P. Parnassiaceae. Pp. 291–296. In: Kubitzki, K. (ed.), *The Families and Genera of Vascular Plants. VI. Flowering Plants. Dicotyledons: Celastrales, Oxalidales, Rosales, Cornales, Ericales*. Berlin, Springer-Verlag (2004).

iii Sharma, V. K. Morphology, floral anatomy and embryology of *Parnassia nubicola*. Wall. *Phytomorphology* 18: 193–204 (1975).

iv Bensel, C. R. & Palser, B. F. Floral anatomy in the Saxifragaceae *sensu lato*. I Introduction, Parnassioideae and Brexoideae. *American J. Bot.* 62: 176–185 (1975).

v Hideux, M. J. & Ferguson, I. K. The stereostructure of the exine and its evolutionary significance in Saxifragaceae *sensu lato*. In: Ferguson, I. K. & Muller, J. (eds), *The Evolutionary Significance of the Exine*. London, Academic Press (1976).

vi Simmons, M. P. *et al.* Phylogeny of the Celastraceae inferred from phytochrome B gene sequence and morphology. *American J. Bot.* 88: 313–325 (2001).

vii Simmons, M. P. *et al.* Phylogeny of the Celastraceae inferred from 26S nuclear ribosomal DNA, phytochrome B, *rbc*L, *atp*B, and morphology. *Molec. Phylogen. Evol.* 19: 353–366 (2001).

viii Matthews, M. L. & Endress, P. K. Comparative floral structure and systematics in Celastrales (Celastraceae, Parnassiaceae, Lepidobotryaceae). *Bot. J. Linn. Soc.* 149: 129–194 (2005).

PASSIFLORACEAE
PASSION FLOWER, GRANADILLO

The many climbers in this family are famous for their extraordinary flowers and edible fruit. The poisonous leaves of *Passiflora* are also well known as a foodstuff for *Heliconius* butterflies.

Distribution The family is found throughout the tropics and subtropics, with some species reaching temperate regions. *Passiflora* (c. 525 spp.) has its center of diversity in wet tropical South America with only c. 20 spp. reaching southern Asia to New Zealand and with none native to Africa. By contrast, *Adenia* (c. 95 spp.) is most diverse in semiarid regions of Africa but extends to tropical Australia. *Basananthe* (25 spp.) occurs in tropical and southern Africa. *Paropsia* (11 spp.) occurs in tropical Africa, Madagascar, and eastern Malaysia. Other genera are small: 8 occur in Africa and Madagascar, 3 in tropical America, and *Paropsiopsis* in Malaysia and Papuasia.

Description Trees or shrubs (e.g., *Androsiphonia, Barteria, Paropsia, Viridivia,* some *Adenia*), woody or herbaceous climbers (e.g., *Passiflora,* some *Adenia*), annual or perennial herbs (e.g., *Basananthe*), sometimes with a swollen tuber or large swollen trunk (e.g., some *Adenia*), sometimes spiny, with alternate leaves. Tendrils derived from modified axillary inflorescences are present on climbing species. The leaves are simple or lobed, rarely compound, strongly 2-lobed in some *Passiflora,* petiolate and with small to large and foliose stipules present. Extrafloral nectaries are often present on the petiole or stem. The inflorescence is usually axillary, rarely terminal or cauliflorous, a 1 to many-flowered cyme,

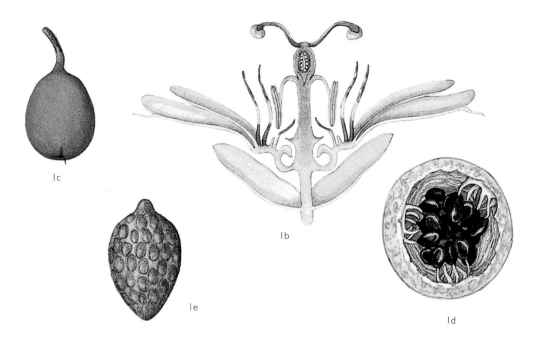

PASSIFLORACEAE. 1 *Passiflora caerulea* (a) twining stem with coiled tendrils, solitary flower with conspicuous filamentous corona and 5-lobed leaves subtended by leafy bracts (x⅔); (b) vertical section of flower with, from the base upward, subtending bracts, hollowed-out receptacle bearing spurred sepals, petals and filaments (the latter forming the corona) and a central stalk (androgynophore) with at the apex the ovary bearing long styles with capitate stigmas and at the base downward curving stamens (x1½); (c) fruit (x⅔); (d) cross section of fruit containing numerous seeds (x1½); (e) seed (x6).

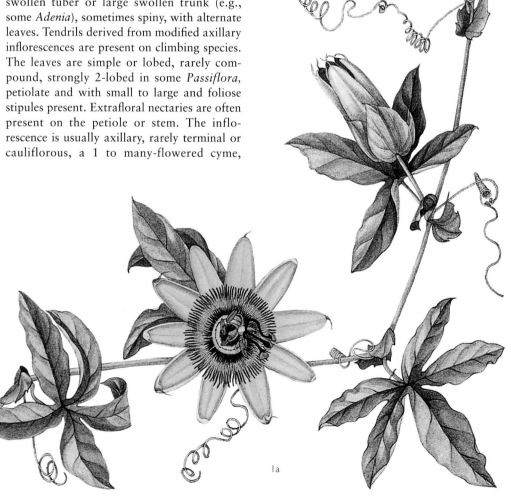

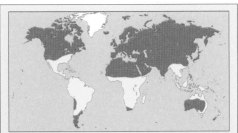

Genera 16   Species c. 700
Economic uses Ornamentals and edible fruit, the sap and leaves of most species are poisonous

growths on the leaves that look like *Heliconius* eggs. *Barteria* from Africa has a mutualistic association with *Pachysima* ants that keep insect predators away[ix].   AC

[i] Davis, C. C. & Chase, M. W. Elatinaceae are sister to Malpighiaceae; Peridiscaceae belong to Saxifragales. *American J. Bot.* 91: 262–273 (2004).
[ii] Chase, M. W. *et al.* When in doubt, put it in *Flacourtiaceae*: a molecular phylogenetic analysis based on plastid *rbc*L DNA sequences. *Kew Bull.* 57: 141–181 (2002).
[iii] Bernhard, A. Floral structure, development, and systematics in Passifloraceae and in *Abatia* (Flacourtiaceae). *Int. J. Plant Sci.* 160: 135–150 (1999).
[iv] Krosnick, S. E. & Freudenstein, J. V. Monophyly and floral character homology of Old World *Passiflora* (subgenus *Decaloba*: Supersection *Disemma*). *Syst. Bot.* 30: 139–152 (2005).
[v] Yockteng, R. & Nadot, S. Phylogenetic relationships among *Passiflora* species based on the glutamine synthetase nuclear gene expresssed in chloroplast (ncpGS). *Molec. Phylog. Evol.* 31: 379–396 (2004).
[vi] Muschner, V. C. *et al.* A first molecular phylogenetic analysis of *Passiflora* (Passifloraceae). *American J. Bot.* 90: 1229–1238 (2003).
[vii] Vanderplank, J. *Passion Flowers and Passion Fruit.* Massachusetts, MIT Press (1991).
[viii] Turner, J. R. G. Adaptation and Evolution in *Heliconius*: A defence of NeoDarwinism. *Ann. Rev. Ecol. Syst.* 12: 99–121 (1981).
[ix] Janzen, D. H. Protection of *Barteria* (Passifloraceae) by *Pachysima* ants (Pseudomyrmecinae) in a Nigerian Rain Forest. *Ecology* 53: 885–892 (1972).

rarely a raceme or fascicle. The flowers are actinomorphic, hermaphrodite, varying in size and shape. The sepals are (3–)5(–8), petals equalling the sepals or rarely absent. A distinctive extrastaminal ring of filaments, the corona, usually forms a bright and showy structure between the petals and stamens. The stamens are (4)5 or 8(–10), on the hypanthium or androgynophore (*Passiflora*). The ovary is superior, of (2)3(–5) carpels, usually forming a many-seeded berry.

Classification Passifloraceae is placed in the Malpighiales (APG II) and closely related to Malesherbiaceae and Turneraceae[i]. The family now includes some genera, on the basis of DNA sequence evidence[ii], that were previously in Flacourtiaceae, a case strengthened by floral structure and developmental similarities[iii]. The published phylogenies of the family focus on *Passiflora* and show *Hollrungia* and *Tetrapathaea* are embedded in it[iv]. Here they are treated as part of that genus. The most extensive study to date[v] shows 3 major clades in *Passiflora,* clades found in a slightly different relationship, and with sometimes weak support, in an earlier publication[vi]. It would appear generally that the more basal genera of the family are African in origin.

Economic uses Many *Passiflora* species are grown for their edible fruit, especially *P. edulis* (including var. *flavicarpa*), the basis of industrially produced juice, and *P. quadrangularis,* the Giant Granadilla, grown for its exceptionally large fruit. Other species grown for their fruit are the Banana Passion Fruit (*P. mollissima*), Fragrant Granadilla (*P. alata*), Red Granadilla (*P. coccinea*), Maypop (*P. incarnata*), Sweet Calabash (*P. maliformis*), Sweet Granadilla (*P. ligularis*), and Yellow Granadilla (*P. laurifolia*)[vii], although even more are utilised locally. Many species of *Passiflora* are also grown as ornamentals, *P. caerulea* being popular in Europe due to its frost hardiness, showy flowers, and colorful fruit. Some species of *Adenia* are grown as ornamental curiosities for their swollen stem (caudiciform) habit. *Passiflora* has a close interrelationship with *Heliconius* butterflies[viii], whose larvae feed on the leaves and are resistant to the numerous toxins the plants produce. Some *Passiflora* deter the butterflies due to out-

PEDALIACEAE
SESAME FAMILY

An Old World family of herbs to shrubs, or in Madagascar trees to 8 m, usually with a showy corolla and often with spines, hooks, or wings on the fruit.

Distribution Tropical and South Africa, Madagascar, tropical Asia, and northern Australia, often in dry, sandy, disturbed habitats and often weedy in other parts of the world. The larger genera are *Sesamum* (19 spp.) in Africa and the Indian subcontinent, *Pterodiscus* (13 spp.) in tropical and South Africa, and *Uncarina* (13 spp.) in Madagascar.

Description Perennial or somewhat robust annual herbs or sometimes shrubs or even (*Uncarina*) trees to 8 m. The leaves are opposite or sometimes alternate, simple and entire to variously lobed or occasionally pinnatifid, rarely digitately compound, usually with glands that produce mucilage when wetted. The flowers are solitary in leaf axils or in few-flowered axillary cymose clusters, usually with nectar glands at the base of the pedicels. They are zygomorphic and often large and conspicuous. The sepals are 5, connate. The corolla is campanulate or trumpet-shaped to cylindrical, often constricted at the base, with 5 lobes. The stamens are 4, 2 long and 2 short, sometimes also with a staminode as well, these inserted in the corolla tube. Pollen is 5- to 12-colpate. There is usually a nectariferous disk below the superior ovary, normally consisting of 2 fused carpels that are usually divided by false septa to give 4 locules, each with 1 to many

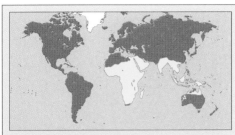

Genera 12–13   Species c. 74
Economic uses Sesame seeds and oil; occasional uses as vegetables; some medicinal products

ovules borne on an axile placenta. The fruit is a capsule or is indehiscent, often conspicuously provided with horns, spines, hooks, or wings.

Classification The irregular gamopetalous corolla and bicarpellate ovary indicate affinity with the Scrophulariaceae group of families, which is confirmed by molecular analyses. Most closely related families have usually been thought to be Martyniaceae and Trapellaceae, both of which are also characterized by conspicuous appendages of various sorts on the fruits. These families have sometimes been included in Pedaliaceae, but it seems preferable to separate Martyniaceae on the basis of its terminal inflorescences, inaperturate pollen, parietal placentation, and fruit structure, and Trapellaceae on the basis of its semi- to fully inferior ovary, probably non-homologous fruiting appendages, and 3-colpate pollen. Molecular cladograms also separate the groups[i], but this needs to be confirmed by extensive sampling.

Three tribes are recognized[ii], one containing only *Sesamothamnus* with petioles converted to spines and pollen in tetrads, the other two separated largely on the anther dehiscence.

Economic uses Sesame (*Sesamum indicum,* syn. *S. orientale*) has been grown as a crop plant since 5000 BCE for its seeds, which are edible and used to coat bread or ground into a meal. An important oil is also extracted from the seeds for use in cooking or to a lesser extent for soap, margarine, and cosmetics, with the residue used as a cattle feed. An extract from tubers of *Harpagophytum* is used to treat rheumatism, arthritis, and digestive disorders.   RKB

[i] Olmstead, R. G. *et al.* Disintegration of the Scrophulariaceae. *American J. Bot.* 88: 348–361 (2001).
[ii] Ihlenfeldt, H.-D. Pedaliaceae. Pp. 307–322. In: Kadereit, J. W. (ed.), *Families and Genera of Vascular Plants. VII. Flowering Plants. Dicotyledons: Lamiales.* Berlin, Springer-Verlag (2004).

PEGANACEAE

A family of dry-country plants, with leaves divided into strap-shaped lobes.

Distribution *Peganum harmala* is widespread from the Mediterranean to the dry areas of Asia (and widely introduced elsewhere), *P. nigellastrum* is in the region of the Gobi Desert, and *P.*

Genera 2 **Species** 4–5
Economic uses Occasional cultivation, toxic properties

mexicanum is in southern Texas and northern Mexico. *P. multisectum* from China may be distinct[i]. *Malacocarpus crithmifolius* occurs from the Caspian Sea to Turkmenistan and Afghanistan. They all occur in arid regions, up to an altitude of 1,850 m or more.

Description Stiff perennial herbs or subshrubs, erect to 1 m high or prostrate to 2 m across. The leaves are alternate, simple but characteristically divided into 2 to several strap-shaped (occasionally spathulate) lobes, each lobe c. 1–2 mm broad. Small stipules are sometimes present. The flowers are solitary on stiff pedicels in leaf axils and are actinomorphic, bisexual, and 4- to 5-merous. The sepals are strap-shaped and occasionally lobed like the leaves. The petals are free, to 1 cm long, white to yellow or greenish. The stamens are 12–15, inserted in 3 whorls on an extrastaminal nectary disk. The ovary is superior, syncarpous with 2–3 lobes and locules, with numerous ovules per locule. The fruit is a loculicidal capsule in *Peganum* and a fleshy berry in *Malacocarpus*.

Classification The 2 genera have usually been referred to the Zygophyllaceae, but they differ markedly in their alternate strap-shaped simple leaves (neither with separate leaflets nor reduced to a single leaflet), in their 2- to 3-carpellate ovary, and in some anatomical characters[ii]. In a molecular analysis[iii], *Peganum* separated together with *Tetradiclis* (see Tetradiclidaceae) and *Nitraria* (see Nitrariaceae) widely from Zygophyllaceae and were placed in the Sapindales. A further anatomical study[iv] has concluded that extensive further work is needed before the family position is clear.

Economic uses *Peganum harmala* is occasionally cultivated as an ornamental or curiosity. It is said to be highly toxic to cattle. RKB

[i] Ma, J., Wang, X., & Zhao, S. A study on the seeds micromorphological characteristics of *Peganum* from China northwest and its taxonomic and ecological significance. *J. Wuhan Bot. Res.* 15: 323–327 (1997).
[ii] Sheahan, M. C. & Cutler, D. F. Contribution of vegetative anatomy to the systematics of the Zygophyllaceae. *Bot. J. Linn. Soc.* 113: 227–262 (1993).
[iii] Sheahan, M. W. & Chase, M. W. A phylogentic analysis of Zygophyllaceae R.Br. based on morphological, anatomical and *rbcL* DNA sequence data. *Bot. J. Linn. Soc.* 122: 279–300 (1996).

[iv] Ronse-Decraene, L. P., De Laet, J., & Smets, E. F. Morphological studies in the Zygophyllaceae: 2. The floral development and vascular anatomy of *Peganum harmala. American J. Bot.* 83: 201–215 (1996).

PELLICIERACEAE

A monotypic family. *Pelliciera rhizophorae* occurs in the mangrove swamps of central and northern South America and the Galápagos Islands. This species forms small, strongly

Genera 1 **Species** 1
Economic uses None

buttressed trees with spirally arranged leaves. The leaves are leathery, simple, elliptical, asymmetric (one side of the lamina notably larger than the other) and entire, without stipules. The inflorescence is of single axillary flowers arising in the last 1–3 leaf axils before the terminal bud[i]. The flowers are sessile, appearing terminal, white, or flushed pink and showy, with 2 large bracts, pentamerous, the sepals short and elliptical but the petals long (4–6 cm), the sides reflexed to give a starlike impression. The stamens are erect, with the anthers reflexed at dehiscence, surrounding the style like those of Tetrameristaceae. The ovary is superior, of 2 fused carpels, each forming a locule, the whole about 6 cm long. The fruit is c. 10 cm long, subglobose with a terminal beak containing 1 seed in a spongy matrix surrounded by a leathery coat.

The genus has usually been placed in the Theaceae (Dahlgren, 1983) or as a separate family in the Theales (Cronquist, 1981) but molecular evidence supports affinity with the Balsaminoid group of Ericales (cf APG II) with close relationship to Marcgraviaceae and Tetrameristaceae[ii,iii] in which family the genus is sometimes included[iv]. AC

[i] Kubitzki, K. Pellicieraceae. Pp. 297–299. In: Kubitzki, K. (ed.), *The Families and Genera of Vascular Plants. VI. Flowering Plants. Dicotyledons: Celastrales, Oxalidales, Rosales, Cornales, Ericales.* Berlin, Springer-Verlag (2004).
[ii] Anderberg, A. A., Rydin, C., & Källersjö, M. Phylogenetic relationships in the order Ericales *s.l.*: analyses of molecular data from five genes from the plastid and mitochondrial genomes. *American J. Bot.* 89: 677–687 (2002).
[iii] Luna, I. & Ochoterena, H. Phylogenetic relationships of the genera of Theaceae based on morphology. *Cladistics* 20: 223–270 (2004).
[iv] Schönenberger, J., Anderberg, A. A., & Sytsma, K. J. Molecular phylogenetics and patterns of floral evolution in the Ericales. *Int. J. Plant Sci.* 166: 265–288 (2005).

PENAEACEAE

A small family of heathlike shrubs from the Cape region of South Africa.

Distribution The family is found only in south and southwestern parts of the Cape Province of South Africa, confined to the fynbos vegetation.

Genera 3
Species c. 20
Economic uses Some ornamental value

Description Small, often ericoid, evergreen shrubs or subshrubs. The leaves are small, opposite, often sessile, entire, with or without minute stipules. The flowers are actinomorphic, hermaphrodite, borne singly, but often crowded in the upper leaf axils; often brightly colored, pink, purple, white, or yellow, and subtended by 2 to 4 leafy and colored bracts. The perianth consists of one whorl of 4 sepals[i] inserted on the rim of the often elongate hypanthial tube; the lobes valvate; the petals are effectively absent. The stamens are 4 alternating with the sepals, the filaments short. The ovary is superior, of 4 carpels alternating with the calyx lobes, and 4 locules, each with 2–4 or more ovules; placentation is axile, basal, or apical. The fruit is a loculicidal capsule with 1 or 2 seeds per locule.

Classification There is general consensus today on the inclusion of Penaeaceae in the Myrtales. Molecular analyses have suggested a sister group relationship with Oliniaceae (q.v.) and that the Rhynchocalycaceae is sister to the Oliniaceae/Penaeaceae[i].

Economic uses Some species, e.g., *Brachysiphon fucatus* and *Endonema retzioides*, are grown as ornamentals. The gum

1a   4a

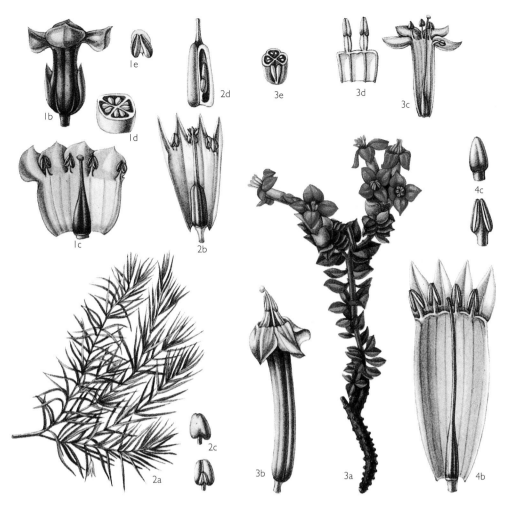

PENAEACEAE. 1 *Brachysiphon fucatus* (a) leafy shoot and flowers (×⅔); (b) flower (×3); (c) flower opened out showing stamens attached to the calyx (×3); (d) cross section of ovary (×6); (e) anthers (×6). 2 *Penaea ericifolia* (a) shoot (×⅔); (b) flower opened out (×4); (c) stamens front (lower) and rear (upper) views (×8); (d) ovary with part of wall removed (×6). 3 *P. squamosa* (a) flowering shoot (×⅔); (b) flower (×2½); (c) flower opened out (×1½); (d) stamens (×3); (e) cross section of ovary with part of vertical wall cut away to show basal ovules (×4). 4 *Glischrocolla formosa* (a) flowering shoot (×⅔); (b) flower opened out (×2); (c) stamens back (upper) and front (lower) views (×3).

obtained from *Penaea mucronata* and *Saltera sarcocolla* (known as Sarcocolla) has been used locally in medicine.   VHH

PENNANTIACEAE

The Pennantiaceae is a small family of trees and shrubs (rarely woody climbers) with alternate, simple entire or dentate leaves, and terminal inflorescences on leafy shoots, or more often ramiflorous or cauliflorous (*P. baylisiana*). It comprises a single genus, *Pennantia*, with 4 spp. from eastern Australia, Norfolk Island, Three Kings Islands (*P. baylisiana*, known in the wild only from a single tree on Great Island and Critically Endangered), and New Zealand[i]. The

flowers tend to be functionally unisexual (plants ± dioecious), actinomorphic, usually 5-merous. The carpels are apparently 3 and 1-locular with a solitary ovule. The fruit is a single-seeded drupe.

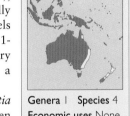

Genera 1 | Species 4
Economic uses None

The genus *Pennantia* has previously been included in the Icacinaceae but according to Kårehed[ii] falls within a broadly defined Apiales with which it shares paracytic stomata.   VHH

PENTADIPLANDRACEAE
J'OUBLIE, BRAZZEIN

A monotypic family of shrubs, sometimes developing into lianas, that grows in tropical western Africa. The leaves are simple, elliptical, with entire margins and rudimentary stipules. The inflorescences are axillary or terminal and can be male, female, hermaphrodite, or mixed on the same plant (plants polygamous). The flowers are 5-merous; the sepals are joined at the base only, saccate and overlapping in bud; the petals are free, thickened, and concave at the base, but

thin and flat at the apex, the basal portions held to each other by fringes of interwoven hairs. The androgynophore is thick, supporting (9)10(–13) stamens (staminodes in female flowers), joined at the base of the filaments. A prominent gynophore is sometimes present, supporting the 4- to 5-locular superior ovary (reduced to a rudiment

Genera 1 | Species 1
Economic uses
Protein-based sweetener

in male flowers), capped with a style and 4- to 5-lobed stigma. The fruit is a many-seeded berry enclosed in a sweet pulp.

Pentadiplandra brazzeana was first placed in the Tiliaceae but transferred to Capparaceae by 1897[i] and then transferred to its own family by Hutchinson (1973). The family is included in Capparales based on DNA sequence evidence, but its precise relationship is somewhat uncertain, the most likely sister group being Tovariaceae[ii]. This monotypic family has come to the fore as a source of a potent, protein based sweetener, making the fruit popular among Gabonese locals where the plant is native. The active constituent, brazzein, was isolated[iii] in the USA and subsequently patented and can now be produced in transgenic maize[iv].   AC

i Bayer, C. & Appel, O. Pentadiplandraceae. Pp. 329–331. In: Kubitzki, K. & Bayer, C. (eds), *The Families and Genera of Vascular Plants. V. Flowering Plants. Dicotyledons: Malvales, Capparales and Non-betalain Caryophyllales*. Berlin, Springer-Verlag (2002).
ii Rodman, J. E. *et al*. Parallel evolution of glucosinolate biosynthesis inferred from congruent nuclear and plastid gene phylogenies. *American J. Bot.* 85: 997–1007 (1998).
iii Ming, D. & Hellekant, G. Brazzein, a new high-potency thermostable sweet protein from *Pentadiplandra brazzeana* Baill. *FEBS Letters* 355: 106–108 (1994).
iv Lamphear B. J. *et al*. Expression of the sweet protein brazzein in maize for production of a new commercial sweetener. *Plant Biotech. J.* 3: 103–114 (2005).

PENTAPETACEAE
DOMBEYA FAMILY

A family of tropical trees and shrubs, rarely herbs, formerly included in a more broadly circumscribed Sterculiaceae (q.v.).

Distribution The family is centered in Africa and neighboring islands of Madagascar, the Mascarenes, and St. Helena (*Trochetiopsis*), with a secondary center (7 genera) in India to southwestern China. It is absent from the New World. *Dombeya* (c. 200 spp.) is found in deciduous thicket and scrub from sea-level to 2,000 m in Africa and, principally, Madagascar. A similar habitat is occupied by many of the other genera of the family.

Description Small trees, less usually medium-sized trees or shrubs, very rarely (*Pentapetes*) herbs. The bark is usually fibrous, and the

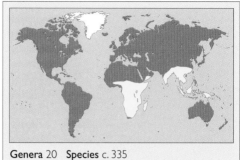

Genera 20 **Species** c. 335
Economic uses Timber; ornamental flowers

indumentum is mainly of stellate hairs, as in most core Malvales. The leaves are alternate, simple, rarely lobed, usually palmately veined and cordate, less usually pinnately nerved and cuneate (*Nesogordonia*); the margins are serrate, less usually entire; the petiole is pulvinate at base and apex; and the linear stipules are moderately persistent to caducous. The inflorescence is usually an axillary thyrse or rarely cymose. An epicalyx of 3 bracts is usually present but often caducous. The calyx is deeply divided into 5 valvate sepals. The petals are 5, non-nectariferous, usually imbricate, pink, white, or yellow, ± distinctly stalked with a flat blade. An androgynophore is present in *Nesogordonia*. The androecium is uniseriate, the filaments united into a ± short cylinder. The fertile stamens are (5–)10–20(–30), joined by 5 linear-ligulate staminodes, each alternating with (1)2–4 fertile stamens and each as long as or longer than them. The pollen is spheroidal, spiny, and often triporate. The ovary is superior, (2–)5(–10) locular, with 2–4 or numerous anatropous ovules per locule and generally ovoid to globose and covered in stellate hairs or scales. Placentation is axile. The single style is usually 5-branched at the apex. The fruit is a loculicidally dehiscent 5-valved capsule, usually leathery or subwoody, with several seeds, which, in a few genera, are winged, and in which the cotyledons in all genera (except *Nesogordonia*, *Pterospermum*, *Schoutenia*, and *Burretiodendron*) are bifid.

Classification Pentapetaceae was formerly included in Sterculiaceae as the Dombeyeae or Dombeyoideae, together with the genera here included in Byttneriaceae, Helicteraceae, and Sterculiaceae *sensu stricto*. These 4 families fall in core Malvales (q.v.) together with Malvaceae, Bombacaceae, Tiliaceae, Sparrmanniaceae, and Brownlowiaceae. Pentapetaceae are recognized by their uniseriate androecium in which fertile stamens usually alternate with longer, ligulate staminodes that are all joined at the base into a ± short tube. Pentapetaceae is also distinctive in the presence of a 3-bracted epicalyx, although this is often caducous, and in the marcescent petals, spiny pollen, and bifid cotyledons. While these characters are shared by the majority of genera, which we call core Pentapetaceae, *Nesogordonia*, *Schoutenia*, *Pterospermum*, and *Burretiodendron* are

anomalous in lacking several of these characters and owe their placement in the family mainly to molecular evidence.

Economic uses *Nesogordonia papavarifera* is a significant timber tree of West Africa, similarly for *Pterospermum* in Southeast Asia, while several species of *Dombeya* are trees cultivated in the tropics on a small scale for their ornamental flowers.   MRC

PENTAPHRAGMATACEAE

A poorly known family of tropical herbs with asymmetrical leaves.

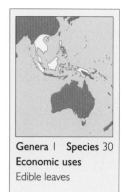

Genera 1 **Species** 30
Economic uses Edible leaves

Distribution The sole genus *Pentaphragma* comprises some 30 spp. confined to Southeast Asia and western Malaysia.

Description Perennial and somewhat fleshy herbs. The leaves are 2-ranked, distinctly asymmetrical at the base, exstipulate, with the margins serrate, dentate, or entire. The flowers are actinomorphic, usually hermaphrodite, in helicoid cymes. The sepals are 5, usually unequal, free. The petals are 4 or 5, usually fleshy or cartilaginous, usually united, deeply lobed, rarely free. The stamens are 4 or 5, adnate to the corolla tube, extrorse, basifixed. The ovary is inferior, syncarpous, with 2–3 carpels; locules 2 or 3 with many ovules per locule; the style is short and thick, with a large stigma. The fruits are baccate with the apical persistent perianth; seeds minute.

Classification The Pentaphragmataceae has been associated by several authors with the Campanulaceae or even included in it. It is placed in the Campanulales by Takhtajan (1997), who regarded it as nearest to the ancestral stock of that order. Phylogenetic analyses of molecular data suggest the Pentaphragmataceae is linked to the Campanulaceae/Stylidiaceae clade[i] in the Asterales while an analysis of phenetic characters and 3 nucleotide sequence data sets of the currently recognized families in the Asterales suggests that Rousseaceae together with Pentaphragmataceae and Campanulaceae is the sister group to the rest of the Asterales[ii], but the support is not strong.

Economic uses The leaves of some species may be eaten.   VHH

[i] Bremer, B. *et al.* Phylogenetics of asterids based on 3 coding and 3 non-coding chloroplast DNA markers and the utility of non-coding DNA at higher taxonomic levels. *Molec. Phylog. Evol.* 24: 274–301 (2002).
[ii] Lundberg, J. & Bremer, K. A Phylogenetic Study of the Order Asterales using one morphological and three molecular data sets. *Int. J. Plant Sci.* 64: 553–578 (2003).

PENTAPHYLACACEAE

A tropical and subtropical family of evergreens, previously included in the Theaceae.

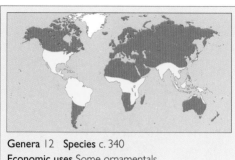

Genera 12 **Species** c. 340
Economic uses Some ornamentals

Distribution *Ternstroemia* is distributed in both the neotropics, palaeotropics, and subtropics and *Cleyera* from Central America to Japan and India. Other genera are American (*Freziera*, *Symplococarpon*) or Asian and Malaysian (*Pentaphylax*, *Anneslea*, *Adinandra*, *Eurya*, *Euryodendron*), African (*Balthasaria*), Macaronesian (*Visnea*), or from New Guinea (*Archboldiodendron*). They grow in subtropical and warm-temperate hill and mountain habitats.

Description Trees or shrubs with unicellular indumentum and usually evergreen leaves. The leaves are simple and often leathery, distichous or spirally arranged, the margins entire to crenate or serrate, stipules absent. The inflorescence is axillary, rarely appearing terminal, of single or fascicled pedicellate flowers, rarely a raceme of up to 15 flowers[i]. The flowers are usually hermaphrodite but sometimes unisexual (*Eurya*, *Freziera*) (plants dioecious), 5-merous, the 5 sepals free or fused, thick and persistent in fruit, the 5 petals free or fused at the base. The stamens are 5 to many, free or joined, sometimes to the petals, the filaments often wide and flat or thickened. The ovary is superior (inferior in *Anneslia* and *Symplococarpon*), of (2)3–5(6) fused carpels, each forming a locule bearing 1 to many ovules. The fruit is usually a berry but sometimes a loculicidal or irregularly dehiscent capsule or drupe.

Classification *Pentaphylax* has been placed in Pentaphylacaceae (e.g., Cronquist 1981, Dahlgren 1983), the Ternstroemiaceae (Bentham 1862), or included in Theaceae (Dalla Torre and Harms 1901). The remaining genera of Pentaphylacaceae are usually placed in Theaceae or in Ternstroemiaceae. Phylogenetic analysis of DNA sequences supported the 4 sampled genera (*Pentaphylax*, *Ternstroemia*, *Eurya*, and *Cleyera*) as forming a monophyletic group[ii] that could be recognized as a single family. There are several morphological differences between *Pentaphylax* and other genera in the group that would support recognition of 2 families rather than one. Whether Pentaphylacaceae or Ternstroemiaceae is the correct name to apply to the broader family concept has been

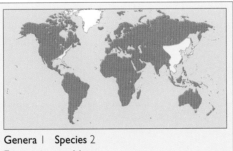

PENTAPHYLACACEAE. 1 *Symplocarpon hintonii* (a) half flower showing inferior ovary (×6); (b) dry indehiscent fruit (×1). 2. *Eurya macartneyi* (a) leafy shoot with female flowers (×⅔); (b) female flower (×6); (c) male flower (×6); (d) epipetalous stamens with lengthwise dehiscence (×12). 3 *E. japonica* fruit—a berry.

a matter of debate and both are used in recent literature. The component genera are divided into 3 tribes[i]: Pentaphylaceae (*Pentaphylax* 1 sp.); Ternstroemieae (*Ternstroemia*, c. 100 spp. and *Anneslea*, 3 spp.); and Freziereae (*Adinandra* 80 spp., *Cleyera* 8 spp., *Archboldiodendron* 1 sp., *Eurya* 50–100 spp., *Freziera* 57 spp., *Symplococarpon* 9 spp., *Euryodendron* 1 sp., *Visnea* 1 sp., and *Balthasaria* 1 sp.). The genera of Sladeniaceae (*Sladenia* and *Ficalhoa*) are also sometimes included in the family[iii].

Economic uses Some species of *Eurya*, *Ternstroemia*, and *Cleyera* are used as ornamentals. Some species are threatened in the wild[iv].   AC

i Weitzman, A. L., Dressler, S., & Stevens, P. F. Ternstroemiaceae. Pp. 450–462. In: Kubitzki, K. (ed.), *The Families and Genera of Vascular Plants. VI. Flowering Plants. Dicotyledons: Celastrales, Oxalidales, Rosales, Cornales, Ericales*. Berlin, Springer-Verlag (2004).
ii Anderberg, A. A., Rydin, C., & Källersjö, M. Phylogenetic relationships in the order Ericales *s.l.*: analyses of molecular data from five genes from the plastid and mitochondrial genomes. *American J. Bot.* 89: 677–687 (2002).
iii Schönenberger, J., Anderberg, A. A., & Sytsma, K. J. Molecular phylogenetics and patterns of floral evolution in the Ericales. *Int. J. Pl. Sci.* 166: 265–288 (2005).
iv Vega, I. L., Ayala, O. A., & Contreras-Medina, R. Patterns of diversity, endemism and conservation: an example with Mexican species of Ternstroemiaceae Mirb. ex DC. (Tricolpates; Ericales). *Biodiversity & Conservation* 13: 2723–2739 (2004).

PENTHORACEAE

Genera 1	Species 2
Economic uses None	

The Penthoraceae comprises a single genus, *Penthorum*, containing 2 morphologically similar species of somewhat fleshy, fibrous, stoloniferous, or rhizomatous herbs from eastern and Southeast Asia (*P. chinense*) and eastern North America (*P. sedoides*). The leaves are alternate, shortly petiolate or sessile, the lamina lanceolate to elliptical, acuminate. The flowers are bisexual, actinomorphic, small and numerous in terminal and axillary scorpioid cymes. The sepals are 5(–8), yellowish or greenish. The petals are 5(–8) or absent. The stamens are 10–16. The ovary is semi-inferior to superior with 5(–8) carpels, partly connate below, white in flower, with free central placentation and numerous ovules. The fruit is a capsule, each carpel rostrate and splitting circumscissilely at the base; seeds are small, numerous.

Penthorum has been regarded as closely related to the Saxifragaceae and sometimes included in it as a subfamily, although recent studies sug-

gest that it should be kept as a separate family within the Saxifragales clade, where it is most closely related to Haloragaceae[i,ii,iii].   VHH

i Soltis, D. E. & Soltis, P. S. Phylogenetic relationships in Saxifragaceae *sensu lato*: A comparison of topologies based on 18S rDNA and *rbc*L sequences. *American J. Bot.* 84: 504–522 (1997).
ii Soltis, D. E. *et al*. Angiosperm phylogeny inferred from 18S ribosomal DNA sequences. *Ann. Missouri Bot. Gard.* 84: 1–49 (1997).
iii Fishbein, M., Hibsch-Jetter, C., Soltis, D. E., & Hufford, L. Phylogeny of Saxifragales (angiosperms, eudicots): analysis of a rapid, ancient radiation. *Syst. Biol.* 50: 817–847 (2001).

PERIDISCACEAE

A poorly understood family comprising 2 monotypic genera from northern South America (*Peridiscus* and *Whittonia*) and, apparently, *Soyauxia* (7 spp.) from tropical West Africa. They are deciduous trees or erect shrubs with alternate simple leaves, the laminas coriaceous and with entire

Genera 3	Species 9
Economic uses None	

margins. The flowers are small, fragrant, actinomorphic, hermaphrodite, in axillary racemose or fasciculate inflorescences. The sepals are 4–5(6) in *Peridiscus* or 7 in *Whittonia*, sometimes the inner ones petaloid. The petals are absent except in *Soyauxia*. The stamens are numerous, inserted on or around the fleshy disk, the filaments free or slightly connate at the base. The ovary is superior, more or less submerged in the disk in *Whittonia*, of 3–4 fused carpels, 1-locular; the placentation is apical with 6–8 pendulous ovules. The fruit is a drupe (*Peridiscus*) or a capsule (*Soyauxia*), with a single seed per fruit.

The affinities of this somewhat heterogeneous family are still somewhat uncertain. *Peridiscus* and the poorly known *Whittonia*, which was later included, have been placed in the Violales, and near the Flacourtiaceae by Takhtajan (1997). Molecular evidence suggested a placement in the Malpighiales (see APG II), although this has been queried[i]. More recent studies not only indicate that it belongs in the Saxifragales[ii] but add *Soyauxia*, which was previously thought to belong to the Medusandraceae.   VHH

i Stevens, P. F. *Angiosperm Phylogeny Website*. Version 6, May (2005).
ii Davis, C. C. & Chase, M. W. Elatinaceae are sister to Malpighiaceae; Peridiscaceae belong to Saxifragales. *American J. Bot.* 91: 262–273 (2004).

PETIVERIACEAE
BLOODBERRY

A small family of tropical, subtropical, and warm temperate largely South American species that is distinguished from Phytolaccaceae *sensu lato* by its single carpel.

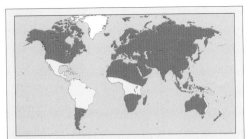

Genera 9 **Species** 36
Economic uses Invasive weed, sometimes ornamental, local medicinal uses

Distribution *Gallesia integrifolia* (1 sp.) is from Bolivia, Brazil, Ecuador, and Peru; *Hilleria* (4 spp.) is endemic to tropical and warm-temperate South America but *H. latifolia* extending to western tropical Africa; *Ledenbergia* (3 spp.) is from Ecuador and Venezuela through to Mexico; *Petiveria alliacea* is widespread and frequent in central, tropical, and warm-temperate South America; *Rivina humilis* (1 sp.) occurs in tropical and subtropical America and is a widespread weed in Africa and Asia; *Schindleria* (6 spp.) is found in Peru and Bolivia; *Seguieria* (16 spp.) is widespread from warm-temperate through tropical South America. *Monococcus echinophorus* (1 sp.) is endemic to rain forests of New South Wales and Queensland in Australia. *Trichostigma* (3 spp.) grows in tropical to temperate Americas.

Description Shrubs, climbers, perennial and annual herbs, plants often smelling of onions. The leaves are entire, the stipules absent or small, sometimes tuberculate or prickly. The flowers are hermaphrodite, rarely unisexual (*Monococcus*), in racemes, panicles, or spikes. The calyx is small, of 4–5 segments, sometimes petaloid or sometimes enlarging in fruit; petals absent. The stamens are 4–25, when isomerous then alternating with the calyx. The ovary is superior with a single carpel, 1-locular, single-seeded. The fruit is a fleshy berry (*Rivina, Trichostigma*), samara-like (*Gallesia, Seguiera*), an achene (*Monococcus, Petiveria*), or an utricle (*Hilleria, Ledenbergia, Schindleria*).

Classification This family is usually placed in Phytolaccaceae *s.l.*[i], but morphological[ii] and molecular[iii] data support its recognition at family rank. The genera form a coherent group both geographically and morphologically, but *Monococcus* stands out as distinct on both counts and may be misplaced here.

Economic uses Some species are weeds. *Rivina humilis* is a problem invasive outside its natural range but is also used as an ornamental in North America. A few species have local medicinal value. AC

[i] Rohwer, J. G. Phytolaccaceae. Pp. 506–515. In: Kubitzki, K., Rohwer, J. G., & Bittrich, V. (eds), *The Families and Genera of Vascular Plants. II. Flowering Plants. Dicotyledons: Magnoliid, Hamamelid and Caryophyllid Families*. Berlin, Springer-Verlag (1993).
[ii] Brown, G. K. & Varadarajan, G. S. Studies in Caryophyllales I: Re-evaluation of classification of Phytolaccaceae *s.l. Syst. Bot.* 10: 49–63 (1985).
[iii] Cuénoud, P. *et al.* Molecular phylogenetics of Caryophyllales based on nuclear 18S and plastid *rbc*L, *atp*B and *mat*K DNA sequences. *American J. Bot.* 89: 132–144 (2002).

PHELLINACEAE

A single genus (*Phelline*) of 10–12 spp. of evergreen trees or shrubs, endemic to New Caledonia, often placed in the Aquifoliaceae. The leaves are alternate, simple, more or less crowded toward the tips of the branches, exstipulate, the lamina margin entire. The flowers are small, unisexual (plants dioecious), in axillary racemes or panicles. The sepals are 4–6, more or less connate at the base; the petals are 4–6, fleshy, free, valvate; in the staminate flowers, there are 4–6 stamens opposite the sepals; in the pistillate flowers, staminodes are present. The ovary is superior, of 2–5 fused carpels and 2–5 locules; placentation apical, with 1 ovule per locule. The fruit is a drupe with 2–5 pyrenes. The seeds have copious endosperm.

The Phellinaceae has been included in the Aquifoliaceae by several authors; Takhtajan (1997) placed it next to the Aquifoliaceae in the Icacinales. Molecular evidence places the family surprisingly in the Asterales as a monophyletic clade along with the Alseuosmiaceae and Argophyllaceae, and this is confirmed by a later combined morphological and molecular analysis[i]. However, the relationships between these 3 families are not yet fully resolved[ii]. VHH

[i] Kårehed, J., Lundberg, J., Bremer, B., & Bremer, K. Evolution of the Australasian families Alseuosmiaceae, Argophyllaceae, and Phellinaceae. *Syst. Bot.* 24: 660–682 (1999) [2000].
[ii] Bremer, B. *et al.* Phylogenetics of asterids based on 3 coding and 3 non-coding chloroplast DNA markers and the utility of non-coding DNA at higher taxonomic levels. *Molec. Phylog. Evol.* 24: 274–301 (2002).

PHRYMACEAE

A family with a single species, *Phryma leptostachya*, with a disjunct distribution between Asia and eastern North America.

Distribution In Asia from Pakistan and Tibet to Japan and Russian Far East (Primorsk), and in North America from southern Canada to Texas and Florida, usually in deciduous forests.

Genera 1 **Species** 1
Economic uses None

Description An erect perennial herb up to 1 m tall, with a cluster of stout, fibrous roots at the base. Stalked glandular hairs occur on the stem and inflorescence, and subsessile glands are found on the leaves. The leaves are opposite, petiolate, more or less ovate, and serrate to crenate. The inflorescence is a slender terminal raceme, the paired flowers being each subtended by a small, linear bract and a pair of bracteoles, with the flowers erect in bud, horizontal at anthesis, and conspicuously deflexed in fruit. The calyx is cylindrical, with 5 teeth, accrescent in fruit, with the 2 posterior teeth becoming hardened and usually hooked at their apex. The corolla is tubular with 5 lobes, white to lilac. The stamens are 4, 2 long and 2 short. The ovary is morphologically 2-carpelled, but the abaxial carpel is suppressed, leaving the mature ovary 1-carpelled and 1-locular. There is a single ovule that is basal or may become sub-basal owing to unequal growth of the carpel wall. The fruit is a single-seeded nutlet that falls enclosed by the accrescent calyx.

The square stem and opposite serrate leaves give the appearance of a member of the Lamiaceae, while the slender terminal inflorescence is much more like that of Verbenaceae. The characteristic strongly deflexed fruit also give the plant a facies similar to *Achyranthes* in the Amaranthaceae. The geographical discontinuity of the species is interesting, and it has been estimated that it has existed for up to 25 million years[i].

Classification The family is part of the Lamiaceae-Verbenaceae-Scrophulariaceae complex of families, but its combination of inflorescence and gynoecium characters exclude it from any of the larger families. A recent molecular study[ii] placed *Phryma* nesting within the genus *Mimulus*, prompting the authors to include 8 genera traditionally placed in Scrophulariaceae in the Phrymaceae despite the obvious morphological differences, such as the single-seeded nutlet. In FGFP, it is indicated that *Phryma* should be included in the Scrophulariaceae[iii], although the monotypic family Phrymaceae is in fact maintained, as here. RKB

[i] Lee, N. S. *et al.* Molecular divergence between disjunct taxa in eastern Asia and eastern North America. *American J. Bot.* 83: 1373–1378 (1996).
[ii] Beardsley, P. M. & Olmstead, R. G. Redefining Phrymaceae: the placement of *Mimulus*, tribe Mimuleae and *Phryma*. *American J. Bot.* 89: 1093–1102 (2002).
[iii] Fischer, E. Phrymaceae. Pp. 401–405. In: Kadereit, J. W. (ed.), *Families and Genera of Vascular Plants. VII. Flowering Plants. Dicotyledons: Lamiales*. Berlin, Springer-Verlag (2004).
[iv] Cantino, P. D. Phrymaceae. Pp. 323–326. In: Kadereit, J. W. (ed.). *Ibid*.

PHYLLANTHACEAE

A diverse cosmopolitan family, similar to Euphorbiaceae but with 2 ovules per locule

Distribution Cosmopolitan except most of the north temperate region, with greatest diversity in the tropics.

Description Trees, shrubs, herbs, rarely climbers, succulents, or aquatics (*Phyllanthus fluitans*). Latex is absent although resinous exudate is rarely present. The indumentum, if present, is simple or rarely lepidote, stellate, or dendritic. Armature is rarely present. Stipules are free or rarely intrapetiolar, entire or rarely laciniate, sometimes foliaceous and rarely spinose. The leaves are simple (compound only in *Bischofia*), rarely absent in mature plants (phyllocladous species of *Phyllanthus*), alternate, spiral, subopposite, more rarely fasciculate, whorled or opposite, the blades symmetrical or rarely asymmetrical, with pinnate venation and usually entire margins. Foliar glands and domatia are rare. Inflorescences are axillary, more rarely cauline or terminal; *Uapaca* has pseudanthia with large, colored bracts. The flowers are actinomorphic, unisexual (plants monoecious or dioecious) except for 4 bisexual *Aporosa* species. The sepals are 3–8. The petals are (2–)4–6 or absent. A disk is often present, and in staminate flowers may be divided, extrastaminal-annular, rarely intrastaminal, or have the stamens inserted in cavities of the disk. The pistillate disk is divided or annular. The stamens are 3–10(–19), the filaments free or variously fused, and the anthers longitudinally dehiscing (rarely appearing oblique or horizontal) or rarely poricidal. The thecae are sometimes separate and the connectives enlarged. A pistillode may be present but staminodes are rare. The ovary is superior, with (1)2–5(–15) carpels and locules, with 2 ovules per locule, often only 1 ovule developing into a seed. The ovules are anatropous or hemitropous, apical-axile, epitropous, and have an obturator. The styles are usually bifid, sometimes entire or rarely multifid, and the stigmas ± terete or sometimes flattened. The fruits are explosively dehiscent (schizocarp), tardily dehiscent or indehiscent (drupes or berries), vividly colored in bird-dispersed taxa. The seeds are ecarunculate (minutely carunculate in *Celianella*) but sometimes with a brightly colored sarcotesta or aril. Endosperm is usually present in mature seeds, and the embryos usually have thin, flat cotyledons. Exalbuminous seeds with plicate or fleshy cotyledons occur in some genera. They are rare in the remainder of Euphorbiaceae *sensu lato,* found only in *Picrodendron* (Picrodendraceae) and in *Elateriospermum, Syndyophyllum,* and *Trigonopleura* (Euphorbiaceae *sensu stricto*).

Classification Phyllanthaceae is the largest segregate from Euphorbiaceae *sensu lato* (largely congruent with Euphorbiaceae-Phyllanthoideae in previous classifications[i,ii]). APG II places it with all other segregate families in the rosid (fabid) order Malpighiales. In a recent 8-loci study[iii], it is strongly supported as sister to Picrodendraceae. The molecular phylogenetics of the family have been comprehensively studied[iv,v]. The circumscription of Phyllanthaceae was modified from Webster's Euphorbiaceae-Phyllanthoideae to include *Croizatia* (from Euphorbiaceae-Oldfieldioideae), *Dicoelia* (from Euphorbiaceae-Acalyphoideae), and *Tacarcuna* (previously *incertae sedis* in Euphorbiaceae). The genera *Drypetes* (including *Sibangea*) and *Putranjiva* are excluded as Putranjivaceae, and monospecific *Centroplacus* is excluded as Centroplacaceae (not treated separately here); *Phyllanoa* belongs to Violaceae[vi]. West African *Centroplacus glaucinus* is a nondescript plant with small flowers in branched inflorescences and carunculate seeds and differs from other biovulate lineages by its lack of an obturator. *Antidesma, Bischofia, Hymenocardia, Martretia,* and *Uapaca,* all of which have been placed in monogeneric families by some authors, are confirmed as members of Phyllanthaceae.

The family falls into 2 major clades now recognized at subfamilial level[vii] and characterized by inflorescence and leaf anatomical features. Several traditional groupings were retrieved with minor modifications but, with the exception of monospecific Bischofieae, none of the tribes from previous classifications are supported by molecular data. Floral morphology is highly homoplasious whereas pollen and seed characters agree well with the molecular results. **SUBFAM. PHYLLANTHOIDEAE** *sensu stricto.* Lacking tanniniferous cells in the leaf epidermis, usually with contracted inflorescence axes and predominantly explosively dehiscent fruits. The following tribes are recognized:

PHYLLANTHACEAE. 1 *Phyllanthus* sp. (a) shoot with flat green phylloclades (modified stems) that bear flowers on their margins (x⅔); (b) female flower with single perianth whorl and 3-lobed stigma (x12).

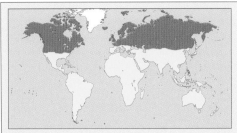

Genera c. 60 Species c. 2,000
Economic uses Edible fruits, timber

Poranthereae Unites genera previously thought to be unrelated, such as the arborescent tropical *Actephila* and *Andrachne*, small-leaved subshrubs of arid regions; disjunct shrubby *Meineckia;* and the ericoid Australian herb *Poranthera*. The latter had been classified in the "Stenolobeae" apart from all other Euphorbiaceae prior to Webster[viii] (see also Picrodendraceae). *Leptopus sensu* Webster is triphyletic (does not include *Chorisandrachne*). *Zimmermannia* and *Zimmermanniopsis* are embedded in *Meineckia*, and *Oreoporanthera* is embedded in *Poranthera*. More genera may need to be recognized in this tribe[ix].

Brideliae The most genus-rich tribe of the family. It contains *Savia*, which is triphyletic in previous circumscriptions (other species now in Phyllantheae and Wielandieae). *Bridelia* and *Cleistanthus* had been separated from all other phyllanthoids on account of their valvate sepals, which have now been shown to be much less significant. *Cleistanthus* is paraphyletic, with 3 embedded genera. Asian and Australian species of *Cleistanthus* group with *Bridelia*, and African *Cleistanthus* appears to form a grade with *Pentabrachion* and *Pseudolachnostylis*.

Wielandieae The smallest tribe. The clades have disjunct distributions: *Astrocasia* and *Chascotheca* occur in central America, *Heywoodia* in southern Africa, *Wielandia* in the western Indian Ocean, and *Chorisandrachne* and *Dicoelia* in Southeast Asia. *Chorisandrachne* was synonymized with *Leptopus* by Webster[i] but has since been shown to be sister to *Dicoelia*[v]. All Malagassian taxa ascribed to *Savia* belong to *Wielandia sensu lato*[x].

Phyllantheae In its composition almost identical to subtribe Flueggeinae in Webster's classification[i], with the addition of *Savia* sect. *Heterosavia* and the geographically disjunct *Lingelsheimia* (previously in Drypeteae/Putranjivaceae)[iv,v]. It includes *Phyllanthus*, the largest genus of the family and one of the largest plant genera. Molecular studies have shown *Phyllanthus* in its traditional circumscription to be non-monophyletic. Monospecific *Plagiocladus* is to be excluded and *Breynia, Glochidion, Reverchonia,* and *Sauropus* are to be included in the genus. These adjustments bring the species number in

Phyllanthus sensu lato to c. 1,200. Similar obligate pollination mutualism has been reported between Asian *Glochidion*[xi], *Breynia*[xii], and New Caledonian *Phyllanthus*[xiii], with the same genus of seed-consuming moths, a phenomenon otherwise only known in *Ficus* (Moraceae) and *Yucca* (Agavaceae). *Richeriella* was found to be congeneric with *Flueggea*[vii].

SUBFAM. ANTIDESMATOIDEAE Tanniniferous cells in the leaf epidermis, usually elongated inflorescence axes and predominantly indehiscent or tardily dehiscent fruits; almost all taxa are dioecious. There are 6 tribes:

Antidesmateae Includes some of the morphologically most highly modified genera in the family, e.g., *Antidesma* and *Martretia*, with unusual fruit morphology, which are found to be sisters to the morphologically more conventional genera *Thecacoris* and *Apodiscus*, respectively. Most genera in Antidesmateae have modified anthers with separate thecae and usually enlarged connectives.

Scepeae Corresponds to Webster's[i] subtribe Scepinae after exclusion of *Jablonskia* and *Apodiscus*. It contains the large, mainly Asian, genera *Aporosa* and *Baccaurea*.

Jablonskieae, Spondiantheae, Uapaceae, and Bischofieae are small tribes, with only Jablonskieae comprising an additional genus—the rare monospecific *Celianella* from the tepuis of northern South America. Monospecific *Spondianthus* is strongly toxic and, besides *Uapaca*, is the only taxon with resinous exudate in Phyllanthaceae. The flowers of African *Uapaca* are arranged in pseudanthia with colored bracts, resembling a single flower, and some species have stilt roots. Finally, monospecific *Bischofia* is the only species in the family with compound, distinctly toothed leaves.

Economic uses Some taxa are regionally cultivated for their fleshy edible fruits (e.g., *Antidesma* spp., *Baccaurea* spp., *Phyllanthus acidus*, *P. emblica*, and *Uapaca* spp.). Some provide timber or fish poison, or show medicinal promise. The fruits of *Phyllanthus emblica* (Emblic, Nelli, or Indian Gooseberry) are one of the richest sources of natural ascorbic acid (vitamin C).   PH

i Webster, G. L. Synopsis of the genera and suprageneric taxa of Euphorbiaceae. *Ann. Missouri Bot. Gard.* 81: 33–144 (1994).
ii Radcliffe-Smith, A. *Genera Euphorbiacearum.* Richmond, Royal Botanic Gardens, Kew (2001).
iii Wurdack, K. J. & Davis, C. C. Unpublished manuscript.
iv Wurdack, K. J. *et al.* Molecular phylogenetic analysis of Phyllanthaceae (Phyllanthoideae pro parte, Euphorbiaceae *sensu lato*) using plastid *rbc*L DNA sequences. *American J. Bot.* 91: 1882–1990 (2004).
v Kathriarachchi, H. *et al.* Molecular phylogenetics of Phyllanthaceae inferred from 5 genes (plastid *atp*B, *mat*K, 3' *ndh*F, *rbc*L) and nuclear PHYC. *Mol. Phyl. Evol.* 36: 112–134 (2005).
vi Hayden, W. J. & Hayden, S. M. Two enigmatic biovulate Euphorbiaceae from the Neotropics: relationships of *Chonocentrum* and the identity of *Phyllanoa. American J. Bot.* 83 (Supplement): 162 (Abstract). (1996).
vii Hoffmann, P., Kathriarachchi, H., & Wurdack, K. J. A phylogenetic classification of Phyllanthaceae (Malpighiales; Euphorbiaceae *sensu lato*). *Kew Bull* 61: 37–53 (2006).
viii Webster, G. L. Conspectus of a new classification of the Euphorbiaceae. *Taxon* 24: 593–601 (1975).
ix Vorontsova, M. S., Hoffmann, P., & Chase, M. W. Evolution of tribe Poranthereae (Phyllanthaceae or Euphorbiaceae *s.l.*). Cardiff, Wales, Abstract: 37, 5th Biennial Meeting of The Systematics Association, August 2005.
x Hoffmann, P. & McPherson, G. Revision of *Wielandia sensu lato* including *Blotia* and *Petalodiscus* (Phyllanthaceae; Euphorbiaceae *sensu lato*). *Ann. Missouri Bot. Gard.* In press.
xi Kato, M., Takimura, A., & Kawakita, A. An obligate pollination mutualism and reciprocal diversification in the tree genus *Glochidion* (Euphorbiaceae). *Proc. Nat. Acad. Sci. USA* 100: 5264–5267 (2003).
xii Kawakita, A. & Kato, M. Obligate pollination mutualism in *Breynia* (Phyllanthaceae): further documentation of pollination mutualism involving *Epicephala* moths (Gracillariidae). *American J. Bot.* 91: 1319–1325 (2004).
xiii Kawakita, A. & Kato, M. Evolution of obligate pollination mutualism in New Caledonian *Phyllanthus* (Euphorbiaceae). *American J. Bot.* 91: 410–415 (2004).

PHYLLONOMACEAE

A monogeneric family (syn. Dulongiaceae) with 4 spp. of glabrous shrubs or small trees, with epiphyllous inflorescences, endemic to the neotropics, in Mexico, Central America, and the Andean region of South America, from Colombia to Bolivia. The leaves are alternate, simple, with small, somewhat fimbriate, stipules. The flowers are actinomorphic, bisexual, small, borne in usually branched inflorescences on the adaxial leaf surface near the apex. The sepals are 4–5, free, and the petals 4–5, free, greenish, yellowish, or reddish. The stamens are 5. The ovary is superior, with 2 fused carpels, incompletely 1-locular, and with parietal placentation; the ovules are numerous. The fruit is a berry with 3–6 seeds.

Genera 1   Species 4
Economic uses None

Phyllonoma has often been included in the Escalloniaceae or Grossulariaceae but molecular evidence places it as sister to Helwingiaceae (q.v.) in the Aquifoliales[i]. A revision was published by Mori and Kallunki[ii].   VHH

i Bremer, B. *et al.* Phylogenetics of asterids based on 3 coding and 3 non-coding chloroplast DNA markers and the utility of non-coding DNA at higher taxonomic levels. *Molec. Phylog. Evol.* 24: 274–301 (2002).
ii Mori, S. A. & Kallunki, J. A. A revision of the genus *Phyllonoma* (Grossulariaceae). *Brittonia* 29: 69–84 (1977).

PHYSENACEAE

A monogeneric family of 2 spp. of small or medium sized dioecious trees or shrubs, endemic to Madagascar: *Physena madagascariensis* occurs largely in the east and center of the island and *P. sessiliflora* in the west and south. The leaves are alternate, simple, with entire margins, coriaceous, exstipulate. The flowers are actinomorphic, small to medium sized, unisexual (plants dioecious), in a loose many-flowered axillary raceme. The perianth is 1 whorl of 5–9 sepals (covered internally with unbranched hairs), free or fused at base, with blunt lobes or teeth. In male flowers, the fertile stamens are (8–)10–14(–25), free or

Genus 1   Species 2
Economic uses None

partially fused in 1 whorl, with thin filaments and long anthers. In female flowers, the ovary is superior, of 2 fused carpels, partially 2-locular, with 2 ovules per locule from placentas on the central septum. The styles are 2, long. The fruit is a single-seeded, indehiscent capsule.

The placement of this genus was uncertain for many decades with relationships to Capparaceae and Flacourtiaceae suggested. These relationships were refuted by Thorne (1992), who listed *Physena* as *incertae sedis*, and Cronquist (1988), who recognized Physenaceae in Urticales. A study of wood anatomy[i] indicated *Physena* and *Asteropeia* to be closely related, and the recognition of Physenaceae argued strongly on morphological and anatomical grounds[ii]. Later studies of morphology, anatomy, and chloroplast DNA sequences[iii] indicated that they are basal to the Caryophyllales, a result further supported by more comprehensive DNA sequence analysis[iv]. Triterpenes present in the genus have cytotoxic properties.   AC

i Miller, R. B. & Dickison, W. C. Wood anatomy of *Asteropeia* (Asteropeicaceae) and *Physena* (Physenaceae): two endemics from Madagascar. *Abstr. American J. Bot.* 79, 6: 41 (1992).
ii Dickison, W. C. & Miller, R. B. Morphology and anatomy of the Malagasy genus *Physena* (Physenaceae), with a discussion of the relationships of the genus. *Bull. Mus. Natl. Hist. Nat. Paris IV, sect B, Adansonia. Botanique, Phytochemie* 15: 85–106 (1993).
iii Morton, C. M., Karol, K. G., & Chase, M. W. Taxonomic affinities of *Physena* (Physenaceae) and *Asteropeia* (Theaceae). *Bot. Rev.* 63: 231–239 (1997).
iv Cuénoud, P. *et al.* Molecular phylogenetics of Caryophyllales based on nuclear 18S and plastid *rbc*L, *atp*B and *mat*K DNA sequences. *American J. Bot.* 89: 132–144 (2002).

PHYTOLACCACEAE
POKEWEED

A small family of trees, shrubs, woody climbers, and herbs.

Distribution Most members of the family are native to tropical America and the West Indies, but some are found in central and temperate South America, the eastern Mediterranean area, in tropical Africa and South Africa, Madagascar, the Indian subcontinent, Malaysia, China, Japan, and Australasia.

Description Trees, climbers, shrubs, and herbs. The leaves are alternate or rarely opposite, petiolate, simple and entire, typically without

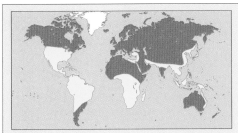

Genera 4 **Species** 31
Economic uses Many medicinal uses; yield red
dyes; used as ornamentals and pot herbs

stipules or with minute stipules, sometimes
as spines. The inflorescences are mostly indeterminate, frequently racemes or spikes, but
sometimes in cymes. The flowers are small,
regular, bisexual, rarely unisexual (plants
dioecious or monoecious). The perianth is a
single whorl of 4 or 5 usually free and persistent segments, usually green or whitish but
sometimes petaloid. The stamens are usually
hypogynous, sometimes united at the base, as
many as the perianth lobes or much more
numerous, often as a result of branching. The
ovary is usually superior, sometimes raised on a
gynophore (*Nowickea*), rarely more or less infe-

PHYTOLACCACEAE. I *Phytolacca clavigera* (a) shoot with
axillary leaf-opposed racemose inflorescence bearing
flowers and fruits (x⅔); (b) flower with numerous stamens
with swollen bases and 7 free carpels (x6); (c) fleshy fruit
(x3); (d) vertical section of fruit (x4).

rior; it comprises 1 to many separate or united
carpels, each carpel with a single basal or
axillary ovule. The styles are as many as the
carpels, short or more or less filiform, or
absent. The fruit is a fleshy berry, dry nut or,
rarely, a loculicidal capsule; the seed has mealy
perisperm and a curved embryo.

Classification This family is consistently placed
in the old Centrospermae and now the Caryophyllales[i]. There is considerable controversy over
the delimitation of this family, some authors
treating it as 6 different families[ii], although the
delimitation of Nowicke in 1968[iii], recognizing 1
family with 6 subfamilies, is usually followed.
Despite the majority of authors treating it as
a broad family, most also comment on the
highly variable morphology and inconsistencies
between the subfamilies. DNA sequence analysis
indicates that a broad Phytolaccaceae is polyphyletic[ii] and that each group is a distinct
lineage. The combined evidence of molecular
and morphological data has led to the treatment
of this family here in a narrow sense, containing
only those genera placed in the subfamily Phyto-
laccoideae in the broader circumscriptions (i.e.,
Anisomeria, Ercilla, Nowickea, Phytolacca).

Lophiocarpus is traditionally placed in
Phytolaccaceae *s.l.*, but both molecular[i] and
morphological[ii] evidence demonstrate it does
not belong in the family *sensu stricto* nor in any
currently circumscribed family so it is mentioned here. The genus comprises 4 spp. of
herbaceous or shrubby plants from southern

Africa, up to 1.2 m tall with terminal inflorescences. *Corbichonia* is believed to be related to
this genus on molecular evidence but is usually
treated in Molluginaceae.

Economic uses *Phytolacca americana* (*P.
decandra*), the Pokeweed or Red Ink Plant, is
cultivated as an ornamental shrub. Similar to
some other *Phytolacca* species, it yields edible
leaves and a red dye from the berries; sometimes it is used as a pot herb. The medicinal
uses are many and varied, including treatment
for rabies, insect bites, lung diseases, and
tumors by species of *Phytolacca,* mainly in
root preparations.   AC

[i] Cuénoud, P. *et al.* Molecular phylogenetics of
Caryophyllales based on nuclear 18S and plastid *rbc*L,
*atp*B and *mat*K DNA sequences. *American J. Bot.* 89:
132–144 (2002).
[ii] Brown, G. K. & Varadarajan, G. S. Studies in
Caryophyllales I: Re-evaluation of classification of
Phytolaccaceae *s.l. Syst. Bot.* 10: 49–63 (1985).
[iii] Nowicke, J. W. Palynotaxonomic study of the
Phytolaccaceae. *Ann. Missouri Bot. Gard.* 55: 294–364
(1968).

PICRAMNIACEAE

A New World segregate from the Simaroubaceae, characterized by long catkin-like
pendent racemes.

Distribution *Picramnia* (41 spp.) and
Alvaradoa (5 spp.) both extend from Florida
and Mexico south to northern Argentina, but
Alvaradoa is absent from all of the Amazon
Basin except the extreme south[i]. Three of the 5
Alvaradoa species are single island endemics
in the Caribbean, on Cuba, Hispaniola, and
Jamaica. *Picramnia* is usually found in rain
forests but occasionally in drier areas of the
Brasilian Planalto, while *Alvaradoa* is characteristic of arid regions. An undescribed third
genus, extending from Panama to Peru, has
been referred to recently[ii].

Description Shrubs and trees up to 12 m. The
leaves are alternate, compound, imparipinnate,
the leaflets alternate to subopposite or (especially the lower ones on each leaf) opposite. Those
of *Alvaradoa* have a mimosoid aspect, oblong
or elliptical, up to 2 cm, usually 8–25 closely
adjacent on each side of the rhachis, while those
of *Picramnia* are usually larger and fewer,
usually ovate-acuminate, up to 15 cm, and 3–8
well spaced on each side of the rhachis. The
inflorescence is a linear, usually pendent, spike
up to 30 cm long, with the flowers sessile or
shortly pedicellate, rarely cauliflorous, the
pedicels often elongating in fruit. The flowers
are small (1–3 mm), actinomorphic, unisexual
(plants dioecious), 3- to 5-merous. The sepals
are free or slightly fused at the base and usually
about as long as the petals, which are caducous
or absent in male flowers. The stamens are as
many as, and opposite, the petals, sometimes
on a column, usually clearly exceeding the

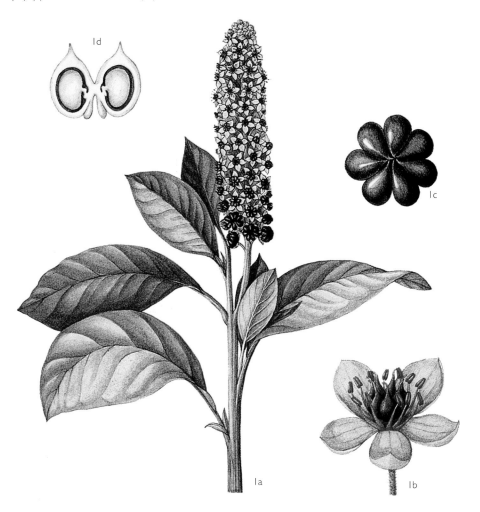

Genera 2 **Species** 46
Economic uses Wood for carpentry; edible fruits; medicinal properties

sepals and petals. The ovary is superior, borne on a short disk or gynophore, with 2–3 sessile divergent stigmas, syncarpous with 2–3 carpels, but in *Alvaradoa* only 1 of them fertile; 1–3 locules, 2 ovules per locule, terminal and pendulous in *Picramnia,* but basal and erect in *Alvaradoa.* The fruit is a berry in *Picramnia* and a samara in *Alvaradoa.*

Classification The 2 genera have been treated as separate subfamilies of the Simaroubaceae since Engler's account in the *Pflanzenfamilien* (1931). With the break-up of the traditional concept of that family on grounds of fruit structure and molecular data (see Simaroubaceae for discussion and references), the genera have been placed together in a separate family[iii]. They closely resemble each other in their long slender inflorescences, fully syncarpous ovary with 2–3 carpels each with 2 ovules, fruit anatomy, wood anatomy, and production of anthraquinones, and molecular evidence places them together. However, they differ markedly from each other in their ovary and fruit and 2 subfamilies are recognized, Picramnioideae and Alvaradooideae. Picramniaceae are unplaced in the Rosids in APG II.

Economic uses *Alvaradoa* produces wood for carpentry and decoctions for digestive and urinary complaints, skin diseases, and rheumatism. *Picramnia* has edible fruits and wood, leaves, and roots used against malaria, stomach complaints, and sexually transmitted diseases[ii]. RKB

[i] Thomas, W. W. The American genera of Simaroubaceae and their distribution. *Acta Bot. Brasil* 4: 11–18 (1990).
[ii] Thomas, W. W. Picramniaceae. Pp. 294–295. In: Smith, N. *et al.* (eds), *Flowering Plants of the Neotropics.* Princeton, Princeton University Press (2004).
[iii] Fernando, E. S. & Quinn, C. J. Picramniaceae, a new family, and a recircumscription of Simaroubaceae. *Taxon* 44: 177–181 (1995).

PICRODENDRACEAE

A small, pantropical, woody family, similar to Euphorbiaceae but with 2 ovules per locule.

Distribution Pantropical, especially in the southern hemisphere. The northernmost genus (*Tetracoccus*) extends to the southern USA. Most endemic genera are found in Australia and New Guinea (10 genera), New Caledonia (3 genera), continental Africa (2 genera), and Madagascar

(2 genera). South Africa, Caribbean, Amazonia, Andes, Sri Lanka, and South India each have 1 endemic genus.

Description Trees, shrubs, or subshrubs with simple indumentum and without exudate. The stipules when present are sometimes intrapetiolar. The leaves are alternate, opposite or whorled, simple or palmately compound (then sometimes secondarily 1-foliate), the margin entire or toothed. The flowers are unisexual (plants monoecious or dioecious) and always apetalous. The sepals range from 2–10+, and stamens can be as few as 2 or too numerous to count. The disk, when present, is situated within the staminate whorl, interspersed between it, or the stamens are inserted in cavities of the disk. Pollen grains are 4- to 8-colporate or pantoporate with up to 60 apertures and have an echinate to verruculose exine. A pistillode is sometimes present. The ovary is superior, of 2–5 carpels forming 2–5 locules, each containing 2 anatropous ovules topped with an obturator. *Scagea* and *Pseudanthus ovalifolius* have secondarily lost an ovule, the latter being the only 1-ovulate species in an otherwise 2-ovulate genus[i]. The fruit is an explosive schizocarp or rarely indehiscent, and the seeds bear a caruncle in about half the genera (thought to be associated with myrmecochorous dispersal). The seeds have copious endosperm except in *Picrodendron* where the strongly folded cotyledons fill the entire seed. An elaborate dispersal mode is found in *Petalostigma pubescens.* The fleshy fruit is first eaten by emus, guaranteeing a wide dispersal. After passing through the bird, only the woody endocarp remains. On drying, this dehisces explosively and expels the seeds up to 2.5 m wide. Eventually, the carunculate seeds are collected by ants and scattered further[ii].

Classification Picrodendraceae was first recognized as a distinct taxon (as subfamily Oldfieldioideae of Euphorbiaceae) on account of its echinate pollen exine[iii]. When its distinctiveness was confirmed by molecular phylogenetics[iv], it was recognized at family rank by APG II within the order Malpighiales (rosids; fabids). It is similar to Phyllanthaceae, sharing the presence of 2 ovules per locule, a nucellar beak, and an obturator. Echinate pollen and pollen with more than 3 apertures is rare in Phyllanthaceae and seems to be a parallel

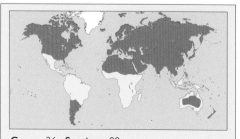

Genera 26 **Species** c. 80
Economic uses Hardwoods, poisons, medicines

development[v,vi]. Phyllanthaceae lacks caruncles (except *Celianella montana*) and frequently has petals, whereas Picrodendraceae is apetalous and often has carunculate seeds. Compound leaves are common in Picrodendraceae but occur only in 1 Phyllanthaceae species. A strongly supported sister relationship between Picrodendraceae and Phyllanthaceae was recovered in an ordinal study[vii] using DNA sequences of 8 loci from 3 genomes. As euphorbiaceous nomenclature has been sufficiently upset recently, the status quo seems preferable to further change. The third biovulate family, Putranjivaceae, shares fewer characters and is more distantly related to Picrodendraceae and Phyllanthaceae.

Pre-Websterian classifications[viii] placed Australian *Micrantheum, Pseudanthus,* and *Stachystemon* with several unrelated Euphorbiaceae *sensu stricto,* and Phyllanthaceae with a similar ericoid habit, in series "Stenolobeae" separate from all other Euphorbiaceae *sensu lato* because of their narrow cotyledons. The distribution of this character suggests that cotyledon shape reflects the environmentally favored needlelike shape of the mature leaves rather than phylogenetic relationships (although the "stenolobous" Picrodendraceae genera form a monophyletic group).

South American monospecific *Podocalyx* is sister to all other Picrodendraceae. The remainder fall into 2 main clades: 1 from America and Africa (plus 1 sp. from Sri Lanka) and 1 from Australia, New Guinea, and New Caledonia (only *Austrobuxus* with wider distribution[ix]). All taxa with compound or secondarily 1-foliate leaves belong to the first clade; narrow cotyledons and secondarily 1-ovulate ovary locules occur only in the latter clade.

Croizatia, previously classified in Oldfieldioideae on account of its echinate pollen, belongs to Phyllanthaceae. It would be aberrant in Picrodendraceae in possessing petals and 3-aperturate pollen. Its spiny pollen exine is not homologous to that in Picrodendraceae. *Paradrypetes* is also excluded from Picrodendraceae. Its opposite leaves with interpetiolar stipules agree well with its new placement as sister to Rhizophoraceae *sensu stricto.*

Economic uses Some species are valuable hardwoods, e.g., *Androstachys johnsonii* in Africa; *Piranhea,* common in central Amazonian seasonally inundated forest. *Hyaenanche globosa* (syn. *Toxicodendrum*) from South Africa contains a toxin; ground seeds are used to poison hyenas and jackals. The bitter bark of *Petalostigma* species (Quinine Bush) was traditionally used by aboriginal Australians for pain relief and against fever. PH

[i] Halford, D. A. & Henderson, J. F. Studies in Euphorbiaceae *sensu lato* 5. A revision of *Pseudanthus* Sieber ex Spreng and *Stachystemon* Planch. (Oldfieldioideae Köhler & Webster, Caletieae Müll. Arg.). *Austrobaileya* 6: 497–532 (2003).
[ii] Clifford, H. T. & Monteith, G. B. A three phase dispersal mechanism in Australian Quinine bush (*Petalostigma pubescens* Domin). *Biotropica* 21: 284–286 (1989).

iii Köhler, E. Die Pollenmorphologie der biovulaten Euphorbiaceae und ihre Bedeutung für die Taxonomie. *Grana Palynol.* 6: 26–120 (1965).
iv Wurdack, K. J. *et al.* Molecular phylogenetic analysis of Phyllanthaceae (Phyllanthoideae pro parte, Euphorbiaceae *sensu lato*) using plastid *rbcL* DNA sequences. *American J. Bot.* 91: 1882–1990 (2004).
v Levin, G. A. & Simpson, M. G. Phylogenetic implications of pollen ultrastructure in the Oldfieldioideae (Euphorbiaceae). *Ann. Missouri Bot. Gard.* 81: 203–238 (1994).
vi Simpson, M. G. & Levin, G. A. Pollen ultrastructure of the biovulate Euphorbiaceae. *Int. J. Pl. Sciences* 155: 313–341 (1994).
vii Wurdack, K. J. & Davis, C. C. Unpublished manuscript.
viii Webster, G. L. Conspectus of a new classification of the Euphorbiaceae. *Taxon* 24: 593–601 (1975).
iv Wurdack, K. J. Molecular systematics and evolution of Euphorbiaceae *sensu lato*. Ph.D. dissertation. Chapel Hill, USA, University of North Carolina, (2002).

PIPERACEAE

PEPPER FAMILY

Trees, shrubs, lianas, and herbs, with spicate inflorescences.

Distribution Tropical and subtropical regions of the world, often colonizing forest clearings. *Macropiper* is confined to the South Pacific; *Zippelia* to Southeast Asia.

Description Small trees or shrubs, lianas (climbing by means of adventitious roots), and herbs, sometimes epiphytic. The leaves are entire, membranous or succulent, alternate, opposite, spirally arranged or basal, from 2 mm to 70 cm. In many *Piper* species, a stipule-like structure called the prophyll is present[i]. The inflorescences are spicate, erect, arched or pendulous, axillary, terminal or leaf-opposed, and composed of minute asepalous apetalous flowers, stalked or sessile, each subtended by a tiny bract. The flowers are actinomorphic, unisexual (plants dioecious or monoecious), or bisexual. The stamens are 2–6, free. The ovary is superior, 1-locular, with a single ovule. The fruits are drupaceous or berrylike, with glochidiate hairs (*Zippelia*), fleshy and coalescent (*Sarcorhachis* and *Macropiper*), beaked with pennicillate stigmas (*Peperomia*), and obovoid or flask-shaped, often with prominent styles (*Piper*). The seeds are starchy. The larger fleshy infructescences are often eaten by bats or birds.

Classification The Piperales, containing Piperaceae and Saururaceae, were thought to be derived from the Magnoliales. However, the char-

Genera 5 **Species** c. 3,000
Economic uses Culinary pepper; ornamentals

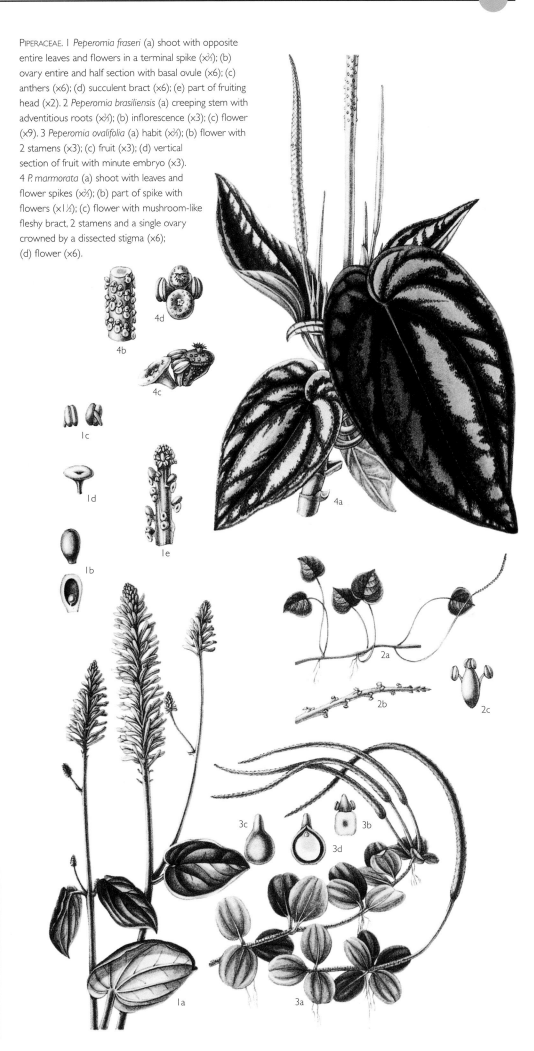

PIPERACEAE. 1 *Peperomia fraseri* (a) shoot with opposite entire leaves and flowers in a terminal spike (x⅔); (b) ovary entire and half section with basal ovule (x6); (c) anthers (x6); (d) succulent bract (x6); (e) part of fruiting head (x2). 2 *Peperomia brasiliensis* (a) creeping stem with adventitious roots (x⅔); (b) inflorescence (x3); (c) flower (x9). 3 *Peperomia ovalifolia* (a) habit (x⅔); (b) flower with 2 stamens (x3); (c) fruit (x3); (d) vertical section of fruit with minute embryo (x3). 4 *P. marmorata* (a) shoot with leaves and flower spikes (x⅔); (b) part of spike with flowers (x1½); (c) flower with mushroom-like fleshy bract, 2 stamens and a single ovary crowned by a dissected stigma (x6); (d) flower (x6).

acters of spicate inflorescences, absence of perianth, bilateral floral symmetry, and monosulcate pollen support the Piperales as a natural group[ii].

Economic uses *Piper nigrum* is the source of commercial peppercorns; *P. longum* is the "Long Pepper" of India; *P. betle* leaves are used to wrap betel nuts (*Areca catechu*); species of *Piper* and *Macropiper* are used in local medicine or as infusions; some species of *Peperomia* are grown as ornamental plants.   MT

[i] Tebbs, M. C. Piperaceae. Pp. 516–520. In: Kubitzki, K., Rohwer, J. G., & Bittrich, V. (eds), *The Families and Genera of Vascular Plants. II. Flowering Plants. Dicotyledons: Magnoliid, Hamamelid and Caryophyllid Families.* Berlin, Springer-Verlag (1993).
[ii] Burger, W. C. The Piperales and the monocots. *Bot. Rev.* 43: 345–393 (1977).

PITTOSPORACEAE

PARCHMENT-BARK

A medium-sized family of evergreen shrubs and trees, often with showy flowers.

Distribution The family is native mostly to the Old World tropics, all but 1 (*Pittosporum*) of the 9 or 10 genera being endemic to Australasia.

Description Shrubs or small trees, sometimes lianas or scramblers, occasionally spiny, often aromatic, the bark traversed by resin canals. The leaves are usually alternate, sometimes opposite or whorled, evergreen and leathery, typically entire, exstipulate. The infloresence is umbellate, corymbose, paniculate, sometimes a solitary flower. The flowers are usually actinomorphic or weakly zygomorphic (*Cheiranthera*), bisexual, rarely unisexual (plants polygamous). The sepals are 5, free or slightly connate. The petals are 5; white, blue, or red; mostly connate below; and often clawed. The stamens are 5, attached to the sepals. The ovary is superior, of 2 fused carpels (sometimes 3–5) and 1 or many locules with placentas in 2 ranks, axillary or parietal. The style is simple. The fruit is a loculicidal capsule or a berry; the seeds are mostly numerous, sometimes winged, often (e.g., *Pittosporum*) smeared with a brownish resin-like mucilage (the Greek word *pittos* meaning "pitch"); there is abundant endosperm.

PITTOSPORACEAE. 1 *Pittosporum crassifolium* (a) leafy shoot and inflorescences of male flowers (x⅔); (b) male flower (x1); (c) male flower with perianth removed showing large stamens and vestigial ovary (x1½); (d) gynoecium from female flower with vestigial stamens (x1½); (e) cross section of ovary (x1½); (f) dehiscing fruit—a capsule (x1½). 2 *Billardiera mutabilis* (a) flowering and fruiting shoot (x⅔); (b) fruit—a berry (x1½). 3 *Billardiera* (*Sollya*) *heterophylla* flowering shoot (x⅔). 4 *Marianthus ringens* (a) twining, leafy stem and inflorescence (x⅔); (b) flower (x1½); (c) androecium (x2); (d) stamen with flattened filament (x2⅔); (e) gynoecium (x2⅔).

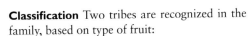

Classification Two tribes are recognized in the family, based on type of fruit:

Pittosporeae Fruit a capsule. *Pittosporum* (including *Citriobatus*) c. 140 spp. from the Canary Islands, through West and East Africa and eastern Asia, to Hawaii, Polynesia, and chiefly Australia and New Zealand, 1 sp. extending to Malaysia. *Auranticarpa* (6 spp., Australia), *Cheiranthera* (4 spp., Australia); *Hymenosporum* (1 sp., *H. flavum*, Australia and New Guinea); *Bursaria* (3 spp., Australia); *Bentleya* (2 spp., western Australia).

Billardiereae Fruit a berry. *Billardiera* (including *Sollya* and *Pronaya*) 11 spp. from Australia); *Rhytidosporum* (5 spp., Australia); *Marianthus* (14 spp., Australia).

A series of cladistic analyses of morphological data suggests that *Billardiera* is

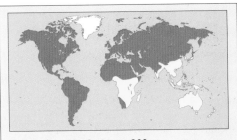

Genera 9 or 10 **Species** c. 200
Economic uses Ornamentals (*Pittosporum* and *Billardiera*) and timber (*Pittosporum*) used locally

monophyletic[i], including *Sollya* and *Pronaya* but excluding *Marianthus* (which is recircumscribed[ii]) and *Rhytidosporum*[iii]. A new genus, *Auranticarpa*, was recently described[iv]. The family had been included in the Rosales and considered closely related to the Escalloniaceae (itself perhaps near the Apiales), but both morphological and molecular data suggest its inclusion in the Apiales[v].

Economic uses Several *Pittosporum* species (Parchment-bark, Australian Laurel) are grown as ornamentals. The following shrub and small tree species are widely grown for their flowers (purple, white or greenish yellow, sometimes fragrant) and attractive foliage and are hardy in warm temperate sheltered sites: *Pittosporum crassifolium, P. ralphii, P. tenuifolium, P. eugenioides* (all from New Zealand), *P. tobira* (Japan and China), and *P. undulatum* (Victoria Box, Australia). The Tasmanian evergreen climber *Billardiera longiflora* (Apple Berry) is grown for its creamy white to purple flowers and blue edible berries. The wood of *Pittosporum* is used locally and for inlay work.　　　　VHH

i Cayzer L. W., Crisp M. D., & Telford I. R. H. Cladistic analysis and revision of *Billardiera* (Pittosporaceae). *Australian Syst. Bot.* 17: 83–125 (2004).
ii Cayzer L. W. & Crisp M. D. Reinstatement and revision of the genus *Marianthus* (Pittosporaceae). *Australian Syst. Bot.* 17: 127–144 (2004).
iii Cayzer, L. W., Crisp M. D., & Telford I. R. H. Revision of *Rhytidosporum* (Pittosporaceae). *Australian Syst. Bot.* 12: 689–708 (1999).
iv Cayzer, L. W., Crisp M. D., & Telford I. R. H. *Auranticarpa*, a new genus of Pittosporaceae from northern Australia. *Australian Syst. Bot.* 13: 903–917 (2000).
v Plunkett, G. M. *et al.* Recent advances in understanding Apiales and a revised classification. *S. African J. Bot.* 70: 371–381 (2004).

PLANTAGINACEAE
PLANTAINS

Distribution Most species belong to the genus *Plantago*, with more than 250 spp. It is native in temperate regions and on mountains in the tropics. The genus *Littorella* has 3 lacustrine

Genera 3　Species c. 260
Economic uses Limited medicinal value

species in Europe, North America, and Argentina to Chile, while *Bougueria* has a single species (*B. nubigena*) native to the Andes of southern Peru, Bolivia, and northern Argentina.

Description The plants are annual to perennial herbs or occasionally subshrubs, mostly herbaceous with a basal rosette of leaves surrounding the base of 1 or more elongate scapes, each with a compact terminal inflorescence, but some species have leafy stems with opposite leaves and many inflorescences. The leaves are entire or variously divided or lobed, and lack stipules. The inflorescence is a compact spike or spherical head or, in *Littorella*, a lax, few-flowered cluster with a single central male flower on a long pedicel. The flowers are small and individually inconspicuous, bisexual in *Plantago*, unisexual in *Littorella* (plants dioecious) and *Bougueria* (plants gynodioecious). The calyx has 4 lobes, usually somewhat unequal. The corolla is scarious and not showy, with a short tube and 4(5) lobes or in *Bougueria* irregularly divided. The 4 stamens alternate with the corolla lobes, or in *Bougueria* only 1(2). The anthers are versatile, dehiscing inward. The ovary is superior, of 2 fused carpels, *Plantago* having 2 locules with 2 to many ovules on axile placentas in each locule, but *Littorella* and *Bougueria* having a single locule with 1 basal ovule. The style is solitary and the stigma simple. The fruit in *Plantago* is a capsule with circumscissile dehiscence, but in the other 2 genera it is a single-seeded nut. The seed has a fleshy endosperm and is mucilaginous when wet; the embryo is erect or in *Bougueria* curved.

Classification The family has been variously considered close to Scrophulariaceae or somewhat isolated among the gamopetalous families. Molecular evidence[i] places it nesting within Scrophulariaceae close to *Aragoa*, a northern Andean genus of shrubby species with small imbricate leaves, white corollas, and a capsule dehiscing longitudinally into 4 valves. In general appearance and in technical characters, however, the 3 genera of Plantaginaceae are rather conspicuously different. The circumscissile capsules characteristic of *Plantago*, and even more so the single-seeded nutlets of the other 2 genera, are unlike anything found in Scrophulariaceae. Some recent authors[i,ii] have sunk the family into part of the Scrophulariaceae, but others[iii,iv,v] have not done so. Bearing in mind other families segregated from the Scrophulariaceae complex, virtually all having a different ovary and fruit structure, it seems preferable to maintain a separate family in the traditional sense. Three genera have long been recognized, but some recent publications[i,ii,iv,v] have sunk them all into *Plantago*. In a morphological and chemical analysis[iii], both *Littorella* and *Bougueria* were nested within *Plantago*, but in molecular analyses[i,ii] *Littorella* appeared more or less basal in the family and *Bougueria* almost basal. The 1845 illustration with the protologue of *Bougueria* is instructive in showing how different *B. nubigena* is from *Plantago*.

Economic uses *Plantago* seeds are used as a laxative and in treatment of dysentery. The economic significance is largely negative through the aggressive weediness of some species, especially *P. major* and *P. lanceolata*.　　RKB

i Bello, M. A. *et al.* The páramo endemic *Aragoa* is the sister genus of *Plantago* (Plantaginaceae; Lamiales): evidence from plastid *rbc*L and nuclear ribosomal ITS sequence data. *Kew Bull.* 57: 585–597 (2002).
ii Rønsted, N. *et al.* Phylogenetic relationships within *Plantago* (Plantaginaceae): evidence from nuclear ribosomal ITS and plastid *trn*L–*trn*F sequence data. *Bot. J. Linn. Soc.* 139: 323–328 (2002).
iii Chavez, F. Plantaginaceae. Pp. 297–298. In: Smith, N. *et al.* (eds), *Flowering Plants of the Neotropics*. Princeton, Princeton University Press (2004).
iv Rahn, K. A phylogenetic study of the Plantaginaceae. *Bot. J. Linn. Soc.* 120: 145–198 (1996).
v Schwarzbach, A. E. Plantaginaceae. Pp. 327–329. In: Kadereit, J. W. (ed.), *Families and Genera of Vascular Plants. VII.* Berlin, Springer-Verlag (2004).

PLATANACEAE
PLANE TREES

Genera 1　Species c. 10
Economic uses Widely cultivated in urban areas (*P. hybrida*, London Plane); wood used in veneers

The family comprises a single genus of some 9 or 10 tree species mostly from temperate North America, with 2 spp. in Mexico and Central America, 1 sp. (*P. orientalis*) in southeastern Europe and southwestern Asia to northern Iran, and 1 sp. (*P. kerrii*) in Indochina. The bark is light and characteristic, often flaking off in large plates. The leaves are alternate, simple, usually palmately 3- to 9-lobed or elliptical-oblong, unlobed (*P. kerrii*), with the petioles swollen at the base and covering the axillary bud (scarious and not covering the bud in *P. kerrii*); the margins are coarsely dentate; stipules are prominent and leaflike. The flowers are actinomorphic, unisexual (plants monoecious), in 1 to many terminal, stalked, globose heads. The sepals in the staminate flowers are 3–4(–7), short, free or connate, and the petals vestigial; the stamens are as many as the sepals. The pistillate flowers have 3–4(–7) short sepals and no petals; the ovary is superior, apocarpous, of (3–)5–8(9) carpels, with elongate stigmas; placentation is apical, with 1 (rarely 2) pendulous ovule per carpel. The fruit is an achene with tufts of hair in globose heads.

The Platanaceae is one of the earliest eudicot lineages to evolve, with only the genus *Platanus* surviving today[i]. It was included in Hamamelidales by Takhtajan (1997) but is regarded today as part of the Proteales, where it is considered sister to the Proteaceae[ii]. The London Plane (*P. acerifolia*) was widely planted in towns and cities for shade and ornament and as a street tree because of resistance to pollution. Its origin is uncertain and may be a hybrid. The wood of several species is used for furniture.　　VHH

i Hoot, S. B., Magallón, S., & Crane, P. R. Phylogeny of basal eudicots based on three molecular data sets: *atp*B, *rbc*L, and 18S nuclear ribosomal DNA sequences. *Ann. Missouri Bot. Gard.* 86: 1–32 (1999).

ii Chase, M. W. *et al.* Phylogenetics of seed plants: An analysis of nucleotide sequences from the plastid gene *rbc*L. *Ann. Missouri Bot. Gard.* 80: 528–580 (1993).

PLOCOSPERMATACEAE

The family is monotypic and comprises *Plocosperma buxifolium,* a shrub or tree to 5 m tall or more, with opposite entire and subsessile leaves, recorded from southern Mexico, Guatemala, Nicaragua, and Costa Rica. The flowers are in axillary dichasia of 1–7 flowers and functionally unisexual.

Genus 1 Species 1
Economic uses None

The corolla is funnel-shaped to campanulate, with 5 rounded lobes, blue to purplish. The stamens are 5, inserted on the corolla tube, and exserted with versatile anthers, those in functionally female flowers producing sterile pollen. The ovary has a twice-bifid stigma and a single locule with 4 erect basal ovules. The fruit is linear and conspicuous, up to 12 cm long or more, much exceeding the leaves, dehiscing by 2 valves, bearing usually 1 seed (rarely up to 4), which is linear and several centimeters long and bears an apical conspicuous tuft of light brown hairs c. 1.5 cm long. The species is variable. Plants from the north of its range tend to have smaller leaves, flowers, and fruits, and 3 spp. have been recognized, but the variation seems to be continuous. Although formerly regarded as a rather aberrant member of the Loganiaceae, it is now thought to be more closely related to Lamiaceae or Verbenaceae.   RKB

PLUMBAGINACEAE

LEADWORT, SEA LAVENDER, AND THRIFT

A cosmopolitan family of herbs, shrubs, and climbers, many of which are halophytes or psammophytes.

Distribution The family is cosmopolitan, although absent from Antarctica, but especially frequent in dry or saline habitats, e.g., sea coasts and salt steppes. Subfamily Plumbaginoideae is predominantly tropical and warm temperate, while subfamily Staticoideae is largely coastal and north-temperate Old World. *Aegialitis* (2 sp.) is restricted to mangroves of Indo-Malaysia and Australia. *Limonium* (c. 350 spp.), by far the largest genus in the family, has many narrow endemics.

Description Shrubs, lianes, or annual and perennial herbs with spirally arranged leaves, often in a basal rosette. The leaves are simple and entire, sometimes with basal auricles but

PLUMBAGINACEAE. 1 *Limonium imbricatum* (a) habit showing part of tap root, rosette of leaves and flowers in branched panicles (x⅓); (b) part of inflorescence (x2). 2 *L. tunetanum* (a) half flower showing stamens inserted at base of corolla tube (x8); (b) cross section of ovary with a single ovule (x40). 3 *L. thouini* vertical section of fruit (x⅔). 4 *Aegialitis annulata* indehiscent fruit with persistent calyx (x1). 5 *Armeria pseudarmeria* habit showing radical leaves and flowers in dense capitulate clusters (x⅔). 6 *A. maritima* half flower with lobed petals, epipetalous samens and gynoecium with simple hairy styles and a single basal ovule (x4). 7 *Plumbago auriculata* shoot bearing simple leaves and inflorescences (x⅔).

without stipules, often with secretory glands exuding water, calcium salts, or mucilage. The flowers are actinomorphic, hermaphrodite, in cymose or racemose (e.g., *Limonium*), spicate (e.g., *Acantholimon*) infloresescences or in dense, capitulate clusters (e.g., *Armeria*). The

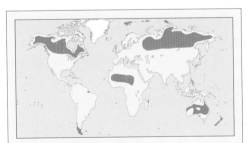

Genera 25 Species c. 840
Economic uses Garden ornamentals (Thrift, *Plumbago*), medicinal uses

bracts are scarious and sometimes form an involucre. The sepals are 5, persistent, fused to form a 5-toothed or 5-lobed tube, which is often membranous, ribbed, and colored. The petals are 5, free, connate only at the base, or fused into a long tube. The stamens are 5, opposite the petals, free or inserted at the base of the corolla. The anthers are 2-locular and split open longitudinally. The ovary is superior, of 5 fused carpels, and with a single locule containing a single basal anatropous ovule; 5 styles or 5 sessile stigmas surmount the ovary. The fruit is usually enclosed by the calyx and is normally indehiscent. The seed contains a straight embryo surrounded by mealy endosperm.

Classification The family has sometimes been placed with Primulales[i], based on floral structure, or with Centrospermae, but the lack of betalains has been used to refute the latter[i]. Cladistic analysis of DNA sequence data support Plumbaginaceae as a monophylum sister to the Polygonaceae[ii], which confirms some previous classifications (e.g., Cronquist 1981). The chief genera are *Limonium* (c. 350 spp.), *Acantholimon* (c. 150 spp.), *Armeria* (c. 80 spp.), *Plumbago* (15 spp.), *Limoniastrum* (10 spp.), and *Ceratostigma* (8 spp.).

Economic uses A number of species yield extracts that are used in medicine, e.g., those from *Plumbago europaea* and *P. scandens* are used to treat dental ailments; those from tropical *P. zeylanica* are used to treat skin diseases; and those from the roots of the European *Limonium vulgare* are used to treat bronchial hemorrhages. Many members of the family are grown in gardens, such as *Armeria* spp. (Sea Pink or Thrift) and *Limonium* spp. (Sea Lavender) whose cut flowers may be dried and used as everlastings. The climbers *Plumbago auriculata* (pale blue flowers) and *P. rosea* (red flowers) are widely grown. *Ceratostigma willmottianum* is a popular ornamental garden shrub.   AC

[i] Kubitzki, K., Plumbaginaceae. Pp. 523–530. In: Kubitzki, K., Rohwer, J. G., & Bittrich, V. (eds), *The Families and Genera of Vascular Plants. II. Flowering Plants. Dicotyledons: Magnoliid, Hamamelid and Caryophyllid Families.* Berlin, Springer-Verlag (1993).
[ii] Cuénoud, P. *et al.* Molecular phylogenetics of Caryophyllales based on nuclear 18S and plastid *rbc*L, *atp*B and *mat*K DNA sequences. *American J. Bot.* 89: 132–144 (2002).

PODOSTEMACEAE

RIVERWEEDS

A family of seasonal aquatics that resemble liverworts, mosses, or algae and grow in torrential water in the tropics and warm-temperate regions of the world.

Distribution The family is widely distributed through the tropics and some warm-temperate regions (eastern Asia and northeastern America) on rocks in seasonally fast-flowing water. The species grow submerged in the wet season

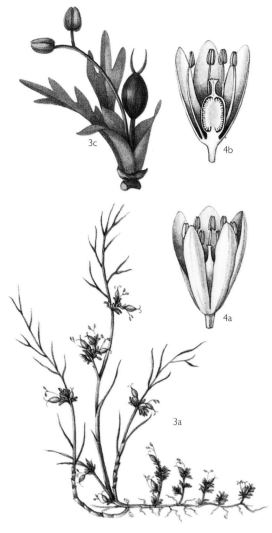

and emerge in the dry season when flowering and fruiting take place. Genera are usually restricted to 1 continent, and species are sometimes endemic to one river system. The major genera are: *Apinagia* (c. 50 spp.), northern and central South America; *Ledermanniella* (c. 44 spp.), tropical Africa, especially West Africa; *Marathrum* (c. 25 spp.), Central America, West Indies, and northwestern South America; *Rhyncholacis* (c. 25 spp.), northern South America; and *Podostemum* (c. 10 spp.), tropical America through to northeastern North America. *Cladopus* (6 spp.) is endemic to Asia, occurring in southern Asia, Southeast Asia, and Japan. The remaining genera are small (fewer than 10 spp., and usually 1–2 spp.) but 1 sp., *Tristicha trifaria*, is very widespread, occurring in tropical America, Africa, Madagascar, and the Mascarenes.

PODOSTEMACEAE. 1 *Dicraeia algiformis* (a) ribbonlike shoot bearing flowers (x⅔); (b) fruit—a capsule (x3); (c) expanded flower with basal involucre (x3); (d) vertical section of ovary showing numerous ovules on axile placentas (x6); (e) portion of ribbonlike stem bearing flowers and fruit (x⅔). 2 *Mourera weddelliana* (a) habit showing basal holdfast and erect inflorescence (x⅔); (b) vertical section of ovary surrounded by persistent stamens (x4). 3 *Podostemum ceratophyllum* (a) habit (x⅔); (b) fruit with wall cut away (x6); (c) flower with 2 divided bracts, 3 green sepals, stalked ovary and 2 stamens with filaments fused at base (x3). 4 *Weddellina squamulosa* (a) entire and (b) half flower (x9).

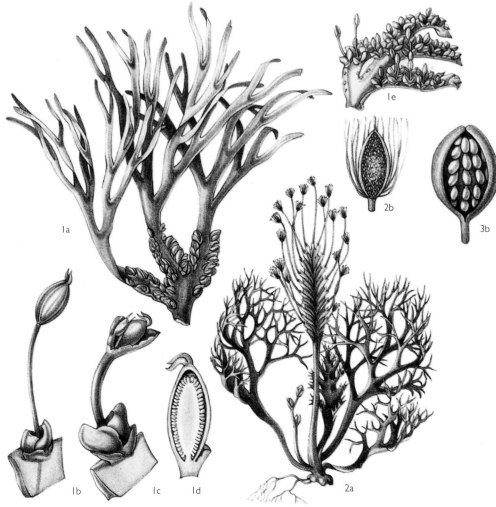

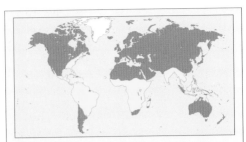

Genera 47 **Species** c. 300
Economic uses Salad leaves, traditional medicine

Description Annual or perennial with photosynthetic roots, forming crusts or mats on rocks of fast-flowing streams, rivers, and waterfalls. The roots are adventitious, arising early in development from the hyopcotyl[i], flattened, ribbonlike or thallose, or rounded and threadlike (especially when young), developing adventitious reproductive (sometimes vegetative) shoots on the upper surface, adhering to rocks by adhesive hairs to sticky microbial films[ii], often with the aid of specialized discoid root outgrowths (hapters[iii]). The stems are sometimes tiny, except when flowering, but sometimes well-developed floating structures to 80 cm long[iv], branches usually occur only on fertile shoots. The leaves are simple, rarely absent, and linear to subulate, or more complex, sometimes divided into many lobes[v], occurring on adventitious shoots; stipules sometimes present. Inflorescences are complex on adventitious shoots arising from the roots, flowers in cymes, spikes or single, usually developing and flowering only when the plants emerge during the dry season; or flowers developing underwater and self-pollinating. The flowers are enclosed in a membrane (spathella) in Podostemoideae or not, actinomorphic, bisexual; the perianth consists of 2–3 or 5 to many free or basally joined tepals, or tepals absent; the stamens are 1 to many, in 1–2 whorls, alternating with the tepals, 1 partial whorl or 2(4) borne on a lateral androphore; the ovary is superior, of 1–3 carpels, each forming a locule, topped by 1–3 stigmas on a short to elongated style. The fruit is a capsule; seeds 2 to many. It is important to note that the structures in this family do not follow the usual ground plan of flowering plants, and there is some debate on the correct identity of roots, leaves, and stems, although the floral morphology fits a more usual ground plan.

Classification The Podostemaceae have a highly unusual morphology that made systematic placement of the family difficult until the easy availability of DNA sequence comparison. Traditionally, the family has been placed in its own order, Podostemales, in most systems[vi], although some place them in Rosales (e.g., Thorne 1983). Early molecular phylogenetic classifications identified the family as a member of the Malpighiales (APG II), although the relationships of the family were not clear within the

order. A more recent analysis of 4 DNA regions sampling all 3 genomic compartments indicates Hypericaceae (here in Clusiaceae) to be the most closely related family based on samples of only 1 sp. per family[vii], although a study of 1 nuclear and 2 plastid genes, using multiple samples within families, has suggested the closest relationship to be with Ochnaceae[viii]. The family is now usually divided into 3 subfamilies based on morphological and DNA sequence evidence[ix]:
SUBFAM. PODOSTEMOIDEAE with spathella and 2 carpels (c. 40 genera, 260 spp.);
SUBFAM. WEDDELLINOIDEAE no spathella and 2 carpels (*Weddellina squamulosa* only);
SUBFAM. TRISTICHOIDEAE no spathella and 3 carpels (5 genera, c. 15 spp.) Tristichoideae has often been treated as a separate family (Tristichaceae) based on a different mode of microsporogenesis, but recent work shows elements of Podostemoideae also share this mode[x]. The extreme morphological oddity of the genera makes it difficult to establish whether generic limits are applied consistently. The 47 genera and 268 spp. recognized by Cook[xi] still represents the most complete treatment, although c. 30 spp. have been described since.

Economic uses *Dicraeanthus* is reportedly used locally in salad. Several species are locally protected under conservation laws due to their limited distribution and the vulnerability of their habitat.   AC

[i] Suzuki, K., Kita, Y., & Kato, M. Comparative developmental anatomy of seedlings in nine species of Podostemaceae (Subfamily Podostemoideae). *Ann. Bot.* 89: 755–765 (2002).
[ii] Jäger-Zürn, I. & Grubert, M. Podostemaceae depend on sticky microbial biofilms with respect to attachment to rocks in waterfalls. *Int. J. Plant Sci.* 161: 599–607 (2000).
[iii] Jäger-Zürn, I. Comparative studies in the morphology of *Crenias weddelliana* and *Maferria indica* with reference to *Sphaerothylax abyssinica* (Podostemaceae: Podostemoideae). *Bot. J. Linn. Soc.* 138: 63–84 (2002).
[iv] Ruithauser, R., Pfeifer, E., & Bernhard, A. Podostemaceae of Africa and Madagascar: keys to genera and species, including generic descriptions, illustration to all species known, synonyms, and literature list. http://www.systbot.unizh.ch/podostemaceae (2004 onwards).
[v] Jäger-Zürn, I. Morphology and morphogenesis of ensiform leaves, syndesmy of shoots and an understanding of the thalloid plant body in species of *Apinagia, Mourera*, and *Marathrum* (Podostemaceae). *Bot. J. Linn. Soc.* 147: 47–71 (2005).
[vi] Brummitt, R. K. *Vascular Plant Families and Genera.* Richmond, Royal Botanic Gardens, Kew (1992).
[vii] Davis, C. C. *et al.* Explosive radiation of Malpighiales supports a mid-Cretaceous origin of tropical rain forests. *American Nat.* 165: E36–E65 (2005).
[viii] Davis, C. C. & Chase, M. W. Elatinaceae are sister to Malpighiaceae; Peridiscaceae belong to Saxifragales. *American J. Bot.* 91: 262–273 (2004).
[ix] Kita, Y. & Kato, M. Infrafamilial phylogenetic relationships of the aquatic angiosperm family Podostemaceae inferred from *mat*K sequence data. *Plant Biology* 3: 156–163 (2001).
[x] Jäger-Zürn, I., Novelo R. A., & Philbrick, C. T. Microspore development in Podostemaceae–Podostemoideae, with implications on the characterization of the subfamilies. *Plant Syst. Evol.* 256: 209–216 (2006).
[xi] Cook, C. D. K. *Aquatic Plant Book.* The Hague, The Netherlands, SPB Academic Publishing (1990).

POLEMONIACEAE
PHLOX FAMILY

Temperate herbs centered in western North America, and woody plants and climbers centered in Mexico and the Andes, often with showy flowers.

Distribution The family is predominantly in the New World, but *Polemonium* is widespread in northern temperate regions, and *Phlox* has 1 sp. in Siberia. The herbaceous genera occur mostly in drier habitats in North America, especially western USA and Mexico, but extend to Chile and Argentina, while the trees and climbers are predominantly in Mexico and the Andes, and sometimes reach cloud forest at up to 3,000 m.

Description Small annual to perennial herbs to shrubs (spiny in *Acanthogilia*) or trees up to 8 m (*Cantua*) or herbaceous to woody tendrilled climbers up to 25 m (*Cobaea*). The leaves are alternate or (*Phlox, Leptodactylon,* and *Linanthus*) opposite or (*Gymnosteris*) in a whorl below the inflorescence, varying from simple (often linear in herbaceous species but broad and coriaceous in woody species) to linear-pinnatisect or (tribe Leptodactyloneae) with linear palmate leaflets, or pinnate with distinct leaflets. In *Cobaea* the leaves usually have about 3 pairs of fairly large distinct leaflets and the rhachis terminated by branched tendrils that each end in a pair of clawlike hooks important in the climbing habit[i]. These hooks closely resemble the apex of the leaflets, suggesting that each tendril branch is a modified leaflet. Stipules are absent (in some *Cobaea* species, the basal pair of leaflets may resemble stipules). The inflorescence is a usually terminal cyme, often aggregated and more or less capitate, in *Cobaea* leaf-opposed and long-peduncled, with 2 opposite bracteoles closely resembling leaves in being pinnate and tendrilled. The flowers are actinomorphic or sometimes bilateral, bisexual, 5-merous, or occasionally 4- to 6-merous in *Linanthus*, often showy. The sepals are 5, usually fused into a tube but in *Cobaea* free and sometimes broad and overlapping. The corolla is usually narrowly tubular, up to 8 cm in *Cantua*, but broadly campanulate in *Cobaea*. The stamens are as many as the corolla lobes and alternating with them. There is usually an annular nectariferous disk surrounding the ovary base. The ovary is superior of (2)3(4) fused

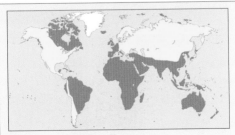

Genera 20 **Species** 350–380
Economic uses Garden ornamentals

carpels, each with a separate locule bearing 1 to many ovules. The fruit is a capsule, dehiscing loculicidally or (*Cobaea*) septicidally, with 1 to many seeds, which may be winged.

Classification The family has been almost universally regarded as being related to the group of families around Solanaceae, Boraginaceae, and Convolvulaceae, but molecular data place it remote from these families and close to Fouqueriaceae, Styracaceae[ii], or the Primulaceae group. In APG II it is accordingly placed in the Ericales. As is clear from the above description, *Cobaea*

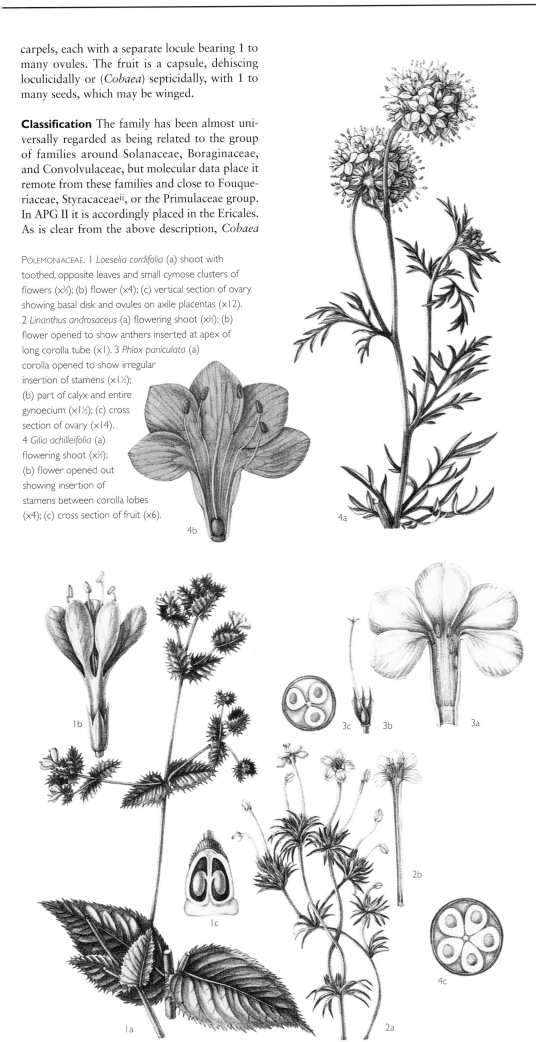

POLEMONIACEAE. 1 *Loeselia cordifolia* (a) shoot with toothed, opposite leaves and small cymose clusters of flowers (x⅔); (b) flower (x4); (c) vertical section of ovary showing basal disk and ovules on axile placentas (x12). 2 *Linanthus androsaceus* (a) flowering shoot (x⅔); (b) flower opened to show anthers inserted at apex of long corolla tube (x1). 3 *Phlox paniculata* (a) corolla opened to show irregular insertion of stamens (x1½); (b) part of calyx and entire gynoecium (x1½); (c) cross section of ovary (x14). 4 *Gilia achilleifolia* (a) flowering shoot (x⅔); (b) flower opened out showing insertion of stamens between corolla lobes (x4); (c) cross section of fruit (x6).

differs markedly from other genera, not least in its climbing habit. It has sometimes been placed in its own family Cobaeaceae, but it has obvious affinities with *Cantua* and other genera and its place in Polemoniaceae is confirmed by all molecular analyses.

Four classifications of the family have recently been proposed[iii,iv,v,vi] with significant differences at subfamily, tribal, and generic levels. A further paper by Grant has discussed differences in approach between his system and 2 of the others and has emphasized the important differences between a phylogeny and a classification[vii]. A later synopsis of his classification[viii] is followed here. The subsequent treatment by Wilken[vi] has followed Porter & Johnson[iv] in separating *Acanthogilia* as its own subfamily, referred Grant's Loeselieae and Gilieae to Polemonioideae, and differed from both previous treatments in accepting or sinking certain genera.

SUBFAM. COBAEOIDEAE Herbs to trees and climbers; pollen pantoporate; seeds usually winged; chromosomes small. There are 4 tribes:

Cantueae Shrubs and trees; leaves entire; calyx herbaceous; corolla sometimes bilateral; seeds many, winged. *Cantua* (including *Huthia*), 12 spp., Andes.

Cobaeae Climbers; leaves pinnate, tendrilled; sepals free, herbaceous; corolla campanulate; seeds flat, winged. *Cobaea*, 18 spp., Mexico to Venezuela and Peru, widely naturalized elsewhere.

Bonplandieae Perennial herbs to subshrubs; leaves entire to pinnatisect; calyx herbaceous, bilateral; corolla bilateral; locules 1-seeded; seeds winged or not. *Bonplandia*, 2 spp., Mexico and Guatemala.

Loeselieae Annuals to subshrubs; leaves entire or serrate; bracts often conspicuously veined; calyx membranous; corolla regular or bilateral; seeds winged or not. *Loeselia*, 15 spp., Texas to northern South America.

SUBFAM. POLEMONIOIDEAE Herbs to subshrubs; pollen pantoporate, except some *Collomia*; seeds lack wings; chromosomes usually medium to large. There are 3 tribes:

Polemonieae Herbs; leaves alternate, simple or pinnately lobed or sometimes entire; locules often single-seeded; seeds brown or black. *Polemonium*, *Collomia*, *Allophyllum*, and *Navarretia*, with a total of c. 78 spp., northern temperate regions.

Phlocideae Herbs and subshrubs; leaves (at least lower) opposite, entire, narrow; locules usually single-seeded; seeds brown. *Phlox*, *Microsteris*, and *Gymnosteris*, with a total of c. 66 spp. in North America and 1 sp. in Asia.

Leptodactyloneae Herbs and subshrubs; leaves usually opposite and palmately linear-lobed; locules rarely 1-seeded; seeds usually sandy colored. *Leptodactylon*, *Linanthus* and *Maculigilia*, with a total of c. 50 spp. in arid western North America.

SUBFAM. ACANTHOGILIOIDEAE Herbs to shrubs; pollen zonocolporate except some *Eriastrum*; seeds winged or not; chromosomes usually medium to large. There are 2 tribes:

Acanthogilieae *Acanthogilia gloriosa*, a *Ulex*-like desert shrub from Baja California, with leaf spines and small herbaceous leaves in their axils; seeds flat with broad wings; chromosomes small. This species has been regarded a basal to the family[v] and is given subfamily status by Porter and Johnson[iv,vi].

Gilieae Nonspiny herbs to subshrubs; leaves not dimorphic; seeds not winged; chromosomes medium to large. Genera include *Gilia, Tintinabulum, Ipomopsis, Eriastrum,* and *Langloisia*, with a total c. 105 spp. in mostly arid western North America, although *Gilia* reaches to Chile and Argentina.

Economic uses Numerous species of *Phlox, Gilia,* and *Polemonium* are grown in temperate gardens as ornamentals. *Cobaea scandens* is a spectacular climber and may be grown as an annual in temperate regions but is widely naturalized in tropical countries.   RKB

i Prather, A. L. Systematics of *Cobaea* (Polemoniaceae). *Syst. Bot. Monogr.* 57: 1–81 (1999).
ii Johnson, L. A., Soltis, D. A., & Soltis, P. A. Phylogenetic relationships of Polemoniaceae inferred from 18S ribosomal DNA sequences. *Pl. Syst. Evol.* 214: 65–89 (1999).
iii Grant, V. Primary classification and phylogeny of the Polemoniaceae, with comments on molecular cladistics. *American J. Bot.* 85: 741–752 (1998).
iv Porter, J. M. & Johnson, L. A. A phylogenetic classification of Polemoniaceae. *Aliso* 19: 55–91 (2000).
v Prather, L. A., Ferguson, C. J., & Jansen, R. K. Polemoniaceae phylogeny and classification: implications of sequence data from the chloroplast gene *ndh*F. *American J. Bot.* 87: 1300–1308 (2000).
vi Wilken, D. H. Polemoniaceae. Pp. 300–312. In: Kubitzki, K. (ed.), *The Families and Genera of Vascular Plants. VI. Flowering Plants. Dicotyledons: Celastrales, Oxalidales, Rosales, Cornales, Ericales.* Berlin, Springer-Verlag (2004).
vii Grant, V. A guide to understanding recent classifications of the family Polemoniaceae. *Lundellia* 4: 12–24 (2001).
viii Grant, V. Taxonomy of the Polemoniaceae: the subfamilies and tribes. *Sida* 20: 1371–1385 (2003).

POLYGALACEAE

MILKWORT FAMILY

Herbs, shrubs, small trees, climbers, and even saprophytes, remarkable for the superficial resemblance of the flowers to the well-known papilionaceous flower of the Leguminosae.

Distribution The Polygalaceae is almost cosmopolitan, being absent only from New Zealand and many of the southern Pacific Islands and the extreme northern parts of the northern hemisphere.

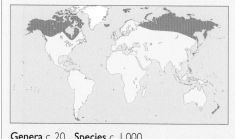

Genera c. 20 **Species** c. 1,000
Economic uses Local medicines

Description Perennial or annual herbs and shrubs, occasionally trees or lianas, rarely saprophytes (*Epirhizanthes*), with usually alternate leaves. The leaves are simple, without stipules, but sometimes with extra-floral nectaries at the junction of the petiole and stem. The inflorescence is axillary or terminal spikes, racemes, solitary or paniculate (*Monnina hirta*). The flowers are hermaphrodite, usually zygomorphic, each subtended by a bract and 2 bracteoles. The calyx of 5 (rarely 4–7) sepals is variously modified, most commonly either with the 2 lowermost united or with the 2 inner (lateral) enlarged and often petaloid. The corolla is usually of 3 petals, with the lowest (median) petal often saucer-shaped and sometimes with a fringed crest. The stamens are usually 8, generally joined to the base of the

POLYGALACEAE. 1 *Xanthophyllum scortechinii* (a) leafy shoot and irregular flowers (x⅔); (b) flower with petals removed showing free stamens (x1½); (c) petal (x1½); (d) gynoecium (x2); (e) cross section of ovary (x2); (f) globose fruit (x⅔). 2 *Polygala apopetala* (a) inflorescence (x⅔); (b) flower with lateral sepals removed (x2); (c) androecium with filaments united in a split sheath (x3); (d) stamens (x4); (e) gynoecium (x3); (f) vertical section of ovary (x8). 3 *Carpolobia lutea* (a) leaves and fruit—a drupe (x⅔); (b) fruit entire and in cross section (x⅔). 4 *Bredemeyera colletioides* flowering shoot (x⅔). 5 *Securidaca longipedunculata* winged fruits—samaras (x⅔).

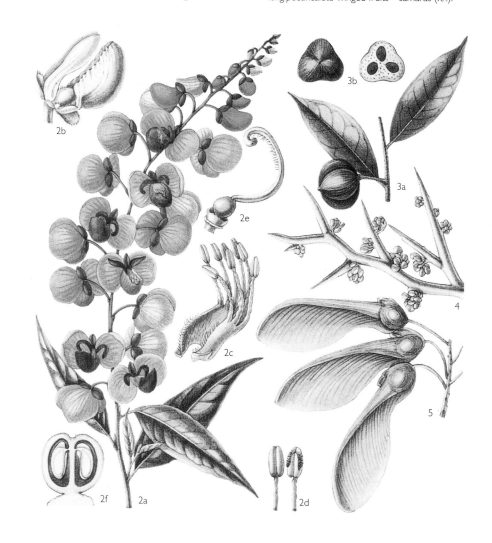

corolla, with their united filaments forming a split sheath; the anthers are basifixed, usually dehiscing by an apical pore; the pollen grains have a distinctive pattern on their outer wall. A ring-shaped disk is sometimes present inside the base of the staminal whorl. The ovary is superior, of (1)2(5) united carpels, with a single pendulous ovule on an axile placenta in each of the 2 locules, although there are various exceptions to this general structure. The style is simple. The fruit is usually a loculicidal capsule (e.g., *Polygala, Muraltia*), septicidal in *Salomonia*, a drupe (e.g., *Nylandtia, Monnina*), samara-like with 1 (*Securidaca*) or 2 (*Ancylotropis, Pteromonnina*) wings, or a nut (*Atroximia*). The seeds, sometimes hairy, generally have an aril and contain a straight embryo and fleshy endosperm (sometimes absent).

Classification The Polygalaceae family is divided into 4 tribes:

Xanthophylleae 1-locular ovary, 4–40 ovules, and comprises the sole genus *Xanthophyllum* (c. 90 spp.) from the Indomalaysian region, treated by some authorities as a distinct family, the Xanthophyllaceae (Cronquist 1981), on account of the almost free stamens and close superficial resemblance to the Caesalpinioideae of the Leguminosae but most commonly included in Polygalaceae (Thorne 1991).

Moutabeae Sepals and petals each forming a tube and usually actinomorphic flowers. Moutabeae is a paraphyletic group; genera include *Balgoya, Barnhartia, Diclidanthera,* and *Moutabea*.

Carpolobieae 5 petals and stamens and a 3-locular ovary. Genera include *Atroxima* and *Carpolobia*.

Polygaleae 2 large petaloid sepals, 3 petals, and usually 6–8 stamens and a 2-locular ovary. There are c. 14 genera, including *Polygala* (c. 325 spp.), *Monnina* (c. 150 spp.), *Muraltia,* (c. 115 spp.), and *Securidaca* (c. 80 spp.).

The large genus, *Polygala*, and the much smaller *Bredemeyera* are highly polyphyletic in one detailed molecular phylogeny[ii]. The flowers of many species resemble those of legumes, but this was long treated as a convergent character, and the family placed in its own order in many systems[iii]. Extensive molecular evidence shows the family to have a close affinity with the legumes[iv], the Quillajaceae, and Surianaceae in a well supported, but poorly resolved, Fabales (APG II).

Economic uses Local medicines are extracted from several species, e.g., snake-root from *Polygala senega* of eastern North America. The constituent glucoside seregin is used by Native Americans to cure snake bites.　　AC

i Persson, C. Polygalaceae. Pp. 306–308. In: Smith, N. *et al.* (eds), *Flowering Plants of the Neotropics*. Princeton, Princeton University Press (2004).
ii Persson, C. Phylogenetic relationships in Polygalaceae based on plastid DNA sequences from the *trnL-trn*F region. *Taxon* 25: 763–779 (2001).
iii Brummitt, R. K. *Vascular Plant Families and Genera*. Richmond, Royal Botanic Gardens, Kew (1992).
iv Soltis, D. E. *et al.* Angiosperm phylogeny inferred from 18S rDNA, *rbc*L, and *atp*B sequences. *Bot. J. Linn. Soc.* 133: 381–461 (2000).

POLYGONACEAE
BUCKWHEAT, RHUBARB, AND DOCK FAMILY

A large cosmopolitan family of herbs, some shrubs, and a few trees.

Distribution Most genera inhabit the temperate northern regions. A few are tropical or subtropical, notably *Antigonon* (Mexico and Central America), *Coccoloba* (tropical America and Jamaica), and *Muehlenbeckia* (Australasia and South America).

Description Annual or perennial herbs, shrubs, trees, scramblers, or climbers, often with swollen nodes. The leaves are usually basal or alternate, opposite, or occasionally whorled (some *Eriogonum*), simple (but highly reduced in *Calligonum* and *Muehlenbeckia*), with or without a petiole and, in subfamily Polygonoideae[i], with a characteristic ochrea, or membranous sheath, around the stem and uniting the stipules. The inflorescence is terminal or axillary, simple or a branched thyrse, appearing paniculate, racemose, spikelike, or umbel-like (*Eriogonum heracleioides*). The flowers are usually hermaphrodite, occasionally unisexual (plants generally dioecious, sometimes monoecious), small, white, greenish, or pinkish, and subtended by a sheathing membranous tube of fused bracteoles. The tepals are 3–6, usually in 2 whorls of 3 or 1 whorl of 5 (by fusion of 2 tepals), often becoming enlarged and membranous in fruit. The stamens are (2–)6–9, with 2-locular anthers opening lengthwise. The ovary is superior, of (2)3(4) fused carpels, with a single locule containing a single basal ovule. The styles are 2–4, usually free. The fruit is a triangular achene or nut that is sometimes attached to, or enclosed in, a fleshy expanded perianth (e.g., *Antigonon, Coccoloba,* and *Muehlenbeckia*).

Classification Polygonaceae is sister to Plumbaginaceae in molecular analysis[ii] in the non-core Caryophyllales. This contrasts with the traditional placement of the family in its own order, Polygonales (e.g., Thorne, 1992; Dahlgren, 1983; Cronquist, 1981). The family has great diversity in its wood anatomy[iii], probably reflecting the diversity of growth forms and habits of Polygonaceae. Two subfamilies are recognized[i]:

SUBFAM. **ERIOGONOIDEAE** Stipules poorly defined, with sympodial growth and a specialized involucre around the cymose inflorescence.

Eriogoneae Involucre tubular or multibracteate. The tribe comprises 15 genera, including: *Eriogonum*, with c. 240 spp. of annual or perennial herbs and subshrubs mostly from the dry, warm regions of western North America; *Chorizanthe*, with c. 50 spp.; and 13 very small genera.

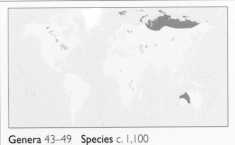

Genera 43–49 **Species** c. 1,100
Economic uses Food plants; several ornamentals

Pterostegieae Single bisaccate and inflated bract. There are 2 genera: *Pterostegia* (1 sp.) and *Harfordia* (1 sp.).

SUBFAM. **POLYGONOIDEAE** Stipules (ochrea) sheathing, growth monopodial, and inflorescences racemose with cymose branches.

Triplareae Usually dioecious trees or shrubs, with 2 whorls of 3 tepals, the outer often enlarged in fruit. 5 genera: *Ruprechtia* (c. 20 spp), *Triplaris* (c. 18 spp.), and 3 genera with 3 or fewer species.

Coccolobeae Usually perennials with a 5-merous perianth. 6 genera: *Coccoloba* (c. 120 spp.) tropical trees, shrubs, and climbers; *Muehlenbeckia* (23 spp.) climbing or woody, hardy or half-hardy plants; *Antigonon* (c. 4 spp.) flamboyant climbers; and 3 other genera with fewer than 10 spp.

Rumiceae Herbs with 2 whorls of 3 tepals. 4 genera: *Rumex* (c. 200 spp.), *Rheum* (c. 30 spp., strong, large-leaved herbs from Siberia, the Himalayas, and eastern Asia, and 2 other genera with 2 spp.

Polygoneae 2 whorls of 5 tepals, often with keels or wings. 7–8 genera: *Calligonum*, c. 80 spp. of broomlike xerophytic shrubs from central Asia; *Oxygonum* (c. 30 spp.); *Atraphaxis* (c. 25 spp.), desert-loving hardy and half-hardy shrubs and subshrubs; *Polygonum* (c. 20 spp.); and 3–4 genera with fewer than 10 spp.

Persicarieae 2 whorls of 5 tepals, not usually winged or keeled. 3 genera: *Persicaria* (c. 150 spp.); *Fagopyrum* (c. 8 spp.), perennial and annual herbaceous species often with succulent stems, native to temperate regions of Eurasia; and *Koenigia* (6 spp.); *K. islandica* is circumpolar.

In a preliminary molecular phylogenetic analysis of the family[iv], including 14 genera of Polygonoideae, but only *Eriogonum* from Eriogonoideae, *Eriogonum* was nested well within a clade comprising genera of tribe Coccolobeae (and *Triplaris*, but with weak support). This group was placed sister to the remaining genera. The other main group included a monophyletic Rumiceae and Polygoneae each nested within a highly divided Persicarieae. At the generic level; *Muehlenbeckia* grouped with tribe Polygoneae and not Coccolobeae, *Koenigia* was derived within *Persicaria*, and *Reynoutria* was not included in *Polygonum* if *Atraphaxis* was to be recognized. There is still much phylogenetic work to be done in this family.

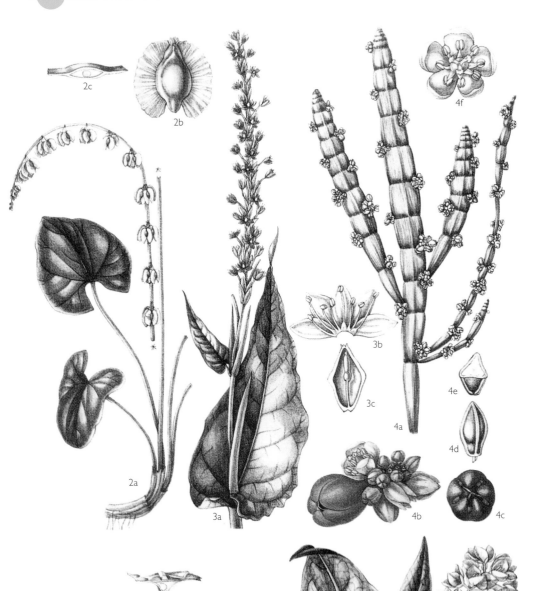

POLYGONACEAE. 1 *Rumex hymenosepalus* (a) leafy shoot with flowers and winged fruit (x⅔); (b) mature fruits showing persistent perianth (x⅔); (c) cross section of fruit (x1); (d) flower (x2). 2 *Oxyria digyna* (a) habit (x⅔); (b) winged fruit (x4); (c) cross section of fruit (x4). 3 *Polygonum amplexicaule* (a) flowering spike showing sheathing stipules or ochreas clasping the stem above the leaf bases (x⅔); (b) perianth opened out to show 8 stamens (x2); (c) vertical section of ovary (x4). 4 *Homalocladium platycladum* (a) flowering shoot (x⅔); (b) flower buds and young fruits (x4); (c) mature fruit (x4); (d) seed (x4); (e) cross section of seed (x4); (f) flower viewed from above (x7).

Economic uses Cultivated ornamentals include *Antigonon leptopus* (Coral Vine or Rosa de Montaña), *Muehlenbeckia axillaris*, *Atraphaxis frutescens*, the rock-garden species of *Eriogonum* (grown for their gray and white foliage), the waterside *Rheum palmatum* and *Rumex hydrolapathum*, and fast-growing border, ground cover, and rock garden species of *Persicaria*. The purple berries of the West Indian Seaside Grape, *Coccoloba uvifera*, are eaten, as are the leaves of the Common Sorrel, *Rumex acetosa*, used as a salad and pot herb,

and the petioles of the Common Rhubarb, *Rheum × hybridum*. *Fagopyrum esculentum* (Common Buckwheat) is widely cultivated for its seeds and as a manure and cover crop. Similar, but less extensive, uses are made for *F. tataricum*, Tartary Buckwheat, although the seeds are not eaten by humans. AC

[i] Brandbyge, J. Polygonaceae. Pp. 531–544. In: Kubitzki, K., Rohwer, J. G., & Bittrich, V. (eds), *The Families and Genera of Vascular Plants. II. Flowering Plants. Dicotyledons: Magnoliid, Hamamelid and Caryophyllid Families*. Berlin, Springer-Verlag (1993).
[ii] Cuénoud, P. *et al*. Molecular phylogenetics of Caryophyllales based on nuclear 18S and plastid *rbc*L, *atp*B and *mat*K DNA sequences. *American J. Bot*. 89: 132–144 (2002).
[iii] Carlquist, S. Wood anatomy of Polygonaceae: Analysis of a family with exceptional wood diversity. *Bot. J. Linn. Soc*. 141: 25–51 (2003).
[iv] Lamb Frye, A. S. & Kron, K. A *rbc*L phylogeny and character evolution in Polygonaceae. *Syst. Bot*. 28: 326–332 (2003).

PORTULACACEAE

PURSLANE AND LEWISIA

This nearly cosmopolitan family is related to Cactaceae and other succulent plants.

Distribution The family is nearly cosmopolitan with the major diversity in the southern hemisphere, with centers of diversity in the Andes and South Africa, but species also occur in North America and Eurasia. *Portulaca* is pantropical. Many genera occur in arid or semiarid conditions (e.g., *Anacampseros*, *Calandrinia*, *Portulaca*, *Talinum*).

Description Trees (*Calyptrotheca*), shrubs, scramblers (*Grahamia*), and herbs, sometimes with a woody base, succulent or tuberous. The leaves are spirally arranged or apparently opposite, usually simple with entire margins, usually glabrous but sometimes with hairs or glandular barbs, often fleshy, the petioles short or almost absent and poorly defined; stipules are lacking but basal scales, spines, or bristles are sometimes present. The flowers are hermaphrodite (unisexual in *Ceraria*; plants dioecious), actinomorphic, rarely slightly zygomorphic, in terminal cymes or panicles, sometimes condensed into a dense head or flowers solitary. The sepals are 2(3)–5 or more in *Lewisia*, and often unequal in size when 2, persistent or deciduous. The petals are (2–)5(–12 or more), sometimes brightly

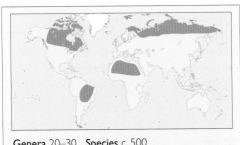

Genera 20–30 Species c. 500
Economic uses Ornamentals and medicinal uses

in the core Caryophyllales in a close-knit group. Within the family, the classification has varied drastically. One classification splits the family into 7 tribes and 19 genera[iii], but a recent account recognizes 4 tribes and 29 genera[iv]. Some genera are ill-defined and can be distinguished only by microscopic characters of the seed.

Economic uses Purslane (*Portulaca oleracea*) is a widely eaten, but not commercialized, vegetable. Several genera are grown as ornamentals because of their bright flowers or as curiosities (*Anacampseros* spp., *Portulaca grandiflora*, *Lewisia tweedyi*, and hybrids). *Portulaca oleracea* is used medicinally.   AC

[i] Cuénoud, P. *et al*. Molecular phylogenetics of Caryophyllales Based on Nuclear 18S and Plastid *rbc*L, *atp*B and *mat*K DNA sequences. *American J. Bot.* 89: 132–144 (2002).
[ii] Downie, S. R., Katz-Downie, D. S., & Cho, K.-J. Relationships in the Caryophyllales as suggested by phylogenetic analysis of partial chloroplast DNA ORF2280 homolog sequences. *American J. Bot.* 84: 253–273 (1997).
[iii] MacNeill, J. Synopsis of a revised classification of the Portulacaceae. *Taxon* 23: 725–728 (1974).
[iv] Carolin, R. C. Portulacaceae. Pp. 544–550. In: Kubitzki, K., Rohwer, J. G., & Bittrich, V. (eds), *The Families and Genera of Vascular Plants. II. Flowering Plants. Dicotyledons: Magnoliid, Hamamelid and Caryophyllid Families*. Berlin, Springer-Verlag (1993).

POTTINGERIACEAE

A family comprising 1 sp. (*Pottingeria acuminate*) of uncertain affinity, known from Assam to northwestern Thailand. It is a slender evergreen tree or shrub with alternate, entire, coriaceous leaves, with very small stipules. It has small actinomorphic, bisexual flowers in axillary cymes,

Genera 1	Species 1
Economic uses None	

with 5 free sepals and 5 free petals, 5 stamens borne on a large fleshy extrastaminal disk. The ovary is superior, of 3 fused carpels, 1-locular, with intrusive parietal placentation and numerous ovules. The fruit is a septicidal capsule.

It was included in the Celastraceae by Airy-Shaw[i] as a separate subfamily and later questionably retained there in some recent treatments[ii]. It is included in the Hydrangeales by Takhtajan (1997) and is listed as an unplaced eudicot I in APG II.   VHH & RKB

[i] Airy Shaw, H. K., Cutler, D. E., & Nilsson, S. *Pottingeria*, its taxonomic position, anatomy and palynology. *Kew Bull.* 28: 97–104 (1973).
[ii] Simmons, M. P. Celastraceae. Pp. 29–64. In: Kubitzki, K. (ed.), *The Families and Genera of Vascular Plants. VI. Flowering Plants. Dicotyledons: Celastrales, Oxalidales, Rosales, Cornales, Ericales*. Berlin, Springer-Verlag (2004).

PORTULACACEAE. 1 *Claytonia perfoliata* (a) habit showing flower stalks erect before, and curved downward after, pollination, and erect when bearing fruit (x⅔); (b) mature capsule with 1 of 2 persistent sepals removed (x6); (c) flower with petals party removed (x12). 2 *Lewisia cotyledon* (a) habit showing basal rosette of leaves and inflorescence (x⅔); (b) vertical section of ovary with basal ovules (x3). 3 *Portulaca grandiflora* (a) flowering shoot showing hairy stipules (x⅔); (b) cross section of 1-locular ovary (x4); (c) half flower showing overlapping petals (x4). 4 *Montia fontana* (a) habit (x⅔); (b) fruit—a capsule dehiscing by 3 valves (x6); (c) flower (x6).

colored, free or connate. The stamens are usually 5, or 1 (in *Monocosmia*), sometimes many, free or connate to the base of the petals. The ovary is superior or inferior (*Portulaca*) of 2–5 fused carpels, 1-locular, with free central placentation at maturity. The style is usually divided. The fruit is capsular, usually dehiscing through terminal valves, or circumscissile, or the fruit is a nut. Seeds with or without an aril.

Classification The family has been undersampled in molecular analyses. In one study based on one sample each from *Portulaca* and *Claytonia* using one nuclear and 3 chloroplast genes, the family is paraphyletic and groups with Cactaceae, Halophytaceae, Didieraceae and Basellaceae[i]. Another study groups 3 genera in a polytomy[ii] again with Cactaceae, Didieraceae, and Basellaceae. While monophyly of the family is questionable, there is no doubt that the component genera belong

PRIMULACEAE
PRIMULA FAMILY

A predominantly north-temperate family of herbs with often showy tubular flowers and stamens opposite the corolla lobes.

Distribution Most genera are in the northern-temperate region, including the Himalayas and China. *Lysimachia* is in all the major regions but is very restricted in Australasia and South America. *Anagallis* is widespread but is not native in most of Malaysia and Australasia. *Ardisiandra* is confined to African mountains, from Bioko to Ethiopia and south to Zimbabwe. *Cyclamen* extends from the Mediterranean to Ethiopia. *Asterolinon* extends from the Mediterranean to Tanzania, while the closely related *Pelletiera* occurs in temperate South America (and is introduced in the Canary Islands). *Primula* has perhaps 400 northern temperate species, 1 in

eastern tropical Africa, 2 in Malaysia to New Guinea, 2–3 in Mexico, and 1–2 in temperate South America, including the Falkland Islands. *Androsace* has more than 150 northern temperate species and 1 in temperate South America.

Description Annual to perennial herbs, sometimes caespitose or mat-forming. The leaves are alternate or opposite to whorled, often forming a basal rosette, simple and linear (*Asterolinon, Pelletiera*) to orbicular, with an entire to serrate margin, except in the aquatic genus *Hottonia*, in which they are finely dissected into linear segments and in *Potamosace*, in which they are pinnatisect. Occasionally (some *Lysimachia*) the leaves are glandular-punctate. The inflorescence may be scapose with a solitary terminal flower, a terminal umbel of several whorls, a terminal raceme, an axillary cluster, or solitary axillary flower. The flowers are actinomorphic, bisexual, and often heterostylous, and usually 5-merous except in *Trientalis*, which is 7-merous, and in *Pelletiera* in which the corolla is 3-merous. The sepals are free or fused into a tubular to campanulate calyx. The corolla is tubular to campanulate, with the tube sometimes very short and the lobes free almost to the base; in *Glaux* the corolla is absent. The stamens are 5, attached to the corolla opposite the lobes. The ovary is superior, of 5 fused carpels, and has a single locule. Placentation is free central, with usually many (up to 400) ovules that are often embedded in the placenta. The fruit is usually a characteristically subspherical capsule, but in some genera of Primuleae it is elongate, especially in *Bryocarpum*, where it is linear and c. 5 cm long. The capsule either dehisces by apical valves or by a circumscissile split (*Anagallis, Potamosace*, some *Primula*), or the upper part lifts off irregularly (*Soldanella, Bryocarpum*, some *Primula*), or the whole capsule breaks

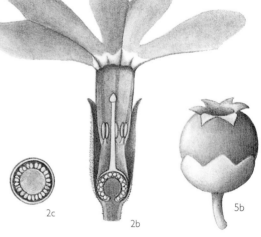

up irregularly to release the seeds. Seeds are angular, and there are few in *Asterolinon* and *Pelletiera* but many in the other genera.

Classification The Primulaceae has generally been grouped with several other families in which the stamens are inserted opposite the corolla lobes, the group having been long known as Primulales. This group has more recently been placed within the broader order Ericales, members of which have a similar disposition of the stamens. There has been only marginal disagreement over the circumscription of the family until recently. A close affinity with the Myrsinaceae has long been apparent, but molecular analysis[i] has found that the latter family nests within one half of the traditional Primulaceae. This has resulted in the accounts in FGVP[ii] placing more than half of the genera traditionally placed in Primulaceae (including *Lysimachia, Anagallis, Cyclamen,* etc.) in Myrsinaceae to avoid paraphyly. Furthermore, no infrafamilial groups such as subfamilies or tribes are recognized in the new concept of Myrsinaceae to accommodate these genera, presumably because that would again create a paraphyletic taxon. Key characters

PRIMULACEAE.

1 *Dodecatheon meadia* habit showing basal rosette of leaves and flowers, with reflexed petals, borne on leafless stalks (x⅔). 2 *Primula veitchii* (a) habit (x⅔); (b) half flower showing epipetalous stamens and ovules on a free central placenta (x4); (c) cross section of 1-locular ovary (x6). 3 *Soldanella alpina* habit showing flowers with deeply divided petals (x⅔). 4 *Primula veris* dehisced fruit (a capsule) with part of persistent calyx removed (x3). 5 *Cyclamen hederifolium* (a) habit showing basal tuber (x⅔); (b) dehisced fruit (x4). 6 *Lysimachia punctata* leafy terminal inflorescence with yellow flowers (x⅔).

epicalyx fused with the calyx, and several other significant features. These distinctive basal genera are best referred to their own families. The recent FGVP accounts[ii] have already accorded separate family status to *Maesa* and *Samolus*, and here the family Coridaceae is also recognized to accommodate *Coris*. Four tribes may be recognized in the Primulaceae:

Primuleae Plants scapose with a basal rosette, mat-forming, or (*Hottonia*) aquatic, the calyx usually tubular or campanulate in the lower part, the corolla aestivation imbricate, and the cells of the corolla epidermis isodiametric, and the ovules not immersed in the placenta except in *Dionysia*. *Primula* (c. 400 spp.), *Androsace* (c. 160 spp.), *Dionysia* (c. 48 spp.), *Omphalogramma*, *Dodecatheon*, and *Soldanella* all with 10–15 spp., and *Hottonia*, *Cortusa*, *Kaufmannia*, *Bryocarpum*, *Vitaliana*, and *Potamosace* each with 1–2 spp.

Ardisiandreae Stems are leafy, elongate, the calyx deeply divided into ovate segments, corolla imbricate but with elongate epidermal cells, the stamens forming a cone, and seeds at first not immersed in the placenta but surrounded by it at maturity[i]. *Ardisiandra*, with 3 spp. in Africa. This genus is basal in the family and intermediate[iii] in most respects between the preceding and following tribes. *Stimpsonia* (1 sp. in Asia) is similarly intermediate but not morphologically similar to *Ardisiandra*[iii] and might be treated as a fifth tribe.

Lysimachieae Stems are more or less elongate and leafy, calyx is usually divided almost to the base into lanceolate to linear segments, corolla with contorted aestivation and elongate epidermal cells, and the ovules immersed in the placenta (as in Myrsinaceae and Theophrastaceae). *Lysimachia* (c. 150 spp.), *Anagallis* (34 spp.), and *Trientalis*, *Asterolinon*, *Pelletiera*, and *Glaux* (all with 1–3 spp.).

Cyclamineae Tuber with a rosette of leaves, only a single cotyledon developing, calyx campanulate with reflexed lobes, corolla with contorted aestivation and elongate epidermal cells, and the ovules unitegmic (not bitegmic as in other tribes) and immersed in the placenta. *Cyclamen* (c. 20 spp.).

According to recent literature, generic limits are being challenged by cladistic approaches. *Douglasia*, *Vitaliana*, and *Potamosace* all make *Androsace* paraphyletic, *Dodecatheon*[iv], *Dionysia*, *Soldanella*, *Omphalogramma*, *Kaufmannia*, and *Cortusa* all make *Primula* paraphyletic, and *Anagallis* and *Glaux* both make *Lysimachia* paraphyletic. It is to be hoped on practical grounds that this will not lead to wholesale renaming in these cases.

used in FGVP to link these genera with Myrsinaceae rather than Primulaceae are ambivalent, since *Ardisiandra* and *Dionysia* are exceptions in the embedding of the ovules in the placenta, and the presence of gland dots is far from constant. On the other hand, the Myrsinaceae *sensu stricto* are largely evergreen tropical trees and shrubs (few herbs) with usually large, thick, leathery leaves and fleshy single-seeded drupes with subglobose seeds, readily separable from Primulaceae, which are temperate herbs with dry, capsular fruits and usually many angular seeds. It is preferred here to retain the herbaceous genera as one family, Primulaceae, which is apparently paraphyletic with respect to the Myrsinaceae. The relationships of the Primulaceae to allied families have been clarified by recent morphological and molecular studies. There are 3 main morphological groups: Theophrastaceae (woody) and Primulaceae (herbaceous) are sister groups, while the Myrsinaceae (woody) are derived from Primulaceae. Each of these major groups, however, has 1 genus associated with it, which stands out in basal or subbasal position in phy-

logenetic analyses and has distinctive characters. The large genus, *Maesa*, has been included in Myrsinaceae until recently but is basal to the whole group and differs in its semi-inferior ovary, many seeds, wood anatomy, and other characters. The genus *Samolus* has generally been placed in Primulaceae but is clearly closer to Theophrastaceae, differing from it, however, in its herbaceous habit, semi-inferior ovary, and numerous other significant characters. The genus *Coris* has also been placed in Primulaceae but has zygomorphic flowers, a spiny

Economic uses *Primula*, *Lysimachia*, and *Cyclamen* are all of major horticultural importance, and *Androsace*, *Dionysia*, *Cortusa*, *Dodecatheon*, and *Soldanella* are all also widely cultivated. Otherwise there is little economic use of the family. Poisonous glycosides are produced in *Cyclamen* and *Anagallis*. RKB

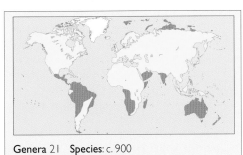

Genera 21   Species: c. 900
Economic uses Many cultivated species

[i] Källersjö, M., Bergqvist, G., & Anderberg, A. A. Generic realignment in primuloid families of the Ericales *s.l.* *American J. Bot.* 87: 1325–1341 (2000).

[ii] Ståhl, B. & Anderberg, A. A. Maesaceae, Myrsinaceae, and Primulaceae. Pp. 255–257, 266–281, & 313–319. In: Kubitzki, K. (ed.), *The Families and Genera of Vascular Plants. VI. Flowering Plants. Dicotyledons: Celastrales, Oxalidales, Rosales, Cornales, Ericales.* Berlin, Springer-Verlag (2004).

[iii] Anderberg, A. A., Peng, C.-I., Trift, I., & Källersjö, M. The *Stimpsonia* problem; evidence from DNA sequences of plastid genes *atp*B, *ndh*F and *rbc*L. *Bot. Jahrb. Syst.* 123: 369–376 (2001).

[iv] Mast, A. R., Feller, D. M. S., Kelso, S., & Conti, E. Buzz-pollinated *Dodecatheon* originated from within heterostylous *Primula* subgenus *Auriculastrum* (Primulaceae). *American J. Bot.* 91: 926–942 (2004).

PROTEACEAE

PROTEA FAMILY

Trees and shrubs, mostly characterized by flowers with a single whorl of 4 tepals, fused to the inside of which are 4 stamens.

Distribution Mainly in the southern hemisphere (but extending into southern and eastern Asia, Central America, and western and northeastern tropical Africa), where the family is almost completely restricted to Gondwanic continental blocks and fragments. They occur in a variety of habitats, ranging from open shrubland or grassland to rain forests, from alpine meadows to tropical lowlands, usually on acidic, well-drained, nutrient-poor soils.

PROTEACEAE. 1 *Leucospermum conocarpodendron* (a) leafy shoot with terminal conelike inflorescence, each flower with a conspicuous arrow-shaped pollen presenter (×⅔); (b) flower with stamens fused to perianth segments and long style with arrow-shaped pollen presenter (×1½); (c) perianth opened out to show fruit (×10); (d) hairy bract (×4). 2 *Grevillea robusta* (a) deeply divided leaf (×⅔); (b) inflorescence each flower with a projecting style (×⅔); (c) young flower with pollen presenter retained in bud (×2); (d) tepal with anthers directly attached (×5½); (e) mature flower with extended style and stigma (×2); (f) fruits (×⅔); (g) winged seed (×⅔).

Description Perennial subshrubs, shrubs, or trees to 40 m. Clusters of short, lateral roots ("proteoid roots") are often produced. The leaves are highly variable but mostly leathery, with pinnate venation. They are compound, deeply dissected or lobed, up to a fourth order of dissection, toothed, lobed and toothed, or simple and entire. The degree to which leaf shape varies between different stages in development is a striking feature of some species, especially many rain forest taxa. The inflorescence varies spectacularly but can be interpreted as variation on 2 basic "structural plans": simple or branching racemes of flower pairs in the subfamily Grevilleoideae, and racemose, simple, or compound inflorescences in the other subfamilies. These have been modified in some taxa by reduction, elaboration, compaction, and thickening to yield conelike, headlike, leafy, and 1-flowered inflorescences. The flowers are usually completely bisexual but sometimes unisexual (plants dioecious or monoecious), usually either radially or bilaterally symmetrical, with a perianth of 4 (rarely 3 or 5) free or variously united tepals. The stamens are 4 (rarely 3 or 5), opposite the tepals, with their filaments partly or wholly fused to them or rarely free, usually all fertile but sometimes 1 or more are sterile, usually with 2-locular anthers, but occasionally the lateral anthers are 1-locular. One to 4 hypogynous nectary glands are usually present, free or fused into a crescentic or annular nectary. The ovary is superior, sessile or stipitate, with a single carpel (rarely 2, free) containing 1 to many ovules inserted on marginal placentae. The style is usually distinct, often with its apex functioning as a pollen presenter. The fruits are dehiscent or indehiscent, dry or succulent, containing 1 to many, sometimes winged, seeds.

Classification Molecular systematic research has shown that the Proteaceae are most closely related to the superficially dissimilar Platanaceae, with which they share characteristic hair bases but no other distinctive morphological characters. Together, these are most closely related to the even more dissimilar aquatic family Nelumbonaceae. The most recent generic monograph[i] recognizes 5 subfamilies, 10 tribes, and 19 subtribes, most of which show fairly wide character variation, and easy morphological diagnosis of some of them is not possible.

SUBFAM. BELLENDENOIDEAE Only 1 sp., *Bellendena montana*, a subalpine shrub from Tasmania, with ebracteate racemes and the stamens free.

SUBFAM. PERSOONIOIDEAE Distinguished by exceptionally large chromosomes and absence of proteoid roots. There are 2 tribes:

Placospermeae With 1 sp., *Placospermum coriaceum,* a large rain forest tree with follicular fruits from northeastern Australia.

Persoonieae (4 genera, 103 spp.), fruits drupaceous, nonarid Australia, New Zealand, and New Caledonia.

SUBFAM. SYMPHIONEMATOIDEAE Proteoid roots absent; fruits dry, indehiscent, (2 genera, 3 spp.), Tasmania and southeastern Australia.

SUBFAM. PROTEOIDEAE Proteoid roots, non-auriculate cotyledons, indehiscent, dry or drupaceous fruits. There are 4 tribes:

Conospermeae (3 genera, 110 spp.), southwestern and southeastern Australia.

Petrophileae (2 genera, 56 spp.), southern Australia and South Africa.

Proteeae (2 genera, 118 spp.), widespread in sub-Saharan Africa and Madagascar.

Leucadendreae (12 genera, 322 spp.), southern Australia and southern Africa. There are 6 genera of Proteoideae, containing 21 spp. from New Caledonia, Australia, and Madagascar, ungrouped at tribal level.

SUBFAM. GREVILLEOIDEAE Proteoid roots, auriculate cotyledons, usually with paired flowers, fruits dry or succulent, follicular or indehiscent. There are 4 tribes:

Roupaleae (13 genera, 171 spp.), Australia, Melanesia, southern and eastern Asia, South and Central America, New Caledonia, and New Zealand.

Banksieae (4 genera, 173 spp.), restricted to non-arid parts of Australia and southern New Guinea.

Embothrieae (12 genera, 565 spp.), Australia, New Caledonia, South America, and Melanesia, including the large genera *Grevillea* (362 spp.) and *Hakea* (149 spp.).

Macadamieae (16 genera, 92 spp.), South and Central America, Australia, Southeast Asia, New Caledonia, Melanesia, South Africa, and Madagascar, including commercially important species of *Macadamia* (9 spp.). Two genera of Grevilleoideae, *Carnarvonia* and *Sphalmium,* are ungrouped at tribal level.

Recent molecular systematic analyses[ii,iii], have clarified the membership of major clades in the Proteaceae, strongly corroborating the membership of many groups that were previously recognized but also revealing some unanticipated relationships. According to these analyses, as well as floral morphology and

Genera 80 **Species** c. 1,700
Economic uses Edible nuts (*Macadamia*), cultivated ornamentals, some fine timber trees, source of traditional medicines

stomatal anatomy, *Bellendena montana* (i.e., subfamily Bellendenoideae) is the sister group to the rest of the family. The Persoonioideae are probably the next most basal subfamily. Symphionematoideae is the sister group to the Proteoideae, the subfamily in which it has previously been included. *Carnarvonia* and *Sphalmium* previously have been treated as separate subfamilies, but molecular systematic analysis, as well as fruit and seedling morphology, strongly places them in the Grevilleoideae.

Economic uses Proteaceae have long been used by various indigenous peoples as sources of food, medicine, tannins and other dyes, firewood, and timber. Many arborescent species have been logged commercially for timber, but this use is declining with the gradual cessation of rain forest logging in the southern hemisphere. The commercial significance of the Proteaceae is now dominated by trade in macadamia nuts (*Macadamia integrifolia, M. tetraphylla,* and their hybrids) and, to a much lesser extent, the sale of ornamentals. Important ornamental genera include *Protea, Leucadendron, Leucospermum, Serruria, Banksia, Grevillea, Telopea,* and *Persoonia.*   PHW

[i] Weston, P. H. Proteaceae. In: Kubitzki, K. (ed), *Families and Genera of Vascular Plants. IX. Flowering Plants. Eudicots.* Berlin, Springer-Verlag (2006).
[ii] Hoot, S. B. & Douglas, A. W. Phylogeny of the Proteaceae based on *atp*B and *atp*B-*rbc*L intergenic spacer region sequences. *Australian Syst. Bot.*11: 301–320 (1998).
[iii] Barker, N. P., Weston, P. H., Rourke, J. P., & Reeves, G. The relationships of the southern African Proteaceae as elucidated by internal transcribed spacer (ITS) DNA sequence data. *Kew Bull.* 57: 867–883 (2002).

PTAEROXYLACEAE
SNEEZEWOOD FAMILY

A pantropical, compound-leaved, woody family here given an expanded circumscription to include genera that have formerly been misfits in Sapindaceae, Simaroubaceae, and Rutaceae.

Distribution *Ptaeroxylon* has 1 sp. in woodland or closed forest, from northeastern Tanzania to the Cape Province of South Africa. *Cedrelopsis* comprises 8 spp. in dry deciduous to subhumid forest in Madagascar. *Bottegoa* has 1 sp. in dry woodland or acacia bush in Ethiopia, southern Somalia, and northeastern Kenya. *Harrisonia* has 3 spp., 1 sp. in dry forest or forest margins from Guinea and Somalia to Angola and Mozambique, 1 sp. from Myanmar and Hainan to Sulawesi, and 1 sp. from the Philippines to northern Australia. *Spathelia* contains c. 20 spp. from Mexico and the West Indies to Peru and southeastern Brazil, in thickets, woodland, and forest. *Dictyoloma* has 2 spp. in dry forests and disturbed areas in Brazil, Peru, and Bolivia.

Description Shrubs or more usually trees, *Spathelia* and *Dictyoloma* being little branched but up to 30 m high (*Spathelia*) with a terminal

Genera 6 **Species** 35
Economic uses Durable mahogany-like timber, some medicinal uses

cluster of large leaves[i]. The distribution and nature of secretory cavities and gland dots need further investigation (see discussion of classification below). The leaves are alternate, pinnate or (*Bottegoa, Dictyoloma*) bipinnate, or occasionally ternate in *Harrisonia,* in *Spathelia* up to 70 cm or more long with 32 or more pairs of leaflets, the leaflets often rhombic and asymmetric, and with entire to serrate margins. Stipules are absent (the recurved spines of *Harrisonia* being apparently stem prickles). The inflorescence is a terminal or subterminal or occasionally axillary panicle, sometimes large and many-flowered (up to 1 m across). The flowers are actinomorphic, bisexual or unisexual (monoecious in *Dictyoloma,* dioecious in *Ptaeroxylon* and *Cedrelopsis*), 4-merous in *Ptaeroxylon,* 4- to 5-merous in *Cedrelopsis, Bottegoa,* and *Harrisonia,* 5-merous in *Spathelia* and *Dictyoloma.* The sepals are 4–5, free or slightly fused. The petals are 4–5, free, well exceeding the sepals, yellow or white. The stamens are usually 4 in *Ptaeroxylon* and *Bottegoa,* 5 in *Cedrelopsis, Spathelia* and *Dictyoloma,* and 8–10 in *Harrisonia,* inserted outside a disk. The ovary is superior, of 2 carpels in *Ptaeroxylon* and *Bottegoa,* 2–3 in *Spathelia,* 3–5 in *Cedrelopsis,* 4–5 in *Harrisonia,* and 5 in *Dictyoloma.* Each carpel has 1 ovule in *Ptaeroxylon, Bottegoa,* and *Harrisonia,* 1–2 in *Spathelia,* 2 in *Cedrelopsis,* and 3–4 in *Dictyoloma,* subspherical to deeply longitudinally ridged. The fruit is indehiscent and conspicuously winged in *Bottegoa* (2 wings) and *Spathelia* (2–3 wings), or a capsule splitting longitudinally to the base in *Ptaeroxylon, Cedrelopsis,* and *Dictyoloma,* or a fleshy berry in *Harrisonia.* The seeds are winged in *Ptaeroxylon, Cedrelopsis, Spathelia,* and *Dictyoloma.*

Classification Bentham (in Bentham and Hooker 1862) placed *Ptaeroxylon* in the Sapindaceae, with which it shares a common ovule structure. In 1890, Radlkofer transferred it to Meliaceae. In 1959, Leroy pointed out the relationship with the Madagascan *Cedrelopsis* and recognized the family Ptaeroxylaceae for the 2 genera, an opinion confirmed by Pennington and Styles in their generic revision of the Meliaceae in 1975. Growing chemical evidence, especially the presence of chromones, for linking *Spathelia* and *Harrisonia* with Ptaeroxylaceae and Cneoraceae was reviewed in 1983[ii]. In 1995, a careful study of *Bottegoa* showed that this genus, hitherto placed in Sapindaceae, is best also referred to Ptaeroxylaceae[iii], and an affinity with *Harrisonia* (a misfit in the Simaroubaceae) was also pointed out. *Bottegoa* was included in Ptaeroxylaceae by Verdcourt in the *Flora of Tropical East Africa* in 1996. Meanwhile, work on Simaroubaceae fruit morphology[iv] and molecular data[v] had clearly shown *Harrisonia* to be misplaced in that family and noted the chemical evidence placing it with *Spathelia*. Both this and *Dictyoloma*, also from South America, have respectively been treated as isolated genera in their own subfamilies in Rutaceae. Interestingly, *Dictyoloma* not only shares an unusual growth habit with *Spathelia* despite floral differences (cf. also *Tetrameles* and *Octomeles* in Datiscaceae) but also shows a marked similarity to *Ptaeroxylon*, especially in its fruit. *Bottegoa* seems to connect to the South American genera in having bipinnate leaves (as in *Dictyoloma*) and broadly winged, indehiscent, 2-seeded fruits (as in many *Spathelia* species). A broader molecular analysis in 2000[vi] has placed 5 of these 6 genera (*Cedrelopsis* was not included in the analysis), plus *Cneorum* (actually *Neochamaelea*), together as a sister group to the main body of Rutaceae and has included all of these in a broad concept of the latter. This has been followed by APG II.

It seems, however, to be unwise to include the group in Rutaceae. The survey by van der Ham *et al*[ii] has quoted leaf anatomy, extrafloral nectaries, solitary oil cells rather than schizolysigenous oil glands, presence of chromones of the ptaeroxylin group, and sometimes solitary apotropous ovules, as characters conflicting with Rutaceae. The solitary oil cells are found in all 4 Old World genera, but the New World genera need investigation. The pollen of all 6 genera is similar, with a reticulate exine, but does not provide any clear separation from Rutaceae (M. M. Harley, *pers. comm.*). Perhaps the most morphologically divergent genus is *Harrisonia*, with stem prickles, small often axillary inflorescences, 8–10 stamens, and a fleshy fruit with rutaceous exocarp, but chemically, and in its oil cells, it falls with the other genera. Although it is difficult to find one character that will now define Ptaeroxylaceae, the reticulate morphological relationships make it desirable to adopt a new concept of the family to include these 6 genera. This may be paraphyletic with respect to Cneoraceae, but the latter is so distinctive (see that family) that it can comfortably be excluded.

Economic uses *Ptaeroxylon obliquum* provides a hard and durable wood resembling mahogany, which is used in construction, for telegraph poles, and for machine bearings, where it lasts longer than brass or iron. When sawn, the dust causes violent sneezing, giving rise to the common name Sneezewood. Madagascan *Cedrelopsis grevei*, "Katrafay," is well known as the source of a medicinal tonic.   RKB

i Pennington, T. D., Reynel, C., & Daza, A. *Illustrated Guide to the Trees of Peru*. Sherborne, D. Hunt (2004).
ii Taylor, D. A. H. & Waterman, P. G. Pp. 353–375 & 377–400. In: Waterman, P. G. & Grundon, M. F. (eds), *Chemistry and Chemical Taxonomy of the Rutales*. London, Academic Press (1983).
iii van der Ham, R. W. J. M. *et al.* *Bottegoa* Chiov. transferred to Ptaeroxylaceae. *Kew Bull.* 50: 243–265 (1995).
iv Fernando, E. S. & Quinn, C. J. Pericarp anatomy and systematics of the Simaroubaceae *sensu lato*. *Australian J. Bot.* 40: 263–289 (1992).
v Fernando, E. S., Gadek, P. A., & Quinn, C. J. Simaroubaceae, an artificial construct: evidence from *rbc*L sequence variation. *American J. Bot.* 82: 92–103 (1995).
vi Savolainen, V. *et al.* Phylogeny of the eudicots: a nearly complete familial analysis based on *rbc*L gene sequences. *Kew Bull.* 55: 257–309 (2000).

PTELEOCARPACEAE

A single species, *Pteleocarpa lamponga,* a tree up to 40 m in dipterocarp forests of Borneo, the Malay Peninsula, Sumatra, and Thailand. The leaves are alternate, up to 14 × 6 cm, elliptical, cuneate, acuminate, thin to slightly coriaceous, glabrous, with a petiole up to 2 cm. The inflorescence is a terminal many-flowered

Genera 1 Species 1
Economic uses Timber

panicle with pedicels up to 8 mm. The flowers are up to 6 mm, actinomorphic, and bisexual. The sepals are 5, fused below. The corolla has a tube c. 2 mm and 5 lobes 3-6 mm, yellow. The stamens are 5, exserted from between the corolla lobes. The ovary is superior, of 2 fused carpels, with a style divided almost to the base, with 2 locules, each including 1 large pendent ovule and 1 small erect ovule, only the former developing in fruit. The fruit is a characteristic, flat, suborbicular samara, up to 5.5 × 5 cm, with a single central seed and broad scarious wings on each side, divided by a suture leading to the retuse apex.

The similarity of the inflorescence to that of *Cordia* has led to its placement by some authors in Boraginaceae, but this affinity and other suggested relationships have been disputed[i] and the fruit is different from anything in that family. Molecular evidence (R. Olmstead, *pers. comm.*) places it well apart from Boraginaceae, and family status seems inevitable. The tree has been well described and illustrated, with local use as timber in Sumatra recorded[ii].   RKB

i Veldkamp, J. F. Notes on *Pteleocarpa, incertae sedis*. *Fl. Males. Bull.* 10: 47–50 (1988).
ii Dayang Awa, A. L. *Pteleocarpa*. In: Soepadmo, E., Wong, K. M., & Saw, L. G. *Tree Flora of Sabah and Sarawak* 2: 103–105 (1996).

PTEROSTEMONACEAE

A monogeneric family of 2 or 3 spp. of shrubs with showy, aromatic, white or pink flowers, endemic to central and southern Mexico. The leaves are alternate, simple, ovate to nearly orbicular, glossy above, hairy beneath, with conical or peltate hairs and minute stipules. The flowers are actinomorphic, bisexual, in few-flowered corymbose cymes. The sepals are 5, connate below into a tube, with erect triangular valvate lobes. The petals are 5, free, imbricate, reflexed. The stamens are 5, with broad winged filaments

Genera 2
Species 2–3
Economic uses
Local medicines

that are toothed near the apex. There are 5 staminodes opposite the petals, the filaments coarsely toothed like the stamens. The ovary is inferior, with 5 fused carpels and locules; the placentation is axile with 4–6 ovules per locule. The fruit is a woody, septicidal, few-seeded capsule crowned by the persistent calyx and petals.

Pterostemon was included in the Grossulariaceae by Cronquist (1981) and as a separate family in the Saxifragales by Takhtajan (1997). Molecular evidence indicates that *Pterostemon* and *Itea* (Iteaceae) have a sister group relationship[i,ii,iii] within the Saxifragales, but they differ in several morphological features.   VHH

i Savolainen, V. *et al.* Phylogeny of the eudicots: a nearly complete familial analysis based on *rbc*L gene sequences. *Kew Bull.* 55: 257–309 (2000).
ii Soltis, D. E. & Soltis, S. Phylogenetic relationships among Saxifragaceae *sensu lato*: a comparison of topologies based on 18S rDNA and *rbc*L sequences. *American J. Bot.* 84: 504–522 (1997).
iii Soltis, P. S. *et al.* Angiosperm phylogeny inferred from 18S rDNA, *rbc*L, and *atp*B sequences. *Bot. J. Linn. Soc.* 133: 381–461 (2000).

PUTRANJIVACEAE

A small pantropical family of shrubs and trees, similar to Euphorbiaceae, but always with drupaceous fruits.

Distribution Pantropical distribution, but mainly in the Old World. Only c. 20 *Drypetes* species occur in the Americas, the remainder in Africa (c. 70 spp.) and Asia and Australasia (c. 100 spp.). *Putranjiva* (4 spp.) is restricted to Asia (India to Ryukyu Islands, in the south to New Guinea).

Description Small to large forest trees or shrubs, with simple leaves arranged in 2 distinct rows (distichous), or rarely opposite or subopposite (*Drypetes oppositifolia*). The indumentum is simple and exudate is absent. Some species contain mustard oil and smell strongly of horseradish. The leaves are penninerved and eglandular, usually have asymmetrical leaf bases and entire to toothed margins; some species have hard, shiny leaves with spiny margins; stipules small. The flowers are unisexual (plants dioecious; rarely monoecious), arranged in axillary fascicles (sometimes cauline), and subtended by inconspicuous bracts. The sepals are 3–7, usually strongly imbricate and often unequal, often caducous in fruit in pistillate flowers. Petals are absent in both sexes. The staminate flowers have a central disk that sometimes envelopes the stamens from the inside (*Drypetes*) or lack a disk (*Putranjiva*). The disk in pistillate flowers is annular (*Drypetes*) or absent (*Putranjiva*). The stamens are 2–20, sometimes up to 50, the filaments free (sometimes subconnate in *Putranjiva*), and anthers dehiscing longitudinally. The pollen is 3-colporate, prolate to prolate-spheroidal, with a semi- to pertectate exine. The pistillode is minute or absent. The ovary is superior, with 1–6 carpels forming as many locules, each containing 2 anatropous ovules with axile placentation that are provided with an obturator. Styles are short and stigmas are usually distinctly flattened, entire, or bifid. The fruits are drupaceous, green to brown (sometimes covered with various indumentum), ellipsoidal, fusiform or globose, sometimes lobed; pericarp usually hard, sometimes fleshy. The seeds are ecarunculate and solitary per locule due to abortion of the second ovule. The endosperm is copious, and the cotyledons are flat. The record of unusual sieve-element plastids (PIcs type) in *Drypetes* and *Bischofia*, but no other biovulate Euphorbiaceae *sensu lato*[i] was erroneous. *Drypetes* has S-type plastids similar to all other examined species of Euphorbiaceae *sensu lato*[ii].

Classification Putranjivaceae are a segregate of Euphorbiaceae *sensu lato*. They were excluded in molecular studies[iii] as a separate lineage and found to be sister to Lophopyxidaceae. They belong to Malpighiales (rosids; fabids), as do all segregates of Euphorbiaceae *sensu lato*. Placement in previous classifications[iv,v] was in Euphorbiaceae-Phyllanthoideae-Drypeteae. The position of Putranjivaceae in the phyllanthoids had not been questioned until the late 20th century[vi], mainly because their morphological characters can easily be accommodated within the diverse Phyllanthaceae. However, the characters shared between Putranjivaceae and Phyllanthaceae can be interpreted as plesiomorphic and Putranjivaceae always have drupes and never the typical euphorbiaceous schizocarps. They also have a central staminate disk, which is very rare in Phyllanthaceae. Another significant difference of Putranjivaceae from the other biovulate euphorbiaceous families Phyllanthaceae and Picrodendraceae is the lack of a nucellar beak[vii]. Putranjivaceae differ in their

chromosome numbers (n=20) from Phyllanthaceae which mostly have n=13[viii]. The presence of mustard oils (glucosinolates) in Putranjivaceae, unique outside Brassicales[ix], is clearly homoplasious. Molecular phylogenetic analyses strongly support a relationship of Putranjivaceae to the monospecific liana Lophopyxidaceae with which it shares leaves with theoid teeth, dioecy, and 2 pendulous apical-axile anatropous ovules per locule (each with an obturator), 1 of which is aborted in the indehiscent fruit. Phylogenetic relationships within Putranjivaceae are still poorly understood, and whether *Sibangea* and *Putranjiva* are distinct from the large genus *Drypetes* has been variously debated. African *Sibangea* (3 spp.) closely resembles the large pantropical genus *Drypetes* and is of questionable generic distinctiveness. The Asian genus *Putranjiva* is sister to *Drypetes*, including *Sibangea* when investigated with *rbcL*[iii], but this relationship lacks support. *Lingelsheimia* (= *Aerisilvaea*, *Danguyodrypetes*), which had been included in *Drypeteae* in previous classifications[iv,v] belongs to Phyllanthaceae.

Economic uses Some members of the family are used for timber.   PH

[i] Behnke, H.-D. Sieve-element characters. *Nordic J. Bot.* 1: 381–400 (1981).
[ii] Behnke, H.-D. Personal communication.
[iii] Wurdack, K. J. *et al.* Molecular phylogenetic analysis of Phyllanthaceae (Phyllanthoideae pro parte, Euphorbiaceae *sensu lato*) using plastid *rbc*L DNA sequences. *American J. Bot.* 91: 1882–1990 (2004).
[iv] Webster, G. L. Synopsis of the genera and suprageneric taxa of Euphorbiaceae. *Ann. Missouri Bot. Gard.* 81: 33–144 (1994).
[v] Radcliffe-Smith, A. *Genera Euphorbiacearum*. Richmond, Royal Botanic Gardens, Kew (2001).
[vi] Meeuse, A. D. J. *The Euphorbiaceae auct. plur., an unnatural taxon.* Delft, The Netherlands, Eburon (1990).
[vii] Tokuoka, T. & Tobe, H. Embryology of tribe Drypeteae, an enigmatic taxon of Euphorbiaceae. *Pl. Syst. Evol.* 215: 189–208 (1999).
[viii] Hans, A. S. Chromosomal conspectus of the Euphorbiaceae. *Taxon* 22: 591–636 (1973).
[ix] Rodman, J., Karol, K. G., Price R. A., & Sytsma K. J. Molecules, morphology, and Dahlgren's expanded order Capparales. *Syst. Bot.* 21: 289–307 (1996).

QUIINACEAE

A family of neotropical trees and shrubs resembling Ochnaceae but with opposite or whorled leaves with interpetiolar stipules

Distribution Lowland forests of the New World, especially the Amazon Basin. *Quiina* has 34 spp. from Belize and Jamaica to Bolivia and southeastern Brazil. *Lacunaria* has 10 spp. from Costa Rica to the Amazon Basin. *Touroulia* has 2 spp. from Venezuela and the Guianas to northern Brazil. *Froesia* has 5 spp. in the northern Amazon Basin.

Description Shrubs or trees up to 20 m or more, rarely lianas. The leaves are opposite in *Quiina* (except *Q. pteridophylla*), *Froesia*, *Touroulia*, and *Lacunaria oppositifolia*, and

in whorls of 3–4(5) in other *Lacunaria* species and *Q. pteridophylla*. In *Froesia*, the leaves are pinnate and up to 2 m long, with long petioles, large opposite or subopposite to alternate leaflets on petiolules up to 5 cm long. In *Touroulia*, they are more or less pinnate with long petioles

Genera 4   **Species** 51
Economic uses None

and some or all of the leaflets sessile and decurrent down the rhachis. In *Quiina* and *Lacunaria*, the leaves are simple (? 1-foliate), with short petioles resembling pulvini (in some species the seedling leaves are said to be pinnate). The leaves or leaflets are up to 50 cm, usually rather shiny, with closely parallel and usually rather obscure tertiary veins and linear cells between the finer veins, the margins undulate or crenate to serrate. The stipules are conspicuous, persistent, linear or sometimes (e.g., *Q. glaziovii*) foliaceous. The inflorescence is a terminal or axillary raceme or panicle. The flowers are actinomorphic, bisexual (*Froesia*), unisexual (plants androdioecious in *Quiina* and *Touroulia*, dioecious in *Lacunaria*[i]), usually 4- or 5-merous, and usually small (up to 3 mm) but larger in *Froesia*. The sepals are usually 4 or 5, rarely 2–3 or 6–8, imbricate, unequal. The petals are usually 4 or 5(–8). The stamens are 10–80, or in *Froesia* 100–200. The ovary is superior, of 3 free carpels in *Froesia*, united in the other genera, and with 2–8 carpels, or in *Lacunaria* up to 14, with as many free styles as carpels, with 2(–4) ovules per carpel. The fruit is 3 fleshy follicles in *Froesia*, a tardily dehiscent capsule in *Lacunaria*, and a fleshy berry in *Quiina* and *Touroulia*.

Classification The family has usually been regarded as in or close to the Theales, as also has Ochnaceae but, surprisingly, a close relationship between the 2 families was not generally appreciated. Molecular evidence, however, has clearly linked Ochnaceae, Quiinaceae, and Medusagynaceae—see the latter for reference. Thus, some authors have sunk the last 2 families[ii]. In APG II, Ochnaceae is listed as optionally including Quiinaceae and Medusagynaceae, all placed in the Malpighiales. However, in most of the numerous analyses available, either Quiinaceae appear as a sister group to Ochnaceae and Medusagynaceae or sometimes *Medusagyne* is sister to the other families. In one case, Quiinaceae is sister to Ochnoideae but *Luxemburgia* of the Sauvagesioideae is basal to the whole group[iii], so that both Quiinaceae and Medusagynaceae nest within Ochnaceae. The topology of the cladograms can be quite different according to which genera are included in the analysis. The sampling of the Ochnaceae has been so limited that no clear view of phylogenetic relationships can be deduced. It will be interesting to see an analysis that includes

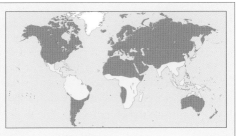

Genera (1)2(3)   **Species** c. 200
Economic uses Timber trees

Froesia, which retains the more primitive features of Quiinaceae (including pinnate leaves), together with the 1 member of the Ochnaceae with pinnate leaves, *Rhytidanthera*, as well as other genera of the Sauvagesioideae.   RKB

i Schneider, J. V., Swenson, U., & Zizka G. Phylogenetic reconstruction of the neotropical family Quiinaceae (Malpighiales) based on morphology with remarks on the evolution of an androdioecious sex distribution. *Ann. Missouri Bot. Gard.* 89: 64–76 (2002).
ii Chase, M. W. *et al.* When in doubt, put it in *Flacourtiaceae*: a molecular phylogenetic analysis based on plastid *rbc*L DNA sequences. *Kew Bull.* 57: 141–181 (2002).
iii Savolainen, V. *et al.* Phylogeny of the eudicots: a nearly complete familial analysis based on *rbc*L gene sequences. *Kew Bull.* 55: 257–309 (2000).

QUILLAJACEAE
SOAP BARK, QUILLAY, PALO JABÓN

A single genus, *Quilla-ja*, of 2 spp. of evergreen trees endemic to South America: *Q. brasiliensis* occurs in Brazil, Argentina, and Uruguay, while *Q. saponaria* is restricted to the Andean coastal zone of central Chile. The leaves are alternate, spirally arranged, dark green, simple, serrate, with tiny stip-

Genera 1 | Species 2
Economic uses Source of saponins

ules. The inflorescence is terminal, often on short side shoots, a few flowered cyme. The flowers are radially symmetrical, bisexual, the sepals are 5, valvate, each with a nectary, the petals are 5, alternating with the sepals, the stamens in a highly distinctive arrangement forming 2 whorls of 5, one whorl borne part way along the sepals, the other between the petals and carpels. The ovary is superior, of 5 basally joined carpels forming a lobed fruit that splits revealing 2 locules per lobe.

The genus has generally been placed in the Rosaceae[i]. Molecular evidence shows the family to have a close affinity with the legumes[ii,iii], the Surianaceae, and Polygalaceae in Fabales (APG II). This relationship is further supported by the presence of styloids in the phloem[iv]. *Quillaja saponaria* is now of major economic importance due to exploitation for triterpenoid saponins used in human vaccines, as a food emulsifier, feed additive, in hair treatment, and in fish farming. The saponins are usually extracted from the bark but are present in other parts of the tree. Currently 60,000 trees each year are harvested from the wild, yielding bark exports worth US$4million per annum[v].   AC

i Brummitt, R. K. *Vascular Plant Families and Genera*. Richmond, Royal Botanic Gardens, Kew (1992).
ii Morgan, D. R., Soltis, D. E., & Robertson, K. R. Systematic and evolutionary implications of *rbc*L sequence variation in Rosaceae. *American J. Bot.* 81: 890–903 (1994).

iii Soltis, D. E. *et al.* Angiosperm phylogeny inferred from 18S rDNA, *rbc*L, and *atp*B sequences. *Bot. J. Linn. Soc.* 133: 381–461 (2000).
iv Lersten, N. R. & Horner, H. T. Macropattern of styloid and druse crystals in *Quillaja* (Quillajaceae) bark and leaves. *Int. J. Plant Sci.* 166: 705–711 (2005).
v http://www.fondef.cl/ingles/resumen/forestal/D9712/D9712010eng.html

QUINTINIACEAE
POSSUMWOOD

A poorly known monogeneric family of 25 spp. of evergreen trees or shrubs from the southwestern Pacific (Philippines, New Zea-land, New Caledonia, and Australia). The leaves are alternate, simple, with peltate scales, the margins entire or serrate; stipules absent. The flowers are actinomorphic, bisexual, in terminal or axillary racemes or panicles. The sepals are 5, connate below. The petals are 5, free or rarely connate. The stamens are 5, the filaments free. The ovary is inferior, of 3–5 fused carpels, with parietal or axile placentation; ovules numerous. The fruit is a septicidal capsule.

Genera 1 | Species 25
Economic uses None

The affinities of *Quintinia* are not yet clear. It is often included in the Saxifragaceae or Escalloniaceae. Lundberg[i] considers *Quintinia* to be closest to *Sphenostemon* and *Paracryphia* and has proposed grouping all 3 genera in the Paracryphiaceae, treated by APG as a Euasterid II but unplaced as to order.   VHH

i Lundberg, J. Phylogenetic studies in the Euasterids II with particular reference to Asterales and Escalloniaceae. Chapter V: A well resolved and supported phylogeny of Euasterids II based on a Bayesian inference, with special emphasis on Escalloniaceae and other incertae sedis. Uppsala, Sweden, *Acta Univ. Ups. Comprehensive Summaries of Uppsala Dissertations from the Faculty of Science and Technology* 676 (2001).

RAFFLESIACEAE

A tropical and subtropical family of stem and root parasites.

Distribution *Rafflesia* (c. 15 spp.) grows in western Malaysia, as does *Rhizanthes* (2 spp.), while *Sapria* (2 spp.) extends from Assam and Bhutan to Myanmar, Cambodia, Vietnam, China, and Thailand.

Description Endoparasitic, achlorophyllous herbs, with a vegetative body similar to a mycelium, residing completely in the roots and stems of the vine *Tetrastigma* (Vitaceae). There are no stems, and the leaves are represented by a variable number of bracts subtending the flowers. The flowers are solitary or in inflorescences and arise from within the host; they are usually unisexual (plants monoecious or dioecious), occasionally also with bisexual flowers (*Rhizanthes*), actinomorphic, ranging in size from 8 cm (*Rhizanthes*) to more than 1 m and weighing up to 7 kg (*Rafflesia*). The flower color varies from pure white to brown (*Rhizanthes*), orange, brick red, with white spots or yellow and black (*Rafflesia*)[i]. They are often malodorous, smelling of rotten flesh (*Rafflesia*) or rotting apples (*Rhizanthes*). The perianth is uniseriate with 5 (*Rafflesia*), 10 (*Sapria*), or 16 (*Rhizanthes*) tepals, connate below and fused to the ovary, forming a perigone tube, the lobes imbricate or valvate (*Rhizanthes*). The perianth is fused distally to the ovary into a perigone tube that forms a centrally located aperture called the diaphragm

RAFFLESIACEAE. 1 *Rafflesia manillana* (a) male flower (x⅓); (b) flower buds (x¼). 2 *R. patma* half male flower bud showing "mycelia" ramifying through host tissue (x¼). 3 *R. rochussenii* vertical section of fruit (x⅓).

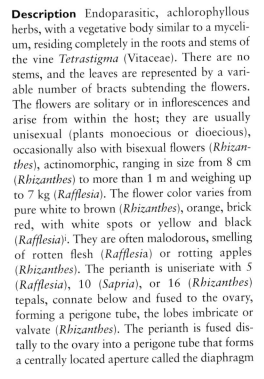

in *Sapria* and *Rafflesia* (but not *Rhizanthes*). In the staminate flowers, the androecium consists of a ring of 12–40 extrorse anthers with 2 (*Rhizanthes, Sapria*) to several (*Rafflesia*) pore-s[i], the connectives fused and adnate to a central column (gynostemium). In the pistillate flowers, the ovary is inferior of 4–8 fused carpels, 1-loc-

Genera 3
Species c. 20
Economic uses None

ular (*Rhizanthes*) or with numerous locules forming a honeycomb-like structure due to intrusive placentas (*Rafflesia, Sapria*). The fruit is baccate, indehiscent, or irregularly dehiscent, with many small seeds. Nickrent's Parasitic Plant Connection[ii] is a valuable source of information and illustrations.

Classification The Rafflesiaceae has been interpreted by some authors in a broad sense so as to include the Cytinaceae, Apodanthaceae, and Mitrastemonaceae as subfamilies or tribes. Others, such as Takhtajan, have interpreted the Rafflesiaceae in a narrow sense and split off these groups as distinct families within the order Rafflesiales. Support for this viewpoint comes from some molecular analyses[iii]. On the other hand, other molecular analyses suggest that that Rafflesiales are not monophyletic but composed of several independent lineages: that formed by *Rafflesia, Rhizanthes,* and *Sapria* (Rafflesiaceae in the narrow sense) should be placed in or near the Malpighiales[iv]; the Apodanthaceae (q.v.) should possibly belong in the Malvales or the Cucurbitales; the Cytinaceae should also be placed in or near the Malvales; and the Mitrastemonaceae should be included in the Ericales[iii,v]. Davis and Wurdack[v] found that other gene sequence data indicated a relationship between Rafflesiaceae and Vitaceae, which they postulated could be caused as a result of horizontal gene transfer from Vitaceae to Rafflesiaceae. The relationships and phyletic position of this group of families are still unclear. *Rafflesia arnoldii* is reputed to be the largest flower in the world, and it and other species are an attraction to ecotourists. VHH

[i] Blarer, A., Nickrent, D. L., & Endress, P. K. Comparative floral structure and systematics in Apodanthaceae (Rafflesiales). *Plant Syst. Evol.* 245: 119–142 (2004).
[ii] Nickrent, D. The Parasitic Plant Connection. http://www.science.siu.edu/parasitic-plants/
[iii] Nickrent, D. L. *et al.* Phylogenetic inference in Rafflesiales: the influence of rate heterogeneity and horizontal gene transfer. *BMC Evol. Biol.* 4: 40 (2004). http://www.biomedcentral.com/1471-2148/4/40
[iv] Barkman, T. J. *et al.* Mitochondrial DNA sequences reveal the photosynthetic relatives of *Rafflesia*, the world's largest flower. *Proc. National Acad. Sci. USA* 101: 787–792 (2004).
[v] Davis, C. C. & Wurdack, K. J. Host-to-parasite gene transfer in flowering plants: Phylogenetic evidence from Malpighiales. *Science* 305: 676–678 (2004).

RANUNCULACEAE

BUTTERCUP, MONKSHOOD, COHOSH, COLUMBINE

One of the few truly cosmopolitan families, spreading across all continents except Antarctica and from sea level to high altitude.

Distribution The family is distributed throughout the world but is centered in temperate and cold regions of the northern and southern hemispheres, with few tropical species. At least 75% of the genera are found in East Asia, and many are endemic; about 45% of genera are found in North America, of which 5 are endemic; and a similar number of genera occur in Europe but

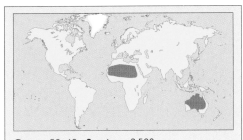

Genera 50–60 **Species** c. 2,500
Economic uses Source of alkaloids, ornamentals

RANUNCULACEAE. 1 *Trollius europeaus* flowering shoot (×⅔).
2 *Helleborus cyclophyllus* flowering shoot (×⅔). 3 *H. niger* (Christmas Rose) half flower showing large sepals, small tubular nectaries (modified petals), numerous stamens, and carpels slightly fused at base, each containing numerous ovules (×⅔).

none are endemic. Most species grow in mesic or wet environments, some are aquatic (*Ranunculus* subg. *Batrachium*).

Description Perennial, or sometimes annual, herbs, rarely shrubby, or lianas (many *Clematis*). The perennial herbaceous species usually persist by means of a condensed sympodial rootstock or rhizome, swelling up into storage tubers in *Aconitum* and some *Ranunculus*. The leaves are opposite or spirally arranged, simple (e.g., *Myosurus, Caltha*), palmately lobed (e.g., *Ranunculus*) or compound, sometimes highly so (e.g., *Thalictrum*), usually with a petiole and without stipules. The aquatic species of *Ranunculus* (*R.* subg. *Batrachium*) usually have special submerged, much-dissected leaves with capilliform segments. In *Clematis*, the petiole is sensitive to touch and supports the stem as a tendril. In *Clematis aphylla*, the whole of the leaf becomes a tendril, photosynthesis being carried out by the stem cortex. The petioles are broadened into a sheathing base. Stipules are absent, except in *Thalictrum, Caltha, Trollius,* and *Ranunculus*. The inflorescence is terminal. In *Eranthis* and some species of *Anemone*, the flowers are solitary; more usually they are in racemes or cymes. In *Anemone*, (including *Pulsatilla*), and *Nigella* there is a protective involucre of leaves beneath the flower, alternate with the calyx segments. The flowers are usually hermaphrodite and regular but irregular in *Aconitum* and *Delphinium*. The flower has its parts typically arranged in spirals along a more or less elongated receptacle, but often the perianth

3 2 1

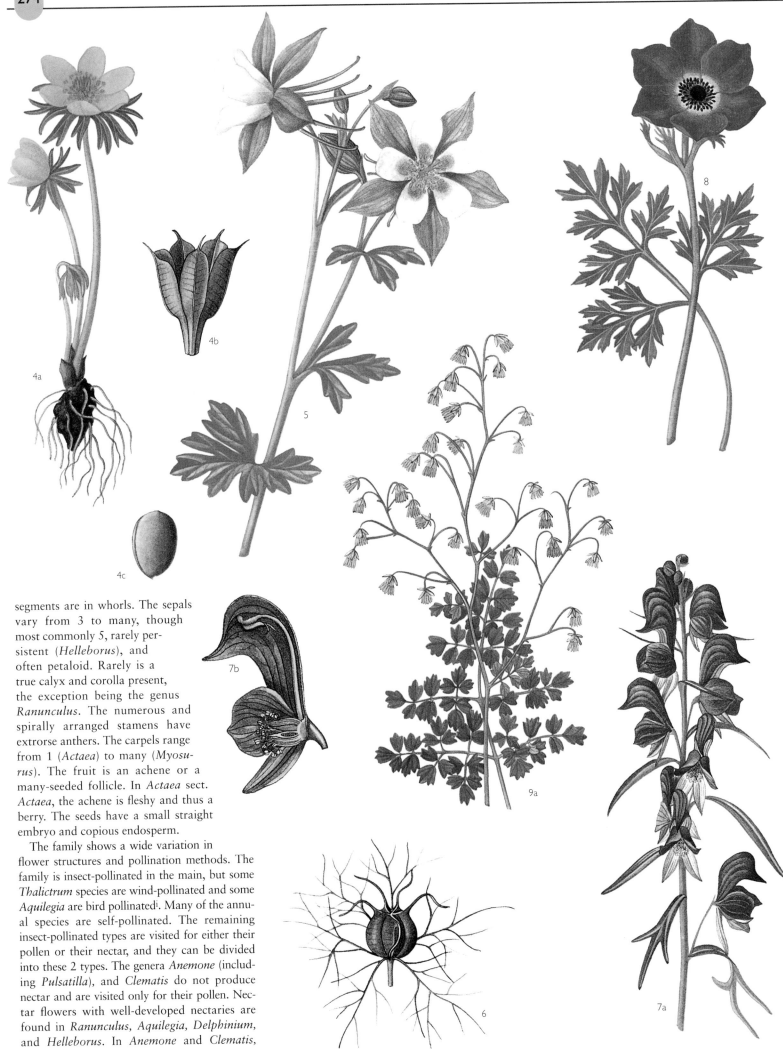

segments are in whorls. The sepals vary from 3 to many, though most commonly 5, rarely persistent (*Helleborus*), and often petaloid. Rarely is a true calyx and corolla present, the exception being the genus *Ranunculus*. The numerous and spirally arranged stamens have extrorse anthers. The carpels range from 1 (*Actaea*) to many (*Myosurus*). The fruit is an achene or a many-seeded follicle. In *Actaea* sect. *Actaea*, the achene is fleshy and thus a berry. The seeds have a small straight embryo and copious endosperm.

The family shows a wide variation in flower structures and pollination methods. The family is insect-pollinated in the main, but some *Thalictrum* species are wind-pollinated and some *Aquilegia* are bird pollinated[i]. Many of the annual species are self-pollinated. The remaining insect-pollinated types are visited for either their pollen or their nectar, and they can be divided into these 2 types. The genera *Anemone* (including *Pulsatilla*), and *Clematis* do not produce nectar and are visited only for their pollen. Nectar flowers with well-developed nectaries are found in *Ranunculus*, *Aquilegia*, *Delphinium*, and *Helleborus*. In *Anemone* and *Clematis*,

RANUNCULACEAE. 4 *Eranthis hyemalis* (Winter Aconite) (a) habit (x⅔); (b) fruit—a group of follicles (x1⅓); (c) seed (x4). 5 *Aquilegia caerulea* shoot bearing regular flowers with spurred perianth segments (x⅔). 6 *Nigella damescena* fruit—a capsule surrounded by feathery bracts (x⅔). 7 *Aconitum napellus* (a) shoot bearing irregular flowers (x⅔); (b) half flower with large hooded upper sepal (x1⅓). 8 *Anemone coronaria* leaf and flower (x⅔). 9 *Thalictrum minus* (a) leaf and flowers in a terminal inflorescence (x⅔); (b) flower with inconspicuous perianth and numerous large stamens (x4). 10 *Clematis alpina* (a) shoot with opposite leaves with long petioles that aid climbing and axillary flowers (x⅔); (b) fruit—a group of achenes, each with a persistent hairy style (x2⅔). 11 *Ranunculus sp.* half flower with free sepals and petals, numerous stamens and carpels, the latter with single ovules (x2). 12 *Myosurus minimus* (a) habit (x⅔); (b) flower with long receptacle bearing numerous carpels (x2⅔). 13 *Ficaria verna* habit showing tuberous roots (x⅔).

insects are attracted by brightly colored sepals, in *Ranunculus* by showy petals (with prominent nectar pouches, known as honey-leaves); in *Aconitum* by showy sepals and petals; and in some *Thalictrum* species by attractive stamen filaments or anthers. The Ranunculaceae are generally protandrous, the stamens shedding their pollen before the ovaries mature, but the reverse case of protogyny (ovary maturing before the stamens) also occurs. However, these characteristics may not be strongly developed, and some anthers may not have shed their pollen before the ovaries start to mature. These processes favor cross-pollination and outbreeding. Seed dispersal is by a variety of agencies. *Clematis* and *Anemone* sect. *Pulsatilla* have styles that lengthen into long feathery structures adapted for wind distribution after pollination. Some species of *Ranunculus* (e.g., *R. arvensis*) have tubercules or hooked spines on the surface for animal dispersal. *Helleborus* species have an elaiosome or oil-containing swelling on the raphe, which attracts ants that then disperse the seeds (myrmecochory).

Classification The family are derived within the Ranunculales[ii], forming the sister group to Berberidaceae. The main division in the family is traditionally between 2 chromosome types[iii]: "T" type in *Thalictrum* and its relatives, and "R" Type in *Ranunculus* and its relatives. Cladistic analysis of morphological[iv] and molecular[v] data suggests that T-type occurs in all basal lineages of the family and R-type occurs later. The widely used classification of Tamura[vi], comprising 2 subfamilies and 5 tribes, later updated[vii] to include 5 subfamilies and 10 tribes, based on chromosome size and morphological data, does not correlate well with the phylogenetic patterns found in the family. The family is now generally treated as comprising 3 or 5 subfamilies based on molecular evidence[ii,v,viii].

SUBFAM. COPTIDOIDEAE *Coptis* (15 spp.) is native to western North America and East Asia. *Xanthorhiza simplicisissima* (Yellowroot) is a small shrub native to eastern North America. Both have T-type chromosomes.

SUBFAM. THALICTROIDEAE Widespread, with 9–11 genera (*Thalictrum*, c. 330 spp; *Aquilegia*, c. 80 spp.). T-type chromomes.

SUBFAM. RANUNCULOIDEAE Worldwide, with 39–45 genera (*Ranunculus*, c. 600 spp.[ix]; *Delphinium*, c. 360 spp.; *Clematis*, c. 325 spp.; *Aconitum*, c. 300 spp.; *Anemone*, c. 200 spp.[x]; *Trollius*, c. 30 spp.; *Adonis*, c. 26 spp.; *Actaea*, c. 25 spp.[xi]), R-type chromosomes. *Asteropyrum*, usually treated in Coptideae[vii], is sister[xii] to tribe Actaeeae[xiii] based on DNA sequence evidence. A recent molecular evaluation of *Ranunculus* showed it to be a well supported and monophyletic group sister to *Ceratocephala* and *Myosurus*[xiv], which excludes *Ficaria* (Lesser Celandine).

Economic uses Many genera have species that provide excellent plants for the garden. Most of the generic names are familiar to nurserymen and gardeners as showy herbaceous border plants (in the case of *Clematis*, climbers): *Trollius*, (Globe Flower), *Aconitum* (Monkshood and Aconites), *Helleborus* (Christmas rose, Lenten Lily, and other Hellebores). Other genera that include ornamental species (many with cultivated varieties) are *Actaea* (Baneberries, Herb Christopher), *Anemone, Aquilegia* (columbines), *Ranunculus* (buttercups, Spearwort and crowfoots), *Caltha* (Marsh Marigold), *Delphinium* (including larkspurs), *Eranthis* (Winter Aconite), *Hepatica, Nigella* (Love-in-a-Mist, Fennel Flower), *Thalictrum*, and *Ficaria*. A number of genera are highly poisonous and have caused deaths. Victorian medical books give lurid details of the symptoms and deaths of gardeners who have inadvertently eaten *Aconitum* tubers, having confused them with Jerusalem artichokes. Many species have traditional medicinal uses, several having been researched for their active constituents although this has led some species close to extinction through overharvesting. AC

[i] Grant, V. Effects of Hybridization and selection on floral isolation. *Proc. Natl. Acad. Sci. USA.* 90: 990–993 (1993).
[ii] Soltis, D. E. *et al.* Angiosperm phylogeny inferred from 18S rDNA, *rbc*L, and *atp*B sequences. *Bot. J. Linn. Soc.* 133: 381–461 (2000).
[iii] Langlet, O. Über Chromosomeverhältnisse und Systematik der Ranunculaceae. *Sven. Bot. Tidskr.* 26: 381–400 (1932).
[iv] Hoot, S. B. Phylogeny of the Ranunculaceae based on epidermal microcharacters and macromorphology. *Syst. Bot.* 16: 741–755 (1991).
[v] Ro, K.-E., Keener, C. S., & McPheron, B. A. Molecular phylogenetic study of the Ranunculaceae: utility of the nuclear 26S ribosomal DNA in inferring intrafamilial relationships. *Molec. Phylogen. and Evol.* 8: 117–127 (1997).
[vi] Tamura, M. A new classification of the family Ranunculaceae L. *Acta Phytotax. et Geobot.* 41: 93–101 (1990).
[vii] Tamura, M. Ranunculaceae. Pp. 563–583. In: Kubitzki, K., Rohwer, J. G., Bittrich, V. (eds), *The Families and Genera of Vascular Plants. II. Flowering Plants. Dicotyledons: Magnoliid, Hamamelid and Caryophyllid Families.* Berlin, Springer-Verlag (1993).
[viii] Jensen, U., Hoot, S. B., Johansson, J. T., & Kosuge, K. Systematics and phylogeny of the Ranunculaceae – A revised family concept on the basis of molecular data. *Plant Syst. Evol.* 9 (Supplement): 273–280 (1995).

[ix] Horandl, E., Paun, O., Johansson, J. T., Lehnebach, C., Armstrong, T., Chen, L. X., & Lockhart, P. Phylogenetic relationships and evolutionary traits in *Ranunculus s.l.* (Ranunculaceae) inferred from ITS sequence analysis. *Molec. Phylogen. Evol.* 36 (2): 305–327 (2005).
[x] Schuettpelz, E., Hoot, S. B., Samuel, R., & Ehrendorfer, F. Multiple origins of Southern Hemisphere *Anemone* (Ranunculaceae) based on plastid and nuclear sequence data. *Plant Syst. Evol.* 231 (1–4): 143–151 (2002).
[xi] Compton, J. A., Culham, A., & Jury, S. L. Reclassification of *Actaea* to include *Cimicifuga* and *Souliea* (Ranunculaceae): phylogeny inferred from morphology, nrDNA ITS, and cPDNA *trn*L-*trn*F sequence variation. *Taxon* 47: 593–634 (1998).
[xii] Wang, W., Li, R.-Q., & Chen, Z. D. Systematic position of *Asteropyrum* (Ranunculaceae) inferred from chloroplast and nuclear sequences. *Plant. Syst. Evol.* 255: 41–54 (2005).
[xiii] Compton, J. A. & Culham, A. Phylogeny and circumscription of tribe Acteeae (Ranunculaceae). *Syst. Bot.* 27: 502–511 (2002).
[xiv] Paun, O., Lehnebach, C., Johansson, J. T., Lockhart, P., & Hörandl, E., Phylogenetic relationships and biogeography of *Ranunculus* and allied genera (Ranunculaceae) in the Mediterranean region and in the European Alpine System. *Taxon* 54: 911–930 (2005).

RESEDACEAE

MIGNONETTE, WELD

A small family of herbs and shrubs mostly of arid places, with some ornamentals.

Distribution Centered on the Mediterranean region, the family extends into parts of northern Europe and eastward to Central Asia and India. *Oligomeris* (3 spp.) is widely distributed, with outliers in South Africa, the Canary Islands, and a single species in the southwestern United States and Mexico. *Caylusea* (3 spp.) spreads from the Cape Verde Islands across northern Africa to India. *Reseda* (55 spp.), by far the largest genus, is restricted to Europe and the Mediterranean region to Central Asia.

Description Herbaceous annuals, perennials, and biennials, occasionally small shrubs (*Ochradenus, Randonia*) or rarely climbing (*Ochradenus*). The leaves are alternate and entire or divided or sometimes absent; glandular stipules are present. The flowers are irregular, usually bisexual, arranged in a bracteate raceme or spike. The sepals are 2–8, usually free, sometimes unequal. The petals are 2–8 (occasionally absent) but are not always equal in number to the sepals; they are mostly broadly clawed, with a scalelike appendage at the base, and usually a more or less deeply cut

Genera 6 Species 70
Economic uses Sources of a yellow dye, perfume oil, and an ornamental (*Reseda odorata*)

Id

Ic

Ia

Ib

lamina. The stamens and ovary are often on a short androgynophore, and there is usually an irregular disk outside the 3–45 stamens. The ovary is superior, of 2–7 more or less fused carpels, which are open above, and 1-locular with the ovules usually on parietal placentas. The fruit is usually an indehiscent capsule, rarely of separate, spreading carpels; *Ochradenus* has a berry.

Classification The family is generally considered to be allied to the Brassicaceae and Capparaceae and has usually been placed in Capparales[i]. DNA sequence data suggests a close relationship with Gyrostemonaceae or Tovariaceae[ii] and with *Emblingia*[iii] and the glucosinolate-containing families in the core Brassicales[iv].

Economic uses *Reseda odorata* (Mignonette) provides a perfume oil and is grown for ornament. *R. luteola* yields an orange-yellow dye (luteolin) used since Roman times. *Oligomeris linifolia* is toxic to grazing cattle. AC

[i] Brummitt, R. K. *Vascular Plant Families and Genera.* Richmond, Royal Botanic Gardens, Kew (1992).
[ii] Rodman, J. E. *et al.* Parallel evolution of glucosinolate biosynthesis inferred from congruent nuclear and plastid gene phylogenies. *American J. Bot.* 85: 997–1007 (1998).
[iii] Chandler, G. T. & Bayer, R. J. Phylogenetic placement of the enigmatic Western Australian genus *Emblingia* based on *rbc*L sequences. *Plant Species Biol.* 15: 67–72 (2000).

iv Hall, J. C., Iltis, H. H., & Sytsma, K. J. Molecular phylogenetics of core Brassicales, placement of orphan genera *Emblingia, Forchhammeria, Tirania,* and character evolution. *Syst. Bot.* 29: 654–669 (2004).

RHABDODENDRACEAE

A single genus of 3 spp. of evergreen trees or shrubs, restricted to Amazonian northeastern Brazil and the Guianas. The leaves are alternate, entire coriaceous, spotted with glands, peltate hairs present on underside. Stipules are absent or minute[i], but the leaf bases bear flanges[ii]. The inflorescences are axillary racemes or raceme-like cymes[iii] sometimes clustered at the shoot tip. The flowers are actinomorphic, hermaphrodite, 5-merous (rarely 4-merous) with distinct calyx and corolla; the sepals are small, imbricate, shrivelling at anthesis; the petals are free, imbricate, and alternating with the sepals; the stamens are 27–53, in 3 whorls; the ovary is superior, with 1 carpel and 2 ovules, only 1 maturing. The fruit is fleshy.

RESEDACEAE. 1 *Randonia africana* (a) shoot bearing flowers and fruit (×⅔); (b) flower showing petals with incised margins (×4); (c) vertical section of flower showing numerous stamens and superior ovary with ovules on axile placentas (×4); (d) fruit—a capsule (×4). 2 *Sesamoides canescens* (a) leafy shoot, inflorescences, and fruits (×⅔); (b) flower with equal green sepals and both incised and linear petals (×10); (c) dehiscing fruit (×8). 3 *Reseda villosa* (a) shoot bearing fruits (×⅔); (b) tip of inflorescence (×⅔); (c) vertical section of flower showing sessile stigma on top of ovary (×3); (d) flower showing small petals (×⅔); (e) dehisced fruit with apical opening (×3).

The relationships of this genus have long been controversial. It was placed in Chrysobalanaceae by Bentham, Rutaceae by Gilg and Pilger, Phytolaccaceae by Record, whereas Metcalfe and Chalk recognized the affinities of the genus with Achatocarpaceae. In 1968, the family was recognized as distinct[iv] by Prance. Only when molecular data became available did its placement as a basal group in the Caryophyllales become established. In fact, it may be sister to the entire order[v].

AC

Genera 1 Species 3
Economic uses None

i Prance, G. T. Rhabdodendraceae. Pp. 321–322. In: Smith, N. *et al.* (eds), *Flowering Plants of the Neotropics.* Princeton, Princeton University Press (2004).
ii Prance, G. T. Rhabdodendraceae. Pp. 339–341.In: Kubitzki, K. & Bayer, C. (eds), *The Families and Genera of Vascular Plants. V. Flowering Plants. Dicotyledons: Malvales, Capparales and Non-betalain Caryophyllales.* Berlin, Springer-Verlag (2002).
iii Watson, L. & Dallwitz, M. J. The families of flowering plants: descriptions, illustrations, identification, information retrieval. Version: 13th January 2005. http://delta-intkey.com
iv Prance, G. T. The systematic position of *Rhabdodendron* Gilg. & Pilg. *Bull. Jard. Bot. Nat. Belg.* 38: 127–146 (1968).
v Fay, M. F. *et al.* Familial relationships of *Rhabdodendron* (Rhabdodendraceae): plastid *rbc*L sequences indicate a caryophyllid placement. *Kew Bull.* 52: 923–932 (1997).

RHAMNACEAE
BUCKTHORN, JUJUBE

A large family of temperate and tropical trees and shrubs with some climbers.

Distribution Cosmopolitan, most common in tropical and subtropical regions.

Description Most species are trees or shrubs (several are adapted to climbing by the use of twining stems, *Berchemia*, tendrils, *Gouania*, or hooks, *Ventilago*), while some are herbs (*Crumenaria*). Spines often present, e.g., *Rhamnus* and *Paliurus* (Christ's crown of thorns was reputedly made from *Paliurus spina-christi*). In some, e.g., *Colletia*, leaf axils have 2 buds, the upper developing into a thorn, the lower into a shoot. The leaves are usually alternate, occasionally opposite, simple, the margins entire or serrate. Stipules usually present, small, spine-like, caducous. The inflorescence is axillary or terminal, usually cymose, rarely racemose. The flowers are small, inconspicuous, actinomorphic, sometimes without petals, bisexual (rarely unisexual). The sepals are 4–5, thick, valvate, often caducous. The petals are (0–)4–5, small, incurved, often closed over the stamens. The stamens are 4–5, opposite the petals, the filaments adnate to base of petals. The ovary is superior to inferior, of 1–3 carpels that are free or sometimes embedded in a prominent disk. The locules are 2–3(–5), or 1 by abortion, each with a single (rarely 2; *Karwinskia*) basal, anatropous ovule. The styles are 2–4, simple or connate. The fruits are fleshy drupes or nuts, dispersed by mammals and birds that eat them, occasionally dry, dehiscent capsules, or schizocarps and wind-dispersed (e.g., *Paliurus*). The seeds are few, the endosperm sparse or absent, and the embryo large, straight, and oily.

Classification The Rhamnaceae were placed in the Rhamnales by Cronquist (1988) but differ from the Celastraceae in having stamens opposite the petals not opposite the sepals. Melchior (1964) related the family to the Vitaceae from which it differs in most not being lianas, the ovary superior, the ovules mostly 1 per locule, and fruits mostly drupaceous. Molecular analyses[i] place the family clearly in the Rosales with Barbeyaceae and Dirachmaceae, forming their sister. The tribes Rhamneae and Zizypheae

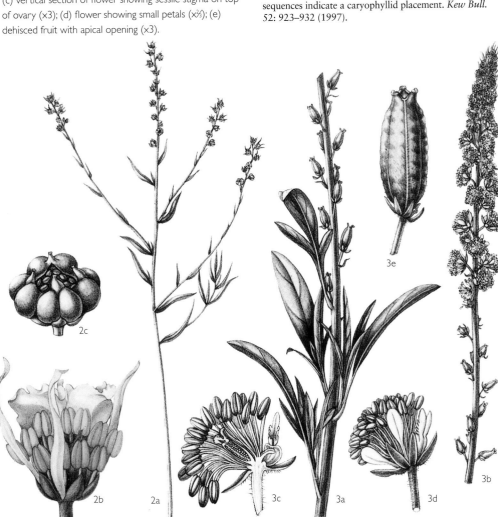

2c 2b 2a 3c 3a 3d 3e 3b

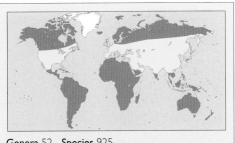

Genera 52 Species 925
Economic uses Ornamentals, medicinal plants, some dyes, charcoal and soap-substitute

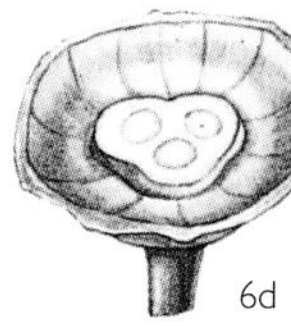

RHAMNACEAE. 1 *Zizyphus jujuba* (a) shoot bearing inconspicuous flowers and leaves with thorny stipules (x⅔); (b) fruits (x⅔); (c) cross section of fruit (x⅔). 2 *Gouania longipetala* (a) inflorescence, leaf, and coiled tendril-like stipule (x⅔); (b) half flower (x6); (c) section of ovary (x12); (d) winged fruit (x4). 3 *Phylica nitida* (a) flowering shoot (x⅔); (b) flower (x6); (c) vertical section of gynoecium and receptacle (x12); (d) cross section of fruit (x3). 4 *Paliurus virgatus* winged fruits (x⅔). 5 *Ceanothus veitchianus* (a) flowering shoot (x⅔); (b) flower (x8). 6 *Colletia cruciata* (a) flowering shoot (x⅔); (b) flower opened out (x4); (c) vertical section of flower base (x4); (d) cross section of flower base (x6).

circumscribed by Suessenguth[ii] are unsatisfactory. Although a molecular approach[iii] to tribal classification has been undertaken, it is now clear that many morphological characters have evolved in parallel. The chief genera include *Rhamnus*, *Hovenia*, *Zizyphus*, *Ceanothus*, *Ventilago*, *Phylica*, and *Frangula*.

Economic uses Sap-green dye is derived from the berries of *Rhamnus cathartica*, yellow dye from the berries of *R. infectoria*, and Chinese green-indigo from the bark of *R. chlorophora*. Other species are used in medicine, notably for their purgative properties, the best-known of these being *R. purshiana*, a North American species whose bark yields cascara sagrada (sacred bark). In the West Indies, the bark of *Gouania domingensis* is chewed as a stimulant. *Hovenia dulcis* (Japan, China) has pink, fleshy flower stalks that are dried and used locally in medicine. Leaf and bark extracts of African *Gouania* species are often applied to sores and wounds. *Ventilago oblongifolia* is similarly used in Malaysia as a poultice to cure cholera. Chemical analysis has shown that many members of the Rhamnaceae contain substances related to quinine, which may account for their wide use in medicine. In the Philippines, the root extract of *Gouania tiliifolia* contains saponin and is used as a soap substitute. *Zizyphus jujuba* is the jujube or Chinese date. *Zizyphus lotus* is believed to be the lotus fruit of antiquity. Several genera are well-known ornamentals, the best-known being *Ceanothus*, which contains many beautiful flowering shrubs. Other genera that are occasionally cultivated include *Pomaderris*, *Phylica*, *Noltea*, *Rhamnus*, and *Colletia*.   SLJ

[i] Richardson, J. E. *et al*. A phylogenetic analysis of Rhamnaceae using rbcL and *trn*L-*trn*F plastid DNA sequences. *American J. Bot.* 87: 1309–1324 (2000).
[ii] Süssenguth, K. Rhamnaceae, Vitaceae, Leeaceae. In: Engler, A. & Prantl, K. (eds), *Die naturlichen Pflanzenfamilien*, 2 aufl. 20d (1953).
[iii] Richardson, J. E. *et al*. A revision of the tribal classification of the Rhamnaceae. *Kew Bull.* 55: 311–340 (2000).

RHIZOPHORACEAE
RED MANGROVE FAMILY

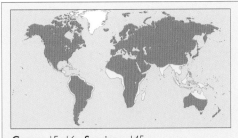

Genera 15–16 **Species** c. 145
Economic uses Sources of timber, fuel, and tanning materials

A tropical family of shrubs or trees, including 4 genera (*Rhizophora*, *Bruguiera*, *Cassipourea*, and *Kandelia*) that are mangroves.

Distribution The family is found throughout the tropics with some extension into the subtropics, mainly in rain forests and mangals but with some species of the non-mangrove genus *Cassipourea* in dry scrub. The main center of distribution is in the Old World tropics, with only 3 genera (*Rhizophora*, *Cassipourea*, and *Sterigmapetalum*) occuring in the neotropics.

Description Shrubs or trees, some with characteristic well-developed aerial stilt roots (*Rhizophorea*, *Carallia*, *Crossostylis*, *Gynotroches*), some with pneumatophores (*Bruguiera*, *Ceriops*). The leaves of most species are usually opposite, sometimes whorled, simple and the margins entire, crenate or dentate, and with conspicuous, interpetiolar, caducous stipules. The flowers are bisexual, rarely unisexual (plants monoecious), actinomorphic, and hypogynous to epigynous, borne in cymes or racemes, rarely solitary in the leaf axils. The sepals are 4–5(–16) persistent, valvate, connate below. The petals are 4–5(–16) or more, usually with a claw and jagged or divided at the tip, rarely entire (*Rhizophora*). The stamens are 8–10 or numerous, free or sometimes connate below, inserted on the edge of a nectarial disk. Staminodes are present in the pistillate flowers, sometimes adnate to the petals. The ovary is semi-inferior to inferior, or superior (Macarisieae), of 2–12 fused carpels, with 2–12 locules; placentation apical or axile, usually with 2 anatropous, pendulous ovules per locule. The style is simple. The fruit is a berry or drupe or dry, leathery, and indehiscent (*Rhizophora*), rarely a dehiscent capsule (*Cassipourea*, *Sterigmapetalum*); the seeds are 1 (Rhizophoreae) or 2, winged (*Sterigmapetalum*) or sometimes with an aril, and germinating on the plant and remaining attached (vivipary) in Rhizophoreae species, with or without fleshy endosperm.

Classification The family may be divided into 3 tribes[i]:
Macarisieae (7 genera, 87 spp.) is distributed throughout the tropics and includes the largest genus in the family, *Cassipourea* (c. 50 spp.), in addition to *Anopyxis* (3 spp.), *Macarisia* (7 spp.), *Blepharistemma* (2 spp.), *Comiphyton* (1 sp.), *Dactylopetalum* (14 spp.), and *Sterigmapetalum* (9 spp.). Fruits capsular.

Gynotrocheae (4 genera, 40 spp.) is distributed in Asia and Madagascar. Genera include *Carallia* (15 spp.), *Crossostylis* (13 spp.), *Gynotroches* (2–4 spp.), and *Pellacalyx* (8 spp.). Fruits berries; aerial stilt roots.

Rhizophoreae (4 genera, 16 spp.) comprises the mangrove genera and is regarded as the most derived. Genera include *Rhizophora* L. (6 spp.) pantropical; *Ceriops* (3 spp.), throughout tropical Asia and Afrca; *Bruguiera* Sav. (6 spp.), tropical Asia and in Africa; and *Kandelia* (1 sp.), confined to Southeast Asia. Fruits indehiscent; seeds viviparous; aerial stilt roots.

The Rhizophoraceae was placed by Cronquist (1981) in its own order situated next to the Myrtales, and Takhtajan (1997) took a similar position. It is placed in the Malpighiales in APG II. Analyses of the affinities of the Rhizophoraceae based on molecular data and morphology identified Erythroxylaceae as sister group[i,ii] within the Malpighiales and rejects previously suggested relationships with the Celastraceae or Elaeocarpaceae.

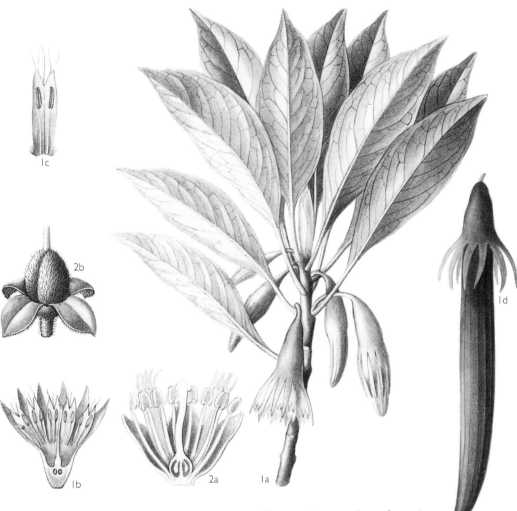

RHIZOPHORACEAE. 1 *Bruguiera gymnorhiza* (a) shoot with alternate leaves and solitary axillary flowers (×⅔); (b) half flower showing notched petals and inferior ovary (×⅔); (c) petal and stamens (×1½); (d) fruit (×⅔). 2 *Cassipourea rowensorensis* (a) half flower with superior ovary and single style (×4); (b) fruit (×2⅔).

Economic uses Several species, especially mangroves, are valuable sources of timber used for local building, fuel, and charcoal. Mangrove bark is also a source of tannins. In addition, mangrove communities are of major economic importance in providing habitats for fish and other marine organisms and in stabilizing coastal areas. VHH

i Schwarzbach, G. & Ricklefs, R. E. Systematic affinities of Rhizophoraceae and Anisophylleaceae, and intergeneric relationships within Rhizophoraceae, based on chloroplast DNA, nuclear ribosomal DNA, and morphology. *American J. Bot.* 87: 547–564 (2000).
ii Setoguchi, H., Kosuge, K., & Tobe, H. Molecular phylogeny of Rhizophoraceae based on *rbc*L gene sequences. *J. Plant Res.* 112: 443–455 (1999).

RHOIPTELIACEAE
HORSE-TAIL TREE FAMILY

Rhoipteleaceae comprises the single species *Rhoiptelea chiliantha*, a small, rare, deciduous tree restricted to southwestern China and northern Vietnam[i]. The leaves are imparipinnate with 9 or 11 pairs of toothed leaflets, covered with simple hairs and resinous gland-scales. Leafy stipules are present but caducous. The flowers

Genera 1 **Species** 1
Economic uses None

are small, inconspicuous in 3-flowered glomerules on large panicles of 16–60 catkins, nodding from the shoot apex (hence Horsetail Tree), and wind pollinated. The flowers are of 2 types, bisexual (functionally male) and female; the plants are functionally monoecious[ii]. The functionally male flowers have 2+2 tepals, 6 stamens with the anthers basifixed on short filaments, a superior 2-carpellate ovary with 2 locules (but only 1 developing), and a single ovule; stigmas 2. The female flowers are similar but smaller and lacking stamens. The fruits developed from the bisexual flowers are wind-dispersed winged nutlets, each containing a single ovate-triangular seed with a hard brown seed coat. The seed is without endosperm, embryo straight, and with 2 thick cotyledons.

The family is closely related to the Juglandaceae (APG II), and earlier placings in the Urticales by Melchior (1964) are erroneous. The family Rhoipteleaceae is regarded to have a number of primitive characteristics compared to the Juglandaceae, including possession of stipules, superior ovary, and bitegmic ovule[iii].　SLJ

[i] Shi-Guo Sun, Yang Lu, & Shuang-Quan Huang. Floral phenology and monoecy in the anemophilous tree, *Rhoiptelea chiliantha* (Rhoipteleaceae). *Bot. J. Linn. Soc.* (2006, in press).
[ii] Wu Cheng-Yih & Kubitzki, K. Rhoipteleaceae. Pp. 584–585. In: Kubitzki, K., Rohwer, J. G., & Bittrich, V. (eds), *The Families and Genera of Vascular Plants. II. Flowering Plants. Dicotyledons: Magnoliid, Hamamelid and Caryophyllid Families.* Berlin, Springer-Verlag (1993).
[iii] Stone, D. E. Biology and evolution of temperate and tropical Juglandaceae. Pp. 117–145. In: Crane, P. R. & Blackmore, S. (eds), *Evolution, Systematics and Fossil History of the Hamamelididae.* Vol. 1. *Introduction and Lower Hamamelididae.* Systematics Association Special vol. no. 40A. Oxford, Clarendon Press (1989).

RHYNCHOCALYCACEAE

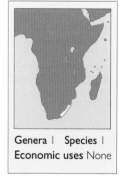

Genera 1　Species 1
Economic uses None

A monotypic family consisting of the single species *Rhynchocalyx lawsonioides*, which is a rare, small evergreen tree with opposite or whorled simple leaves. The species is confined to the Natal and eastern Cape provinces of South Africa. The flowers are actinomorphic, bisexual in terminal and axillary panicles. The sepals are (5)6(–8), inserted on the rim of a short hypanthium. The petals are (5)6(–8). The 6 stamens alternate with the sepals. The ovary is superior with 2(3) fused carpels with axile placentation; styles 1 with a capitate stigma. The fruit is a loculicidal capsule.

Rhynchocalyx was first described in the Lythraceae and then included in the Crypteroniaceae. It was subsequently described as a distinct family because of its unusual embryology. Its separate family status within the Myrtales has been confirmed by molecular

phylogenetic analyses[i,ii] and a study of perianth development and structure in both it and related families, which identified the Rhynchocalycaceae as sister to the clade formed by Oliniaceae and Penaeaceae[iii].　VHH

[i] Conti, E. *et al.* Interfamilial relationships in Myrtales: molecular phylogeny and patterns of morphological evolution. *Syst. Bot.* 22: 629–647 (1997).
[ii] Clausing, G. S. & Renner, S. Molecular phylogenetics of Melastomataceae and Memecylaceae: implications for character evolution. *American J. Bot.* 88: 486–498 (2001).
[iii] Schönenberger, J. & Conti, E. Molecular phylogeny and floral evolution of Penaeaceae, Oliniaceae, Rhynchocalycaceae, and Alzateaceae (Myrtales). *American J. Bot.* 90: 293–309 (2003).

RORIDULACEAE

VLIEËBOS

Genera 1　Species 2
Economic uses
Ornamental curiosity

The family comprises 2 species of *Roridula*: *R. gorgonias* and *R. dentata*. The species are semicarnivorous, trapping insects but gaining nutrients from the accumulated detritus on the ground rather than through direct digestion. The genus is endemic to the Cape province of South Africa, *R. dentata* in the Cedarberg, the Kouebokkeveld, and the Ceres Mountains, and *R. gorgonias* in the western Cape Mountains around Hermanus, Caledon, and Betty's Bay in marshes and along riverbanks. Both species are evergreen perennial shrubs with a taproot, woody stems, and leaves crowded at the ends of the sparse branches. The alternate leaves are linear-lanceolate, with distinct multicellular narrow teeth in *R. dentata*, tapering to a point, without stipules, and bearing numerous glandular hairs that trap insects. The flowers are borne in few-flowered terminal racemes. The flowers are 5-merous, regular, and hermaphrodite. The sepals are linear-lanceolate, green or tinged red and bearing glandular hairs on the outer surface. The 5 free petals are bright pink and overlapping at the base. The stamens are 5, free, and the ovary is superior, of 3 fused carpels each forming a chamber with 1 (*R. dentata*) or many (*R. gorgonias*) seeds. The fruit is a loculicidal capsule.

Roridula has been included in Droseraceae (Bentham and Hooker 1864) and Byblidaceae (Cronquist 1988), Ochnaceae and Clethraceae[i] based on a variety of morphological characters. Molecular studies have placed *Roridula* with *Euonymus* nested in Celastranae[i] or more recently as a sister group of *Actinidia*[ii] with Sarraceniaceae within Ericales. The family has limited value as an ornamental curiosity.　AC

[i] Conran, J. G. & Dowd, J. M. The phylogenetic relationships of *Byblis* and *Roridula* (Byblidaceae-Roridulaceae) inferred from partial 18S ribosomal RNA sequences. *Pl. Syst. Evol.* 188: 73–86 (1993).
[ii] Anderberg, A. A., Rydin, C., & Källersjö, M. Phylogenetic relationships in the order Ericales *s.l.*: analyses of molecular data from five genes from the plastid and mitochondrial genomes. *American J. Bot.* 89: 677–687 (2002).

ROSACEAE

ROSE FAMILY

A large family of woody trees, shrubs, climbers, and herbaceous plants. The family is valued both for its genera of bush and tree fruits and for many popular horticultural ornamentals.

Distribution The family is worldwide but with maximum development in the temperate to subtropical zones of the northern hemisphere.

Description Deciduous or evergreen trees, shrubs, shrublets, or mostly perennial herbs; few climbers but no aquatics. Branch thorns occur in *Crataegus, Prunus,* and other genera, with emergences (surface prickles) in *Rosa* and *Rubus.* The leaves are alternate (rarely opposite), simple, or compound (pinnate or palmate); stipules are present on twigs or at the petiole base (but not always obvious), absent in a few genera (*Exochorda, Spiraea* and allies), persistent or caducous; glands are commonly present, often paired at the top of the petiole. The inflorescences are terminal or axillary, racemose, cymose, paniculate, or flowers solitary. The flowers are actinomorphic, usually bisexual, rarely unisexual (plants dioecious), frequently large and showy (insect-pollinated, rarely small and wind-pollinated), usually (4-)5-merous, showing a series from hypogyny through perigyny to epigyny, the carpels appearing as if swallowed up by the hypanthium. *Rosa* is unique in retaining free carpels, and although the flower appears epigynous, it is perigynous. An epicalyx is usually present as a second, smaller whorl of 5 sepal-like organs below and alternating with the calyx lobes. The sepals and petals are (3–)5(–10), rarely absent (*Acaena*), although petal multiplication is common in cultivars developed for ornament (*Kerria, Prunus, Rosa,* etc.) by replacement of the numerous stamens (sometimes also styles) by petaloid organs. (In extreme cases, e.g., Japanese cherries and old-fashioned roses, the center of the bloom may be green, where the carpels have been transformed into leaflike appendages, when the flower is usually sterile.) The range of petal color is wide but blue is almost completely absent. The stamens are typically numerous, whorled, 2, 3, or more times as many as the petals; anthers are 2-locular, dehiscing longitudinally. The ovary is superior, semi-inferior to inferior, the carpels are normally numerous, free, with varying degrees of fusion occur; in Prunoideae reduced to 1, with 2 usually anatropous ovules. The fruits are diverse: fleshy drupes (*Prunus*), pomes (*Malus, Crataegus*), drupelets (*Rubus*), or dry capsules or follicles, dehiscent or not; seeds 1 to several, testa firm; endosperm usually absent; cotyledons flat.

Woody members of Rosaceae may propagate vegetatively by suckers, as in *Rubus,* which also tip-roots in the brambles. Runners (stolons) are

Genera 85–100 **Species** c. 2,000 sexual in addition to many obligate or facultative apomictic microspecies (1,300–1,500)
Economic uses Bush, tree, and soft fruits of temperate regions and many valued ornamentals

ROSACEAE. 1 *Rubus ulmifolius* (a) flowering shoot (x⅔); (b) fleshy fruits (x⅔). 2 *Rubus occidentalis* half flower showing hypogynous arrangement of parts (x2). 3 *Fragaria* sp. vertical section of false fruit—a fleshy receptacle with the true fruits embedded in it (x1). 4 *Sanguisorba minor* leafy shoot and fruit (x⅔).

a characteristic of some herbaceous genera (*Fragaria*). The flowers of the Rosaceae are mostly among the simplest and least specialized for pollination, relying on a large and wasteful production of pollen that attracts a wide range of insects, large and small. Some genera, such as *Rosa*, produce pollen only, but most also secrete nectar from a disk surrounding the carpels. This disk may be freely exposed (*Rubus*) or more or less screened by the filaments (*Geum*). The latter flowers are regarded as more highly evolved, eliminating short-tongued visitors and attracting only the longer-tongued flies and bees. Protandry is the general rule, and self-compatibility is exceptional. Several genera are characterized by agamospermy, combined with polyploidy and hybridization. The dog roses (*Rosa* section *Caninae*) are said to be "subsexual" because half or more of the chromosomes remain unpaired and are lost prior to gamete formation. *Alchemilla*, *Sorbus*, and the brambles (*Rubus*) are more or less completely apomictic. A marked departure from insect pollination is found in the Sanguisorbeae, notably in the genera *Acaena* and *Poterium*. These rely on wind pollination. The flowers are much reduced, in part unisexual, lacking petals and nectar, and are massed together in capitate or spicate heads.

Classification The family has traditionally been placed in the Rosales, which in the light of recent morphological and molecular data now includes Barbeyaceae, Cannabaceae, Dirachmaceae, Elaeagnaceae, Moraceae, Rhamnaceae, Ulmaceae, and Urticaceae (AGP II). The Chrysobalanaceae (q.v.) and Neuradaceae (q.v.) have both been associated with the Rosaceae or included within it but are segregated here as separate families. The taxonomy and phylogenetic relationships of the Rosaceae are complex and difficult, largely as a result of the size of the family, its range of morphological diversity, and the role of hybridization, polyploidy, and agamospermy. This diversity has been recognized by its division into 4 subfamilies: Spiraeoideae, Rosoideae, Maloideae (Pomoideae), and Prunoideae (Amygdaloideae), sometimes recognized as separate families, and from 8 to 18

tribes. Molecular analyses support the recognition and monophyly of Rosoideae[i] and Maloideae[ii], although recirumscribed, while the Amygdaloideae and Spiraeoideae are polyphyletic and lack coherence. Within the Rosoideae several clades may be recognized[iii] but the definition of clades in the Maloideae remains unclear. The origin of the Maloideae has long been considered to have involved ancient wide hybridization between amygdaloid and spiraeoid ancestors, but some recent work falsifies this hypothesis[ii] although such an origin cannot yet be discounted. A satisfactory subdivision of the Rosaceae is not yet available and for reference the traditional tribes and their principal genera are listed below:

Exochordeae Capsules with winged seeds (*Exochorda*).

Spiraeeae Follicles with wingless seeds; stipules almost or quite absent (*Aruncus, Holodiscus, Sibiraea, Spiraea*).

Neillieae Hypanthium open at top; epicalyx absent (*Neillia, Physocarpus, Stephanandra*).

Gillenieae Stipules on junction of twig and leaf; follicles present (*Sorbaria, Spiraeanthus, Gillenia*)

Kerrieae Hypanthium flat or convex; carpels few, whorled; stamens many, narrowed above from a broad base (*Kerria, Neviusia, Rhodotypos*).

Dryadeae Hypanthium open at top, dry at maturity, epicalyx absent (*Dryas*).

Ulmarieae Receptacle flat or weakly concave; filaments almost club-shaped, soon becoming deciduous (*Filipendula*).

Sanguisorbeae Receptacle urn-shaped, usually hard, enclosing 2 or more achenes (*Acaena, Agrimonia, Poterium, Bencomia, Sanguisorba*).

Potentilleae As Kerrieae, but carpels usually many, on a convex gynophore (*Alchemilla, Fragaria, Geum, Potentilla*).

Rubeae Epicalyx absent, carpels many, free on elevated torus (*Rubus*).

Roseae Receptacle urn-shaped or cylindrical, soft in fruit, enclosing many free carpels (*Hulthemia, Rosa*).

Pruneae Carpels 1 (rarely up to 5), free, with terminal styles and pendulous ovules; fruit a drupe (*Prunus*, including *Amygdalus, Armeniaca, Cerasus, Laurocerasus, Padus, Persica, Maddenia*).

Maleae Carpels 2–5, usually fused with the inner wall of the concave receptacle which together with the calyx enlarges to enclose the fruits as a pome (*Amelanchier, Aronia, Cydonia, Eriobotrya, Malus, Photinia, Pyrus, Raphiolepis, Sorbus, Stranvaesia*).

Crataegeae Stipules on petiole base, hypanthium open becoming a fleshy pyrene with 1–5 pyrenes (*Cotoneaster, Pyracantha, Crataegus, Chaenomeles, Mespilus*).

The delimitation of genera is unclear in some groups such as *Potentilla, Sorbus,* and *Prunus,* notably those of economic importance. The subgenera within *Prunus* are raised to generic level by some botanists, as *Amygdalus* (for the Almonds), *Cerasus* (Cherries), *Padus* (Bird Cherries), and *Laurocerasus* (Portugal and Cherry Laurels), etc. As regards species, greatest problems concern those genera in which subsexual or asexual reproduction is normal. These do not display the normal patterns of disjunct populations in the wild, and there is no agreement amongst taxonomists on how to treat them. In such groups, e.g., *Rubus, Rosa,* and *Alchemilla,* species circumscription is complicated by hybridization, polyploidy, agamospermy, and lack of a universal species concept[iv]. For example, the single species *Rubus fruticosus* has been split up into many hundreds of self-perpetuating microspecies (agamospecies). A molecular phylogeny of *Rubus* has recently been proposed[iv].

Economic uses Most of the important bush and tree fruits of temperate regions are found in the Rosaceae. By far the most important economically is the apple (*Malus*), now grown in numerous hybrid cultivars of complex origin. They are grown mainly for dessert, but are also used for making cider. The next most important genus is *Prunus*, which produces almonds, apricots, cherries, damsons, nectarines, peaches, and plums, all of which are grown extensively for consumption as fresh fruit and for canning and making into jams, conserves, and liqueurs. Other major rosaceous fruits are blackberries, loganberries, and raspberries (*Rubus*); loquats (*Eriobotrya*); medlars (*Mespilus*); pears (*Pyrus*); quinces (*Cydonia*); and strawberries (*Fragaria*).

Many *Prunus* species are also cultivated as ornamentals, notably the Japanese flowering cherries. However, it is the rose, the "queen of flowers," that overshadows all the other ornamentals, being probably the most popular and widely cultivated garden flower in the world, valued since ancient times for its beauty and fragrance. Modern roses are complex hybrids descended from about 9 of the wild species. Rose-growing is now a large industry, with some 5,000 named cultivars estimated to be in cultivation.

Among other popular cultivated genera are herbaceous perennials such as *Alchemilla* (Lady's Mantle), *Geum* (Avens), *Filipendula* (Meadowsweet), and *Potentilla* (Cinquefoil); and trees and shrubs such as *Amelanchier, Chaenomeles* (flowering quinces, including *C. lagenaria,* better known as the Japonica), *Cotoneaster, Exochorda* (Pearl Bush), *Sorbus* (Rowan, Mountain Ash), *Photinia,* and *Pyracantha* (Fire Thorn).

Attar or Otto of roses, a volatile fragrant oil, is extracted or distilled from fresh flowers of *Rosa damascena, R. gallica, R. centifolia,* and other species; its production is a major industry in Bulgaria and parts of western Asia.

[GDR] SLJ

[i] Morgan, D. R., Soltis, D. E., & Robertson, K. R. Systematic and evolutionary implications of *rbc*L sequence variation in Rosaceae. *American J. Bot.* 81: 890–903 (1994).
[ii] Evans, R. C. & Campbell, C. S. The origin of the apple subfamily (Maloideae; Rosaceae) is clarified by DNA sequence data from duplicated GBSSI genes. *American J. Bot.* 89: 1478–1484 (2002).
[iii] Eriksson, T. *et al.* The phylogeny of Rosoideae (Rosaceae) based on sequences of the internal transcribed spacers (ITS) of nuclear ribosomal DNA and the *trn*L-*trn*F region of chloroplast DNA. *Int. J. Plant Sci.* 164: 197–211 (2003).
[iv] Alice, L. A. & Campbell, C. A. Phylogeny of *Rubus* (Rosaceae) based on nuclear ribosomal DNA internal transcribed spacer region sequences. *American J. Bot.* 86: 91–97 (1999).

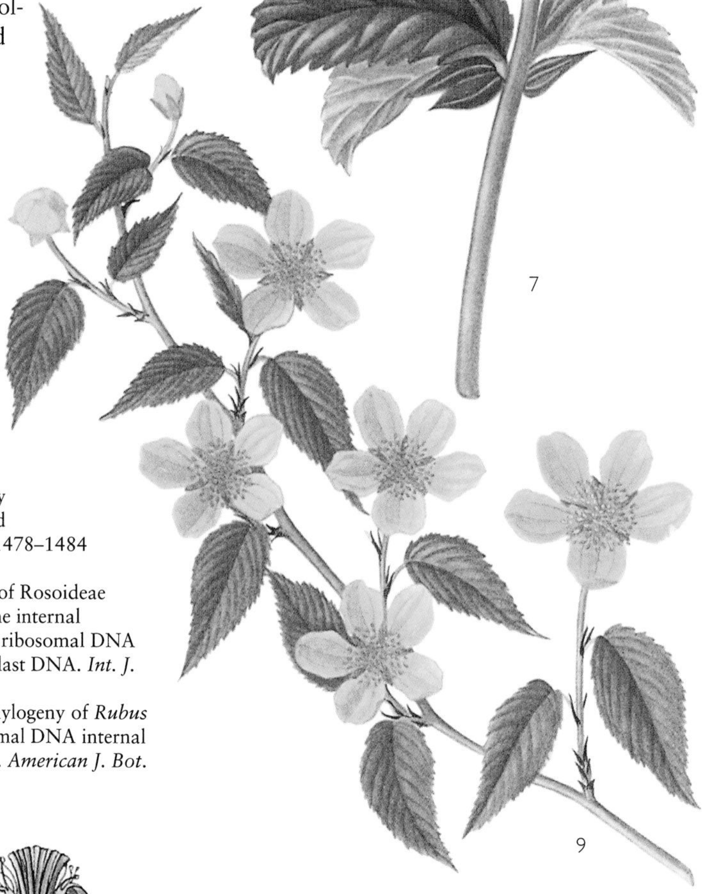

ROSACEAE. 5 *Agrimonia odorata* fruit comprising a receptacle covered with hooks enclosing the achenes (not visible) (×4). 6 *Rosa* sp, vertical section of hep showing urn-shaped receptacle enclosing the achenes (×1⅓). 7 *Potentilla agyrophylla* var. *atrosanguinea* flowering shoot (×⅔). 8 *Rosa pendulina* flowering shoot clearly showing the stipules at the base of the leaf stalks (×⅔). 9 *Kerria japonica* flowering shoot (×⅔). 10 *Chaenomeles speciosa* (a) flowering shoot (×⅔); (b) vertical section of fruit comprising swollen receptacle and calyx enclosing the true fruits (×⅔); (c) cross section of ovary (×6); (d) vertical section of flower showing epigynous arrangement of parts (×2). 11 *Cotoneaster salicifolius* (a) leafy shoot with fruit (×⅔); (b) vertical section of fruit (×2⅔). 12 *Sorbus aria* flowering shoot (×⅔). 13 *Prunus insititia* leafy shoot and fruits—drupes (×⅔). 14 *Prunus* sp. vertical section of fruit (×1⅓). 15 *Prunus avium* vertical section of flower with petals removed showing perigynous arrangement of parts (×2⅔). 16 *Spiraea cantoniensis* vertical section of flower (×4).

16

10d

10b

10c

11b

10a

14

15

11a

5

13

12

ROUSSEACEAE

A small family of tropical climbers or trees of uncertain affinities.

Distribution Subfamily Rousseoideae comprises a single species, *Roussea simplex*, which is endemic to Mauritius, while the 3 genera of subfamily Carpodetoideae (*Carpodetus*, *Cutsia,* and *Abrophyllum*) range from eastern Australia and New Zealand to New Guinea.

Genera 4 Species 13
Economic uses None

Description Evergreen trees or lianas with alternate or opposite (*Roussea*), simple, serrate leaves. The flowers are actinomorphic, bisexual, large and solitary (*Roussea*) or small and in panicles, 4–5(6)-merous. The sepals are valvate, the petals thickish, connate or free. The stamens are 5–6, the filaments free. The ovary is superior of 3–7 carpels, with 3–7 locules. The fruit is a berry (*Roussea*) or dry indehiscent or capsular and dehiscent.

Classification The affinities of *Roussea* have been uncertain. It was previously placed along with *Brexia* and *Ixerba* in the Saxifragaceae, Grossulariaceae, or Brexiaceae, but molecular data indicates a placement in the Asterales[i], where it is sister to *Carpodetus*[ii], a genus formerly placed in or near the Escalloniaceae. Both have been included in an expanded Rousseaceae in the Asterales[iii], along with *Cutsia* and *Abrophyllum*, also formerly placed in the Saxifragaceae, although they differ in many features. Further work is needed to clarify the position of these genera.   VHH

[i] Lundberg, J. The Asteralean affinity of the Mauritian *Roussea* (Rousseaceae). *Bot. J. Linn. Soc.* 137: 267–276 (2001).
[ii] Bremer, B. *et al.* Phylogenetics of Asterids based on 3 coding and 3 non-coding chloroplast DNA markers and the utility of non-coding DNA at higher taxonomic levels. *Molec. Phylogen. Evol.* 24p: 274–301 (2002).
[iii] Lundberg, J. Phylogenetic studies in the Euasterids II with particular reference to Asterales and Escalloniaceae. Uppsala, Sweden, *Acta Univ. Upsal.* (2001).

RUBIACEAE

COFFEE OR MADDER FAMILY

The fourth largest flowering plant family after the Orchidaceae, Asteraceae, and Fabaceae, usually recognized by the presence of simple, opposite, or whorled, entire leaves, interpetiolar stipules and an inferior ovary.

Distribution The Rubiaceae has a cosmopolitan distribution, but species diversity and biomass is distinctly concentrated in the tropics and the subtropics. In humid tropical forests, the Rubiaceae is often the most species abundant woody plant family. The family is less frequent and less diverse, but still widespread, in the temperate regions. It is also found in the subpolar regions of the Artic and Antarctic.

Description Mostly small trees or shrubs (sometimes armed), but nearly all life-forms are found, including large trees, annual and perennial herbaceous plants, woody monocaul dwarfs, lianas, epiphytes, geofrutices (± herbaceous stems from a woody rootstock), and rarely succulent or aquatic; occasionally associated with ants (hollow stems or special chambered tubers e.g., *Myrmecodia*). The leaves are opposite or sometimes whorled, simple and nearly always entire, a few genera with bacterial nodules; domatia (pits or hairy tufts in the axils) often present. Colleters, which are modified hairs secreting exudate, are found on the inside surface of the stipule or otherwise associated with the stipule and sometimes on inflorescence and floral parts. Stipules are always present, most commonly interpetiolar, often fused above axils, sometimes forming a sheath or less often entirely free, or exactly leaflike (e.g., *Galium*), rarely intrapetiolar. The inflorescence is highly variable. The flowers are usually bisexual but also unisexual (plants usually dioecious), often heterostylous or with pollen presentation, and commonly 4- or 5-merous. The calyx is adnate to the ovary, extending to a free limb-tube and/or lobes, occasionally reduced to a rim, sometimes with calycophylls (or semaphylls: large, leaflike, white or colored calyx lobes) present. The corolla is always tubular (gamopetalous, although in a few taxa the tube is almost obsolete or rarely absent [*Dialypetalanthus*]), with radial symmetry (actinomorphic), although some species are secondarily zygomorphic. The stamens are usually as many as the corolla lobes, alternate to them and fixed to the internal surface of the corolla tube (epipetalous), occasionally the filaments are free to the base, sometimes fused in a ring (monadelphous, e.g., Chiococceae); the anthers are usually dorsifixed, but occasionally basifixed; dehiscence is commonly introrse or infrequently porate. A disk usually functioning as a nectary is positioned above the ovary inside the calyx limb, most commonly cushionlike or annular. The ovary is inferior, except for the genera *Gaertnera* and *Pagamea*, which have secondarily

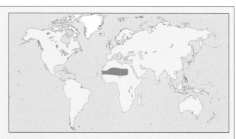

Genera 615 Species 13,150
Economic uses Coffee, quinine, ipecacuanha, some ornamentals

derived superior ovaries, the carpels usually 2, sometimes 5 or more, with axile placentation, or sometimes parietal, or a mixture of axile and parietal (some members of tribe Gardenieae), and ovules are 1 to numerous per locule, anatropous. The style is simple, and the stigma lobed or capitate. Morphologically elaborate stigmas are present in those groups with distinct secondary pollen presentation mechanisms, e.g., in the tribe Vanguerieae. The fruit is small or quite large, indehiscent (berries or drupes with usually 2 or more single-seeded pyrenes, less frequently a multiseeded stone) or dehiscent (capsules, mericarps), sometimes united into syncarps. The seeds may be small or large, sometimes winged. Endosperm is present and usually conspicuous, sometimes ruminate, absent in tribe Guettardeae.

Classification The Rubiaceae is a member of the Gentianales and shares many features common to other families of the order, particularly leaf and floral morphology, the presence of colleters, and lack of internal phloem. Recently, 3 small families, Dialypetalanthaceae[i], Henriqueziaceae[ii], and Theligonaceae[iii], have been confirmed as members of the Rubiaceae. Systematic data convincingly support the recognition of the Rubiaceae as a monophyletic group[iv,v,vi], and the morphological circumscription is relatively straightforward[vii,viii]. Only 1 genus (*Mastixiodendron*) is regarded as a doubtful member. Classification into subfamilies and, to a lesser extent, tribes is problematic[iv], and there is still some way to go before a robust, natural classification can be produced. Recent systematic studies divide the family into 3 subfamilies, the Rubioideae, Cinchonoideae, and Ixoroideae, which are monophyletic on the basis of molecular sequence data[i,ii,iii,ix,x]. Separation between subfamilies Cinchonoideae and Ixoroideae from Rubioideae is relatively easy. In the Rubioideae, for example, raphides are almost universally present, and in the Cinchonoideae they are almost entirely absent[iv,viii,xi]. There are 2 main exceptions in the Cinchonoideae, the closely related tribes Hamelieae and Hillieae, although there are a few other miscellaneous genera in the Cinchonoideae that possess raphides[ix]. Other than presence of raphides, morphological delimitation of subfamilies is not as clear cut, and the recognition of subfamilies currently depends on combinations of morphological characters rather than synapomorphies. Previous concepts of Cinchonoideae and Ixoroideae [v,xi] separated these subfamilies primarily on the basis of aestivation (mostly valvate or imbricate in Cinchonoideae, always contorted in Ixoroideae). Due to the transfer of several genera from Cinchonoideae to Ixoroideae, however, this is no longer possible. For this reason Cinchonoideae and Ixoroideae are often referred to as Cinchonoideae *sensu stricto* and Ixoroideae *sensu lato*[i,iii], respectively, especially in informal divisions of the family. Since these 2 subfamilies are now difficult to separate by morphological characters, an alternative and more

RUBIACEAE. 1 *Ixora chinensis* (a) flowering shoot showing stipules between the petioles (×⅔); (b) tubular corolla opened out to show epipetalous stamens and simple style with a lobed stigma and vertical section of ovary (×1½). 2 *Asperula suberosa* flowering shoot (×⅔). 3 *Mussaenda* sp. flower with 1 calyx lobe much enlarged (×1½). 4 *Coffea arabica* (a) fruits (×⅔); (b) cross section of fruit (×1). 5 *Sarcocephalus pobeguinii* vertical section of fruit (×½). 6 *Sherbournia calycina* (a) flowering shoot (×⅔); (b) vertical section of ovary (×2); (c) cross section of ovary (×3).

practical approach is to combine them under Cinchonoideae, giving a classification of the Rubiaceae into 2 subfamilies[iv,xii]: Rubioideae and Cinchonoideae. Previous classifications involving the division of the Rubiaceae into 4[v,xiii] or more[xiv] subfamilies have now been abandoned in the light of new systematic data.

SUBFAM. RUBIOIDEAE Raphides present, rarely absent; trees, shrubs, climbers, and numerous herbs; external indumentum (hairs) clearly septate or sometimes non-septate; stipules interpetiolar, entire, divided or fimbriate; corolla lobes valvate (often with thickened margins or induplicate); heterostyly frequent, homostyly infrequent; wind pollination occurs in a few genera; secondary pollen presentation absent; stamens attached in upper part of corolla tube, anthers dorsifixed; style usually divided into 2 linear lobes; ovules solitary to numerous in each locule; ovules erect from the base, on a placenta

attached to the septum or pendulous from the apex; placentation axile; fruit dry or succulent, dehiscent or indehiscent; syncarps sometimes present; seeds sometimes with a small radial marginal wing; endosperm present. Around 24 tribes with cosmopolitan distribution.

SUBFAM. CINCHONOIDEAE Raphides absent, rarely present (Hillieae, Hamelieae only, and a few other genera); trees, shrubs, woody climbers; external indumentum (hairs) non-septate or incompletely septate; stipules interpetiolar or rarely intrapetiolar (Henriquezieae), entire, divided, rarely fimbriate; corolla lobes valvate (simple, reduplicate, never with thickened margins), imbricate, rarely contorted; stamens attached in upper part of corolla tube or sometimes filaments free near the base or very rarely fused into an androecial ring above the ovary (*Dialypetalanthus*); anthers dorsifixed or basifixed; heterostyly infrequent (*Guettarda*), homostyly more or less ubiquitous; secondary pollen presentation present or absent; style divided into 2 lobes, capitate, fusiform, or clavate; ovules solitary or numerous in each locule; ovules attached to the septum, more or less at the apex, rarely from the base, often imbricate along the placenta; placentation axile; fruit dry or succulent, dehiscent or indehiscent, syncarps often present; seeds often winged (wings radial or strongly bipolar, rarely as hairlike outgrowths in Hillieae); endosperm present or ± absent (Guettardeae only). Around 12 tribes with pantropical distribution but strongly centered in the New World.

SUBFAM. IXOROIDEAE Raphides absent; trees, shrubs, woody climbers, rarely herbs; external indumentum (hairs) non-septate or incompletely septate; stipules interpetiolar, sometimes intrapetiolar (e.g., *Capironia*), entire, rarely laciniate or fimbriate; corolla lobes contorted or valvate (simple, induplicate or reduplicate, or with thickened margin) or infrequently imbricate (*Heinsia*); heterostyly infrequent (Sabiceeae, Mussaendeae), homostyly common; secondary pollen presentation present, rarely absent; style divided into 2 (or more) lobes, capitate, fusiform, or clavate; ovules solitary or numerous in each locule; ovules attached to the septum, often near the middle, or more or less at the apex, rarely from the base, or embedded in 1 to several placentas; placentation axile, rarely parietal, or rarely axile and parietal (within the same ovary); fruit dry or succulent, sometimes indehiscent or rarely dehiscent (*Crossopteryx*), syncarps not known; seeds rarely winged (e.g., *Crossopteryx*), ± radial and fringed; endosperm present. Around 16 tribes with pantropical distribution.

Altogether around 52 tribes are currently recognized, and generally they are relatively easy to characterize morphologically. Compared to other large flowering plant families, the Rubiaceae is understudied. Thousands of species still need formal scientific description, and several large generic complexes (e.g., *Hedyotis*) still need thorough assessment.

Economic uses Coffee (*Coffea*) is by far the most important economic product of the Rubiaceae, with some 100 million people worldwide depending on it for their livelihood[xv]. Indeed, it has become one of the world's most important commodities. Two main species are used in the production of coffee: *C. arabica* (arabica coffee) and *C. canephora* (robusta coffee), with a small amount of production for *C. liberica* (Liberian or Liberica coffee, and excelsa coffee). *Coffea arabica* is by far the most important commercial species, making up more than 95% of the total marketable crop. Quinine (*Cinchona* spp.) was an historically important malarial prophylaxis and curative. Modern synthetic counterparts and alternatives are far more widely used, although quinine is still used as a food and drink flavoring, e.g., in tonic water. Other economically important Rubiaceae include Ipecacuanha (*Carapichea ipecacuanha*), an amoebicide, emetic, and expectorant; and Gambir or White Cutch (*Uncaria* spp.), which is used as a tanning agent for leather and to a lesser extent as a dyestuff and in medicine. Of lesser importance are the timber species (e.g., *Adina* spp., *Anthocephalus* spp.) and the dye commonly known as madder, which is derived from the root of *Rubia tinctorum*. Numerous other taxa have ethnic use as red or black dyes. The alkaloid Yohimbine (*Paucinystalia johimbe*) is used as an aphrodisiac both for humans and in veterinary practice. The Rubiaceae are rich in alkaloids and other chemicals, and numerous taxa have local medicinal uses; a few taxa are poisonous to cattle; *Catunaregam* spp. are used as a fish poison. *Gardenia* is used in the perfume industry. *Galium verum* has been used in cheese making as "vegetarian rennet." Dried plants of *Galium odoratum* and *G. verum* were historically placed among linen to give it a pleasant freshly mown hay scent (coumarin). A number of species are used as ornamentals in the tropics, such as *Bouvardia, Gardenia, Hamelia, Ixora, Manettia, Mussaenda, Rondeletia, Warszewicia,* and there are a few garden plants of the temperate zones, including *Asperula, Galium, Houstonia,* and *Leptodermis*. House plants include *Gardenia, Nertera, Pentas,* and *Serissa* (as a bonsai).

APD & DMB

[i] Fay, M. F. *et al*. Plastid *rbc*L sequence data show *Dialypetalanthus* to be a member of Rubiaceae. *Kew Bull.* 55: 853–864 (2000).
[ii] Rogers, G. K. *Gleasonia, Henriquezia* and *Platycarpum* (Rubiaceae). *Flora Neotropica* 39: 1–135 (1984).
[iii] Wunderlich, R. Die systematische Stellung von *Theligonum. Öst. Bot. Zeitschr.* 119: 329–394 (1971).
[iv] Andersson, L. & Rova, J. H. E. The *rps*16 intron and the phylogeny of the Rubioideae (Rubiaceae). *Pl. Syst. Evol.* 214: 161–186 (1999).
[v] Bremer, B. *et al*. Subfamilial and tribal relationships in the Rubiaceae based on *rbc*L sequence data. *Ann. Missouri Bot. Gard.* 82: 383–397 (1995).
[vi] Bremer, B. & Thulin, M. Collapse of Isertieae, re-establishment of Mussaendeae, and a new genus of Sabiceeae (Rubiaceae); phylogenetic relationships based on *rbc*L data. *Pl. Syst. Evol.* 211: 71–92 (1998).
[vii] Bridson, D. M. & Verdcourt, B. Rubiaceae. Pp. 379–720. In: Pope, G. V. (ed.), *Flora Zambesiaca 5*, part 2. Richmond, Royal Botanic Gardens, Kew (2003).

[viii] Robbrecht, E. Tropical woody Rubiaceae. *Opera Bot. Belg.* 1: 1–271 (1988).
[ix] Bremer, B. *et al*. More characters or more taxa for a robust phylogeny—case study from the coffee family (Rubiaceae). *Syst. Biol.* 48: 413–435 (1999).
[x] Bremer, B. & Manen, J. F. Phylogeny and classification of the subfamily Rubioideae (Rubiaceae). *Pl. Syst. Evol.* 225: 43–72 (2000).
[xi] Jansen, S. *et al*. A survey of the systematic wood anatomy of the Rubiaceae. *IAWA J.* 23: 1–67 (2002).
[xii] Verdcourt, B. Remarks on the classification of the Rubiaceae. *Bull. Jard. Bot. État Brux.* 28: 209–281 (1958).
[xiii] Robbrecht, E. Supplement to the 1988 outline of the classification of the Rubiaceae. *Opera Bot. Belg.* 6: 173–196 [1993] (1994).
[xiv] Bremekamp, C. E. B. Remarks on the position, the delimitation and the subdivision of the Rubiaceae. *Acta Bot. Neerland.* 15: 1–33 (1966).
[xv] Vega, F. E. *et al*. Global project needed to tackle coffee crisis. *Nature* 425: 343 (2003).

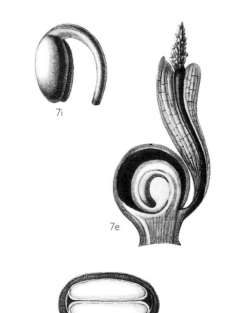

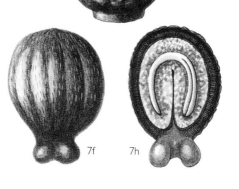

RUBIACEAE. 7 *Theligonum cynocrambe* (a) habit showing opposite leaves (appearing alternate at the apex; ×⅔); (b) united membranous stipules (×6); (c) male flower with bilobed perianth and numerous stamens with short, filiform filaments and long anthers (×10); (d) male flower bud (×10); (e) half female flower with lateral tubular perianth and style that arises from the base of the ovary (×10); (f) fruit—a nutlike drupe (×10); (g) cross section fruit (×10); (h) vertical section of fruit showing single seed with a curved embryo embedded in fleshy endosperm (×10); (i) embryo with 2 large, globose cotyledons (×12).

RUTACEAE

RUE AND CITRUS FAMILY

A variable family with usually pinnately compound leaves, characterized by secretory cavities with aromatic oils and superficial pellucid gland dots.

Distribution Tropical and southern hemisphere, with proliferation of many endemic genera in Australia and South Africa. Rather few genera extend into the northern temperate regions, but *Phellodendron, Skimmia,* and *Dictamnus* can be found in the Russian Far East, *Haplophyllum* has more than 30 spp. from the Mediterranean to Mongolia and *Zanthoxylum* and *Ptelea* reach southern Canada. In the tropics, many species are in the understory of evergreen forest, but in more temperate regions they are often in dry country.

Description Trees (usually rather small and evergreen), shrubs, or woody climbers, occasionally herbs, sometimes with prickles or spines. Most species are aromatic and have secretory cavities in most tissues and superficial gland-dots. The leaves are usually alternate, rarely opposite or whorled, usually compound with pinnate, 3-foliate or 1-foliate (with a joint and pulvinus at base of lamina, e.g., *Citrus*), or perhaps rarely simple and then entire to bipinnatisect (perhaps *Calodendrum, Thamnosma, Ruta,* etc.), characteristically showing pellucid gland dots on their surface or at least at their margins. The inflorescence is variable, terminal, or axillary. The flowers are usually bisexual, rarely unisexual (plants occasionally dioecious), actinomorphic or occasionally somewhat zygomorphic with unequal petals or some stamens sterile, usually 5-merous but sometimes (2–3)4-merous, usually hypogynous. The sepals are usually 5, free or united. The petals are usually 5, free or sometimes united at the base. The stamens are as many as the petals (sometimes 2 fertile and 3 staminodes) or more commonly twice as many in 2 whorls (or 1 whorl reduced to staminodes), occasionally up to 4 times as many, variously united at the base and inserted at the base of a disk or gynophore. The ovary is superior, with (1–2–)4–5(–10) carpels, which are fully united or more or less free and united only by their connate styles, each carpel with a

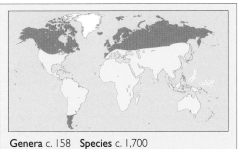

Genera c. 158 **Species** c. 1,700
Economic uses Citrus for edible fruits and drinks; some cultivation for essential oils; numerous cultivated ornamentals

RUTACEAE. 1 *Ruta graveolens* (a) shoot with bipinnate leaves and cymose inflorescences (x⅔); (b) flower with 4 sepals, 4 petals, 8 stamens, and a superior, lobed ovary with a basal disk and crowned by a single style (x2⅔); (c) vertical section of ovary (x6); (d) cross section of ovary showing 4 locules and ovules on axile placentas (x4). 2 *Citrus aurantium* (sweet orange) (a) half flower with numerous stamens and prominent disk at the base of the ovary (x2); (b) fruit—a hesperidium (x⅔).

separate locule
or sometimes the
partitions are incom-
plete, each carpel with
usually 2 ovules on an axile placenta. The fruit
is a capsule, schizocarp, drupe, or berry, in
Citrus and related genera a hesperidium with
the locules filled by many juicy hairs.

Classification The family is well established as
closely related to Meliaceae, Simaroubaceae,
and Ptaeroxylaceae, which are included in
Sapindales by Cronquist (1981) and APG II.
The classification within the family, however,
is far from clear and seems to be in a state of
flux. Engler's account in the *Pflanzenfamilien*
in 1931, largely repeated by Scholz in the *Syllabus* in 1964, recognized 7 subfamilies. His
Rhabdodendroideae has long been treated
as a separate and unrelated family, and his
highly anomalous monogeneric subfamilies
Spathelioideae and Dictyolomatoideae are here
referred to Ptaeroxylaceae in accordance with
molecular evidence. The group that has
remained most stable in recent classifications
based on chemical[i] and molecular [ii]data is the
subfamily Aurantioideae (syn. Citroideae) in
which the fruit is a relatively large, fleshy berry
or hesperidium (such as *Citrus* fruits), sometimes with a hard epicarp. The Rutoideae
of Engler have dehiscent fruits, with usually
2 ovules per carpel, while his Toddalioideae
were defined as having an indehiscent drupe or
samara with one or 2 ovules per carpel. This
separation has been considered artificial,
and the subfamilies have been combined as
subfamily Rutoideae *sensu lato*, the chemical
evidence suggesting 17 tribes[i]. Engler's subfamily Flindersioideae included only *Flindersia* and
Chloroxylum, which have capsules with 1 to
several winged seeds, both genera having been
referred in the past to Meliaceae or a separate
family Flindersiaceae. These were kept together
as a subfamily by the chemical data[i], but the
molecular data have not supported a close relationship, *Chloroxylum* grouping with *Ruta* in a
rather isolated position, with *Flindersia* being
sister to *Lunasia* in the rest of the former
Rutoideae, but in a rather meagre sampling of
genera[ii]. Much further work is needed to produce an acceptable classification of the family.

Economic uses The genus *Citrus* is of major
economic importance. The fruits of its species,
hybrids, and backcrosses include the lemon, cit-
ron, sweet orange, Seville orange, tangerine,
satsuma, mandarin, clementine, lime, grapefruit,
and kumquat. Other species are cultivated for
their essential oils, such as bergamot oil. Among
many cultivated ornamentals are *Choisya terna-
ta* and *Skimmia japonica*.   RKB

[i] da Silva, M. F., Das G. F., Gottlieb, O. R., &
Ehrendorfer, F. Chemosystematics of the Rutaceae:
suggestions for a more natural taxonomy and
evolutionary interpretation of the family. *Pl. Syst. Evol.*
161: 97–134 (1988).
[ii] Chase, M. W., Morton, C. M., & Kallunki, J.
Phylogenetic relationships of Rutaceae: a cladistic
analysis of the subfamilies using evidence from *rbc*L
and *atp*B sequence variation. *American J. Bot.* 86:
1191–1199 (1999).

RUTACEAE. 3 *Ptelea trifoliata* winged fruit—an unusual feature
for the family (x1½). 4 *Citrus limon* (lemon) flowering shoot
(x⅔). 5 *Crowea saligna* flowering shoot (x⅔).

SABIACEAE

A family of woody climbers from tropical Asia.

Distribution *Sabia* extends from western
Himalayas to Japan, Indonesia, and the
Solomon Islands, especially in eastern
Himalaya, northern Indochina, and southwest-
ern China, in thickets and forests up to 3,300m
in altitude.

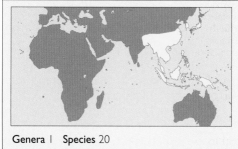

Genera 1 **Species** 20
Economic uses Occasional medicinal applications

Description Evergreen or deciduous scandent
shrubs to lianes up to 12(–20) m high, sometimes erect in juvenile stage. Bifurcated spines
below the nodes are found in *S. japonica* only.
The leaves are alternate, not clustered, simple,
ovate to elliptical, up to 25 × 10 cm but usually
smaller, usually acuminate, with entire to
minutely toothed margins, petiolate, without
stipules. The inflorescence is an axillary cyme
or aggregations of these into a thyrse, or flowers solitary, with long pedicels often thickened
above in fruit, with often conspicuous bracts
and bracteoles. The flowers are actinomorphic,
bisexual, 5- to 7-merous, up to 15 mm. The
sepals are sometimes unequal, connate at the
base. The petals are lanceolate to suborbicular.
The stamens are 5, epipetalous. There is a
crownlike disk below the ovary. The ovary is
superior, of 2 carpels united at the base, with 2
ovules per carpel. In fruit, either 1 or more
commonly both of the carpels develop into a 2-
seeded laterally compressed drupe with a pulpy
mesocarp and thin, crustaceous, ribbed endocarp, the whole fruit often appearing markedly
bipartite, white to red or blue at maturity.

Classification The Meliosmaceae (*Meliosma,
Ophiocaryon*) are often included in Sabiaceae
but differ markedly in their erect non-climbing
habit, leaves with a basal pulvinus, paniculate
inflorescence, often strongly zygomorphic flowers with only 2 fertile stamens and usually 3
staminodes that form a hood over the fertile stamens and ovary, and globose drupes with a
hard, bony, unribbed endocarp. It is preferred

here to maintain 2 separate families. A good argument for uniting them might be that they are much closer to each other than either is to the other families that have been included with them on molecular grounds in the Proteales, which are Proteaceae, Platanaceae, and Nelumbonaceae[i]. In APG II, the Sabiaceae are placed in eudicots unplaced as to order. A relationship with Menispermaceae has been previously postulated based on evidence of pollen morphology and embryology. A placement in the Rutales has been advocated based on wood anatomy, but Cronquist placed the family in Ranunculales.

Economic uses In India, the leaves may be applied to relieve pains or swellings of ankles and wrists, while in New Guinea they are eaten to control fever[ii].   RKB

[i] Savolainen, V. *et al.* Phylogeny of the eudicots: a nearly complete familial analysis based on *rbc*L gene sequences. *Kew Bull.* 55: 257–309 (2000).
[ii] van de Water, T. P. M. A taxonomic revision of the genus *Sabia* (Sabiaceae). *Blumea* 26: 1–64 (1980).

SALICACEAE

WILLOW AND POPLAR FAMILY

Trees and shrubs, sometimes dwarf, with catkins of flowers lacking a normal perianth.

Distribution *Salix* has c. 350 spp. widespread in the northern hemisphere from the Arctic to subtropics, especially in China, but few in the southern hemisphere (1 each in South America, Africa, and tropical Asia, but absent from New Guinea and Australasia). *Populus* has 30–40 spp. in the northern hemisphere but is absent from the southern hemisphere apart from 1 sp. in East Africa. The species occupy a wide range of habitats, from arctic tundra to riverbanks, other wet places, and mountains.

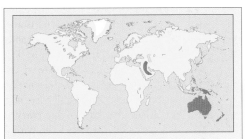

Genera 2 **Species** c. 385
Economic uses Fast-growing, low-quality timber; flexible branches for fencing and baskets; many cultivated trees and shrubs; bark used in tanning; medicinally the origin of aspirin

Description Dwarf prostrate subshrubs (almost herbaceous in *S. herbacea*) to shrubs and trees up to 30 m tall. The leaves are simple, linear to suborbicular, entire to serrate, or in *Populus* sometimes somewhat lobed, and usually deciduous. The stipules are usually conspicuous but often caducous. The flowers are unisexual (plants dioecious), reduced, borne in compact spikes (catkins) up to several cen-

timeters long, which are held upright in nearly all *Salix* (see below) but are pendent in *Populus*, these often appearing before the leaves develop. Each flower is subtended by a bract that is small and entire in *Salix* but obtriangular and toothed to fimbriate in *Populus*. In *Salix*, the perianth is replaced by 1–3 small nectariferous glands, while in *Populus* it is represented by a raised cuplike glandular rim, and normal sepals and petals are lacking. The male flowers in *Salix* consist of (1)2(–5) stamens, and in *Populus* of 5–60 stamens. The female flowers consist of a superior ovary derived from 2(–4) carpels with a single locule with 2–4 parietal placentas, each with numerous ovules and narrowed above to a short style with 2 often bifid stigmas. The fruit is a 2- to 4-valved capsule dehiscing from the top. The seeds are small and enveloped by a plume of long hairs arising from their base. Dehiscence of all the capsules of a catkin of a female plant releases a mass of wind-borne plumose seeds, or the whole dehisced catkin may fall to the ground.

SALICACEAE. 1 *Populus sieboldii* (a) leafy shoot and pendulous fruiting catkins (x⅔); (b) young female catkin (x⅔); (c) female flower with cuplike disk (x6); (d) ovary (x6); (e) stigmas (x6); (f) shoot with young male catkin (x⅔); (g) male flower (x6); (h) mature male catkins together with remains of 1 from the previous year (x⅔). 2 *P. nigra* "Italica" (Lombardy Poplar) habit. 3 *Salix caprea* (a) leaves (x⅔); (b) young female catkins (x⅔); (c) female flower and bract (x6); (d) vertical section of female flower (x6); (e) cross section of ovary (x8); (f) mature female catkins (x⅔); (g) male catkin; (h) male flower (x6).

Classification Although in early times the family was usually placed with other catkinate families with flowers lacking a perianth, it has been widely recognized that its affinities are with the Parietales, especially the Flacourtiaceae with which it shares many vegetative features. Recent molecular evidence[i] has placed it nested within traditional Flacourtiaceae, which has been divided into 2 parts. To avoid a paraphyletic family, the authors have combined the Salicaceae *sensu stricto* with the larger part of the Flacourtiaceae, and since the name Salicaceae has priority over Flacourtiaceae, the enlarged group has been called Salicaceae. This has resulted in more than half of the genera formerly referred to Flacourtiaceae being referred instead to Salicaceae, causing much confusion (cf. also Achariaceae). The traditional Salicaceae have then become the tribe Saliceae. In such a classification, the Saliceae nest within the Flacourtieae, so paraphyly is still present—changing the rank of taxa does not remove it. This new taxonomy is not followed here. It is considered that the catkinate inflorescences of reduced flowers lacking a normal calyx and corolla, and the very characteristic plumed seeds, justify recognition of Salicaceae in the traditional sense. For further discussion see Flacourtiaceae. Two segregate genera from eastern Asia, *Chosenia* and *Toisusu*, with 1 and 2 spp. respectively, have been separated from *Salix,* differing from it in having the catkins pendent (like *Populus*) and the upper part of the stigma caducous. Hybrids of both with *Salix* spp. occur. *Chosenia* also differs from other *Salix* in not having nectariferous glands in the flowers, and is thought to be wind pollinated similar to *Populus*. On the other hand, *Toisusu* has lateral glands in the female flowers and dorsiventral glands in the male flowers, and is thought to be insect pollinated like *Salix*. These differences are slight, and it is here preferred to retain them within *Salix* as anomalous species.

Economic uses The wood is fast growing but is usually of poor quality as timber and is often used for making small objects. A specialist use of *Salix* is in making cricket bats, for which female plants of *S. alba* var. *caerulea* are the most prized. The Osier, *S. viminalis*, is harvested in juvenile growth for its pliant stems, which have long been woven into baskets or used for hedging. Many species are planted for ornamental or achitectural effect, especially the Weeping Willow, aptly named as *S.* × *sepulcralis* (*S. alba* × *babylonica*). The fastigiate Lombardy Poplar, *P. nigra* var. *italica,* is widely planted as an avenue tree or wind break and is a conspicuous feature of many French roadsides. The bark has been used for tanning leather. Aspirin was originally derived from the bark of *Salix* but is now manufactured synthetically. RKB

[i] Chase, M. W. *et al.* When in doubt, put it in Flacourtiaceae: a molecular phylogenetic analysis based on plastid *rbc*L DNA sequences. *Kew Bull.* 57: 141–181 (2002).

SALVADORACEAE
TOOTHBRUSH TREE

A small family of trees and shrubs of the Old World arid tropics.

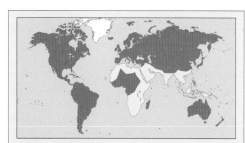

Genera 3 **Species** 8–10
Economic uses Shoots, leaves used as salad; local medicinal uses; fat for candles; Kegr salt; perfume oil

Distribution The family is native to arid, often saline, areas in Africa, Madagascar, Arabia, India, Asia, including New Guinea.

Description Shrubs to small trees, with (*Azima*) or without thorns, and with opposite leaves. The leaves are leathery, entire, and simple, with minute stipules, *Azima* with 2 or more thorns in the axil. The inflorescence is a terminal or axillary, simple or compound raceme, with numerous small flowers. The flowers are actinomorphic, hermaphrodite (*Dobera, Salvadora*) or unisexual (plants polygamodioecious in *Dobera,* or dioecious in *Azima*). The sepals are 2–4(5) joined in a tube with imbricate or valvate lobes. The petals are 4(5) free or fused at the base (*Salvadora*), with teeth or glands on the inner side. A disk in the form of separate glands may be present between the filaments. The stamens are 4(5), alternating with the petals, free in *Azima,* fused into a tube in *Dobera,* or joined to the base of the petals in *Salvadora*. The ovary is superior, of 2 fused carpels, forming 1 (*Dobera, Salvadora*) or 2 (*Azima*) locules, and bearing a short to almost lacking style and 2-lobed stigma. The fruit is a berry or drupe and contains a single seed without endosperm and an embryo with thick cotyledons.

Classification The 3 genera differ in habit and floral structure: *Azima* (3–4 spp.) are thorny shrubs whose flowers possess free petals and stamens and a 2-locular ovary, while *Dobera* (2 spp.) and *Salvadora* (c. 4 spp.) are unarmed trees and shrubs. The family is often placed in Celastrales or Olealesi[i]. DNA sequence evidence suggests a relationship with Gyrostemonaceae or Tovariaceae[ii] in one study but with Bataceae[iii] in another and with *Forchammeria* and *Tirania*[iv] in a third due to differences in sampling. The family is now placed in core Brassicales/Capparales.

Economic uses The shoots and leaves of *Salvadora persica* (Toothbrush Tree or Salt Bush) are used in salads or as food for camels and the sweet fruit are eaten. Kegr salt is obtained from the plant ash. The sweet fruit of *S. oleioides* is eaten dried or raw when fully ripe. The flowers of *Dobera roxburghii* provide an essential oil used by Sudanese women as perfume and the fruit can be eaten. The chewed stems and roots are used to clean and brush teeth. AC

[i] Brummitt, R. K. *Vascular Plant Families and Genera.* Richmond, Royal Botanic Gardens, Kew (1992).
[ii] Rodman, J. E. *et al.* Parallel evolution of glucosinolate biosynthesis inferred from congruent nuclear and plastid gene phylogenies. *American J. Bot.* 85: 997–1007 (1998).
[iii] Chandler, G. T. & Bayer, R. J. Phylogenetic placement of the enigmatic Western Australian genus *Emblingia* based on *rbc*L sequences. *Plant Species Biol.* 15: 67–72 (2000).
[iv] Hall, J. C., Iltis, H. H., & Sytsma, K. J. Molecular phylogenetics of core Brassicales, placement of orphan genera *Emblingia, Forchhammeria, Tirania,* and character evolution. *Syst. Bot.* 29(3) 654–669 (2004).

SAMBUCACEAE
ELDER FAMILY

One widespread genus of robust herbs to small trees with pinnate leaves and broad corymbs of white to pink flowers.

Distribution *Sambucus* occurs in all continents. It is widespread in the northern-temperate region (*S. nigra, S. racemosa,* and *S. ebulus*), extending south in the Old World to the Canaries, North Africa, the Middle East, and China, and in the New World through Central America to the Andes. *S. adnata* and *S. wightiana* occur in the Himalayan region, and *S. javanica* extends from the eastern Himalayas to Japan and New Guinea. *S. australis* is native in northeastern Argentina, Uruguay, and southern Brazil. In tropical Africa, the European *S. ebulus* has an isolated disjunct subsp. on the high mountains of East Africa. The other 2 spp. are endemic to Australia from the Cape York Peninsula to Tasmania.

Description Robust somewhat shrubby rhizomatous herbs to many-stemmed shrubs or small deciduous to semievergreen trees up to 10 m tall. The stems of herbaceous species are grooved, those of woody species conspicuously lenticillate, with a well-developed white or brown pith. Stipule-like outgrowths that occur in whorls of 3–4(–6) at the nodes, which may

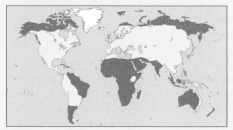

Genera 1 **Species** 9
Economic uses Many medicinal properties; fruit used for food and wine; shrubs often cultivated as ornamentals

be peg-shaped or stipitate-glandular or folia-ceous and up to 3 cm, are found irregularly in all species, but are sometimes absent, their development being sometimes dependent on growth conditions. The leaves are opposite, pinnate with up to 7 pairs of leaflets, or occasionally bipinnate, sometimes with stipels on the midrib; the leaflets are up to 15 cm long, ovate to lanceolate or oblanceolate, with serrate margins and usually an acute apex. The inflorescence is a broad flat or rounded terminal panicle or corymb up to 30 cm across, with the primary rays umbellate or alternate. Bracts and bracteoles are usually small. The flowers are sessile or pedicillate, actinomorphic, unisexual in *S. australis*, usually 5-merous but 3- or 4-merous respectively in the Australian species *S. australasica* and *S. gaudichaudiana*, often fragrant. The calyx is small and inconspicuous. The corolla has a short tube and (3–)5 broad rounded lobes, and is white or cream to greenish-yellow or pinkish. The stamens are as many as corolla lobes, inserted near the base of the tube, sometimes with purple anthers. The ovary is three-quarters inferior, with 3–5 locules, each with a single pendulous ovule. The stigma is sessile, 3- to 5-lobed. The fruit is a spherical to ovoid berry, black or blue to purple or red, including 3–5 hard pyrenes.

Classification *Sambucus* has often been placed in Caprifoliaceae but differs from it in many ways, including the compound leaves, corolla without nectaries, only partially inferior ovary, and sessile stigma. Its affinities lie on one hand with the delicate herbaceous Adoxaceae and on the other with the woody *Viburnum*, which is here also excluded from Caprifoliaceae as Viburnaceae. Some authors have therefore included all 3 in a broad Adoxaceae. Despite these undoubted affinities, however, the argument for a monogeneric family Sambucaceae based on numerous morphological characters has been compellingly argued in the recent comprehensive monograph[i]. The same family concept had been adopted by Airy Shaw, Dahlgren and Takhtajan (see Adoxaceae and Viburnaceae for further discussion and references.) These 3 families together form a sister group to the other Dipsacales of APG II, which includes Caprifoliaceae. The monograph cited has convincingly reduced the number of species recognized in *Sambucus* from previous estimates of 20–40 to 9. *S. nigra* is treated as comprising 6 subspp., including the North American subsp. *canadensis* (American Elder) and subsp. *caerulea* (Blue Elder), formerly *S. canadensis* and *S. caerulea* respectively, the former extending broadly from Canada to Panama and the other from British Columbia to California, while subsp. *peruviana* extends from Panama to northern Argentina. Subsp. *nigra*, the Common Elder of Europe, occurs eastward to the Volga River, and other subspp. are endemic in Madeira and the Canaries, respectively. The Red Elder, *S. racemosa*, is pannorthern temperate.

Economic uses The monograph cited describes *Sambucus* as a multipurpose chemist's shop. The flowers of European *S. nigra* are said to be diuretic, diaphoretic, and antipyretic, while subsp. *canadensis* is used to treat gastrointestinal and dermatophytic disorders in Guatemala and to have antibacterial properties. *S. javanica* has antihepatotoxic properties and helps cure rheumatic pain and fractures. The flowers and fruits of *S. nigra* are used to make wine, champagne, liqueurs (sambuca), herbal tea, and syrup, or are made into fritters, jam, or pie-fillings. Several species are cultivated as ornamental shrubs, with many cultivars, and have sometimes escaped as widespread weeds. RKB

i Bolli, R. Revision of the genus *Sambucus*. *Dissert. Bot.* 223: 1–227, pl. 1–29 (1994).

SAMOLACEAE

Herbaceous relatives of the Theophrastaceae.

Genera 1 Species c. 10[i]
Economic importance Occasional aquarium plants

Distribution *Samolus* is widespread, but rare in the tropics. The main centers of diversity are in the southern USA and adjacent Mexico and in Australia. All species generally prefer wet places, sometimes beside brackish water.

Description Annual to perennial herbs, erect or trailing, rarely slightly woody at the base. The leaves are usually in a basal rosette and with few to many alternate, linear to spathulate leaves on the stems, but reduced in size or lacking in some species. The inflorescence is a lax terminal raceme with rather stiffly spreading pedicels usually much longer than the flower. The lower nodes of the inflorescence may bear foliage leaves, but most of the raceme is leafless. Each pedicel is either subtended by a small bract or bears the bract somewhere along its length so that the pedicel may appear jointed. The flowers are bisexual, 5-merous, actinomorphic, and perigynous. The sepals are small, triangular, not overlapping, inserted near the top of the ovary, and persistent. The corolla is campanulate or sometimes urceolate, white or pink to purple, with broadly ovate rotate lobes, often scarcely exceeding the calyx. There are usually 5 staminodes inserted near the sinus between the corolla lobes. The fertile stamens are inserted on the corolla tube opposite the lobes. The ovary is three-quarters inferior, more or less globose, with a single

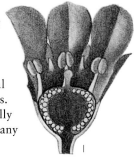

locule and a free central placenta with many ovules. The fruit is a loculicidally dehiscent capsule with many small angular seeds.

Classification *Samolus* has usually been included in the Primulaceae, where it has always been anomalous in its inferior ovary. Morphological and molecular characters place it much nearer to Theophrastaceae, in which it has been included in the APG classifications. The similarity of the inflorescences seems to suggest an obvious relationship to Theophrastaceae (compare, e.g., the inflorescences of a robust *Samolus* with that of *Jacquinia macrocarpa*), and is supported by the ring of staminodes in both. However, there are many characters that keep *Samolus* out of Theophrastaceae, notably the herbaceous habit, inferior ovary, dehiscent capsule, lack of the placental pulp surrounding the seeds, absence of calcium oxalate crystals in the anthers, and the worldwide distribution. The recognition of the family Samolaceae in FGVP[ii] is followed here. For further discussion and references, see Primulaceae.

Economic uses Some species are cultivated in aquaria or terraria. RKB

i Crusio, W. E. Notes on the genus *Samolus* (Primulaceae). *Mededel. W. A. P.* 2: 13–25 (1982) and 6: 13–16 (1984).
ii Ståhl, B. Samolaceae. Pp. 387–389. In: Kubitzki, K. (ed.), *The Families and Genera of Vascular Plants. VI. Flowering Plants. Dicotyledons: Celastrales, Oxalidales, Rosales, Cornales, Ericales*. Berlin, Springer-Verlag (2004).

SANTALACEAE
SANDALWOOD

Tropical and temperate herbs, shrubs, and trees, mostly semiparasites.

Distribution The family is widely distributed in tropical, subtropical, and temperate regions, being concentrated in relatively dry areas.

Genera c. 40 Species c. 490
Economic uses Sandalwood; oil; edible fruits

Description Mostly root-parasites, but a few are epiphytic branch-parasites. The leaves are usually alternate or in some genera (e.g.

Santalum, Buckleya) opposite, without stipules. Scalelike leaves are found in several genera (e.g., *Anthobolus, Exocarpos,* and *Phacellaria*), and the plants then superficially resemble a *Cytisus* or *Cupressus,* while other species have flattened branches (cladodes) that imitate true leaves. The flowers are borne in spikes, racemes, or heads and are actinomorphic, bisexual, or unisexual (plants monoecious or dioecious), generally small and inconspicuous. The perianth comprises a single united whorl of 3–6 segments (tepals) that may be greenish or colored. The stamens are as many as, and opposite, the tepals and adnate to them at the base. The ovary is inferior or semi-inferior, of 2–5 carpels, with a single locule containing 1–5 naked ovules suspended from a central placental column. The style is more or less simple. The fruit is a nut or drupe containing a single seed that has no testa and copious endosperm. The peduncle becomes fleshy in *Exocarpus.*

Classification The Santalaceae has previously been divided into 3 or 4 tribes: Amphorogyneae, Anthoboleae, Santaleae, and Thesieae. However, recent work indicates[i,ii] that the family is paraphyletic with respect to Viscaceae (q.v.) and Eremolepidaceae (q.v.), which were included in a broadly defined family by APG II. Seven well-supported clades are recognized in molecular phylogenetic analyses of Santalaceae, and a new classification has been proposed[ii] in which Viscaceae, Amphorogynaceae, Santalaceae, Nanodeaceae, Pyrulariaceae, Thesiaceae, and Comandraceae are recognized while *Anthobolus* is excluded from Santalaceae and allied with Opiliaceae, but as full details of this new arrangement are not yet available, this account of the family is provisional.

Economic uses *Santalum album,* the Sandalwood Tree, from southern India yields a fragrant white timber suitable for fine carvings and carpentry, while sandal oil, used in eastern countries for anointing the body and in the manufacture of soap and perfumes, is distilled from the yellow heartwood and roots. Sandalwood is also used as incense in Hindu, Buddhist, and Muslim ceremonies and used as joss sticks. *Exocarpos cupressiformis,* Australian Cherry, is one of the few species with edible fruits. The fruit of *Acanthosyris falcata* is also edible.   VHH

[i] Nickrent, D. L. & Malécot, V. A molecular phylogeny of Santalales. Pp. 69–74. In: Fer, A., Thalouarn, P., Joel, D. M., Musselman, L. J., Parker, C., & Verkleij, J. A. C. (eds), *Proceedings of the 7th International Parasitic Weed Symposium.* Nantes, Faculté des Sciences, Université de Nantes (2001).

[ii] Der, J. & Nickrent, D. Molecular systematics of Santalaceae: phylogeny and classification of a paraphyletic family of hemiparasitic plants. Systematics Section /ASPT Abstracts, Botanical Society of America (2005). http://www.2005.botanyconference.org/engine/search/index.php?func=detail&aid=213

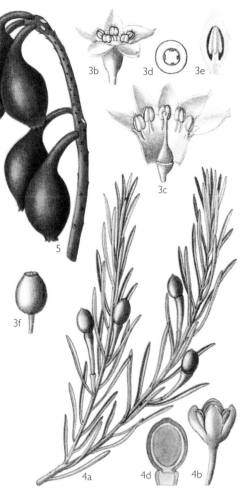

SANTALACEAE. 1 *Quinchamalium majus* (a) leafy shoot and terminal inflorescence (×⅔); (b) flower (×3); (c) stamen attached to perianth-segment (×4); (d) vertical section of ovary (×12). 2 *Thesium lacinulatum* (a) flowering shoot (×⅔); (b) shoot tip showing scalelike leaves (×3); (c) flower (×6); (d) flower with 2 perianth segments removed (×6); (e) fruit (×6). 3 *Santalum yasi* (a) flowering shoot (×⅔); (b) flower (×6); (c) flower opened out (×8); (d) cross section of ovary (×18); (e) vertical section of ovary (×18); (f) fruit (×⅔). 4 *Anthobolus foveolatus* (a) leafy shoot and fruits (×⅔); (b) partly open flower (×4); (c) flower (×4); (d) vertical section of fruit (×1½). 5 *Scleropyrum wallichianum* fruits (×⅔).

SAPINDACEAE
LITCHI, MAPLE, AND HORSE CHESTNUT FAMILY

Mainly tropical and southern trees and woody climbers with pinnate leaves and unisexual flowers, here also including the more temperate Aceraceae and Hippocastanaceae.

Distribution Worldwide except in the extreme north and south. The northern-temperate distribution is largely due to the genera *Acer* and *Aesculus.* The rest of the family is in the tropics and southern hemisphere, occurring commonly in wet forests up to 3,600 m altitude but sometimes in open drier habitats, extending north sparingly to southern USA and temperate eastern Asia. Highest numbers of species are in the New World (tropical America 968 spp., tropical Africa 234 spp., South Africa 27 spp., India 37 spp., *Flora Malesiana* region 235 spp., Australia 190 spp., northern temperate c. 120 spp.).

Genera c. 145 **Species** c. 1,900
Economic uses Edible fruit arils (Lychee, Akee, Rambutan), timber, ornamentals

Description Trees up to 40 m or more, woody lianes often with tendrils (modified inflorescences), or shrubs, occasionally herbaceous climbers (*Cardiospermum*) or rarely rhizomatous herbs (*Diplopeltis* in Australia). The leaves are alternate or sometimes opposite (especially Aceroideae and Hippocastanoideae, see below), usually paripinnate or imparipinnate, sometimes bipinnate or ternate to biternate or palmately compound, or sometimes 1-foliate or simple, and the leaflets of compound leaves are alternate to opposite or digitate, entire to serrate, and often large. Small stipules or pseudostipules are present in some climbing genera. The inflorescence is usually a panicle or thyrse, or rarely the flowers are solitary or fascicled in leaf axils. The flowers are actinomorphic or sometimes weakly zygomorphic (petals crowded to one side in *Chytranthus* etc.) to strongly zygomorphic in the Hippocastanoideae, usually 4- to 5-merous but sometimes with multiple parts, usually small and not showy except in *Aesculus,* nearly always functionally unisexual (staminodes or pistillodes often present), the plants then monoecious or dioecious, or the flowers rarely bisexual in *Dodonaea* and unisexual and the plants andromonoecious in *Aesculus.* The sepals are (3–)5(–7), free or fused into a tube, which is sometimes urceolate, occasionally petaloid. The petals are (2–)4–5(6), free, often clawed, or sometimes absent. A complete or 1-sided extrastaminal disk is present, or this may be reduced to a pair of glands. The stamens are 5–10(–74) but most commonly 8 in a single whorl. The ovary is superior, with 1–3(–8) locules, with 1 to several ovules in each locule. The fruit may be a capsule, drupe, berry, or double samara, often strongly lobed or variously ornamented. The seeds often have an aril or sarcotesta, as in the fleshy part of lychee fruit.

Classification The concept of the Sapindales has varied widely from author to author, and many families formerly associated with Sapindaceae are now placed elsewhere. According to molecular evidence[i], the major families related are the Rutaceae, Meliaceae, and Anacardiaceae. The only question over the delimitation of the family is whether the Aceraceae and Hippocastanaceae should be included or not. Both have opposite leaves with palmate lobes or leaflets, and characteristic fruits. Opposite leaves are apparently otherwise only found in Sapindaceae in *Matayba,* especially *M. oppositifolia* in the West Indies and in the simple-leaved *Guindilia* in Chile, while digitate leaves are found rarely in *Allophylus.* The large fruits with large seeds of *Aesculus* are not out of place in Sapindaceae (see the temperate Asian *Xanthoceras*), while the double samara of *Atalaya* in eastern Malaysia and Australia is such an exact match for the fruit of *Acer* that this cannot be seen as a family character either. The occurrence of 5 petals with 8 stamens in both *Acer* and *Aesculus* agrees well with many Sapindaceae. A broad sampling of genera for molecular analysis is still awaited, but available evidence[i] from a few genera has placed both *Acer* and *Aesculus* near the base of Sapindaceae but with *Xanthoceras* sister to the whole group. There seems to be no case for maintaining these 2 families distinct from Sapindaceae, but in the absence of a modern reassessment of the tribes recognized by Radlkofer in his *Pflanzenreich* monograph (1930–1933), it is preferred here to recognize them as 2 subfamilies. Radlkofer's division of the remainder of the family into 2

SAPINDACEAE. **I** *Acer platanoides* (a) shoot with opposite, palmately lobed leaves and fruits comprising pairs of winged samaras (x⅔); (b) shoot with terminal inflorescence, young leaves, and bud scales (x⅔); (c) male flower with 4 sepals, 4 petals, 8 stamens, and central vestige of the ovary (x3); (d) half bisexual flower showing winged ovary with forked style and short stamens on a lobed disk (x3); (e) silhouette of a leafless tree showing the much branched habit.

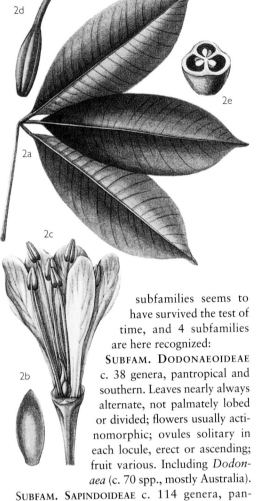

2d

2e

2a

2c

2b

from South America produce a drink with a very high caffeine content. Soap has been made from fruits of many species, especially *Sapindus saponaria*. Many trees yield valuable timber, and the stems of the climbers are used as rope. The stems of the climbers are also ground up as fish poison. *Schleicheria trijuga* produces macassar oil used in ointments. Many species and cultivars of *Acer* (Maples) are grown in temperate regions. *Aesculus hippocastanum* (Horse Chestnut) is commonly planted, and several other species are grown as ornamentals. *Koelreuteria paniculta* (Golden Rain Tree) is widely planted and produces panicles of large papery capsules at the end of a hot summer. RKB

[i] Savolainen, V. *et al.* Phylogeny of the eudicots: a nearly complete familial analysis based on *rbc*L gene sequences. *Kew Bull.* 55: 257–309 (2000).
[ii] Xiang, Q.-Y. *et al.* Origin and biogeography of *Aesculus* (Hippocastanaceae), a molecular phylogenetivc perspective. *Evolution* 52: 988–997 (1998).

8b

8a

6

8c

8e

8d

7b

7c

7d

3c

3a

3b

3g

3d

3e

3f

subfamilies seems to have survived the test of time, and 4 subfamilies are here recognized:

SUBFAM. DODONAEOIDEAE c. 38 genera, pantropical and southern. Leaves nearly always alternate, not palmately lobed or divided; flowers usually actinomorphic; ovules solitary in each locule, erect or ascending; fruit various. Including *Dodonaea* (c. 70 spp., mostly Australia).

SUBFAM. SAPINDOIDEAE c. 114 genera, pantropical and southern. Leaves nearly always alternate, rarely palmately lobed or divided; flowers usually actinomorphic; ovules 2 to several per locule or rarely 1 pendulous; fruit various. Genera include *Serjania* (c. 220 spp.), *Allophylus* (c. 200 spp., famously reduced to 1 in *Fl. Malesiana*), and *Paullinia* (c. 180 spp.).

SUBFAM. HIPPOCASTANOIDEAE 3 genera, northern warm-temperate to tropical. Leaves opposite, palmately divided into 3–5 leaflets; flowers strongly zygomorphic, andromonoecious; ovules 2 per locule; fruits spherical with one large seed. *Aesculus* (13 spp., northern temperate), *Billia* (2 spp., Mexico to northern South America, leaves 3-foliate), and *Handeliodendron* (1 sp., China)[ii].

SUBFAM. ACEROIDEAE 2 genera, widespread north-temperate, extending sparingly into Southeast Asia and Central America. Leaves opposite, palmately lobed to 3-foliate or pinnate, or entire with 3 veins from the base in Asia; flowers actinomorphic, functionally unisexual; ovules 2 per locule; fruit a double unilateral samara or a circular samara. *Acer* (c. 110 spp., distribution as subfamily) and *Dipteronia* (2 spp., China).

Economic uses Fruits known for their edible aril occur in the Asiatic *Litchi chinensis* (Lychee) and *Nephelium lappaceum* (Rambutan), the African *Blighia sapida* (Akee, poisonous if not cooked), and the South American *Melicoccus bijuga* (Memoncillo). Seeds of *Paullinia cupana*

SAPINDACEAE. 2 *Billia hippocastanum* (a) leaf (x⅔); (b) sepal
(x2); (c) flower with sepals removed to show 5 slightly
unequal petals and 6 stamens (x1½); (d) gynoecium with
curved style and simple stigma (x4½); (e) cross section
of 3-locular ovary with ovules on axile placentas (x10).
3 *Aesculus hippocastanum* (a) leafless mature tree; (b)
digitate leaf and inflorescence (x⅔); (c) half flower with
fused sepals and unequal petals (x1); (d) dehiscing fruit
(a capsule) exposing seeds (x⅔); (e) seed with large
hilum (x⅔); (f) cross section of ovary (x4); (g) vertical
section of ovary (x4). 4 *Dodonaea bursarifolia* (a) flowering
shoot (x⅔); (b) male flower (x4); (c) female flower (x4);
(d) fruit (x1½); (e) cross section of fruit (x1½); (f) vertical
section of part of fruit (x4). 5 *Litchi chinensis* fruits, entire,
and in section (x⅔). 6 *Serjania exarata* shoot tip with coiled
tendrils, winged fruits, and inflorescences (x⅔). 7 *Paullinia
thalictrifolia* (a) flowering shoot (x⅔); (b) flower (x4); (c)
flower with 2 petals removed (x4); (d) gynoecium (x6).
8 *Cupaniopsis anacardioides* (a) pinnate leaf (x⅔); (b)
inflorescence (x⅔); (c) flower viewed from above (x2);
(d) gynoecium surrounded by disk and calyx (x2);
(e) cross section of ovary (x4).

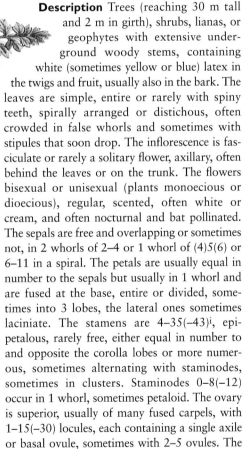

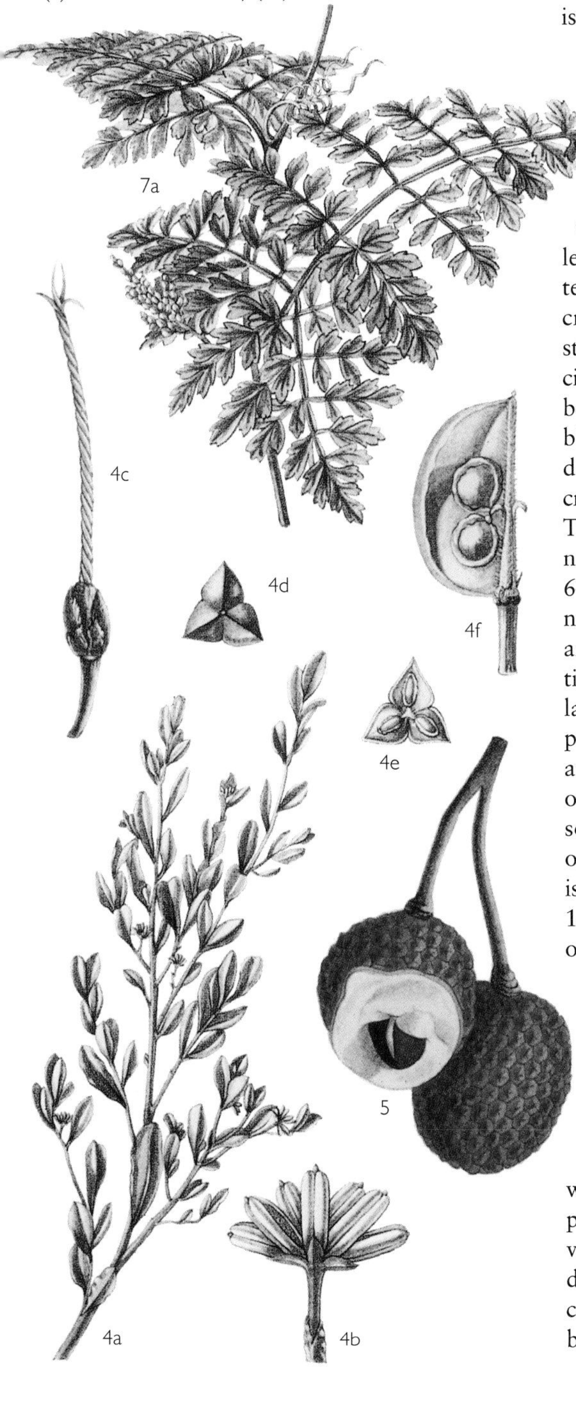

SAPOTACEAE

CHICLE, GUTTA-PERCHA, MASTIC, AND SAPODILLA

A large family of pantropical trees containing
latex in the twigs and fruit.

Distribution The Sapotaceae occurs pantropi-
cally, mainly in lowland and lower montane rain
forest. The largest genera are *Pouteria* (c. 330
spp.), c. 200 spp. in Tropical America, 120 spp.
in Asia and the Pacific, and 5 spp. in Africa;
Palanquium (120 spp.), Taiwan and Indo-
Malaysia; *Madhuca* (110 spp.), Indo-Malaysia;
Chrysophyllum (80 spp.), 43 spp. in Tropical
America, 15 spp. in Africa, 10 spp. in Madagascar, and 2 or 3 spp. in Indo-Malaysia and
Australia; *Manilkara* (80 spp.), 30 spp. in tropical America, 20 spp. in Africa and Madagascar,
and 15 spp. in Asia and the Pacific; and *Sideroxy-
ylum* (75 spp.), 49 spp. in tropical America, 6
spp. in Africa, 6 spp. in Madagascar, 8 spp. in
the Mascarenes, 4 spp. in Asia, and 1 sp. in Pakistan through to Ethiopia.

Description Trees (reaching 30 m tall
and 2 m in girth), shrubs, lianas, or
geophytes with extensive underground woody stems, containing
white (sometimes yellow or blue) latex in
the twigs and fruit, usually also in the bark. The
leaves are simple, entire or rarely with spiny
teeth, spirally arranged or distichous, often
crowded in false whorls and sometimes with
stipules that soon drop. The inflorescence is fasciculate or rarely a solitary flower, axillary, often
behind the leaves or on the trunk. The flowers
bisexual or unisexual (plants monoecious or
dioecious), regular, scented, often white or
cream, and often nocturnal and bat pollinated.
The sepals are free and overlapping or sometimes
not, in 2 whorls of 2–4 or 1 whorl of (4)5(6) or
6–11 in a spiral. The petals are usually equal in
number to the sepals but usually in 1 whorl and
are fused at the base, entire or divided, sometimes into 3 lobes, the lateral ones sometimes
laciniate. The stamens are 4–35(–43)[i], epipetalous, rarely free, either equal in number to
and opposite the corolla lobes or more numerous, sometimes alternating with staminodes,
sometimes in clusters. Staminodes 0–8(–12)
occur in 1 whorl, sometimes petaloid. The ovary
is superior, usually of many fused carpels, with
1–15(–30) locules, each containing a single axile
or basal ovule, sometimes with 2–5 ovules. The
style is simple. The fruit is a berry, not articulated. The seeds are 1 or few, with an oily
endosperm, bony testa, and a large embryo.

Classification The Sapotaceae is a member
of Ericales[ii], sister to Lecythidaceae. It is one
of the families (another is the Lauraceae)
within which generic limits are difficult to
perceive, and opinions on subdivision
vary considerably. From 35 to 75 ill-
defined genera can be distinguished;
currently 53 genera are accepted[iii],
but many of these are not natural

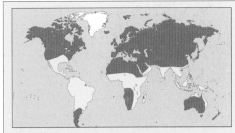

Genera 35–75 **Species** c. 800
Economic uses Latex; oil; timber; fruit

groups[iv], and molecular evidence is sparse so
far[v]. A detailed study of New Caledonian
species indicates many genera to be paraphyletic[vi]. Some authors recognize 5 tribes[iii] but
others 3[iv]. *Sarcosperma*, sometimes considered
a mono-generic family (Sarcospermataceae), is
sister to the remaining species sampled in an
analysis of chloroplast DNA sequence[iv].

Economic uses The Sapotaceae are an important component of many tropical rain forests
(as in Malaysia and Borneo). Some species
have heavy timber, which is hard and naturally
durable, but often siliceous; others have lighter
timber, some without silica. Gutta-percha,
obtained from the latex of *Palaquium* species (especially *Palaquium gutta* of Sumatra,

SAPOTACEAE. 1 *Sideroxylon costatum* (a) flowering shoot
(x⅔); (b) flower (x3); (c) corolla opened out (x3);
(d) gynoecium (x4); (e) vertical section of ovary (x4).
2 *Manilkara zapota* (*Achras zapota*) (a) fruit—the sapodilla
plum (x⅔); (b) cross section of fruit (x⅔).

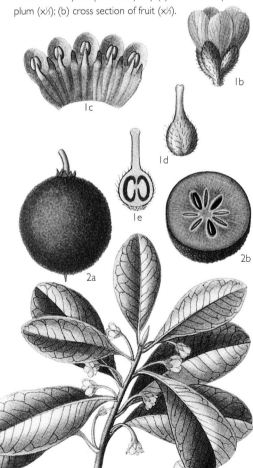

SAPOTACEAE. 3 *Madhuca parkii* (a) tip of leafy
flowering shoot with flowers in fascicles (×⅔);
(b) cross section of ovary showing 8 locules each
with an ovule on an axile placenta (×3). 4
Mimusops zeyheri var. *laurifolia* (a) leafy shoot with
axillary fascicles of flowers (×⅔); (b) perianth opened out
(×3); (c) petal with appendages (×3); (d) stamen (×4); (e)
staminode (×4); (f) vertical section of ovary (×3).

solitary. The perianth is absent in male flowers and sepaline in the female flowers, with 2 calyx elements that are accrescent and fleshy, later winged and enclosing the fruit. The stamens are 1–4, free. The ovary is partly inferior, with 2 carpels.

The family was not described until 1997[i], but the one component genus had been recognized as an anomalous element of the Chenopodiaceae gaining subfamily rank[ii]. *Sarcobatus* is now distinguished from Chenopodiaceae by its different sieve element plastids[i]. Molecular evidence from *rbc*L and *mat*K sequence data suggests the family is most closely related to Agdestidaceae in the "core" Caryophyllales[iii]. Greasewoods are used traditionally as high quality firewood and for tools[iv]. AC

[i] Behnke, H.-D. Sarcobataceae – a new family of Caryophyllales. *Taxon* 46: 495–507 (1997).
[ii] Ulbrich, E. In: Engler, A. & Prantl, K. (eds), *Die natürlichen Pflanzenfamilien*, ed. 2, 16c. Leipzig (1934).
[iii] Cuénoud, P. *et al.* Molecular phylogenetics of Caryophyllales based on nuclear 18S and plastid *rbc*L, *atp*B and *mat*K DNA sequences. *American J. Bot.* 89: 132-144 (2002).
[iv] http://www.desertusa.com/mag00/oct/papr/gwood.html

[ii] Anderberg, A. A., Rydin, C., & Källersjö, M. Phylogenetic relationships in the order Ericales *s.l.*: analyses of molecular data from five genes from the plastid and mitochondrial genomes. *American J. Bot.* 89: 677–687 (2002).
[iii] Pennington, T. D. *The Genera of Sapotaceae.* Richmond, Royal Botanic Gardens, Kew (1991).
[iv] Swenson, U. & Anderberg, A. A. Phylogeny, character evolution, and classification of Sapotaceae (Ericales). *Cladistics* 21: 101–130 (2005).
[v] Anderberg, A. A. & Swenson, U. Evolutionary lineages in Sapotaceae (Ericales): a cladistic analysis based on *ndh*F sequence data. *Int. J. Plant Sci.* 164: 763–773 (2003).
[vi] Bartish, I. V. *et al.* Phylogenetic relationships among New Caledonian Sapotaceae (Ericales): molecular evidence for generic polyphyly and repeated dispersal. *American J. Bot.* 92: 667–673 (2005).

SARCOLAENACEAE

Genera 8
Species c. 60
Economic uses
The timber of some species is used locally.

A small family of often handsome, mainly evergreen, small to medium trees and shrubs, endemic to Madagascar, growing in littoral to montane forests. The leaves are alternate, simple, entire, ovate, oblong or elliptical to circular, with caducous stipules. The flowers are actinomorphic, hermaphrodite, often showy, borne singly or in pairs, enclosed or surrounded by involucres of bracts of various forms and degrees of development, often accrescent. The inflorescences are various, terminal, or axillary. The sepals are 3 or 5 (the outer 2 smaller) and the petals 5(6). The stamens are 6–12 or many (20+), free or partially connate. An extrastaminal disk is usually present. The ovary is superior, with (1–)3–4(5) united carpels, with axile placentation; ovules 1 or few to numerous. The fruit is a loculicidal capsule or an indehiscent nut, with accrescent bracts or a cupule. The seeds are 1 to several. The timber of some species is used locally.

The Sarcolaenaceae was placed in the Theales by Cronquist, but recent phylogenetic analyses based on molecular evidence suggest that the family is sister to Dipterocarpaceae within an expanded Malvales[i,ii]. VHH

Malaysia, Java, and Borneo) is a polymer of isoprene, differing from rubber by having *trans-* instead of *cis-* isomerization, being almost nonelastic, a better insulator of heat and electricity, becoming plastic on heating, and on cooling retaining any shape imparted while hot. *Manilkara bidentata* of northern South America also yields a latex, balata, formerly of considerable importance. Chicle, the elastic component of earlier chewing gum, is produced from the latex of *Manilkara zapota* (*Achras zapota*). *Manilkara zapota* also yields the popular edible fruit Chiku or Sapodilla Plum, and *Chrysophyllum cainito* the Star Apple; both are of American origin but now planted elsewhere in the humid tropics. Some of the fruits termed Sapote are the product of *Calocarpum* species, notably *Calocarpum sapota*. The seeds of the north tropical African *Butyrospermum paradoxum*, the Shea Butter Tree, yield an edible oil. AC

[i] Pennington, T. D. Sapotaceae. Pp. 390–421. In: Kubitzki, K. (ed.), *The Families and Genera of Vascular Plants. VI. Flowering Plants. Dicotyledons: Celastrales, Oxalidales, Rosales, Cornales, Ericales.* Berlin, Springer-Verlag (2004).

SARCOBATACEAE
GREASEWOOD

A family consisting of a 1 genus of spiny succulent shrubs, with 2 spp.: *Sarcobatus vermiculatus* restricted to western North America and *S. baileyi* endemic to Nevada, recognizable by the thick fleshy leaves borne on spiny branches. Greasewoods grow in arid or saline habitats in the Great Basin, where they can dominate plant communities, and in southwestern deserts south to the Sonoran Desert. Leaves are alternate, fleshy, without petioles or stipules, the lamina simple, entire. The flowers are unisexual (plants monoecious or monoecious and dioecious): male flowers are aggregated into catkinlike inflorescences, the female flowers are

Genera 1 Species 2
Economic uses Tools; high-quality firewood

SARGENTODOXACEAE

i Alverson, W. S. *et al.* Circumscription of the Malvales and relationships to other Rosidae: evidence from *rbc*L sequence data. *American J. Bot.* 85: 876–877 (1998).
ii Ducousso, M. *et al.* The last common ancestor of Sarcolaenaceae and Asian dipterocarp trees was ectomycorrhizal before the India–Madagascar separation, about 88 million years ago. *Molec. Ecol.* 13: 231–236 (2004).

A monotypic family, comprising *Sargentodoxa cuneata*, a dioecious woody climber from China (including Hainan), Laos, and Vietnam, similar in appearance to Lardizabalaceae and sometimes included there. Leaves are 3-foliate, the inflorescence is a lax raceme, and there are 6 elongate sepals and small nectariferous petals, giving an appearance similar to Lardizabalaceae. It also has many stipitate fleshy uniovulate carpels spirally arranged on a swollen fleshy receptacle, very unlike the ring of 3 leathery follicles or fleshy multiovulate carpels in Lardizabalaceae. The reference by Takhtajan (1997) to a "deep-seated corm" is apparently a misprint for "deep-seated cork," but this is a significant anatomical difference, and leaf anatomy is also different. Molecular and other studies have placed *Sargentodoxa* outside Lardizabalaceae (see that family for references especially the monograph

Genera 1 | Species 1
Economic uses None

of that family by Qin) but sister to it. Family status is retained here even though it has been sunk in FGVP into Lardizabalaceae.   RKB

SARGENTDOXACEAE. 1 *Sargentodoxa cuneata* (a) shoot bearing 3-foliate leaf and male inflorescence (x⅔); (b) stamens (x4); (c) half female flower (x2); (d) fruits (x⅔); (e) vertical section of fruit containing a single seed (x3).

SARRACENIACEAE

NORTH AMERICAN PITCHER PLANT, COBRA LILY, AND MARSH-PITCHER

An American family of carnivorous pitcher plants where each trap is formed from a single rolled and modified leaf.

Distribution *Sarracenia* (8–11 spp.) has its center of diversity in southeastern USA and occurs predominantly on wet sandy coastal plains but extends north around the Great Lakes and into Canada. *Darlingtonia* (1 sp.) is restricted to the mountains of California and Oregon in wet flushes and by streams. *Heliamphora* (8–11 spp.) grows in marshes on sandstone outcrops among the tepuis of the Guyana Highlands.

Description Rhizomatous or rarely woody (*H. tatei*) perennials with distinctive rolled leaves forming insect-trapping pitchers. Each leaf is rolled into a tube with a ventral wing, variously with a flat or expanded, swollen-globose or small cup-shaped lid, sometimes water filled (*Heliamphora* and *S. purpurea* agg.), sometimes green but often with bright red markings and sometimes clear windows (*S. minor, S. psittacina,* and *D. californica*). Sometimes heterophyllous (*Sarracenia*) with noncarnivorous laterally

SARRACENIACEAE. 1 *Heliamphora nutans* (a) pitchers (leaves) and flowers on leafless stalks (x⅔); (b) stamens, ovary, and style (x2); (c) section of 3-locular ovary with ovules on axile placentas (x3). 2 *Sarracenia purpurea* (a) pitchers and flower stalk (x2); (b) flower with green sepals and reddish petals (x1); (c) style and stigma (x1); (d) cross section of ovary with ovules on inrolled carpel walls (x2). 3 *Darlingtonia californica* (a) pitchers and atypical flowers (x⅔); (b) vertical section of gynoecium with stamens attached at the base (x3).

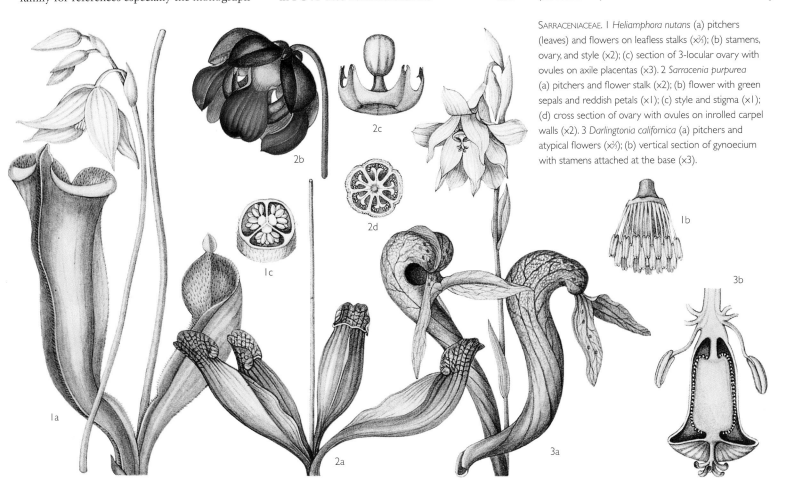

flattened leaves produced in winter (*S. flava* and *S. oreophila*). The inflorescence is terminal from the rhizome on an erect scape, of 1 flower (*Sarracenia* and *Darlingtonia*) or a few-flowered raceme (*Heliamphora*). The flowers actinomorphic, hermaphrodite, 5-merous, with distinct calyx and corolla (*Sarracenia* and *Darlingtonia*) or with

Genera 3
Species c. 20
Economic uses
Ornamental curiosities

overlapping petaloid segments (*Heliamphora*). The sepals, where present, are in 1 whorl of 5 overlapping (*Sarracenia*) or non-overlapping (*Darlingtonia*) segments, sometimes dark red, usually green; the petals are 5, hanging below sepals and bending outward, red or yellow (*Sarracenia*) or translucent with red veining and adpressed to form a chamber (*Darlingtonia*). The petal-like structures of *Heliamphora* are 4–6 overlapping, usually white, sometimes red or pink tinged. The stamens are 10–20 (*Heliamphora*) or many (*Darlingtonia* and *Sarracenia*). The ovary is superior, 1-locular formed from 3 (*Heliamphora*) or 5 (*Darlingtonia* and *Sarracenia*) fused carpels. The style in *Sarracenia* is distinctive, forming an umbrella-like outgrowth from the tip of the ovary, curving out to project between the petals. The fruit is a longitudinally splitting globose capsule containing many seeds.

Classification This family is placed in the Ericales by DNA sequence data[i] in a group with *Roridula*, a semicarnivorous genus, and *Actinidia*, the Kiwi fruit, although the 3 show no obvious morphological similarities. The family has been placed with Theales (Thorne 1983) or usually in its own order (e.g., Takhtajan 1983) or with other carnivorous plants in Nepenthales (Cronquist 1988). The 3 component genera are clearly defined on morphological grounds and have non-overlapping distributions. This distinctness does not extend to the species, as there is pronounced interspecies fertility within each genus. A molecular phylogenetic study showed almost no resolution within *Sarracenia*, and *Heliamphora* was insufficiently sampled to establish the relationships among species[ii].

Economic uses There is amateur interest in the family as an ornamental curiosity. Many species are listed by CITES[iii]. AC

[i] Anderberg, A. A., Rydin, C., & Källersjö, M. Phylogenetic relationships in the order Ericales *s.l.*: analyses of molecular data from five genes from the plastid and mitochondrial genomes. *American J. Bot.* 89: 677–687 (2002).
[ii] Bayer, R. Phylogenetic relationships in Sarraceniaceae based on *rbc*L and ITS sequences. *Syst. Bot.* 21: 121–134 (1996).
[iii] Simpson, R. B. *Pitchers in Trade: a Conservation Review of the Carnivorous Genera of Sarracenia, Darlingtonia, and Heliamphora*. Richmond, Royal Botanic Gardens, Kew (1994).

SAURURACEAE
LIZARD'S TAIL FAMILY

A small family of herbs with cordate leaves and axillary spicate inflorescences.

Distribution The family is disjunct between North America and eastern Asia: *Saururus cernuus* extends from southeastern Canada to Texas and Florida, while *S. chinensis* is in China, Korea, Japan, and the Philippines. *Gymnotheca* has 2 spp. in China, *G. chinensis* extending into northern Vietnam. *Anemopsis californica* occurs from Washington State to Texas and southwestern Mexico. *Houttuynia cordata* extends from central Himalaya to Korea, Japan, and western Java and is cultivated and perhaps naturalized elsewhere (Madagascar, New Caledonia, etc.).

SAURURACEAE. 1 *Anemopsis californica* (a) shoot with upper leaves and inflorescence surrounded by petal-like involucre of bracts (×⅔); (b) shoot with fruiting head (×⅔); (c) leaf (×⅔); (d) flower lacking a perianth and with an ovary that is sunk into the receptacle (×4); (e) cross section of inflorescence (×1½). 2 *Gymnotheca chinensis* (a) flowering shoot with cordate leaves (×⅔); (b) flowers (×2); (c) gynoecium (×4); (d) ovary opened out to show ovules on parietal placentas (×6). 3 *Saururus cernuus* (a) flower and bract (×2); (b) half flower with free carpels (×3). 4 *Houttuynia cordata* (a) flowering shoot (×⅔); (b) flower (×4); (c) fruit dehiscing from apex (×4); (d) fruit opened out to show seeds (×4).

SAXIFRAGACEAE. I *Darmera peltata* (a) leaf (x⅔); (b) inflorescence (x⅔); (c) flower with 2 petals removed to show stamens (x2); (d) fruits (x2). 2 *Bergenia crassifolia* (a) tip of shoot and inflorescence (x⅔); (b) flower opened out showing central bicarpellate ovary (x2); (c) vertical section of ovary (x2).

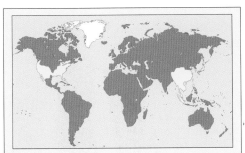

Genera 4 Species 6
Economic uses Medicinal uses, edible leaves or shoots and rhizomes , some horticultural value

Description Rhizomatous or stoloniferous, usually aromatic, perennial herbs, procumbent or up to 1.2 m high. The leaves are alternate, simple, cordate at the base, with entire margins, the petioles with membranous stipules along their lower margin, these connate and encircling the stem at their base. The inflorescence is an axillary dense spike or narrow raceme of 10–350 flowers, sometimes (*Anemopsis, Houttuynia, Gymnotheca involucrata*) with an involucre of petaloid bracts at its base and so superficially resembling a single flower. The flowers lack a perianth and are actinomorphic and bisexual. The stamens are 3 in *Houttuynia*, and (3–)6(–8) in other genera, free in *Saururus* but elsewhere adnate to the base of the ovary. The ovary is susperior in *Saururus* with (1–)4(–7) carpels, free or basally connate, and in *Anemopsis* with 1–3 fused carpels, or semi-inferior in *Houttuynia* and *Gymnotheca*, with 3 or 4 carpels respectively and 1-locular. Placentation is parietal, with 6–13 ovules per locule. The fruit in *Saururus* is a schizocarp with 4 single-seeded mericarps, but a capsule with 8–40 seeds in other genera.

Classification The family is closely related to the Piperaceae, differing from it mainly in its parietal placentation with several ovules and its dry fruit (a single basal ovule and fleshy berry or drupe in Piperaceae). For further details and references, see the recent formal revision[i].

Economic uses Several species are used medicinally, e.g., the rhizomes of *Saururus chinensis* are boiled and eaten in India, and the leaves and shoots of *Houttuynia* are eaten as a vegetable. *Houttuynia* is also grown as an ornamental or curiosity ground cover plant, with several variegated cultivars.   RKB

[i] Brach, A. R. & Xia Nian-he. Saururaceae. *Species Plantarum* 11: 1-12. Canberra, ABRS (2005).

SAXIFRAGACEAE
SAXIFRAGES, ASTILBES, ELEPHANT'S EARS

A family of north temperate and high altitude species including many plants of garden interest.

Distribution The family occurs predominantly in the northern hemisphere, reaching high latitudes and altitudes. In the southern hemisphere,

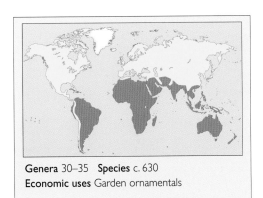

Genera 30–35 Species c. 630
Economic uses Garden ornamentals

the family occurs along the Andes and south into Tierra del Fuego (*Saxifraga, Saxifragodes, Chrysosplenium*).

Description Herbaceous perennials, with alternate, rarely opposite (e.g., *S. oppositifolia, Chrysosplenium*) leaves and often with multicellular, sometimes glandular, hairs. The leaves are evergreen or deciduous, simple, sometimes petiolate, sometimes pinnately or palmately lobed, peltate, or rarely compound, sometimes with stipules. The inflorescence is terminal or axillary, a raceme, cyme, spike, dense head, panicle, or solitary flower. The flowers are usually actinomorphic, rarely slightly zygomorphic, bisexual or rarely unisexual (plants monoecious, e.g., *S. eschscholtzii*). The sepals are usually (3–)5, free. The petals are as many as, and usually longer than, the sepals, free,

sometimes clawed or dissected. The stamens are 3–10, usually equal or double the number of sepals, free. The ovary is superior or inferior, of 2(3) partially joined carpels, each forming a lobe elongated to a stylar beak[i]. The fruit is usually a dry dehiscent capsule.

Classification Saxifragaceae in the narrow sense has usually been placed in the Rosales[ii], but APG II states that its position is still uncertain: one analysis places it sister to the eurosid group[iii], while another places it sister to all eudicots except Gunnerales[iv], but it is consistently sister to Grossulariaceae. The composition of the family has changed dramatically since the introduction of DNA sequence data when genera treated in a broad Saxifragaceae[v] were found to be spread across 4 (or 6, Cronquist 1981) subclasses of flowering plants[vi]. *Saxifraga* (c. 350 spp.) accounts for more than half the family even after splitting off *Micranthes*[vii] (c. 70 spp.). Other major genera include *Chrysosplenium* (c. 57 spp.); *Heuchera* (c. 50 spp.); *Astilbe* (22 spp.), and *Mitella* (20 spp.). The genera fall into 2 major groups: *Saxifraga* s.s. and the Heucheroid clade[viii] containing all the other genera.

Economic uses The separation of Grossulariaceae from Saxifragaceae leaves no food plants in the family but several genera are of ornamental value. *Saxifraga* spp. are commonly grown in rock gardens and borders, and

Astilbe, Bergenia, Heuchera, Mitella, Rodgersia, and *Tiarella* are widely grown as ornamental plants. AC

i Soltis, D. E. Saxifragaceae. Pp. 346–348. In: Smith, N. *et al.* (eds), *Flowering Plants of the Neotropics*. Princeton, Princeton University Press (2004).
ii Brummitt, R. K. *Vascular Plant Families and Genera*. Richmond, Royal Botanic Gardens, Kew (1992).
iii Soltis, D. E. *et al.* Angiosperm phylogeny inferred from 18S rDNA, *rbc*L, and *atp*B sequences. *Bot. J. Linn. Soc.* 133: 381–461 (2000).
iv Soltis, D. E. *et al.* Gunnerales are sister to other core eudicots: Implications for the evolution of pentamery. *American J. Bot.* 90: 461–470 (2003).
v Schulze-Menz, G. K. Saxifragaceae. Pp. 201–206. In: Melchior, H. & Engler, A. (eds.), *Syllabus der Pflanzenfamilien*. Berlin, Gebrüder Borntraeger, (1964).
vi Soltis, D. E. & Soltis, P. S., Phylogenetic relationships in Saxifragaceae *sensu lato:* a comparison of topologies based on 18S rDNA and *rbc*L sequences. *American J. Bot.* 84: 504–522 (1997).
vii Soltis, D. E. *et al. mat*K and *rbc*L gene sequence data indicate that *Saxifraga* (Saxifragaceae) is polyphyletic. *American J. Bot.* 83: 371–382 (1996).
viii Soltis D. E. *et al.* Elucidating deep-level phylogenetic relationships in Saxifragaceae using sequences for six chloroplastic and nuclear DNA regions. *Ann. Missouri Bot. Gard.* 88: 669–693 (2001).

SCHISANDRACEAE

Genera 2   **Species** 50
Economic uses Local medicines and tonics

A small tropical family of aromatic twining woody climbers. Most species occur in eastern and Southeast Asia, ranging from east India and Sri Lanka to Indochina, China, and Siberia south to western Malaysia A single species of *Schisandra, S. glabra,* is found in the southeastern USA and has recently been recorded from Mexico[i]. The leaves are alternate, spirally arranged, simple, often dotted with pellucid glands, the margins denticulate or shallowly serrate, sometimes entire. The flowers are unisexual (plants monoecious or dioecious), actinomorphic, smallish, whitish or yellowish to pink or orange, solitary or paired, sometimes in tight clusters. The tepals are 5–24 in 1 or more series, free, the outer and innermost modified. The stamens are 4–60 (up to 80 in *Kadsura*), free to connate, the filaments short, connate below or fully fused. The ovary is superior, 1-locular, of 12 to numerous (up to 300 in *Kadsura*) free carpels, spirally arranged on the receptacle, which is obovoid to ellipsoid (*Kadsura*) or conical-terete to cylindrical (*Schisandra*); the ovules are 2–3 (*Schisandra*) or 2–5(–11) (*Kadsura*). The fruit is an aggregate of berries, closely adpressed (*Kadsura*) or separated at maturity (*Schisandra*)[ii].

The 2 genera have recently been revised by Saunders[iii,iv], who has also produced an account of the family[v]. The family is closely related to the Illiciaceae in which it is often included and forms part of the basal angiosperm group known as the ANITA grade.

Species of *Schisandra*, especially *S. chinensis,* are used in traditional medicine and as a phytoadaptogen. Some species are cultivated as ornamental plants. VHH

i Panero, J. L. & Dávila Aranda, P. The family Schisandraceae: a new record for the flora of Mexico. *Brittonia* 50: 87–90 (1998).
ii Hao, G., Chye, M.-L., & Saunders, R. M. K. A phylogenetic analysis of the Schisandraceae based on morphology and nuclear ribosomal ITS sequences. *Bot. J. Linn. Soc.* 135: 401–411 [Erratum: 136: 449–450] (2001).
iii Saunders, R. M. K. Monograph of *Kadsura* (Schisandraceae). [Systematic Botany Monographs 54]. Ann Arbor, Michigan, The American Society of Plant Taxonomists (1998).
iv Saunders, R. M. K. Monograph of *Schisandra* (Schisandraceae). [Systematic Botany Monographs 58]. Ann Arbor, Michigan, The American Society of Plant Taxonomists (2000).
v Saunders, R. M. K. Schisandraceae. *Species Plantarum: Flora of the World 4*. Canberra, Australian Biological Resources Study (2001).

SCHLEGELIACEAE

A family of woody climbers, epiphytic shrubs and forest trees, often previously referred to either Bignoniaceae or Scrophulariaceae but differing from both in its fleshy berry fruits.

Distribution *Schlegelia* has 15 spp. from southeastern Mexico and the West Indies to Peru and Brazil, *Gibsoniothamnus* has 8 spp. from Mexico to Panama, *Exarata* has 1 sp. in Colombia and Ecuador, and *Synapsis* has 1 sp. in Cuba. All occur in wet forests.

Genera 4   **Species** 25
Economic uses None

Description Shrubs (often epiphytic) to woody lianes up to 20 m or more or (*Exarata*) a buttressed canopy tree to 30 m. The leaves are opposite, simple, usually shortly petiolate, the lamina typically thick and leathery and broadly elliptical (up to 30 × 15 cm) and with entire margins, or in *Synapsis* lobed and strongly spiny and mimicking leaves of holly (*Ilex aquifolium*). In *Gibsoniothamnus*, the leaves are usually markedly anisophyllous. The inflorescence is a short terminal or axillary cyme or raceme, and the flowers are usually distinctly pedicellate. The flowers are more or less actinomorphic, bisexual. The calyx is campanulate and 3- to 5-lobed, usually persistent as a cup round the base of the fruit. The corolla is 5-lobed, tubular, but slightly 2-lipped. The stamens are 4, included within the corolla tube. The ovary is superior

and develops into a fleshy spherical berry. The seeds are numerous, flattened and usually tending to be winged.

Classification Gentry[i,ii] placed this group as a tribe of Bignoniaceae and was followed by Cronquist, albeit commenting on characters transitional to Scrophulariaceae. Others have preferred to place it within Scrophulariaceae[iii]. Molecular evidence based on limited sampling has placed *Schlegelia* basal to the Bignoniaceae and near Verbenaceae[iv]. The fleshy fruit is characteristic of none of these families, and the habit and general appearance is different from most Scrophulariaceae. It is preferred here to recognize it as a separate family. RKB

i Gentry, A. H. Bignoniaceae – part 1. *Flora Neotropica Monographs 25.* New York Botanical Garden (1980).
ii Gentry, A. H. *Exarata* (Bignoniaceae), a new genus from the Chocó region of Ecuador and Colombia. *Syst. Bot.* 17: 503–507 (1992).
iii Fischer, E. Scrophulariaceae. Pp. 333–432. In: Kadereit, J. W. (ed.), *Families and Genera of Vascular Plants. VII.* Berlin, Springer-Verlag (2004).
iv Olmstead, R. G. *et al.* Disintegration of the Scrophulariaceae. *American J. Bot.* 88: 348–361 (2001).

SCROPHULARIACEAE
FIGWORT AND FOXGLOVE FAMILY

A family in the throes of dismemberment and reassimilation, with consequent major disruption of internal parts.

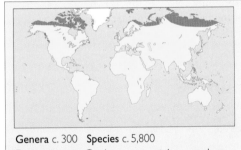

Genera c. 300   **Species** c. 5,800
Economic uses Garden ornamentals, some drugs, parasitic weeds

Distribution Cosmopolitan and in a wide range of habitats, commonly in open grassy places, rarely (*Wightia* etc.) in tropical forests.

Description Annual to perennial herbs, occasionally climbing (*Asarina*), shrubs, or rarely 30 m lianes (*Wightia*) or trees (e.g., *Paulownia, Halleria*), often holoparasitic or hemiparasitic, rarely epiphytic (*Dermatobotrys*), sometimes aquatic (*Limosella*), sometimes resurrection plants (*Craterostigma*). The leaves are alternate or opposite or occasionally whorled, simple to pinnatisect, sometimes reduced (aquatic *Limosella*). The inflorescence is a raceme or thyrse, or the flowers are solitary. The flowers are strongly zygomorphic to occasionally subactinomorphic, sometimes resupinate (*Nemesia, Pedicularis resupinata,* etc.), bisexual, usually 4- to 5-merous. The sepals are free or fused, regular or irregular. The corolla is usually strongly zygomorphic and 2-lipped, usually with the upper lip 2-lobed and lower 3-lobed, often gibbous to spurred on the abaxial side, with paired spurs in *Angelonia* and *Diascia*. The stamens are usually 4, sometimes 5, inserted on the corolla tube. The ovary is superior, usually of 2 free carpels with a terminal style and 2 locules, each with 2 to many ovules on axile placentae, rarely the ovules reduced to 1 (*Globulariopsis*), but sometimes in Orobancheae with 3 carpels and parietal placentae. The fruit is a capsule or rarely a berry (*Dermatobotrys, Teedia*), sometimes ballistic (*Lathraea*), and the seeds are usually many and often winged.

Classification The traditional Scrophulariaceae has been affected more than most other families by the application of cladistics in the last decade. It has long been considered to be at the base of a group of gamopetalous families that seem to have radiated out from it, but has retained considerable morphological cohesion (despite the wide range of morphological variation one would expect in a family of 300 genera). The circumscription adopted here differs markedly from some recent literature but coincides largely with that adopted in separate recent accounts by long-time specialists in the family N. Holmgren[i] and E. Fischer[ii] (though it may be noted that the latter has subsequently adopted a narrower circumscription). This coincides largely with tradition over the last 150 years.

In the light of molecular analyses, it has repeatedly been said that the traditional concept of the family is polyphyletic, but this has yet to

SCROPHULARIACEAE. I *Erinus alpinus* habit showing rosette of leaves and terminal inflorescence of irregular flowers (x⅔). 2 *Verbascum betonicifolium* shoot with alternate leaves and inflorescence of irregular flowers (x⅔). 3 *Rhinanthus minor* shoot with opposite leaves and inflorescence (x⅔).

be substantiated. While it is likely that many of the gamopetalous families such as Bignoniaceae, Lentibulariaceae, Gesneriaceae, Pedaliaceae, Martyniaceae, Myoporaceae, Verbenaceae, etc., plus Acanthaceae and Avicenniaceae, and also the apetalous families Hippuridaceae and Callitrichaceae (i.e., most of the order Lamiales of APG II), have been descended from ancestors that possessed the characters of the present Scrophulariaceae, this would make the latter paraphyletic rather than polyphyletic.

If we consider the characters by which some of the families involved have been defined up to now, in the case of the Hippuridaceae, the vegetative structure is highly modified for its aquatic habitat while the flowers lack a perianth and consist only of an inferior ovary surmounted by a single stamen, and it would make no sense to put it in the Scrophulariaceae. In the Callitrichaceae, another aquatic group with

simplified morphology and no perianth, the ovary is superior, but its pendulous apical ovules and fruit breaking up into 4 mericarps together indicate sufficient divergence from Scrophulariaceae to merit its own family. Similarly with the Lentibulariaceae, where the vegetative parts are much modified for its carnivorous habit, and although the flower resembles that of Scrophulariaceae, the free central placentation is different.

Holmgren[i] has noted that the unifying characters of Scrophulariaceae are a bilateral corolla, a bicarpellate ovary with axile placentation, many-seeded capsules, and seeds with well-developed endosperm. The bicarpellate ovary with axile placentation is particularly characteristic of nearly all the family, and it is noticeable that when segregate families have been recognized the ovary characters are usually different. Thus the Myoporaceae, in which *Eremophila* has bilateral flowers similar to Scrophulariaceae but *Myoporum* is different, have a drupaceous fruit, unlike the capsule of Scrophulariaceae. Although clear distinctions in all the characters are often not possible, by considering the combination of characters it seems possible to recognize the more distinctive of the derivatives of Scrophulariaceae as separate families. Of the segregates recognized here, one of the most difficult decisions concerned the Plantaginaceae, in which the 3 genera do have a calyx and corolla, albeit non-colorful and reduced, but the circumscissile capsule of *Plantago* and 1-locular ovary with basal ovules developing into single-seeded nutlets in the other genera are markedly at variance with Scrophulariaceae. However, there has been a tendency recently to sink the Plantaginaceae, and, through the unfortunate nomenclatural priority of that name, apply it to a part of the traditional Scrophulariaceae, which includes *Veronica*. Perhaps even more marginal is the decision to maintain Globulariaceae on the basis largely of the 1-locular ovary and single pendulous ovule, and the family has in fact been earlier considered a tribe of Scrophulariaceae[iv].

It has been noted recently in a Scrophulariaceous context[v] that if one requires only monophyletic taxa, one may have to either sink taxa that appear quite distinct or split the main body where there are no good characters at all, neither of which is an attractive option. The alternative, increasingly recognized as the only logical solution by a broad body of the taxonomic community[iii], is to accept paraphyletic taxa.

In a major disintegration of the traditional Scrophulariaceae[vi], 5 families were recognized—Scrophulariaceae, Orobanchaceae (including the hemiparasitic genera), Calceolariaceae, Stilbaceae, and Veronicaceae (the latter later renamed as Plantaginaceae)—but no family descriptions or keys to the families were given. Furthermore, the number of genera sampled was small (about 38 genera out of about 300 in the traditional Scrophulariaceae). Later a further family was segregated as Phrymaceae[vii] to include *Mimulus*, in which *Phryma* (treated as a separate family here and in VPFG) had been found to nest. Support for such splitting has been forthcoming from further molecular studies[viii]. More recently 2 further families, Linderniaceae and Gratiolaceae, have been recognized, again without descriptions or diagnoses from other segregates[ix]. Much further evidence is needed, however, before a clear picture emerges. While considerable thought has been given here to the delimitation of the Scrophulariaceae, it is clearly quite beyond the scope of the present work to offer a modern synthesis of the classification within the family. The traditional division of the family into 3 subfamilies by Bentham was modified by Pennell in 1935 who combined 2 of these to leave 2 subfamilies separated by the aestivation of the corolla lobes. Others have opted to divide the family into parasitic and the non-parasitic subfamilies. For a modern assessment of the classification within the family, see Fischer in FGVP[ii], who has recognized 8 "families," 1 of which was divided into 3 subfamilies, and a total of 40 or more tribes. The present account differs from this only in recognising Schlegeliaceae as a distinct family. Recognition of Orobanchaceae for parasitic genera with 3 carpels and parietal placentation might have been another option.

Economic uses The family is of major importance in horticulture, including such genera as *Antirrhinum*, *Calceolaria*, *Cymbalaria*, *Digitalis*, *Hebe*, *Linaria*, *Mimulus*, *Nemesia*, *Parahebe*, *Paulownia*, *Pedicularis*, *Penstemon*, *Scrophularia*, *Verbascum*, and *Veronica*. The drugs digoxin and digitalin are extracted from *Digitalis* for heart treatment. The parasitic genera may be negatively important, and *Striga* is a major scourge in tropical agriculture.   RKB

i Holmgren, N. Scrophulariaceae. Pp. 348–350. In: Smith, N. *et al.* (eds), *Flowering Plants of the Neotropics*. Princeton, Princeton University Press (2004).
ii Fischer, E. Scrophulariaceae. Pp. 333–432. In: Kadereit, J. W. (ed.), *Families and Genera of Vascular Plants. VII*. Berlin, Springer-Verlag (2004).
iii Nordal, I. & Stedje, B. Paraphyletic taxa should be accepted. *Taxon* 54: 5–6 (2005).
iv Barringer, K. Five new tribes in the Scrophulariaceae. *Novon* 3: 15–17 (1993).
v Albach, D. C., Martinez-Ortega, M. M., Fischer, M. A., & Chase, M. W. A new classification of the tribe Veroniceae – problems and a possible solution. *Taxon* 53: 429-452 (2004).
vi Olmstead, R. G. *et al.* Disintegration of the Scrophulariaceae. *American J. Bot.* 88: 348–361 (2001).
vii Beardsley, P. M. & Olmstead, R. G. Redefining Phrymaceae: the placement of *Mimulus*, tribe Mimuleae and *Phryma*. *American J. Bot.* 89: 1093–1102 (2002).
viii Oxelman, B., Kornhall, P., Olmstead, R. G., & Bremer, B. Further disintegration of Scrophulariaceae. *Taxon* 54: 411–425 (2005).
ix Rahmanzadeh, R. *et al.* The Linderniaceae and Gratiolaceae are further lineages distinct from the Scrophulariaceae (Lamiales). *Plant Biol.* 7: 67–78 (2005).

SCROPHULARIACEAE. **4** *Linaria vulgaris* (a) shoot with linear leaves and inflorescence (x⅔); (b) half flower with spurred corolla and stamens of 2 lengths (x3). **5** *Digitalis obscura* leafy shoot and inflorescence (x⅔). **6** (OPPOSITE PAGE) *Veronica fruticans* leafy shoot and inflorescence (x⅔). **7** *Scrophularia macrantha* (a) lower lip of corolla opened out showing 4 stamens with anthers linked in pairs and a central, small staminode (x4⅔); (b) cross section of ovary showing 2 locules and axile placentas (x6). **8** (BELOW) *Sibthorpia europeae* dehiscing fruit—a capsule (x10). **9** *Penstemon lyallii* leafy shoot with irregular flowers and young fruits (x⅔).

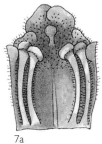

7b

7a

6

SCYPHOSTEGIACEAE

Scyphostegia borneensis is a dioecious large shrub or tree to 20 m found in primary and secondary forests in Borneo (Sarawak, Sabah, eastern Kalimantan) to an altitude of 900 m. The leaves are elliptic to oblong, shortly petiolate, with crenate margins. In male trees, the inflorescences are laxly clustered on leafless terminal parts of branches[i], each with a short peduncle. Each inflorescence is up to 5 cm long and 2–3 mm across, oblong to linear in outline, and consists of up to c. 23 concentric shortly tubular bracts, each bract projecting slightly from within the bract below and having a single flower in its axil or 2 in the uppermost axil. In female trees, the inflorescences are solitary on long peduncles in the leaf axils, and each inflorescence has 1 or 2 concentric tubular bracts and a single flower in the axil of the upper one. The male flowers are inconspicuous and have a greenish tubular perianth with 6 short lobes (3 inner smaller than the outer), 3 nectary glands within it, and 3 stamens united into a column. The female flowers are exserted from the upper bract and have a greenish tubular perianth with 6 larger subequal lobes, no glands, and a superior ovary. The ovary is urceolate with a reflexed margin, 9–12 longitudinal grooves, and has a single locule with many elongate ovules crowded together on its floor. The fruit is accrescent, subspherical, to 1.5 cm across, fleshy, dehiscing into 9–12 valves, which reflex to reveal a mass

of vertical seeds nesting on a spongy tissue. For more detail, see van Steenis's account[ii], especially the 1972 supplement.

This extraordinary plant was first referred to Monimiaceae but later referred to its own family by many authors, with much speculation about its affinities. Molecular analysis[iii] has confirmed suspicions that it is allied to Flacourtiaceae, placing it within that family but only one step from the base of the family. Scyphostegiaceae was therefore reduced to a tribe of Flacourtiaceae. The advice given in the title of the molecular paper is not followed here. The unique inflorescences, the tubular 6-lobed green perianth, the 3 stamens united into a column and the unusual ovary, fruit, and seeds all argue for maintaining Scyphostegiaceae as a family. Van Steenis said, when discussing the affinities of the plant, "All recent authors agree that the genus represents a distinct family," and there seems no good reason to disagree with his approach. RKB

i Baehni, C. Note sur les inflorescences mâle et femelle du *Scyphostegia borneensis* Stapf. *Bull. Soc. Bot. Suisse* 48: 22–28 (1938).
ii van Steenis, C. G. G. J. Scyphostegiaceae. *Flora Malesiana*, ser. 1, 5: 297–299 (1957) and 6: 967–968 (1972).
iii Chase, M. W. *et al*. When in doubt, put it in Flacourtiaceae: a molecular phylogenetic analysis based on plastid *rbc*L DNA sequences. *Kew Bull.* 57: 141–181 (2002).

SETCHELLANTHACEAE

The Setchellanthaceae comprises 1 sp., *Setchellanthus caeruleus*, endemic to the Chihuahuan and Tehuacán deserts of Mexico[i]. The pungent-smelling shrub has tiny leaves on long and short shoots and is covered in stiff hairs. The leaves are alternate, with petioles but without stipules. Only the long shoots bear flowers, these solitary in the leaf axils. The flowers are actinomorphic and bisexual, the (5)6(7) sepals are fused for their entire length and thickened at the base, splitting into 1 or 2 flaps when flowering. The petals are (5)6(7), clawed and overlapping, blue-lilac. The stamens are (40–)60–76, in (5)6(7) paired rows. A short gynophore (stalk) supports the superior ovary of 3 fused carpels, each forming a locule, the locules containing 10–14 ovules. The short style bears 3 stigmatic stylodia. The fruit is a 3- to 10-seeded capsule dehiscing by the outer surfaces of the carpel peeling away from the central column.

The family was described as distinct from Capparaceae in 1999[i] but its close affinity to other Brassicales was still recognized on anatomical and vegetative characters[ii]. DNA sequence evidence suggested a relationship with Capparales[iii] close to Caricaceae, Limnanthaceae, and the "core" Brassicales but support for a fully resolved relationship was weak[iv]. No economic uses are reported. AC

i Iltis, H. H. Setchellanthaceae (Capparales), a new family for a relictual, glucosinolate-producing endemic of the Mexican deserts. *Taxon* 48: 257–275 (1999).

ii Carlquist, S. & Miller, R. B. Vegetative anatomy and relationships of *Setchellanthus caeruleus* (Setchellanthaceae). *Taxon* 48: 289–302 (1999).
iii Karol, K. G., Rodman, J. E., Conti, E., & Sytsma, K. J. Nucletide sequence of *rbc*L and phylogenetic relationships of *Setchellanthus caeruleus* (Setchellanthaceae). *Taxon* 48: 303–315 (1999).
iv Rodman, J. E. *et al*. Parallel evolution of glucosinolate biosynthesis inferred from congruent nuclear and plastid gene phylogenies. *American J. Bot.* 85: 997–1007 (1998).

SIMAROUBACEAE

TREE OF HEAVEN AND QUASSIA FAMILY

A family of essentially pinnate-leaved shrubs and trees that has been much reduced in size recently by the segregation of other families.

Distribution Pantropical but with most genera in the Old World, sparingly extending outside the tropics to northern Japan (*Picrasma quassioides*), New South Wales (*Ailanthus triphysa*), Florida (*Simarouba glauca*), Texas (*Castela erecta, C. stewartii*), and northern Argentina (*C. alternifolia*). Most members of the family occur in lowland forest.

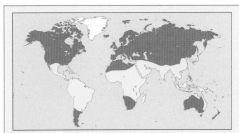

Genera c. 18 **Species** c. 115
Economic uses Many medicinal properties; limited use as timber and wood pulp; some cultivated species

Description Shrubs to large trees. *Castela* has conspicuous stem spines. The leaves are alternate or rarely opposite, usually pinnate with opposite or subopposite leaflets, sometimes 3-foliate, occasionally (*Castela, Samadera, Simaba, Soulamea*) 1-foliate, in *Quassia* with a winged and articulated rhachis, without stipules or rarely (*Picrasma*) stipulate. The inflorescence is a terminal or axillary thyrse, cyme, or raceme. The flowers are actinomorphic, bisexual or more often unisexual (plants dioecious or andromonoecious) with rudimentary organs of the other sex, 3- to 8-merous, usually small but occasionally larger and showy (*Quassia*). The sepals are most often 5, slightly united at the base. The petals are 5 or rarely absent, free. There are usually twice as many stamens as petals, rarely more than twice as many, but sometimes 5 are reduced to staminodes. There is usually a disk within the stamens, which may be extended as a gynophore or androgynophore. The ovary is superior, of 2–5(–8) carpels, weakly to strongly united, sometimes free (*Picrolemma, Ailanthus* spp.) or united only by the styles, with 1 ovule per locule. The fruit is a capsule, samara (*Ailanthus*), drupe, or berry.

Classification The Simaroubaceae has formerly been interpreted in a much wider sense than here. Engler in the *Pflanzenfamilien* (1931), recognized 6 subfamilies, 5 of which are now separated as the families Surianaceae and Irvingiaceae (both with simple leaves), Kirkiaceae and Picramniaceae (the latter including Engler's subfamily Alvaradooideae as well as *Picramnia*), and also included *Harrisonia*, which is here referred to Ptaeroxylaceae. Fruit anatomy[i] and molecular analysis[ii] have clarified family relationships and established a position in Sapindales near Rutaceae and Ptaeroxylaceae. It is preferred here not to include Leitneriaceae, with simple leaves, in Simaroubaceae as has been proposed. Members of Simaroubaceae with apparently simple leaves usually have a distinct pulvinus below the lamina and are presumably merely 1-foliate by reduction from a compound leaf. While the family concept seems clear, a new generic assessment is needed, particularly to determine whether it is preferable to adopt the broad concept of *Quassia*, which includes *Hannoa*, *Odyendea*, *Pierreodendron*, *Samadera*, *Simaba*, and *Simarouba*[iii], or to recognize these as separate genera, leaving *Quassia* as only 1 sp. (as is favored more recently[ii] and as is accepted here).

Economic uses Many species are known for their medicinal properties or are used as insecticides or a source of incense. Some trees provide timber. *Ailanthus altissima*, the Tree of Heaven, from China is one of the fastest-growing trees known and is widely planted in temperate regions, often becoming naturalized, and used for wood pulp. *Quassia amara*, Bitterwood, from Brazil is a spectacularly red-flowered shrub or small tree grown in the tropics, and *Samadera indica* is also widely grown as a tropical ornamental. RKB

i Fernando, E. S. & Quinn, C. J. Pericarp anatomy and systematics of the Simaroubaceae *sensu lato*. *Austral. J. Bot.* 40: 263–289 (1992).
ii Fernando, E. S., Gadek, P. A., & Quinn, C. J. Simaroubaceae, an artifical construct: evidence from *rbc*L sequence variation. *American J. Bot.* 82: 92–103 (1995).
iii Nooteboom, H. P. The generic delimitation in Simaroubaceae tribus Simaroubeae and conspectus of the genus *Quassia*. *Blumea* 11: 509–528 (1962).

SIMAROUBACEAE. | *Quassia amara* (a) shoot with 3-foliate leaves and inflorescence (x⅔); (b) flower with petals removed to show numerous stamens (x1½); (c) calyx and gynoecium with disk at base (x1½); (d) fruits (x⅔); (e) cross section of fruit (x⅔). 2 *Ailanthus excelsa* (a) part of pinnate leaf (x⅔); (b) fruits— twisted samaras (x⅔); (c) half section of fruit showing single seed (x⅔).

SIMMONDSIACEAE

JOJOBA OR GOAT NUT

Genus 1 Species 1
Economic uses Extract used to make jojoba oil

The family comprises a single species, *Simmondsia chinensis*, of small evergreen xerophytic trees or shrubs from southwestern North America, California and offshore islands, New Mexico and Arizona, and northern Mexico. The leaves are opposite, bluish green and leathery, elliptical-oblong, opposite, almost sessile and without stipules, pubescent. The flowers are actinomorphic, unisexual (plants dioecious) in axillary inflorescences, the male inflorescence a capitate cluster of short pedicellate flowers; the female flower solitary. The flowers are apetalous, the male ones usually with 10–12 stamens on a flat receptacle; the female flowers larger, soft, and hairy, with 5 greenish sepals and a superior ovary of 3(4) fused carpels with axile placentation. The fruit is a loculicidally dehiscent capsule, with 3(4) valves, containing a single dark brown seed.

The family is often treated in the Buxaceae and sometimes in Euphorbiales, but a wide accumulation of evidence from seed morphology and anatomy[i], phytochemistry, and plastid type[ii] refutes these relationships. DNA sequence analysis places the family at the border of the core and non-core Caryophyllales[iii].

Jojoba oil is a wax with similar properties to whale oil[iv] and is widely used in cosmetics and in pesticides. Extensive domestication of jojoba occurred in the 1970s due to tax incentives in the USA and the increased pressure on the international whaling industry. AC

i Wunderlich, R. Some remarks on the taxonomic significance of the seed coat. *Phytomorphology* 17: 301–311 (1967).
ii Behnke H.-D. Distribution and evolution of forms and types of sieve-element plastids in the dicotyledons. *Aliso* 3: 167–182 (1991).
iii Cuénoud, P. *et al.* Molecular phylogenetics of Caryophyllales based on nuclear 18S and plastid *rbc*L, *atp*B and *mat*K DNA sequences. *American J. Bot.* 89: 132-144 (2002).
iv Undersander, D. J. *et al.* Jojoba. In: *Alternative Field Crops Manual.* http://corn.agronomy.wisc.edu/ AlternativeCrops/Jojoba.htm (1990).

SIPARUNACEAE

A small family of tropical evergreen shrubs or trees with opposite or whorled simple leaves and unisexual flowers.

Distribution The family comprises 2 genera, *Glossocalyx*, with a single species in West Africa, and *Siparuna* (60–70 spp.), which grows in the neotropics, from Mexico and Central America, the Caribbean and northern South America

(Ecuador and Colombia) down to Peru, Bolivia, and Paraguay. It occurs mainly in humid tropical forests.

Genera 2
Species c. 70
Economic uses some medicinal uses

Description Evergreen shrubs or trees whose bark, leaves, and fruits emit a pungent lemony smell when cut or crushed. The leaves are opposite or in whorls of 2–3, simple, the margins entire or serrate, without stipules. The flowers are in axillary cymes or sometimes fascicles, or cauliflorous on older wood, unisexual (plants monoecious or dioecious), usually actinomorphic, small and inconspicuous, the fertile flowers functionally male, or functionally female, without organs of the other sex; receptacle cup-shaped or globose. The perianth has 4–6(7) distinct or basally connate cream to yellowish tepals (rarely 1 tepal longer than the others in *Glossocalyx* or perianth calyptrate in *Siparuna decipiens*). The male flowers have 1–72 stamens, usually numerous, arranged inside the cupular receptacle, the filaments usually free, without nectariferous appendages. The female flowers have a superior ovary of 3–30 free carpels that are immersed in the receptacle, the styles distinct; ovules are 1 per locule. A characteristic feature of *Siparuna* is the membranous roof (velum) that covers the sexual organs but with a hole at the center through which the styles or anthers emerge[ii]. The fruit is composed of drupelets (1–)3–25 in a fleshy receptacle which, when mature, usually opens to reveal the drupelets; drupelets with a red or orange-colored fleshy appendage (aril) in the dioecious taxa.

Classfication The neotropical genus *Bracteanthus* is sometimes recognized as a third genus but is best included in *Siparuna*. The family belongs in the Laurales and is sometimes included in the Monimiaceae, although on morphological and molecular grounds it is closer to other families in the order such as Gomortegaceae and Atherospermataceae[i,iii]. Molecular phylogenetic studies support its monophyly and indicate that the dioecious *Glossocalyx* is sister to the remaining members of the family. It appears that dioecism evolved from monoecy more than once in *Siparuna*[iv].

Economic uses Some species of *Siparuna* are widely used locally for their medicinal properties in the neotropics.   VHH

i Renner, S. S. Siparunaceae. Pp. 351–353. In: Smith, N. *et al.* (eds), *Flowering Plants of the Neotropics*. Princeton, Princeton University Press (2004).
ii Renner, S. S., Schwarzbach, A. E., & Lohmann, L. Phylogenetic position and floral function of *Siparuna* (Siparunaceae: Laurales). *Int. J. Plant Sci.* 158 (6 Suppl.): S89–98 (1997).

iii Renner, S. S. Circumscription and phylogeny of the Laurales: evidence from molecular and morphological data. *American. J. Bot.* 86: 1301–1315 (1999).
iv Renner, S. S. & Won, H. Repeated evolution of dioecy from monoecy in *Siparuna* (Siparunaceae, Laurales). *Syst. Biol.* 50: 700–712 (2001).

SLADENIACEAE

A monogeneric family of 2 spp. of evergreen trees to 15 m. *Sladenia celastrifolia* grows in southwestern China, northern Thailand, and northern Myanmar, and the recently described *S. integrifolia*, is endemic to secondary evergreen forests in southeastern Yunnan, China. The leaves are alternate,

Genera 1 **Species** 2
Economic uses Timber used locally

exstipulate, petiolate, simple, ovate or lanceolate to oblong-elliptical, the margins entire or serrate. The flowers are in axillary dichasial cymes, hermaphrodite. The sepals are 5, free, persistent in fruit. The petals are 5, partly connate at the base, white. The stamens usually 10, rarely up to 13, with dilated filaments. The ovary is superior, of 3 fused carpels, 3-locular with axial placentas. The fruit is a dry, 8- to 10-ribbed schizocarp, with persistent sepals. The seeds are 2 per locule.

The position of *Sladenia* continues to be debatable. After originally being included in the Theaceae, it has also placed in the Dilleniaceae and later the Actinidiaceae. It was recognized as a separate family, Sladeniaceae, in 1965, and molecular studies have suggested either an uncertain affinity or inclusion in the Ternstroemiaceae[i,ii]. Studies of spore and gamete production support the recognition of *Sladenia* as a separate family[iii]. Molecular studies[ii] have indicated that the tropical East African *Ficalhoa laurifolia*, usually included in the Theaceae, forms a sister group with *Sladenia* but further work is needed to clarify its position. The wood of *Ficalhoa* is used locally.   VHH

i Savolainen, V. *et al.* Phylogeny of the eudicots: a nearly complete familial analysis based on *rbc*L gene sequences. *Kew Bull.* 55: 257–309 (2000).
ii Anderberg, A. A., Rydin, C., & Källersjö, M. Phylogenetic relationships in the order Ericales *s.l.*: analyses of molecular data from five genes from the plastid and mitochondrial genomes. *American J. Bot.* 89: 677–687 (2002).
iii Li, L., Liang, H.-X., Peng, H., & Lei, L.-G. Sporogenesis and gametogenesis in *Sladenia* and their systematic implication. *Bot. J. Linn. Soc.* 143: 305–314 (2003).

SOLANACEAE
POTATO FAMILY

A cosmopolitan, economically important family of herbs, shrubs, trees, and lianas, including the potato, tomato, eggplant (aubergine), and many alkaloid-containing poisonous species.

Distribution The family has a near worldwide distribution but is mainly tropical and subtropical, especially in Central and South America with c. 63 genera and 1,200 spp.[ii], and Australia in which the subfamily ANTHOCERCIDEAE is endemic. There is also a considerable representation in tropical Africa. It occurs in a range of habitats from deserts to tropical rain forests.

Description Trees (some *Cyphomandra, Solanum*) or shrubs (*Cestrum, Lycium*) to vines, lianas, epiphytes, and perennial or annual herbs. The leaves are alternate, usually simple, lamina entire or dissected, exstipulate. The flowers are bisexual or rarely unisexual (plants andromonoecious), usually actinomorphic, sometimes slightly to markedly zygomorphic, in terminal or axillary, sometimes complex, cymose infloresecences or reduced to a single flower (*Solanum, Nierembergia, Mandragora*). The sepals are (4)5(–7), in 1 whorl, partly fused, usually regular, usually persistent, and sometimes enlarged around the fruit (e.g., *Physalis, Nicandra*). The petals are (4)5(–10), usually contorted and plicate, imbricate or valvate, variously fused, making the corolla round and flat (e.g., *Solanum, Lycopersicon*), campanulate (e.g., *Nicandra, Withania, Mandragora*) or tubular (e.g., *Cestrum, Nicotiana*); rarely 2-lipped as in *Schizanthus*, where the abaxial petals form a keel. The stamens are usually 4 or 5, 2+2, or up to 8 or 10. The ovary is superior, of 2 fused carpels (3–5 in *Nicandra* or more in some *Nolana*) with a single style, and usually with 2 locules, sometimes 3–5 (*Nicandra, Datura*), generally with numerous axile ovules. The fruits show considerable diversity[iii] but are usually bicarpellate and a septicidal capsule or a berry, sometimes a drupe or pyrene (*Goetzea* group, *Capsicum, Lycium*), or a schizocarpic with up to 15(–30) single-seeded mericarps (*Nolana*).

Classification The subdivision of the Solanaceae remains unclear. It is traditionally divided into 2 subfamilies: Cestroideae and Solanoideae. Along with 19 tribes, these are maintained in a new recent classification by Hunziker[iv]. On the other hand, a provisional phylogenetic classification based on molecular data by Olmstead *et al*[v] recognizes 7 subfamilies and 20 tribal-level groupings and includes

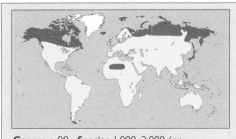

Genera c. 90 **Species** 1,000–2,000 (or 3,000–4,000)[i]
Economic uses Food crops such as aubergine, chilli peppers, potato, and tomato; some ornamentals; source of alkaloids and tobacco

SOLANACEAE. 1 *Salpiglossis atropurpurea* (a) flowering shoot (×⅔); (b) part of flower showing 2 pairs of unequal stamens and a single infertile reduced stamen (×1½); (c) fruit (×2). 2 *Datura stramonium* var. *tatula* (a) flowering shoot (×⅔); (b) fruit—a capsule (×⅔). 3 *Solanum rostratum* (a) flowering shoot (×⅔); (b) flower with 2 petals and 2 stamens removed (×2). 4 *Physalis alkekengi* (a) shoot showing fruits enclosed in a persistent orange-red calyx (×⅔); (b) calyx removed to show the fruit (×⅔). 5 *Nicotiana tabacum* cross section of ovary showing 2 locules and axile placentas (×6).

groups that have previously been recognized as separate families, namely the Nolanaceae (*Nolana*), Duckeodendronaceae (*Duckeodendron*), and Goetzeaceae. The Goetzeaceae is recognized by Olmstead *et al* as a subfamily and is confined to the Greater Antilles (*Coeloneurum* on Hispaniola, *Espadaea* and *Henoonia* endemic to Cuba, all monotypic, whereas *Goetzea* has 2 spp., 1 endemic to Hispaniola, and the other endemic to Puerto Rico). A recent study[vi] of the Goetzeoideae suggests that monotypic Brazilian *Metternichia* and *Duckeodendron* should also be included in it, although the former has dry, capsular fruits, and the latter has a pyriform drupe with an orange mesocarp and an unusually thick fibrous mesocarp, unlike the drupes of the Antillean members, which lack this covering. The Nolanaceae are included here in the Solanaceae, although they differ by their unusual schizocarpic fruits. The Solanaceae are included in the Solanales along with the Convolvulaceae, Hydroleaceae, Montiniaceae, and Sphenocleaceae. Molecular data suggest that they are closest to the Convolvulaceae[vii]. It is clear that much further work and sampling is needed before an acceptable circumscription and subdivision of the Solanaceae can be attained.

Economic uses The Solanaceae contains many species of major economic importance. These include the potato (*Solanum tuberosum*) from the Peruvian Andes and Bolivia, the fourth most important food crop in terms of tonnage produced, Eggplant or Aubergine (*S. melongena*), tomato (*Lycopersicon esculentum*), and the Peppers (various *Capsicum* spp.), including Paprika, Chillies, Cayenne, and Sweet or Bell peppers). Others, popular in tropical America but relatively little known outside this area, include the Husk Tomato (*Physalis pubescens*), the Tomatillo (*P. ixocarpa*), the Cape Gooseberry (*P. peruviana*), the Tree Tomato (*Cyphomandra betacea*), the Pepino (*Solanum muricatum*), the Cocona (*S. topiro*), the Lulita (*S. hirsutissimum*), and the Naranjilla or Lulo (*S. quitoense*). The family also includes various species that are grown as ornamentals, especially in the genera *Browallia*, *Brugmansia*, *Brunfelsia*, *Cestrum*, *Datura*, *Nicotiana*, *Nierembergia*, *Petunia*, *Salpiglossis*, *Schizanthus*, *Solanum*, and *Solandra*, which have showy flowers. Some *Capsicum* and *Solanum* species are widely grown for their colorful fruits, while certain *Cestrum*, *Lycium*, *Solanum*, and *Strep-*

tosolen species are popular shrubs. The Chinese Lantern Plant (*Physalis alkekengi*) is extensively used in dried floral arrangements. Many members of the Solanaceae are famed for their alkaloid content and have been used since historic times for their medicinal, poisonous, or psychotropic properties. The most important example is tobacco (*Nicotiana tabacum*), which contains the addictive and highly toxic alkaloid nicotine and is grown extensively for smoking, chewing, and snuff manufacture. It is a major world commodity, but the most harmful plant in the world through the diseases it causes. Other *Nicotiana* species also contain nicotine and are used for preparing insecticides. Several genera contain steroid alkaloids that have been used medicinally for centuries, including Deadly

Nightshade (*Atropa belladonna*), Thorn-Apple (*Datura* spp.), and Black Henbane (*Hyoscyamus niger*). They and many other solanaceous species are highly poisonous. [JME] VHH

i Solanaceae Source. A global taxonomic resource for the nightshade family. http://internt.nhm.ac.uk/jdsml/botany/databases/solanum/about_solanaceae/diversity.dsml
ii Nee, M. Solanaceae. Pp. 355–357. In: Smith, N. *et al.* (eds), *Flowering Plants of the Neotropics*. Princeton, Princeton University Press (2004).
iii Knapp, S. Tobacco to tomatoes: a phylogenetic perspective on fruit diversity in the Solanaceae. *J. Experimental Bot.* 53, No. 377, Fruit Development and Ripening Special Issue, 2001–2022 (2002).
iv Hunziker, A. T. *Genera Solanacearum: the Genera of Solanaceae Illustrated, Arranged According to a New System*. Ruggell, LIE: Koenigstein, Koeltz Scientific Books (2001).

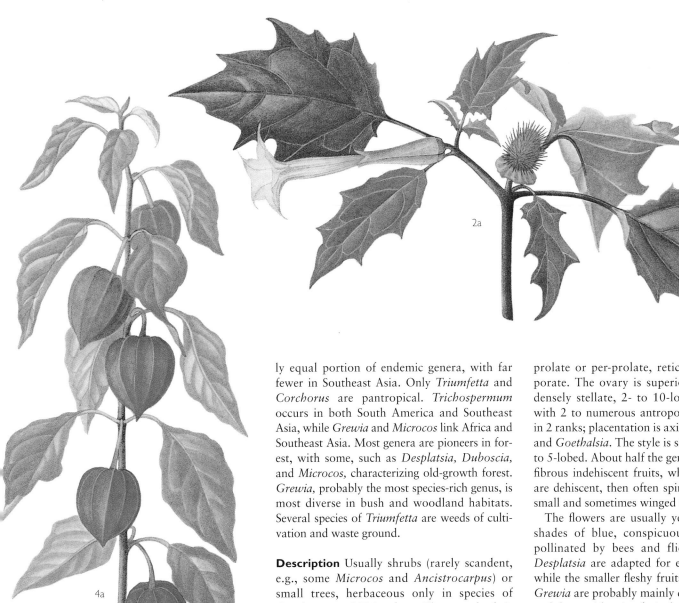

5

v Olmstead, R.G. *et al*. Phylogeny and provisional classification of the Solanaceae based on chloroplast DNA. Pp. 111–137. In: Nee, M., Symon, D., Lester, R. N., & Jessop, J. P. (eds), *Solanaceae IV: Advances in Biology and Utilization*. Richmond, Royal Botanic Gardens, Kew (1999).

vi Santiago-Valentin, E. & Olmstead, R. G. Phylogenetics of the Antillean Goetzeoideae (Solanaceae) and their relationships within the Solanaceae based on chloroplast and ITS DNA sequence data. *Syst. Bot.* 28: 452–460 (2003).

vii Savolainen, V. *et al*. Phylogeny of the eudicots: a nearly complete familial analysis based on *rbc*L gene sequences. *Kew Bull.* 55: 257–309 (2000).

SPARRMANNIACEAE

JUTE, SPARRMANNIA, GREWIA

A family of tropical shrubs, generally small trees and herbs, comprising most of the genera formerly included in Tiliaceae.

Distribution The family is pantropical with extensions to South Africa and New Zealand. Tropical America and Africa each have a rough-ly equal portion of endemic genera, with far fewer in Southeast Asia. Only *Triumfetta* and *Corchorus* are pantropical. *Trichospermum* occurs in both South America and Southeast Asia, while *Grewia* and *Microcos* link Africa and Southeast Asia. Most genera are pioneers in forest, with some, such as *Desplatsia, Duboscia,* and *Microcos,* characterizing old-growth forest. *Grewia,* probably the most species-rich genus, is most diverse in bush and woodland habitats. Several species of *Triumfetta* are weeds of cultivation and waste ground.

Description Usually shrubs (rarely scandent, e.g., some *Microcos* and *Ancistrocarpus*) or small trees, herbaceous only in species of *Corchorus* and *Triumfetta*. The stem bark is particularly fibrous and often mucilaginous, and the indumentum is stellate or bushy, often with some simple hairs. The leaves are simple, rarely lobed, alternate, often serrate and triplinerved, sometimes palmately nerved. The petioles are pulvinate, with the stipules often caducous. The inflorescences are usually leaf-opposed and modified cymes. Epicalycular bracts are usually absent or at least caducous. The calyx is divided to the base into 5 equal, valvate sepals. The petals are usually 5, generally clawed, often with a hairy basal nectary pocket. The andro-gynophore is usually nectariferous if the petals are not; it is shorter than the ovary and usually angular. The stamens usually by far exceed 15 in number, and the filaments are usually free to the base. In exceptional cases, however, only 4 or 5 stamens occur (*Tetralix* and e.g., *Triumfet-ta pentandra*). In a few cases (*Mollia* and *Ancistrocarpus*), the stamens are grouped in antesepalous bundles, the filaments rarely unit-ed below in a tube that envelops the ovary (*Desplatsia*). Staminodes, in the form of sterile, but otherwise scarcely modified outermost stamens, occur in several genera, such as *Sparrmannia, Clappertonia,* and *Apeiba*. The 2 anther thecae are collateral, usually dorsifixed, and often versatile and introrse. The pollen is prolate or per-prolate, reticulate, and tricol-porate. The ovary is superior, ovoid, usually densely stellate, 2- to 10-locular, the locules with 2 to numerous antropous ovules, usually in 2 ranks; placentation is axile, except in *Sicrea* and *Goethalsia*. The style is simple and capitate to 5-lobed. About half the genera have fleshy or fibrous indehiscent fruits, while the remainder are dehiscent, then often spiny with numerous small and sometimes winged seeds.

The flowers are usually yellow, less usually shades of blue, conspicuous, and probably pollinated by bees and flies. The fruits of *Desplatsia* are adapted for elephant dispersal, while the smaller fleshy fruits of *Microcos* and *Grewia* are probably mainly dispersed by birds, and those with spiny fruits by mammals.

Classification Sparrmanniaceae formed the major part of traditional Tiliaceae (q.v.), but molecular evidence supported by morphology[i] has unequivocally shown them to be most closely related to Byttneriaceae (formerly included in Sterculiaceae). They are more distantly related to the other families of core Malvales (see Malvaceae), such as Tiliaceae *sensu stricto,* Sterculiaceae *sensu stricto,* Bombacaceae, and Malvaceae. They are distinguished from Byttneriaceae by their almost always more numerous (15+) stamens (versus usually 5[–10]), which are free to the base

Genera 26 **Species** c. 780
Economic uses Sources of fibers and tropical ornamental plants

SPARRMANNIACEAE. I *Triumfetta subpalmata* leafy shoot with flowers and fruits (×⅔). 2 *Corchorus bullatus* (a) leafy shoot with narrow stipules, flowers, and fruits (×⅔); (b) fruit (a capsule) with wall removed to show seeds (×⅔). 3 *Grewia parvifolia* (a) flowering shoot (×⅔); (b) flower with sepals fused into a tube and small free petals (×5); (c) cross section of ovary with 4 locules (×10); (d) vertical section of gynoecium (×15).

(versus often united into a tube) and, if clustered in bundles, these are antesepalous and not antepetalous. Staminodes are always outermost and scarcely modified, versus inner and often fleshy, petaloid, or in other ways modified. Within the Malvales, they are also unusual in the hairy nectaries at the base of the petals and in their mainly leaf-opposed inflorescences. No satisfactory infrafamilial classification is available.

Economic uses Throughout the tropics, the stem bark of this family is peeled from live treelets and used as rope at a local level (e.g., "Tsely" in Madagascar). Some *Triumfetta* have been cultivated for this purpose in the Mascarenes. *Sparrmannia africana* and a few *Grewia* have been cultivated for their flowers, those of the former being remarkable for their touch-sensitive stamens. Some *Grewia* have medicinal properties. One species in Baluchistan is so sought for its medicinal roots that it is now locally extinct.   MRC

i Bayer, C. *et al.* Support for an expanded family concept of Malvaceae within a recircumscribed order Malvales: A combined analysis of plastid *atp*B and *rbc*L DNA sequences. *Bot. J. Linn. Soc.* 129: 267–303 (1999).

SPHAEROSEPALACEAE

Distribution The family is confined to Madagascar. *Rhopalocarpus* (15 spp., syn. *Sphaerosepalum*) is widely distributed through-out the island in both subhumid evergreen forest up to 1,000 m and subarid deciduous thicket, while *Dialyceras* (3 spp.) grows at low altitude in the northeast[i].

Description Deciduous trees up to 25 m or more. The leaves are alternate, simple, entire, coriaceous, with closely parallel pinnate venation or (some *Rhopalocarpus*) with 3 prominent veins from the base to leaf apex, or some species with an intermediate condition, with sheathing but caducous stipules, the petiole slender to pulvinus-like. The inflorescence is a regularly few to many branched axillary raceme with conspicuous pedicels. The flowers are actinomorphic, bisexual, 4–12 mm in diameter, usually 4-merous. The sepals are 4, in pairs, broad, and concave. The petals are 4, free, caducous. The stamens are 25–160 in 2 to many whorls inserted around the base of the gynophore. The ovary is superior, borne on an annular disk at the top of a distinct gynophore, with 2–4(5) fused carpels with a terminal style (*Rhopalocarpus*), or 4 carpels apparently separate from each other with a gynobasic style (*Dialyceras*). The ovules are 7–9 per locule in *Rhopalocarpus* and 2–6 in *Dialyceras*. The fruit consists of 1–2(–4) pyriform to fusiform single-seeded mericarps up to 4 cm long, with a pronounced beak in *Dialyceras*, or a subspherical or 2- to 4-lobed fleshy to woody 1- to 4-seeded berry, which is often strongly verrucose and resembles a lychee fruit (*Rhopalocarpus*)[ii].

Classification Various affinities for the family have been suggested, but molecular data[iii] have placed it close to Bixaceae, Cochlospermaceae, Diegodendra-ceae and Thymelae-aceae in a broad Malvales. Marked similarities between *Dialyceras* and another endemic Madagascan family, Diegodenraceae, in habit, inflorescence, gynobasic style, and single-seeded mericarps, as well as nearly identical specialized wood structure and sheathing idioblasts, may suggest that these 2 families should be united as one. The molecular data[iii] gave a cladogram topology with rather low bootstrap values and a later analysis[iv] has given a slightly different topology in which a close relationship with Diegoden-draceae is more plausible. A recent intensive study of the group[v] has pointed to significant differences between Sphaerosepalaceae and Diegodendraceae and has suggested the similarities are the result of parallel evolution, but the matter might deserve further consideration. RKB

Genera 2 Species 18
Economic uses None

i Schatz, G. E., Lowry, P. P., & Wolf, A.-E. Endemic fasmilies of Madagascar, 2. A synoptic revision of Sphaerosepalaceae. *Adansonia* ser. 3, 21: 107–123 (1999).
ii Schatz, G. Sphaerosepalaceae. Pp. 391–392. In: *Generic Tree flora of Madagascar*. Kew, Richmond, UK, and St. Louis, USA (2001).
iii Fay, M. F. *et al.* Plastic *rbc*L sequence data indicate a close affinity between *Diegodendron* and *Bixa*. *Taxon* 47: 43–50 (1998).
iv Savolainen, V. *et al.* Phylogeny of the eudicots: a nearly complete familial analysis based on *rbc*L gene sequences. *Kew Bull.* 55: 257–309 (2000).
v Horn, J. W. The morphology and relationships of the Sphaerosepalaceae (Malvales). *Bot. J. Linn. Soc.* 144: 1–40 (2004).

SPHENOCLEACEAE

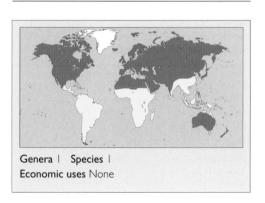

Genera I Species I
Economic uses None

Sphenoclea zeylanica is a glabrous annual herb with a pantropical distribution but possibly only introduced in the New World. The stems are up to 1.5 m, hollow. The leaves are alternate, simple, entire, linear-elliptical. The inflorescence is a dense terminal spike up to 8(–12) cm, with spirally arranged flowers, each subtended by a bract and 2 bracteoles. The flowers are 2–3 mm diameter, sessile, actinomorphic, bisexual, 5-merous, epigynous. The sepals are attached midway up the ovary. The corolla is shortly campanulate to urceolate, white to greenish or pink, the lobes folded over the top of the ovary in bud. The stamens are inserted on the upper corolla tube, alternating with the corolla lobes, with very short filaments. The ovary is semi-inferior, of 2 fused carpels, with a short glabrous style and a capitate to slightly 2-lobed stigma. The locules are 2, with numerous ovules inserted on swollen spongy axile placentas. The fruit is a circumscissile capsule.

The semi-inferior ovary has suggested to many that the affinities lie in the Campanulaceae, but the habit and inflorescence have also suggested affinity with Phytolaccaceae. The molecular evidence does not support either case and places it with Hydroleaceae and Vahliaceae[i,ii] in the Solanales. RKB

i Savolainen, V. *et al.* Phylogeny of the eudicots: a nearly complete familial analysis based on *rbc*L gene sequences. *Kew Bull.* 55: 257–309 (2000).
ii Bremer, K. *et al.* A phylogenetic analysis of 100+ genera and 50+ families of euasterids based on morphological and molecular data. *Pl. Syst. Evol.* 229: 137–169 (2001).

SPHENOSTEMONACEAE

A poorly understood monogeneric family of evergreen trees and shrubs from Australia, New Guinea, and New Caledonia.

Geography *Sphenostemon* has 1 sp. in Queensland, Australia. The remaining species occur from central Malaysia to New Caledonia.

Genera 1 Species 10
Economic uses Wood of some species is used as timber

Description Evergreen trees, usually small, or shrubs. The leaves are alternate or subopposite, sometimes whorled, simple, exstipulate, entire, or dentate. The flowers are actinomorphic, small, in terminal racemose inflorescences. The sepals are 4, free, imbricate, caducous. The petals are 4 or absent, free, fleshy, caducous. The stamens are 4–12, the filaments short and stout, the anthers sessile with a thick connective. The ovary is superior, of 2 fused carpels, with 1 ovule per locule; the style is absent and the stigma sessile. The fruit is drupaceous with 1–3 pyrenes.

Classification *Sphenostemon* has been placed near to the Aquifoliaceae in the Celastrales of Cronquist or Icacinales of Takhtajan (1997). It is treated as a Euasterid II but unplaced as to order by APG II . Lundberg[i] found that *Sphenostemon* is closely related to *Paracryphia*, also unplaced by APG II, a relationship also suggested by an earlier *rbc*L analysis[ii], and considered that these 2 genera, along with *Quintinia,* belonged in the same clade and should be merged in a single family Paracryphiaceae.

Economic uses The timber of some trees is used locally. VHH

i Lundberg, J. Phylogenetic studies in the Euasterids II with particular reference to Asterales and Escalloniaceae. Chapter V: A well resolved and supported phylogeny of Euasterids II based on a Bayesian inference, with special emphasis on Escalloniaceae and other incertae sedis. Uppsala, Sweden, *Acta Univ. Ups. Comprehensive Summaries of Uppsala Dissertations from the Faculty of Science and Technology* 676 (2001).
ii Savolainen, V. *et al.* Phylogeny of the eudicots: a nearly complete familial analysis based on *rbc*L gene sequences. *Kew Bull.* 55: 257–309 (2000).

STACHYURACEAE

Distribution From the Himalayas (Nepal) through China to northern Indochina, Taiwan, Bonin Island, and Japan north to Hokkaido. They occur in thickets, broad-leaf woodland, and occasionally rain forests, from sea level to more than 3,000 m.

Genera 1 Species c. 15
Economic uses Some cultivated ornamental shrubs

Description Deciduous or evergreen shrubs and small trees, sometimes scandent. The leaves are alternate, simple, suborbicular-acuminate to linear, petiolate, the margins finely to strongly serrate. Stipules are small, deciduous. The inflorescence is an erect or pendent spike or raceme up to 12 cm, each flower subtended by a broad bract and 2 bracteoles. The flowers are bisexual or functionally unisexual (plants dioecious), actinomorphic, hypogynous, 4-merous. The sepals are 4, free. The petals are 4, free, yellow. The stamens are 8, in 2 whorls. The ovary is superior, of 4 fused carpels, with a terminal style and capitate to peltate stigma, which is slightly 4-lobed. There are 4 intrusive parietal placentae that are free of each other in the upper part of the ovary but coalesce in the lower part to divide the locule incompletely into 4, bearing many ovules. The fruit is a leathery berry with 4 locules and many seeds, which are small and with an aril.

Classification *Stachyurus* has been placed in various orders, particularly Violales and Theales, but is now convincingly associated with Crossosomataceae and Staphyleaceae (for references, see Crossosomataceae), which also have arillate seeds. However, morphological similarities to the families Aphloiaceae, Ixerbaceae, Geissolomataceae and Strasburgeriaceae, which are also referred to Crossosomatales, are difficult to discern.

Economic uses Several species are cultivated as ornamental shrubs, especially *S. praecox*, which is well known for its pendent spikes of yellow flowers in the spring. RKB

STAPHYLEACEAE
BLADDER NUT FAMILY

Distribution *Staphylea* (c. 10 spp.) grows mostly in more southerly parts of the northern temperate regions, from southern Europe to the Caucasus, eastern Afghanistan to China (south to Yunnan) and Japan (Hokkaido southward), and North America south to northern Mexico. *Euscaphis* (1–2 spp.) occurs in China, northern Vietnam, Taiwan, Japan, and Korea. Both of these genera occur in broad-leaf woodland and open habitats. *Turpinia* (c. 35 spp.) has 23 spp. in tropical Asia from India and China to New Guinea and southern Japan, and 12 in the Neotropics from Mexico and West Indies to Peru, mostly found in tropical rain forests.

Description *Staphylea* and *Euscaphis* are deciduous large shrubs to small trees up to c. 6 m, while *Turpinia* consists of evergreen trees up to 25(–35) m high. The leaves are opposite, usually pinnately or 3-foliately compound but occasionally in some *Turpinia* species 1-foliate (with a basal pulvinus), the leaflets usually ovate or elliptical and acuminate with crenate or serrate margins, usually with stipellae at their base, and small and deciduous stipules. The inflorescence is a terminal or axillary panicle, often pendent. The flowers are usually bisexual but sometimes unisexual (plants dioecious), actinomorphic, hypogynous, 5-merous, up to c. 1 cm long. The sepals are 5, free, often petaloid, and almost equalling the petals. The petals are 5, free, usually white or greenish. The stamens are 5, alternating with the petals. There is a well developed annular disk below the ovary. The ovary is superior, with 2–3(4) carpels, which in *Euscaphis* are free or connate at their apex only, fully united in the other genera, each carpel having 2–12 ovules on an axile placenta. The fruit in *Euscaphis* consists of 1–3 leathery asymmetrically elliptic follicles 1–2 cm long, dehiscing ventrally; in *Staphylea*, an inflated papery capsule up to 6 × 4 cm with 2–3 locules that break open at the tip; and in *Turpinia*, a leathery to fleshy ovoid or spherical indehiscent berry. The seeds are hard and shiny, yellowish brown to black, with an aril only in *Euscaphis*.

Classification After a history of being passed from order to order (most often Sapindales and Celastrales), the family has now apparently

Genera 3 Species c. 46
Economic uses Cultivated ornamentals; timber

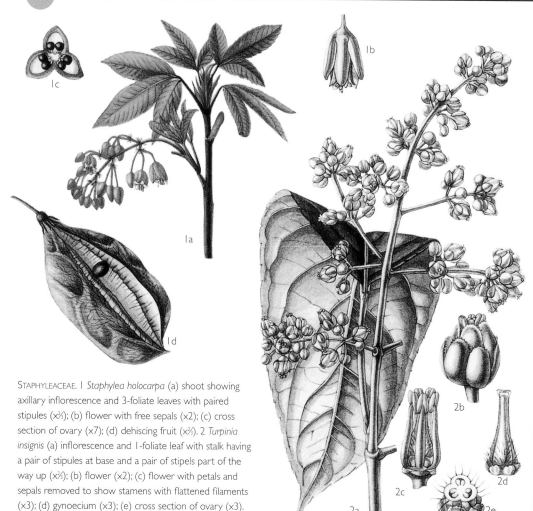

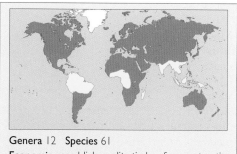

STAPHYLEACEAE. I *Staphylea holocarpa* (a) shoot showing axillary inflorescence and 3-foliate leaves with paired stipules (×⅔); (b) flower with free sepals (×2); (c) cross section of ovary (×7); (d) dehiscing fruit (×⅔). 2 *Turpinia insignis* (a) inflorescence and 1-foliate leaf with stalk having a pair of stipules at base and a pair of stipels part of the way up (×⅔); (b) flower (×2); (c) flower with petals and sepals removed to show stamens with flattened filaments (×3); (d) gynoecium (×3); (e) cross section of ovary (×3).

come to rest in the Crossosomatales, close to the northern hemisphere families Stachyuraceae and Crossosomataceae with Aphloiaceae, Geissolomataceae, Ixerbaceae, and Strasburgeriaceae from the southern hemisphere more distantly related in the same group (for references see Crossosomataceae). Arillate seeds, often characteristic of this group of families, are found only in *Euscaphis* in Staphyleaceae. *Tapiscia* (1 sp., China) and *Huertea* (4 spp., West Indies to Peru) have commonly been included in Staphyleaceae and are vegetatively similar (*Tapiscia* may sometimes have opposite leaves) but are here separated as Tapisciaceae (q.v.) on molecular grounds despite the difficulty of quoting key characters. In *Flora Malesiana* (1960), van Steenis drew attention to a close vegetative similarity also to *Sambucus* (Sambucaceae), even down to the stipellae on the rhachis, but the modern methods he awaited have not confirmed any relationship.

Economic uses Several species of *Staphylea* are cultivated as ornamentals for their flowers and striking inflated papery fruits. Some *Turpinia* species provide timber.   RKB

STEGNOSPERMATACEAE

This monogeneric family of 3 spp. of hairless shrubs or lianas, sometimes small trees, usually with spreading branches, grows in arid or coastal regions of Central America, from Nicaragua to northwestern Mexico and the Antilles. The leaves are fleshy, alternate, petiolate and entire, without stipules. The inflorescence is a thyrse borne either terminally or in the leaf axils. The flowers are actinomorphic, hypogynous, hermaphrodite, with 5 free sepals and 5 white petals with green or red tinge. The 5(–10) stamens are united at the base. The carpels are 3–5, fused into a 3- to 5- then 1-locular ovary with a central column; ovules epitropous; styles absent or short; stigmas recurved and free. The fruit is capsular, the seeds nearly covered with a red aril.

This family is often included in Phytolaccaceae but differs in wood anatomy, sieve plastid ultrastructure, and floral morphology and is securely placed in the core Caryophyllales but not closely related to any individual family[i,ii]. No economic uses are recorded.   AC

Genera 1   Species 3
Economic uses None

[i] Downie, S. R., Katz-Downie, D. S., & Cho, K.-J. Relationships in the Caryophyllales as suggested by phylogenetic analysis of partial chloroplast DNA ORF2280 homolog sequences. *American J. Bot.* 84: 253–273 (1997).
[ii] Cuénoud, P. *et al.* Molecular phylogenetics of Caryophyllales based on nuclear 18S and plastid *rbc*L, *atp*B and *mat*K DNA sequences. *American J. Bot.* 89: 132–144 (2002).

STEMONURACEAE

A family of tropical trees recently segregated from Icacinaceae.

Distribution The family is in all major tropical wet forest regions. *Lasianthera* occurs in Guineo-Congolan Africa, *Grisollea* in Madagascar and the Seychelles, *Irvingbaileya* in Queensland, and *Discophora* from Panama to Brazil, while the other 8 genera are in tropical and subtropical Asia to the Pacific. *Gomphandra* (33 spp.) grows from India to China and the Solomons, *Stemonurus* (12 spp.) from Sri Lanka to the Solomons; *Medusanthera* (4 spp.) from Malaya to Samoa; and *Cantleya, Codiocarpus, Gastrolepis, Hartleya,* and *Whitmorea* each have 1–2 spp. with narrower distributions.

Genera 12   Species 61
Economic uses High-quality timber for construction

Description Evergreen shrubs and trees. The leaves are alternate, simple, and entire. The inflorescence is an axillary or terminal cyme or panicle, sometimes ramiflorous. The flowers are actinomorphic, bisexual or often unisexual (plants dioecious), usually 5-merous, usually small (3–6 mm), but *Whitmorea* up to 1.5 cm. The calyx is cuplike with (4)5(6) short broad lobes. The petals are (4)5(–7), free or sometimes united in *Gomphandra,* often inflexed at the apex, absent in male *Grisollea*. The stamens are as many as the petals or corolla lobes, and the filaments are usually broadened above, bearing many club-shaped hairs that are conspicuously exserted in the open flowers. The ovary is superior, often with a fleshy appendage on one side (absent in *Gomphandra*) and with a short broad sessile stigma, and with a single locule with 2 pendent apical ovules. The fruit is a single-seeded drupe up to several cm, and the appendage (if present) usually brightly colored.

Classification This family was formerly included in Icacinaceae, but as early as 1941 it had emerged as a distinct group on evidence of wood anatomy[i]. Recent morphological and molecular analysis[ii] has maintained the distinctness of the group and has found it be be widely separated from Icacinaceae *sensu stricto,* and a new family Stemonuraceae has been circumscribed. Further discussion has been given by the same author[iii]. Morphologically, the family is easily recognized by the always cymose inflorescence, leaf arrangement on the stem, distinctive cigar-shaped buds, club-shaped hairs on the filament apex protruding from the opened flowers, and the flattened drupe often with a colored appendage.

Economic uses The Malaysian *Cantleya corniculata* produces highly prized timber often used in house- or ship-building or as a substitute for sandalwood. *Codiocarpus* and *Whitmorea* have minor use in construction. TMAU & RKB

i Bailey, I. W. & Howard, R. A. The comparative morphology of the Icacinaceae, 2. Vessels. J. *Arnold Arbor.* 22: 171–187 (1941).
ii Kårehed, J. Multiple origin of the tropical forest tree family Icacinaceae. *American J. Bot.* 88: 2259–2274 (2001).
iii Kårehed, J. Not just hollies – the expansion of Aquifoliales. In: *Evolutionary Studies in Asterids emphasising Euasterids II.* Uppsala, Sweden, Acta Univ. Uppsala, thesis 761.

STERCULIACEAE
COLAS AND STERCULIAS

A family of tropical trees and shrubs placed in the core Malvales but not morphologically similar to any other family.

Distribution The family is pantropical. The largest genus *Cola* is confined to continental sub-Saharan Africa with *Octolobus*, all other genera being restricted to Southeast Asia from China to Australia except *Heritiera*, which extends to Africa, and *Hildegardia*, *Pterygota*, and *Sterculia*, which are pantropical. Most species are lowland rain forest trees, either pioneer trees (many *Sterculia*) or characterizing undisturbed forest (most *Cola* spp.). However, some African *Sterculia* characterize arid bush, as do most species of predominantly Australian *Brachychiton*. The species of *Hildegardia*, with one exception, are unusual in being rare, threatened with extinction, and preferring dry, rocky slopes. *Heritiera littoralis* occurs in Indian Ocean mangrove communities.

Description Trees or shrubs, both evergreen and deciduous, all with particularly fibrous bark and producing mucilage when wounded. Some *Brachychiton* and *Hildegardia* have inflated trunks ("bottle trees"). Stellate or bushy hairs characterize the family but are sometimes only found on shoot apices or flowers. The leaves are usually spiral, simple, entire and palmately nerved; some are palmately lobed, fewer palmately compound; some species of *Cola* with oblong-elliptic leaves are pinnately nerved. Pulvinate petioles are typical in the family. The stipules are usually inconspicuous and caducous. The inflorescences are usually axillary and thyrsoid with female flowers usually distal and rarer than the males. In many *Cola* species, however, flowers are cauliflorous and fasciculate. The flowers are actinomorphic or slightly zygomorphic, unisexual (plants monoecious), or appearing bisexual but when studied in detail are functionally female. Epicalycular bracts are absent or inconspicuous. The perianth is a single whorl, apparently by loss of the corolla, usually

uniform, cuplike, with 5 lobes, rarely nearly divided to the base, or with up to 8 lobes (*Octolobus*). The male flowers have a conspicuous androgynophore terminated by a globose or discoid head of extrorse sessile anther cells numbering 5–c. 20; the anther cells are so equally divided that it is usually difficult to decide where 1 anther ends and another begins; vestigial carpels are detectable within the anther head. The functionally female flowers (appearing bisexual) have anthers that appear fully formed but do not dehisce, surrounding the superior ovary, which is usually sessile, but in many *Sterculia* sits with the anthers on a well-developed androgynophore. The carpels are 5 erect, hairy, in a whorl (numerous in some *Cola*, numerous and spiralled in *Octolobus*) but are free, although they often appear ± united in a single ovary. Each carpel tapers into a single apical unbranched style. The numerous ovules, orthotropous or atropous, are inserted along a ventral placenta. In the fruits, the carpels are clearly separated, each usually developing into a fruitlet or mericarp. In *Sterculia* and *Brachychiton*, the carpels dehisce ventrally along the placental line, displaying the seeds to dispersers, believed usually to be birds, and are sometimes arillate. In *Pterocymbium*, the carpels dehisce before the seeds are developed; when mature, the seeds remain attached to the winged mericarp and are dispersed by wind. Winged mericarps also occur in *Scaphium*, *Hildegardia*, *Franciscodendron*, and many *Heritiera*, while in *Pterygota* it is the seeds that are winged, the carpels dehiscing. Indehiscent fleshy or papery fruits, usually with flesh-coated seeds, are usual in *Cola* and *Octolobus*.

Classification Formerly Sterculiaceae was more broadly circumscribed to include species treated here as Byttneriaceae, Helicteraceae, and Pentapetaceae (q.v.). Molecular analysis supported by morphology[i] has shown these 4 groups to be disparate, with the Byttneriaceae, for example, more closely related to part of the former Tiliaceae than to any of the other families once included in traditional Sterculiaceae. Recent combined molecular analysis[ii] has shown that the genera fall into 4 clades: the *Cola* clade, with *Cola*, *Octolobus*, *Pterygota*, *Hildegardia*,

STERCULIACEAE. I *Cola acuminata* (a) flowering shoot (x⅔); (b) male flower (x⅔); (c) stamens united into a column (x2); (d) gynoecium (x2); (e) cross section of ovary (x2). 2 *Sterculia rupestris* tree in leaf.

Firmiana, *Pterocymbium*, and *Scaphium*; the *Brachychiton* clade, with *Brachychiton*, *Argyrodendron*, *Acropogon*, and *Franciscodendron*; the *Sterculia* clade, with *Sterculia* alone; and the *Heritiera* clade, with *Heritiera* alone. Should further work prove their robustness and morphological support be found, these clades will probably prove to be the basis for a future infrafamilial classification of the family.

Economic uses The seeds of the west African trees *Cola nitida* and *Cola acuminata* (Cola nuts), together with extracts of *Coca* leaves, provided the key ingredients of the original Coca Cola drink. Cola drinks are now popular throughout the world, although the extent to which extracts of *Cola* nuts are still employed is shrouded in secrecy. Several species of *Cola*, including principally those mentioned above, are cultivated for the stimulant effect of their bitter seed embryos, which have been important in West African culture and internationally traded there for many hundreds of years. Several other species, such as *Cola lepidota*, are harvested and sometimes cultivated for the sweet seed coats on a local scale. Several species of *Sterculia*, *Cola*, *Heritiera*, and *Pterygota* provide internationally traded timber, although none are of the most prized categories. Species of *Brachychiton*, *Hildegardia*, and *Firmiana* are cultivated on a small-scale throughout the tropics as ornamental plants. *Hildegardia barteri* is encouraged on farms in Nigeria and adjacent countries as a multipurpose tree: its seeds are eaten as "nuts," its leaves can be edible if treated correctly, its bark-fiber employed for basket making, and the pollarded tree can produce poles for yam cultivation. MRC

i Bayer, C. *et al.* Support for an expanded family concept of Malvaceae within a recircumscribed order Malvales: a combined analysis of plastid *atp*B and *rbc*L DNA sequences. *Bot. J. Linn. Soc.* 129: 267–303 (1999).

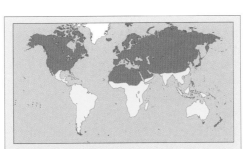

Genera 13 **Species** c. 415
Economic uses Timber and ornamental trees, cola-drink flavoring, cola nuts

ii Wilkie, P. *et al.* Phylogenetic relationships within the subfamily Sterculioideae (Malvaceae/Sterculiceae–Sterculieae) using the chloroplast gene *ndh*F. *Syst. Bot.* 31: 160–170 (2006).

STILBACEAE

A family of ericoid shrubs confined to the Western Cape area of South Africa.

Distribution All 6 genera are endemic to the mountains and plains of the Western Cape area of South Africa. *Stilbe* has 7 spp., *Kogelbergia* 2 spp., and the other genera 1 sp. each.

Description Either a single-stemmed branching shrub or a ligno-tuber with many erect simple stems, always with characteristic narrow and closely imbricate leaves and flowers in compact terminal inflorescences. The leaves are linear with recurved margins, arranged in whorls of 3–7, the whorls over-

Genera 6 **Species** 13
Economic uses Limited use as cut flowers

lapping. Inflorescences are globose to cylindrical heads borne terminally on main or axillary stems, and each flower is subtended by a pair of linear bracteoles. The sepals are 5, free or fused. The corolla is tubular with 4 or 5 narrow lobes and a ring of hairs at the throat and is either regular or strongly 2-lipped, not showy and up to only 12 mm long, and whitish to light mauve or pink, except in *Retzia* where it is up to 5.5 cm long and orange-red with black tips. The stamens are 4 with the posterior stamen lacking, or 5 (*Retzia*). The ovary is superior, of 2 fused carpels, usually with 2 locules and a single basal ovule in each, sometimes (*Kogelbergia*) 1-locular with 2 basal ovules, or with 2 locules with 2–3 ovules in each (*Retzia*). The fruit is a dehiscent capsule.

Classification Although the family has often been included in Verbenaceae, molecular evidence[i,ii] places it nesting in Scrophulariaceae as here defined. Furthermore, it is placed close to *Nuxia,* which has usually been referred to Loganiaceae (a family that is now widely scattered taxonomically). A recent generic account[iii] follows the molecular evidence in including *Nuxia* in Stilbaceae, thus greatly expanding the family in morphology and distribution. *Nuxia* has about 16 spp. extending across Africa and Madagascar, including large shrubs and trees with broad leaves, and it has a 2-locular ovary with many seeds in each locule on a peltate-axile placenta. Its leaves are usually in whorls of 3 or 4, similar to Stilbaceae, and its corolla has a ring of hairs similar to that found in Stilbaceae. Other morphological and chemical characters also support a close relationship. The available molecular evidence is based on a small sample of genera and

species and moreover suggests that the montane tropical African tree genus *Halleria,* traditionally placed in Scrophulariaceae, is also closely related to Stilbaceae, and it has been included in the family[ii] but there seems to be no morphological support for this[iii]. As in other cases, the problem is where to draw the line between 2 families, and here it is preferred to confine Stilbaceae to the ericoid genera in the Western Cape, with their distinctive habit and different ovary, in agreement with the family specialist[iv,v]. However, the matter should be revisited when much wider sampling has been done. *Retzia* was for a long time referred to the Buddlejaceae or placed in its own family Retziaceae, but is accommodated easily in Stilbaceae having an ericoid habit and distribution in the Western Cape. It is, however, referred to a separate subfamily Retzioideae distinguished by its larger and showy flowers, 5 stamens, and several ovules in the ovary.

Economic uses Plants may be used as ornamental cut flowers, along with other South Africa ericoid plants. RKB

i Oxelman, B., Backlund, M., & Bremer, B. Relationships of the Buddlejaceae *s.l.* investigated using parsimony jackknife and branch support analysis of chloroplast *ndh*F and *rbc*L sequence data. *Syst. Bot.* 24: 164–182 (1999).
ii Olmstead, R. G. *et al.* Disintegration of the Scrophulariaceae. *American J. Bot.* 88: 348–361 (2001).
iii Linder, H. P. Stilbaceae. Pp. 433–440. In: Kadereit, J. W. (ed.), *Families and Genera of Vascular Plants. VII.* Berlin, Springer-Verlag (2004).
iv Rourke, J. P. A review of generic concepts in the Stilbaceae. *Bothalia* 30: 9–15 (2000).
v Rourke, J. P. Stilbaceae. Pp. 541–543. In: Leistner, O.A. (ed.), *Seed Plants of Southern Africa: Families and Genera* (*Strelitzia* 10). Pretoria, South Africa, National Botanical Institute, (2001).

STRASBURGERIACEAE

Strasburgeria robusta is a tree up to 10 m or more high, confined to New Caledonia. The leaves are alternate, mostly clustered toward the ends of branches, to 25 cm or more long, obovate, coriaceous, with sinuate margins or shallowly and sparsely toothed toward the apex, tapered to a short winged petiole, with conspicuous ovate-connate intrapetiolar stipules. The inflorescence is a solitary axillary pedicellate flower among the clustered leaves. The flowers are bisexual, actinomorphic, hypogynous, large (to 5.5 × 2.5 cm) and showy. The sepals are 8–11, spirally arranged, imbricate, ovate to suborbicular, leathery, the inner ones progressively larger. The petals are usually 5, imbricate, decreasing in size toward the center of the flower, yellowish. There are 10 stamens in a single whorl. There is a disk below the ovary, which is lobed between the stamens. The ovary is superior, of (4)5(–7) fused carpels, terminated by a prominent style, and there is a single ovule in each locule. The fruit is large, woody and indehiscent, and the seeds are hard but without an aril.

The family has had many different affinities suggested for it, especially Brexiaceae, Theaceae, and Ochnaceae[i]. Recent molecular[ii]

and floral anatomical studies have placed it in a new concept of Crossosomatales, closest to Ixerbaceae from New Zealand and Geissolomataceae from South Africa (see Crossosomataceae for additional references). Fossil pollen identifiable with *Strasburgeria* has been found in Australia and New Zealand. RKB

i Dickison, W. C. Contributions to the morphology and anatomy of *Strasburgeria* and a discussion of the taxonomic position of the Strasburgeriaceae. *Brittonia* 33: 564–580 (1981).
ii Cameron, K. M. On the phylogenetic position of the New Caledonian endemic families Paracryphiaceae, Oncothecaceae and Strasburgeriaceae: a comparison of molecules and morphology. *Bot. Rev.* 68: 428–443 (2003).

STYLIDIACEAE
TRIGGERPLANT AND STYLEWORT FAMILY

Distribution Stylidiaceae is predominantly from temperate and tropical Australia, with a single species known from southern South America, and a small number of taxa distributed in New Zealand and across Southeast Asia. The south-western part of Western Australia is a center of diversity, with 70% of all species, most from the genus *Stylidium*. The family is commonly represented in wet or seasonally wet habitats.

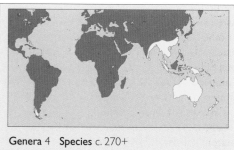

Genera 4 **Species** c. 270+
Economic uses Limited cultivation

Description Annual or perennial herbs, often with a rosette or tussock habit, more rarely creepers, climbers, subshrubs, or cushion plants, commonly bearing glandular and/or eglandular hairs, seldom completely glabrous. The stems are shortly elongated, or compact and prominently thickened, occasionally modified into ligno-tubers, corms, or rhizomes, with adventitious roots common. Anomalous secondary growth is present within numerous species of *Stylidium*. The leaves are commonly arranged in a basal rosette, or alternate with a terminal rosette, occasionally whorled or imbricate, rarely absent altogether in mature plants. The leaves are simple but varied in shape and size, often fibrous, sessile, or petiolate, and exstipulate. A hydathode is usually present in *Forstera* and *Phyllachne*. The inflorescences are terminal, with flowers arranged in racemes, panicles, and cymes, more rarely in pseudoheads and corymbs or with flowers solitary. The flowers are bisexual, rarely unisexual by abortion of parts, commonly zygomorphic, occasionally actinomorphic. The calyx tube is adnate to the ovary with 5(–9) free or partly fused, equal or unequal lobes, occasionally 2-lipped. The corolla is sympetalous with (4)5(–9) lobes, the anterior lobe highly modified in *Levenhookia* and *Stylidium* to form a labellum. The corolla varies in color and shape, commonly adorned with appendages and/or hairs. Two stamens are fused to the style to form a floral column that bears the anthers and stigma at its tip. The anthers are extrorse, opening by longitudinal slits, and typically develop before the stigma. Extrastaminal nectaries are present near the base of the column. The ovary is inferior and imperfectly 2-locular, rarely with 1 locule sterile, occasionally becoming 1-locular by loss of septum. Placentation is axile, with 2–850 ovules per flower. The fruit is a capsule, typically dehiscent. The seeds are small, smooth, or variously ornate, with a minute embryo and copious endosperm. In *Stylidium* (the triggerplants), the column is touch sensitive. It rotates rapidly when triggered so as to deposit or retrieve pollen from visiting insects, after which it gradually resets. In *Levenhookia* (the styleworts), the column is insensitive, but some movement is generated when it is released from the hooded labellum.

Classification Stylidiaceae is an unusual family that possesses a number of apomorphies including a floral column. While various systematic positions for Stylidiaceae have been proposed, the family is now confidently placed within the Asterales[i,ii], although its precise position relative to other taxa remains somewhat ambiguous. A sister group relationship with Donatiaceae is well supported[iii]. However, an association with Campanulaceae has also been suggested[iv]. The monospecific New Zealand genus *Oreostylidium* has previously been recognized within Stylidiaceae[v]; however, molecular, morphological, and anatomical data clearly support the placement of this taxon within *Stylidium*[vi]. *Donatia* has sometimes been included within Stylidiaceae (see under Donatiaceae). A molecular dataset is required to improve our understanding of phylogenetic relationships within *Stylidium*.

Economic uses A small number of species are grown as ornamentals. JAW

i Erbar, C. Floral development of two species of *Stylidium* (Stylidiaceae) and some remarks on the systematic position of the family Stylidiaceae. *Can. J. Bot.* 70: 258–271 (1992).
ii Gustafsson, M. H. G. & Bremer, K. Morphology and phylogenetic interrelationships of the Asteraceae, Calyceraceae, Campanulaceae, Goodeniaceae and related families (Asterales). *American J. Bot.* 82: 250–265 (1995).
iii Lundberg, J. & Bremer, K. A phylogenetic study of the order Asterales using one morphological and three molecular data sets. *Int. J. Plant Sci.* 164: 553–578 (2003).
iv Bremer, B. *et al.* Phylogenetics of Asterids based on 3 coding and 3 non-coding chloroplast DNA markers and the utility of non-coding DNA at higher taxonomic levels. *Molec. Phylogen. Evol.* 24(2): 274–301 (2002).
v Mildbraed G. W. J. Stylidiaceae. In: Engler, A. (ed.) Das Pflanzenreich IV, 278. Berlin, Wilhelm Engelmann (1908).
vi Wagstaff, S. J. & Wege, J. Patterns of diversification in New Zealand Stylidiaceae. *American J. Bot.* 89: 865–874 (2002).

STYRACACEAE
SILVERBELL AND SNOWBELL TREES

A family of shrubs and trees best known as the source of benzoin (gum bejamin).

Distribution The greatest number of genera is found in Eastern Asia to Malaysia and New Guinea, especially China, with all but 2 of them endemic to the area: *Huodendron* (4 spp.), *Bruinsmia* (2 spp.), *Alniphyllum* (3 spp.), *Changiostyrax* (1 sp.), *Sinojackia* (5 spp.), *Parastyrax* (2 spp.), *Pterostyrax* (4 spp.), *Rehderodendron* (5 spp.), and *Melliodendron* (1 sp.). *Halesia* (3 spp.) occurs in China and eastern USA, while *Styrax* (130 spp.) occurs in all areas above as well as southeastern North America to South America, with 1 sp. (*Styrax officinale*) in the Mediterranean.

Description Evergreen or deciduous, pubescent, scaly or rarely hairless, trees or shrubs usually with alternate leaves. The leaves are simple, without stipules (unless minute) and usually entire. The inflorescence is a terminal or axillary raceme, cyme, or panicle, rarely single-flowered or fascicled. The flowers are actinomorphic, usually bisexual, or female (plants gynodiecious). The calyx is tubular with 4–5(–9) persistent lobes. The corolla is usually white, tubular at the base, but often only very shortly so, with (4)5(–7) valvate, rarely overlapping, lobes. The stamens are double the number, or equal in number, to and alternate with, the corolla lobes, usually joined to the corolla tube or united as a tube. The ovary is superior (*Alniphyllum*, *Bruinsmia*, and *Styrax*) semi-inferior (*Huodendron*), or inferior or usually so (*Mellidodendron*, *Rehderodendron*, *Sinojackia*, *Halesia*, *Parastyrax*, and *Pterostyrax*), of 3 (*Styrax*) to 5 fused carpels with 3 to 5 locules, each containing 1 to many anatropous ovules on axile placentas. The style is simple with a capitate or lobed stigma. The fruit is a drupe or capsule, with the calyx persistent. The 1 to few seeds with copious endosperm are eaten and dispersed by animals.

Classification The family is placed in Ericales on molecular evidence[i], and sister to Diapensiaceae[ii]. The family has traditionally, and consistently, been placed in Ebenales (reviewed in Brummitt 1992) based on morphological and anatomical characters and is now included with many of the old Ebenales in the expanded Ericales of APG II. For full accounts of all the genera see the *Flora of China*[iii]. Phylogenetic relationships within the family are established for 9 genera[iv] and for sections within *Styrax*[v] based on molecular evidence.

Economic uses The chief product is resin, with the tropical resins (chiefly from *Styrax benzoin*) traded as benzoin (corrupted as gum bejamin, sometimes benjamin), and used medicinally (in friar's balsam) and in incense. *Styrax officinale* (Mediterranean) is often

Genera 11 **Species** c. 150
Economic uses Pharmaceutical, fragrance and confectionary use of the resin benzoin; some ornamentals

reported to yield the resin storax, used as an antiseptic, inhalant, and expectorant, but this is produced by *Liquidambar*[vi]. *Halesia* includes the Silverbell or Snowdrop Trees, *Styrax* the Snowbell Trees; both groups are beautiful, distinctive ornamentals. AC

i Anderberg, A. A., Rydin, C., & Källersjö, M. Phylogenetic relationships in the order Ericales *s.l.*: analyses of molecular data from five genes from the plastid and mitochondrial genomes. *American J. Bot.* 89: 677–687 (2002).
ii Bremer, B. *et al.* Phylogenetics of asterids based on 3 coding and 3 non-coding chloroplast DNA markers and the utility of non-coding DNA at higher taxonomic levels. *Molec. Phylog. Evol.* 24: 274–301 (2002).
iii Huang, S. & Grimes, J. W. Styracaceae. Pp. 253–271. In: Wu, Z.-Y. & Raven, P. H. (eds), *Flora of China*. Vol. 15. *Myrsinaceae through Loganiaceae*. Beijing, Science Press & St. Louis, Missouri Botanical Garden (1996).
iv Fritsch, P. W., Morton, C. M., Chen, T., & Meldrum, C. Phylogeny and biogeography of the Styracaceae. *Int. J. Plant Sci.* 162 (6 Supplement): S95–S116 (2001).
v Fritsch, P. W. Phylogeny and biogeography of the flowering plant genus *Styrax* (Styracaceae) based on chloroplast DNA restriction sites and DNA sequences of the internal transcribed spacer region. *Molec. Phylog. Evol.* 19: 387–408 (2001).
vi Fritsch, P. W. Styracaceae. Pp. 434–442. In: Kubitzki, K. (ed.), *The Families and Genera of Vascular Plants. VI. Flowering Plants. Dicotyledons: Celastrales, Oxalidales, Rosales, Cornales, Ericales*. Berlin, Springer-Verlag (2004).

SURIANACEAE
BAY CEDAR, OOLINE, PEBBLE BUSH

A small but widespread family related to the legume family.

Distribution *Suriana maritima* is distributed along tropical and subtropical coasts in the Gulf of Mexico and Caribbean, across the Indian Ocean and into the Pacific. *Recchia* (3–4 spp.) is restricted to Mexico and Costa Rica. *Cadellia pentastylis* and *Guilfoylia monostylis* are endemic to eastern regions of Australia. *Stylobasium* (3 spp.) occurs throughout Australia, except Victoria and Tasmania.

Description Trees and shrubs with alternate leaves. The leaves are simple, elliptical, or pinnate (some *Recchia*), sometimes fleshy (*Suriana*), sometimes with minute stipules (*Recchia*)[i]. The inflorescence is axillary or terminal, a panicle, or of 1–2 flowers, sometimes in axillary clusters. The flowers are radially symmetrical, bisexual,

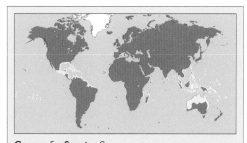

Genera 5 **Species** 9
Economic uses Limited use of wood for carving

with 5 free or fused sepals alternating with 5 free or 0 (*Stylobasium*[ii]) petals, 5 or 10 stamens, and 5 or 0 staminodes. The ovary is superior, of 1, 2, or 5 free carpels each having 2 ovules. The fruit is a drupe, berry, or nut.

Classification The family has often been placed within the Simaroubaceae[iii]. Extensive molecular evidence shows the family to have a close affinity with the legumes[iv,v,vi], the Quillajaceae and Polygalaceae in a well-supported, but poorly resolved, Fabales (APG II). Monophyly of the family was demonstrated in an extensive study of all genera and most species in the family[vii] that showed *Recchia* to be sister to *Cadellia* and *Suriana* to be sister to *Stylobasium* and *Guilfoylia*.

Economic uses The heartwood of *Suriana* is used for small decorative carvings.   AC & RKB

i Thomas, W. W. Surianaceae. P. 364. In: Smith, N. *et al.* (eds), *Flowering Plants of the Neotropics.* Princeton, Princeton University Press (2004).
ii Macfarlane, T. D., Watson L., & Marchant N. G. (eds), Western Australian Genera and Families of Flowering Plants. Western Australian Herbarium (Version: August 2002). http://florabase.calm.wa.gov.au/
iii Dahlgren, R. M. T. General aspects of angiosperm evolution and macrosystematics. *Nordic J. Bot.* 3: 119–149 (1983).
iv Fernando, E. S. *et al.* Rosid affinities of Surianaceae: molecular evidence. *Mol. Phylogenet. Evol.* 2: 344–350 (1993).
v Fernando, E. S., Gadek, P. A., & Quinn, C. J. Simaroubaceae, an artificial construct: Evidence from *rbc*L sequence variation. *American J. Bot.* 82: 92–103 (1995).
vi Soltis, D. E. *et al.* Angiosperm phylogeny inferred from 18S rDNA, *rbc*L, and *atp*B sequences. *Bot. J. Linn. Soc.* 133: 381–461 (2000).
vii Crayn, D. M., Fernando, E. S., Gadek, P. A., & Quinn, C. J. A reassessment of the familial affinity of the Mexican genus *Recchia* Mociño & Sessé ex DC. *Brittonia* 47: 397–402 (1995).

SYMPHOREMATACEAE

A family of high-climbing woody lianes from tropical Asia.

Distribution *Symphorema* (3 spp.) occurs from western and northern India to Thailand and the Malay Peninsula, with outliers in northern Philippines and Tanimbar Islands. *Sphenodesme* (14 spp.) extends from India and southwestern China to Borneo. *Congea* (10 spp.), occurs from northern India and southwestern China through Indochina to Sumatra.

Description Massive woody lianes climbing to 15 m or more and with stems up to 10 cm thick or more. The leaves are opposite, entire or toothed. The flowers are small and sessile but borne in capitate cymose clusters of 3–7, each cluster terminal on axillary simple or branching peduncles and surrounded by a conspicuous involucre of 3–4 (*Congea*) or 6,

Genera 3 **Species** 27
Economic uses Occasional cultivation as ornamentals in the tropics

often brightly colored, elliptical to spathulate bracts, much exceeding the flowers, each cluster terminal. The calyx is actinomorphic or in *Sphenodesme* rarely 2-lipped, with 5–10 nerves and 3–8 lobes, which are sometimes bifid. The corolla is actinomorphic with 5–6(7) lobes in *Sphenodesme* and 6–16 lobes in *Symphorema*, or 2-lipped with 5 lobes in *Congea*. The stamens are usually about as many as the corolla lobes, 4, and didynamous in *Congea* but up to 18 in *Symphorema*, inserted on the corolla tube and exserted from or included in the tube. The ovary is superior, of 2 carpels, imperfectly 2-locular, the septum being incomplete above, and bears 2 pendulous ovules at the top of each locule. The fruit is slightly fleshy or dry and indehiscent and bears 1(2) seeds.

Classification The family was long included in a heterogeneous Verbenaceae, although it has been separated out by several authors. Munir[i] considered it distinct from the then Verbenaceae. Molecular evidence places it basal to the genera formerly in Verbenaceae that are now transferred to Lamiaceae, and in an analysis of combined data sets[ii] Symphoremataceae has again been maintained as a separate family. A more recent revision of the genera of Lamiaceae[iii] has included it as subfamily Symphorema-toideae, but the capitate inflorescence with conspicuous involucral bracts is unmatched in Lamiaceae and characteristic of all 3 genera, providing an immediate means of recognition of the family. Other characters conflicting with Lamiaceae are the normally actinomorphic calyx and corolla, the many stamens in *Symphorema,* and particularly the incompletely 2-celled ovary with ovules pendulous from the apex.

Economic uses *Congea griffithiana*, *C. tomentosa*, and *C. velutina* are cultivated in the tropics as ornamentals or botanical curiosities.   RKB

i Munir, A. A. Revision of *Congea*. *Gard. Bull Singapore* 21: 259–314 (1966). Revision of *Sphenodesme*. *Ibid*. 21: 315–378 (1966). Revision of *Symphorema*. *Ibid*. 22: 153–171 (1967).
ii Wagstaff, S. J. *et al.* Phylogeny inferred in Labiatae *s.l.* from *cp*DNA sequences. *Pl. Syst. Evol.* 209: 265–274 (1998).
iii Harley, R. M. *et al.* [12 additional authors are cited] Lamiaceae. Pp. 167–275. In: Kadereit, J. W. (ed.), *Families and Genera of Vascular Plants. VII.* Berlin, Springer-Verlag (2004).

SYMPLOCACEAE
SWEETLEAF

Genera 1 **Species** c. 320
Economic uses Source of yellow dy and of maté in South America

One genus, *Symplocos*, with c. 320 spp. of evergreen (sometimes deciduous) trees or shrubs with astringent bark, with a trans-Pacific distribution from southeastern USA to southern Brazil and from northern China to subtropical Australia but found primarily in humid tropical montane forests. Two deciduous species, *S. tinctoria* and *S. paniculata*, are warm temperate occurring in southeastern USA and eastern Asia respectively. The leaves are spirally or distichously arranged, usually sweet tasting, simple, the margin dentate, glandular-dentate, or entire; petiolate but without stipules. The inflorescence is usually axillary, near the stem tips but sometimes terminal, and is a spike, raceme, compact cyme or thyrse, panicle or of 1 flower. The flowers are actinomorphic, hermaphrodite, usually 5-merous. The sepals are (3–)5, persistent in fruit and valvate or overlapping. The petals are (3–)5(–11), white (sometimes yellow), overlapping, are fused at the base (subgenus *Hopea*) or to the middle (subgenus *Symplocos*). The stamens are (4)5–60, joined to the base of the petals, and fused into a tube (subgenus *Symplocos*), fused at the base (some subgenus *Hopea*) or in 5 clusters (some subgenus *Hopea*). The ovary is inferior or semi-inferior, of 2–5 fused carpels, each forming a locule containing 2–4 ovules. The fruit is a drupe, the seeds with copious endosperm.

The family is placed in Ericales by APG II and in molecular analyses[i] and is probably sister to Diapensiaceae and Styracaceae[ii]. Studies of floral ontogeny failed to resolve the position of the family[iii] but provide an insight into the complex development of the fused stamens and petals. DNA sequence analysis of 111 spp. of *Symplocos*[iv] refutes the 2 subgenera recognized on morphological grounds[v] due to several misplaced species and sections. The bark of some species provides a yellow dye for batik and a mordant for other pigments. South American species are used for maté.   AC

i Anderberg, A. A., Rydin, C., & Källersjö, M. Phylogenetic relationships in the order Ericales *s.l.*: analyses of molecular data from five genes from the plastid and mitochondrial genomes. *American J. Bot.* 89: 677–687 (2002).
ii Schönenberger, J., Anderberg, A. A., & Sytsma, K. J. Molecular Phylogenetics and patterns of floral evolution in the Ericales. *Int. J. Plant Sci.* 166: 265–288 (2005).

iii Caris, P., Ronse Decraene, L. P., Smets, E., & Clinckemaillie, D. The uncertain systematic position of *Symplocos* (Symplocaceae): evidence from a floral ontogenetic study. *Int. J. Plant Sci.* 163: 67–74 (2002).
iv Wang, Y. *et al.* Phylogeny and infrageneric classification of *Symplocos* (Symplocaceae) inferred from DNA sequence data. *American J. Bot.* 91: 1901–1914 (2004).
v Nooteboom, H. P. Revision of the Symplocaceae of the old world New Caledonia excepted. *Leiden Botanical Series* NR.1. Leiden, Leiden University Press (1975).

TAMARICACEAE

TAMARISKS

A family of small heathlike shrubs and small trees, usually with tiny pink or white flowers.

Distribution Temperate and subtropical, growing in maritime or sandy places from the Mediterranean through North Africa and southeastern Europe via central Asia to India and China. It also occurs in Norway and in southwestern Africa. The family is halophytic, xerophytic, or rheophytic.

Description Small trees, shrubs, or subshrubs with woody stems bearing slender branches. The leaves are alternate, small, tapering or scalelike, without stipules. The inflorescence consists of one (*Hololachna, Reaumuria*) to many flowers in dense spikes or racemes (*Myricaria, Myrtama, Tamarix*). The flowers are minute, regular, bisexual, without bracts. The sepals and petals are 4 or 5, free. The petals and the 5–10(–many) stamens are inserted on a fleshy, nectar-secreting disk. The stamens are free or slightly fused at the base. The carpels are 2, 4, or 5, fused into a superior ovary with 1 locule; ovules are few to numerous on parietal or basal placentas. The styles are usually free, absent in some species, and the stigmas sessile (*Myricaria*). The fruit is a capsule containing seeds with or without endosperm. The seeds are sometimes winged but usually covered with long hairs either all over or in a tuft at the apex.

Classification The family is a member of the non-core Caryophyllales[i] and is sister to the Frankeniaceae[ii] with which it shares many morphological and ecological features. There are 2 tribes:

Reaumurieae Includes the genus *Reaumuria* (20 spp., halophytic shrubs or undershrubs native to the eastern Mediterranean and central Asia) with solitary terminal flowers with showy petals, each bearing 2 longitudinal appendages on the inside, the stamens numerous and either free or more or less fused into 5 bundles opposite the 5 petals, and the ovary with 5 styles and few seeds; and *Hololachna* (2 spp., sometimes treated as a synonym of *Reaumuria*[iii], native to central Asia), with axillary solitary flowers whose petals are devoid of appendages, the stamens only 5–10, free or shortly connate at the base, and the ovary with 2–4 styles and few seeds.

TAMARICACEAE. 1 *Tamarix aphylla* (a) shoot with minute leaves and dense raceme of flowers (×⅔); (b) part of inflorescence (×4); (c) stamens and gynoecium (×6); (d) 4-lobed gynoecium (×8); (e) vertical section of ovary with basal ovules (×10). 2 *Reaumuria linifolia* (a) leafy shoot with solitary flowers (×⅔); (b) gynoecium (×2); (c) petal (×1½). 3 *Tamarix africana* habit (×⅟₂₀). 4 *Myricaria germanica* (a) flowering shoot (×⅔); (b) flower (×2); (c) stamens united at the base (×4); (d) fruit—a capsule (×2); (e) dehiscing capsule showing cluster of hairy seeds (×2); (f) seed, which is hairy at the apex only (×4).

Tamarisceae Comprises *Tamarix* (55 spp., western Europe, the Mediterranean, North Africa, northeast China, and India.), *Myricaria* (10 spp., Europe, China, and central Asia), and *Myrtama* (1 sp., Pakistan, Kashmir, and Tibet), all characterized by their numerous flowers in a spikelike or racemose inflorescence and numerous non-endospermic seeds. In *Tamarix,* the inflorescence is borne either on the woody lateral branches or on the terminal young shoots. In *Myricaria*, the inflorescence is a long terminal raceme. A phylogenetic study based on ITS sequence data[iv] has suggested *Myricaria* is paraphyletic to *Tamarix*.

Economic uses The twigs of the shrub *T. mannifera* (from Egypt to Afghanistan) yield the white sweet gummy substance manna as a result of puncture by the insect *Coccus maniparus.* Insect galls on species of *Tamarix* (*T. articulata* and *T. gallica*) are a source of tannin, dyes and medicinal extracts. *Tamarix gallica*

and *T. africana* are often grown as ornamental shrubs for their rather feathery appearance and their catkin-like inflorescences. *T. pentandra* is sometimes grown as a hedge or windbreak, as is *T. gallica* in Mediterranean coastal regions. Species of *Tamarix* have become widely naturalized in North and South America and in Australia and some have become serious invasives, particularly *T. parviflora* and *T. ramosissima* in the USA[v], and *T. aphylla* in Australia[vi].　　[SRC] AC

Genera 5
Species c. 80
Economic uses Some ornamental species and others yielding dyes, medicines, and tannins from galls

[i] Cuénoud, P. *et al*. Molecular phylogenetics of Caryophyllales based on nuclear 18S and plastid *rbc*L, *atp*B and *mat*K DNA sequences. *American J. Bot*. 89: 132–144 (2002).
[ii] Gaskin, J. F. *et al*. A systematic overview of Frankeniaceae and Tamaricaceae from nuclear rDNA and plastid sequence data. *Ann. Missouri Bot. Gard*. 91: 401–409. (2004).
[iii] Gaskin, J. F. Tamaricaceae. Pp. 363–368. In: Kubitzki, K. & Bayer, C. (eds), *The Families and Genera of Vascular Plants. V. Flowering Plants. Dicotyledons: Malvales, Capparales and Non-betalain Caryophyllales*. Berlin, Springer-Verlag (2002).
[iv] Zhang, D.-Y. *et al*. Systematic studies on some questions of Tamaricaceae based on ITS sequence. *Acta. Bot. Boreal.-Occident. Sin*. 20: 421–431 (2000).
[v] Carpenter, A. T. Element Stewardship Abstract For *Tamarix*. Arlington, VA, USA, The Nature Conservancy (1998).
[vi] Anon. Athel pine or tamarisk (*Tamarix aphylla*): *Weeds of National Significance: Weed Management Guide*. Department of the Environment and Heritage and the CRC for Australian Weed Management (2003).

TAPISCIACEAE

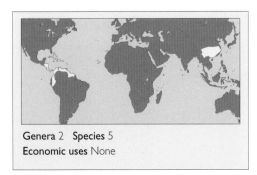

Genera 2　**Species** 5
Economic uses None

A family of 2 genera, *Tapiscia* with a single species, *T. sinensis,* endemic to China, and *Huertea,* comprising 4 spp., restricted to the neotropics, from Honduras, Hispaniola, and Cuba to northern South America. They are sometimes included in the Staphyleaceae as a subfamily, but they differ from it by their alternate leaves (sometimes opposite), connate sepals, nectary disk less well developed or absent, and seeds with a well-developed fibrous exotegmen. They are shrubs or small trees, with imparipinnate or 3-foliate leaves, the leaflets

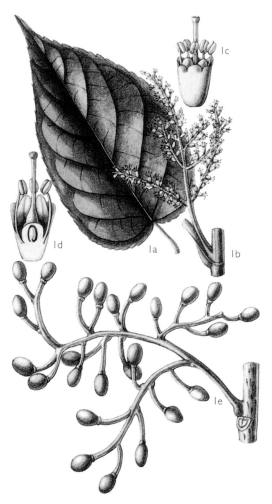

TAPISCIACEAE. I *Tapiscia sinensis* (a) leaflet (x⅔); (b) inflorescence (x⅔); (c) flower showing fused sepals (x4); (d) vertical section of flower (x6); (e) indehiscent fruits (x⅔).

with dentate margins. The flowers are small, actinomorphic, hermaphrodite. The ovary is superior, of 2 fused carpels, with 1 or 2 ovules per locule. The fruit is drupaceous or baccate. The family is included in the Sapindales by Takhtajan (1997). Molecular data indicate a close relationship to the Dipentodontaceae (q.v. for references) and to the Malvales and Sapindales. Syn. Huerteaceae.　　VHH

TEPUIANTHACEAE

A family of mainly trees comprising the genus *Tepuianthus,* with 6 spp. endemic to the Guayana highlands region of Venezuela and neighbouring Colombia and Brazil, all but one of them confined to the sandstone mountain summits, known as tepuis, the exception being *T. savannensis,* which grows on the lowland savannas of southwestern Venezuela[i]. They are trees or less frequently shrubs (*T. auyantepuiensis*). The leaves are usually alternate, sometimes opposite, simple, with entire margins. The

Genera I　**Species** 6
Economic uses None

flowers are actinomorphic, normally unisexual (plants androdioecious), rarely bisexual, in terminal or axillary cymose inflorescences. The sepals are 5, free, covered with dense, silky hairs on lower surface. The petals are 5, free, imbricate, yellowish. There is an extrastaminal disk of 5–10 fleshy glands. The stamens are 5 to 12(–16). The ovary is superior, with 3 carpels and usually 3 locules, with parietal placentation. The fruit is a loculicidal capsule.

Affinities suggested for the Tepuianthaceae include the Celastrales and Rutales, but morphological and molecular data have indicated its placement in the Malvales as sister to the Thymelaeaceae and as a subfamily of it[ii]. Until the genera of the Thymelaeaceae have been more thoroughly studied, it is preferred here to keep *Tepuianthus* as a separate family.　　VHH

[i] Boom, B. & Stevenson, D. W. Tepuianthaceae. Pp. 366–367. In: Smith, N. *et al*. (eds), *Flowering Plants of the Neotropics*. Princeton, Princeton University Press (2004).
[ii] Wurdack, K. J. & Horn, J. W. A reevaluation of the affinities of the Tepuianthaceae: molecular and morphological evidence for placement in the Malvales: 151. In: *Botany 2001: Plants and People*, Abstracts [Albuquerque] (2001).

TETRACARPAEACEAE

Tetracarpaea tasmannica is an erect subshrub or shrub 15–60 (100) cm high confined to the mountains of Tasmania. The leaves are alternate, evergreen, up to 25(–30) × 8(–10) mm, oblanceolate or elliptical, cuneate and entire in the lower part, rounded and crenate to serrate in the upper part, with a petiole up to 2(3) mm. The inflorescence is a fairly dense erect terminal raceme 1–5 cm long, with up to perhaps 20 shortly pedicellate flowers. The flowers are c. 5 mm, actinomorphic, bisexual, 4-merous. The sepals are 4, small. The petals are 4, erect, shortly clawed, white or cream. The stamens are 8, in one series. The ovary is superior, of 4 free carpels, each one shortly stipitate and with a short style and notched stigma and with many ovules attached to the ventral margin. The fruit is a group of 4 erect follicles up to 5 mm long, each dehiscing along its ventral margin to release the many tiny seeds.

The genus has been variously considered related to Cunoniaceae, Dilleniaceae, Saxifragaceae, Hydrangeaceae, and Crassulaceae. Molecular evidence[i] has placed it with Penthoraceae of East Asia and eastern North America and Haloragaceae, which is predominantly Australian, this group of families being sister to Crassulaceae. There are some similarities with each of these, and APG II has suggested that Haloragaceae might be defined to include both *Penthorum* and *Tetracarpaea*. However, the ovary of *Tetracarpaea* comprises free carpels with many small seeds, as in Crassulaceae, but in *Penthorum* the ovary is semi-inferior with many seeds, and in Haloragaceae it is inferior with few seeds. The habit and infructescence are similar to those of *Penthorum,* but the 4-merous flowers with clawed petals and 8 stamens perhaps point to further similarity with Halor-agaceae. Tetracarpaeaceae may perhaps be seen as representing a more plesiomorphic condition. However, all these families seem sufficiently distinctive to be easily maintained separately. RKB

i Morgan, D. R. & Soltis, D. E. Phylogenetic relationships among members of Saxifragaceae *sensu lato* based on *rbc*L sequence data. *Ann. Missouri Bot. Gard.* 80: 631–660 (1993).

TETRACENTRACEAE

Genera 1 Species 1
Economic uses
Cultivated ornamental

Tetracentron sinense is a deciduous tree up to 30 m found from eastern Nepal[i] to northern Myanmar and southwestern and central China[ii], usually in mixed deciduous and evergreen forest up to 3,000 m. It produces long shoots and short shoots, the latter bearing a single leaf and inflorescence. Leaves are *Tilia*-like, broadly ovate, cordate at the base, with palmate venation, the margins with many acute to rounded crenations. The inflorescence is a many-flowered, linear, pendulous catkinlike spike 6–18 × 0.3–0.5 cm. The flowers are actinomorphic, bisexual, 4-merous, small (1–2 mm) and sessile. The sepals are 4 but there are no petals. The stamens are 4, opposite the sepals. The carpels are 4, alternating with the stamens in a ring, connate at the base but separate distally, the styles at first erect but soon reflexing. Each carpel has 5–6 ovules, pendulous, in 2 rows from the marginal placenta. The fruit comprises 4 laterally connate follicles.

A close relationship with the Trochodendraceae has always been suggested by the similar ovary and fruit, and *Tetracentron* has been included in that family by some authors. Its general appearance, however, is distinctive, and there are many obvious characters to separate them and allow family recognition. *Tetracentron* differs from *Trochodendron* in its branching, with long and short shoots, its solitary leaves with palmate venation, its catkinlike inflorescence of small sessile flowers, its flat receptacle, its 4-merous flowers, the presence of sepals, and its 4 stamens opposite the sepals. The embryology is so different that a separate order Tetracentrales has been suggested[iii]. Trees from the Himalayan part of the distribution area have been treated as taxonomically separable, as var. *himalense*[ii], but this was not recognized in *Flora of Bhutan* (1984). RKB

i Noshiro, S. Westernmost *Tetracentron sinense. News Lett. Himalayan Bot.* 5: 11–13 (1989).
ii Hara, H. *Flora of the Eastern Himalaya.* Pp. 85–86 (1966).
iii Pan, K. Y. *et al.* The embryology of *Tetracentron sinense* Oliver and its systematic signifiacance. *Cathaya* 5: 49–58 (1993).

TETRACHONDRACEAE

A family of low-growing herbs recently expanded from 1 genus to 2, now showing remarkable geographical disjunctions.

Distribution *Tetrachondra* has 1 sp. in New Zealand (North Island, South Island, and Stewart Island), where it grows in wet places, and 1 sp. in southern Argentina, also in damp places. *Polypremum* occurs in the eastern and southern USA, through Central America and the West Indies to Surinam, usually occurring in dry sandy habitats, and is introduced elsewhere in North America, in Paraguay, and on Hawaii and other islands in the Pacific[i].

Description Low-growing perennial herbs not exceeding 20 cm, prostrate and mat-forming with roots at the nodes (*Tetrachondra*) or stiffly ascending from a tap root (*Polypremum*). The leaves are opposite with a short interpetiolar ridge between the pairs, ovate or elliptical (*Tetrachondra*) to linear (*Polypremum*). The flowers are borne in leaf axils or short cymes and are small and inconspicuous, regular, and 4-merous. The sepals are 4, slightly fused at the base, rotate and ovate in *Tetrachondra*, stiffly erect and narrowly lanceolate with scarious margins and an acute apex in *Polypremum*. The corolla is tubular and rotate, with a ring of hairs in the throat in *Polypremum* only. The stamens are 4, with short filaments attached near the mouth of the corolla. The ovary in *Tetrachondra* is superior, 4-lobed with a gynobasic style, with a single erect ovule in each lobe, while in *Polypremum* it is semi-inferior, syncarpous with a terminal style, and with many ovules. The fruit breaks up into 4 single-seeded nutlets (*Tetrachondra*) or is a many-seeded dehiscent capsule (*Polypremum*).

Classification The affinities of *Tetrachondra* are uncertain. The 4-lobed ovary with gynobasic style has suggested that it should be placed in or near the Lamiaceae. Since 1924, it has been optionally placed in its own family, while *Polypremum* has had a varied history of placement in the Loganiaceae or Buddlejaceae[ii], or even Rubiaceae. Molecular data[iii] have placed the 2 genera close to each other, and they have recently formally been described as a single family[ii]. However, apart from being both low-growing herbs with opposite leaves, 4-merous

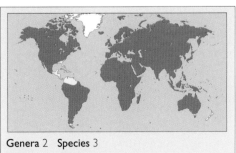

Genera 2 Species 3
Economic uses None

flowers, and a gamopetalous corolla, they have little in common to support such a close relationship. In particular, the ovary and fruit characters are different. The exact nature of the placentation of *Polypremum* appears somewhat different in published illustrations, but the ovary certainly has many ovules and develops into a dehiscent capsule, very unlike that of *Tetrachondra*. The concept of a single family based on the molecular evidence is accepted here only with considerable hesitation. Classification within *Tetrachondra* may be controversial[iv].

Economic uses None known; *Polypremum* is a widespread weed. RKB

i Rogers, R. K. The genera of Loganiaceae in the southeastern United States. *J. Arnold Arbor.* 67: 143–185 (1986).
ii Norman, E. N. Buddlejaceae. Pp. 66–67. In: Smith, N. *et al.* (eds), *Flowering Plants of the Neotropics.* Princeton, Princeton University Press (2004).
iii Oxelman, B., Backlund, M., & Bremer, B. Relationships of the Buddlejaceae *s.l.* investigated using parsimony jackknife and branch support analysis of chloroplast *ndh*F and *rbc*L sequence data. *Syst. Bot.* 24: 164–182 (1999).
iv Rossow, R. A. Tetrachondraceae. In: *Flora Patagonica* 6: 221–222 (1999).

TETRADICLIDACEAE

Genera 1 Species 2
Economic uses None

A family of fleshy annuals comprising a single genus with 2 spp.: *Tetradiclis tenella,* which occurs in saline pans and sandy places from south-central European Russia, Ukraine, and Egypt eastward to Pakistan, and *T. corniculata,* which has been described recently from the eastern end of Kazakhstan near the Chinese border. They are much-branched, decumbent-ascending annual herbs with stems to 15(20) cm. The leaves are alternate, fleshy, strap-shaped to spathulate, with usually 1 or 2 similarly shaped lateral lobes, up to c. 1.2 cm long. The inflorescence is a terminal one-sided cyme with small (1 mm) sessile flowers in the axils of small bracts. The flowers are actinomorphic, bisexual, and (3-) 4-merous. The sepals are connate at the base. The petals are free and shortly clawed. The stamens are attached to an intrastaminal nectary disk. The ovary is superior with 4 carpels basally fused to present a flat disk (*T. tenella*) or a cross-shape with 2-horned arms (*T. corniculata*), with the style arising from a central depression. The locules are 4, each with 6 seeds in a complicated arrangement. The fruit is at first fleshy and red but later dry and capsule-like, dehiscing its seeds in 2 stages. The unique structure of the ovary and dispersal of the seeds are fully described by Takhtajan (1997). The embryology is also said to be unique.

The family has been variously associated with Crassulaceae, Elatinaceae, Rutaceae, and particularly Zygophyllaceae. Molecular evidence places it close to Peganaceae and Nitrariaceae which, like Tetradiclidaceae, are separated widely from Zygophyllaceae and placed in the Sapindales. The lobed strap-shaped leaves of *Tetradiclis* show some similarity to those of *Peganum* and *Malacocarpus* of the Peganaceae. No economic uses are known. RKB

[i] Savolainen, V. *et al.* Phylogeny of the eudicots: a nearly complete familial analysis based on *rbc*L gene sequences. *Kew Bull.* 55: 257–309 (2000).

TETRAMERISTACEAE

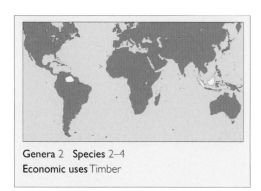

Genera 2 **Species** 2–4
Economic uses Timber

A family of 2 monotypic genera of large trees or shrubs with brown bark and alternate, spirally arranged leaves: *Pentamerista neotropica* in the Guyana highlands, and *Tetramerista glabra* from the Malay peninsula, Borneo, and Sumatra. *Tetramerista* is sometimes treated as having 3 spp. The leaves are leathery, simple, ovate, and entire, with brown indumentum below and without stipules. The inflorescence is axillary, pedunculate, with a crowded, umbel-like, condensed raceme of hermaphrodite pedicellate flowers at the apex. The flowers are 4-merous (*Tetramerista*) or 5-merous (*Pentamerista*) throughout, with distinct ovate sepals alternating with narrower petals bearing pits on the upper surface[i]. The stamens are erect, with anthers reflexed, surrounding the style. The ovary is superior, of 4–5 fused carpels, each forming a locule with a single apical style. The fruit is a berry containing 4 or 5 seeds.

The genera have usually been placed in the Theaceae (Dahlgren, 1983) or as a separate family in the Theales (Cronquist, 1981), but molecular evidence supports affinity with the Balsaminoid group of Ericales (APG II), with close relationship to Marcgraviaceae and Pellicieraceae[ii], which is sometimes included in this family[iii]. The wood of *Tetramerista glabra* is used in Malaysia and Indonesia for general construction work[iv]. AC

[i] Kubitzki, K. Pellicieraceae. Pp 297–299. In: Kubitzki, K. (ed.), *The Families and Genera of Vascular Plants. VI. Flowering Plants: Dicotyledons. Celastrales, Oxalidales, Rosales, Cornales, Ericales.* Berlin, Springer-Verlag (2004).
[ii] Anderberg, A. A., Rydin, C., & Källersjö, M. Phylogenetic relationships in the order Ericales *s.l.*: analyses of molecular data from five genes from the

plastid and mitochondrial genomes. *American J. Bot.* 89: 677–687 (2002).
[iii] Schönenberger, J., Anderberg, A. A., & Sytsma, K. J. Molecular phylogenetics and patterns of floral evolution in the Ericales. *Int. J. Plant Sci.* 166: 265–288 (2005).
[iv] Chudnoff, M. *Tropical Timbers of the World.* Agriculture Handbook #607. Madison, WI, Forest Products Laboratory, Forest Service, U.S. Dept. of Agriculture (1984).

THEACEAE
TEA, CAMELLIA, AND FRANKLINIA

A small family of trees and shrubs, usually with simple evergreen leaves, including the economically important *Camellia sinensis*, the leaves of which are brewed to make tea.

Distribution The family is mainly restricted to tropical and subtropical regions and centered chiefly in America and Asia. There are close links between North American and Southeast Asian species[i] dating from the Late Miocene.

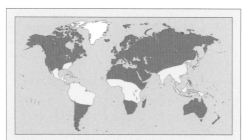

Genera 7 **Species** c. 200 (sometimes to 400)
Economic uses Tea and ornamentals

Description Usually evergreen trees or shrubs, with unicellular indumentum and alternate leaves arranged spirally or distichously. The leaves are simple and coriaceous with a toothed (rarely entire) margin and no stipules. The flowers are solitary, borne in the leaf axil, hermaphrodite, actinomorphic, large and showy, the outer bracts grading into the calyx of 5(6, rarely more) free, or basally connate, imbricate, usually thick, sepals that grade in to the 5 (rarely many) free or basally connate, imbricate petals. The stamens are 20 or more, free, rarely connate, frequently joined to the corolla base. The ovary is superior, of (3–)5(–10) fused carpels, each with 2 to few ovules usually on an axile placenta; the styles are simple or branched with a lobed stigma. The fruit is usually a loculicidal capsule, but occasionally a drupe or with irregular dehiscence.

Classification Theaceae is treated as a core element of the Ericales (APG II) based on DNA sequence analysis. The circumscription of the family has narrowed considerably since Cronquist (1981) included some 40 genera and 600 spp. In a recent treatment[ii], only 7 genera remain and 200–400 spp. Families previously included in the Theaceae are: Asteropeiaceae, now in Caryophyllales; Pellicieraceae, Tetrameristicaceae, Symplocaceae, Pentaphylacaceae, and Ternstroemiaceae (as tribe Ternstroemieae of Pentaphylacaceae) all now in

Ericales; Caryocaraceae, Bonnetiaceae, and Medusagynaceae now in Malpigiales; and Stachyuraceae in Crossosomatales. A study of chloroplast DNA resulted in the remaining Theaceae being divided into 3 tribes[iii]: Theeae includes the genera *Camellia*, *Pyrenaria*, and *Apterosperma*; Gordonieae includes the genera *Gordonia*, *Franklinia*, and *Schima*). Stewartieae includes the single genus *Stewartia*. Some of these larger genera have been subdivided; *Camellia* into as many as 9 genera: *Camellia*, *Camelliastrum*, *Dankia*, *Glyptocarpa*, *Parapiquetia*, *Piquetia*, *Stereocarpus*, *Theopsis*, and *Yunnanea*, but the characters used in these generic divisions vary from author to author and appear to form a continuum. A similar situation is found in *Pyrenaria*, in which three segregate genera are sometimes recognized. These include a broadly circumscribed *Pyrenaria*[iv] and *Camellia*[v]. A cladistic study of morphology of Theaceae *s.l.* indicated that Sladeniaceae and Ternstroemiaceae might be placed within Theaceae *s.s.*, but this was not supported by bootstrap analysis[vi]. In contrast, other authors state that there are no morphological synapomorphies for Ternstroemiaceae and Theaceae[vii].

Economic uses The genus *Camellia* accounts for almost all economic use of the family. *Camellia sinensis* yields tea from plantations primarily in India, Sri Lanka, China, and East Africa. Tea is drunk either black or green by about one half of the world's population. Tea seed oil is extracted from the seeds of *Camellia sasanqua*. *Camellia japonica*, *C. reticulata*, and *C. sasanqua* are the parents of the roughly 30,000 cultivars of the garden camellia valued for its large flowers and glossy evergreen leaves. *Franklinia alatamaha*, originally distributed over a small area near Fort Barrington in Georgia in the southern USA, but now extinct in the wild, is still cultivated in North America and Europe. AC

[i] Li, J. *et al.* Phylogenetic relationships and biogeography of *Stewartia* (Camellioideae, Theaceae) inferred from nuclear ribosomal DNA ITS sequences. *Rhodora* 104: 117–133 (2002).
[ii] Stevens, P. F., Dressler, S., & Weitzman, A. L. Theaceae. Pp 463–471. In: Kubitzki, K. (ed.), *The Families and Genera of Vascular Plants. VI. Flowering Plants: Dicotyledons. Celastrales, Oxalidales, Rosales, Cornales, Ericales.* Berlin, Springer-Verlag (2004).
[iii] Prince, L. M. & Parks, C. R. Phylogenetic relationships of Theaceae inferred from chloroplast DNA sequence data. *American J. Bot.* 88: 2309–2320 (2001).
[iv] Yang, S. X. A discussion on relationships among the genera in Theioideae (Theaceae). *Acta Sci. Nat. Univ. Sunyatseni* 29: 74–81 (1998).
[v] Sealy, J. R. *A revision of the genus* Camellia. London, Royal Horticultural Society (1958).
[vi] Luna, I. & Ochoterena, H. Phylogenetic relationships of the genera of Theaceae based on morphology. *Cladistics* 20: 223–270 (2004).
[vii] Anderberg, A. A., Rydin, C., & Källersjö, M. Phylogenetic relationships in the order Ericales *s.l.*: analyses of molecular data from five genes from the plastid and mitochondrial genomes. *American J. Bot.* 89: 677–687 (2002).

THEACEAE. 1 *Camellia rosiflora* flowering shoot (x⅔). 2 *C. japonica* "Kimberley" half flower with semi-inferior ovary and numerous stamens, united at their bases (x⅔). 3 *C. salicifolia* (a) gynoecium with 3-lobed stigma (x4); (b) cross section of 3-locular ovary (x6); (c) stamens with fused filaments (x3); (d) stamen (x4); (e) fruit—a capsule (x2).

THEOPHRASTACEAE

A distinctive family of pachycaul long-leaved or branching short-leaved shrubs and trees.

Distribution Tropical America, especially the West Indies and Central America. *Clavija* (55 spp.[i]) occurs from Nicaragua to Bolivia and Paraguay, especially in the Andes, with 1 isolated species on Hispaniola, and *Jacquinia* (32 spp.[ii]) from Mexico and Florida to Peru and northern Brazil, especially in the West Indies. By contrast, 4 genera are extremely restricted: *Theophrasta* (2 spp.) on Hispaniola, *Deherainia* (2 spp.) in southeastern Mexico and Honduras, *Neomezia* (1 sp.) in Cuba, and *Votschia* (1 sp.) in northeastern Panama. They usually grow in lowland seasonally dry deciduous or semideciduous forest or thorn scrub.

Genera 6
Species c. 93
Economic uses Fish poisons; soap; some edible fruits; cultivated ornamentals

Description Either little-branched pachycaul shrubs to small trees up to 3(–8) m (*Clavija*, *Theophrasta*, *Neomezia*) or branching shrubs and trees up to 8(–15) m (*Jacquinia*, *Deherainia*, *Votschia*). The leaves of the pachycaul genera are usually conspicuously long, 15–60 cm, narrowly oblanceolate, with entire to sharply toothed margins, while those of the branching genera are smaller, 2–10(–15) cm, lanceolate to elliptic, with entire margins, and usually a spinose tip. The leaves are alternate but are often aggregated into pseudowhorls at the branch tips or nodes. The leaves are also stiff and coriaceous and glandular-punctate. The inflorescences are terminal in the pachycaul genera and axillary in the others and are often rather stiff lax racemes, with rigid pedicels longer than the flowers, but more crowded and with short pedicels in *Clavija*, or sometimes umbellate in *Jacquinia*, or reduced to a single flower in *Deherainia*. The flowers are actinomorphic, bisexual, or in *Clavija* usually unisexual (plants polygamodioecious), 5-merous or sometimes 4-merous in *Clavija*. The calyx lobes are free, overlapping, and persistent. The corolla is broadly campanulate to urceolate with rounded lobes, waxy and very often orange, but sometimes white or yellow in *Jacquinia* or green in *Deherainia*. Staminodes are present, alternating with the corolla lobes, usually fused to each other and inserted on the corolla, sometimes appearing as a second ring of corolla lobes. The fertile stamens are opposite the corolla lobes, the filaments flattened and often connate. The upper and lower parts of the anthers are filled with a characteristic meal of calcium oxalate crystals. The ovary is superior, and syncarpous and 1-locular, with free central placentation and usually 40–250 ovules, but sometimes fewer in *Clavija*. The fruit is an indehiscent berry with a hard pericarp, usually subglobose but elongate to up to 6 cm in *Deherainia*, with 1–10 seeds embedded in a sweet orange placental pulp.

Classification The family falls into the complex around Primulaceae in the Ericales of APG II. The Samolaceae are closely allied in morphological and molecular characters. For discussion and other references, see Primulaceae. The family falls clearly into 2 tribes: Theophrasteae, including the 3 pachycaul genera (see description above); and Jacquinieae, including the other 3 genera.

Economic uses Roots of *Jacquinia* have been used as a fish poison (barbasco), and the bark and stem cortex used as a soap. Some *Clavija* species have edible fruits. Some species are widely cultivated in the tropics as ornamentals. RKB

i Ståhl, B. A revision of *Clavija* (Theophrastaceae). *Opera Botanica* 107: 1–78 (1991).
ii Ståhl, B. A synopsis of *Jacquinia* (Theophrastaceae) in the Antilles and South America. *Nordic J. Bot.* 15: 493–511 (1995).

THOMANDERSIACEAE

Thomandersia comprises shrubs and small trees of the Guineo-Congolan forests from Nigeria to Gabon and the Democratic Republic of Congo. The leaves are opposite, entire, mostly elliptic-acuminate, with short to long petioles. The inflorescence is a narrow, terminal or axillary, compact to usually elongate, spike up to 20 cm with flowers 0.8–2 cm long. The flowers are zygomorphic and bisexual. The calyx is shortly campanulate, 5-lobed, with 1 or 2 nectariferous glands at its base. The corolla is tubular, gibbous at the base, bilabiate, with the upper lip 2-lobed and the lower 3-lobed. There are 4 fertile stamens and 1 tiny staminode, all attached to the corolla tube. The ovary is superior, thick-walled, of 2 fused carpels, and the separate locules each have 1–3 ovules on axile placentas. The fruit is a somewhat woody globose to ovoid capsule up to 1.5 cm long, with up to 6 seeds that are each subtended by a broad, flat retinaculum up to 5 × 1.5 mm.

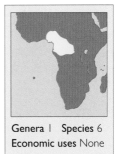

Genera 1 **Species** 6
Economic uses None

The genus has long been referred to Acanthaceae because of its hooklike retinacula subtending the seeds, but was—perhaps surprisingly at the time—placed in its own family in 1977[i]. However, recent work has shown that the retinacula of *Thomandersia* are not homologous with those of Acanthaceae, and molecular analysis has shown that it is remote from that family[ii]. The molecular data place it nearest to Schlegeliaceae, a family of large trees from tropical America, but it seems to have rather little in common with that within the context of the Lamiales. It appears to be an isolated genus which should be placed in its own family. No economic uses are known. RKB

[i] Sreemadhaven, C. C. Diagnoses of some new taxa and some new combinations in Bignoniales. *Phytologia* 37: 413–416 (1977).
[ii] Wortley, A. H., Scotland, R. W., & Rudall, P. J. Floral anatomy of *Thomandersia* (Lamiales) with particular reference to the nature of the retinaculum and extranuptial nectaries. *Bot. J. Linn. Soc.* 149: 469–482 (2005).

THYMELAEACEAE
DAPHNE FAMILY

A medium-sized cosmopolitan family of mainly shrubs and trees.

Distribution The family is cosmopolitan in distribution, occurring in both temperate and tropical regions, and is especially well represented in Africa and Australia. It shows more diversity in the southern hemisphere than in the northern hemisphere.

Description Trees, shrubs, or more rarely lianas or herbs, with long fibers in the bark and phloem giving the bark, stems, and leaves a certain toughness and flexibility. The leaves are simple, exstipulate, alternate or opposite, sometimes with the internodes closely spaced, making the alternate leaves appear opposite or whorled. The lamina is entire with parallel venation, prominent on the upper surface. The flowers are actinomorphic, bisexual or unisexual, sometimes polygamous (plants usually dioecious, sometimes monoecious), in racemes, capitula, or fascicles. The flowers have a hollowed out receptacle, forming a deep tube (hypanthium) from the rim of which the floral parts are borne. The sepals are (3)4–5(6), united or rarely free, commonly petaloid, greenish white, cream, red, pink, or brown. The petals are as many as the sepals or twice as many, reduced to scales or absent, free. The stamens are 3–5 or 8–10 or up to 100 (1–2 in *Pimelea*), attached in the hypanthium or on the rim, staminodial or absent in pistillate flowers. The ovary is superior, situated at the base of the receptacular cup, sometimes with a short gynophore, and comprises 2–5(–8–12) fused carpels, containing as many locules, each with 1 axial or parietal, pendulous ovule. The style is simple. The fruit is variable, usually a drupe, an achene, berry, or occasionally a capsule (*Aquilaria*); the seed has little or no endosperm, and the embryo is straight.

Classification The Thymelaeaceae has been variously divided into subfamilies, some of which are also recognized by some authors as separate families, and there is currently no agreement on what to include or exclude. Cronquist's treatment is followed here; it recognizes the subfamilies Thymelaeoideae, Aquilarioideae, Gonostyloideae, and Gilgiodaphnoideae (=Synandrodaphnoideae), and this is supported in a recent molecular phylogenetic analysis[i], which also considered the family

THYMELAEACEAE. **1** *Octolepis flamignii* (a) leafy shoot with flowers and flower buds (×⅔); (b) flower (×3); (c) half flower (×4); (d) hypanthial cup with stamens, and style and stigma (×4); (e) gynoecium (×5); (f) stamen (×5); (g) dehiscing fruit (×1½). **2** *Pimelea buxifolia* (a) leafy shoot with terminal inflorescences (×⅔); (b) flower (×3); (c) flower opened out (×3); (d) vertical section of ovary (×4). **3** *Daphne mezereum* (a) flowering shoot (×⅔); (b) leafy shoot with fruits (×⅔); (c) flower opened (×2); (d) fruit (×2). **4** *Gonystylus augescens* flower with 2 sepals removed (×4).

monophyletic. A small group of Australasian genera previously included in the Gonostyloideae together with the African genus *Octolepis* is separated out as subfamily Octolepidioideae by Wurdack and Horn[ii], who follow its circumscription as given by Herber[iii] based on pollen morphology. Stevens[iv] keeps *Gonostylus* and *Octolepis* together in the same subfamily, which he calls Octolepioideae, and following Horn[v] he also includes the northern South American *Tepuianthus* in the family although it is maintained here as a separate family, Tepuianthaceae (q.v.). Until a thorough morphological review is made of the whole family and more comprehensive molecular sampling undertaken, any treatment of the Thymelaeaceae must remain tentative, with the placement of many genera uncertain.

SUBFAM. THYMELAEOIDEAE Worldwide, but mainly in Africa and Australia, including *Gnidia* (140 spp.) from Africa to Madagascar and through India to Sri Lanka; *Pimelea* (110 spp.) in Australia and New Zealand; *Wikstroemia* (50 spp.) in Australasia, reaching South China; *Daphne* (50 spp.) in Australasia, extending across Asia to Europe and North Africa; and *Daphnopsis* (50 spp.) in the neotropics.

SUBFAM. AQUILARIOIDEAE Seven small genera from the Pacific area and Africa, including *Aquilaria* (15 spp.) and *Gyrinops* (8 spp.).

SUBFAM. GILGIODAPHNOIDEAE A monotypic genus (*Synandrodaphne*) that grows in western tropical Africa.

SUBFAM. GONYSTYLOIDEAE Includes the genera *Gonostylus* (20 spp.), *Octolepis* (6 spp.) *Microsemma* (= *Lethedon*, 10 spp.)

The Thymelaeaceae has usually been situated in the Myrtales, but recent molecular studies have placed it within an expanded Malvales clade[vi].

Economic uses Some tree species (*Gonostylus*) provide timber. Several species of *Daphne* are cultivated as ornamental shrubs, often with fragrant flowers. The bark of several genera,

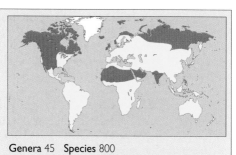

Genera 45 **Species** 800
Economic uses Ornamental shrubs (*Daphne*)

Genera 1 **Species** 1
Economic uses None

particularly *Wikstroemia,* yields fibers that are used locally in paper manufacture. The inner bark of the West Indian Lace-bark, *Lagetta lagetto,* is stretched to yield an ornamental textile in Jamaica. VHH

i van der Bank, M., Fay, M. F., & Chase, M. W. Molecular phylogenetics of Thymelaeaceae with particular reference to African and Australian genera. *Taxon* 51: 329-339 (2002).
ii Wurdack, K. J. & Horn, J. W. A reevaluation of the affinities of the Tepuianthaceae: molecular and morphological evidence for placement in the Malvales: 151. In: *Botany 2001: Plants and People,* Abstracts [Albuquerque] (2001).
iii Herber, B. E. Pollen morphology of the Thymelaeaceae in relation to its taxonomy. *Pl. Syst. Evol.* 232:107–121 (2002); Herber, B. E. Thymelaeaceae. Pp. 373–396. In: Kubitzki, K. (ed.), *The Families and Genera of Vascular Plants. V. Flowering Plants. Dicotyledons. Malvales, Capparales and Non-betalain Caryophyllales.* Berlin, Springer-Verlag (2002).
iv Stevens, P. F. Angiosperm Phylogeny Website. Version 6, May 2005 [and more or less continuously updated since]. http://www.mobot.org/MOBOT/research/APweb/
v Horn, J. W. The morphology and relationships of Sphaerosepalaceae (Malvales). *Bot. J. Linn. Soc.* 144: 1–40 (2004).
vi Alverson, W. S. *et al.* Circumscription of the Malvales and relationships to other Rosidae: Evidence from *rbcL* sequence data. *American J. Bot.* 85: 876–887 (1998).

TICODENDRACEAE

Ticodendron incognitum was probably first collected in 1925 in Costa Rica but was not described until 1989, after it had become well known there. The name is derived from the local diminutive "tico." It was referred to a new family 2 years later by the same authors[i]. It is now known to extend from southern Mexico (Oaxaca) to central Panama, where it occurs in wet montane forests from 500m to 2,400m and is thought to be a relict of the Tertiary Laurasian flora[ii]. It is an evergreen tree to 25 m or more, sometimes locally dominant[iii]. The leaves are alternate, simple, ovate to elliptic, with serrate margins in the upper part, with a short petiole. The male inflorescences are axillary, to 4 cm long, simple or branched, with dense clusters of sessile flowers at 1 to several nodes, occasionally with a terminal female flower. The female inflorescences are axillary with a solitary terminal flower. The flowers are actinomorphic, unisexual (plants dioecious or polygamodioecious), and lacking a perianth. The male flowers consist of 8–10 stamens subtended by 3 bracts. The female flowers consist of a single superior ovary of 2 carpels, terminated by 2 long pubescent styles, and with 2 locules per carpel, with 1 marginal ovule per locule. The fruit is a single seeded asymmetrical drupe up to 7 × 4 cm with a hard endocarp. The seed is large and oily.

The family seems to be clearly referable to the Fagales and is closest to Betulaceae. Interestingly, the vernacular names used in Costa Rica suggest affinity with *Alnus.* RKB

i Gómez-Laurito, J. & Gómez P. L.D. Ticodendraceae, a new family of flowering plants. *Ann. Missouri Bot. Gard.* 78: 87–88 (1991).
ii Hammel, B. & Burger, W. G. Neither oak nor alder, but nearly: the history of Ticodendraceae. *Ann. Missouri Bot. Gard.* 78: 89-95 (1991).
iii Meave, J., Gallardo, C., & Rincón, A. Plantas leñosas raras del bosque mesófilo de montaña, 2. *Ticodendron incognitum* Gómez-Laurito, J. & Gómez P. *Bol. Soc. Bot. Mexico* 59: 149–152 (1996).

TILIACEAE
LIME OR LINDEN FAMILY

A family of trees. Formerly much more broadly circumscribed to include species here treated as Sparrmanniaceae and Brownlowiaceae.

Distribution *Tilia* is circumboreal in warm-temperate conditions on fertile well-drained, lowland, soils. *Craigia* is confined to temperate southwestern China and northern Vietnam. *Mortoniodendron* is Central American, extending from Mexico to northern Colombia.

Description Trees with fibrous bark and leaves alternate, usually distichous, simple, often slightly cordate, triplinerved, serrate, petioles pulvinate, stipules caducous and inconspicuous. The hairs are stellate, often mixed with simple and/or tufted hairs. The inflorescences are thyrsoid, axillary, those of *Tilia* with a bractlike modified leaf toward the base. The flowers are (4-)5-merous, hypogynous, and actinomorphic. The sepals are valvate, free, and nearly equal, secreting nectar in *Tilia* (uninvestigated in other genera). The petals are free (absent in some *Mortoniodendron*), white or yellow, slightly clawed, and lacking nectaries. The stamens are grouped in bundles opposite the petals (but free in some *Mortoniodendron* and appearing so in some *Tilia*), each bundle consisting of 4–12 extrorse stamens, the innermost of which is a petaloid staminode in *Craigia* and some *Tilia*. The 2 anther thecae dehisce longitudinally producing oblate, tricolporate, finely reticulate pollen. The ovary is 4- to 5-angled, stellate hairy, of 4–5 carpels, with an apically 5-branched style. Placentation is axile, the ovules 2–18 per locule, in 2 rows, ascending (descending in *Mortoniodendron*). The rugulose capsular fruits are 1- to 5-seeded, the sutures raised to form longitudinal ridges (extended into wings in *Craigia*). The ellipsoid seeds have foliose cotyledons and abundant endosperm.

Classification Tiliaceae belong to the Malvales *sensu stricto,* also referred to as core Malvales and formerly including Tiliaceae, Malvaceae, Bombaceae, and Sterculiaceae. Molecular evidence has shown indisputably that some parts of traditional Tiliaceae and Sterculiaceae are more closely related to each other than to the other parts of their respective families. A change in classification has been unavoidable. Within Malvales there are 10 groups that are morphologically distinct and, on molecular evidence, monophyletic in all cases but 1. One approach is to lump these all into a "Super Malvaceae," recognizing them as subfamilies (FGVP[i]). Here they are treated as separate families, and traditional Tiliaceae is divided into: Sparrmanniaceae (about 90 percent of the traditional Tiliaceae) most closely related to Byttneriaceae of former Sterculiaceae; Brownlowiaceae (q.v.); and Tiliaceae *sensu stricto.*

Tiliaceae *sensu stricto* is distinguished from Sparrmanniaceae by having staminal bundles opposite the petals (not the sepals); 1- to 5-seeded, rugulose fruits, which when capsular have raised rims to the loculicidal valves (mostly fleshy or many-seeded fruits in Sparrmanniaceae); oblate, not prolate pollen; and distichous not spiral phyllotaxy.

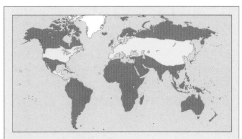

Genera 3 **Species** c. 45
Economic uses Ornamental trees in temperate areas; wood; bark fiber; honey sources

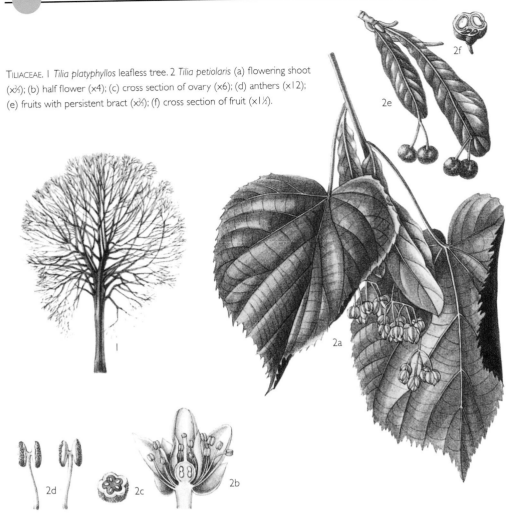

TILIACEAE. I *Tilia platyphyllos* leafless tree. 2 *Tilia petiolaris* (a) flowering shoot (x⅔); (b) half flower (x4); (c) cross section of ovary (x6); (d) anthers (x12); (e) fruits with persistent bract (x⅔); (f) cross section of fruit (x1½).

Brownlowiaceae is probably closely related to Tiliaceae, sharing for example, the same pollen type, but differs in having an incompletely divided calyx (divided to the base into 5 sepals in Tiliaceae *sensu stricto*) and stamens with the 2 thecae divergent and touching each other only at the apex.

Tiliaceae *sensu stricto* consists of 3 distinct genera. *Mortoniodendron* is the most morphologically distinct of the 3. Formerly placed on molecular evidence in Brownlowiaceae[ii], it has since been reallocated to Tiliaceae on more recent evidence (Baum, pers. comm. 2004). No infrafamilial classification is available. Both *Craigia* and *Mortoniodendron* are poorly studied and have not been revised.

Economic uses The main use of *Tilia* today is probably as a street tree, although declining due to the antipathy of car owners to the unsightly black mold that grows on the sugary secretion deposited on cars parked below aphid-inhabited *Tilia* trees. *Tilia* is an important source of honey in areas such as China and was formerly an important wood for ornamental carving (e.g., Grinling Gibbons) and for e.g., shields, platters, toys, piano keys, house and furniture frames, and canoes. The bark (bass) fiber was an important source of nets, brushes, footwear, and rope, and trees were once planted throughout Europe for avenues or, earlier still, as symbols to mark events[iii]. No information is available on the economic importance of *Craigia* and *Mortoniodendron*.   MRC

[i] Bayer, C. & Kubitzki, K. Malvaceae. P. 225. In: Kubitzki, K. & Bayer, C. (eds), *The Families and Genera of Vascular Plants. V. Flowering Plants. Dicotyledons: Malvales, Capparales and Non-betalain Caryophyllales.* Berlin, Springer-Verlag (2002).
[ii] Bayer, C. *et al.* Support for an expanded family concept of Malvaceae within a recircumscribed order Malvales: A combined analysis of plastid *atp*B and *rbc*L DNA sequences. *Bot. J. Linn. Soc.* 129: 267–303 (1999).
[iii] Piggot, C. D. *Limes.* UK, Sage Press (2005).

TORRICELLIACEAE

The Torricelliaceae is a small family of shrubs to small trees, with alternate simple, entire, dentate, or dissected leaves and small flowers in terminal or subterminal panicles or racemes. The family comprises two genera: *Torricellia* (endemic to the eastern Himalayas and western China, 3 spp.) and *Aralidium* (endemic to western Malaysia, 1 sp.). The flowers are largely unisexual (plants either monoecious in *Torricellia* or dioecious in *Aralidium*), with 5 sepals or 3–5 (in female flowers of *Torricellia*); the stamens are 5; the ovary is inferior, of 3–4 joined carpels, 2-, 3-, or 4-locular. The fruit is drupaceous.

Both of the genera recognized in this family have previously been placed in the Cornaceae,

Genera 2 Species 4
Economic uses None

along with *Melanophylla*, and each has also at some stage been placed in an eponymous family. Recent molecular sequence data suggest that both genera cluster together and have been placed in the expanded Torricelliaceae[i,ii]. Some authors place *Melanophylla* within this cluster, but here the genus has been placed in Melanophyllaceae (q.v.).   VHH

[i] Chandler, G. T. & Plunkett, G. M. Evolution in Apiales: nuclear and chloroplast markers together in (almost) perfect harmony. *Bot. J. Linn. Soc.* 144: 123–147 (2004).
[ii] Plunkett, G. M. *et al.* Recent advances in understanding Apiales and a revised classification. *South African J. Bot.* 70: 371–381 (2004).

TOVARIACEAE

The Tovariaceae comprises the single genus *Tovaria*. The 2 spp. are native to montane tropical America (*T. pendula* in Bolivia, Mexico, and Veracruz) and the Caribbean (*T. diffusa* in Jamaica), colonizing openings caused by land disturbance[i] and in dense wet thickets. Both

Genera 1 Species 2
Economic uses None

species are green-barked herbs or shrubs that sometimes behave like annuals. The plants are pungent-smelling, with alternate, 3-foliate leaves and without (or with minute[ii]) stipules. The flowers are regular, hermaphrodite, hypogynous in lax terminal racemes, and consist of 8 narrow sepals with overlapping edges, 8 petals, and 8 free stamens with hairy filaments dilated near the base. The ovary is superior and has 6–8 fused carpels, with 6–8 locules containing numerous ovules borne on axile placentas; the locules are formed by membranous dividing walls. The style is short with a lobed stigma. The fruits are berries, mucilaginous when young, but with a membranous outer coat, and about 1 cm in diameter. The many small, shiny seeds have curved embryos and a sparse endosperm. In *Tovaria pendula*, the flowers and fruits are greenish, the anthers brown or yellow. *T. diffusa* has pale green or yellow flowers and long, sparsely flowered racemes.

Tovariaceae has been consistently placed with Capparales/Brassicales[iii] in the major systems of classification. DNA sequence data indicate a close relationship to Pentadiplandraceae[iv], to Emblingiaceae and Resedaceae[v].   AC

[i] Nee, M., Tovariaceae. Pp. 376–377. In: Smith, N. *et al.* (eds), *Flowering Plants of the Neotropics.* Princeton, Princeton University Press (2004).
[ii] Appel, O. & Bayer, C. Tovariaceae. Pp. 397–399. In: Kubitzki, K. & Bayer, C. (eds), *The Families and Genera of Vascular Plants. V. Flowering Plants. Dicotyledons: Malvales, Capparales and Non-betalain Caryophyllales.* Berlin, Springer-Verlag (2002).
[iii] Brummitt, R. K. *Vascular Plant Families and Genera.* Richmond, Royal Botanic Gardens, Kew (1992).
[iv] Rodman, J. E. *et al.* Parallel evolution of glucosinolate biosynthesis inferred from congruent nuclear and plastid gene phylogenies. *American J. Bot.* 85: 997–1007 (1998).

v Chandler, G. T., & Bayer, R. J. Phylogenetic placement of the enigmatic Western Australian genus *Emblingia* based on *rbc*L sequences. *Plant Species Biol.* 15: 67–72 (2000).

TRAPELLACEAE

A family of 2 spp. of floating aquatics from eastern Asia, with a similarity in habit and leaf shape to *Trapa* but apparently referable to the complex around Scrophulariaceae.

Distribution Central and eastern China to Korea, Japan, and Khabarovsk in the Russian Far East, where it is found in shallow still water.

Genera 1   Species 2
Economic uses None

Description Aquatic herbs with a rhizome rooting underwater and stems reaching to the water surface, with dimorphic leaves and flowers. The leaves are in opposite pairs, the submerged ones narrowly elliptical, attenuate to a narrow base, and with serrate margins, those floating on the surface of the water broadly deltoid, shortly cuneate to a distinct petiole and with crenate margins. The flowers are solitary in leaf axils, those below water small (to 4 mm), cleistogamous and with reduced corolla and stamens[i], those at water level or above[ii] opening normally and up to 2 cm long. In normal flowers, the calyx is attached from just above half way up the ovary to more or less at the top of the ovary and is shortly tubular with 5 teeth, with rudiments of the fruiting appendages inserted immediately below the calyx and alternating with the teeth. The corolla is trumpet-shaped and slightly irregular, with 5 rounded lobes. There are 2 fertile stamens and 2 staminodes. The ovary is half to fully inferior, of 2 carpels, of which the abaxial one is abortive and sterile, the adaxial locule bearing 2 pendulous ovules at the top, of which 1 is sessile and the other funiculate. The fruit is single-seeded, accrescent up to 2 cm, oblong (*T. sinensis*) to obconical (*T. antennifera*), and hard-walled, developing 5 conspicuous filiform appendages from near the point of attachment of the sepals, each usually hooked at the apex.

Classification The trumpet-shaped corolla and bicarpellary ovary immediately suggest close affinity with Scrophulariaceae, and the hooked fruiting appendages are reminiscent of Pedaliaceae. It was originally included tentatively within Pedaliaceae, although several significant differences were noted. More recent authors either place it in that family or recognize the family Trapellaceae[iii]. Anatomical evidence suggests that the fruiting appendages may not be homologous with the appendages found in Pedaliaceae (q.v.).

Although it is often regarded as monospecific, 2 spp. may be recognized[i]: *T. antennifera*, with a somewhat obconic fruit, with irregular raised warty ridges, and 5 equal long appendages, found in central Japan (Honshu) and the Lake Hanka area of Primorsk in Russia; and *T. sinensis*, with an oblong, smooth fruit and unequal fruiting appendages occuring from central China to southern Japan (Kyushu) and north to Khabarovsk in Russia.   RKB

i Glück, H. Die Gattung *Trapella*. *Bot. Jahrb.* 71: 267–336 (1940).
ii Satake, T. *et al.* (eds), *Wild Flowers of Japan: Sympetalae* 3: t. 108, 5 (1981).
iii Ihlenfeldt, H.-D. Trapellaceae. Pp. 445–448. In: Kadereit, J. W. (ed.), *Families and Genera of Vascular Plants. VII.* Berlin, Springer-Verlag (2004). See also Ihlenfeldt *loc. cit.* 315 (2004).

TRIGONIACEAE

A small family of evergreen tropical trees or lianas with papilionoid flowers.

Distribution The largest genus *Trigonia* (c. 27 spp.) occurs in Central and South America. *Trigoniodendrum spiritusanctense* is endemic to southeastern Brazil, and the genus *Isidodendron*, with a single species, *I. tripterocarpum*, was recently described from Colombia[i]. In the eastern hemisphere, *Humbertiodendrum* (1 sp.) is endemic to Madagascar, and *Trigoniastrum* (1 sp.) is confined to Southeast Asia.

Description Evergreen trees or lianas. The leaves are usually opposite, sometimes alternate (*Trigoniodendrum*), simple, stipulate, densely covered with whitish hairs beneath. The flowers are zygomorphic, somewhat papilionoid, herma-phrodite, in terminal or axillary racemes or panicles. The sepals are 5, usually connate at the base. The petals are 5, unequal, the dorsal petal largest and forming a saccate standard, the laterals spathulate, the anterior petals sometimes forming a saccate keel. The stamens are variable in number, 5–7(–12), some of them as staminodes, with the filaments connate at the base. The ovary is superior, of 3 carpels, syncarpous, with 3 locules or rarely 1. The style is terminal, capitate. The fruit is a septicidal capsule or 3-winged samara (*Humbertiodendrum* and *Trigoniastrum*). The seeds are often hairy, without endosperm.

Classification The relationships of the Trigoniaceae remain somewhat uncertain. It has been placed in the Polygales and later the Vochysiales by Cronquist. Molecular evidence indicates a placement in the Rosidae[ii], together with the Dichopetalaceae in the order Malpighiales. Later molecular studies suggest a close relationship with the Chrysobalanaceae and Dichopetalaceae[iv], but Prance[iii] notes that all these are separated by a series of morphological, anatomical, and chemical differences that would favor maintaining them as separate families. The Euphroniaceae, which is sister to the

Genera 5   Species c. 31
Economic uses Wood used locally as timber

Chrysobalanaceae in the same clade as the Dichapetalaceae and Trigoniaceae according to a further molecular study[v], is sometimes included in the Trigoniaceae (q.v.) but differs from it in several features, and on balance is better retained as a separate family.

Economic uses The family has little economic interest apart from local use of the wood of some species for timber.   VHH

i Fernández-Alonso, J. L., Perez-Zabala, J. A., & Idárraga-Piedrahita, A. *Isidodendron*, un nuevo género neotropical de árboles de la familia Trigoniaceae. *Revista Acad. Colomb. Ci. Exact.* 24(92): 347–357 (2000).
ii Savolainen, V. *et al.* Phylogeny of the eudicots: a nearly complete familial analysis based on *rbc*L gene sequences. *Kew Bull.* 55: 257–309 (2000).
iii Prance, G. T. & Sothers, C. A. *Species plantarum. Flora of the World.* Part 9. Chrysobalanaceae 1: *Chrysobalanus* to *Parinari*. Part 10. Chrysobalanaceae 1: *Acioa* to *Magnistipula*. Canberra, Australian Biological Resources, (2003).
iv Chase, M. W. *et al.* When in doubt, put it in Flacourtiaceae: a molecular phylogenetic analysis based on plastid *rbc*L DNA sequences. *Kew Bull.* 57: 141–181 (2002).
v Litt, A. J. & Chase, M. W. The systematic position of *Euphronia*, with comments on the position of *Balanops*: An analysis based on *rbc*L sequence data. *Syst. Bot.* 23: 401–409 (1999).

TRIMENIACEAE

A small tropical family of aromatic trees, shrubs, or lianas, growing from central Malaysia and eastern Australia to the south-western Pacific (Fiji, Samoa, Marquesas). The leaves are opposite, petiolate, exstipulate, simple, ovate to lanceolate, entire or serrate, dotted with pellucid glands. The flowers are

Genera 1–2
Species 8
Economic uses None

unisexual (plants andromonoecious) or bisexual, small, actinomorphic, deciduous, with a small slightly convex receptacle, in axillary or terminal cymose, racemose, or paniculate inflorescences. The perianth is composed of 2–40 or more spirally arranged free sepaline tepals, the outer ones merging into the bracteoles. The stamens are 7–25, spirally arranged, free. The ovary is superior, consisting of a solitary carpel, rarely 2, with a single locule. The fruit is a small, fleshy berry.

There is a single genus, *Trimenia*, with 8[i] spp., although a second genus, *Piptocalyx*, is sometimes recognized. The Trimeniaceae has been previously included in the Monimiaceae but are now generally recognized on morphological and molecular grounds to be allied the Illiciaceae, Schisandraceae, and Austrobaileyaceae, with which it forms a clade[i]. No economic uses are reported.   VHH

[i] Philipson, W. R. Trimeniaceae. Pp. 596–599. In: Kubitzki, K., Rohwer, J. G., & Bittrich, V. (eds), *The Families and Genera of Vascular Plants. II. Flowering Plants: Dicotyledons, Magnoliid, Hamamelid and Caryophyllid Families*. Berlin, Springer-Verlag (1993).
[ii] Qiu, Y.-L. *et al.* The earliest angiosperms: Evidence from mitochondrial, plastid and nuclear genes. *Nature* 402: 404–407 (1999).

TRIPLOSTEGIACEAE

Rhizomatous herbs intermediate between Valerianaceae and Dipsacaceae, distinguished from them by their double epicalyx.

Distribution *Triplostegia glandulifera* occurs from western Himalayas to central and southwestern China, Taiwan, Sulawesi, and New Guinea. *T. grandiflora* occurs in the eastern Himalayas and southwestern China. Both grow in high-altitude grasslands, usually between 2,500 and 4,000 m.

Description Slender herbs with erect or ascending stems 20–50 cm high, with long and slender rhizomes. The leaves are opposite and often crowded at the base of the stem, sometimes more or less rosette-forming. The lower leaves are usually petiolate and lyrate to deeply pinnatisect, with serrate margins on the lobes, but upper leaves are usually smaller and sessile and unlobed but with serrate margins. The upper stem and inflorescence are densely clothed in gland-tipped hairs. The upper part of the plant forms a lax trichotomous cyme with few-flowered clusters of flowers on the terminal branches. The ovary is surrounded by an outer epicalyx of 4 bracts, which are fused at their base, covered with gland-tipped hairs, and each with a small, terminal, incurved hook, and inside that an urceolate 8-ribbed inner epicalyx (involucel) with a terminal rim with 8 teeth. The calyx is a short 5-toothed rim. The corolla is 3–8 mm long, funnel-shaped, and more or less regular. The stamens are 4, small, at the mouth of the corolla. The ovary is inferior, with a single locule and 1 apical pendulous ovule. The fruit is a dry cypsela with a single seed, which falls enclosed by the double epicalyx and is probably dispersed by means of the hooks on the tips of the outer epicalyx lobes.

Genera 1 | Species 2
Economic uses None

Classification *Triplostegia* has the habit of Valerianaceae but an involucel like the Dipsacaceae, and has been included variously in either family. A position as a subfamily of the former has been suggested on palynological evidence[i], or of the Dipsacaceae on molecular evidence[ii]. In lacking a capitate inflorescence it differs markedly from Dipsacaceae, but in the presence of an involucel it differs also from Valerianaceae. In its double epicalyx and densely glandular inflorescence it differs from both of these, and in recent years has been increasingly placed in its own family[iii]. Analyses of morphological characters[iv] and molecular data[v] consistently place *Triplostegia* outside both Valerianaceae and Dipsacaceae and in a position between them both[vi].   RKB

[i] Backlund, A. & Nilsson, S. Pollen morphology and the systematic position of *Triplostegia* (Dipsacales). *Taxon* 46: 21–31 (1997).
[ii] Zhang, H. W., Chen, Z. D., Chen, H. B., & Tang, Y. C. Phylogenetic relationships of the disputed genus *Triplostegia* based on *trn*L-*trn*F sequences. *Acta Phytotax. Sin.* 39: 337–344 (2001).
[iii] Hofmann, U. & Göttmann, J. *Morina* L. und *Triplostegia* Wall. ex DC. im Vergleich mit Valerianceae und Dipsacaceae. *Bot. Jahrb. Syst.* 111: 499–553 (1990).
[iv] Caputo, P. & Cozzolino, S. A cladistic analysis of Dipsacaceae (Dipsacales). *Pl. Syst. Evol.* 189: 41–61 (1994).
[v] Bell, C. D. Preliminary phylogeny of Valerianaceae (Dipsacales) inferred from nuclear and chloroplast DNA sequence data. *Molec. Phylogen. Evol.* 31: 340–350 (2004).
[vi] Pyck N. & Smets, E. On the systematic position of *Triplostegia* (Dipsacales): a combined molecular and morphological approach. *Belg. J. Bot.* 137: 125–139 (2004).

TROCHODENDRACEAE

Trochodendron aralioides is an evergreen tree up to 25 m in height and 5 m diameter, found in southern Korea, Japan (not in Hokkaido), Nansei-shoto (Ryukyu Islands), and Taiwan. It is usually in evergreen forests, and in Taiwan it may grow in pure stands at altitudes of 2,000–3,000 m. The leaves are alternate, but usually densely clustered or almost whorled at the ends of branches, long petiolate, elliptical to obovate, cuneate to rounded at the base, crenate to minutely serrate in the distal part. The inflorescence is an erect terminal raceme with long ascending pedicels. The flowers are actinomorphic, bisexual or unisexual (plants androdioecious), lacking sepals and petals, up to 1 cm diameter. The stamens are usually 40–70, spirally inserted on the sides of a swollen disk or receptacle similar to pins on a pincushion. The ovary is superior or slightly sunk into the receptacle, consisting of a whorl of 4–11(–17) carpels that are connate in their lower parts and

Genera 1 | Species 1
Economic uses Timber; adhesive resin; cultivated ornamental

free distally, the styles at first forming a cone in the middle of the flower but soon reflexing. Each carpel has 20–30 ovules that are pendulous from the upper margin of the carpel. The fruit is a ring of connate follicles.

The family is closely related to the Tetracentraceae, which is also Asian and monotypic and has been included by some authors in Trochodendraceae. For characters separating them, see Tetracentraceae. It has been sometimes associated with Magnoliales, or more recently placed in the Hamamelididae, but molecular evidence[i] places it near Didymelaceae and Buxaceae. In APG II, it is unplaced to order in the eudicots. Several fossil genera have been referred to Trochodendraceae, suggesting a former much wider distribution of the family, but their taxonomic placement has been queried in the light of seed strucure[ii].   RKB

[i] Savolainen, V. *et al.* Phylogeny of the eudicots: a nearly complete familial analysis based on *rbc*L gene sequeunces. *Kew Bull.* 257–309 (2000).
[ii] Doweld, A. B. Carpology, seed anatomy and taxonomic relations of *Tetracentron* (Tetracentraceae) and *Trochodendron* (Trochodendraceae). *Ann. Bot.* 82: 413–443 (1998).

TROPAEOLACEAE
NASTURTIUM, CANARY CREEPER, MASHUA

Climbing, fleshy herbs, including cultivated ornamentals and an important food crop.

Distribution The family is native mainly to the mountains from Mexico to southern Chile and across to eastern Argentina[i].

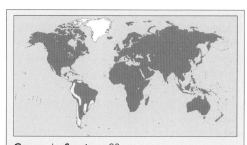

Genera 1 | Species c. 90
Economic uses Ornamentals; food crops; flavorings

Description Twining fleshy herbs with an acrid mustard oil present in the sap, sometimes with root tubers. The leaves are alternate, peltate, sometimes deeply lobed, the petioles sometimes twining around supports and without stipules. The showy flowers are bisexual, zygomorphic and spurred and usually borne singly in the axils of leaves. The sepals are 5, free, 1 modified to form a long nectar spur. The petals are 5, free, usually clawed, the upper 2 smaller than the lower 3. The stamens are 8, free. The ovary is superior, of 3 fused carpels, with 3 locules each containing 1 axile pendulous ovule; the single apical style has 3 stigmas. The fruit is a 3-seeded schizocarp, each mericarp separating to become an indehiscent "seed" lacking endosperm. The embryo is straight and has thick fleshy cotyledons.

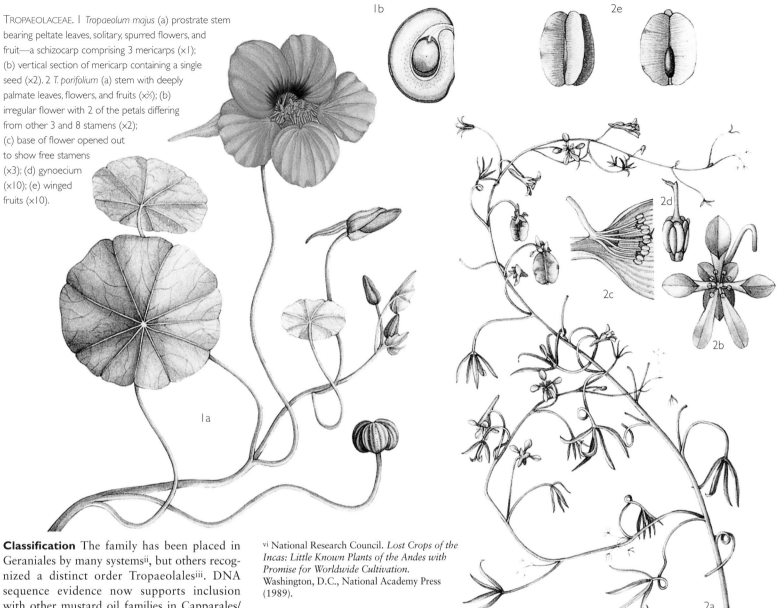

TROPAEOLACEAE. 1 *Tropaeolum majus* (a) prostrate stem bearing peltate leaves, solitary, spurred flowers, and fruit—a schizocarp comprising 3 mericarps (×1); (b) vertical section of mericarp containing a single seed (×2). 2 *T. porifolium* (a) stem with deeply palmate leaves, flowers, and fruits (×⅔); (b) irregular flower with 2 of the petals differing from other 3 and 8 stamens (×2); (c) base of flower opened out to show free stamens (×3); (d) gynoecium (×10); (e) winged fruits (×10).

Classification The family has been placed in Geraniales by many systems[ii], but others recognized a distinct order Tropaeolales[iii]. DNA sequence evidence now supports inclusion with other mustard oil families in Capparales/ Brassicales[iv] grouped with Akaniaceae and Bretschneideraceae. Phylogenetic study of the species using chloroplast DNA sequence[v] data gave evidence for sinking segregate genera *Magallana* and *Tropheastrum* into *Tropaeolum*, leaving the family with its single genus.

Economic uses About 8 spp. are cultivated for ornament; most commonly encountered are *Tropaeolum majus,* the Garden Nasturtium, and *T. peregrinum* (*T. canariense*), Canary Creeper or Canary-bird Flower. The unripe seeds of *T. majus* are occasionally pickled and used like capers. The tubers of *T. tuberosum* are the fourth most important indigenous root crop in the Andean region[vi].   [SLJ] AC

[i] Sparre, B. & Andersson, L. A taxonomic revsion of the Tropaeolaceae. *Opera Bot.* 108: 1–139 (1991).
[ii] Brummitt, R. K. *Vascular Plant Families and Genera.* Richmond, Royal Botanic Gardens, Kew (1992).
[iii] Dahlgren, R. M. T. General aspects of angiosperm evolution and macrosystematics. *Nordic J. Bot.* 3: 119–149 (1983).
[iv] Rodman, J. E. *et al.* Parallel evolution of glucosinolate biosynthesis inferred from congruent nuclear and plastid gene phylogenies. *American J. Bot.* 85: 997–1007 (1998).
[v] Andersson, L. & Andersson, S. A molecular phylogeny of Tropaeolaceae and its systematic implications. *Taxon* 49: 721–736 (2000).

[vi] National Research Council. *Lost Crops of the Incas: Little Known Plants of the Andes with Promise for Worldwide Cultivation.* Washington, D.C., National Academy Press (1989).

TURNERACEAE

Mainly tropical and subtropical shrubs and herbs, with often showy yellow flowers.

Distribution The family is found in the tropics and subtropics of the New World, tropical and southern Africa, Madagascar, and the Mascarene Islands. The largest genera are *Turnera* (120 spp.) and *Piriqueta* (41 spp.), both of which are widespread in tropical America and with 2 and 1 sp. in Africa, respectively. The monospecific *Adenoa* is endemic to Cuba.

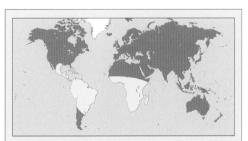

Genera 8 Species c. 190
Economic uses Local medicinal use; some ornamental value

Description Herbs, shrubs, or rarely trees. The leaves are alternate, simple, often with trichomes or glandular hairs of various types, and often extrafloral nectaries at the base of the lamina; the stipules are usually small or absent, developed in *Erblichia* and some *Turnera*; the leaf margins are entire, dentate, crenate, or rarely pinnatifid. The flowers are actinomorphic, hermaphrodite, often showy, homostylous or heterostylous, usually solitary in the leaf axils or in cymose or racemose axillary or terminal inflorescences. The heterostylous species tend to have long- and short-styled flowers, although some have long, short, and homostylous flowers. The sepals are 5, connate into a tube, with imbricate lobes (*Adenoa, Piriqueta, Turnera*), or sepals more or less free (*Erblichia*). The petals are 5, distinct, clawed, adnate below to the calyx, and forming a tube. The stamens are 5, inserted at the base of the floral tube, or the filaments adnate to the tube up to the throat, and with 5 nectariferous pockets[i]. The ovary is superior to semi-inferior, of 3 fused carpels, 1-locular, the placentation parietal, with 3 to many ovules; styles 3, distinct, the stigmas

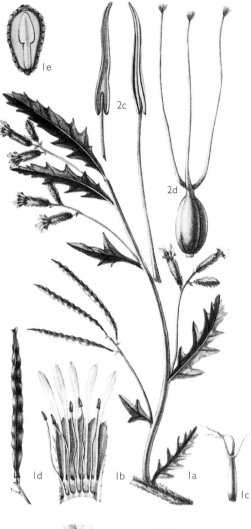

usually brushlike. The fruit is a loculicidal capsule with 1 to many arillate, often finely reticulate seeds.

Classification The Turneraceae is usually considered to be close to the Passifloraceae and Malesherbiaceae in the Violales, and the limited molecular data available support this position[ii], especially the association with the former family, although all 3 families are now placed in the Malpighiales.

Economic uses Some species of *Turnera*, e.g., *T. diffusa* and *T. aphrodisiaca* (*Damiana*) are used locally as medicines and as an aphrodisiac and are available commericially. Some species are grown as ornamentals, such as *Erblichia odorata* and several *Turnera* species. *Turnera subulata*, *T. ulmifolia,* and others are weedy or invasive in some parts of the neotropics.   VHH

TURNERACEAE. I *Tricliceras heterophylla* (a) shoot with alternate leaves, flowers in racemes, and elongate fruits (x⅔); (b) perianth opened out (x2); (c) gynoecium (x2); (d) fruit—a capsule (x1); (e) vertical section of seed (x4). 2 *Turnera berneriana* (a) shoot with leaves and solitary flowers (x⅔); (b) half flower (x3); (c) 2-locular anthers (x8); (d) gynoecium showing fringed stigmas (x8); (e) vertical section of ovary (x15); (f) cross section ovary with ovules on 3 parietal placentas (x15). 3 *Turnera angustifolia* (a) flowering shoot (x⅔); (b) perianth opened out (x⅔); (c) gynoecium (x1); (d) fruit dehiscing by 3 valves (x1½).

i Arbo, M. M. Turneraceae. Pp. 380–382. In: Smith, N. *et al.* (eds), *Flowering Plants of the Neotropics.* Princeton, Princeton University Press (2004).
ii Chase, M. W. *et al.* When in doubt, put it in Flacourtiaceae: a molecular phylogenetic analysis based on plastid *rbc*L DNA sequences. *Kew Bull.* 57: 141–181 (2002).

ULMACEAE
ELM FAMILY

Trees with small apetalous flowers and winged or drupaceous fruits, related to Urticaceae.

Distribution Widespread in tropical and temperate regions, from tropical forests to temperate woodland, but absent from some cool-temperate countries.

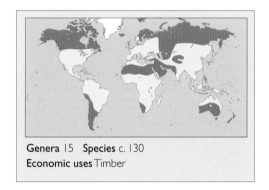

Genera 15   **Species** c. 130
Economic uses Timber

Description Trees up to 40(–60) m high or occasionally shrubs or rarely lianes, armed with axillary spines in *Chaetacme* and some *Celtis*. The leaves are alternate or rarely (*Lozanella*) opposite, simple and unlobed, often asymmetrical, either with 3 main veins from the base or with pinnate venation, the margins crenate, toothed, or rarely entire. Stipules are inconspicuous and caducous. Dotlike cystoliths are sometimes present. The inflorescence is a terminal or axillary cyme, often aggregaed into loose clusters, or the female flowers sometimes solitary. The flowers are unisexual (plants monoecious) or sometimes some are bisexual, actinomorphic, with a small and usually greenish, 4- to 5-merous, uniseriate perianth. The tepals are 4–5, free or shortly united. The stamens are as many or sometimes twice as many as the tepals, not inflexed in bud. The ovary is superior, of 2 united carpels, with 2 divergent styles, with 1(2) locules and a single ovule pendent from the apex. The fruit is a samara or fleshy drupe.

Classification The family is clearly referable to the group long known as Urticales, now included by APG II in Rosales—see Urticaceae for other references. Two subfamilies are recognized[i].
SUBFAM. ULMOIDEAE Includes 6 genera, c. 35 spp., mostly northern temperate. Leaves pinnately veined; fruits flattened and winged except in *Planera* and *Zelkova*, where it is a leathery drupe. *Ulmus* has up to 25 spp. in conservative taxonomy but establishes widespread clones that are treated as many species by some authors. Other genera are *Hemiptelea*, *Holoptelea,* and *Phyllostylon*.

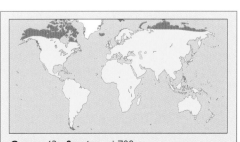

SUBFAM. CELTIDOIDEAE Includes 9 genera, c. 95 spp., mostly tropical. Leaves with 3 major veins from the base or rarely pinnately veined; fruit a fleshy drupe. Including *Celtis*, with c. 50 spp., mostly in the tropics and southern hemisphere.

Some authors[ii] have preferred to separate Celtidoideae as a family Celtidaceae. However, the characters to separate them are not clear-cut, and *Ampeloceras* and *Aphananthe* in particular are somewhat intermediate. *Zelkova* has drupaceous fruits more like the Celtidoideae but other characters are similar to those of Ulmoideae and has been put into either subfamily. The broader concept of the family seems preferable and is readily recognizable by the combination of regular flowers with paired divergent styles, straight stamens unlike those of Urticaceae, caducous stipules, and lack of latex.

Economic uses *Ulmus* was formerly a plentiful source of good, decay-resistant timber and has been much planted in hedgerows and avenues but has been depleted in recent decades by Dutch Elm disease. *Zelkova* is planted for its colorful leaves in the fall and is more resistant to the disease. Other genera provide timber in the tropics, and bark has been used for fiber, while *Celtis* sometimes has edible fruits (Hackberry) and is planted as a street tree.   CMW-D & RKB

ULMACEAE. 1 *Ulmus campestris* (a) habit; (b) leafy shoot (x⅔); (c) flowering shoot (x⅔); (d) flower with calyx, no petals, 5 stamens, and 2 styles (x8); (e) anther posterior (right) and anterior (left) view (x24); (f) gynoecium crowned by 2 styles with stigmas on inner faces (x12); (g) winged fruit (x1). 2 *Trema orientalis* (a) leafy shoot with flowers (x⅔); (b) fruit (x6). 3 *Celtis integrifolia* (a) male flower with 5 stamens and hairy vestige of ovary (x6); (b) bisexual flower with 5 stamens and gynoecium crowned by 2 styles forked at apices (x4); (c) vertical section of ovary with single pendulous ovule (x4).

[i] Manchester, S. P. Systematics and fossil history of the Ulmaceae. Pp. 221–251. In: Crane, P. R. & Blackmore, S. *Evolution, Systematics and Fossil History of the Hamamelidae, vol. 2.* Systematics Association Special vol. no. 40A. Oxford, Clarendon Press (1989).
[ii] Ueda, K., Kosuge, K., & Tobe, H. A molecular phylogeny of the Celtidaceae and Ulmaceae. *J. Plant Res.* 110: 171–178 (1997).

URTICACEAE
NETTLE AND RAMIE FIBER FAMILY

Herbs, shrubs, and trees with small greenish, usually unisexual, flowers, some species bearing stinging hairs.

Distribution Worldwide distribution in a wide range of habitats, with plants ranging from ruderal temperate weeds to rain forest climbers, predominantly in the tropics, usually in moist places, often epiphytic.

Description Annual or perennial herbs (sometimes with succulent stems) to soft-wooded trees up to 15 m or high-climbing lianas. Dot-like or elongate cystoliths are found on leaves and stems throughout the family, and stinging hairs are found on most parts in the tribe Urticeae. The leaves are alternate or opposite, simple, occasionally deeply lobed, often asymmetrical, often with 3 major veins from the base, the margins entire or toothed. Stipules are usually present and often conspicuous but are absent in most Parietarieae. The inflorescence is a terminal or usually axillary variously

Genera 43   **Species** c. 1,700
Economic uses Fibers; edible leaves; some indoor plants

URTICACEAE. 1 *Pilea microphylla* (a) bud of male flower (×10); (b) male flower (×12); (c) perianth of female flower (×24); (d) gynoecium (×24); (e) vertical section of ovary (×24). 2 *Parietaria judaica* (a) bisexual flower (×4); (b) perianth opened out to show single and hairy capitate stigma (×6). 3 *Forsskaolea angustifolia* male flower (×6). 4 *Urtica magellanica* (a) flowering shoot (×⅔); (b) bud of male flower (×10); (c) male flower (×12); (d) vertical section of male flower with cuplike vestige of ovary (×12); (e) female flower (×12).

elaborated cyme, often condensed into heads or aggregated into large panicles, sometimes reduced to 1 or a few flowers. The flowers are small (usually less than 2 mm), usually greenish, unisexual (plants monoecious or dioecious) or rarely (some Parietarieae) some flowers bisexual, and actinomorphic or female flowers sometimes zygomorphic, with the perianth uniseriate or sometimes absent in female flowers. Tepals in male flowers are (1–2)3–5. Tepals in female flowers are 3–5 or apparently absent, sometimes unequal, often united into a tube. The stamens are equal in number to, and opposite, the tepals, inflexed in bud, and reflexing abruptly to eject pollen. The ovary is superior, sometimes adnate to the perianth tube, with a single style, originating as 2 carpels but only 1 developing, with 1 locule and a single basal ovule. The fruit is a small achene, often enclosed in an accrescent perianth and sometimes also persistent involucral bracts.

Classification From the nineteenth century onward, the family has been grouped with the families here recognized as Cannabaceae, Moraceae, Cecropiaceae, and Ulmaceae as the order Urticales[i], but the place of this order in the flowering plants has been disputed. Different authors have placed it with Euphorbiales, Fagales, Malvales, and Hamamelidales, but molecular evidence places all 5 families in the Rosales of APG II. Five tribes are recognized[ii]:

Urticeae (10 genera, c. 180 spp.). Herbs to trees (*Dendrocnide, Obetia*); cystoliths dotlike or elongate; stinging hairs usually present; female flowers with 4 tepals, usually free. Including *Urtica* (c. 60 spp.).

Elatostemeae (7 genera, c. 1,250 spp.). Usually succulent-stemmed herbs; often anisophyllous; cystoliths elongate; an involucre often present; female flowers with 3–5 tepals, usually free; stigma penicillate-capitate. Including *Elatostema* (c. 500 spp.) and *Pilea* (c. 720 spp.).

Boehmerieae (17 genera, c. 240 spp.). Herbs to trees (several genera); cystoliths dotlike; often anisophyllous; leaves often 3-veined, often discolorous; involucre absent; female flowers with perianth tubular or absent. Including *Boehmeria* (c. 70 spp.).

Parietarieae (5 genera, c. 20 spp.). Herbs to shrubs (*Hemistylus*); cystoliths dotlike; leaves never opposite; involucre present; stipules usually absent; flowers sometimes bisexual.

Forsskaoleeae (4 genera, 16 spp.). Herbs to subshrubs; cystoliths dotlike; leaves sonetimes discolorous; involucre often present; male flowers with 1 tepal and 1 stamen; stigma filiform or subcapitate.

Economic uses *Boehmeria nivea* (Ramie Fiber) and *Laportea canadensis* provide fibers that are made into clothes, and other species have fibers that are made into amate paper. Some *Urtica* and *Laportea* species are eaten as a vegetable. Some genera are cultivated, often as indoor plants, especially *Pilea* spp. *Soleirolia soleirolii* is often an abundant carpeting weed in greenhouses. Urticeae stings can be very painful and persistent, especially from *Dendrocnide. Urtica* pollen is a serious contributor to hayfever. CMW-D & RKB

[i] Berg, C. C. Systematics and phylogeny of Urticales. Pp. 193–217. In: Crane, P. R. & Blackmore, S. *Evolution, Systematics and Fossil History of the Hamamelidae, vol. 2.* Systematics Association Special vol. no. 40A. Oxford, Clarendon Press (1989).
[ii] Friis, I. The Urticaceae, a systematic review. Pp. 287–308. In: Crane, P. R. & Blackmore, S. *Evolution, Systematics and Fossil History of the Hamamelidae, vol. 2.* Systematics Association Special vol. no. 40A. Oxford, Clarendon Press (1989).

VAHLIACEAE

The family comprises a single genus *Vahlia*, of 5 spp. of annual to perennial herbs to subshrubs, from mesic parts of Africa from Senegal and Egypt south to the Cape, Madagascar, and from Iraq to India, Sri Lanka, and Phu Quoc Island of Vietnam, apparently absent from the Arabian peninsula[i], growing in open places, often on river banks, cultivated ground or saline flats, sometimes in grassland or woodland. The leaves are opposite, simple, entire, ovate to linear, sessile to subsessile, subglabrous or with multicellular and often gland-tipped hairs. The inflorescence is an axillary, crowded cyme of paired flowers that are sessile or shortly pedicellate. The flowers are actinomorphic, bisexual, usually small, 1–3(–7) mm, 5-merous. The sepals are 5, inserted on the top of the ovary surrounding the styles, persistent in fruit. The petals are 5, about as long as the sepals, lanceolate to rounded, white to yellow. The stamens are 5, inserted on a small disk on the top of the ovary. The ovary is inferior, with a single locule, with 2 swollen placentas suspended from its apex each bearing many ovules. Styles 2, broad-based divergent, each with a capitate stig-

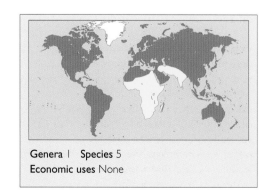

Genera 1 Species 5
Economic uses None

ma. The fruit is a subglobose capsule, dehiscing by 2 valves between the style bases to release many tiny seeds[ii].

Often mistaken superficially for Rubiaceae, Vahliaceae has traditionally been regarded as referable to Saxifragales. However, molecular evidence places it well removed from Saxifragaceae and close to Hydroleaceae[iii], perhaps best referred to Solanales, but left uplaced to order in the Euasterids 1 of APG II. RKB

[i] Bridson, D. M. A revision of the family Vahliaceae. *Kew Bull.* 30: 163-182 (1975).
[ii] Bridson, D. M. *Flora of Tropical East Africa: Vahliaceae.* 6 pp. Rotterdam, Balkema and Bentham-Moxom Trust (1975).
[iii] Savolainen, V. *et al.* Phylogeny of the eudicots: a nearly complete familial analysis based on *rbc*L gene sequences. *Kew. Bull.* 55: 257–309 (2000).

VALERIANACEAE

VALERIAN FAMILY

A herbaceous family related to Caprifoliaceae and Dipsacaceae, showing great variation in South America.

Distribution Widespread distribution, but with centers of generic diversity in the Mediterranean and temperate Asia, and with greatest morphological diversity in the Andes, but poorly represented in Africa and tropical Asia and absent from Australasia.

Description Mostly annual to perennial herbs, sometimes woody at the base, densely caespitose (*Phyllactis, Belonanthus, Stangea,* and *Aretiastrum* in the high Andes), microphyllous with imbricate leaves (*Aretiastrum*), or a herb climbing to several meters (*Valeriana scandens,* widespread in tropical America). The leaves are opposite or rarely alternate, simple and entire, to 3-foliate or pinnately compound. The inflorescence is a terminal dichasial cyme, usually lax and pyramidal, sometimes compact and subcapitate but without an involucre, the bracts at each node usually small but large and scarious in *Nardostachys*. In *Fedia* and *Astrephia* the inflorescence axes are markedly swollen. *Patrinia* has 2 bracteoles, often scarious and accrescent, and persistent in fruit. The flowers are regular to weakly zygomorphic, rarely unisexual (plants polygamomonoecious). The calyx is usually either absent or reduced to small teeth, or forms a pappus persisting on the fruit, but in *Nardostachys* it has 5

VALERIANACEAE. 1 *Valeriana officinalis* (a) pinnatisect leaf and cymose inflorescence (×⅔); (b) flower with small corolla spur (×6); (c) fruit crowned by plumose calyx (×6). 2 *Centranthus lecoqii* flower with distinct spur and single stamen (×4). 3 *Patrinia villosa* fruit (×4). 4 *Nardostachys jatamansi* (a) leafy shoot and inflorescence (×⅔); (b) flower opened out showing 4 stamens and 1 style (×4); (c) fruit (×4). 5 *Valerianella* species have varied fruits due to the growth of the calyx; shown here are those of (a) *V. echinata,* (b) *V. vesicaria,* and (c) *V. tuberculata.*

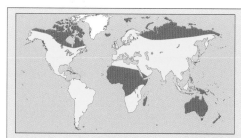

Genera 13 **Species** c. 350
Economic uses Few medicinal plants; essential oils for perfumes; salad vegetables; garden ornamentals

ii Bell, C. D. Preliminary phylogeny of Valerianaceae (Dipscacales) inferred from nuclear and chloroplast DNA sequence data. *Molec. Phylogen. Evol.* 31: 340–350 (1994).
iii Eriksen, B. Notes on generic and infrageneric delimitation in the Valerianaceae. *Nordic J. Bot.* 9: 179–187 (1989).

VERBENACEAE

VERVAIN FAMILY

A family of predominantly New World tropical herbs to forest trees, recently greatly reduced in size by the transfer of many genera to Lamiaceae.

Distribution Nearly all genera have at least some representatives in the New World, many being confined to South America, but others extending to the Old World tropics and becoming less common in the northern-temperate zone. Habitats range from open ground, deserts, and high alpine places to evergreen forests.

Description Annual or perennial herbs to shrubs, climbers, and small trees. The leaves are usually opposite, occasionally alternate or whorled, with margins entire or variously toothed to very deeply divided, sometimes fleshy or reduced almost to scales. The inflorescence is racemose and usually with many flowers arranged in a long terminal or axillary spike, or contracted into a dense head that may have an involucre of bracts below it, rarely only 2–6 flowers in an inflorescence. The calyx consists of (2–)4–5 fused sepals and is tubular to campanulate and persistent in fruit. The corolla is irregular, tubular to trumpet-shaped, often 2-lipped, with 5 usually rather short lobes. The

VERBENACEAE. 1. *Glandularia peruviana* (a) shoot bearing opposite leaves and a terminal raceme (x⅔); (b) flower with irregular corolla (x2); (c) corolla opened out to show epipetalous stamens (x6); (d) gynoecium showing lobed ovary surmounted by a single style with a lobed stigma (x3).

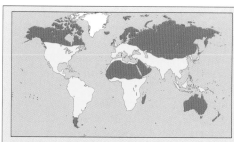

Genera 34 **Species** c. 1,150
Economic uses Garden ornamentals; small trees used for timber; edible tubers; infusions; essential oils; medicinal herbs

stamens are (2–)4(5), attached in the corolla tube, and not exserted. The ovary is superior, unlobed, of 2 carpels (in *Duranta* arguably 4), each carpel divided into 2 locules by inrolling of the margin, with 1 or 2 ovules per locule attached basally or, rarely, in the upper part of the locule; the style is terminal. The fruit is either a dry or fleshy schizocarp splitting into 1- to 2-seeded mericarps or drupaceous with 1–2 single-seeded pyrenes or 2–4, 2-seeded pyrenes, usually retained within the persistent calyx.

Classification The family is part of the complex close to Scrophulariaceae. The close relationship with the Lamiaceae, formerly assumed because of the fruit dehiscence into mericarps, seems to be sometimes supported by molecular evidence[i], although a closer relationship with the Acanthaceae and Pedaliaceae may be suggested[ii]. No clear idea of relationships will be obtained until many more genera in this and related families are analyzed. However, the removal of many genera traditionally placed in Verbenaceae (including *Vitex, Clerodendrum, Gmelina, Tectona,* etc.), and their transfer to Lamiaceae, proposed recently on morphological and molecular evidence[iii,iv], seems incontrovertible. The new slimmed-down Verbenaceae forms a natu-

well-developed, scarious, rounded lobes. The corolla is tubular to trumpet-shaped, almost regular, but sometimes weakly 2-lipped, with 3, 4, or 5 lobes, spurred in *Centranthus*. The stamens are 1 in *Centranthus* and *Patrinia monandra*, 2 in *Fedia*, 3 in most genera, but 4 in *Nardostachys* and other *Patrinia* spp. The ovary is inferior, 3-carpellate, but 2 of the carpels abort and leave a single locule with a single pendulous ovule. The style is slender with a 2- to 3-lobed stigma. The fruit is a dry cypsela, often surmounted by an accrescent or pappus-forming calyx.

Classification The family shows obvious affinity with the Dipsacaceae, Morinaceae, and Triplostegiaceae, from which it differs notably in the lack of an involucel surrounding the ovary. Like those families, it apparently has an ancestry in Caprifoliaceae. More than half of the family is referable to *Valeriana,* which has perhaps 250 spp. more or less throughout the distribution of the family. The next largest genus is *Valerianella* with c. 50 spp. in the northern-temperate region. Analyses based on morphological characters[i] and on molecular data[ii] place the Asian genera *Patrinia* and *Nardostachys* basal to the rest of the family. The great morphological diversity in the Andes is apparently derived from within the broad genus *Valeriana.* Some generic limits in the family are controversial. The sinking into *Valeriana* on cladistic grounds[iii] of the 4 cushion-forming Andean genera mentioned above, which also have technical characters to justify them, is unacceptable for many taxonomists and fails to recognize the importance of the major evolutionary diversification that has taken place in South America.

Economic uses Roots and leaves of some species have medicinal properties for relieving stress. *Nardostachys jatamansi* (*N. grandiflora*) (Spikenard) produces essential oils used in perfumery. *Valerianella* (Lambs' Lettuce) is eaten as a salad vegetable. Cultivated ornamental plants include *Centranthus ruber* (Red Valerian) from the Mediterranean, *Patrinia triloba* (*P. palmata*) from Japan, and several species of *Valeriana* (Valerian). RKB

i Caputo, P. & Cozzolino, S. A cladisitc analysis of Dipsacaceae (Dipsacales). *Pl. Syst. Evol.* 189: 41–61 (1994).

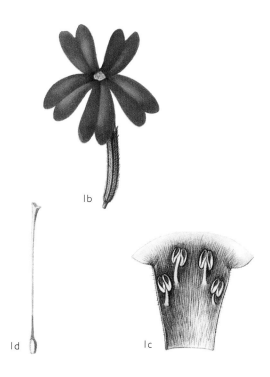

ral family with a characteristic facies deriving largely from its racemose inflorescence and trumpet-shaped corolla with stamens included in the tube, whereas the traditional concept included a wide range of different genera including some big trees. It can readily be distinguished from Lamiaceae by its racemose rather than cymose inflorescence, trumpet-shaped corolla with stamens not exserted from the tube, and unlobed ovary with a terminal style. Verbenaceae is now divided into 6 tribes[v] based largely on characters of the ovary and fruit dehiscence. The main genera are *Verbena* (more than 200 spp.), *Glandularia* (c. 100 spp., often included in *Verbena*, but separated by their compact inflorescence and long style), *Lantana* (c. 150 spp.), *Lippia* (c. 200 spp.), and *Citharexylum* (c. 130 spp., mostly small trees). Other families recently segregated from the traditional Verbenaceae are Cyclocheilaceae, Nesogenaceae, and Symphoremataceae. *Nyctanthes*, also formerly included in Verbenaceae, is now referred to Oleaceae.

Economic uses The family is well known for ornamental herbs (especially *Verbena* and *Glandularia*) and shrubs (*Lantana, Duranta, Citharexylum*), and *Phyla nodiflora* is widely grown as a ground-cover plant. Essential oils are extracted from *Lantana, Lippia, Phyla, Aloysia,* and *Acantholippia,* and infusions and medicinal remedies come from *Verbena, Lampayo, Aloysia, Lippia,* and *Stachytarpheta,* while tubers of *Pitraea* are edible. Wood from *Citharexylum* and *Aloysia* is of commercial importance, but the reputed use of the former for making musical instruments ("zither wood") is an apparently fictitious corruption of the generic name. The polyploid *Lantana camara*[vi] is recognized as one of the worst invasive weeds in the world. RKB

i Savolainen, V. *et al.* Phylogeny of the eudicots: a nearly complete familial analysis based on *rbc*L gene sequences. *Kew Bull.* 55: 257–309 (2000).
ii Olmstead, R. G., Kim, K.-J., Jansen, R. K., & Wagstaff, S. J. The phylogeny of the Asteridae *sensu lato* based on chloroplast *ndh*F gene sequences. *Molec. Phylogen. Evol.* 16: 96–112 (2000).
iii Cantino, P. D. Evidence for a polyphyletic origin of the Labiatae. *Ann. Missouri Bot. Gard.* 79: 361–379 (1992).
iv Wagstaff, S. J. & Olmstead, R. G. Phylogeny of Labiatae and Verbenaceae inferred from *rbc*L sequences. *Syst. Bot.* 22: 165–169 (1997).
v Atkins, S. Verbenaceae. Pp. 449–468. In: Kadereit, J. W. (ed.), *Families and Genera of Vascular Plants. VII.* Berlin, Springer-Verlag (2004).
vi Stirton, C. H. Some thoughts on the polyploid complex *Lantana camara* L. (Verbenaceae). Pp. 321–340. In: Annecke, D. P., *Proceedings of the 2nd National Weeds Conference in South Africa* (1977).

VIBURNACEAE
WAYFARING TREE AND GUELDER ROSE FAMILY

Shrubs and small trees with flat or rounded inflorescences as Sambucaceae but with simple leaves.

Distribution Widespread in northern temperate and tropical regions, but absent from tropical and southern Africa, New Guinea,

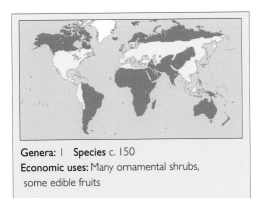

Genera: 1   Species c. 150
Economic uses: Many ornamental shrubs, some edible fruits

Australasia, the Pacific, most of Brazil, and southern South America. The main centers of distribution are China and the Americas. The species occur in open places, woodland, and wet tropical forests.

Description Shrubs or small to medium trees, deciduous or evergreen. The leaves are opposite or rarely in whorls of 3, petiolate, simple, sometimes 3- to 5-lobed, the margins entire to crenate or serrate. Stipules may be present or absent. The inflorescence is a terminal corymb or panicle, with the primary rays usually umbellate. In many species, the outermost flowers of the inflorescence are highly modified, sterile, with a much enlarged corolla (similar to *Hydrangea* of the Hydrangeaceae). The flowers are shortly pedicellate, actinomorphic (except the sterile marginal ones), bisexual, and 5-merous. The calyx is small and inconspicuous. The corolla is tubular to funnel-shaped or campanulate, white to pink. The stamens are inserted on the corolla tube. The ovary is semi-inferior, of 3 carpels, but 2 of these abort leaving a single fertile locule with a single pendulous ovule. The stigma is subsessile and 3-lobed. The fruit is a drupe with a single 1-seeded pyrene.

Classification *Viburnum* has often been included in the Caprifoliaceae, but its closer affinities are with *Sambucus* and *Adoxa* (see Adoxaceae for discussion and references). While *Viburnum* can be readily separated from Adoxaceae by the same characters of habit and inflorescence that separate *Sambucus,* it does show considerable similarity to *Sambucus* and one questions whether these 2 genera should not be placed together in 1 family. The most obvious difference is that 1 has simple leaves while the other has pinnate leaves. There are other morphological characters, such as the single-seeded fruits of *Viburnum,* while the often quoted embryological characters put *Sambucus* with *Adoxa* rather than *Viburnum.* It seems marginally preferable to maintain separate families. *Viburnum* has often been regarded as the most basal member of the Dipsacales *sensu* APG II. Further discussion and a key to 10 sections recognized in the genus have been given by Hara[i], who placed *Viburnum* and *Sambucus* as separate subfamilies of a broad Caprifoliaceae.

VIBURNACEAE. 1 *Viburnum tinus* (a) leafy shoot and inflorescences (×⅔); (b) corolla and stamens (×3); (c) vertical section of gynoecium showing capitate stigma and pendulous ovule (×3); (d) fruits (×⅔).

Economic uses Many species are cultivated in temperate regions as ornamental shrubs or small trees. *V. tinus* (Laurustinus) is an evergreen, winter-flowering shrub much used in urban plantings, but with rather dull, white flowers. *V. farreri, V. grandiflora,* and their hybrid *V. × bodnantense* are grown for their fragrance. *V. plicatum* has horizontal inflorescences of bright white flowers radiating at the margins and has many cultivars. *V. rhytidophyllum* has leaves up to 20 cm long. *V. lantana* (Wayfaring Tree) and *V. opulus* (Guelder Rose) are often planted for their colorful leaves during the fall. In North America, the fruits of *V. lentago* have been used locally as food. RKB

i Hara, H. A revision of Caprifoliaceae of Japan with reference to allied plants in other districts and the Adoxaceae. *Ginkgoana* 5: 1–336, t. 1–55 (1983).

VIOLACEAE
VIOLET FAMILY

Tropical trees, shrubs, vines, and lowland herbs, and temperate montane and alpine herbs (*Viola*).

Distribution The family is predominantly pantropical in distribution, excepting 1 *Hybanthus* species (in temperate North America) and

Viola, which is cosmopolitan and has centers of diversity in the mountains of western South America, western North America, eastern Asia, and southern Europe.

Description Trees or treelets, shrubs or subshrubs, vines, and herbs. The leaves are alternate except in a few *Hybanthus* and some *Rinorea*, simple (infrequently dissected or lobed in *Viola*), linear to reniform, with entire to serrate margins, and bearing stipules. The inflorescence is mostly thyrsoid, dichasial, cymose or racemose in woody genera, and mostly fasciculate or solitary in herbaceous ones, with flower pedicels usually bearing a pair of

Genera 22 **Species** c. 890
Economic uses Bird lime; local medicines and food; cultivated ornamentals

bractlets. The flowers are actinomorphic in approximately half of the genera (tribe Rinoreeae) or zygomorphic in the other half (tribe Violeae), bisexual or unisexual (some dioecious *Melicytus*), hypogynous or slightly perigynous. The sepals are 5 and small to conspicuous. The petals are 5 and linear to orbicular. The stamens are 5 (3 in 1 *Leonia* sp.), alternating with petals and free to variously fused, bearing separate nectariferous glands on 2 or 5 stamens or fused to the filament tube as an extended "collar," each anther connective tipped by a conspicuous to rudimentary dorsal scale (absent in *Leonia*, ventral only in *Fusispermum*). The ovary is superior, of 3 fused carpels (5 in some *Leonia*, 2–5 in *Melicytus*), each with 1 to many ovules with parietal placentation. The fruit is a 3-valved capsule in most genera (6-valved in one *Agatea*), sometimes somewhat woody, rarely a fleshy berry (*Gloeospermum* and *Melicytus*), nut (*Leonia*), follicle (*Hybanthopsis*), or papery bladder (*Anchietea*). The seeds are mostly globose to ovoid (flattened in *Corynostylis*, winged in *Anchietea* and *Agatea*, with basal projections in *Hybanthopsis*), bearing an elaiosome in most genera.

Classification Relationships formerly assumed in the subclass Dilleniidae of, for example, Cronquist have not been borne out by molecular data. Violaceae is placed in the Higher Rosids in APG

II, most closely related to Krameriaceae and more distantly to Passifloraceae, Turneraceae, and Malesherbiaceae. Traditionally, the family has been split into monotypic subfamilies Fusispermoideae and Leonioideae, as well as subfamily Violoideae with tribes Rinoreeae and Violeae. Current molecular and traditional systematic evidence rejects the Leonioideae (placing *Leonia* in the Violeae), reveals a complex series of relationships among actinomorphic and zygomorphic genera, and urges segregation of 6 additional genera from a polyphyletic *Hybanthus*[i]. Because current intrafamilial studies are not yet sufficient to confidently remodel higher-level groupings for all genera, the higher taxa in Hekking's most recent classification[ii] are presented here.

SUBFAM. FUSISPERMOIDEAE Flowers small, actinomorphic, in pseudoracemose or narrowly thyrsoid inflorescences; petal aestivation convolute; filaments partially fused into a low tube; anther connectives each with a ventral scale and lacking a dorsal one; fruit a trivalvate capsule; seeds of 2 types, discoid or fusiform. Comprises the single genus *Fusispermum* (3 spp., Panama and northern South America).

SUBFAM. LEONIOIDEAE Flowers moderate to large, actinomorphic, in loose pseudoracemose thyrses or cymes; petal aestivation quincuncial; stamens fused into a tall tube with apical anthers; connective scale lacking; fruit a nut (one with thin edible pulp); seeds globose. Comprises the single genus *Leonia* (6 spp., Panama and northern South America).

SUBFAM. VIOLOIDEAE Flowers small to large, actinomorphic to zygomorphic, in thyrses or more commonly racemes, cymes, fascicles, or solitary; petal aestivation typically descending (rarely quincuncial, in some *Gloeospermum* and few *Rinorea*); filaments free to partially fused (stamens fully fused to a tall tube in *Hekkingia* and *Paypayrola*); anther connectives with a conspicuous to rudimentary dorsal scale; fruit mostly a trivalvate capsule (6-valved in 1 *Agatea*, follicle in *Hybanthopsis*, papery bladder in *Anchietea*, fleshy berry in *Gloeospermum* and *Melicytus*); seeds globose to long-ovoid, winged, or flattened in a few genera. Two tribes are recognized:

Rinoreeae Flowers actinomorphic or slightly zygomorphic; filaments commonly partially fused or fully fused into a low tube, rarely "free;" glands 5 and free on each filament or base of anther, or fused to the filament tube as extensions of it; dorsal connective appendage conspicuous (rarely rudimentary, in *Isodendri-*

on, *Hekkingia*, and *Paypayrola*); fruits commonly large and somewhat woody, mostly trivalvate capsules (fleshy berries in *Gloeospermum* and *Melicytus*); seeds globose, smooth, hairy, or glabrous, rarely bearing an elaiosome (*Allexis* and *Hekkingia*, sometimes *Paypayrola*). Represented by *Allexis* (western Africa, 4 spp.), *Amphirrhox* (Latin America, 1–3 spp.), *Decorsella* (western Africa, 1 sp.), *Gloeospermum* (Latin America, c. 12 spp.), *Hekkingia*[iii] (South America, 1 sp.), *Isodendrion* (central Pacific, 4 spp.), *Melicytus* (Australia and New Zealand, c. 10 spp.), *Paypayrola* (Latin America, 8 spp.), *Rinorea* (pantropical, c. 150 spp.), *Rinoreocarpus* (South America, 1 sp.).

Violeae Flowers strongly zygomorphic; filaments free or intermittently weakly fused (rarely fully fused in a low tube); glands 2, free on pair of filaments, alternate with the saccate or spurred "bottom" petal; dorsal connective appendage conspicuous; fruits small to moderate sized, typically thin and elastic-walled trivalvate capsules (papery bladders in *Anchietea*); seeds mostly globose, infrequently flattened (*Corynostylis*), winged (*Anchietea*, *Agatea*), or with basal projections (*Hybanthopsis*), often with surficial patterning, commonly glabrous. Represented by *Agatea* (South Pacific, c. 10 spp.), *Anchietea* (South America, 6 spp.), *Corynostylis* (Latin America, 3–4 spp.), *Hybanthopsis*[iv] (South America, 1 sp.), *Hybanthus* (pantropical and temperate North America, c. 110 spp.), *Mayanaea* (Central America, 1 sp.), *Noisettia* (South America, 1 sp.), *Orthion* (Latin America, 6 spp.), *Schweiggeria* (South America, 1 sp.), *Viola* (worldwide, c. 550 spp.).

Economic uses A few *Hybanthus* species are used for bird lime, while certain larger *Rinorea* species provide occasional timber. Some *Rinorea* species in Southeast Asia take up heavy metals and may prove useful in bioremediation. Both *Rinorea* and *Viola* are used in

traditional medicine owing to methyl salicylate; leaves of the latter are utilized occasionally for a potherb, and the flowers as a jelly, syrup, or confection. Numerous *Viola* species are grown as ornamentals (Pansies, Violets, Violas).　　HEB

i Ballard, Jr., H. E., de Paula-Souza, J., & Feng, M. Dismantling the polyphyletic genus *Hybanthus* Jacq. (Violaceae). *Botany Abstracts* Systematics Section/ASPT. Botanical Society of America. http://www.2005.botanyconference.org/engine/search/index.php?func=detail&aid=286
ii Hekking, W. H. A. Violaceae: Part I, *Rinorea* and *Rinoreocarpus*. *Flora Neotropica Monographs* 46: 1–207 (1988).
iii Munzinger, J. K. & Ballard, Jr., H. E. *Hekkingia* (Violaceae), a new arborescent violet from French Guiana, with a key to genera in the family. *Syst. Bot.* 28: 345–351 (2003).
iv de Paula-Souza, J. & Souza, V. C. *Hybanthopsis*, a new genus of Violaceae from Eastern Brazil. *Brittonia* 55: 209–213 (2003).

VISCACEAE
MISTLETOE FAMILY

A small family of shrubby to herbaceous parasitic plants, with or without leaves and with small flowers.

Distribution Widespread in Europe, south Asia, Malaysia, Pacific region, North and South America, tropical Africa, and Madagascar, infecting both gymnosperms and angiosperms. The family is centered in south-

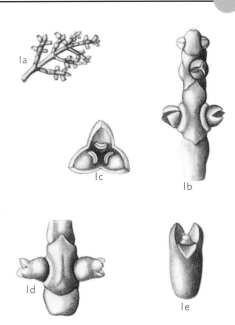

VISCACEAE. I *Dendrophthora cupressoides* (a) flowering branch (x⅔); (b) male inflorescence (x4); (c) male flower from above (x8); (d) part of female inflorescence (x4); (e) female flower with sessile stigma (x8).

ern Asia and may have diversified from there[i], the few genera having partially overlapping and partially disparate ranges. Two genera are restricted to the New World and *Arceuthobium* is widely distributed in northern temperate and tropical mountains.

Description Shrubby or small herbaceous plants that grow on the branches of gymnosperms and generally dicotyledonous woody angiosperms, attached to the vascular system of the host by sinkers and with strands in the bark of the host adjacent to the cambium. The leaves are opposite, simple, entire, often rather fleshy, sometimes reduced to scales, without stipules. The flowers are small, unisexual (often monoecious, rarely dioecious), solitary, clustered or in spikes, often in 3s enfolded by a pair of joined bracts. The perianth segments (tepals) are 3–4(–6), valvate. The stamens are as many as the perianth segments and often attached basally, the anthers occasionally fused. The ovary is inferior without chambers or ovules, the several embryo sacs originating from a short placental column. The fruit is generally a single-seeded berry. The seeds lack a testa but may be enclosed in a thin, hardened endocarp and are either sticky and dispersed by birds or squirted out explosively.

Classification The Viscaceae were formerly generally included as a subfamily of the Loranthaceae, but are now considered to be of separate origin, derived from Santalaceae[ii].

Economic uses *Viscum album* is used in Europe for Christmas festivities and also in the palliative treatment of cancer. *Phoradendron serotinum* ("*P. flavescens*") is used as a Christmas decoration substitute in North America. *Arceuthobium* is a serious pest of

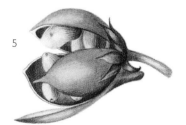

VIOLACEAE. I *Anchietea salutaris* shoot with alternate leaves and fruit (x⅔). 2 *Rinorea* sp. half flower showing stamen filaments fused at the base and the anthers with a membranous extension to the connective (x10). 3 *Viola hederacea* (a) habit (x⅔); (b) vertical section of flower showing irregular petals and anthers in a close ring around the ovary (x4). 4 *Corynostylis arborea* (a) shoot with leaf and inflorescence (x⅔); (b) cross section of ovary with 1 locule containing numerous ovules on parietal placentas (x4). 5 *Hybanthus enneaspermus* var. *latifolius* dehiscing fruit—a capsule (x4). 6 *Melicytus obovatus* leafy shoot with fruit (x⅔).

Genera 7 **Species** c. 550
Economic uses Curious plants with some use in folk customs and medicine; pests of timber trees

gymnosperms in north temperate regions[iii], and *Phoradendron* attacks hardwoods and conifers in tropical America to a limited extent.   RMP

[i] Barlow, B. A. Biogeography of the Loranthaceae and Viscaceae. Pp. 19–46. In: Calder, M. & Bernhardt, P. (eds), *The Biology of Mistletoes.* Sydney, Academic Press, (1983).
[ii] Nickrent, D. L. & Malécot, V. A molecular phylogeny of Santalales. Pp. 69–74. In: Fer, A. *et al.* (eds), *Proceedings of the 7th International Parasitic Weed Symposium.* Nantes, Faculté des Sciences, Université de Nantes (2001).
[iii] Hawksworth, F. G. & Wiens, D. *Dwarf mistletoes: Biology, Pathology and Systematics. Agriculture Handbook #709.* Pp. 1–410. Washington, D.C., Forest Service, U.S. Dept. of Agriculture (1996).

VITACEAE

GRAPEVINE AND VIRGINIA CREEPER

Vitaceae is a family of mainly lianas with tendrils, rarely shrubs, renowned for the grapevine, *Vitis vinifera.*

Distribution The Vitaceae is mainly found in the tropics and subtropics and widely distributed in both the northern and southern hemispsheres. *Cissus* (200–350 spp.) is entirely

VITACEAE (ABOVE). I *Vitis thunbergii* (a) inflorescence (x⅔); (b) flower bud (x3); (c) flower with petals removed showing cuplike calyx and 5 stamens (x3); (d) gynoecium (x3); (e) part of shoot with leaf and immature and mature fruits—berries (x⅔). 2 *Tetrastigma obtectum* (a) leafy shoot with axillary inflorescences (x⅔); (b) flower bud (x4); (c) flower viewed from above showing 4 petals and 4 stamens (x3); (d) stamens (x6).

VISCACEAE (BELOW). 2 *Viscum album* (Mistletoe) (a) shoot with berries (x⅔); (b) leaf base (x2); (c) male flowers (x6); (d) stamen adnate to perianth segment (x8); (e) female flowers (x6); (f) ovary (x10); (g) fleshy fruit (x2).

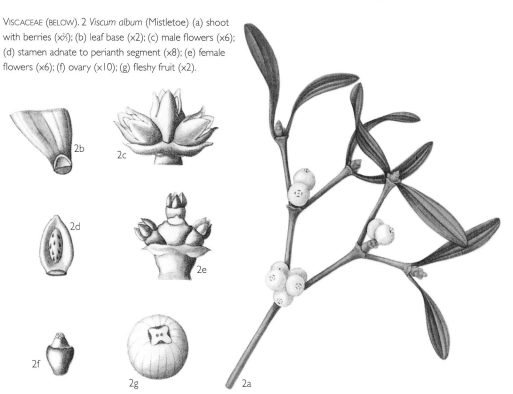

tropical, *Cyphostemma* (150–250 spp.) is warm temperate and tropical, *Ampelocissus* (100 spp.) is widespread in the tropics, *Vitis* (c. 70 spp.) is found mainly in the northern hemisphere, and *Leea* (c. 70 spp.) ranges from Africa to northern Australia and the Pacific Islands[i].

Description Mostly lianas climbing by simple or branched leaf-opposed tendrils, but some are shrubs, small trees, or herbs. The leaves are alternate, simple (the lamina palmately lobed or veined, often coarsely dentate, rarely entire, with pellucid glands), or compound (3-foliate, palmate, 1- to 3-pinnate). Stipules are usually small, caducous, sometimes persistent. The inflorescence is usually leaf-opposed or terminal, cymose or racemose. The flowers are actinomorphic, bisexual, rarely unisexual (plants usually monoecious, dioecious in *Tetrastigma*), small. The sepals are (3)4–5, connate, forming a collar or tube. The petals are (3)4–5, free, valvate, often deciduous. The stamens are as many as and opposite the petals, very small. A ringlike or lobed intrastaminal disk is usually present, absent in *Parthenocissus*. The ovary is superior, of 2 carpels and 2 locules, with 2 ovules per locule. The style is simple with a minute stigma. The fruit is a berry with 1(–4) seeds.

Classification The Vitaceae were included in the Rhamnales (Cronquist 1981) or in the adjacent Vitales (Takhtajan 1997). Recent molecular analyses have suggested a number of different

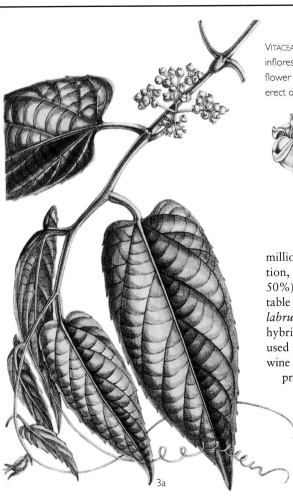

VITACEAE. 3 *Cissus velutinus* (a) leafy shoot with axillary inflorescence and unbranched, coiled tendrils (×⅔); (b) flower (×4); (c) vertical section of gynoecium showing erect ovules and short style with a discoid stigma (×4).

3b

3c

3a

Genera 3–7
Species 18
Economic uses None

million hectares are devoted to grape cultivation, the main producers being in Europe (c. 50%), Asia (22%), and the USA (18.5%). Most table wine is produced from *Vitis vinifera* but *V. labrusca*, *V. rotundifolia*, *V. amurensis*, and hybrids with these and other species are also used on a smaller scale. World production of wine is nearly 3,000 milllion hl, 79% of which is produced in Europe. Some *Cissus* species are used in traditional medicines. Species cultivated as ornamental climbers include *Parthenocissus quinquefolia* (Virginia Creeper), *P. tricuspidata* (Boston Ivy), *Vitis amurensis*, *V. davidii*, *Ampelopsis arborea*, and *Cissus* spp. VHH

i Ingrouille, M. J. *et al.* Systematics of Vitaceae from the viewpoint of plastid *rbc*L sequence data. *Bot. J. Linn. Soc.* 138: 421–432 (2002).
ii Soltis, D. E. *et al. Phylogeny and Evolution of Angiosperms.* Sunderland, Massachusetts, Sinauer Associates (2005).
iii Rossetto, M. *et al.* Is the genus *Cissus* (Vitaceae) monophyletic? Evidence from plastid and nuclear ribosomal DNA. *Syst. Bot.* 27: 522–533 (2002).

placements, including as sister to Caryophyllales, the asterid clade, the rosid clade, and Dilleniales (see discussion in Soltis *et al*[ii]). The inclusion here of *Leea*—a genus of shrubs or trees, lacking tendrils, often treated a separate although closely related family, Leeaceae—is supported by molecular analyses that indicate that *Leea* and *Ampelopsis* are ancestral to the rest of Vitaceae[i].

Generic limits and the delimitation of species from the family are in a state of flux, and much further work is needed to resolve these uncertainties and the taxonomic and cladistic relationships between species[iii].

Economic uses The grapevine is widely cultivated across the world for the production of table grapes, wine, fruit juice, and dried fruit (raisins, currants, sultanas). Globally over 6.5

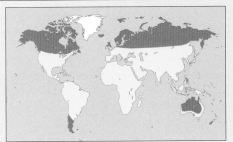

Genera 12–14 Species c. 800
Economic uses Wine; grapes; dried grapes (raisins, sultanas, currants); ornamentals

VIVIANIACEAE

A small family of herbs and shrubs from South America, often included in the Geraniaceae.

Distribution The family is restricted to temperate and tropical South America. *Balbisia* (11 spp.) and *Viviania* (6 spp.) occur along the Andes of Chile and Argentina[i], and along the east coast from São Paulo to Buenos Aires, whereas the monotypic *Rhynchotheca spinosa* is restricted to valleys in the Andes of Cotopaxi province in Ecuador[ii].

Description Shrubs or herbs, sometimes basally woody, rarely annual (*Araeoandra tenuicaulis*), with opposite leaves. The leaves are simple or deeply 3 lobed (some *Balbisia*), entire, crenate or toothed, sessile or shortly petiolate, exstipulate. The inflorescences are cymose. The flowers[iii] are actinomorphic, 4- or 5-merous (rarely 8- or 10-merous), with distinct sepals and petals or petals absent (*Rhynchotheca*). The sepals are tapering, and entire or emarginate. The petals are as many as, and equal to, or longer than, the sepals, overlapping in bud, yellow, white, pink, or violet. An extrastaminal hypogynous disk is present. The

stamens are 8–10, those opposite sepals long, and those opposite petals short, free. The ovary is superior, of 3–5 fused carpels, is sessile, ovoid to globose, and densely pubescent. The style is short, but there are 3–5 elongated stigmatic lobes with a dry (*Balbisia*, *Viviania*) or mucilaginous (*Rhynchotheca*) papillate surface. The fruit is a dehiscent capsule with 1–2 seeds per locule.

Classification The component genera of the family are members of the Geraniales[iv] and were included in Geraniaceae by Cronquist (1981). The family has been placed in Caryophyllales (Tahktajan, 1987) or even Pittosporales (Hutchinson, 1973). The limits of Vivianiaceae have been controversial, as have the limits of the genera within. Up to 3 separate familes and 7 genera have been recognized. In the last revision of Vivianiaceae[v], some 4 genera were recognized on the basis of floral and vegetative characters, all of which are all now included in *Viviania*. In more recent floristic accounts, Vivianiaceae[vi] has been treated as including 2 genera, both now treated in *Viviania*, and Ledocarpaceae[vii] containing only the genus *Balbisia*. *Rhynchotheca* has sometimes been treated in Rhynchothecaceae or included in either Vivianiaceae or Ledocarpaceae. Despite the differences of opinion at generic and family levels the number of species recognized has been consistent at c. 18. AC

i Brako, L. & Zarruchi, J. L. Catalogue of the flowering plants and gymnosperms of Peru. *Monogr. Syst. Bot.* 45. St. Louis, Missouri Botanical Garden (1993).
ii Jørgensen, P. M. & León, S. Y. Checklist of the vascular plants of Ecuador. *Monogr. Syst. Bot.* 75. St. Louis, Missouri Botanical Garden (1999).
iii Weigend, M. Notes on the floral morphology in Vivianiaceae (Geraniales). *Pl. Syst. Evol.* 253: 125–131 (2005).
iv Price, R. A. & Palmer, J. D. Phylogenetic relationships of the Geraniaceae and Geraniales from *rbc*L sequence comparisons. *Ann. Missouri Bot. Gard.* 80: 661–671 (1993).
v Lefor, M. W. A taxonomic revision of the Vivianiaceae. *University of Connecticut Occasional Papers in Biological Science, Series 2.* 15: 225–255 (1975).
vi Ariza Espinar, L. *Flora Fanerogámica Argentina* Fasciculo 8, 129b. Vivianiaceae. CONICET, Cordoba (Argentina) (1995).
vii Ariza Espinar, L. *Flora Fanerogamica Argentina* Fasciculo 18, 129a. Ledocarpaceae. CONICET, Cordoba (Argentina) (1995).

VOCHYSIACEAE

A small mainly neotropical family of trees with attractive and distinctive flowers.

Distribution The family is mainly confined to tropical Central and South America, with a few species found in West Africa (*Erismadelphus* and the recently discovered *Korupodendron*[i]). They grow in lowland rain forests and savannas.

Description Medium to large trees, sometimes shrubby, and an occasional climber. The leaves are opposite or whorled, simple, with small, paired, usually deciduous stipules and sometimes with pairs of stipular extrafloral nectaries at the base of the petiole. The inflorescence is terminal or axillary, usually racemose, or reduced to a few axillary flowers in (some *Qualea, Callisthene*). The flowers are zygomorphic, bisexual. The calyx is often showy, the sepals 5, connate at the base, the fourth spurred and sometimes elongate. The petals are usually 3, rarely 5 (*Salvertia* and *Erismadelphus*), sometimes 1 (*Callisthene, Erisma, Qualea*, some *Vochysia*), or absent (some *Vochysia*). The stamens are only 1 fertile, antepetalous except in *Callisthene* and *Qualea*. The ovary is superior, 3-locular (Vochysieae) or inferior, 1-locular (Erismeae), with marginal (Erismeae) or axile placentation, the ovules 2 to many per locule; the style is simple, the stigma usually terminal. The fruit is a loculicidal capsule (Vochysieae) or indehiscent and usually winged (Erismeae). The seeds are 1-numerous, winged or not.

Classification The Vochysiaceae has been placed in the Polygalales (e.g., Cronquist 1981) and in the Vochysiales along with Malpighi-

VOCHYSIACEAE. 1 *Vochysia divergens* (a) leafy shoot and inflorescence (x⅔); (b) winged fruit (x1⅓). 2 *V. guatemalensis* (a) vertical section of flower (x1⅓); (b) stamen (x2); (c) staminode (x4); (d) cross section of ovary (x2); (e) vertical section of ovary (x2). 3 *V. obscura* winged seed (x1). 4 *Salvertia convallariodora* part of inflorescence (x⅔). 5 *S. convallariodora* flower showing single fertile stamen (x⅔). 6 *Erismadelphus exsul* var. *platiphyllus* (a) flower (x4); (b) vertical section of flower base (x6); (c) winged fruit (x⅔).

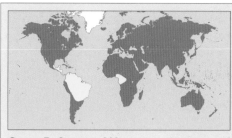

Genera 7 **Species** c. 200
Economic uses Locally used timber; medicine; gums

aceae and Trigoniaceae by Takhtajan (1997). However, morphological, anatomical, and molecular[ii,iii,iv] studies indicate inclusion in the Myrtales despite the family's unusual floral structure. Two distinct tribes may be recognized, the Vochysieae comprising *Vochysia* (100 or more spp.), *Qualea* (50 spp.), *Callisthene* (8 spp.), and *Salvertia* (1 sp.); and the Erismeae, containing *Erisma* (20 spp.) and the West African *Erismadelphus* (2 spp.) and *Korupodendron* (1 sp.), separated by the characters noted above. Analysis of the relationships suggests that there are 3 groups, the Erismeae (probably monophyletic), *Vochysia/Salvertia* and *Qualea/Callisthene*[v]. The position of *Korupodendron* is not yet resolved.

Economic uses The timber of some species is used locally. Some species are used medicinally or as a source of gums or infusions. The seeds of several species of *Erisma* are used as food or a source of oil, including Jaboty Butter from *E. calcaratum*, which used for making candles and soap, as well as *E. japura* and *E. splendens*.   VHH

i Litt, A. & Cheek, M. *Korupodendron songweanum*, a new genus and species of Vochysiaceae from West-Central Africa. *Brittonia* 54: 3–17 (2002).
ii Litt A. Floral morphology and phylogeny of Vochysiaceae. Ph.D. dissertation. New York, City University of New York (1999).
iii Litt, A. & Stevenson, D. W. Floral development and morphology of Vochysiaceae I. The structure of the gynoecium. *American J. Bot.* 90: 1533–1547 (2003).
iv Litt, A. & Stevenson, D. W. Floral development and morphology of Vochysiaceae II. The position of the single fertile stamen. *American J. Bot.* 90: 1548–1559 (2003).
v Litt, A. Vochysiaceae. Pp. 396–398. In: Smith, N. *et al.* (eds), *Flowering Plants of the Neotropics*. Princeton, Princeton University Press (2004).

WINTERACEAE

WINTER'S BARK

A family of aromatic evergreen trees or shrubs, whose wood lacks vessels.

Distribution Generally, the family is found in montane or cool temperate rain forests. In the Americas, it is represented by the genus *Drimys* (4 spp.), which ranges from Mexico south to Guyana, Brazil, Tierra del Fuego, and the Juan Fernandez Islands. Other genera grow in New Zealand and Tasmania, and then north through eastern Australia and New Caledonia to the Solomon Islands and then

1b   3   2c   2b   2d   2e   2a   6c   1a

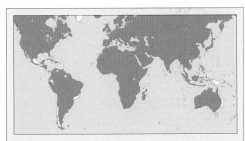

Genera 4–9 **Species** 90–100

Economic uses Some species are valued as ornamentals; Winter's Bark (*Drymis winteri*) formerly used as an antiscorbutic

to Borneo and the Philippines. A monotypic genus, *Takhtajania*, occurs in Madagascar, but the family is not represented on mainland Africa. A summary of the geographical history of the family is given by Doyle[i].

Description Evergreen trees or shrubs, the bark and leaves having a peppery taste. The leaves are alternate, simple, entire, often leathery, without stipules, usually glaucous beneath thanks to the waxy deposits that plug the stomata. The flowers are actinomorphic, hermaphrodite (unisexual and dioecious in *Tasmannia*), in terminal or axillary cymes, with the sepals and petals weakly differentiated. The sepals are 2–3(–6), valvate, or calyptrate. The petals are (0–)2–25 or more, imbricate. The stamens are few to numerous, with short, broad filaments. The ovary is superior, apocarpous, with 1 to numerous carpels, with the partly free margins of the ventral suture bearing papillae and forming a double stylelike stigmatic crest; the ovules are numerous to few, in 2 rows on marginal or submarginal placentas. The fruits are usually indehiscent, monocarpic, berrylike, or dry follicles; the seeds few to numerous, with abundant endosperm.

Classification The Winteraceae was previously considered to be closely allied to the Magnoliaceae, but recent research and phylogenetic studies have indicated that it is close to the Canellaceae, nested among Magnoliales, Laurales, and Piperales. The rare monotypic *Takhtajania perrieri*, which is endemic to Madagascar, was originally described in the genus *Bubbia*, but on the basis of its apparently paracarpous bicarpellate gynoecium and anomocytic stomata it was placed in a new genus and later recognized as a separate family, Takhtajanaceae. It has been the subject of many multidisciplinary studies recently[ii] and is now retained in the Winteraceae as a genus or subfamily. Recent molecular analysis indicated that *Takhtajania* is basal in the family and that *Tasmannia* is sister to the group of *Drimys*, *Pseudowintera*, and *Zygogynum sensu lato*[iii] (including *Bubbia* and *Exospermum*). Morphological and molecular studies suggest that *Tasmannia* and *Drimys* form a monophyletic group and that *Takhtajania* forms part of a clade including *Pseudowintera* and *Zygogynum sensu lato*[iv], but later floral developmental and molecular studies support the separate generic status of *Tasmannia*[v]. The wood of the Winteraceae is without vessels, a feature that it shares with a few other angiosperm families and the Monocotyledons, and has been interpreted, along with its simple floral structure, to suggest its archaic nature. However, the vessel-less wood may be a secondary acquisition.

Economic uses Some species are cultivated as ornamental plants. The bark of *Drimys winteri*, Winter's Bark, was used as an antiscorbutic by seafarers in the sixteenth century and subsequently. VHH

WINTERACEAE. 1 *Pseudowintera axillaris* leafy shoot with small axillary flowers (x⅔). 2 *Drimys winteri* (a) leafy shoot with bundles of conspicuous flowers (x⅔); (b) half flower showing free sepals and petals, numerous short stamens, and free carpels containing ovules on marginal placentas (x2); (c) fruits—berries (x2).

i Doyle, J. A. Paleobotany, relationships, and geographic history of Winteraceae. *Ann. Missouri Bot. Garden* 87: 303–316 (2000).
ii Investigations into the systematic botany and phylogenetic relatonships of *Takhtajania perrieri* (Capuron). Baranova & Leroy, J.-F. (Winteraceae), The "Takhtajania issue." *Ann. Missouri Bot. Gard.* 87 (3): 297–434 (2000).
iii Karol, K. G., Suh, Y., Schatz, G. E., & Zimmer, E. A. Molecular evidence for the phylogenetic position of *Takhtajania* in the Winteraceae: inference from nuclear ribosomal and chloroplast gene space sequences. *Ann. Missouri Bot. Gard.* 87: 414–414 (2000).
iv Endress, P. K. Igersheim A., Sampson, F. B., & Schatz, G. E. Floral structure of *Takhtajania* and its systematic position in Winteraceae. *Ann. Missouri Bot. Gard.* 87: 347–347 (2000).
v Doust, A. N. & Drinnan, A. N. Development and molecular phylogeny support the generic status of Tasmannia (Winteraceae). *American. J. Bot.* 91: 321–331 (2004).

ZYGOPHYLLACEAE

CREOSOTE BUSH AND LIGNUM-VITAE FAMILY

Distribution Mostly in drier parts of the tropics and warm-temperate regions, absent or poorly represented (*Tribulus* only) in much of tropical Asia and wetter parts of Africa and South America. Many species occur in desert or subdesert areas. Some New World genera include evergreen shrubs and trees, which may reach an altitude of 3,000 m in the Andes, while some *Balanites* spp. are large trees of lowland forests in Africa.

Description Annual or perennial herbs or subshrubs to deciduous or evergreen shrubs and trees, in *Balanites* occasionally a large, evergreen

tree up to 50 m. *Balanites* has supraaxillary branch spines. The leaves are opposite or alternate or fascicled and may be (especially *Tribulus*) markedly anisophyllous. They are essentially pinnately compound, and may have up to 10 pairs of leaflets, but are often reduced to 1 pair, with or without a single terminal leaflet, occasionally reduced to a single terminal leaflet and so appearing simple, or (*Larrea*) with a pair of leaflets fused at the base. The leaflets are sessile, entire to occasionally pinnatisect (*Larrea*), usually markedly oblique, sometimes very fleshy (especially *Augea*). Stipules are present and sometimes developed into needlelike spines up to 2 cm long (especially *Fagonia*). The inflorescence is an axillary or terminal cyme, or the flowers are paired or solitary in leaf axils. The flowers are actinomorphic, bisexual or unisexual (plants dioecious, *Neoluederitzia*), 4- or more usually 5-merous, usually up to 1(2) cm but up to 5 cm across in *Morkillia*. The sepals are free or connate at the base. The petals are free, or absent in *Seetzenia* and rarely so in *Zygophyllum*. The stamens are twice as many as the petals. The ovary is syncarpous, with (4)5 carpels and as many locules, or in *Augea* and *Kallstroemia* with usually 10 locules, with axile placentation, and 1 to many pendulous ovules per locule. The ovary is often deeply lobed or winged, or with paired spines on each carpel, attenuate above into a simple style (or 5 styles in *Seetzenia*), with lobed or capitate stigma. It may be glabrous or pubescent, or in some genera conspicuously villous. The fruit is a capsule or schizocarp, or in *Balanites* a fleshy drupe.

Classification In the past, the family has usually been referred to the Geraniales or Sapindales and has usually included the Nitrariaceae, Peganaceae, and Tetradiclidaceae, while *Balanites* has often been placed in a separate family Balanitaceae. Segregation of the 3 families has been increasingly accepted in the last 30 years and has been confirmed in morphological and molecular analysis[i], resulting in their placement in the Sapindales of APG II while the remaining body of Zygophyllaceae is merely referred to Eurosids 1. *Balanites* falls within Zygophyllaceae in the molecular analysis and is here included in the family.

Engler (*Pflanzenfamilien*, 1931) divided the main part of the family as here defined into 2 main groups, based on whether the seeds are endospermous or not, those without endosperm (tribe Tribuleae) differing also in having usually alternate leaves (and those in this group with opposite leaves are often anisophyllous). Later, El Hadidi[ii] separated the same group of genera as the family Tribulaceae. The molecular data, including a further analysis[iii], have divided the family into more or less the same 2 groups, but without any morphological diagnosis of them. It seems tempting to take these 2 main groups as subfamilies and leave the lesser groupings of the molecular cladograms as tribes, but further analysis of the morphological characters is needed. The first 4 tribes below would fall into the

Genera 24 **Species** c. 275
Economic uses Timber; pharmaceutical products; edible buds and fruits

Tribuloideae (but see comments on first and third tribes below) and the other 3 into the Zygophylloideae if this division were shown to be justifiable.

Morkillieae (3 genera, 4 spp.). All endemic to Mexico. Shrubs with alternate or fascicled leaves, which are imparipinnate or reduced to a terminal single leaflet. *Sericodes* differs in its fascicled leaves and villous ovary (similar to that of the next tribe), breaking up into single-seeded mericarps, and may be misplaced here. The 2 other genera are anomalous in this part of the family in having endospermous seeds, which breaks down the suggested subfamily distinction. *Morkillia* (1 sp.), *Viscainoa* (2 spp.), *Sericodes* (1 sp.).

Sisynditeae 2 monotypic genera from Namibia and the northern Western Cape. Shrubs with alternate pinnate leaves and villous ovary. *Sisyndite* (1 sp.), *Neoluederitzia* (1 sp.).

Balaniteae (1 genus, 9 spp.). From Africa to Myanmar. Shrubs to large trees, sometimes semievergreen, with simple or branching supraaxillary stem-spines, alternate leaves, and single-seeded drupes. Molecular analyses group this with the previous tribe, in the first paper as sister to it, but in the later one sandwiched between 2 genera, but it differs in many ways and has often been thought to be unrelated to Zygophyllaceae. An anatomical study[iv] made before the molecular data were available concluded that the considerable discontinuities from the Zygophyllaceae supported the view that it might constitute its own family. Takhtajan has noted that its wood anatomy, pollen morphology, and drupaceous fruit all argue for family status. There may be a good case for recognizing this genus as a third subfamily. *Balanites* (9 spp.).

Tribuleae (4 genera, 41 spp.). Widespread distribution. Herbs to subshrubs with opposite pinnate leaves, often anisophyllous, with spiny or schizocarpic fruits and distinctive anatomy[iv]. *Tribulus* (25 spp., widespread), *Tribulopis* (5 spp., Australia), *Kelleronia* (3 spp., Arabia and Somalia), *Kallstroemia* (c. 8 spp., USA to Argentina).

Seetzenieae (1 genus, 2 spp.). Decumbent herbs to subshrubs with opposite digitately 3-foliate leaves, lacking petals. *Seetzenia* (2 spp., 1 North Africa to India; 1 in South Africa).

Larreeae (c. 7 genera, 25 spp.). All New World. A rather diverse group of tropical American shrubs to large trees, sometimes evergreen, with leaves opposite and pinnate, or with 2

leaflets that are sometimes fused at the base. *Larrea* (5 spp., southern USA to the Andes and Argentina), *Pintoa* (1 sp., Chile), *Porlieria* (3–4 spp., Texas to Argentina), *Guaiacum* (5–6 spp., southern USA to North and South America), and *Bulnesia* (9 spp., South America). *Metharme* (1 sp., Chile) and *Plectrocarpa* (1 sp., Argentina) were not included in the molecular analyses but are probably referable here.

Zygophylleae (6 genera, 187 spp.). Confined to the Old World. This tribe of herbs and shrubs with opposite leaves has been reorganized generically recently[v]. *Tetraena mongolica* was previously given subfamily status by Takhtajan but is now congeneric with other species formerly in *Zygophyllum*. *Augea* (1 sp., Namibia and northern Western Cape), *Fagonia* (34 spp., Cape Verde to India and South Africa), *Melocarpum* (2 spp., northeastern tropical Africa), *Tetraena* (40 spp., Canaries and Greece to South Africa, Madagascar, and China), *Zygophyllum* (50 spp., southeastern Europe to China, especially central Asia), *Roepera* (60 spp., Australia and southern Africa).

Economic uses The Desert Date, *Balanites aegyptiaca*, has been cultivated widely in Africa for 4,000 years for its edible fruits, which yield oils, saponins, and the steroid diosgenin, and may be important in controlling bilharzia[vi]. *Guaiacum officinale* produces one of the hardest woods known, Lignum-vitae, which has been extensively used in ship construction and marine engineering because of its resistance to saltwater. The same species produces many pharmaceutically important products, such as guaiac, guaicol, guaiazulene etc., as well as guaiac gum, which was formerly used as a cure for syphilis. Other timber trees are the Maracaibo Lignum-vitae (*Bulnesia arborea*) and Paraguay Lignum-vitae (*B. sarmienti*). Leaves of *Larrea tridentata*, the abundant and ecologically important Creosote Bush of southern USA and Mexico, are boiled for an antiseptic for people and animals, while the flower buds are pickled and eaten as capers. The plant also produces anti-oxidants, which are used to preserve rubber and chocolate confectionary. The buds of *Zygophyllum fabago* are also eaten. RKB

i Sheahan, M. C. & Chase, M. W. A phylogenetic analysis of Zygophyllaceae R.Br. based on morphological, anatomical and *rbc*L sequence data. *Bot. J. Linn. Soc.* 122: 279–300 (1996).
ii El Hadidi, N. Tribulaceae as a distinct family. *Publ. Cairo Univ.* 7–8: 103–108 (1977).
iii Sheahan, M. C. & Chase, M. W. Phylogentic relationships within Zygophyllaceae based on DNA sequences of the plastid regions, with special emphasis on Zygophylloideae. *Syst. Bot.* 25: 371–384 (2000).
iv Sheahan, M. C. & Cutler, D. F. Contribution of vegetative anatomy to the systematics of the Zygophyllaceae R.Br. *Bot. J. Linn. Soc.* 113: 227–262 (1993).
v Beier, B.-A., Chase, M. W., & Thulin, M. Phylogenetic relationships and taxonomy of subfamily Zygophylloideae (Zygophyllaceae) based on molecular and morphological data. *Pl. Syst. Evol.* 240: 11–39 (2003).
vi Sands, M. J. S. The Desert Date and its relatives: a revision of the genus *Balanites*. *Kew Bull.* 56: 1–128 (2001).

MONOCOTYLEDONS

ACORACEAE

CALAMUS OR SWEET FLAG

Genera 1 Species 2–4
Economic uses Dried rhizomes and Calamus oil have been used for centuries as medicaments

A single genus of 2 to 4 spp. of sweet-smelling rhizomatous herbs widely distributed in the northern hemisphere, excluding the Arctic and dry areas. They have distichous, ensiform leaves with intravaginal scales. The inflorescence is borne laterally on a leaflike stem and is dense, spicate overtopped by a leaflike spathe. The flowers are bisexual, usually 3-merous with 3+3 tepals and 3+3 stamens, and the ovary is 2- to 3-carpellate, superior. The fruit is a capsule or berry.

Traditionally, the family has been included in the Araceae[i] but has recently been segregated as a separate family[ii,iii]. Most molecular analyses placed it as sister group to the remaining monocots[iv,v], but there are exceptions[vi]. The rhizomes of *Acorus calamus* have been used since biblical times as a tonic and medicine, and the oil (Calamus oil) has been used in perfumery and in the pharmaceutical industry. Oil produced in Asian countries from tetraploid and hexaploid strains has been shown to be toxic and carcinogenic due to beta-asarone calamus while the oil produced from diploid strains in North America are asarone free and triploid strains in Europe contain up to 8% beta-asarone and are thus acceptable. OS

[i] Dahlgren, R. M. T., Clifford, H. T., & Yeo, P. F. *The Families of the Monocotyledons*. Berlin, Springer-Verlag (1985).
[ii] Grayum, M. H. A summary of evidence and arguments supporting the removal of *Acorus* from the *Araceae*. *Taxon* 36: 723–729 (1987).
[iii] Mayo, S. J., Bogner, J., & Boyce, P. C. *The genera of Araceae*. Richmond, Royal Botanic Gardens, Kew (1997).
[iv] Duvall, M. R., Learn Jr., G. H., Eguiarte, L. E., & Clegg, M. T. Phylogenetic analysis of *rbc*L sequences identifies *Acorus calamus* as the primal extant monocotyledon. *Proc. Nat. Acad. Sci., USA* 90: 4611–4644 (1993).
[v] Chase, M. W. *et al.* Multi-gene analysis of monocot relationships: A summary. Pp. 63–75. In: Columbus, J. T. *et al.* (eds), *Monocots: Comparative Biology and Evolution*. Claremont, California, Rancho Santa Ana Botanic Garden (2006).
[vi] Davis, J. I. *et al.* A phylogeny of the monocots, as inferred from *rbc*L and *atp*A sequence variation, and a comparison of methods for calculating jackknife and bootstrap values. *Syst. Bot.* 29: 467–510 (2004).

AGAPANTHACEAE

A small, monogeneric family of perennial, tuberous herbs, with fleshy roots, restricted to southern South Africa. The leaves are basal, distichous, and linear. The inflorescence is scapose, umbel-like, with 2 involucral bracts. The bisexual, slightly zygomorphic flowers consist of 5, 3-merous whorls. The 3+3 tepals are more or less equal, blue or white, and fused basally into a tubular structure. The 3+3 stamens are fused at the base with the tube. The ovary is superior, 3-locular with numerous semi-campylotropous ovules, and a single, hollow style, and septal nectaries. The fruit is a loculicidal capsule with many seeds, with a large embryo embedded in an endosperm of aleurone, oil, and cellulose.

Genera 1 Species 9
Economic uses
Several species are of horticultural value, e.g., *Agapanthus praecox*

The Agapanthaceae has previously been included in the Alliaceae (q.v.)[i], but is now considered either sister group to Alliaceae plus Amaryllidaceae (q.v.)[ii], or to an even larger assemblage of families[iii]. OS

[i] Dahlgren, R. M. T., Clifford, H. T., & Yeo, P. F. *The Families of the Monocotyledons*. Berlin, Springer-Verlag (1985).
[ii] Pires, J. C. *et al.* Phylogeny, genome size, and chromosomal evolution of Asparagales. Pp. 287–304. In: Columbus, J. T. *et al.* (eds), *Monocots: Comparative Biology and Evolution*. Claremont, California, Rancho Santa Ana Botanic Garden (2006).
[iii] Meerow, A. W., Fay, M. F., Guy, C. L., Li, Q.-B., Zaman, F. Q., & Chase, M. W. Systematics of Amaryllidaceae based on cladistic analysis of Plastid *rbc*L and *trn*L-*trn*F sequence data. *American J. Bot.* 86: 1325–1345 (1999).

AGAVACEAE

AGAVE FAMILY

The Agavaceae are herbs, shrubs, or trees, with rosulate, often succulent, leaves, with usually large, many-flowered inflorescences.

Distribution Restricted to the New World in arid regions from the temperate to the tropical zone (USA and Canada to Bolivia), where it occurs in dry, rocky slopes and dry woodland.

Description Perennial, small to gigantic rosette herbs, up to 4 m in width, or shrubs or trees, with anomalous secondary growth. Some species are monocarpic. The leaves are rosulate, spirally arranged, either annual or perennial, and occasionally succulent, often terminated by a soft or sharp point, and with entire or prickly margins. The inflorescence is sometimes enormous, reaching a length of 8–12 m in some species of *Agave* and is a terminal or axillary, bracteate, spike, raceme, or panicle. The flowers are actinomorphic or zygomorphic, with 2 whorls of petaloid, slightly succulent tepals usually green, yellow, or white, either free or united into a tube. The 3+3 stamens are free, sometimes exserted, and bent in bud. The ovary is inferior or superior, consisting of 3 carpels and is 3-locular, with numerous anatropous ovules arranged in 2 rows in each locule, and a single style or 3 styles that may be exserted or included. The fruit is a loculicidal or septicidal capsule with flat, black seeds, with a cylindrical, straight, or somewhat curved embryo in an endosperm of lipids and proteins.

Classification The Agavaceae is here defined in a narrow sense[i] and includes the highly deviant species *Hesperocallis undulata*, which is sometimes, doubtfully, placed in Hostaceae (q.v.)[ii]. The 3 taxa share an unusual basic chromosome complement consisting of 5 long and 25 short chromosomes[iii]. However, molecular data place *Hesperocallis* as the sister group of Agavaceae[iv]. In the delimitation used here Agavaceae is closely related to a clade consisting of Anthericaceae (q.v.), Behniaceae (q.v.), and Herreriaceae (q.v.). Broadly defined, the Agavaceae could be extended to include these families plus Anemarrhenaceae (q.v.)[v] or even more of the families recognized here[vi], and even subsumed in a enlarged Asparagaceae[vi]. The pollination syndrome of *Yucca* is often considered as a classical example of coevolution[vii].

Economic uses Species of Agavaceae have been used by aboriginal populations in the New World for fibers, foods, medicine, and ornaments. The largest genus is *Agave* (c. 200 spp.), with hapaxanthic shoots. *A. americana* (Century Plant) is native to eastern Mexico and has been cultivated in Europe since 1561. Several *Agave* species are used to make fibers e.g., sisal (*A. sisalana*) or henequen (*A. fourcroydes*) and for the production of alcoholic drinks, e.g., tequila, mescal, and pulque. *Yucca*, *Agave*, and *Polianthes* are used as ornamentals or cut flowers. OS

Genera 9 Species c. 300
Economic uses
Species of the Agavaceae are rich source of fibers, food, beverages, and local medicine

i APG II. An update of the Angiosperm Phylogeny Group classification for the orders and families of flowering plants: APG II. *Bot. J. Linn. Soc.* 141: 399–436 (2003).
ii Kubitzki, K. Hostaceae. Pp. 256–260. In: Kubitzki, K. (ed.), *The Families and Genera of Vascular Plants. III. Flowering Plants. Monocotyledons: Lilianae (except Orchidaceae).* Berlin, Springer-Verlag (1998).
iii Gómez-Pompa, A., Villalobos-Pietrini, R., & Chimal, A. Studies in the Agavaceae. I. Chromosome morphology and number of seven species. *Madroño* 21: 208–221 (1991).
iv Pires, J. C., Maureira, J., Rebman, J. P., Salazar, G. A., Cabrera, L. I., Fay, M. F., & Chase, M. W. Molecular data confirm the phylogenetic placement of the enigmatic *Hesperocallis* (Hesperocallidaceae) with agave. *Madroño* 51: 307–311 (2004).
v Pires, J. C. *et al.* Phylogeny, genome size, and chromosomal evolution of Asparagales. Pp. 287–304. In: Columbus, J. T., Friar, E. A., Hamilton, C. W., Porter, J. M., Prince, L. M., & Simpson, M. G. (eds), *Monocots: Comparative Biology and Evolution.* Claremont, California, Rancho Santa Ana Botanic Garden (2006).
vi Chase, M. W. *et al.* Multi-gene analysis of monocot relationships: A summary. Pp. 63–75. In: Columbus, J. T. *et al.* (eds), *Monocots: Comparative Biology and Evolution.* Claremont, California, Rancho Santa Ana Botanic Garden (2006).
vii Pellmyr, O. Yuccas, yucca moths, and coevolution: A review. *Ann. Missouri Bot. Gard.* 90: 35–55 (2003).

ALISMATACEAE

WATER PLANTAINS

This small family of aquatics or marsh plants

Genera 11 **Species** c. 80
Economic uses A few species are tropical water weeds and some are important food for wildlife

has petiolate leaves, and flowers with white petals and many stamens and carpels.

Distribution The family is cosmopolitan but absent from Arctic and arid zones.

Description Mostly robust perennials but a few may be annual or perennial depending on the water regime. The stems are cormlike or stoloniferous. Most species have simple, entire leaves with a well-developed, often sagittate or hastate blade and distinct petiole with an expanded, sheathing base and intervaginal scales. Some species have 2 forms of leaf: narrow submerged and broader floating or emerged leaves, or both. Secretory ducts containing white latex are present. The inflorescence is usually terminal on a leafless scape, paniculate, or racemose, often complex with whorls of branches, but some species have umbel-like inflorescences or solitary flowers. Flowers regular, bisexual, or unisexual (monoecious or dioecious). The 3 sepals usually persist

in fruit, and the 3 petals are deciduous, crumpled in bud, white or pink in *Burnatia* and *Wiesneria,* minute or occasionally absent in *Burnatia*. The stamens are 3, 6, 9, or numerous. The ovary is superior, comprising 3 to numerous free carpels in 1 whorl or in a clustered head; each carpel contains 1 (rarely 2) anatropous ovules. The fruit is a head of nutlets (except in *Damasonium,* which has 6 to 10 dehiscent or semidehiscent, several-seeded follicles in 1 whorl united at the base or adnate to the elongated receptacle and spread starlike in fruit). The seeds without endosperm; embryo usually strongly curved.

Classification Generic circumscriptions in the Alismataceae are somewhat unsatisfactory and at least 1 of its largest genera, *Echinodorus,* with 27 spp., is polyphyletic[i]. Due to the numerous stamens and carpels, the Alismataceae have often been considered as primitive Monocots and related to the dicotyledonous family Ranunculaceae (Cronquist 1981). However, the resemblance of the families is superficial[ii]. The Alismataceae are closely related to Limnocharitaceae (q.v.).

Economic uses *Sagittaria sagittifolia* is cultivated in Asia for its edible corms, and the roots of *S. latifolia* were used as food by Native Americans in North America. Species of *Echinodorus* and *Sagittaria* are decorative and used as poolside or aquarium plants.   [CDC] OS

i Soros, C. L. & Les, D. H. Phylogenetic relationships in the Alismataceae. *American J. Bot.* 89, Suppl.: 152 (2002).
ii Les, D. H. & Haynes, R. R. Systematics of subclass Alismatidae: a synthesis of approaches. Pp. 353–377. In: Rudall, P. J., Cribb, P. J., Cutler, D. F., & Humphries, C. J. (eds), *Monocotyledons: Systematics and Evolution.* Richmond, Royal Botanic Gardens, Kew (1995).

ALLIACEAE

ONION FAMILY

The Alliaceae are bulbous or rhizomatous herbs with a characteristic smell and scapose umbel-like inflorescence.

Distribution The family is mainly distributed in areas with Mediterranean type climate in the western hemisphere, but *Allium* reaches far into Central Asia. Several genera are restricted to southern South America, and only few species are found in humid areas or in the tropics.

Description Mostly bulbous or rhizomatous, perennial or biennial geophytes, usually with a characteristic smell of onions. The leaves are linear, filiform, or lanceolate; rarely ovate; often terete and hollow; and fleshy with closed basal sheaths that occasionally form a pseudostem. The inflorescence is terminal, scapose, and usually umbel-like; rarely reduced to a single flower; and often subtended by 1 or 2 spathelike bracts, which enclose the young

Genera 13 **Species** 600–750
Economic uses Several species especially of *Allium* are widely used to flavor food, as vegetables, or as ornamentals

inflorescence. The flowers are bisexual and mostly actinomorphic, rarely zygomorphic, with 5, 3-merous whorls. The tepals are petaloid, either united at the base sometimes forming a tube or almost free, and occasionally with appendages or scales between the tepals and stamens. There are 3+3 stamens, rarely less; the filaments are often basally fused with the tepals and occasionally united into a tube, sometimes with apical appendages. The ovary is

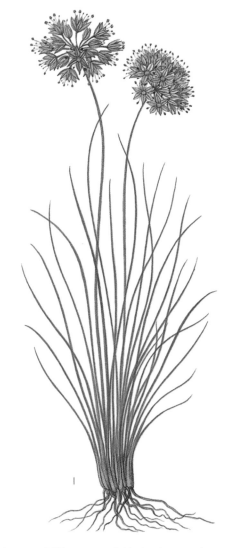

ALLIACEAE. 1 *Allium cyaneum* habit showing terete leaves, flowers, and umbel-like inflorescence (x⅔).

almost always superior, consisting of 3 carpels, and is 3-locular, with septal nectaries and 2 to many anatropous or campylotropous ovules in each locule. The style is single, apical, or gynobasic. The fruit is a loculicidal capsule with few to many seeds; a short, straight or long, curved embryo; and an endosperm of aleurone and fatty oils.

Classification As circumscribed here, Alliaceae appears to be monophyletic[i,ii], and several genera often included are treated in families of their own[iii]. Some are included in Themidaceae (q.v.)[i,iv,v] others in Agapanthaceae[vi]. The Alliaceae is sister to Amaryllidaceae (q.v.)[vii] or a broadly defined Asparagaceae (q.v.)[viii].

Economic uses *Allium* is of great economic importance and several species are widely cultivated: *A. cepa* (cepa) is used to flavor food and as a vegetable, *A. schoenoprasum* (chives) and *A. tuberosum* (Chinese chive) to flavor food, as are the bulbs of *A. sativum*. *A. ampeloprasum* (leek) is used as a vegetable. Species of several genera, e.g., *Allium*, *Ipheion*, *Nothoscordum*, are of horticultural importance. OS

i Fay, M. F. and Chase, M. W. Resurrection of Themidaceae for the *Brodiaea* alliance, and recircumscription of Alliaceae, Amaryllidaceae and Agapanthoideae. *Taxon* 45: 441–451 (1996).
ii Meerow, A. W., Fay, M. F., Guy, C. L., Li, Q.-B., Zaman, F. Q., & Chase, M. W. Systematics of Amaryllidaceae based on cladistic analysis of plastid *rbc*L and *trn*L-*trn*F sequence data. *American J. Bot.* 86: 1325–1345 (1999).
iii Dahlgren, R. M. T., Clifford, H. T., & Yeo, P. F. *The Families of the Monocotyledons*. Berlin, Springer-Verlag (1985).
iv Rahn, K. Themidaceae. Pp. 436–441. In: Kubitzki, K. (ed.), *The Families and Genera of Vascular Plants. III. Flowering Plants. Monocotyledons: Lilianae (except Orchidaceae)*. Berlin, Springer-Verlag (1998).
v Pires, J. C., Fay, M. F., Davis, S., Hufford, L., Rova, J., Chase, M. W., & Sytsma, K. J. Molecular and morphological phylogenetic analyses of the Themidaceae (Asparagales). *Kew Bull.* 56: 601–626 (2001).
vi Kubitzki, K. Agapanthaceae. Pp. 58–60. In: Kubitzki, K. (ed.), *The Families and Genera of Vascular Plants. III. Flowering Plants. Monocotyledons: Lilianae (except Orchidaceae)*. Berlin, Springer-Verlag (1998).
vii Pires, J. C. *et al.* Phylogeny, genome size, and chromosomal evolution of Asparagales. Pp. 287–304. In: Columbus, J. T. *et al.* (eds), *Monocots: Comparative Biology and Evolution*. Claremont, California, Rancho Santa Ana Botanic Garden (2006).
viii Chase, M. W. *et al.* Multi-gene analysis of monocot relationships: A summary. Pp. 63–75. In: Columbus, J. T. *et al.* (eds), *Monocots: Comparative Biology and Evolution*. Claremont, California, Rancho Santa Ana Botanic Garden (2006).

ALSTROEMERIACEAE

ALSTROEMERIA FAMILY

A family of rhizomatous herbs or vines often with swollen roots.

Distribution The family is restricted to the New World, from Mexico along the Andes to Chile and around the Paraná River and into adjacent Brazil. It grows in a variety of habitats in both the temperate and tropical zones.

Description Almost always erect or twining perennials (*Alstroemeria graminea* is annual), glabrous, with rhizomes and swollen storage roots. The simple leaves are alternate, dispersed along the stem or gathered in a basal rosette, often more or less twisted at the base, so the blade becomes inverted. The inflorescence is either terminal bracteate, with leaflike bracts, few to many-flowered, and often umbel-like, or rarely the flowers are solitary. Flowers are bisexual, actinomorphic or zygomorphic (in *Alstroemeria*), with 2 whorls of 3 petaloid tepals, often the inner whorl nectariferous at the base, and 2 whorls of 3 free stamens. The ovary is inferior with 3 carpels and is 1-(*Leontochir*) or 3-locular, with parietal or axile placentas. Ovules are numerous and anatropous, and the style is filiform with 3 stigmatic branches. The fruit is usually a dehiscent, dry or leathery, loculicidal capsule, which in *Alstroemeria* is explosive. The seeds are globose, often with an orange-red aril, and contain a cylindrical embryo in an endosperm of aleurone and oil.

Classification The genera *Taltalia* and *Schickendantzia* are often segregated from *Alstroemeria*, but their exclusion makes *Alstroemeria* paraphyletic[i]. The Alstroemeriaceae is undoubtedly monophyletic[ii,iii] and has been considered closely related to the Orchidaceae[iv]. Recent molecular analysis points at a sister-group relationship to Luzuriagaceae[ii,v,vi] (q.v.).

Economic uses Several species and hybrids of *Alstroemeria* (Inca Lily) are in cultivation and sold as cut flowers or grown in gardens. Species of *Bomarea*, especially *B. edulis*, are cultured for their starchy, edible roots. OS

Genera 3
Species c. 160
Economic uses Of great importance in horticulture

i Aagesen, L. & Sanso, A. M. The phylogeny of Alstroemeriaceae, based on morphology, *rps*16 intron, and *rbc*L sequence data. *Syst. Bot.* 28: 47–69 (2003).
ii Davis, J. I. *et al.* A phylogeny of the monocots, as inferred from *rbc*L and *atp*A sequence variation, and a comparison of methods for calculating jackknife and bootstrap values. *Syst. Bot.* 29: 467–510 (2004).
iii Fay, M. F. *et al.* Phylogenetics of Liliales: Summarized evidence from combined analyses of five plastid and one mitochondrial loci. Pp. 559–565. In: Columbus, J. T., Friar, E. A., Hamilton, C. W., Porter, J. M., Prince, L. M., & Simpson, M. G. (eds), *Monocots: Comparative Biology and Evolution*. Claremont, California, Rancho Santa Ana Botanic Garden (2006).
iv Dahlgren, R. M. T., Clifford, H. T., & Yeo, P. F. *The Families of the Monocotyledons*. Berlin, Springer-Verlag (1985).
v Chase, M. W. *et al.* Multi-gene analysis of monocot relationships: A summary. Pp. 63–75. In: Columbus, J. T. *et al.* (eds), *Monocots: Comparative Biology and Evolution*. Claremont, California, Rancho Santa Ana Botanic Garden (2006).
vi Vinnersten, A. & Bremer, K. Age and biogeography of major clades in Liliales. *American J. Bot.* 88: 1695–1703 (2001).

AMARYLLIDACEAE

DAFFODIL FAMILY

A family of usually bulbous, perennial herbs frequently with distichous, simple, often linear leaves and scapose, umbel-like inflorescences.

Distribution The family is cosmopolitan but occurs mainly in the tropics and subtropics, although there are several representatives in temperate areas in Europe and a few in Asia. Most species are adapted to seasonally dry habitats.

Description Mostly bulbous, perennial geophytes, but sometimes aquatics (e.g., *Crinum*) or epiphytes, with distichous or spirally arranged, linear or elliptic leaves, which occasionally have a sheathing base that forms a pseudostem. The inflorescence is often scapose and umbel-like, but sometimes almost sessile (e.g., *Gethyllis*, *Sternbergia*) with 2 bracts that enclose the flowers in bud. There are 1 to many, often large and showy, sessile or pedicellate flowers. The flowers are actinomorphic or zygomorphic, with 3+3 petaloid tepals, the inner smaller than the outer, usually fused into a tube at the base (e.g., *Crinum*, *Zephyranthes*, *Sternbergia*, *Cyrtanthus*, *Stenomesson*) or free (e.g., *Galanthus*, *Leucojum*, *Amaryllis*, *Nerine*). In some genera, there is a conspicuous (e.g.,

Genera c. 60   Species 850
Economic uses A rich source of ornamentals

Narcissus) or inconspicuous corona (e.g., *Hippeastrum*, *Lycoris*). Generally, there are 3+3 more or less equal stamens, rarely fewer or more, inserted at the base of the tepals. The stamens are occasionally connate into a cup or paracorona (e.g., *Pancratium*, *Nerine*). The ovary is inferior, 3-carpellate and usually 3-locular, with septal nectaries and axile or basal placentas with anatropous ovules. The style is filiform with a capitate, 3-lobed or 3-fid stigma. The fruit is a loculicidal capsule, rarely baccate (e.g., *Haemanthus*). The seeds are frequently numerous, black or dark brown, and have a small, straight embryo and an endosperm consisting of hemicellulose and lipids or starch.

Classification The circumscription of the Amaryllidaceae has varied tremendously, but the present delimitation of the family has gained widespread acceptance[i,ii,iii]. There is little doubt that the family is monophyletic[iv,v], and closely related to Alliaceae (q.v.)[iv,vi].

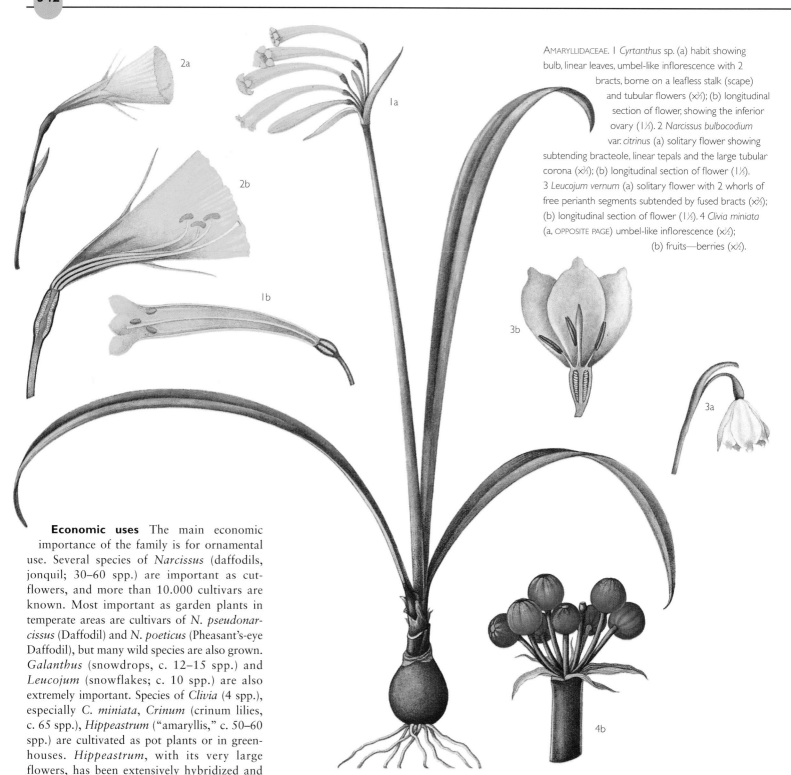

AMARYLLIDACEAE. I *Cyrtanthus* sp. (a) habit showing bulb, linear leaves, umbel-like inflorescence with 2 bracts, borne on a leafless stalk (scape) and tubular flowers (x⅔); (b) longitudinal section of flower, showing the inferior ovary (1⅓). 2 *Narcissus bulbocodium* var. *citrinus* (a) solitary flower showing subtending bracteole, linear tepals and the large tubular corona (x⅔); (b) longitudinal section of flower (1⅓). 3 *Leucojum vernum* (a) solitary flower with 2 whorls of free perianth segments subtended by fused bracts (x⅔); (b) longitudinal section of flower (1⅓). 4 *Clivia miniata* (a, OPPOSITE PAGE) umbel-like inflorescence (x⅓); (b) fruits—berries (x⅓).

Economic uses The main economic importance of the family is for ornamental use. Several species of *Narcissus* (daffodils, jonquil; 30–60 spp.) are important as cut-flowers, and more than 10.000 cultivars are known. Most important as garden plants in temperate areas are cultivars of *N. pseudonarcissus* (Daffodil) and *N. poeticus* (Pheasant's-eye Daffodil), but many wild species are also grown. *Galanthus* (snowdrops, c. 12–15 spp.) and *Leucojum* (snowflakes; c. 10 spp.) are also extremely important. Species of *Clivia* (4 spp.), especially *C. miniata*, *Crinum* (crinum lilies, c. 65 spp.), *Hippeastrum* ("amaryllis," c. 50–60 spp.) are cultivated as pot plants or in greenhouses. *Hippeastrum*, with its very large flowers, has been extensively hybridized and is an important bulb crop in the Netherlands. Some species of *Crinum* are sold as aquarium plants. Several genera are of local medical importance and have attracted attention for their possible anticancer alkaloids.   [BFM] OS

i Cronquist, A. *An Integrated System of Classification of Flowering Plants*. New York, Columbia University Press (1981).

ii Dahlgren, R. M. T., Clifford, H. T., & Yeo, P. F. *The Families of the Monocotyledons*. Berlin, Springer-Verlag (1985).

iii Meerow, A. & Snijman, D. A. Amaryllidaceae. Pp. 83–110. In: Kubitzki, K. (ed.), *The Families and Genera of Vascular Plants. III. Flowering Plants. Monocotyledons: Lilianae (except Orchidaceae)*. Berlin, Springer-Verlag (1998).

iv Fay, M. F. *et al*. Phylogenetic studies of Asparagales based on four plastid DNA regions. Pp. 360–371. In: Wilson, K. L. & Morrison, D. A. (eds), *Monocots. Systematics and Evolution*. Melbourne, CSIRO Publishing (2000).

v Meerow, A. W., Fay, M. F., Guy, C. L., Li, Q.-B., Zaman, F. Q., & Chase, M. W. Systematics of Amaryllidaceae based on cladistic analysis of plastid *rbc*L and *trn*L-*trn*F sequence data. *American J. Bot.* 86: 1325–1345 (1999).

vi Pires, J. C. *et al*. Phylogeny, genome size, and chromosomal evolution of Asparagales. Pp. 287–304. In: Columbus, J. T. *et al*. (eds), *Monocots: Comparative Biology and Evolution*. Claremont, California, Rancho Santa Ana Botanic Garden (2006).

ANARTHRIACEAE

A small family of evergreen, caespitose, dioecious herbs, endemic to southwestern Australia, with either well-developed, longitudinally folded, linear leaves, with a split, sheathing base (*Anarthria*) or only the sheaths well-developed, the leaves rudimentary (*Hopkinsia*) or lacking (*Lyginia*). The inflorescence is racemose with few or many sessile or pedi-cellate flowers. The flowers consists of 6 tough or membranaceous tepals in 2 whorls. The male flowers have 3+3 stamens opposite the inner tepals. The female flowers have a superior 3-locular ovary, with the 3 styles free to the base, and a single, hanging, pendulous, atropous ovule. The fruit is a capsule.

Molecular evidence suggests that the Anarthriaceae genera are

Genera 3 **Species** 11
Economic uses None

monophyletic[i]. Previously, the genera have all been placed in the Restionaceae (q.v.)[ii], a position that may still be tenable as molecular analysis either places them as sister group to the remaining genera of Restionaceae[iii] or as related to Restionaceae plus Centrolepidaceae[iv,v,vi]. However, the genera have also all been placed in 3 separate families[v]. OS

[i] Doyle, J. J., Davis, J. I., Soreng, R., Garvin, D., & Andersson, M. J. Chloroplast DNA inversions and the origin of the grass family (Gramineae). *Proc. Nat. Acad. Sci., USA*. 89: 7722–7726 (1992).
[ii] Cutler, D. F. & Shaw, H. K. A. Anarthriaceae and Ecdeiocoleaceae: two new monocotyledonous families, separated from Restionaceae. *Kew Bull.* 19: 160–168 (1965).
[iii] Davis, J. I. *et al.* A phylogeny of the monocots, as inferred from *rbc*L and *atp*A sequence variation, and a comparison of methods for calculating jackknife and bootstrap values. *Syst. Bot.* 29: 467–510 (2004).
[iv] Chase, M. W. *et al.* Multi-gene analysis of monocot relationships: A summary. Pp. 63–75. In: Columbus, J. T. *et al.* (eds), *Monocots: Comparative Biology and Evolution*. Claremont, California, Rancho Santa Ana Botanic Garden (2006).
[v] Briggs, B. G. & Johnson, L. A. S. Hopkinsiaceae and Lyginiaceae, two new families of Poales in Western Australia, with revisions of *Hopkinsia* and *Lyginia*. *Telopea* 8: 477–502 (2000).
[vi] Briggs, B. G., Marchant, A. D., Gilmore, S., & Porter, C. L. A molecular phylogeny of Restionaceae and allies. In: Wilson, K. L. & Morrison, D. A. (eds), *Monocots. Systematics and Evolution*. Melbourne, CSIRO Publishing (2000).

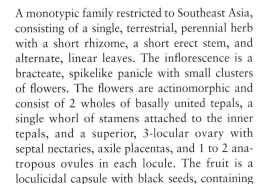

4a

ANEMARRHENACEAE

A monotypic family restricted to Southeast Asia, consisting of a single, terrestrial, perennial herb with a short rhizome, a short erect stem, and alternate, linear leaves. The inflorescence is a bracteate, spikelike panicle with small clusters of flowers. The flowers are actinomorphic and consist of 2 wholes of basally united tepals, a single whorl of stamens attached to the inner tepals, and a superior, 3-locular ovary with septal nectaries, axile placentas, and 1 to 2 anatropous ovules in each locule. The fruit is a loculicidal capsule with black seeds, containing a linear embryo in a fleshy endosperm.

The position of *Anemarrhena asphodeloides* has been problematic. Previously, it has been placed in the Anthericaceae (q.v.)[i], but mo-lecular evidence now includes it either in a broadly circumscribed Agavaceae[ii] or an even more broadly defined Asparagaceae[iii] or to be a sister group of a clade that contains Agavaceae in a narrow sense, plus Antheri-caceae and some smaller families[iv]. OS

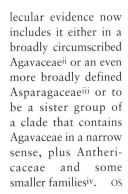
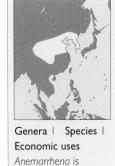

[i] Dahlgren, R. M. T., Clifford, H. T., & Yeo, P. F. *The Families of the Monocotyledons*. Berlin, Springer-Verlag (1985).
[ii] APG II. An update of the Angiosperm Phylogeny Group classification for the orders and families of flowering plants: APG II. *Bot. J. Linn. Soc.* 141: 399-436 (2003).
[iii] Chase, M. W. *et al.* Multi-gene analysis of monocot relationships: A summary. Pp. 63–75. In: Columbus, J. T. *et al.* (eds), *Monocots: Comparative Biology and Evolution*. Claremont, California, Rancho Santa Ana Botanic Garden (2006).
[iv] Pires, J. C. *et al.* Phylogeny, genome size, and chromosomal evolution of Asparagales. Pp. 287–304. In: Columbus, J. T. *et al.* (eds), *Monocots: Comparative Biology and Evolution*. Claremont, California, Rancho Santa Ana Botanic Garden (2006).

ANTHERICACEAE

SPIDER PLANT FAMILY

A small family of terrestrial herbs, often with spirally arranged or distichous, linear basal leaves.

Distribution The Anthericaceae has a subcosmopolitan distribution, however, it is absent from extremely cold climates. The family occurs mostly in open habitats and woodlands, but very rarely in rain forest.

Description Perennial, mostly rhizomatous herbs, rarely woody, with spirally arranged or distichous, more or less sheathing, linear to oblong/lanceolate, often rosulate leaves. The inflorescence is either terminal, usually a scapose, bracteate raceme or panicle, or umbel-like arising directly from the rhizomes (e.g., *Leucocrinum*). There are 1 to several bisexual, actinomorphic or zygomorphic, rather small flowers at each node. The flowers

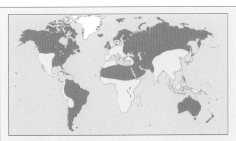

have 2, 3-merous whorls of more or less equal tepals that are either free or fused at the base into a long tube (e.g., *Diora*). The stamens are 3+3, united with the tepals at the base, or rarely inserted near the throat or fused at the base. The ovary is superior and 3-locular, with 2 to many, nearly always anatropous or campylotropous, ovules in 2 rows in each locule on axile placentas, septal nectaries, and a filiform capitate or 3-lobed style. The fruit is a loculicidal capsule, which in *Leucocrinum* is subterranean, containing 2 to many black seeds, with a cylindrical embryo and a fleshy endosperm of fat, oil, and starch.

Classification Parts of Laxmanniaceae (q.v.), Hemerocallidaceae (q,v.), and the whole of Boryaceae (q.v.) and Johnsoniaceae (q.v.) are occasionally included in Antheriaceae[i], but here the family is recognized in a narrow sense, excluding these taxa[ii]. In this narrow sense, Anthericaceae is sister group to Herreri-aceae (q.v.)[iii,iv], and both families are, together with several other families, sometimes included in broadly defined Agavaceae (q.v.)[iii] or a very broadly defined Asparagaceae (q.v.)[v]. The relationship between *Echeandia* (c. 65 spp.) and New World *Anthericum* (c. 65 spp.) species is unclear. Most American species of *Anthericum* and *Chlorophytum* (150–200 spp.) have been transferred to either *Echeandia* or *Hagenbachia* (5–6 spp.).

Economic uses Species of *Anthericum* and *Chlorophytum*, especially *C. comosum* (Spider Plant), are widely cultivated as ornamentals. OS

[i] Dahlgren, R. M. T., Clifford, H. T., & Yeo, P. F. *The Families of the Monocotyledons*. Berlin, Springer-Verlag (1985).
[ii] Conran, J. G. Anthericaceae. Pp. 114–121. In: Kubitzki, K. (ed.), *The Families and Genera of Vascular Plants. III. Flowering Plants. Monocotyledons: Lilianae (except Orchidaceae)*. Berlin, Springer-Verlag (1998).
[iii] Pires, J. C. *et al.* Phylogeny, genome size, and chromosomal evolution of Asparagales. Pp. 287–304. In: Columbus, J. T. *et al.* (eds), *Monocots: Comparative Biology and Evolution*. Claremont, California, Rancho Santa Ana Botanic Garden (2006).
[iv] Chase, M. W. *et al.* Multi-gene analysis of monocot relationships: A summary. Pp. 63–75. In: Columbus, J. T. *et al.* (eds), *Monocots: Comparative Biology and Evolution*. Claremont, California, Rancho Santa Ana Botanic Garden (2006).

APHYLLANTHACEAE

Aphyllanthus monspeliensis is a caespitose, perennial herb that occurs in rocky soils in southern France, northern Spain, and Morocco. The leaves are alternate, distichous, and sheathing with a strongly reduce lamina. The inflorescence is scapose, capitate, with scarious bracts, and 1 to 2, or rarely 3, actinomorphic, sessile flowers. The flowers have 3+3, blue, free, overlapping, equal tepals, 3+3 stamens inserted at the base of the tepals, and an inferior, 3-locular ovary, with septal nectaries and 3 stigmatic lobes. Each locule contains 1 anatropous ovule. The fruit is a loculicidal capsule with 1 to 3 black seeds and a linear embryo in an endosperm of aleurone and lipids.

Classification Due to its deviant morphology Aphyllanthaceae has often been placed in a family of its own with supposedly obscure relationships[i,ii]. Molecular evidence places the family as sister group to Laxmanniaceae (q.v.)[iii], which is part of an enlarged Asparagaceae[iv]. This position has also been suggested by morphology[v,vi].   OS

Genera 1 | Species 1
Economic uses None

[i] Dahlgren, R. M. T., Clifford, H. T., & Yeo, P. F. *The Families of the Monocotyledons*. Berlin, Springer-Verlag (1985).
[ii] Conran, J. G. Aphyllanthaceae. Pp. 122–124. In: Kubitzki, K. (ed.), *The Families and Genera of Vascular Plants. III. Flowering Plants. Monocotyledons: Lilianae (except Orchidaceae)*. Berlin, Springer-Verlag (1998).
[iii] Pires, J. C. *et al.* Phylogeny, genome size, and chromosomal evolution of Asparagales. Pp. 287–304. In: Columbus, J. T. *et al.* (eds), *Monocots: Comparative Biology and Evolution*. Claremont, California, Rancho Santa Ana Botanic (2006).
[iv] Chase, M. W. *et al.* Multi-gene analysis of monocot relationships: A summary. Pp. 63–75. In: Columbus, J. T. *et al.* (eds), *Monocots: Comparative Biology and Evolution*. Claremont, California, Rancho Santa Ana Botanic Garden (2006).
[v] Huber, H. Die Samenmerkmale und Verwandtschaftsverhältnisse der Liliiflore. *Mitt. Bot. Staats.* München 8: 219–538 (1969).
[vi] Chandra, S. & Gosh, K. Pollen morphology and its evolutionary significance. In: Ferguson, I. K. & Muller, J. *The Evolutionary Significance of the Exine*. London, Academic Press (1976).

APONOGETONACEAE
WATER HAWTHORN

This family comprises a single genus of rhizomatous or aquatics with a spicate inflorescence and more or less conspicuous tepals.

Distribution Restricted to warm and tropical parts of the Old World and northern Australia, almost all species are exclusively aquatics, although a few occur in marshy habitats. Most species are found in Africa and Madagascar. The South African *Aponogeton distachyos* (Water Hawthorn or Cape Pondweed) is naturalized in southern Australia, western South America, and Western Europe.

Description Perennial aquatics with short, tuberlike corms or, rarely, elongated and branched rhizomes. Many species survive dry periods as dormant tubers. The usually distinctly petiolate leaves have intervaginal scales and are spirally arranged at the base of the stems. Submerged leaves are linear to oblong/elliptic, often undulate or contorted, whereas floating leaves are ovate to lanceolate, mostly with a distinct midrib and 1 or more pairs of parallel main nerves connected by numerous cross veins. Both leaves and stems have oil or tannin-containing articulated laticifers. In the Madagascan Lace Plant, *A. madagascariensis* (*A. fenestralis*), the whole blade is fenestrated and consists of a lattice of veins and nerves. Several species develop thick, leathery, floating leaves. In some species, such as *A. junceus*, the leaves are reduced to elongated midribs and resemble rushes. Flowers are usually bisexual or occasionally unisexual; some species are agamospermous. The inflorescences are spikelike and borne on long stalks, which emerge above the water surface. In bud, each spike is enveloped in a spathe. In all the Asian and most Australian species the spikes are single; in most African species they are bifurcating and in some Madagascan species there are up to 10 spikes on 1 stalk. The flowers are usually small and zygomorphic. There are usually 2 tepals (only 1

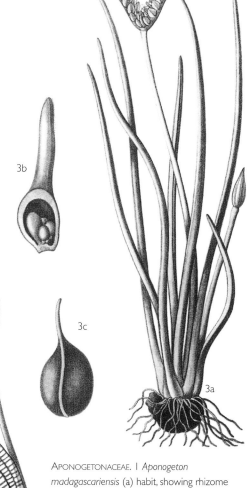

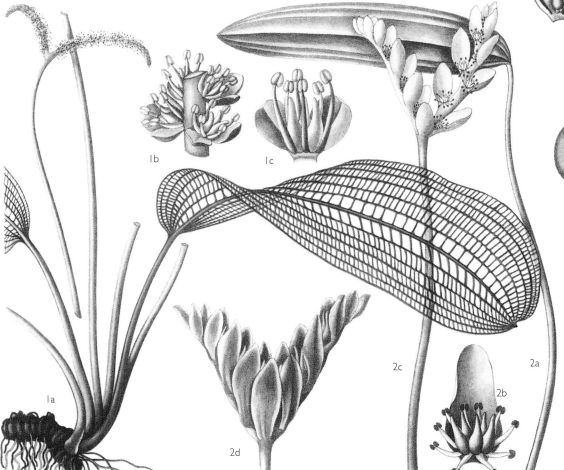

APONOGETONACEAE. 1 *Aponogeton madagascariensis* (a) habit, showing rhizome bearing leaves in which the blade is merely a lattice of veins and nerves, and a bifurcating inflorescence born on a long stalk (×⅔); (b) portion of inflorescence (×4); (c) flower, consisting of 2 tepals, 6 stamens, and 3 sessile carpels (×6). 2 *Aponogeton distachyos* (a) aerial leaf (×⅔); (b) flower with a single tepal (×2); (c) inflorescence (×⅔); (d) infructescence with persistent tepals (×⅔). 3 *Aponogeton junceus* (a) habit showing tuberlike corms, tufts of straplike leaves and bifurcating inflorescence (×⅔); (b) vertical section of carpel showing sessile ovules (×16); (c) fruit—a leathery follicle (×6).

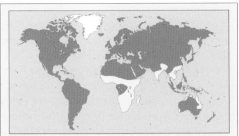

Genera 1 **Species** c. 45
Economic uses Some species are edible and others are grown as aquarium plants

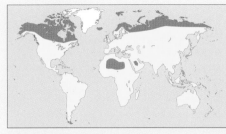

Genera c. 110 **Species** c. 3,200
Economic uses Many species with edible tubers, a subsistence food crop of importance locally and throughout the tropics. Some species are medically useful. Some are economically valuable ornamentals

large tepal in *A. distachyos* and 6 in *A. hexatepalus*) often white, persistent, or caducous. The 6 stamens are in 2 or more whorls of 3 each, but there are up to 16 stamens in *A. distachyos*. Between 2 and 9 ovaries are free, sessile, and superior. The usually 3 free carpels have 2 to numerous anatropous ovules on a basal placenta. The carpels may mature under water into free, leathery follicles with 2 to numerous seeds. The seeds have a straight embryo and no endosperm; the testa is usually single but can be split in two.

Classification The presence of lactifers suggests a close relationship with Butomaceae and Alsimataceae[i], but this is contradicted by molecular data[ii].

Economic uses Tubers and leaves are eaten locally by humans and livestock. Several species are popular aquarium plants. Due to large-scale trade the Madagascan Lace Plant (*A. madagascariensis*) has become extinct in many localities. [CDC] OS

[i] Les, D. H., Cleland, M. A., & Waycott, M. Phylogenetic studies in Alismatidae, II: Evolution of marine angiosperms (seagrasses) and hydrophily. *Syst. Bot.* 22: 443–463 (1997).
[ii] van Bruggen, H. W. E. Aponogetonaceae. Pp. 21–25. In: Kubitzki, K. (ed.), *The Families and Genera of Vascular Plants. IV. Flowering Plants. Monocotyledons: Alismatanae and Commelinanae (except Gramineae)*. Berlin, Springer-Verlag (1998).

ARACEAE
AROIDS AND DUCKWEEDS

A family of terrestrial herbs with flowers in a spadix (aroids) and minute floating aquatic herbs with reduced flowers (duckweeds).

Distribution Cosmopolitan, with most species in tropical Southeast Asia, Africa, and America, fewer in temperate and boreal zones.

Description Perennial, mainly terrestrial, herbs with a variety of lifeforms, including geophytes, epiphytes, and climbers, rarely free-floating aquatics, with either rosulate leaves (*Pistia*) or platelike fronds (*Lemna*). Both raphides and laticifers, with a milky or watery sap (latex), are common. The roots are adventitious, often thickened, rhizomes or subglobose tubers. Most

ARACEAE. 1 *Spirodela polyrhiza* (a) fronds with adventitious roots and prominent root caps (×2⅔); (b) underside of frond with 2 daughter fronds (×2⅔); (c) vertical section of frond with multilayered air spaces (×5⅓). 2 *Wolffia arrhiza* (a) frond with several daughter fronds (×26); (b) section of frond with a daughter frond budding of from a lateral pouch and a flower (sometimes interpreted as a reduced inflorescence) in a single cavity (×52). 3 *Lemna gibba* (a) fronds (×12); (b) vertical section of frond with a single layered air space (×8); (c) inflorescence of 2 male flowers and 1 female flower enclosed by a scalelike leaf (×20); (d) opened fruit showing several seeds (×20); (e) cross section of fruit with 4 seeds (×20); (f) vertical section of seed with embryo at the top (×40). 4 *L. minor* opened fruit with a single seed (×20). 5 *L. trisulca* (a) frond with several daughter fronds (×2⅓); (b) seedling, showing first frond and suspensor (×6); (c) fruit enclosed in scalelike leaf (×26).

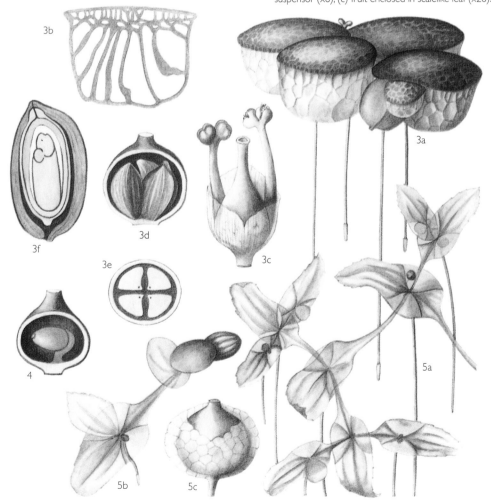

climbers and epiphytes have root dimorphism e.g., some roots are absorbent, growing downward toward the soil, while the others, not influenced by gravity, grow away from light and clasp firmly as they grow into crevices in the branches of the supporting tree. Many aerial roots develop an outer water-absorbing tissue similar to the velamen of orchids.

A few genera e.g., *Anthurium* and *Philodendron*, include species that are truly epiphytic and collect water and mineral salts in their leaf rosettes. The aquatics either lack roots (*Pistia, Wolffia*) or develop simple roots (*Lemna*). The leaves are usually differentiated in petiole sheath, petiole, and an expanded leaf blade of variable shape. Some species e.g., the Swiss Cheese Plant (*Monstera deliciosa*) develops fenestrated or holed leaves during development.

The inflorescence usually consists of a dense spadix with numerous, ebracteate flowers and is usually subtended by a conspicuous spathe that is often green but sometimes elaborately colored. In the aquatics (e.g., *Lemna*), the inflorescence is reduced to a single flower more or less hidden in a cavity of the fronds. The flowers are sessile and bi- or unisexual. With a few exceptions (e.g., *Arisaema*), the inflorescence is bisexual, with male flowers borne above female flowers. The tepals are often more or less connate in bisexual flowers and absent in the unisexual flowers. There are usually 2 whorls of 2 or 3 stamens more or less opposite the tepals, if present. The ovary is superior or embedded in the spadix and generally composed of 1 to 3, rarely more, carpels with a comparable number of locules with 1 to numerous anatropous, amphitropous, or atropous ovules on basal, apical, axile, or parietal placentas. The style is usually inconspicuous with lobed stigma. The fruit is a berry, sometimes a utricle, a drupe, or

6

nutlike, with 1 to many seeds, which usually have endosperm. The embryo is generally straight, although some species have a curved embryo with or without endosperm.

Nearly all species, except some of the aquatics (e.g., *Lemna*), are insect pollinated. The inflorescence of many species emits a nauseous, foetid odor that attracts carrion flies to effect pollination, but some have less offensive odors (e.g., *Spathiphyllum* and *Philodendron* spp.). The production of odor may be accompanied by a rise in inflorescence temperature.

Classification Traditionally, the Lemnaceae has been considered a separate family[i,ii], closely related to the Araceae, but is here included in the Araceae as phylogenetic analyses have unanimously shown that the Araceae are paraphyletic if Lemnaceae are excluded[iii,iv]. A close relationship between the Araceae and various aquatic Monocot families (e.g., Alismataceae) has previously been suggested mainly on morphological evidence[i] and has now been supported by molecular data[v]. In the current circumscription, the Araceae consists of 8 subfamilies:

SUBFAM. GYMNOSTACHYDOIDEAE Monotypic. *Gymnostachys anceps,* an acaulescent herb with linear leaves, which are not differentiated into blade and petiole, and have parallel venation; restricted to forest in eastern Australia; lacking laticifers; inflorescence scapose, usually consisting of 3 to 7 partial inflorescences separated from each other by a peduncular axis, each partial inflorescence with several spadices and inconspicuous spathes; flowers bisexual, with 4 tepals, 4 stamens, 2-carpellate, with a single atropous ovule.

SUBFAM. ORONTIOIDEAE 3 genera, North America and East Asia, all found in swampy habitats; lacking laticifers; rhizomes erect, branching sympodially; leaves entire, petiolated, oblong/lanceolate to cordate, with reticulate higher order venation; 1 to 2 erect inflorescences with

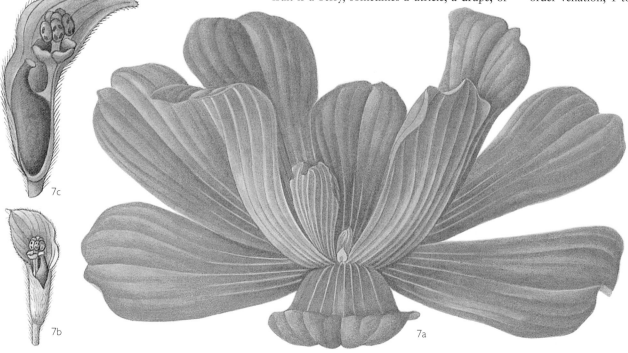

ARACEAE. 6 *Philodendron verrucosum* epiphytic stem with clinging, adventitious roots (x⅓). 7 *Pistia stratiotes* (a) habit, showing floating rosette of leaves and a single inflorescence (x⅔); (b) inflorescence with subtending, basally connate spathe (x2); (c) vertical section of inflorescence with spathe, a single basal female flower and above a whorl of male flowers each with pairs of fused stamens (x4).

7c

7b

7a

suppressed peduncles, with or without conspicuous spathe; flowers bisexual, 2- or 3-merous, with tepals, and the ovary is immersed in the spadix axis. The 2 spp. of *Lysichiton*, *L. camtschatcensis* from E Asia and *L. americanus* from western North America, both have conspicuous spathes (white and yellow, respectively) are grown as ornamentals.

SUBFAM. POTHOIDEAE This subfamily consists of 4 genera, 3 of which are found in Southeast Asia, and 1, *Anthurium*, in Central and South America. Their growth is monopodial, and they are mostly aerial herbs without laticifers. The leaves are differentiated into blade and petiole and have reticulate higher-order venation. The petiole is geniculate near the apex. The spathe does not enclose the spadix, and the flowers are bisexual, 3-merous, or rarely 2-merous, with tepals. The largest genus is the largely epiphytic *Anthurium* (c. 800 spp.), the leaf shape is variable and some species have leaves up to 2 m long. The inclusion of *Anthurium* in this subfamily is uncertain[iv].

SUBFAM. MONSTEROIDEAE 12 genera, largely pantropical; mostly terrestrial or climbing, rarely epiphytes; lacking laticifers; leaves sometimes pinnatifid or perforated; spathe is expanded or boat-shaped and the spadix fertile to the apex; flowers nearly always bisexual, 2-merous, with or without tepals.

SUBFAM. LASIOIDEAE 10 genera of terrestrial or aquatic plants from tropical parts of South America, Africa, and Southeast Asia; with laticifers; leaves have both blade and a petiole that is geniculate at the apex, often either armed with prickles or warty; with a striking coloration and reticulate higher-order venation; spadix flowering and fruiting in a basipetal sequence; flowers bisexual. *Dracontium gigas* has only 1 leaf, up to 3 m long, and a single inflorescence, up to 1.5 m tall.

SUBFAM. CALLOIDEAE Monotypic. The aquatic *Calla palustris*, circumboreal in distribution; lacking laticifers; stem rhizomatous with distichously arranged long-petiolated cordate leaves with parallel-pinnate venation; spathe is expanded, white within and green on the outside at anthesis; flowers are bisexual and 3-merous without tepals.

SUBFAM. AROIDEAE The largest subfamily, 72 genera, cosmopolitan, most genera and species tropical; with laticifers; mostly tuberous or rhizomatous climbers or epiphytes; few are truly aquatics (*Pistia*); leaves differentiated into blade and petiole, rarely geniculate at the apex, and often with reticulate higher-order venation; spathe usually with a basally convolute tube and an upper, expanded blade; flowers unisexual (plant monoecious), lacking tepals. The genus *Amorphophallus* (c. 150 spp.) is restricted to the Old World tropics. *A. titanium* is said to have the largest inflorescence of any herbaceous plant, weighing up to 50 kg but is mostly likely surpassed by *A. brooksii* with an inflorescence weighing up to 70 kg. The monotypic *Pistia stratiotes* is a pantropical, free-floating, evergreen aquatic. The genus *Arum* has highly specialized pollination mechanisms.

SUBFAM. LEMNOIDEAE Worldwide distribution, but not in dry and arctic regions. Morphologically, the 5 genera of the Lemnoideae deviate from the remaining species in the family. All are free-floating herbs reduced to small, poorly differentiated fronds, which float on or just below the surface, either single or several cohering together, with up to 21 roots on a frond or only 1. The fertile fronds have 1 to 2 lateral reproductive pouches, which have 1 to 2 bisexual flowers, surrounded by a scalelike leaf or utricular structure at the base. The ovary is 1-locular and bottle-shaped and the 1 to 7 ovules are anatropous, hemianatropous, or atropous. The fruit is an indehiscent utricle, although flowering and fruiting is rare. The interpretation of the reproductive structures of Lemnoideae is controversial, being considered either as a reduced inflorescence with 1 to 2 male flowers and 1 female flower[vi] or as a single flower[vii]. *Wolffia anusta* and *W. globosa*

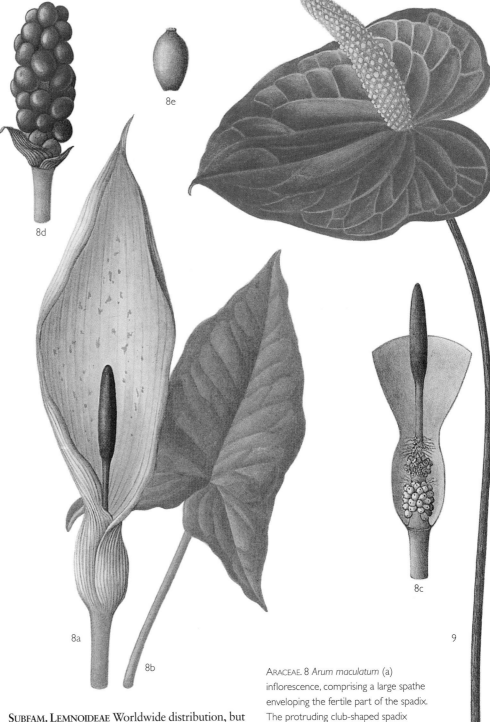

ARACEAE. **8** *Arum maculatum* (a) inflorescence, comprising a large spathe enveloping the fertile part of the spadix. The protruding club-shaped spadix appendix is sterile (x⅔); (b) leaf (x⅔); (c) vertical section of inflorescence showing, from the base, female flowers, male flowers, sterile hairlike female flowers, and spadix appendix (x⅔); (d) fruiting spike (x⅔); (e) ovary (x4). **9** *Anthurium andraeanum* spathe and spadix with bisexual flowers (x⅔).

are the smallest known flowering plants, scarcely visible as scum on the surface of water. *Wolffia* is rich in protein and is eaten in Southeast Asia. Especially *Lemna minor* is eaten by wildfowl.

Economic uses The family is of considerable importance as it includes several species with edible tubers (e.g., taro, cocoyam, dasheen, *Colocasia esculenta*) or stems (tannia, *Xanthosoma sagittifolium*), which are a subsistence food crop locally and throughout the tropics. In some countries, cultivation has reached a commercial scale. More locally grown are species of *Amorphophallus* (e.g., elephant yam, *A. paeoniifolius*, in tropical Asia and konjaku, *A. konjac*, in Japan), *Cyrtosperma*, and *Alocasia* (e.g., swamp taro, babai, *C. merkusii*, gigant taro, cunjevoi, *A. macrorrhizos* both in Southeast Asia and Oceania). The inflorescence of *Monstera deliciosa* (cheeseplant) is sometimes used fresh as food. Many aroids are poisonous and must be cooked before they are eaten. Some species are of medicinal use or are used as arrow poisons. Many genera contain species widely grown as garden ornamentals, house or aquarium plants, or as cut flowers. The genera *Arum* and *Arisaema* are frequently grown in the garden, while *Aglaonema*, *Anthurium*, *Dieffenbachia* (poisonous and containing highly irritant raphides), *Monstera*, *Philodendron*, *Spathiphyllum*, *Zantedeschia*, and *Zamioculcas* are grown as house plants. Species of *Dieffenbachia* are a frequent cause of poisoning. The genus *Cryptocoryne* is widely used in aquaria, and *Zanthedeschia aethiopica* (Arum Lily), *Anthurium andraeanum,* and *A. scherzerianum* (Flamingo Flowers) and their hybrids are important cut flowers. Members of the Lemnoideae are used to remove nutrients and heavy metals from waste water.   OS

i Dahlgren, R. M. T., Clifford, H. T., & Yeo, P. F. *The Families of the Monocotyledons*. Berlin, Springer-Verlag (1985).
ii Landolt, E. Lemnaceae. Pp. 264–270. In: Kubitzki, K. (ed.), *The Families and Genera of Vascular Plants. IV. Flowering Plants. Monocotyledons: Alismatanae and Commelinanae (except Gramineae)*. Berlin, Springer-Verlag (1998).
iii Rothwell, G. W., Van Atta, M. R., Ballard, Jr., H. E., & Stockey, R. A. Molecular phylogenetic relationships among Lemnaceae and Araceae using the chloroplast *trn*L-*trn*F intergenic spacer. *Molec. Phyl. Evol.* 30: 378–385 (2004).
iv Mayo, S. J., Bogner, J., & Boyce, P. C. *The Genera of Araceae*. Richmond, Royal Botanic Gardens, Kew (1997).
v Chase, M. W. *et al.* Multi-gene analysis of monocot relationships: A summary. Pp. 63–75. In: Columbus, J. T. *et al.* (eds), *Monocots: Comparative Biology and Evolution*. Claremont, California, Rancho Santa Ana Botanic Garden (2006).
vi Den Hartog, C., van der Plas, F. A synopsis of the Lemnaceae. *Blumea* 18: 355–368 (1970).
vii Hillman, W. S. The Lemnaceae or duckweeds. A review of the descriptive and experimental literature. *Bot. Rev.* 27: 221–287 (1961).

ARECACEAE (PALMAE)

PALM FAMILY

Arecaceae consists of tree- or shrublike plants, or lianas, with woody stems and often large, tough, plicate, split leaves; axillary inflores-

Genera c. 190 **Species** c. 2,000
Economic uses Numerous uses, as important crop plants, used for house construction, in the household, and many are important ornamentals

cences with a characteristic prophyll; and numerous usually 3-merous flowers and single-seed fruits.

Distribution Palms are mainly distributed in the tropics, with many species in Central America, northern South America, and East Asia, while Africa has a surprisingly low number of species compared to its size. Palms occur in many habitats, from mangrove swamps to lowland rain forest but, with the exception of East Asian mangrove swamps (*Nypa*) and wetlands (*Mauritia*, *Copernicia*) in northern South America, palms are rarely the dominant or most conspicuous part of any vegetation type. There are a few subtropical and temperate outliers, including the European *Chamaerops humilis*, which attains latitude 44°N. All palms have a single apical bud; if killed by frost the stem dies. Few species have overcome this limitation. *Trachycarpus* reaches 2,400 m altitude at 32°N in the Himalayas, where the land is under snow from November to March, and *Serenoa* reaches 30°N in North America. On Borneo, *Calamus gibbianus* reaches an elevation of more than 3,000 m but *Ceroxylon utile* has been recorded at more than 4,000 m in Colombia.

Description The life form of palms is highly variable. Some are tree- or shrublike, others are nearly stem-less or climbing, but in principle all have woody stems that terminate in a crown of

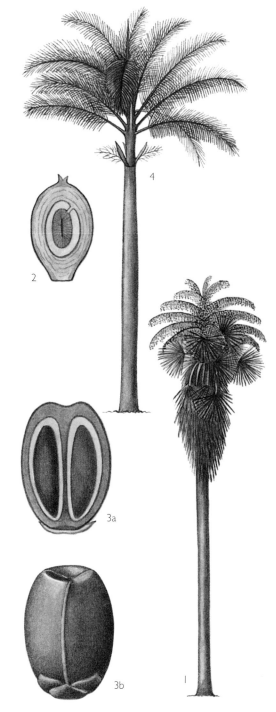

ARECACEAE. (ABOVE:) 1 *Corypha umbraculifera* habit, showing massive terminal inflorescence. 2 *Elaeis guineensis* vertical section of the fruit, containing a single seed in a hollow cavity surrounded by a thick mesocarp. 3 *Arenga westerhoutii* (a) vertical section of fruit with 2 seeds, surrounded by a fleshy mesocarp (x⅔); (b) fruit, with perianth remains (x⅔). 4 *Roystonea regia* habit, showing the inflorescences appearing below the crown. (LEFT:) 5 *Caryota cumingii* (a) male flower with 3 sepals and 3 petals (x6); (b) open male flower, showing enlarged petals and 6 stamens (x5). 6 *Corypha umbraculifera* bisexual flower (x6). 7 *Hyphaene thebaica* habit showing the dichotomous branching, an unusual feature in the palms. 8 *Phoenix dactylifera* carpel in cross section (x6). (OPPOSITE PAGE): 9 *Livistona rotundifolia* part of branch bearing indehiscent fruit (x⅔). 10 *Chamaedorea geonomiformis* habit showing bifid leaves and male inflorescences. 11 *Caryota mitis* part of bipinnate leaf (x⅔). 12 *Raphia vinifera* fruit covered with shiny scaly fruit, a distinctive feature of the Calamoideae (x⅔). 13 *Chamaedorea fragrans* male inflorescence (x⅔).

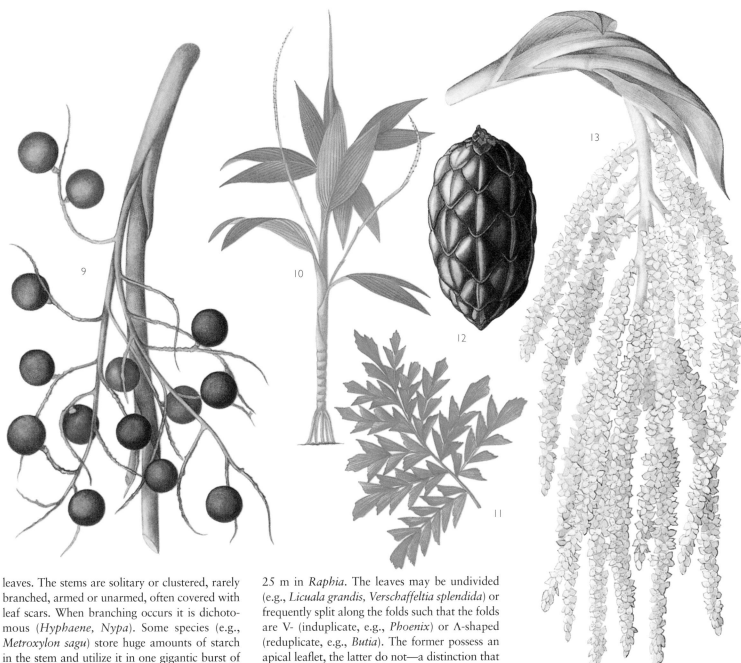

leaves. The stems are solitary or clustered, rarely branched, armed or unarmed, often covered with leaf scars. When branching occurs it is dichotomous (*Hyphaene, Nypa*). Some species (e.g., *Metroxylon sagu*) store huge amounts of starch in the stem and utilize it in one gigantic burst of reproduction after which they die back (hapaxanthic). Despite the often massive size of the stem, the diameter increases solely by primary growth, once fully formed its width is maintained. The leaves are usually spirally arranged, only rarely di- or tristichous, initially with a tubular sheath, which may later split open and may be armed with spines or prickles, and generally have a distinct petiole, which is unarmed or armed with spines or teeth. The blade is palmate, pinnate, rarely bipinnate (*Caryota*), bifid (*Asterogyne*), or entire (*Johannesteijsmannia*), tough, and usually plicate (folded like a fan). The folds are either arranged on either side of an elongated central rachis (feather palms), or arise crowded from a very short central rachis or costa (fan palm). Large fan leaves are intermediate between the feather and fan types and have a massive central costa (costapalmate).The folds develop as the leaf grows in bud before it is unfurled and, as it expands from the bud, swellings (pulvini) develop that push the folded leaflets out to their mature position. Palm leaves vary enormously in size from less then 15 cm in species of *Chamaedora* to

25 m in *Raphia*. The leaves may be undivided (e.g., *Licuala grandis, Verschaffeltia splendida*) or frequently split along the folds such that the folds are V- (induplicate, e.g., *Phoenix*) or Λ-shaped (reduplicate, e.g., *Butia*). The former possess an apical leaflet, the latter do not—a distinction that arises early in development. The "leaflets" may contain 1 fold or several and be linear, fishtail-shaped (*Ceratolobus pseudoconcolor*), acute, or praemorse (looking as if bitten off). Before final expansion of the leaf, superficial cells often divide to form hairs or scales that are often of great complexity and beauty. The loose surface layer so formed acts as a lubricant during the opening of the leaf. Palm inflorescences are axillary, simple or huge panicles, with up to 6 orders of branching, often emerging among the leaves, but sometimes below or above the crown (*Corypha, Metroxylon*), usually solitary, but occasionally several appear together. The inflorescence invariably has a more or less prominent prophyll. Species of *Corypha* have treelike inflorescences that carry up to 10 million flowers. The flowers occur individually, 2 or 3 together, in cincinnate clusters or are 2-ranked. Sometimes the flowers are borne in pits in the inflorescence braches. The flowers may be bisexual or unisexual (the plants then monoecious, andromonoecious or dioecious), basically 3-merous. There are 3, somewhat imbricate, sepals; 3 slightly imbricate

petals; 3+3 stamens; and a superior ovary with 3 carpels, each with a single ovule. The carpels are either free (apocarpous) or wholly or partly fused (syncarpous), and the ovules are erect or pendulous anatropous (rarely semianatropous or atropous). Male flowers occasionally have both 4 sepals and petals and the female flower may have up to 10 carpels. More than 6 stamens per flower are not uncommon and more than 1,000 have been counted in *Ammandra*. Unisexual flowers often develop by abortion. Palms are wind- or animal-pollinated (by ants, bees, beetles, or flies), or both. The latter may in fact be the prevailing method of pollination. The fruits are mostly one-seeded berries or drupes with a single seed, or rarely 2 to 10. They vary in size from 4.5 mm in diameter to a length of 50 cm and a weight of 18 kg found in *Lodoicea maldivica* (double coconut)—the world's largest fruit (and seed). The fruit surface is most often smooth but may be covered with exquisite, geometrically arranged scales, prickles, hairs, or warts. The mesocarp is fleshy or dry and variously fibrous. The endocarp if distinct is mostly thin. Many fruits are brightly colored, and virtually all are indehiscent. The storage tissue is endosperm, which is oily or fatty rather than starchy; it is sometimes extremely hard, in which case it is known as vegetable ivory (*Phytelephas*) and used for carvings or (formerly) buttons. In a few genera (e.g., *Cocos*), the seed is hollow. Upon germination, the cotyledon remains as a haustorium within the testa.

Classification The monophyly of the Arecaceae has never seriously been challenged. Molecular analyses of the relationships of the Arecaceae to other groups of Monocots and of the relationships within the family have been hampered by unusually slow sequence evolution[i]. Recent morphological and molecular analyses point to a close relationship between the Arecaceae and families related to grasses, zingibers, and commelinoids[ii],[iii],[iv]. The classification used here divides the family into 6 subfamilies[v]. However, recent investigations have shown that several of these subfamilies are not monophyeletic and proposals are to be made to reduce the number of subfamilies to 5 and move part of Arecoideae into an enlarged Coryphoideae, subsume the Phytelephantoideae into a reduced Ceroxyloideae, as part of the latter will be included in Arecoideae[vi].

SUBFAM. CORYPHOIDEAE The members of this subfamily have palmate or costapalmate, usually induplicate, leaves, and solitary flowers or flowers in cincinate clusters. The inflorescences emerge among or above the leaves and are highly branched. The flowers are bi- or less commonly unisexual, 3-merous in all whorls, and apocarpous or more or less syncarpous. This large subfamily of palms is restricted to the Old World and consists of 31 genera and c. 300 spp. The genera *Phoenix*, *Chamaerops*, and *Lodoicea* belong to the Coryphoideae. *Phoenix* occurs in Africa, Asia, and the Middle East, mainly in dry habitats, though several species occur in swamps, including 2 in mangrove forest; 1 sp. (*P. theophrasti*) occurs in eastern Crete and southern coast of Turkey. The monotypic *Chamaerops humilis* (European Fan Palm) is native to the western Mediterranean in Europe and Africa. *Lodoicea maldivica* (Double Doconut) endemic to the Seychelles Islands.

SUBFAM. CALAMOIDEAE Strongly armed, with pinnate, bipinnate, or palmate, reduplicate leaves and a wide diversity of inflorescence types, although the flowers nearly always arranged in pairs and hermaphrodite or unisexual (plants polygamous, monoecious, or dioecious). Ovary is syncarpous, and the fruits are covered in beautifully arranged, reflexed scales in vertical rows. The majority of the 21 genera and c. 650 spp. occur in the Old World, while 4 are restricted to the New World. Most climbing palms belong here, including *Calamus* (rattan palms)—the largest palm genus with c. 400 spp.. Another important member of the group is *Metroxylon sagu* (Sago Palm), which is cultivated in most of Asia and is most likely native to New Guinea.

SUBFAM. NYPOIDEAE Comprises only the monotypic species, *Nypa fruticans* (Mangrove Palm), which is distributed from East Asia to Australia, with outliers in the Pacific. The species has a subterranean or dichotomously branching, creeping stem and pinnate, reduplicate leaves. The inflorescences are solitary, monoecious and appear among the leaves. The female inflorescence is headlike and surrounded by several catkinlike male inflorescences at the base. The flowers are 3-merous in all whorls, and the infrutescence globose. *Nypa* is sometimes place as sister group to the remaining palms[vii].

SUBFAM. CEROXYLOIDEAE Leaves pinnate, reduplicate or entire, and flowers unisexual, only rarely bisexual, solitary, or flowers borne in 2-ranked rows. The unisexual flowers are rarely dimorphic. There are 9 genera and 125–150 spp.. The largest genus is *Chamaedorea*, which occurs in South and Central America. Other genera occur on Madagascar, Mascarene Islands, and northern Australia. Molecular data suggest that the subfamily is not monophyletic[vi].

SUBFAM. ARECOIDEAE Leaves pinnate or bipinnate (*Caryota*), re- or induplicate, and bisexual flowers borne in pairs or 3 together and with the perianth in 2 whorls and a usually 1- to 3-locular ovary. Many species are hapaxanthic. By far the largest subfamily of palms, comprising 6 separate groups with a total of 112 genera and c. 1,400 spp.. The fruits of *Cocos* are unusual, having a thick, fibrous pericarp, and inner thick, hard endocarp with 3 germination pits, 2 of which are sealed and only the largest hides the embryo. Within the endocarp is a single seed, with its minute embryo embedded in a peripheral layer of endosperm rich in oil. The huge, hollow inner part of the fruit is filled with liquid endosperm.

SUBFAM. PHYTELEPHANTOIDEAE Dioecious palms, with pinnate, reduplicate leaves and dimorphic inflorescences. Female flowers are borne in large heads; the male flowers in spikelike or racemose inflorescences. The female flowers have 5 to 10 connate carpels. The perianth is reduced in the male flowers, which have numerous, occasionally more than 1,000 stamens. Restricted to the Amazonian Basin and the costal ranges of northern South America, consisting of 3 genera and c. 10 spp. It is most likely monophyletic[viii].

Economic uses The economic importance of palms, both locally and globally, is immense, and several books[ix] have been written on the topic, hence only a limited range of well-known uses are mentioned below. Coconuts are the main products of the Coconut Palm (*Cocos nucifera*), which also yield copra (the dried kernels which yield coconut oil) as an important by-product. Oil is also extracted from *Elaeis guineensis* (Oil Palm) and *Orbigyna* species. Dates, the fruits of *Phoenix dactylifera*, are central to the economy of many producing countries and are a major crop in Middle East, where they have been cultivated for at least 5,000 years; other species of *Phoenix* are used as source of fibers, sugar, and a multiplicity of other things, e.g., fuel, or used as pot plants (*P. roebelenii*). A major source of carbohydrate for many people living in the tropics is sago, which is processed from the pith of palms of the genus *Metroxylon sagu* (Sago Palm) and some species of other genera (e.g., *Arenga* and *Caryota*). Palm wine (toddy) is another useful product of many palms including *Borassus* and *Caryota*. This is evaporated to produce palm sugar (jaggery) or distilled to form the base for the liquor called arrack. However, the term *arrack* is most often applied to the alcoholic spirit obtained from the sap of the Coconut Palm. Palms also produce a variety of useful fibers. These include coir, which comes from the husk of the coconut, raffia fiber obtained by stripping the surface of young leaflets of the genus *Raphia*, and piassava fiber, from the leaf sheaths or fibrous stems of South American *Leopoldinia piassaba* and *Attalea funifera*. *Caryota urens* yields a black bristle fiber (kitul fiber) used for ropes and broom heads. For canework, the much-used rattan cane is invaluable; this comes mainly from *Calamus* spp., which also provides malacca cane—a stout cane used for walking sticks and baskets. Waxes are obtained from *Copernicia* (carnauba wax) and *Ceroxylon*. Vegetable ivory, from *Phytelephas* (Ivory Nut Palm) and others, was once an important commodity, and used for buttons and as a substitute for real ivory. The betel nut is rich in alkaloids and is used as a masicatory in betel-chewing. It comes from *Areca catechu* and is widely used in India, Malaysia, and tropical Africa. Many palms, with tall slender trunks and dense crown of leaves, are widespread ornamentals in the Tropics. Notable examples are Cuban Royal Palm (*Roystonea regia*), Chinese Windmill Palm (*Trachycarpus fortunei*), Coquitos Palm (*Jubaea chilensis*)), and *Washingtonia filifera*. OS

i Wilson, M. A., Gaut, B. S., & Clegg, M. T. Chloroplast DNA evolves slowly in the palm family (Arecaceae). *Molec. Biol. Evol.* 74: 303–314 (1990).
ii Stevenson, D. W. & Loconte, H. Cladistic analysis of monocot families. Pp. 543–578. In: Rudall, P. J., Cribb, P. J., Cutler, D. F., & Humphries, C. J. (eds), *Monocotyledons: Systematics and Evolution.* Vol. 2. Richmond, Royal Botanic Gardens, Kew (1995).
iii Chase, M. W. *et al.* Multi-gene analysis of monocot relationships: A summary. Pp. 63–75. In: Columbus, J. T. *et al.* (eds), *Monocots: Comparative Biology and Evolution.* Claremont, California, Rancho Santa Ana Botanic Garden (2006).
iv Davis, J. I. *et al.* A phylogeny of the monocots, as inferred from *rbc*L and *atp*A sequence variation, and a comparison of methods for calculating jackknife and bootstrap values. *Syst. Bot.* 29: 467–510 (2004).
v Uhl, N. W. & Dransfield, J. *Genera Palmarum, a Classification of Palms Based on the Work of Harold E. Moore, Jr.* Lawrence, Kansas, Allen Press (1987).
vi Dransfield, J., Uhl, N. W., Asmussen, C. B., Baker, W. J., Harley, M. M., & Lewis, C. E. A new phylogenetic classification of the palm family, Arecaceae. *Kew Bull.* 60: 559–569 (2005).
vii Uhl, N. W., Dransfield, J., Davis, J. I., Louckow, M. A., Hansen, K. S., & Doyle, J. J. Phylogenetic relationships among palms: cladistic analyses of morphological and chloroplast DNA restriction site variation. Pp. 623–661. In: Rudall, P. J., Cribb, P. J., Cutler, D. F., & Humphries, C. J. (eds), *Monocotyledons: Systematics and Evolution.* Vol. 2. Richmond, Royal Botanic Gardens, Kew (1995).
viii Barfod, A. S., Ervik, F., & Bernal, R. Recent evidence on the evolution of phytelephantoid palms. *Mem. New York Bot. Gard.* 83: 265–277 (1998).
ix Balick, M. J., Beck, H. T. *et al. Useful Palms of the World.* New York, Columbia University Press (1990).

ASPARAGACEAE

ASPARAGUS FAMILY

A family of perennial herbs or subshrubs that lacks true leaves but have phylloclades as their main photosynthetic organs.

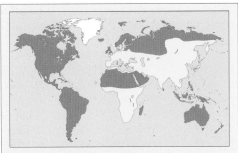

Genera 2 **Species** 150–300
Economic uses Mainly used as vegetables and ornamentals

Distribution *Asparagus* is restricted to the Old World in areas with arid, semiarid, or Mediterranean climate, while *Hemiphylacus* is restricted to Mexico.

Description Perennial herbs or subshrubs with rhizomes, rarely tubers, and erect or scandent stems. The leaves are scalelike, spurred at the base, and subtending solitary or clustered, leaflike phylloclades that may be spiny. The flowers are bi- or unisexual (plants monoecious or dioecious), pendulous or erect and axillary, and occurring singly or in small clusters. There are 3+3 free or basally fused petaloid tepals, 3+3 often free stamens, but sometimes fused with tepals basally, occasionally the outer whorl in female flowers staminodial and adnate to the tepals. The ovary is superior, 3-locular and 3-carpellate, with 2 to 12 hemianatropous or atropous ovules per locule and septal nectaries. The style has 3 short stigmatic branches or is capitate. The fruit is a berry or rarely a nut, often conspicuously colored, and the seeds are black and have slightly curved embryo in an endosperm of fat and hemicellulose.

Classification The placement of the highly deviant genus *Hemiphylacus* in Asparagaceae is controversial[i,ii,iii,iv], but recent molecular studies unanimously point at a sister group relationship with *Asparagus*[iii,iv,v]. The Asparagaceae is defined here in a narrow sense[v] and is most likely sister to Ruscaceae (q.v.)[v,vi] or Convallariaceae (q.v.)[iv].

Economic uses *Asparagus officinalis* (Garden Asparagus) has been cultivated since ancient Greek times for its young shoots, which are now widely eaten as a vegetable. *A. setaceus* is often used in bouquets of cut flowers. OS

i Mabberley, D. J. *The Plant Book*. 2nd edn. Cambridge, Cambridge University Press (1997).
ii Kubitzki, K. & Rudall, P. J. Asparagaceae. Pp. 21–25. In: Kubitzki, K. (ed.), *The Families and Genera of Vascular Plants. III. Flowering Plants. Monocotyledons: Lilianae (except Orchidaceae)*. Berlin, Springer-Verlag (1998).
iii Rudall, P. J., Engelman, E. M., Hanson, L., & Chase, M. W. Systematics of *Hemiphylacus, Asparagus*, and *Anemarrhena*. *Pl. Syst. Evol.* 211: 181–199 (1998).
iv Fay, M. F. *et al.* Phylogenetic studies of Asparagales based on four plastid DNA regions. Pp. 360–371. In: Wilson, K. L. & Morrison, D. A. (eds), *Monocots. Systematics and Evolution*. Melbourne, CSIRO Publishing (2000).
v Pires, J. C. *et al.* Phylogeny, genome size, and chromosomal evolution of Asparagales. Pp. 287–304. In: Columbus, J. T. *et al.* (eds), *Monocots: Comparative Biology and Evolution*. Claremont, California, Rancho Santa Ana Botanic 2006).
vi Judd, W. S. *et al. Plant Systematics. A Phylogenetic Approach.* 2nd edn. Sunderland, Massachusetts, Sinauer Associates (2002).

ASPHODELACEAE

ALOE FAMILY

A small family, predominantly herbs with succulent leaves in basal or terminal rosettes.

Distribution Asphodelaceae is widespread in temperate, subtropical, and tropical regions of the Old World, primarily South Africa. Many species prefer arid habitats.

Description Often succulent, geophytic herbs, climbers, or pachycaul trees with leaves in basal or terminal rosettes and characteristic succulent roots. The leaves are distichous or spirally arranged, lanceolate/acuminate, linear or subulate, terete, often succulent, and with toothed margins. The inflorescence is a raceme or panicle usually with a well-developed peduncle. The flowers are bisexual and actinomorphic, rarely zygomorphic, and bilabiate (*Haworthia*), with 2 whorls of 3, sometimes fleshy, petaloid, brightly colored tepals that are free or fused into at tube at the base. There are 3+3 free stamens, often the same color as the tepals. The ovary is superior, 3-carpellate, 3-locular with 1 to many mostly anatropous, atropous or campylotropous ovules in each locule, and a simple style. The fruit is a loculicidal capsule, rarely berrylike, with brown or black seeds that are covered by an outgrowth from their funicule. The embryo is straight with an endosperm that stores fat and aleurone.

ASPHODELACEAE.

I *Aloe jucunda* habit, showing basal rosette of spiny, fleshy leaves (×⅔).

Genera 15 **Species** c. 780
Economic uses *Aloe* spp. used for medicine and in cosmetic industry; species of several genera are grown as ornamentals and for cut flowers

Classification The monophyly of Asphodelaceae is supported both by morphological[i,ii] and molecular data[iii,iv,v]. The family is usually divided into 2 subfamilies Asphodeloideae and Alooideae, but only the latter is monophyletic[iii]. Molecular data place the Asphodelaceae in a clade that also contains Xanthorrhoeaceae (q.v.) and Hemerocallidaceae (q.v.). All 3 families are sometimes combined (as in APG II) in an enlarged Xanthorrhoeaceae[iv]. In a narrow sense, the Asphodelaceae is either sister group to Xanthorrhoeaceae plus Hemerocallidaceae[iii,iv], to Xanthorrhoeaceae alone[vi], to Hemerocallidaceae alone[iv], or the relationship is unresolved[vii].

Economic uses Of particular economic importance are *Aloe vera* and *A. ferox*. *A. vera* is widely cultivated in North America and the West Indies, and the bitter leaf sap is used to make Curaçao aloes. *A. ferox* is harvested in South Africa, and the leaf sap is called Cape aloe. Both include the same purgative substances. The sap of many *Aloe* spp. is used to treat burns. The central part of leaves produces a sap free from bitter ingredients, and it is widely used in the cosmetics and health industries. Many genera include species that are used as ornamentals, e.g., *Aloe* (c. 400 spp.), *Asphodeolus* (Asphodel, 12–16 spp.), *Bulbine* (50–60 spp.), *Haworthia* (70–90 spp.), and *Kniphofia* (Red-hot Pokers, Torch Lily, c. 70 spp.). The latter is also cultivated for cut flowers. OS

i Smith, G. F. & Van Wyk, B.-E. Asphodelaceae. Pp. 130–140. In: Kubitzki, K. (ed.), *The Families and Genera of Vascular Plants. III. Flowering Plants. Monocotyledons: Lilianae (except Orchidaceae)*. Berlin, Springer-Verlag (1998).
ii Dahlgren, R. M. T., Clifford, H. T., & Yeo, P. F. *The Families of the Monocotyledons*. Berlin, Springer-Verlag (1985).
iii Chase, M. W. *et al.* Phylogenetics of Asphodelaceae (Asparagales): An analysis of plastid *rbc*L and *trn*L-*trn*F DNA sequences. *Ann. Bot.* 86: 935–991 (2000).
iv Pires, J. C. *et al.* Phylogeny, genome size, and chromosomal evolution of Asparagales. Pp. 287–304. In: Columbus, J. T. *et al.* (eds), *Monocots: Comparative Biology and Evolution*. Claremont, California, Rancho Santa Ana Botanic Garden (2006).

v Chase, M. W. *et al.* Multi-gene analysis of monocot relationships: A summary. Pp. 63–75. In: Columbus, J. T. *et al.* (eds), *Monocots: Comparative Biology and Evolution.* Claremont, California, Rancho Santa Ana Botanic Garden (2006).

vi Fay, M. F. *et al.* Phylogenetic studies of Asparagales based on four plastid DNA regions. Pp. 360–371. In: Wilson, K. L. & Morrison, D. A. (eds), *Monocots. Systematics and Evolution.* Melbourne, CSIRO Publishing (2000).

vii McPherson, M. A., Fay, M. F., Chase, M. W., & Graham, S. W. Parallel loss of slowly evolving intron from two closely related families in Asparagales. *Syst. Bot.* 29: 296–307 (2004).

ASTELIACEAE

Distribution Found mainly in the southern hemisphere, though a few species occur on Hawaii, from sea level to alpine areas on rocks, in bogs, or in forests.

Description Perennial, rhizomatous herbs with short stems. The leaves are linear to lanceolate, spirally arranged, often densely, with an open or closed sheathing base and silvery hairy when young. The inflorescence is terminal, consisting of few-flowered racemes and may be densely hairy. The flowers are actinomorphic, unisexual (plants usually dioecious or gynodioecious), or bisexual, with 2 whorls of between 3 and 7 free or basally fused tepals, and 6 often free stamens. The ovary is superior, 3-carpellate, or of up to 7 carpels, 3- or 1-locular, with a short style, and septal nectaries. There are several anatropous ovules in each locule. The fruit is a berry or capsule (*Milligania*). The seeds are black, with a straight or slightly curved embryo in an endosperm of oil and aleurone.

Classification The Asteliaceae are most likely monophyletic and sister to Hypoxidaceae (q.v.), plus Lanariaceae (q.v.) and Blandfordiaceae (q.v.)[i,ii,iii,iv], or only to the former 2 families[v]. OS

i Fay, M. F. *et al.* Phylogenetic studies of Asparagales based on four plastid DNA regions. Pp. 360–371. In: Wilson, K. L. & Morrison, D. A. (eds), *Monocots. Systematics and Evolution.* Melbourne, CSIRO Publishing (2000).

ii Pires, J. C. *et al.* Phylogeny, genome size, and chromosomal evolution of Asparagales. Pp. 287–304. In: Columbus, J. T. *et al.* (eds), *Monocots: Comparative Biology and Evolution.* Claremont, California, Rancho Santa Ana Botanic Garden (2006).

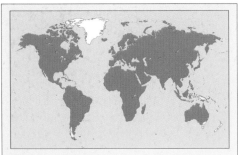

Genera 4 Species c. 35
Economic uses None

iii Rudall, P. J., Chase, M. W., Cutler, D. F., Rusby, J., & de Bruijn A. Y. Anatomical and molecular systematics of Asteliaceae and Hypoxidaceae. *Bot. J. Linn. Soc.* 127: 1–42 (1998).

iv Davis, J. I. *et al.* A phylogeny of the monocots, as inferred from *rbc*L and *atp*A sequence variation, and a comparison of methods for calculating jackknife and bootstrap values. *Syst. Bot.* 29: 467–510 (2004).

v Chase, M. W. *et al.* Higher-level systematics of the monocotyledons: An assessment of current knowledge and a new classification. In: Wilson, K. L. & Morrison, D. A. (eds), *Monocots. Systematics and Evolution.* Melbourne, CSIRO Publishing (2000).

BEHNIACEAE

A monotypic family that is restricted to forested areas of southeastern Africa. Behniaceae comprises a single perennial shrublike herb, with thin, branching, erect, or twining stems with distichous, petiolate leaves. The inflorescence is axillary and consists of a single or a few flowers. The

Genera I Species I
Economic uses None

flowers are white or green, actinomorphic, with two 3-merous whorls of tepals, united for two-thirds of their length, 2 whorls of stamens adnate to the tepals, and a 3-carpellate, 3-locular, superior ovary, with septal nectaries, and a few anatropous ovules arranged in 2 rows in each locule. The fruit is a berry with few to numerous seeds, which turn black when drying. The embryo is club shaped, and the endosperm contains aleurones and oil.

Classification The position of *Behnia reticulata* has been controversial. Its status as a family of its own was suggested by Dahlgren in 1989 but first formally made in 1997. Previously, the species has been placed in the Luzuriagaceae (q.v.)[i], but molecular evidence now appears either to include it in a broadly circumscribed Agavaceae[ii,iii] or to be sister group of a clade that contains Anthericaceae (q.v.) and Herreriaceae (q.v.)[iv]. Behniaceae is sometimes included in a greatly enlarged Asparagaceae[ii,iii]. OS

i Dahlgren, R. M. T., Clifford, H. T., & Yeo, P. F. *The Families of the Monocotyledons.* Berlin, Springer-Verlag (1985).

ii APG II. An update of the Angiosperm Phylogeny Group classification for the orders and families of flowering plants: APG II. *Bot. J. Linn. Soc.* 141: 399–436 (2003).

iii Chase, M. W. *et al.* Multi-gene analysis of monocot relationships: A summary. Pp. 63–75. In: Columbus, J. T. *et al.* (eds), *Monocots: Comparative Biology and Evolution.* Claremont, California, Rancho Santa Ana Botanic Garden (2006).

iv Pires, J. C. *et al.* Phylogeny, genome size, and chromosomal evolution of Asparagales. Pp. 287–304. In: Columbus, J. T. *et al.* (eds), *Monocots: Comparative Biology and Evolution.* Claremont, California, Rancho Santa Ana Botanic Garden (2006).

BLANDFORDIACEAE

Distribution A monogeneric family of 4 spp. of perennial, caespitose herbs with short stems, and linear alternate, distichous leaves with sheathing base, found only in eastern Australia from the subtropics to temperate areas, mostly in coastal heathlands. The inflorescence is a bracteate raceme, with bisexual, actinomorphic flowers. There are 6 tepals, petaloid fused into a campanulate corolla with broad lobes, and 2, 3-merous whorls of stamens fused to the corolla tube for up to half its length. The ovary is superior, 3-carpellate and 3-locular, with many anatropous ovules in each locule and septal nectaries. The fruit is a septicidal capsule, and the seeds brown, with a linear or slightly curved embryo in a starchless endosperm. The Blandfordiaceae are monophyletic and sister to Hypoxidaceae (q.v.), plus Lanariaceae (q.v.) and Asteliaceae (q.v.)[i,ii,iii,iv,v]. OS

Genera I Species 4
Economic uses
Cultivated as ornamentals

i Fay, M. F. *et al.* Phylogenetic studies of Asparagales based on four plastid DNA regions. Pp. 360–371. In: Wilson, K. L. & Morrison, D. A. (eds), *Monocots. Systematics and Evolution.* Melbourne, CSIRO Publishing (2000).

ii Pires, J. C. *et al.* Phylogeny, genome size, and chromosomal evolution of Asparagales. Pp. 287–304. In: Columbus, J. T. *et al.* (eds), *Monocots: Comparative Biology and Evolution.* Claremont, California, Rancho Santa Ana Botanic Garden (2006).

iii Rudall, P. J., Chase, M. W., Cutler, D. F., Rusby, J., & de Bruijn A. Y. Anatomical and molecular systematics of Asteliaceae and Hypoxidaceae. *Bot. J. Linn. Soc.* 127: 1–42 (1998).

iv Davis, J. I. *et al.* A phylogeny of the monocots, as inferred from *rbc*L and *atp*A sequence variation, and a comparison of methods for calculating jackknife and bootstrap values. *Syst. Bot.* 29: 467–510 (2004).

v Chase, M. W. *et al.* Higher-level systematics of the monocotyledons: An assessment of current knowledge and a new classification. In: Wilson, K. L. & Morrison, D. A. (eds), *Monocots. Systematics and Evolution.* Melbourne, CSIRO Publishing (2000).

BORYACEAE

Description This family comprises 2 genera (*Alania* and *Borya*) of xeromorphic perennial, caespitose, or shrubby herbs that are endemic to rocky areas in Australia, with short stems and linear spirally arranged leaves with a strongly, sheathing base. The inflorescence is a terminal, scapose, involucrate spike, or raceme. The flowers are bisexual and actinomorphic, with two

Genera 2 Species 12
Economic uses None

3-merous, petaloid whorls of similar, white tepals united into a tube at the base, or free. The stamens are in 2 whorls. The ovary is superior, 3-carpellate, 3-locular, with numerous anatropous ovules in each locule and septal nectaries. The fruit is a loculicidal capsule, containing black seeds with small, ovoid embryos and a starchless endosperm.

Boryaceae was included in the Anthericaceae by Dahlgren *et al.* (1985), but is now generally accepted as a separate family[i]. The Boryaceae is monophyletic and either sister to Hypoxidaceae, Blandfordiaceae, Lanariaceae, and Asteliaceae (q.v.)[ii,iii,iv,v] to Blandfordiaceae alone[vi], or to a large clade that excludes all these taxa[vii].   OS

[i] Conran, J. G. Boryaceae. Pp. 151–154. In: Kubitzki, K. (ed.), *The Families and Genera of Vascular Plants. III.* Berlin, Springer-Verlag (1998).

[ii] Rudall, P. J., Chase, M. W., Cutler, D. F., Rusby, J., & de Bruijn A. Y. Anatomical and molecular systematics of Asteliaceae and Hypoxidaceae. *Bot. J. Linn. Soc.* 127: 1–42 (1998).

[iii] Fay, M. F. *et al.* Phylogenetic studies of Asparagales based on four plastid DNA regions. Pp. 360–371. In: Wilson, K. L. & Morrison, D. A. (eds), *Monocots. Systematics and Evolution.* Melbourne, CSIRO Publishing (2000).

[iv] Graham, S. *et al.* Robust inference of monocot deep phylogeny using an expanded multigene plastid data set. In: Columbus, J. T. *et al.* (eds), *Monocots: Comparative Biology and Evolution.* Claremont, California, Rancho Santa Ana Botanic Garden (2006).

[v] Chase, M. W. *et al.* Higher-level systematics of the monocotyledons: An assessment of current knowledge and a new classification. In: Wilson, K. L. & Morrison, D. A. (eds), *Monocots. Systematics and Evolution.* Melbourne, CSIRO Publishing (2000).

BROMELIACEAE
PINEAPPLE OR BROMELIAD FAMILY

A family of frequently epiphytic, rosette plants with spirally arranged leaves often with serrate margins and inflorescences usually comprising showy bracts.

Distribution Almost exclusively native to the Americas, occuring from Virginia in the USA to Patagonia in South America. A single species (*Pitcairnia feliciana*) is native to Africa. A few species are found on islands in the Pacific e.g., *Racinaea insularis* of the Galápagos Islands and *Greigia berteroi* and *Ochagavia elegans* of the Juan Fernandez Islands. *Tillandsia usneoides* is supposed to be one of the world's widest-ranging plants[i]. Bromeliads occur in a variety of different habitats and are a characteristic element in Neotropical forests, but they are also common in arid areas. They occur at altitudes of up to 4,500 m in the Andes mountain range.

Description Most bromeliads are herbaceous epiphytes, but a few have elongated stems or are rosette trees. They vary in size from few centimeters in some epiphytic species of *Tillandsia* to 3 m in *Puya raimondii*. The main function of the roots in epiphytic species is to anchor the plant, whereas the terrestrial species (e.g., *Pitcairnia, Puya*) have fully developed root systems. The most specialized epiphytic genera (e.g., *Tillandsia*) lack roots (except as young seedlings) and leaf-base tanks and absorb water from the atmosphere by means of multicellular, peltate hairs. The leaves are spirally arranged, often creating a basal rosette of stiff, usually evergreen, leaves, with spiny margins. There is seamless transition between the sheath and the blade, but often a characteristic difference in color. In the terrestrial species, the leaves have unexpanded leaf-bases and leaf hairs that serve only to reduce transpiration. In some tank-forming species (e.g., *Ananas*), the overlapping leaf-bases act as reservoirs for water and humus, which are utilized by adventitious roots growing up between the leaf-bases. The majority of genera have larger leaf-base tanks, and absorption from the tanks is mainly carried out by specialized peltate hairs (trichomes) composed of a foot, a stalk of living cells, and a shield of dead cells. These hairs are often positioned at the bottom of a pit and expand when wetted, so water is drawn into the dead cells of the hair and thence osmotically through the living cells of the stalk of the trichome into the leaf. The scales collapse when dry, closing the pit like a lid and permitting gas exchange through the stomata (almost entirely

Genera 56   Species c. 2,600
Economic uses The pineapple is the most important cash crop in the family, but other species have edible fruits, some are sources of fibers, and several are important ornamentals

confined to the underside of the leaves) but reducing water loss from the surface of the plant. The plants can thus survive in extremely arid habitats, but not in very humid habitats such as rain forests. The tanks may hold up to 5 liters of water and serve as a water reservoir for canopy-dwelling animals. The tanks contain a highly specialized flora and fauna, including species of *Utricularia* (bladderworts), tree frogs (*Syncope antenori*), and various insects[ii].

The inflorescence is terminal, produced out of the center of the tank in tank-forming species and may be a spike, raceme, or panicle, often with brightly colored, conspicuous bracts, rarely with solitary flowers (*Tillandsia*). Many bromeliads die after flowering, including some of the genera cultivated for their inflorescence, but these produce suckers and can be readily propagated. The flowers are bisexual or rarely

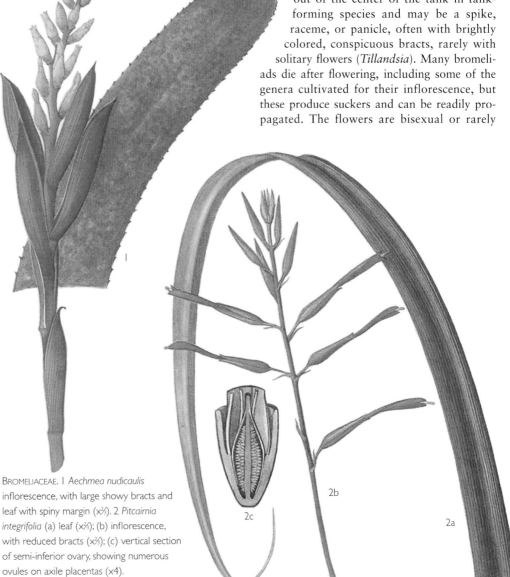

BROMELIACEAE. I *Aechmea nudicaulis* inflorescence, with large showy bracts and leaf with spiny margin (x⅔). 2 *Pitcairnia integrifolia* (a) leaf (x⅔); (b) inflorescence, with reduced bracts (x⅔); (c) vertical section of semi-inferior ovary, showing numerous ovules on axile placentas (x4).

functionally unisexual (*Hectia*), usually actinomorphic, although slightly zygomorphic in some genera (*Pitcairnia*). The flowers are 3-merous, with a greenish calyx and a showy petaloid corolla. The 6, free stamens are often attached to the base of the perianth or more or less fused. The ovary is superior to inferior and consists of 3 fused carpels. There are 3 locules, each with numerous ovules on axile placentas. The stigmas are 3-lobed and borne on a slender style. The fruit is a berry or capsule. The seeds contain a small embryo and a starchy endosperm. In several genera, the seeds have wings, are plumose, or have tailed appendages. In most genera, the showy inflorescences and the nectaries on the septa of the ovaries are adaptations to pollination by birds (hummingbirds and honeycreepers), bats, or insects.

Classification The Bromeliaceae family is undoubtedly monophyletic[iii,iv,v]. Molecular analyses indicate that the family is closely related to a large number of other families, including among others, the Poaceae[iv,v]. The family is often divided into 3 subfamilies[vi] which, based on morphology, are considered monophyletic[vii], but the monophyly of the Pitcairnioideae is doubtful[viii].

SUBFAM. PITCAIRNIOIDEAE Largely composed of terrestrial bromeliads, with well-developed roots and usually superior ovary. The fruit is a capsule. The subfamily includes 16 genera. Among them are *Brocchinia*, with 21 species restricted to the Guayana Highlands. *B. reducta* has been reported to be passive carnivorous. Other notable genera are *Pitcairnia* and *Puya*.

SUBFAM. TILLANDSIOIDEAE Largely composed of epiphytes, with or without anchoring roots. The ovary is superior, and the fruit a capsule with plumose seeds. It includes 9 genera of which *Tillandsia* is by far the largest with more than 500 spp..

SUBFAM. BROMELIOIDEAE The largest subfamily with 30 genera, mostly epiphytes, with anchor roots, usually with inferior ovary and a berrylike fruit. In *Ananas* and the related *Pseudananas*, the individual fruits fuse, and the inflorescence swells to form a multiple fruit.

Economic uses *Ananas comosus* (Pineapple) is an extremely important edible fruit of tropical and subtropical regions. In 2005, the world production exceeded 16 million tonnes, accounting for the largest part of the world's export in tropical fruits. Most commercially grown pineapples are canned or made into juice rich in vitamins A and B. Pineapple stems and fruits are a possible commercial source of enzymes. Bromelain, which has anti-inflammatory properties, is a mixture of 5 different proteolytic enzymes. Other species with edible fruits are found in the genera *Aechmea*, *Bromelia*, and *Greigia*. They are consumed locally. A range of species produce fibers used locally for clothmaking and cordage, notably the pineapple in the Philippines, *Aechmea magdalenae* (Pita) in Colombia, and *Neoglaziovia variegata* (Caroa) in Brazil. *Tillandsia usneoides* (Spanish Moss) was previously widely used as a substitute for horsehair in upholstery. Various genera are grown as ornamentals in the open in frostfree regions, under glass, or as house plants in temperate regions. The foliage alone may be attractive, as in variegated forms of the pineapple, the striped leaves of certain species of *Billbergia*, *Cryptanthus*, and *Guzmania*, and the dense rosettes of *Dyckia*, *Nidularium*, and *Aregelia*. Additionally, several genera include specie which produce inflorescences with showy bracts, for example

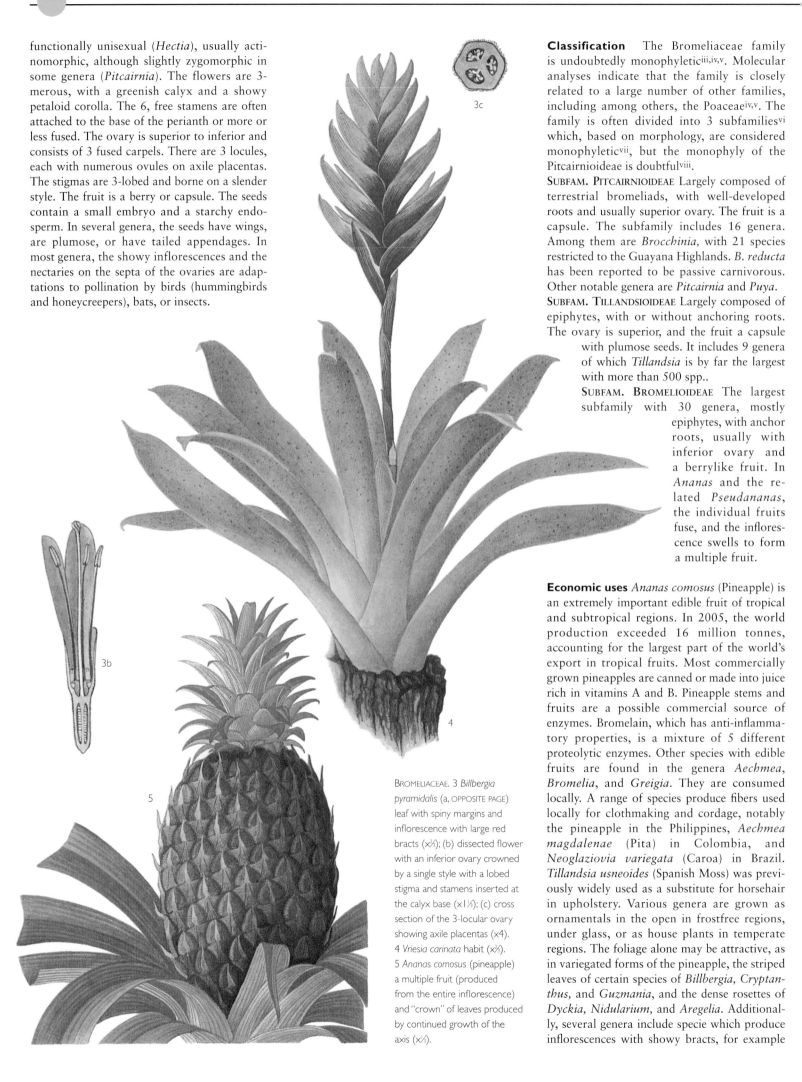

3c

3b

5

4

BROMELIACEAE. 3 *Billbergia pyramidalis* (a, OPPOSITE PAGE) leaf with spiny margins and inflorescence with large red bracts (x⅔); (b) dissected flower with an inferior ovary crowned by a single style with a lobed stigma and stamens inserted at the calyx base (x1½); (c) cross section of the 3-locular ovary showing axile placentas (x4). 4 *Vriesia carinata* habit (x⅔). 5 *Ananas comosus* (pineapple) a multiple fruit (produced from the entire inflorescence) and "crown" of leaves produced by continued growth of the axis (x⅓).

Pitcairnia, *Billbergia*, *Aechmea*, and *Vriesea*. In parts of the dry tropics, bromeliads have hampered malaria control since the water retained in the leaf tanks of certain native species may serve as a breeding ground for malaria-carrying *Anopheles* mosquitos. The tanks cannot easily be sprayed for mosquito control.   [BP] OS

i Holst, B. K. & Luther, H. E. Bromeliaceae (Bromeliad family). Pp. 418–421. In: Smith, N. *et al.* (eds), *Flowering Plants of the Neotropics*. Princeton/Oxford, Princeton University Press (2004).
ii Laessle, A. M. A micro-limnological study of Jamaican bromeliads. *Ecology* (1961).
iii Gilmartin, A. J. & Brown, G. K. Bromeliales, related monocots, and resolution of relationships among Bromeliaceae subfamilies. *Syst. Bot.* 12: 493–500 (1987).
iv Davis, J. I. *et al.* A phylogeny of the monocots, as inferred from *rbc*L and *atp*A sequence variation, and a comparison of methods for calculating jackknife and bootstrap values. *Syst. Bot.* 29: 467–510 (2004).
v Chase, M. W. *et al.* Multi-gene analysis of monocot relationships: A summary. Pp. 63–75. In: Columbus, J. T. *et al.* (eds), *Monocots: Comparative Biology and Evolution*. Claremont, California, Rancho Santa Ana Botanic Garden (2006).
vi Dahlgren, R. M. T., Clifford, H. T., & Yeo, P. F. *The Families of the Monocotyledons*. Berlin, Springer-Verlag (1985).
vii Judd, W. S. *et al. Plant Systematics. A Phylogenetic Approach*. 2nd edn. Sunderland, Massachusetts, Sinauer Associates (2002).
viii Crayn, D., Winter, K., & Smith, J. A. C. Multiple origins of crassulacean acid metabolism and the epiphytic habit in the neotropical family Bromeliaceae. *Proc. Nat. Acad. Sci. USA* 3703–3708 (2004).

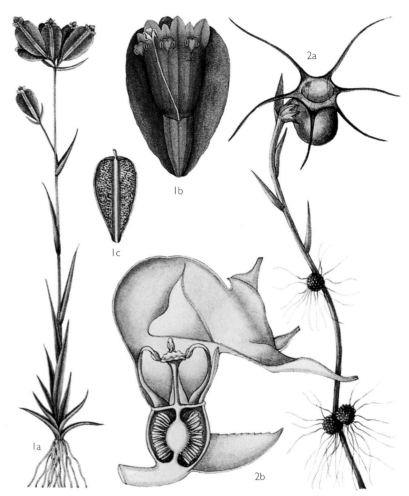

BURMANNIACEAE.
1 *Burmannia coelestis* (a) is a hemisaprophytic species with both basally aggregated and stem leaves, unbranched stem, and terminal inflorescence of flowers with winged perianth tubes (×1); (b) part of perianth opened out to show the 6 tepals and the enlarged perianth tube expanded below the ovary, 3 anthers, and style with 3 stigmas (×4); (c) capsule (×4). 2 *Afrothismia winkleri* (a) saprophytic species, with scalelike leaves, rhizome with tubers, flower with nearly equal tepals, and an urnlike perianth tube (×⅔); (b) longitudinal section of flower showing urnlike perianth tube with internal flange, inferior ovary containing numerous ovules on parietal placentas, and a sterile column in the middle (×8).

3a

BURMANNIACEAE
BURMANNIA FAMILY

The Burmanniaceae is a family of mostly small, achlorophyllous, saprophytic or hemisaprophytic herbs, which are often tinted reddish, yellowish, or white.

Distribution The Burmanniaceae is largely pantropical but also reaches warm-temperate regions. Most species occur in the moist understory of lowland and montane forests, but some inhabit open areas such as savannas and swampland. Several species are extremely rare.

Description Mostly small, saprophytic or hemisaprophytic, annual or perennial herbs with slender, upright, unbranched stems produced from often tuberous rhizomes. The saprophytic species are achlorophyllous, often colored white, yellow, or red, with alternate, sessile, and scalelike leaves. The hemisaprophytic species (*Burmannia* spp.) are green, often with well-developed, basally aggregated, alternate leaves. The inflorescence is a terminal many-flowered cyme, often a bifurcate, bracteate circinnus, rarely the flowers are solitary. The flowers are bisexual, actinomorphic, usually pedicellate, and often white, bright blue, or rarely yellow. There are usually 6 tepals in 2 whorls, rarely only 3 or 1. They are basally fused into a tube, which often has longitudinal wings or ribs. The outermost whorl encloses and protects the inner whorl in the bud. The stamens are 6 or 3 on short fila-ments, often fused with the tube and opposite the inner tepals. The thecae open transversely or longitudinally and often bear various appendages or are fused into an anther tube (*Thismia*). The ovary is inferior, consisting of 3 carpels and 3- or 1-locular. The placentation is axile, parietal, or free, with numerous small, anatropous ovules. The style is cylindrical with 3 apical branches or the stigma is capitate. The fruit is a capsule, rarely fleshy, often winged, and variously dehiscent. The seeds are tiny with endosperm comprising just a few cells and containing a few starch grains that are replaced by protein and fat.

Classification The achlorophyllous members of the Burmanniaceae live symbiotically with mycorrhizal fungi, mostly Vesicular Arbuscular Mycorrhiza (VAM). The hemisaprophytic species represent an intermediary stage between autotrophy and heterotrophy. The Corsiaceae

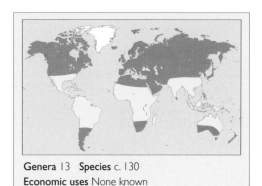

Genera 13   Species c. 130
Economic uses None known

(q.v.) is occasionally included in Burmanniaceae, and the latter is in its current circumscription often divided into 2 separate families, Burmanniaceae and Thismiaceae. However, Corsiaceae has been shown to be only distantly related to Burmanniaceae[i]. Due to its tiny seeds, the Burmanniaceae has been considered closely allied to the Orchidaceae, but this is not supported by morphology[i,ii]. The sister group of the Burmanniaceae is most likely the Dioscoreaceae (q.v.)[iii].   OS

[i] Davis, J. I. *et al.* A phylogeny of the monocots, as inferred from *rbc*L and *atp*A sequence variation, and a comparison of methods for calculating jackknife and bootstrap values. *Syst. Bot.* 29: 467–510 (2004).
[ii] Rasmussen, F. N. Relationships of Burmanniales and Orchidales. Pp. 353–377. In: Rudall, P. J., Cribb, P. J., Cutler, D. F., & Humphries, C. J. (eds), *Monocotyledons: Systematics and Evolution.* Richmond, Royal Botanic Gardens, Kew (1995).
[iii] Caddick, L. R., Rudall, P. J., Wilkin, P., Hedderson, T. A. J., & Chase, M. W. Phylogenetics of Dioscoreales based on combined analyses of morphological and molecular data. *Bot. J. Linn. Soc.* 138: 123–144 (2002).

BUTOMACEAE
FLOWERING RUSH

The Butomaceae comprises a single herbaceous aquatic species, easily recognizable by its 3-angled leaves and its umbellate inflorescence.

Distribution Widespread in Europe and temperate Asia, with scattered occurrences in Eastern China, the single representative of this family prefers aquatic and marshy habitats.

Description *Butomus* is a rhizomatous perennial with linear leaves up to 1 m long or more. The leaves are triangular in transverse section, more or less spirally twisted, in 2 rows along the rhizome, with intervaginal scales. The umbel-like inflorescence terminates at a leafless scape and consists of a single terminal flower surrounded by 3 cymes. The flowers are regular and bisexual. The 3 sepals are petal-like, pink, and persist in fruit. The 3 petals are larger but similar to sepals. Stamens usually 3+6. The ovary is superior, with usually 6 carpels, in 2 alternating whorls, and slightly united at the base. The ovules are numerous, anatropous, and scattered over the inner surface of the carpel wall. The fruit is a follicle crowned by a persistent style. Seeds without endosperm; embryo straight.

Classification The Butomaceae (together with e.g., Hydrocharitaceae and Limnocharitaceae) has been considered closely related to Cabombaceae[i,ii]. Most molecular analyses indicate that Butomaceae is closely related to Hydrocharitaceae[iii,iv,v].

Economic uses Becoming a weed in parts of North America and in ricefields in the Danube Delta in Romania. The rhizomes are rich in starch and eaten mixed with grain starch or as vegetables in parts of Russia.   [CDC] OS

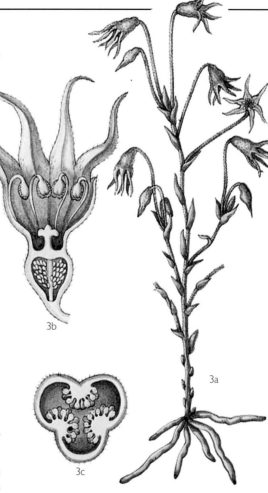

BURMANNIACEAE. 3 *Haplothismia exannulata* (a) saprophytic species, with scalelike leaves and 3–6-flowered monochasial inflorescences, with nodding flowers (x⅔); (b) longitudinal section of flower showing pendant stamens and 1-locular ovary (c) cross section of ovary with numerous ovules on 3 parietal placentas (x6).

Genera 1   Species 1
Economic uses Widely cultivated as an ornamental

[i] Dahlgren, R. M. T. & Clifford, H. T. Some conclusions from a comparative study of the monocotyledons and related diccotyledons. *Ber. Deutsch. Bot Gesel.* 94: 203–227 (1981).
[ii] Doyle, J. A. & Endress, P. K. Morphological phylogenetic analysis of basal angiosperms: comparison and combination with molecular data. *Int. J. Pl. Sci.* 161, Suppl. S121–S153 (2000).
[iii] Les, D. H., Cleland, M. A., & Waycott, M. Phylogenetic Studies in Alismatidae, II: Evolution of marine angiosperms (seagrasses) and hydrophily. *Syst. Bot.* 22: 443–463 (1997).
[iv] Davis, J. I. *et al.* A phylogeny of the monocots, as inferred from *rbc*L and *atp*A sequence variation, and a comparison of methods for calculating jackknife and bootstrap values. *Syst. Bot.* 29: 467–510 (2004).
[v] Chase, M. W. *et al.* Higher-level systematics of the monocotyledons: An assessment of current knowledge and a new classification. In: Wilson, K. L. & Morrison, D. A. (eds), *Monocots. Systematics and Evolution.* Melbourne, CSIRO Publishing (2000).

CAMPYNEMATACEAE

Distribution The family consist of 2 genera: 1 native to New Caledonia and 1 native to Tasmania.

Description Small, terrestrial herbs with a single basal leaf or several basal clusters of leaves, which are linear to elliptical with a 3-dentate apex. The inflorescence is bracteate, paniculate, or umbel-like, with small, sometimes solitary, flowers. The flowers have 5, 3-merous whorls, and the tepals are greenish, and at least in *Campynemanthe*, with nectaries. The ovary is inferior, 3- or 1-locular, with free styles and axile or parietal placentas and few to many anatropous ovules. The fruit is a capsule, containing few to many seeds, with a tiny embryo and an endosperm of hemicellulose and oil.

Genera 2   Species 4
Economic uses None

Campynematace was considered closely related to Melanthiaceae (q.v.) and the families around Burmanniaceae (q.v.)[i]. Molecular evidence either points at Campynematace as sister group to a larger group of families including, among others, Alstroemeriaceae, Colchicaceae, Liliaceae, and Melanthiaceae[ii,iii,iv], but excluding Corsiaceae (q.v.), or the same group of families but with the exclusion of Alstroemeriaceae, Colchicaceae, and Luzuriagaceae[v].   OS

[i] Dahlgren, R. M. T., Clifford, H. T., & Yeo, P. F. *The Families of the Monocotyledons.* Berlin, Springer-Verlag (1985).
[ii] Davis, J. I. *et al.* A phylogeny of the monocots, as inferred from *rbc*L and *atp*A sequence variation, and a comparison of methods for calculating jackknife and bootstrap values. *Syst. Bot.* 29: 467–510 (2004).
[iii] Fay, M. F. *et al.* Phylogenetics of Liliales: Summarized evidence from combined analyses of five plastid and one mitochondrial loci. Pp. 559–565. In: Columbus, J. T. *et al.* (eds), *Monocots: Comparative Biology and Evolution.* Claremont, California, Rancho Santa Ana Botanic Garden (2006).
[iv] Chase, M. W. *et al.* Multi-gene analysis of monocot relationships: A summary. Pp. 63–75. In: Columbus, J. T. *et al.* (eds), *Monocots: Comparative Biology and Evolution.* Claremont, California, Rancho Santa Ana Botanic Garden (2006).
[v] Vinnersten, A. & Bremer, K. Age and biogeography of major clades in Liliales. *American J. Bot.* 88: 1635–1703 (2001).

CANNACEAE
CANNA FAMILY

A family of large, rhizomatous herbs with alternate, petiolate, simple leaves and large, asymmetric flowers with petal-like staminodes.

Distribution Restricted to the tropical parts of the Americas, where it grows on riverbanks.

Description Large, erect plants, with starchy, tuberous rhizomes and distichously or spirally arranged leaves with sheathing bases that gradually taper into the petiole, no ligule, and a lamina that is rolled up from one side and with a distinct compound midrib and pinnately arranged lateral veins. The inflorescence is terminal with 3-ranked bracts, each subtending either few-flowered cymes or single flowers. The flowers are short-lived, large, and showy, bisexual with 5 whorls, but strongly asymmetric. The perianth comprises 3, imbricate, basally connate sepals, which are usually green or purple, much smaller than the petals, and persistent in fruit. The 3 petals, 1 of which is usually smaller than the other, are often yellow or white and basally fused. The 6 stamens are petal-like and brightly colored, 1 of which

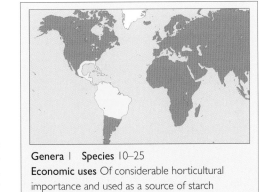

Genera 1 **Species** 10–25
Economic uses Of considerable horticultural importance and used as a source of starch

bears a bithecal anther along the edge, the remaining 1–5 being staminodes. The staminodes are all of different form and size, the 1 opposite the functional stamen, which is reflexed and rolled back on itself, is called the labellum, and the others called wings. The ovary is inferior, of 3 fused carpels, and has 3 locules each containing 2 rows of numerous ovules on axile placentas. The fruit is usually a hard, tuberculate or bristly, irregularly splitting capsule, containing many small seeds with a tuft of hairs and straight embryos surrounded by a thin starchy and a copious, extremely hard, starchy endosperm.

Classification The family is monophyletic and, on the basis of both morphological and molecular data, the sister group to the Marantaceae[i,ii,iii] (q.v.) with which it shares

such features as asymmetric flowers, reduction in the number of functional stamens, and free staminodes. Both families belong to a large monophyletic group that also includes, among other families, Zingiberaceae (q.v.) and Musaceae (q.v.)[i,iii].

Economic uses *Canna edulis*, the source of purple or Queensland arrowroot, has been cultivated in the Andes for around 4,500 years and is now grown all over the tropics for its starchy rhizomes, although it is most important in Australia and India. The starch is easily digestible, the grains of which are among the largest known and are visible to the naked eye. More than 1,000 horticultural varieties of cannas are known, often with showy leaves and large flowers that vary from yellow (*Canna* × *generalis*) to red and orange (*C.* × *orchidoides*) and are widely grown as tropical ornamentals or in greenhouses. The seeds are used locally in the production of necklaces.   OS

[i] Chase, M. W. *et al.* Multi-gene analysis of monocot relationships: A summary. Pp. 63–75. In: Columbus, J. T. *et al.* (eds), *Monocots: Comparative Biology and Evolution.* Claremont, California, Rancho Santa Ana Botanic Garden (2006).
[ii] Johansen, L. B. Phylogeny of *Orchidantha* (Lowiaceae) and the Zingiberales based on six DNA regions. *Syst. Bot.* 30: 106–117 (2005).
[iii] Kress, J. W., Prince, L. M., Hahn, W. J., & Zimmer, E. A. Unraveling the evolutionary radiation of the families of the Zingiberales using morphological and molecular evidence. *Syst. Bot.* 50: 926–944 (2001).

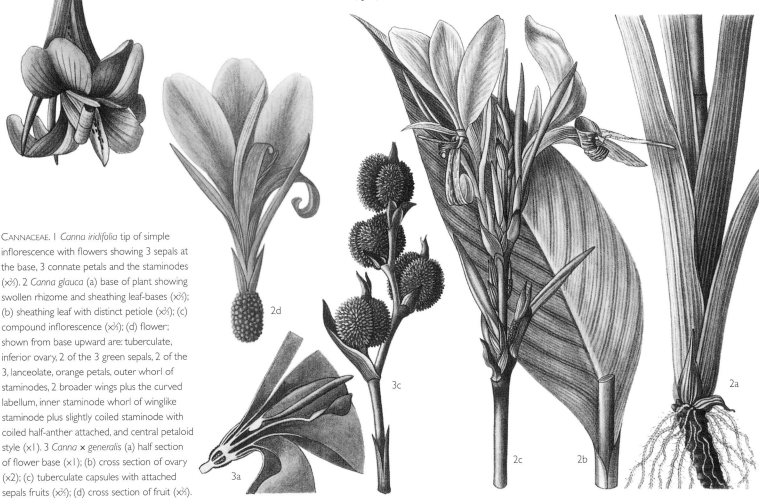

CANNACEAE. 1 *Canna iridifolia* tip of simple inflorescence with flowers showing 3 sepals at the base, 3 connate petals and the staminodes (×⅔). 2 *Canna glauca* (a) base of plant showing swollen rhizome and sheathing leaf-bases (×⅔); (b) sheathing leaf with distinct petiole (×⅔); (c) compound inflorescence (×⅔); (d) flower; shown from base upward are: tuberculate, inferior ovary, 2 of the 3 green sepals, 2 of the 3, lanceolate, orange petals, outer whorl of staminodes, 2 broader wings plus the curved labellum, inner staminode whorl of winglike staminode plus slightly coiled staminode with coiled half-anther attached, and central petaloid style (×1). 3 *Canna* × *generalis* (a) half section of flower base (×1); (b) cross section of ovary (×2); (c) tuberculate capsules with attached sepals fruits (×⅔); (d) cross section of fruit (×⅔).

CENTROLEPIDACEAE

The Centrolepidaceae is a family of small, cushion-forming, annual or perennial, grass-like, rushlike, or even mosslike herbs, with scapose, spicate, or capitate inflorescences and conspicuous bracts.

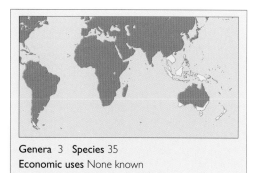

Genera 3 **Species** 35
Economic uses None known

Distribution Disjunct distribution, mainly found in Australasia, Southeast Asia, and in the southernmost tip of South America. They grow in heath, scrub, and woodland.

Description Members of the family are cushionlike herbs with bristlelike, linear leaves that are basal in annuals and imbricate and crowded along the stems in perennials. The inflorescence is terminal, usually scapose, either condensed and capitate, or spikelike, with 2 or more distichous, glumelike bracts. The flowers are minute, unisexual, and naked, often combined into a "pseudanthium." The male flower has a single stamen, and the female flower has a solitary carpel, which terminates in a slender, undivided style. The flowers are grouped into uni- or bisexual "pseudanthia," which are composed either of 1 or 2 male flowers or 1 to many female flowers or, more frequently, a combination of male and female flowers grouped together in the capitate or spicate inflorescences. The ovary is 1-locular, with a single, hanging, atropous ovule. The fruit is a 1-seeded follicle, and the seeds have a starchy endosperm.

Classification The Centrolepidaceae are often considered an offshoot of Restionaceae[i] (q.v.) and a sister group relationship to the latter has been noted[ii,iii,iv]. *Aphelia* is restricted to southern Australia, while *Centrolepis* is limited to Southeast Asia, Australia, and New Zealand. *Gaimardia* species occur in South America, Southeast Asia, Tasmania, and New Zealand; *G. australis* is the only species that reaches southern South America. OS

[i] Dahlgren, R. M. T., Clifford, H. T., & Yeo, P. F. *The Families of the Monocotyledons.* Berlin, Springer-Verlag (1985).
[ii] Linder, H. P. and Caddick, L. R. Restoniaceae seedlings: Morphology, anatomy and systematic implications. *Feddes Repertorium* 112: 59–80 (2001).
[iii] Chase, M. W. *et al.* Multi-gene analysis of monocot relationships: A summary. Pp. 63–75. In: Columbus, J. T. *et al.* (eds), *Monocots: Comparative Biology and Evolution.* Claremont, California, Rancho Santa Ana Botanic Garden (2006).
[iv] Brigg, B. G., Marchant, A. D., Gilmore, S., & Porter, C. L. A molecular phylogeny of Restionaceae and allies. Pp. 661–671. In: Wilson, K. L. & Morrison, D. A. (eds), *Monocots. Systematics and Evolution.* Melbourne, CSIRO Publishing (2000).

COLCHICACEAE
COLCHICUM FAMILY

A small family of perennials with underground corms or rhizomes and often large flowers.

Distribution Representatives of the family Colchicaceae are typically found in areas with a Mediterranean climate, especially in Africa, Asia, and the Mediterranean countries. The only genus in the New World is *Uvularia*.

Description Members of this family are perennial herbs, with underground rhizomes or corms, and erect, or sometimes scandent, stems, with distichous, rarely verticillate, often sheathing leaves. Occasionally, the roots are tuberous (*Gloriosa*). The blades are ovate, lanceolate or linear, rarely cuspidate or cirrhose. The inflorescences are axillary, few- to many-flowered, spikes, racemes, or more or less condensed cymes, rarely the flowers are solitary. The flowers are usually bisexual, actinomorphic, sessile, or pedicellate. The 6 tepals are more or less equal, generally nectariferous, and connate at the base or free. The 6 stamens sometimes have nectar glands at the base. The ovary is superior, 3-carpelllate, completely or partially syncarpous, 3-locular with few or many anatropous or campylotropous ovules and parietal placentas, often with free or partly to completely united styles. The fruit is a septicidal or loculicidal capsule, which is occasionally fleshy and rarely becomes baccate. The seeds have straight, linear embryos, and endosperms with or without starch.

Classification The circumscription of the family varies and occasionally members of the Liliaceae (q.v.) and Convallariaceae (q.v.) are included[i]. However, as circumscribed here, the monophyly of the family is well supported[ii,iii,iv], although part of the family (including *Uvularia*) occasionally is separated as the Uvulariaceae[v,vi], justified by its position as sister group to the remaining Colchicaceae[iii,iv]. The sister group of Colchicaceae is Alstroemeriaceae (q.v.) plus Luzuriagaceae (q.v.)[iii,iv,vii].

Economic uses Species of *Colchicum* (Naked Ladies, c. 90 spp.) that flower in the fall are widely cultivated in gardens in temperate areas, while *Gloriosa superba* (Flame Lily) is often grown in greenhouses. Many members of the family store highly toxic alkaloids and are a serious hazard to live-stock. Due to their toxic contents several species are also of medical importance. Colchicine, an alkaloid from *Colchicum autumnale* and related species, is well known for it ability to interfere with normal cell division, by disorganising the spindle-mechanism at mitosis, and is used in plant breeding to create polyploids. OS

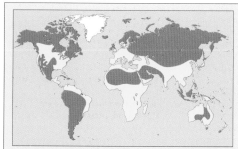

Genera c. 18 **Species** c. 225
Economic uses Several important ornamentals or medicinal uses; source of colchicine

[i] Nordenstam, B. Colchicaceae. Pp. 175–185. In: Kubitzki, K. (ed.), *The Families and Genera of Vascular Plants. III. Flowering Plants. Monocotyledons: Lilianae (except Orchidaceae).* Berlin, Springer-Verlag (1998).
[ii] Rudall, P. A., Stobart, K. L., Hong, W.-P., Conran, J. G., Furness, C. A., Kite, G. C., & Chase, M. W. Consider the lilies: Systematics of Liliales. Pp. 347–359. In: Wilson, K. L. & Morrison, D. A. (eds), *Monocots. Systematics and Evolution.* Melbourne, CSIRO Publishing (2000).
[iii] Fay, M. F. *et al.* Phylogenetics of Liliales: Summarized evidence from combined analyses of five plastid and one mitochondrial loci. Pp. 559–565. In: Columbus, J. T. *et al.* (eds), *Monocots: Comparative Biology and Evolution.* Claremont, California, Rancho Santa Ana Botanic Garden (2006).

COLCHICACEAE. I *Colchicum callicymbium* habit showing basal corm, emerging leaves, and flowers with 6 tepals, 6 yellow stamens, and trifid style (x⅔).

iv Chase, M. W. *et al.* Multi-gene analysis of monocot relationships: A summary. Pp. 63–75. In: Columbus, J. T. *et al.* (eds), *Monocots: Comparative Biology and Evolution.* Claremont, California, Rancho Santa Ana Botanic Garden (2006).

v Dahlgren, R. M. T., Clifford, H. T., & Yeo, P. F. *The Families of the Monocotyledons.* Berlin, Springer-Verlag (1985).

vi Judd, W. S. *et al. Plant Systematics. A Phylogenetic Approach.* 2nd edn. Sunderland, Massachusetts, Sinauer Associates (2002).

vii Vinnersten, A. & Bremer, K. Age and biogeography of major clades in Liliales. *American J. Bot.* 88: 1685–1703 (2001).

COMMELINACEAE

SPIDERWORT FAMILY

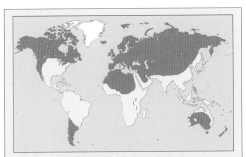

Genera c. 40 **Species** 650
Economic uses A number of genera include species that are used as garden ornamentals or pot plants. Some local medicinal uses

A family of herbs, often somewhat succulent, with a mucilaginous sap and alternate, entire leaves with a sheathing base.

Distribution The family is found throughout warm-temperate, subtropical, and tropical regions. In temperate regions, the family is best represented in North America and Asia. No species are native to Europe. The Commelinaceae is found in most habitats but prefer humid conditions and are typically found in grassland and forests.

Description Although most species are stoloniferous, rhizomatous, terrestrial perennial herbs, some species are annual (*Commelina*) or epiphytic (*Cochliostema*), and a few are climbers (e.g., *Dichorisandra*). Usually they have jointed, succulent aerial stems, though sometimes almost stemless, with alternate leaves. The leaves are distichous or spirally arranged, often somewhat succulent, entire, frequently narrowing into a pseudopetiole, and with a closed basal sheath. The inflorescence is essentially a cyme with cincinate components and is borne either at the end of the stem or in the leaf axils; rarely the flowers are solitary (*Sauvallea*). The flowers are bi- or unisexual, usually actinomorphic, but occasionally zygomorphic. The perianth consists of a calyx of 3 free or partly fused sepals and a corolla of 3 free or basally fused, white or colored petals. When fused, the corolla may have a basal tube (*Coletrype, Cyanotis, Tradescantia*), while rarely 1 of the 3 petals is reduced in size (e.g.,

Aneilema). The flowers often are of a short duration, typically no more than a day. The stamens are in 2 whorls of 3, usually free, but in some genera only 2 or 3 of the stamens are fertile, the others being reduced to staminodes. *Callisia* has only 1 functional stamen. Several genera have brightly colored, often beadlike hairs on the filaments. The ovary is superior and consists of 3 fused carpels with 3 locules (rarely 2), each containing 1 to many atropous ovules on axile placentas. The style is terminal and simple, terminating in 3 branches or capitate. The flowers are strongly scented in some genera (e.g., *Palisota, Tripogandra*) but always without nectar. The fruit is usually a dehiscent capsule, rarely indehiscent and berrylike. The seeds have a distinct, caplike enlargement (embryostega) that covers the embryo. They contain a copious, starchy endosperm.

COMMELINACEAE. 1 *Commelina erecta* shoot, showing sheathing leaf-bases, flowers with 3 petals, 3 stamens, and 3 staminodes (x⅔). 2 *Gibasis graminifolia* (a) leafy shoot and inflorescence, each flower with 6 stamens (x⅔); (b) capsule (x3). 3 *Tradescantia zebrina* (= *Zebrina pendula*) leafy stem and solitary flower (x⅔).

ii Evans, T. M., Sytsma, K. J., Faden, R. B., & Givnish, T. J. Phylogenetic relationships in the Commelinaceae: II. A cladistic analysis of *rbc*L and morphological data. *Syst. Bot.* 28: 270–292 (2003).

iii Davis, J. I. *et al.* A phylogeny of the monocots, as inferred from *rbc*L and *atp*A sequence variation, and a comparison of methods for calculating jackknife and bootstrap values. *Syst. Bot.* 29: 467–510 (2004).

iv Kellogg, E. A. & Linder, H. P. Phylogeny of Poales. Pp. 353–377. In: Rudall, P. J., Cribb, P. J., Cutler, D. F., & Humphries, C. J. (eds). *Monocotyledons: systematics and evolution.* Richmond, Royal Botanic Gardens, Kew (1995).

v Chase, M. W. *et al.* Multi-gene analysis of monocot relationships: A summary. Pp. 63–75. In: Columbus, J. T. *et al.* (eds), *Monocots: Comparative Biology and Evolution.* Claremont, California, Rancho Santa Ana Botanic Garden (2006).

CONVALLARIACEAE

A small family of rhizomatous herbs, with alternating, distichous, opposite or whorled leaves, and flowers in a spike or raceme.

Distribution Restricted to the northern hemisphere in Asia, Europe, and North America. Most species are understory in forests, but a few are epiphytic.

Description Mostly terrestrial, rhizomatous herbs with perennial or annual stems up to 1 m high, usually carrying alternating, distichous, opposite or whorled, sessile or shortly petiolate entire, ovate to linear leaves, or rarely the plants stemless and the leaves are in a basal rosette. The inflorescence is terminal or axillary, usually a spike or raceme, rarely reduced to a single flower. The flowers are bisexual, rarely unisexual, and usually actinomorphic with 2 whorls of 3, rarely 4 to 5, or only 2, petaloid tepals, which are often fused into a tube at the base. There are mostly 2 whorls of 3 stamens, but sometimes only 4, or 8 to 12. The ovary is superior or perigynous with septal nectaries and usually 3-carpellate and 3-locular, but sometimes with fewer or more carpels and locules. The anatropous, campylotropous, or more or less atropous ovules are

Classification Both morphological[i] and molecular data[ii,iii] support the monophyly of the Commelinaceae. Molecular analyses point at a close relationship between Commelianaceae and Pontederiaceae[iii,iv] or Hanguanaceae[v]. The family is often divided into to the subfamilies Cartonematoideae and Commelinoideae. The first has yellow, actinomorphic flowers; the second, rarely yellow flowers that are either actinomorphic or zygomorphic. Incongruence between different data sets has questioned this division[ii].

Economic uses A number of genera are important pot or garden plants, e.g., *Commelina* (170 spp.), *Tradescantia* (including both *Rhoeo* and *Zebrina*, 70 spp.), *Cyanotis* (50 spp.), and *Dichorisandra* (25 spp.). Different species of *Tradescantia* are sold under the common names Tradescantia, Wandering Jew, or Wandering Sailor. *Tradescantia virginiana* is known as the Spiderwort. An extract of leaves and stems of the tropical African perennial herb *Aneilema beninense* is used as a laxative. The sap of several species is used to treat eye inflammation (e.g., *Floscopa scandens*, *Commelina* spp.) in Africa, Asia, and South America. Some species are weeds.   OS

i Evans, T. M., Faden, R. B., Simpson, M. G., & Sytsma, K.J. Phylogenetic relationships in the Commelinaceae: I. A cladistic analysis of morphological data. *Syst. Bot.* 25: 668–691 (2000).

COMMELINACEAE. 4 (ABOVE, LEFT) *Tradescantia sillamontana* leafy shoot with inflorescence subtended by boat-shaped, leafy bracts (x⅔). 5 *Tradescantia* (= *Rhoeo*) *spathacea* shoot showing rosette of bromeliad-like leaves and inflorescence partly hidden by boat-shaped bracts (x⅓). 6 *Tradescantia navicularis* (a) juvenile plant and (b) adult shoot (x⅔).

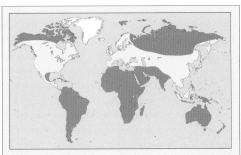

Genera 17 Species c. 130
Economic uses Several species are ornamentals; few are used for food or have medical uses

borne in 2 rows in each locule. The fruit is an often brightly colored berry, a capsule, or drupe, or is indehiscent and ruptures during development, with 1 to many black or colored seeds. The embryo is linear in a starchless endosperm, consisting of aleurones and oil.

Classification The Convallariaceae is here accepted in a narrow sense[i,ii]. Molecular data indicate that a larger group of families, Convallariaceae (q.v.), Dracaenaceae, Eriospermaceae (q.v.), Nolinaceae (q.v.), and Ruscaceae (q.v.), here treated as separate taxa constitute a monophyletic group, sister to the Asparagaceae[iii,iv,v] or including it in a strict sense (q.v.)[vi]. This is also supported by a combined morphological and molecular analysis[vii]. The relationship of these families and parts of them accepted here, within Ruscaceae in a broad sense, and inclusion in it, is not convincing. Acceptance of a broadly defined Ruscaceae would create a morphologically extremely heterogeneous assemblage[viii]. However, acceptance of the other families in Ruscaceae makes Convallariaceae paraphyletic.

Economic uses Several species of *Convallaria* (Lily-of-the-valley), *Lirope* (Border Plant), *Ophiopogon* (Mondo Grass), and *Polygonatum* (Solomon's Seal) and are used as ornamentals. Different variegated forms of *Aspidistra elatior* (Cast-iron plant, Aspidistra) were very popular as house plants in the 19th century. The rhizomes of several genera are eaten locally, e.g., *Maianthemum* (May Lily) and *Polygonatum*. *Convarllaria majalis* is poisonous and has medical properties. *Maianthemum dilatatum* is a weed in swamps in North America. OS

[i] Dahlgren, R. M. T., Clifford, H. T., & Yeo, P. F. *The Families of the Monocotyledons.* Berlin, Springer-Verlag (1985).
[ii] Conran, J. G. & Tamura, M. N. Convallariaceae. Pp. 186–198. In: Kubitzki, K. (ed.), *The Families and Genera of Vascular Plants. III. Flowering Plants. Monocotyledons: Lilianae (except Orchidaceae).* Berlin, Springer-Verlag (1998).
[iii] Yamashita, J. & Tamura, M. N. Molecular phylogeny of the Convallariaceae (Asparagales). Pp. 387–400. In: Wilson, K. L. & Morrison, D. A. (eds), *Monocots. Systematics and Evolution.* Melbourne, CSIRO Publishing (2000).
[iv] Pires, J. C. *et al.* Phylogeny, genome size, and chromosomal evolution of Asparagales. Pp. 287–304. In: Columbus, J. T. *et al.* (eds), *Monocots: Comparative Biology and Evolution.* Claremont, California, Rancho Santa Ana Botanic Garden (2006).

[v] Tamura, M. N., Yamashita, J., Fuse, S., & Haraguchi, M. Molecular phylogeny of monocotyledons inferred from combined analysis of plastid *mat*K and *rbc*L gene sequences. *J. Plant Res.* 117: 109–120 (2004).
[vi] Fay, M. F. *et al.* Phylogenetic studies of Asparagales based on four plastid DNA regions. Pp. 360–371. In: Wilson, K. L. & Morrison, D. A. (eds), *Monocots. Systematics and Evolution.* Melbourne, CSIRO Publishing (2000).
[vii] Rudall, P. J., Conran, J. G., & Chase, M. W. Systematics of Ruscaceae/Convallariaceae: a combined morphological and molecular investigation, *Bot. J. Linn. Soc.* 134: 73–92 (2000).
[viii] Judd, W. S. *et al. Plant Systematics. A Phylogenetic Approach.* 2nd edn. Sunderland, Massachusetts, Sinauer Associates (2002).

CORSIACEAE

Distribution The family has a highly disjunct distribution in the southern hemisphere, with 1 genus (*Arachnitis*) occurring in southern Andean South America (including the Falkland Islands and Bolivia); 1 genus (*Corsia*) on New Guinea, the Solomon Islands, and northern Australia; and 1 genus in southern China (*Corsiopsis*). They occur mostly in humid forests.

Description Small, achlorophyllous, saprophytic, perennial, unbranched herbs with simple, ovate, more or less distichous leaves. The bisexual flowers are terminal, solitary, and zygomorphic with a perianth of 6 tepals, the 2 outer and 3 inner tepals are linear, ovate, or lanceolate, whereas the third, outer, median tepal is broader and upright or reflexed. The stamens are 6; the ovary is inferior, syncarpous, and consists of 3 carpels with free or connate styles and parietal placentas. The ovules are numerous, with tiny embryos. The fruit is a capsule with dustlike seeds.

Classification The completely or largely completely achlorophyllous families Corsiaceae, Burmanninaceae (q.v.), and Thismiaceae (here included in Burmanniaceae) have been considered closely related[i], and together related to the Orchidaceae (q.v.)[ii], but the latter relationship is largely unsupported[iii,iv]. There is, however, some evidence that the family is polyphyletic, and that *Arachnitis* is sister group to Burmanniaceae (q.v.) and *Corsia* to Campynemataceae (q.v.)[v]. *Arachnitis* is sometimes considered related to a large clade including, inter alia, Liliaceae and Colchicaceae [vi,vii]. OS

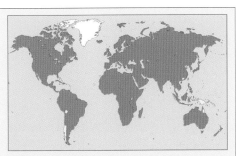

Genera 3 Species 30
Economic uses None known

[i] Dahlgren, R. M. T., Clifford, H. T., & Yeo, P. F. *The Families of the Monocotyledons.* Berlin, Springer-Verlag (1985).
[ii] Rübsamen, T. Morphologische, embryologische und systematische Untersuchungen an Burmanniaceae und Corsiaceae (mit Ausblick auf die Orchidaceae Apostasioideae). *Dissert. Bot.* 92: 1–310 (1986).
[iii] Rasmussen, F. N. Relationships of Burmanniales and Orchidales. Pp. 227–241. In: Rudall, P. J., Cribb, P. J., Cutler, D. F., & Humphries, C. J. (eds), *Monocotyledons: systematics and evolution.* Richmond, Royal Botanic Gardens, Kew (1995).
[iv] Neyland, R. & Hennigan, M. A phylogenetic analysis of large-subunit (26S) ribosome DNA sequences suggests that the Corsiaceae are polyphyletic. *N. Z. J. Bot.* 41: 1–11 (2003).
[v] Davis, J. I. *et al.* A phylogeny of the monocots, as inferred from *rbc*L and *atp*A sequence variation, and a comparison of methods for calculating jackknife and bootstrap values. *Syst. Bot.* 29: 467–510 (2004).
[vi] Fay, M. F. *et al.* Phylogenetics of Liliales: Summarized evidence from combined analyses of five plastid and one mitochondrial loci. Pp. 559–565. In: Columbus, J. T. *et al.* (eds), *Monocots: Comparative Biology and Evolution.* Claremont, California, Rancho Santa Ana Botanic Garden (2006).

COSTACEAE
COSTUS FAMILY

A small family of rhizomatous herbs, with simple leaves with sheathing base and ligule and a terminal, globose, or conelike spike, often with brightly colored bracts and flowers.

Distribution The family is pantropical, most species occurring in humid, lowland, or montane forest, with some species found along rivers or in swampy or disturbed areas.

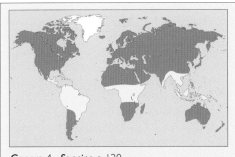

Genera 4 Species c. 120
Economic uses Source of cut flowers

Description Perennial, non-aromatic, rhizomatous, terrestrial, rarely epiphytic herbs, usually with unbranched stems, and spirally arranged, more or less elliptical leaves, with a short petiole, closed tubular sheath, and ligulate. The inflorescence is a condensed, globose, or conelike spike, terminating a leafy shoot or appearing on a separate leafless shoot, rarely with solitary flower. The bracts are persistent and imbricate, supporting 1 to 2 flowers. The flowers are zygomorphic, with a tubular 2- to 3-lobed calyx and a 3-lobed corolla. There is only 1, often petaloid, stamen with 2 bisporangiate thecae. The remaining 5 staminodes are fused into an often 3-lobed labellum of the same color as the corolla. The ovary is usually inferior, consisting of 3 carpels, 3-locular with

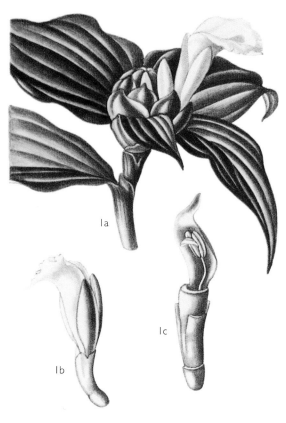

COSTACEAE. 1 *Costus afer* (a) globose, bracteate inflorescence showing upper leaves with basal ligule (×⅔); (b) flower, with 3-lobed calyx and corolla and protruding labellum (×1); (c) flower with calyx and corolla tubes removed showing labellum, fertile stamen and hoodlike stigma, protruding from the thecae (×½).

axile placentas bearing numerous, anatropous ovules. The style is filiform and frequently surrounded by the thecae. Usually, the fruit is a 3-locular capsule with a persistent calyx, dehiscing loculicdally. The seeds have a white or yellow aril and a straight embryo with a copious, starchy endosperm.

Classification All recent classifications points to a sister group relationship between Zingiberaceae (q.v.) and Costaceae [i,ii,iii], and the 2 families have previously been combined. The largest genus in the family, *Costus,* is most likely not monophyletic[iv].

Economic uses Of limited economic importance, a few species are grown in greenhouses and sold as cut flowers. OS

[i] Chase, M. W. *et al.* Multi-gene analysis of monocot relationships: A summary. Pp. 63–75. In: Columbus, J. T. *et al.* (eds), *Monocots: Comparative Biology and Evolution.* Claremont, California, Rancho Santa Ana Botanic Garden (2006).
[ii] Kress, J. W., Prince, L. M., Hahn, W. J., & Zimmer, E. A. Unraveling the evolutionary radiation of the families of the Zingiberales using morphological and molecular evidence. *Syst. Biol.* 50: 926–944 (2001).
[iii] Johansen, L. B. Phylogeny of *Orchidantha* (Lowiaceae) and the Zingiberales based on six DNA regions. *Syst. Bot.* 30: 106–117 (2005).
[iv] Specht, C. D., Kress, W. J., Stevenson, D. W., & DeSalle, R. A. Molecular phylogeny of Costaceae (Zingiberales). *Mole. Phyl. Evol.* 21: 333–345 (2001).

CYCLANTHACEAE
PANAMA HAT FAMILY

The Cyclanthaceae is a small family of perennial epiphytes, root-climbers, or terrestrial herbs, with alternate, simple leaves, usually with a bifid apex or, rarely, palmlike.

Distribution The family occurs in lowland and montane rain forests (up to 3,000 m), ranging from southern Mexico through Central America, the West Indies, the Amazonian part of South America, to subtropical coastal areas of Brazil.

Genera 12
Species c. 225
Economic uses Panama hats are produced from young petioles and leaves of *Carludovica palmata.* Other species are used for thatching and medicine

Description Monoecious, perennial, rhizomatous plants that may be terrestrial, epiphytic, or climbing lianas, with or without, more or less lignified, aerial stems. The leaves are distichous or spirally arranged, sheathing at the bases, and in most species, petiolate, frequently with a bifid apex, rarely fan- or palmlike. The inflorescence is an unbranched, axillary, or terminal, usually cylindrical to ellipsoid, spadix, subtended by 2 to 11, mostly 3 to 4, bracts or spathes. The flowers are densely crowded, either arranged in 5-flowered groups, with 1 female flower surrounded by 4 male flowers, in a spiral along the spadix, or the individual flowers are not discernable. The flowers are unisexual (plants monoecious). The ovary has apical or subapical placentas with numerous anatropous ovules. The fruit is usually fleshy, with numerous seeds, comprising minute, straight embryos surrounded by copious endosperm of fat.
SUBFAM. CARLUDOVICOIDEAE Male flowers with an indefinite number of stamens, connate at the base, and surrounded by 1 to 2 whorls of perianth segments, which may be missing.

CYCLANTHACEAE. 1 *Asplundia vagans* habit, with bifid leaf apices and lateral inflorescence (×⅔). 2 *Stelestylis stylaris* female flower with 1 perianth segment removed (×4½). 3 *Evodianthus funifer* (a) half male flower in bud, with numerous stamens, and asymmetric position of the pedicel (×4½); (b) male flower (×6); (c) young fruit, with 1 perianth segment (×4). 4 *Cyclanthus bipartitus* (a) tip of spadix with alternating whorls of male and female flowers (×1½); (b) section of spadix showing male and female flowers (×4). 5 *Sphaeradenia chiriquensis* portion of spadix showing the characteristic arrangement of male and female flowers (×2).

Usually there are fewer than 10, but up to 150 have been counted in male flowers of *Asplundia.* Female flowers free or connate, with 4 tepals, mostly epigynous or perigynouos, and with 4 filiform staminodes opposite the tepals. Ovary of 4 carpels, 1-locular, more or less embedded in the rachis, and with 1 to 4 spreading stigmas that are stalkless or mounted on a short style. Fruits are united into a syncarp.
SUBFAM. CYCLANTHOIDEAE Monotypic, including only *Cyclanthus bipartitus.* The individual flowers cannot be distinguished but are arranged in alternating male and female cycles or spirals along the spadix. Stamens in 4 rows alternating with 2 rows of pistillate flowers surrounded by staminodes. Each cycle consists of 2 rows of pistillate flowers with confluent loci that make up an ovarian chamber, with numerous parietal placentae. The ring dries up in fruit to leave a hollow ring filled with seeds.

Classification The 2 subfamilies are most likely monophyletic[i]. The family has for long been considered closely related to the Pandanaceae (q.v.), a relationship that is supported by both morphological[ii,iii] and molecular[iv,v,vi] evidence.

Economic uses This family is economically important for *Carludovica palmata* (the Panama Hat Plant). The young petioles and leaves are made into fibers (paja toquilla) from which Panama hats are weaved. These hats have been

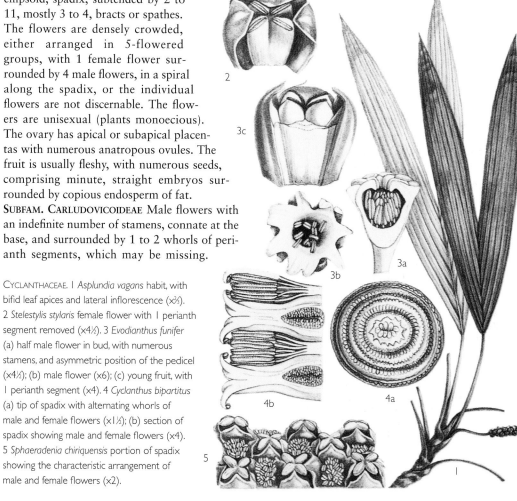

CYCLANTHACEAE. 6 *Carludovica rotundifolia* (a) habit, with spadices and fanlike leaves (½); (b) female flower, with a long staminodes (the others removed) (×2); (c) young fruit, inside the tepals, and with the staminodes removed (×4½); (d) male flower, with numerous stamens (×2); (e) inflorescence with connate female flowers, cut open to show the stalk (pink) into which bases of the fleshy (orange) fruits are partially imbedded (×⅔).

produced for centuries and were sold and shipped from Panama in the 1800s, giving rise to their name. Ecuador is the main producer of these hats, and alone exports more than 1 million Panama hats every year. About 6 young leaves are required to make a single hat. Older, coarser leaves are used to make mats and baskets. The leaves of species of *Carludovica*, *Asplundia*, and some other genera are used as thatching material. Species of *Asplundia* are used to treat snake bites, and *Virola theiodora* contains a hallucinogen.   SRC/OS

i Erikson, R. Phylogeny of the Cyclanthaceae. *Pl. Syst. Evol.* 190: 31–47 (1994).
ii Dahlgren, R. M. T., Clifford, H. T., & Yeo, P. F. *The Families of the Monocotyledons.* Berlin, Springer-Verlag (1985).
iii Stevenson, D. W. & Loconte, H. Cladistic analysis of monocot families. Pp. 543–578. In: Rudall, P. J., Cribb, P. J., Cutler, D. F., & Humphries, C. J. (eds). *Monocotyledons: Systematics and Evolution.* Richmond, Royal Botanic Gardens, Kew (1995).
iv Caddick, L. R., Rudall, P. J., Wilkin, P., Hedderson, T. A. J., & Chase, M. W. Phylogenetics of Dioscoreales based on combined analyses of morphological and molecular data. *Bot. J. Linn. Soc.* 138: 123–144 (2002).
v Davis, J. I. *et al.* A phylogeny of the monocots, as inferred from *rbc*L and *atp*A sequence variation, and a comparison of methods for calculating jackknife and bootstrap values. *Syst. Bot.* 29: 467–510 (2004).
vi Chase, M. W. *et al.* Multi-gene analysis of monocot relationships: A summary. Pp. 63–75. In: Columbus, J. T. *et al.* (eds), *Monocots: Comparative Biology and Evolution.* Claremont, California, Rancho Santa Ana Botanic Garden (2006).

CYMODOCEACEAE

The Cymodoceaceae is small family of sea-grasses, growing submerged in seawater where they provide both food and shelter for fish.

Distribution The Cymodoceaceae are a family of marine plants that live in tropical and subtropical seas; a few species are found in warm temperate waters. The genera *Amphibolis* and *Thalassodendron* are restricted to the temperate seas of Australia, and *Cymodocea nodosa* is found in the Mediterranean.

Description Rhizomatous, creeping monopodial, herbaceous and leafy, or more or less sympodial and woody plants with scale-leaves.

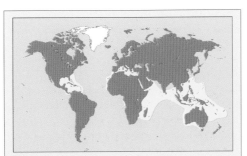

Genera 5 Species 16
Economic uses No direct use to humans but they provide food for fish stocks

The leaves are distichous, linear, or terete with a persistent sheathing base and intervaginal scales. Flowers unisexual (plants dioecious), naked, and either terminal on a short branch or, in *Syringodium*, arranged in a cymose inflorescence. The male flower is pedunculate or nearly sessile, with 2 laterally fused stamens prolonged apically (some authorities consider the male flower to be an inflorescence of 2 flowers each consisting of 1 stamen[i]). The pollen is filiform, and pollination takes place underwater. The female flower is sessile, with 2 free carpels with a simple, long (*Halodule*) or short style and usually 2, rarely 3, stigmas. Each carpel contains a solitary, straight, pendulous ovule. The fruit is indehiscent and 1-seeded, either with a hard (e.g., *Halodule*) or fleshy exocarp (e.g., *Amphibolis*), or consisting of a fleshy bract enclosing the fertilized ovaries (e.g., *Thalassodendron*). The seeds have no endosperm. They occasionally germinate while attached to the parent plant (e.g., *Amphibolis*, *Thalassodendron*).

Classification The genera are *Cymodocea* (4 spp.), *Syringodium* (2 spp.), *Halodule* (6 spp.), *Amphibolis* (2 spp.), and *Thalassodendron* (2 spp.). The latter 2 genera both have woody rhizomes and sessile male flowers and are occasionally included in their own subfamily[i]. Cymodoceaceae is separated from Ruppiaceae with doubt, and both families are closely related to Posidoniaceae[ii].

Economic uses Beds of Cymodoceaceae sea-grasses are used by many fish and other marine organisms as important sources of food and as spawning grounds.   [CDC] OS

i Dahlgren, R. M. T., Clifford, H. T., & Yeo, P. F. *The Families of the Monocotyledons.* Berlin, Springer-Verlag (1985).
ii Les, D. H., Cleland, M. A., & Waycott, M. Phylogenetic studies in Alismatidae, II: Evolution of marine angiosperms (seagrasses) and hydrophily. *Syst. Bot.* 22: 443–463 (1997).

CYPERACEAE
SEDGE FAMILY

A large family of mainly perennial, grasslike herbs, with trigonous, solid stems and 3-ranked leaves, and the tepals reduced to bristles or scales, or are entirely absent.

Distribution The Cyperaceae has a worldwide distribution and is only absent from Antarctica. Members of the family are especially abundant in damp, wet, or marshy regions of the temperate and subarctic zones, where species can dominate the vegetation entirely.

CYPERACEAE. 1 *Carex decurtata* (a) habit (x⅔); (b) male and female spikelet (x3); (c) 3 stigma-branches emerging through the apical opening in the utricle (x6); (d) utricle split open showing the naked ovary (x8); (e) naked, male flower with 3 stamens and supporting glume (x6). 2 *Cladium tetraquetrum* (a) habit showing sheathing leaf-bases (x⅔); (b) inflorescence (x6); (c) spikelet showing a single, bisexual flower (x6); (d) naked, bisexual flower and subtending bract (glume) (x6). 3 *Cyperus compressus* (a) habit (x⅔); (b) flower showing 3-fid style and 3 stamens (x18); (c) spikelet of bisexual flowers—only the stigmas are visible (x3); (d) naked flower with 3 stamens and an ovary crowned by a style with 3 stigma-branches (x12).

Description Mostly perennial, rhizomatous, or caespitose herbs, rarely annuals, shrubs (e.g., *Microdracoides*), or lianas (e.g., in *Scleria*), generally with solid culms that are trigonous in cross section, but hollow and terete culms are not uncommon. The leaves are usually tristichous, rarely distichous (e.g., *Oreobolus*), and basal, occasionally absent, with closed sheaths (rarely open) and with or without a ligule at the junction with the linear blade. The leaf epidermis has cells with 1 or more, conical silica bodies with their base resting on the inner surface of the cell. The inflorescence is terminal and composed of spikelets that are arranged in a large variety of compound, usually open inflorescences, often with erect branches, e.g., spikes, panicles, or corymbs, but may be condensed, e.g., headlike, or even reduced to a single spikelet. The spikelets consists of 1 to many, spirally or distichously arranged glumes (bracts), the lateral spikelets often with a sterile glume (prophyll) at the base, and the remaining glumes support small, inconspicuous bisexual or unisexual flowers (plants usually monoecious). The perianth is typically represented by 3 to 6 scales, bristles, or hairs but may be absent in some genera (e.g., *Carex*). There are 3 stamens, rarely only 1 or numerous. The ovary is superior, 2- or 3-carpellate, with a single locule and a single, basal, anatropous ovule. The style is divided into 2 or 3 branches and often persistent on the ripe fruit, which is an achene, or rarely a drupe. In *Carex* and related genera, the fruit is completely enclosed by a prophyll (utricle). The seed contains a small embryo surrounded by copious mealy or oily endosperm.

Classification Their unique epidermal silica bodies and the presence of pollen pseudomonads (3 of the 4 pollen in a tetrad degenerate and are incorporated into the wall of the remain pollen grain—a type of pollen also known from the Epacridaceae [q.v.]) leave virtually no doubt that the Cyperaceae is monophyletic[i,ii,iii], a fact that is also supported by molecular analyses[iv]. The sister group to the Cyperaceae is most likely Juncaceae[ii,iv], although the latter family may be paraphyletic with respect to Cyperaceae[iii]. The classification of the genera within the family is somewhat controversial. The most recent classifications recognized 2[v] or 4[i] subfamilies, respectively, but in neither case are the subfamilies monophyletic[vi]. With around 2,000 spp., *Carex* is one of the largest genera of flowering plants.

Economic uses The Cyperaceae is of limited economic importance, but several species are used for human food, notably *Eleocharis dulcis* (Chinese Water Chestnut) and *Cyperus esculen-*tus (Tigernut, Chufa), which are cultivated in Southeast Asia. In Africa, especially, species are used as fodder (e.g., *C. involucratus*) for livestock. Several species are noxious weeds in ricefields e.g., *C. rotundus* (Nut Sedge) and *C. esculentus*. The stems of a number of other *Cyperus* spp., such as *C. malacopsis* and *C. tegetiformis* (Chinese Mat Grass), are used to make mats as are the stems of the *Eleocharis austrocaledonica*. *C. dispalatha* is cultivated in Japan for its leaves, which are used to make hats. Stems of *Cladium mariscus* are used for thatching material for houses in Europe and parts of North Africa. The stems of *Scirpus totara* are used to make canoes and rafts in tropical South America, and those of *S. lacus-*

Genera c. 100 **Species** c. 4,500
Economic uses Some used to make mats, hats, baskets, paper, and fodder for animals. Some edible tubers or stems used locally as medicines; a few sold as pot plants and water-garden ornamentals

tris (Bulrush) of North and Central America in basketwork, mats and chair seats. *Cyperus papyrus (*Papyrus or Paper Reed) was used by the ancient Egyptians, and later by the Greeks, to make papyrus more than 5,000 years ago. The word *paper* is derived from the Egyptian "papyrus." Today, *C. papyrus* is gaining importance as a fuel. Some species of *Carex, Caustis, Cyperus,* and *Scirpus* are also cultivated as pot plants and water-garden ornamentals.   SRC/OS

i Goetghebeur, P. Cyperaceae. Pp. 141–190. In: Kubitzki, K. (ed.), *The Families and Genera of Vascular Plants. IV. Flowering Plants. Monocotyledons: Alismatanae and Commelinanae (except Gramineae).* Berlin, Springer-Verlag (1998).
ii Simpson, D. Relationships within Cyperales. Pp. 497–509. In: Rudall, P. J., Cribb, P. J., Cutler, D. F., & Humphries, C. J. (eds), *Monocotyledons: Systematics and Evolution.* Richmond, Royal Botanic Gardens, Kew (1995).
iii Munro, S. L. & Linder, H. P. The phylogenetic position of *Prionium* (Juncaceae) within the order Juncales based on morphological and *rbc*L sequence data. *Syst. Bot.* 23: 43–45 (1998).
iv Chase, M. W. *et al.* Multi-gene analysis of monocot relationships: A summary. Pp. 63–75. In: Columbus, J. T. *et al.* (eds), *Monocots: Comparative Biology and Evolution.* Claremont, California, Rancho Santa Ana Botanic Garden (2006).
v Buhl, J. J. Sedge genera of the world: relationships and a new classification of the Cyperaceae. *Australian Syst. Bot.* 8: 125–305 (1995).
vi Muasya, A. M., Bruhl, J. J., Simpson, D. A., Culham, A., & Chase, M. W., Suprageneric phylogeny of Cyperaceae: A combined analysis. Pp. 593–601. In: Wilson, K. L. & Morrison, D. A. (eds), *Monocots. Systematics and Evolution.* Melbourne, CSIRO Publishing (2000).

DASYPOGONACEAE

A family of 4 genera and between 8 and 16 spp. of shrubby or arborescent perennial plants restricted to southwestern Australia and western Victoria, inhabiting dry woodlands and heaths. The leaves are spirally arranged, V- or U-shaped in transection. The flowers are bisexual, with

Genera 4
Species 8–16
Economic uses
Some ornamentals

2 whorls of persistent, colored, free or united tepals; 6 stamens; and a 1- to 3-locular ovary, solitary or in globular heads. The fruit is usually dry and indehiscent, enclosed in the perianth. The family sometimes has a wider circumscription[i] and is also occasionally included in Xanthorrhoeaceae (q.v.). The monotypic *Baxteria australis* has been transferred to its own family Baxteriaceae by Takhtajan (1997). Current evidence supports the monophyly of Dasypogonaceae[ii,iii,iv].   OS

i Dahlgren, R. M. T., Clifford, H. T., & Yeo, P. F. *The Families of the Monocotyledons.* Berlin, Springer-Verlag (1985).
ii Rudall, P. & Chase, M. W. Systematics of the Xanthorrhoeaceae *sensu lato:* evidence for polyphyly. *Telopea* 6: 629–647 (1996).
iii Chase, M. W. *et al.* Multi-gene analysis of monocot relationships: A summary. Pp. 63–75. In: Columbus, J. T. *et al.* (eds), *Monocots: Comparative Biology and Evolution.* Claremont, California, Rancho Santa Ana Botanic Garden (2006).
iv Davis, J. I. *et al.* A phylogeny of the monocots, as inferred from *rbc*L and *atp*A sequence variation, and a comparison of methods for calculating jackknife and bootstrap values. *Syst. Bot.* 29: 467–510 (2004).

DIOSCOREACEAE
DIOSCOREA FAMILY

A small family of mainly tropical vines or lianas, with usually cordate, simple, or compound leaves.

Distribution Pantropical, with a few species reaching the subtropical zone of the northern hemisphere and the temperate zone of the southern hemisphere. Most species are found in tropical forest but some occur in grassland and semideserts.

Description Perennial geophytes, usually with tubers or rhizomes. The tubers are either annual and renewed each year or persistent with secondary growth sometimes attaining a massive size. The stems are short-lived, twining, rarely erect, or procumbent. The leaves are alternate, rarely opposite or whorled, and usually cordate, entire or variously lobed, and long-petioled. The secondary veins in the lamina anastomose by means of lateral, small veins. The inflorescence is axillary, usually a panicle, raceme, or spike. The flowers are arctinomorphic, small, inconspicuous, unisexual (dioecious) or bisexual (*Stenomeris*), pedicellate or sessile. There are 2 whorls of 3 similar tepals, often fused at the base, and 1 (the inner may be wanting or replaced by staminodes) or 2 whorls of 3 stamens. The ovary is inferior, consisting of 3 fused carpels and is 3-locular, with axile placentation and 1 to 2, rarely many, anatropous ovules per locule. The style is 3-fid or 3-lobed. The fruit is often a 3-winged capsule, a berry, or a samara, in which 2 of the locules may be aborted, and the seeds are usually winged or flattened, with endosperm and a small embryo.

Classification The Dioscoreaceae is divided into 2 subfamilies, Stenomerioideae and Dioscoreoideae[i], the former is recognized by having

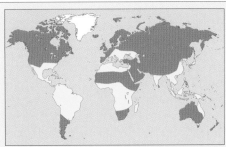

Genera 3–4 **Species** c. 880
Economic uses *Dioscorea* (yams) are a staple food in large parts of the tropics and a rich source of steroidal saponins, e.g., diosgenin; also used locally to make arrow poison

bisexual flowers and numerous seeds in each locule, the latter by having unisexual flowers and usually 1 to 2 ovules in each locule. The family is largely tropical, but *Tamus* is restricted to the Mediterranean reaching from Macronesia to Iran and Iraq. The taxonomy of largest genus, *Dioscorea*, with c. 800 spp., is extremely complicated, with many infrageneric taxa often given the rank of genera[i]. *Stenomeris* is sometimes considered a family of its own and is occasionally included in Dioscoreaceae but with misgivings[i]. Additionally, Taccaceae (q.v.) is occasionally included in Dioscoreaceae, but the relationships between *Stenomeris*, Taccaceae, and Dioscoreaceae do not appear well resolved[ii,iii].

Economic uses Yams (*Dioscorea* spp.) are a staple food in large parts of the tropics, and about 60 spp. are cultivated as a subsistence crop in 3 main centers: Southeast Asia, West Africa, and Central and South America. It has been suggested that cultivation took place 11,000 BCE[iv]. The rhizomes are a rich source of steroidal saponins. These compounds are important for the pharmaceutical industry since they are used in the manufacture of hormonal medicines, e.g., diosgenin, which is used to make oral contraceptives. Several species are used to treat rheumatism. Many species also have local uses, e.g., in the production of arrow poison and in extracts to stupefy fish.   [CJH] OS

i Huber, H. Dioscoreaceae. Pp. 216–235. In: Kubitzki, K. (ed.), *The Families and Genera of Vascular Plants. III. Flowering Plants. Monocotyledons: Lilianae (except Orchidaceae).* Berlin, Springer-Verlag (1998).
ii Caddick, L. R., Rudall, P. J., Wilkin, P., Hedderson, T. A. J., & Chase, M. W. Phylogenetics of Dioscoreales based on combined analyses of morphological and molecular data. *Bot. J. Linn. Soc.* 138: 123–144 (2002).
iii Chase, M. W. *et al.* Multi-gene analysis of monocot relationships: A summary. Pp. 63–75. In: Columbus, J. T. *et al.* (eds), *Monocots: Comparative Biology and Evolution.* Claremont, California, Rancho Santa Ana Botanic Garden (2006).
iv Coursey, D. G. The origins and domestication of yams in Africa. Pp. 383–408. In: Harlan, J. R., De Wit, J. M. J., & Stemler, A. B. L. (eds), *Origins of African Plant Domestication.* The Hague/Paris, Mouton Publications (1976).

DORYANTHACEAE

Distribution Endemic to eastern Australia in rocky areas.

Description Enormous, perennial, caespitose herbs that measure up to 5.5 m high, with linear to narrowly lanceolate, spirally arranged leaves, that reach up to 2.5 m in length, and with a thin sheathing base encircling the axis. The inflorescence is a terminal thyrse with numerous smaller leaves and red bracts.

Genera 1 Species 2
Economic uses
Doryanthes excelsa
is widely cultivated

The flowers are large, actinomorphic, and bisexual; with 3+3 similar tepals fused at the base and 3+3 stamens. The ovary is inferior, 3-carpellate and 3-locular, with numerous, anatropous ovules in 2 rows in each locule, and septal nectaries. The fruit is a loculicidal capsule with numerous, usually winged seeds. The embryo is straight and embedded in an endosperm rich in oil and fat. The Doryanthaceae is sister group to a large clade including among other families, Agavaceae (q.v.), Amaryllidaceae (q.v.), and Iridaceae (q.v.)[i,ii,iii,iv].   OS

[i] Fay, M. F. *et al.* Phylogenetic studies of Asparagales based on four plastid DNA regions. Pp. 360–371. In: Wilson, K. L. & Morrison, D. A. (eds), *Monocots. Systematics and Evolution.* Melbourne, CSIRO Publishing (2000).
[ii] Davis, J. I. *et al.* A phylogeny of the monocots, as inferred from *rbc*L and *atp*A sequence variation, and a comparison of methods for calculating jackknife and bootstrap values. *Syst. Bot.* 29: 467–510 (2004).
[iii] Pires, J. C. *et al.* Phylogeny, genome size, and chromosomal evolution of Asparagales. Pp. 287–304. In: Columbus, J. T. *et al.* (eds), *Monocots: Comparative Biology and Evolution.* Claremont, California, Rancho Santa Ana Botanic Garden (2006).
[iv] Chase, M. W. *et al.* Higher-level systematics of the monocotyledons: An assessment of current knowledge and a new classification. In: Wilson, K. L. & Morrison, D. A. (eds), *Monocots. Systematics and Evolution.* Melbourne, CSIRO Publishing (2000).

DRACAENACEAE

DRAGON-TREE FAMILY

Distribution The family is almost entirely tropical, occurring worldwide in semiarid areas or rain forests but mainly found in the Old World; 1 sp. of *Dracaena* occurs in Mesoamerica but is not found in South America. A few species, e.g., *Dracaena draco* and *D. tamaranae*, grow in Macronesia (Canary Islands, Cape Verde Islands, and Madeira) and northwestern Africa (Morocco).

Description Up to 40 m high trees, but some species are shrubs or geophytes. The leaves are alternate, distichous or spirally arranged, entire and ovate, linear or sword-shaped,

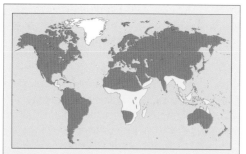

Genera 1 Species c. 100
Economic uses Several members of the family are cultivated as indoor and outdoor ornamentals; a red resin called dragon's blood is extracted from *Dracaena cinnabari*

up to 2 m long, with a sheathing base. The inflorescence is terminal pedunculate, paniculate, racemelike, or capitate. The flowers are actinomorphic, with 2 whorls of equal tepals fused into a tube at the base, and 2 whorls of 6 stamens. The ovary is superior, 3-carpellate, 3-locular with septal nectaries, and 1 anatropous ovule per locule. The fruit is a berry with 3 seeds, and the endosperm stores aleurone and oil.

Sansevieria (including *S. trifasciata*, Mother-in-law's Tongue) is included here in *Dracaena*. The Dracaenaceae is interpreted here in a narrow sense[i,ii]. For a full discussion of its relationships, see under Convallariaceae. As accepted here, the Dracaenaceae is monophyletic in several analyses, but its recognition makes Convallariaceae paraphyletic.

Economic uses Some *Dracaena* species (e.g., *D. draco*, Dragon Tree) are cultivated as house plants (e.g., in temperate countries or grown outdoors as ornamentals in subtropical and tropical regions. A red resin called dragon's blood is extracted from *D. cinnabari*, which is endemic to Soqotra.   OS

[i] Dahlgren, R. M. T., Clifford, H. T., & Yeo, P. F. *The Families of the Monocotyledons.* Berlin, Springer-Verlag (1985).
[ii] Bos, J. J. Dracaenaceae. Pp. 238–241. In: Kubitzki, K. (ed.), *The Families and Genera of Vascular Plants. III. Flowering Plants. Monocotyledons: Lilianae (except Orchidaceae).* Berlin, Springer-Verlag (1998).

ECDEIOCOLEACEAE

A small family endemic to southwestern Australia and composed of 2 monotypic genera of rushlike, caespitose, monoecious herbs, with creeping rhizomes. The leaves are reduced to sheaths split to the base. The inflorescence consists of 1 to few spikelets, each with several

Genera 2 Species 2
Economic uses None

glumes. The lowest glumes may be male or female, the sexes alternating once or twice along the spike. The flowers are flattened, 3-merous, with 2 whorls of 3 tepals, and either 2 sets of 3 stamens, or 2 to 3 superior carpels with frees styles. There is 1 pendulous, atropous ovule in each locule. The fruit is a capsule or nut. The family has previously been placed in the Restionaceae (q.v.) but was separated from the latter on the basis of anatomical differences[i]. Molecular evidence points at close relationships among Ecdeiocoleaceae, Joinvilleaceae, and Poaceae, with Joinvilleaceae as sister group to Ecdeiocoleaceae plus Poaceae[ii,iii,iv].   OS

[i] Cutler, D. F. & Shaw, H. K. A. Anarthriaceae and Ecdeiocoleaceae: two new monocotyledonous families, separated from Restionaceae. *Kew Bull.* 19: 488–497 (1965).
[ii] Davis, J. I. *et al.* A phylogeny of the monocots, as inferred from *rbc*L and *atp*A sequence variation, and a comparison of methods for calculating jackknife and bootstrap values. *Syst. Bot.* 29: 467–510 (2004).
[iii] Chase, M. W. *et al.* Multi-gene analysis of monocot relationships: A summary. Pp. 63–75. In: Columbus, J. T. *et al.* (eds), *Monocots: Comparative Biology and Evolution.* Claremont, California, Rancho Santa Ana Botanic Garden (2006).
[iv] Michelangeli, F. A., Davis, J. I., & Stevenson, D. W. Phylogenetic relationships among Poaceae and related families as inferred from morphology, inversions in the plastid genome, and sequence data from the mitochondrial and plastid genomes. *American J. Bot.* 90: 93–106 (2003).

ERIOCAULACEAE

PIPEWORT FAMILY

A small family of mainly tropical herbs, often with the spirally arranged leaves forming dense rosettes and bearing characteristic head- or buttonlike inflorescences at the ends of long, leafless peduncles.

Distribution Mainly tropical and subtropical, with species concentrations in South America and Africa, but with a few outliners in North America and Europe, members of the family often occur in seasonally wet, marshy, and boggy habitats. A few genera include true aquatics (e.g., *Eriocaulon, Tonina*) and some even grow in temporarily dry habitats.

Description Rhizomatous, perennial, or annual herbs, usually with rosettes of spirally, rarely distichously arranged, usually linear grasslike leaves or filiform when submerged in aquatic species, rarely thick and coriaceous. Some species have large trunks covered in adventitious roots or large (up to 4 m high) leafy stems (*Paepalanthus*), others, especially the aquatics, have leafy, floating stems. The inflorescence is a single, indeterminate, headlike spike borne at the end of a leafless peduncle or composed of up to 100 such heads in an umbel-like inflorescence. The peduncles usually extend beyond the leaves, which may sheathe them at the base. Each head is subtended by an involucre of bracts and is composed of 10 to more than 1,000 flowers. The bracts subtending the individual flowers are very different from the involucral bracts. The flowers are usually unisexual, rarely bisexual (e.g., in *Rondonanthus* and *Syngonanthus*), and within each head the male and female flowers are mixed, or the male flowers are in the center surrounded by female flowers (plants monoecious); occasionally, the male and female flowers are on separate plants (plants dioecious). The flowers are usually small, actinomorphic, 3-merous (most genera) or 2-merous (e.g., *Eriocaulon*, *Syngonanthus*). The sepals are free or fused at the base, while the petals are commonly fused into a tube. There are 1 or 2 whorls of 3 or 2 stamens, rarely only 1. The ovary is superior, with 2 or 3 fused carpels and 2 or 3 locules and a single terminal style bearing 2 or 3 elongate stigmas, occasionally with appendices that may be longer than the stigmas. There is a solitary, atropous, pendulous

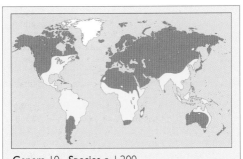

Genera 10 **Species** c. 1,200
Economic uses Source of everlasting flowers

ovule in each locule. The fruit is a membranous, loculicidal capsule. The seeds contain copious starchy endosperm and a small embryo.

Classification The family has it greatest diversity in South America. *Eriocaulon* (400 spp.), *Syngonanthus* (200 spp.), and *Paepalanthus* (c. 400 spp.) are the largest genera in the family. However, species delimitation is difficult in *Syngonanthus* and *Paepalanthus* is poorly defined. The family is undoubtedly monophyletic[i] and several attempts have been made to subdivide it into monophyletic subgroups[i], but without success[ii]. The Eriocaulaceae has often been considered closely allied to the Xyridaceae (q.v.)[iii], a fact that is supported by recent molecular analyses[iv,v].

Economic uses Apart from their use as everlasting flowers, which has made some species endangered, there are no reported economic uses. However, *Syngonanthus chrysanthus* has within the last 3 to 4 years become a common pot plant in Europe and elsewhere. Species of *Eriocaulon* occur as weeds in ricefields, e.g., *E. aquaticum* in Europe and North America and *E. cinereum* in northern Italy, but are not overly troublesome. [CDC] OS

i Stützel, T. Eriocaulaceae. Pp. 197–207. In: Kubitzki, K. (ed.), *The Families and Genera of Vascular Plants. IV. Flowering Plants. Monocotyledons: Alismatanae and Commelinanae (except Gramineae)*. Berlin, Springer-Verlag (1998).

ii Giulietti, A. M. *et al*. Multipisciplinary studies on neotropical Eriocaulaceae. Pp. 580–589. In: Wilson, K. L. & Morrison, D. A. (eds), *Monocots. Systematics and Evolution*. Melbourne, CSIRO Publishing (2000).
iii Dahlgren, R. M. T., Clifford, H. T., & Yeo, P. F. *The Families of the Monocotyledons*. Berlin, Springer-Verlag (1985).
iv Davis, J. I. *et al*. A phylogeny of the monocots, as inferred from *rbc*L and *atp*A sequence variation, and a comparison of methods for calculating jackknife and bootstrap values. *Syst. Bot.* 29: 467–510 (2004).
v Chase, M. W. *et al*. Multi-gene analysis of monocot relationships: A summary. Pp. 63–75. In: Columbus, J. T. *et al*. (eds), *Monocots: Comparative Biology and Evolution*. Claremont, California, Rancho Santa Ana Botanic Garden (2006).

ERIOCAULACEAE. 1 *Eriocaulon aquaticum* (a) habit, showing dense head- or buttonlike inflorescence and basal rosette of leaves (x⅔); (b) male flower with 2 free sepals and 2 fused petals, and 4 stamens (x8); (c) inner perianth segment from male flower, showing rudimentary stamen and fertile stamen with gland behind (x12); (d) female flower with 2 free sepals and petals (x8); (e) head- or buttonlike inflorescence (x3); (f) vertical section of fruit (a capsule), showing pendulous seeds (x12). 2 *Syngonanthus laricifolius* (a) habit; note the internodes between the rosettes from different growth seasons (x⅔); (b) male flower with 2 bracts and 3 free sepals and 3 fused petals (x15). 3 *Paepalanthus riedelianus* (a) habit, each inflorescence supported by a leaf (x⅔); (b) gynoecium with 3 brushlike stigmas and elongated appendices (x16); (c) female flower with free sepals and petals (x8); (d) the fused petals of a male flower opened out to show 3 stamens and rudimentary ovary (x10); (e) male flower with free sepals (x8).

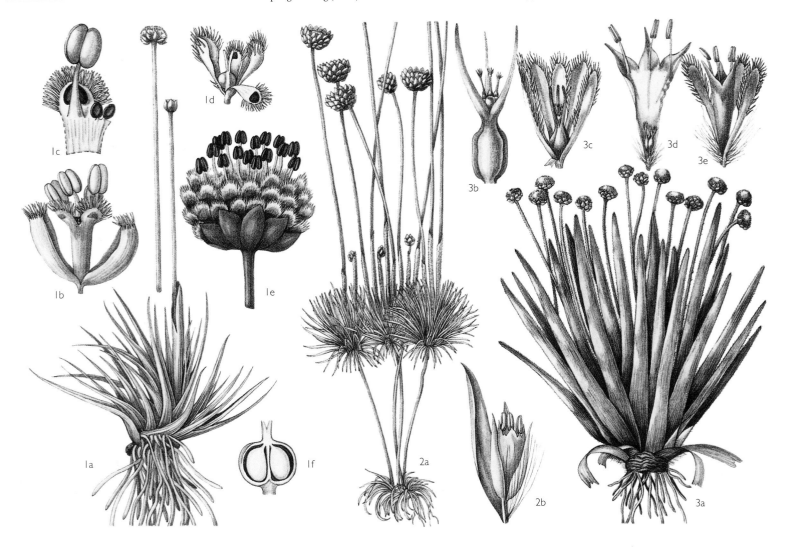

ERIOSPERMACEAE

A monogeneric family from sub-Saharan Africa in semiarid areas and grassland, with c. 100 spp. of perennial, tuberous geophytes, rarely reaching a height of 40 cm. In the Cape region, flowers are produced during the dry summer and the leaves in the rainy winter, elsewhere in Africa the leaves develop shortly after the flowers, which appear at the end of the dry winter season.

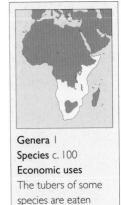

Genera 1
Species c. 100
Economic uses
The tubers of some species are eaten as vegetables

The tuber varies in size from the size of a pea to more than 10 cm in length. The leaves, or the single leaf found in many species, are linear or ovate to orbicular and have a petiole-like sheathing base, which connects the tuber and the blade, and persists as a fibrous or membranous sheath when the blade wither. The new leaf and the inflorescence emerge through this protective cover. The inflorescence is an erect scapose, few to many-flowered raceme. The flowers are bisexual, with 6 tepals in 2 whorls that are fused at the base, either similar or the outer spreading and the inner erect. The 6 stamens are adnate to the tepals. The ovary is superior, 3-carpellate, and 3-locular, with basal septal nectaries and 3 to 6 ovules in each locule. The fruit is a loculicidal capsule with 6 to 12 seeds. The embryo is large and straight. The endosperm is largely replaced by a perisperm of aleurone and oil.

The Eriospermaceae is interpreted here in a narrow sense[i,ii]. It is often placed as sister group to the remaining taxa in a broadly defined Ruscaceae. For further discussion of its affinities, see under Convallariaceae. OS

[i] Dahlgren, R. M. T., Clifford, H. T., & Yeo, P. F. *The Families of the Monocotyledons.* Berlin, Springer-Verlag (1985).
[ii] Bogler, D. Nolinaceae. Pp. 392–397. In: Kubitzki, K. (ed.), *The Families and Genera of Vascular Plants. III. Flowering Plants. Monocotyledons: Lilianae (except Orchidaceae).* Berlin, Springer-Verlag (1998).

FLAGELLARIACEAE

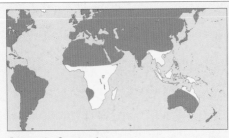

Genera 1 **Species** 4
Economic uses Basket-making and local medicines

A single genus of 4 spp. of climbing or scrambling, tropical lianas that arise from sympodial creeping rhizomes. They are strictly Paleotropical, extending to northeastern Australia and the Pacific Islands, in humid lowland forest and mangroves. The leaves are grasslike, distichously arranged, ending in a sensitive tendril and with a sheathing base that terminates in 2 lateral lobes. The inflorescence is terminal and paniculate and carries many, sessile, 3-merous, and bisexual flowers. The flowers have whitish, membranous tepals in 2 whorls and are slightly fused at the base. There are 2 whorls of stamens and a superior, syncarpous, 3-locular ovary with axile placentation. There are 3 linear stigmatic lobes, and a single, nearly atropous ovule in each locule. The fruit

FLAGELLARIACEAE. 1 *Flagellaria guineensis* (a) shoot bearing a terminal paniculate inflorescence and leaves with sheathing bases and tips formed into a coiled tendril (x⅔); (b) tip of inflorescence bearing drupaceous fruits (x⅔); (c) lower surface of leaf showing parallel veins (x2); (d) bisexual, 3-merous flower (x3½); (e) tepal (x7½); (f) tepal and stamen (x5); (g) gynoecium with 3, narrow, hairy stigmatic surfaces (x5); (h) cross section of ovary (x5); (i) fleshy indehiscent fruit (x2⅔); (j) cross section of fruit (x1½).

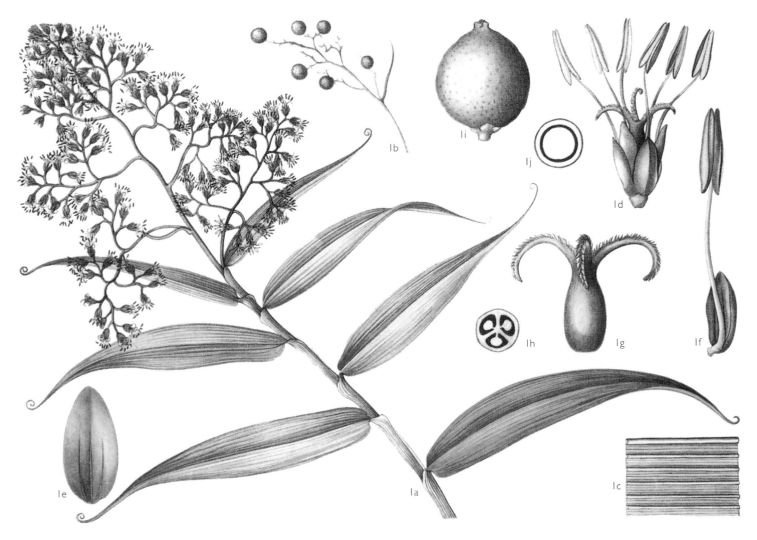

HAEMODORACEAE. 1 *Anigozanthos flavidus* (a) habit (x⅔); (b) inflorescence (x⅔); (c) flower showing curved green perianth tube and 6 stamens (x1); (d) stamen front (lower) and back view (upper) (x3); (e) flower dissected showing epipetalous stamens (x⅔); (f) cross section of 3-locular ovary with 3 axile placentas (x6); (g) vertical section of ovary (x6). 2 *Phlebocarya ciliata* (a) habit (x⅔); (b) flower showing perianth in 2 whorls and 6 stamens (x6); (c) flower dissected showing epipetalous stamens (x6); (d) vertical section of ovary (x14).

is drupaceous, red or black, usually with a single, lens-shaped seed, with a minute embryo and a starchy endosperm.

The Flagellariaceae have often been considered closely related to Joinvilleaceae (q.v.), which is generally accepted as a separate family, and both families are closely related to the Poaceae (q.v.). The resemblance of these families to Hanguanaceae is superficial[i] as most recent analyses support. Their mutual relationships are different in individual analyses and involve other families such as Restionaceae and Ecdeiocolaceae[ii,iii,iv]. *Flagellaria indica* is used in Thailand and Malaysia to make baskets and also has local medical use.   OS

[i] Dahlgren, R. M. T., Clifford, H. T., & Yeo, P. F. *The Families of the Monocotyledons.* Berlin, Springer-Verlag (1985).
[ii] Davis, J. I. *et al.* A phylogeny of the monocots, as inferred from *rbc*L and *atp*A sequence variation, and a comparison of methods for calculating jackknife and bootstrap values. *Syst. Bot.* 29: 467–510 (2004).
[iii] Chase, M. W. *et al.* Multi-gene analysis of monocot relationships: A summary. Pp. 63–75. In: Columbus, J. T. *et al.* (eds), *Monocots: Comparative Biology and Evolution.* Claremont, California, Rancho Santa Ana Botanic Garden (2006).
[iv] Michelangeli, F. A., Davis, J. I., & Stevenson, D. W. Phylogenetic relationships among Poaceae and related families as inferred from morphology, inversions in the plastid genome, and sequence data from the mitochondrial and plastid genomes. *American J. Bot.* 90: 93–106 (2003).

HAEMODORACEAE

BLOODWORT OR KANGAROO PAW FAMILY

A small herbaceous family with distinctive red colored roots and ensiform leaves.

Genera 13 Species c. 100
Economic uses Some cultivated as ornamentals

Distribution The Haemodoraceae is distributed in temperate to tropical Australia, northern South America, and Africa, although a few species are found along the Atlantic coast of North America and on New Guinea.

Description Perennial herbs with rhizomes, corms or bulbs, which are often—as are the roots—red or reddish (e.g., *Haemodorum*). The inflorescences, flowers, and bracts are often more or less densely covered in simple, stellate

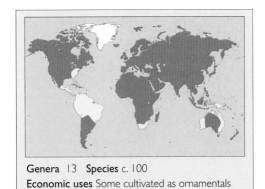

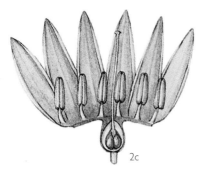

or dendritic hairs. The leaves are 2-ranked ensiform, narrowly linear or acicular, glabrous or hairy, with a sheathing base. The inflorescences are terminal, variable, but often a raceme or a panicle, with bracts and composed either of 1- to 3-flowered clusters or bi- or trifurcate helicoid cymes. The flowers are bisexual, actinomorphic or slightly zygomorphic, often conspicuously colored, consisting either of 3+3 free (*Phlebocarya*), or basally fused, tepals, or the tepals united into a straight or curved tube with 6 free, valvate lobes. The flowers vary in length from 5 mm in *Barberetta* to more than 9 cm long in *Anigozanthos*. The stamens are 3 or 6, rarely only 1, free or adnate to the tepals or tube. Species with 3 stamens may have staminodes. The ovary is either superior or inferior, of 3 fused carpels, with 3 locules, each locule containing 1 to many anatropus or atropous ovules usually on axile placentas. The style is usually filiform, with a capitate stigma or 3-lobed. The fruit is a capsule, opening by 3 valves. The seeds have a tiny embryo and a starchy endosperm.

Classification The family is most likely monophyletic[i,ii] and often divided into 2 subfamilies, Haemodoroideae and Conostylidoideae[iii], which are also monophyletic. The former has 6 stamens and red pigmentation of the subterranean parts (except in *Barberetta*), the latter 1 or 3 stamens and lack red pigment and is restricted to southwestern Australia. Dahlgren[iv] grouped the family with the Philydraceae (q.v.) and Pontederiaceae (q.v.). Molecular analyses points at a close relationship between Haemodoraceae and Commelinaceae (q.v.) and Pontederiaceae[iv,v].

Economic uses Several species of Haemodoraceae, especially *Anigozanthos* (Kangaroo Paw), but also species of *Blancoa, Conostylis, Haemodorum, Tribonanthes,* and *Xiphidium,* are cultivated as garden and pot plants. *Anigozanthos manglesii* is the State emblem of Western Australia. *Lachnanthes caroliniana* is an agricultural pest in commercial cranberry bogs. Some species have been used locally as medicines or dyes e.g., *Haemodorum corymbosum* produces a red pigment with antitumor and antibacterial qualities.   OS

i Simpson, M. G. Phylogeny and classification of the Haemodoraceae. *Ann. Missouri Bot. Gard.* 77: 722–784 (1990).
ii Hopper, S. D., Fay, M. F., Rossetto, M., & Chase, M. W. A molecular phylogenetic analysis of the bloodroot and kangaroo paw family, Haemodoraceae: Taxonomic, biogeographic and conservation implications. *Bot. J. Linn. Soc.* 131: 2.
iii Simpson, M. G. Haemodoraceae. Pp. 212–222. In: Kubitzki, K. (ed.), *The Families and Genera of Vascular Plants. IV. Flowering Plants. Monocotyledons: Alismatanae and Commelinanae (except Gramineae).* Berlin, Springer-Verlag (1998).
iv Dahlgren, R. M. T. A revised system of classification of the angiosperms. *Bot. J. Linn. Soc.* 80: 91–124 (1990).
v Davis, J. I. *et al.* A phylogeny of the monocots, as inferred from *rbc*L and *atp*A sequence variation, and a comparison of methods for calculating jackknife and bootstrap values. *Syst. Bot.* 29: 467–510 (2004).

vi Chase, M. W. *et al.* Multi-gene analysis of monocot relationships: A summary. Pp. 63–75. In: Columbus, J. T. *et al.* (eds), *Monocots: Comparative Biology and Evolution.* Claremont, California, Rancho Santa Ana Botanic Garden (2006).

HANGUANACEAE

The Hanguanaceae is a monogeneric family comprising 5 spp. of dioecious herbs, with spirally arranged, linear to lanceolate leaves, and a sheathing base. They are found in tropical Southeast Asia, Micronesia, and northern Australia in lowland swamps and humid forest. The inflorescence is terminal,

Genera 1 **Species** 5
Economic Used as traditional medicines

paniculate, or whorled, and the flowers are unisexual (plants dioecious), actinomorphic, and 3-merous. The male flowers have 6 stamens and a rudimentary ovary, while the female flowers have a superior, 3-locular ovary and 6 staminodes. The fruit is a 1- or, rarely, 3-seeded berry. The seeds are bowl-shaped.

The Hanguanaceae was previously often included in the Flagellariaceae, but it is now generally accepted as a distinct family. Its relationships have been highly controversial, however. The current evidence points at sister group relationships to either the Commelinaceae[i] alone or indeed the Commelinaceae plus the Pontederiaceae[ii].   OS

i Chase, M. W. *et al.* Multi-gene analysis of monocot relationships: A summary. Pp. 63–75. In: Columbus, J. T. *et al.* (eds), *Monocots: Comparative Biology and Evolution.* Claremont, California, Rancho Santa Ana Botanic Garden (2006).
ii Davis, J. I. *et al.* A phylogeny of the monocots, as inferred from *rbc*L and *atp*A sequence variation, and a comparison of methods for calculating jackknife and bootstrap values. *Syst. Bot.* 29: 467–510 (2004).

HELICONIACEAE
HELICONIA FAMILY

The Heliconiaceae comprises a single genus, *Heliconia*, which is primarily found in the Neotropics with a few species in Southeast

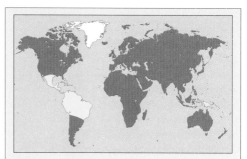

Genera 1 **Species** 100–200
Economic uses Tropical ornamentals and some used as cut flowers

Asia. They are rhizomatous herbs, either with pseudostems formed by the overlapping leaf sheaths, or with aerial shoots. The leaves are distichous, simple, and have open sheaths, but both petiole and ligule may be absent. The inflorescence is large and terminal with conspicuous, colored, carinate, bracts, which support cincinnate clusters of flowers. The flowers are bisexual and zygomorphic, with 5 whorls. The tepals and stamen filaments are basally fused into a short tube. Five of the tepals remain fused into a sheathlike structure, whereas the median tepal and the filaments are free for most of their remaining length. Five of the stamens are fertile; 1 is a scalelike staminodes. The ovary is inferior, of 3 fused carpels, 3-locular with axile placentas, and with a single anatropous ovule per locule. The fruit is drupaceous, and the seeds have no aril but a copious starchy and oily endosperm.

Molecular and morphological evidence suggest that Heliconiaceae is monophyletic[i,ii] and sister group to Musaceae[i] (q.v.). This position is in agreement with the previous inclusion of Heliconiaceae in Musaceae[iii]. However, some molecular data point at a relationship with other taxa e.g., Maranthaceae (q.v.) and Zingiberaceae (q.v.)[ii], or indicate that Musaceae plus Heliconiaceae is paraphyletic[iv].   OS

i Johansen, L. B. Phylogeny of *Orchidantha* (Lowiaceae) and the Zingiberales based on six DNA regions. *Syst. Bot.* 30: 106–117 (2005).
ii Kress, W. J., Prince, L. M., Hahn, W. J., & Zimmer, E. A. Unraveling the evolutionary radiation of the families of the Zingiberales using morphological and molecular evidence. *Syst. Biol.* 50: 926–944 (2001).
iii Stevenson, D. W. & Stevenson, J. W. Heliconiaceae. Pp. 442–443. In: Smith, N. *et al.* (eds), *Flowering Plants of the Neotropics.* Princeton/Oxford, Princeton University Press (2004).
iv Chase, M. W. *et al.* Multi-gene analysis of monocot relationships: A summary. Pp. 63–75. In: Columbus, J. T. *et al.* (eds), *Monocots: Comparative Biology and Evolution.* Claremont, California, Rancho Santa Ana Botanic Garden (2006).

HEMEROCALLIDACEAE
DAYLILY FAMILY

The Hemerocallidaceae is a small family of rhizomatous or caespitose herbs, with distichous leaves and paniculate inflorescences or with solitary, 3-merous flowers.

Distribution The family has a wide distribution in tropical and temperate parts of Eurasia, Australia, New Zealand, the Pacific Islands, the Andes, and Madagascar, but it is not found on mainland Africa.

Description Perennial, caespitose, or rhizomatous herbs, up to 2 m high. The leaves are distichous, linear, or strap-shaped, measuring from a few mm up to 2 m long, with a sheathing base and cauline or gathered in tufts. The leaves are swordlike in *Dianella* and *Phormium*. The inflorescence is either a branched panicle or reduced to a solitary flower. The flowers are bisexual, actinomorphic or zygomorphic (e.g., *Hemerocallis, Phormium*), with 3+3, more or less-

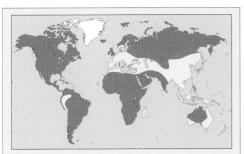

Genera 7–8 Species c. 40
Economic uses Mainly ornamental use but a few
have medicinal use or are a source of fibers

similar, often persistent tepals that are free or
basally united in a tube. The stamens are 3+3,
free, or connate at the base. The ovary is superior
or rarely semi-inferior, of 3 carpels, and is
1-locular or 3-locular, with 1 to numerous, anatropous ovules on axile or parietal placentas, and a
single style sometimes terminating in a tuft of
hairs (*Hemerocallis*). Most genera have septal nectaries. The fruit is a berry or capsule with black
seeds. The embryo is linear and the endosperm
does not contain starch.

Classification The Hemerocallidaceae is sometimes considered monotypic and restricted to
the genus *Hemerocallis*[i]. In the broader circumscription used here, the family is monophyletic
and often includes the Johnsoniaceae (q.v.)[ii].
However, the relationship between these 2 families is not clear, and the Jonnsoniaceae is here
treated as a separate family, sister to the
Hemerocallidaceae. Molecular data places
Hemerocallidaceae in a clade that also contains
Asphodelaceae (q.v.) and Xanthorrhoeaceae
(q.v.) and all 3 families are sometimes combined
into an enlarged Xanthorrhoeaceae (q.v.)[iii]. In
a narrow sense (including the Johnsoniaceae),
the Hemerocallidaceae are either sister group
to Asphodelaceae[iv], to Xanthorroeaceae (q.v.)[v],
to Asphodelaceae plus Xanthorroeaceae[iii,vi], or
the relationship is unresolved[ii].

Economic uses Species and hybrids of *Hemerocallis* (Daylillies, Spiderlilies), especially *H.
fulva*, are cultivated; other species of the genus
have medicinal qualities, e.g., *H. citrina*, which
is used to inhibit fibroblast proliferation.
Phormium tenax (New Zealand Flax) is a
source of fibers used to manufacture cloth in
New Zealand. Some species of *Dianella* and
Stypandra are poisonous and therefore haz-
ardous to livestock.   OS

i Dahlgren, R. M. T., Clifford, H. T., & Yeo, P. F. *The
Families of the Monocotyledons*. Berlin, Springer-Verlag
(1985).
ii McPherson, M. A., Fay, M. F., Chase, M. W., &
Graham, S. W. Parallel loss of slowly evolving intron
from two closely related families in Asparagales.
Syst. Bot. 29: 296–307 (2004).
iii Chase, M. W. *et al.* Multi-gene analysis of monocot
relationships: A summary. Pp. 63–75. In: Columbus,
J. T. *et al.* (eds), *Monocots: Comparative Biology
and Evolution*. Claremont, California, Rancho Santa
Ana Botanic Garden (2006).

iv Chase, M. W. *et al.* Phylogenetics of Asphodelaceae
(Asparagales): An analysis of plastid *rbc*L and *trn*L-*trn*F
DNA sequences. *Ann. Bot.* 86: 935–991 (2000).
v Pires, J. C. *et al.* Phylogeny, genome size, and
chromosomal evolution of Asparagales. Pp. 287–304.
In: Columbus, J. T. *et al.* (eds), *Monocots: Comparative
Biology and Evolution*. Claremont, California, Rancho
Santa Ana Botanic Garden (2006).
vi Fay, M. F. *et al.* Phylogenetic studies of Asparagales
based on four plastid DNA regions. Pp. 360–371.
In: Wilson, K. L. & Morrison, D. A. (eds), *Monocots.
Systematics and Evolution*. Melbourne, CSIRO
Publishing (2000).

HERRERIACEAE

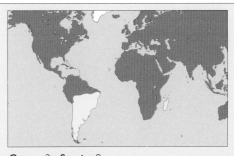

Genera 2 Species 9
Economic uses None

The Herreriaceae consists of 2 genera of perennial herbs with tuberous rhizomes and
branching, spiny, and sometimes twining stems
or rosette plants. The family is restricted to
moist temperate and subtropical forests in South
America (*Herreria*) and Madagascar (*Herreriopsis*). The leaves are spirally arranged, clustered,
ovate-lanceolate, or linear. The inflorescence is
an axillary or terminal raceme or panicle. The
flowers are bisexual, actinomorphic, with 2
whorls of 3 more or less equal, colored tepals,
with nectaries basally (*Herreriopsis*), 3+3 free
stamens, and a 3-merous, 3-locular superior
ovary, with septal nectaries (*Herreria*), a single
capitate style, and 2 rows of anatropous ovules
on axile placentas in each locule. The fruit is a
septicidal capsule, with few to numerous black
seeds, each seed having a small straight embryo
in an endosperm of aleurone and oil.

The position of the family has been controversial[i,ii], but molecular evidence includes it in a
broadly circumscribed Agavaceae[iii] (see also
APG II) or as sister group to Anthericaceae[iv].
Herreriaceae is sometimes included in a greatly
enlarged Asparagaceae[iii]   OS

i Dahlgren, R. M. T., Clifford, H. T., & Yeo, P. F.
The Families of the Monocotyledons. Berlin,
Springer-Verlag (1985).
ii Conran, J. G. Herreriaceae. Pp. 253–255. In:
Kubitzki, K. (ed.), *The Families and Genera of
Vascular Plants. III. Flowering Plants. Monocotyledons:
Lilianae (except Orchidaceae)*. Berlin, Springer-Verlag
(1998).
iii Chase, M. W. *et al.* Multi-gene analysis of monocot
relationships: A summary. Pp. 63–75. In: Columbus,
J. T. *et al.* (eds), *Monocots: Comparative Biology
and Evolution*. Claremont, California, Rancho Santa
Ana Botanic Garden (2006).
iv Pires, J. C. *et al.* Phylogeny, genome size, and
chromosomal evolution of Asparagales. Pp. 287–304.
In: Columbus, J. T. *et al.* (eds), *Monocots: Comparative*

Biology and Evolution. Claremont, California, Rancho
Santa Ana Botanic Garden (2006).

HOSTACEAE
HOSTA FAMILY

This monogeneric fam-
ily comprises around
25 spp. of perennial
rhizomatous herbs,
with basal, spirally
arranged, petiolated
and linear-lanceolate
to ovate leaves. The
family is restricted to
Japan, China, and
Korea in shady forests
and rocky outcrops.
The inflorescence is
a bracteate or leafless
raceme arising directly
from the rhizome, and
the blue or white flow-

Genera 1
Species c. 25
Economic uses *Hosta*
is an important
ornamental in
temperate regions

ers are bisexual, funnel-shaped, or campanulate,
with 2, 3-merous whorls of similar tepals, and
3+3 stamens attached to the base of the tepals or
to the top of the ovary. The ovary is superior,
consisting of 3 carpels, and is 3-locular, with septal nectaries, a thin filiform style, with a 3-lobed,
capitate stigma, and numerous anatropous
ovules on an axile placenta in each locule. The
fruit is a pendant capsule, dehiscing from the
apex, with numerous black seeds.

The genera *Leucocrinum* and *Hemerocallis*
are often included in the Hostaceae, but are
here treated in the Anthericaceae (q.v.) and in a
family of their own, Hemerocallidaceae (q.v.),
respectively. The Hostaceae are often erroneously called Funkiaceae. Even defined in a
narrow sense[i] (see also APG II), Agavaceae
(q.v.) could be enlarged to include both *Hesperocallis undulata* and Hostaceae[ii]. The 3 taxa
share a unique basic chromosome complement
of 5 long and 25 short chromosomes. However,
molecular evidence places Hostaceae in an
unresolved relationship with Agavaceae[iii,iv].
Broadly defined, the Agavaceae could be
extended to include the above-mentioned families plus Anemarrhenaceae (q.v.), Anthericaceae
(q.v.), Behniaceae (q.v.), and Herreriaceae
(q.v.). The Hostaceae is sometimes included in a
broadly defined Asparagaceae[v].   OS

i Kubitzki, K. Hostaceae. Pp. 256–260. In: Kubitzki, K.
(ed.), *The Families and Genera of Vascular Plants. III.
Flowering Plants. Monocotyledons: Lilianae*. Berlin,
Springer-Verlag (1998).
ii Gómez-Pompa, A., Villalobos-Pietrini, R., &
Chimal, A. Studies in the Agavaceae. I. Chromosome
morphology and number of seven species.
Madroño 21: 208–221 (1991).
iii Bogler, D. J. & Simpson, B. B. A Chloroplast DNA
study of the Agavaceae. *Syst. Bot.* 20: 191–205 (1995).
iv Bogler, D. J. & Simpson, B. B. Phylogeny of
Agavaceae based on ITS rDNA sequence variation.
American J. Bot. 83: 1225–1235 (1996).
v Chase, M. W. *et al.* Multi-gene analysis of monocot
relationships: A summary. Pp. 63–75. In: Columbus,
J. T. *et al.* (eds), *Monocots: Comparative Biology
and Evolution*. Claremont, California, Rancho Santa
Ana Botanic Garden (2006).

HYACINTHACEAE

HYACINTH FAMILY

A family of usually bulbous herbs, with basal, often filiform to elliptic, leaves and a simple raceme of often brightly colored flowers.

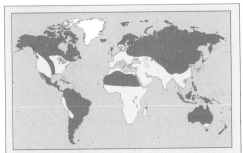

Genera c. 70 **Species** c. 900
Economic uses Several species are cultivated as ornamentals, and a few have been used medically

Distribution The Hyacinthaceae is widely distributed in areas with Mediterranean climate but reaches the tropics. The greatest diversity is in the Mediterranean and southern Africa, extending to Central and East Asia; a few species occur in North America, and a single species is found in South America. They prefer open, sunny habitats but are also found in swamps.

Description Bulbous, rarely rhizomatous plants, the bulbs consisting of cataphylls, leaf-bases, or both. Leaves are basal, spirally arranged, or distichous, rarely lacking (e.g., *Bowiea*), usually without petiole, mostly filiform to elliptical, and flat or terete. The inflorescence is a scapose, occasionally bracteate, raceme, rarely a compound raceme (e.g., *Camassia*) or a spike, with 1 to hundreds of flowers. The flowers are bisexual, actinomorphic, or rarely zygomorphic, with 3+3 tepals, which are either free or more or less fused at the base, more or less equal, often brightly colored, and occasionally with tufts of hair at the apex (*Albuca*). The stamens are 3+3, occasionally fused at the base; rarely the outer whorl is sterile. The ovary is superior, rarely semi-inferior (*Bowiea*), 3-carpellate, and 3-locular, with septal nectaries; a simple, capitate or 3-lobed style; and 1 to many anatropous ovules in 1 or 2 rows in each locule. The fruit is a loculicidal capsule that may remain fleshy. The seeds are colored, often black; the embryo is straight, usually in an endosperm of aleurone or oil, but rarely starch.

Classification The monophyly of the Hyacinthaceae is in doubt, and several genera included here, including *Camassia*, have been transferred to a family of their own (Camassiaceae[i]) or are included in still narrowly defined Agavaceae (q.v.) as defined here[ii]. In the present circumscription, Hyacinthaceae is sometimes part of an enlarged Asparagaceae (q.v.)[iii]. The Hyacinthaceae is sister to the Aphyllanthaceae (q.v.)[iv] or more likely the Themidaceae (q.v.)[v,vi].

Economic uses Many genera includes species that are widely cultivated in gardens and parks in temperate areas in the northern hemisphere, e.g., *Scilla* (c. 30 spp.), *Muscari* (c. 50 spp.), *Hyacinthus* (3 spp.), *Puschkinia* (3-4 spp.), and *Camassia* (5 spp.). Recently, it has been suggested that the genus *Ornithogalum* (c. 50 spp.) should be enlarged to include several other genera, e.g., *Albuca*, *Dipcardi*, and *Pseudogaltonia*[vii]. Some species, such as *O. thyrsoides* (Chincherinchee), are grown as cut flowers, and *O. umbellatum* (Star of Bethlehem) and *Lachenalia* (c. 110 spp.) are also cultivated. *Drimia* (*Urginea*) *maritima* (Sea Onion) and *Thurarthos indicum* both contain cardiac glycosides, and *D. maritima* has been used medically for centuries. 					OS

i Pfosser, M. & Speta, F. Phylogenetics of Hyacinthaceae Based on Plastid DNA sequences. *Ann. Missouri Bot. Gard.* 86: 852–875 (1999).
ii Judd, W. S. *et al.* *Plant Systematics. A Phylogenetic Approach.* 2nd edn. Sunderland, Massachusetts, Sinauer Associates (2002).
iii Chase, M. W. *et al.* Multi-gene analysis of monocot relationships: A summary. Pp. 63–75. In: Columbus, J. T. *et al.* (eds), *Monocots: Comparative Biology and Evolution.* Claremont, California, Rancho Santa Ana Botanic Garden (2006).
iv Fay, M. F. *et al.* Phylogenetic studies of Asparagales based on four plastid DNA regions. Pp. 360–371. In: Wilson, K. L. & Morrison, D. A. (eds), *Monocots. Systematics and Evolution.* Melbourne, CSIRO Publishing (2000).
v Pires, J. C. *et al.* Phylogeny, genome size, and chromosomal evolution of Asparagales. Pp. 287–304. In: Columbus, J. T. *et al.* (eds), *Monocots: Comparative Biology and Evolution.* Claremont, California, Rancho Santa Ana Botanic Garden (2006).
vi Speta, F. Hyacinthaceae. Pp. 261–285. In: Kubitzki, K. (ed.), *The Families and Genera of Vascular Plants. III. Flowering Plants. Monocotyledons: Lilianae (except Orchidaceae).* Berlin, Springer-Verlag (1998).
vii Manning, J., Goldblatt, P., & Fay, M. F. A revised generic synopsis of Hyacinthaceae in sub-Saharan Africa, including new combinations and the new tribe Pseudprospereae. *Edinburgh J. Bot.* 60: 533–568 (2004).

HYDATELLACEAE

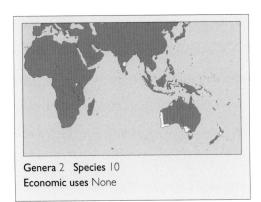

Genera 2 **Species** 10
Economic uses None

A small family of minute, mostly annual, aquatics that grow in leafy tufts. The family is restricted to Australia, New Zealand, and India and are found in swamps, along lake-shores, and in seasonally wet habitats. The leaves are basal, 1-veined, without sheath or ligule, and linear to filiform. The inflorescence is a capitulum on short scape, which is hidden among the leaves while flowering, and extends beyond them while in fruit. The capitulum usually has 2 to 4 bracts, with numerous unisexual flowers (plants dioecious or monoecious), the male consisting of a single stamen, and the female consisting of a stalked ovary. The ovary is 1-locular with a single anatropous ovule and a bearded, terminal, sessile stigma with between 2 and 10 hairs in a single row. The fruit is either dehiscent or indehiscent, membranaceous, and with a single seed enclosing a minute embryo in a starchy perisperm.

The Hydatellaceae has previously been included in the Centrolepidaceae (q.v.) but is now widely recognised as a separate family[i]. It is with doubt placed as sister group to Xyridaceae (q.v.)[ii]. 					OS

i Hamann, U. Hydatellaceae. Pp. 231–234. In: Kubitzki, K. (ed.), *The Families and Genera of Vascular Plants. IV. Flowering Plants. Monocotyledons: Alismatanae and Commelinanae (except Gramineae).* Berlin, Springer-Verlag (1998).
ii Davis, J. I. *et al.* A phylogeny of the monocots, as inferred from *rbc*L and *atp*A sequence variation, and a comparison of methods for calculating jackknife and bootstrap values. *Syst. Bot.* 29: 467–510 (2004).

HYDROCHARITACEAE

CANADIAN WATERWEED AND FROG-BIT

The Hydrocharitaceae is a family of marine and freshwater aquatics with an inflorescence that often has 2 fused bracts at the base.

Distribution Cosmopolitan but mainly tropical, the family is found in a wide range of marine and freshwater habitats.

Description Perennial or annual aquatic herbs, having either a creeping rhizome with leaves arranged in rosettes, or an erect main stem with roots at the base and spirally arranged or whorled leaves. The leaves are simple, highly variable in shape, usually submerged, sometimes floating, rarely emergent, from linear to orbicular, with or without a petiole, with intervaginal scales, and with or without sheathing bases. The inflorescence is scapose or sessile, initially enclosed in 1 to 2 free or fused bracts, with between 1 and 100 or more flowers. Generally, the flowers are unisexual (plants dioecious), usually actinomorphic, rarely slightly irregular (e.g., *Vallisneria*), and either bisexual or unisexual (male and female then being borne on separate plants). The perianth segments are in 1 or 2 series of 3 (rarely 2) free segments; the inner series when present are usually showy and petal-like. Stamens usually numerous, in up to 6 or more series; the inner stamens are sometimes sterile (in *Lagarosiphon* the staminodes function as sails for the free-floating male flowers). The pollen is inaperturate, globular, and free or in tetrads. In the marine genera *Thalassia* and *Halophila*, the pollen grains are liberated in chains. The ovary is inferior with between 2 and 20 incompletely united carpels, and with a long hypanthium in many genera. The ovules are

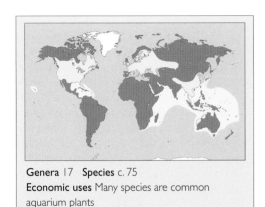

Genera 17 Species c. 75
Economic uses Many species are common
aquarium plants

numerous and scattered over the inner surface of the carpel walls, with as many styles as there are carpels. The fruits are globular to linear, dry or pulpy, usually indehiscent, and splitting up irregularly. Usually, the seeds have straight embryos and lack endosperm.

The family has a spectacular variety of pollination mechanisms. Some genera, such as *Hydrocharis* (Frogsbit), *Ottelia*, and *Stratiotes,* have relatively large showy flowers that are insect-pollinated. Several genera are pollinated at the water surface. In *Elodea,* the pollen is liberated at or below the water surface and floats to the stigmas of the female flowers. In *Lagarosiphon, Maidenia, Nechamandra,* and *Vallisneria,* the male flowers become detached as buds from the parent plant and rise to the surface, where they

expand and drift or sail to the female flowers, where pollination takes place above the water surface. In *Hydrilla,* male flowers are also released, but pollen is liberated from the anthers by an explosive mechanism; only pollen that reaches the stigmas directly pollinates the ovules. In *Limnobium,* the male flowers are held above the female flowers, and pollen is liberated from the anthers, and usually drifts through the air to the stigma. In *Halophila* and *Thalassia,* pollination takes place below water.

Classification The Hydrocharitaceae is divided into 5 or 6 groups[i,ii], occasionally recognized as subfamilies[iii]. However, the groups circumscribed by different authors are different, and the relationships between different genera are unclear. Molecular analyses indicate that Hydrocharitaceae is closely related to Butom-aceae[i,iv] and clearly place Najadaceae within Hydrocharitacae[i]. However, the inclusion of Najadaceae in Hydrocharitaceae is not well-supported by morphology[ii].

Economic uses Several species are used in aquaria, and a number of introduced species have become pernicious weeds, e.g., *Egeria densa* and *Hydrilla verticillata* in southern USA, *Hydrocharis morsus-ranae* (Frog-bit) in eastern

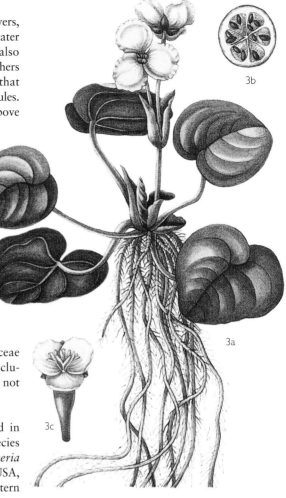

Canada, *Elodea canadensis* (Canadian Waterweed) in Europe (where it is now being replaced by *E. nuttallii*), and *Lagarosiphon major* in New Zealand. [CDC] OS

i Les, D. H., Cleland, M. A., & Waycott, M. Phylogenetic studies in Alismatidae, II: Evolution of marine angiosperms (seagrasses) and hydrophily. *Syst. Bot.* 22: 443–463 (1997).
ii Cook, C. D. K. Hydrocharitaceae. Pp. 234–248. In: Kubitzki, K. (ed.), *The Families and Genera of Vascular Plants. IV. Flowering Plants. Monocotyledons: Alismatanae and Commelinanae (except Gramineae).* Berlin, Springer-Verlag (1998).
iii Dahlgren, R. M. T., Clifford, H. T., & Yeo, P. F. *The Families of the Monocotyledons.* Berlin, Springer-Verlag (1985).
iv Davis, J. I. *et al.* A phylogeny of the monocots, as inferred from *rbc*L and *atp*A sequence variation, and a comparison of methods for calculating jackknife and bootstrap values. *Syst. Bot.* 29: 467–510 (2004).

HYDROCHARITACEAE. 1 *Vallisneria spiralis* (a) habit, showing stolons bearing new plants, ribbon-shaped leaves, and the long-stalked female flowers that reach the water surface (x⅔); (b) male flower, which separates from the parent plant and floats to the surface (x12); ovary (c) in cross section (x8) and (d) in vertical section (x4); (e) female flower, at pollination it remains attached but floats horizontally on the surface (x4). 2 *Elodea canadensis* (a) habit showing female flowers on long hypanthia (x⅔); (b) vertical section of ovary (x5); (c) female flower with 3 forked styles (x5). 3 (ABOVE) *Hydrocharis morsus-ranae* (a) general habit of this free-floating plant, shown here with male flowers (x⅔); (b) cross section of fruit (x2); (c) female flower (x⅔).

HYPOXIDACEAE

A small family of geophytes with linear, rosulate, 3-ranked, often plicate or folded leaves and inflorescences on hairy scapes.

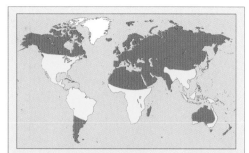

Genera 9 **Species** c. 100–200
Economic uses A few species are cultivated as ornamentals

Distribution The family is largely restricted to the southern hemisphere in tropical and subtropical regions of Africa, Asia, Australia, and South America, but a few genera reach into the northern hemisphere. Prefer seasonally wet habitats but is also find in tropical forests.

Description Perennial, geophytic herbs with a rhizome or a tunicate corm and linear to lanceolate, rosulate, 3-ranked leaves with more or less sheathing bases, usually sessile but occasionally with a pseudo-petiole and often plicate or folded when young. The inflorescences are axillary on hairy scapes, bracteate, and usually spicate, corymbose, or rarely capitate, umbel-like or reduced to a solitary flower. The flowers are usually actinomorphic and bisexual, with 2, 3-merous, rarely 2-merous, whorls of similar, usually free tepals, though fused into a tube in some genera (e.g., *Rhodohypoxis*), often colored on the inside and green and hairy on the outside. The stamens are 3+3, rarely 2+2 or only 3 (*Pauridia*), with short filaments arising from the base of the free tepals or from the mouth of the corolla tube, or lacking completely. The ovary is inferior, often separated from the perianth by a structure of unknown origin (perianth tube or ovary extension), with 3 carpels, 3-lobed with axile placentas, or 1-locular with parietal placentas usually with numerous, anatropous ovules. The style is single with 3 to 6 stigmas, with the ovary occasionally prolonged into a beak (e.g., *Curculigo*). The fruit is often a loculicidal capsule, but sometimes opening with a lid, or fleshy and indehiscent or irregularly dehiscent. The seeds are black with a small embryo in a non-starchy endosperm, storing aleurone and oil.

Classification The number of species in the Hypoxidaceae is difficult to ascertain mainly because of the apomictic nature of *Hypoxis*, which make species delimitation extremely difficult. The family has been considered to be closely related to the Orchidaceae[i,ii], but recent molecular and morphological analyses point to the Hypoxidaceae as the most likely sister group to Lanariaceae (q.v.) [iii,iv,v,vi].

Economic uses A few species of *Curculigo* are cultivated as ornamentals.   OS

[i] Stevenson, D. W. Hypoxidaceae. Pp. 418–421. In: Smith, N. *et al.* (eds), *Flowering Plants of the Neotropics*. Princeton/Oxford, Princeton University Press (2004).
[ii] Nordal, I. Hypoxidaceae. Pp. 286–294. In: Kubitzki, K. (ed.), *The Families and Genera of Vascular Plants. III. Flowering Plants. Monocotyledons: Lilianae (except Orchidaceae)*. Berlin, Springer-Verlag (1998).
[iii] Fay, M. F. *et al.* Phylogenetic studies of Asparagales based on four plastid DNA regions. Pp. 360–371. In: Wilson, K. L. & Morrison, D. A. (eds), *Monocots. Systematics and Evolution*. Melbourne, CSIRO Publishing (2000).
[iv] Rudall, P. J. *et al.* Anatomical and molecular systematics of Asteliaceae and Hypoxidaceae. *Bot. J. Linn. Soc.* 127: 1–42 (1998).
[v] Pires, J. C. *et al.* Phylogeny, genome size, and chromosomal evolution of Asparagales. Pp. 287–304. In: Columbus, J. T. *et al.* (eds), *Monocots: Comparative Biology and Evolution*. Claremont, California, Rancho Santa Ana Botanic Garden (2006).
[vi] Davis, J. I. *et al.* A phylogeny of the monocots, as inferred from *rbc*L and *atp*A sequence variation, and a comparison of methods for calculating jackknife and bootstrap values. *Syst. Bot.* 29: 467–510 (2004).

IRIDACEAE
IRIS FAMILY

The Iridaceae is a family of herbs with rhizomes, corms, or bulbs, and flowers with conspicuous, petaloid tepals.

Distribution The family is cosmopolitan and occurs in both tropical and temperate regions, but South Africa and Central and South America are especially rich in species. Most species occur in open scrub, deserts, and grassland, but the family is conspicuously absent from the Indian subcontinent, Sahara, and the interior of Australia, and few species occur in forests.

Description Perennial, rarely annual (e.g., *Sisyrinchium*) herbs, with rhizomes, corms, or bulbs. Most species are deciduous, although a few are evergreen. *Geosiris* is achlorophyllous. The leaves are mostly distichous, sometimes ensiform, basal, or cauline, with an imbricate, open or closed sheathing base, usually appearing with the flowers, or scalelike in *Geosiris*. The inflorescence varies considerably but is usually bracteate and either composed of terminal, 1 to many flowered, umbellate, monochasial cymes, or spicate, more rarely paniculate, or reduced to a single, almost sessile, flower (e.g., *Crocus*). The flowers are bisexual, actinomorphic, or zygomorphic (e.g., *Gladiolus*) and frequently with a large conspicuous perianth of 2 whorls of 3 petaloid tepals, although the inner whorl is sometimes reduced (e.g., *Iris*) or missing; when zygomorphic, the flowers are more or less bilabiate. The tepals are free or united at the base into a straight or curved, tubular or funnel-shaped tube. The 3, rarely only 2 (*Diplarrhena*), stamens are opposite the outer tepals. The ovary is inferior, very rarely superior (*Isophysis*), with septal nectaries, and consists of 3 fused carpels, with 3 locules and axile placentas, or more rarely 1 locule with 3 parietal placentas. The ovules are usually numerous, rarely 1 or few, anatropous or campylotropous, in 1 or 2 rows in each locule. The style is terminal and usually 3-branched or 3-lobed. The branches are filiform or expanded and petaloid, with the stigmatic surface facing the tepals. In several genera (e.g., *Iris, Moraea*), the style is a 3-branched, flattened, petaloid structure, which has enlarged showy "crests" overtopping the stigmatic surface. In *Iris*, these style branches curve outward, away from the axis

Genera c. 70 **Species** c. 1,800
Economic uses The family is of great horticultural importance and includes numerous garden and indoor ornamentals and cut flowers

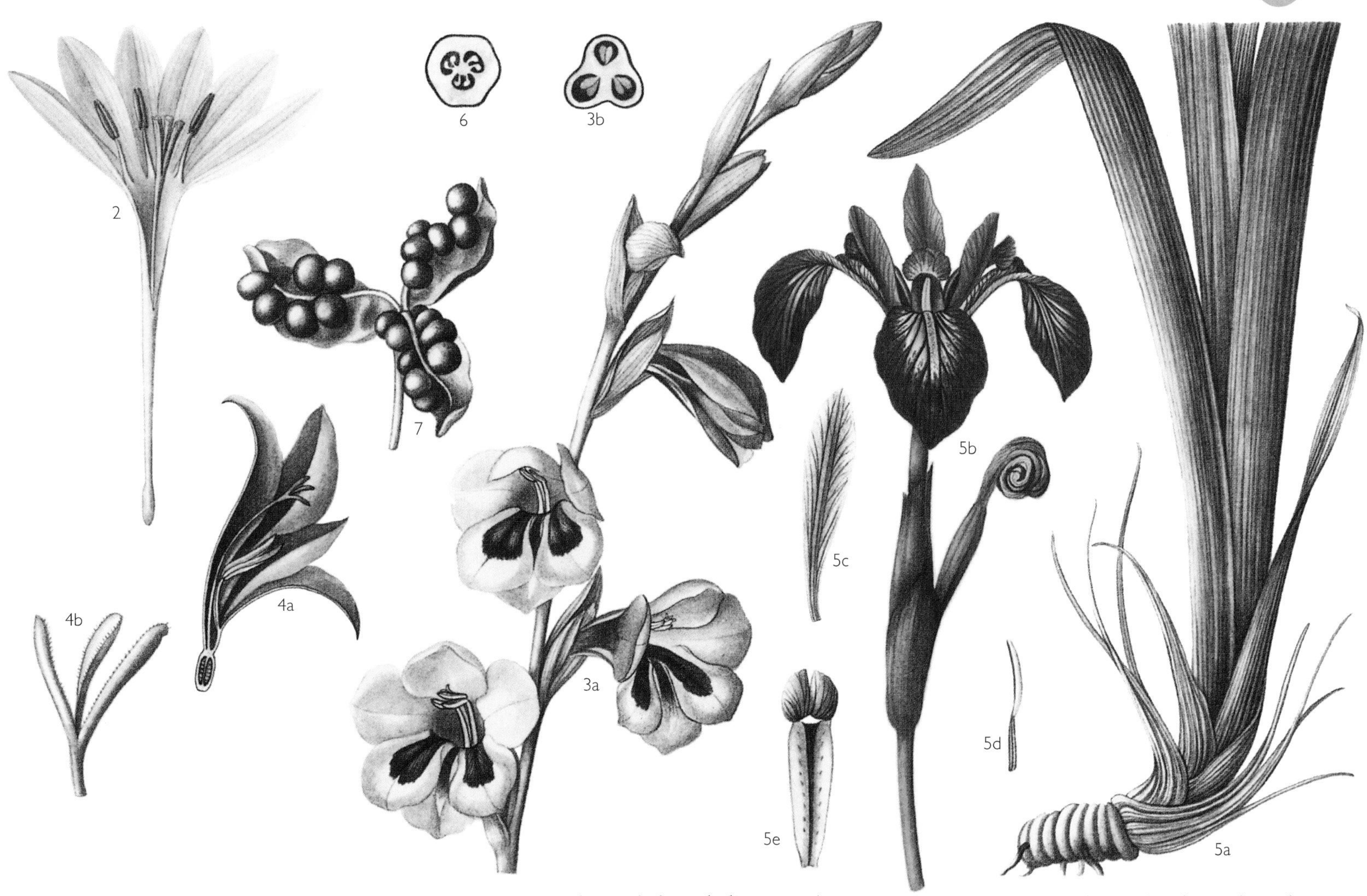

IRIDACEAE. 1 (OPPOSITE PAGE) *Crocus flavus* (a) habit, showing tunicated corm and sessile flowers 3-lobed style (x4); (b) capsule (x⅔); (c) tip of trilobed style (x4). (ABOVE:) 2 *Crocus* sp. flower with perianth opened (x⅔). 3 *Gladiolus papilio* (a) spicate inflorescence, with bract and large prophyll (x⅔); (b) cross section of 3-locular ovary with 2 rows of ovules in each locule (x2⅔). 4 *G. melleri* (a) longitudinal section through "bilabiate" flower (x⅔); (b) tip of style (x3). 5 *Iris laevigata* (a) apex of rhizome and leaf-bases of the ensiform leaves (x⅔); (b) inflorescence with fully opened flower consisting of three reflexed "falls," three erect inner "standards," and three petaloid style branches behind the stamens (x⅔); (c) bearded standard (x1); (d) stamen (x1); (e) petaloid style branch, the stigmatic surface is on the small triangular crest (x1). 6 *I. germanica* cross section of ovary with 2 rows of ovules in each locule (x2⅔). 7 *I. foetidissima* loculicidally dehiscing capsule, with seeds (x⅔).

of the flower, and form, with 3 of the perianth segments, a protective tunnel-like organ over each anther. The fruit is a loculicidal capsule, usually with many brown seeds, with a small embryo in an endosperm that contains hemicellulose, oil, and protein but rarely starch.

Classification *Geosiris aphylla* is sometimes considered as a separate family, Geosiridaceae[i], but undoubtedly belongs in the Iridaceae[ii,iii]. The family is often divided into a number of subfamilies[iv,v], which were found to be monophyletic, except the Nivenioideae, in a multigene analysis[vi]. In the present circumscription, the Iridaceae is monophyletic and sister group to a huge clade including e.g., Asparagaceae (q.v.) in the broad sense [ii,iii,vi,vii].

Economic uses The family is of great horticultural importance and includes numerous garden and indoor ornamentals. Genera cultivated in temperate areas include *Crocus* (85 spp.), *Iris* (225 spp.), *Sisyrinchium* (60–80 spp.), and *Tigridia* (35 spp.). *Crocosmia* (Montbretia; 9 spp.), *Gladiolus* (255 spp.), *Iris,* and *Freesia* (14 spp.) are important as cut flowers The stigmas of the triploid *Crocus sativus* provide the important spice saffron, grown commercially in Europe, the Middle East, and India. Harvesting is done manually, and the spice is consequently exceedingly expensive. Its use was known in ancient Greece. The powdered, fragment rhizome of *Iris germanica* (German Iris) is a source of orris, which is used in perfumes and potpourris. A few South African species are serious weeds in Australia, e.g., *Sparaxis, Romulea,* and *Watsonia.* [BFM] OS

i Dahlgren, R. M. T., Clifford, H. T., & Yeo, P. F. *The Families of the Monocotyledons.* Berlin, Springer-Verlag (1985).
ii Fay, M. F. *et al.* Phylogenetic studies of Asparagales based on four plastid DNA regions. Pp. 360–371. In: Wilson, K. L. & Morrison, D. A. (eds), *Monocots. Systematics and Evolution.* Melbourne, CSIRO Publishing (2000).
iii Pires, J. C. *et al.* Phylogeny, genome size, and chromosomal evolution of Asparagales. Pp. 287–304. In: Columbus, J. T. *et al.* (eds), *Monocots: Comparative Biology and Evolution.* Claremont, California, Rancho Santa Ana Botanic Garden (2006).
iv Goldblatt, P., Manning, J. C., & Rudall, P. Iridaceae. Pp. 295–333. In: Kubitzki, K. (ed.), *The Families and Genera of Vascular Plants. III. Flowering Plants. Monocotyledons: Lilianae (except Orchidaceae).* Berlin, Springer-Verlag (1998).
v Judd, W. S. *et al. Plant Systematics. A Phylogenetic Approach.* 2nd edn. Sunderland, Massachusetts, Sinauer Associates (2002).
vi Reeves, G. *et al.* Molecular systematics of Iridaceae: Evidence from four plastid regions. *American J. Bot.* 88: 2074–2097 (2001).
vii Chase, M. W. *et al.* Multi-gene analysis of monocot relationships: A summary. Pp. 63–75. In: Columbus, J. T. *et al.* (eds), *Monocots: Comparative Biology and Evolution.* Claremont, California, Rancho Santa Ana Botanic Garden (2006).

IXIOLIRIACEAE

The Ixioliriaceae is a monogeneric family of geophytes with bulblike corms and alternate, basally sheathing, linear leaves on erect stems. Species are found in semiarid areas throughout Egypt

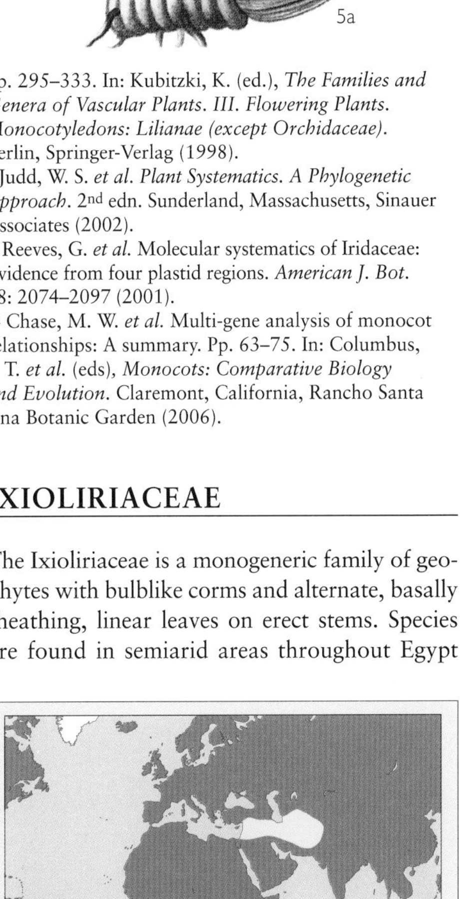

Genera 1 **Species** c. 3
Economic uses Cultivated as ornamentals

and southwestern and Central Asia. The inflorescence is a few to many flowered thyrse or is umbel-like. The flowers are bisexual and actinomorphic, with 3+3 free, petaloid tepals, equal, with the outer tepal ending in pointed apices; 3+3 stamens; an inferior 3-carpellate, 3-locular ovary, containing numerous anatropous ovules on axile placentas; and a single style with 3 stigmatic lobes. The fruit is a loculicidal capsule, with numerous black seeds, each one containing a long embryo in a starchless endosperm.

The Ixioliriaceae has been included in the Amaryllidaceae, but in recent years it has been treated as a separate family, a fact based on morphological evidence. Recent molecular studies support this latter view[i,ii,iii] but disagree on its closer relationships, pointing to either Tecophilaeaceae (q.v.)[i,iv] or Iridaceae (q.v.)[iii,v] as its sister group. OS

i Fay, M. F. *et al.* Phylogenetic studies of Asparagales based on four plastid DNA regions. Pp. 360–371. In: Wilson, K. L. & Morrison, D. A. (eds), *Monocots. Systematics and Evolution*. Melbourne, CSIRO Publishing (2000).
ii Pires, J. C. *et al.* Phylogeny, genome size, and chromosomal evolution of Asparagales. Pp. 287–304. In: Columbus, J. T. *et al.* (eds), *Monocots: Comparative Biology and Evolution*. Claremont, California, Rancho Santa Ana Botanic Garden (2006).
iii Chase, M. W. *et al.* Multi-gene analysis of monocot relationships: A summary. Pp. 63–75. In: Columbus, J. T. *et al.* (eds), *Monocots: Comparative Biology and Evolution*. Claremont, California, Rancho Santa Ana Botanic Garden (2006).
iv Chase, M. W. *et al.* Phylogenetics of Asphodelaceae (Asparagales): An analysis of plastid *rbc*L and *trn*L-*trn*F DNA sequences. *Ann. Bot.* 86: 935–991 (2000).
v Davis, J. I. *et al.* A phylogeny of the monocots, as inferred from *rbc*L and *atp*A sequence variation, and a comparison of methods for calculating jackknife and bootstrap values. *Syst. Bot.* 29: 467–510 (2004).

JOHNSONIACEAE

Distribution The Johnsoniaceae is largely restricted to Australia and New Guinea, although a single genus (*Caesia*) reaches southern Africa and Madagascar.

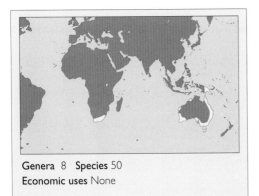

Genera 8 Species 50
Economic uses None

Description Perennial, tufted or rhizomatous herbs, rarely with stilt-roots (*Hensmania, Stawellia*), sometimes shrublike. The leaves are distichous and linear, and the inflorescence is usually a spike or raceme or is umbel-like. The flowers are bisexual and actinomorphic with 3+3 free or basally united tepals and 3+3 sta-

mens that are basally adnate to the tepals or with 3 staminodes. The ovary is superior, 3-carpellate, 3-locular, usually with 2 anatropous or campylotropous ovules in each locule, and a single, terminal, rarely gynobasic style. The fruit is a capsule, nut, or schizocarp with black seeds. The embryo is linear. The Johnsoniaceae has been included in Anthericaceae (q.v.)[i] but is here treated as a separate family, although it is sometimes included in the Hemerocallidaceae (q.v.)[ii]. However, the relationship between the 2 families as treated here is not well understood, although the Johansoniaceae appears to be a clade embedded in the Hemerocallidaceae[2]. The Johnsoniaceae, Hemerocallidaceae, Asphodelaceae (q.v.), and Xanthorrhoeaceae (q.v.) are occasionally included in an enlarged Xanthorrhoeaceae[iii] (cf. APG II). OS

i Dahlgren, R. M. T., Clifford, H. T., & Yeo, P. F. *The Families of the Monocotyledons.* Berlin, Springer-Verlag (1985).
ii McPherson, M. A., Fay, M. F., Chase, M. W., & Graham, S. W. Parallel loss of slowly evolving intron from two closely related families in Asparagales. *Syst. Bot.* 29: 296–307 (2004).
iii Chase, M. W. *et al.* Multi-gene analysis of monocot relationships: A summary. Pp. 63–75. In: Columbus, J. T. *et al.* (eds), *Monocots: Comparative Biology and Evolution*. Claremont, California, Rancho Santa Ana Botanic Garden (2006).

JOINVILLEACEAE

A family comprising a single genus of 2 spp. of perennial, erect herbs, which are restricted to the Malay Peninsula, northern Borneo, Sumatra, and the Pacific Islands. The leaves are distichous, linear to lanceolate, with open sheaths, and the inflorescence terminal and paniculate. The flowers are bisexual, actinomorphic, and 3-merous in all whorls. The ovary is superior, 3-locular, with a single, hanging, atropous ovule in each locule. The fruit is a drupe, usually containing 3 seeds, with a starchy endosperm and a disk-shaped embryo.

Genera 1 Species 2
Economic uses None

The Joinvilleaceae have often been considered to be closely related to Flagellariaceae, and both families closely related to the Poaceae, whereas their resemblance to Hanguanaceae is merely superficial. This fact has been supported by nearly all recent analyses. The Joinvillaceae is very likely sister group to Ecdeiocolaceae as well as Poaceae. For literature see the account under Flagellariaceae. OS

JUNCACEAE
RUSH FAMILY

A small family of grasslike, predominantly terrestrial herbs, usually with tristichous, rarely distichous, simple sheathing leaves.

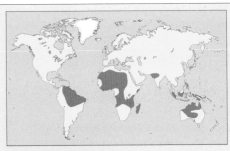

Genera 7 Species c. 440
Economic uses Few species are of commercial value: some are used to make baskets or mats, and others are aquatic ornamentals; cushions of *Distichia* are used as fuel in Peru

Distribution The family is cosmopolitan but mainly found in cold temperate or montane regions in wet or damp, occasionally saline, habitats. A few genera (e.g., *Distichia, Oxychloe,* and *Patosia*) consist of cushion-forming species adapted to the harsh diurnal freezing and thawing in the High Andes of South America and reaching the limit of vegetation.

Description Most species are perennial, rarely annuals (e.g., in *Juncus* and *Luzula*), with erect or horizontal rhizomes and erect or ascending stems. The leaves are usually tristichous, but some species have distichous leaves, linear or filiform, with open or closed sheaths, usually flat, dorsiventral but occasionally terete, or the blade is reduced. The inflorescence is terminal or apparently lateral, many-flowered (*Juncus* and *Luzula*), compound, and consist of open panicles and head- or spikelike inflorescences. In the remaining genera, the flowers are solitary and either terminal (*Rostkovia*) or lateral (*Oxychloe*). The flowers are actinomorphic, typically bisexual, or rarely female. The tepals are small, similar, glumaceous or herbaceous, arranged in 2 whorls of 3. They are dull in color, often green, brown, or black but sometimes also white or yellowish. There are usually 6 stamens in 2 whorls, alternating with the tepals. Occasionally, the inner whorl is missing. The ovary is superior, terminated by a style with 3 branches, consisting of 3 carpels, and usually 3-locular, rarely 1-locular. There are 3 (e.g., *Luzula*) to many anatropous ovules on central, axile, or parietal placentas. The fruit is a loculicidal capsule, and the seeds have a copious starchy endosperm and a straight embryo.

Classification The genera *Juncus* and *Luzula*, with c. 300 and c. 115 spp., respectively, make up most of the family. *Juncus* spp., which are probably not monophyletic[i], usually have open leaf sheaths, glabrous leaves, and capsules with 15 to many seeds, whereas *Luzula* spp. have closed leaf sheaths, more or less ciliate leaf margins, and capsules with only 3 seeds. The remaining 5 genera have solitary flowers and are small, with only 1 to 5 spp. each. They are restricted to the southern hemisphere, where *Distichia, Patosia,* and *Oxychloe* all are confined to the Andes mountains, while *Marsippospermum* and *Rostkovia* also occur in New Zealand. *Rostkovia* is even found on some of the subantartic islands. The family is closely related to the Cyperaceae[ii,iii] (q.v.), and the 2 families share a couple of specialized features, e.g., diffuse centromeres and pollen in tetrads (in Cyperaceae further reduced to monads)[iv], but Juncaceae might be paraphyletic[v] with respect to Cyperaceae. The genus *Prionium* has traditionally been included in the Juncaceae[vi] but is now recognized as a separate family, Prioniaceae[vii,viii] (q.v.). Prioniaceae plus Thurniaceae (q,v.) are sister to Juncaceae plus Cyperaceae. Both *Thurnia* and *Prionium* share pollen tetrads with Juncaceae, but only *Thurnia* and Cyperaceae deposit silica.

Economic uses Generally, the family is of very limited commercial value. However, some species of *Juncus* and *Luzula* are important components in pastures. *Distichia muscoides* is used as fuel in the treeless Andes Mountains. Split rushes used in basket making and the manufacture of chair bottoms are taken from the stems of *Juncus effusus* (Soft Rush) and *J. squarrosus* (Heath Rush). [CJH] OS

i Drábková, L., Kirschner, J., Seberg, O., Petersen, G., & Vlcek, C. Phylogeny of Juncaceae based on *rbc*L sequences, with special emphasis on *Luzula* L. and *Juncus*. L. *Pl. Syst. Evol.* 240: 133–147 (2003).
ii Roalson, E. H. Phylogenetic relationships in the Juncaceae inferred from nuclear ribosomal DNA internat transcribed spacer sequence data. *Int. J. Pl. Sci.* 166: 397–413 (2005).
iii Chase, M. W. *et al.* Multi-gene analysis of monocot relationships: A summary. Pp. 63–75. In: Columbus, J. T. *et al.* (eds), *Monocots: Comparative Biology and Evolution.* Claremont, California, Rancho Santa Ana Botanic Garden (2006).
iv Davis, J. I. *et al.* A phylogeny of the monocots, as inferred from *rbc*L and *atp*A sequence variation, and a comparison of methods for calculating jackknife and bootstrap values. *Syst. Bot.* 29: 467–510 (2004).
v Judd, W. S. *et al. Plant Systematics. A Phylogenetic Approach.* 2nd edn. Sunderland, Massachusetts, Sinauer Associates (2002).
vi Simpson, D. Relationships within Cyperales. Pp. 497–509. In: Rudall, P. J., Cribb, P. J., Cutler, D. F., & Humphries, C. J. (eds). *Monocotyledons: systematics and evolution.* Richmond, Royal Botanic Gardens, Kew (1995).
vii Balslev, H. Juncaeae. Pp. 252–260. In: Kubitzki, K. (ed.), *The Families and Genera of Vascular Plants. IV. Flowering Plants. Monocotyledons: Alismatanae and Commelinanae (except Gramineae).* Berlin, Springer-Verlag (1998).
viii Munro, S. L. & Linder, H. P. The phylogenetic position of *Prionium* (Juncaceae) within the order Juncales based on morphological and *rbc*L sequence data. *Syst. Bot.* 23: 43–45 (1998).
ix Kirschner, J. *Species Plantarum.* Flora of the World. Part 6. Juncaceae 1: Rostkovia to Luzula. Canberra, Australian Biological Resources Study (2002).

JUNCAEAE. 1 *Luzula nodulosa* habit showing, ciliate leaf margins and sheathing leaf-bases (×¼). 2 *L. alpinopilosa* (a) a perfect flower, with 5 whorls (×20); (b) cross section of 3-locular ovary (×100). 3 *Distichia muscoides* a low-growing cushion plant with distichous leaves and solitary flowers (×⅔). 4 *Juncus bufonius* habit showing erect linear leaves with loosely sheathing bases and flowers in dense cymose heads subtended by leaflike bracts (×⅔). 5 *J. acutiflorus* inflorescence (×⅔). 6 *J. bulbosus* half flower (×14). 7 *J. capitatus* dehiscing capsule consisting of 3 valves (×18).

JUNCAGINACEAE
ARROWGRASS

A small family of marsh herbs with basal leaves and inconspicuous flowers.

Genera 4 **Species** 15
Economic uses A few species are edible

Distribution The family is subcosmopolitan and occurs in marshy habitats in temperate and cold areas.

Description Annual or perennial, rhizomatous, tufted herbs, with basal sheathing, usually emergent but sometimes floating, leaves with intervaginal scales. The inflorescence is a scapose, terminal spike or raceme. The flowers

are usually regular, without bracts (except in *Lilaea*), bisexual or unisexual (plants monoecious or dioecious, or polygamous). The tepals are 1 to 4, or 6 in 2 whorls, green, rarely absent. Stamens 1, 4, or 6, with mostly sessile anthers, sometimes the outer whorl adnate to the tepals. The ovary is superior and consists of 1 (*Lilaea*), 3, or 6 free or partly united carpels, each containing a single or a few basal, anatropous ovules (rarely apical and orthotropous). The styles are short or absent. The fruit is schizocarp, separating into achenes, and the seeds have a straight embryo with no endosperm. The flowers are mainly wind-pollinated.

Classification The family includes 4 genera: *Maundia* (1 sp.), *Lilaea* (1 sp.), *Tetroncium* (1 sp.), and *Triglochin* (12 spp.). *Lilaea* is sometimes recognized as a separate family Lilaeaceae[i], but it is usually included in the Jucaginaceae[ii,iii]. The tepals in *Triglochin* are occasionally interpreted as extensions from the stamens[ii], and the individual flowers in Juncaginaceae as synanthia[iv].

Economic uses The leaves and rhizomes of some species such as *Triglochin procerum* are used as food by native Australian aborigines. [CDC] OS

i Tomlinson, P. B. *Anatomy of the Monocotyledons. VII. Helobiae (Alismatidae).* Oxford, Clarendon Press (1982).
ii Dahlgren, R. M. T., Clifford, H. T., & Yeo, P. F. *The Families of the Monocotyledons.* Berlin, Springer-Verlag (1985).
iii Les, D. H., Cleland, M. A., & Waycott, M. Phylogenetic studies in Alismatidae, II: Evolution of marine angiosperms (seagrasses) and hydrophily. *Syst. Bot.* 22: 443–463 (1997).
iv Burger, W. C. The Piperales and the monocots. Alternative hypotheses for the origin of the monocotyledon flower. *Bot. Rev.* 43: 345–393 (1977).

LANARIACEAE

A monotypic family of perennial herbs endemic to the fynbos in the Cape Province of South Africa. The species is densely covered with branched, woolly hairs and has spirally arranged or distichous, linear, basally sheathing leaves. The inflorescence is a dense, corymbose panicle with actinomorphic, bisexual flowers, which have 3+3 equal, basally fused tepals; 3+3 stamens fused with the tepal tube; and an inferior ovary with septal nectaries. The ovary is 3-carpellate and 3-locular, with 2 anatropous ovules in each locule. The fruit is a single seed capsule, with black seeds containing a slightly curved embryo in an endosperm, which initially comprises starch, but later proteins and oil.

The Lanariaceae is occasionally included in the Tecophilaceae (q.v.)[i] but is now generally placed in a family of its own[ii]. The Lanariaceae is sister to Hypoxidaceae (q.v.)[iii,iv]. OS

Genera 1 Species 1
Economic uses None

i Dahlgren, R. M. T., Clifford, H. T., & Yeo, P. F. *The Families of the Monocotyledons.* Berlin, Springer-Verlag (1985).
ii Rudall, P. J. Lanariaceae. Pp. 340–342. In: Kubitzki, K. (ed.), *The Families and Genera of Vascular Plants. III. Flowering Plants. Monocotyledons: Lilianae (except Orchidaceae).* Berlin, Springer-Verlag (1998).
iii Fay, M. F. *et al.* Phylogenetic studies of Asparagales based on four plastid DNA regions. Pp. 360-371. In: Wilson, K. L. and Morrison, D. A. (eds), Monocots: systematics and evolution. Melbourne, CSIRO Publishing (2000).
iv Fay, M. F. *et al.* Phylogenetics of Liliales: Summarized evidence from combined analyses of five plastid and one mitochondrial loci. Pp. 559–565. In: Columbus, J. T. *et al.* (eds), *Monocots: Comparative Biology and Evolution.* Claremont, California, Rancho Santa Ana Botanic Garden (2006).

LAXMANNIACEAE

Laxmanniaceae is a small family of shrubby or treelike perennials of unresolved circumscription and affinity.

Distribution Mainly distributed in the southern hemisphere, with several genera confined to southwestern Australia but also known from New Zealand, New Caledonia, Madagascar, Southeast Asia, and South and North America.

Description Rhizomatous or tufted, shrubby or treelike perennials (*Cordyline*) up to 10 m high, sometimes with rhizomes that resemble stilt roots (e.g., *Laxmannia*). The leaves are spirally arranged or distichous, more or less linear, occasionally terete, with sheathing base and entire, scarious, or spiny margins. The inflorescence is an erect, simple, or compound spike or raceme, occasionally bracteate, umbellike, or cymose. The flowers are unisexual or bisexual and actinomorphic, with 2, 3-merous whorls of usually similar tepals, which may be united at the base and have fimbriate apices (e.g., *Thysanotus*). The stamens are 3+3, or only the inner 3, which may have hairy appendages, and all or only the inner are attached to the tepals. The ovary is inferior, consists of 3 carpels, and is 3-locular. There are several anatropous or campylotropous ovules on an axile placenta in each locule, septal nectaries, and a single style with 1 to 3 lobes. The fruit is a berry, a drupe, a loculicidal capsule, or by reduction a nutlet, with few to numerous colored, occasionally black, seeds, with a linear embryo in an endosperm of aleurone, hemicellulose, and fat.

Classification The circumscription of the Laxmanniaceae (sometimes erroneously named Lomandraceae[i]) has been controversial and is likely to change in the future[i]. One group of genera, the Lomandra-group (e.g., *Lomandra, Acanthocarpos, Chamaexeros*) is characterized by their distichous leaves and by embryological characters and has been placed in the Dasypogonaceae (q.v.)[ii] or Xanthorrhoeaceae (q.v.)[iii]. The remaining genera are grouped in a very heterogeneous Cordyline-group (e.g., *Cordyline, Arthropodium, Eustrephus*) and have been placed in different families, e.g., *Cordyline* in Asteliaceae (q.v.)[ii] and *Arthropodium* in Anthericaceae (q.v.)[ii]. *Eustrephus* has been placed in Luzuriagaceae (q.v.)[ii], in the Smilaceaeae (q.v.) by Cronquist (1981), or in the Geitonoplesiaceae[iii], which is a family not recognized here. *Sowerbaea* has been placed as a separate tribe in the Liliaceae[iv]. Members of the Lomandra-group and Cordyline-group, including *Eustrephus*, are monophyletic in several analyses[v,vi,vii] and sister group to Convallariaceae plus a narrowly circumscribed Asparagaceae(q.v.)[v], to Aphyllanthaceae (q.v.)[vi], or Anthericaceae[vii].

Economic uses Several species of e.g., *Cordyline* and *Lomandra* have been used for human consumption and in the production of fibers, while others are cultivated as ornamentals e.g., *Cordyline, Thysanotus,* and *Arthropodium*. OS

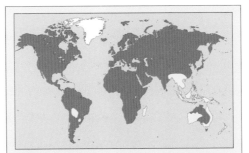

Genera 14 Species c. 180
Economic uses Locally used as a source of fibers and as ornamentals

i Conran, J. G. Lomandraceae. Pp. 354–356. In: Kubitzki, K. (ed.), *The Families and Genera of Vascular Plants. III. Flowering Plants. Monocotyledons: Lilianae (except Orchidaceae).* Berlin, Springer-Verlag (1998).
ii Dahlgren, R. M. T., Clifford, H. T., & Yeo, P. F. *The Families of the Monocotyledons.* Berlin, Springer-Verlag (1985).
iii Conran, J. G. The Geitonoplesiaceae Dahlgren ex Conran (Liliiflorae: Asparagales). A new family of monocotyledons. *Telopea* 6: 39 (1994).
iv Keighery, G. J. The Johnsonieae (Liliaceae): biology and classification. *Flora* 175: 103–108 (1984).
v Fay, M. F. *et al.* Phylogenetic studies of Asparagales based on four plastid DNA regions. Pp. 360–371. In: Wilson, K. L. and Morrison, D. A. (eds), *Monocots: Systematics and Evolution.* Melbourne, CSIRO Publishing (2000).
vi Pires, J. C. *et al.* Phylogeny, genome size, and chromosomal evolution of Asparagales. Pp. 287–304. In: Columbus, J. T. *et al.* (eds), *Monocots: Comparative Biology and Evolution.* Claremont, California, Rancho Santa Ana Botanic Garden (2006).
vii Davis, J. I. *et al.* A phylogeny of the monocots, as inferred from *rbc*L and *atp*A sequence variation, and a comparison of methods for calculating jackknife and bootstrap values. *Syst. Bot.* 29: 467–510 (2004).

LILIACEAE

LILY FAMILY

A family of geophytes with large, often conspicuously colored, typical monocot flowers, which was formerly interpreted in a much wider sense.

Distribution Mainly distributed in the temperate zone of the northern hemisphere, although a few genera reach the subtropics. They occur mostly in forests, grassland, and alpine meadows, reaching 4300 m, but rarely in deserts (*Calochortus*) or alpine tundra (*Lloydia*).

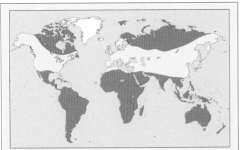

Genera 16 **Species** c. 640
Economic uses Many genera include important ornamentals; the bulbs are also used as drugs in medicine and for food

Description Perennial, unbranched, or rarely branched geophytes, with bulbs or rhizomes. The bulb is often tunicate and composed of a single or numerous fleshy scales. The leaves are alternate but may appear opposite or verticillate, usually sessile, and sometimes sheathing. The inflorescence is a raceme, occasionally umbel-like (e.g., *Gagea*, *Fritillaria*), but often the flowers are solitary. The flowers are often large, bisexual, actinomorphic, or slightly zygomorphic, with 2 whorls of 3, usually similar, free, and conspicuously colored tepals that are nectariferous. There are 3+3, rarely only 3, free stamens. The ovary is superior, consists of 3 carpels, and is usually 3-locular, with 3 free, or more or less fused, styles and numerous, anatropous ovules on axile placentas. The fruit is a loculicidal capsule, rarely a septicidal capsule (*Calochortus*) or a berry (e.g., *Clintonia*), usually with numerous, often flattened seeds in each locule, each with a minute embryo.

Classification The circumscription of Liliaceae has varied enormously. In the classical broad sense, Liliaceae includes between 200 and 300 genera, which here are distributed into different families. Even as presently defined, a small group of genera, including *Calochortus*, are sometimes recognized as a family of their own, Calochortaceae[i,ii], which, however, may or may not be monophyletic[ii,iii]. Other genera, including *Clintonia*, have been transferred to Uvulariaceae (= Colchicaceae)[ii]. The family is often divided into 2 subfamilies, Lilioideae with loculicidal, and Calochortoideae with septicidal, capsules. The sister group of the Liliaceae is most likely Smilacaceae (q.v.)[iv,v].

Economic uses The family includes some of the oldest known ornamentals of the western world. *Lilium candidum* (Madonna Lily, Bourbon Lily) is depicted on a 5,000-year-old tablet from Sumeria, and today many *Lilium* species (lily; 150 spp.) and hybrids are in cultivation in

LILIACEAE. **1** *Lilium martagon* (a) racemose inflorescence (x⅔); (b) loculicidal capsule (x⅔). **2** *L. canadense* longitudinal section of flower showing petaloid tepals, stamens, and superior ovary containing numerous, stacked ovules and crowned by a single style with a lobed stigma (x½).

gardens and as cut flowers. *Tulipa* (tulip; 150 spp.) was introduced to Vienna in 1554 via Turkey, and to Holland in 1571, where it created a veritable "tulipomania." Today, more than 2,600 cultivars are in cultivation and around 300 cultivars are grown on a commercial scale. *Fritillaria* (130 spp.) is also widely cultivated, and in China more 18 spp. are known to possess medicinal properties. Species of all 3 genera have edible bulbs.   OS

[i] Tamura, M. N. Calochortaceae. Pp. 164–172. In: Kubitzki, K. (ed.), *The Families and Genera of Vascular Plants. III. Flowering Plants. Monocotyledons: Lilianae*

(except Orchidaceae). Berlin, Springer-Verlag (1998).
[ii] Rudall, P. A., Stobart, K. L., Hong, W.-P., Conran, J. G., Furness, C. A., Kite, G. C., & Chase, M. W. Consider the lilies: Systematics of Liliales. Pp. 347–359. In: Wilson, K. L. & Morrison, D. A. (eds), *Monocots. Systematics and Evolution*. Melbourne, CSIRO Publishing (2000).
[iii] Fay, M. F. *et al.* Phylogenetics of Liliales: Summarized evidence from combined analyses of five plastid and one mitochondrial loci. Pp. 559–565. In: Columbus, J. T. *et al.* (eds), *Monocots: Comparative Biology and Evolution*. Claremont, California, Rancho Santa Ana Botanic Garden (2006).
[iv] Davis, J. I. *et al.* A phylogeny of the monocots, as inferred from *rbc*L and *atp*A sequence variation, and a comparison of methods for calculating jackknife and bootstrap values. *Syst. Bot.* 29: 467–510 (2004).
[v] Chase, M. W. *et al.* Multi-gene analysis of monocot relationships: A summary. Pp. 63–75. In: Columbus, J. T. *et al.* (eds), *Monocots: Comparative Biology and Evolution*. Claremont, California, Rancho Santa Ana Botanic Garden (2006).

LIMNOCHARITACEAE
WATER POPPY FAMILY

A small family of perennial, aquatic, laticiferous herbs with ephemeral petals.

Distribution Pantropical, preferably in stag- nant water. The monotypic genus *Butomopsis* (*Tenagocharis*) is found in tropical Africa, Southeast Asia, and northern Australia. *Hydrocleys* and *Limnocharis* are found in the neotropics. *L. flava* has become naturalized in India and Southeast Asia.

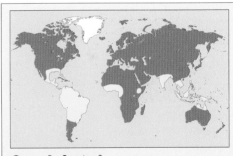

Genera 3 **Species** 8
Economic uses Several species are pernicious weeds

Description Perennial, emergent, submersed, or floating herbs, with erect, unbranched stems. The leaves are basal and pseudodistichous, ovate to cordate or lanceolate, with distinct, curved, parallel nerves, intervaginal scales, and petioles. All species have secretory ducts containing latex. The inflorescence is scapose, usually umbel-like, with bracts subtending each flower. The flowers are borne on long pedicels and are showy, regular, and bisexual. The 3 sepals are green and persistent in fruit. The 3 petals are white or yellow, usually delicate and ephemeral. *Butomopsis* has relatively small white petals, and the New World genera have large, yellow petals. Stamens are 3 to many, and occasionally staminodes occur. The pollen usually has 3 pores or none. The carpels are superior, 3 to 20, free or basally fused, in 1 or, rarely, 2 whorls. The ovules are numerous and scattered over the inner surface of the carpel wall. The fruit is a follicle with numerous seeds with a curved embryo and without endosperm.

Classification Species are notoriously difficult to identify due to great phenotypic plasticity. The relationships between Alismataceae (q.v.) and Limnocharitaceae are not clear. Limnocharitaceae was with some hesitation recognized as distinct from Alismataceae by Dahlgren *et al*[i], but merged by others[ii,iii]. Recent analyses[iii] show Limnocharitaceae as being monophyletic and arising from a paraphyletic Alismataceae. However, this hypothesis is poorly supported. The Limoncharitaceae differs from the Alismataceae by having numerous ovules per carpel, laminar placentation, and dehiscents fruits.

Economic uses *Limnocharis flava* is cultivated for food in India and Southeast Asia, the leaves being eaten as an alternative to spinach or endive, as well as providing fodder for pigs. Both *Hydrocleys* and *Limnocharis* are gaining added importance as ornamentals and aquaria plants. *Hydrocleys nymphoides* (Water Poppy) has large, decorative, shining yellow blossoms with a reddish brown center. It has been grown in greenhouses in Europe since 1830.   [CDC] OS

i Dahlgren, R. M. T., Clifford, H. T., & Yeo, P. F. *The Families of the Monocotyledons*. Berlin, Springer-Verlag (1985).
ii Judd, W. S. *et al*. *Plant Systematics. A Phylogenetic Approach*. 2nd edn. Sunderland, Massachusetts, Sinauer Associates (2002).
iii Les, D. H., Cleland, M. A., & Waycott, M. Phylogenetic studies in Alismatidae, II: Evolution of marine angiosperms (seagrasses) and hydrophily. *Syst. Bot*. 22: 443–463 (1997).

LOWIACEAE

A small monogeneric family of around 20 spp. of perennial, rhizomatous herbs, with distichous, entire, lanceolate leaves, sheathing at the base, without ligule, but with a distinct petiole. They are restricted to Southeast Asia, occurring from southern China, through Vietnam, to Malaysia and

Genera 1
Species c. 20
Economic uses None

Borneo. The inflorescence is branched but spikelike. The flowers are bisexual and zygomorphic, with 3 basally fused sepals and 3 petals of unequal size, 2 narrow and lateral, and 1 large, forming a labellum. There are 5 free stamens and the ovary is inferior, of 3 fused carpels with numerous anatropous ovules on axile placentas, and with a long, solid extension that persists in the fruit. There are 3 large stigmatic lobes. The fruit is a loculicidal capsule, and the seeds have a hairlike aril.

Molecular and morphological evidence suggest that Lowiaceae is monophyletic[i,ii] and sister group to Streliziaceae (q.v.)[ii,iii]. However, some molecular data point at another position[i].   OS

i Johansen, L. B. Phylogeny of *Orchidantha* (Lowiaceae) and the Zingiberales based on six DNA regions. *Syst. Bot*. 30: 106–117 (2005).
ii Kress, W. J., Prince, L. M., Hahn, W. J., & Zimmer, E. A. Unraveling the evolutionary radiation of the families of the Zingiberales using morphological and molecular evidence. *Syst. Biol*. 50: 926–944 (2001).
iii Chase, M. W. *et al*. Multi-gene analysis of monocot relationships: A summary. Pp. 63–75. In: Columbus, J. T. *et al*. (eds), *Monocots: Comparative Biology and Evolution*. Claremont, California, Rancho Santa Ana Botanic Garden (2006).

LUZURIAGACEAE

Distribution The family is restricted to the southern hemisphere, where it is found in southern Chile, Argentina, and the Falklands Islands, and in southeastern Australia, including Tasmania, and New Zealand. It grows in moist, shady, temperate forests.

Genera 2   Species 4–6
Economic uses Several species are grown as ornamentals

Description Perennial, rhizomatous, terrestrial or epiphytic, climbing shrublets, with distichous twisted, ovate to lanceolate leaves on more or less woody stems. The inflorescence is an axillary circinnus, or the flowers are solitary. The flowers are small, actinomorphic, with 2 whorls of 3, rarely 4 (*Drymophila*), more or less similar, free tepals that are nectariferous at the base. The ovary is superior, consists of 3 or 4 carpels and is usually 3- or 4-locular, with a simple capitate or trifid style. The ovules are hemianatropous on axile or subparietal placentas. The fruit is a white, blue, or orange berry, with between 1 and 10 seeds. The endosperm is copious and consists of aleurone and oil. It has been considered that Luzuriagaceae is closely related to Philesiaceae (q.v.) and Alstromeriaceae (q,v.)[i]. Recent molecular evidence convincingly points at a sister group relationship to Alstroemeriaceae (q.v.)[ii,iii,iv].   OS

i Dahlgren, R. M. T., Clifford, H. T., & Yeo, P. F. *The Families of the Monocotyledons*. Berlin, Springer-Verlag (1985).
ii Davis, J. I. *et al*. A phylogeny of the monocots, as inferred from *rbc*L and *atp*A sequence variation, and a comparison of methods for calculating jackknife and bootstrap values. *Syst. Bot*. 29: 467–510 (2004).

iii Fay, M. F. *et al*. Phylogenetics of Liliales: Summarized evidence from combined analyses of five plastid and one mitochondrial loci. Pp. 559–565. In: Columbus, J. T. *et al*. (eds), *Monocots: Comparative Biology and Evolution*. Claremont, California, Rancho Santa Ana Botanic Garden (2006).
iv Chase, M. W. *et al*. Multi-gene analysis of monocot relationships: A summary. Pp. 63–75. In: Columbus, J. T. *et al*. (eds), *Monocots: Comparative Biology and Evolution*. Claremont, California, Rancho Santa Ana Botanic Garden (2006).

MARANTACEAE
PRAYER-PLANT FAMILY

A small tropical family of herbaceous perennials, with distichous, simple, often asymmetric leaves and flowers.

Distribution The Marantaceae is pantropical but absent from Australia. With the exception of 2 genera (*Halopegia* and *Thalia*), the family is restricted to either Africa, Asia, or the neotropics. They occur mainly in disturbed areas in tropical rain forests, usually below 1,000 m.

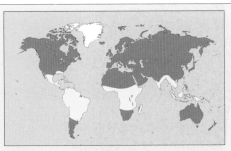

Genera 31 Species c. 550
Economic uses Only West Indian Arrowroot (*Maranta arundinacea*) is of economic importance, as a source of high-quality starch. Several spp. used as ornamentals; some cultivated locally

Description Terrestrial, perennial herbs, with underground rhizomes or tubers and distichous, sheathing leaves, with or without a petiole, and the lamina with a strong midrib and closely set, parallel, more or less sigmoid lateral veins, which are united near the margin. The junction between the blade and the petiole is pulvinate. The inflorescence is lateral or terminal, simple, or consisting of spikelike or capitate partial inflorescences, with distichously or spirally arranged bracts. The flowers are not very conspicuous, bisexual, and asymmetric, usually in pairs and mirror images of each other. The 3 free sepals are rarely connate at the base (*Megaphrynium*) and distinct. The 3 petals, stamens, and style are fused below into a corolla tube of variable length. The outer whorl of the androecium usually consists of between 1 and 2 petaloid staminodes but may be missing; the inner whorl has 1 fertile often petaloid and monothecal staminode, 1 hoodlike staminode, and 1 firm and fleshy staminode. The ovary is inferior, of 3 fused carpels, 3-locular, but usually 2 of the locules are aborted. The fertile locule contains 1 basal, anatropous ovule. There is a single style fused to the corolla tube at the base.

MARANTACEAE. I *Calathea villosa* (a) leaf, showing basal sheath, petiole, blade, and characteristic venation (x⅔); (b) a simple inflorescence with flowers subtended by green bracts (x⅔). 2 *C. concolor* (a) flower comprising free sepals, 3 irregular petals, I outer petaloid staminodes, and I petaloid stamen, with a fertile anther, and the hooded and fleshy staminode (x1); (b) the opened corolla tube, with fertile stamen and staminodes, I of which is hooded (x1). 3 *Stromanthe sanguinea* (a) upper rolled leaf and complex inflorescence (x⅔); (b) flower the corolla tube almost hidden (x2); (c) open corolla tube (x3). 4 *Maranta arundinacea* (a) shoot with leaves and inflorescence (x⅔); (b) tuber (x⅔); (c) flower (x1); (d) hooded petaloid staminodes, dorsal view of fertile stamen and style (x1x⅓); (e) cross section of 1-locular ovary (x3); (f) fruit with style (x2).

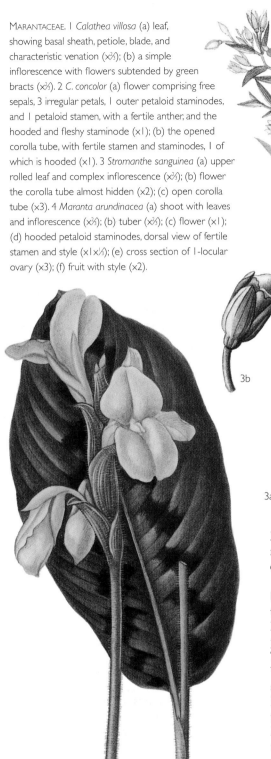

Until pollination, the style is enclosed in the hood-shaped staminode and bent backward under tension, but when touched it curls up. The fruit is usually a loculicidal capsule, more rarely berrylike or a caryopsis. The seeds have an aril and abundant starchy endosperm surrounding a horseshoe-shaped embryo.

Classification The family is monophyletic and according to both morphological and molecular data, the sister group to the Cannaceae[i,ii,iii] (q.v.) with which it shares such features as asymmetric flowers, reduction in the number of functional stamens, and free staminodes. The

Maranthaceae and Cannaceae belong to a large monophyletic group that also includes Zingiberaceae (q.v.) and Musaceae (q.v.)[i,iii].

Economic uses Economically, the most important crop plant is *Maranta arundinacea* (West Indian Arrowroot), which has rhizomes that produce a high-quality, easily digestible starch. The species is cultivated commercially in the West Indies and the tropical Americas. Several species are used in temperate zones as greenhouse ornamentals and houseplants mainly for their attractive foliage, e.g., *Calathea lancifolia*, *Maranta bicolor*, and *M. leuconeura*. The tough, durable leaves of *Ischnosiphon* are used to make basketry, and those of some species of *Calathea* for roofing, lining baskets, and food wraps. The inflorescences of some species of *Calathea* are cooked and eaten as a vegetable, and the tubers of *C. allouia* (Topee-tampoo) are eaten as potatoes in the West Indies.   SRC/OS

i Chase, M. W. *et al.* Multi-gene analysis of monocot relationships: A summary. Pp. 63–75. In: Columbus, J. T. *et al.* (eds), *Monocots: Comparative Biology and Evolution.* Claremont, California, Rancho Santa Ana Botanic Garden (2006).
ii Johansen, L. B. Phylogeny of *Orchidantha* (Lowiaceae) and the Zingiberales based on six DNA regions. *Syst. Bot.* 30: 106–117 (2005).
iii Kress, W. J., Prince, L. M., Hahn, W. J., & Zimmer, E. A. Unraveling the evolutionary radiation of the families of the Zingiberales using morphological and molecular evidence. *Syst. Biol.* 50: 926–944 (2001).

MAYACACEAE

A monotypic family of small perennial herbs, with 1 sp., *Mayaca baumii,* being native to western Africa, and the remaining 3 to 9 spp. native to the Americas, extending from southeastern USA to Paraguay, growing on open, sandy soil along rivers, often seasonally submerged. The leaves are simple, spirally arranged, sessile, usually with a bidentate apex. The flowers are solitary in leaf axils, actinomorphic, and 3-merous. The sepals and petals are free, green and white, respectively. The stamens are 3, and the ovary is superior, composed of 3 fused carpels, 1-locular, with few to many atropous, parietal ovules. The fruit is a capsule.

The Mayacaceae has been considered closely allied to the Commelinaceae (q.v.)[i,ii]. However,

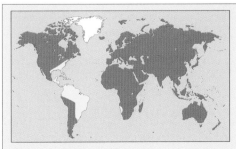

Genera 1   Species 4–10
Economic uses Aquarium plants

recent molecular evidence places the family as sister group to Hydatellaceae (q.v.) plus *Xyris*, a possible affinity suggested previously by Cronquist (1981) and Takhtajan (1987). Some species are used as aquarium plants, but difficult to cultivate. [CDC] OS

i Dahlgren, R. M. T., Clifford, H. T., & Yeo, P. F. *The Families of the Monocotyledons.* Berlin, Springer-Verlag (1985).
ii Michelangeli, F. A., Davis, J. I., & Stevenson, D. W. Phylogenetic relationships among Poaceae and related families as inferred from morphology, inversions in the plastid genome, and sequence data from the mitochondrial and plastid genomes. *American J. Bot.* 90: 93–106 (2003).

MELANTHIACEAE

DEATH CAMAS FAMILY

A small family of bulbous or rhizomatous, erect, unbranched herbs, previously included in the Liliaceae.

Distribution Mainly distributed in the temperate zone of the northern hemisphere, where it occurs in forests and in dry, rocky habitats.

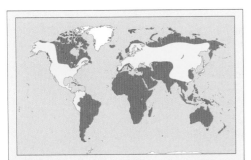

Genera 16 Species c. 170
Economic uses A few species, mainly *Veratrum* spp., are of medicinal use

Description Perennial herbs, usually with rhizomes or bulbs, rarely with corms. The stems are erect and unbranched, sometimes with fibrous leaf sheaths at the base. The leaves are cauline or in a basal rosette, spirally arranged, and occasionally sheathing, or otherwise gathered in pseudowhorls at the shoot apex. The inflorescence is frequently a raceme, rarely a panicle, a spike or umbel-like, or the flowers are solitary. The flowers are sessile or pedicellate and usually actinomorphic, unisexual or bisexual (plants monoecious, dioecious, or polygamous). The tepals are 3+3, similar, petaloid, free or basally connate, or 3 sepaloid and 3 petaloid, although the whorls occasionally have 4 to 10 tepals or 1 is missing. When present, the nectaries are either perigonial or septal. The stamens are 3+3, free, rarely up to 24 in 6 whorls, occasionally with elongated connectives. The ovary consists of 3 carpels, occasionally up to 10, which are more or less syncarpous and usually superior, rarely inferior or half-inferior. The styles are 3 or united into a single style. The ovule are 2 to numerous per locule, pleurotropus, or anatropous. The fruit is baccate or a septicidal, loculicidal, or ventri-

cidal capsule. The embryo is ovoid or globose, and the endosperm composed of aleurone, fatty acids, and starch.

Classification The circumscription of Melanthiaceae has varied considerably in the past, and several families recognized here as distinct have previously been included in it, e.g., Nartheciaceae (q.v.), Peterosaviaceae (q.v.), and Tofieldiaceae (q.v.).[i] On the other hand, several genera included here in Melanthiaceae may also be separated in their own family, e.g., Trilliaceae (q.v.),[ii,iii] but their exclusion makes Melanthiaceae paraphyletic[iv,v]. Melanthiaceae is sister to a large clade including e.g., Liliaceae (q.v.) and Colchicaceae (q.v.)[vi].

Economic uses Several *Veratrum* spp. have some medicinal importance, e.g., *V. viride*, which is poisonous to both livestock and humans but has been used as a treatment for hypertension. Some members of the family are grown as ornamentals. OS

i Dahlgren, R. M. T., Clifford, H. T., & Yeo, P. F. *The Families of the Monocotyledons.* Berlin, Springer-Verlag (1985).
ii Judd, W. S. *et al. Plant Systematics. A Phylogenetic Approach.* 2nd edn. Sunderland, Massachusetts, Sinauer Associates (2002).
iii Tamura, M. N. Trilliaceae. Pp. 444–452. In: Kubitzki, K. (ed.), *The Families and Genera of Vascular Plants. III. Flowering Plants. Monocotyledons: Lilianae (except Orchidaceae).* Berlin, Springer-Verlag (1998).
iv Fuse, S. & Tamura, M. N. A phylogentic analysis of the plastid *matK* gene with emphasis on Melanthiaceae *sensu lato. Plant Biol. (Stüttg.)* 2: 415–427 (2000).
v Tamura, M. N., Yamashita, J., Fuse, S., & Haraguchi, M. Molecular phylogeny of monocotyledons inferred from combined analysis of plastid *matK* and *rbcL* gene sequences. *J. Plant Res.* 117: 109–120 (2004).
vi Fay, M. F. *et al.* Phylogenetic studies of Asparagales based on four plastid DNA regions. Pp. 360–371. In: Wilson, K. L. & Morrison, D. A. (eds), *Monocots. Systematics and Evolution.* Melbourne, CSIRO Publishing (2000).

MUSACEAE

BANANA FAMILY

A small family of large to gigantic herbs, with pseudostems formed of overlapping sheaths; very large, simple, entire leaves; and huge pendant, bracteate inflorescences.

Distribution *Musa* occurs from the Himalayas, through southern China, to northern Australia and the Philippines; *Ensete* from Central Africa across Southeast Asia, to New Guinea and Java. Most species live in wet, tropical lowlands, with outliers in cooler, hilly country.

Description Large to gigantic, monocarpic or suckering herbs with pseudostems formed by the leaf sheaths of the enormous leaves. *Musa* has lacticifers sprouting from the rhizome, whereas *Ensete* lacks lacticifers, is unbranched, and monocarpic. The leaves are spirally arranged, very large, entire, and sheathing with a distinct petiole; they have a thick midrib with

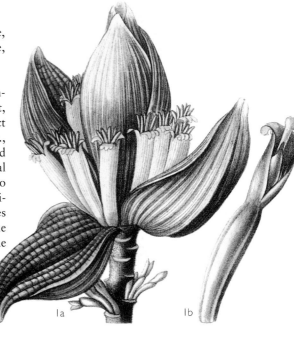

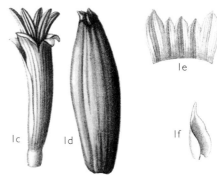

MUSACEAE. 1 *Musa rubra* (a) inflorescence with functionally female flowers below and male above subtended by large bracts (x⅔); (b) female flower (x1); (c) male flower, both with 5 fused and 1 free tepal (x1); (d) fruit—a fleshy berry (x⅔); (e) apex of the 5 fused tepals (x1); (f) single free tepal (x1). 2 (OPPOSITE PAGE) *Ensete ventricosum* (a) large herb (up to 10 m high) with pseudostem formed by sheathing leaf-bases; and terminal inflorescence (x⅔) (b) bract subtending numerous flowers (x⅓); (c) male flower (x (x⅔)); (d) bisexual flower (x⅔); (e) cross section of ovary (x1); (f) seed with window cut out to show embryo (x1).

parallel, slightly sigmoid veins running from it and fusing near the margin, and are usually split along the lateral veins by wind. The inflorescences are derived from the growing points of the basal corms and are terminal with spathaceous bracts, each supporting a cincinnate cluster of flowers. The flowers are zygomorphic and usually unisexual, female in basal and male in terminal clusters on the same plant, rarely with bisexual flowers in the proximal part. There are 2 whorls of 3 petal-like perianth segments, 3 outer and 2 of the inner tepals are fused, the third inner tepal is free. The stamens are 5 plus a small staminode, rarely 6. The pollen is sticky and pollination is commonly by macroglossine bats, but pollination by tree shrews and sunbirds has been observed. The ovary is inferior, of 3 fused carpels, and 3-locular, with each locule containing many anatropous ovules on

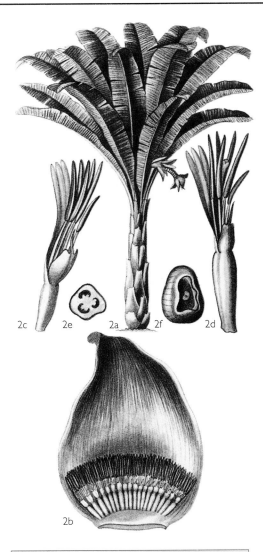

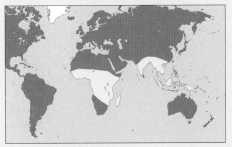

an axile placenta. The style is filiform and often expanded at the apex. The fruit is a fleshy berry containing numerous stony seeds, and the fruits finally form compact infructescences. The seed has copious, starchy endosperm and a straight embryo.

Classification. *Musa* has 30 to 40 spp. and *Ensete* about 6. The monotypic genus *Heliconia* is sometimes included in the Musaceae[i], and some investigations place the Heliconiaceae (q.v.) as sister group to Musaceae[ii], while others place Musaceae as sister to Strelitziaceae (q.v.) plus Lowiaceae (q.v.)[iii], but neither of these placements are supported by more comprehensive molecular analysis[iv].

Economic use The family provides a major food crop, the banana, which is comparable only to *Citrus* on the world's fruit trade. Most of the commercially grown bananas are triplod and originated in Southeast Asia from 2 wild species, either from *Musa acuminata* alone or from its hybrids with *M. balbisiana*. The plants are grown from suckers, and banana varieties are therefore clones. Being triploid, the fruits develop without forming seeds and become filled with the characteristic sweet, slightly acidic, aromatic, parenchymatous pulp, which is the edible part. Bananas are of enormous importance as food (locally staple) in the tropics and an important item in international trade. Other cultivars are grown locally but are of no commercial value. Fibers from *M. textilis* (Abaca or Manila Hemp) were previously widely used in the production of robes and coarse textiles mainly in the Philippines and the South Pacific Islands. *Ensete ventricosa* (Abyssinian Banana) is cultivated for its fiber and for food— the stem pulp and young shoots are cooked and eaten. Cultivars of both *Musa* and *Ensete* are also grown as ornamentals.   OS

i Lane, L. E. Genera and generic relationships in Musaceae. *Mitt. Bot. Staatsamml. München*, 13: 114–131. (1955).
ii Johansen, L. B. Phylogeny of *Orchidantha* (Lowiaceae) and the Zingiberales based on six DNA regions. *Syst. Bot.* 30: 106–117 (2005).
iii Kress, W. J., Prince, L. M., Hahn, W. J., & Zimmer, E. A. Unraveling the evolutionary radiation of the families of the Zingiberales using morphological and molecular evidence. *Syst. Biol.* 50: 926–944 (2001).
iv Chase, M. W. *et al.* Multi-gene analysis of monocot relationships: A summary. Pp. 63–75. In: Columbus, J. T. *et al.* (eds), *Monocots: Comparative Biology and Evolution*. Claremont, California, Rancho Santa Ana Botanic Garden (2006).

NAJADACEAE

The Najadaceae contains a single genus, *Najas*, consisting of annual or perennial aquatics, growing entirely submerged.

Distribution Subcosmopolitan, this family occurs throughout temperate and warm regions of the world in aquatic habitats.

Description The stem is slender and usually much-branched, with linear leaves that appear to be opposite but are crowded in leaf axils and may be described as pseudo-whorled or in bunches. The leaves are sessile, usually toothed at the margin with a sheathing base, and with intervaginal scales. The flowers are solitary or rarely in small clusters, unisexual (plants monoecious, dioecious, or polygamous) and reduced. The male flowers are enclosed in 1 to 2 membranous sheathing, occasionally purple, envelopes that completely surround the solitary stamen. Pollination takes place underwater, and the ellipsoid to globose pollen grains germinate prior to dehiscence, enhancing the likelihood of contact with the stigma. The female flowers are either naked or, rarely, surrounded by a purple envelope. The ovary consists of a single carpel ending

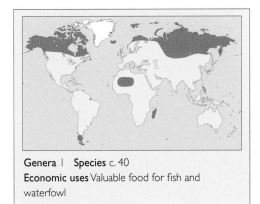

in a short style with 3 to 4 stigmas. There is a single basal, anatropous ovule. The fruit is a single-seeded nutlet. The seeds usually have a straight embryo and lack endosperm.

Classification About 40 spp. are recognized, but identification is difficult. Structural reductions have made the placement of Najadaceae difficult. Recent molecular investigations point to a close relationship with Hydrocharitaceae and recognition of the family consequently renders the former paraphyletic[i,ii]. The interpretation of the floral envelopes is controversial—the inner envelope is sometimes interpreted as a perianth[i].

Economic uses *Najas* is valuable food for waterfowl and fish. In warmer regions, *Najas* species may become aggressive weeds in ricefields, lakes, and waterways in tropical regions and the USA.   [CDC] OS

i Les, D. H., Cleland, M. A., & Waycott, M. Phylogenetic studies in Alismatidae, II: Evolution of marine angiosperms (seagrasses) and hydrophily. *Syst. Bot.* 22: 443–463 (1997).
ii Haynes, R. R., Holm-Nielsen, L. B., & Les, D. H. Najadaceae. Pp. 301–306. In: Kubitzki, K. (ed.), *The Families and Genera of Vascular Plants. IV. Flowering Plants. Monocotyledons: Alismatanae and Commelinanae (except Gramineae)*. Berlin, Springer-Verlag (1998).

NARTHECIACEAE
BOG ASPHODEL FAMILY

A small family of perennial herbs with basal, ensiform, or lanceolate leaves, and bractate inflorescences with erect, typical monocot flowers.

Distribution The family grows in temperate regions with a disjunct distribution on the eastern and western sides of North America, in northwestern Europe, and Southeast Asia, including Japan and the Himalayas.

Description Perennial herbs with creeping or erect rhizomes and basal, distichous or spirally arranged, linear, ensiform, or lanceolate leaves (*Aletris*). The inflorescence is a terminal, bracteate spike, raceme, or terminal, often many-flowered corymb with erect flowers. The flowers are bisexual, actinomorphic with 5, 3-merous whorls. The 3+3 tepals are petaloid, free, or connate at the base, persistent in fruit. The 3+3 stamens are inserted at the base of, or sometimes adnate to, the tepals for most of their length. The ovary is superior, rarely slightly or completely inferior, composed of 3 carpels, 3-locular, with numerous anatropous or campylotropous ovules. There is a single style or 3 styli. The fruit is a loculicidal capsule, and the seeds occasionally have long appendages at both ends.

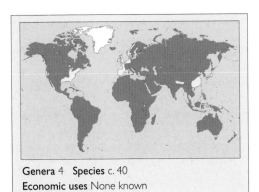

Genera 4 **Species** c. 40
Economic uses None known

Classification Nartheciaceae is in its present circumscription most likely monophyletic[i]. It has been segregated from a broadly defined, rather heterogenous Melanthiaceae[ii], following further segregation[iii] of Toefieldiaceae (q.v.) and Petrosaviaceae (q.v.), which is sometimes treated as a tribe or subfamily[iii] within Nartheciaceae. The relationships of Petrosaviaceae to other members of the monocotyledons is uncertain[iv,v], but the family is most likely the sister group of Burmanniaceae (q.v.) plus Dioscoreaceae (q.v.)[i].   OS

[i] Caddick, L. R., Rudall, P. J., Wilkin, P., Hedderson, T. A. J., & Chase, M. W. Phylogenetics of Dioscoreales based on combined analyses of morphological and molecular data. *Bot. J. Linn. Soc.* 138: 123–144 (2002).
[ii] Dahlgren, R. M. T., Clifford, H. T., & Yeo, P. F. *The Families of the Monocotyledons.* Berlin, Springer-Verlag (1985).
[iii] Tamura, M. N. Nartheciaceae. Pp. 381–392. In: Kubitzki, K. (ed.), *The Families and Genera of Vascular Plants. III. Flowering Plants. Monocotyledons: Lilianae (except Orchidaceae).* Berlin, Springer-Verlag (1998).
[iv] Chase, M. W. *et al.* Multi-gene analysis of monocot relationships: A summary. Pp. 63–75. In: Columbus, J. T. *et al.* (eds), *Monocots: Comparative Biology and Evolution.* Claremont, California, Rancho Santa Ana Botanic Garden (2006).
[v] Davis, J. I. *et al.* A phylogeny of the monocots, as inferred from *rbc*L and *atp*A sequence variation, and a comparison of methods for calculating jackknife and bootstrap values. *Syst. Bot.* 29: 467–510 (2004).

NOLINACEAE

Distribution Restricted to the dry regions of southern USA and Central America, especially in mountainous areas.

Description Perennial, treelike, or shortly caulescent plants, with unbranched or branched stems, covered by persistent leaves or scales, and linear leaves with prickly margins aggregated in terminal rosettes. The inflorescence is a many-flowered, bracteate panicle. The flowers are usually unisexual, rarely bisexual (plants

Genera 4 **Species** c. 50
Economic uses The leaves of some species are used in basketry

dioecious or polygamodioecious) and actinomorphic. The 6, free tepals are papery, and the 6, free stamens are in 2 whorls; staminodes are present in the female flowers. The ovary consists of 3 carpels and is superior with 1 or 3 locules and septal nectaries, with 2 ovules per locule or a total of 6. The fruit is dry and indehiscent, a capsule or a nut or breaking up in winged nutlets. The embryo is straight in a starchless endosperm.

The Nolinaceae is here accepted in a narrow sense[i,ii]. In such a circumscription it is monophyletic in several analyses but that makes Convallariaceae (q.v. for further discussions on relationships) paraphyletic.   OS

[i] Dahlgren, R. M. T., Clifford, H. T., & Yeo, P. F. *The Families of the Monocotyledons.* Berlin, Springer-Verlag (1985).
[ii] Bogler, D. Nolinaceae. Pp. 392–397. In: Kubitzki, K. (ed.), *The Families and Genera of Vascular Plants. III. Flowering Plants. Monocotyledons: Lilianae (except Orchidaceae).* Berlin, Springer-Verlag (1998).

ORCHIDACEAE. 1 *Bulbophyllum barbigerum* habit, showing pseudobulbs, single leaf, inflorescence, and flowers with strongly dissected labellum (x⅔). 2 *Dendrobium pulchellum* habit showing creeping stem rooting at the nodes, distichous, spotted leaves, and single flowers (x⅔). 3 *Sophronitis coccinea* habit (x⅔).

ORCHIDACEAE
ORCHID FAMILY

The largest family of flowering plants, characterized by their often showy, strongly zygomorphic flowers and numerous dustlike seeds.

Distribution The family has a cosmopolitan distribution, and orchids may be found under nearly all conditions—as understory plants in dark, tropical lowland forests; at the top of tall

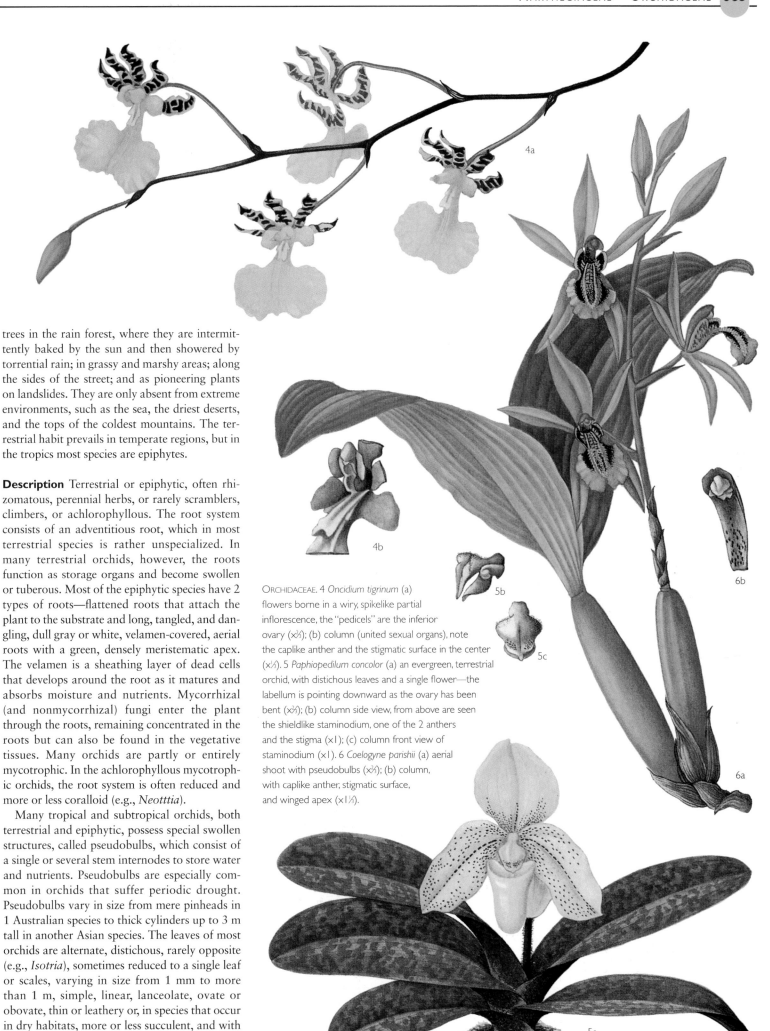

trees in the rain forest, where they are intermittently baked by the sun and then showered by torrential rain; in grassy and marshy areas; along the sides of the street; and as pioneering plants on landslides. They are only absent from extreme environments, such as the sea, the driest deserts, and the tops of the coldest mountains. The terrestrial habit prevails in temperate regions, but in the tropics most species are epiphytes.

Description Terrestrial or epiphytic, often rhizomatous, perennial herbs, or rarely scramblers, climbers, or achlorophyllous. The root system consists of an adventitious root, which in most terrestrial species is rather unspecialized. In many terrestrial orchids, however, the roots function as storage organs and become swollen or tuberous. Most of the epiphytic species have 2 types of roots—flattened roots that attach the plant to the substrate and long, tangled, and dangling, dull gray or white, velamen-covered, aerial roots with a green, densely meristematic apex. The velamen is a sheathing layer of dead cells that develops around the root as it matures and absorbs moisture and nutrients. Mycorrhizal (and nonmycorrhizal) fungi enter the plant through the roots, remaining concentrated in the roots but can also be found in the vegetative tissues. Many orchids are partly or entirely mycotrophic. In the achlorophyllous mycotrophic orchids, the root system is often reduced and more or less coralloid (e.g., *Neotttia*).

Many tropical and subtropical orchids, both terrestrial and epiphytic, possess special swollen structures, called pseudobulbs, which consist of a single or several stem internodes to store water and nutrients. Pseudobulbs are especially common in orchids that suffer periodic drought. Pseudobulbs vary in size from mere pinheads in 1 Australian species to thick cylinders up to 3 m tall in another Asian species. The leaves of most orchids are alternate, distichous, rarely opposite (e.g., *Isotria*), sometimes reduced to a single leaf or scales, varying in size from 1 mm to more than 1 m, simple, linear, lanceolate, ovate or obovate, thin or leathery or, in species that occur in dry habitats, more or less succulent, and with a sheathing base, which usually clasps the stem.

ORCHIDACEAE. 4 *Oncidium tigrinum* (a) flowers borne in a wiry, spikelike partial inflorescence, the "pedicels" are the inferior ovary (x⅔); (b) column (united sexual organs), note the caplike anther and the stigmatic surface in the center (x⅓). 5 *Paphiopedilum concolor* (a) an evergreen, terrestrial orchid, with distichous leaves and a single flower—the labellum is pointing downward as the ovary has been bent (x⅔); (b) column side view, from above are seen the shieldlike staminodium, one of the 2 anthers and the stigma (x1); (c) column front view of staminodium (x1). 6 *Coelogyne parishii* (a) aerial shoot with pseudobulbs (x⅔); (b) column, with caplike anther, stigmatic surface, and winged apex (x1½).

The inflorescence is either lateral or terminal, and frequently a bracteate raceme, but sometimes a spike, a panicle, or the flowers are solitary. Basically the flowers are bisexual, though unisexual flowers are sometimes found (plants moneocious or dioecious, e.g., *Catasetum*), and strongly zygomorphic. They have 2 whorls of 3 tepals, 3 sepals, and 3 petals. The sepals are usually rather similar, but the 2 laterals or the single dorsal may be elongated or bear a longitudinal crest. In some genera, the lateral sepals are partially or completely united (e.g., *Cypripedium*). However, the petals are usually clearly different with the 2 laterals being quite similar to the sepals but distinct from the dorsal, the lip, or labellum. The labellum is highly modified, variously lobed or dived, often with characteristic raised ridges, lamellae, or with hairs and glands, and often extended at the base to form a spur, which may contain nectar. The spur, which may be bilobed apically, can be up to 30 cm long in *Angraecum sesquipedale* from Madagascar. Although technically seen dorsally in the flower, the labellum is very often facing downward as the whole flower is turned or bent over, because the inferior ovary is twisted lengthwise or bent downward at the apex. As a result, the flower is turned 180° i.e., upside down, in most terrestrial orchids. The flowers are ephemeral to very long-lasting, often fragrant, the scent varying from the smell of rotting carrion through sickly sweet vanillalike odor to unquestionably very pleasing perfumes. Opposite the labellum in the floral whorl are the sexual organs, which are reduced and always united to form a single structure called the column (gynostemium). In its simplest, basic form, the column is surmounted by the anther with the receptive stigmatic surface just underneath, usually separated from each other by a flap of sterile tissue (the rostellum). The column is often winged subapically and with a foot that is often fused to the sepals. There are usually 1, rarely 2 or 3, stamens with 2-locular anthers and 3 stigmas, one of which is usually sterile and transformed into the rostellum. The anther many be parallel to the column or variously bent, lying as a cap at the end of the column. The huge variations of columnar structures provide the basis for much orchid classification. The pollen are often agglutinated into 2, 4, 6, 8 discrete masses (pollinia) that are sessile or on short, sticky stalks, grouped 1 or 2 together.

The simplest form of pollinating mechanism is the accidental removal, by a bee, of 1 or more, or all, of the pollinia, which attach themselves to the bee's head or thorax during its search for nectar. Bees, wasps, flies, fungus gnats, ants, beetles, hummingbirds, bats, and frogs have all been observed as pollinating agents for orchids. The attachment of the pollinia to the insect is helped by a variety of additional mechanisms such as a quick-setting glue on the pollinia stalks or an explosive device that can project the pollinia up to 60 cm from the flower (e.g., *Catasetum*), and by the development of features, both visual and olfactory, that attract the insect to the right position on the right flower. A well-known mechanism is called pseudocopulation (e.g., *Ophrys*).

The ovary is inferior, usually 3-carpellate, 1-locular with parietal placentas, or rarely 3-locular with axile placentas. There are

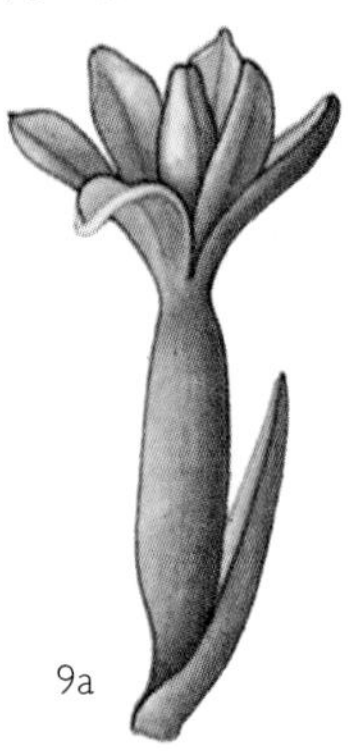

ORCHIDACEAE. (LEFT:) 7 *Neottia nidus-avis* (Bird's Nest Orchid), an achlorophyllous, mycotrophic plant, with scale leaves and coralloid roors (x⅔). 8 *Anoectochilus roxburghii* net-veined leaf and flowering spike, with flowers with a clearly twisted inferior ovary making the labellum point downward (x⅔). 9 *Apostasia nuda* (a) almost symmetrical flower (x4); (b) column with stigma and 2 anthers (x8). (RIGHT:) 10 *Disa hamatopetala* spike (x⅔). 11 *Ophrys bertolonii* habit, with the characteristic, egg-shaped tuberous roots (x⅔). (OPPOSITE PAGE, ABOVE:) 12 *Cypripedium calceolus* (Lady's Slipper) flower section, showing the staminodium, the stigmatic surface, and the inner hairy part of the labellum (x1). 13 *Cypripedium irapeanum* solitary flower and leaves (x⅔). (OPPOSITE PAGE, FAR RIGHT:) 14 *Orchis purpurea* (Lady Orchid) leaves and inflorescence (x⅔). 15 *Corybas aconitifolius* habit, with a single leaf, tuberous root, and a single flower composed of the labellum, maintain in a dorsal position, and the dorsal sepal, the remaining tepals being reduced, filiform (x⅔). 16 *Dactylorhiza fuchsii* (Common Spotted-Orchid) flower (a) side view, labellum with spur (x2⅔) and (b) front view; note the anther with 2 pollinia (x2).

12   13

numerous anatropous ovules in each locule. The fruit is a capsule, which opens by 6 lateral, longitudinal slits. The seeds are dustlike, without endosperm. A single fertilized flower spike may produce 1 million or more seeds. In most species, they contain a more or less undifferentiated embryos. After association with the appropriate fungus, the seeds develop an ephemeral tubercle, a protocorm, and only later is the first seedling leaf and roots developed. This symbiosis appears to persist in terrestrial, but not in epiphytic, orchids.

Classification As described here, the Orchidaceae is beyond doubt monophyletic[i] and sister group to a large clade consisting of e.g., Asparagaceae (q.v.), Alliaceae (q.v.), Hypoxidaceae (q.v.), Iridaceae (q.v.), etc. [ii,iii,iv]. The family is

Genera c. 800   **Species** 18,000–20,000
Economic uses The Orchidaceae is the basis of a huge floriculture industry, a source of flavoring essence (vanillin), and has local medicinal uses

currently divided into 5 subfamilies[v,vi,vii], 2 of which are occasionally treated as separate family, Apostasiaceae and Cypripediaceae[viii]. The relationship between some of the subfamilies is in need of further investigation[ix].

SUBFAM. APOSTASIOIDEAE Comprising only 2 genera, *Apostasia* and *Neuwiedia*, restricted to the Southeast Asia and northeastern Australia. They are easily recognized by their 2 to 3 anthers, the fingerlike staminodium, and by their lack of pollinia. Apostasioideae is the sister group of the remaining subfamilies[x,xi].

SUBFAM. CYPRIPEDIOIDEAE A small subfamily widespread in the temperate regions of the northern hemisphere, in Southeast Asia, and northern South America. It consists of 5 genera that all have 2 anthers and a characteristic slipper- or urn-shaped labellum, and nearly always lack pollinia. The column is stalked and ends in a shield-shaped, entire staminodium. It may be grouped with the Vanilloideae[xii] or with the remaining Orchidaceae, except Apostasioideae[xiii]. The genus *Cypripedium* (slipper orchid) belongs to this subfamily.

SUBFAM. VANILLOIDEAE A small subfamily of around 15 genera with a largely pantropical distribution but also found in eastern North America and eastern Australia. Its members have only 1 anther, and the pollen is not usually aggregated in pollinia but spread singly or in tetrads. It includes vines (e.g., *Vanilla*) and achlorophyllous, mycotrophic herbs (e.g., *Erythrorchis*).

SUBFAM. ORCHIDOIDEAE Easily recognizable by having a single anther, which is erect or bent backward with an acute apex, the pollen in pollinia, the convolute leaves, and fleshy roots or tubers. It includes around 200 mostly terrestrial genera, including many of those found throughout Europe; commonly encountered genera are *Aceras* (Man Orchid), *Anacamptis* (Pyramidal Orchid), *Coeloglossum* (Frog Orchid), *Dactylorhiza* (Marsh Orchids), *Gymnadenia* (Fragrant Orchid), *Habenaria* (Musk Orchid), *Himantoglossum* (Lizard Orchid), *Ophrys* (Spider Orchid), *Orchis*, and the mycotrophic *Neottia*.

SUBFAM. EPIDENDROIDEAE Containing around 600 genera, this is the largest orchid subfamily. Like the Orchidoideae, it has only a single, but beaked, anther and the pollen in pollinia. The leaves are either convolute or plicate, but the roots are not fleshy. It contains the majority of tropical epiphytic orchids but also some terrestrial species. Genera commonly encountered include *Cattleya*, *Cymbidium*, *Dendrobium*, *Epidendrum*, *Masdevallia*, *Oncidium*, *Phalaenopsis*, and *Vanda*.

Economic uses Excepting the flavoring essence vanillin, orchids have little direct economic importance other than as the basis for a vast floricultural industry. Vanilla beans are the cured, unripe fruits of *Vanilla planifolia*

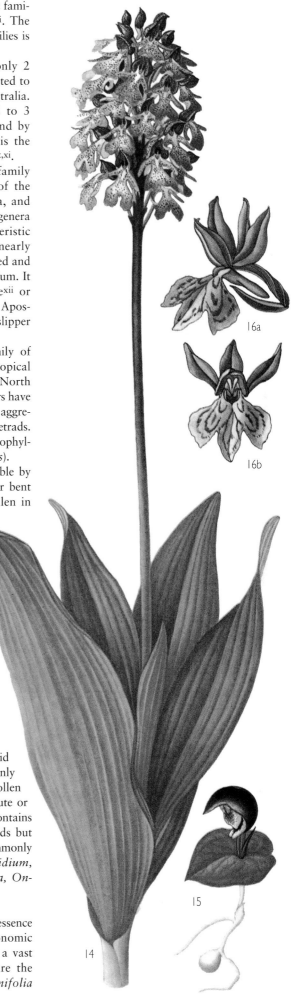

16a

16b

15

14

(Bourbon Vanilla) or *V. tahitensis* (Tahiti Vanilla). The main source of vanilla is the islands of Madagascar, the Comores, and Réunion, which together account for about 75 % of the world production. The characteristic aroma is developed by enzymatic actions during curing, and the cured capsules contain 2% vanillin. Due to their alkaloid content, several orchid species are used medicinally in China and India.

The modern orchid industry is based in the USA, but it is also a major export earner in countries such as Australia, Britain, Germany, Hawaii, Holland, Indonesia, Malaysia, and Singapore, and Thailand. Orchids are cultivated in the controlled environment of greenhouses in the temperate regions of the world; elsewhere they are grown outdoors in the same way as other plants. The last decade has witnessed a vast increase in the production of orchids as pot plants. Many orchids are now endangered, and avid collecting of rare species has brought many to the brink of extinction, as growers are willing to pay exorbitant amounts for such species. However, orchid trade is among the most controlled and regulated[xiv].   [PFH] OS

[i] Rasmussen, F. N. The development of orchid classification. Pp. 3–12. In: Pridgeon, A. M., Cribb, P. J., Chase, M. W., & Rasmussen, F. N. (eds), *Genera Orchidacearum*. Vol. 1. *General Introduction, Apostasiaceae, Cypripedioideae*. Oxford, Oxford University Press (1999).
[ii] Davis, J. I. *et al.* A phylogeny of the monocots, as inferred from *rbc*L and *atp*A sequence variation, and a comparison of methods for calculating jackknife and bootstrap values. *Syst. Bot.* 29: 467–510 (2004).
[iii] Chase, M. W. *et al.* Multi-gene analysis of monocot relationships: A summary. Pp. 63–75. In: Columbus, J. T. *et al.* (eds), *Monocots: Comparative Biology and Evolution*. Claremont, California, Rancho Santa Ana Botanic Garden (2006).
[iv] Pires, J. C. *et al.* Phylogeny, genome size, and chromosomal evolution of Asparagales. Pp. 287–304. In: Columbus, J. T. *et al.* (eds), *Monocots: Comparative Biology and Evolution*. Claremont, California, Rancho Santa Ana Botanic (2006).
[v] Pridgeon, A. M. *et al.* (eds), *Genera Orchidacearum*. Vol. 1. *General Introduction, Apostasiaceae, Cypripedioideae*. Oxford, Oxford University Press (1999).
[vi] Chase, M. W. Classification of Orchidaceae in the age of DNA data. *Curtis's Bot. Mag.* 22: 2–7 (2005).
[vii] Judd, W. S. *et al. Plant Systematics. A Phylogenetic Approach*. 2nd edn. Sunderland, Massachusetts, Sinauer Associates (2002).
[viii] Rasmussen, F. N. Orchids. Pp. 249–274. In: Dahlgren, R. M. T., Clifford, H. T., & Yeo, P. F. *The Families of the Monocotyledons*. Berlin, Springer-Verlag (1985).
[ix] Freudenstein, J. V., van den Berg, C., Goldman, D. H., Kores, P. J., Molvray, M., & Chase, M. W. An expanded plastid DNA phylogeny of Orchidaceae and analysis of jackknife branch support strategy. *American J. Bot.* 91: 149–157 (2004).
[x] Freudenstein, J. V. & Rasmussen, F. N. What does morphology tell us about orchid relationships? – A cladistic analysis. *American J. Bot.* 86: 225–248 (1999).
[xi] Cameron, K. M. *et al.* A phylogenetic analysis of the Orchidaceae: Evidence from *rbc*L nucleotide sequences. *American J. Bot.* 86: 208–224 (1999).
[xii] Freudenstein, J. V. & Chase, M. W. Analysis of mitochondrial *nad*1b-c intron sequences in Orchidaceae: Utility and coding of length-change characters. *Syst. Bot.* 26: 643–657 (2001).
[xiii] Kocyan, A., Qiu, Y.-L., Endress, P. K., & Conti, E. A phylogenetic analysis of Apostasioideae (Orchidaceae) based on ITS, *trn*L-*trn*F and *mat*K sequences. *Plant Syst. Evol.* 247: 203–213 (2004).
[xiv] Roberts, J. A., Beale, C. R., Bensler, J. C., McGough, H. N., & Zappl, D. C. *CITES Orchid Checklist I–III*. Richmond, Royal Botanic Gardens, Kew (1995–2002).

PANDANACEAE

SCREWPINE FAMILY

The Pandanaceae is a small tropical family of woody plants, with long, spiny, ensiform leaves and usually naked, unisexual flowers.

Distribution The Pandanaceae grow in thickets or as single individuals from sea level to altitudes of 3,000 m or more in the tropics and subtropics, from western Africa to the Pacific Ocean, including New Zealand.

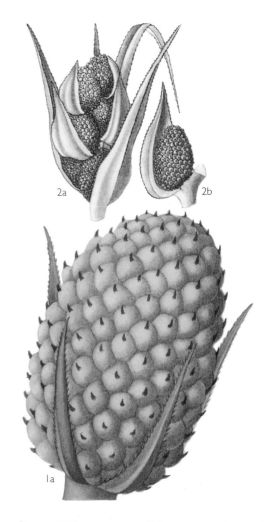

PANDANACEAE. 1 *Pandanus minor* (a) conelike fruit (x⅔); (b) 2 segments of the fruit (x1). 2 *P. pygmaeus* (a) female inflorescences (x⅔); (b) female inflorescence and subtending bract (x⅔); (c) cross section of female inflorescence (x1); (d) female flowers showing sessile stigmas and with part of wall cut away to show basal ovules (x4). 3 *P. houlletii* (a) male inflorescences and bracts (x⅔); (b) male flower comprising a whorl of stamens with fused filaments (x6); (c) dehiscing stamen (x8). 4 *P. kirkii* (a) habit showing aerial roots at base of trunk (x⅟₅₀); (b) fruit (x⅕). 5 *Freycinetia angustifolia* (a) flowering shoot with swordlike leaves (x⅔); (b) ovary and base of style (x16).

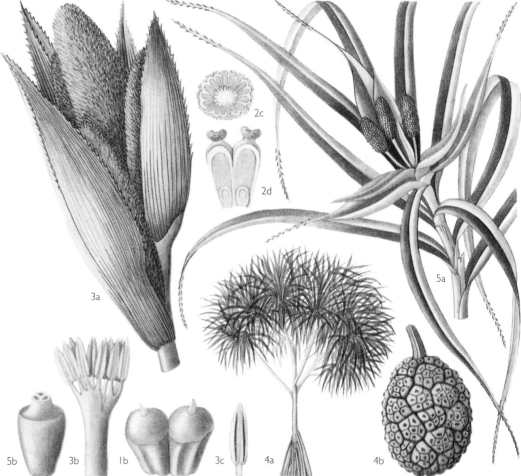

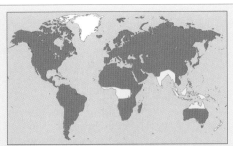

Genera 3 Species c. 800–900
Economic uses Species of *Pandanus* and
Freycinetia are used locally for thatching and
weaving, and some are edible

Description Tall, usually branched trees, shrubs, or climbers, with stems that are annulated by leaf scars and often supported by adventitious roots, which serve as prop or clasp roots. The long, narrow, sometimes stiff and ensiform leaves are tristichous or arranged in 4 rows, which may be spirally twisted due to the spirally growth of the stems and are often aggregated at the end of the stems. The leaves are usually armed on the margins and occasionally on the upper or both sides of the lamina. The inflorescences are unisexual, terminal or less frequently lateral, initially erect but later nodding, consisting of racemes or spikes, which are gathered into large spikes or umbel-like inflorescences, or into repeatedly branched panicles, often with green lower bracts and colored upper bracts. The flowers are naked or with a vestigial perianth, unisexual (plants dioecious), occasionally with remnants of the other sex. The flowers are distinct (*Sararanga*) or more or less indistinct (*Freycinetia*, *Pandanus*). The staminate flowers have a few to several hundred, sometimes basally fused, stamens. The female flowers are composed of 1 to several, superior carpels that are partially or completely fused, with 1 to many locules, depending on the degree of fusion between the carpels, and 1 to numerous anatropous ovules per locule, on basal or parietal placentas. The fruit is a multilocular drupe or a berry with small seeds, a minute straight embryo, and a starchy or oily endosperm.

Classification Traditionally, the family has been considered a close relation to Cyclanthaceae (q.v.), and this relationship is supported by both morphological evidence[i,ii] and molecular evidence[iii,iv,v].

Economic uses Species of the genus *Pandanus* (screw pines) and *Freycinetia* are used locally as sources of food—the fruit pulp, endosperm, and bracts of species being eaten. Other species produce spice or perfumes. The leaves of many species are used for thatching and weaving. Several species are ornamentals, notably *Freycinetia banksii* and *Pandanus veitchii*. OS

i Dahlgren, R. M. T., Clifford, H. T., & Yeo, P. F. *The Families of the Monocotyledons*. Berlin, Springer-Verlag (1985).

ii Stevenson, D. W. & Loconte, H. Cladistic analysis of monocot families. Pp. 543–578. In: Rudall, P. J., Cribb, P. J., Cutler, D. F., & Humphries, C. J. (eds). *Monocotyledons: Systematics and Evolution*. Vol. 2. Richmond, Royal Botanic Gardens, Kew (1995).
iii Caddick, L. R., Rudall, P. J., Wilkin, P., Hedderson, T. A. J., & Chase, M. W. Phylogenetics of Dioscoreales based on combined analyses of morphological and molecular data. *Bot. J. Linn. Soc.* 138: 123–144 (2002).
iv Davis, J. I. *et al.* A phylogeny of the monocots, as inferred from *rbc*L and *atp*A sequence variation, and a comparison of methods for calculating jackknife and bootstrap values. *Syst. Bot.* 29: 467–510 (2004).
v Chase, M. W. *et al.* Multi-gene analysis of monocot relationships: A summary. Pp. 63–75. In: Columbus, J. T. *et al.* (eds), *Monocots: Comparative Biology and Evolution*. Claremont, California, Rancho Santa Ana Botanic Garden (2006).

PETERMANNIACEAE

This monotypic family is restricted to the rain forests in the northern temperate part of eastern Australia. The single representative of the group is a perennial, rhizomatous, densely spiny climber (*Petermannia cirrhosa*), reaching up to 6 m high, with alternate petiolate, entire, more or less lanceolate leaves. The inflorescence is a terminal cyme, which is overtopped by the lateral axis below, lacking bracts, and partly modified into tendrils. The flowers are small, actinomorphic, and bisexual, with 3+3 tepals with basal nectaries, 3+3 stamens, and a 3-carpellate, inferior, 1-locular, rarely becoming 3-locular ovary, with parietal placentas and with many anatropous ovules in 2 rows. The fruit is a red berry with brown seeds, a linear embryo, and a starchless endosperm.

The Petermaniaceae has been included in Smilaceaeae (q.v.)[i] and has also been considered closely related to Phile-

Genera 1 Species 1
Economic uses None

siaceae (q.v.) and Rhipogonaceae (q.v.)[ii], and Smilaceaeae is sometimes enlarged to include both Petermaniaceae and Philesiaceae. More recently, it has been included in Colchicaceae (q.v.)[iii] (see APG II), but here it is recognized as a separate family. It is most likely the sister group of a clade consisting of Alsteromeriaceae (q.v.), Colchicaceae (q.v.), and Luzuriagaceae (q.v.)[iv]. OS

i Dahlgren, R. M. T., Clifford, H. T., & Yeo, P. F. *The Families of the Monocotyledons*. Berlin, Springer-Verlag (1985).
ii Conran, J. G. & Clifford, H. T. The taxonomic affinities of the genus *Ripogonum*. *Nordic J. Bot.* 6: 215–219 (1985).
iii Rudall, P. J. *et al.* Consider the Lilies. Systematics of Liliales. Pp. 347–359. In: Wilson, K. L. & Morrison, D. A. (eds), *Monocots. Systematics and Evolution*. Melbourne, CSIRO Publishing (2000).
iv Fay, M. F. *et al.* Phylogenetics of Liliales: Summarized evidence from combined analyses of five plastid and one mitochondrial loci. Pp. 559–565. In: Columbus, J. T. *et al.* (eds), *Monocots: Comparative Biology and Evolution*. Claremont, California, Rancho Santa Ana Botanic Garden (2006).

PETROSAVIACEAE

Petrosaviaceae is a small family, consisting of only 2 genera with a patchy distribution in Southeast Asia. Both occur in montane areas, and *Petrosavia* is known from Japan through southeastern China to the Celebes, whereas *Japanolirion* is restricted to Japan. The 3 spp. of *Petrosavia* are leafless, achlorophyllous saprophytes, while *Japanolirion* has linear, basal, and green leaves. The inflorescence is usually a bracteate raceme; the flowers are regular, erect, with 5, 3-merous whorls, a usually superior ovary, and with numerous ovules; and the fruit consists of follicles.

Genera 2 Species 4
Economic uses
None known

The family has been difficult to circumscribe[i,ii] and place in a taxonomic context (cf. APG II), and it is only recently that *Japanolirion* has been included based both on morphological and molecular evidence [iii,iv,v]. Both genera have previously been placed in a very broadly defined Melanthiaceae[i], but recently, Petrosaviaceae (q.v.), Toefieldiaceae (q.v.), Nartheciaceae, and (q.v.) Melianthaceae (q.v.) have all been recognized as separate families as is done here. The relationships between the Petrosaviaceae and the remaining monocotyledons vary considerably among different analyses[iii,iv,v,vi]. OS

i Dahlgren, R. M. T., Clifford, H. T., & Yeo, P. F. *The Families of the Monocotyledons*. Berlin, Springer-Verlag (1985).
ii Tamura, M. N. Nartheciaceae. Pp. 381–392. In: Kubitzki, K. (ed.), *The Families and Genera of Vascular Plants. III. Flowering Plants. Monocotyledons: Lilianae (except Orchidaceae)*. Berlin, Springer-Verlag (1998).
iii Cameron, K. M., Chase, M. W., & Rudall, P. J. Recircumscription of the monocotyledonous family Petrosaviaceae to include Japonolirion. *Brittonia* 55: 215–225 (2003).
iv Chase, M. W. *et al.* Multi-gene analysis of monocot relationships: A summary. Pp. 63–75. In: Columbus, J. T. *et al.* (eds), *Monocots: Comparative Biology and Evolution*. Claremont, California, Rancho Santa Ana Botanic Garden (2006).
v Davis, J. I. *et al.* A phylogeny of the monocots, as inferred from *rbc*L and *atp*A sequence variation, and a comparison of methods for calculating jackknife and bootstrap values. *Syst. Bot.* 29: 467–510 (2004).
vi Caddick, L. R., Rudall, P. J., Wilkin, P., Hedderson, T. A. J., & Chase, M. W. Phylogenetics of Dioscoreales based on combined analyses of morphological and molecular data. *Bot. J. Linn. Soc.* 138: 123–144 (2002).

PHILESIACEAE

This family comprises 2 monotypic genera of erect or twining shrubs that occur in the forests of southern Chile. The leaves are spirally arranged or distichous, sessile or shortly petiolate, and basally sheathing. The inflorescence is axillary or terminal, with 1 to 3, bisexual, actinomorphic, pendulous flowers. The tepals are nectariferous and in 2 whorls of 3—either the outer sepal-like and shorter than

PHILESIACEAE. I *Lapageria rosea* leafy, twining shoot with solitary axillary flowers, with spotted tepals (x⅔).

the inner (*Philesia*), or the outer petaloid, red or white, and of the same length as the inner (*Lapageria*). The inner tepals are occasionally spotted white. There are 3+3 stamens, and the ovary is superior, 3-carpellate, with an erect, filiform style, and 1-locular, with parietal placentas. The fruit is a red berry. The Philesiaceae is considered to be sister to the Rhipogonaceae (q.v.)[i,ii]. Both species are cultivated as ornamentals.   OS

Genera 2   **Species** 2
Economic uses None

[i] Fay, M. F. *et al.* Phylogenetics of Liliales: Summarized evidence from combined analyses of five plastid and one mitochondrial loci. Pp. 559–565. In: Columbus, J. T. *et al.* (eds), *Monocots: Comparative Biology and Evolution.* Claremont, California, Rancho Santa Ana Botanic Garden (2006).
[ii] Chase, M. W. *et al.* Multi-gene analysis of monocot relationships: A summary. Pp. 63–75. In: Columbus, J. T. *et al.* (eds), *Monocots: Comparative Biology and Evolution.* Claremont, California, Rancho Santa Ana Botanic Garden (2006).

PHILYDRACEAE

A small family of perennial herbs from tropical East Asia and Australia, with rhizomes or corms and basally crowded, distichous, linear, ensiform, or terete leaves. The inflorescence is a simple or compound spike, often with hairy axes, and the flowers are solitary, sessile, bisexual, and zygomorphic. The perianth is petaloid and consists of 2 whorls, each of 2 free or basally connate tepals. The 2 lateral inner and the median upper tepals are often smaller than the median lower tepals. There is only 1 stamen, with a flattened filament. The ovary is superior, of 3 carpels, and is usually 1-locular, with parietal placenta, rarely more or less 3-locular with axile placenta and numerous often atropous or anatropous ovules per locule. The style is simple. The fruit is a capsule containing numerous seeds, with a copious endosperm. Morphologically, Philydraceae is closely related to Pontederiaceae (q.v.)[i]. Recent molecular analyses place it either as sister group to Pontederiaceae and Haemodoraceae (q.v.)[ii] or to Haemodoraceae alone[iii].   OS

Genera 4   **Species** c. 5
Economic uses
Ornamentals

[i] Hamann, U. Philydraceae. Pp. 389–394. In: Kubitzki, K. (ed.), *The Families and Genera of Vascular Plants. IV. Flowering Plants. Monocotyledons: Alismatanae and Commelinanae (except Gramineae).* Berlin, Springer-Verlag (1998).
[ii] Chase, M. W. *et al.* Multi-gene analysis of monocot relationships: A summary. Pp. 63–75. In: Columbus, J. T. *et al.* (eds), *Monocots: Comparative Biology and Evolution.* Claremont, California, Rancho Santa Ana Botanic Garden (2006).
[iii] Davis, J. I. *et al.* A phylogeny of the monocots, as inferred from *rbc*L and *atp*A sequence variation, and a comparison of methods for calculating jackknife and bootstrap values. *Syst. Bot.* 29: 467–510 (2004).

POACEAE (GRAMINEAE)

GRASS FAMILY

A large and, economically, the most important family of flowering plants.

Distribution The family is truly cosmopolitan, ranging from the polar circles to the equator (1 of the 2 flowering plants on Antarctica is a grass, *Deschampsia antarctica*), and from mountain

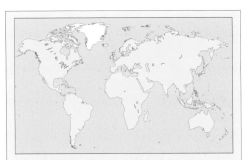

Genera c. 675   **Species** 10,000
Economic uses Cereal crops (barley, maize, millet, oats, rice, sorghum, and wheat), many species used for fodder, sugarcane, bamboos and canes used for building materials and basketry, and ornamental and lawn grasses

summits to the sea itself. It has been estimated that grasses are the main component in some 25% of Earth's vegetation cover. Native grassland has developed on most continents; prairie and plains in North America, pampas and llanos in South America, veldt in South Africa, and steppe in Eurasia. Savannah, another community type with an extensive grass cover, but interspersed with trees and bushes, has developed in Africa, Australia, India, and South America.

Description Mostly rhizomatous, stoloniferous, or caespitose perennial or annual herbs, although sometimes woody in the tropics. The roots are fibrous and often supplemented by adventitious roots from the lower nodes of the stem such that each new shoot has its own root system. It has been estimated that a single plant of *Festuca rubra*, which spreads by rhizomes, may be some 250 m in diameter and up to 400 years old, and that a large tussock of *F. ovina*, some 8 m across, could be 1,000 years old. The upright stems are cylindrical, usually hollow but sometimes solid, pithy, and mostly unbranched. A few species develop bulbs or corms. The leaves are usually distichous, emerge from the nodes, and are composed of 2 parts—sheath and blade. The node has an intercalary meristem that extrudes the stem from the sheath and remains active. Differential growth at this meristem enables the stem to bend upright again after being flattened (lodging) by rain or trampling. The sheath encircles the stem tightly and has overlapping margins, or the margins are occasionally fused to form a tube around the stem. The blade is long and narrow but may be broad, lanceolate to ovate. In a few genera, it is deciduous from the sheath, and occasionally it is narrowed at the base into a pseudopetiole. The base of the blade also has a meristematic zone, permitting the blade to continue growth despite the removal of its distal parts by grazing or cutting. In size, leaves can vary from the bladeless sheaths of the Australian *Spartochloa scirpoidea*, to the enormous blades (up to 5 m long) of the South American bamboo *Neurolepis nobilis*. There is usually a membranous or ciliate rim (ligule) at the junction between sheath and blade.

The inflorescence is a specialized leafless branch system, which usually surmounts the stem and may be an open or contracted panicle, sometimes to the extent that it becomes a spike or raceme. The inflorescence develops inside the uppermost leaf sheath, emerging only when almost mature. Extremes of size range from the spectacular 2 m plume of *Gynerium sagittatum*, to the solitary single-flowered spikelet of *Aciachne pulvinata*. The basic unit of the inflorescence is the spikelet, which consists of an often elongated axis (rhachilla) that may extend beyond the last floret and have 2-ranked scales. The 2 lowest scales (glumes) are empty, but the remainder (lemmas) each form part of a floret, whose floral parts are enclosed between the lemma on the outside and a delicate membranous scale (palea) on the inside. Glumes or lemmas are often produced into 1 or more long, stiff, straight, or bent bristles (awns). Infrequently, either 1 or both of the glumes are absent. The floral parts consist of 2 (rarely 0 or 3) tiny, membranaceous scales (lodicules), 3 (rarely 1 to 6 or more) stamens, and 2 styles and (rarely 1 or 3) feathery stigmas surmounting the ovary. The interpretation of the spikelet remains controversial, but it is often related to a branch system in which the glumes and lemmas represent modified leaf sheaths, the palea a prophyll, and the lodicules supposed to represent vestigial perianth members[i,ii]. The basic pattern is remarkably uniform throughout the family, but an extraordinary number of variations have been developed from this simple arrangement by changes in the size, shape, ornamentation, or sex of the parts, or by reduction in their number. Bisexual spikelets are the rule, although some of the florets are often unisexual, mostly male, or barren (plants monoecious or rarely dioecious). The pollen is viable for less than a day—the shortest-lived of all angiosperm pollens—and of a distinctive type, monoporate, operculate, with a very finely granular surface. Both apomixis and cleistogamy are found in the family and often associated with special spikelets hidden in the axils of the leaf sheath or even beneath the soil surface. In some mountain or arctic species, the spikelets are transformed into bulbils and have become a regular means of propagation. Vivipary (the germination of the seed while still attached to the parent) has been recorded in *Melocanna*.

The ovary is superior and has a single locule containing 1 ovule, usually adnate to the adaxial side of the carpel. The most peculiar features of the embryo is the transformation of the cotyledon into a haustorial organ (scutellum) attached to the starchy (rarely liquid) endosperm. The fruit is a caryopsis (fruit wall fused to the seed), although some bamboos have a berrylike fruit (*Melocanna*) and a few genera (notably *Sporobolus*) have a nut. At germination, a special organ (coleoptile) protects the shoot meristem while it penetrates the surface. Part of the inflorescence is frequently incorporated in the dispersal unit, ranging from an adherent palea to the complete inflorescence functioning as tumbleweed. Awns, wind-borne plumes of various kinds, hooks and barbs, and occasionally an adhesive, often play a part in seed dispersal.

Classification Both morphological and molecular data unanimously support the monophyly of the Poaceae[iii,iv,v]. The classification of the family presented below is based on a huge variety of morphological, biochemical, anatomical, and molecular characters. There are 12 recognized subfamilies, but none of these can be distinguished on morphological characters alone, and 5 genera are unplaced[iv]. There are 2 huge clades in this system, the BEP clade—consisting of Pooideae, Ehrhartoideae, and Bambusoideae—and the PACCAD clade—consisting of the Panicoideae, Centothecoideae, Aristidoideae, Aanthonioideae, Arundinoideae,

POACEAE. 1 *Arundinaria japonica* tip of shoot, showing leaves with pseudopetiole and spicate inflorescence arranged in a panicle (x⅔). 2 *Phleum pratense* spikelike inflorescence (x⅔). 3 *Stipa capillata* dispersal unit—lemma with enclosed caryopsis and long, twisted, feathery awn (x⅔). 4 *Aristida kerstingii* dispersal unit—lemma with enclosed caryopsis and apically 3-branched awn (x⅔). 5 *Tristachya decora* paniculate inflorescence, with the spikelets arranged in groups of 3 mimicking a single spikelet (x⅔). 6 *Poa annua* junction between blade and sheath, a long membranaceous ligule inserted where the blade and base meet (x6).

and Chlorioideae[v]. The most likely sister group of the Poaceae is the Joinvilleaceae (q.v.) plus the Ecdeiocoleaceae (q.v.)[vi,vii] or the Joinvilleaceae alone[v].

SUBFAM. ANOMOCHLOOIDEAE Perennial, rhizomatous, herbaceous plants with hollow or solid stems and broad, spirally arranged or distichous leaves, which have a pseudopetiole with characteristic swellings and ligulate or not. The inflorescence is spicate with a highly complex branching pattern. The ultimate structure on the branches is usually considered homologous to the spikelet found in the remaining grasses. The bracts within the spikelets are sometimes awned. The spikelet equivalent is bisexual, 1-flowered with 4 or 6 stamens, and an ovary with a single style with 1 or 3 stigmas and no lodicules. The subfamily includes 2 tribes and 2 genera (*Anomochloa,* 1 sp., and *Streptochaeta,* 3 spp.), and all species are found in the understory of tropical forests. The monophyly of this subfamily is in doubt[vi].

SUBFAM. PHAROIDEAE Perennial, rhizomatous, monoecious, herbaceous, with hollow or solid stems and broad leaves with twisted pseudopetiole, turning the leave upside down, and ligule. The inflorescence is paniculate, with unisexual, 1-flowered spikelets, and usually male and female spikelets are found in pairs on small branchlets. Female spikelets have 2 glumes; a tubular, awnless lemma; a well-developed palea; and no lodicules. The ovary has a single style, 3 stigmas, and lacks appendages. Male spikelets have 2 glumes, 0 or 3 lodicules, and 6 stamens. The single tribe in this subfamily includes 12 spp. of understory plants in tropical and warm temperate forests. In *Leptaspis,* the female lemma is inflated, cockleshell, or urnlike, and the panicle may be shed intact.

SUBFAM. PUELIOIDEAE Perennial, rhizomatous, herbaceous, with hollow stems and broad leaves with a pseudopetiole and ligule. The inflorescence is racemose or paniculate, and the spikelets have 2 glumes and several florets. Either the 1 to 3 basal florets in the spikelets are male, the next female-fertile, and the uppermost sterile or the 3 to 6 basal florets are male or sterile and uppermost female. The lemma has no awn, and the palea is well developed. There are 3 lodicules, 6 stamens, and an ovary with 2 to 3 styles and 2 to 3 stigmas, with or without an apical appen-dage. The 14 spp. in this subfamily are divided into 2 tribes and occur in the shaded understory of rain forests. The genus *Puelia* has a characteristic rachilla extension and *P. schumanniana* only develops a single leaf.

SUBFAM. BAMBUSOIDEAE Usually perennial, rhizomatous, herbaceous, or woody, with hollow or solid stems and broad leaves with a pseudopetiole and ligule. The inflorescence is spicate, racemose or paniculate, and the spikelets are bisexual or unisexual, with or without 1 to 2 or more glumes, single or numerous

POACEAE. 7 *Avena sativa* (a) paniculate inflorescence (x⅔); (b) fruit (x6). 8 *Lolium perenne* habit showing adventitious roots, leaves with sheathing bases, and spicate inflorescence (x⅔).

florets, a sometimes awned lemma, and a well-developed palea. The lodicules are usually 3, rarely 6 or many, and the stamens 2, 3, or 6, rarely many. The ovary has 2 to 3 styles, 2 to 3 stigmas, and sometimes an apical appendage. This subfamily is divided into 2 tribes, and most of the approximately 1,200 spp. are found in temperate and tropical forests, high montane grassland, along riverbanks, and in savannahs. Species of this clade are of great economical importance. Species of *Bambusa* can grow to a height of c. 40 m and have daily growth rate of up to almost 1 m.

SUBFAM. EHRHARTOIDEAE Perennial or annual, rhizomatous or stoloniferous, herbaceous or slightly woody, usually with narrow leaves with a ligule and, occasionally, pseudopetiole. The inflorescence is paniculate or racemose, and the spikelets are unisexual or bisexual, usually with 2 glumes, although glumes may be absent. The spikelets have 0 to 2 sterile florets and 1 female-fertile, the lemmas are sometimes awned, and the palea is well developed. The lodicules are 2 (often), the stamens 3 or 6 (rarely fewer), and the ovary lacks an appendage and has 2 free styles or a more or less fused style with 2 stigmas. This subfamily is divided into 3 tribes, and the 120 spp. mainly occur in forests, on hillsides, and in aquatic habitats. The small tribe includes both *Oryza* (rice) and *Zizania,* e.g., *Z. aquatica* (Canadian wild rice), a traditional food of many Native Americans.

SUBFAM. POOIDEAE Perennial or annual, sometimes rhizomatous or stoloniferous, herbaceous, usually with hollow culms, and broad or narrow leaves with ligule, but rarely with pseudopetiole. The inflorescence is spicate, racemose, or paniculate, and the spikelets are mostly bisexual, with 2 glumes, rarely without, or 1 glume absent. The spikelets have 1 to many female-fertile florets, the lemma with or without awn, and a usually well-developed palea. The lodicules are 2, rarely 0 or 3, the stamens 1 to 3, and the ovary has 2 styles and stigmas, and usually no appendage. The 3,300 spp. in this subfamily are divided into 13 tribes, and most are found in cold temperate and boreal areas, extending into the mountains in the tropics. This subfamily includes most of the grasses found in temperate areas, including the cereals barley (*Hordeum*), oat (*Avena*), rye (*Secale*), and wheat (*Triticum*).

SUBFAM. ARISTIOIDEAE Perennial or annual, caespitose, herbaceous, with hollow or solid culms, and narrow leaves with ligule, but without a pseudopetiole. The inflorescence is paniculate, with bisexual spikelets with 2 glumes, and the spikelets have a single female-fertile floret, lemma with 3 awns, and a short palea. There lodicules are usually 2, the stamens 1 to 3, and the ovary with 2 styles and stigmas, and usually no appendage. This subfamily includes a single tribe with 350 spp. that live in open habitats from temperate to tropical areas. All the species in this subfamily have very efficient dispersal mechanisms, and entanglement of fruits of *Aristida contorta* around the legs of sheep can immobilize them.

POACEAE 9 *Andropogon fastigiatus* inflorescences, consisting of single racemes (x⅔). 10 *Imperata cylindrica* inflorescence, the silky appearance of stems from hairs that arise from the glumes (x⅔). 11 *Cynodon dactylon* habit, showing creeping stem bearing adventitious roots and branching (tillering) at ground level. The inflorescence is composed of spikes (x⅔). 12 *Brachiaria brizantha* inflorescence with spikelets in spikes (x⅔).

SUBFAM. ARUNDINOIDEAE Usually perennial, rhizomatous, stoloniferous or caespitose, herbaceous, with hollow culms. The leaves are relatively broad to narrow, with ligule, and lacking a pseudopetiole. The inflorescence is generally paniculate, with bisexual spikelets with 2 glumes, and occasionally a sterile lemma. The spikelets have 1 to several female-fertile florets, lemma with a single (rarely 3) awns, and a well-developed palea. The lodicules are 2, the stamens 1 to 3, and the ovary has 2 styles and stigmas, and no appendage. The single tribe has c. 40 spp. found in temperate and tropical areas. *Molinia caerulea* (Purple Moor Grass), *Phragmitis australis* (Reed), and *Arundo donax* all belong to this subfamily.

SUBFAM. DANTHONIOIDEAE Usually perennial, rhizomatous, stoloniferous or caespitose, herbaceous or slightly woody, frequently with solid culms. The leaves are narrow, with a ligule and no pseudopetiole. The inflorescence is paniculate, rarely racemose or spicate, with bisexual or unisexual spikelets with 2 glumes and 1 to 6, rarely up to 20, female-fertile florets, the lemma with a single awn, and a well-developed palea. The lodicules are 2, the stamens 3, and the ovary has 2 styles and stigmas but no appendage. The single tribe includes 250 spp., which occur in grasslands, heathlands, and open woodlands.

SUBFAM. CENTOTHECOIDEAE Perennial or annual, rhizomatous or stoloniferous, herbaceous or reedlike, with hollow or solid culms. The leaves are rather broad or narrow with a ligule and, often, pseudopetiole. The inflorescence is racemose or paniculate, with bisexual or unisexual spikelets with 2 glumes and 2- to many-flowered spikelets. The lemma is occasionally awned, and the palea is well developed. The lodicules are 2 or sometimes absent, the stamens 1 to 2 to 3, and the ovary has 2 styles and stigmas, without appendage. There are about 45 spp. in the subfamily, divided into 2 tribes, most species occuring in temperate woodlands and tropical forests. The lemma bristles of *Centotheca* serve as an effective dispersal mechanism.

SUBFAM. PANICOIDEAE Perennial or annual, rhizomatous, stoloniferous or caespitose, mostly herbaceous, frequently with solid culms. The leaves are relatively broad or narrow, with or without a ligule, and occasionally with pseudopetiole. The inflorescence is a panicle, a raceme, or a spike with bisexual or unisexual (plants dioecious or monoecious) spikelets. The spikelets are paired on a long and short stalk, respectively, with 1 female-fertile floret, and have 2 glumes, a sterile lemma and a lemma that is sometimes awn, and a well-developed palea. The lodicules are 2, the stamens 3, and the ovary has 2 styles and stigmas, lacking an apical appendage. This subfamily includes 6 tribes, and the approximately 3,700 spp. are found in the temperate, subtropic, and tropic zones. Sugarcane (*Saccharum*), *Miscanthus*, and *Sorghum* belong to this tribe.

SUBFAM. CHLORIDOIDEAE Perennial or annual, rhizomatous, stoloniferous or caespitose, mostly herbaceous, with hollow or solid culms. The leaves are rather narrow with ligule, and without pseudopetiole. The inflorescence is paniculate (or paniculate with spicate branches), racemose, or spicate, usually with bisexual spikelets. The spikelets have 2 glumes, 1 to many female-fertile flowers, a lemma with 1 or more awns, and a well-developed palea. The lodicules are 0 or 2, the stamens 1 to 3, and the ovary has 2 styles and stigmas and no appendage. The 1,400 spp. in this subfamily are divided into 5 tribes, most species being found in dry habitats in the tropics and subtropics. Species of *Eragrostis* are cultivated for their edible seeds. The Buffalo Grass (*Buchloe dactyloides*) from the North American plains also belong here.

Economic uses The adoption of the grasses as a principal source of food was a milestone in human development, with many of the great civilization being founded on the cultivation of grass crops. In many areas of the world, however, there was a shift in human endeavour from foraging to farming around 8,000–10,000 years ago, both in the Old World and New World[viii]. Recent evidence suggests that Stone Age people in Israel collected the seeds of wild grasses about 23,000 years ago. These grasses included wild emmer wheat and barley—forerunners of the crops grown today.

In the Near East, in the area known as the "Fertile crescent," which includes parts of Iran, Iraq, Turkey, Syria, Lebanon, and Israel, wild species of *Triticum* and *Hordeum* yielded the cereals wheat and barley. *Hordeum vulgare* subsp. *vulgare* (barley) is one of the oldest crops to be domesticated in this region, from its wild progenitor *H. vulgare* subsp. *spontaneum*, which is widely distributed in the Eastern Mediterranean and reaching all the way to Afghanistan. Domestication of barley is likely to have taken place around 9,000 years ago, although barley grains have been recovered from Egyptian tombs dating back 15,000 years, although it is difficult to tell which subsp. they belong to. From the Fertile crescent, domestication spread both westward in the Mediterranean area and eastward through Tibet to China. *Triticum aestivum* (Common Wheat) is an allohexaploid, arising from hybridization between the diploid *Aegilops tauschii* and the tetraploid *Triticum turgidum* (Wild Emmer), which in turn is a hybrid between the 2 diploid species *Aegilops speltoides* and *Triticum urartu*[ix]. The domestication of common wheat took place later than barley, around 8,000 years ago.

Secale cereale (Rye) is most likely a domesticated form of *S. strictum* (Mountain Rye), a perennial species found from Morocco to Iran

and Iraq, which occurs as a weed in arable land. Cultivated rye is known from the Neolithic age in Austria, but cultivation seems to have become widespread in Europe only after the Bronze Age. *Avena sativa* (Oat) is a domesticated form of *A. sterilis* (Wild Oat), which is a perpetual invader of cereal fields and, like rye, became widespread in Europe after the Bronze Age. In the last decades, hybrids between *S. cereale* and *Triticum aestivum* have resulted in Triticale, which has resistance to several diseases and is now widely grown, especially in Europe.

The origins of *Oryza sativa* (Rice) have been debated for some time, but the plant is of such antiquity that the precise time and place of its first development will perhaps never be known. However, the earliest evidence for domestication of rice in Southeast Asia is from remains found in Thailand, dating back at least 4,000 years. Signs of early cultivation of rice have been found in the Yangtze valley dating back to about 8500 BCE. *Oryza glaberrima* (African Rice) has been cultivated for 3,500 years. Between 1500 and 800 BCE, African Rice was propagated from its original center, the Niger River delta, and extended to Senegal. However, the crop never developed far from its original region, and cultivation even declined in favor of the Asian species, possibly brought to the African continent by Arabs coming from the east coast between the 7th and 11th centuries. The cultivation of *Zea mays* (Maize, Corn) is thought to have started from 7500 to 12,000 BCE. Archaeological remains of the earliest maize cob, found in the Oaxaca Valley of Mexico, date back around 6,250 years. Many other species of grasses are cultivated. The most important are Foxtail Millet (*Setaria italica*) and Proso (*Panicum miliaceum*) in Asia and Sorghum (*Sorghum bicolor*) and Bulrush or Pearl Millet (*Pennisetum glaucum*) in Africa. This is supplemented by a number of rather local minor grains, including Finger Millet (*Eleusine coracana*), Fundi (*Digitaria exilis*), and Tef (*Eragrostis tef*).

Saccharum officinarum (Sugarcane) originated in New Guinea, where it has been known since about 6000 BCE. From about 1000 BCE, sugarcane spread with human migrations to Southeast Asia, India, and east into the Pacific. It has probably hybridized with wild sugarcanes of India and China to produce "thin" canes. It spread further westward to the Mediterranean between 600 and 1400 CE, and the crop was subsequently taken to Central and South America from the 1520s onward, and later to the West Indies. Probably only a single cultivar made the journey westward from India and was carried to the New World. It was the only sugarcane grown in the New World for more than 250 years, until it was replaced by the noble cane "Otaheite" ("Bourbon") at the end of the 18th century.

The second facet of humankind's dependence on the grasses springs from the domestication of animals, which was roughly contemporaneous with the advent of agriculture. Until recent times, livestock rearing was based upon the exploitation of natural grasslands, although the preservation of fodder as hay had been introduced by the

POACEAE. 13 *Bromus commutatus* (a) exploded view of a floret showing 2 scales (the awned lemma and palea with hairy margins) and flower with 3 stamens, ovary with appendage and 2 feathery styles (×4); (b) spikelet showing the 2 lowermost scales without awns (glumes) and several upper scales with awns (lemmas). 14 *Olyra ciliatifolia* tip of shoot showing broad leaves with pseudopetiole and a paniculate inflorescence (×⅔).

PONTEDERIACEAE. 1 *Pontederia cordata* (a) leaf and inflorescence with subtending lower spathe visible (x⅔); (b) 2-lipped flower, showing 3 short upper and 3 longer lower stamens (x3); (c) gynoecium (x4); (d) cross section of 1-locular ovary (x5); (e) vertical section of ovary with a single ovule (x5). 2 *Heteranthera limosa* (a) leafy shoot and solitary flowers (x⅔); (b) flower with most of the perianth removed showing 2 stamens with smaller anthers than the other (x3); (c) cross section of ovary (x3). 3 *Eichhornia paniculata* (a) base of plant with sheathing leaf-bases and swollen petioles (x⅔); (b) inflorescence, with both bracts visible, flowers slightly zygomorphic (x⅔); (c) flower opened out, glandular hairs are clearly visible (x1½); (d) gynoecium, style with glandular hairs, and entire stigma (x3); (e, f) upper, long (x4) and lower, short stamen, both have filaments with glandular hairs (x6).

Roman era. Sown pastures, based on rye grass, date from the 12th century in northern Italy and from the late 16th century in northern Europe.

Bamboos provide an excellent raw material for people in many parts of the world. On a local scale, bamboos are used for housing and furnishings, for tools, hunting weapons and utensils, in basketry, for fuel, and even for food. Young bamboo shoots, especially of *Phyllostachys*, are eaten as vegetables, either cooked or pickled. On an industrial scale, bamboos are used as timber for construction and as a source of pulp for the production of paper and cardboard.

An aromatic oil is distilled from the leaves of lemon grass (*Cymbopogon* spp.), imparting a cit-

ronella scent to soaps and other perfumery and it is also used as a spice. Among a host of minor uses are necklace beads (*Coix* involucres), brush bristles (*Sorghum* inflorescence branches), pipe bowls (*Zea* cobs), clarinet reeds (*Arundo donax* stems), and corn dollies or various garishly dyed inflorescences sold as house decorations. Lawns grasses (especially species of *Festuca* and *Cynodon*) have an honored place in horticulture, and several are used as ornamentals e.g., *Festuca* spp. (Fescue) and *Arundo donax* for their foliage, *Cortaderia selloana* (Pampas Grass) and *Miscanthus* spp. and hybrids for their impressive foliage and silvery inflorescences, *Briza* spp. (Doddering-dillies, Jiggle-joggies) for their characteristic nodding spikelets, a large variety of *Pennisetum* spp. for their showy inflorescences, and varieties of *Phalaris arundinacea* for the often variegated leaves.

The obnoxious properties of grasses lie mainly in their success as weeds of cultivation. Indeed, many species are serious weeds, including *Elytrigia repens* (Common Couch), *Imperata cylindrica*, *Avena fatua* (Wild Oat), and *Poa annua* (Annual Meadowgrass). [WOC] OS

i Tran, V. N. Sur la valeur morphologique des lemmes de Graminées. *Bull. Mus. Nat. Hist. Natur.* sér. 3, 128: 33–57 (1973).
ii Soreng, R. J. & Davis, J. I. Phylogenetic and character evolution in the grass family (Poaceae); simultaneous analysis of morphological and chloroplast restriction site character sets. *Bot. Rev.* 64: 1–85 (1998).
iii Clayton, W. D. & Renvoize, S. A. *Genera Graminum. Grasses of the World. Kew Bull.* Additional Series XIII (1986).
iv Watson, I. & Dallwitz, M. J. *The Grass Genera of the World.* Wallingford, UK, CAB International (1992).
v GPWG (The Grass Phylogeny Working Group). A phylogeny of the grass family (Poaceae) as inferred from eight character sets. Pp. 3–7. In: Jacobs, S. W. L. & Everett, J. (eds), *Grasses: Systematics and Evolution.* Melbourne, CSIRO Publishing (1998).
vi GPWG. Phylogeny and subfamilial classification of the Poaceae. *Ann. Missouri Bot. Gard.* 88: 373–457 (2001).
vii Chase, M. W. *et al.* Multi-gene analysis of monocot relationships: A summary. Pp. 63–75. In: Columbus, J. T. *et al.* (eds), *Monocots: Comparative Biology and Evolution.* Claremont, California, Rancho Santa Ana Botanic Garden (2006).
viii Davis, J. I. *et al.* A phylogeny of the monocots, as inferred from *rbc*L and *atp*A sequence variation, and a comparison of methods for calculating jackknife and bootstrap values. *Syst. Bot.* 29: 467–510 (2004).
ix Damania, A. B., Valkoun, W. G., & Qualset, C. O. (eds), *The Origins of Agriculture and Crop Domestication.* Aleppo, Syria, ICARDA, (1998).
x Cox, T. S. Deepening the Wheat Gene Pool. *J. Crop Production* 1: 11–25 (1998).

PONTEDERIACEAE
WATER HYACINTH AND PICKEREL WEED

A relatively small family of freshwater aquatics or marsh plants.

Distribution The family is widespread in tropical and subtropical regions, with a few temperate outliers. It occurs, often as dominating component, in all types of freshwater habitats,

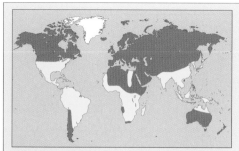

Genera 9   Species 33
Economic uses Many species are cultivated as aquatic ornamentals; some have leaves that are used as vegetables or fodder; some are weeds

marshlands, low-lying pastures, and ricefields. Some species of *Monochoria* may occur in salt or brackish water.

Description Annual and perennial, submerged, free-floating, or emergent species with indeterminate vegetative, leafy stems and determinate, leafless flowering stems. Leaves are distichous, rarely apparently whorled (*Hydrothrix*), with a sheathing base, with or without stipules that may envelop the stems or appear as a ligule, either linear or differentiated into a petiole (occasionally inflated, e.g., *Eichhornia*), and simple blade. The highly plastic blade is orbicular or sagittate, emergent, floating, or sub-merged. The inflorescence is a terminal spike, umbel-like panicle, or the flowers are solitary, subtended by 2 bracts (spathes), the lower often leaflike. The flowers are typically short-lived, lasting only a day. The flowers are bisexual, more or less zygomorphic, 2-lipped in species with large flowers (*Pontederia*), usually with 6, rarely 3 or 4, free or basally connate, showy, petaloid tepals. The stamens are 3 or 6, often dimorphic, rarely 1 or 4. The ovary is superior, of 3 fused carpels, often 3-locular, with axile placentas that may appear parietal, and an entire or 2-lobed stigmas. The ovules are anatropous, usually numerous, but occasionally only one. The fruit is a capsule or a nutlet. The seeds have a straight embryo and starchy endosperm.

The showy flowers are insect-pollinated, although visitors may be rare and seed set low. In some species (*Heteranthera multiflora*), different individuals produce left- and right-handed flowers (1 type of flower may be blue and have 1 anther, the other being yellow and having 5 anthers). Trimorphic heterostyly is known in *Eichhornia*, *Pontederia*, and *Reussia*. Some species have cleistogamous flowers developing on underwater stems (e.g., *Hydrothrix*, *Zosterella*). After flowering, species of *Eichhornia*, *Pontederia*, *Reussia*, and *Monochoria* exhibit a downward curvature of the floral axis (hydrocarpy), and their fruits mature below the water surface.

Classification Monophyly of the Pontederiaceae is supported by both morphology[i] and molecular data[ii,iii]. The family is sister group to either Haemodoraceae (q.v.)[iv] or Commelinaceae (q.v.)[v].

Economic uses *Eichhornia crassipes* (Water Hyacinth) and *Pontederia cordata* (Pickerel Weed) are widely grown as aquatic ornamentals, and a few genera include aquarium plants (e.g., *Heteranthera* and *Zosterella*). *Eichhornia crassipes* is one of the world's most noxious aquatic weeds and other species e.g., *Heteranthera limosa* and *H. reniformis* (USA), *Reussia rotundifolia* and *Pontederia cordata* (South America), *Eichhornia natans* (Africa), and *Monochoria vaginalis* (Asia) are serious weeds in ricefields. The leaves of many genera are used as fodder or as a green vegetable.   [SCHB] OS

[i] Eckenwalder, J. E. & Barrett, S. C. H. Phylogenetic systematics of Pontederiaceae. *Syst. Bot.* 11: 373–391 (1986).
[ii] Barrett, S. C. H. & Graham, S. W. Adaptive radiation in the aquatic plant family Pontederiaceae: insights from phylogenetic analysis. Pp. 225–258. In: Givnish, T. J. & Sytsma, K. J. (eds), *Molecular Evolution and Adaptive Radiation*. Cambridge, Cambridge University Press (1997).
[iii] Graham, S. W., Kron, J. R., Morton, B. R., Eckenwalder, J. E., & Barrett, S. C. H. Phylogenetic congruence and discordance among one morphological and three molecular data sets from Pontederiaceae. *Syst. Biol.* 47: 545–567 (1998).
[iv] Chase, M. W. *et al.* Multi-gene analysis of monocot relationships: A summary. Pp. 63–75. In: Columbus, J. T. *et al.* (eds), *Monocots: Comparative Biology and Evolution*. Claremont, California, Rancho Santa Ana Botanic Garden (2006).
[v] Davis, J. I. *et al.* A phylogeny of the monocots, as inferred from *rbc*L and *atp*A sequence variation, and a comparison of methods for calculating jackknife and bootstrap values. *Syst. Bot.* 29: 467–510 (2004).

POSIDONIACEAE

A family of submerged perennial "seagrasses" with a single genus, *Posidonia*.

Distribution A remarkable disjunct distribution with 1 species in the Mediterranean, where it often forms extensive sea meadows, and the other along the southern Australian coast.

Description The stem is a creeping, monopodial rhizome, which is covered with the fibrous remains of old leaf sheaths. The leaves are linear to filiform, with distinct persistent leaf sheaths that surround the stem and with intravaginal scales. The leaf blade and sheath have many scattered dark spots due to buildup of tannin. The inflorescence is racemose and spike-like. The flowers are actinomorphic, usually

Genera 1   Species 9
Economic uses Food source for marine organisms

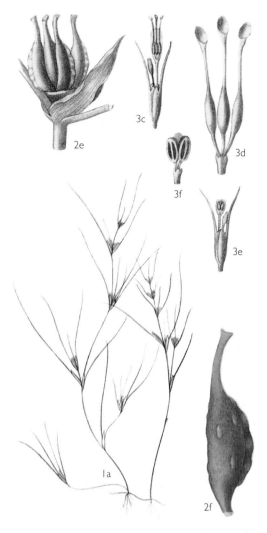

bisexual, and lacking tepals. The 3 stamens are sessile, the anthers with a broad shieldlike connective part that ends in a prominent tip overtopping the locules. The pollen is filiform, swollen at center, and pollination occurs underwater. The ovary is superior, with a single, naked carpel, containing a single, straight, pendulous ovule. The stigma is sessile and irregularly lobed. The fruit is best described as a buoyant follicle with a fleshy pericarp, and the seed is filled by the embryo. The tangled fibrous remains of old leaf sheaths are aggregated by the sea currents into so-called marine, or *Posidonia*, balls or aegagropyles.

Classification The Mediterranean *Posidiona oceanica* differs from the Australian species in that the leaf sheaths are incompletely wrapped around the stem, the ligule is very short, the bracts of the inflorescence are larger than its sheaths, and the seed coat lacks a wing. The Australian species fall in 2 groups, the *australis* and *ostenfeldii* complexes, but the exact number of species is debatable[i,ii]. Posidoniaceae is well-defined and closely related to either Ruppiacae (q.v.) or Cymodoceaceae (q.v.)[iii].

Economic uses *Posidonia* meadows are the largest littoral source of primary production and biomass and house many species of algae, fish, and marine invertebrates for which they provide a food source. *Posidonia australis* is the

POTAMOGETONACEAE (LEFT & RIGHT). I *Althenia filiformis* (a) habit, showing narrow leaves with sheaths joined to their bases and rhizomatous, erect stems (x⅔); (b) female flowers, showing pelate stigmas (x4). 2 *Zannichellia palustris* (a) habit, showing free leaf sheaths and floral complexes (x⅔); (b) male flower of 1 stamen and female flower of 4 carpels (x6); (c) carpel with funnel-shaped stigma (x16); (d) vertical section of carpel (x16); (e) fruits (x4); (f) fruit (x8). 3 *Lepilaena preissii* (a) shoot with female flowers (x⅔); (b) shoot with male flowers (x⅔); (c) female inflorescence (x2); (d) gynoecium of 3 free carpels with funnel-shaped stigmas (x6); (e) male inflorescence (x2); (f) stamen (x6).

source of posidonia fibers, which are used for textiles, paper, and insulation.   [CDC] OS

i Kuo, J. & Cambridge, M. L. A taxonomy study of the *Posidonia ostenfeldii* complex (Posidoniaceae) with description of four new Australian seagrasses. *Aquatic Botany* 20: 267–295 (1984).
ii Campey, M. L., Waycott, M., & Kendrick, G.A. Re-evaluating species boundaries among members of the *Posidonia ostenfeldii* species complex (Posidoniaceae)—morphological and genetic variation. *Aquatic Botany* 66: 41–56 (2000).
iii Les, D. H., Cleland, M. A., & Waycott, M. Phylogenetic studies in Alismatidae, II: Evolution of marine angiosperms (seagrasses) and hydrophily. *Syst. Bot.* 22: 443–463 (1997).

POTAMOGETONACEAE

PONDWEEDS

This family of herbs that grow in aquatic habitats throughout the world.

Distribution Cosmopolitan, but most occur in temperate regions of the northern hemisphere. All species grow in fresh- or brackish water. *Potamogeton* and *Zannichellia* are cosmopolitan, and *Groenlandia* is found in western Europe and North Africa to southwestern Asia. The monotypic, *Pseudalthenia* (*Vleisia*) *aschersoniana* is endemic to the Cape Province, South Africa.

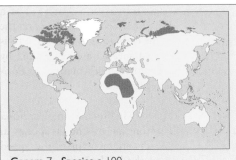

Genera 7 **Species** c. 100
Economic uses Species of *Potamogeton* are important food plants for many waterfowl

Description Submerged or floating herbs, usually perennial but some annual. The lower parts of the stems are creeping and rhizomatous, the upper parts elongate, flexible, erect, or floating. Some species develop specialized winter buds (turions). The leaves are alternate, opposite, or in bunches, with sheathing bases and intravaginal scales. The submerged leaves are simple, often linear to orbicular, when present the floating leaves are broader, lanceolate to ovate, entire, or serrulate. The inflorescence is usually a stalked spike or an axillary cluster of flowers. Flowers bisexual, rarely unisexual (plants monoecious or dioecious), either regular with 2 or 4 free, bract-like, clawed scales opposite each stamen or a small cuplike, 3-lobed sheath, or absent. Most species have 1, 2, or 4 stamens. Pollen is more or less spherical; pollination occurs above or underwater. The ovary is superior and usually consists either of 4 (rarely fewer or more) free or partly united carpels, each containing a single basal, anatropous or campylotropous ovule, or of 1 to 8 free carpels, each with a single, pendulous, anatropous ovule. In the latter case, the stigmas are conspicuous, enlarged, funnel-shaped, feathery, or peltate. The fruit is a drupe or berry. The seeds have a coiled embryo and no endosperm.

Classification The Zannichelliaceae is here included in Potamogetonaceae. This is supported by molecular evidence[i,ii]. Morphology points at a relationship with Cymodoraceae (q.v.) and Posidoniaceae (q.v.), but there is no phylogenetic evidence[ii]. The monoecious *Zanichellia palustris* (Horned Pondweed) has the leaf sheath free from the leaf; leaves are usually opposite or in false whorls. The dioecious *Althenia* and *Lepilaena* have leaf sheaths joined to the leaves, the leaves being alternate. The monotypic *Groenlandia densa* differs from *Potamogeton* in having opposite or whorled leaves without sheathing bases and a nutlet with a thin pericarp as fruit.

Potamogeton (c. 80 spp.) is the largest exclusively aquatic genus of flowering plants. Most of the species have aerial pollination.

Economic uses *Potamogeton* spp. are food sources for waterfowl. Other members of the family are invasive and block waterways.   [CDC] OS

i Les, D. H., Cleland, M. A., & Waycott, M. Phylogenetic studies in Alismatidae, II: Evolution of marine angiosperms (seagrasses) and hydrophily. *Syst. Bot.* 22: 443–463 (1997).
ii Haynes, R. R., Holm-Nielsen, L. B., & Les, D. H. Zannichelliaceae. Pp. 470–474. In: Kubitzki, K. (ed.), *The Families and Genera of Vascular Plants. IV. Flowering Plants. Monocotyledons: Alismatanae and Commelinanae (except Gramineae).* Berlin, Springer-Verlag (1998).

PRIONIACEAE

A monotypic family growing along permanent rivers and streams in South Africa. The only species, *Prionium serratum*, has a basally wood trunk more than 1 m high and 10 cm in diameter, clothed in withered leaf-bases. The leaves are borne in terminal rosettes, tristichous, parallel-sided, and serrate, with an open sheathing base. The inflorescence is a terminal, profusely branched panicle up to 1 m tall. The flowers are bisexual, with 5, 3-merous whorls, 3+3 stamens, and chaffy tepals. The pollen is dispersed in tetrads. The ovary is superior, 3-locular with erect, anatropous ovules. The fruit is a capsule. The seeds have a starchy endosperm.

The family is closely related to the Junca-ceae (q.v.) and has traditionally been included in that family[i], but it has recently been segregated into a family of its own, Prinon-iaceae[ii]. Recent molecular analyses place *Thurnia* and *Prionium* as sister groups related to Cyperaceae plus Juncaceae[iii,iv]. *Prioni-um* share pollen tetrads with Juncaceae, but only *Thurnia* and Cyperaceae deposit silica[i]. A strong fiber, palmite is made from the serrate leaves of the palmlet, *Prionium serratum (P. palmitum)*, but the fibers are of limited use.   os

Genera | Species |
Economic uses Fibers have limited use

[i] Balslev, H. Juncaeae. Pp. 252–260. In: Kubitzki, K. (ed.), *The Families and Genera of Vascular Plants. IV. Flowering Plants. Monocotyledons: Alismatanae and Commelinanae (except Gramineae).* Berlin, Springer-Verlag (1998).
[ii] Munro, S. L. & Linder, H. P. The phylogenetic position of *Prionium* (Juncaceae) within the order Juncales based on morphological and *rbc*L sequence data. *Syst. Bot.* 23: 43–45 (1998).
[iii] Chase, M. W. *et al.* Multi-gene analysis of monocot relationships: A summary. Pp. 63–75. In: Columbus, J. T. *et al.* (eds), *Monocots: Comparative Biology and Evolution.* Claremont, California, Rancho Santa Ana Botanic Garden (2006).
[iv] Davis, J. I. *et al.* A phylogeny of the monocots, as inferred from *rbc*L and *atp*A sequence variation, and a comparison of methods for calculating jackknife and bootstrap values. *Syst. Bot.* 29: 467–510 (2004).

RAPATEACEAE. I *Rapatea pandanoides* habit, showing entire leaves with large sheathing bases, and petiole with short spines (x⅓). 2 *R. paludosa* (a) inflorescence with 2 subtending bracts (x⅔); (b) longitudinal section of flower, showing free, stiff outer sepals; inner basally fused petals with adnate stamens; and the superior 3-locular ovary crowned by a simple style and each locule with a single ovule (x5⅔); (c) capsule dehiscing by 3 valves to reveal I seed in each locule (x5⅔). 3 *Schoenocephalium arthro-phyllum* (a) habit, showing linear, sheathing leaves and inflorescences with 2 subtending bracts (x⅓); (b) dehiscing capsule with 2 seeds in each locule (x4).

RAPATEACEAE

A small tropical family of perennial herbs.

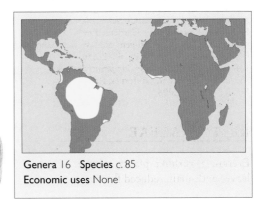

Genera 16   Species c. 85
Economic uses None

Distribution The Rapateaceae is almost exclusively neotropical, with the highest diversity ocurring in the lowland Guayana–Amazon region, a few outliers in Colombia and Panama, and 1 monotypic genus (*Maschalocephalus*) in West Africa. Members of the family are largely confined to bogs but may also be found in forests and shrubland.

Description Mostly terrestrial, perennial herbs with thick rhizomes. The leaves are distichous, simple, rarely petiolate, and often ensiform, with an open asymmetric sheath. The inflorescences are terminal or lateral, crowed or solitary, borne on long leafless scapes, often subtended by 2 or more bracts, and globose or flat, composed of 1 to 3 (*Monotrema, Stegolepis*) or up to 70 spikelets (*Saxofridericia, Spathanthus*). The flowers are 3-merous, bisexual, and actinomorphic, with 3 sepals, 3 petals, and 2 whorls of stamens, which are basally fused or fused with the base of the petals and with porose dehiscence. The ovary is superior, 3-locular, rarely 1-locular, with a capitate stigma, and between 1 and 8 anatropous ovules on basal or axile placentae. The fruit is a capsule, and the seeds have a starchy endosperm.

Classification It is doubtful whether the term *spikelet* is applied correctly when describing the partial inflorescences of Rapateaceae, but it is conventionally used, as here. The partial inflorescences may be derived from helicoid cymes[i]. The family is most likely monophyletic[ii,iii], but there is disagreement on the exact relationships of the family. It is usually placed in a huge clade (the Poales) that also includes the Bromeliaceae, Cyperaceae, Juncaceae, and Poaceae[i,iv,v]. OS

[i] Stevenson, D. W., Colella, M., & Boom, B. Rapateaceae. Pp. 415–424. In: Kubitzki, K. (ed.), *The Families and Genera of Vascular Plants. IV. Flowering Plants. Monocotyledons: Alismatanae and Commelinanae (except Gramineae).* Berlin, Springer-Verlag (1998).
[ii] Givnish, T. J. *et al.* Ancient vicariance or recent long-distance dispersal? Inferences about phylogeny and South American-African disjunctions in Rapateaceae and Bromeliaceae based on *ndh*F sequence data. *Int. J. Plant Sci.* 165 (4 Suppl.): S35–S54 (2004).
[iii] Davis, J. I. *et al.* A phylogeny of the monocots, as inferred from *rbc*L and *atp*A sequence variation, and a comparison of methods for calculating jackknife and bootstrap values. *Syst. Bot.* 29: 467–510 (2004).
[iv] Stevenson, D. W. & Loconte, H. Cladistic analysis of monocot families. Pp. 543–578. In: Rudall, P. J., Cribb, P. J., Cutler, D. F., & Humphries, C. J. (eds), *Monocotyledons: Systematics and Evolution.* Vol. 2. Richmond, Royal Botanic Gardens, Kew (1995).
[v] Chase, M. W. *et al.* Multi-gene analysis of monocot relationships: A summary. Pp. 63–75. In: Columbus, J. T. *et al.* (eds), *Monocots: Comparative Biology and Evolution.* Claremont, California, Rancho Santa Ana Botanic Garden (2006).

RESTIONACEAE

Evergreen, rushlike plants, with much reduced leaves and small, reduced flowers in spikelets.

Distribution The family is almost exclusively distributed in the southern hemisphere, with

Genera c. 50 **Species** c. 500
Economic uses Members of a few genera are used in horticulture and for thatching in South Africa

most species in southwestern Africa and southwestern Australia, but individuals also occur in Chile, Madagascar, New Zealand, and Malaysia. Most genera are restricted to a single continent, but *Apodasmia,* with 4 spp., occurs in Australia, New Zealand, and Chile, and 1 sp., *Dapsilanthus disjunctus* (= *Leptocarpus d.*), occurs widely in Southeast Asia and on the Chinese island of Hainan. The majority of species are found in marshes and swamps, often seasonally dry, fire-prone habitats, from sea level up to altitudes of 2,100 m, where they are often locally dominant.

Description Evergreen, up to 3.5 m, rushlike, caespitose, rhizomatous or stoloniferous, usually dioecious plants, with photosynthetic stems,

which branch dichotomously or have the branches arranged in whorls. The stems are solid or hollow, often variously ribbed and may be circular, semicircular, or more or less square in cross section. The stem anatomy is very characteristic for the family, with 1 to 2 layers of chlorenchymatous, palisadelike cells inside the epidermis, and often with protective cells lining the substomatal chambers. Additionally, the chlorenchyma is separated from the cortex by a ring of undifferentiated cells, 1 to several layers

RESTIONACEAE. 1 *Willdenowia lucaeana* (a) paniculate male inflorescence, with single male flowers each subtended by a bract (x⅔); (b) racemose female inflorescence with spikelets, each consisting of single flowers and numerous sterile bracts (x⅔); (c) male flower with subtending bract (x2); (d) a woody nut (x2). 2 *Acion (Restio) monocephalus* (a) habit (x⅔); (b) male flower with 3 stamens (x4); (c) female flower with 2 styles and 3 staminodes (x4); (d) cross section of fruit (x6); (e) entire capsule (x4). 3 *Thamnochortus insignis* (a) female spikelets, the flowers are hidden by the bracts (x⅔); (b) vertical section of female spikelet (x1½); (c) female flower, somewhat flattened, the outer tepals keeled, making the flower winged, and the subtending bract (x3). 4 *Leptocarpus simplex* (a) shoot with erect male spikelets (x⅔); (b) male flower (x4); (c) female flower, with keeled outer tepals and 3 styles (x4); (d) vertical section of ovary (x4). 5 *Elegia juncea* (a) shoot with exserted male spikelets, and flowers, and prominent spathes (x⅔); (b) shoot with hidden female spikelets hidden by the prominent spathes (x⅔); (c) indehiscent fruit (x6).

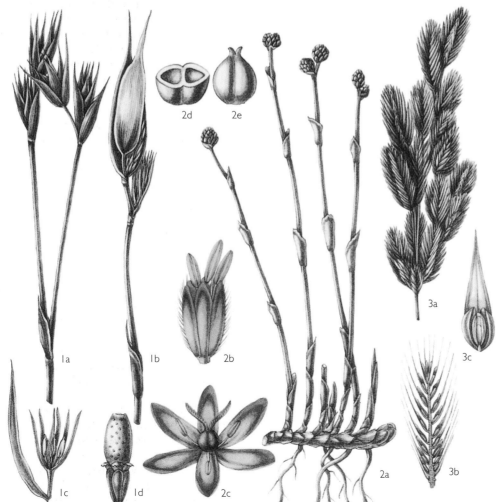

thick. The leaves are often strongly dimorphic, in adult plants consisting only of a sheath, often split to the base, and in juvenile plants with a distinct leaf blade (e.g., *Lepyrodia*) or a large, leaflike apex. The inflorescence is a spicate, racemose, or paniculate, occasionally with conspicuous spathes. The flowers are solitary, subtended by 1, or rarely 2, bracts or arranged in 1 to many flowered spikelets. The flowers are actinomorphic, generally unisexual (plants usually dioecious), often showing sexual dimorphism, or bisexual, with a perianth of 6 tepals in 2 whorls, the outer whorl often keeled and the flowers appearing winged. The male flowers have 3 stamens opposite the inner tepals. In female flowers, the stamens are absent or present as staminodes, and the ovary is superior, with 3 locules, each with a single, pendulous, atropous ovule. There are 1 to 3 filaments, or these are united into a style. The fruit is a capsule or a nut. The seeds have a copious, starchy endosperm.

Classification Both morphology[i] and molecular data[ii,iii,iv] support the Restionaceae as monophyletic. In the morphological analysis, however, the family is paraphyletic unless Centrolepidaceae (q.v.) and part of Anarthriaceae (e.g., the genera *Hopkinsia*, *Lyginia*) are included. The relationship between Centrolepidaceae and Restionaceae in particular is supported by unique morphological characters[i]. Molecular data point at a close relationship between Restionaceae and Anarthriaceae[iii] or Centrolepidaceae[iv]. Members of family fall in either of 2 clades, 1 containing the African genera and the other the Australian genera. Morphologically, the African clade may be separated from the Australian clade by the presence of a raised or swollen rim of the germination pores of the pollen and well-developed layers of protective cells, while the Australian genera have longer pubescent rhizomes[v] than the African genera. Presently, the Restionaceae is subject of extensive investigations and the generic circumscriptions are likely to change.

All species are wind-pollinated. In the Australian genus *Alexgeorgia*, the female flowers are solitary and develop directly from the rhizomes. Only the styles and the tip of the bracts reach above ground level at maturity.

Economic uses Members of a few genera are used in horticulture and for thatching in South Africa. [DC] OS

i Linder, H. P., Briggs, B. G., & Johnson, L. A. S. Restionaceae: A morphological phylogeny. Pp. 653–660. In: Wilson, K. L. & Morrison, D. A. (eds), *Monocots. Systematics and Evolution*. Melbourne, CSIRO Publishing (2000).
ii Briggs, B. G., Marchant, A. D., Gilmore, S., & Porter, C. L. A molecular phylogeny of Restionaceae and allies. Pp. 661–671. In: Wilson, K. L. & Morrison, D. A. (eds), *Monocots. Systematics and Evolution*. Melbourne, CSIRO Publishing (2000).
iii Davis, J. I. *et al.* A phylogeny of the monocots, as inferred from *rbc*L and *atp*A sequence variation, and a comparison of methods for calculating jackknife and bootstrap values. *Syst. Bot.* 29: 467–510 (2004).

iv Chase, M. W. *et al.* Multi-gene analysis of monocot relationships: A summary. Pp. 63–75. In: Columbus, J. T. *et al.* (eds), *Monocots: Comparative Biology and Evolution*. Claremont, California, Rancho Santa Ana Botanic Garden (2006).
v Linder, H. P., Briggs, B. G., & Johnson, L. A. S. Restionaceae. Pp. 425–445. In: Kubitzki, K. (ed.), *The Families and Genera of Vascular Plants. IV. Flowering Plants. Monocotyledons: Alismatanae and Commelinanae (except Gramineae)*. Berlin, Springer-Verlag (1998).

RHIPOGONACEAE

This family comprises a single genus of 6 spp. of shrubs or woody twiners with prickly stems, which grow in forests from New Guinea, along the eastern coast of Australia to New Zealand. The leaves are alternate, distichous, opposite or verticillate, petiolate, often with a well-developed drip tip and 3 prominent veins. The inflorescence is an axillary or terminal spike, raceme, or panicle with sessile or pedicellate flowers. The flowers are rather small, bisexual, actinomorphic, with free tepals in 2, 3-merous whorls and basally nectariferous. There are 6 free stamens, and the ovary is superior, of 3 fused carpels, 3-locular, with 2 anatropous ovules in each locule, and with a single style or the style absent. The fruit is a black or red berry. The seeds contain an endosperm of aleurone, fatty acids, and starch.

Genera 1 Species 6
Economic uses Some species used locally to make baskets

The Rhipogonaceae are often included in Smilacaceae, but appear to be sister to the Philesiaceae (q.v.)[i,ii,iii]. Some species are used locally for basket making. OS

i Davis, J. I. *et al.* A phylogeny of the monocots, as inferred from *rbc*L and *atp*A sequence variation, and a comparison of methods for calculating jackknife and bootstrap values. *Syst. Bot.* 29: 467–510 (2004).
ii Fay, M. F. *et al.* Phylogenetics of Liliales: Summarized evidence from combined analyses of five plastid and one mitochondrial loci. Pp. 559–565. In: Columbus, J. T. *et al.* (eds), *Monocots: Comparative Biology and Evolution*. Claremont, California, Rancho Santa Ana Botanic Garden (2006).
iii Chase, M. W. *et al.* Multi-gene analysis of monocot relationships: A summary. Pp. 63–75. In: Columbus, J. T. *et al.* (eds), *Monocots: Comparative Biology and Evolution*. Claremont, California, Rancho Santa Ana Botanic Garden (2006).

RUPPIACEAE

DITCH GRASSES

A monogeneric family of submerged aquatic herbs with 2-ranked, apically serrulate leaves.

Distribution Cosmopolitan, members of the family occur in freshwater or saltwater. A few species are found inland in freshwater in South America and New Zealand. Plants have been collected at altitudes of 4,000 m in the Andes.

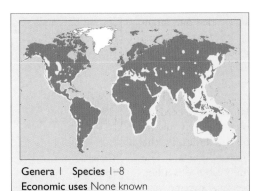

Genera 1 Species 1–8
Economic uses None known

Description Usually annual, submerged herbs with slender stems, erect or floating in the upper parts, rhizomatous in the lower, with alternate or opposite leaves, which are simple, linear, and somewhat serrulate at the apex. The leaf-bases are enlarged and sheathing with intravaginal scales. The inflorescence is short with 1 to few flowers, capitate, and long-stalked axillary or terminal spike. The axillary stalks are at first short and enveloped in the sheathlike leaf-base. Following pollination, the inflorescence stalks elongate and sometimes become coiled. Flowers actinomorphic, bisexual, small, and lacking tepals. There are 2 stamens. The ovary is superior, usually with 4 free carpels, without a style, each with a single, pendulous, campylotropous ovule. The peduncle of the inflorescence elongates prior to flowering, lifting the flowers to the water surface where pollination takes place. The carpels are sessile or long-stalked in fruit. The fruit is usually a single-seeded drupe with a somewhat spongy exocarp. The seeds have no endosperm.

Classification Ruppiaceae is often considered related to Potamogetonaceae and occasionally included in the latter[i]. Recent molecular analysis indicates that it is closely related to Cymodoceaceae (q.v.) and Posidoniaceae (q.v.) and even doubtfully distinct from the former[ii]. The number of accepted species varies from 1 polymorphic species to 8. [CDC] OS

i Dahlgren, R. M. T., Clifford, H. T., & Yeo, P. F. *The Families of the Monocotyledons*. Berlin, Springer-Verlag (1985).
ii Les, D. H., Cleland, M. A., & Waycott, M. Phylogenetic studies in Alismatidae, II: Evolution of marine angiosperms (seagrasses) and hydrophily. *Syst. Bot.* 22: 443–463 (1997).

RUSCACEAE

BUTCHER'S BROOM FAMILY

The Ruscaceae is a small family of shrubs or lianas with leaflike phylloclades.

Distribution In the Mediterranean, extending to northern Iran in the east and Madeira in the west, mostly found in woodland habitats.

Description Perennial shrubs or lianas, with distichous, spirally arranged or whorled, leathery or horny leaves, which are caducous or soon dried out, largely present only as inflorescence bracts. The stems are green, simple or branched, carrying leaflike, lanceolate to cordate-ovate, nearly sessile phylloclades in the upper parts. The inflorescence is a lax, terminal raceme, or a condensed raceme or fascicle, of flowers on the margin or surface of the phylloclades.

Genera 3
Species c. 10
Economic uses
Some species grown as ornamentals

The flowers are actinomorphic, bisexual (*Danae*) or unisexual (plants monoecious, e.g., *Danae, Semele,* and some *Ruscus* or dioecious, e.g., some *Ruscus*), with 6 free, or partly fused, nearly equal, small, petaloid tepals in 2 whorls. The androecium consist of 3 or 6 anthers, with the filaments fused into a tube or the anthers reduced to staminodes. The ovary is superior, 3-carpellate, either 3-locular with 2 ovules in each locule, or 1-locular with 1 to 4 ovules, with or without a style, or with a pistillodium. The ovules are anatropous, and the placentation variable, apical, or parietal. The fruit is a red or orange berry, the seeds with a small embryo and a starchless endosperm, storing aleurone, lipids, and hemicellulose.

Classification The Ruscaceae is here accepted in a narrow sense[i,ii] and is monophyletic in several analyses, but its acceptance makes Convallariaceae paraphyletic. For a discussion of its affinities see under Convallariaceae.

Economic uses Several species of *Ruscus, Danae racemosa* and *Semele androgyna* are occasionally grown as ornamentals.   OS

i Dahlgren, R. M. T., Clifford, H. T., & Yeo, P. F. *The Families of the Monocotyledons.* Berlin, Springer-Verlag (1985).
ii Yeo, P. F. Ruscaceae. Pp. 412–416. In: Kubitzki, K. (ed.), *The Families and Genera of Vascular Plants. III. Flowering Plants. Monocotyledons: Lilianae (except Orchidaceae).* Berlin, Springer-Verlag (1998).

SCHEUCHZERIACEAE
RANNOCH-RUSH FAMILY

The Scheuchzeriaceae consists of a single species (*Scheuchzeria palustris*), which is a grasslike, slender, rhizomatous perennial herb, growing in bogs in the northern hemisphere. The leaves are 2-rowed, linear, with a sheathing base and intravaginal hairs. The flowers are regular, bisexual, and borne in simple, terminal racemes with bracts. The tepals are inconspicuous, free, and yellow-green. The stamens are in 2 whorls of 3, and the ovary is superior, of usually 3, rarely 6, carpels, united at the base only, each locule containing 1 or 2 anatropous ovules. The fruit is a 1- or 2-seeded follicle, and the seeds have no endosperm. The embryo is

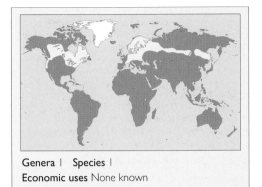

Genera I   Species I
Economic uses None known

chlorophyllous and has a rounded cotyledon and a small plumule.

The family has previously been included in the Juncaginaceae (q.v.), but both morphology[i,ii] and molecular data[iii] justify its positioning as a separate family.   [CDC] OS

i Haynes, R. R., Holm-Nielsen, L. B., & Les, D. H. Scheuchzeriaceae. Pp. 449–451. In: Kubitzki, K. (ed.), *The Families and Genera of Vascular Plants. IV. Flowering Plants. Monocotyledons: Alismatanae and Commelinanae (except Gramineae).* Berlin, Springer-Verlag (1998).
i Dahlgren, R. M. T., Clifford, H. T., & Yeo, P. F. *The Families of the Monocotyledons.* Berlin, Springer-Verlag (1985).
ii Les, D. H., Cleland, M. A., & Waycott, M. Phylogenetic studies in Alismatidae, II: Evolution of marine angiosperms (seagrasses) and hydrophily. *Syst. Bot.* 22: 443–463 (1997).

SMILACACEAE
SARSAPARILLA FAMILY

A small family of usually twining vines with prickly stems and leaves often with paired tendrils at the base.

Distribution The family is widespread in tropical, subtropical, and temperate zones in both wet and dry habitats, such as moist forest and savannahs. Most species occur in South America, Africa, and Southeast Asia.

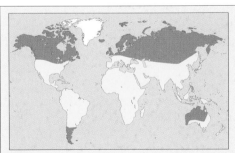

Genera 2   Species c. 320
Economic uses Several *Smilax* spp. are edible or have medical uses and are a source of "Sarsaparilla," in the past used to treat syphilis

Description Usually twining vines with long shoots and prickly stems; less commonly perennial shrubs, climbing by tendrils; and rarely annuals. The leaves are distichous, opposite or verticillate, petiolate and sheathing, with paired

tendrils from the sheath. The lamina is entire, leathery, with 3 to 7 prominent main veins and reticulate venation; the leaf margins and the main vein are often armed with prickles. The inflorescence is a simple or compound raceme or is umbel-like. The flowers are unisexual, erect, and actinomorphic, with 2, 3-merous whorls of free or fused, greenish, white, or cream-colored tepals with basal nectaries. The male and female flowers sometimes carry vestiges of the opposite sex. The stamens are 3 to 6, free of fused (*Heterosmilax*), rarely more. The ovary is superior and consists of 3 fused carpels, rarely only 1, and is 3-locular with a single atropous ovule in each locule. The fruit is a red or black berry. The seeds have a large embryo embedded in a endosperm with aleurone and fatty acids.

Classification *Rhiponogum,* which here is treated in its own family, Rhipogonaceae (q.v.) is often included in Smilacaceae[i], occasionally together with Philydraceae (q.v.)[ii]. However, Smilacaceae appears to be sister group to a larger clade that excludes Rhipogonaceae and includes, among other families, Liliaceae and Colchicaceae[iii,iv,v]. As circumscribed here, the Smilacaceae only include 2 genera, *Smilax* (c. 300 spp.) and *Heterosmilax* (10–12 spp.), but the species of *Heterosmilax* with 9–12 stamens are sometimes segregated in their own genus *Pseudosmilax*[vi].

Economic uses Various species of *Smilax* are edible, whereas others, including S. *aristolochiaefolia,* S. *regelii,* and S. *febrifuga,* are sources of commercial "sarsaparilla," which in the past was used for the treatment of syphilis but is now used as a digestive tonic and in the treatment of a large variety of maladies, such as anemia, fevers, rheumatism, and sexually transmitted diseases. In addition, the dried rhizome of *Smilax china* (China Root) yields an extract used in the treatment of gout and has stimulant qualities. Some species are used in horticulture.   OS

i Mitchell, J. D. Smilacaceae. Pp. 480–481. In: Smith, N. *et al.* (eds), *Flowering Plants of the Neotropics.* Princeton/Oxford, Princeton University Press (2004).
ii Cronquist, A. *An Integrated System of Classification of Flowering Plants.* New York, Columbia University Press (1981).
iii Davis, J. I. *et al.* A phylogeny of the monocots, as inferred from *rbc*L and *atp*A sequence variation, and a comparison of methods for calculating jackknife and bootstrap values. *Syst. Bot.* 29: 467–510 (2004).
iv Fay, M. F. *et al.* Phylogenetics of Liliales: Summarized evidence from combined analyses of five plastid and one mitochondrial loci. Pp. 559–565. In: Columbus, J. T. *et al.* (eds), *Monocots: Comparative Biology and Evolution.* Claremont, California, Rancho Santa Ana Botanic Garden (2006).
v Chase, M. W. *et al.* Multi-gene analysis of monocot relationships: A summary. Pp. 63–75. In: Columbus, J. T. *et al.* (eds), *Monocots: Comparative Biology and Evolution.* Claremont, California, Rancho Santa Ana Botanic Garden (2006).
vi Dahlgren, R. M. T., Clifford, H. T., & Yeo, P. F. *The Families of the Monocotyledons.* Berlin, Springer-Verlag (1985).

STEMONACEAE

The Stemonaceae is a small family predominantly found in Southeast Asia, from southern China to northern Australia, but *Croomia* also occurs in Japan and southeastern USA. It is found in regions with seasonal climates in rather dry vegetation or in rain forests. They are twinning, creeping, or erect perennials, with alternate, opposite or in whorled leaves, and laminas that have parallel main veins connected by transverse veins. The inflorescence is an axillary, simple, or compound raceme or a 1- to few-flowered lax cymose cluster, or occasionally several clusters are united in an umbel-like inflorescence. The flowers are actinomorphic or rarely zygomorphic, 4-merous (5-merous in *Pentastemona*), usually bisexual, but sometimes functionally unisexual (plants monoecious). The tepals are 4, in 2 whorls or 1 whorl of 5. The stamens are in 2 whorls of 2, stalked, or 1 whorl of 5, sessile. In the 4-merous flowers, the stamens are united with the tepals, sometimes with a sterile, apical appendages that may be fused in the tips. In the 5-merous flowers, the stamens are fused into a conspicuous, fleshy ring and become partly hidden in pouches made by the hypanthium and ovary. The ovary is inferior, semi-inferior, or superior, of 2 carpels, 1-locular, with few to many anatropous ovules on parietal, basal, or apical placentas. The fruit is berrylike or sometimes a 2-valved capsule. The seed contains a small embryo surrounded by copious endosperm.

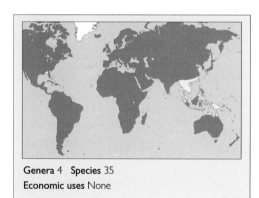

Genera 4 **Species** 35
Economic uses None

Pentastemona is included here in Stemonaceae[i] but is occasionally considered a family of it own, Pentastemonaceae and, indeed, deviates morphologically from the 3 other genera in the family (*Stemona, Stichoneuron, Croomia*)[ii]. In molecular analyses, *Pentastemona* comes out as the sister group to the remaining members of the Stemonaceae[iii]. Stemonaceae has often been considered closely related to the families around Dioscoreaceae[i] (q.v.), however, it is now considered to be sister group to Cyclanthaceae (q.v.) plus Pandanaceae (q.v.)[iii,iv,v]. OS

[i] Dahlgren, R. M. T., Clifford, H. T., & Yeo, P. F. *The Families of the Monocotyledons.* Berlin, Springer-Verlag (1985).

[ii] Kubitzki, K. Pentastemonaceae. Pp. 404–406. In: Kubitzki, K. (ed.), *The Families and Genera of Vascular Plants. III. Flowering Plants. Monocotyledons: Lilianae (except Orchidaceae).* Berlin, Springer-Verlag (1998).
[iii] Caddick, L. R. *et al.* Phylogenetics of Dioscoreales based on combined analyses of morphological and molecular data. *Bot. J. Linn. Soc.* 138: 123–144 (2002).
[iv] Davis, J. I. *et al.* A phylogeny of the monocots, as inferred from *rbc*L and *atp*A sequence variation, and a comparison of methods for calculating jackknife and bootstrap values. *Syst. Bot.* 29: 467–510 (2004).
[v] Chase, M. W. *et al.* Multi-gene analysis of monocot relationships: A summary. Pp. 63–75. In: Columbus, J. T. *et al.* (eds), *Monocots: Comparative Biology and Evolution.* Claremont, California, Rancho Santa Ana Botanic Garden (2006).

STRELITZIACEAE
BIRDS-OF-PARADISE FLOWER FAMILY

A family of large herbs with large, entire, distichous leaves and inflorescences with tough, spathaceous bracts and distinctive flowers.

Distribution The 3 genera are endemic to tropical America (*Phenakospermum*), South Africa (*Strelizia*), and Madagascar (*Ravenala*). They are found along rivers, or in swampy or disturbed areas.

Description Perennials with a short and massive sympoidal rhizome, either arborescent, with the pseudostems formed by the sheathing leaf-bases, or acaulescent. The leaves are alternate and distichous, sheathing, with or without an indistinct petiole. The lamina is entire, the midrib thick, with parallel, slightly sigmoid veins running from it and fusing near the margin. The inflorescence is terminal or lateral with tough, spathaceous bracts, each supporting a cincinnate cluster of flowers. The flowers are zygomorphic and bisexual. The tepals are 3, the outer free and more or less similar, the inner connate at the base, and the 2 anterior are distinctly larger than the posterior. In *Strelitzia*, the 2 anterior tepals are arrow-shaped with a central groove, which enclose the filaments and style. The stamens are 6 (*Ravenala*) or 5 (*Phenakospermum, Strelitizia*). The ovary is inferior, of 3 fused carpels, 3-locular, with numerous anatropous ovules per locule on axile placentas. The style is filiform and often expanded at the apex. The fruit is a woody capsule, dehiscing loculicidally by 3 valves. The seeds have an orange, red, or blue aril and abundant starchy endosperm surrounding a straight embryo.

Classification The Strelitziaceae were originally treated as part of the Musaceae (q.v.). In a recent analysis based on morphological and molecular data, however, it is sister to Lowiaceae (q.v.),

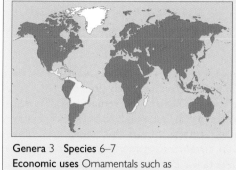

Genera 3 **Species** 6–7
Economic uses Ornamentals such as bird-of-paradise flower (*Strelitizia*) and Traveller's Tree (*Ravenala madagascariensis*)

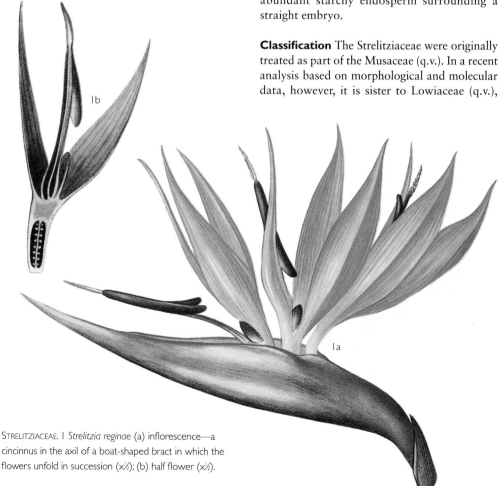

STRELITZIACEAE. I *Strelitzia reginae* (a) inflorescence—a cincinnus in the axil of a boat-shaped bract in which the flowers unfold in succession (x⅓); (b) half flower (x⅔).

and Lowiaceae plus Sterlitziaceae are sister to Musaceae[i]. The close relationship between Lowiaceae and Strelitziaceae is supported by larger molecular analyses[ii].

Economic uses *Ravenala madagascariensis* (Traveller's Tree) is widely grown in the tropics as an ornamental. The vernacular name refers to the belief that the distichous leaves always align north-south and thus indicate the direction. *Strelitzia reginae* and *S. nicolai* (bird-of-paradise flowers) are widely cultivated, and the latter is an important cut flower.   OS

[i] Kress, W. J., Prince, L. M., Hahn, W. J., & Zimmer, E. A. Unraveling the evolutionary radiation of the families of the Zingiberales using morphological and molecular evidence. *Syst. Biol.* 50: 926–944 (2001).
[ii] Chase, M. W. *et al.* Multi-gene analysis of monocot relationships: A summary. Pp. 63–75. In: J. T. *et al.* (eds), *Monocots: comparative biology and evolution.* Rancho Santa Ana Botanic Garden, Claremont, California, USA (2006).

TACCACEAE
TACCA FAMILY

A monogeneric family of perennial, tropical, rhizomatous herbs, with a cymose inflorescence subtended by 2 conspicuous bracts.

Distribution Pantropical, reaching into China and Australia, usually growing in the understory of humid and seasonal tropical or subtropical rain forests. The only species outside Asia is the tropical American *T. sprucei.*

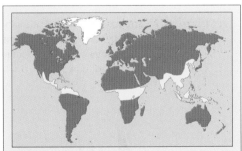

Genera 1   Species c. 15
Economic uses The tubers of *Tacca leontopetaloides* are a rich source of starch

Description Perennial rhizomatous herbs with elongated, vertical or horizontal, rhizomes and basal, long-petiolate, entire or variously dissected, leaves. The cymose, umbellate inflorescence is borne on a radical scape and subtended by usually 4, but occasionally only 2 or up to 12, involucral bracts. The flowers are actinomorphic and bisexual, with 2 whorls of 6 petaloid tepals fused into a short cuplike tube at the base and often dull or dark-colored. The stamens are 6, slightly petaloid, in 2 whorls attached to the tepals or the perianth tube and with expanded connectives, which in the form of a hood, hides the anthers. The ovary is inferior, consisting of 3 carpels and is 1-locular with 3 protruding placentas, with numerous pendulous, anatropous ovules. Several species have septal nectaries. The style is short and terminates in 3 reflexed stigmas, which are often petaloid. The fruit is a berry or rarely a loculicidal capsule, containing numerous ribbed, reniform seeds, each with a small embryo surrounded by copious endosperm of protein and fat.

Classification Taccaceae is occasionally included in Dioscoreaceae (q.v.), but the relationships between Taccaceae, *Stenomeris*, and the remaining Dioscoreaceae do not appear well resolved[i,ii].

Economic uses The starch-rich rhizomatous tubers of *Tacca leontopetaloides* are used as a source of arrowroot and falsely traded as East Indian arrowroot; the stems and scapes used as fibers in the manufacture of hats.   OS

[i] Caddick, L. R., Rudall, P. J., Wilkin, P., Hedderson, T. A. J., & Chase, M. W. Phylogenetics of Dioscoreales based on combined analyses of morphological and molecular data. *Bot. J. Linn. Soc.* 138: 123–144 (2002).
[ii] Chase, M. W. *et al.* Multi-gene analysis of monocot relationships: A summary. Pp. 63–75. In: Columbus, J. T. *et al.* (eds), *Monocots: Comparative Biology and Evolution.* Claremont, California, Rancho Santa Ana Botanic Garden (2006).

TECOPHILAEACEAE
TECOPHILA FAMILY

A small family of herbs with corms, and simple, usually basal leaves.

Distribution There are 3 genera endemic to Chile, 3 or 4 occur only in Africa, and 1 is endemic to California, USA.

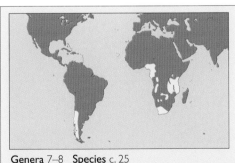

Genera 7–8   Species c. 25
Economic uses A few species are cultivated as ornamentals

Description Perennial, terrestrial herbs with corms that are often covered with a fibrous tunic. The leaves are parallel-veined, sessile, or more or less petiolate, spirally arranged, and generally basal, except in *Walleria,* where they are cauline. The leaf-base is occasionally sheathing, and the blade entire, glabrous, narrowly linear to lanceolate-ovate, and occasionally with an undulate margin. The inflorescence is a panicle, a raceme, or the flowers are solitary or in small groups, sometimes in the axile of the cauline leaves. The flowers are large, bisexual, and zygomorphic or actinomorphic,

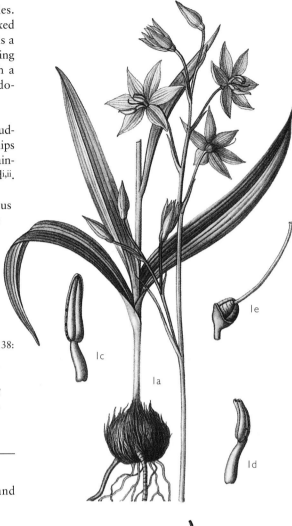

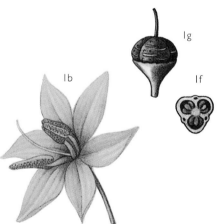

TECOPHILAEACEAE. I *Cyanella lutea* (a) habit, showing tunicate corm, linear, sheathing leaves, and paniculate inflorescence (×⅔); (b) flower, showing dimorphic stamens: five smaller anterior, and 1 large, anterior (×1½); (c) large anther (×2); (d) small anther dehiscing by apical pores (×2); (e) semi-inferior ovary with "sunken" style (×2); (f) cross section of ovary (×3); (g) fruit—a loculicidal capsule (×2⅔).

with 3+3 membranous tepals basally fused into a tube and sometimes with appendages between the adnate stamens. The tepals of the outer whorl have a mucronate or aristate apical extension. The stamens are 6, in 2 whorls, all fertile or some are staminodes. The anthers of most species have poricidal dehiscence. The ovary is semi-inferior or inferior, 3-carpellate, and

TECOPHILAEACEAE. 2 *Cyanastrum cordifolium* (a) habit, showing nontunicated corm, cordate leaves, and racemose inflorescence (x⅔); (b) part of flower, showing 3 stamens, semi-inferior ovary with "sunken" style (x2); (c) stamen dehisced by pores (x4). 3 *Conanthera campanulata* (a) leafy shoot and paniculate inflorescence with slightly zygomorphic flowers (x⅔); (b) part of flower, showing 3 of the 6 fertile anthers (x2⅔); (c) anther with extended connective (x4).

3-locular with axile placentas carrying 2 to numerous ovules in 2 rows in each locule. The fruit is a loculicidal capsule, with numerous seeds, each with a large embryo, surrounded by fleshy endosperm or lacking an endosperm.

Classification As defined here, the Tecophilaeaceae is often divided into 2 separate families, Tecophilaeaceae and Cyanastraceae[i], and was previously considered related to Haemodoraceae (q.v.)[ii], a family which is now considered related to Commelinaceae (q.v.) and Pontederiaceae (q.v.). However, recent molecular evidence points to the family being monophyletic as circumscribed here and sister group to the Ixioliriaceae[iii], Doryanthaceae (q.v.)[iv], or to a very large clade that includes e.g., Asparagaceae (q.v.), Hyacinthaceae (q.v.), and Asparagaceae in a narrow sense (q.v.)[iv].

Economic uses Some species of *Cyanella* e.g., *C. hyacinthoides*, and *Tecophilaea* e.g., *T. cyanocrocus* (Chilean Crocus), are cultivated as ornamentals. OS

i Dahlgren, R. M. T., Clifford, H. T., & Yeo, P. F. *The Families of the Monocotyledons*. Berlin, Springer-Verlag (1985).
ii Simpson, M. G. & Rudall, P. J. Tecophilaeaceae. Pp. 429–436. In: Kubitzki, K. (ed.), *The Families and Genera of Vascular Plants. III. Flowering Plants. Monocotyledons: Lilianae (except Orchidaceae)*. Berlin, Springer-Verlag (1998).
iii Fay, M. F. *et al*. Phylogenetic studies of Asparagales based on four plastid DNA regions. Pp. 360–371. In: Wilson, K. L. & Morrison, D. A. (eds), *Monocots. Systematics and Evolution*. Melbourne, CSIRO Publishing (2000).
iv Chase, M. W. *et al*. Multi-gene analysis of monocot relationships: A summary. Pp. 63–75. In: Columbus, J. T. *et al*. (eds), *Monocots: Comparative Biology and Evolution*. Claremont, California, Rancho Santa Ana Botanic Garden (2006).

THEMIDACEAE

A small family of geophytes with corms, until recently included in the Alliaceae.

Distribution Apart from 1 sp. that occurs in Guatemala, the family is only found in western North America, from sea level to 3,000 m.

Description Perennial, stemless or short stemmed geophytes, with corms that are renewed every year and are enveloped in the leaf-bases from the previous year, creating

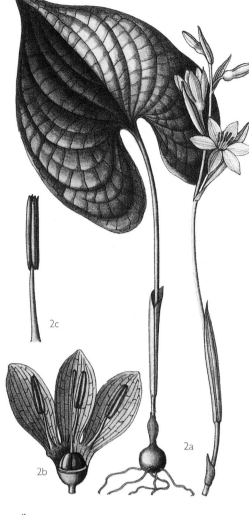

a characteristic tunic around them. The leaves are linear, spirally arranged, flat or terete, and sometimes hollow. The inflorescence is a scapose umbel, rarely reduced to a single flower. The flowers are bisexual, actinomorphic, with 2 whorls of 3 tepals fused at the base, and often a corona or scales between the 2 whorls of 3 stamens, with 1 whorl occasionally reduced to staminodes. The filaments are free or frequently united with apical appendages. The ovary is superior, often stalked, 3-carpellate, and 3-locular, with 2 to several anatropous ovules in each locule. The style is solitary, capitate, or 3-lobed. The fruit is a loculicidal capsule with 2 or more seeds in each locule. The seeds have short straight embryos and starchless endosperms of aleurone and oil.

Genera 12
Species c. 60
Economic uses Mainly used as ornamentals

Classification The Themidaceae has been segregated from the Alliaceae (q.v.)[i] and appears to be monophyletic[i,ii,iii] and sister to the Hyacinthaceae (q.v.)[ii], Hyacinthaceae plus Aphyllanthaceae (q.v.)[iii], or Amaryllidaceae plus Alliaceae (q.v.)[iv].

Economic uses Several genera, e.g., *Brodiaea* and *Dichlostemma*, include species that are used as ornamentals. OS

i Fay, M. F. & Chase, M. W. Resurrection of *Themidaceae* Salisb. for the *Brodiaea* alliance, and recircumscription of *Alliaceae, Amaryllidaceae* and Agapanthoideae. *Taxon* 45: 441–451 (1996).
ii Pires, J. C. *et al*. Phylogeny, genome size, and chromosomal evolution of Asparagales. Pp. 287–304. In: Columbus, J. T. *et al*. (eds), *Monocots: Comparative Biology and Evolution*. Claremont, California, Rancho Santa Ana Botanic Garden (2006).
iii Fay, M. F. *et al*. Phylogenetic studies of Asparagales based on four plastid DNA regions. Pp. 360–371. In: Wilson, K. L. & Morrison, D. A. (eds), *Monocots. Systematics and Evolution*. Melbourne, CSIRO Publishing (2000).
iv Tamura, M. N., Yamashita, J., Fuse, S., & Haraguchi, M. Molecular phylogeny of monocotyledons inferred from combined analysis of plastid *mat*K and *rbc*L gene sequences. *J. Plant Res*. 117: 109–120 (2004).

THURNIACEAE

A small, little-known family, comprising a single genus (*Thurnia*) and 3 spp., all growing in flooded areas along rivers and streams and restricted to northeastern South America. All species are perennial herbs with basal, tristichous, parallel-sided and some-times serrate leaves, V-shaped in cross section, with a sheathing base. The inflorescence is terminal, racemose, and consists of 1 or more globose to ellipsoid heads, supported by leafy bracts. The flowers are bisexual, with 5, 3-merous whorls, with 3+3 stamens, and chaffy tepals. The pollen is dispersed in tetrads. The ovary is superior, 3-locular, with 3 or more erect, anatropous ovules. The fruit is a loculicidal capsule, with 3 spindle-seeds and a starchy endosperm.

Genera 1 Species 3
Economic uses None known

The family is closely related to the Juncaceae (q.v.) but often recognized as a separate family, Thurniaceae[i,ii], which is occasionally considered related to the Rapateaceae (q.v.) or Xyridaceae (q.v.). However, recent molecular analyses place *Thurnia* and *Prionium* (here recognized as a separate family, Prioniaceae [q.v.]) as sister groups, related to Cyperaceae (q.v.) plus Juncaceae[iii,iv]. Diffuse centromeres are known from Thurniaceae, Juncaceae, and Cyperaceae[ii,iii]. Both *Thurnia* and *Prionium* share pollen tetrads with Juncaceae, but only Thurnia and Cyperaceae deposit silica[i]. OS

i Stevenson, D. W. Thurniaceae. Pp. 486–487. In: Smith, N. *et al.* (eds), *Flowering Plants of the Neotropics.* Princeton/Oxford, Princeton University Press (2004).
ii Kubitzki, K. Thurniaceae. Pp. 455–457. In: Kubitzki, K. (ed.), *The Families and Genera of Vascular Plants. III. Flowering Plants. Monocotyledons: Lilianae (except Orchidaceae).* Berlin, Springer-Verlag (1998).
iii Dahlgren, R. M. T., Clifford, H. T., & Yeo, P. F. *The Families of the Monocotyledons.* Berlin, Springer-Verlag, (1985).
iv Chase, M. W. *et al.* Multi-gene analysis of monocot relationships: A summary. Pp. 63–75. In: Columbus, J. T. *et al.* (eds), *Monocots: Comparative Biology and Evolution.* Claremont, California, Rancho Santa Ana Botanic Garden (2006).
v Davis, J. I. *et al.* A phylogeny of the monocots, as inferred from *rbc*L and *atp*A sequence variation, and a comparison of methods for calculating jackknife and bootstrap values. *Syst. Bot.* 29: 467–510 (2004).

TOFIELDIACEAE

A small family of herbs, which is widespread in the northern temperate and subartic zone, in the European Alps and in northwestern South America. Members of this family have basal, ensiform, distichous leaves, and a racemose or rarely spicate inflorescence with 3 bracteoles (calyculus), occasionally reduced to a single

Genera 3–4 Species c. 25
Economic uses Some use as ornamentals

flower (*Harperocallis*). The flowers are usually 3-merous with 3+3 tepals and 3+3 stamens, and a 3-carpellate, septicidal capsule.

The Tofieldiaceae is often included in either the Melianthaceae (q.v.)[i] or the Nartheciaceae (q.v.)[ii]. However, recent molecular analyses[iii] place in these plants in a separate family, with an unresolved relationship to the Araceae (q.v.) and a clade consisting predominantly of aquatics (e.g., Alismataceae, Hydrocharitaceae, Potamogetonaceae). OS

i Frame, D. Melianthaceae. Pp. 460–462. In: Smith, N. *et al.* (eds), *Flowering Plants of the Neotropics.* Princeton/Oxford, Princeton University Press (2004).
ii Tamura, M. N. Nartheciaceae. Pp. 381–392. In: Kubitzki, K. (ed.), *The Families and Genera of Vascular Plants. III. Flowering Plants. Monocotyledons: Lilianae (except Orchidaceae).* Berlin, Springer-Verlag (1998).
iii Tamura, M. N., Fuse, S., Azuma, H., & Hasebe, M. Biosystematic studies on the family Tofieldiaceae I. Phylogeny and circumscription of the family inferred from DNA sequences of *mat*K and *rbc*L. *Plant Biol. (Stuttg.)* 6: 562–567 (2004).

TRIURIDACEAE
TRIURID FAMILY

The Triuridaceae is small family of saprophytic, achlorophyllous herbs with scalelike leaves.

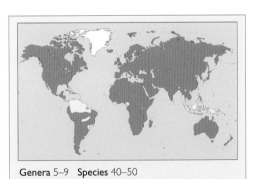

Genera 5–9 Species 40–50
Economic uses None known

Distribution The Triuridaceae is found mainly in tropical America, West Africa, Madagascar, and Asia, but a few genera have representatives in subtropical South America. Most species live in humid, dense forest.

Description Small, either colorless, yellow, or purple, achlorophyllous, mycotrophic, perennial herbs with tiny, alternate, scalelike leaves. The inflorescence is a terminal, bracteate, few- to many-flowered raceme. The flowers are small, actinomorphic, and usually unisexual (plants monoecious or dioecious) and colored, rarely bisexual (e.g., in *Sciaphila*). The perianth consists of between 3 and 6, rarely more, basally connate tepals in a single whorl. The tepals are more or less triangular, often covered with hairs, papillae, or glands, and sometimes with apical appendages. The bisexual flowers have between 2 and 6 sessile stamens, mostly opposite the tepals, and many free carpels. In the male flowers, the stamens are borne on an androphore. The stamens may have connective

appendages, and both male and female flowers may have staminodes. In the female flowers, the ovary is superior, of 10 to many free carpels immersed in the receptacle, each with a single basal, anatropous ovule. The stigmatic part of the style is glabrous or brush shaped. The fruit is a follicle or an achene. The seeds are minute with a small, undifferentiated embryo.

Classification The position of the Triuridaceae has been difficult to resolve and is still in need of further investigation[i], and there is evidence that it is not monophyletic[ii]. The family has been considered closely related to Petrosaviaceae (q.v.) and the Alismataceae (q.v.)[iii,iv]. However, analyses based on mitochondrial data and morphological have placed them as sister group to Pandanaceae (q.v.), Stemonaceae (q.v.), and Cyclanthaceae (q.v.)[i,v]. OS

i Gandolfo, M. A. Triuridaceae. Pp. 487–488. In: Smith, N. *et al.* (eds), *Flowering Plants of the Neotropics.* Princeton/Oxford, Princeton University Press (2004).
ii Gandolfo, M. A., Nixon, K. C., & Crepet, W. L. A phylogenetic analysis of modern and Cretaceous Triuridaceae (Monocotyledonae). *American J. Bot.* 85: 131 (1998).
iii Cameron, K. M., Chase, M. W., & Rudall, P. J. Recircumscription of the monocotyledonous family Petrosaviaceae to include *Japonolirion*. *Brittonia* 55: 215–225 (2003).
iv Rudall, P. J. Monocot pseudanthia revisited: Floral structure of themycoheterotrophic family Triuridaceae. *Int. J. Plant Sci.* 164 (5 suppl.): S307–320 (2003).
v Davis, J. I. *et al.* A phylogeny of the monocots, as inferred from *rbc*L and *atp*A sequence variation, and a comparison of methods for calculating jackknife and bootstrap values. *Syst. Bot.* 29: 467–510 (2004).

TYPHACEAE
CATTAIL FAMILY

A small family of rhizomatous herbs with linear leaves and unisexual flowers either in very dense, clublike terminal inflorescences or in racemose inflorescences of globose heads.

Distribution The genus *Typha* is cosmopolitan, but absent in the Arctic region and arid areas, whereas *Sparganium* (Bur-reed) occurs in the temperate and Arctic parts of the northern hemisphere, with outliers in Australia and New Zealand. Both usually grow in shallow, freshwater habitats.

Description Perennial, rhizomatous plants, with starchy rhizomes, usually emerging from shallow water or growing in wet soil, with stems up to 2 m. The leaves are often distichous, with a sheathing base forming a false stem, linear, keeled, or flat with a concave abaxial surface, and spongy, occasionally floating. The inflorescence terminates the stem; in *Typha*, it is spikelike, consisting of 2 extremely dense, cylindrical parts, a lower female and an upper male section, separated by a short piece of stem or contiguous; in *Sparganium*, it is racemose, consisting of unisexual, globose heads, the female heads below the male. The perianth consists of 1 to several inconspicuous tepals or

Genera 2 **Species** c. 25
Economic uses Some species, especially *Typha* (cattail), may be used for weaving material, e.g., basket-making, and the rhizomes are edible

numerous scales or bristlelike hairs. The male flowers have usually 3 stamens, rarely fewer or more. The female flowers have a superior, pseudomonomerous, rarely 2- or 3-carpellate ovary, and in *Typha* an elongated stipe densely covered with long hairs. The single ovule is apotropus and pendulous. The fruit is a spongy drupe (*Spraganium*) or an achenelike follicle (*Typha*), and the embryo is straight, with a copious, mealy endosperm, containing starch, protein, and oil.

Classification The Sparganiaceae and the Typhaceae are occasionally considered separate families[i], but morphological[ii] and recent molecular evidence[iii,iv] makes it appropriate to combine them as is done here. The female inflorescences are most likely contracted branching systems[ii].

Economic uses *Typha* spp. play a significant ecological role as important constituents of the reed belt. The leaves, especially those of the Bulrush (*Typha latifolia*), are used as a weaving material in the manufacture of baskets, chair bottoms, and mats. The starchy rhizomes of some species are edible, and the plants have limited ornamental value. [CJH] OS

i Dahlgren, R. M. T., Clifford, H. T., & Yeo, P. F. *The Families of the Monocotyledons.* Berlin, Springer-Verlag (1985).
ii Müller-Doblies, U. Über die Verwandtschaft von *Typha* und *Sparganium* im Infloreszenz- und Blütenbau. *Bot. Jahrb. Syst.* 89: 451–562 (1970).
iii Davis, J. I. *et al.* A phylogeny of the monocots, as inferred from *rbc*L and *atp*A sequence variation, and a comparison of methods for calculating jackknife and bootstrap values. *Syst. Bot.* 29: 467–510 (2004).
iv Chase, M. W. *et al.* Multi-gene analysis of monocot relationships: A summary. Pp. 63–75. In: Columbus, J. T. *et al.* (eds), *Monocots: Comparative Biology and Evolution.* Claremont, California, Rancho Santa Ana Botanic Garden (2006).

VELLOZIACEAE
VELLOZIA FAMILY

The Velloziaceae is a small family of herbs or shrubs, with stems often covered by persistent leaf-bases and adventitious roots and linear leaves aggregated at the stem tips.

Distribution The family grows in South America in a broad zone south of the Amazonian rain forest and in the foothills of the Northern Andes, with single species reaching as far north as Panama. In the Old World, the family occurs on Madagascar and in Africa, largely south of the Sahara, but reaching as far north as the southwestern corner of the Arabian Peninsula. Members of the family are almost always found in highly arid habitats.

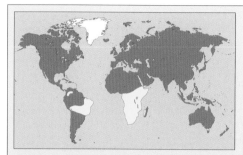

Genera 8 **Species** c. 250
Economic uses Some species are used locally as fire lighters and brushes

Description Herbs or shrubby perennials, ranging in height from a few centimeters to 6 m, with woody and simple or dichotomously branched stems, often covered with persistent leaves or fibrous leaf-bases. The leaves are tristichous, usually clustered at the end of the stems, linear with a sheathing base, and entire or denticulate at the margins. Adventitious roots, which can rapidly absorb any available water, alternate with the leaves. The inflorescence is 1- to many-flowered, terminal with caducous scape, rarely a few-flowered capitulum (*Acanthochlamys*). The flowers are mostly bisexual, only rarely unisexual (*Barbaceoniopsis*) plants dioecious, and actinomorphic. The perianth is frequently brightly colored and consists of 2 whorls of 3 similar tepals. The stamens are 6, free or variously fused to the perianth, in 2 whorls of 3, or numerous stamens in 6 bundles. The ovary is inferior or semi-inferior, consists of 3 carpels, and is 3-locular, with a style that is enlarged near the 3 stigmas. The ovary walls are often conspicuously prolonged into a hypanthium, which may carry a corona between the tepals and the stamens. The ovules are numerous on 2 axile placentas per locule. The fruit is a loculicidal capsule, with numerous seeds. The seed contains a small embryo surrounded by hard, starchy endosperm.

Classification *Acanthochlamys bracteata,* which is restricted to southwestern China, is occasionally considered to make up a separate family, Acanthochlamydaceae[i], and in recent molecular analyses it is sister to the remaining genera of the Velloziaceae[ii,iii,iv]. Velloziaceae is most likely monophyletic and sister group to a clade composed of Cyclanthaceae (q.v.), Pandanaceae (q.v.), and Stemonaceae (q.v.)[iv,v].

Economic uses Some species are used locally as fire lighters and as paint brushes. They have limited use as ornamental plants. OS

i Kao, P.-C. Acanthochlamydaceae — a new monocotyledonous family. Pp. 483–507. In: Kao, P.-C. & Tan, Z.-M. (eds), *Flora sichuanica.* Vol. 9. Chengdu, Sichuan Science and Technology Press.
ii Kubitzki, K. Acanthochlamydaceae. Pp. 55–58. In: Kubitzki, K. (ed.), *The Families and Genera of Vascular Plants. III. Flowering Plants. Monocotyledons: Lilianae (except Orchidaceae).* Berlin, Springer-Verlag (1998).
iii Behnke, H.-D. *et al.* Systematics and evolution of Velloziaceae, with special reference to sieve-element plastids and *rbc*L sequence data. *Bot. J. Linn. Soc.* 134: 93–129 (2000).
iv Salatino, A., Salatino, M. L. F., De Mello-Silva, R., Giannasi, D. E., & Price, R. A. Phylogenetic Inference in Velloziaceae Using Chloroplast *trn*L-*trn*F sequences. *Syst. Bot.* 26: 92–103 (2001).
v Davis, J. I. *et al.* A phylogeny of the monocots, as inferred from *rbc*L and *atp*A sequence variation, and a comparison of methods for calculating jackknife and bootstrap values. *Syst. Bot.* 29: 467–510 (2004).
vi Chase, M. W. *et al.* Multi-gene analysis of monocot relationships: A summary. Pp. 63–75. In: Columbus, J. T. *et al.* (eds), *Monocots: Comparative Biology and Evolution.* Claremont, California, Rancho Santa Ana Botanic Garden (2006).

XANTHORRHOEACEAE
GRASS TREE FAMILY

A monogeneric family of perennials, with woody stems, either arborescent or subterranean, with tufts of linear leaves terminating the branches, and an erect, massive spike of flowers and capsular fruit.

Distribution Restricted to subtropical and tropical Australia in open *Eucalyptus* forests and heathlands.

Genera 1 **Species** 30
Economic uses Previously used in the production of varnishes

Description Perennial, woody plants, with branched or unbranched stems, which are either arborescent, up to 3 m high, or subterranean. The leaves are arranged in tufts at the end of branches and are erect at first, later spreading or recurving, simple, linear, from only 3–5 cm to up to 1 m long, and usually sheathing, with resiniferous leaf-bases that persist on the stems. The inflorescence is an erect, massive, scapose spike, up to 4 m high, with spirally arranged bisexual, actinomorphic flowers supported by bracts. The 2 persistent tepal whorls are differentiated into 3 petaloid and 3 sepaloid tepals. The sepals are scarious, and the tepals are white or yellow membranacous. The stamens are 3+3, free, and the ovary superior, 3-carpellate, 3-locular, with a simple style, and 2 rows of anatropous ovules per locule. The fruit is a loculicidal capsule with 1 to 2 black seeds in each locule. The embryo is straight in a starchless endosperm.

Classification *Xanthorrhoea* was considered by Cronquist (1981) as part of a larger family, including for example members of the Dasypogonaceae (q.v.) and Laxmanniaceae (q.v.), but it was later considered as constituting a monotypic family morphological grounds[i]. Molecular data place the Xanthorrhoeaceae in a clade that also contains Asphodelaceae (q.v.) and Hemerocallidaceae (q.v.), and all 3 families are sometimes combined into an enlarged Xanthorrhoeaceae (see APG II). In a narrow sense, the Xanthorrhoeaceae is either sister group to Asphodelaceae (q.v.) plus Hemerocallidaceae (q.v.)[ii], to Asphodelaceae (q.v.) alone[iii], to Hemerocallidaceae alone[iv], or the relationship is unresolved[v].

Economic uses *Xanthorrhoea* spp. e.g., *X. australis* and *X. hastilis,* yield a gum, Grass Tree Gum or Red Acaroid Gum, once used to make varnishes. *X. preissii* (Black Boy) is a characteristic component of Australian heathlands. OS

i Dahlgren, R. M. T., Clifford, H. T., & Yeo, P. F. *The Families of the Monocotyledons.* Berlin, Springer-Verlag (1985).
ii Chase, M. W. *et al.* Phylogenetics of Asphodelaceae (Asparagales): An analysis of plastid *rbc*L and *trn*L-*trn*F DNA sequences. *Ann. Bot.* 86: 935–991 (2000).
iii Fay, M. F. *et al.* Phylogenetic studies of Asparagales based on four plastid DNA regions. Pp. 360–371. In: Wilson, K. L. & Morrison, D. A. (eds), *Monocots. Systematics and Evolution.* Melbourne, CSIRO Publishing (2000).
iv Pires, J. C. *et al.* Phylogeny, genome size, and chromosomal evolution of Asparagales. Pp. 287–304. In: Columbus, J. T. *et al.* (eds), *Monocots: Comparative Biology and Evolution.* Claremont, California, Rancho Santa Ana Botanic Garden (2006).
v McPherson, M. A., Fay, M. F., Chase, M. W., & Graham, S. W. Parallel loss of slowly evolving intron from two closely related families in Asparagales. *Syst. Bot.* 29: 296–307 (2004).

XERONEMATACEAE

The Xeronemataceae is a small family of shrublike, rhizomatous perennials with 1 sp. found in New Zealand and 1 sp. in New Caledonia. They have basal, distichous, ensiform leaves with sheathing base. The inflorescence is a brushlike spike, with bisexual, actinomorphic, upward-facing, large flowers; with 3+3 more or less equal, bright red tepals fused at the base forming a curved tube; and 3+3 free, strongly exserted stamens. The ovary is superior, 3-carpellate, 3-locular, on a small stalk, and with septal nectaries. The fruit is a loculicidal capsule with black seeds, with a linear embryo in a starchless endosperm.

Genera 1   Species 2
Economic uses None

Xeronema has been included in Phormiaceae[i] (which is part of Hemerocallidaceae) or in Hemerocallidaceae as an anomalous genus[ii]. In

most molecular analyses, however, it occupies a position as sister group to a huge clade including Agavaceae (q.v.), Hyacinthaceae (q.v.), Hemerocallidaceae (q.v.) etc.[iii,iv,v]. OS

i Dahlgren, R. M. T., Clifford, H. T., & Yeo, P. F. *The Families of the Monocotyledons.* Berlin, Springer-Verlag (1985).
ii Clifford, H. T., Henderson, R. J. F., & Conran, J. G. Hemerocallidaceae. Pp. 245–253. In: Kubitzki, K. (ed.), *The Families and Genera of Vascular Plants. III. Flowering Plants. Monocotyledons: Lilianae (except Orchidaceae).* Berlin, Springer-Verlag (1998).
iii Chase, M. W., Rudall, P. J., Fay, M. F., & Stobart, K. L. Xeronemataceae, a new family of asparagoid lilies from New Caledonia and New Zealand. *Kew Bull.* 55: 865–870 (2000).
iv Fay, M. F. *et al.* Phylogenetic studies of Asparagales based on four plastid DNA regions. Pp. 360–371. In: Wilson, K. L. & Morrison, A. (eds), *Monocots. Systematics and Evolution.* Melbourne, CSIRO Publishing (2000).
v Chase, M. W. *et al.* Multi-gene analysis of monocot relationships: A summary. Pp. 63–75. In: Columbus, J. T. *et al.* (eds), *Monocots: Comparative Biology and Evolution.* Claremont, California, Rancho Santa Ana Botanic Garden (2006).

XYRIDACEAE
YELLOW-EYED GRASSES FAMILY

The Xyridaceae is a small family of herbaceous plants, with basal, equitant leaves; scapose inflorescences; and ephemeral flowers.

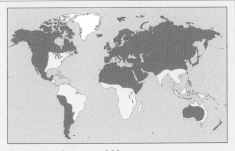

Genera 5   Species c. 300
Economic uses A few species are aquarium ornamentals or of local medicinal use

Distribution The family is pantropical with 4 genera restricted to South America. Only *Xyris* has a wider distribution, reaching East Asia, Australia, western North America, and Canada. Nearly all species occur in wetands, and a few are truly aquatics (e.g., *X. aquatica*).

Description Mostly perennial, rarely annual, rushlike herbs, with distichous or spirally arranged leaves; a sheathing, often keeled, base; and occasionally ligulate. The leaves are often basal and equitant, less often cauline or terete or angulate. The lateral or terminal inflorescence is mostly scapose, often with a pair of bracts on the scape, and terminated by a spike or head or a panicle of spikes. The spikes or heads have 1 to many, solitary and subsessile, rarely pedicellate, bisexual, slight bilateral or zygomorphic, ephemeral flowers supported by stiff, papery, or leathery bracts. The perianth has 2 whorls, usually with 3 unequal sepals in

the outer whorl and 3 petals in the inner whorl. The inner sepal is membranous and wrapped around the petals, falling when the flower opens, while the 2 lateral sepals are small and often keel-shaped, persistent on the ripe fruit. The petals are more or less equal, free or united with a basal tube, often yellow, white, or blue. The stamens are 3, attached to the corolla tube, opposite the petals, and with 3 staminodes with characteristic moniliform hairs, rarely the stamens 6 (*Xyris*). The ovary is superior, 3-carpellate, with a single locule or incompletely 3-locular with a 3-branched style, occasionally with lateral appendages. The placentation is marginal, parietal, basal, free-central or axile, with numerous anatropous or atropous ovules. The fruit is a loculicidal capsule, often with numerous prominent longitudinal ridges. The embryo is small with a mealy or sometimes oily endosperm.

Classification Two of the genera here included in the Xyridaceae, *Abolboda* and *Orectanthe*, have been assigned to a family of their own, Abolbodaceae (Takhtajan, 1980), a point of view that largely has been abandoned[i,ii]. However, there is some indication that the Xyridaceae is not monophyletic[iii]. The family may be related to the Eriocaulaceae[ii]. The genus *Xyris* (yellow-eyed grasses) is the largest and most variable in the family and contains about 240 spp., whereas 3 of the genera have only 1 or 2 spp.: *Achlyphila*, *Aratitiyopea*, and *Orectanthe*.

Economic uses A few species in the genus *Xyris* are used as aquarium plants. The leaves and roots of 2 North American species, *Xyris ambigua* and *X. caroliniana*, have been used as domestic remedies in the treatment of colds and skin diseases, respectively. OS

i Judd, W. S. *et al. Plant Systematics. A Phylogenetic Approach.* 2nd edn. Sunderland, Massachusetts, Sinauer Associates (2002).
ii Chase, M. W. *et al.* Multi-gene analysis of monocot relationships: A summary. Pp. 63–75. In: Columbus, J. T. *et al.* (eds), *Monocots: Comparative Biology and Evolution.* Claremont, California, Rancho Santa Ana Botanic Garden (2006).
iii Davis, J. I. *et al.* A phylogeny of the monocots, as inferred from *rbc*L and *atp*A sequence variation, and a comparison of methods for calculating jackknife and bootstrap values. *Syst. Bot.* 29: 467–510 (2004).

ZINGIBERACEAE
GINGER FAMILY

A tropical family of aromatic, rhizomatous herbs, with distichous, simple, sheathing leaves, and often showy flowers.

Distribution Zingiberaceae has a pantropical distribution, chiefly occurring in Indomalaysia. The majority of species are found in the humid, tropical lowlands, but several occur in montane forests or drier dipterocarp forests, and a few more are epiphytes.

ZINGIBERACEAE. 1 *Aframomum melegueta* (a) flowering, leafless shoots (x⅔); (b) opened flower, with inferior ovary, with 2 glands, the style freed from the embracing thecae (x⅔); (c) style in place between the thecae, stigma protuding (x1); (d) fruit (x⅔); (e) cross section of fruit, with many seeds (x⅔). 2 *Zingiber officinale* (Ginger) (a) flowering leafless shoot (x½); (b) flower, with clearly visible labellum and anther with style in place between the thecae (x1); (c) longitudinal section of flower (x1); (d) thecae, with extended connective folded around the style (x10). 3 *Alpinia officinarium* (a) leafy shoot and inflorescence (x⅔); (b) flower, with clearly visible labellum and anther (x1); (c) gynoecium, with glands at base of the style (x1½); (d) cup-shaped stigma (x4); (e) anther (x1½).

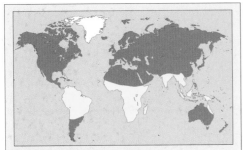

Genera 50 Species c. 1,300

Economic uses Zingiberaceae is an extremely important source of spices (e.g., cardamom, ginger, turmeric), perfumes, medicines, dyes, and tropical and greenhouse ornamentals

Description Aromatic, perennial herbs, with branched, thick, and fleshy rhizomes, frequently possessing tuberous roots. The aerial stems are largely unbranched pseudostems, may vary considerably in size (up to 8 m high), but usually the stems are short. The leaves are distichous or tufted, often without blades at the bases. The leaf sheath is often open, with or without a ligule, and a petiole of varying length, sometimes absent. The lamina is usually entire; elliptical; with a prominent midrib; and parallel, pinnate, lateral veins, diverging obliquely from the midrib. The inflorescence terminates a leafy shoot or emerges from a separate, sheath-covered, leafless shoot directly from the rhizome. The inflorescence is often more or less cylindrical or globose and usually a thyrse in which each bract subtends a short cincinnate partial inflorescence, rarely only a single flower, a spike, or raceme. The flowers are zygomorphic and bisexual and last only a day. The calyx is 3-lobed or 3-dentate, and the corolla tubular at the base, 3-lobed, the lobes varying in size and shape, the median, posterior lobe often larger than the others. Only the median, posterior stamen of the inner whorl is fertile; the 2 others of the inner whorl are petaloid staminodes that are fused into a conspicuous 2- or 3-lobed labellum. The 2 lateral stamens of the outer whorl are petaloid or absent, whereas the median, anterior stamen of the outer whorl is always reduced. The fertile stamen has a long or short filament but is occasionally sessile. The ovary is inferior, of 3 carpels, and either 3-locular with axile placentas or 1-locular with parietal or basal, many, anatropous ovules. The style is often extremely thin, almost always embedded in a furrow in the filament of the fertile stamens, emerging between the thecae. The fruit is a dry or fleshy capsule, usually dehiscing loculicidally, with few to many seeds, with a distinctive white or red aril, straight embryos, and scanty, often starchy endosperm.

Classification Most recent classifications have a distinct, monophyletic group of families that include Zingiberaceae, Costaceae (q.v.), Maranthaceae (q.v.), Cannaceae (q.v.), Lowiaceae (q.v.), Strelitziaceae (q.v.), Heliconiaceae (q.v.), and Musaceae (q.v.)[i,ii,iii]. The relationships between the families vary, however, research sometimes pointing to a sister group relationship between Zingiberaceae and Costaceae[iv,v,vi], and the 2 families have previously been combined. The family is tentatively divided into 4 tribes: Alpinieae, Hedychieae, Globbeae, and Zingibereae[vii].

Economic uses The Zingiberaceae are rich in aromatic, volatile oils and are widely used as condiments, herbs, dyes, and medicinal plants. The rhizomes of *Zingiber officinale* (Ginger) and *Curcuma* spp. are important on the world market. Bombay or East Indian

Arrowroot is derived from the tubers of *Curcuma angustifolia*, whereas *C. longa* yields turmeric, one of the main coloring and aromatic ingredients of curry powder and also used as a yellow dye. *Elettaria cardomomum* from Indonesia yields the important eastern spice cardamom, while the seeds of *Aframomum melegueta* are marketed as Melegueta pepper or grains of paradise. Other important products include abir—a perfumed powder obtained from the rhizome of *Hedychium spicatum*—and zedoary—a spice, tonic, and perfume extracted from the rhizomes of *C. zedoaria*. *Alpinia officinale* and *A. galanga* yield the medicinal and flavoring rhizome galangal. Many species have beautiful flowers and are cultivated in the tropics and as greenhouse ornamentals in temperate countries, including *Hedychium coronarium* (Ginger Lily), *Alpinia purpurata* (Red Ginger) and *A. zerumbet* (Shell Ginger). *Roscoea* is cultivated as a garden plant in temperate areas. [CJH] OS

i Dahlgren, R. M. T., Clifford, H. T., & Yeo, P. F. *The Families of the Monocotyledons*. Berlin, Springer-Verlag (1985).
ii Judd, W. S. *et al. Plant Systematics. A Phylogenetic Approach*. 2nd edn. Sunderland, Massachusetts, Sinauer Associates (2002).
iii Stevenson, D. W. & Stevenson, J. W. Zingiberaceae. Pp. 494–495. In: Smith, N. *et al.* (eds), *Flowering Plants of the Neotropics*. Princeton/Oxford, Princeton University Press (2004).
iv Chase, M. W. *et al.* Multi-gene analysis of monocot relationships: A summary. Pp. 63–75. In: Columbus, J. T. *et al.* (eds), *Monocots: Comparative Biology and Evolution*. Claremont, California, Rancho Santa Ana Botanic Garden (2006).
v Kress, J., Prince, L. M., Hahn, W. J., & Zimmer, E. A. Unraveling the evolutionary radiation of the families of the Zingiberales using morphological and molecular evidence. *Syst. Biol.* 50: 926–944 (2001).
vi Johansen, L. B. Phylogeny of *Orchidantha* (Lowiaceae) and the Zingiberales based on six DNA regions. *Syst. Bot.* 30: 106–117 (2005).
vii Larsen, K., Lock, J. M., Maas, H., & Maas, P. J. M. Zingiberaceae. Pp. 474–495. In: Kubitzki, K. (ed.), *The Families and Genera of Vascular Plants. IV. Flowering Plants. Monocotyledons: Alismatanae and Commelinanae (except Gramineae)*. Berlin, Springer-Verlag (1998).

ZOSTERACEAE
EEL GRASSES

A small family of marine grasslike herbs, with characteristic flattened inflorescences, which live entirely submerged.

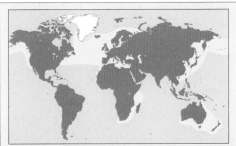

Genera 2–4 **Species** 18
Economic uses Food source for marine organisms and packing material when dried

Distribution Marine aquatics, rarely in the intertidal zone, found mainly in temperate seas of the northern and southern hemispheres. A few species extend into tropical seas.

Description Usually perennial aquatics, with creeping, monopodial rhizomes. The leaves are distichous, linear, without stomata, and grasslike, with distinctly sheathing bases and intravaginal scales. The inflorescence is a flattened spadix, enclosed by a modified leaf sheath; it may be branched (*Heterozostera*, *Zostera*) or unbranched (most *Phyllospadix* spp.). Flowers unisexual (plants monoecious or dioecious), naked, and arranged in 2 rows on 1 side of the spadix. In monoecious plants, the male and female flowers are arranged alternately along the spadix. The flowers lack tepals, and the male flowers have 2 stamens joined by a ridgelike connective. The pollen grains are filamentous, of the same buoyancy as seawater, and adhere to each other in a network, floating freely. In most species, a curious hooklike

process (retinaculum) of unknown origin is found beside each stamen or in the female spadixes of *Phyllospadix*, where they alternate with the female flowers. The female flower consists of a single, naked carpel, with a short style and 2 relatively long stigmas. The ovule is solitary, straight, and pendulous. The fruit is ovoid or ellipsoidal achene. Seeds without endosperm.

Classification The number of recognized genera in Zosteraceae is controversial and varies from 2 (*Phyllospadix, Zostera*)[i] to 4 (*Heterozostera, Nanozostera, Phyllospadix,* and *Zostera*)[ii]. The Zosteraceae are closely related to Potamogetonaceae[iii].

Economic uses The dried leaves and stems of *Zostera marina* are used as a packing material, particularly for Venetian glass. *Zostera* species are primary producers in coastal habitats and an important source of food for waterfowl and fish. As a result, the "wasting disease" that hit *Z. marina* in Western Europe in the 1930s and 1940s led to a decline in the waterfowl and fish populations. [CDC] OS

i Les, D. H., Moody, M. L., Jacobs, S. W.-L., & Bayer, R. J. Systematics of seagrasses (Zosteraceae) in Australia and New Zealand. *Syst. Bot.* 27: 468–484 (2002).
ii Tomlinson, P. B. & Posluzny, U. Generic limits in the seagrass family Zosteraceae. *Taxon* 50: 429–437 (2001).
iii Les, D. H., Cleland, M. A., & Waycott, M. Phylogenetic studies in Alismatidae, II: Evolution of marine angiosperms (seagrasses) and hydrophily. *Syst. Bot.* 22: 443–463 (1997).

INDEX

This index includes Latin names of genera as well as English names mentioned in the book. When a Latin name is used as a popular name for a different plant, for example—*Amaryllis* or *Nasturtium*—an entry for each use of the name will be found. When the Latin name is also the name commonly used in English, the whole entry is under the Latin form; for example, *Asparagus, Crocus, Dahlia.* Page numbers in *italics* refer to illustration captions and those in **boldface** to families with their own entries.